Air Bag Development and Performance
New Perspectives from Industry, Government and Academia

Editor
Richard Kent

Associate Editors
Charles Strother
Jeff Crandall
Hugo Mellander
Charles Griswold

Published by
Society of Automotive Engineers, Inc.
400 Commonwealth Drive
Warrendale, PA 15096-0001
U.S.A.
Phone (724) 776-4841
Fax: (724) 776-5760
www.sae.org

All rights reserved. No part of this publication may be reproduced, stored in a retrieval system, or transmitted, in any form or by any means, electronic, mechanical, photocopying, recording, or otherwise, without the prior written permission of SAE.

    For permission and licensing requests contact:

    SAE Permissions
    400 Commonwealth Drive
    Warrendale, PA 15096-0001-USA
    Email: permissions@sae.org
    Fax: 724-772-4891
    Tel: 724-772-4028

All SAE papers, standards, and selected books are abstracted and indexed in the Global Mobility Database.

For multiple print copies contact:

    SAE Customer Service
    Tel: 877-606-7323 (inside USA and Canada)
    Tel: 724-776-4970 (outside USA)
    Fax: 724-776-1615
    Email: CustomerService@sae.org

ISBN 0-7680-1119-1
Library of Congress Catalog Card Number: 2003101391
SAE/PT-88
Copyright © 2003 SAE International

Positions and opinions advanced in this publication are those of the author(s) and not necessarily those of SAE. The author is solely responsible for the content of the book.

SAE Order No. PT-88

Printed in USA

Preface

The first U.S. patents related to deploying devices for occupant protection in a crash were filed before World War II. These precursors to the air bags of today included sprung instrument panels (e.g., U.S. Patents 2,070,760, 1935 and 2,091,057, 1937) like the one shown below. This system would displace toward an occupant in a frontal crash, providing "a relatively compressible protective device in connection with the part or parts of an automobile body most subject to impact by passengers" (Straith, "Safety Device for Automobiles" U.S. Patent 2,070,760, 1935). In the 1950s, patents for inflatable bags were filed, including a single-bag system by John Hetrick in 1953 (U.S. Patent 2,649,311) and the multiple-bag system by H. A. Bertrand in 1958 (2,834,606) shown below. Since these early patents, activity related to air bags has exploded – especially in recent years. A U.S. patent search as of this writing will yield over 3,000 patents related to the air bag, 2,500 of which have been filed since 1996. The evolutionary journey of these systems is a fascinating one, and is the subject of this book.

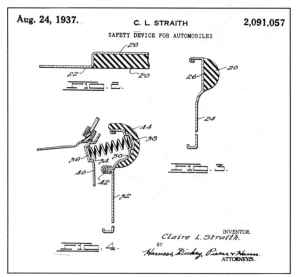 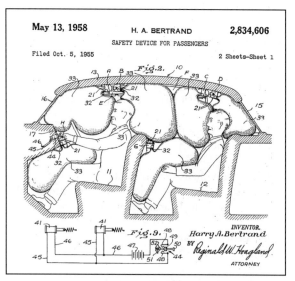

SAE originally approached me about editing a book on air bags several years ago. I was given freedom regarding the structure and content of the book; they asked only that it would deal with the topic of air bags and that it would be primarily a compendium of papers published since the 1987 compendium edited by Viano (SAE PT-31). As I considered this offer, I decided that a well-constructed effort with a limited scope, focusing on a significant amount of new material augmented by existing SAE and non-SAE literature, would be a valuable contribution to the field. This is what I set out to do.

The first question was how to go about limiting the scope in a manner that would be most useful and practicable. The nature of air bag design and performance is so transient that a book lends itself better to a historical accounting rather than to a discussion of contemporary design issues. With this in mind, I recruited some of the pioneering engineers who developed these systems and solicited their input regarding the content and scope of the project. I tried to find people to represent four perspectives: U.S. industry, European industry, U.S. government, and academia. I was fortunate enough to convince some truly excellent and influential engineers to get involved. They include the four Associate Editors listed on the title page. Charles Strother spent many years at NHTSA and as a NHTSA contractor during the 1970s. Hugo Mellander is one of Europe's top automobile safety experts and was influential at Volvo during their early air bag development efforts. Charles Griswold was one of the first to work on the GM air bag systems. Dr. Jeff Crandall has written over 100 papers dealing with occupant protection in automobile crashes. Together, we decided to approach this project in the following manner.

The Associate Editors were primarily responsible for two aspects of this book. First, they contributed an original work that gives their perspective on the history, development, or performance of air bags. Their four chapters and mine make up the first section of this book. Second, each Associate Editor was to recruit input and use his knowledge of the air bag literature to identify papers that represent his perspective or that reveal the current thinking at important points during the air bag's history. Our starting point was a broad search of the literature using the SAE archives, Medline, Ovid, the historical proceedings of the Stapp, AAAM, ESV and IRCOBI conferences, the federal docket, and other databases. From these sources, we identified over 2000 papers dealing with the subject of air bags. We reduced this to approximately 1000 that dealt with historical, development, or performance issues. Approximately 800 of these are included in some form in this book. Approximately 50 selected SAE, government, and, where copyright permission could be obtained, other papers are included in their entirety. About 90 other papers, listed as recommended reading, are included by title and abstract. Finally, some 600 papers (SAE and other) are listed for further reading in a bibliography at the end of the book.

While the Associate Editors represent certain interests and characteristics, no single person could completely define the perspectives of groups as large and diverse as the four listed above. We therefore consulted with many individuals in an effort to incorporate a range of perspectives. The following people contributed substantially as co-authors and also provided input into the paper selection process:

David Viano, Ph.D., M.D.
ProBiomechanics, LLC

Hansjürgen Scholz
Daimler-Benz (ret.)

Cameron R. "Dale" Bass, Ph.D.
University of Virginia

Michael U. Fitzpatrick, M.S.
Fitzpatrick Engineering

Joseph N. Kanianthra, Ph.D.
National Highway Traffic Safety Administration

Donald E. Struble, Ph.D.
Struble-Welsh Engineering

Richard M. Morgan, M.S.
National Highway Traffic Safety Administration

Sherman Henson
Consultant

Karl Erik Nilsson
Autoliv Inc. (ret.)

Other individuals did not write new material, but contributed their expertise in other ways. The following individuals provided reviews of the new material, input regarding the papers to include in the compendium, or other technical consultation:

Dr. Leonard Evans
Science Serving Society

Michael B. James, M.S.
Collision Safety Engineering, LLC

Maria Segui-Gomez, Ph.D., M.D.
The Johns Hopkins Bloomberg School of Public Health

Donald F. Heulke, Ph.D.
University of Michigan (ret.)

Dr. rer. nat. Dimitrios Kallieris
University of Heidelberg

Dr. Guy S. Nusholtz
DaimlerChrysler

Uwe Meissner, M.S.
Volkswagen

Chris Sherwood, M.S.
University of Virginia

Finally, the following people are acknowledged for their efforts in the production of this book. Acquiring print-quality versions of these documents required numerous hours of searching through old archives, scanning, facilitating copyright transfer, and even tracking down retired authors. I also acknowledge and thank the Association for the Advancement of Automotive Medicine for allowing us access to their historic archives. Without the efforts of these dedicated individuals, this book would not exist.

Nathan Benson
Amy Vellinga
Di Loveless
Nanette Strother (ret.)
Collision Safety Engineering, LLC

Irene Herzau
Association for the Advancement of Automotive Medicine

Martha Swiss
Tracy Fedkoe
SAE

Tim Gillespie
University of Virginia

While many people contributed to the various chapters in this book, the chapters should not be considered as representing the consensus view of any group. The opinions expressed are those of the authors and are not necessarily the views of SAE, the editor, or the authors' current or previous employers. I simply tried to recruit excellent engineers having diverse perspectives. It is up to the reader to decide how well I accomplished my task.

We followed a paper selection methodology similar to that outlined by Stanley Backaitis in his excellent SAE compendia on biomechanics and trauma, but I acknowledge that the method for selecting the papers is not an entirely scientific one. Because of our limited scope, we did not include many papers discussing detailed air bag design issues. We focused on papers that reveal historical conceptions or misconceptions about air bags or that reveal their fundamental characteristics. We also targeted papers that have long term reference value. Furthermore, SAE papers were necessarily given greater weight for full inclusion since they could be obtained easily and without copyright transfer. The reprinted papers should therefore not be considered as the only worthy ones. I do feel, however, that the selected papers, combined with those compiled by David Viano in the earlier compendium, describe well the historical development of air bags and their performance. They also represent a subset of particularly outstanding or influential papers published over the years. In fact, despite the papers being selected by people representing a diversity of perspectives, I found a surprising amount of agreement regarding the papers to include. As a result, it became inappropriate to categorize a particular paper as representing "the industry perspective", "the European perspective", or "the government perspective" and the papers are grouped by more descriptive topics.

It is not necessary for the reader to proceed linearly through the book. In particular, the first five chapters often assume a certain level of knowledge regarding terminology or air bag design in general. I would therefore encourage the reader to refer, for example, to the included Struble paper (980648) and its clear description of air bags if this background information is needed. Similarly, the first five chapters cite many of the seminal works that are included in this compendium; so the reader may refer directly to the cited document for additional detail.

Finally, I would like to acknowledge and thank the many outstanding researchers who contributed to this history. Many of the documents compiled here were written by individuals who dedicated their careers to increasing occupant protection in crashes. This has been an educational and personally enriching journey for me, and I hope that this book provides you with a valuable reference as we build on this legacy.

Richard Kent
Research Assistant Professor
Department of Mechanical and Aerospace Engineering
Center for Applied Biomechanics
University of Virginia

Table of Contents

Preface iii

I. History, Development, and Performance from Five Perspectives

Chapter 1. A Brief History of the Air Bag Development and Implementation Activities of the
United States Government 3
      Charles E. Strother, Joseph N. Kanianthra, Richard M. Morgan, Michael U. Fitzpatrick,
      Donald E. Struble

Chapter 2. U.S. Industrial Air Bag Development 35
Charles Griswold, Sherman Henson

Chapter 3. European Air Bag Development 49
      Karl Erik Nilsson, Hansjürgen Scholz, Hugo Mellander

Chapter 4. The Biomechanics of Inflatable Restraints – Occupant Protection and Induced Injury 69
      Jeff Crandall, Richard Kent, David Viano, Cameron R. "Dale" Bass

Chapter 5. The Field Performance of Frontal Air bags 111
      Richard Kent, David Viano, Jeff Crandall

II. Federal Air Bag Legislation

The National Traffic and Motor Vehicle Safety Act of 1966 149

Abstract to Analysis of Effects of Proposed Changes to Passenger Car Requirements
of MVSS 208, 1974 157

Final Regulatory Impact Analysis, Amendment to FMVSS 208, July 11, 1984 159

SAE Paper 950865 Regulatory History of Automatic Crash Protection in FMVSS 208
Kratzke 175

Executive Summary of Fourth Report to Congress, Effectiveness of Occupant Protection
Systems and Their Use, May 1999 185

Executive Summary of Final Economic Assessment, FMVSS No. 208, Advanced Air Bags,
May 2000 291

SAE Paper 2001-01-0157 Advanced Air Bag Systems and Occupant Protection: Recent
Modifications to FMVSS 208 205
Hinger and Clyde

SAE Paper 2001-01-0165 Theoretical Evaluation of the Requirements of the 1999 Advanced
Airbag SNPRM – Part One: Design Space Constraint Analysis 211
Laituri, Sriram, Kachnowski, Scheidel, Prasad

III. Air Bag Development

Development of Advanced Restraint Systems for Minicars RSV, Proc. 6th ESV, 1976 233
Strother, Fitzpatrick, Egbert231

The Daimler-Benz Development of a Final Production Air Bag System for the U.S.A.,
Proc. 8th ESV, 1980 251
Scholz

The Development of an Advanced Airbag Concept, Proc. 12th ESV, 1989.....................................261
Johansson, Billig, Mellander, Werner, Hora

SAE Paper 950347 Investigation of Sensor Requirements and Expected Benefits of Predictive
Crash Sensing 267
Swihart and Lawrence

SAE Paper 960226 The BMW Seat Occupancy Monitoring System – A Step Towards "Situation
Appropriate Air Bag Deployment"...279
Kompaß, Witte

SAE Paper 980646 An Innovative Approach to Adaptive Airbag Modules...................................289
Ryan

SAE Paper 980648 Airbag Technology: What it is and How it Came to be...................................295
Struble

SAE Paper 982293 Direct Thermal Detection for Front Passenger Seat Airbag Suppression.......315
Lambert

SAE Paper 1999-01-0761 Occupant Classification System for Smart Restraint Systems............319
Billen, Federspiel, Schockmel, Serban, Sherrill

SAE Paper 2001-01-0164 Influence of Air Bag Folding Pattern on OOP-Injury Potential...........325
Mao and Appel

IV. Laboratory Performance of Air Bags – Occupant Restraint

SAE Paper 740578 Human Volunteer and Anthropomorphic Dummy Tests of General Motors
Driver Air Cushion System...335
Smith.

SAE Paper 933121 Thoracic Biomechanics with Air Bag Restraint 351
Yoganandan, Pintar, Skrade, Chmiel, Reinartz, Sances

SAE Paper 942216 The performance of active and passive driver restraint systems in
simulated frontal collisions 363
Kallieris, Stein, Mattern, Morgan, Eppinger

SAE Paper 950886 The Effect of Limiting Shoulder Belt Load with Air Bag Restraint................375
Mertz, Williamson, Lugt

ESV Paper 98-S5-O-07 Assessment of Air Bag Performance Based on the 5th Percentile
Female Hybrid III Crash Test Dummy, Proc. ESV, 1998..383
Dalmotas

SAE Paper 2001-22-0008 The Influence of Superficial Soft Tissues and Restraint Condition on
Thoracic Skeletal Injury Prediction 401
Kent, Crandall, Bolton, Prasad, Nusholtz, Mertz

SAE Paper 2002-22-0002 Laboratory Reconstructions of Real World Frontal Crash Configurations
Using the Hybrid III and THOR Dummies and PMHS 423
Petitjean, Lebarbe, Potier, Trosseille, Lassau

V. Laboratory Performance of Air Bags – Occupant Injuries

Possible effect of air bag inflation on a standing child, Proc. AAAM, 1974 453
Aldman, Andersson, Saxmark

SAE Paper 841656 A Biomechanical Analysis of Head, Neck, and Torso Injuries to Child
Surrogates Due to Sudden Torso Acceleration 469
Prasad and Daniel

SAE Paper 902324 Assessment of Air Bag Deployment Loads 485
Horsch, Lau, Andrzejak, Viano

SAE Paper 922510 Investigation of airbag-induced skin abrasions 507
Reed, Schneider, Burney

SAE Paper 933119 Assessment of Air Bag Deployment Loads with the Small Female
Hybrid III Dummy 519
Melvin, Horsch, McCleary, Wideman, Jensen, Wolanin.

Thoracic Response and Trauma of Out-of-Position Drivers Resulting from Air Bag
Deployment, Proc. AAAM, 1997 531
Crandall, Duma, Bass, Pilkey, Kuppa, Khaewpong, Eppinger

SAE Paper 980636 Evaluation of 5th Percentile Female Hybrid III Thoracic Biofidelity
during Out-of-position Tests with a Driver Airbag 549
Crandall, Bass, Duma, Kuppa

SAE Paper 982325 Dual Stage Inflators and OOP Occupants – A Performance Study 557
Malczyk, Franke, Adomeit

SAE Paper 983162 Investigation into the Noise Associated with Air Bag Deployment:
Part II - Injury Risk Study Using a Mathematical Model of the Human Ear 565
Rouhana, Webb, Dunn

SAE Paper 1999-01-0764 Deployment of Air Bags into the Thorax of an Out-of-Position Dummy 585
Bass, Crandall, Bolton, Pilkey, Khaewpong, Sun

SAE Paper 2001-01-0179 Air Bag Loading on In-Position Hybrid III Dummy Neck 601
Kang, Agaram, Nusholtz, Kostyniuk

VI. Field Performance of Air Bags

SAE Paper 880400 Restraint Performance of the 1973-76 GM Air Cushion Restraint System 619
Mertz

SAE Paper 910901 Effectiveness of Safety Belts and Airbags in Preventing Fatal Injury 631
Viano

SAE Paper 940714 Airbag Protected Crash Victims – The Challenge of Identifying Occult Injuries 645
Augenstein, Perdeck, Digges, Stratton, Lombardo, Malliaris, Byers, Nunez, Zych, Andron,
Craythorne, Verga

SAE Paper 940716 Upper Extremity Injuries Related to Air Bag Deployments 655
Huelke, Moore, Compton, Samuels, Levine

SAE Paper 940802 Survey of Airbag Involved Accidents - An Analysis of Collision Characteristics,
System Effectiveness and Injuries 665
Werner and Sorenson

SAE Paper 950866 An Overview of Airbag Deployments and Related Injuries. Case Studies and a Review of Literature...685
Huelke

SAE Paper 960659 Air Bag Field Performance and Injury Patterns...693
Malliaris, DeBlois, Digges

SAE Paper 960664 Air Bag Deployment Frequency and Injury Risks...717
Werner, Roberson, Ferguson, Digges

Supplemental Restraint Systems: Friend or Foe to Belted Occupants? Proc. AAAM, 1996.......................729
Dalmotas, Hurley, German

SAE Paper 970491 Injury Risks in Cars with Different Air Bag Deployment Rates 743
Werner, Roberson, Ferguson, Digges

Mechanisms of Injuries for Adults and Children Resulting from Airbag Interaction,
Proc. AAAM, 1997...753
Kleinberger and Summers

Driver air bag effectiveness by severity of the crash. American Journal of Public
Health 90:1575-1581. Segui-Gomez...769

SAE Paper 2001-01-0156 Analysis of Driver Fatalities in Frontal Crashes of Airbag-Equipped
Vehicles in 1990-98 NASS/CDS 777
Zuby, Ferguson, Cammisa

Air bag crash investigations, Proc. ESV, 2001 789
Chidester and Roston

SAE Paper 2002-01-0186 Performance of Depowered Air Bags in Real World Crashes 801
Augenstein, Perdeck, Stratton, Digges, Steps

VII. Abstracts for Further Reading...807

VIII. Bibliography for Further Reading 827

* * *

About the Contributors 857

I. HISTORY, DEVELOPMENT, AND PERFORMANCE FROM FIVE PERSPECTIVES

Chapter 1

A Brief History of the Air Bag Development and Implementation Activities of the United States Government

Charles E. Strother, Joseph N. Kanianthra, Richard M. Morgan, Michael U. Fitzpatrick, Donald E. Struble

The chapter reviews the 35-year history of the efforts of the U.S. Department of Transportation to develop and evaluate the life-saving potential of air bag systems, as well as to mandate their installation in vehicles. In producing this review, the authors rely upon their personal experiences as well as the technical literature documenting this history. The paper is conceptually divided into parallel discussions of the research and rule-making efforts. Each of these discussions, in turn, is chronologically divided into periods before and after 1982-83. The choice of year reflects the announcement of the passive restraint rule that finally came into effect, and the redirection of federal research efforts away from outside contractors and toward in-house facilities. While preparing the paper, the authors revisited the existing air bag literature, resulting in a bibliography of the more significant publications in this field.

Introduction

Although the U.S. federal government neither originated the concept nor ever installed such a device in a production automobile, it did nevertheless have a significant role in the evolution of air bag restraint systems. This role resulted from two of the primary activities conducted by the National Highway Traffic Safety Administration (NHTSA), namely rulemaking and safety research. In the first instance, the issuance of safety regulations, or the announced intention to do so, provided a powerful impetus for the automobile industry to continue development of the air bag. In the second instance, the research conducted by the Agency, or under its sponsorship, contributed to the body of knowledge about the potentialities and problems associated with trying to develop this most complex of automotive safety systems.

This paper is an attempt to summarize these activities from the advent of the NHTSA to the present. In writing this work, the authors have drawn upon their own experiences in this area in addition to making use of the available literature. Three of the authors of this paper either are or were NHTSA researchers. Some began working at the Agency in the very early 1970s and were involved in the planning and execution of federal research projects conducted in support of the anticipated passive restraint mandate. Three of the authors worked as independent contractors, executing air bag research and development efforts under NHTSA sponsorship. Thus, this paper not only reflects what one could obtain from a reading of the air bag literature, but includes the authors' recollections concerning activities during this period.

The paper is divided chronologically into two periods, the dividing line between them being the early 1980s. This date is considered significant for two reasons. First, the final rule mandating air bags in passenger vehicles was proposed in 1983. Secondly, it was at about this time that the NHTSA started to conduct virtually all of its own research in-house (primarily at the Vehicle Research and Test Center in Ohio) and thus, in the main, stopped sponsoring large research and development programs with non-government research and development companies as it had done from its inception. Prior to these discussions, however, it seems appropriate to first briefly discuss air bag development prior to the creation of the federal safety agency in 1967.

Evolution of the Airbag Concept Prior to Federal Safety Involvement

The first hint that perhaps the newly invented automobile had some unforeseen side effects that would profoundly affect peoples lives became apparent when, in the year 1899, Mr. H.H. Bliss stepped off a streetcar in San Francisco into the path of one of these early automobiles and was immediately killed. No, air bags were not yet even the germ of an idea, but this was the first indication that the protection of people both inside and outside the car would eventually have to be considered. This first automobile-induced death in the U.S. would prove to be the precursor to many decades of research and development sponsored first by industry and later mandated by a government agency that would have as its objective the protection and the safety of people both inside and, like Mr. Bliss, outside the boundaries of a hurtling automobile.

Those of us who were there at the early stages of automobile safety efforts were witness to the awakening concern of the need to protect people who could be killed and maimed when interacting with energies in the hundreds of thousands of Newton-meters. We can thus look back these many years later with the hindsight of experience and see certain key events and/or factors that affected the safety of persons associated with automobiles.

One of these key factors was the evolution of the air bag concept, which has an extensive and convoluted history.

The concept of using a gas bag to protect occupants in a frontal collision apparently goes back to the 1920s, perhaps earlier. A number of U.S. patents for automotive air bags can be found, starting in the early 1950s. Several years ago, the Washington Automotive Press Association reviewed the history of the air bag and awarded its prestigious "Golden Gear Award" to Mr. John W. Hetrick as the original inventor of the automobile air bag (Reed 1999). Hetrick's thinking apparently began when, during a Sunday drive, he suddenly had to veer off the road to avoid a large rock. In doing so, he impacted a relatively soft embankment and escaped serious injury, but he and his wife had to physically restrain their daughter from impacting the windshield. This experience was an epiphany for Mr. Hetrick. He couldn't get his mind off the idea that, with a little thought, some sketches and a bit of testing, he might be able to develop an inflatable cushion that could prevent the type of impact his daughter only narrowly missed. While there were others with similar concepts during this time frame, Mr. Hetrick's concept was apparently the first to envision the basic type of system in cars today. His invention, see Figure 1, consisted of a tank of compressed air and inflatable cushions on the steering wheel, the middle of the dashboard, and the glove compartment to protect front seat occupants. The "sensor" in Hetrick's design consisted of a sliding, spring-loaded weight which, when propelled forward by crash forces, would open the tank valve to send air into the bags (Hetrick 1953).

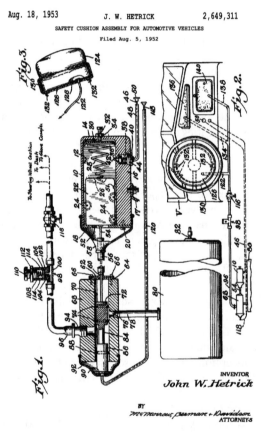

Figure 1. Illustration accompanying the Hetrick airbag patent (adapted from Hetrick 1953).

Another early pioneer in the air bag field was Dr. Carl Clark. Dr. Clark, working at Martin-Marietta in the early 1960s, conducted experiments with his "air-stop" restraint system (Clark et al. 1964). This concept was initially aimed at protecting space-capsule passengers during landing, as seen in Figure 2. His NASA-sponsored research suggested that the air-stop system might also be applied to protect aircraft and ground vehicle passengers in severe collisions. The system envisioned by Dr. Clark was one in which a set of unvented bags would be inflated relatively slowly, either by the driver or by anticipatory sensors that would detect unusual pre-impact vehicle motion.

Figure 2. Dr. Clark demonstrating his "Airstop" system via a pendulum swing impact (adapted from Clark et al. 1964).

Ford and General Motors also experimented with inflatable restraints in the late 1950s. The air bag research and development efforts of the U.S. automobile industry are chronicled in detail elsewhere in this book, but selected activities are mentioned here as a foundation for subsequent work by the federal government. Work at Ford using rubber life rafts to clock inflation times for compressed gas inflation systems were disappointing, casting some doubt that such a system could be inflated quickly enough to be effective (Sherman 1995). One other factor dampening the enthusiasm of the Ford engineers for an air bag system was the idea that the "Lifeguard" safety equipment, introduced on 1956 production models, would prove so popular and effective that the relatively complex air bag system would not be necessary. The Lifeguard package consisted of seat belts, a padded dashboard, a deep-dished steering wheel, and padded sun visors (Ford Motor Company 1957). Unfortunately, early studies indicated that seat belt usage was very low (e.g., MacVean 1966). Indeed, if seat belt usage with this system had been significant, air bag development would have probably come to a halt at this point in time. Later studies, for example, have shown that the life-saving potential of lap and shoulder belts exceeds that of air bags, given a belt usage rate of at least 54 percent (Evans 1989).

The next significant contributor to air bag development appears to have been David Peter Haas. Around 1960, Haas wrote about the concept of a gas-filled air cushion for his master's thesis at the Chrysler Institute of Engineering. Like Hetrick, Haas apparently got his original inspiration from a personal experience - a frontal impact involving his family when he was a child. After spending a number of years at Chrysler in a non safety-related capacity, in 1964 Haas moved to Eaton, Yale, and Towne (EYT - later to become the Eaton Corporation), where he began conducting research on the air bag concept. In 1966, Haas and his fellow researchers at EYT successfully tested a pre-inflated air bag system of their own design. Also that year, a new detonating valve was created by the U.S. Army at the White Sands Proving Ground (Sherman 1995). This valve gave new hope to the idea that a compressed gas system could inflate an air bag with sufficient speed to be effective in a frontal crash. Haas was given the responsibility for a multimillion-dollar air bag feasibility study at EYT. Soon thereafter, EYT and Ford Motor Company joined forces in air bag development, wherein Ford specified the design parameters and conducted crash tests.

The Ford/Eaton air bag prototypes were initially focused on protection for the right front passenger. As illustrated in Figure 3, upper body restraint in a frontal collision was provided by a 9-10 cubic-foot bag that was inflated with nitrogen gas stored in a 3500 lb/in^2 (gage) tank. An acceleration sensor sent a signal to the tank valve, where a detonator would fire, opening the valve. The high-pressure nitrogen then rushed into a gas diffuser and from there, into the air bag, which was initially attached to and folded against the diffuser. The air bag inflated in about 40 milliseconds to a pressure of about 2.2 lb/in^2 (gage). Lower body restraint, and protection in rollovers and multiple collisions, was provided by a lap belt. Thus, this system was not truly passive since some action was required by the occupant to fully utilize its protective potential. In addition to the front passenger air bag efforts, driver upper body protection was attempted via the steering column, which retracted toward the instrument panel to make room for a lateral extension of the passenger bag. This technique proved impractical, however, and later work employed a separate bag housed in the hub of the steering wheel (Kemmerer et al. 1968, Slack 1968).

An initial series of forty-two crash tests was conducted using, first, pre-inflated and then sensor-triggered unvented air bags (Frey 1970). The results were viewed by the investigators as encouraging. Lap belt loads were reduced by 50 to 70 percent in the 30-mph barrier tests. Head and chest deceleration levels were decreased by about 40-50 percent. The crash sensors used provided fast and reliable system activation in the barrier tests. The researchers concluded that additional sensor work was needed as well as the development of a means to deflate the bags quickly after the collisions. Also, the need for the development of a driver air bag system was established.

In connection with this study, sensor issues were investigated. That is, the Ford engineers recognized that a simple acceleration sensor (g-switch) for bag deployment was in no way satisfactory. For the air bag system to be practical, a more sophisticated sensor would have to be developed that could, with very high reliability, discriminate between deployment and non-deployment situations. To this end, vehicle tests were conducted to measure vehicle acceleration levels in a limited number of on-road situations. These included a vehicle running over railroad tracks and through potholes and chuckholes, as well as impacts into barriers at low (around 5 mph) speed, 'threshold' speeds (around 15 mph), and 'protection-limit' speeds (30 mph).

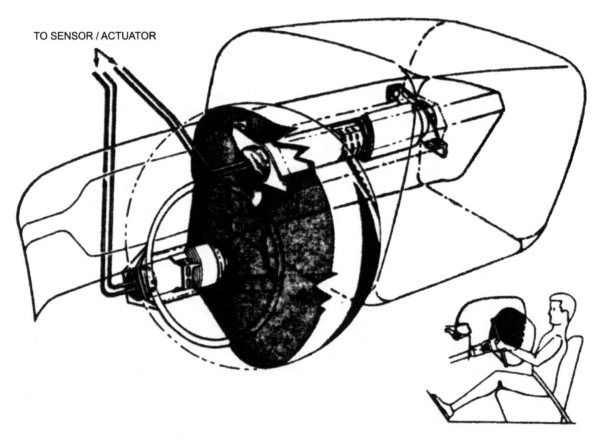

Figure 3. The Ford/Eaton airbag system (adapted from Kemmerer et al. 1968).

Airbag Regulatory Efforts, 1967-1982

In 1966, Congress passed the National Traffic and Motor Vehicle Safety Act, establishing a federal automotive safety agency. This agency was initially named the National Highway Safety Bureau, a division of the Federal Highway Administration, but a few years thereafter the Agency was promoted to administration level and renamed the National Highway Traffic Safety Administration (NHTSA). The enabling legislation called for the safety agency to promulgate motor vehicle safety standards that would be "reasonable, practicable, and meet the need for highway safety" (U.S. Congress 1966).

One of the original 19 Federal Motor Vehicle Safety Standards (FMVSS) promulgated was FMVSS 208, the Occupant Protection Standard (NHTSA 1967). This standard required seatbelts in all passenger cars manufactured after January 1, 1968 and specifically called for lap and shoulder belts at the front seat outboard positions. Like many of the first set of standards, the original FMVSS 208 was a 'design standard' (i.e., a standard that specified particular hardware).

During the very early history of the Agency, it became apparent that frontal impacts were responsible for a significant percentage of motor vehicle fatalities and serious injuries. Two strategies emerged for greatly reducing the numbers of these frontal casualties. The first, the "active" strategy, so named because the occupant would have to take some action to be restrained, involved drastically increasing the usage of the existing seat belts mandated by FMVSS 208. Studies conducted at this time showed that seat belt use was, at most, 15-20 percent (MacVean 1966). This usage rate indicated that a significant reduction in serious injuries could be achieved if occupants could be either convinced or required to wear not only the lap belt, but the shoulder belt as well. A second strategy called for a "passive" approach to occupant protection - that is, protection that would not require any action on the part of the occupant. The idea of "passively" protecting vehicle occupants was, in fact, advanced in Ralph Nader's book, *UnSafe at Any Speed*, one of the factors leading to the Safety Act (Nader 1965). At this time, the air bag was the only promising method of providing passive protection in frontal impacts with a velocity change (delta-V) in the range of 30 mph.

Encouraged by the results of the air bag research and testing under the Ford/Eaton program, the NHTSA issued, on July 2, 1969, an Advanced Notice of Proposed Rulemaking (ANPRM), announcing its intention to mandate passive protection for front seat occupants starting with Model Years (MY) 1972 for passenger cars and 1974 for light trucks and vans (LTV). Unlike the original FMVSS 208, this proposed revision would essentially convert the standard from a design standard to a 'performance standard' (i.e., a standard that specified a level of performance in a compliance test without dictating what hardware would be used to achieve that performance). The preference for performance standards existed from the day the Agency was established. Thus, although the U.S. Congress certainly may have contemplated and eventually mandated (in 1991) an 'air bag standard', no standard specifically requiring air bags was ever contemplated by the NHTSA until dictated to do so by the legislative branch.

This ANPRM was the first action in a series of occupant restraint regulatory, legislative and judicial events that would take place over the next 25-30 years. These events would see delays in the scheduled implementation of passive restraints, the rescinding and reinstating of passive protection requirements, court battles over the appropriateness of NHTSA regulation, and the assumption of regulatory duties by several Secretaries of Transportation and even the Congress of the United States. To detail all of these activities would unduly burden the current effort. Fortunately, such an undertaking is not required, as the legislative history of FMVSS 208 has been documented in detail by two previous efforts. The first deals with the history of the occupant protection standard from the inception of the safety agency until the mid-1990s (Kratzke 1995). The second paper details the history of the passive restraint standard during the period 1996-1998 (Stocke 1998). Without attempting to detail the tortuous history of FMVSS 208 during the 1970s and early 1980s, several highlights will be briefly discussed as they influenced the course of air bag development and implementation during this time frame.

When the efforts to mandate passive restraints were initiated in 1969, it seemed clear to all concerned that the only method of complying with this proposed performance standard was with air bag systems. This is not to say that concepts for converting seat belts to a passive system were nonexistent, at least in patent form. These concepts, however, were viewed by the vast majority of safety engineers as being impracticable. This view changed around 1974 with the advent of the Volkswagen Restraint Automatic (VWRA) system (Schimkat et al. 1974, States and Rosenau 1982). The VWRA, see Figure 4, consisted of a two-point shoulder belt, the outboard anchor of which was attached to the upper rear corner of the door, and a knee bolster system for lower body restraint. With the introduction of the VWRA system into the FMVSS 208 picture, provisions (such as an emergency shoulder belt release) had to be made to minimize the probability that an occupant so restrained would not be trapped inside the vehicle after a collision. Moreover, the existence of such a device could have detrimental effects on the usage of the system, particularly since the NHTSA, as of the fall of 1974, was expressly prohibited by an act of Congress from promulgating any safety standard that would make it impossible to operate the vehicle without the belts being worn (this, as a result of the belt interlock 'fiasco'[1]). As time passed and more passive belt systems came into being to compete with the air bag as a viable passive restraint, the NHTSA was placed in a difficult position with regard to FMVSS 208. On the one hand, the Agency wanted to retain the standard as a performance standard to allow vehicle manufacturers the freedom to evolve innovative means of occupant protection. On the other hand, if passive belts that could be easily circumvented were the result of a 'passive' FMVSS 208, then the original intent of the standard would be compromised. This conundrum existed all during the history of air bag regulation from the mid-seventies until air bag systems started to became the obvious consumer preference in the mid-1980s.

During the early 1970s, from the first ANPRM of July 2, 1969 until the middle of 1975, the posture of the NHTSA was clearly to press forward with the implementation of passive restraints. These efforts were interrupted in 1972 when several automobile manufacturers took the NHTSA to court, challenging the agency's authority to require air bags in vehicles, given the fact that the air bag systems were still in development and the fact that the Anthropomorphic Test Device (ATD) proposed for the compliance test was not adequately described in the proposed standard. The U.S. Court of Appeals, in the case entitled Chrysler v. DOT, sided with the NHTSA in the main, but did agree with the petitioners with regard to the need for a more adequately defined test dummy. About a year and a half later, the NHTSA satisfied the court's concerns by essentially specifying the GM Hybrid II dummy as the test device (reference?). Starting in the mid-1970s, however, the aggressive posture of the NHTSA in pursuing the implementation of passive restraints began to be interrupted with periods where it was not clear just what course the federal government would take. In 1976, Secretary of Transportation William Coleman issued a ruling (this after indicating that he, rather than the Administrator of the NHTSA, would determine the government's course of action) calling for an industry/NHTSA sponsored air bag demonstration program of 500,000 vehicles over a two-year period (MY 1979 and 1980) while, in the interim, the NHTSA would redouble its efforts to promote seat belt use. Given the rather unspectacular results of the 1974-1975 GM ACRS vehicle introduction (wherein GM consumers were given an option of purchasing air-bag equipped GM large cars – see Chapters X and Y for more detail), there was considerable skepticism with regard to the realism of this proposed demonstration. About six months later, under a new Administration and a new DOT Secretary (Brock Adams), the passive restraint mandate in FMVSS 208 was reinstated, calling for a phase-in of passive-restraint-equipped passenger vehicles, starting with large vehicles (i.e., vehicles with a wheelbase of over 114 inches) in MY 1982, vehicles with a wheelbase of greater than 100 inches then becoming passively-equipped in MY 1983, with all passenger vehicles complying in MY 1984. This ruling was also challenged in court in a case called Pacific Legal Foundation v. DOT, but the court found in favor of the Agency. This change in policy preempted implementation of the large-scale air bag demonstration program mentioned above.

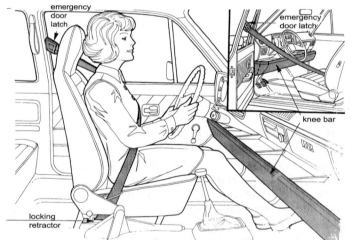

Figure 4. The VW Restraint Automatic (VWRA) (adapted from States and Rosenau 1982).

---

[1] In 1972, manufacturers had the choice to use either passive restraints or manual restraints with an interlock device that prevented the car from starting unless the seat belt was fastened. In response to public outcry regarding seatbelt interlock devices, Congress passed a law (Motor Vehicle and School Bus Safety Amendments of 1974) forbidding manufacturers to meet FMVSS 208 using them.

In 1981, following another change in administration, the NHTSA reversed itself again, rescinding the passive restraint requirements proposed under FMVSS 208, citing concerns over the public's acceptance of passive belts, which were understood to be installed in vehicles instead of the air bags originally envisioned by the Agency. Since the NHTSA could not promulgate any rule requiring passive belts to be ignition interlocked and since some sort of emergency belt release was necessary, the NHTSA position was that passive belt usage would not be high enough to justify the added complexity of these systems. Once again, the courts became involved, and in a case entitled State Farm v. DOT, the U.S. Court of Appeals overturned the NHTSA rescission and ordered the Agency to reinstate the passive restraint requirements. This decision was later upheld by the U.S. Supreme Court. Thus, as the year 1982 came to an end, it appeared that passive restraints would indeed become mandatory under the provisions of FMVSS 208, although the timetable for such implementation was unclear. There were some curves yet to come on the road to mandatory air bags - a story that we'll return to later.

Airbag Research and Development Efforts, 1967-1982

Starting in the late 1960s, there were essentially three forces that molded the early air bag development. Namely, the government, through the fledgling NHTSA, the independent research organizations, whose task it was to try to achieve the design goals outlined by the NHTSA research programs, and finally the motor vehicle manufacturers themselves, who would have to bear the final responsibility for the systems that they chose to install in their vehicles. These three entities were each to form a large role in the air bag development that would proceed over the next thirty-odd years. This paper will concentrate on describing the activities of the first two entities, while a companion paper will deal with the industry efforts.

NHTSA's role was two-fold. First, they had the responsibility to set vehicle safety standards. The Agency's second role was to see to it that whatever safety performance was being mandated would be 'feasible, practical, and meet the need for highway safety.' That is, the Agency was legally obligated to establish (1) that a significant safety problem existed, (2) that means existed to ameliorate that problem, and (3) that the cost of the 'means' would be more than offset by its benefits. It was to establish these latter two points that research organizations such as UMTRI, the Cornell Aeronautical Laboratory (later to become Calspan), and others became involved as government-sponsored researchers. That is, for the first years of its existence, the NHTSA did not have laboratory test facilities of its own. As stated earlier, however, the NHTSA is required by the enabling legislation to support any proposed safety regulation by establishing the existence of practical means for compliance. Thus, during this period, these private organizations and others would bid on Requests for Proposals (RFP) and perform research projects outlined by the Agency. To this end, one of the first investigations of frontal impact protection systems undertaken by the Agency was performed under contract to the University of Michigan's automotive safety laboratory (first called the Highway Safety Research Institute, later to be re-named the University of Michigan Transportation Research Institute - UMTRI). Under this research program, investigators performed frontal impact sled tests with both a passenger air bag system and an integrated (i.e., seat-attached), multi-point seat belt system. The purpose of these tests was to explore the protection limits of these two competing restraint system types, passive and active, when used by various 'normally seated' adult occupants. The air bag system used in this program was basically the Ford/Eaton system discussed above. The integrated belt system tested consisted of a 5-point seat belt array, anchored to an essentially rigid seat structure. Using the Hybrid II ATDs available at that time and injury criteria based on chest and head CG accelerations, it appeared to the researchers that both systems could provide frontal impact protection at delta-Vs of 30 mph and beyond.

Over the first few years of the NHTSA's existence, its first Director, Dr. William Haddon, and his colleagues became convinced that Americans were just not going to 'buckle up' in sufficient numbers and, thus, that the air bag concept offered the better potential to reduce greatly the large numbers of serious injuries and fatalities occurring each year in the United States. In 1969, persuaded that the air bag concept was both feasible and practicable, the Agency issued its first Advanced Notice of Proposed Rulemaking (ANPRM), announcing its intention to revise FMVSS 208 to require passive front seat occupant protection in a 30 mph frontal rigid flat barrier compliance test (NHTSA 1969). To support this proposed rule and to provide research to justify an anticipated 'upgrading' of the rule to higher delta-Vs, a number of contracted research programs were begun in the late 1960s, aimed at either trying to allay criticisms or fears of some of the perceived undesirable 'side effects' of air bags or to better understand their protective capabilities.

As an example of the first type of research, an investigative effort was initiated to squelch concerns that air bag inflations could cause significant auditory problems. This Agency-sponsored effort consisted of a human volunteer study at the Wright-Patterson Aerospace Medical Research Laboratory. In this study, volunteers were exposed to the deployment noise of the Ford/Eaton compressed gas system (Nixon 1969). These systems had been observed to create sound levels in the range of about 160 decibels during the early phases of deployment. The set-up for these tests is illustrated in Figure 5. While the volunteers did experience some temporary threshold shifts, they suffered no significant injuries or hearing loss. As another example, in a later study at Southwest Research Institute (SWRI), human volunteers were subjected to surprise driver inflations to establish that they would not lose control of their vehicle, should an inadvertent deployment occur (Ziperman and Smith 1975).

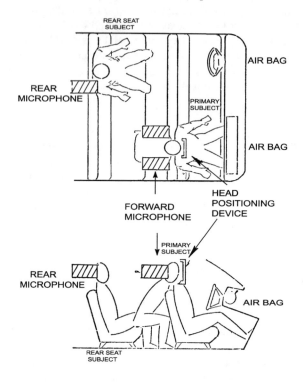

Figure 5. Test set-up for the airbag noise experiments with human volunteers (adapted from Nixon 1969).

In 1970, an Agency-sponsored research program was initiated that involved static and dynamic (sled) tests of a General Motors passenger air bag system with human volunteers (i.e., military personnel) (Smith et al. 1972). These tests, performed at Holloman Air Force Base in New Mexico, were done in cooperation with GM, who supplied the sled buck as well as the air bag systems. Forty-one 'statically deployed' experiments were followed by thirty-five dynamic tests at severities in the range of 15.1 mph delta-V (8.6 g) to 31.5 mph delta-V (21.7 g). The objective of this research was to determine the air bag characteristics that could improve the safety performance for protecting human beings. In some of the tests, the volunteers were pre-positioned with their torsos pitched forward in a simulation of a pre-braking attitude. One design feature that was changed after static deployment testing was to reduce the bag inlet diameter, slowing down the inflation rate and thus reducing the "aggressiviness" of the bag deployment. This change produced desirable results in the dynamic tests. No severe injuries other than abrasions, contusions, blisters, and erythema were sustained by the volunteers. The results from this research, together with the primate (baboon) tests conducted at Wayne State University, enabled GM to improve their air bag designs.

Early (1967-1978) Out-of-Position Occupant Concerns

Among the potential problems presented by the air bag, the possibility of producing injuries during bag deployment was viewed both by the industry and the NHTSA as the most serious. It was recognized early on by the Agency that not all occupants could be counted upon to be in the so-called "Normal Seated Position" (NSP) at the time of bag deployment. Indeed, the term Normal Seated Position had long been recognized as being a misnomer. That is, it was known that there exists a wide range of driving/riding positions under normal driving conditions. Further, it was realized that many frontal crashes are preceded by vehicle motions (e.g., swerving, panic braking) that afford opportunities for occupants - particularly unrestrained children - to be thrown into further so-called Out-of-Position (OOP) configurations at bag deployment. Indeed, to highlight this reality, in 1971 Ford submitted to federal docket 69-11 (the FMVSS 208 docket) a film from a study of panic-braking exercises with human volunteers (NHTSA 1971).

One of the earliest efforts by the NHTSA to address the problem of deployment-induced injuries was their research on aspirated-flow air bag inflators. This concept, initially evolved for use with inflatable aircraft escape chutes, involved the idea of entraining compartment air into the flow of gases to fill the bag (Southerland 1970). The NHTSA contracted with Rocket Research Corporation to explore the merits of this idea (NHTSA 1975). Elaborate inflator housings, provided with one-way flapper valves, were evolved in an effort to allow ambient air into the bag in the early stages of inflation, but then to prevent its escape later when bag pressures would rise. The hope was that this entrainment of air would provide an inflation system that would at least partially 'stall' if an OOP occupant were in the path of deployment. Unfortunately, the aspiration concept was never demonstrated to be of sufficient benefit to justify its added complexity.

The thrust of the air bag research sponsored by the NHTSA up until the late 1970s, however, was concerned not so much with deployment injury issues but rather with extending the barrier impact performance capabilities of air bag systems in protecting the NSP adult occupant beyond the initially-proposed 30 mph requirement. That is, the NHTSA was primarily interested in demonstrating that air bag systems could keep dummy injury measures below the proposed FMVSS 208 levels (head accelerations below 80 g - later HIC levels below 1000, chest accelerations below 60 g, and femur loads below 2250 pounds) at delta-Vs of 45-50 mph. This was felt by the research engineers at the Agency to be the next logical step in occupant protection since field crash studies were showing that the elimination of fatalities at impact severities up to a 30 mph delta-V would 'only' address about one-half of the deaths in frontal impacts. This emphasis on high-speed impact performance was both reflected in Experimental Safety Vehicle specifications and memorialized in an NHTSA-written paper entitled "Passive Protection at 50 MPH," authored in 1972 by several members of the Crashworthiness Research Section of the NHTSA (Carter et al. 1972).

Both the NHTSA and the researchers who would be developing these 'advanced' air bag systems were well aware that these systems, by their very nature, would be 'aggressive air bag designs' in terms of their deployment energies. Simple physics indicates that the occupant kinetic energy to be absorbed in a 50-mph delta-V crash is almost three times that to be absorbed at 30 mph. Thus, because of the view that 'exotic' restraint systems would not be necessary at the FMVSS 208 severity level, these researchers were focusing on the harder problem (higher speed protection) first and would only later be allowed to simplify the problem (down to 30 mph protection). This procedure runs counter to the normal design practice, wherein one solves the relatively simple problem first and then later evolves the design as technology improves, more is learned, and more performance is desired.

The primary reason that the NHTSA did not focus more early attention on the problem of deployment-induced injuries was that the emphasis within the Agency at this time was on evolving a passive protection system that would save the most lives with as high a benefit/cost ratio as possible. This mind set, together with a lack of appreciation for the negative public reaction to instances where occupants (particularly child occupants) were seriously injured or killed in otherwise survivable crashes, led to the early concentration on higher delta-V protection rather than on deployment problems. To fully appreciate this point, we must revisit the role of the Congress in the establishment and oversight of the Agency.

In the early 1970s, Congress made it clear to the NHTSA that any mandated safety feature, such as air bags, would have to be shown to have benefits to society that significantly outweighed their costs to consumers (i.e., the 'practicable' requirement for rule making was more rigorously defined). Benefit/Cost analyses and Societal Loss computations thus became the order of the day. A key element of these analyses was, of course, the assignment of dollar amounts to a wide range of injury levels, including the cost to society for a death. Insurance companies, of course, had been making such calculations for years. Within the context of the NHTSA mission, this meant that a total dollar amount (the 'Benefit') could be assigned to protect occupants to various levels of protection. This, in turn, could be compared to the estimated costs incurred in implementing these protection levels into vehicles. Moreover, the benefit could be computed for each occupant size for a given set of crash conditions. The Benefit/Cost methodology thus pointed the way toward higher delta-V frontal crash protection, since the relatively small numbers of OOP occupants who might represent a safety loss were overwhelmed in importance by the larger numbers of NSP occupants who would benefit from this 'enhanced' capability. Thus, the NHTSA was armed with a perfectly logical decision-making tool to resolve the OOP conundrum. The selected path was basically "do whatever you can to minimize the OOP problem without compromising the high-end delta-V protection." Stated another way, the NHTSA was making the assumption that the industry could somehow resolve the air bag aggressiveness problems, be it with FMVSS 208 (i.e., 30 mph) systems or the more aggressive 45-50 mph systems. Establishing a more analytical approach to policy making was in keeping with the trend at that time towards systems analysis, championed by Robert McNamara at Ford and then at the Department of Defense. Of course, the public does not necessarily react to a serious injury or death in an otherwise survivable crash (particularly that of a child) in terms of Benefit/Cost. The failure to fully appreciate public reaction would have significant consequences down the road.

In any event, to bid on these NHTSA contracts to develop "advanced" air bag systems, the bidding organization had to determine how they would conduct the specified tests to both develop these high delta-V systems and to demonstrate their success. This caused many organizations to invest their own and some government money to develop new sled and car crash facilities. Since full scale crash tests were expensive and buying complete HYGE sled assemblies from the Bendix Corporation (Skeels and Falzon 1962) could soon exhaust development budgets, a few brave companies decided to design their own sled and car crash facilities. Notable among these were Dynamic Science, Minicars Inc., and the Southwest Research Institute (SWRI).

A great deal of innovation was going on in these areas - all revolving around the air bag and the need to test it. Producing a reasonable crash pulse (vehicle deceleration time history) was a primary goal. This was not a big problem with car crashing since the crash pulse was part and parcel of the experience. However, in the case of sled testing, it was realized that one must design a method to generate the characteristic crash signature for a variety of vehicles. This led to a wide variety of innovative designs: deforming metal bands, bungee cords, hydraulic buffers, and the controlled rearward acceleration of the mounted sled carriage by the metered flow of air through a variable orifice. In accomplishing all these things in a tight time frame, some inevitable misadventures occurred that became a part of the air bag development legend. Sleds blew up, sled bucks flipped upside down during their 'crash' and cars became detached from their towing carriages. In an attempt to conduct angled car-to-car crashes, the vehicles occasionally actually totally missed each other. In this latter case, it became obvious in hindsight that, over a quarter of a mile, cable stretch can reach distances well over a vehicle length in dimension. Failure to take stretching into account thus caused more than one "whiff."

During the early 1970s, the NHTSA funded private contractors like Calspan of Buffalo, N.Y. and Minicars Inc. of Goleta, California to evolve developmental systems that were aimed at minimizing deployment times and maximizing system "stroke efficiency" (a measure of the uniformity of restraint loads during the occupant's forward translation) (Michie and Bronstad 1973, Fitzpatrick 1974, Shoemaker and Biss 1974, Fitzpatrick 1975). Unfortunately, during these early efforts only grudging attention was paid to two aspects of the air bag which, in great part, affect its feasibility - namely (1) crash sensing (i.e., the ability to accurately and quickly determine the necessity of deployment) and (2) deployment-induced injuries - particularly to the OOP occupant - including, of course, children. An overview of crash sensing issues and the sensor systems developed by the auto manufacturers and their suppliers during this period was presented by Struble in 1998. By way of contrast, virtually all of the NHTSA's air bag investigation programs employed either (1) simple, delayed-action switches (strip switches) to simulate what was then believed to be technically-feasible sensing times, or (2) sensors taken from early GM or Ford air bag vehicles for use in flat barrier vehicle crash testing. Thus, the variations in sensing times in single-vehicle pole/tree impacts or in vehicle-vehicle crashes involving offsets and/or over-ride were largely ignored. Also ignored were the potential benefits to be gained from what we would call today "Smart Sensors". This lack of consideration stemmed primarily from that fact that these devices were not technically feasible at this time.

That is, during the mid-1970s and into the 1980s, solid state electronics was in its infancy. This was amply demonstrated in 1974 with the introduction of the first electronic systems - seat belt interlocks - into production vehicles. Rampant system failures added to the widespread public disgust with these devices, and undoubtedly contributed to their being outlawed by Congress. They did portend, however, the widespread use of electronics that we see today. But at that time, sensors capable of detecting occupant sizes and positions and reacting to this information by making adjustments in the deployment mechanics of the air bag were simply not available to air bag researchers. This lack of sensor sophistication meant that these researchers had limited ability to make system adjustments to accommodate the OOP child occupant circumstance without unduly compromising the effectiveness of the systems in protecting the more numerous NSP adult passengers. In hindsight, perhaps if the NHTSA had included tests for OOP occupants in its early proposed FMVSS 208, the industry would have reacted to this action by initiating sensor development programs that, in turn, could have accelerated the development of solid state electronic sensors to assist in handling the OOP situation.

The single exception to the NHTSA's lack of sensor research was a program initiated in the early-mid 1970s to look at the feasibility of so-called "anticipatory sensors" - sensors that could detect an impending crash before vehicle contact so that the air bag systems could be deployed more slowly. This look into the feasibility of such systems was conducted by the DOT's Transportation Systems Center (now the Volpe National Transportation Systems Center) in Boston (Hopkins 1973). Unfortunately, the program was well ahead of its time and no reliable systems were uncovered.

Deployment concerns were not totally ignored by the rule makers and researchers at the NHTSA during the early 1970s. It was in this time frame that several engineers within the Agency, concerned that the OOP problem might prove to require years of research to solve, argued that a passive restraint requirement should logically begin with a drivers-only rule. Such a strategy, they reasoned, would yield about 2/3 of the benefits of an all-front-seated-positions rule (since vehicle occupancy was about 1.5 persons/vehicle) and would avoid virtually all of the concerns with deployment issues (the deployment problems with small adult drivers was not appreciated at that time). Unfortunately, a position within the legal department of the NHTSA was advanced that front seat passengers could not be denied the benefits being offered to drivers, and so the idea of phasing-in passive restraints in this manner did not gain any widespread support within the Agency until years later. This insistence on including front passenger protection in the early proposed passive restraint rules, in the opinion of the authors, probably delayed the introduction of air bag systems in production vehicles by ten years or more.

In keeping with the main thrust of the NHTSA's air bag research efforts during the mid to late 1970s, so-called 'advanced' driver and front seat air bag systems were evolved under the Agency's sponsored research that were capable of producing dummy injury measures within FMVSS 208 limits at barrier impact speeds approaching 50 mph. Researchers at Minicars Inc. evolved an experimental driver air bag system for a structurally modified 1974 Ford Pinto that was able to produce satisfactory (NSP) dummy injury measures in sled tests and car crashes (into a flat, rigid barrier) involving delta-Vs of 45-50 mph (Fitzpatrick 1974), see Figure 6. This driver restraint system later evolved into the system employed to protect the driver of the Minicars Research Safety Vehicle (RSV) at similar impact severities (Strother et al. 1976). The Minicars system, see Figure 7, employed a pyrotechnic inflator (capable of providing about 30 percent more gas than would be required for a system meeting the proposed FMVSS 208 requirements), a 'bag within a bag' system for chest and then head load distribution and energy dissipation, and a unique steering column system that possessed increased energy absorption capabilities and was oriented optimally for driver upper body restraint.

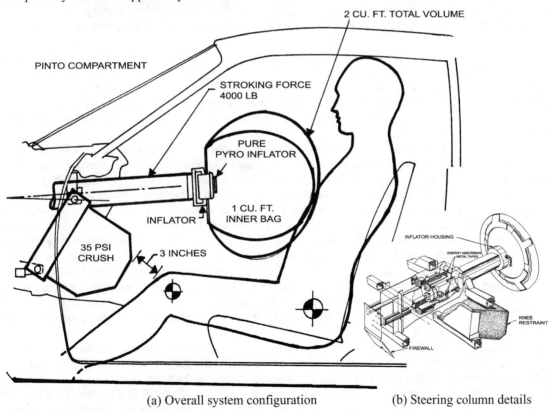

(a) Overall system configuration　　　　(b) Steering column details

Figure 6. The Minicars Pinto driver airbag system (ca. 1974).

Figure 7. The Minicars Research Safety Vehicle (RSV) and front passenger airbag systems (ca. 1976).

Another RSV project was conducted at Calspan Corporation in Buffalo, New York in cooperation with the Chrysler Corporation (NHTSA 1980). Unlike the Minicars RSV, the Calspan/Chrysler RSV was not a new vehicle but rather a structurally modified production car, namely a Simca 1308. The Calspan/Chrysler RSV was designed with a driver air bag system that had a similar capability to that evolved by Minicars. That is, in tests conducted under FMVSS 208 test conditions, but at speeds of about 50 mph, the compliance dummies in the NSP were able to pass the FMVSS 208 injury criteria.

The Minicars researchers also evolved a unique front passenger air bag system, first for integration into the modified Ford Pinto (Fitzpatrick 1975) and later for the Minicars RSV (Fitzpatrick 1977, Strother et al. 1976). This system, also shown in Figure 7, was comprised of a so-called 'mid-mount' air bag system (i.e., the bag was mounted relatively high on the vertical face of the dashboard and was deployed virtually straight rearward), a two-chamber air bag system (a lower chamber for chest protection and an upper chamber, which was inflated solely with gas vented from the lower chamber, for head protection), and a 'stroking' dash element that contained both the air bag and mechanical knee restraint systems. The stroking dash was intended to function like the steering column of the driver restraint system, allowing the air bag system to translate forward during the impact, providing additional 'stroking' distance, accomplished at a higher level of energy absorbing efficiency. The combination of the separately tailored head and chest air bags plus the mechanical efficiency of the stroking dash combined to allow the design engineer to obtain great latitude in tailoring each for achieving minimum injury levels. Later, this system was adapted to the Minicars RSV, wherein the stroking dash feature was found to be unnecessary and thus eliminated (Strother et al. 1976).

Following the end of the RSV air bag work in the late 1970s, the NHTSA sponsored further air bag development work at Minicars Inc. to adapt the systems developed for the RSV to several late-1970s small cars. The first of these programs was focused upon using the insights gained in the RSV work to develop driver air bag systems for three late-1970s compact cars that would meet the proposed FMVSS 208 requirements (Strother and Broadhead 1978). Typically, the program was focused exclusively on the fixed rigid barrier (FRB) environment and presupposed the feasibility of bumper-mounted sensors. Figure 8 is an illustration of one of the driver air bag systems evolved. It was during this first small-car driver air bag program that attention was first directed to the protection of an out-of-position (OOP) occupant. That is, a contract modification was instituted to explore the feasibility of developing a front passenger air bag system for one of these small cars (the Vega) wherein the protection of an OOP child was considered. The system evolved for the Vega was based, in part, upon the earlier RSV passenger air bag system and is illustrated in Figure 9. The OOP tests used to develop the Vega system were designated OOP Cases "A" and "C", see Figure 10a. Position "A" employed a 6-year-old dummy, seated upright on the front edge of the seat. Position "C" was similar to "A", except that the torso was pitched forward. The results of this first attempt to reconcile OOP child protection with protection under the proposed FMVSS 208 conditions were not successful (Strother and Zinke 1978).

Following this initial attempt at addressing the OOP child problem, further, more successful, efforts were pursued. Under a subsequent contract, researchers at Minicars, Inc. began working on passenger air bag systems for small cars that could reconcile these two diverse restraint conditions (Zinke 1980). The first of these efforts concerned the Chevette and the Dodge Omni and were targeted for 30-35 mph protection for NSP adult occupants as well as a measure of OOP child protection. Again, the focus of the program was exclusively on the FRB environment and again, the feasibility of bumper-mounted sensors with their very fast response times was presumed. The OOP child testing at first focused on the 6 yr-old dummy in OOP Positions "A" and "C" as defined earlier, but later the focus was shifted to a test configuration in which a 3-year-old dummy was standing on the floor in front of the seat, see Figure 10b (Zinke 1981a).

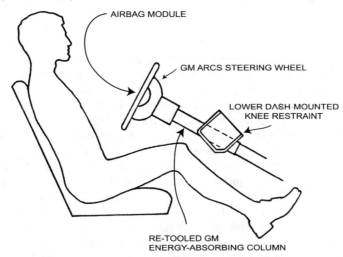

Figure 8. The Chevette driver airbag system designed by Minicars under NHTSA contract (adapted from Strother and Broadhead 1978).

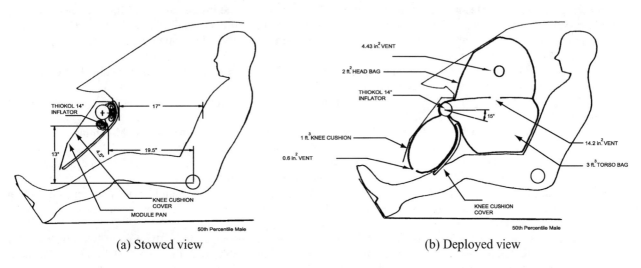

(a) Stowed view (b) Deployed view

Figure 9. Front passenger airbag system evolved for the Vega by Minicars under NHTSA contract (adapted from Strother and Zinke 1978)

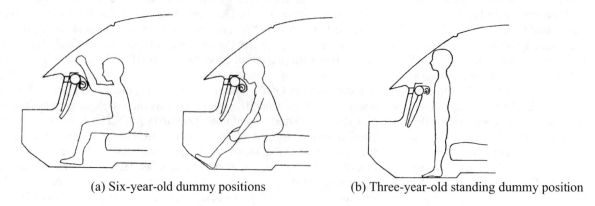

(a) Six-year-old dummy positions   (b) Three-year-old standing dummy position

Figure 10. The OOP child positions tested under the Minicars/NHTSA front passenger airbag investigations (adapted from Zinke 1980, 1981).

Although the program was declared a success by the researchers, an honest appraisal of the work done would conclude that, although progress had been made, the issue of OOP child protection was still unresolved. That is, the researchers were laboring under constraints that did not allow them to totally solve this problem. First, because budget constraints required the program to deal exclusively with the flat rigid barrier environment, the problems associated with collisions with objects that could produce disadvantageous crash signatures and sensor response times had to be ignored. In fact, as was the case in the prior Vega program, no real sensor work was accomplished, sensor investigation being outside the scope of these efforts. Rather, what were believed to be reasonable sensing times in the flat barrier crash environment (i.e., sensing times of 8-10 ms) were simulated in the programs via delayed-action strip switches. Thus, the air bags in this program were always triggered relatively early in the crash sequence; a situation entirely appropriate in terms of the barrier environment but one not to be anticipated in the real world. Secondly, there was very little available in the way of analytical tools (i.e., computer models of the deploying air bag and occupant) and thus virtually all of the exploration had to be done using relatively expensive dynamic testing. Because of this total dependence upon experimental methods, deployment-induced injuries with the evolved passenger air bag systems were only partially explored due to funding limitations. That is, only a limited number of child OOP situations were investigated, see Figure 10, with an incomplete understanding as to whether or not any of these selected situations truly represented the 'worst case' scenario. Further, due to the state of the art in child injury criteria and dummy development, these experiments had to be conducted with tools and techniques of dubious validity. Nevertheless, the investigation did find that changing bag folding patterns, inflator mass flow rates, and diffuser orientations had beneficial effects for the OOP situations investigated while, at the same time, not unduly compromising the protective capabilities of the system in the standard FMVSS 208 environment.

The NHTSA-sponsored air bag development efforts discussed above (i.e., those accomplished during the period up to the very late 1970s) were accomplished almost exclusively by experimental methods. That is, the typical program would begin with an initial series of developmental sled tests. During this testing, the results of prior sled experiments would be studied by the investigator(s) and they would then make an educated guess at how the system could be improved by altering one or two system parameters. This process would continue until some satisfactory level of performance was achieved, whereupon a series of evaluation tests - sled tests and a limited number of FMVSS 208-type vehicle barrier tests (at speeds higher than 30 mph) - would suffice to define the performance characteristics of the system. A few crude computer simulation models of the interaction of an occupant and air bag were available (e.g., the ABAG19 Program - a model of a single mass occupant interacting with a deployed cylindrical air bag), but they were viewed as insufficiently realistic to be of any great value in the development effort (Carter et al. 1972).

Starting around 1978-79, however, computer simulation models began to evolve into truly useful tools for this research. The first of these tools developed under NHTSA sponsorship were the DRAC (DRiver Air Cushion) and DEPLOY (Passenger air cushion model) programs developed by Fitzpatrick Engineering (Fitzpatrick 1980). The DRAC program, shown schematically in Figure 11, was a two-dimensional simulation of a three-mass driver interacting with a deploying elliptical air bag mounted on a stroking (and perhaps binding) steering column. The DEPLOY program, shown schematically in Figure 12, was another essentially two-dimensional program of a two-mass passenger chest interacting with a deploying cylindrical air bag mounted on a stroking or crushing dashboard. Once programs such as this were available to the engineers, they could 'tweak' variables to their heart's content in an effort to better understand the relative importance of and interactions among the system parameters. This, in turn, led to the identification of 'optimized' sets of system parameters that would be predicted to result in the lowest dummy injury measures over the spectrum of test conditions. With the insights and guidance from these simulations, sled testing became much more efficient. And with feedback from the sled testing, the models became more and more realistic. In addition, this type of analysis led to some of the early warnings about the OOP child problem.

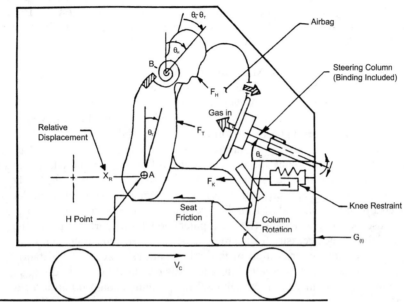

Figure 11. Schematic of "DRAC" computer model (adapted from Biss et al. 1980).

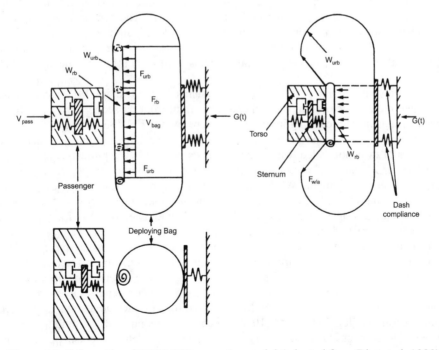

Figure 12. Schematic of DEPLOY computer model (adapted from Biss et al. 1980).

In the late 1970s, the NHTSA embarked upon an experimental safety vehicle program titled the Modified Integrated Vehicle Program (MIV). The objective of the MIV program was to evolve a safety vehicle that would be more practical, nearer-term than the predecessor RSV. Unlike the RSV, the MIV would start off being a structurally unmodified production vehicle. This vehicle would then be equipped with various safety features, principal among them being a side impact protection system meeting the contemplated FMVSS 214 requirements and an 'advanced' air bag system for the front seat passengers. The significance of the MIV program in the context of NHTSA air bag history is that the frontal impact protection goals differed from those of the RSV in two important aspects. First, the high-speed impact protection goal for the MIV was 35-40 mph delta-V, not the 45-50 mph of the earlier experimental vehicle. Secondly, the passenger air bag system was to give increased importance to the OOP child occupant. Thus, the MIV program signaled a shift in the NHTSA focus, moving away from high speed protection for the NSP occupant (although a 40 mph system still represented a system with about 75-80% more energy capability than a FMVSS 208 system) and toward the protection of the OOP child occupant (the latter not being a requirement in the RSV program).

To support the air bag work on the MIV program, the Small Car Front Seat Passenger Inflatable Restraint System project was further modified to include work on driver and front passenger restraint systems for the Chevrolet Citation, the vehicle selected for MIV work by Minicars, one of the two MIV contractors (the other being VW, which evolved a safety vehicle using the VWRA-equipped Rabbit as the base vehicle) (Zinke 1981b). The Citation was selected primarily based upon its performance in 35 and 40 mph FRB tests done under the New Car Assessment Program (NCAP). At the 8th International ESV Conference in 1980, the NHTSA and its contracted researchers presented a report summarizing the work done in evolving driver and passenger air bag systems for the Chevrolet Citation (Biss et al. 1980). This summary report is considered significant by the authors of the current paper for the following reasons. First, it presents a clear shift in emphasis from 50-mph delta-V protection to OOP child protection. Secondly, it presents a shift from the previous, purely experimental air bag development efforts, to investigation making use of analytical tools (i.e., computer models) to perform parameter studies, guide system development, and to minimize the necessity for the more expensive sled and car crash testing.

The system evolved for the protection of the right front passenger of the Citation MIV was a dual level inflation system wherein the so-called 'low-level' inflation was provided by a cylindrical sodium azide inflator and the 'high-level' inflation was provided by the additional (and simultaneous) inflation of a solid propellant driver inflator. That is, the investigators clearly saw problems with a single-stage "40-mph" passenger air bag system in terms of OOP child protection. In the evaluation tests, all of the inflations were of the 'high-level' type, save the OOP test. The decision whether to deploy in the 'high' or 'low' mode was (conceptually) made by the sensor system (again, no actual sensor work was done under this program). The sensor system was envisioned to be comprised of a relatively sensitive sensor or set of sensors that would initiate the cylindrical inflator (i.e. the 'low' mode) with a threshold delta-V of about 12 mph, and a relatively insensitive sensor(s) (located forward of the first sensor(s)), which would always simultaneously initiate the booster (driver) inflator at delta Vs of about 28-29 mph or more.

The efforts to evolve driver and passenger air bag systems for the Citation MIV were considered successful by both the NHTSA and the contractors. In the context of the right front passenger air bag system, this meant that the restraint system was capable of producing injury measures below the FMVSS 208 injury criteria with a NSP 50th percentile male dummy during a 40 mph FRB impact while, with the same system, producing acceptable injury measures during a series of evaluation sled tests. The sled tests included 30 mph tests with all three adult dummy sizes (5th percentile female, 50th and 95th percentile male) and one child dummy (6 yr. old) in the NSP, and a single OOP test (standing 3 yr. old) at 12 mph. While the Biss et al. report is considered significant for the historical reasons identified above, again it would be inappropriate to conclude that the NHTSA or its contractors "solved" the problem of how to design a passenger air bag system that would simultaneously provide high-speed protection for adult NSP occupants and not create deployment-induced injuries for OOP children; the reason being simply that the researchers were operating under significant handicaps. First, as was the case in the prior NHTSA programs, the program focused exclusively on the flat barrier environment. And again, no real sensor work was conducted (although it was indicated in the ESV paper as being planned[2]) to explore whether the optimistic characteristics of the hypothetical sensor system could ever be achieved in practice. This deficiency is even more critical in the MIV program, due to the more demanding sensor system hypothesized (one that would always simultaneously trigger the main and booster inflators in the 'high' mode at 8-10 milliseconds). Second, realistic funding limitations necessarily limited the crash conditions investigated to a very small number of delta-V, crash pulse, sensing time, crash direction, occupant size, and occupant position combinations. Third, due to the timing of the study, the state of the

---

[2] However, no such sensor work was ever conducted, either by the NHTSA or its researchers.

art with regard to child dummy construction, instrumentation, and injury criteria was such that drawing conclusions about OOP child dummy test results would be risky at best.

Nevertheless, these researchers do deserve credit for their work both in evolving better analytical tools for air bag development and in providing valuable early insights into what air bag parameters would be critical in achieving satisfactory performance. Among these findings was the conclusion that the early gas mass flow rate was an important factor in determining OOP protection (and hence that the standard tank test was inadequate to define inflator output). Other important findings included the conclusion (perhaps obvious) that the initial occupant/bag standoff was a critical factor over the first 12 inches, the finding from the analytical study confirming the minimal effect of aspiration, and the interesting conclusion that bag fabric weight (and hence bag folding technique) was only important if the early gas flow rate was relatively high. Perhaps most importantly, it was discovered that there were important interactive effects among the air bag parameters that were as important as the individual parameter effects themselves.

Airbag Regulatory Efforts, 1983-Present

In the fall of 1983, as a result of the rulings by the judiciary, the NHTSA issued an NPRM that outlined three alternative courses of action on frontal occupant protection being considered (NHTSA 1984). First, the Agency might reinstate the passive requirements of FMVSS 208, establishing a new implementation timetable. Second, it might amend the passive restraint requirements of the proposed standard to preclude detachable passive belts (although just how this would be accomplished in view of the 'interlock' law was not clear). Lastly, the NHTSA was considering rescinding the passive requirements altogether. This notice received thousands of comments and eventually led to the issuance of a Final Rule by Secretary of Transportation Elizabeth Dole. The Dole decision confronted some harsh realities with respect to seat belts. On the one hand, they offered the greatest protection of any single restraint system (when used) at little or no additional cost. On the other hand, belt usage levels in the early 1980s were low, and the Agency could not forecast either widespread usage of, or refusal to use, automatic, detachable belts. A mandatory belt use law (MUL) (with "teeth," and enforced accordingly) was an obvious remedy, as had been demonstrated in many other countries around the world. But the federal government could not enact and enforce such a statute. The requisite authority rested in the various states, only a few of which had enacted MULs, and weak ones at that.

Adding to the conundrum was the uncertainty whether the American public would accept (or emphatically reject) air bags or automatic belts. Again, the "interlock" experience gave the NHTSA good cause to take full account of public and Congressional reaction. In what seemed like a bizarre linkage at the time, Ms. Dole reinstated the passive restraint requirement but gave the country the "out" of avoiding passive restraints if the States were willing to pass MULs meeting certain conditions that cover two-thirds of the U. S. population.

This action was the strongest statement yet from the federal government regarding the benefits of belt use. It led to the passage of MULs in Puerto Rico, the District of Columbia, and a number of states. As reported to Congress in 1993, by January of that year, the number of states with safety belt use laws on the books had risen from zero in 1983 to 42 states, the District of Columbia and Puerto Rico. As of the writing of this paper, this number has risen to 49 (New Hampshire being the only hold-out). Figure 13, based upon data from the NHTSA, shows how seat belt usage climbed from 1983, when Secretary Dole made her decision, to 1998. A couple of interesting points need to be made about this Figure. First, in 1983, the year of the decision, the use of lap and shoulder belts by front-seat occupants was only about 14 percent. Thus the potential benefits to be accrued from passive protection were high. By the time the passive restraint phase-in began, MY 1987, this usage rate had risen to over 40 percent, and by the time it was complete, MY 1990, belt usage was close to 60 percent. Thus, the potential benefits of passive restraints were dropping dramatically. Although the rise in belt usage has apparently leveled off, recent seat belt usage figures are in the range of just over 70 percent. The population of unbelted occupants has thus dropped by a factor of about 3 since the Dole decision.

With the reinstated passive restraint requirements came a new phase-in schedule for their implementation. The phase-in was based upon production percentages rather than vehicle size, thus giving manufacturers much more flexibility to account for major model changes, production volumes, and marketability. It was specified to begin in MY 1987 and to be completed in MY 1990. To encourage manufacturers to install air bags rather than passive belts, Dole created a credit program during the phase-in period. Passive-restraint vehicles equipped with driver air bag systems during the phase-in counted as 1.5 vehicles. About a year later, an additional air bag credit incentive was adopted wherein a phase-in vehicle with a driver air bag system and a manual belt system for the front passenger counted as one whole passive restraint-equipped vehicle.

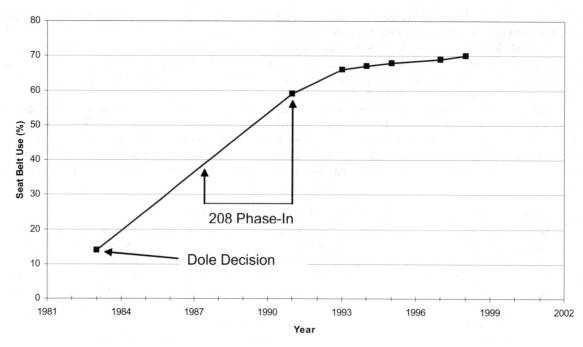

Figure 13. The increase in seat belt usage, 1984-1998 (based on data from NHTSA 1993a, 1996a, 1999).

Once again, the federal court was asked to rule upon the legality of the new FMVSS 208. And once again, in a case called State Farm v. Dole, the government decision was upheld. The Dole decision stood and by MY 1990 several new passenger vehicle models offered for sale in the United States came equipped with air bag systems for front seat occupants.

Along the way, public acceptance came into play, although perhaps not in the way anyone in Washington anticipated. This occurred when Lee Iacocca, a vigorous opponent of air bags when he was with Ford, did a crisp about-face and announced in 1988 that his then-employer, Chrysler, would have driver-side air bags in all its domestically-produced MY 1990 cars. Mr. Iacocca, an astute observer of the car-buying public, had apparently decided that people wouldn't like passive belts but would enthusiastically endorse air bags. In fact, the public went beyond endorsement and virtually equated inflatable devices to crash safety, despite the technical problems that would eventually surface. None of this could have been predicted in 1983.

The Intermodal Surface Transportation Efficiency Act (ISTEA) was enacted by Congress on December 18, 1991. The Act called for all passenger car and light trucks to have driver and passenger air bags besides lap and shoulder safety belts. Thus, the federal government would require not just "passive restraints" but specifically air bags. For passenger cars, this requirement was to start in MY 1998; for light trucks, MY 1999. Congress also charged the Department of Transportation to report the effectiveness of occupant protection systems based on their actual usage. The report was to cover lap and shoulder belt usage by the public.

During the early 1990s, reports of injuries and fatalities to front seat occupants from air bag deployments began to appear in the news. These incidents were mainly concerned with unbelted front seat passengers, including small adults and children. As the numbers of vehicles equipped with air bags climbed, so did these reports of deployment injuries. As its first regulatory action in response to this problem, the NHTSA, in 1991, issued a Final Rule amending FMVSS 201 to facilitate the installation of top-mounted, vertically-deploying front passenger air bags (NHTSA 1991a). This rule change was in response to a petition by Chrysler, which wanted to install this type of air bag system to reduce the risks for OOP occupants. Ford and GM supported the Chrysler position. It was also during this time frame that the NHTSA became acutely aware of the potential problems encountered when parents put a rear-facing child seat in the right-front seat of a vehicle with a passenger-side air bag. That is, it became clear that a deploying air bag could very well injure an infant severely by impacting the child seat and the child's head. This was in addition to the already-recognized concerns about protecting unbelted or improperly belted children in the front seats of air bag-equipped vehicles. In one of the Agency's first actions to address this problem, the NHTSA issued a Consumer Advisory, warning owners of rear-facing child seats not to place these seats in the front passenger position of air bag-equipped vehicles (NHTSA 1991b).

In January of 1993, the NHTSA made its first ISTEA report to Congress (NHTSA 1993a). One of the principal findings in the report was that the combination of increased belt use (from 14 to over 59 percent) and air bags reduced the fatality risk by about 23 percent. Also in 1993, the NHTSA issued its Final Rule implementing the air bag mandate required by ISTEA (NHTSA 1993b). As part of this rule, vehicles were required to attach warning labels on the sun visors providing four specific cautions. These cautions included a statement not to install rearward-facing seats in the front passenger position and directed the owner to read the owner's manual for further information and explanations. Early in the following year, the NHTSA amended FMVSS 213, Child Restraint Systems, to require rear-facing child seats to bear a warning against using this device in the front seat of an air bag equipped vehicle (NHTSA 1994).

Concerned that these actions were insufficient, the Alliance of Automobile Manufacturers (the Alliance) met with the NHTSA on January 24, 1994 to signal a need to do more to avoid a possibly dangerous situation. The Alliance asked for the meeting to explore the prospect of allowing a cutoff switch for the front passenger air bag. The industry suggested using manual switches because they assumed automatic cutoff switches were not ready for production. After considering these industry concerns, the NHTSA issued an NPRM in October 1994 proposing to permit such manual switches in vehicles with no rear seats. A final rule was published in May of 1995 (NHTSA 1995a). The final rule allowed manufacturers, for a limited time (up to September 1, 1997), the choice of installing a manual device in certain vehicles. Motorists could use the device to turn off the passenger air bag when an infant restraint had to be placed in this position. The capability to deactivate the passenger air bag would thus enable parents to safely take advantage of rear-facing infant seats in these vehicles. Later that year, in response to several fatalities to improperly-restrained children in air bag equipped vehicles, the NHTSA issued a strong warning in a press release (NHTSA 1995b). This release broadened the previous agency warnings about young children to apply to older children and even adults who were not using their seat belts.

In February of 1996, NHTSA published its second ISTEA report to Congress concerning the effectiveness of occupant restraint systems. This report gave four major findings (NHTSA 1996a). The first three of these findings concerned the Agency's estimates of the fatality and injury-reduction effectiveness of the air bags. The fourth major finding was that seat belt usage was continuing to climb and had reached about 67 percent by the end of 1994. Thus, none of the major findings concerned deployment-induced injury.

On August 8, 1996, the NHTSA again reacted to the reports of deployment-induced injuries by issuing an NPRM to further modify FMVSS 208. These modifications were (1) to include the requirement for warning labels, and (2) to allow manufacturers to install manual cutoff switches for the passenger side air bag in virtually any passenger vehicle (NHTSA 1996b). The purpose of these modifications was "to reduce the adverse effects of air bags, especially those on children." The warning labels would be placed on sun visors and instrument panels, and better labeling would be placed on child restraint systems (CRS). The manual cutoff switch was proposed for all vehicles mainly to allow the accommodation of rear-facing infant seats in the front of any vehicle. It was made clear in this notice that the need for these measures was viewed as temporary by the NHTSA - temporary because the future use of so-called "smart air bags" would make the labels unnecessary. The definition of smart air bags was given in the notice several times. Basically, as defined by the Agency, a smart bag was one that automatically prevents it from injuring either infants in rear-facing infant seats or unbelted or improperly-belted OOP children. This would be accomplished by either not deploying at all in such situations or by deploying at a reduced energy level. On November 27, 1996, the Agency issued its final rule concerning the proposed modifications. The requirements for labeling were basically as specified in the NPRM, but the requirement for extending the cutoff switch option to all vehicles was dropped (NHTSA 1996c). Thus cutoff switches continued under the earlier rule to apply only to vehicles with no rear seat.

At the end of 1996, it was time again to report to Congress on the effectiveness of occupant restraint systems (NHTSA 1996d). The NHTSA covered many topics in this report, including the issuance of the proposed NPRM to amend FMVSS 208 to include warning labels (see above). The major findings in this report dealt with the fatality-reducing effectiveness of driver air bags, but also, for the first time, dealt with deployment injuries to children in the front passenger position. The NHTSA concluded that these injuries (31 as of the report), with one exception, were either (1) sustained by infants that, despite all of the warning to the contrary, were in rear-facing CRS or (2) unrestrained or improperly restrained older children. The following month, January of 1997, the Agency issued two separate NPRMs and one final rule, all aimed at reducing the incidences of deployment-induced injuries (NHTSA 1997a, NHTSA 1997b, NHTSA 1997c). The final rule extended for three years the time period during which manual cutoff switches for passenger bags would be allowed from September 1, 1997 to September 1, 2000. The idea of extending the cutoff switches to all vehicles, not just vehicles without an effective rear seat, was considered. Indeed, the idea was supported by several parties, including two vehicle manufacturers, two air bag sensor suppliers, and a CRS manufacturer. More vehicle manufacturers, as well as the NHTSA, however, were opposed to this idea, feeling that the switches would be misused and thus their presence would do more harm than good. Hence, the proposal to extend cutoff switches to all vehicles was not incorporated into the Final Rule.

That same day an NPRM was issued proposing to depower future air bags. The Agency studied the consequence of two different strategies to allow the automotive manufacturers to design less-aggressive air bags under FMVSS 208. The first possible approach was to increase the allowable acceleration in the chest of unbelted dummies from 60 g to 80 g. The second possible approach was to relax the test requirement for the unbelted dummy of FMVSS 208 through the use of a generic 30 mph sled test with a relatively mild crash pulse of 125 milliseconds duration. On the one hand, it was recognized that belt usage in 1997 was markedly improved over that in 1984 when the Dole decision to phase in air bags was made. This was important because deployment aggressivity is dictated by the need to arrest the forward motion of the unbelted 165 pound dummy in the 30 mph barrier impact. As we have discussed earlier, concentrating efforts on unbelted occupants made much more sense in 1984 when usage levels were only around 14 percent. On the other hand, crash statistics indicate that, in 1998, almost two-thirds of those occupants killed in frontal collisions were belted (Stocke 1998).

The third action taken in January of 1997 was a second NPRM proposing to allow the deactivation of manufacturer-installed air bags under specific circumstances. Under this proposal, dealers and repair businesses, upon written authorization from a vehicle owner, would deactivate either or both of the vehicle's air bags. Deactivation would only be allowed if the vehicle did not have a cutoff switch or if the vehicle did not have 'smart' air bags. Warning labels were also proposed to be required as a condition of deactivation. The vehicle manufacturer, under this proposed rule, would have to be informed of this action by the deactivating party, who would forward the owner's authorization to them.

On March 19, 1997, the NHTSA issued a Final Rule, in effect calling for the depowering of manufacturer-installed air bags (NHTSA 1997d). The 125-millisecond duration sled test approach was adopted as the means for achieving this end. The rule change was indicated to expire on September 1, 2001; the idea being that, by this time, 'smart' air bags would be available to make depowering unnecessary. In issuing the rule, the Agency went on record as establishing a policy that it was not acceptable for a safety device to cause fatal injuries in an otherwise survivable crash, particularly if the occupants being killed were children. Thus the NHTSA was fully prepared to accept the fact that some unbelted teenage and adult occupants would not be saved in the effort to avoid fatally injuring children. Having some indications that some vehicles might be able to pass the more lenient requirements of the sled test without the use of air bags, the Agency added a neck injury criterion to ensure that this did not happen.

On November 18, 1997, the NHTSA issued the final rule on the deactivation proposal (NHTSA 1997e). This rule stated that, as of January 19, 1998, manual air bag cutoff switches, which can be retrofitted into older vehicles, will be available to qualifying consumers (e.g., those with a medical condition, those whose size and vehicle prevent them from distancing themselves at least 10 inches from the steering wheel, or whose vehicle has no effective rear seating area). The rule also exempted dealers and repair shops from the statutory prohibition against deactivating federally-mandated safety equipment. Unlike the proposed rule, the Final Rule called for the installation of cutoff switches rather than permanent deactivation.

About 1997, the industry began to discuss new restraint designs that could possibly alleviate trauma to an out-of-position occupant. These discussions typically suggested that current restraint design focused on the mid-size male dummy in a rigid wall crash at 30 mph. The new approach would be to combine the restraint hardware with acceptable sensing technologies and control techniques directed by a decision-making procedure. In particular, these intelligent restraint systems might detect an out-of-position occupant and lower the rate of air bag inflation in a vehicle crash (e.g., Musiol et al. 1997).

On June 8, 1998, the September 1, 2001 sunset date for the sled test option (i.e., the 'depowering' provision) of FMVSS 208 was superseded by a provision in the Transportation Efficiency Act (TEA - 21, the successor to ISTEA enacted by Congress). In a paragraph titled "Coordination of Effective Dates," TEA - 21 provided that the unbelted sled test option "shall remain in effect unless and until changed" by the final rule for advanced air bags.

The NHTSA's fourth ISTEA report to Congress came out in May 1999. As of December 1997, the national safety belt use rate was approximated as 69 percent. Many major findings were similar to those in previous reports to Congress. Two new findings discussed the protection offered by the different combinations of occupant restraints and the tradeoff between the decreased risk of neck injury versus the increased risk of arm trauma that accompanies air bag use (NHTSA 1999).

On September 18, 1998, the agency published yet another NPRM in the Federal Register (63 FR 49958) proposing to upgrade FMVSS 208 to require advanced air bags (NHTSA 1998). To reduce the risk to small children and others positioned too close to a deploying air bag, new tests would be required. The proposed new tests use the 12-month-old infant, 3-year-old, 6-year-old, and 5th percentile female dummies. Several tests would be used to assess the protection afforded these smaller occupants. The purpose of the proposed additional tests was to have the air bag either not deploy or else deploy at a low enough level to present a minimal probability of trauma.

Three alternative sets of six high speed tests were offered to define air bag protection, all but two of which would be full-frontal tests. Each alternative was divided into three tests to be run with unbelted ATDs and three tests with belted dummies. The first set of "unbelted" tests would be run at delta-Vs either in the range of 20-25 mph (Alternatives 1 and 3) or at 20-30 mph (Alternative 2). The first "unbelted" test would employ the small, 5th percentile female dummy, the second the mid-sized, 50th percentile male. The third and last "unbelted" test would be under similar conditions to the second test, but would be a 30-degree oblique impact. The second set of "belted" tests would start out with two tests using the 5th percentile female ATD. The first test would be a full frontal impact at a delta-V up to 30 mph, and the second test would be an offset frontal impact at a delta-V up to 25 mph. Finally, the third and last "belted" test would use the 50th percentile male ATD in a full frontal impact at a delta-V up to 30 mph (Alternatives 1 and 2) or 0-35 mph (Alternative 3). Alternative 3 was chosen for the Final Rule, which was issued in May of 2000 (NHTSA 2000a, see also Hinger and Clyde 2001 for a summary of the requirements in this version of FMVSS 208). Additionally, the final rule established new injury criteria for the mid-size male, small female, 6-year-old child, 3-year-old child, and the 12-month-old infant. The NHTSA estimated that the adopted Alternative 3 test requirements would reduce fatalities by 117 to 215 per year (NHTSA 2000b).

NHTSA Research Efforts, 1982-Present

As discussed previously, starting around 1982-1983, the NHTSA began conducting the vast majority of its research in-house. The facilities available to conduct this work existed in basically two locales. First, there was the NHTSA headquarters in Washington, D.C. This facility was staffed by a number of engineers, statistical analysts and the like, armed with some computer equipment and library resources, but it contained no laboratory facilities. However, the Agency did acquire these capabilities when, in the late 1970s, it relocated the Safety Research Laboratory (which had been in Riverdale, Maryland). This entity was combined with the Engineering Testing Laboratory (a small group doing special defect investigations) to create the Vehicle Research and Test Center (VRTC) in East Liberty, Ohio, on the grounds of the newly developed Transportation Research Center (TRC), a state-run facility. That is, in addition to a staff of engineers, most of whom were engineers transferred from NHTSA headquarters to staff the new operation, VRTC had access to the shop, sled, and car crash facilities at TRC. Starting around 1982 or 1983, these two entities, VRTC in Ohio and DOT headquarters in D.C., conducted virtually all of the Agency's air bag research.

Also, from the early 1980s onwards, the air bag development research conducted by the NHTSA took on a different character from that of the previous efforts. Instead of the research being directed toward a clearer understanding of the possibilities and limitations of air bag designs, the Agency's research efforts, starting in the early 1980s, were focused primarily towards the evaluation of production air cushion systems. These investigations involved monitoring industry air bag research and development programs, conducting tests of production vehicles with retrofitted air bags, and analyzing the performance of air bag systems using computer models.

Three investigations to look at the performance of retrofit air bag systems were conducted by the NHTSA in the early 1980s. The first of these programs was carried out in 1982, when the Agency conducted a series of crash tests to evaluate the performance of a driver air bag system that had been retrofitted into selected police vehicles (NHTSA 1983). The system, developed by Romeo Kojyo Company Ltd., consisted of an air bag module, steering wheel, and auxiliary knee bolster. Five FRB impacts and one pole impact were conducted under the program, all of which utilized the 50th percentile male compliance ATD in the compliance posture (i.e., the NSP). The second of the programs looking at retrofit air bag systems was conducted in 1983. This test program was comprised of 29 sled tests using two sled bucks, one representing a 1982 Plymouth Gran Fury/Dodge Diplomat and the other a 1978 Ford LTD (Esser 1983). All three adult-sized ATDs, all postured in the NSP, were used in the test program and both the belted and unbelted condition were tested. Unfortunately, this test program was never written up in a report or paper in the open literature. In 1984-85 the NHTSA conducted a third retrofit "demonstration program," this time using the Breed retrofit driver air bag system (Brantman 1984). The objective was to establish the feasibility of developing viable driver air bag systems that would operate using a mechanical sensor contained within the gas generator. The testing portion of the program utilized three Ford LTDs, a Dodge Diplomat and a Chevrolet Impala, all of which had been provided with the Breed retrofit system. Testing consisted of rough road driving tests, impacts into dirt ditches, and a series of crash tests. The analytical portion of the program was comprised of a computer-assisted investigation to evaluate the predictability and timeliness of air bag deployments.

The NHTSA's Vehicle Research and Test Center conducted a program in the early 1990s to look at the ability of production passenger air bag systems to provide frontal impact protection to CRS-restrained children (Sullivan 1992). This study utilized the HYGE sled programmed with the FMVSS 213 deceleration pulse. The air bags tested included mid-mounted and top-mounted systems. The four types of CRS used included car beds, rear-facing infant seats, convertible infant/toddler seats, and booster seats. Child dummies representing a new born infant, 9-month-old and a 3-year-old were employed as test subjects. The results indicated that, while the 3-year-old in the booster seat was receiving some benefit from the air bag, infants in both rear-facing and forward-facing CRS were subjected to high HIC and chest accelerations. It was on the basis of these test results that the Agency issued its December 1991 consumer advisory warning referred to earlier about placing infant seats in the front seat position of air bag equipped vehicles.

In August of 1996, Dr. Charles Kahane, an Agency researcher working at NHTSA headquarters, analyzed the difference in performance for air bag and non-air bag vehicles (Kahane 1996). The statistical analysis was based on the Fatal Accident Reporting System (FARS) data during the years 1986 to 1996. He compared the fatality risk to front-seat occupants in passenger cars and light trucks fitted with air bags to the analogous risk in like vehicles without air bags. The fatality risk reduction of air bags for all drivers was found to be about 11 percent. Airbags were credited with providing roughly the same benefit for passengers, age 13 or older, as for drivers. Kahane found the overall fatality risk reduction was similar in light-weight, medium-weight, and heavy cars. Based on the data available in those eleven years, air bags appeared to have about the same lifesaving benefit in light trucks and vans as in passenger cars.

In November of 1996, NHTSA published its analysis of the safety trade-off associated with the use of cutoff switches. For the analysis, the Agency considered the operational lifetime of one year's production of vehicles that would be eligible for a cutoff device. The analysis approximated that 32,680 air bags would deploy with a rear-facing infant or a child age 1-12 occupying the front seat. Some of these children were predicted to be badly traumatized. About 1.1 percent of the vehicles with a manual cutoff device were estimated to have an infant and 7.8 percent a child aged 12 years or less. If roughly 10 percent of the older occupants forgot to turn the air bag back on, then the analysts estimated that between 0-2 fatalities might occur to the older occupants. Similarly, if 10 percent forget to turn the air bag back on, 5 to 38 AIS 2-5 injuries were estimated to occur to the older occupants (NHTSA 1996e).

In February of 1997, NHTSA researchers published an analysis of the effects of the two depowering alternatives previously discussed (increasing the chest acceleration requirement to 80 g or incorporating a milder, generic, sled acceleration pulse) (NHTSA 1997f). For the passenger side, the Agency estimated that about 140 children would be killed over the lifetime of the fleet for one model year, assuming no change to FMVSS No. 208. With the 80 g approach, they predicted 14 of these children could be saved. For the generic sled test approach (the approach eventually adopted), a predicted 47 children's lives would be saved. For the passenger side, the Agency estimated an estimated 714 adults (12-years-old and older) would be saved over the lifetime of the fleet for one model year assuming no change to FMVSS No. 208. Using the 80 g approach, the agency estimated 11 to 87 more fatalities to unbelted adult passengers. For the generic sled approach, the Agency estimated 34 to 280 more fatalities to adult passengers who failed to buckle up. For the driver side, the Agency approximated that 1600 to 2700 arm injuries (at AIS injury level 2 and 3) could be avoided under the 80 g approach. Using the eventually-adopted generic sled approach, approximately 5,100 to 8,800 arm injuries (at AIS injury level 2 and 3) were estimated to be avoided.

Later in 1997, in connection with the deactivation proposal issued in January of that year, the NHTSA researchers published the results of an analysis of the potential consequence of allowing the installation of on-off switches in vehicles (NHTSA 1997g). This analysis focused on the consequences that might occur during the years 1998 to 2001. In prefacing the study, the Agency noted that (1) vehicle manufacturers were expected to install many advanced air bags in the fleet by model year 2002 and (2) that education and labeling efforts were expected to result in fewer children being in the front seat. For the driver side, if on-off switches were installed and used by all drivers actually at risk, about 45 deaths were estimated to be prevented during 1998 to 2001. For each 1 percent of the drivers who were not at risk and turned off the air bag, about 42 deaths were predicted over this same period. For the passenger side, if on-off switches were installed and used for all children 0-12 years of age, Agency analysts estimated 177 deaths would be avoided. For each 1 percent of the passengers above 12 years of age who were not at risk and turned off the air bag, about 9 deaths were projected.

In the late 1990s, as changes in air bag designs were occurring, the Agency sought to team up with suppliers to conduct research through cooperative agreements. However, because of limited resources and the difficulty in finding first-tier suppliers willing to participate, the NHTSA efforts were not entirely successful. One company who has been conducting air bag research jointly with the Agency for a number of years, however, has been ASL/Takata. In April, 1993, the NHTSA entered into an agreement with ASL Inc., the organization within Takata responsible for advanced air bag research. Initially, the focus of the research was on the improvement of air bag performance in high severity crashes. However, the problem of air bag-induced injuries occurring in low and moderate severity crashes was receiving the attention of the public, the government, and the industry. By 1995, a change in the direction of research efforts was made to identify the critical parameters that could reduce the risk of injury to OOP occupants while still being able to meet the 30 mph FRB requirements of FMVSS 208 with unbelted dummies. As a major producer of air bags, Takata has been exploring a number of advances in air bag technology. Research has included the development of inflators with unconventional gas flow rates, dual stage inflators, modified deployment patterns and sophisticated sensors and algorithms for making "smart" deployment decisions. Evaluations of the systems developed under those research projects were undertaken with small stature driver dummies initially, followed by tests of 'advanced' passenger air bag systems with child dummies. Some of those research projects are still continuing (NHTSA 1997h). A cooperative research program with another air bag supplier, ATI, Inc. is also ongoing as this report is being written.

In searching for approaches to reduce fatalities of children in moderate-speed crashes, the Agency entered into an Interagency Agreement with the National Aeronautics and Space Administration (NASA). Specifically, the objective was to have NASA help the NHTSA focus on advanced air bag technology that could proceed at a fast pace. The Jet Propulsion Laboratory (JPL), working under this Agreement, assessed advanced air bag technology and published their report in April 1998 (Phen et al. 1998). The report went into myriad surveys of advanced technology, and JPL suggested a more systematic investigation of air bag performance. Airbag performance, they said, should be determined based on four criteria. First, determine how the air bag should deploy for occupants of different sizes sitting at different distances from the stored air bag. Second, an air bag system should handle vehicle crashes varying from low-speed vehicle-to-vehicle crashes to high-speed rigid wall crashes. Third, JPL suggested that ambient temperature be considered. Finally, they recommended that one needed to determine how the air bag should deploy for unbelted occupants and how should it deploy for belted occupants. JPL identified what they thought were five key technology development needs. First, they felt that air bag deployment time variability needed to be reduced by improvements in the vehicle crush/crash sensor system. Second, they thought that inflator variability had to be reduced so that dual-stage inflators could be applied effectively. Third, they indicated that system and component reliability needed diligent attention to achieve the high level required under field conditions. Fourth, they saw a need for occupant sensors to be developed that could distinguish with high accuracy between small, medium, and large adults, children, and infant seats. And last, they urged that there was a necessity for position sensors to measure occupant proximity to the air bag module with the required response time and accuracy.

In June of 1998, two members of the NHTSA staff presented a paper at the 16th International Technical Conference on the Enhanced Safety of Vehicles (Chidester and Rutland 1998). This paper presented the results of an investigation into the field performance of air bag equipped vehicles using the NASS (National Automotive Sampling System) and Special Crash Investigations (SCI) files. The study found that air bags were highly effective in reducing fatalities, and estimated that, as of March 1, 1998, about 2,920 lives had been saved by inflatable restraints. At the same time, 54 children were indicated as having been fatally injured by the deploying air bag. The vast majority of these fatalities were found to be sustained by children OOP or not properly restrained at the time of deployment. At that same ESV Conference, the NHTSA presented the results of another field study, this time tracking the performance of air bag systems using the NASS files for the years 1988-1994 (Summers and Hollowell 1998). The results indicated that, when comparing drivers in air bag-equipped vehicles to those in vehicles without air bags, moderate and serious injury risk, as well as the fatality risk, was about the same for belted drivers. The fatality risk in air bag vehicles, however, was found to be significantly lower for unbelted drivers.

In 1999, concerned that depowered air bag systems might fail to protect occupants in high severity crashes, the NHTSA conducted a series of 30-mph FMVSS 208-type FRB impacts of 13 MY 1998-1999 production vehicles (NHTSA 2000c). Unbelted 50th percentile male dummies occupied the driver and front passenger positions, positioned in the compliance (i.e., NSP) posture. With one minor exception (the femur loads for one dummy), the driver dummies in all 13 tests passed the FMVSS 208 injury criteria with most values being below 80 percent of the limits. For the passenger position the requirements of FMVSS 208 were met in almost every instance (in the one exception, the chest acceleration slightly exceed the criterion) and most were below 80 percent of the limits. Thus, with minor exceptions, the tested vehicles, although certified to the sled test, also passed the pre-existing 30 mph barrier test with unbelted adult dummies.

At the 17th ESV Conference, held in Amsterdam in June of 2001, a paper was presented, summarizing the development of an occupant position sensor by Automotive Technologies International (ATI) (Breed et al. 2001). This project was partially funded by the NHTSA (and Autoliv), starting in the late 1990s. The position sensor developed by ATI used ultrasonic transducers in conjunction with pattern recognition algorithms to determine the existence of an OOP occupant or a rear-facing CRS in the front seat. The system was reported to have been incorporated into a production vehicle. At that same ESV Conference, the NHTSA analysts presented the results of their latest study of air bag field performance (Chidester and Roston 2001). This study, which used SCI data, concluded that the rulemaking changes of March 1997 to allow air bag depowering had had a positive effect on reducing air bag-related fatalities. The study found that, as of January 1, 2001, there were no adult driver or adult passenger air bag-related fatalities in MY 1999 and newer vehicles. In addition, the study found that there were no MY 2000 or 2001 vehicles involved in a child passenger air bag-related fatality. These reductions were recognized as being the result of two factors. First, there was the behavior change in the motoring public. As a result of public education programs, improved labeling and media coverage, the public was concluded to have become more aware of the dangers air bags pose to children in the front seat and to have more frequently heeded the NHTSA's call for placing children in the rear. Additionally, greater numbers of drivers were assumed to be consciously avoiding sitting too close to the steering wheel. Secondly, there were technological changes as air bag systems evolved. Airbag inflator outputs (i.e., the air bag pressure rise rates and levels attained), for example, were reduced significantly in many MY 1998 and later vehicles. Additional design changes included a recessing of the driver module in the steering wheel, as well as passenger air bag folding patterns and tethering.

The NHTSA has been conducting a Special Crash Investigation for several years now to, among other things, assess the rate of air bag-induced fatalities for various MY vehicles. Figure 14 is based upon these SCI data, as reported on the NHTSA website in January 2002, and summarizes the findings for the two deployment situations of most interest: passenger air bag deployment-caused child fatalities and driver air bag deployment-caused adult fatalities (NHTSA 2002). The findings indicate that deployment-caused child fatalities peaked around MY 1995 and have sharply declined since. Deployment-caused driver fatalities are seen to have been generally lower and to have followed a similar trend with a peak around MY 91 and a steady decline since. Again, these positive results were accredited to the increased public awareness as well as the improved air bag designs.

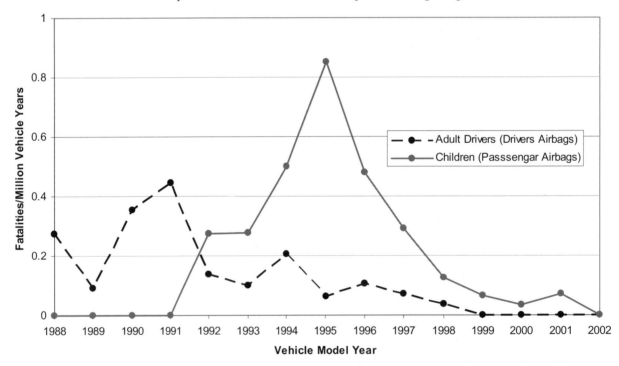

Figure 14. Distribution of airbag-induced fatalities by vehicle model year (data from NHTSA 2002).

Summary

The federal government, by virtue of the activities of either the Secretary of Transportation or the NHTSA, played a significant role in the evolution of the air bag. The NHTSA's contribution to air bag development came via their two primary activities: rulemaking and research. In the area of rulemaking, the issuance of either passive restraint rules or the intention to issue or modify such rules provided a powerful impetus to the automotive community to continue the development, evaluation, and implementation of air bags. This rulemaking activity started in 1969 with the first issuance of an NPRM announcing the Agency's intention to require passive restraints in all production vehicles, and it continues to this day.

For the first 15 years or so of the NHTSA's existence, research was conducted by independent contractors working under programs outlined by the Agency and overseen by an NHTSA Contract Technical Manager (generally a staff engineer who was involved in generating the program and selecting the contractor). The work during this period can be characterized as follows:

1. Due mainly to funding limitations, the programs addressed a limited scope of frontal impact situations. Virtually all simulated and staged impacts, for example, were single vehicle collisions with the FMVSS 208 barrier. Also, the vast majority of impacts were conducted at the upper limit of crash severity or delta-V with comparatively few tests at lesser severities, say, near the sensor trigger threshold. This was particularly true of the early efforts, which concentrated exclusively upon the development of systems capable of protecting normally-seated adult occupants at extreme impact severities (i.e,. delta-Vs of 40-50 mph). Occupant postures at impact were mainly in the so-called Normal Seated Position of FMVSS 208. Even later, when tests with "out-of-position" child dummies were incorporated into the programs, these tests were very limited in terms of the positions, dummy sizes, and crash severities investigated.

2. The efforts were predominantly experimental rather than analytical in nature, typically relying almost exclusively upon sled testing for system development and evaluation followed by a few FMVSS 208-type car-to-barrier crashes for additional system evaluation. This was mainly due to the lack of viable vehicle/occupant/restraint models during this time frame. Federal programs could have been initiated during this time frame to accelerate the development of computer models, but this tactic was not employed. Nevertheless, during the last few years of this period, useful computer models were evolved and employed to make the investigation process more effective.

3. Development and evaluation efforts did not include any significant crash sensor work. That is, these programs either employed delayed-action strip switches or production sensors taken from early production air bag-equipped vehicles, to trigger the systems being developed or evaluated.

The air bag research conducted by the NHTSA during the period from about 1982 to the present has been performed by Agency personnel, working either at the NHTSA's facilities in Ohio (the Vehicle Research and Test Center) or at Agency headquarters in Washington, D.C. The work during this period can be characterized as follows:

1. The research efforts were mainly directed at evaluating or monitoring the safety performance of production air bag systems. This work was of three types. First, there were programs that involved the full-scale testing of production air bag systems. Second, there were programs that employed computer modeling to investigate the performance of such systems. Last and more typically, crash investigations were conducted to monitor or analyze the field performance of production systems.

2. The full scale programs carried out during this period were initially concerned with evaluating the performance of retrofit air bag systems. These programs, conducted in the early 1980s, used either sled tests or fixed-object vehicle crash tests and employed adult dummies in the normal seated position.

3. The three full scale testing programs carried out during the 1990s were mainly concerned with the aggressivity of bag deployments, either directly or indirectly. Two of these programs involved testing with child dummy occupants. In the first instance, the focus of a 1990 study was on the interaction of air bags with child restraint systems. In the second instance, out-of-position occupant tests were conducted in 1999 with production air bag systems produced under the relaxed requirements of FMVSS 208. Finally, in 1999, the NHTSA conducted FMVSS 208-type vehicle impacts with depowered air bag vehicles to see if these systems could provide protection under the conditions of the original compliance test.

4.   The research efforts that involved the collection and analysis of field crash data have mainly been directed at investigating the performance of production air bag systems. The first of these studies, conducted in 1996, focused on the performance of such systems relative to active, belt-only systems. Later, the focus shifted to the investigation of the effect of the regulatory modifications adopted to deal with air bag deployment problems (e.g., cutoff switches, depowering).

The air bag regulatory efforts began with an Advanced Notice of Proposed Rulemaking in 1969, with the latest modification to FMVSS 208 being issued in May of 2000. The period between 1969 and 1982 saw many starts and stops to the passive restraint requirement involving a number of court challenges and changes in direction by the federal government. The Dole decision, in 1983, came at a time when air bag systems were beginning to be accepted by the public and industry and this rule finally resulted in mandatory passive restraints in new vehicles.

During the 1990s a new round of regulatory activity was begun in response to the emergence of air bag deployment injuries and fatalities. The initial efforts during this era consisted of temporizing measures to make the air bag systems less aggressive without unduly compromising the high-speed capabilities of the systems. The end result of this second period of regulation has been the establishment of a more long-term, albeit a much more complicated rule, addressing various occupant sizes and positions in an effort to minimize the probability of an air bag caused serious injury or fatality and to retain all of the system's high-speed protection capability.

Conclusions

1.   The early (1969-1979) research efforts of the NHTSA were generally narrowly focused programs designed to demonstrate the ability of air bag systems to provide protection in high speed, FMVSS 208-type crashes with normally seated adult occupants. As such, these programs did not address the aggressive deployments of the developed systems.

2.   The later (1979-1982) research efforts of the Agency saw a shift in focus, reducing the high speed performance goals of the air bag systems being developed and evaluated, and introducing, for passenger air bag systems, a limited number of out-of-position tests with child dummies. For a variety of reasons, including state-of-the-art considerations and funding limitations, the child dummy testing was only able to identify promising directions for system modifications to reconcile the disparate requirement to protect small, otherwise unrestrained occupants while, with the same system, providing high-speed protection to otherwise unrestrained, normally seated adult occupants.

3.   The lack of adequate attention to sensor issues in the contracted research efforts limited the ability of the researchers to fully understand and investigate the role of this important element in overall system performance.

4.   Rulemaking in the area of passive frontal protection provided a valuable stimulus to the development of production air bag systems. With 20/20 hindsight, some would say that the federal government moved too fast to implement air bag restraints. But that implementation strategy, although not perfect, nevertheless had the net effect of saving lives that would have been lost with a slower, more cautious approach.

5.   The increase in seat belt use from the time that the passive restraint rule was promulgated has undoubtedly dramatically decreased the benefit/cost ratio achievable solely by the addition of an air bag. The original benefits were derived based upon the projection that the air bag would be the primary means of protection in frontal impacts for the vast majority of (unbelted) users. With current usage rates for seat belts, this benefit is now mainly in terms of head protection for the majority of (belted) users. This decrease in the benefit/cost ratio is such that, had an air bag requirement been proposed in the present environment, with seat belt usage in the range of 70 percent or more, the rule could not have been justified on purely economic grounds. On the other hand, having a situation where the vast majority of occupants are belted would allow the possibility of an air bag and seat belt combination with a very high efficiency. That is, with the quicker occupant coupling afforded by the seat belt (particularly contemporary pre-tensioned seatbelts), designers could force-limit the belts and soften the air bags to result in a system that could make

more efficient use of the available compartment space while further reducing the probability of a deployment-induced injury. Such a strategy, of course, would require future modifications of the existing federal regulations.

6.   A failure to fully appreciate the negative public reaction to air bag deployment injuries resulted in a passive restraint rule that, until relatively recently, did not address the situation of an out-of-position, small occupant. Had the NHTSA attempted to incorporate out-of-position provisions in FMVSS 208 earlier, the history of air bag adoption would undoubtedly been quite different. Early difficulties in formulating an objective rule would have been encountered by the Agency, due to difficulties in identifying the "worst case scenario," and due to the lack of development of child dummies and child injury criteria. Also, manufacturers would have had difficulties meeting these requirements due to the state of the art in air bag technology. The effect of proposed out-of-position compliance testing, however, would probably have been to accelerate efforts by the auto industry and other research groups in these areas.

References

Biss, D., Fitzpatrick, M., Zinke, T., Strother, C., Kirchoff, G. (1980) A systems analysis approach to air bag design and development. Proc. 8th ESV Conference, Wolfsburg, Germany.

Brantman, R. (1984) "Breed Driver Airbag Fleet Demonstration Program," NHTSA Contract DTNH-22-84-C-07077, Copy in Docket No. 74-14-GR-505, NHTSA, US DOT, Washington, DC.

Breed, D., Summers, L., Carlson, J., Koyzreff, M. (2001) Development of an occupant position sensor system to improve frontal crash protection. Paper No. 325, Proc. 17th Technical Conference on the Enhanced Safety of Vehicles, Amsterdam, The Netherlands.

Carter, R., Hofferberth, J., Strother, C., et al. (1972) "Passive Protection at 50 mph", U. S. Department of Transportation Report DOT-HS-810-197, NHTSA, US DOT, Washington, DC.

Childester, A. Rutland, K. (1998) Airbag crash investigations. Paper No. 98-S6-0-02, Proc. 16th International Technical Conference on the Enhanced Safety of Vehicles, Windsor, Canada.

Chidester, A. and Roston, T. (2001) Airbag crash investigations. Paper No. 246, Proc. 17th International Technical Conference on the Enhanced Safety of Vehicles, Amsterdam, The Netherlands.

Clark, C., Blechschmidt, C., Gordon, F. (1964) Impact protection with the 'AirStop' restraint system. Proc. 8th Stapp Car Crash Conference.

Esser, R. (1983) "Airbag Demonstration Program," Report RD 583-2, NHTSA, US DOT, Washington, DC.

Evans, L. (1989) Passive compared to active approaches to reducing occupant fatalities. Proc. of the 12th International Conference on Experimental Safety Vehicles, Washington, DC.

Fitzpatrick, M. (1974) "Development of Advanced Passive Restraint System for Sub-Compact Car Drivers," Final Report under NHTSA Contract DOT-HS-113-3-742, NHTSA, US DOT, Washington, DC.

Fitzpatrick, M. (1975) "Pinto Front Passenger Airbag Final Report," NHTSA, US DOT, Washington, DC.

Fitzpatrick, M. (1977) "RSV - Phase II, Comprehensive Technical Results, Section 4.3.3 System Operation - Out of Position Child," November, NHTSA, US DOT, Washington, DC.

Fitzpatrick, M. (1980) "Development of the DEPLOY Computerized Math Model of a Deploying Airbag" Final Report for Office of Passenger Vehicle Research, Contract No. DTNH22-80-C-07120, October, NHTSA, US DOT, Washington, DC.

Ford Motor Company (1957) "Ford Reports – Safety Features Save Lives," Traffic Safety and Highway Improvement Department, Ford Motor Company, Dearborn, MI.

Frey, S. M. (1970) "History of Airbag Development," Paper Presented at the International Conference on Passive Restraints.

Hetrick, J. (1953) "Safety Cushion Assembly for Automotive Vehicles," U.S. Patent No. 2,649,311. John W. Hetrick, Inventor, August 18.

Hopkins, J.B. (1973) "Anticipatory Sensors for Collision Avoidance and Crash Safety Prediction as Applied to Vehicle Safety Research," 1973

Hinger, J. and Clyde, H. (2001) Advanced air bag systems and occupant protection: Recent modifications to FMVSS 208. Paper 2001-01-0157, Society of Automotive Engineers, Warrendale, PA.

Kahane, C. J. (1996) "Fatality Reduction by Airbags, Analysis of Accident Data through Early 1996," Department of Transportation Report DOT HS 808 470, August, NHTSA, US DOT, Washington, DC.

Kemmerer, R., Chute, R., Haas, P., Slack, W. (1968) Automotive inflatable occupant restraint system, Parts I and II. Paper No. 680033, Society of Automotive Engineers, Warrendale, PA.

Kratzke, S. (1995) Regulatory history of automatic crash protection in FMVSS 208. Paper No. 950865, Society of Automotive Engineers, Warrendale, PA.

MacVean, S. (1966) "Seat Belt Usage and Vehicle Occupancy Data", Ford Motor Company Technical Memo PRM 66-26, August, Ford Motor Company, Dearborn, MI.

Michie, V. and Bronstad, M. (1973) "Evaluation of Airbag Restraints for Subcompact Car Passengers," Final Report under Department of Transportation Contract DOT-024-1-165, May, NHTSA, US DOT, Washington, DC.

Musiol, J., Norgan-Curtiss, L., Wilkins, M. (1997) Control and application of intelligent restraint systems. Paper No. 971052, Society of Automotive Engineers, Warrendale, PA.

NHTSA (1967) Federal Motor Vehicle Safety Standard 208 (the first version, requiring lap and shoulder belts at all outboard front seating positions), NHTSA, US DOT, Washington, DC.

NHTSA (1969) Advanced Notice of Proposed RuleMaking on FMVSS 208, 34 Federal Register 11148, July 2, NHTSA, US DOT, Washington, DC.

NHTSA (1971) Film submitted to the FMVSS 208 Docket showing unrestrained NHTSA representative in the right-front passenger seat subjected to panic braking at 5, 10, 20, and 40 mph - various occupant pre-braking situations. NHTSA, US DOT, Washington, DC.

NHTSA (1975) "Development of Improved Inflation Techniques," prepared by Rocket Research Corporation, Final Report DOT HS-801 724, August, NHTSA, US DOT, Washington, DC.

NHTSA (1980) "Calspan/Chrysler Research Safety Vehicle: Final Technical Report, Volume 1, Executive Summary," Report No. DOT HS 805-322, April, NHTSA, US DOT, Washington, DC.

NHTSA (1983) "Airbag Fleet Retrofit Program Crash Tests," NHTSA Report DTNH22-82-A-17148, September, NHTSA, US DOT, Washington, DC.

NHTSA (1984) "Final Regulatory Impact Analysis, Amendment to FMVSS No. 208, Passenger Car Front Seat Occupant Protection," July 11, 1984. NHTSA, Plans and Programs, Office of Planning and Analysis, US DOT, Washington, DC.

NHTSA (1991a) Federal Register, Volume 56, Final Rule to Amend FMVSS 201 to Allow Top-Mounted Passenger Airbag Systems, June, NHTSA, US DOT, Washington, DC.

NHTSA (1991b) Consumer warning issued by the NHTSA regarding the use of rear-facing CRS in the front seat of air bag-equipped vehicles, NHTSA, US DOT, Washington, DC.

NHTSA (1993a) "Effectiveness of Occupant Protection Systems and Their Use," Report to the Congress, Department of Transportation Report DOT HS 808 019, January, NHTSA, US DOT, Washington, DC.

NHTSA (1993b) Federal Register, Volume 58, Final Rule to Mandate Airbags (rather than passive restraints) in Vehicles, September 2, NHTSA, US DOT, Washington, DC.

NHTSA (1994) Federal Register, Volume 59, Final Rule to Amend FMVSS 213 to Require Warning Labels on Rear-Facing CRS Regarding Placement in the Front Seat of Airbag Equipped Vehicles, February 16, NHTSA, US DOT, Washington, DC..

NHTSA (1995a) Federal Register, Volume 60, Number 99, Final Rule Allowing Cutoff Switches for Certain Vehicles, May 23, NHTSA, US DOT, Washington, DC.

NHTSA (1995b) Press Release, "Warning of Dangers to Children and Unbelted Adults Riding in the Front Passenger Seat of Airbag Equipped Vehicles", October 27, NHTSA, US DOT, Washington, DC.

NHTSA (1996a) "Effectiveness of Occupant Protection Systems and Their Use," Second Report to Congress, Department of Transportation Report DOT HS 808 389, February, NHTSA, US DOT, Washington, DC.

NHTSA (1996b) Federal Register, Volume 61, Number 152. Docket 74-14, Notice 100. Warning Label NPRM for Airbags, August 8, NHTSA, US DOT, Washington, DC.

NHTSA (1996c) Federal Register, Volume 61, Number 230. Docket 74-14, Notice 103. Warning Label Final Rule, November 27, NHTSA, US DOT, Washington, DC.

NHTSA (1996d) "Effectiveness of Occupant Protection Systems and Their Use," Third Report to Congress, Department of Transportation Report DOT HS 808 537, December, NHTSA, US DOT, Washington, DC.

NHTSA (1996e) "FMVSS No. 208 Airbag Cutoff Device," Final Regulatory Evaluation, Office of Regulatory Analysis, Plans and Policy, November, NHTSA, US DOT, Washington, DC.

NHTSA (1997a) Federal Register, Volume 62, Number 3. Docket 74-14, Notice 107. NPRM, Airbag Deactivation, January 6, NHTSA, US DOT, Washington, DC.

NHTSA (1997b) Federal Register, Volume 62, Number 3. Docket 74-14, Notice 108, January 6, NHTSA, US DOT, Washington, DC.

NHTSA (1997c) Federal Register, Volume 62, Number 3. Docket 74-14, Notice 109, January 6, NHTSA, US DOT, Washington, DC.

NHTSA (1997d) Federal Register, Volume 62, Number 53. Docket 74-14, Depowering Final Rule, March 19, NHTSA, US DOT, Washington, DC.

NHTSA (1997e) Federal Register, Volume 62, Docket 74-14, Final Rule on Cutoff Switches, November 18, NHTSA, US DOT, Washington, DC.

NHTSA (1997f) "Actions to Reduce the Adverse Effects of Airbags, FMVSS No. 208 Depowering," Final Regulatory Evaluation, Office of Regulatory Analysis, Plans and Policy, February, NHTSA, US DOT, Washington, DC.

NHTSA (1997g) "FMVSS No. 208, Airbag On-Off Switches," Final Regulatory Evaluation, Office of Regulatory Analysis, Plans and Policy, November, NHTSA, US DOT, Washington, DC.

NHTSA (1997h) Report on ASL/Takata Cooperative Research with NHTSA, September, NHTSA, US DOT, Washington, DC.

NHTSA (1998) Federal Register, Docket 74-14. NPRM on Advanced Airbags, NHTSA, US DOT, Washington, DC.

NHTSA (1999) "Effectiveness of Occupant Protection Systems and Their Use," Fourth Report to Congress, Department of Transportation Report DOT HS 808 537, May, NHTSA, US DOT, Washington, DC.

NHTSA (2000a) Federal Register, Docket 2000-7013-1. Final Rule on Advanced Airbags, May 5, NHTSA, US DOT, Washington, DC.

NHTSA (2000b) "FMVSS No. 208, Advanced Airbags", Final Economic Assessment, Office of Regulatory Analysis & Evaluation, Plans and Policy, May, NHTSA, US DOT, Washington, DC.

NHTSA (2000c) Submission to Docket No. NHTSA-2000-7013-16 concerning the NHTSA testing of 13 late-1990s production air bag-equipped vehicles to determine if depowering adversely affected high-speed protection. June 20, NHTSA, US DOT, Washington, DC.

NHTSA (2002) National Center for Statistics and Analysis, Special Crash Investigation Data on Airbag Fatalities. www-nrd.nhtsa.dot.gov/pdf/nrd-30/NCSA/SCI/4Q.htm. January, NHTSA, US DOT, Washington, DC.

Nader, R. (1965) "UnSafe At Any Speed - The Designed-In Dangers of the American Automobile" Pocket Books, New York, NY.

Nixon, C. (1969) "Human Auditory Response to an Airbag Inflation Noise", Final Report, DOT Contract No. P.O. 9-1-1151, NHTSA, US DOT, Washington, DC.

Phen, R., Dowdy, M., Ebbeler, D., Kim, E., Moore, N., VanZandt, T. (1998) "Advanced Airbag Technology Assessment," Final Report, Jet Propulsion Laboratory Publication 98-3, April, NHTSA, US DOT, Washington, DC.

Reed, D. (1999) Father of the air bag. Automotive Engineering, Volume 99, No. 2, February, Society of Automotive Engineers, Warrendale, PA.

Schimkat, H., Weissner, R., Schmidt, G. (1974) A comparison between Volkswagen automatic restraint and three-point automatic belt on the basis of dummy and cadaver tests. Paper No. 741183, Society of Automotive Engineers, Warrendale, PA.

Sherman, D. (1995) The rough road to air bags.Invention and Technology, Vol. 2, No. 1.

Shoemaker, N. and Biss, D. (1974) The development of an air bag on collapsible dashpanel restraint system for right front seat occupants. Paper No. 740576, Society of Automotive Engineers, Warrendale, PA.

Skeels and Falzon (1962) A new laboratory device for simulating vehicle crash conditions. Proc. 6th Stapp Car Crash Conference, November 7-9.

Slack, W. (1968) Automotive inflatable occupant restraint system, Part II. Paper No. 680033, Society of Automotive Engineers, Warrendale, PA.

Smith, G., Hurite, S., Yanik, A. (1972) Human volunteer testing of GM air cushions. Paper No. 720443, Society of Automotive Engineers, Warrendale, PA.

Southerland, G. (1970) "Self-deployed, air induction inflation systems. Proc. International Conference on Passive Restraints.

Stocke, J. (1998) Recent regulatory history of air bags. Paper No. 980650, Society of Automotive Engineers, Warrendale, PA.

States, J., Rosenau, W. (1982) "Field Performance of Volkswagen Automatic Restraint System, Progress Report," November, NHTSA, US DOT, Washington, DC.

Strother, C., Fitzpatrick, M., Egbert, T. (1976) Development of advanced restraint systems for Minicars RSV. 6th International Conference on Experimental Safety Vehicles, Washington, D.C.

Strother, C., Broadhead, W. (1978) Small car driver inflatable restraint system evaluation. Final Report, Contract DOT-HS-6-01412, NHTSA, US DOT, Washington, DC.

Strother, C., Zinke, T. (1978) "Small Car Driver Inflatable Restraint System Evaluation Vol. 4: Evolving a Low Mount Passenger Air Cushion Restraint System (ACRS) for the Vega Subcompact Vehicle", Final Report, Contract DOT-HS-6-01412, July, NHTSA, US DOT, Washington, DC.

Struble, D. E. (1998) Airbag technology: What it is and how it came to be. Paper No. 980648, Society of Automotive Engineers, Warrendale, PA.

Sullivan, L. (1992) "Child Restraint/Passenger Airbag Interaction Analysis," NHTSA Report VRTC-87-0074, October, NHTSA, US DOT, Washington, DC.

Summers, L., Hollowell, W. (1998) Status of NHTSA research into the field performance of air bags. Proc. 16th ESV Conference, Windsor Canada.

U.S. Congress (1966) The National Traffic and Motor Vehicle Safety Act of 1966. U.S. Congress, Washington, DC.

Zinke, T. (1980) The development of air cushion restraint systems for small car front seat occupants. Paper No. 800294, Society of Automotive Engineers, Warrendale, PA.

Zinke, T. (1981a) "Small Car Front Seat Passenger Inflatable Restraint System - Volume I - Interim Results", Final Report, Contract DOT-HS-8-01809, April, NHTSA, US DOT, Washington, DC.

Zinke, T. (1981b) "Small Car Front Seat Passenger Inflatble Restraint System - Volume II - Citation Airbag Systems", Final Report, Contract DOT-HS-01809, April, NHTSA, US DOT, Washington, DC.

Ziperman, H., Smith, G. (1975) Startle reaction to air-bag restraints. Journal of the American Medical Association, 223(5):436-440.

Chapter 2

U.S. Industrial Air Bag Development

Charles J. Griswold, Jr., and Sherman E. Henson

Preface

The research and development of automotive inflatable air bag systems has spanned over fifty years, or roughly half as long as the automobile itself. The first United States Patent for an inflatable cloth device that would absorb the impact energy of a person in an automobile crash was issued in 1952, followed by several others in 1953 and 1955. In 1956, the first annual Stapp Car Crash Conference was held in Detroit at Wayne State University, with a presentation by Dr. John Paul Stapp describing his human tolerance studies on the rocket sled at Holloman Air Force Base. By 1957, both Ford and General Motors Research were studying inflatable devices as a means of absorbing the "second collision" in car crashes. To further stimulate new thinking and development, General Motors Research Laboratories built a Safety Concept Car starting with a production 1959 Buick. This car incorporated safety designs already in production, such as door latches designed to resist crash loads and padded instrument panels. Some of the research concepts shown below were energy absorbing steering columns and steering wheels, windshields to absorb head impact, and an instrument panel incorporating an inflatable restraint system.

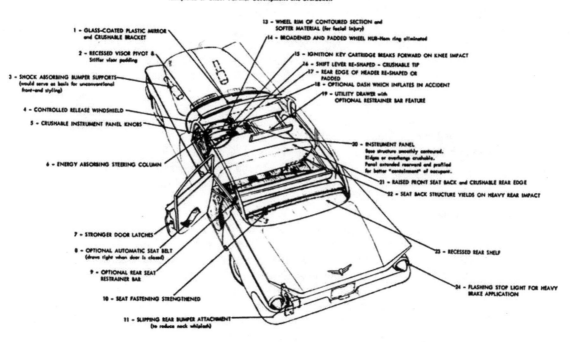

Diagram of 1959 Buick Safety Concept Car[1]

The Fundamental Problem of Deployment of Air Bags

All of these early air bag efforts dealt primarily with the energy absorbing characteristics of deployable devices. In order to provide protection in a crash situation, a system must deploy without adding additional hazard to the restraint of the occupants. This became the most difficult challenge for the researcher and designer, since a deployable air bag inherently adds additional energy to the energy present in a crash. A deployable system should inflate the bags fast enough to protect occupants in the relatively infrequent life threatening crashes without inducing serious or fatal injuries in the more frequent minor events. This problem is compounded by the random character of injuries. Fatal injuries are rare in minor accidents, but have been known to occur. Inflation should occur when

---

[1] Personal communication – C.J.Griswold with Charles W. Gadd and Charles E. Kroell

serious injury is possible, not just when it is probable. Some occupants emerge from serious accidents in non-airbag cars with minor injuries. Bag inflation in serious accidents brings with it the possibility of increasing injury severity. This fundamental problem bears on the capability of crash sensors to know when it is necessary and when it is not desirable to activate inflation. It is also desirable to control the forces created by inflation. The design of the system should recognize many other limiting factors such as the tolerance of children to inflation forces and the various positions and postures an occupant may be in at bag contact. The objective is that the inflation energy should not be the cause of significant injuries that might not occur in the absence of the air bag.

Largely because of these fundamental issues, the air bag history has been remarkable by the wide range of public, professional and industry controversy. In addition, air bag design and development after 1969 has been driven by a regulatory requirement for meeting certain anthropometric test device measurements in a frontal crash test. The speed of the crash test was first proposed to be 50 miles per hour in the Experimental Safety Vehicle Programs of the late 1960's. The proposed FMVSS 208 regulation, to be effective in 1973, designated a crash test speed, with an unrestrained occupant, of 30 miles per hour into a fixed barrier.

Even today, the problem of balancing the benefits of air bags against possible harm is exceedingly complex. The many technical papers reproduced and cited in this book, as well as in an earlier SAE compendium,[2] illustrate the efforts of the engineering and biomechanical community to find answers and develop systems to address these considerations. As this is being written, the problem has still not been resolved, and a search for a practicable industry standard still continues fifty years after the concept was advanced.

First Full-Vehicle Component Development

In the mid to late 1960's, a number of potential air bag suppliers began product research and development of individual components to support the research efforts of General Motors and Ford Motor Company. The principle system developers were Eaton Yale and Towne and Allied Chemical. The earliest inflator systems used stored inert gas under pressure as the inflating medium, and later moved to stored gas combined with internal chemical gas generators. The augmented systems were smaller and more adaptable to the limited space behind the instrument panels where the systems needed to be placed. The NASA space program of the 1960's provided the rocket motor technology that could be adapted for air bag inflators by Talley Defense Systems, Morton Thiokol, Olin Chemical and Rocket Research. Even smaller gas generators were developed for systems that would be mounted in the steering wheel hub. Another spin-off from the space program was technology that could be used for sensing the severity of the crash. Part of the General Motors team in the Delco Electronics Division that developed the Moon Rover vehicle put their expertise into sensor design. Other new companies emerged to develop both electronic and mechanical designs. Eaton developed a demonstration system that was shown to the NHSB in 1968, and about four years later persuaded a stunt driver and a passenger to drive a large size Ford Sedan equipped with a driver in-the-wheel and passenger system into a barrier at about 24 miles per hour. [3]

Early Deployment Hazard Concerns

The early crash tests at 30 miles per hour demonstrated the relative simplicity of systems that could apparently provide restraint at high crash speeds. In fact, General Motors and others demonstrated air bag restraint in experimental safety vehicles at barrier speeds up to 50 miles per hour. The test dummies available at the time, however, were not capable of evaluating the biomechanical and potential injury effect of the inflation forces necessary to deploy the air bags at the higher speeds. This led a GM spokesman to say that the system was not yet ready for humans.[4] In 1971, Wayne State University air bag experiments with baboons showed broken teeth and dislocated bones.[5] Subsequent research and development, using human and animal surrogates and more sophisticated dummies has shown that there is an upper limit to deployment energy that can safely be tolerated in a passenger vehicle. Tragically, an unyielding regulatory requirement that air bags must restrain unbelted occupants in a frontal barrier crash at 30 miles per hour regardless of the size, weight or crash pulse of the vehicle led to this upper limit being proven in public by fatalities experienced in 1990 era vehicles.

---

[2] Passenger Car Inflatable Restraint Systems – Edited by D.C.Viano - SAE PT-31 - 1987
[3] "A Movie for Air Bag Devotees", Washington Report, Automotive Industries, March 15, 1973
[4] Associated Press, Phoenix 1972
[5] Auto Air Bags a Potential Hazard to Children, WSU Study, Detroit Free Press, July 7, 1971

General Motors Air Cushion Restraint System Development[6]

In November of 1967, Edward N. Cole was named President of General Motors. He saw both a business potential and the occupant protection possibility of an air bag system that would not require the use of seat belts, In 1968, the GM research effort was moved to the Fisher Body Division as Project Manager with a goal to develop a production system to meet the new NHTSA's 1973 target for "Passive Protection".

By 1969, GM had developed a passenger system that inflated quickly enough to restrain a dummy in a 30 mile per hour frontal barrier test. This system utilized a 160 cubic inch inflator using Nitrogen pressurized at 3500 psi. Concerns about the effect of this inflation led to surrogate tests using baboons at Wayne State University in Detroit and, with the cooperation of the NHTSA, a human volunteer test program using the sled facility at Holloman Air Force Base in New Mexico. One of the baboons in the Wayne State tests was fatally injured. One of the Holloman volunteers suffered a mild concussion in a static deployment while leaning forward before deployment.[7] These development events showed that an inflatable restraint could potentially cause serious injury if an occupant was out of position and too close to the bag. The early experimental passenger system was redesigned completely. The inflation energy and subsequent occupant rebound out of the bag was reduced, and the bag redesigned to improve energy absorption.

By the fall of 1972, the second generation GM passenger air cushion restraint system had been evaluated with human volunteers on the Holloman Air Force Daisy sled at speeds up to 30 miles per hour. The results were satisfactory.[8]

The final inflator for the 1974 production passenger air bag was a hybrid using stored gas and an internal gas generator. There were two bags, one outer and one inner. The outer bag consisted of a porous nylon bag that absorbed the energy of the occupant impact, with a non-porous inner bag for knee restraint. The internal gas generator inflator design permitted a further refinement. The development of a two stage sensor system by the Delco Electronics division allowed the passenger air bag to be initially inflated by the stored gas and only one of the generators. If the crash severity was sensed to be at a higher level, the later signal from the high level sensor triggered the second generator to provide additional gas for the greater energy required. Further development showed that the placement of the sensors provided a variable inflation rate that increased as impact speed increased. The variable rate system was used only on the passenger side; the driver air bag was deployed fully as soon as the low level threshold was detected.[9] .

The driver system was engineered by the Oldsmobile Division, along with Saginaw Steering, Delco Products and Inland, and presented several unique problems somewhat different from the passenger system.[10] The requirement to meet test requirements without seat belts presented a major challenge in providing restraint for large male and small female drivers. This required the design and development of a new knee restraint in the lower instrument panel which would not interfere with the energy absorbing steering column. The EA column unit itself was redesigned from the production mesh-type element to a sliding ball design that would accommodate the higher bending forces created by the air bag system. A major concern for the driver system was the effect on driver control when the system deployed. A major volunteer test program was undertaken to evaluate this possibility and was found not to be a major concern.[11] The driver system was also evaluated with human volunteers up to 30 miles per hour at Southwest Research Institute, Texas.[12]

The GM test and development program from 1969 through 1974 consisted of over two thousand Hy-G sled tests and several hundred full scale crash tests utilizing dummies ranging in size from the 95% male to a 3 year old child. The dummy technology was primitive by today's standards.

---

[6] Lundstrom Louis, "Relating Air Cushion Performance to Human Factors and Human Tolerance Levels", SAE 746031 – ESV Conference 1974 (and SAE PT-31)

[7] Mertz, H.J., Marquardt, J.F., "Small Car Air Cushion Performance Considerations". SAE 851199, Government-Industry Meeting, Washington, DC, 1985 (and SAE PT-31)

[8] G.R.Smith, etal, "Human Volunteer Testing of GM Air Cushions, SAE 720443 (and SAE - PT-31)

[9] Campbell, David D. "Air Cushion Restraint System and Vehicle Development", SAE 720407 1972 2nd Passive Conference. (and SAE PT-31)

[10] T.N.Louckes, R.J.Slifka, T.C.Powell & S.G.Dunford – "General Motors Driver Air Cushion Restraint System" – SAE 730605, Detroit, Michigan 1973

[11] L.Lundstrom, 1974 ESV Conference ibid.

[12] G.R.Smith, E.C.Gulash & R.G.Baker – "Human Volunteer and Anthropometric Dummy Tests of General Motors Driver Air Cushion System" SAE 740578, 3rd International Conference on Occupant Protection, Troy, Michigan 1974

Field Trial Programs

In order to gain exposure to the real-world environment, General Motors built test vehicles to be driven by engineering personnel on public roads in all kinds of driving conditions, and on the proving grounds in more severe conditions. Some of these were equipped with sensor systems, most had full airbag systems installed. Since the deployment reactions with humans was still largely unknown, these first systems were non-deployable. In fact, many local, state and federal regulation issues involving pressure vessels and explosive valve devices had to be resolved before the full system cars could be driven on public roads.[13]

After the controlled human volunteer test experience, 1000 Chevrolet Impala sedans with production design driver and passenger systems were built to serve as a field evaluation fleet. These cars were leased to police and other selected fleet users with the agreement that any accidents that would occur would be reported for analysis by a field investigation team. The NHTSA also participated in this field evaluation. These Chevrolets did not have belts installed for the front seat occupants.

Ford Motor Company had a similar program involving 831 Mercury Montereys with passenger systems and lap belts installed.[14]

General Motors Production – 1974 to 1976

General Motors invested over $62 million dollars ($246 million in year 2001 Dollars) in the development of the Driver and Passenger system that was offered as an option in 1974 through 1976 full-size Oldsmobiles, Buicks and Cadillacs.[15] An extensive advertising campaign was undertaken by General Motors and others, including educational movie machines placed in dealerships. In the Chicago Auto Show in February 1974, an air bag inflation demonstration car was part of the GM new model exhibit. There was, however, wide-spread distrust and opposition to air bags in cars. The AAA Motor News ran ads calling air bags "probably the greatest deception in the history of auto safety". On the other hand, Allstate Insurance supported the concept by offering small insurance discounts for air bag owners.[16] General Motors production tooling capability was in place to produce 100,000 cars a year equipped with air bags, but only slightly over 10,000 cars were purchased by GM customers. The remainder of the approximately 33,000 passenger systems that were built was left in the factory inventory when production was terminated in 1976.

Injuries Related to Air Bags

With the introduction of the 1973 Chevrolet Field Trial Fleet and the 1974 Optional systems, General Motors put an in-depth accident investigation and data collection system in place. Individual accidents were investigated by engineering teams who reconstructed events and analyzed the performance of the systems.[17] As air bag-equipped cars accumulated miles on the highways, attempts to measure the injury-reducing effectiveness of the restraints were made. It became apparent that while air bags were reducing injuries, in some cases they were actually causing injury. These injuries increased the concern regarding inflation hazard.

In a study of crashes involving the 1973-1976 GM cars reported in a 1978 SAE paper[18], air bag effectiveness was found to be less than expected, especially when AIS 2 level injuries were included. In fact, a negative 34% effectiveness was found for the passenger side bag. In a 1988 SAE paper[19], Mertz reported that a case-by-case study of the GM files found some injuries greater than expected in 1973-1976 GM cars. The greater than expected injuries were found for the Head/Neck, Arm/Hand, Thorax, and Leg/Pelvis. In the Head/Neck area, three drivers were concussed, one reported to be unconscious for 25 minutes. In the Arm/Hand area, nine drivers and six passengers sustained fractures, In the Thorax area, one driver was fatally injured, although there was no autopsy to allow a determination of the cause. Five other drivers suffered rib fractures. There were two drivers and nine passengers with fractures in the Leg/Pelvis area.

---

[13] "Law Change Needed for Air Bag Tests" Detroit Free Press, May 9, 1971

[14] "Bag-Belt Test Cars Delivered" – The State Journal, August 28, 1972

[15] "GM Trims Goal for Air Bag Sales", Detroit Free Press, July 12, 1974

[16] "AAA Blasts Allstate's ads for Airbags", The Detroit News, February 2, 1974

[17] Smith, G.R. "Air-Bag Update, Recent Crash Case Histories" – SAE 770155, 1977

[18] Pursel, H.D., Bryant, R.W., Scheel, J.W. and Yanik, A.G., "Matching Case Methodology for Measuring Restraint Effectiveness," SAE 780415, February 1978.

[19] Mertz, Harold J., "Restraint performance of the 1973-76 GM air cushion restraint system", January 1988, SAE 880400

Later tests with the new GM Hybrid III dummies and animals showed that the injuries being caused by air bags were consistent with the occupants being close to the bag at the time of deployment.

Early Actions in 1989 and Later Systems

It soon became apparent that children in rear-facing infant seats could be vulnerable to injuries from a deploying passenger-side air bag. NHTSA issued a consumer bulletin[20] advising parents to put children in rear-facing infant seats in the rear seat of passenger-side air bag equipped vehicles. The SAE formed a Child Restraint /Air Bag Interaction [CRABI] task force to recommend a test procedure.

As exposure to air bags continued to grow, so did injuries. At first, the injuries were mostly minor abrasions. Then there were fatalities to short-stature, elderly women. Injuries to the eye and upper extremities also appeared. Most alarming were fatalities to children riding in the right front seat. Fatalities to children began to be reported in minor collisions where injuries would not have occurred without an air bag deployment.

These injuries occurred in spite of manufacturers' efforts to minimize them by developing systems with lower inflation pressures, internal tethers and larger vents. Since it was clear that the deployment force of the bag was creating injuries, the manufacturers told NHTSA that further reductions in air bag power could not be made without a change to FMVSS 208. The 48 km/h (30 mi/h) unbelted male dummy barrier crash, they said, was driving the power levels of air bags, and the bags could not be made less powerful with lower inflation forces without changes to this test.

Depowering

In March 1997, NHTSA temporarily permitted manufacturers to certify vehicles using a sled test with a prescribed pulse (125 ms) and air bag deployment time (20ms) instead of a 48 km/h (30 mi/h) fixed barrier crash test with an unbelted dummy. Manufacturers had persuaded NHTSA that the sled test would lead to reduced power air bags in the shortest time, lessening the risk of injury. In the 1997 rule, the sled test option was set to expire on September 1, 2001[21]. However, this termination date was superceded by the NHTSA

Reauthorization Act of 1998 part of TEA 21 to be discussed below, which kept the option in effect until superceded by the final rule for advanced air bags.

Congressional Action

In view of the mounting number of child deaths related to air bag deployments, and after considerable deliberation, Congress included an advanced air bag provision in the Transportation Equity Act for the 21st Century[22], also known as TEA 21 in June 1998. This law directed DOT to take the following actions:

1. **Notice of proposed rulemaking** - By September 1, 1998, issue a notice of proposed rulemaking to improve occupant protection for occupants of different sizes, belted and unbelted, while minimizing the risk to infants, children, and other occupants from injuries and deaths caused by air bags, by means that include advanced air bags.

2. **Final rule** – Issue a final rule by September 1, 1999. The DOT Secretary was allowed to extend the date to March 1, 2000 if the earlier date could not be met.

3. **Effective date** – Provide for a three-year phase-in of the rule beginning sometime between September 1, 2002 and September 1, 2003.

4. **Credit for early compliance** – Grant credits toward the phase-in percentages for vehicles certified before the beginning of the phase-in or in excess of the required percentages.

While the law directed DOT to require "Advanced Airbags," the law did not define such an air bag, giving NHTSA wide latitude in setting the technical requirements, if not the timing, of the new rule. However, it was unusual for Congress to dictate the timing of the entire rulemaking process for a safety standard.

---

[20] Avenessian, Ridella et al, "An analytical model to study the infant seat/airbag interaction", January 1992, SAE 920126

[21] 62 FR 12968; March 19, 1997

[22] A part of Public Law 105-178

Advanced Air Bag NPRM

PROVISIONS OF THE PROPOSAL - On September 18, 1998, NHTSA issued the Notice of Proposed Rulemaking called for in TEA 21. This sixty five-page document laid the groundwork for what was to become the most complex FMVSS to date. In the NPRM, NHTSA noted that as of June 1, 1998 its Special Crash Investigations Unit had confirmed 105 crashes in which deployment of an air bag had resulted in a fatal injury. Sixty-one of those fatalities involved children, four were adult passengers and forty were drivers.

The mandate to improve protection for all occupants was accomplished, in NHTSA's view, by:
- o adding 48 km/h (30 mi/h) fixed barrier crash tests using belted and unbelted 5th percentile female dummies
- o adding a 40 km/h (25 mi/h) 40% offset deformable barrier crash test using a 5th percentile female dummies
- o deleting the sled test option that had been used to expedite reduced power air bags
- o introducing new injury criteria for the neck and chest

To minimize risk to infants, children and other occupants, the Agency proposed that:
- o manufacturers either automatically suppress the air bag for normally seated children, children in rear-facing infant seats small female is drivers, or
- o demonstrate a "low-risk" deployment for those same occupants
- o Also, the NPRM invited comment on a "Dynamic Out-of-Position" test in which the air bag would be automatically suppressed during a crash whenever an occupant movement indicates a risk of injury from the air bag.

At the same time, the Agency proposed injury criteria for the new 3 and 6 yr old child, and 5th percentile female dummies, which, had not yet been used in a Federal Motor Vehicle Safety Standard.

In the NPRM, the Agency estimated the implementation of its requirements would prevent 563-644 fatalities annually when compared to vehicles designed to the "sled test" option. Annual costs were projected to be as high as $2.51 billion depending on the options chosen by manufacturers. The Agency assumed that technologies such as dual level inflators, weight sensors and "low-risk deployment" air bags, all unproven, would be ready for production in time to be used to meet the rule, and would work reliably in accidents.

PUBLIC COMMENTS - Most of those who formally commented to NHTSA on the proposal, including manufacturers, favored the inclusion of the small female and child dummies, improved injury criteria and additional tests for the "out of position" occupants. The most controversial issue in the NHTSA proposal was the elimination of the sled test option that had led to the depowering of air bags.

Most auto manufacturers, restraint system suppliers, The Insurance Institute for Highway Safety, The American Trauma Society, the National Transportation Safety Board and others opposed the elimination of the sled test option and the consequent return to the 48 km/h (30 mi/h) unbelted barrier test. These organizations expressed concern that the severity of the 48 km/h (30 mi/h) unbelted barrier test would lead to increased air bag power levels and therefore, an increase in injuries and fatalities when compared to depowered bags. Furthermore, the commenters questioned the technical feasibility of meeting the proposed requirements in the broad range vehicles preferred by consumers. The Harvard Center for Risk Analysis pointed to flaws in NHTSA's projected benefits for the proposed regulation, and recommended priority be given to children, belted adults, and short-statured drivers. The Center cited the engineering reality of design tradeoffs that would need to be made in order to meet the conflicting requirements of the proposed rule.

Opposition to the "sled test" option came primarily from Consumers Union and the Center for Auto Safety. These groups stated their belief that the sled test was inadequate to certify vehicles with air bags, because the test does not reproduce the crash pulse and air bag deployment timing of each individual vehicle, the vehicle pitch and occupant compartment deformation, and does not simulate oblique collisions.

Manufacturers expressed their concerns that unproven advanced technologies needed to meet the proposed rule would not be ready for mass production in time to meet the requirements of the proposed rule. Of particular concern were "Occupant Classification" technologies that would discriminate between children and small adults in order to deploy or suppress the air bag. Comments to the NPRM stated that technologies under development could misclassify occupants, allowing an air bag deployment that could harm a child, or suppressing an air bag deployment needed to lower the injury risk of a small adult. Also, commenters feared that mandating flawed Occupant Classification sensors would encourage parents to allow children to ride in the front seat, reversing gains made by "kids in the back" public information campaigns.

Supplemental NPRM

In response to comments received, NHTSA issued a Supplemental Notice of Proposed Rulemaking on November 5, 1999. In the SNPRM, The Agency announced that it was considering two alternative tests for enhancing the protection of unbelted occupants in high-speed crashes:

o   An unbelted rigid barrier test in the range of 40-48 km/h (25-30 mi/h). If 40 km/h (25 mi/h) were chosen, the speed of the belted rigid barrier test would be increased to 56 km/h (35 mi/h).

o   An offset deformable barrier test in the range of 48-56 km/h (30-35 mi/h).

In comments to NHTSA on the SNPRM, car companies and others continued to urge NHTSA not to reinstate the 48 km/h (30 mi/h) unbelted barrier crash test. They pointed out that an air bag designed for a 5th percentile female driver would be quite different from one designed for a 50th percentile male driver not only because the female is smaller and lighter, but because the female sits considerably closer to the air bag. As a result of these differences in anthropometry, which result in different proximities to the inflating bag, an air bag that is designed for optimal performance for the larger male would be injurious to the smaller, closer female in a 48 km/h (30 mi/h) rigid barrier crash. Conversely, a bag designed with lower power to protect the female, would not meet the proposed requirements for the larger male at 48 km/h (30 mi/h). Simply increasing the power of the second stage of the dual stage inflator to better restrain the larger male would also increase the power of the first stage, because the dual stage air bags then under development used a common bag and venting for both stages. For this reason, the manufacturers favored the 40 km/h (25 mi/h) unbelted rigid barrier test.

The Alliance of Automobile Manufacturers[23] (AAM) also submitted a study of air bag performance that used computer simulations to predict the performance of various advanced air bag designs in the proposed FMVSS 208 tests as well as the overall occupant injury risk in real world crashes. The study found that design tradeoffs necessitated by the 48 km/h (30 mi/h) unbelted test made it doubtful that any air bag system using technologies available concurrent with the effective date of the proposed rule could meet all the requirements. Meeting the requirements of the rule would, however, be feasible with the 40 km/h (25 mi/h) rigid barrier test. Additionally, the study found that restraint systems designed to the 48 km/h (30 mi/h) test, if possible, would result in more overall risk to occupants than the 40 km/h (25 mi/h) rigid barrier test.

In addition, the Insurance Institute for Highway Safety continued to oppose the return to a 48 km/h (30 mi/h) unbelted barrier test, saying that the test resulted in air bags with too much power. The Institute submitted a study of air bag accidents, which found that 15% of deaths in high-speed crashes in vehicles with air bags certified to the 48 km/h (30 mi/h) barrier crash requirement were most likely caused by the air bag itself. Many of these deaths were due to occupants that were no longer in a "normal" position at the time of air bag deployment. Previous studies by NHTSA and others had not identified any deaths caused by the air bag in high-speed crashes.

A second issue that was again stressed in the comments was the inability of occupant classifications systems to reliably distinguish between the 5th percentile female dummy and the 6 yr old child dummy. The systems could be "fooled" by subtle shifts in seating posture that would make an occupant appear to weigh more or less than the actual weight. These errors in classification could result in an air bag being deployed for a child and causing an injury, or not being deployed for an adult when needed to prevent an injury. The weight sensors could be further tricked during normal installation of a rear-facing infant seat when the seat belt is tightened around the seat in accordance with recommended practice. This belt "cinching" places an additional load on the seat cushion, appearing to the sensor as a heavier occupant that would require an air bag. While it might be feasible to develop sensors that account for belt "cinching," these devices may not meet the timing of the rule.

Interim Final Rule

On May 12, 2000, NHTSA issued an Interim Final Rule on Advanced Air Bags. An outline of the rule is shown in Figures 1 and 2. Although many technical changes had been made since the 1998 and 1999 proposals, the most significant change was the lowering of the unbelted, fixed barrier test speed from 48 km/h (30 mi/h) to 40 km/h (25 mi/h). NHTSA explained that "…After further examination of the issues and information before us and an assessment of the areas of uncertainty about simultaneously improving protection and minimizing risk, we have concluded that the adoption of a 48 km/h (30 mph) unbelted requirement would not be in the best overall interest of safety."

To compensate for the predicted reduction in benefits resulting from the lowered speed of the unbelted test, the speed of the belted, fixed barrier test would be gradually raised from 48 km/h (30 mi/h) to 56 km/h (35 mi/h)

---

[23] The Alliance member companies are BMW Group, DaimlerChrysler, Fiat, Ford Motor Company, General Motors, Isuzu, Mazda, Mitsubishi Motors, Nissan, Porsche, Toyota, Volkswagen and Volvo.

beginning with the 2008 model year. The sled test was gone. The Agency announced in the Rule that it would seek additional data on the appropriate speed for the unbelted test, and consider raising the speed at a later date.

It is evident from figures 1 and 2 that several tests had been added to FMVSS 208. To improve occupant protection for different size occupants, a set of fixed barrier crash tests using the 5th percentile female dummy, both with and without belts, was added. In addition, the rule added a new test mode, the 40% offset deformable barrier test using the belted 5th percentile female dummy in the forward most seating position. The rule also requires testing with a large number of child seats with a 1-year-old test dummy. In addition, new injury criteria for neck loads and chest deflection have been added. These new tests will require air bag systems to balance performance for the large and small dummies, belted and unbelted, and in rigid and offset barrier tests.

The rule specified a phase-in schedule that is to begin September 1, 2003, and all vehicles must meet the rule by September 1, 2006. Beginning September 1, 2007, the belted barrier crash test speed is increased to 56 km/h (35 mi/h) in a phase-in that is to be completed by 8/31/2010.

Rulemaking Petitions

On October 30, 2001, The Alliance of Automobile Manufacturers submitted a petition to NHTSA requesting changes in the interim final rule. AAM stated that the development of prototype occupant classification systems had not advanced as rapidly as expected, resulting in serious challenges for the industry. Specifically, occupant classification systems could not yet reliably distinguish between a 5th percentile female and a 6 yr old child. The Alliance referred to studies by the United States General Accounting Office and NASA's Jet Propulsion Laboratory to support its prediction that these systems would not be ready for production in time to meet the rule. The AAM asked for four changes to the rule:

- Postpone for three years the static out of position requirement using the 6-year-old child dummy.
- Shorten the data acquisition time for out of position testing.
- Allow an air bag override switch in vehicles with 3 position front seating.
- Reduce the percentage of vehicles required to meet the regulation in the first year of the phase-in from 35% to 10%.

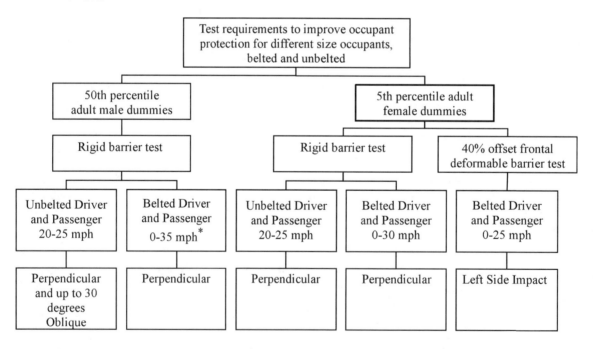

*Range is 0-30 mph during first stage of the phase-in of the final rule

Figure 1. Test requirements to improve occupant protection for different size occupants, belted and unbelted

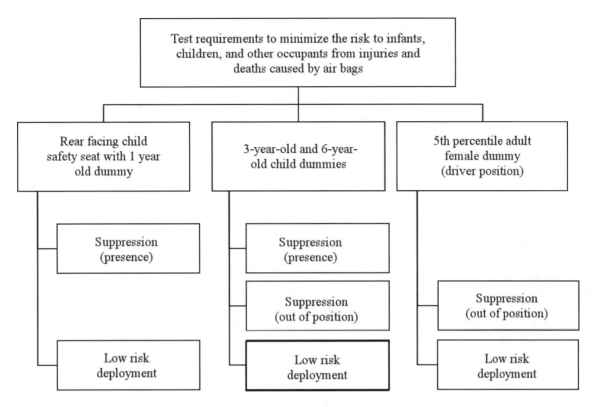

Figure 2. Test requirements to minimize the risk to infants, children, and other occupants from injuries and deaths caused by airbags

In responding to the AAM petition on September 24, 2002, NHTSA proposed to reduce the first year phase-in from 35% to 20%, and requested comments on this proposal within 30 days. The agency also noted that the 300 ms data acquisition time had been reduced in another rulemaking action, and it denied the requests related to the 6 year old child dummy and the air bag override switch.

Advanced Air Bag Hardware

There is no single definition of an "advanced air bag." Congress directed NHTSA to mandate them, but did not define them. However, there are several components that could make up today's advanced air bag concept:

- **Dual Level Inflators**. These inflators can deploy at either a high or a low pressure level. The low level is intended for low severity accidents or for children and small adults.
- **Crash Severity Sensors**. In addition to sensing a crash, these sensors detect the severity of a crash so that a deployment decision can be made.
- **Occupant Sensors**. These sensors are intended to detect the presence, weight, size, position or other characteristics of the occupants in order to make a deployment decision.
- **Safety Belt Buckle Switches**. Sensors in the safety belt buckles would allow the system to alter deployment based on belt usage. For example, a higher speed threshold could be used for bag deployment when an occupant is belted.
- **Belt Force Limiters**. These devices limit the peak loads on the belts in severe crashes.
- **Belt Webbing Spool-out Sensors**. By measuring the amount of shoulder belt webbing unreeled from the retractor, these sensors may give an indirect indication of a belted out of position occupant.
- **Control Module**. A central computer receives signals from the crash severity and occupant sensors and controls the deployment of the dual level inflators and belt pretensioners in order to minimize occupant injury risk.

Some components of advanced systems have been in production already. As described previously in this book, GM introduced a dual level passenger side air bag inflator with crash severity sensing in 1973. BMW included an occupant sensor in its 1994 vehicles to suppress air bag deployment for small occupants. NHTSA has described a Mercedes dual threshold system based on belt usage. Dual speed thresholds for deploying air bags have been used in Mercedes-Benz vehicles produced for the U.S. market for several years. In these vehicles, the lower threshold for air bag deployment is approximately 19 km/h (12 mph) when an occupant is unbelted and a higher threshold of approximately 29 km/h (18 mph) is utilized when the an occupant is belted. A belt buckle switch provides the information to allow the selection between these two thresholds."[24]

In advance of a "smart" air bag rule, manufacturers continued to introduce advanced restraint technologies. For example, in the fall of 1999, Ford introduced a "smart" restraint system on the 2000 model year Ford Taurus and Mercury Sable. That system included all the technologies listed above except for the passenger occupant sensor and spool-out sensor. As noted above, these "occupant classification sensors" were not yet reliable enough for production.

Engineering Challenges

The sheer number of additional test requirements will compel manufacturers to employ ever more efficient engineering tools, such as finite element structural analysis and occupant simulation models to keep development costs and lead times practical. Some other challenges are:

Suppression: To minimize the injury risk from inflating air bags, tests with one, three, and six-year-old child dummies on the passenger side, and tests with a 5th percentile female dummy on the driver side were added. In meeting these requirements, manufacturers can choose between "suppression" and "low-risk" deployment. "Suppression" means that the air bag would not deploy when the seat is occupied by one of the child or small adult dummies, but would deploy when occupied by a larger dummy. Another choice for the manufacturer for the child dummy suppression certification is to use a sensor that detects when the dummy is too close to the air bag. There is no provision for air bag protection when the system sensors suppress inflation.

Low risk deployment – This certification option would require a static air bag deployment test with a dummy against the steering wheel or the instrument panel. Injury criteria must be below specified levels to pass this test. To meet this test, first stage air bag power would need to be reduced to a level that would not exceed injury criteria in the child and small adult dummies used in test. The dilemma for the air bag designer is that a first stage power level that will not harm a small child could be ineffective for an adult in a low-speed collision. Therefore, it will be very difficult to certify a vehicle on the passenger side using the low-risk deployment strategy.

As noted in the manufacturers' comments to the rulemaking, the "suppression" systems have developed slower than expected. Nor has a successful "low-risk" deployment air bag been developed for the passenger side. It is not the intent here to exhaustively list all types of occupant classification and out-of-position sensors. The following list includes some examples of different types of these sensors that have been under development and some of the challenges in making these systems work reliably:

o **Seat weight sensors** – these use load cells in the seat cushion or on the seat track to estimate the occupant weight, and classify the occupant as either a 6-year-old child or smaller, or as an adult. These sensors have had problems reliably classifying all occupants under some real world conditions due to large variations in weight seen by the sensors for the same size occupant.
o **Occupant pressure distribution** – these sensors use a matrix of strain gages or load cells in the seat cushion to measure the pressure profile caused by the occupant's buttocks and legs pressing on the seat. By relating the pressure distributions to those of humans, it may be possible to classify adult and child occupants. In addition, the pressure distribution sensors may be able to recognize a child seat. The Alliance of Automobile Manufacturers and its contractor, First Technology Safety Systems, have developed special dummies (OCATD) with anatomically correct buttocks to be used as standard test devices for these systems. AAM has petitioned NHTSA to allow these test devices for compliance testing in lieu of Hybrid III dummies or humans.
o **Infrared, Sonar, Capacitance** – these sensors can estimate an object's presence, size and distance from the air bag and potentially sense an out-of-position occupant.
o **Video** – using a video camera and image recognition software, these systems could potentially classify occupants by their images.

---

[24] Federal Register/Vol. 65, No. 93/Friday, May 12, 2000

As noted in manufacturers' comments to the rule, there is no technology currently available that is reliable enough to classify occupants flawlessly when the advanced air bag regulation takes effect. This lack of reliability could result in a child getting the higher power air bag deployment or an adult getting the lower power when a high power is needed. Both scenarios could result in serious injury or death. In its response to comments on this issue, the NHTSA has suggested that passenger seat occupants should observe the required "air bag deactivated" warning light on the instrument panel, and adjust their seating position accordingly. This, of course, pre-supposes that all occupants will have a sufficient understanding of the system to make a correct decision.

On the driver's side, Ford has used a seat position switch that suppresses the high level of the dual-level inflator when the seat is toward the front of its adjustment, which would be typical of a small driver.

The Future

More technologies are being researched to try to make air bag deployments less risky. Among these are:

o   Continuously variable inflators – instead of deploying at two or more distinct power levels, these inflators could inflate at any power level within its range as requested by the control module.
o   Pre-crash sensing – sensors that use, for example radar or video, to detect a collision a few feet before impact. The extra distance would allow air bags to inflate less rapidly, reducing the potential for injury.

Although greatly reduced by air bag depowering, "kids in back" public information campaigns and other measures, air bag related fatalities still occur (Figure 3). As of April 1, 2002, NHTSA has confirmed a total of 208 air bag related fatalities in the US and Puerto Rico, including 129 children.

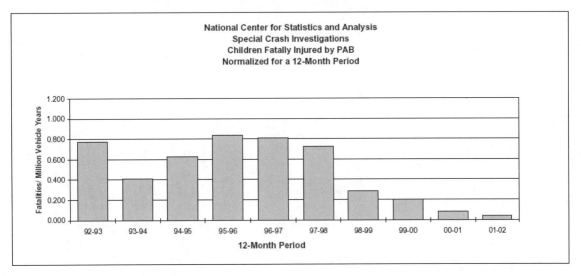

Figure 3. Children fatally injured by the passenger air bag. Source NHTSA SCI website.

Summary

The air restraint system was conceived with high hopes many years ago and has had more basic research and development effort, as evidenced by the almost 800 technical papers cited in this book, than any other occupant protection device. As this history is written, new standards are in place and are being proposed that challenge the ingenuity of researchers and engineers to satisfy the necessary requirements of a system that is not passive, but very active. The most rigorous challenge, as always, is to develop devices that will survive the durability demands of automobiles that will be used in all kinds of field conditions for many years. As an example, the Insurance Institute for Highway Safety found one of the original 1973 General Motors Field Trial Chevrolets after fifteen years in regular use and used it as the test vehicle in one of the first public demonstrations of their new crash test facility. The air bag system functioned reliably as designed.

In simple words, the field and test history, over the almost thirty years that air bag systems have been on the public roads, has shown that a sensibly powered and designed system can provide injury reduction, but that the potential still exists for the systems to create serious injuries.   The challenge is to establish performance standards which take advantage of proven technologies to minimize injury risks.

In an October 7, 2002 editorial *USA Today* notes that NHTSA's "…hurried pace (to introduce smart air bags) is particularly misguided given that changes NHTSA already has adopted are working remarkably well."

Chapter 3

European Air Bag Development

Karl Erik Nilsson, Hansjürgen Scholz, Hugo Mellander

Over the last 35 years, no specific issue in automotive development has been debated and reported like the air bag. This matter of public health was a technological challenge with significant costs, risks, and expectations involved. It was not only a concern of the broad car driving public, but a responsibility of manufacturers and of society. After two decades of US legislation and "learning-by-doing" efforts, the air bag experienced enormous success in the early 1990s. Today, not one new car – whether in the USA or in Europe – is sold without at least two air bags; with some cars including eight of them.

One could think that everything has already been written about air bags. In retrospect, however, different approaches to the air bag in the USA and Europe deserve some historical illumination. In addition, much of the basic development and industrialization effort that led to a realization of the air bag was never published.

Seat belts experienced faster development and acceptance in Europe, while the low "buckle up" rates in the USA called for air bags as "passive" restraints. The US legislation forced European manufacturers, who exported cars to the USA, to consider air bags. Early industrial, institutional, and legislative efforts in the USA provided basic guidelines, but the difficult way to convert and apply some of the needed military technologies into commercial use was anything but given. The US legislation experienced massive resistance, especially from the US car manufacturers, and became a matter of decade-long controversial debates and adjustments.

While preparing for this US legislation, Daimler-Benz (today DaimlerChrysler) believed that the air bag would have the best safety potential as a supplement to the seat belt rather than as a stand-alone passive restraint. Unlike their contemporaries in the US, Europeans never expected the air bag to replace the seat belt. They recognized that the seat belt is the primary life saver and that the air bag is best suited as a supplement to the belt. In fact, Volvo proposed a smaller version of the air bag for the largely belted European population. The European air bag development, as demonstrated by Daimler-Benz, was driven by an early dedication to apply appropriate technologies and to test these thoroughly over sufficient time. The goal was to prove a level of reliability that would remove any doubt about the efficiency of this restraint system. In December of 1980, Daimler-Benz's European introduction of an optional driver air bag (including a belt pretensioner for the passenger) initiated a decade of real-world verification to show the air bag system to be technically reliable, efficient, understood, and accepted by car customers. From that point in time, the air bag became primarily a market-driven issue. In contrast to the seat belt, air bags never became required by law in Europe.

In the following, the experiences from two development sides will focus on the industrial participation in the European air bag history. The perspectives are from the car manufacturers' side, mainly from Daimler-Benz as the driving force for the realization of the air bag, and from the supply industry side with editorial coordination. This historic outlook contains documentation and personal impressions gained not only from involvement in the development of specific air bag systems, but also from many years of pioneering occupant safety in Europe. Opinions and judgments are those of the authors. Because this text is drawn mainly from personal experience, references are made to only a small fraction of the air bag literature available. This historic research can be further studied in the selected references presented at the end of the chapter and in the bibliography presented at the end of the book.

Introduction

The first known air bag patent (DP 896312) was applied for by Mr. Walter Linderer, Munich, in October of 1951. It described an inflatable, elastic container for protection of vehicle occupants. At that time, when even seat belts were not considered standard equipment, Mr. Linderer was much too early to get a serious response to his visionary idea. The American, Mr. John W. Hetrick, did not get better response regarding his patent (US 2649311) for a "safety cushion assembly for automotive vehicles", for which he applied in 1952. Despite his active presentation to car manufacturers, he could not generate interest for this idea.

New visions and serious attempts from automotive manufacturers to improve occupant safety in crashes came up in the early 1950s. In January of 1951, Daimler-Benz AG applied for a patent (DP 854157) for a safety car body with a stiff occupant compartment and softer crash zones in the front and rear. Other safety-related patents soon followed, including one in 1952 for an air cushion which automatically inflates during a car deceleration, one in 1953 for an inflatable bag which is installed in the dash board, and one in 1954 for a safety steering wheel with energy absorption. First investigations on air bags, performed by Ford and General Motors in the US in the later 1950s, indicated that the development of such an inflatable system, based on pressurized gas vessels, would be too difficult to become practicable.

For that decade, the introduction of seat belts was enough of a challenge. The 1950s ended relatively successfully in Europe, with 3-point seat belts being introduced as standard equipment in 1959 Volvos. However, in the USA – the big automobile nation – the lack of seat belt acceptance and the rapidly growing number of victims on the roads forced Congress to act. In 1966, The National Traffic and Motor Vehicle Safety Act was signed into law and a federal agency was established to "reduce traffic accidents and deaths and injuries to persons resulting from traffic accidents."

Meanwhile, the air bag idea had been taken up more seriously in the USA and was supported by surgeons and consumer groups, led by the young lawyer, Ralph Nader, who wrote the legendary book Unsafe at Any Speed in 1965. In the industry, Eaton, Yale and Towne (later Eaton Corporation) started development of air bags based on a pressurized gas vessel technique, a so-called stored gas system, in the mid 1960s. Further investigations by Ford/Eaton, General Motors/Allied Chemical, and others provided information to the safety agency for considering the air bag as a means to reduce the consequences of traffic crashes.

In July 1969, The National Highway Traffic Safety *Bureau* (later *Administration*, NHTSA) published an Advance Notice of Proposed Rulemaking for installation of air bags in all new passenger cars, effective January 1, 1972. NHTSA's rulemaking was based on their own research in cooperation with the industry. The notice was followed by an abrupt end to the US car manufacturers' support of the air bag. Only General Motors continued their plans for some time, while warning that a small child standing close to the instrument panel of a car could be seriously injured or even killed by an air bag.

The upcoming air bag requirements also challenged European manufacturers, which exported cars into the US market. Air bags would require "explosives" and a rapid application of military technologies with quite new aspects regarding mass production and approval for commercial use. As a result of these challenges, air bags would later become recognized as one of the most – if not the most – extensively developed products in automotive history.

The "Learning-by-Doing" Way of Daimler-Benz

Daimler-Benz ("DB") started their first air bag principle tests in 1966, based on a fluid Freon and solid propellant ("hybrid") principle developed by Yanase, Japan. Over the next two years DB tried to improve this technology, including a switch in 1967 to solid propellants produced by Dynamit Nobel, Fürth. No crash simulation equipment was available on the market, however, so for dynamic tests a strong engine pulled a sled via a wire or, for crash tests, a steam rocket was used to accelerate the car.

The primary purpose of this experimental phase was to learn whether a bag could be inflated fast enough to provide a basic level of protection. The gas-generation principle was soon considered insufficient and necessitated a follow-up study. This was primarily due to the dangerous nature of the pyrotechnic propellant. Therefore, the employees directly involved were sent to a two-week training course in order to get an official permit to handle explosives.

Logically, DB contacted US companies such as Eaton, Ohlin, and Allied Chemical, which had started the development of air bag technology based on stored gas. Thiokol Corporation, which initiated research based on solid propellant at the request of Chrysler in 1968, and Rocket Research were also contacted.

In 1970, basic investigations were initiated with different potential suppliers and institutions. These efforts spanned a large range, from tests of an Eaton crash sensor to establish threshold criteria to sound pressure tests to judge human tolerances. Discussions at the "Materialprüfanstalt" (MPA, later BAM) in Berlin included such topics as the legal aspects of introducing pyrotechnics into the manufacturing process and into publicly marketed cars. Thiokol visited DB in May of 1970 to discuss solid propellant gas generators. Throughout 1970, a large number of tests with components from different companies, mainly from the USA, were carried out or planned.

In October of 1970, DB started investigation of air bag modules with Thiokol driver and Olin passenger gas generators by sled tests performed at Calspan in Buffalo, New York. Instead of a crash sensor, a time switch was used for ignition. The tests were considered to be successful, as they showed that NHTSA's injury criteria for the head (HIC <1000), for chest acceleration (<60 g), and for femur load (10 kN) could basically be fulfilled without a

3-point belt in a 48 km/h head-on collision. Integration constrains were found, however, in particular with the bulky Olin pressurized gas vessel.

DB established an important new partnership in 1970. An engineer, who had left DB at the end of 1969 to assume a new position at Bayern-Chemie Gesellschaft für flugchemische Antriebe mbH (BC), a subsidiary of Messerschmitt-Bölkow-Blohm GmbH (MBB), Ottobrunn by Munich (and at that time still Wasag Chemie), was contacted from his earlier colleagues regarding gas generation based on solid propellant. After a preliminary positive response, DB sent a letter to BC in July 1970 as well as to some other potential partners regarding the urgent issue of gas generation for air bags in passenger cars. BC confirmed its basic development capabilities and interest in a letter to DB in August 1970.

DB invited BC for a first meeting in December of 1970. Meanwhile, the US legislation had been delayed, but DB still demanded air bags by mid-1973. BC proposed a solid propellant gas generator, small enough to be integrated in DB's concept for an energy absorbing steering wheel, with air bag module capable of inflating a 60-liter bag. From some given bag inflation data, BC estimated that the sound pressure would reach a 170 decibel level, caused by the rapid displacement of the air. As a result of this meeting, the parties agreed to start a basic air bag development program immediately.

In March of 1971, the US regulation, FMVSS 208, required all cars to be equipped with "passive" restraints (either "automatic" seat belts or air bags) by 1974. This provoked immediate reactions, mainly from the US car manufacturers. Their primary argument, which would be bolstered by experiences in the coming years, was that no air bag system could have been ready in time to provide a sufficient reliability and liability base. The 1971 NHTSA ruling was followed by more than twenty years of manufacturer-government controversy before the air bag mandate was finally signed into US law at the end of 1991.

As a push for air bags and other technologies, US Department of Transportation (DOT) requested car manufacturers in 1970 to participate in the development of experimental safety vehicles (ESV). These vehicles would be designed to meet the established injury criteria at a crash test speed of 80 km/h. The concept cars required reinforcements of the bodies, elongation of the front-end and the consideration of four air bags (for the front and rear seats). The steam rocket at DB was not powerful enough for sufficient acceleration of the two-ton experimental car. Pragmatically, a BC solid propellant rocket was selected to add thrust and solve that problem.

General Motors continued its air bag program by offering driver and passenger air bags, with subsidized costs, in 1974 to 1976 model year vehicles. Despite the financial incentives, only approximately 10,000 cars were equipped with an air bag as a result of this program. The program was much smaller than planned and did not meet the initial expectations. General Motors concluded that there was not only no demand for air bags, but that they were actively resisted by the customers.

In parallel, Volvo was developing a stored gas air bag system with Eaton for a Volvo Experimental Safety Car (VESC) as part of the ESV program. Seventy-five cars of the 240 model were also equipped with driver and passenger air bags and tested on the American roads in the mid-seventies. Of these cars, Allstate Insurance was driving 50, Eaton 20 and Volvo 5 cars. The experience was analyzed, but did not lead to a continuation of the program.

In early 1974, DB started an investigation of the air bag's effect on a standing child. Beginning in the same year, Volvo and Chalmers University of Technology, Gothenburg, made similar investigations and published a report on "Possible Effects of Air bag Inflation on a Standing Child" in 1978. It focused on the occupant "out-of-position" (OOP) problem, which would eventually be observed in some field cases with unbelted occupants.

DB did not favor the option of automatic seat belts in the US regulation of 1971, but continued their air bag development efforts. The concentration at BC on one dedicated partner certainly created the best conditions for a successful development in the decade to follow. On the other hand, the general attitude of car manufacturers to the air bag, the recognized air bag risks, and the uncertain outcome of the US legislation often reduced the motivation to continue the challenge to a few enthusiastic individuals. The positive side of the long list of predicted risks was a thorough consideration of related matters in the development.

As the cooperative work of DB and BC continued, early cylindrical bags had to be redesigned since the fabric was penetrated by the hot gas as the bag inflated. Thicker nylon fabric and a shield portion were used to address this problem. The shield or "parachute" also restricted the air bag's extension in the direction of the driver's face and provided a better bag shape. In May 1971, DB visited Ottobrunn for the first feasibility tests. These confirmed realistic bag inflation and duration times in an "S-class, W108" body, which justified a decision for further evaluations in dynamic tests. Furthermore, the sound pressure reached only 158 dB (reflected 163 dB), which was far below the earlier prediction. It was agreed that BC/MBB would investigate the medical aspects of sound pressure, gas toxicity, and gas temperature. The transformation of military standards to simulate real-world long-term aging of the propellant and the establishment of test specifications were brought up as matters for further investigation. To deal with these issues, BC proposed a cooperative effort with the "Institut der Chemie für Explosiv - und Treibstoffe, ITC" in Pfinztal-Berghausen.

DB confirmed that the tested gas generator, the lightest and by far the smallest of all proposals, was acceptable for integration in the planned steering wheel. BC even expected that the size (diameter 103 mm, height 54 mm) and weight (550 g) could be further reduced – not yet knowing that the nitrocellulose propellant would not fulfill DB's requirements.

As a pragmatic way to evaluate the immediate environment with a sensitive toxicity "instrument", a mechanic at BC brought a cage of canary birds to his work place. The couple soon hatched a new generation, providing evidence that the long-term toxicity risk was low. As no reliable data were available on the auditory effects of sound pressures above 130 dB, the influence from air bags started to be thoroughly investigated in tests with volunteers in 1972. No damages to the hearing were found and – as a sign of enthusiasm – one engineer even claimed to hear better after the tests.

A new crash facility was in construction at DB in Sindelfingen, but it would not be ready until the end of 1972. Initial dynamic tests with the BC gas generator were therefore performed on a Bendix sled at Porsche in Weissach in July 1971. They showed encouraging protection performance. In a follow-up meeting between MBB/BC and DB including top management further cooperation was discussed and agreed upon. The topics included proprietary and financial terms and the system coordination, which would be the responsibility of DB.

Primarily, BC had brought in a patent application for a crash protection system with a rocket booster type of gas generator. This generator had a fast burning charge, a centrally arranged ignition unit, and a solid cooling ring with a circular layout of combustion chamber and "Laval" nozzle. These patent claims were soon found less relevant for an ideal gas generator design. As mutual ideas, DB and BC had applied for patents for a cushion shape and also for a knee protection system based on aluminum "honeycomb" structures. While this knee protection system was later supplanted by a thin transversal tube, the honeycomb aircraft design element became widely used for energy absorption in dynamic tests.

The development of an appropriate air bag gas propellant was particularly challenging. Nothing available at that time could fulfill DB's high requirements on non-toxicity and long   term chemical stability at high temperatures. The preferred production of nitrogen and avoidance of carbon monoxide led to the decision of BC to change to an inorganic system based on sodium azide.

Almost simultaneously, BC, Canadian Industries (CIL), Thiokol, and Talley filed patent applications for sodium azide-based gas propellants with different oxygen carriers. Thiokol used molybdenum disulfide (plus sulfur); Talley used copper oxide; and BC and CIL used silicon dioxide plus potassium nitrate. It was later recognized that the patents of BC and CIL did not only cover the same claims but also had the same priority date of July 24, 1972. CIL became the major supplier of sodium azide to the gas generator manufacturers. Other suppliers were Dynamit Nobel, Troisdorf, and Daicel, Japan.

In early 1973, BC started the development of a prototype gas generator for the driver side, based on the sodium azide propellant that had been in development for almost a year. The propellant delivered non-toxic, almost pure nitrogen gas and was chemically stable in regard to the high temperature and long-time aging requirements. Known as a toxic chemical, sodium azide had been processed and used as fertilizer, for cleaning, etc. for a long time. Precautions due to toxicity risks were nevertheless required and the manufacturing process had to be free from heavy metals to avoid explosive reactions.

BC suffered a fatal accident when a person misused a steel rod to support the material flow in a grinding/mixing operation. This lead to the development of a safer wet process. Positive cooperative arrangements with the chemical union and "Gewerbeaufsichtsamt" in Munich supported the development of safe processes and storage conditions. For independent chemical judgments and approval, investigations were carried out by ICT, later re-named "Frauenhofer-Institut für Treib- und Explosivstoffe". The "Materialprüfanstalt" (MPA) in Berlin, meanwhile twice re-named and today "Bundesanstalt für Materialforschung und –prüfung" (BAM) was responsible for testing to reach national transportation and storage approvals.

A main disadvantage of sodium azide propellants was the production of some 60 percent slag, which required voluminous and expensive filtering to avoid inhalation of alkaline dust and heat transferring particles. These particles could damage the bag and possibly burn the exposed skin of occupants. Regardless, it was believed for many years that the established specifications would not allow a substitute to the inorganic sodium azide propellants.

The wet propellant powder was pressed into pellets and dried. Size and relation between total surface and weight of the ca. 75-gram load provided defined burning rates in the combustion chamber and pressure-time diagrams in a 60 liter test tank at +20, −35 and +85°C. The pellets were ordered in screen layers in the combustion chamber. BC applied for a patent for an all-side gas-permeable container for fixation of the pellets. The purpose was to prevent pellets from cracks, which would increase the burning rate and pressure.

Similar to other military appliances, the gas generator ignition chain started with the electrical signal (from the sensor) to a squib bridge wire, surrounded by a small drop of initial explosive and some black powder. This was placed together with a transmission charge (ca. 2 grams) in a central screw with nozzles, providing hot particles for a defined combustion of the pellets. The gas passed an axial filter chamber for cooling and slag/particle absorption.

The gas generator was defined as a prototype to be qualified primarily for internal DB tests. Main steel housing parts were expensively machined, chrome-plated against corrosion, and screwed together with seals for hermetic protection of the pyrotechnics. Even the axial gas outlet nozzles were expensive to make. With a 110-mm diameter and a 40-mm height, the weight of the "F12" prototype reached 1.6 kg (Figure 1). However, it needed additional attachment parts and the axial gas outlet required an external secondary 0.2 kg filter. All this was tolerable for this development stage and the basic function was the best known so far, thus serving the internal test purpose well.

Figure 1. F12 prototype gas generator (Bayern-Chemie).

A first proposal for an air bag specification was developed among DB, BC, and ITC at the end of 1971. After a year of F12 improvements and some adjustments to the requirements, a preliminary DB specification could be issued in 1974. This specified various requirements, including toxicity data and gas performance at +20, -35 and +85°C temperature, weight, size, igniter electrical data, burst and fire tests, and safety instructions. It focused on the environmental test program, including temperature storage and cycling, humidity, vibration, shock and corrosion tests, which simulated functional reliability after 10 years of real world application. This was deemed necessary in order for the system to qualify for production, transportation, and commercial use.

Test drives with electromechanical "spring-mass" crash sensors since the beginning of 1970 did not convince DB that this principle would provide sufficient discrimination of different crash situations. Therefore, at the recommendation of BC, the development of an electronic crash sensor and a diagnostic unit was initiated between DB and MBB's division for missile electronics in Schrobenhausen in January of 1972. After only one month, a first functional sample, based on the piezoelectric principle, was delivered for sled tests. Two months later, an optimized version was used in worst-case road test drives to determine threshold limits. Beside a 4 g threshold definition, the sensor could provide a reaction time of 6 ms, by far the best of the few known alternatives.

The sensor and diagnostic units were designed to detect the car body crash pulse at a central place within the compartment, supported by a relatively stiff body structure. Thus, it was different than the electromechanical spring-mass principle first used by Eaton and later by Technar (later TRW), or the ball-in-tube (Breed) principle, later widely used in the USA. The electromechanical sensors needed a dual front application with wiring through the engine compartment. The crash sensor development proved to become an extraordinary difficult matter, requiring thorough and time consuming attention.

The first sensor/diagnostic specification was issued by DB in August of 1973. These inflator/module and crash sensor/diagnostic specifications were further refined over the next five years. The basics of this pioneering work were eventually accepted throughout the industry and are, with product specific adjustments, still considered as standards.

Other air bag research was being performed concurrently, including investigations of drivers' reaction to an inadvertent deployment while driving. All categories of volunteers managed the tests without any failures when the air bag was activated suddenly while driving. The bag deflated after 100 ms and disappeared fast enough not to hinder the sight. Neither the bag itself, nor the sound – like from a gun – caused any disturbing reaction.

BC/MBB's experience from missiles and its wiring to create a potential risk as flying "antennas" gave rise to electromagnetic interference concerns for the air bag. Pragmatic test drives close to the strongest radio stations caused a few electrical functions to fail; though these were not of major concern at that test condition. Still, in order to prevent inadvertent ignition in an increasing electromagnetic environment, a high frequency filter was integrated in a specially developed connector at Amphenol-Tuchel in Heilbronn. To secure a reliable electric connection, strict attention was paid to the pull and push forces between the connector and the gas generator igniter.

Through these development achievements, DB became convinced of the air bag system's potential and, in August of 1974, decided internally to continue these development efforts for introduction as soon as possible – regardless of the uncertain outcome of the US legislation.

At BC, 1,000 F12 prototypes were built and 500 units were environmentally tested as specified. The qualification was completed in mid-1975. Conditional to this qualification, DB had ordered 1,300 units a few months earlier, of which the majority was a somewhat higher version for a dual arrangement in a 160-liter passenger air bag module. Fleet tests of these units were planned in preparation of a US demonstration program.

However, a few F12 units were found to rupture – the worst case scenario – due to high temperature overpressure and screw connection design and tolerance problems. After process and design improvements, the gas generator passed a re-qualification in May of 1977. The ordered gas generators could be delivered and, with some delay, DB started the planned fleet tests program with 250 cars.

Despite the technological lead MBB had in the sensor development, time and cost constraints caused DB to interrupt this development in early 1977 and continue to volume production with Robert Bosch GmbH in Stuttgart-Schwieberdingen.

In October 1977, DB decided internally to offer an air bag option in the new S-class and to announce this decision at the International Automobile Exhibition (IAA) in Frankfurt in the late summer of 1979. For the US demonstration program, DB requested alternative quotations for 2,500 and 4,500 units.

Meanwhile, general confidence regarding the future of air bags had reached a low point at BC. The outcome of the US legislation seemed uncertain. DB put massive pressure on BC to continue the development with more resources, but was the only car manufacturer seriously working on air bags. In November of 1977, BC informed DB of its intention to stop this development and spend the resources on military programs. BC would be ready to quote only 2,500 F12 gas generators.

DB immediately intensified its contacts with the US manufacturers Eaton, Rocket Research, Talley, and Thiokol. The pioneer Eaton then stopped its air bag program in 1978. Thiokol was awarded the gas generator supply for the planned US demonstration program, which had to be delayed two years until 1982, and manufactured 17,000 inflators for this program.

In 1977, NHTSA had instigated a demonstration program, which was to include 500,000 cars. General Motors and Ford had agreed to support this program, but would supply only 60,000 cars each. DB had agreed to supply 2,250 cars. A new situation came up in December 1977, however, as NHTSA required a three-year phase-in of passive restraints, starting in September of 1981. Four months after this announcement, NHTSA cancelled the demonstration program. Nevertheless, DB carried out extensive system-specific tests in 1978 to investigate potential problems like inadvertent deployment, pipe smoking, and the wearing of glasses. A broader fleet investigation program was initiated in mid-1978 using an internal test fleet of 700 cars fitted also with crash sensor prototypes from Bosch.

Meanwhile, concerns about the safe handling of a separate gas generator with axial gas outlets came up. When firing an F12 with the exhaust aimed at the floor, the generator could exceed 200 km/h when it struck a 2.4 m high ceiling, leaving a dent there before returning to its original position. This rocket function was impressive, but less useful in regard to commercial handling and approval.

It also became quite clear, at least to the BC project manager, who had production experience in the automotive supply industry, that the complicated design of the F12 prototype was not a realistic base for higher-volume production. Design improvements were proposed, but the management of BC was not ready to spend money on further development. This critical situation was reported from the BC project side to the head of MBB in December of 1977. Suddenly, the air bag interest within BC was back again and hectic plans for a high volume manufacturing of the F12 took place – without participation of the project manager. A replacement of the general management followed. The two new general managers left their positions as assistants to the head of MBB and started the BC appointments in the autumn of 1978.

Unofficially, the BC project manager had shown ideas for a more realistic sheet metal design for a gas generator to the air bag project manager of DB, who strongly supported it. The new gas generators, or "inflators", which Thiokol and Talley presented to European car manufacturers, contained all filters and had radial exhausts, thus being thrust neutral. The complete weight of these inflators was about 1,200 grams. DB revised its specification for the gas generator to require a weight not exceeding 1,200 grams and a nearly thrust neutral function, defined as a maximum jump height of 1 meter.

In February of 1979, DB invited BC to a major meeting, asking for a new gas generator compatible with the new specification. DB requested a high volume design and serial production until September 1980. One week later, BC confirmed in a letter that this would be possible.

Petri AG in Aschaffenburg was the partner of DB for the steering wheel, including assembly and integration of the driver air bag module. BC intensified its direct contacts with Petri and found in this supplier a qualified partner for the manufacturing of some inflator parts. BC supported Petri´s establishment of its own equipment and procedures for the acceptance testing.

The "GG4" (Figure 2) development, qualification, and approvals were completed, and new supply relations, processes, and production equipment were established within the required 18 months. On September 1, 1980, a first serial lot of 30 inflators was supplied for initial DB street approval.

Figure 2. GG4 production driver inflator (Bayern-Chemie).

The housing of the small GG4 was of stainless steel sheet parts, creating a geometrically optimized and homogenous welded container structure. The use of an electron beam technique allowed minimal heat transfer to the pyrotechnics and prevented tension cracks and stress corrosion. Without seal additions (except for normal nozzle foil), it was hermetically tight, corrosion- and tamper-proof. The pyrotechnics were taken over from F12 and provided a very clean gas consisting mainly of nitrogen (Figure 3). Gas performance deviations in the temperature range (Figure 4) could be reduced by mechanical means.

Nitrogen	N_2	96%
Hydrogen	H_2	2%
Oxygen	O_2	1%
Carbon dioxide	CO_2	1%
Hydrocarbons	C_XH_Y	< 900 ppm
Carbon monoxide	CO	< 160 ppm
Ammonia	NH_3	< 85 ppm
Nitrous oxides	NO, $(NO)_X$	< 65 ppm
Hydrogen chloride	HCl	< 2 ppm

Figure 3. GG4 gas emission (ITC, February 1980).

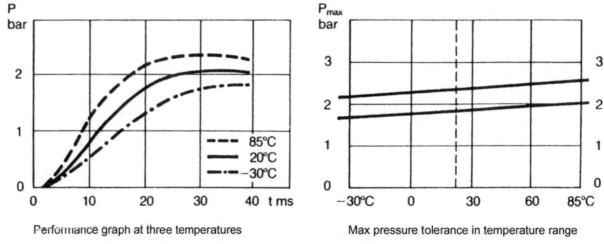

Figure 4. GG4 gas performance (60-liter tank), tolerance in temperature range.

Similar to the US inflators, the radial exhaust allowed sufficient filter volume to eliminate the need for a secondary filter. The principal function is shown in Figure 5. At a weight of 890 grams (including flange), the burst safety factor exceeded twice the required value. Thus, the sheet thickness of the housing was not optimal. First, however, the tight schedule did not allow a safe optimization of the thickness and, second, for a later redesign a change to aluminum with the same sheet thickness already took shape. This would allow a further use of most of the GG4 parts, tooling, and equipment.

With the rescission of FMVSS 208 in October of 1981, the DB's plans to introduce a driver air bag option in the USA were postponed until September of 1984. When this US option became available, DB used the inflators already produced for the US program by Thiokol, which differed from the European program.

The electronic sensor of Bosch was an analog device with a strain gauge accelerometer. It was centrally placed on the mid tunnel. To this, Bosch delivered a separate control/diagnostic unit. The function was not only to detect the crash severity and provide a correctly timed electrical current to the gas generator igniter, but also to check the system via a control current. It also stored functional information and provided reserve energy in case of an interruption of the connection from the battery. For ten seconds, a flashing light on the dash board confirmed the system readiness at every start. In case of continuous (or no) light, service was requested.

For optimal protection and prevention of a "blow into the face", a flatter and broader shape of the inflated bag was essential. Therefore, tethers were integrated between the bag mouth and the central front area. To avoid a possible burn-through of the bag by hot particles, DB replaced the early shield by neoprene coating of the fabric inside. Vent holes close to the bag mouth provided a defined gas evacuation for appropriate energy absorption when the occupant loaded the bag. Another difficulty was to prevent material damage of the bag after being folded inside the steering wheel cover for at least ten years. Even after this long, the bag had to be ready to be filled within one 25th of a second also in the difficult condition of $-35°C$ temperature. The fabric was woven from a fine polyamide 6.6 yarn into a 970 dtex specific mass/strength. DB's bag partners were Enka Glanzstoff, Obernburg (later Akzo) for the fabric, Phoenix AG, Hildesheim for confection, and Petri for the assembly of the module.

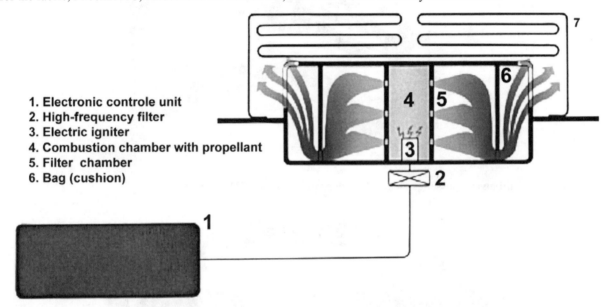

Figure 5. Inflator function principle.

One challenge in DB's development with Petri was the cover of the air bag steering wheel. It should withstand the static pressure of the folded bag over ten years and be opened from the rapidly inflating bag at temperatures between –35 and +85°C in a defined process without fragmentation. Moreover, it had to harmonize well in design, color, and structure with the steering wheel in the direct view of the driver. As used in the steering wheel rim, a polyurethane foam with a reinforcement fabric, connected to a circulating belt, was finally found, tested, and released.

The electrical supply to the steering wheel required a reliable and noise-free electrical connection. Initially, coal brushes and slip rings, protected against humidity, were used. Later, DB together with Petri developed a clock-spring for the contact transfer.

In September 1980, DB started a street test program. This program included 600 cars equipped with a driver-side air bag and a passenger-side pyrotechnic retractor pretensioner, also triggered by the air bag crash sensor. The pretensioner eliminated the webbing slack, thus utilizing the available space to the dash board for improved energy absorption and seat belt efficiency.

Ten Years of Market-Driven Air Bag Introduction

DB introduced the driver air bag and passenger belt pretensioner as an option (Figure 6 and Figure 7) in Europe at the end of 1980, initially for the S-class and later for other models. This was the first air bag offered in Europe and was rapidly recognized as an important milestone for the future of air bags, not only by providing direction for later fulfillment of the US passive requirement, but also for improved occupant safety in general. To avoid the risk that the air bag may be seen as a substitute for seat belts, DB named it a "Supplemental Restraint System, SRS".

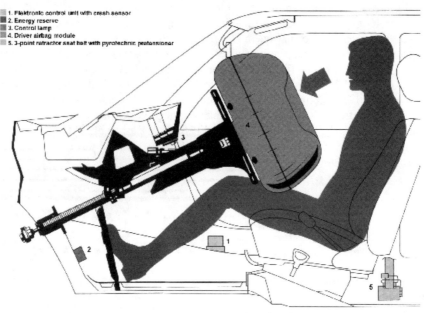

Figure 6. Driver airbag and passenger seat belt pretensioner option of Daimler-Benz.

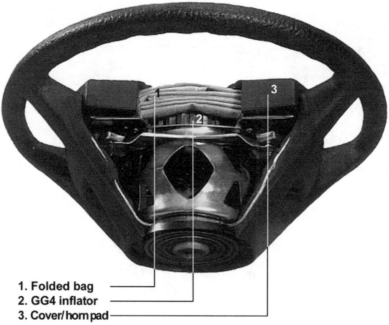

Figure 7. Daimler-Benz airbag steering wheel.

This initiative of DB was published in many newspapers and magazines. It showed the air bag function sequences in sled tests (Figure 8 through Figure 11) and how the bag inflates out of the steering wheel (Figure 12 through Figure 15).

Figure 8 and 9. Driver airbag sled test (Daimler-Benz) 25 ms and 33 ms after impact

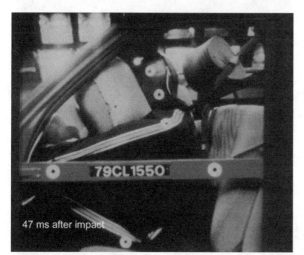 

Figure 10 and 11. Driver airbag sled test (Daimler-Benz) 47 ms and 90 ms after impact

The high development costs for DB and its suppliers justified an initial price for the option package of DM 1,800. Despite intensive advertisement and publications, the projected sales were lower than expected. Therefore, BC initiated a marketing study during 1981 and presented the results to DB and other potential customers. It showed that the air bag was already widely known and understood, but that a price level of 3.5 percent of the total car price was seen as the limit for ordering the air bag option. This meant that the air bag price must decrease in order to reach higher volumes.

Many European car manufacturers, such as BMW, Porsche, Volvo and Audi, which exported significant volumes into the US market, were also convinced that the air bag would provide a better alternative to the passive requirement than "automatic" seat belts. Especially in combination with a high use of the convenient 3-piont retractor belts, the air bag promised superior protection. DB´s initiative pushed these manufacturers to speed up their air bag plans.

Among those active in the US market, VW was initially the only European manufacturer to opt for a belt to satisfy the passive restraint requirement. To meet the US standard, VW provided a passive shoulder seat belt combined with a stationary knee bolster. Other volume manufacturers, like Fiat, Ford of Europe, Opel, PSA, and Renault had no US exports and were not forced to take an immediate position.

BC intensified its redesign efforts for an aluminum inflator in 1981, using the GG4 tools. DB became very supportive in switching over to an inflator, allowing further weight reduction to an incredible 490 grams. This would bring important benefits, including the reduction of steering wheel vibration. After a delivery of around 40,000 units of the GG4 steel version for the European option program of DB, the "GG5" aluminum version was qualified for production at the end of 1984. The GG5 dimensions, which are the same as those of the GG4, are shown in Figure 16.

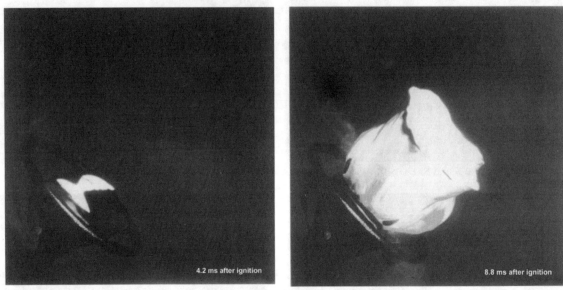

Figure 12 and 13. Driver airbag inflation (Daimler-Benz) 4.2 ms and 8.8 ms after ignition

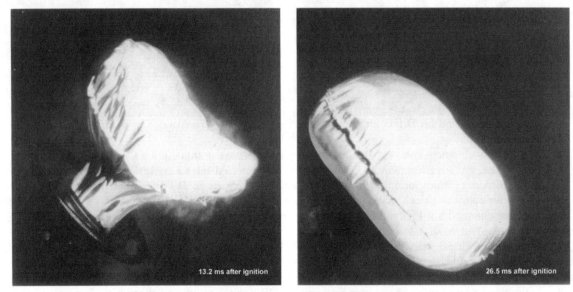

Figure 14 and 15. Driver airbag inflation (Daimler-Benz) 13.2 ms and 26.5 ms after ignition

This was followed in 1988 by a simplified "GG6" version in connection with a completely new air bag facility in Aschau am Inn. Since tests showed that a bulk load of the pellets had no negative influence – rather positive – the expensive process of screening the pellets was eliminated, thus further reducing the investment and manufacturing costs. The GG6 became the first high-volume inflator for most of the European air bag introduction programs.

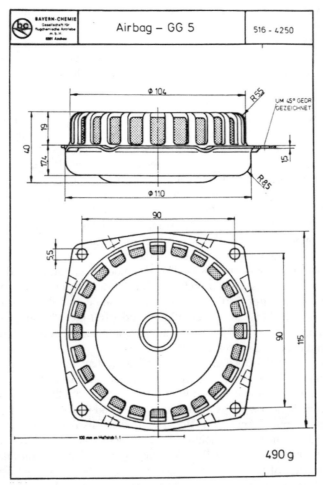

Figure 16. GG5 driver inflator dimensions (Bayern-Chemie).

Bosch introduced a second generation of the electronic sensor in 1984. With the change to digital technology, the sensor and diagnostics could be integrated into one central unit and the weight reduced to one third or 0.35 kg total.

Thiokol was awarded the inflators for DB's US demonstration program, rescheduled for 1982, and manufactured 17,000 units. After the October 1981 rescission of FMVSS 208, DB decided to use these inflators for a US driver option program in 1984. DB therefore asked Thiokol – in 1983 Morton-Thiokol (MT) – to develop a smaller and lighter aluminum inflator to better match the GG5 of BC. This "MTG3" was inertia welded and had a weight of 600 grams. The inflator design was considered to be excellent for high volume production.

For Mercedes' standard introduction of driver air bags in the US market in September of 1985, both of these technologically leading inflators were used. The MTG3 got a higher share than the GG5 for the US market, while DB continued only with GG5 in Europe.

BMW introduced driver air bags in the 5, 6 and 7 series as a European option in April of 1985 and as a US option two years later. For the US market, the long version of the 7 model ("L7") got the driver air bag as standard in September of 1985. BC supplied the inflator and the Japanese seat belt manufacturer, Takata Corporation, supplied a silicon-coated bag to the steering wheel and module supplier, Petri. BMW used sensors from Technar, Arcada (CA), and a diagnostic unit from Siemens AG, Regensburg.

In September of 1986, Porsche was the first manufacturer to introduce driver and passenger air bags, first for the US market as standard in the 944 Turbo model and as an option in the 944 and 944S models. Three BC driver inflators were used, two of them in a dual arrangement for the passenger module. The first inflator was ignited simultaneously with the driver inflator and the second one a few milliseconds later, thus creating a softer bag pressure onset. Phoenix delivered the bags to the steering wheel and module supplier, Kolbenschmidt AG (KS), in Aschaffenburg. Technar supplied the sensors and Siemens the diagnostic unit.

Volvo installed driver air bags for the US market from December 1986 on, as standard in the 780 model and as an option in the 740 model. The 760 model was equipped standard with the driver air bag for the USA in April of 1987. BC, Phoenix, Petri and Bosch were suppliers.

Audi developed a "procon-ten" system until 1986. During a crash, the engine displacement pulled seat belts tight and the steering wheel away from the driver via steel wires. This system was considered a substitute to the air bag, but it could not fulfill the US passive requirement. Starting in the European market, Audi introduced a driver air bag option for the 100 and 200 models at the end of 1987. Suppliers were BC, Phoenix, KS and Bosch.

As part of the electrical system, sensors and diagnostics were handled separately by the car manufacturers. The inflators, as module heart pieces, had taken a central place in the air bag development, which was carried out directly with the car manufacturers in all technical and commercial aspects. As a result of the interactions among different car manufacturers, a performance standardization was developed in addition to mutual specifications. A German working group for air bags ("Arbeitskreis") was formed with the participation of Audi, BMW, DB, Porsche, and VW and their related suppliers. With some exceptions for sensor/diagnostics and inflators, module suppliers started to take over more project coordination at the end of the 1980s.

Saab introduced driver air bags in 1987. The KS module contained a Morton-Thiokol inflator and a Phoenix bag. The sensor supplier was TRW (Technar) and the diagnostic supplier was Siemens. In 1988, DB added a passenger air bag as an option in the S-class. The module was developed together with the seat belt supplier, TRW Repa, Alfdorf. It had two BC driver inflators and a 160-liter Phoenix bag.

Alfa Romeo introduced driver air bags as optional equipment in the 164 model in 1990. The module supplier was KS with BC and Phoenix as sub suppliers.

Porsche made application developments of its air bag system for GM Corvette and Rolls Royce, to be equipped in 1990. The Rolls-Royce was initially offered without passenger air bags. The air bag system for the VW Golf Cabrio (convertible) was also developed with Porsche. The suppliers were the same as for Porsche. In September of 1991, VW offered driver air bags in the Golf Cabrio and Passat models in the US market.

In 1990, Breed Automotive Corporation (Breed) introduced its first "self contained" driver air bag system as an option on the Jaguar XJ-S. The module included Breed's own inflator and contained a mechanical ball-in-tube sensor with a percussion fuse and propellant pellets, supplied by BC. This "all-mechanical" module did not reach expected volumes, partly because of the required diagnostics and the benefit of one central electronic sensor for at least two air bags and additional belt pretensioners.

Audi and BMW developed passenger air bags with TRW Repa based on dual driver inflators from BC. These options were first offered in the top-line models in 1990.

Greater acceptance of DB's European driver air bag option followed a mid-1980s period of stagnation at around 25,000 units per year. Air bags started to become a competition issue among the growing number of car models that offered air bags. In 1990, driver air bags could be ordered at a price level of DM 2,500 at, for example, DB, Volvo or Saab, while BMW charged about DM 1,000 less. For both driver and passenger air bags, the option was over DM 5,000 at Porsche, DM 4,700 at DB, and around DM 3,000 for the BMW 7 model. Toyota Lexus and the exotic Honda NSX had driver air bags as standard equipment, but did not offer passenger side air bags.

In the first ten years of market-driven air bag introduction – until the end of 1990 – nine car manufacturers, including GM Corvette, installed the European air bag systems in 1.3 million cars, primarily on the driver side. During that decade, these cars traveled an estimated 35 billion km.

While the US legislative activities played a basic role for the development of air bags, the early air bag success in Europe – without legislation – was driven largely by education and competition. The public interest for new safety technologies was recognized and supported by the mass media. German motor magazines initiated crash tests together with the official technical expert organization, TÜV, and reported safety ratings of new cars beginning in the mid-1980s. This contributed strongly to customers' understanding of the importance of occupant safety, not only with respect to air bags. Safety equipment and performance became the #1 competitive issue.

The excellent reliability and performance experience created a base for introducing air bags in larger volumes. The air bag business had so far been dominated by the European car manufacturers. From 1983 on, some 450 US police cars were retrofitted with driver air bags in a NHTSA demonstration program. The system of Romeo Koyo had BC inflators, bags and steering wheels from Takata, Japan, and crash sensors from Technar, USA. Since

60

1985, Ford offered driver air bags for the company (GSA) fleet and as an option to the public in the Tempo and Topaz models. The module and inflator were supplied by Talley (today TRW), the steering wheel by Sheller, the sensor by Breed, and the diagnostic by Toshiba.

The US legislation required "automatic occupant protection". This was either "passive" seat belts or air bags on at least the driver side in all cars, and became effective September 1, 1989 (MY 1990). Except for the Europeans, almost all car manufacturers had offered more or less inexpensive belt solutions to fulfill the US legislation. Successively, US car manufacturers recognized that air bags sell and, in the 1990s, US air bag systems replaced the passive (or automatic) seat belts in large volumes, in many cases far ahead of the legal requirement.

The "Eurobag" – a Volume Promoter

In 1986, Volvo initiated a "Eurobag" project, giving the system development responsibility to BC. The idea was to have only a driver air bag, and to make it about half the passive US size. This bag was intended for the European market, where the seat belt use rate had reached values of around 80 percent. In order to reduce system costs, it was determined that the sensor should be integrated into the air bag module (Figure 17). MBB became involved as developer of the electronic sensor. The system was ready for a production decision in 1987 but, for several reasons, the decision was delayed and, finally, Volvo cancelled the project.

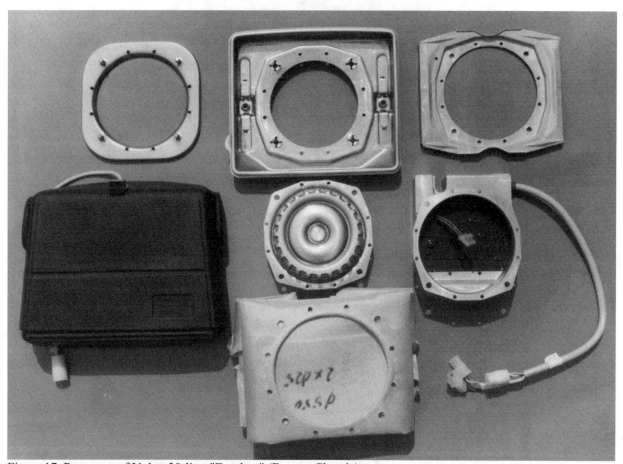

Figure 17. Prototype of Volvo 30-liter "Eurobag" (Bayern-Chemie).

The Swedish company Autoliv (Electrolux Autoliv 1984-1994) was a pioneer not only of regular seat belts, but of "automatic" belts and manufactured these as "passive" alternatives to air bags in the later half of the 1980s. However, the European market trend rapidly made it clear that air bags would become a huge business of the future.

Autoliv initiated air bag development programs with BC inflators in France (first for Peugeot) and in Sweden. This led to them supplying 60-liter driver modules to Saab (Figure 18) and Volvo in 1991. A passenger air bag was also developed for Volvo, using a tube inflator from BC and a bag from Svensk Air bag AB, Kungälv. Volvo introduced the passenger air bag (Figure 19) as an option in 1992. These modules are not depicted at relative scale in Figures 18 and 19.

Figure 18. 60-liter driver airbag module for Saab (Autoliv).

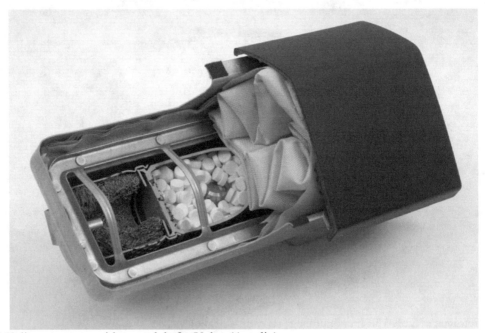

Figure 19. 160-liter passenger airbag module for Volvo (Autoliv).

In parallel, Autoliv decided to further evaluate Volvo's Eurobag idea and employed key people from Volvo and BC. Autoliv had strong seat belt positions not only at Volvo, but also at volume manufacturers like Ford of Europe, PSA (Peugeot, Citroen), Renault, and Rover. This opened an excellent opportunity for Autoliv to offer an all-restraint partnership with seat belt and air bag know-how, engineering and test support. These manufacturers had no US export, but recognized an urgent marketing need for air bags in Europe.

Autoliv intensively tried to secure beneficial purchase conditions, especially for inflators, but failed its price targets, which were clearly lower than those of their established competitors. Thus, to create an independent component position for the Eurobag, Autoliv evaluated the use of nitrocellulose propellant. This type of propellant had been used in a broad spectrum of military appliances since Alfred Nobel. Its primary benefit is a high specific gas yield without particles.

For use in cars, the established specifications restricted the carbon monoxide (CO) production and required a certain level of aging stability at high temperatures. This was the reason for using inorganic sodium azide in all air bag inflators. Seat belt pretensioners involved small amounts of nitrocellulose and their locations were not exposed to the sun like the air bags in the steering wheel or dash board.

The possible use of nitrocellulose in air bags of only half the "normal" sizes had to be thoroughly investigated. French SNPE and BC supplied micro gas generators with a load of less than 1 gram nitrocellulose for seat belt pretensioners. Autoliv formed a "Livbag" joint venture with SNPE, Paris, in 1990 for the development and manufacturing of inflators, initially based on nitrocellulose. With less than 10 grams of propellant, an inflator could be build considerably smaller and at lower cost than a comparable sodium azide inflator. Despite the freedom from particles, a "filter" package was required for cooling due to the high gas temperature of the "Euroflator" (Figure 20).

For the electronic crash sensors, Autoliv established cooperation with Cipro, Sweden, SensorNor, Norway, and later Nokia. For bags, cooperation started with Marling, England, who had developed a technique for "one-piece-woven" bags. After an initial ownership participation, Autoliv acquired all of Marling in 1994.

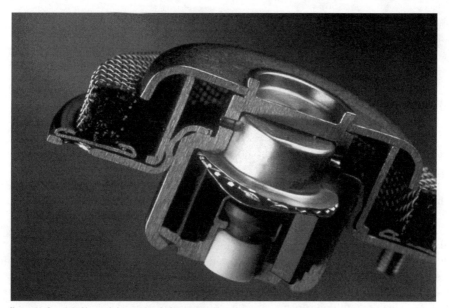

Figure 20. "Euroflator" for 30-liter driver module (Autoliv).

VW also took up the idea of smaller air bags for markets outside the USA and announced a European driver and passenger air bag option (Figure 21) to be available at a price of 1,200 DM in June 1992. The steering wheel and both modules were supplied by KS and contained, for the 30-liter driver side, a smaller BC inflator developed for Volvo and, for the 60-liter passenger side, a Dynamit Nobel tube inflator, both based on sodium azide propellant. Phoenix delivered the bags and Siemens an electronic control unit (ECU) for central placement.

The VW announcement indicated that the air bag would soon be requested as a supplement to the frequently used seat belts in lower-priced, higher-volume cars instead of only in higher-priced vehicles. This set a new milestone in the market-driven air bag expansion.

Autoliv had to ensure that – with the high seat belt use rates in Europe – a 30-liter driver air bag would be sufficient for the main purpose of protecting the face, that the established specifications for high temperature storage were unrealistically high, and that the CO concentration from the smaller air bags would be without toxic risks to car occupants. The insertion of a catalytic "Hopcalite" material in the cooling screens reduced the CO amount. Tests with humans showed that concentrations far above specified values did not lead to critical absorption in the blood hemoglobin within realistic exposure times. Real-world tests showed that specified high temperature levels for storage (e.g., when cars were parking in the sun or painted in the repair shop) could be clearly decreased.

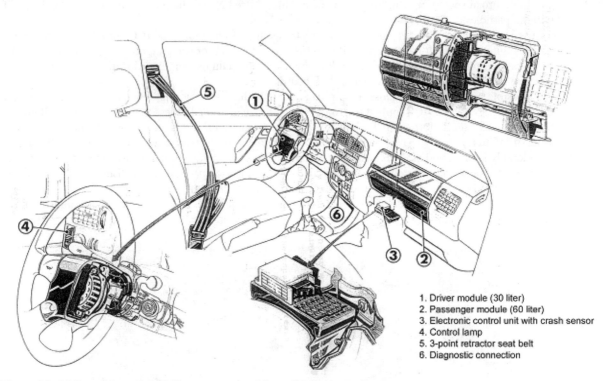

Figure 21. 30-liter driver and 60-liter passenger airbags (Volkswagen).

BMW supported the efforts and arguments of Autoliv and started a project with the nearby engineers at Autoliv in Dachau. This led to BMW's introduction of the Eurobag, including an electronic crash sensor (Figure 22) for a smaller "sport" steering wheel, in the beginning of 1993. A parallel development effort with Ford of Europe provided the final impetus and Autoliv released a Eurobag system at the end of 1991. The package contained a 30-liter driver and a 60-liter passenger module, using three Euroflators, and a centrally located electronic crash sensor, supplied by Nokia in Motala, Sweden. The driver air bag was standard and the passenger-side initially as an option. Supply for the Mondeo model started from Dachau at the end of 1992, for the Escort and Scorpio in mid-1993 and for the Fiesta in the beginning of 1994, followed by the Transit. In total, a yearly equipment rate of more than 1.5 million cars was reached.

Similar systems were developed in France for Renault and PSA and in England for Rover. By 1994, Autoliv was suddenly the number one supplier of air bags and, including seat belts, the largest supplier of occupant restraint systems in Europe. The number one seat belt position (before TRW Repa) followed the acquisition of Autoflug, Rellingen, in 1992.

After 1 million air bag installations, DB announced in July of 1992 that it would start to equip all cars with driver air bags (60 liter) in two months. In April of 1992, BMW internally made a similar decision to install driver air bags universally by September of 1992, but would still offer the "sport" steering wheel option including Eurobag.

The steering wheel manufacturers Petri and KS shared the high volume supply of "full-size" driver air bag modules to DB and BMW. Management discontinuations at BC became one reason for some customers changing inflator suppliers from BC to Morton (until 1989 Morton-Thiokol). Bosch or Siemens supplied the ECUs.

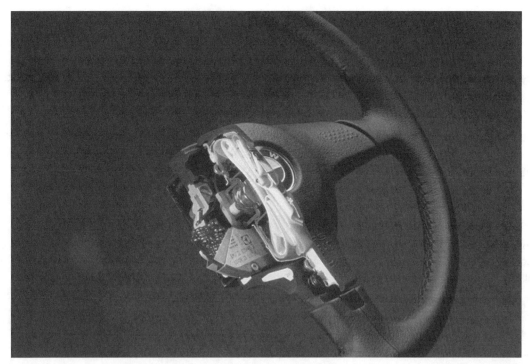

Figure 22. BMW 30-liter Eurobag including electronic sensor (Autoliv).

From 1991 on, DB changed the passenger module for the new S-class, and later for other models, to a tube inflator from Morton. This allowed more space for the glove compartment and non-coated bags. In 1992, BMW also changed the passenger module to a Morton tube inflator.

Opel introduced a driver air bag option for the Omega model in 1992, based on a Porsche system development. Siemens delivered the ECU and Petri the module, using a 60-liter Phoenix bag and a Temic inflator. (In 1990, DB formed Deutsche Aerospace, DASA, including MBB. The automotive activities of AEG and DASA were integrated into Telefunken microelectronic GmbH, "Temic" at the end of 1992 and BC changed to "Temic Bayern-Chemie Air bag GmbH".)

In an advertisement campaign, Opel stressed – as VW´s main competitor – that Opel cars were equipped with "full-size" air bags. For the consumer in general, it seemed obvious that a bag of half the size would likely provide less protection than a full-size bag. Opel continued installations of full-size air bags for both driver and passenger sides, beginning with the Astra and Corsa models in 1994. With the Vectra, all models got standard driver and passenger air bags in 1995. Perti delivered the Astra and Corsa driver modules with Temic (former BC) inflators and Phoenix bags, and Morton supplied the complete passenger modules. The Vectra modules were also delivered by Petri, but included Allied-Bendix stored gas ("hybrid") inflators.

Meanwhile, the rapidly increasing equipment of Eurobags as well as full size air bags had taken dimensions far above any expectation in Europe. In the USA, the "automatic" seat belts were phasing out and being replaced by air bags in an enormous expansion – much faster than the December 1991 air bag legislation required.

VW – Europes´ largest car manufacturer – decided in 1993 to unify air bags into all US type versions, expecting to realize cost benefits by providing uniform air bags as standard equipment in all models. This led in mid-1995 to standard front air bags for the Passat model. Petri delivered the driver module with Morton inflators, and TRW provided the passenger modules, containing Temic inflators.

For the "A4" platform including Golf, Vento, Audi, Seat, and Skoda, the demand for driver and passenger modules exceeded 1.5 million units of each per year. The driver module supply – always with the steering wheel – was shared between KS and Petri with inflators from Temic and Morton, respectively. The complete passenger modules (including a stored gas inflator) for all the A4 platform and the Audi A6 was awarded to Morton. For this supply, a new assembly plant was opened in Braunschweig by Gerhard Schröder, at that time "Ministerpräsident" (governor) of the state of Niedersachsen (Lower Saxony). The standard equipment started with the Audi A3 in 1996 and followed with Audi A6 and all other models of the A4 platform in 1997.

Following the example of VW, European manufacturers began phasing out Eurobags. All cars that were equipped with Eurobags got the larger air bags installed as standard with the next model change. Nitrocellulose was not used anymore, but adjusted specifications opened the opportunity for other organic alternatives to replace sodium azide propellants. Because of large amounts of poisonous sodium azide that were predicted to become exposed to the environment at later shredding, good arguments for alternative propellants came up. This opened the opportunity for other suppliers, who were not able to provide the traditional sodium azide propellant, to enter into the air bag business. The old idea of using stored argon gas with a pyrotechnic charge for release and heating up the gas was further developed. Improved materials and processes allowed some size and weight reductions compared to the designs of the early 1970s. These "hybrid" inflators became widely used for passenger air bags.

Safety Achievement Conclusions

Air bags had become very common by 1993 (Figure 23). Statistics from Germany, a central European transit area including higher "Autobahn" speeds, may be relevant for some conclusions regarding the safety contribution of frontal air bags. It should be underlined, however, that too many factors are involved for making many concrete statements. Therefore, only some general achievements in the total safety development should be referred to.

Model Year	1974	1975	1976	1977	1978	1979	1980	1981	1982	1983	1984	1985	1986	1987	1988	1989	1990	1991	1992	1993
GM, 10,321 DPop.	---	---	---	---																
DB, Dop. Europe								---	---	---	---	---	---	---	---	---	---	---	---	---
DB, Dop. US												---								
Ford, Dop. 2 models												---	---							
DB, Dst. US													---	---	---	---	---	---	---	---
DB, Pop.																		---	---	---
DB, Dst. all																				---
BMW, Dst. L7 US													---	---	---	---	---	---	---	---
BMW, Dop. Europe														---	---	---	---	---	---	---
BMW, Dop. US																---	---	---	---	---
BMW, Pop.																		---	---	---
BMW, Dst. all																				---
Porsche, DPst.* US														---	---	---	---	---	---	---
Porsche, DPst.																			---	---
Volvo, Dst.** US													---	---	---	---	---	---	---	---
Volvo, Dop. Europe																---	---	---	---	---
Saab, Dst. US															---	---	---	---	---	---
Saab, Dst. Europe																	---	---	---	---
Audi, Dop. Europe														---	---	---	---	---	---	---
Audi, Dst. US																		---	---	---
Chrysler, Dop. some															---	---	---			
Lincoln Cont. DPst.																	---	---	---	---
Chrysler, Dst. 97%																	---	---	---	---
Ford, Dst. 50%																	---	---	---	---
GM, Dst. 14%																	---	---	---	---
Corvette, DPst.																		---	---	---
Jaguar XJ-S, Dst.																		---	---	---
Rolls Royce Dst. all																		---	---	---
VW, Dst. 2 US mod.																		---	---	---
Alfa R.164, Dst. US																		---	---	---
Toyota. Dst. US																		---	---	---
Honda, Dst. US																			---	---

D = driver side, P = passenger side, op. = option, st. = standard. */ 944 Turbo; 944 and 944S op. **/ 780; 740 op.

Models are introduction models (may have been followed by others than shown). Percentages apply to first year.

Figure 23. Airbag Introductions – 20 Years of Public Offering.

In 1955, at which time practically no seat belts were used, the number of car crashes with occupant injuries exceeded 4,500 per 1 billion km driven and caused nearly 200 deaths per 1 billion km. The corresponding numbers in 1984, at which time a seat belt law with a 40 DM fine resulted in a 94 percent seat belt use rate, were 934 and

26.5. With the substantially increasing traffic on the West German roads, a peak of almost 9,000 total occupant fatalities occurred in 1970 (corresponding to 35.4 per 1 billion km). This decreased to about 5,100 in 1984.

Of course, increased seat belt use is not the only factor improving the statistics. Simultaneously, a number of "active" safety factors like better roads, improved car handling, and brakes (including ABS) contributed strongly as well. Counteracting these improvements, cars became faster and traffic became more crowded. With the seat belt and the development of safer car body structures, evaluation methods and improved safety awareness followed. All of these achievements made real sense only with the consequent use of the occupant restraints.

At the end of 1974, the German motor magazine "ADAC motorwelt" published the judgment that half of those killed in car crashes would still be alive if they had used the seat belt. Since the early seventies, comfort improvements of the 3-point retractor seat belts with convenient one-hand activation have contributed to a high acceptance. There is hardly any doubt in the general opinion in Europe that the seat belt will remain as "the # 1 life saver".

In 1984, driver air bags and seat belt pretensioners for the passenger had been installed in relatively low numbers in Mercedes cars since the end of 1980. These safety supplements to the seat belts, including passenger air bags, approached full equipment rates for new cars in Germany in 1997. This is certainly reflected in the decrease of car crashes with occupant injuries to 616 per billion km traveled and a decreased fatality rate of 13.8 per 1 billion km in 1997 (including higher accident statistics for the new states after the German unification in 1990).

Autoliv introduced the first side impact protection system (SIPS) in the Volvo 850 in 1994, followed by an inflatable tube system (ITS) at BMW in 1997 and inflatable curtains (IC) for Mercedes and Volvo in 1998. These pelvic/torso and head air bags for side protection have already reached an installation rate of 62 percent (2001) for new cars in Germany. The corresponding rates for frontal air bags have reached 99 percent for the driver and 96 percent for the passenger side.

By 2000, the number of accidents with occupant injuries reached 603 per billion km and the number of fatal injuries 11.8 (Autobahn 4.5) per 1 billion km. This is one fifth and one tenth, respectively, of the numbers 40 years ago – but still too high for the future!

From the beginning, the conversion of military technologies into another purpose – to save lives – has been a most fascinating and finally successful history. It took half of the automobile's history before seat belts started to be recognized as necessary safety devices.

The air bag development was more difficult than anyone could have imagined; it took ten years of intensive development efforts to reach a sufficient reliability level for market introduction and another ten years to reach general acceptance. Since Daimler-Benz launched the air bag at the end of 1980, approximately one million frontal air bags were installed during the 80s (almost all by European car manufacturers). The worldwide number of frontal air bag installations is estimated to reach 750 million by the end of 2002.

Some rough estimates indicate that the lives of more than two million passenger car occupants have been saved by the seat belt and that the number the injuries prevented is an order of magnitude greater than that. The air bag has proven to be an efficient supplement to the seat belt, especially at higher speeds and for avoidance of head and face injuries, which tragic outcomes can not be expressed in numbers.

A recent analysis of some 700 accidents with air bag equipped cars in Germany has shown that 97 percent of all occupants of the cars were using the seat belts. It also showed that in severe frontal collisions with DoD 3-4 (degree of damage to the car in the 1 to 5 scale), the serious to fatal injuries (AIS 3 or more in the abbreviated injury scale of 0 to 6) were 21 percent lower for the driver and 28 percent lower for the front passenger than in comparable accidents without air bags.

Another recent investigation of frontal crashes in the United Kingdom showed that, for 1,942 belted and 253 unbelted drivers, the AIS 2 or higher injury rate for belted drivers was reduced from 32 percent without air bag to 24 percent with air bag. For unbelted drivers this improvement was only from 40 percent to 37 percent. A strong air bag improvement in the severity of head injuries was reported for both belted and unbelted drivers. At the AIS 2 or higher level, the air bag effectiveness was 58 percent for belted and 43 percent for unbelted drivers.

In Europe, as in the US, the air bag has been the sole restraint system in too many cases. Concomitant use of the seat belt would have improved the protection considerably. In general, injuries and even fatal outcomes to children in some extreme situations, caused by a powerful air bag inflation too close to the occupant, would have been avoided if seat belts or child seats had been correctly used.

Further improvements of seat belts, frontal, side and other air bags in combination with each other and with the car itself have led to enormous achievements in the complete safety function of modern cars. In Europe, the seat belt will certainly remain the #1 life saver and injury prevention device, with different air bag applications as excellent safety supplements.

Meanwhile, the "active" and "passive" safety of modern automobiles has shifted the higher fatality rates from frontal impacts to side impacts, where the deformation zones are simply shorter.

The restraints of today will become adapted into more "intelligent" systems for further consideration of different risk situations and individual occupant conditions. New technologies in "pre crash sensing" will allow earlier precautions, including the option of softer inflation of the bags. Most likely, these technologies will also prevent some crashes from occurring at all.

References

Barényi (1951) German Patent No. 854157, applied January 23, 1951, issued August 28, 1952: "Kraftfahrzeug, insbesondere zur Beförderung von Personen".

Linderer (1951) German Patent No. 896312, applied October 6, 1951, issued October 1, 1953: "Einrichtung zum Schutze von in Fahrzeugen befindlichen Personen gegen Verletzungen bei Zusammenstößen".

Patzelt et al. (1971) German Patent No. 2152902, applied October 23, 1971, issued September 2, 1982: "Aufprallschutzvorrichtung für den Insassenschutz in Kraftfahrzeugen".

Passauer et al. (1972) U. S. Patent 3,947,300 filed July 9, 1973, issued March 30, 1976: Fuel for Generation of Nontoxic Propellant Gas. Priority date of July 24, 1972 (Germany).

Aldman et al. (1974) AAAM: Possible Effects of Air bag Inflation on a Standing Child.

Scholz (1975) ATZ ("Automobiltechnische Zeitschrift"): The Airbag as Improvement of the Future Occupant Protection.

Patzelt at al. (1979) German Patent No. 2905618, applied February 14, 1979, issued May 11, 1988: "Auf eine Grundplatte aufklipsbare kappenartige Abdeckung".

Scholz (1980) ESV: The Development of a Final Production Daimler-Benz Airbag System for the USA Market.

Nilsson and Zeuner (1985) Frauenhofer ICT: "Airbag-Insassenshutz für Automobile. lungsmerkmale und praktische Ausführung des pyrotechnischen Gasgenerators".

Johansson et al. (1989) 12th ESV: The Development of an Advanced Airbag Concept.

Roselt, Köster, Langwieder, Hummel (2002) IRCOBI: Injury Patterns of Front Seat Occupants in Frontal Car Collisions with Air bags. Effectivity and Optimasation Potential of Airbags

Kirk, Thomas, Frampton (2002) IRCOBI: An Evaluation of Airbag Benefits/Disbenefits in European Vehicles – a Combined Statistical and Case Study Approach

Chapter 4

The Biomechanics of Inflatable Restraints – Occupant Protection and Induced Injury

Jeff Crandall, Richard Kent, David Viano, Cameron R. "Dale" Bass

The fundamentals of occupant protection in a crash involve vehicle crashworthiness and occupant restraint. Crashworthiness refers to implementing a strong occupant compartment that resists intrusion and to designing crushable front and rear structures that deform and perform work to dissipate the kinetic energy of the crash. This combination provides controlled vehicle deceleration and survival space in the occupant compartment. Occupant restraint refers to the use of lap-shoulder belts, air bags, and other systems to provide ride-down of the vehicle deceleration, containment on the seat, and distribution of forces on the pelvis, shoulder, chest, and other designated anatomical structures to decelerate the occupant. The effectiveness of current restraint designs for protecting the occupant and reducing the risk of serious injury and death in a crash has been thoroughly researched and documented. Lap-shoulder belts are 42% effective in preventing death, with the highest effectiveness of 77% in rollovers and the lowest of 27% in near-side impacts (Viano 1991). In frontal crashes, the addition of the air bag to the three-point belt system raises the effectiveness level to approximately 50%.

This chapter reviews the biomechanics of air bag restraint. It necessarily starts by addressing occupant restraint in general before covering the role air bags play in restraining occupants involved in a frontal impact. The discussion includes a description of occupant kinematics for belted and unbelted occupants in the presence of an air bag. In addition, some indirect benefits of air bags are mentioned, including the air bag's role in facilitating force-limiting belts with lower limits on the shoulder belt loads than would be possible without air bags. The chapter then moves to a discussion of the biomechanics of air bag-induced injury and summarizes the mechanisms for the generally minor trauma sustained by adults exposed to air bag deployments. Finally, details are provided on the few cases of children and small adults who have sustained severe and fatal injuries from air bag loading.

1. Introduction

Air bags and safety belts serve complementary functions for occupant protection since a system of technologies is needed to provide maximum safety in the wide variety of real-world crashes (Reidelbach 1985, Viano 1988). For example, safety belts can prevent fatal injuries in many crash modes, but they must be worn by the occupant to be effective. Conversely, air bags require no action by the occupant, but they do not provide protection in all impact scenarios (e.g., rollover). The belt and air bag system of complementary technologies therefore embraces the concepts of passive and active protection and provides a viable approach to occupant safety. Even as many thousands of lives are being saved by belt use and supplemental air bags, however, serious injury and fatalities still occur in severe crashes. Likewise, any restraint system can have a negative effect in certain instances. Seat belts can cause injurious thoracic, abdominal and, more rarely, neck loading, particularly if they are misused. In particular, injury can result from or be exacerbated by placement of the shoulder harness under the arm and wearing of the lap-belt high on the abdomen with poor seating posture. An air bag, which necessarily "adds" energy to a crash during the inflation process, can also cause injury if the occupant is in the path of deployment. Therefore, the air bag design requires a balance between a long fill time to reduce the risk of inflation injury and a rapid inflation to quickly fill the space between the occupant and the interior (e.g., Mertz and Marquardt 1985). The design, implementation, and legislation of these restraint systems therefore require consideration of a complex web of biomechanical and other tradeoffs.

2. Brief Review of Seat Belt Biomechanics

Although much of the development work on restraints has involved frontal barrier and sled testing, the primary role of the lap belt is to retain an occupant inside the vehicle during a rollover. These types of crashes were recognized through crash investigations to be the leading cause of fatality in the 1950s. In rollover crashes, safety belt use nearly eliminates the risk of paralyzing cervical injury and ejection. Another early use of the lap belt was to complement crash protection of the chest provided by a driver air bag. The lap belt assured that the driver was properly positioned and aligned to utilize the air bag as intended. In the absence of the lap belt, loads are applied through the knees and seat pan to restrain the occupant's pelvis and control the upright posture of the driver, but the use of a lap belt provides a greater level of control.

The combination of a lap and diagonal shoulder belt provides occupant restraint by routing safety belt loads over the bony structures of the pelvis and shoulder. This takes advantage of the relatively high tolerance to impact force for these regions of the skeleton and avoids concentrating load on the more complaint abdominal and thoracic regions. By adding the convenience of an inertial locking retractor to the lap-shoulder belt, Bohlin (1967) proved the system would be worn in the field and would be protective of injury in crashes. This set in motion a series of international efforts to require safety belt use by the motoring public. Among the first countries to adopt safety belt wearing requirements were Sweden and Australia, and this necessitated the development of a better scientific understanding of the principals of belt restraints. The fundamentals of a high quality belt restraint system involve occupant kinematic controls (Adomeit and Heger 1975, Adomeit 1977, Adomeit 1979), which maintain the lap belt low on the pelvis through adequate seat cushion support. This minimizes pelvic rotation and reduces the tendency for the lap-belt to slide off the ileum and directly load the abdomen. Forward rotation of the upper torso, to slightly greater than 90° upright posture, directs a major portion of the upper torso restraint into the shoulder (Figure 1). In a frontal impact, a snug-fitting lap-shoulder belt ties the occupant directly to the passenger compartment and allows the occupant to "ride-down" the crash as the vehicle front-end crushes. This coupling and ride-down decelerates the occupant more gradually than is possible with energy-absorbing interiors. Even with shoulder belt restraint, however, there is forward excursion of the torso and, in particular, movement of the head and neck toward the steering wheel or windshield. Figure 2 shows three time-points in a frontal crash of a lap-shoulder belted passenger dummy and highlights the forward lean of the upper torso, the flexion of the neck, and the excursion of the head. While this kinematic is more favorable than the consequence of an unrestrained occupant striking the interior surfaces at high speed, there is still the potential for head and face contact with the vehicle interior and for inertial injuries of the neck. In fact, an increase in whiplash-type injuries of the neck was observed after the wide-spread use of lap-shoulder belts in the United Kingdom (Rutherford 1985). Figure 3 shows the same type of crash sequence as Figure 2 with a combined driver air bag and lap-shoulder belt restraint. Early in the crash, sensors detect the severity of the crash and activate the inflation of the air bag if the collision severity is above a set threshold. This causes a rapid filling of the bag as it deploys out of the steering wheel hub. The bag then loads the head, neck, and torso to restrain the upper body in ways not possible with only the shoulder belt. Vent holes in the back of the air bag relieve pressure and absorb energy such that, even after occupant rebound, the air bag continues to deflate. This type of restraint provides a benefit to the occupant via several mechanisms, as described in the following section.

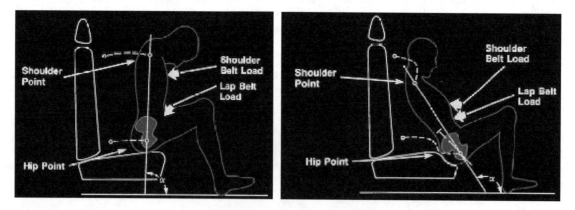

Figure 1. "High quality" (left) and "low quality" (right) belt restraint (Adomeit 1977) in a frontal crash.

3. Inflatable Restraint Biomechanics

In the early days of air bag development, belt use was low and showed no signs of increasing in the near term (e.g., NHTSA 1984). Low belt use rate was the original impetus behind the development of air bag systems since they overcome the primary weakness of belt systems: to be effective the occupant must fasten the belts in advance of the crash. Using a pyrotechnic device to generate nitrogen gas, a bag can be rapidly inflated during the early phase of vehicle frontal crush without action by the occupant. The bag then "fills" some of the space between the occupant and the interior, which couples the occupant to the passenger compartment and achieves the safety benefits of ride-down and load distribution. Energy dissipation is achieved via venting since a non-vented bag acts essentially as an elastic spring once it is inflated (Patrick and Nyquist 1972, Romeo 1975).

For unbelted occupants, the air bag's principle benefit is load distribution and attenuation on the thorax. Despite low belt use being the original motivation, however, the air bag also provides a safety benefit for a belted occupant, albeit via a different mechanism. For belted occupants, the principle benefits of an air bag are due to mitigation of head contact and neck loading. The primary mechanisms by which air bags protect occupants are outlined below.

Figure 2. Kinematics of a belted dummy without an air bag in a frontal collision. Note the neck loading via head inertia and the potential for head/face contact with the vehicle interior.

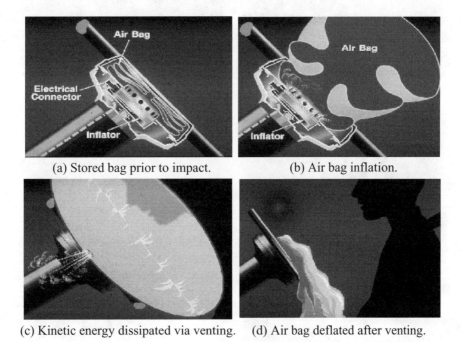

(a) Stored bag prior to impact.   (b) Air bag inflation.

(c) Kinetic energy dissipated via venting.   (d) Air bag deflated after venting.

Figure 3. Occupant interaction with an air bag in a frontal collision. Inertial neck loading and the potential for head/face contact are mitigated and work is performed to dissipate kinetic energy.

3.1 Head and Face Contact Mitigation

First, the air bag provides primary restraint to the head and face so that contact with the steering wheel is mitigated (Figure 1, Figure 2, and Figure 3). This benefit has been documented thoroughly in frontal impact tests. For example, Patrick et al. (1972) compared no belt, lap-belt only, shoulder-lap belt, and shoulder-lap belt plus air bag restraint conditions. They noted that the restraints without the supplemental air bag resulted in either abdominal or facial contact with the steering wheel, while the air bag effectively mitigated these contacts. This is not to say that an air bag precludes any risk of head or face contract, especially in the absence of concomitant belt use. Crandall et al. (1994) and others have observed that head and face injuries are over-represented in unbelted occupants with an air bag compared to belted occupants with no air bag. Berg et al. (1998) presented a series of seven cadaver tests, where an air bag restraint with no belt was used. They noted head impacts into the upper region of the windscreen or the roof in all tests. Depending on the size of the occupant, the severity of the collision, the geometry of the steering column and the vehicle interior, and the characteristics of the air bag, it is possible for an occupant to translate vertically over the air bag and sustain head or face loading via the steering wheel or the windshield/windshield header (Figure 4). This injury mechanism emphasizes, again, the importance of the belt as a primary restraint and the air bag as a supplemental restraint. With a properly positioned belt restraint, the kinematics shown in Figure 4 do not occur.

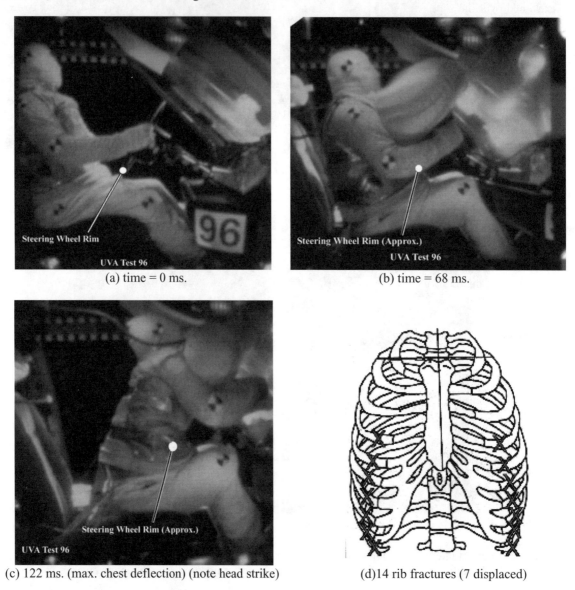

Figure 4. Unbelted cadaver translating vertically over the air bag and sustaining a substantial head strike on the header (~ 175 g acceleration spike recorded at the first thoracic vertebra). Also note the steering wheel loading on the inferior anterior thorax, which resulted in multiple bilateral rib fractures, as shown. Adapted from Kent et al. (2000).

3.2 Load Sharing, Force Distribution, and Work

Second, the air bag distributes forces on the chest and shares loading with the belt system. Numerous studies have shown that cadavers restrained by a belt and an air bag sustain lower levels of injury than cadavers restrained by a belt alone or by an air bag alone (e.g., Kallieris et al. 1982, Yoganandan et al. 1993, Crandall et al. 1997a, Kent et al. 2001a). Shoulder injuries related to belt loading also seem to be reduced with the addition of an air bag restraint (Werner and Sorenson 1994) (Figure 5). These injury reductions are primarily due to the load sharing and force distributing effects of an air bag. It is well established that a person's tolerance to a force applied on the thorax is highly dependent upon the area over which that force is applied. For example, Patrick et al. (1965) performed a series of sled tests with embalmed cadavers impacting padded load cells anteriorly. They found that a 3.3 kN hub load to the sternum resulted in minor trauma, while similar injuries required approximately 8.8 kN if the load was distributed over the shoulders and chest. Bierman et al. (1946) tested volunteers with a drop device that loaded a 4-point belt harness 490 cm^2 in area. Painful reactions and some minor injuries occurred when loads exceeded 8.9 kN. When the load area was increased to 1,006 cm^2, loads up to 13.3 kN were sustained without pain or injury. The potential exists, therefore, to increase substantially the work done on an occupant and to perform this work at a non-injurious force level if the force can be distributed over the entire anterior thorax.

This work can be increased even further, without increasing thoracic injury risk, if the occupant is tensed during the impact. In a series of human volunteer sled tests performed at 48 km/h, loading through the steering wheel via tensed arms and through the floorpan via tensed legs had a pronounced effect in terms of controlling occupant kinematics and minimizing restraining loads through the belt and air bag. The role of arm bracing has been studied by Horsch and Culver (1983) and shows a considerable ability to restrain the upper torso away from the steering wheel when the driver is alert of the pending crash. Similar effects can be seen with bracing by volunteers in simulated crashes (Hendler et al. 1974). However, the proportion of supplemental restraint by bracing decreases with increasing crash severity, simply because the occupant's strength becomes negligible compared to the inertial forces exerted on the upper body. In these cases, the air bag plays a greater role. Nonetheless, comparison of the tensed human volunteer in Figure 6 with the atonic cadaver in Figure 4 indicates the substantial effect of musculature in a crash as severe as 48 km/h. Armstrong et al. (1968) estimated that as much as 55% of an occupant's kinetic energy in a tolerable frontal impact may be dissipated via the work done by "propriotonic" restraint.

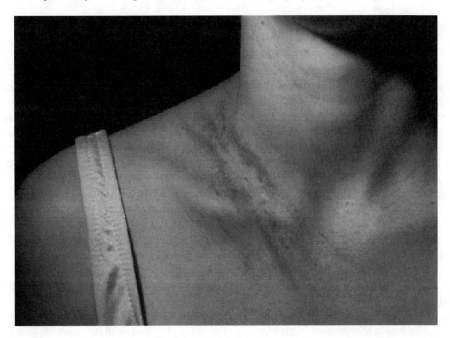

Figure 5. Abrasion from shoulder belt loading. Belt injuries may include fractured or dislocated clavicle, rib fractures, or others. The presence of an air bag may reduce the frequency and severity of belt-related injuries.

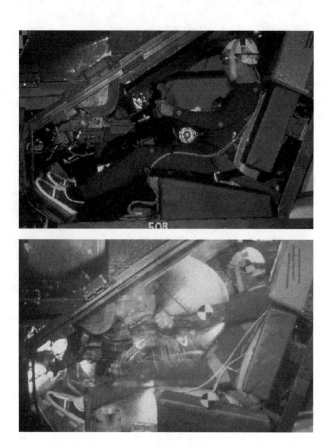

Figure 6. Human volunteer frontal sled test (48 km/h) showing air bag system feasibility and pronounced effect of musculature, which reduces restraint loading relative to a dummy or cadaver test (from Holloman Air Force Base tests of the GM driver air cushion described in Smith et al. 1974).

The load-sharing characteristics of air bags have led to the observation that thoracic acceleration, a widely used injury indicator, may not reflect an air bag's benefit to the occupant. By distributing forces on the chest, the total force on the chest can be higher than the force generated by belt loading, thereby resulting in higher chest acceleration. Furthermore, the fundamental of upper torso lean is an important attribute of belt restraint and loading through the bony structures of the body. Forces concentrated on the shoulder or upper thorax are more tolerable than forces concentrated lower on the thorax or abdomen. With this background, it is necessary to evaluate chest acceleration critically when comparing belt-dominated loading with air bag-dominated loading since a higher level of force, and therefore acceleration, can be tolerated if the force is distributed or if the torso is pitched forward. Grosch (1985) found that the expected benefit due to a supplemental air bag was not reflected by the acceleration levels measured in laboratory tests and concluded that it is necessary to consider other indicators of injury risk. Others have observed that the chest acceleration peak due to bag slap during deployment can be the global maximum, though the magnitude of this peak (approximately 50 g) is insufficient to cause acceleration-induced injury (Enouen et al. 1984). An acceleration peak such as this, having a different cause than a comparable sled test with belt loading, can lead to misinterpretation of the relative benefits of different restraint conditions. In the Grosch test series, chest acceleration and chest deflection exhibited contradictory trends, with acceleration increasing when an air bag was added, while deflection decreased. Further complicating the assessment is the fact that this phasing is different for drivers and passengers. Driver-side air bags typically deploy rearward nearly far enough to load the occupant prior to any forward translation of the driver in the vehicle. The result of the driver's proximity to the deploying bag is that the air bag loading begins sooner after the bag deploys (approximately 50 ms after impact in a 48-km/h full-frontal sled test) than it does for the passenger. The driver's chest acceleration therefore typically exhibits a single peak, which occurs during combined belt and air bag loading. A passenger-side occupant, on the other hand, sustains primarily belt loading over the first approximately 75 ms after impact, which results in peak chest acceleration due primarily to belt loading (Figure 7). The later air bag loading then typically generates a second chest acceleration peak. Depending on the characteristics of the belt, air bag, vehicle interior, and occupant, a passenger's global acceleration maxima may occur either under belt loading or under combined loading, while a normally seated driver's maximum acceleration occurs under combined loading. This phenomenon was observed by (Vezin et al., 2002) and its importance for thoracic injury prediction in the laboratory is discussed in more detail by Kent et al. (2000). The combinatorial nature of belt and air bag loading also necessitates appropriate phasing of belt and air bag loading so that the combined effect does not exceed the acceleration tolerance of the human body, though cases exceeding the acceleration tolerance of a properly restrained occupant in an otherwise survivable frontal crash are assumed to be extremely rare (by inference from Melvin et al. 1998).

 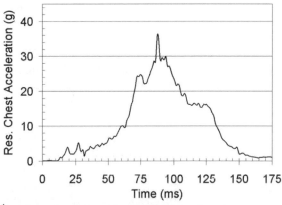

(a) Driver-side test showing Hybrid III 50th male at time of maximum chest acceleration.

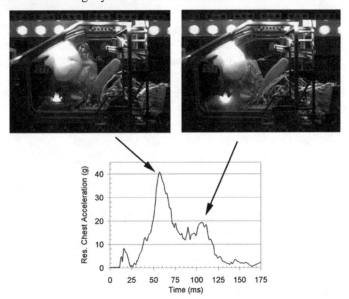

(b) Passenger- side showing Hybrid III 50th male at time of global and local maximum chest acceleration.

Figure 7. Driver-side and Passenger-side occupant kinematics and chest acceleration in a 30-mph frontal sled test with a force-limiting belt and air bag restraint system (see Kent et al. 2000).

While the air bag does share loading with the seat belt and thereby reduces the level of concentrated belt force exerted on the chest, it is typically an insufficient restraint by itself. As shown in Figure 4, injurious thoracic loading can occur from steering wheel contact, even with an air bag restraint. This behavior was observed by Smith et al. (1974) in volunteer tests and produced injuries in cadaver tests conducted by Yoganandan et al. (1993). Because air bags neither remain inflated nor provide lateral restraint, seat belts are needed to adequately control occupant kinematics over the range of crash types, including rollovers and side impacts.

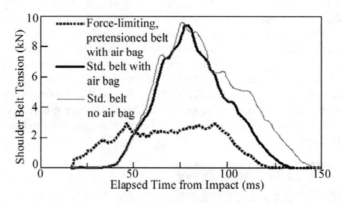

(a) Shoulder belt tension. Note that the air bag reduces the magnitude and the duration of belt loading with a standard belt. Note also the substantial reduction in belt force that can be achieved with a force-limiting belt and air bag (the belt is also pretensioned in this case).

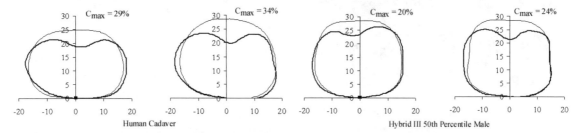

(b) Thoracic profiles at maximum sternal deflection (C_{max}) with 4-kN force-limiting belt (left) and standard belt (right) with air bag. Note the increased C_{max} (expressed as a percent of undeformed chest depth) and decreased radius of curvature for the standard belt (Kent and Crandall 2001) (units are cm).

Figure 8. Shoulder belt tension and thoracic cross-sectional deformation in a 48 km/h frontal impact with three different restraint conditions. Tests performed at the University of Virginia.

3.3 Restraints System Design Flexibility

Third, an air bag facilitates the use of force-limiting belts, and allows for the use of lower force limits than would be possible without a supplemental restraint. While numerous experimental studies have documented the benefits in the laboratory, the value of lower belt force limits (e.g., 4 kN vs. 6 kN) has also been observed in the field (Foret-Bruno et al. 2001). By distributing loads, mitigating head and face contact, and sharing loading with the belt system, an air bag allows for lower levels of belt force to be applied to the occupant. This has two primary benefits. First, a force-limiting belt can be designed to yield before injurious belt forces are developed (Figure 8). Second, a force-limiting belt promotes forward rotation of the torso, which generates a "higher quality" biomechanical restraint condition (Figure 1, Figure 9) (e.g., Adomeit 1977). While general reduction of belt forces seems desirable, there is a limit on how low the shoulder belt force should be reduced without compromising protection in crashes involving long-duration deceleration or multiple impacts. If the belt has yielded in a previous collision, there can be a loss of upper torso control and (with a sliding latch plate design) lap control in subsequent collisions. Since the air bag will deflate after the primary impact, it cannot be relied upon to provide restraint in secondary impacts. In this case, the belt restraint system must be available and functioning to provide retention and restraint of the occupant.

 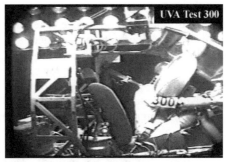

(a) Standard belt with an air bag, at impact (left) and maximum occupant excursion (right).

(b) Standard belt with no air bag, at impact (left) and maximum occupant excursion (right).

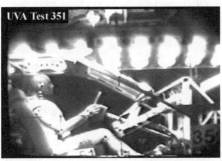 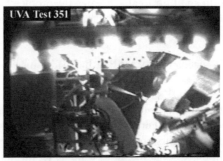

(c) 4-kN force-limiting belt with an air bag, at impact (left) and maximum occupant excursion (right). Note the increased occupant excursion and torso pitch relative to (a).

 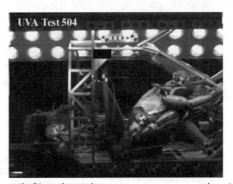

(d) 3.5-kN force-limiting belt with no air bag, at impact (left) and maximum occupant excursion (right). Note that this is a different dummy (THOR) than the Hybrid III used in (a), (b), and (c).

Figure 9. 57-km/h frontal impact restraint scenarios (50th percentile male) illustrating the benefit of an air bag. Comparing (a) and (b) shows the bag's benefit for mitigating head/face contact and for load sharing. The inadequacy in this test of a 4-kN force-limiting belt without an air bag is shown in (d). Tests performed at the University of Virginia

The discussion above describes the mechanisms by which air bags supplement seat belts in frontal collisions, and also illustrates how some benefit can be realized by unbelted occupants. The discussion also implies, however, the biomechanical tradeoffs associated with an inflatable restraint. First, occupant loading should occur as early as possible in a crash, so that the maximum ride-down benefit may be achieved. This emphasizes early deployment of a relatively large air bag and also the use of belt pretensioning with early activation and fast retraction to initiate early belt restraint. Second, the magnitude of restraint loading applied to the occupant should be the maximum tolerable so that injurious contact with interior components can be avoided. These requirements emphasize an air bag with relatively high internal pressure so that large forces can be developed. In contrast, the work performed on the occupant should occur over as much distance as possible prior to contact with the vehicle interior in order to minimize the magnitude of force that must be applied. Depending on the crash severity and occupant size, this emphasizes an air bag with lower internal pressures. Furthermore, occupants of different mass and geometry must be considered. It is a substantial design challenge to optimize the air bag system's performance given the range of conditions in which it must perform, including a single crash that may have both belted and unbelted occupants.

Air bags have followed an evolutionary design path in an effort to reach this level of optimization. Overall, air bags have been shown to be beneficial, but this evolutionary path has resulted in some cases where the air bag generated injuries, sometimes more severe than the injuries that would have been expected without an air bag deployment. While the discussion above outlined the mechanisms by which the air bag provides a benefit, the following section addresses some of the negative consequences of air bag deployment and the biomechanics of air bag-related injuries. The reader should bear in mind that, while the following section describes a wide array of injuries that can be caused by a deploying air bag, the systems overall have saved lives. We dedicate a large section of this chapter to air bag-induced injuries, however, in an effort to characterize the problem and to understand what efforts have been done or need to be done to eliminate the problem. In terms of the overall consequences of these injuries relative to the benefits provided by air bags, the reader is referred to the next chapter in this book.

4. The Biomechanics of Air Bag-Induced Injury

Section 3 showed that air bags can be effective both as a primary safety device for protecting unbelted occupants and as a supplemental restraint for diminishing the risk of head and neck injuries for belted occupants. Furthermore, the air bag facilitated development of force-limiting belt systems since more forward excursion of the occupant could be provided without concern of the head striking the steering wheel. In these scenarios, it is desirable to introduce the air bag early in the event to initiate occupant ride-down with the decelerating vehicle and to utilize all available space for energy absorption. In addition, the air bag must have sufficient pressure to prevent full compression of the air bag during loading by unbelted occupants in severe crashes. Initially, the design direction for the air bag was to have zero punch-through (i.e., impact of the occupant with the underlying surfaces) during restraint of a 95th percentile male in a full-rear seating position. While this situation was later relaxed somewhat as the requirements tended to balance the needs of small occupants and a broader range of real-world crashes, the concern for the occupant not to overload the air bag and interact with the inflator and inflator components led to relatively aggressive requirements for inflation pressures. Since the air bag is not inflated during normal driving, this suggests that the inflatable restraint should be actuated quickly, thereby generating these inflation pressures in a very short period of time. For instance, air bags designed to reduce the risk to unbelted occupants in 48 km/h rigid barrier type crashes are designed to be fully inflated in less than 35 ms. A critical performance issue in the development of any air cushion system is the dichotomy that exists between the air bag characteristics required to meet restraint performance requirements for higher severity full-frontal rigid barrier and the effect that these air bag characteristics have on increasing the risk of deployment-induced injuries to occupants (Mertz et al. 1985). If the occupant is in the path of the deploying air bag, substantial forces can be applied by the air bag that may result in injury to the occupant. Thus, while the estimated benefits of air bags are considerable, some residual risks remain. Huelke (1994) examined crashes involving air bag deployments and determined that nearly 30 percent of drivers reported sustaining some sort of an air bag-related injury. While not uncommon, these injuries are nearly always minor in severity and typically involve erythema or minor skin abrasions. When reading this chapter, it is essential for the reader to realize that while unintended interactions with air bags can produce injuries, an extremely small percentage of these interactions will be severe or fatal. Furthermore, the net protective benefits of an air bag in reducing the overall level of trauma may be present even for those sustaining air bag-induced injuries. Finally, caution must be used in interpreting the significance of the cadaver, dummy, and animal studies relative to the expected performance of inflatable restraint systems in general. Since the objective of these studies is frequently to evaluate the injury potential of air bag systems in "worst-case" loading environments, interpretation of these studies should be limited to injury potential rather than injury risk.

Most air bag-induced injuries occur to occupants who are out-of-position (OOP) during bag deployment. In this chapter, the term out-of-position is used to describe any situation where the occupant's position lies within the deployment space of the air bag and, thereby, generates unintended interactions between the occupant and the deploying air bag. Out-of-position situations for vehicle occupants are, in contrast to situations with a normal belted seating position, fundamentally undefined and numerous (Berg et al. 1998). Obviously, many positions are not uncommon during normal driving or riding (e.g., leaning forward to adjust the radio) but certainly differ from those used as standardized positions for restraint effectiveness evaluation in laboratory settings. In the field, proximity to the air bag at the time of deployment depends on a variety of occupant, vehicle, crash, and restraint factors. Perhaps the most frequent scenario for an occupant being OOP involves unbelted occupants who translate forward during pre-impact braking (Karlow et al. 1994). Even without braking, however, pre-crash position can considerably influence the likelihood of interaction with the deploying air bag. Extreme cases involve medical conditions that prior to the crash cause the occupant to lean toward the steering wheel or instrument panel at time of deployment. In less extreme circumstances, the occupant's proximity to the air bag at the time of deployment is influenced by ergonomic factors such as the seat adjustment, steering column, or wheel tilt adjustment. In particular, smaller drivers have a propensity to sit close to the steering wheel and upright in order to facilitate operation of the pedal controls or to improve visibility.

In addition to the occupant's initial position, crash parameters can influence the likelihood of an occupant sustaining an air bag-induced injury. Since one of the air bag's design objectives was to improve performance of unbelted occupants in FMVSS 208 testing, air bags have been evaluated extensively in rigid barrier tests. Therefore, it is less likely that an occupant will sustain severe air bag-induced injuries when the air bag fires in full-frontal crashes than in other collision scenarios. Real world crashes often involve striking another vehicle and the recorded accelerations are generally longer in duration and lower in amplitude than barrier crashes of the same change in velocity (Delta-V). This complicates the evaluation of triggering and deployment of air bags and setting of system requirements, which aim to link laboratory tests to field crashes and which aim to optimize the occupant restraint system. Other common crash types are the offset frontal crash and tree or pole impacts, which present challenges to crash sensing and the early deployment of the air bag. In frontal crashes where the object struck by the vehicle does not resemble a barrier, the resulting crash pulse may exhibit lower deceleration levels than would occur in a full-frontal impact. This softer pulse can affect the sensing of the air bag system and can produce later deployment of the air bag (Nusholtz et al. 1998c, Dalmotas 1998). While the softer pulse will also tend to delay the occupant's motion relative to the vehicle, the occupant may have translated sufficiently forward in the vehicle to be very close to the air bag module at the time the sensor initiates deployment. Similarly, multiple collisions may allow the occupant to move forward prior to the occurrence of a sufficiently severe impact to deploy the air bag. These crash scenarios can result in the head and neck of even belted occupants being very close to the air bag module at the time of eventual air bag deployment (Nusholtz et al. 1998b).

While air bags are highly effective in preventing serious or fatal head injury in higher severity crashes, concerns have been expressed that these gains may be partially offset by air bag-induced injuries in low severity crashes, particularly for belted occupants (Dalmotas 1998). Since the energy released by an air bag inflator has historically been independent of collision severity or occupant size, the concern is that in low-severity deployments the bag forces may more seriously injure the occupant than would have otherwise occurred (Melvin and Mertz 2002). As such, fatal bag-related injuries can occur at all collision severities. With increasing collision severity, however, uncertainty regarding the injury outcome that would have been sustained in the absence of air bag deployment increases. Consequently, counts of air bag fatalities are generally limited to lower speed crashes, where, in the absence of deployment, the occupant would have been expected to survive the crash.

Even during the initial development efforts for air bags, concerns were raised about air bag-induced injuries. Initial studies in the 1970s on prototype systems were followed by more sophisticated test procedures developed concurrently with widespread implementation of air bags into production vehicles in the late 1980s. Soon thereafter, crashes with air bag-equipped vehicles provided more real-world case studies. In general, the types of injuries caused by air bags can be grouped into those occurring with adults, which may be serious but are usually not life threatening, and those occurring with children and infants.

4.1 Injuries to Adults

There are currently documented cases of 63 adult drivers and 7 adult passengers killed by air bags (Chidester and Roston 2001). In addition, there are numerous studies documenting the increased risk of minor and moderate injuries that accompany an air bag deployment. Many studies of this issue were performed prior to the introduction of air bags into the fleet, and indicated the potential for injury. As air bag-equipped vehicles became prevalent and cases of patients exposed to air bag deployments became more common clinically, case studies began to appear in the medical literature and laboratory tests began to simulate the injury mechanisms observed in the field. By definition, these studies did not attempt to determine air bag performance on a population scale, but they did give an indication of some typical and atypical injury patterns that were observed in the presence of an air bag deployment and, in retrospect, some give an indication of common conceptions and misconceptions regarding air bag performance.

4.1.1 Early Studies – Prototype Air Bag Systems

Along with the perceived benefits from air bags, unintended injuries have been a concern of engineers, researchers and regulators even prior to the implementation of production air bag systems. An early experimental General Motors system was tested with human volunteers at the Holloman Air Force Base in Allamagordo, New Mexico, with the cooperation of the NHTSA. This system had been tested with dummies up to a 30 mile per hour impact with no apparent problem. These tests were to be conducted both statically and dynamically, although the dynamic tests were not run as a result of the injuries in the static tests. (Smith et al. 1972, Lundstrom 1974) In the static deployments, the volunteers were seated so that the bag would inflate fully before contacting the subject. In addition to the normal seating position, it was intended to evaluate unconventional body positions that an occupant might be in during a crash event, such as legs crossed with both heels on the floor, upper extremities raised above the torso so that the bag deployed against the forearms and with the occupant leaning forward into the bag. While the extremity position trauma was limited to minor erythema, abrasions, contusions and blisters, the test in which the volunteer leaned forward into the deploying bag resulted in a short period of unconsciousness diagnosed as mild concussion. The test program was terminated at that point to allow a lower-energy system to be developed. While the lower-energy system was tested dynamically the following year, the unusual position evaluation was moved to Wayne State University with animal surrogates. This Holloman test was significant in that it was the first to show that a bag system that performed well in dynamic tests could cause injury in static deployments if the human subject was close to the deploying bag.

Southwest Research Institute conducted a subsequent series of stepped-severity tests in which dummies and human volunteers in the driver position were exposed to increasingly severe air bag loading environments (Smith 1974, Figure 6). The unbelted occupants were restrained only by the driver air bag. Tests were conducted at eight different impact severity levels with the volunteers nominally in-position. During the deployment process the air bag "wiped" across the face and forearms of the majority of subjects regardless of test severity. The goggles on the volunteer in Figure 6 were a result of an earlier eye injury in testing. Only minor trauma was experienced by the volunteers, however, as erythema, abrasions, and ecchymosis were produced by interactions between the air bag and the arms, face (mouth, nose and chin) and chest. In describing the test experience, many volunteers expressed that they felt a "stinging" sensation when the air bag was "wiping" across their face, hands, and forearms as it deployed.

4.1.2 Production Air Bag Systems

Initial implementation of air bags into fleet vehicles in the 1970s provides feedback on the injury potential of production air bag systems. As early as 1976, Mohan et al. showed that the mean injury level in lower-severity crashes was higher in a fleet of air bag-equipped vehicles than in a similar fleet without air bags. Mertz (1988) presented case reviews of deployment crashes involving 1973-1976 GM air bag systems in which the occupant sustained injury. The passenger air bag included an upper body bag and an internal inflatable knee restraint; and it was a dual level system. Since many of the injuries in this study occurred in crashes of minor to moderate severity, the authors concluded that the injuries could have been caused by air bag deployment. In a related study of this fleet, Pursel et al. (1978) found passenger air bag effectiveness for AIS 2+ injuries to be -34% (i.e., the passengers in air bag-equipped vehicles sustained more AIS 2+ injuries than the passengers in vehicles without air bags).

The adult population at greatest risk for air bag-induced fatal injuries appears to be small stature females with a height of less than 1.6 m (Kleinberger and Summers, 1997). Dalmotas (1998) attributes this susceptibility to their tendency to sit relatively closer to the steering wheel-air bag and to their greater likelihood of having slender, lower density bones than men. In addition, considerations for restraining the 95% male often dictated an aggressive air bag inflation to prevent punch-through and the unintended consequence for small occupants only became obvious later. Consequently, most experimental efforts to assess air bag-induced injuries have focused on small female cadavers and the small female Hybrid III dummy. While not wearing a belt restraint certainly increases the likelihood of being out-of-position, fourteen of the small stature drivers killed by air bag deployments are believed to have been belted, indicating that belt use in this population does not always protect drivers from air bag inflation injury (C. Morris et al. 1998). Dalmotas (1998) demonstrated the significance of seating position and highlighted the fact that wearing a belt does not completely remove the risk of injury to smaller stature persons in a forward seating position. Rigid barrier tests were conducted at 48 km/h using the 50th percentile male Hybrid III in a normal seating position and the 5th percentile female Hybrid III in a full-forward seating position. Both dummies were belted and the vehicles included both first generation and second generation (i.e., "depowered") air bags. For the 50th percentile male dummy, the initial dummy-to-module space allowed sufficient time for full air bag inflation prior to dummy contact, which resulted in well controlled head and neck kinematics. The 5th percentile female dummy's proximity to the steering wheel resulted in dummy interaction while the bag was expanding. This typically resulted in the head being forced upward and rearward as the bag expanded under the chin, producing an extension-tension neck response. While the average response of the second generation systems was lower than that for the first generation, the neck extension IARV for the 5th female dummy was exceeded in 59% of the first generation and 45% of the second generation air bag tests.

For adults mildly out-of-position, the difference between injuries induced by the air bag and those that would be expected to have been sustained in a crash of a given severity can be difficult to discern. Berg et al. (1997) used cadaver tests to demonstrate that contact with the inflating air bag in non-extreme OOP situations does not necessarily lead to injuries or life-threatening loads. In fact, they found that for belted passenger dummies in a forward-leaning posture at impact speeds of 55 km/h, the protective effect of an air bag would merit its deployment in nearly all circumstances. The residual protective effect of air bags for unbelted mildly out-of-position occupants could not be established with equal clarity.

4.1.3 Air bag Loading Mechanisms

For the adult population, loading mechanisms from air bag deployment can be classified into three varieties: bag slap, punch-out forces, and membrane forces. Bag slap occurs when the occupant is struck by a small but rapidly moving portion of the air bag during deployment (Figure 10). Air bag slap generally involves a bag strike to either the face or the thorax of the occupant. Even though the mass of the air bag is comparatively low, the impact from the high velocities can produce at least superficial injuries such as skin abrasions or eye injuries (Spiess et al., 1997). The unique nature of these abrasions justified their classification as a new trauma pattern that was not observed prior to air bag implementation into the vehicle fleets (Dalmotas 1998). Powell and Lund (1995) examined production air bag systems and estimated the maximum leading edge speeds of the deploying fabric ranged from 171 km/h to 328 km/h. Using an approximate reference speed at which abrasions may occur, the tests showed that many production air bags were still traveling fast enough to cause abrasions as far as 28 cm from the steering wheel. To minimize this effect, many air bags present a flat segment of the bag against the occupant and unfold the bag from behind. This directs the unfolding towards the steering wheel or instrument panel, reducing the occupant interaction with the unfolding bag material (see Figure 3b).

If multiple layers of air bag fabric are involved and the brunt of the impact is taken by the sternum or head rather than the entire torso, the increased effective mass of the bag and the reduced effective mass of the occupant can combine to produce high occupant accelerations (Partridge 1979). In at least one study, a passenger-side air bag exhibiting considerable bag slap produced higher chest accelerations from the bag deployment than occurred during the 48 km/h frontal impact event (Enouen et al. 1984). Although the bag slap event is relatively short in duration, heart arrhythmia has been reported with short duration, high sternal accelerations (Viano and Artinian 1978). Prasad and Daniel (1984) conducted air bag interaction tests with piglets and determined that transient heart arrhythmia was correlated with mid-sternal short duration accelerations greater than 1000 g to 1200 g during the bag slap.

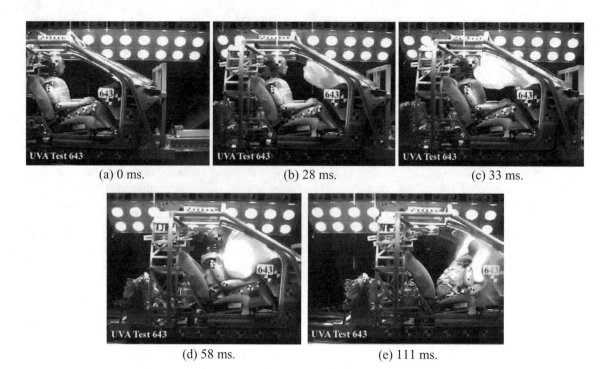

Figure 10. Passenger-side air bag deployment sequence in one frontal impact scenario. Note bag slap at approximately 33 ms, followed by primary air bag loading after approximately 58 ms. Bag slap may be exacerbated if the occupant is OOP. Test performed at the University of Virginia.

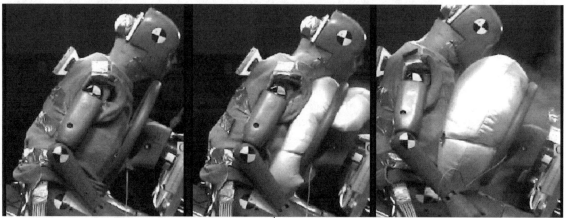

Figure 11. Punch-out loading on a Hybrid III 5th female interacting with a deploying air bag (tests performed at the University of Virginia).

Horsch et al. (1990) developed the term "punch-out forces" to describe the situation when the still-folded air bag attempts to escape from the module and strikes the occupant with a concentrated impact (Figure 11). When the deployment is obstructed by the occupant, the gas pressure inside the air bag becomes greater than that normally required to break the module cover and the resulting high force is concentrated on that part of the occupant blocking the air bag's deployment path. Although generally exhibiting durations less than 10 ms, resulting forces can be as high as 20 kN and can far exceed the injury thresholds of any body region (Lau et al. 1993). The forces generated on an occupant in this situation are related to internal air bag/module pressure, the projected packaged area of the folded bag, and the distance through which the bag must travel before it can unfold circumferentially (Melvin and Mertz 2002). Generally, the risk of injury from punch-out forces is significantly reduced with even a small separation between the air bag module and the occupant. If the person is close to but not in direct contact with the module, the cushion will break out from the cover and surround the occupant, exerting a rearward force on the occupant.

Following punch-out, the subject's interaction with the inflating bag is characterized by the pressurized bag acting as a membrane. Horsch et al. (1990) described the membrane-loading phase as the period after the air bag has deployed from the module but the occupant is sufficiently close to interact with the bag before full inflation. Although the bag pressure is not unusually high at that point in the inflation process, the large area involved and the angles achieved during membrane loading can result in large, injurious forces (Patrick and Nyquist 1972). These forces result from a combination of the air bag's internal pressure and the tension forces arising from the inflating air bag wrapping around the occupant in its path. Whereas punch-out forces produce local injuries, membrane forces act on a much larger area and can induce injuries away from the module (Lau et al. 1993). This is particularly true of the passenger side because the air bag tends to be wider and deeper (Partridge 1979). Unlike punch-out forces, membrane forces can be high even with some separation between the occupant and the module cover (Horsch et al. 1990, Spiess et al. 1997). As the bag wraps around the occupant while continuing to inflate, the membrane forces are greater than the bag inertia forces plus the internal bag pressure on the occupant over the area in contact (Patrick and Nyquist 1972). If wrap-around occurs on both sides of the head, it can propel the head rearward at high velocity, potentially causing a secondary interaction with the interior of the vehicle or primary injury. On the other hand, if only a portion of the head is loaded, the bag may deflect laterally, relieving the membrane tension and forces on the occupant. In tests with porcine surrogates, Lau et al. (1993) believed membrane forces induced abdominal injury, although portions of the abdomen were loaded during punch-out.

4.1.4 Air bag-Induced Injury Patterns

An overview of the mechanisms for fatal air bag-induced injuries has been provided by Kleinberger and Summers (1997) and Summers et al. (1998). The first mechanism involves air bag contact with the face or chin of the driver causing basilar skull fracture with associated brainstem lacerations and/or subdural and subarachnoid hemorrhages. The second mechanism involves air bag-torso interactions and can produce multiple rib fractures, usually bilaterally, with associated lacerations of the underlying thoracic and abdominal organs or other soft tissues. Lacerated organs or tissues can include the heart, spleen, liver, and aorta. In addition to lacerations, fatal cardiac arrhythmias and traumatic apnea may occur. The third injury mechanism is not as common as the first two, but involves cardiac and pulmonary contusions and hemorrhages without rib fractures subsequent to air bag loading of the thorax. Each injury mechanism is discussed in more detail below.

Head and Face Injuries

As previously discussed, injuries to the head and face of adults are most often erythema (i.e., skin redness), abrasions, and contusions caused by contact with the air bag fabric. At least one case study alleging chemical burns on the chin and anterior neck due to the air bag system (Polk and Thomas 1994) may be a misinterpretation of the common air bag abrasions. On occasion, thermal burns to the face due to high-temperature air bag exhaust gas have been observed (Dalmotas 1998) although most of the air bag-induced burns are relatively minor (Reed et al. 1994). In rare cases, entrapment of the driver between the seat and steering wheel during air bag inflation has resulted in third-degree burns when the driver was pinned against the module for some time. For friction-induced abrasions, there remains some controversy as to the precise mechanism of abrasion. Early studies with volunteers attributed abrasions to "wiping" of the air bag across the surface of the skin (Smith et al. 1974) that produced a "stinging" sensation (Lundstrom 1974). Reed et al. (1994) conducted static air bag deployments with volunteers and concluded the patterns of abrasion severity were not sensitive to fabric weave and, therefore, were not a result of friction at the skin surface. Reed et al. (1994) determined that the air bag-induced abrasions are caused by high contact pressures developed as the air bag fabric strikes the skin with the contact pressure at least partially dependent on the speed of the impacting fabric.

Ocular injuries also occur, but much less frequently than skin injuries. The eye injuries range from mild corneal abrasions and hyphema (Figure 12) to retinal detachment, though severe injury is rare. While passenger-side air bags can produce eye injuries, most eye injuries result from interactions with the driver system (Roselt et al. 2002). Isolated cases of air bag perforation from a pipe stem or from eyeglasses have also been reported (Huelke et al. 1995). Examining field cases, Duma et al. (1996) concluded that the eye can be injured by contact with the fully deployed air bag, but that more severe injuries resulted from contact with the actively deploying air bag (see Figure 10). Increased likelihood of eye injury due to deployment of the air bag, rather than contact with a deployed air bag, was supported by the experimental findings of Kikuchi et al. (1975). For corneal injury, they demonstrated a minimum required fabric velocity of 41 m/s, much faster than the velocity expected due to the occupant translating into a fully inflated bag. Using an approximate reference value chosen to represent a speed at which abrasions may occur, tests of model year 1993 air bags showed that many of the air bags were traveling at or above the threshold speed as far as 286 mm from the steering wheel (Powell and Lund 1995).

 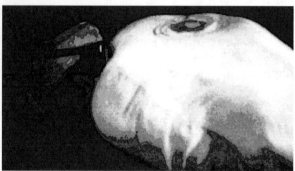

(a) Eye interaction with a deploying air bag.

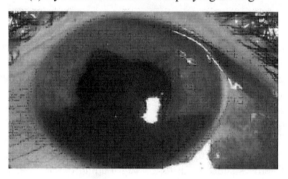

(b) Hyphema (blood in the anterior chamber of the eye, AIS 1).

Figure 12. Images from high-speed video of an air bag interaction with the face. Hyphema, a possible result of air bag loading, is illustrated in the lower photograph (images courtesy of Dr. Tyler Kress of the University of Tennessee and Dr. Stefan Duma of the Impact Biomechanics Laboratory at Virginia Tech).

Facial injuries are not limited to skin abrasions and eye injuries. Injury to the temporomandibular joint has also been attributed to air bag loading (Garcia 1994). In addition to direct loading by the air bag, the face has also injured when the upper extremity is loaded by the air bag and the hand strikes the face causing nasal and anterior face fractures (Huelke, 1995).

Hearing loss from air bag deployment has also been reported (e.g., Yaremchuk and Dobie 2001). Rouhana et al. (1994, 1998) published studies on air bag noise and the assessment of temporary hearing loss based on military standards for impulse noise. Based on these studies, the potential exists for threshold shifts to occur when an air bag (ca. early 1990s) deploys. Richter et al. (1974) used squirrel monkeys to investigate direct slap of the air bag against the ear. While they found no permanent hearing damage or eardrum perforation, post-impact erythema of the eardrum and bruising of the auricle were observed. Based on their results, the authors concluded that slight temporary hearing impairment in humans might be expected from direct air bag slap. Nixon (1970) subjected ninety-one volunteers to an acoustic impulse characteristic of air bag deployment. Similar to the findings of Richter et al. (1974), otologic effects on the drum membrane were negative and no discomfort or aural pain was reported. However, 50% of the subjects and 35% of the ears experienced some temporary threshold shift with 65% of the ears recovering within 24 hours. Many subjects reported short duration tinnitus or ringing following air bag deployment. Roselt et al. (2002) reported on 57 cases of impaired hearing subsequent to air bag deployment in relatively minor crashes. Permanent hearing loss occurred in 11 cases and 9 of these involved deployment of both passenger and driver frontal air bags.

More severe injuries to the head have been associated with being in very close proximity to the air bag at the time of deployment. In the cases of adult fatalities due to air bag loading, short stature occupants have sustained fatal diffuse axonal and brain stem injuries. Confirming this risk, high linear and angular accelerations of the dummy head have been observed in laboratory tests. Saul (1998) conducted passenger air bag deployments with the 5th percentile female Hybrid III dummy and found head injury criterion (HIC) values in excess of 3000 for the out-of-position dummy. Horsch et al. (1990) conducted tests with the Hybrid III dummy and observed that the severity of head loading by the inflating air bag depends on the alignment and separation of the dummy with respect to the air bag modules. In particular, the highest head loading and HIC values were associated with the head directly centered on the module.

Neck Injuries

Neck injuries are not uncommon in crashes involving air bag deployments, but they are usually minor and it is not clear that the commonly reported cervical strains are related to the air bag. In fact, an air bag in combination with a pretensioned belt has been shown to reduce the risk of AIS 1 neck injuries to belted occupants in frontal impacts (Kullgren et al. 2000). Anterior neck abrasions or erythema from the air bag are not rare, but they appear to occur less frequently than they occur to the face. Some of the shorter occupants killed by an air bag have sustained neck extension fractures at the atlanto-occipital joint, with or without spinal cord involvement.

In general, three modes of direct and indirect neck loading have been identified in laboratory out-of-position tests using anthropometric dummies: air bag directly loading the head first, air bag directly loading the neck first, and the air bag becoming entrapped in the jaw-neck region (Kang et al 2001). However, some of these observed interactions may be artifacts of the dummy design. While additional discussion on limitations of current test surrogates will be provided later, cadaver tests have shown that either direct loading of the neck or indirect loading of the neck via head contact can produce extreme neck forces and moments via membrane loading by the air bag (Figure 13). In static deployment tests, Berg et al. (1998) noted that asymmetric lateral loading between the unfolding air bag and the occupant's face produced asymmetric contact leading to rotational loading of the neck vertebrae.

There exists some disagreement as to what pre-deployment positions put the occupant's neck at greater risk. Horsch et al. (1990) found the highest neck loads and moments were observed with the neck located directly over the air bag module. Contrarily, other studies have shown peak extension moments at initial positions where there exists an offset between the dummy head and neck complex and the air bag module. These studies have been able to maximize membrane loading by allowing some initial deployment before radial expansion of the air bag loads the occupant. Malczyk et al. (1998) conducted static deployment tests with an out-of-position dummy and found that an initial distance of 50 mm from the module increased the extension bending moment relative to the 0 mm configuration. In tests with the initial offset, film analysis revealed that the bag initially inflated in front of the dummy's face with increases in extension moment building considerably when the cushion achieved full radial expansion during wrap-around. With initial direct contact of the cover, the air bag spread towards the thorax and distributed more of the deployment loads on the chest.

Schroeder and Eidam (1997) conducted out-of-position tests with cadavers and showed that neck injuries were possible in a static environment even when no contact existed between the head/neck complex and the air bag. In isolated cases, injuries to the ligamentous cervical spine were observed as a result of neck flexion when the chest was loaded by the air bag and the head lagged due to its inertia. Malczyk et al. (1998) explained this phenomenon as the chest being accelerated rearwards while the head remained in place due to its inertia. Unlike most situations in which critical neck loads result from membrane loading, the cause in this case was attributed to the punch-out effect.

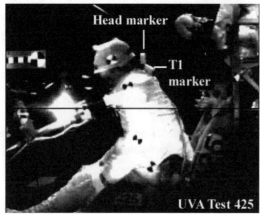 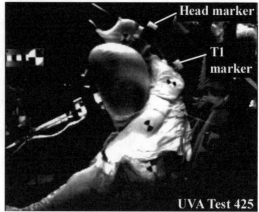

(a) Human cadaver response at start of test and at maximum elongation of the cervical spine.

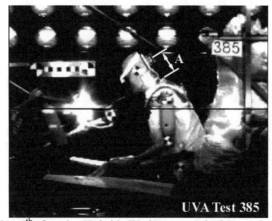 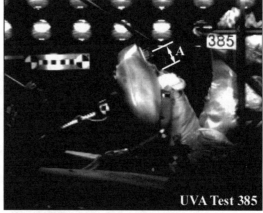

(b) 5th female Hybrid III in a matched test does not exhibit this elongation (dimension A, from instrumentation cube on head to cube on T1, is essentially unchanged during loading).

Figure 13. Images from high-speed video of an OOP deployment into the upper thorax of a small female adult cadaver and Hybrid III (ISO-1 chin-on-module position). Note the markers on the head (occiput) and on the first thoracic vertebra (T1) showing the cadaver's neck elongation during loading. Test performed at the University of Virginia using a ca. early 1990s driver airbag.

Torso Injuries

Except in low-severity crashes, injuries to the thorax and abdomen are often difficult to assign a cause. In many cases, the seatbelt, steering wheel, and air bag are all candidate injury sources. As with the head and neck, the most common thoracic injuries attributed to the air bag are minor abrasions and contusions. For occupants against the air bag module at the time of bag deployment, however, severe loading conditions can be applied to the chest. Separation between the air bag module and occupant's chest has been demonstrated in numerous studies to be the principal factor in determining the injury potential (Musiol et al., 1997; Horsch et al., 1990; Bass et al., 1998; Horsch and Culver 1979; Digges et al., 1998).

Horsch and Culver (1979) showed that the peak interaction force between the air bag and human surrogate varied inversely as separation from the chest and the air bag module increased. Further studies involving both the mid-sized males and small female Hybrid III dummies in various configuration relative the driver module support the conclusions that occupants in close proximity to the air bag can be severely injured in the thorax. Using cadavers in the driver seating position, Schroeder and Eidam (1997) conducted static deployment tests with the chest to air bag distance set to 0, 50, and 100 mm. In cases of 50 mm separation, portions of the module cover made contact with the anterior thorax and produced both rib and sternal fractures. While the severity of injury decreased with separation distance, skin lesions of the unclothed cadaver torso were still observed when the chest was 100 mm away from the air bag module.

High levels of chest deformation have been observed in deployments with the chest-on-module occupant position. In addition to high levels of deformation, Horsch et al. (1990) estimated that the Viscous Criterion (VC_{max}) for test conditions with the chest initially against the module was several times the recommended assessment value of 1.0 m/s. Several experimental studies examining the severity of the this position (Figure 14) have noted the limitations of current dummies to record chest deformations due to either stroke limiting in the dummy (Horsch and Culver 1979) or failures of the instrumentation at high rates of loading (Crandall et al. 1998). In tests with the Hybrid III dummy, the loading severity exceeded the measurement capability when the dummy chest was within 25 mm of the air bag at the time of deployment initiation (Horsch and Culver 1979).

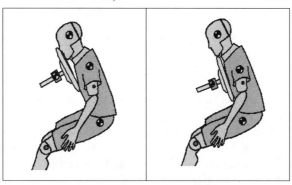

Figure 14. Schematic depictions of the chin-on-module position (ISO-1) (left) and the chest-on-module position (ISO-2) (right).

In crashes with documented air bag-induced thoracic trauma, the fatal injuries attributed to the air bag include multiple rib fractures, flail chest, lung contusions, atrial ruptures, and lacerations of the myocardium, pericardium, or aorta (Huelke et al. 1995, Chidester and Roston 2001). Air bag-related injuries to the abdomen have also been documented but do not occur with the frequency of thoracic trauma (Kleinberger and Summers 1997). Crandall et al (1997b) examined the Special Crash Investigation cases of NHTSA and found that 63% of the cases had thoracic trauma as the most severe trauma. In particular, aortic transactions were identified in 36% of the cases. Whereas there remains much uncertainty as to whether acceleration or compression produces aortic injuries, laboratory tests have confirmed high levels of both parameters when an air bag deploys into a proximate thorax. Saul et al. (1998) conducted static deployments with the small female dummy and observed chest accelerations as high as 170 g for the driver and 358 g for the passenger. As previously mentioned, deformation measurements in the dummies have exceeded 84 mm and were limited only by the range of displacement allowed in the ATD design.

For those occupants against the air bag module at the time of deployment, the amount of coverage of the module by the occupant strongly influences the severity of loading and the magnitude of response and injury. Melvin et al. (1993) compared tests with the 50th percentile male and 5th percentile female dummies to show that the larger male covers the module more completely and exhibits correspondingly higher punch-out forces. Thus, the larger dummy experiences larger VC_{max} and deflections than the smaller dummy, despite the large male being stiffer than the small female. Relative to humans, less complete coverage of the module in tests with swine and dummies has resulted from differing thoracic aspect ratios in the swine (Lau et al. 1993) and decreased degrees of freedom of the Hybrid III anterior chest (Crandall et al. 1998). This suggests the potential for more severe conditions for human occupants assuming the chest-on-module orientation than for surrogates used for out-of-position tests in laboratory experiments. In some cases with low-output inflators or single-stage inflators, the body's inertia may block the deployment such that only a limited portion of the bag escapes through the lower door (i.e., upper door does not open). This can cause a smaller bag volume to escape since a portion of the fabric remains folded in the module. Consequently, the peak pressures in the bag with blocked deployment can be significantly higher than those observed with initial occupant-module separation (Malczyk et al. 1998).

Lau et al. (1993) used domestic swine as human surrogates in air bag tests with modified production air bag systems. Severe and extensive cardiac injuries were observed in 7 of 17 tests with heart perforation as the single, most life-threatening injury. Furthermore, heart perforation was the sole injury induced by deployment of the lowest output inflator. While the majority of tests with cardiac trauma had concomitant skeletal injury, cases of heart perforation were observed without multiple rib fractures. In addition to heart perforations, cardiac contusions occurred as did various arrhythmias. No major lung injuries were identified but spleen lacerations and liver injuries were produced. Peak VC occurred at the end of the punch-out phase and there was apparent coincidence between high VC_{max} values and the punch-out injuries. While the authors state that membrane loading occurred later and may have corresponded to maximum compression, the timing of injury cannot be determined and may have occurred prior to maximum compression. The high risk of injury associated with the chest-on-module position was confirmed by the authors' conclusion that that even the lowest inflator, which may not be sufficient to protect an occupant under normal deployment in a severe crash, produced heart rupture.

In the cadaver tests by Crandall et al., chest deformation exhibited the highest correlation with the severity of injury. While peak VC exceeded 1 m/s in nearly all tests, the lack of observed soft-tissue trauma produced a poor correlation between VC_{max} and the observed injury severity. The air bag characteristic that exhibited the strongest correlation with the extent of injury was the pressure onset rate rather than the peak pressure. All injuries were judged to result from punch-out rather than membrane forces, but the timing of fracture could not be estimated.

In addition to the punch-out injury mechanism, Schroeder and Eidam (1997) hypothesized an indirect mechanism of air bag-induced loading for the torso. In ten experiments with cadavers positioned near the air bag module, the upper body and head were accelerated by the expanding air bag in the direction of the seat back. The air bag static deployments resulted in the cadavers contacting the seat back at impact speeds of 21 km/h to 35 km/h. While potentially a concern for OOP children, the risk of injury via this mechanism for unbelted adult occupants in a dynamic environment remains unknown.

Upper Extremity Injuries

Perhaps the most common non-skin injuries caused by air bag deployment are injuries to the upper extremities (Werner et al., 1996) (Figure 15). Werner and Sorenson (1994) found that drivers were turning, swerving, or changing lanes in 16% of crashes leading to a deployment, which may contribute to upper extremity injuries from the air bag. In particular, deviating from the 3 o'clock and 9 o'clock hand orientation on the wheel can produce wrist pronation. Crossing of the radius and ulna during pronation decreases the overall bending tolerance of the forearm and thereby increases vulnerability of the forearm (Duma et al. 1998). NHTSA (1996) noted that the risk of serious (AIS 3) upper extremity injury to a belted driver may increase by 40 percent with air bags whereas the risk of any upper extremity injury (AIS 2+) among belted drivers was found to increase by as much as a factor of four given air bag deployment in a crash. Dalmotas et al. (1996) showed that harm to the upper extremities and face increases by 71.1% when an air bag deployment occurs.

Huelke (1997) investigated 540 crashes involving air bag-equipped vehicles and found that 38% of drivers sustained some type of upper extremity injury, though most were minor (AIS 1). Contusions, abrasions, and sprains were common consequences of upper limb interaction with air bags during the deployment phase. Hand and wrist burns produced by the venting of the air bag gas are being noted less often with current air bag modules than with earlier modules (Jerrigan et al. 2002). Infrequently, forearm, hand, and digit fractures have been sustained. In 18 of the cases investigate by Huelke et al., the driver sustained an AIS 2 or AIS 3 upper extremity injury, with six due to direct contact with the air bag or module cover and 12 due to impact of the upper limb with the roof rail or other interior structures subsequent to contact with the air bag (i.e., indirect contact). Upper limb injuries resulting from direct contact are related to the punch-out phase of the deployment where the air bags deployment is resisted by the inertia of the forearm which generates large forearm bending moments (Figure 16). Indirect contact injuries are produced when the upper extremity is not injured by the initial interaction with the air bag but is subsequently accelerated by the air bag's membrane forces. After completing the upper extremity and air bag interaction, portions of the upper limb can contact the vehicle interior or even the head and neck of the occupant with sufficient severity to produce injury. The direct and indirect injury mechanisms manifest themselves as different resulting injury patterns. Atkinson et al. (2002) suggested that direct air bag loading tends to result in mid-shaft fractures of the tibia and fibula, while indirect loading is associated with injuries to the wrist and elbow. The associations of injury patterns and loading mechanism are not absolute, however. For example, Huelke et al. (1994) found that hand and forearm injuries could be sustained from indirect loading when the upper limb was accelerated by the air bag into the instrument panel, rearview mirror, or windshield, as indicated by contact scuffs, tissue debris, or windshield damage.

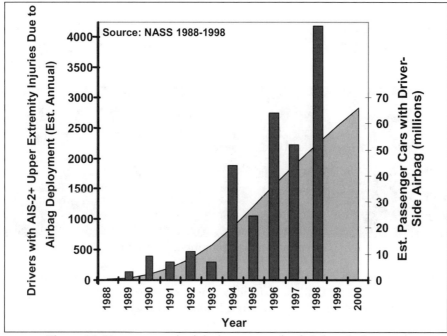

Figure 15. Upper extremity injuries as a result of air bag deployment.

Using multi-body simulations, cadaver tests, instrumented dummy limbs, and simplified component test devices, researchers have shown that module cover characteristics (e.g., stiffness, size, weight, and tear pattern), inflator attributes, distance from the forearm to the module, and orientation of the forearm relative to the steering wheel are all significant factors in determining the outcome of air bag and upper extremity interactions (Sieveka et al. 1997, Kuppa et al. 1997, Atkinson et al. 2002, Bass et al. 1997, Bass et al. 1998, Saul et al. 1996, Johnston et al. 1997, Hardy et al. 1997). These investigations have primarily focused on direct contact injuries. Presumably, decreasing the inflator power or implementing module modifications to reduce the risk of direct contact injuries would likely reduce the risk of indirect contact injuries as well. Changes to the inflator are important for air bag-induced upper limb injuries since the authors are unaware of any sensor technology being developed to detect out-of-position upper extremities at the time of deployment.

In studies of the direct contact injury mechanism, various upper limb orientations have been evaluated with the most common high-risk position being the forearm against the air bag module with a perpendicular orientation to the seam of the module cover. Observed injuries in these cadaveric experiments have been largely transverse, oblique, and wedge fractures of the ulna or radius, or both, and have been similar to those reported in the field investigations (Hardy et al. 1997, Bass et al. 1998). For tests with the forearm initially against the module, forearm fractures occurred early in the deployment (i.e., when in contact with the module cover) before significant upper limb motion had developed. In a limited number of cases, peak forearm moments and by inference injury occurred after the time of primary module cover/forearm interaction (Bass et al. 1998). This suggests that while module cover interaction may be sufficient, it is not necessary and injuries may result from interaction with the air bag itself. Hardy et al. (1997) and Bass et al. (1998) both observed in their experimental studies that female cadavers were more likely than males to sustain forearm fractions during air bag interactions. This was attributed to both geometric as well as bone mineral density differences between the genders.

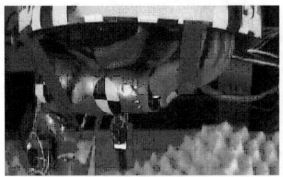 

(a) Immediately before deployment. (b) Arm loaded by deploying air bag.

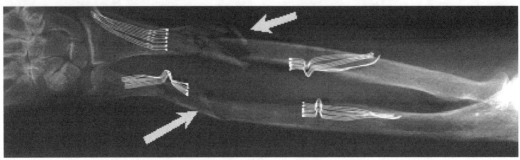

(c) Typical injury pattern for direct loading.

Figure 16. Forearm radiograph showing cadaveric mid-shaft radius and ulna fractures from air bag loading of an arm placed over the module. Radiograph also depicts strain gages mounted on the bones. Dummy arm illustrates interaction on the left. Tests performed at the University of Virginia.

4.2 Injuries to Children

As early as the 1970s, the significance of air bags potentially injuring out-of-position children in the front passenger seat was observed by researchers: "a system acting as if an unbelted 160 pound (70 kg) occupant is present is obviously too powerful to be optimal for a child weighing a fifth as much" (Partridge 1979). The aggressivity of the air bag is particularly concerning when the likelihood of a child being out-of-position is considered. Based on a million car-years of exposure, Montalvo et al. (1982) estimated that there would be 3,732 cars involved in collisions of sufficient severity to deploy an inflatable restraint. In these crashes, there would be 149 unrestrained children of the ages 0 to 4 years, of which 51 would be close enough to the instrument panel at the time of collision due to initial ride position or pre-impact braking to be contacted by a deploying air bag. While this number represents a relatively low exposure, the high risk of injury for these children has generated considerable debate and research.

For older children, the greater risk for air bag-induced injuries relative to adult occupants can be attributed to their propensity not to assume "normal" seating positions, their tendency both to have the seat forward and to sit on the forward edge of their seat, and their diminutive stature, which prevents them from bracing since their lower limbs generally cannot reach the floor or toepan (Aldman et al. 1974, Romeo 1975). Based on an observational study, Mertz et al. (1982) reported that only 45% of children sit in a "normal" position, while the majority is OOP. The propensity for a child to be OOP, coupled with the fact that nearly 70% (Mertz et al. 1982) of frontal crashes involve some form of pre-impact braking, highlights the potential for unintended air bag-child interactions and necessitate the development of air bag systems that do not injure an OOP child.

Child fatalities (normalized for exposure) reached a peak in the 1995-1996 sales period, and have been declining steadily since (see Chapter 5). Even before the change in FMVSS 208, which allowed depowering in model year 1998 vehicles, child fatalities had begun to decline largely due to a successful education campaign. Child fatalities are currently highest for model year 1995 vehicles, decreasing by over half for 1996 vehicles, with continued declines in subsequent model years. While the number of air bag-induced injuries has declined, the authors believe it is important to describe and understand the mechanisms of air bag-induced injuries until all are eliminated. To this end, the following section expands on the injuries to infant and older children that have resulted from air bag interactions.

4.2.1 Infants

As of Jan. 1, 2001, the NHTSA Special Crash Investigations (SCI) program (Chidester and Roston 2001) has confirmed 172 air bag-related fatalities, with an additional 60 cases under investigation. Of the confirmed cases, 102 have been children under 13, and 19 have been infants in a rear-facing child safety seat (RFCSS). Crashes have not been severe and typically have involved Delta-V less than 36 km/h, with adult drivers sustaining either no or very minor injury.

The severity of interaction between an infant and a deploying air bag depends on a number of vehicle, crash, and occupant position factors. In terms of air bag module location, forward deploying air bag can produced large motions and more significant interactions with the RFCSS than top-mounted systems (Avanessian et al. 1982). However, even a top-deploying passenger-side air bag can significantly load a rear-facing infant seat. Neck injuries and scalp and facial abrasions have been observed when a top-deploying bag loads the top of an infant's head (Huelke 1994). The scenario for infants killed in a RFCSS typically involves infants who are properly belted into an appropriately sized rear-facing infant seat, and all but one of these seats has been secured to the vehicle. In general, the air bag deploys into the back of the infant seat, typically with sufficient force to break the plastic shell. The most likely mechanism is that the air bag strikes the RFCSS, which in turn loads the posterior (i.e., occipital) region of the child's head. This interaction typically causes skull fractures bilaterally in the parietal region, and potentially associated subdural and subarachnoid hemorrhages. While cervical injuries have been noted, Kleinberger and Summers (1997) contend that in most cases only a cursory pathologic evaluation has been performed in the upper cervical spine region since fatal brain injuries had been already documented. Therefore, although brain injury is typically assumed to be the cause of death in these RFCSS cases, it is possible that cervical injuries existed but were not reported.

4.2.2 Older Children

Children injured by air bag interventions who are not in a RFCSS are usually older children who are unbelted or improperly belted. As previously noted, many of these cases involve pre-impact braking that places the child near the air bag at the time of deployment. Several experimental investigations have been performed with child dummies and animal surrogates to estimate the out-of-position likelihood during pre-impact braking. Stalnaker et al. (1982) conducted braking tests using 3-year-old and 6-year-old dummies. The child dummies slid forward on the seat in response to hard braking, maintaining their upright seated postures and remaining against the panel until the braking ended. Concern was expressed that the dummy kinematics may not be representative of a child's motion due to the dummy's limited biofidelity. To address this, Stalnaker et al. developed trajectories for unrestrained, anesthetized baboons during hard braking for various initial sitting, kneeling, and standing positions. In a similar study, Mertz et al. (1982) confirmed that when 3-year-old and 6-year-old dummies were not in the normal seating position during 30 km/h braking tests, they moved forward such that their heads and chest touched the instrument panel. Based on hard-braking tests, Montalvo et al. (1982) identified as many as 13 distinct positions that would place an unrestrained child in the path of a deploying air bag.

Examination of current field data confirms earlier concerns and confirms that almost all injury cases involve unrestrained children with pre-impact braking that results in the child's close proximity to the air bag prior to deployment (Kleinberger and Summers 1997). Aldman et al. (1974) suggested three possible mechanisms of injury to the out-of-position child: momentum transfer from the unopened bag to the child in close proximity, acceleration and deformation of the child during air bag inflation and expansion, and the impact of the body on interior car structures following interaction with the air bag. Depending on the orientation and location of the child at the time of deployment, the air bag can contact the chest, neck, or face of the out-of-position child. The small stature of the child makes the upper torso, head, and neck particularly vulnerable. Rapid translation and rotation of the skull following bag contact can cause a number of cervical spine injuries; most notable is a dislocation of the atlanto-occipital joint with contusion or laceration of the brain stem or rostral (upper) spinal cord (Kleinberger and Summers 1997) (Figure 17). Other observed cervical injuries include C2-C3 fractures and subluxations that can result in fatal spinal cord injuries. Resulting closed head injuries, such as subdural and subarachnoid hemorrhages and intraventricular bleeding have been consistent with large and rapid rotations of the head produced by the deploying air bag. Mandibular fractures and avulsed teeth have also been reported but skull fractures were not typically observed due to the relatively distributed loads applied by the air bag. In addition to the head trauma, thoracic injuries such as lung contusions and atrial hemorrhages have been reported.

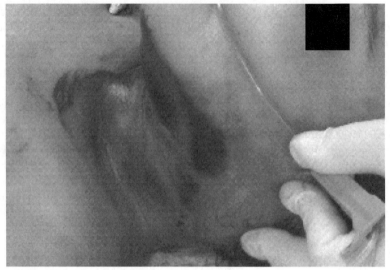

(a) Air bag membrane loading causes bruising and/or abrasions on the anterior neck.

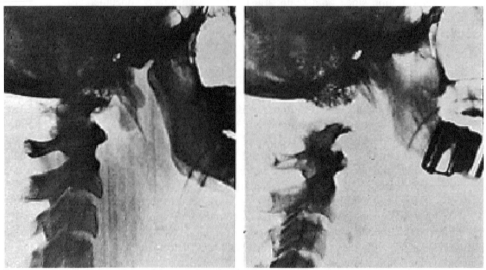

(b) Atlanto-occipital dislocation with associated spinal cord transaction is a cause of death. Basilar skull fractures are sometimes noted.

Figure 17. One scenario for an OOP child fatality due to air bag interaction.

4.2.3 Experimental Investigations

While a number of pediatric injuries and deaths have been observed in the past decade, the number would have been considerably greater if substantial testing had not been performed over the past 30 years to highlight and address the problems of children in close proximity to a deploying passenger air bag. Due to limitations in the biofidelity of child anthropomorphic dummies and the public's sensitivity to child cadaver research, most investigations examining the risk of air bag inflation-related injury for child occupants have involved animal surrogates. One of the earliest studies was performed by Patrick and Nyquist (1972) using baboons to represent the out-of-position child. Superficial soft tissue injuries such as abrasions, contusions, and lacerations resulted from direct contact with the air bag in nearly all tests. In addition, the researchers identified the potential for serious head, neck, chest and abdominal injuries. If the baboon's head was in the path of the air bag, the incidence of concussion, as indicated by a loss of eyelid reflex, was large. Furthermore, the authors state that the tests with the baboon surrogate probably underestimated the likelihood of a child to suffer concussion since the primates possess heavier musculature on the head and a proportionately smaller brain. Fracture of the bones was not a major problem with the animals, but the severity of the interactions and geometric differences between the baboon and a human child led the authors to suggest that the potential for fractures of the child cervical spine might be greater than observed for the baboons. The study indicated that injuries to the thoracic and abdominal organs are typical when the air bag strikes the torso. In addition to direct contact injuries, Patrick and Nyquist noted that in several of the test positions the baboon was propelled over the front seat. While there were no internal vehicle components in the experimental set-up, the kinematics confirm the potential for secondary injuries from interior car structures.

Mertz et al. (1982) and Prasad and Daniel (1984) conducted more comprehensive testing to assess deploying passenger air bag interactions with animals. Mertz et al. (1982) positioned piglets and baboons against the instrument panel with varying vertical heights to examine air bag contact on different body regions, thereby exposing the head, neck, and thorax. Tests were conducted with passenger-side air bag systems installed in low-dash mount, top-dash mount, and high-dash mount positions to compare injury patterns produced with an OOP child. The low-dash mount air bag injured the chest seriously while the top-dash mount injured the head. The high mount bag injured both the head and the chest. Cases with thoracic injuries had concomitant injuries ranged from none to fatal for the head, neck, and abdominal regions. The most significant pig injuries were subdural brain hemorrhages, partial transactions of the cervical cord, cervical and lumbar spine fractures, lung and heart contusions, persistent abnormal ECG readings, extensive rib fractures and liver tears, and lumbar spine fractures.

Prasad and Daniel (1984) conducted piglet tests similar to those performed by Mertz et al. (1982) but used different air bag systems. Observed head injuries ranged from minimal extra-dural hemorrhage to fatal injuries. In all but one case, significant brain injuries were accompanied by severe to fatal neck injuries. Based on film analysis, the head injury without neck injury was due to a severe head impact against the vehicle during rebound. The neck trauma observed in the study was initiated by the rupturing of small blood vessels of the membrane that encase the occipital condylar joints. As the impact severity increased, the damage progressed to rupture of the alar ligament, damage to the spinal cord and brainstem, and finally death. The authors noted combined tensile loads, shear loads, and bending moments were applied to the neck as the air bag deployed. The presence of high tensile and bending stresses at the atlanto-occipital and C1-C2 joints was confirmed by the predominance of brain stem hemorrhages without significant injury elsewhere in the brain. While it was impossible to discern the precise combination of shear, tensile, bending, and torsional forces in the neck, the existence of torn apical ligaments in the neck and cervical spinal cord hemorrhages suggested the existence of high neck tensile loads. The hypothesis of tension-induced failure was also supported by inferior-superior accelerations of the head.

Despite high levels of thoracic compression, Prasad and Daniel (1984) identified no rib fractures. Presumably, this relates to the immature state of ossification in the adolescent animals. The most severe injures in the thoracic region were to the lungs, involving parenchymal hemorrhages and scattered multiple contusions over the entire lung. Since the authors estimated that the rates of compression exceeded 12 m/s in some tests, they postulated that the piglet lung injuries could be associated in part with "air blast" effects. No apparent trends were found between piglet spinal accelerations and thoracic injury.

In addition to the anesthetized animals, the Mertz et al. and Prasad and Daniel studies used an instrumented 3-year-old child dummy to evaluate injury potential. The animal injury severity ratings were paired with the dummy responses measured for each body region in a matched test. The results showed that neck tension was the best indicator for the presence of serious (AIS = 3) neck injuries. Prasad and Daniel did note, however, that combined bending and tensile loading was likely during air bag interactions and suggested that both factors should be considered in the assessment of injury risk. For thoracic trauma, the studies found that HIC and chest acceleration correlated with the presence of injury. In addition, Mertz et al. identified the rate of sternal deformation as a predictor of thoracic injury.

Chalmers University and Volvo conducted static and dynamics tests with porcine subjects intended to represent the out-of-position child (Aldman et al. 1974). From a distance of 10 cm to 15 cm, air bags were deployed into the lateral aspect of the pigs in order to account for aspect differences between the pigs and a child. According to the authors, a side-facing pig was thought to be a better representation of the forward-facing human child for studies of air bag interactions due to the shape of the thoracic cross-section. The vertical height of the pigs was adjusted to focus the deployment at the level of the heart. Correspondingly, numerous cardiac injuries were produced; these injuries were listed as the cause of death in 3 of the 8 fatality cases. In addition to the cardiac injuries, substantial liver injuries were observed. The injuries to the liver were mainly ruptures at the sites of attachment and were considered to be related to displacement of the entire organ rather than local deformation. Similar to the findings of Prasad and Daniel (1984), lung injuries were associated with hemorrhagic spots on the inside of the thoracic cage under the pleura and were compared to air blast injuries.

Subsequent tests with child dummies have tried to provide evaluations of the air bag factors influencing the risk of injury. NHTSA conducted static and dynamic deployment tests with the 3-year-old and 6-year-old child dummies to assess air bag injury potential for the out-of-position child (Summers et al., 1998, Rains et al. 1998). Using depowered air bag systems, they determined that the injury assessment reference values for the dummy were exceeded even when only the primary stage of a multi-stage inflator was used. These tests suggest that air bag suppression may be the only viable option for certain systems and certain child positions, though limitations in the number of tests, the systems evaluated, and the ability of the dummy to identify true injury risk in this complex environment limit the certainty of this conclusion.

4.3 Surrogates and Other Practical Considerations for Interpretation of Injury Risk

There exist considerable practical problems associated with extrapolating laboratory tests with an out-of-position surrogate to real-world performance and injury risks. Significant issues that confound extrapolation include the low but unknown frequency for field OOP exposures, the broad range of potential crash conditions and occupant alignments, the relevance of the laboratory test environment to possible real world conditions, the appropriateness of the injury assessment surrogates, and the lack of injury criteria for OOP conditions (Horsch et al. 1990). Human volunteers, human cadavers, animals, and mathematical models have all been used to evaluate the air bag system. While direct comparison of injury risk among these surrogates is difficult, the advantages and limitations of each surrogate have been explored in the literature. An overview of surrogate applicability is provided to assist in interpretation of the air bag-induced injuries from experimental investigations.

4.3.1 Adult Dummy Designs

Since the field data and air bag deployments with OOP dummies indicate that smaller drivers are at greater risk than larger drivers (Melvin 1993), most efforts to study the adult out-of-position occupant injuries have focused on the small female Hybrid III dummy. The Hybrid III 5th percentile female dummy is essentially a scaled version of the Hybrid III 50th percentile male. Differences in loading between the frontal crash condition for which the dummy was originally designed and the much more complex and potentially more severe out-of-position occupant loading environment with an air bag have necessitated investigations into the applicability of the Hybrid III dummy. While limitations of the dummy design have been noted for the OOP loading environment, it is important to stress that this dummy has been the most advanced dummy available for testing over the past 25 years and has enabled considerable advances in air bag safety.

Thorax

The Hybrid III thorax was developed for frontal blunt loading up to 6.7 m/s impact velocity (Horsch et al. 1990), with design and calibration procedures based on inertial loading during crash events. Crandall et al. (1998) evaluated the biofidelity of the Hybrid III 5th female relative to cadavers in out-of-position tests with a chest-on-module position. Thoracic accelerometers and chestbands were used to compare chest compression, velocity, VC, and acceleration. Despite differences in the amount of chest coverage for the air bag module between the surrogates, differences in measured dummy and cadaver responses were not statistically significant. These findings suggested that the dummy is an adequate surrogate for evaluating air bag aggressivity and predicting thoracic injury in some instances. Subsequent studies with the small female dummy, however, have shown large response sensitivity to initial test conditions. Bass et al. (1999) performed static deployment tests using two different, but reasonable, interpretations of the chest-on-module (ISO-2) position with a 5th percentile female Hybrid III. A large variation in occupant response was found with relatively small changes in distance from the sternum to the air bag module. One particular source of inherent variability in chest response involved the anterior breast skin on the small female dummy. Since the anterior breast of the dummy is not intended to be biofidelic, defining the initial position of the dummy relative to the module becomes crucial. Lin et al. (1995) used a finite element model of the dummy and passenger air bag to demonstrate that skin compliance has a softening effect on the impact response of the dummy. In experiments with and without the breast skin,

Bass et al. (1998) confirmed that the breasts have an effect similar to initial chest-to-module offset. Nearly 50% variation in peak chest acceleration could be observed with as little as a 2 cm gap between the chest and module. While this gap is important in terms of punch-out injury potential, this small change is not thought to reduce injury potential to the degree indicated by the dummy. To eliminate the effective gap introduced by the breasts, Bass et al. applied a preload to the torso and forced the dummy against the module.

In addition to the variability introduced by the breasts, large variations in dummy response have been found with small vertical and horizontal displacements of the occupant. Chest deformations and accelerations dropped by nearly 15% for a 2 cm change in vertical displacement while a 4 cm change altered the values by 40% to 50% Bass et al. (1998). The Viscous Criterion was even more sensitive to initial position with variations of 70% observed with a 4 cm lateral or vertical shift of the dummy chest. Furthermore, for variations approaching 4 cm, qualitative changes in the air bag deployment patterns were observed. Roychoudhury et al. (2000) confirmed the results of Bass et al. using computer simulations. The authors varied dummy positions in three dimensions to show that large variations in dummy response could be achieved with relatively small changes in position. Again, these large sensitivities are not thought to represent a realistic change in injury potential but may suggest a regional response sensitivity based on the dummy structure. In porcine tests, Lau et al. (1993) determined that the location of the module only marginally influenced the risk of thoracic injury. More studies on chest position relative to the air bag are needed to compare sensitivities between dummy responses and thoracic injury outcome in cadaver or animal studies.

Neck

Researchers have identified shortcomings with the Hybrid III neck design in reliably assessing potential interactions between the head-neck and the deploying air bag. The Hybrid III dummy's neck is inherently much smaller in diameter than the human's neck. In addition, there is a hollow area between the dummy's chin and the neck that provides an unrealistically large reaction surface for air bag membrane loading. These anthropometric differences between the dummy and human confound the interpretation of injury measures recorded in the dummy during air bag loading. In general, three modes of air bag-neck interaction have been observed in dummy investigations: air bag directly loading the head, air bag trapped under the chin during the deployment process, and air bag trapped behind the jaw of the dummy head. C. Morris et al. (1998) conducted static air bag deployment tests in vehicles with the 5th female Hybrid III outfitted with several modifications of the dummy's head skin and neck shield. None of the modifications tested was found to provide a reliably biofidelic indication of air bag inflation injury risk. In a similar study, Kang et al (2001) conducted static deployment tests with a passenger air bag and found that the exposed horizontal surface in the chin-jaw region and the near-vertical cavity between the jaw and the neck of the Hybrid III provide unrealistic reaction surfaces for loading, which can result in unrealistic neck deformation. A modified temporomandibular joint head skin and a custom neck shield were unable to impede air bags from being trapped behind the jaw or under the chin. Modifications of the dummy itself helped decrease entrapment of the air bag but were not considered generally applicable for all test conditions.

Even if artifactual interactions between the air bag and the dummy neck can be avoided, unrealistic moment readings can arise from the manner in which the Hybrid III head loads the neck. The original Hybrid III design and calibration procedures were based on inertial loading during crash events. Contact loading resulting from an air bag was not considered when the Hybrid III family was designed (Kang et al. 2001). Air bag loading is significantly different from inertial loading in that there is more than one load path on the head. In the Hybrid III neck design, the nodding block structures transfer moment from the head directly to the neck structure. This design works well for the inertial loading encountered in frontal crash conditions but causes complications during the more complex loading conditions associated with air bag deployment. In particular, the neck experiences second mode bending during air bag loading as opposed to the first mode bending seen in seat belt loading. These complex bending modes can permit little head rotation relative to the neck but provide high bending moment readings. In static deployment tests, higher order bending moments in the dummy have resulted in maximum extension bending moments occurring before virtually any rearward extension. Therefore, even if interactions due to the air bag being trapped behind the jaw or under the chin are prevented, the dummy head and neck interaction may be different from a live human. Based on the higher order bending modes, Kang et al (2001) concluded that the global neck moments in the Hybrid III were not necessarily good insights into the local moments at the occipital condyle for conditions other than the first mode of bending. Furthermore, the influence of neck vertebrae, ligaments, and muscle tone on the bending of the neck is unknown. In humans, the neck muscles partially shield the ligamentous cervical spine structure by transferring moments to the head through an alternate load path (Van Ee et al. 2000).

As previously noted, air bag loading for the OOP occupant frequently generates high tensile loads in addition to bending moments in the neck. In chin-on-module tests (Figure 14) conducted with the Hybrid III 5th percentile female dummy and a small female cadaver, considerable differences have been noted in the extent of axial extension permitted by the dummy (Figure 13). These differences have been attributed to limitations of axial extension by the central steel cable used in the child and adult Hybrid III dummy neck to control flexion and extension. This cable is considerably stiffer in tension than the ligamentous and muscular resistance provided in the human. In addition to influencing injury measures involving tensile forces, extensibility of the neck provides more deployment space for the air bag and influences both occupant and air bag kinematics. Thus, two limitations may exist for evaluating OOP air bag loading with the Hybrid III dummy: the Hybrid III dummy neck may not react to applied tensile loading in a biofidelic manner and this nonbiofidelic response may preclude the dummy from being exposed to the same loads a human would experience in the same air bag loading environment.

Upper Extremity

Numerous upper limb surrogates have been used to evaluate the aggressivity of air bags. They vary considerably in complexity and include the Research Arm Injury Device (RAID) (Kuppa et al. 1997), the 5th percentile female upper limb more commonly known as the SAE arm (Bass et al. 1998), and a customized instrumented version of the Hybrid III 50th percentile male upper extremity (Saul et al. 1996). The RAID is a simple beam test device used to record moments and accelerations during air bag interactions. Saul et al. (1996) used a 50th percentile male Hybrid III upper extremity outfitted with strain gages and accelerometers to record forearm loads. The SAE arm is an anthropomorphic representation of a small female upper extremity instrumented with load cells, potentiometers, and accelerometers. Forearm bending tolerances determined from measurements recorded in the instrumented SAE arm have shown similar injury risk functions when compared with bending tests conducted on cadaveric forearm specimens (Duma, 2000).

In studies using cadaver extremities as the objective injury indicator, all three devices have shown similar abilities to rank the relative injury potential of tested air bag systems. However, significant kinematic differences existed between the RAID and the dummy limbs (Johnston et al. 1997, Bass et al. 1998). Following initial air bag-limb interactions, RAID kinematics were sufficiently different from cadaveric results to make the device unsuitable for studies in which the timing and motion of the arm, cover, and air bag interaction are significant (Bass et al. 1998). This implies that use of the RAID is limited to evaluating initial interactions between the air bag and limb only and is not appropriate for studying indirect loading (i.e., fling) injuries. For the Hybrid III limbs, initial kinematics appear similar to those of the cadaver until considerable shoulder motion is involved.

4.3.2 Child Dummy Designs

The above review of field and experimental data demonstrated that children are arguably the most vulnerable population for air bag-induced injury. Child dummies with improved biofidelity and extended measuring capability and capacity have been developed specifically to address air bag interactions. In particular, the child restraint air bag interaction (CRABI) dummies were developed to represent the 6-month-old, 12-month-old, and 18-old month infant (Mertz 2002). The Hybrid III 3-year-old child was designed to be used for testing child restraints as well as for assessing the injury risks associated with air bag interactions. The dummy's final design was based on a combination of designs from the 3-year-old air bag dummy (Wolanin et al., 1982), a scaled down version of the Hybrid III 50th percentile male, and scaled-up versions of the CRABI dummies. Similarly, the Hybrid III 6-year-old child was designed for use in frontal impact testing and scaled from the Hybrid III 50th percentile male shape and biofidelity response. While studies have been done to confirm the repeatability and durability of these dummies (Saul 1998), limited biomechanical test data for children prevents evaluation of the injury predictive capability of the child dummies. In particular, no child cadaver information exists in the out-of-position loading environment and only very limited data exist for child-restraint loading with forward-facing child safety seats (FFCSS). Given these limitations, Mertz et al. (1982) and Prasad and Daniel (1984) developed paired comparisons between animal surrogates and the dummy response. This enabled dummy-specific injury criteria to be developed for a specific range of test conditions. Comparison of the Hybrid III response against field data and the limited child cadaver for the FFCSS, however, suggests differences in neck kinematics attributed to concentration of thoracic and spinal compliance into only the neck structure (Sherwood et al. 2002). Furthermore, similar to the adult dummy design, examination of neck properties of the Hybrid III three-year-old child dummy indicates that the neck tensile properties are considerably stiffer than equivalent human properties and could influence both air bag kinematics and injury response (Crandall et al. 1999a).

4.3.3 Animal Surrogates

Animals have been used to complement cadaver and dummy tests in air bag out-of-position research. In the absence of child cadaver tests, animal research provides geometric similitude and equivalent developmental ages for direct evaluation of injury potential. Supplemental to what might be discerned from child cadaver research, tests with animals have the added benefit of reproducing some of the injury sequelae that exist in the living population. Subsequent correlations have been made between the observed animal injuries and child dummy responses measured under comparable test conditions. While injury criteria for the dummies have been based largely on the animal research, anatomical, geometric, and developmental differences between the animals and the child population they are intended to represent need to be considered when evaluating injury risks. All authors utilizing animal surrogates in their studies have thoroughly documented the differences between the animals tested and human children. A summary of these differences are provided here to facilitate interpretation of injury and response from air bag studies that have used animals as human surrogates.

Studies have generally involved baboons and pigs of similar weight and height to human child. The pig exhibits more favorable anthropometric characteristics of the chest and abdomen than the baboon and appeared to be more susceptible to injury than the baboon (Mertz et al. 1982). Both the pig and the baboon have a major anthropometric deficiency as a child thoracic injury model, however, in that their chest depth to width ratios are the inverse of the human child. Because of this geometric difference, the fore-aft chest stiffness of both animals is much greater than the human child's. This difference results in greater force levels required to produce compression-type injuries in thoracic organs (Mertz et al. 1982). Of particular importance for assessing air bag-induced head and neck injuries, the pig has no chin protuberance for interacting with the deploying air bag. The neck of the pig attaches to the rear of its skull and its snout is aligned with its cervical spine. While cervical vertebrae of the pig and the child are of similar size, the pig's neck circumference is nearly double that of the human child due to its large dorsal neck muscles. In addition, the functional fore-aft motion of the pig's head-neck structure is much less than the child's, which may result in a smaller degree of rearward motion required to produce a hyperextension neck injury. In terms of brain injury, the thick skull and small brain of the pig would tend to underestimate the injury potential of a child.

4.3.4 Cadavers

The cadaver as a surrogate for the living human exhibits the correct anthropometry for reproducing realistic occupant and air bag interactions in OOP evaluations. The lack of physiologic response, however, limits experimentally-induced injuries to mechanical failures of the tissue and prevents the study of trauma requiring a functional response from the body (e.g., diffuse axonal injury). In addition, the cadaver's ante-mortem pathologies coupled with post-mortem tissue changes can substantially influence the measured biomechanical responses and injury outcomes. Finally, variations in the duration and methods of storage and preservation can create considerable experimental differences that are not easily quantified. For cadavers that have been frozen and thawed prior to testing, minimal changes are generally observed in the material properties of hard tissue structures but the results for material property tests of frozen soft tissues generate uncertainty regarding the validity of visceral responses and injury criteria. This may partially explain the differences in the extent of soft-tissue trauma observed between field studies (Kleinberger and Summers 1997) and cadaver tests (Crandall et al. 1999b) examining air bag-induced thoracic trauma. While the frozen and thawed cadavers exhibited massive numbers of rib fractures in excess of the field cases, the cadaver experiments were unable to recreate the severe visceral injuries identified in the field despite the fact that the cadaver experiments utilized a worst-case scenario with the chest positioned against the module at the time of bag deployment.

While pressurization of the cadaver cardiovascular system and other attempts have been made to model the *in-vivo* condition, another potential difference between cadavers and living humans in a crash is the ability of the occupant to tense during the impact. While the cadaver may suitably represent a portion of the severely injured occupants who have been unconscious or out-of-position due to medical conditions, muscle tensing can significantly influence the kinematics and injury outcome provided the occupant is aware of the impending crash. For the out-of-position occupant, the intercostal muscles likely play a smaller role in influencing thoracic injury than do the neck muscles in determining cervical spine and head injury. Further characterization of muscle response for all body regions is necessary before correlations can be developed between the observed cadaver responses and injury with that of a living, tensed human.

4.4 Air Bag Parameters that Affect Injury Risk

While more than 30 years of research have been devoted to studying air bag-induced injury, continued improvements in air bag technology are still being investigated in the effort to reduce further the potential for air bag-induced injuries. The continued modification of air bag systems involves a multi-stage effort with potential alterations of the air bag materials, sensors, inflators, module designs, as well as the air bag folding and deployment patterns. Coupled with increased public awareness and education campaigns, the effectiveness of current air bag redesign efforts is becoming apparent in the most recent trends in air bag-induced fatalities (Chidester and Roston 2001). Ultimately, "smart" air bag systems may be capable of detecting crash severity, occupant, restraint, and other factors and changing their deployment characteristics in response accordingly (Struble 1998). In the meantime, this chapter examines some of the established relationships between air bag design features and injury potential.

4.4.1 Inflators

The goal in determining inflator output is to reduce occupant loads during air bag inflation with OOP conditions without lowering the occupant protection during in-position conditions. As early as 1972, Patrick and Nyquist recommended that pressure and rate of inflation be investigated as a means of minimizing injuries. While studies have shown that lower energy inflators reduce the contact forces and corresponding risk of air bag-induced injury, decreases in injury risk for OOP occupants relative to potential compromises in protection for unbelted occupants in severe crashes is less clear.

The performance of an inflator is typically described by the final pressure and the maximum rate of pressure rise when the inflator is deployed inside a one-cubic-foot or 60 liter tank. Additionally, the mass loss from the inflator can be measured and used as an approximation of the total gas supplied by the inflator. While the peak pressure and rise rate specifications have proven to be informative with respect to in-position occupants, supplemental information on the inflator thrust has been suggested as a better characterization of the air bag's injury potential (Prasad and Laituri 1996).

While a general relationship between inflator output and injury potential exists, the tank test may have limited relevance to some out-of-position situations. In porcine tests, Lau et al. (1993) found that injury severity was only marginally related to the gas output performance of an inflator. For occupants blocking deployment of the air bag, they concluded that pressure had to build up to a similar level before punch-out regardless of the inflator rating. Similarly, Horsch et al. (1990) showed poor correlation between the inflator's pressure rise rate and the VC_{max} recorded in chest-on-module tests. Several studies have shown, however, that decreasing inflator output may significantly reduce the potential for deployment-related injuries (Musiol et al. 1997). For example, Prasad and Laituri (1996) tested and modeled modified air bag systems to show that maximal reductions in air bag-induced injury potential are achieved through decreasing inflator output. Digges et al. (1998) used a finite element model to generate static deployment conditions for the chest-on-module condition. The model predicted that milder inflation curves that supply the same amount of energy but take longer to peak would result in reduced inflation-related injuries. In dynamic and static deployment tests, Hollowell et al. (2001) concluded that diminished inflator output generally produced lower HICs and chest accelerations for a given vent hole diameter. In cadaver tests with the chest-on-module position, Crandall (1999b) observed a correlation among the inflator's pressure rate, the measured injury predictors (e.g., chest acceleration and compression), and the injury severity.

Horsch's studies in 1979 and 1990 found that when the module cover was blocked by an out-of-position occupant, the internal pressure rose to more than five times the normal inflation pressure. This was a source of high injury risks by punch out forces. In fact, this was probably the cause of the first reported air bag fatality, which occurred with evidence that the driver was slumped against the steering wheel at the time of inflation and blocked the normal path of deployment. Some efforts have been undertaken to design systems with alternative venting, and today dynamic venting is being considered.

Preferable to reducing the overall inflator output or depowering the inflator, multi-stage inflators have demonstrated greater flexibility in tailoring inflator output (Ryan 1998, Reidelbach and Scholz (1979). Ideally, dual-stage air bag systems should be capable of performance equivalent to single stage systems in terms of providing protection for in-position occupants while reducing aggressivity for the OOP occupant. Sensing of impact severity with subsequent dual-stage control of air bag inflation characteristics was examined as early as the initial prototype evaluations (Lundstrom 1974). The inflator consisted of two internal generators capable of producing and releasing two levels of inflation energy. In dual-stage air bag systems, the air bag deploys at low inflation levels during impacts of low severity while impacts of higher severity trigger a faster and higher pressure deployment of the air cushion. Interestingly, the initial GM air bag fleet involved a dual stage passenger air bag system as was the original Mercedes system (see Chapters 2 and 3).

Due to limitations in sensing the occupant's position, thresholds for deployment of each stage of a multi-stage air bag system are based primarily on crash conditions. Deployment of the first stage of the air bag system provides protection for occupants in low speed crashes. In more severe crashes, initial deployment of the first stage assures a base level of protection for the occupant while affording more time for additional sensor information to be gathered on the crash event and for determining the merits of deploying the second stage. As for all air bag systems, benefits achieved for in-position occupants need to be evaluated relative to the risk of producing injury for the out-of-position occupant. Fortunately, the diminished inflator output of the first stage has shown considerable benefits for the OOP occupant. OOP dummy studies have shown considerable reductions in neck and thoracic forces relative to first generation or even second generation (i.e., depowered) systems (Hollowell et al. 2001, Dalmotas 1998). Cadaver investigations of the chest-on-module loading demonstrated significantly fewer rib fractures with single stage deployments than with conventional depowered systems (Crandall et al. 1999b). In addition, investigation of air bag-induced upper limb injuries using cadavers showed single stage deployments did not produce any forearm fractures even in the designated worst-case condition while depowered air bags and dual stage deployments generated injury in each case (Atkinson et al. 2002).

Despite these promising results for adult injuries, however, children may still be at risk for air bag-induced injury with single stage deployments (Dalmotas 1998). Dalmotas noted the lowest dual-stage system capable of passing the 50th male unbelted test at 48 km/h still exceeded the child dummy's injury assessment reference values during single stage deployments in OOP tests. This suggests that further improvements in sensor technology may be required to discriminate the occupant's position and to consider suppression as an option for certain circumstances.

4.4.2 Module Cover

Numerous studies indicate that the severity of loading by the inflating air bag depends on the alignment and separation of the dummy with respect to the air bag module. In general, the highest chest responses have been observed with the cushion centered on the sternum and the chest fully covering the air bag module. For cases where the chest is against the module, incomplete coverage of the module can still exist because the occupant is vertically or laterally offset from the center of the module. A direct relationship exists between the amount of coverage of the module by the test subject, the severity of loading, and the magnitude of response and injury (Horsch et al. 1990, Melvin et al. 1993).

Module cover modifications that may be considered in an effort to reduce OOP injury potential include the tear seam patterns, tear seam area, flap material, and the extent of module recess in the steering wheel. Horsch et al. (1990) conducted static experiments with anesthetized swine, which suggested that break pressure was determined by the module design rather than the inflator. In OOP tests with porcine surrogates, Lau et al. (1993) found module modification to be the most influential independent variable when examining inflator output, module alignment, module tear configuration, and steering wheel characteristics. While orientations of the tear seam were considered, the most effective modification involved alternative deployment paths for the bag if the primary path was blocked. These findings were confirmed in cadaver tests in which severity of the chest interactions were inversely related to the tear seam coverage for the chest-on-module condition (Crandall et al. 1999b). Upper extremity interactions have similarly shown the benefits of providing alternate deployment paths (Kuppa et al. 1997, Bass 1997, Hardy et al. 1997).

Prasad and Laituri (1996) performed computer simulations and experiments to evaluate the occupant's response sensitivity to module cover stiffness. For identical inflators and air bags, stiffer covers generally lead to slightly higher VC_{max} for the chest-on-module condition. In spite of the clear trend of increasing deformation and VC_{max} with inflator cover stiffness, thoracic responses were still likely to produce serious injury when the cover was completely removed from the module.

The sensitivity of thoracic response to initial chest-to-module spacing can be influenced by the relative distance from the module cover to the plane of the steering wheel. For conditions in which the chest is in close proximity to the air bag, recessing the module behind the plane of the wheel effectively allows a portion of the occupant to be supported by the steering wheel rim with a gap between the chest and module. This gap shows similar effects to those achieved by increasing the initial occupant to steering wheel offset (Bass et al. 1997, Roychodbury et al. 2000). On the other hand, while this system may reduce punch-out forces applied to the thorax, neck extension resulting from membrane loads may actually be increased (Spiess et al. 1997).

4.4.3 Folding Patterns

Using the air bag folding pattern as a way of reducing injury has been extensively studied. Four different air bag folding patterns which have been investigated in an effort to reduce air bag-induced injuries were described by Mao (2001). The Leporello or conventional pattern folds the air bag in accordion-type layers into a package that is generally located directly over the inflator. The Raff or Petri (P-fold) pattern is folded into a concentric structure around the inflator and then pressed radially into a package in the air bag module. The Stochastical pattern represents a type of folding in which the folding lines are not regular and the flat air bag is simply pressed from the outside to the center and packed into the module. Finally, the Z-folding pattern folds the air bag vertically like an accordion around the gas inflator and above it.

The effects of folding pattern on injury outcome need to be considered in view of the OOP condition being examined. In particular, air bag folding patterns appear to have the greatest influence on membrane-loading and bag slap interactions. Based on contradictory findings in studies, the potential benefits for reducing punch-out phenomena are less clear.

In early studies, volunteer abrasions were reduced by adjusting the folding pattern (Lundstrom 1974). Romeo (1975) found that the responses of a child dummy during dynamic OOP tests were highly dependent on folding of the air bag. He noted that reductions in the loads on the head and torso could be achieved by adjusting the folding pattern to modify deployment kinematics. In simulation studies of adult dummies, Wang (1991) showed the air bag folding pattern could influence the extent of membrane loading and thereby influence the injury risk.

For the chest-on-module condition, Prasad and Laturi (1996) conducted Hybrid III dummy tests and determined that the folding pattern tested (Leporello and Petri) did not influence a system's performance. The authors hypothesized that during the deployment initiation phase of the inflation process, membrane loading is nearly nonexistent and that peak VC values occur before the air bag is even visible. Conversely, Malczyk et al. (1998) showed that when there is at least a small gap between the module cover and the dummy's body, a P-fold bag can unfold radially, thereby reducing chest deformations, accelerations, and VC_{max} relative to the Leporello pattern. Unlike the benefits observed for the chest response, however, the neck response had a tendency to increase with the P-fold system. For the forehead positioned within 20 cm of the module, the neck had higher tensile forces and extension moments with P-fold.

Similar to the findings of Romeo (1975), Zinke (1980) demonstrated the significance of the folding pattern in determining the injury measures in OOP tests with a child dummy. If the bag is rolled under, it tended to deploy down into the child's lap minimizing high angular rotations of the torso about the hip. If the rolled portion of the bag was placed adjacent to the diffuser slots, only a single layer of material made initial contact with the child's torso. Thus, the effective bags mass was minimized which, in turn, reduced the chest acceleration due to bag slap.

A biased flap is a design innovation used to re-direct the path of air bag inflation for out-of-position occupants. With this approach, an addition layer (or flap) of material is stored with the air bag. During deployment the flap can contact the occupant's head and neck, and present a smooth surface to re-direct the air bag inflation away from the occupant. This reduces the possibility of wrap-around and can have a dramatic effect in lowering forces on the occupant.

The authors of this chapter have attempted to summarize a few of the established relations between air bag design parameters and OOP occupant loading and injury risk. A review of the patent activity on air bags attests to an ever growing number of ideas for future innovations in air bag design.

5.0 Conclusions

A review of the field data shows that the air bag has had a net protective benefit in terms of reducing fatalities and serious injuries in car crashes. The primary mechanisms of this benefit are the air bag's ability to mitigate head and face contact with the steering wheel and other structures in the vehicle, its load-distributing and load-sharing properties, and its ability to perform work and dissipate the occupant's kinetic energy via venting. As a result of these characteristics, air bags facilitate restraint system design flexibility that would not otherwise be possible. For example, 4-kN force-limiting belts, which have shown great promise for reducing thoracic injuries related to belt loading (Foret-Bruno et al. 2001), would not be possible with out a supplemental air bag's ability to restrain the head and assist in decelerating the thorax.

Despite its life-saving benefits, however, the air bag can induce injuries that would otherwise not exist to occupants located within the deployment space of the air bag. While most of these injuries are minor in nature (e.g., abrasions or contusions), close proximity to the air bag has resulted in fatalities, particularly for small adult drivers and children in the passenger position. The nature of these injuries is highly dependent on the occupant's distance from the air bag module at the time of deployment. Punch-out forces result when the occupant blocks the module during deployment, whereas membrane forces develop during subsequent radial expansion of the bag. Serious thoracic trauma includes rib fractures and cardiac injuries is generally associated with punch-out phenomena; though bag slap can produce chest injuries of lesser severity. Closed head injuries can be produced by either membrane or punch-out loads; whereas neck injuries are almost always associated with the former mechanism.

Extensive analytical, computational, and experimental investigations have been performed to document the risk of injury for OOP occupants. Despite the fact that experimental investigations are dependent on surrogates with noted deficiencies, many injury trends predicted by these studies correlate strongly with field data findings. This lends confidence to current investigations that rely on these surrogates to evaluate potential modifications to the air bag system, though any surrogate's limitations must always be considered. Contemporary state-of-the art designs such as multi-stage inflators, folding pattern modifications, and new sensor technology appear to be able to maintain adequate protection for the unbelted occupant in frontal crash conditions while significantly reducing injury risk for OOP occupants.

6.0 References

Adomeit D (1979) Seat design – A significant factor for safety belt effectiveness. Paper Number 791004, Society of Automotive Engineers, Warrendale, PA.

Adomeit D, Heger A. (1975) Motion sequence criteria and design proposals for restraint devices in order to avoid unfavorable biomechanic conditions and submarining. Paper Number 751146, Society of Automotive Engineers, Warrendale, PA.

Adomeit, D. (1977) Evaluation methods for the biomechanical quality of restraint systems during frontal loading. Paper Number 770936, Society of Automotive Engineers, Warrendale, PA.

Aldman, B., Andersson, A., Saxmark, O. (1974) Possible Effects of Air bag Inflation on a Standing Child, Proc. 18th Annual Scientific Conference of the Association of the Advancement of Automotive Medicine, Toronto, Ontario Canada.

Armstrong, R., Waters, H., Stapp, J. (1968) Human muscular restraint during sled deceleration. Paper Number 680793, Society of Automotive Engineers, Warrendale, PA.

Atkinson, P., Hariaharan, P., Mari-Gowda, S., Telehowski, P., Martin, S., Van Hoof, J., Bir, C., Atkinson, T. (2002) An Under-Hand Steering Wheel Grasp Produces Significant Injury Risk to the Upper Extremity During Air bag Deployment. Proc. 46th Annual Scientific Conference of the Association of the Advancement of Automotive Medicine, Tempe, AZ.

Avanessian, H., Ridella, S., Mani, S., Krishnaswamy, P. (1982) An Analytical Model to Study the Infant/Seat Air bag Interaction. Paper Number 920126, Society of Automotive Engineers, Warrendale, Pennsylvania.

Bass, C., Crandall, J., Bolton, J., Pilkey, W., Khaewpong, N., Sun, E. (1999) Deployment of Air Bags into the Thorax of an Out-of-Position Dummy. Paper Number 1999-01-0764, Society of Automotive Engineers, Warrendale, Pennsylvania.

Bass, C., Duma, S., Crandall, J., George, S. (1998) The Interaction of Air bags with Upper Extremity Test Devices. Paper Number 98-S7-O-12, Proc. 16th ESV Conference.

Bass, C., Duma, S., Crandall, J., Morris, R. (1997) The Interaction of Air Bags with Upper Extremities. Paper Number 973324, Society of Automotive Engineers, Warrendale, Pennsylvania.

Berg, F., Grandel, J., Grzelak, R., Schmall, G. (1996) Crash Tests using Passenger Cars fitted with Air bags and a Simulated Out of Position Passenger. Proc. International Research Conference on the Biomechanics of Impact (IRCOBI), Dublin, Ireland.

Berg, F., Schmitt, B., Epple, J., Mattern, J. (1997) Dummy-Loadings Caused by an Air bag in Simulated Out-of-Position Situations. Proc. International Research Conference on the Biomechanics of Impact (IRCOBI). Hanover, Germany.

Berg, F., Schmitt, B., Epple, J., Mattern, R., Kallieris, D. (1998) Results of Full-scale Crash Tests, Stationary Tests and Sled Tests to Analyze the Effects of Air bags on Passengers with or without Seat Belts in the Standard Seating Position and OOP. Paper Number 98-S5-O-10, Proc. 16th ESV Conference.

Bierman, H.R., Wilder, R.M., Hellems, H.K. (1946) The physiological effects of compressive forces on the torso. Report #8, Naval Medical Research Institute Project X-630, Bethesda, MD.

Bohlin, N., (1967) Statistical analysis of 28,000 vehicle accident cases, with emphasis on occupant restraint value. SAE 670925, 11th Stapp Car Crash Conference, Society of Automotive Engineers, Inc., Warrendale, Pennsylvania, USA.

Breed, D., Sanders, W., Castelli, V. (1992) A Critique of Single Point Sensing. Paper Number 920124, Society of Automotive Engineers, Warrendale, Pennsylvania.

Chidester, A. and Roston, T. (2001) Air bag crash investigations. Paper Number 246, Conference on the Enhanced Safety of Vehicles (ESV), National Highway Traffic Safety Administration, Washington, DC.

Clark, C., Blechschmidt, C. (1965) Human Transportation Fatalities and Protection Against Rear and Side Crash Loads by the Airstop Restraint. Proc. 9th Stapp Car Crash Conference, Society of Automotive Engineers, Warrendale, Pennsylvania.

Crandall, J. R., Kuhlmann, T., Martin, P., Pilkey, W.D. (1994) Differing Patterns of Head Injury with Airbag and/or Belt Restrained Drivers in Frontal Collisions," Proc. Advances in Occupant Restraint Technologies, AAAM-IRCOBI, Lyon, France.

Crandall, J., Bass, C., Pilkey, W., Morgan, R., Eppinger, R., Miller, H., Sikorski, J. (1997a) Thoracic response and injury with belt, driver side air bag, and constant force retractor restraints. International Journal of Crashworthiness 2(1):119-132.

Crandall, J., Duma, S., Bass, C., Pilkey, W., Kuppa, S., Khaewpong, N., Eppinger, R. (1997b) Thoracic Response and Trauma of Out-of-Position Drivers Resulting from Air Bag Deployment. Proc. 41st Annual Scientific Conference of the Association of the Advancement of Automotive Medicine, Orlando, FL.

Crandall, J., Bass, C., Duma, S., Kuppa, S. (1998) Evaluation of 5th Percentile Female Hybrid III Thoracic Biofidelity during Out-of-position Tests with a Driver Air bag. Paper Number 980636, Society of Automotive Engineers, Warrendale, Pennsylvania.

Crandall, J.R., Matthews, B., Bass, C.R., Bolton, J. R. (1999a) A Comparison of the Viscoelastic Response of the 3 Year Old Child Dummy Neck Designs, Joint AAAM-IRCOBI Session on Child Occupant Protection, Sitges, Spain.

Crandall, J., Duma, S., Bass, C., Pilkey, W. (1999b) Thoracic Response and Trauma in Air Bag Deployment Tests with Our of Position Small Female Surrogates. Journal of Crash Prevention and Injury Control 1:101-112

Dalmotas, D., Hurley, R., German, A. (1996) Supplemental Restraint Systems: Friend or Foe to Belted Occupants? Proc. 40th Annual Scientific Conference of the Association for the Advancement of Automotive Medicine, Vancouver, British Columbia.

Dalmotas, D. (1998) Assessment of Air Bag Performance Based on the 5th Percentile Female Hybrid III Crash Test Dummy. Paper Number 98-S5-O-07, Proc. 16th ESV Conference.

Digges, K., Noureddine, A., Eskandarian, A., Bedewi, N. (1998) Effect of Occupant Position and Air Bag Inflation Parameters on Driver Injury Measures. Paper Number 980637, Society of Automotive Engineers, Warrendale, Pennsylvania.

Duma, S., Kress, T., Porta, D., Woods, C. (1996) Air bag-Induced Eye Injuries: A Report of 25 Cases. The Journal of Trauma 41:114-119.

Duma, S. M., Schreiber, P., McMaster, J., Crandall, J. R., Bass, C. R., Pilkey, W. D. (1998) Dynamic Injury Tolerances for Long Bones of the Female Upper Extremity, Proc. IRCOBI Conference, Gothenburg, Sweden.

Duma, S. M. (2000), Injury Criteria for the Small Female Upper Extremity, PhD Dissertation, University of Virginia.

Enouen, S., Guenther, D., Saul, R., MacLaughlin, T. (1984) Comparison of models simulating occupant response with air bags. Paper Number 840451, Society of Automotive Engineers, Warrendale, PA.

Foret-Bruno, J-Y., Trosseille, X., Page, Y., Huere, J-F, Le Coz, J-Y., Bendjellal, F., Diboine, A., Phalempin, T., Villeforceix, D., Baudrit, P., Guillemot, H., Coltat, J-C (2001) Comparison of thoracic injury risk in frontal car crashes for occupants restrained without belt load limiters and those restrained with 6 kN and 4 kN belt load limiters. Stapp Car Crash Journal 45:205-224.

Garcia, R. (1994) Air Bag Implicated in Temporomandibular Joint Injury. Cranio 12:125-127.

Grosch, L. (1985) Chest injury criteria for combined restraint systems. Paper Number 851247, Society of Automotive Engineers, Warrendale, PA.

Hardy, W., Schneider, L., Reed, M., Ricci, L. (1997) Biomechanical Investigation of Air bag-Induced Upper Extremity Injuries. Paper Number 973325, Society of Automotive Engineers, Warrendale, Pennsylvania.

Hayano, K., Ono, K., Matuoka, F. (1994) Test Procedures for Evaluating Out-of-Position Vehicle Occupant Interactions with Deployed Air bags. Paper Number 94-S1-O-19, Proc. 14th ESV Conference.

Hendler, E., O'Rourke, J., Schulman, M., et al. (1974) Effect of head and body position and muscular tensing on response to impact. Paper Number 741184, Society of Automotive Engineers, Inc., Warrendale, Pennsylvania, USA.

Hinger, J. and Clyde, H., Advanced air bag systems and occupant protection: recent modifications to FMVSS 208. Paper 2001-01-0157, Society of Automotive Engineers, Warrendale, Pennsylvania, USA.

Hollowell, W., Summers, L., Prasad, A., Narwani, G., Ato, T. (2001) Performance evaluation of dual stage passenger air bag systems. Paper Number 234, 17th Conference on the Enhanced Safety of Vehicles (ESV), National Highway Traffic Safety Administration, Washington, DC.

Horsch, J.D., Culver, C.C. (1979) A Study of Driver Interactions With an Inflating Air Cushion. Paper Number 791029, Society of Automotive Engineers, Warrendale, Pennsylvania.

Horsch, J., Culver, C., (1983) The role of steering wheel structure in the performance of energy absorbing steering systems. Paper Number 831607, Society of Automotive Engineers, Warrendale, Pennsylvania.

Horsch, J., Lau, I., Andrzejak, D., Viano, D. (1990) Assessment of Air Bag Deployment Loads. Paper Number 902324, Society of Automotive Engineers, Warrendale, Pennsylvania.

Horsch JD, Horn, G, McCleary JD. (1991) Investigation of inflatable restraints. Paper Number 912905, Society of Automotive Engineers, Warrendale, Pennsylvania, USA, 1991.

Huelke, D., Moore, J., Compton, T., Samuels, J., Levine, R. (1994) Upper Extremity Injuries Related to Air bag Deployments. Proc. 38th Annual Scientific Conference of the Association of the Advancement of Automotive Medicine, Lyon France.

Huelke, D., Moore, J., Compton, T., Samuels, J. (1995) Upper Extremity Injuries Related to Air bag Deployments. The Journal of Trauma 38:482-488.

Huelke, D. (1997) Children as Front Seat Passengers Exposed to Air bag Deployments. Paper Number 973295, Society of Automotive Engineers, Warrendale, Pennsylvania.

Huelke, D. (1998) Does Stature Influence Driver Injuries in Air bag Deployment Crashes? - Analysis of UMTRI Crash Investigations. Paper Number 980640, Society of Automotive Engineers, Warrendale, Pennsylvania.

Insurance Special Report. (1991) Driver Injury Experience in 1990 Models Equipped with Air Bags or Automatic Belts, Highway Loss Data Institute, Arlington VA, Report No. 38.

Jerrigan, M., Duma, S., Herring, I., Stitzel, J. (2002) Analysis of Burn Injuries in Frontal Automobile Crashes, Proc. 46th AAAM Conference, Tempe, AZ.

Johnston, K., Klinich, K., Rhule, D. (1997) Assessing Arm Injury Potential from Deploying Air bags. Paper Number 970400, Society of Automotive Engineers, Warrendale, Pennsylvania.

Kallieris, D., Mattern, R., Schmidt, G., Klaus, G. (1982) Comparison of three-point belt and air bag-knee bolster systems. Injury criteria and injury severity at simulated frontal collisions. Proc. International Research Council on Biokinetics of Impact (IRCOBI), pp 166-183.

Kang, J., Agaram, V., Nusholtz, G., Kostyniuk, G. (2001) Air Bag Loading on In-Postion Hybrid III Dummy Neck. Paper Number 2001-01-0179, Society of Automotive Engineers, Warrendale, Pennsylvania.

Karlow, J., Jakovski, J., Seymour, B. (1994) Development of a New Downsized Air bag System for Use in Passenger Vehicles. Paper Number 940804, Society of Automotive Engineers, Warrendale, Pennsylvania.

Kent, R. W., Crandall, J. R., Bolton, J. R., Nusholtz, G.S., Prasad, P., Mertz, H. (2001a) The Influence of Superficial Soft Tissues and Restraint Condition on Thoracic Skeletal Injury Prediction, Stapp Car Crash Journal, Vol. 45, pp. 183-204.

Kent, R., Bolton, J., Crandall, J., Prasad, P., Nusholtz, G., Mertz, H., Kallieris, D. (2001) Restrained Hybrid III dummy-based criteria for thoracic hard tissue injury prediction. 2001 Conference of the International Research Council on the Biomechanics of Impact (IRCOBI), Isle of Man.

Kent, R., Crandall, J., Bolton, J., Duma, S. (2000) Driver and right-front passenger restraint system interaction, injury potential, and thoracic injury prediction. Proc. 44th Scientific Conference of the Association for the Advancement of Automotive Medicine, AAAM, Des Plaines, IL.

Kent, R., Crandall, J., Patrie, J., Fertile, J. (2002) Radiographic detection of rib fractures: a restraint-based study of occupants in car crashes. Traffic Injury and Prevention, 3(1):49-57.

Kikuchi, A., Horii, M., Kawai, A., Kawai, S., Kamaki, Y., Matsuno, M. (1975) Injury to Eye and Facila Skin (Rabbit) on Impact with Inflating Air Bag. Proc. International Research Conference on the Biomechanics of Impact (IRCOBI), Birmingham, Alabama.

Kleinberger, M., Summers, L. (1997) Mechanisms of Injuries for Adults and Children Resulting from Air bag Interaction. Proc. 46th Annual Scientific Conference of the Association of the Advancement of Automotive Medicine, Orlando, Florida.

Kullgren, A., Krafft, M., Malm, S., Ydenius, A., Tingvall, C. (2000) Influence of air bags and seatbelt pretensioners on AIS1 neck injuries for belted occupants in frontal impacts. Stapp Car Crash Journal 44:117-125.

Kuppa, S., Yeiser, C., Oslon, M., Taylor, L. (1997) RAID - An Investigative Tool to Study Air Bag/ Upper Extremity Interactions. Paper Number 970399, Society of Automotive Engineers, Warrendale, Pennsylvania.

Lau, I., Horsch, J., Viano, D., Andrzejak, D. (1993) Mechanisms of Injury From Air Bag Deployment Loads. Accident Analysis & Prevention 25(1):29-45.

Lin, T., Wawa, C., Khalil, T. (1995) Evaluation of the Hybrid III Dummy Interactions with Air Bag in Frontal Crash by Finite Element Simulation. Paper Number 952705, Society of Automotive Engineers, Warrendale, Pennsylvania.

Lundstrom, L. (1974) Relating Air Cushion Performance to Human Factors and Tolerance Levels. Paper Number 746031, Proc. 5th ESV Conference.

Malczyk, A., Adomeit, H. (1995) The Air bag Folding Pattern as a Means for Injury Reduction of Out-of-Position. Paper Number 952704, Society of Automotive Engineers, Warrendale, Pennsylvania.

Malczyk, A., Franke, D., Adomeit, H. (1998) A Study On the Benefits of Dual-Stage Inflators Under Out-Of-Position Conditions. Proc. International Research Conference on the Biomechanics of Impact (IRCOBI), Goteborg, Sweden.

Mao, Y. and Appel, H. (2001) Influence of air bag folding pattern on OOP-injury potential. Paper 2001-01-0164, Society of Automotive Engineers, Warrendale, PA.

McElhaney, J., Myers, B. (1993) Biomechanical Aspects of Cervical Trauma. In: Nahum, A., Melvin, J. (eds) Accidental Injury, pp. 311-360. Springer-Verlag, New York, NY.

Melvin, J. W., Horsch, J. D., McCleary, J. D., Wideman, L. C., Jensen, J. L., Wolanin, M. J. (1993) Assessment of Air Bag Deployment Loads with the Small Female Hybrid III Dummy, Paper 933119, Proc. 37th Stapp Car Crash Conference, Society of Automotive Engineers, Warrendale, PA.

Melvin, J., Baron, K., Little, W., Gideon, T., Pierce, J. (1998) Biomechanical analysis of Indy race car crashes. Paper 983161, Society of Automotive Engineers, Warrendale, PA.

Melvin, J., Mertz, H. (2002) Occupant Restraint Systems. Air Bag Inflation-Induced Injury Biomechanics. In: Nahum, A. and Melvin, J. (editors) Accidental Injury: Biomechanics and Prevention. Springer-Verlag, New York, NY.

Mertz, H. (1988) Restraint Performance of the 1973-76 GM Air Cushion Restraint System. Paper Number 880400, Society of Automotive Engineers, Warrendale, PA.

Mertz, H., Driscoll, G., Lenox, J., Nyquist, G., Weber, D. (1982) Responses of Animals Exposed to Deployment of Various Passenger Inflatable Restraint System Concepts for a Variety of Collision Severities and Animal Positions. Paper Number 826047, Proc. 9th ESV Conference.

Mertz, H., Marquardt, J. (1985) Small car air cushion performance considerations. Paper Number 851199, Society of Automotive Engineers, Warrendale, PA.

Mertz, H. (2002) Anthropomorphic Test Devices in: Nahum, A. and Melvin, J. (editors) Accidental Injury: Biomechanics and Prevention. Springer-Verlag, New York, NY.

Mohan, D., Zador, P., O'Neill, B., Ginsburg, M. (1976) Air bags and lap/shoulder belts—a comparison of their effectiveness in real world, frontal crashes. Proc. of AAAM, Des Plaines, IL.

Montalvo, F., Bryant, R., Mertz, H. (1982) Possible Positions and Postures of Unrestrained Front-Seat Children at Instant of Collision. Paper Number 826405, Proc. 9th ESV Conference.

Morris, A., Fildes, B. (1998) Preliminary Experience of Passenger Air bag Deployments in Australia. Paper Number 98-S5-W-17, Proc. 16th ESV Conference.-NEED AN A AND A B

Morris, C., Zuby, D., Lund, A. (1998) Measuring Air bag Injury Risk to Out-of-Position Occupants. Paper Number 98-S5-O-08, Proc. 16th ESV Conference.

Musiol, J., Norgan-Curtiss, L., Wilkins, M. (1997) Control and Application of Intelligent Restraint Systems. Paper Number 971052, Society of Automotive Engineers, Warrendale, Pennsylvania.

National Highway Traffic Safety Administration. (1996) Third Report to Congress: Effectiveness of Occupant Protection Systems and Their Use. National Highway Traffic Safety Administration, U.S. Department of Transportation, Washington DC.

NHTSA (1984) FMVSS 208 regulatory impact analysis. National Highway Traffic Safety Administration, U.S. Department of Transportation, Washington DC.

Nixon, C. (1970) The Human Ear in an Air Bag Noise Environs. Proc. 14th Annual Scientific Conference of the Association of the Advancement of Automotive Medicine, Tempe, AZ.

Nusholtz, G., Wang, D., Wylie, E. (1998a) An Evaluation of Air bag Tank-test Results. Paper Number 980864, Society of Automotive Engineers, Warrendale, Pennsylvania.

Nusholtz, G., Xu, L., Kostyniuk, G. (1998b) Estimation of Occupant Position from Probability Manifolds of Air Bag Fire-times. Paper Number 980643, Society of Automotive Engineers, Warrendale, Pennsylvania.

Nusholtz, G., Xu, L., Mosier, R., Kostyniuk, G. (1998c) Estimation of OOP from Conditional Probabilities of Air bag Fire Times and Vehicle Response. Paper Number 98-S5-O-16, Proc. 16th ESV Conference.

Partridge, L., Young, G. (1979) An Investigation of the Potential Human and Environmental Impacts Associated With Motor Vehicle Air Bag Restraint Systems. Paper Number 790641, Society of Automotive Engineers, Warrendale, PA.

Patrick, L, Bohlin, N., Anderson, A. (1974) Three-point harness accident data and laboratory data comparison. Paper Number 741181, Society of Automotive Engineers, Warrendale, PA.

Patrick, L., Nyquist, G. (1972) Air bag Effects on the Out-of-Position Child. Paper Number 720442, Society of Automotive Engineers, Warrendale, Pennsylvania.

Patrick, L.M., Kroell, C.K., Mertz, H.J. (1965) Forces on the human body in simulated crashes. Proc. 9th Stapp Car Crash Conference, pp. 237-260.

Polk, J. and Thomas, H. (1994) Automotive Air Gag-Induced Second-Degree Chemical burn Resulting in Staphylococcus Aureus Infection. JAOA 94(9):741-743.

Powell, M., Lund, A. (1995) Leading Edge Deployment Speed of Production Air Bags. Paper Number 950870, Society of Automotive Engineers, Warrendale, Pennsylvania.

Prasad, P., Daniel, R. (1984) A Biomechanical Analysis of Head, Neck, and Torso Injuries to Child Surrogates Due to Sudden Torso Acceleration. Paper Number 841656, Society of Automotive Engineers, Warrendale, Pennsylvania.

Prasad, P., Laituri, T. (1996) Consideration for belted FMVSS 208 Testing. Paper Number 86-S3-O-03, Proc. 15th ESV Conference.

Pursel, H., Bryant, R., Scheel, J., Yanik, J. (1978) Matching Case Methodology for Measuring Restraint Effectiveness. Paper Number 780415, Society of Automotive Engineers, Warrendale, PA.

Rains, G., Prasad, A., Summers, L., Terrell, M. (1998) Assessment of Advanced Air bag Technology and Less Aggressive Air Bag Designs through Performance Testing. Paper Number 98-S5-O-06, 16th ESV Conference.

Reed, M., Schneider, L., Burney, R. (1994) Laboratory Investigations and Mathematical Modeling of Air bag-Induced Skin Burns. Paper Number 942217 Society of Automotive Engineers, Warrendale, Pennsylvania.

Reidelbach, W., Scholz, H. (1979) Advanced restraint system concepts. Paper Number 790321, Society of Automotive Engineers, Warrendale, Pennsylvania.

Reidelbach, W., (1985) The Daimler-Benz Supplemental Restraint System. SAE 856016, Tenth International Technical Conference on Experimental Safety Vehicles, Oxford, England, Highway Traffic Safety Administration, Washington, D.C.

Richter, H., Stalnaker, R., Pugh, J. (1974) Otologic Hazards of Air bag Restraint System. Paper Number 741185, Society of Automotive Engineers, Warrendale, Pennsylvania.

Romeo, D. (1975) Front Passenger Passive Restraint for Small Car, High Speed, Frontal Impacts. Paper Number 751170, Society of Automotive Engineers, Warrendale, Pennsylvania.

Romeo, D. (1975) Front passenger passive restraint for small car, high speed, frontal impacts. Paper Number 751170, Society of Automotive Engineers, Warrendale, PA.

Roselt, T., Langwieder, K., Hummel, T., Koster, H. W. (2002) Injury Patterns of Front Seat Occupants in Frontal Car Collisions with Air bags. Effectivity and Optimisation Potential of Air bags. Proc. International Research Conference on the Biomechanics of Impact (IRCOBI), Munich, Germany.

Rouhana, S.W., Webb, S.R., Dunn, V.C. (1998) Investigation into the noise associated with air bag deployment: Part II - Injury risk study using a mathematical model of the human ear. Paper No. 983162, Society of Automotive Engineers, Inc., Warrendale, Pennsylvania, USA.

Rouhana, S.W., Webb, S.R., Wooley, R.G., McCleary, J.D., et al. (1994) Investigation into the noise associated with air bag deployment: Part I – Measurement technique and parameter study. Paper Number 942218, Society of Automotive Engineers, Warrendale, Pennsylvania.

Roychoudhury, T., Sun, S., Hamid, M., Hanson, C. (2000) Fifth Percentile Driver Out of Position Computer Simulation. Paper Number 2000-01-1006, Society of Automotive Engineers, Warrendale, Pennsylvania.

Rutherford, W., H. (1985) The medical effects of seat-belt legislation in the United Kingdom: a critical review of the findings. Archives of Emergency Medicine 2(4):221-3.

Ryan, S. (1998) An Innovative Approach to Adapative Air bag Modules. Paper Number 980646, Society of Automotive Engineers, Warrendale, Pennsylvania.

Saul, R., Backaitis, S., Beebe, M., Ore, L. (1996) Hybrid III Dummy Instrumentation and Assessment of Arm Injuries During Air Bag Deployment. Paper Number 962417, Society of Automotive Engineers, Warrendale, Pennsylvania.

Saul, R., Pritz, H., McFadden, J., Backaitis, S., Hallenback, H., Rhule, D. (1998) Description and Performance of the Hybrid III Three Year Old, Six Year Old and Small Female Test Dummies in Restraint System and Out-Of-Position Air Bag Environments. Paper Number 98-S7-O-01, 16th ESV Conference.

Schroeder, G., Eidam, J. (1997) Typical Injuries Caused by Air-Bag in Out-of-Position Situations– An Experimental Study. Proc. International Research Conference on the Biomechanics of Impact (IRCOBI), Hannover, Germany.

Sherwood, C. P., Crandall, J. R., Shaw, C. G., Gupta, P., Orzechowski, K.,Eichelberger, M., Kallieris, D. (2002) Prediction of Cervical Spine Injury Risk for the 6-year Old Child Involved in Frontal Crashes, Proc. 46th AAAM Conference, Phoenix, AZ.

Sieveka, E., Duma, S., Pellettiere, J., Crandall, J. (1997) Multi-Body Model of Upper Extremity Interaction with Deploying Air bag. Paper Number 970398, Society of Automotive Engineers, Warrendale, Pennsylvania.

Smith, G., Hurite, S., Yanik, A., Greer, C. (1972) Human Volunteer Testing of GM Air Cushions, Paper Number 720443, National Automobile Engineering Meeting, Detroit, Michigan.

Smith GR, Gullash EC, Baker RG. (1974) Human Volunteer and Anthropomorphic Dummy Tests of the General Motors Driver Air Cushion System. Paper Number 740578, Society of Automotive Engineers, Warrendale, PA, USA.

Spiess, O., Marotzke, T., Zahn, M. (1997) Development Methodology of an Air bag Integrated Steering Wheel in Order to Optimize Occupant Protection Balanced Against Out-of-Position Risks. Paper Number 970777, Society of Automotive Engineers, Warrendale, Pennsylvania.

Stalnaker, R., Klusmeyer, L., Peel, H., White, C. (1982) Unrestrained, Front Seat, Child Surrogate Trajectories Produced by Hard Braking. Paper Number 821165, Society of Automotive Engineers, Warrendale, Pennsylvania.

Struble, D. (1998) Air bag Technology: What it is and How it Came to Be. Paper Number 980648, Society of Automotive Engineers, Warrendale, Pennsylvania.

Summers, L., Hollowell, W., Rains, G. (1998) NHTSA's Advanced Air Bag Technology Research Programs. Paper Number 98-S5-W-29, 16th ESV Conference.

Van Ee, C., Nightingale, R., Camacho, D., Chancey, V., Knaub, K., Sun, E., Myers, B. (2000) Tensile properties of the human muscular and ligamentous cervical spine. Paper Number 2000-01-SC07, Society of Automotive Engineers, Warrendale, Pennsylvania.

Vezin, P., Bruyere-Garnier, K., Bermond, F., Verriest, J. (2002) Comparison of hybrid III, thor- and PMHS response in frontal sled tests. Paper Number 2002-22-0001, Stapp Car Crash Journal 46: 1-26.

Viano DC (1988) Cause and control of automotive trauma. Bulletin of the New York Academy of Medicine, Second Series 64(5):376-421.

Viano, D. (1991) Effectiveness of safety belts and air bags in preventing fatal injury. Paper Number 910901, Society of Automotive Engineers, Warrendale, PA.

Viano, D., Artinian, C. (1978) Myocardial Conducting System Dysfunctions for Thoracic Impact. The Journal of Trauma, 18(6):452-459.

Wang, J. (1991) Are Tank Pressure Curves Sufficient to Discriminate Air bag Inflators. Paper Number 910808, Society of Automotive Engineers, Warrendale, Pennsylvania.

Werner, J. and Sorenson, W. (1994) Survey of Air Bag Involved Accidents an Analysis of collision characteristics, System effectiveness and Injuries. Paper Number 940802, Society of Automotive Engineers, Warrendale, PA.

Werner, J. and Sorenson, W. (1994) Survey of air bag involved accidents an analysis of collision characteristics, system effectiveness and injuries. Paper Number 940802, Society of Automotive Engineers, Warrendale, PA.

Werner, J., Roberson, S., Ferguson, S., Digges, K. (1996) Air Bag Deployment Frequency and Injury Risks. Paper Number 960664, Society of Automotive Engineers, Warrendale, PA.

Wolanin, M. J., Mertz, H. J., Nyznyk, R. S., Vincent, H. J. (1982), Description and basis of a 3-year-old child dummy for evaluating passenger inflatable restraint concepts, Proc. 9th ESV Conference, Kyoto, Japan.

Yaremchuk, K. and Dobie, R. (1994) Otologic Injuries From Air Bag Deployment. Otolaryngology Head and Neck Surgery 125(3):130-134.

Yoganandan, N., Pintar, F., Skrade, D., Chmiel, W., Reinartz, J., Sances, A. (1993) Thoracic biomechanics with air bag restraint. Paper Number 933121, Society of Automotive Engineers, Warrendale, PA.

Zinke, D. (1980) The Development of Air Cushion Restraint Systems for Small Car Front Seat Structures. Paper Number 800924, Society of Automotive Engineers, Warrendale, Pennsylvania.

Chapter 5

The Field Performance of Frontal Air bags

Richard Kent, David Viano, Jeff Crandall

Introduction

In 1996, the U.S. vehicle fleet included approximately 89 million cars. Just over 42 million of these, or 48%, were equipped with either a driver or dual frontal air bag system. Essentially all of the 10 million new passenger cars registered each year in the U.S. contain a dual air bag system, and the penetration of air bag-equipped vehicles into the U.S. passenger car fleet will reach essentially 100% by 2010. This chapter discusses the performance of these systems in the field. It is intended to supplement, expand, and place into perspective the material presented in the previous chapter. As a result, we will necessarily reiterate some points for the sake of completeness and to maintain independence between the two discussions.

Air bag legislation and public opinion regarding air bags, and therefore air bag design, are currently in a state of flux and have been since the early 1970s. The result is that the field performance of air bag systems has changed over time, will continue to change as the vehicle fleet evolves, and universal truths regarding air bag performance are few. Some of the performance metrics presented in this chapter seem to have equilibrated and will likely not change substantially if, for example, the details of federal motor vehicle safety standard (FMVSS) 208 (frontal impact protection) continue to change. The fatality effectiveness of the driver-side air bag in frontal collisions is one performance metric that seems to have reached at least a temporarily steady state. Despite the interim change in 1998 to allow certification with a sled test, and the resulting change in air bag design, driver air bag effectiveness at preventing fatalities in purely frontal collisions has remained at approximately 31%[1] (NHTSA 1999). Another universal truth appears to be that the safety benefits of air bags are much smaller than the benefit to be realized if all occupants would simply wear their seatbelts (e.g., Barry et al. 1999). In contrast, performance measures such as the passenger-side air bag's cost-benefit are sensitive to public education and to air bag technology that is beginning to penetrate the fleet. For example, air bag depowering[2], occupant sensing systems that suppress deployment if the passenger seat is unoccupied, and public health campaigns to move children to the rear seat will all tend to improve passenger air bag cost-benefit. The reader should be aware; therefore, that the long-term applicability of some findings presented in this chapter are limited and that some findings provide only a "snap shot" of the current status and a baseline for comparison with air bag performance in the future. We would therefore discourage readers from considering only isolated portions of this chapter. Instead, the chapter should be considered in a holistic sense since that is the only means by which to evaluate the performance of air bags.

The performance of an air bag system can be assessed using a multitude of metrics, which are sensitive to many factors and are influenced by many confounders. No one metric is sufficient to completely define the performance of the air bag, so this chapter includes as many relevant metrics as practicable, interprets the findings, and provides references so the interested reader can further evaluate the limitations, confounders, and utility of each metric. The evaluations presented here range from the very specific (individual case studies) to the general (statistical analyses of large databases). The metrics used to evaluate air bag performance include fatality reduction or increase; serious, moderate, and minor injury reduction or increase; harm[3] reduction or increase; and cost analyses, including insurance costs and the cost of life years saved for various air bag systems and design philosophies. Performance is most often

---

[1] This does not imply that this value will not change in time. As seatbelt usage continues to increase, for example, overall air bag effectiveness will decrease since air bags are more effective for unbelted occupants than for belted (Viano 1995).

[2] "Depowering" is a term used to describe multiple changes in air bag design that many manufacturers have incorporated to reduce air bag-induced injuries. These redesigns potentially involve many changes, including bag material and folding pattern, bag volume, and inflator characteristics. The term "second generation" is also often used to describe air bags that were installed after the change to allow sled testing for FMVSS 208 certification.

[3] Harm (Malliaris et al. 1985) is a metric used to evaluate overall societal costs from injuries of different levels. Harm is sometimes expressed in terms of effective fatalities.

expressed in terms of the when-used effectiveness, which is the reduction in the number of injuries of a certain severity if the entire population being considered changed from not having an air bag to having an air bag, assuming that all other factors such as seatbelt use and the types of vehicles in the fleet are unchanged (Evans 1991). Recognizing that this assumption may not be acceptable in some cases, several factors and factor interactions that influence air bag performance are assessed, including concomitant seatbelt use, impact direction, collision severity, vehicle model year, and occupant age, gender, seating position, and size. The chapter begins with the benefits of air bags. Fatality and injury reductions attributable to the air bag are presented. The chapter then describes the negative consequences of air bag deployment. Injuries to adults and children and the current trends in air bag injury rates are discussed, as are the few documented instances of inadvertent deployments or non-deployment in severe crashes. In the third section, an attempt is made to quantify the influence of the many confounding factors that affect air bag performance. The negative and positive characteristics of air bags are then discussed within the context of societal costs and benefits. Finally, some special topics, including risk homeostasis and the performance of side air bags, are discussed.

1. Fatalities and Injuries Prevented by Air Bags

In August of 1974, the U.S. National Highway Traffic Safety Administration (NHTSA) analyzed the impact of proposed changes to MVSS 208 (NHTSA 1974), which included evaluation of an "air cushion" system. While acknowledging the difficulty in estimating effectiveness for a system that had essentially no field exposure at the time, NHTSA concluded that 15,600 fewer fatalities and 1,000,000 fewer injuries would occur annually in the U.S. if the entire passenger car population were equipped with "air cushion-lap belt" systems with 100% usage. They estimated that the fatality-reducing effectiveness of the air cushion system would be 57% in frontal crashes, 20% higher than lap-shoulder belt systems (Table 1) and substantially higher than the 18% that had been estimated earlier by GM researchers Wilson and Savage (1973). The NHTSA concluded that "passive systems, which do not depend on the knowledge or cooperation of the driving public, are practicable and will provide substantially greater levels of life- and injury-saving protection than active belt systems." NHTSA's optimistic outlook regarding air bags stemmed largely from the feeling that "active" systems (such as seatbelts), which would require action by the occupant, would never reach high levels of use.

Table 1 – Estimates of Restraint System Effectiveness by the NHTSA (1974)

| System | Fatality Effectiveness | | | |
	Frontal Impact	Side Impact	Rear Impact	Rollover
Lap and Shoulder Belt	37	30	0	50
Lap Belt Only	20	30	0	40
Air Cushion and Lap Belt	57	45	0	50
Air Cushion Only	57	20	0	15

| System | Injury Effectiveness | | | |
	Frontal Impact	Side Impact	Rear Impact	Rollover
Lap and Shoulder Belt	27	30	0	50
Lap Belt Only	0	20	0	40
Air Cushion and Lap Belt	64	40	0	50
Air Cushion Only	64	25	0	15

In response to this report, Patrick (1975) estimated injury and fatality effectiveness for several restraint strategies that were currently under consideration: 1) legislated mandatory 3-pt harness use, 2) an air bag plus 20% lap belt use, 3) an air bag only, 4) a passive 3-pt harness, and 5) a torso belt with knee bar. He concluded that the mandatory harness would have the highest effectiveness in terms of fatality and injury reduction (30%, 55%), with the air bag-only system having the lowest of the 5 systems considered (25%, 28%) (Table 2). He estimated that 99,945 lives would be saved from 1977-1986 if a three-point harness was mandated. His estimates for lives saved by mandated air bags over this period were more modest than NHTSA's: 53,433 for the air bag plus 20% lap belt usage and 46,425 for the air bag alone, lower even than NHTSA's revised (Federal Register 1977) estimate that driver and passenger air bags would prevent 12,100 deaths annually (120,000 per decade).

Table 2 – Estimates of Restraint System Effectiveness by Patrick (1975)

System	Fatality Effectiveness	Injury Effectiveness
Mandatory 3-pt harness	30	55
Airbag + 20% lap belt	29	30
Airbag only	25	28
3-pt. harness	30	55
Torso belt and knee bar	25	40

As crashes involving special fleets of air bag-equipped vehicles began to occur, various attempts were made to evaluate the air bag's performance retrospectively rather than by expert opinion or projections based on a fleet of non-air bag-equipped vehicles. Smith and Moffatt (1975) published a summary of 75 air bag-deployment crashes from an 18-month NHTSA study of 800 Mercury vehicles, 1,000 Chevrolet vehicles, and 75 Volvo vehicles equipped with air bags and placed into the fleet from 1972-1975. The lack of data from a control group of non-deployment crashes and an underestimation of the market for air bag-equipped vehicles, which limited the size of the database, precluded a robust assessment of effectiveness. Instead, the authors presented case studies of all 75 cases. They identified three crashes of air bag-equipped vehicles in which the air bag did not deploy but which resulted in moderate or greater injury, and five cases of inadvertent or non-collision deployments, but could not determine the level of benefit provided by the air bag. Backaitis and Roberts (1987) published a summary of 112 crashes involving NHTSA-sponsored fleets of air bag-equipped vehicles. These vehicles were driven by state or city police, armed services personnel, and personnel from various federal agencies. Of these 112 drivers, 103 sustained no serious injuries and none sustained injuries greater than 3 on the Abbreviated Injury Scale (AIS) (AAAM 1990). This study considered no control group and had a small number of cases, so the authors made no attempt to evaluate effectiveness other than to state that both original equipment and retrofit air bag systems deployed as expected and seemed to provide enhanced crash protection. Similarly, Smith (1977) presented case studies from a special fleet of 11,000 General Motors air bag-equipped vehicles that had accumulated over 400 million miles of travel, but determined that the number of collisions was too few to show a clear benefit from the air bag. These early studies, while optimistic in their outlook, were unable to quantify a clear benefit due to the air bag, and were starting to provide evidence that the air bag, while beneficial in many cases, was not a safety panacea.

In perhaps the most comprehensive study performed in the 1970s, and one of the first to quantify air bag effectiveness in a fleet of air bag-equipped vehicles, Mohan et al. (1976) combined data from the NHTSA study and a program of field investigations performed by the University of Michigan. They were able to estimate air bag effectiveness in frontal crashes by comparing similar vehicles with and without air bags. In higher severity crashes, they found that the mean injury severity score (ISS) (Baker et al. 1974) was lower for the cases with air bag deployment (2.7) than for cases with lap/shoulder belts (3.3), but that this difference was not statistically significant. They also found that, while these two restraint systems were not significantly different, they were both significantly better than no restraint (mean ISS 8.0). Their finding that an air bag without a seatbelt does not provide significantly more protection than a seatbelt without an air bag would eventually be corroborated using much larger sets of data (see Section 3a).

A GM study by Pursel et al. (1978) evaluated air bag effectiveness for non-fatal injuries by comparing deployments in similar frontal collisions involving similar vehicles with and without an air bag installed. They estimated combined effectiveness of a driver and passenger system for mitigating AIS 2+ injuries at 6% and for mitigating AIS 3+ injuries at 18%[4]. They also evaluated the driver and passenger systems singly, and estimated passenger air bag effectiveness for AIS 2+ injuries to be -34% (i.e., the passengers in air bag-equipped vehicles sustained more AIS 2+ injuries than the passengers in vehicles without air bags). While air bag-induced injuries had been observed in the laboratory (e.g., Aldman et al. 1974), this unexpected finding was one of the first to confirm the potential for air bag-induced injuries in the field and foreshadowed the passenger-side air bag debate that continues to this day. At the heart of this

---

[4] Viano (1991) noted that the effectiveness figures published by Pursel et al. and by Mertz consider only deployment crashes, which make up only 45% of all crashes. Viano estimated that the driver air bag effectiveness would decrease from 18% to 7% if non-deployment crashes were considered, and that other effectiveness estimates would decrease similarly.

debate is the question of installing an air bag, which has the potential to injure or even to kill, in a seating position that, at least in the U.S., is frequently not occupied by an occupant who would benefit from its deployment (in 1990, for example, the Texas Transportation Institute estimated for urban work trips an average of only about 1.2 persons in each vehicle). This becomes a question of cost/benefit (in terms of deployment into unoccupied space) and risk/reward (in terms of injuries caused vs. injuries averted), which is addressed in Section 4b.

Some of the earliest field performance-based analyses of air bag performance were published by the insurance industry, which was able to accumulate large claims databases relatively quickly after the widespread introduction of air bags into the fleet. In 1994, before large numbers of air bag cases had been investigated by state or federal agencies, the State Farm policyholder group incurred approximately 30 crashes per day in which air bags deployed (Werner and Sorenson 1994). These cases could be compiled and analyzed earlier than federal or state databases, and provided valuable early assessments of air bag performance in the fleet. Werner and Sorenson (1994) presented an analysis of insurance claims resulting from crashes involving vehicles with and without an air bag. Table 3 lists their injury rates for vehicles with and without an air bag by vehicle size, air bag availability, and injury severity. Since air bags do not deploy in low-severity crashes, only non-air bag cases with damage over $5,000 (1994) are included in these rates. Regardless of vehicle class, the frequency of overall injury was slightly greater for crashes involving air bag deployments, but drivers in air bag-deployment crashes were 35% less likely to suffer moderate and severe injuries. This trend was supported by a matched comparison of drivers (with an air bag) and passengers (without an air bag) in the same crash, which showed that the drivers were 16% less likely to sustain moderate to severe injuries than the passengers (standardized for differences in gender).

Table 3 – Injury Experience of Drivers in Frontal Crashes with > $5,000 Damage, by Vehicle Size (Werner and Sorenson, 1994)

	Small		Midsize		Large		Standardized	
	No Airbag	Airbag	No Airbag	Airbag	No Airbag	Airbag	No Airbag	Airbag
Crashes	30	434	152	721	68	499	250	1,654
Avg. Damage Amount	$8,313	$8,564	$8,263	$8,630	$10,882	$9,981	--	--
Drivers injured per 1,000 crashes	600	654	592	634	632	655	606	645
Drivers with moderate and severe injuries per 1,000 crashes	100	83	119	61	74	60	101	66

In addition to evaluating the overall effect of air bags on injury outcome, Werner and Sorenson (1994) had compiled sufficient data to make a preliminary evaluation of injury rates by body region for air bag and non-air bag crashes. They found that drivers exposed to a deployment have lower rates of more serious head and chest injuries, but higher rates of abrasion, laceration, and contusion injuries to the face and upper extremity. They also observed that the air bag seemed to limit belt-induced injuries (neck and shoulder strains and contusions).

Evans (1988) estimated that air bags reduce fatality risk to unbelted drivers by 18% in all crashes (including crash types, such as rear impacts, where an air bag has virtually no effectiveness). Lund and Ferguson (1995) studied a population of both belted and unbelted occupants and found that the air bag reduced fatalities by 24% in frontal crashes and 16% in all crashes. The slightly lower effectiveness for all crashes is consistent with Evans' 18% for unbelted drivers since a population with higher belt use will tend to exhibit lower air bag effectiveness (Viano 1991) (see Section 3a). In a study of 3.36 million Chrysler vehicles, including both belted and unbelted occupants, Edwards (1994) found a fleet-average fatality reduction due to air bags of 22% in frontal crashes, consistent with the 24% found by Lund and Ferguson.

Comparing these studies directly is not possible, however, since it is possible that the belt usage was not equivalent in the two study populations.

For the current U.S. distribution of crash-involved occupants, vehicles, and belt usage, the latest report by NHTSA (1999) estimates driver fatality risk reduction from air bags to be 11% in all crashes. When only purely frontal and near-frontal crashes were considered, the reduction increased to 31% and 19%. All these reductions were significantly greater than zero. These reductions correspond to 2,263 lives saved by air bags from 1987-1997, including 842 lives saved in 1997. The serious injury effectiveness of the air bag for unbelted occupants was large (42%), but, due to the small number of cases, not significantly different than zero (Figure 1). As more field data are compiled this metric will stabilize and attain significance, but the preliminary trend is encouraging.

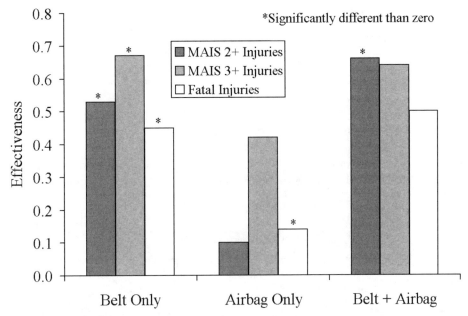

Figure 1. Injury and fatality effectiveness for different restraints in tow-away crashes (data from NHTSA 1999).

The early estimates of air bag effectiveness were over-estimated almost universally. The original estimates by NHTSA of 57% fatality effectiveness in frontal crashes and 20% in side impacts, and by Patrick (25% in all crashes), are actually 31% in frontals, essentially 0% in side impacts, and 11% in all crashes based on the most current field data. While the increase in seatbelt use in the years since those initial estimates explains some of this discrepancy, Dr. John D. Graham's statement before the U.S. Transportation Safety Board Supplemental Restraints Panel that "we oversold the benefits of the…air bag" cannot be denied (Graham 1997a). Likewise, the air bag's benefit in many instances cannot be denied. Recent projections (Martin et al. 2000) based on current effectiveness estimates, market turnover, and belt use trends estimate that the number of lives saved annually by air bags in frontal impacts will approach 3,300 by 2007, much lower than the nearly 15,000 lives saved annually by seatbelts in frontal impacts (Figure 2), but certainly a substantial safety benefit. The number of serious injuries to the trunk and head prevented by air bags is also projected to continue to increase (Figure 3). The lower extremity injuries brought to bear by air bags, as shown in Figure 3, is primarily due to a change in risk to other body regions, which results in lower extremity injuries being used more often to characterize the crash. This phenomenon is discussed in more detail in Section 2a.

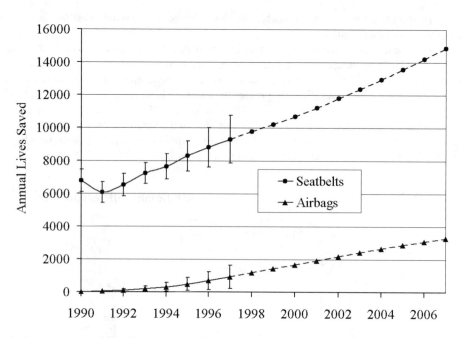

Figure 2. Passenger car driver lives saved (with 68% confidence intervals) by seatbelts and airbags in frontal crashes (dashed line denotes projections) (modified from Martin et al. 2000).

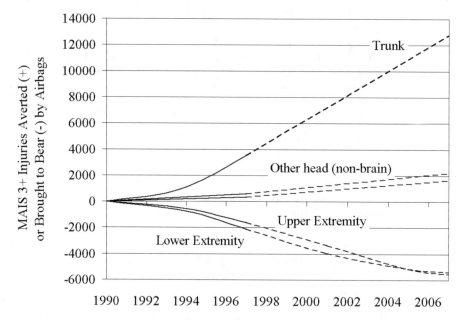

Figure 3. Passenger car driver MAIS 3+ injuries prevented or brought to bear by airbags in frontal crashes, by body region (dashed lines denote projections) (modified from Martin et al. 2000).

While the field experience with air bags is still insufficient to assign much confidence to the projections shown in Figure 2 and Figure 3, the projections do reflect some characteristics of air bags that are reasonably consistent in the literature. First, while air bags provide a clear safety benefit in many instances, they are clearly not a safety panacea and, in fact, do not provide as much safety benefit as a seatbelt when the belt is used (Barry et al. 1999). Second, air bags are effective at preventing fatal injuries in frontal crashes, primarily for unbelted occupants. Third, air bags do not prevent all types of injuries and can actually increase the incidence of some types. The following section discusses the types of injuries that can be caused by air bags and how air bags have changed the types of injuries that occur in the field.

2. Inadvertent Deployment, Non-deployment, and Injuries Caused by Air bags

Much of the impetus behind air bag development was the concept of "extending protection" for a largely unbelted occupant population in severe frontal crashes. This extended protection necessitated air bag systems that deployed quickly and with sufficient internal pressure to restrain a large unbelted occupant. In a 56 km/hr frontal barrier test, for example, accounting for sensing and actuation time, the bag must inflate completely in approximately 30 ms (Struble 1998). Reed et al. (1994) presented a detailed analysis of the inflation velocity and temperature of air bag deployments. If one assumes the chest-to-instrument panel distance of a typical passenger car of 450 mm, the average velocity of the air bag's leading edge is about 15 m/s (34 mi/hr), with a maximum velocity perhaps as high as 90 m/s (200 mi/hr) for ca. early 1990s systems (Powell and Lund 1995). The peak internal pressure in the bag (ca. 1995) will be on the order of 150-300 kPa. Two examples help to put this pressure in perspective: peak air bag pressure (ca. 1995) is approximately half the inflation pressure of a typical automobile tire and about twice the inflation pressure of a regulation basketball. Air bags' peak pressure and leading edge velocity have decreased overall since the 1997 revision of FMVSS 208, but the bag is still not the billowing pillow that it appears in high-speed crash test films.

During development of prototype and research air bag systems, laboratory experiments had shown the potential for minor, moderate, and severe injuries from air bag deployment. Human volunteers had sustained mild concussions, abrasions, and other minor injuries (e.g., Smith et al. 1972). Baboons, chimpanzees, pigs, and human cadavers had shown the potential for severe neck and thoracic trauma (e.g., Patrick and Nyquist 1972, Mertz 1988). It is not surprising, therefore, that the early investigations of field performance indicated air bag-related trauma. A study of air bag inflation loads in out-of-position deployments was initiated by GM after the first suspected death by the air bag. In this case, there was evidence that the driver was against the steering wheel air bag during inflation. In 1979, Horsch and Culver showed that the force of inflation increased dramatically if the torso was against the module preventing free inflation of the air bag. The load in a body block with a separation of 178 mm from the driver air bag was 4.23 kN, whereas the force increased to 11.57 kN with 67 mm and 20.0 kN with 0 mm separation. The increase in load is associated with the constrained inflation of the air bag module. If the inflation path is blocked, the internal pressure of the air bag increases. In a follow-on study, Horsch et al. (1990) showed that the inflation pressure tripled when the inflation path was blocked by an occupant. This increased the pressure from a nominal 180 kPa to nearly 600 kPa when the torso blocked the path of inflation. In test with dummies and animals, the viscous response of the chest exceeded VC = 3.0 m/s and regularly produced lethal injury in experimental animals exposed to inflation (Lau et al. 1993).

As early as 1976, Mohan et al. showed that mean injury level in lower-severity crashes was higher in a fleet of air bag-equipped vehicles than in a similar fleet without air bags. Unlike later studies, however, these authors attributed this finding to under-reporting of minor injuries in non-air bag cases and other factors rather than the increase in some injuries that was later found to accompany air bag deployment.

Mertz (1988) presented case reviews of deployment crashes involving 1973-1976 GM air bag systems in which the occupant received an AIS 3 or higher injury. Many of these injuries occurred in crashes of minor to moderate severity and, based on animal and dummy testing of air bag injury potential, Mertz concluded that the injuries could have been caused by air bag deployment. In a related study of this early GM fleet, Pursel et al. (1978) had found passenger air bag effectiveness for AIS 2 (and some more severe) injuries to be -34% (i.e., the passengers in air bag-equipped vehicles sustained more AIS 2 injuries than the passengers in vehicles without air bags). Mertz (1988) concluded that air bag-induced injuries had a pronounced effect on air bag effectiveness, especially for the front passenger. This effect was so substantial that the elimination of AIS 3 injuries to out-of-position occupants would increase the AIS 3+ effectiveness of the air bag system from 18% to 50% overall, from 21% to 41% for the driver, and from 16% to 71% for the passenger.

The types of injuries caused by air bags can be grouped into those to adults, which may be serious but are usually not life threatening, and those to children and infants. The injuries to children are also mostly not life-threatening, but the bulk of our discussion will focus on the alarming fatal and serious injuries that have been sustained by children and by infants in rear-facing child safety seats when they are placed in the path of a deploying passenger air bag. The reader should bear in mind that the following studies focus on the initial air bag systems introduced on the market. These were full-powered systems sufficient to restrain an unbelted large occupant in a severe crash. In later years, it was found that these systems produced too much inflation force in low-severity crashes with small occupants close to the module. This resulted in changes to FMVSS 208, which facilitated the introduction of depowered air bags and, more recently, some dual and multi-staged inflation air bags that increase the inflation pressure in proportion to crash severity.

2a. Injuries to Adults

NHTSA has currently documented cases of 63 adult drivers and 7 adult passengers killed by air bags (Chidester and Roston 2001) (Figure 4). In addition, there are numerous studies documenting the risk of minor and moderate injuries that accompanies an air bag deployment. Historically, as air bag-equipped vehicles became prevalent and cases of patients exposed to air bag deployments became more common clinically, case studies began to appear in the medical literature. By definition, these studies did not attempt to determine air bag performance on a population scale, but they did give an indication of some typical and atypical injury patterns that were observed in the presence of an air bag deployment and, in retrospect, some give an indication of common conceptions and misconceptions regarding air bag performance. Huelke (1995) provided a detailed overview of several case studies documenting various types of air bag-induced injuries, including injuries to the head and face, neck, thorax, abdomen, back, and upper extremity. Some of these injuries are described below.

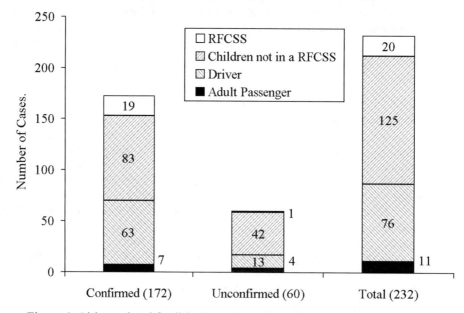

Figure 4. Airbag-related fatalities investigated by NHTSA as of January 1, 2001 (modified from Chidester 2001).

Since the early 1990s, there have been over 200 English language journal articles on injuries associated with air bag deployments in field crashes. Some of these studies are summarized here, but the reader is encouraged to evaluate the validity of all injury claims associated with air bag deployment. It is quite possible that the dramatic increase in public awareness of air bag-induced injuries that accompanied the early occurrences of these injuries resulted in an over-assignment of injuries to an air bag-related cause or in an increased likelihood that an injury of previously unknown source would be attributed to the air bag. Even those injuries caused by the air bag may be misclassified or the exact source misidentified. One early study, for example, alleged chemical burns on the chin and anterior neck due to the air bag system (Polk and Thomas 1994), but gave no indication that this was not the common air bag abrasions due to the fast unfolding of the air bag fabric during deployment or the less common burns associated with venting.

Reported injuries to the head and face of adults are most often erythema (skin redness), abrasions, and contusions. Ocular injuries also occur, but much less frequently than skin injuries. There have been cases of chemical burns of the eye due to contact with the byproducts of sodium azide deployment (Smally et al. 1992). Isolated cases of air bag perforation from a pipe stem or from eyeglasses have also been reported (Huelke 1995, Walz et al. 1995, Gault et al. 1995). Duma et al. (1996) concluded that the eye can be injured by contact with the fully deployed air bag, but that more severe injuries resulted from contact with the actively deploying air bag. Pearlman et al. (2001) recently summarized the range of eye injuries and risk factors with air bag deployments, which can range from abrasions to retinal tears, hyphema, eyeball ruptures, and enucleation (O'Halloran et al. 1998, Ruiz-Moreno 1998, Baker et al. 1996, Goldberg, et al. 1995, Han 1993). The face and eyes can also be injured when the upper extremity is loaded by the air bag and the hand strikes the face. These cases may result in nasal and anterior face fractures, and occur most often when the driver's hand is at 12 o'clock on the steering wheel during an attempted turn at the moment of the crash. The force of deployment thrusts the hand rearward. Wrist and hand injuries can occur when the hand hits the roof rail or pillars in the vehicle. Hearing loss and tinnitus has been reported from air bag deployment (e.g., Buckley et al. 1999, Yaremchuk and Dobie 2001), and there have been cases of ruptured eardrums (McFeely et al. 1999). Rouhana et al. (1994, 1998) published studies on air bag noise and the assessment of temporary hearing loss based on military standards for impulse noise. Based on these studies, the potential exists for threshold shifts to occur when an air bag (ca. early 1990s) deploys. Other studies have found no ear injury, even with air bags deployed directly onto the ear (Richter et al. 1974). A wide range of other injuries has been described associated with air bags, including, for example, basilar skull fractures (Bandstra and Carbone 2001) arterial ruptures (Gottesman et al. 2002), temporomandibular joint and mandibular injuries (Garcia 1994, Van der Linden 2002, Levy et al. 1998), maxillofacial fractures (Roccia et al. 1999), esophogeal ruptures (Cullinan and Merriman 2001), induced asthmatic attacks (Gross et al. 1994, 1995), and dental injuries (Brown 1998). In the cases of adult fatalities due to air bag loading, shorter occupants have sustained fatal diffuse axonal or brain stem injuries as well as cervical fractures (Perez and Palmatier 1996, Hollands et al. 1996).

Neck injuries are not uncommon in crashes involving air bag deployments, but they are almost always minor and it is not clear that the commonly reported cervical strains are related to the air bag. In fact, an air bag in combination with a pretensioned belt has been shown to reduce the risk of AIS 1 neck injuries to belted occupants in frontal impacts (Kullgren et al. 2000). Anterior neck abrasions or erythema from the air bag are not rare, but they appear to occur less frequently than they occur to the face. Some of the shorter occupants killed by an air bag have sustained neck extension fractures at the atlanto-occipital joint, with or without spinal cord involvement.

Thoracic injuries are often difficult to assign a cause. In many cases, the seatbelt, steering wheel, and air bag are all candidates. As with the head and neck, the most common thoracic injuries that can be attributed to the air bag are minor abrasions and contusions. Many of the adults killed by air bags were out-of-position due to a preimpact event (such as braking), and the taller occupants typically sustain thoracic loading rather than the neck loading described above. In these cases, the fatal injuries include multiple rib fractures, flail chest, lung contusions, atrial ruptures, and lacerations of the myocardium, pericardium, or aorta (Huelke 1995, Chidester and Roston 2001). Air bag-related injuries to the abdomen and back are rare.

Perhaps the most common non-skin injuries caused by air bag deployment are injuries to the upper extremities (Werner et al. 1996). Huelke et al. (1997) investigated 540 crashes involving air bag-equipped vehicles and found that 38% of drivers sustained some type of upper extremity injury, though most were AIS 1. In 18 of the cases, the driver sustained an AIS 2 or AIS 3 upper extremity injury, with six due to direct contact with the air bag or module cover and 12 due to "flinging". These two injury mechanisms manifest themselves as different resulting injury patterns. Atkinson et al. (2002) showed that direct air bag loading tends to result in mid-shaft fractures of the tibia and fibula, while flinging is associated with injuries to the wrist and elbow. Dalmotas et al. (1996) showed that harm to the face and the upper extremities increases by 71.1% when an air bag deployment occurs. Werner and Sorenson (1994) found that drivers were turning, swerving, or changing lanes in 16% of crashes leading to a deployment, which may contribute to upper extremity injuries from the air bag.

The introduction of air bags into the fleet has increased the frequency of some types of injuries not because the air bag causes the injury, but because the presence of an air bag changes the type of injury that is used to characterize the crash. For example, an occupant may sustain an AIS 4 head injury and an AIS 2 lower extremity injury in a crash without an air bag. This case would be considered an MAIS 4 head injury. Now assume this occupant had an air bag, which had reduced the head injury to an AIS 1. This is now an MAIS 2 lower extremity case. The field database has a resulting increase in lower extremity injury rate, which is attributable to the air bag despite the air bag's lack of involvement with the lower extremity. Similarly, Crandall et al. (1994) hypothesized different head injury potential for belted (no air bag) occupants vs. air bag (no belt) occupants, based on kinematic differences observed in laboratory tests. Specifically, they observed that, with a seatbelt and no air bag, the occupant often contacts the steering wheel rim while, with an air bag and no seatbelt, the occupant can ride over the steering wheel and air bag and sustain facial and/or head injuries from contact with the front header or windshield. They confirmed this hypothesis using the field data, which showed that brain injuries and facial injuries are both significantly more probable with an air bag only than with a belt (either with or without an air bag). The probability of contacting the windshield was also higher in the case of an air bag with no belt. This finding was supported by Malliaris et al. (1995b), who found that the air bag-only restraint condition results in disproportionately higher percentage of injury to the head/neck compared to other restraint conditions, and McKay et al. (1999), who found that drivers with an air bag in the U.S. have higher rates of facial injury and non-air bag head contact than drivers in Australia, which has a much higher belt use rate. The change in injury outcome, due to air bag-induced injuries and to a changing injury distribution due to air bags, has been reported extensively in the literature and is reflected in the injury projections shown in Figure 3. The reader should keep in mind, however, that, compared to an unbelted occupant without an airbag, injuries to the head and other body regions decrease for both belted occupants and for those restrained by an air bag without a belt (e.g., Mouzakes et al. 2001).

2b. Injuries to Children
    While the increased risk for injuries to adult drivers is of concern, the experience of children exposed to an air bag deployment is cause for alarm. In the 1970s, Volvo (Aldman et al. 1974) raised the concern for child injury by inflation forces with the passenger side air bag. Similar studies in the US (Patrick and Nyquist 1972) showed risks for a child in the path of an inflating air bag. These laboratory findings were supported by the early field evaluation of air bags, which included cases of child deaths in the passenger seat (Mertz 1988). This followed studies with animals and child dummies (Mertz et al. 1982, Mertz and Weber 1982) demonstrating the potential for life-threatening injuries to children against the module during inflation. The out-of-position exposures were found not only with the child initially against the module at deployment, but also secondary to hard braking prior to a crash with the child displaced forward at the point of air bag activation (Stalnaker et al. 1982, Montalvo et al. 1982).
    Ferguson et al. estimated that 74 lives had been saved by front passenger air bags in 1992-1995 model year cars in single and two-vehicle crashes from 1992-1995. During that same time, however, 3 infant and 11 child deaths were attributed to the air bag. Studies have also identified non-fatal injuries to children as a result of air bag loading (Gotschall et al. 1997, Durbin et al. 2002). In their latest report to Congress, the NHTSA (1999) concluded that right-front passengers under the age of 13 have a higher fatality risk with an air bag than without, but was unable to quantify this increased risk. Other researchers have estimated the increased risk at 33% (Ferguson et al. 1996), 67% (Graham 1998b), and 31% (restrained) to 84% (unrestrained) (Glass et al. 2001). In other words, even among properly restrained children on the passenger side, it appears that air bags cause a net increase in fatality risk (cf. Graham 1997a). In terms of overall safety benefit of the passenger-side air bag, Graham et al. (1998b) found that 5-10 adult lives have been saved for every child that has been killed. In contrast, the ratio for the driver-side air bag is approximately 75 adults saved for every adult killed. If years of life saved is used as the metric rather than the number of lives saved, the benefit-to-risk ratio for the passenger air bags currently in the fleet is no greater than 5 to 1. Graham et al. (1998b) state that they are "aware of no mandatory measure in the history of preventative medicine that has been preserved" with a ratio that low, but public education campaigns to move children to the rear seat, changes in legislation that allow for depowered air bags, and advancements in air bag technology are expected to reduce injuries from passenger air bags, thereby improving this ratio.
    As of Jan. 1, 2001 the NHTSA Special Crash Investigations (SCI) program (Chidester and Roston 2001) has confirmed 172 air bag-related fatalities, with an additional 60 cases under investigation. Of the confirmed cases, 102 are children under 13, and 19 are infants in a rear-facing child safety seat (RFCSS). The scenario for children killed in a RFCSS typically involves the air bag deploying into the back of the infant seat, often with sufficient force to break the plastic shell (Augenstein et al. 1997). Skull and brain injuries are the typical cause of death. Children killed while not in a RFCSS are usually unbelted or improperly belted and the cases often involve preimpact braking that places the child near the air bag at the time of deployment. Contact with the instrument panel prior to deployment has been confirmed in some cases. In most cases, the air bag deploys into the child's chest, neck, and face, placing the neck in tension and bending. Typical injuries include atlanto-occipital fracture, often with accompanying spinal cord transaction, and brain stem injuries. Diffuse axonal brain injuries, consistent with extreme linear and angular acceleration of the head, are also common. Skull fractures are not typically found, though they may not be coded in many cases due to the immediately apparent spinal injuries.
    As these alarming fatalities began to occur in the field and the increase in certain types of non-fatal injuries from air bags became understood, air bags underwent a metamorphosis in legislation and design philosophy. This metamorphosis involved two primary components: depowering (Werner et al. 1996) and the development of "smart" systems (Miller 1995). Depowering refers to various design

modifications intended to reduce the severity of injuries sustained during interaction with an air bag. In contrast to depowering, which is primarily intended to minimize injury once an air bag deploys, "smart" systems suppress or change the deployment characteristics in response to a particular condition. Conditions that may affect the deployment of a "smart" system include the presence or absence of a passenger, the presence of a child restraint in the passenger seat, belt use, and many others (e.g., Miller 1995). The effectiveness of these air bag redesign efforts (in addition to public education campaigns to move children to the rear seat and other factors) is becoming apparent in the most recent NHTSA publication on air bag-induced fatalities (Chidester and Roston 2001). Child fatalities (normalized for exposure) reached a peak in the 1995-1996 sales period, and have been declining steadily since (Figure 5, top plot). Even before the change in FMVSS 208, which allowed depowering in model year 1998 vehicles, child fatalities had begun to decline. Child fatalities are currently highest for model year 1995 vehicles, decreasing by over half for 1996 vehicles, with continued declines in subsequent model years (Figure 5, bottom plot). Similar reductions in adult fatalities, both for the driver and for the front passenger, have occurred since the mid- to late-1990s (Chidester and Roston 2001).

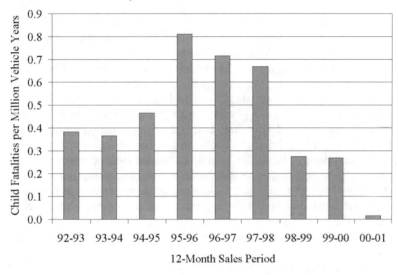

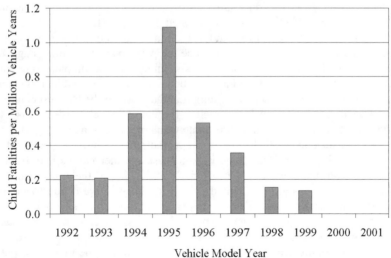

Figure 5. Normalized child fatalities by 12-month sales period and by vehicle model year (modified from Chidester 2001).

2c. Inadvertent Deployment or Non-Deployment

They system reliability of air bags is between 0.9999 and 0.99999 (Phen et al. 1998), indicating a relatively low number on unintended deployments and non-deployments (on the order of tens per year in the U.S.). Despite this excellent reliability, the literature contains isolated references to cases of an air bag deploying when it is not intended to or not deploying when it might be expected to. Virtually all of these cases involve very early air bag designs. Essentially all of the cases of air bag-induced injury discussed above are not considered to fall within this category, since they occurred when the air bag deployed in a crash. Given the number of air bag-equipped vehicles currently on the road the air bag system has proven to be an extremely reliable device, but the documented cases of apparently inappropriate deployment or non-deployment deserve discussion.

One of the earliest studies to identify such cases was by Smith and Moffatt (1975). In a fleet of research vehicles having cumulative exposure of over 100,000 car months, they identified three crashes that did not result in deployment, but did result in moderate or greater injury. Based on the injuries sustained, the implication is that the air bag should have deployed in these cases, but there is no direct evidence given that the air bag system malfunctioned. Smith and Moffatt also observed four cases of non-collision deployments, none of which resulted in serious (AIS 3+) injuries. The non-collision deployments resulted from 1) improper wiring of a wire harness, which allowed the bumper sensor wiring to come into contact with the rotating power steering pump 2) a hammer blow to the bumper while the engine was running 3) overpressure from a propane gas explosion in the trunk and 4) an unknown cause resulted in the passenger air bag deploying while the driver air bag did not (this case is described in detail by Smith (1973)). Backaitis and Roberts (1987) summarized 112 crashes involving NHTSA-sponsored fleets of air bag-equipped vehicles. The authors concluded that both original equipment and retrofit air bag systems deployed as expected. Werner and Sorenson (1994) studied 2,818 air bag deployments and found "very few instances of what might be considered inadvertent deployments." One case involved a single-point sensor that received multiple strikes from a loose drive shaft hitting the floor pan. Langweider and Hummel (1998) published perhaps the most comprehensive list of what they termed "problem" air bag cases. These included cases where the air bag activated while the vehicle was stopped (7 cases), the air bag activated while driving without a collision (7 cases, often associated with an impact to the undercarriage), and 5 cases of no air bag activation despite high crash severity (including a case where the driver air bag deployed and the passenger air bag did not). These cases involved multiple causes, including improperly retrofitted components and damage to wiring from an attempted theft or a previous collision. There have been cases of extreme front deformation of the vehicle involving disruption of the battery and loss of electrical power to the air bag. Air bags are fit with a capacitance charge to handle these situations, but some extraordinary crashes have damaged the system before deployment could occur. Given the large number of air bags on the road and their exposure to diverse environments, the isolated nature of these "problem" cases is testament to the overall excellent technical reliability of these systems.

3. Factors Affecting Air bag Effectiveness

The studies presented in section 1 are almost exclusively evaluations of "as used" effectiveness. In other words, the air bag effectiveness is calculated for the distribution of occupants, crashes, and vehicles that were current at the time of the study. From a research or air bag assessment standpoint, it is desirable to understand how various factors influence the performance of an air bag, so that the air bag's performance may be predicted when these factors change. For example, some of the comparisons presented above contain the inherent assumption that belt use rates are equivalent in the air bag-deployment crashes and the belt-only vehicle crashes, an assumption that may not be true, especially in the early 1990s when air bags were available only on luxury vehicles. If the air bag's effectiveness as a function of seatbelt use could be established, then a change in belt use rate would not confound air bag assessment.

Some factors, such as seatbelt use or impact direction, have a profound effect on air bag effectiveness, and this effect can be quantified in some cases. Other factors may be small, or their effect may be impossible to determine precisely. One example of the latter is the effect of the occupant's position at the time of impact. Obviously, an unbelted driver displaced forward due to preimpact braking will experience a different interaction with the air bag than a driver that remains back in the seat. While general conclusions about the effect of the occupant's pre-impact position can be drawn based on engineering judgement, forensic investigation of crashes, crash testing, or computer modeling, robustly quantifying its effect in the field is not possible for several reasons. First, defining the myriad of potential positions an occupant may assume requires too many parameters to be practicable. Second, this information is often not possible to determine in a post-crash investigation. There are, however, many factors that can be quantified using the current data available on air bag performance in the field. This section presents several of these factors, including the role of seatbelt use and the influence of crash, occupant, and vehicle characteristics.

3a. Seatbelts and Air bags

In a summary of research performed by Evans (1991) and others, Viano (1991) estimated the fatality effectiveness of several different safety systems, divided into the effectiveness due to protection against ejection and the effectiveness due to mitigation of interior impact (Table 4). He found that the fatality effectiveness of a lap-shoulder belt system is split approximately equally between these two components. The potential effectiveness of air bags is limited to mitigating interior contact, with no effectiveness in limiting ejection, which prevents even an optimally designed air bag from matching the effectiveness of a seat belt. Using the finding of Huelke et al. (1979) that approximately 50% of unbelted crash fatalities are unpreventable by any restraint system, Viano concluded that an air bag could provide only a maximum of 4% (1988) to 7% (1991) additional fatality prevention beyond that provided by a worn lap-shoulder belt system; an estimate that was in agreement with NHTSA's (1984) estimate of 5%. This is consistent with the findings of Evans (1986), who showed that lap-shoulder belts are most effective in rollovers, and Huelke et al. (1985), who determined that safety belt use virtually eliminated the risk of paralyzing cervical injury and ejection. In crashes where a rollover is the first harmful event, lap-shoulder belts are 82±5% effective in preventing driver fatalities, a value that cannot be matched by the interior impact mitigating benefit of an air bag when all collision types are considered. When only frontal collision are considered, however, Zador and Ciccone (1991) found that, relative to comparable cars with manual belts only, driver fatalities in frontal crashes were reduced by 28% in cars with both manual belts and air bags. Similarly, their colleagues at the Highway Loss Data Institute (1991) concluded that drivers in air bag-equipped cars experienced 28% lower severe injury rates and 24% lower hospital inpatient rates than drivers in cars with automatic belts.

Table 4 – Fatality Effectiveness of Occupant Protection Systems (Evans 1991 in Viano 1991)

| | Driver | | |
Safety system	Impact mitigation	Ejection prevention	Overall
Lap-shoulder belt	23±4%	19±1%	42±4%
Airbag	18±4%	-	18±4%
Lap-shoulder belt plus airbag	27%	19%	46%

| | Right front passenger | | |
Safety system	Impact mitigation	Ejection prevention	Overall
Lap-shoulder belt	22±4%	17±1%	39±4%
Airbag	13±4%	-	13±4%
Lap-shoulder belt plus airbag	26%	17%	43%

Werner and Sorenson (1994) presented one of the earliest assessments of belt use in a large population of air bag-equipped vehicles, but have an apparent problem with belt use over-reporting (usage reported at 89%)[5]. As a result, their injury reduction rates appear to be biased. For example, they find that the reduction in moderate and severe injuries is equal for the belt (125.5% reduction in cars with no air bag) and the air bag (125.5% reduction for unbelted occupants) (Table 5), a finding that does not agree with the bulk of the previous or subsequent literature.

Table 5 – Moderate and Severe Injury Rates and Reductions by Restraint Conditions
(modified from Werner and Sorenson 1994)

Drivers with moderate and severe injuries per 1,000 frontal crashes	
Airbag vehicles with deployment	
Belted	98
Unbelted	75
Vehicles without an airbag	
Belted	221
Unbelted	98

Change in moderate and severe injuries (negative indicates reduction)	
Effect of belt in airbag-equipped vehicles	(75-98)/75 = -30.7%
Effect of belt in non-airbag-equipped vehicles	(98-221)/98 = -125.5%
Effect of airbag for non-belted occupants	(98-221)/98 = -125.5%
Effect of airbag for belted occupants	(75-98)/75 = -30.7%
Effect of airbag and belt	(75-221)/75 = -194.7%

---

[5] Belt use is often over-reported, particularly if it is reported by the occupant. This is a significant problem in insurance claim-based studies. Databases in which belt use is reported by crash investigators have more reliable belt use rates, but are still subject to over-reporting. See Malliaris and DeBlois (1998) for a discussion of belt use over-reporting and its profound effect on restraint effectiveness studies

Table 6 – Projection of Lives Saved by Restraint Use and Availability
(Assumes 10% Airbag Penetration into Fleet) (modified from Viano 1995)

Belt Use	Total Deaths	Lives Saved	Saved by airbag	Saved by belt
		Driver		
0%	24,475	525	2.1%	0.0%
10%	23,465	1,535	1.9%	4.2%
20%	22,455	2,545	1.8%	8.4%
30%	21,445	3,555	1.6%	12.6%
40%	20,435	4,565	1.5%	16.8%
50%	19,425	5,575	1.3%	21.0%
60%	18,415	6,585	1.1%	25.2%
70%	17,405	7,595	1.0%	29.4%
80%	16,395	8,605	0.8%	33.6%
90%	15,385	9,815	0.7%	37.8%
100%	14,375	10,825	0.5%	42.0%
		Right Front Passenger		
0%	7,896	104	1.3%	0.0%
10%	7,588	412	1.3%	3.9%
20%	7,280	720	1.2%	7.8%
30%	6,972	1,028	1.2%	11.7%
40%	6,664	1,336	1.1%	15.6%
50%	6,358	1,644	1.1%	19.5%
60%	6,048	1,952	1.0%	23.4%
70%	5,740	2,260	1.0%	27.3%
80%	5,432	2,568	0.9%	31.2%
90%	5,124	2,876	0.9%	35.1%
100%	4,816	3,184	0.8%	39.0%

Recognizing that a restraint system is only effective when it is used, Viano (1995) (Table 6) compared fatality effectiveness of occupant protection systems by seating position, taking into account occupancy rates for the different seating positions, penetration of air bags into the fleet, and the effect of belt usage. He cites reported safety belt benefits of, depending on the study, 40%-50% in preventing fatalities and 45%-55% in reducing serious injury (NHTSA 1984), and states that the air bag is providing a 24% lower hospitalization rate and a 25%-29% lower incidence of serious injuries for the then-current safety belt usage of approximately 65%. Viano projected lives saved by restraint use and availability (Table 6). The analysis shows that the safety benefit that can be realized by increased belt use is about two times greater than the safety benefit from a similar increase in air bag penetration into the fleet. For example, a 10% increase in air bag availability produces a safety gain (in terms of fatality reduction) similar to that produced by a 5% increase in belt use. Viano also found that the incremental safety benefit of air bags decreases as belt use rate increases since air bags provide only 5% additional fatality effectiveness in combination with belts but 21% fatality effectiveness alone (i.e., if belt use rate was 0% air bag effectiveness would be 21% and if belt use rate was 100% air bag effectiveness would be 5%). So, for example, equivalent reduction in fatalities is achieved by 20% belt use and 47% penetration of air bags, or by 36% belt use and 100% air bag availability. This finding is supported by Evans (1999), who estimated that, as belt use rates increase from 0% to 100%, deaths prevented by air bags decline from 13/100 to 5/100.

Lund and Ferguson (1995) found that the air bag reduced fatalities by 24% in frontal crashes and 16% in all crashes for the distribution of belt use in their dataset. In contrast to the findings of Viano and others, these values did not change appreciably when the cases were divided into belted and unbelted (21% frontal belted, 23% frontal unbelted), again indicating potential bias from self-reporting of belt use.

Malliaris et al. (1995a, 1996) expanded on these fatality-based analyses by considering injuries in addition to fatalities and by considering the entire population of tow-away crashes instead of only frontal crashes. Recognizing that a single injury metric facilitates the interpretation of air bag performance and design prioritization, Malliaris et al. also evaluated air bags using societal harm, expressed in terms of an effective fatality rate. They found that the rate of harm was highest for drivers with no restraint, followed by drivers restrained by an air bag and no belt, but that the difference between these two restraint conditions was not significant (Figure 6, for non-normalized rates the reader is referred to the original paper). The harm rates for these two restraint conditions were significantly higher, however, than the rate for drivers restrained by a belt. The effect of an air bag for belted drivers was not significant in terms of harm reduction. In fact, the safety benefit of an air bag for either belted or unbelted occupants is not clear in this study. For example, maximum AIS (MAIS) 3 and MAIS 4 injury rates for unbelted drivers are non-significantly greater when an air bag deploys. The significant trends associated with air bag deployment include a decrease in MAIS 2 rate, MAIS 5 rate, and fatality rate for unbelted occupants, an increase in MAIS 1 rate for belted occupants, and a decrease in MAIS 4 rate for belted occupants. In contrast, belt use is associated with significant decreases for all metrics except MAIS 5 injury rate regardless of air bag availability (lack of significance due to the small number of cases with an MAIS 5 injury). The reader should bear in mind that these findings are based on a limited number of cases and are biased by several factors discussed in a companion paper (Malliaris et al. 1995b). For example, air bag availability is higher in larger luxury cars than in smaller economy cars. Likewise, the distribution of belt use is different for different size vehicles. Since restraint system effectiveness will generally be lower, all else being equal, for a more crashworthy car (i.e., there will be fewer fatalities or injuries without the restraint simply because the car is more crashworthy), the inherent increased safety of a larger car will manifest itself as a decrease in air bag effectiveness relative to a lap/shoulder belt and relative to no restraint. With air bag availability in the fleet of 100% this bias disappears, but air bag penetration at the time of the studies by Malliaris et al. was insufficient to remove this confounding effect.

Early studies attempted to evaluate restraint effectiveness by body region (e.g., Blower and Campbell 1994), but there was an insufficient amount of data to draw firm conclusions. By 1999, there were a sufficient number of air bag deployments in the field to make preliminary estimates of restraint effectiveness by injury severity for different body regions (Table 7) (NHTSA 1999). Air bags were shown to be effective at reducing moderate and serious injuries to the head for both belted and unbelted drivers, though drivers gain greater head protection from a belt without an air bag than from an air bag alone. Serious chest injuries were reduced by an air bag for unbelted drivers, but the air bag did not provide large serious-injury benefit for belted occupants (Table 7). In contrast, moderate injuries to the chest of belted occupants were substantially reduced with the addition of an air bag. The findings of Frampton et al. (2000) and Foret-Bruno et al. (2001) indicate that the greatest air bag-related benefit to the chest may be that air bags, by limiting head contacts for belted occupants, allow for the use of force-limiting seatbelts. Force-limiting belts reduce the concentrated belt loads on the occupant's thorax by allowing the belt or retractor mechanism to deform at a specified force level (typically 3 kN to 6 kN). Force-limiting belts provide other benefits to the occupant, including increased torso pitch, which loads the shoulder rather than the vulnerable lower thorax and abdomen. They also shift the restraining load so it does not coincide with the maximum load from the air bag (see Chapter 4).

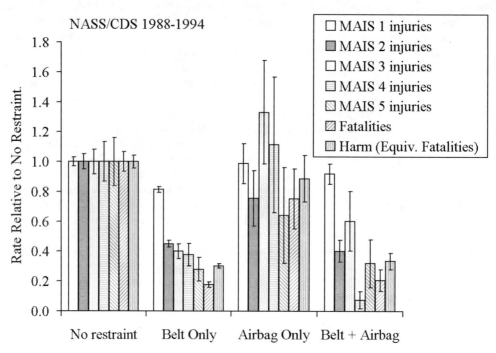

Figure 6. Injury, fatality, and harm rates for different restraints relative to no restraint in tow-away crashes (modified from Malliaris et al. 1996).

Table 7 – Estimated Effectiveness of Restraint Systems by Injury Severity for the Head, Chest, Upper Extremity, and Lower Extremity (NHTSA 1999)

			Major Body Region	
Restraint System	Head	Chest	Upper Extremity	Lower Extremity
Moderate (MAIS 2+) Injury Reduction
Airbag + Belt	**82%**	**64%**	**45%**	47%
Airbag Only	**54%**	17%	32%	-51%
Belt Only	**62%**	25%	**57%**	44%
Serious (MAIS 3+) Injury Reduction
Airbag + Belt	**81%**	52%	20%	**78%**
Airbag Only	58%	17%	57%	36%
Belt Only	**60%**	**50%**	58%	**83%**

Bold indicates statistically significant difference from the risk of unrestrained drivers

NHTSA (1999) also found that air bags reduced both moderate and severe upper extremity injuries for unbelted occupants, but increased upper extremity injuries for belted occupants. The only significant finding for lower extremity injury reduction was that both a seatbelt and a combined belt-and-air bag restraint provided reductions in serious injury. The effect of an air bag on lower extremity injuries for unbelted drivers is not clear, with a negative effectiveness for moderate injuries and a positive effectiveness for serious injuries.

For the current distribution of occupants and vehicles, the latest report by the U.S. government (NHTSA 1999) found that air bags provided approximately 9% fatality risk reduction for belted drivers and 14% reduction for unbelted drivers in all crashes. The fatality effectiveness for combined lap-shoulder belt with an air bag was estimated to be 50%. For serious injury, the lap-shoulder belt provided approximately 64% reduction in injury risk regardless of the presence of an air bag (i.e., an air bag did not improve serious injury protection over a belt alone). The serious injury effectiveness of the air bag alone was large (42%), but not significantly different than zero (Figure 1).

The primary reason that the belt is more effective than the air bag is because air bags and other interior components do not restrain the occupant in a seating position and have minimal effectiveness in reducing ejection. This and potentially other factors result in a lower overall effectiveness for air bags only. Their principal contribution to crash protection is by excellent interior impact mitigation and load distribution, which provides about half the overall benefit of lap-shoulder belt use. The fact that a lap-shoulder belt and air bag is only approximately 50% effective in preventing fatalities underscores that absolute protection is not achievable by occupant restraints, and that injury and fatality will continue to occur to belt wearers even with an overall net safety gain as active restraint usage increases and airbags have become widely available (Viano 1988). The limit reflects the severity of many fatal crashes, which may involve extreme vehicle damage and forces on the passenger compartment, unusual crash configurations and causes of death, unique situations associated with particular seating positions, crash dynamics and human tolerance limits. Efforts continue to increase the overall effectiveness of occupant restraints above the 50% level. This has included the use of features to increase the comfort and restraint performance of the lap-shoulder belts by pre-tensioning, load-limiting, and height adjusting the shoulder belt attachment to the B-pillar.

3b. Occupant Characteristics, Vehicle Characteristics, Crash Type, and Air bag Effectiveness

Seatbelt fatality effectiveness is known to decrease with adult age, from 58.0% for age 16-24 to 31.6% for age over 65 (Evans 1989). Air bag effectiveness may also be related to adult age, though the quantification of this effect is complex. Evans (1989), for example, showed that air bag fatality effectiveness is approximately 18% regardless of adult age when all collisions are considered. This finding is primarily due to the behavioral aspects of aging rather than the biomechanical aspects. For example, an older person is less likely to be involved in a rollover collision, where air bag effectiveness is essentially zero. The over-representation of the young in this type of collision decreases their overall air bag effectiveness relative to the older population. Lund and Ferguson (1995) estimated air bag effectiveness for two different age groups (under 30 and over 30), taking into account the differences in collision involvement and removing rollover collisions from consideration. In frontal impacts, they found the fatality reduction due to air bags to be 31% for the younger group and 20% for the older group. Similarly, in all crashes (rollovers excluded) the fatality reduction was greater for the younger group (20% vs. 13%). The NHTSA (1999) further subdivided air bag fatality effectiveness into four age groups (Table 8). They found the fatality reducing effects of air bags in purely frontal crashes to be almost constant up to 69 years of age, but to decrease for drivers over 70 (though this difference is not statistically significant). It is unclear why air bag effectiveness should be lower for this older population, since an increase in load distribution, while beneficial for all occupants, is generally thought to be particularly beneficial for older occupants (see Zhou et al. 1996, for example). Perhaps as more data become available this age dependency will become better understood.

In addition to age, an occupant's size or gender, or both, may affect air bag performance for that occupant. Previous sections discussed how the injury patterns changed for the shorter or taller adult occupants killed by air bag loading. Shorter occupants also seem to be over-represented in the set of adult fatalities. It is logical that the injury-reducing potential of an air bag also changes for occupants of different height, weight, and body mass index. At this point, we are aware of no studies that have quantified this change using field data. The effect of gender, which is strongly correlated with size, has been studied more extensively and will be discussed within the context of air bag deployment threshold (see Section 3c).

Seating position within the vehicle may also affect air bag effectiveness. Non-drivers represent a substantial portion of the fatalities and severe injuries that occur in automobile collisions. In 1998, drivers accounted for 14,895, or approximately 74% of all U.S. passenger-vehicle frontal impact fatalities, while other occupants, mostly in the right-front seat position (i.e., front passengers, in the U.S.), accounted for 5,248 fatalities (Kent et al. 2000). This uneven fatality distribution is due largely to the higher crash exposure of drivers, though, as mentioned above, impact direction, offset, intrusion, and population age, size, and gender differences also influence the relative fatality risk. Not only are front passengers typically younger, smaller, and more often female than drivers, they also have a different crash experience than drivers. Offset crashes in the U.S. are more often offset to the driver's side, so the crash pulse and the structural intrusion affect the interaction with the vehicle differently for the driver than for the passenger. Furthermore, drivers have less variation in their seating position and gain a potential benefit from a stroking steering column and an earlier engagement with the air bag.

Table 8 – Fatality-Reducing Effects of Airbags in Purely Frontal Crashes for

Different Occupants (NHTSA 1999)

Group	Fatality Reduction
Drivers of passenger cars	**31%**
Belted	**21%**
Unbelted	**34%**
Drivers age 29 and younger	**30%**
Drivers age 30-55	**34%**
Drivers age 56-69	**34%**
Drivers age 70 and older	16%
Drivers of light cars (up to 1,263 kg)	**31%**
Drivers of medium cars (1,263-1,417 kg)	**25%**
Drivers of heavy cars (above 1,417 kg)	**39%**
Drivers of light trucks (pickups, vans, utility vehicles)	**36%**
Right-front passengers age 13 and older	**32%**

*Bold indicates that the estimate is significantly different than zero

There are also inherent differences in the driver and front passenger environments. For example, the proximity of the steering wheel limits the distance available for a driver's forward excursion in a frontal impact. In contrast, a front passenger generally has more distance available over which to reach a common velocity with the decelerating vehicle in a frontal impact. Despite the potential benefit of this additional space, belt effectiveness in vehicles without air bags has generally been found to be slightly lower for front passengers than for drivers, perhaps due to the lower degree of variability in driver seating position. Findings regarding air bag effectiveness for drivers and adult passengers have been mixed. Viano (1995) estimated approximately equal (47%) driver and front passenger fatality effectiveness for combined belt-and-air bag restraint systems. The NHTSA (1999) estimated that, for frontal impacts, passenger air bags are approximately as effective for fatality reduction (32 percent) for all front passengers age 13 and older as driver air bags (31 percent) are for all drivers (Table 8). Ferguson et al. (1996) estimated that air bags decrease front passenger fatalities by approximately 18% in frontal crashes and 11% in all crashes, slightly less effective than the reductions estimated for drivers in frontal crashes (24%) and in all crashes (16%) (Lund and Ferguson 1995). The estimates by NHTSA and Ferguson et al. do not control for differences in injury tolerance between drivers and front passengers, however, or for differences in the collision experience of drivers and passengers. Kent et al. (2000) attempted to remove the effect of these potential confounders by calculating fatality ratios. All frontal impacts in 1998 FARS were analyzed to evaluate the ratio of front-passenger fatalities to total fatalities (FP ratio) for various restraint configurations (Figure 7). Though the greater potential (in the U.S.) for driver-side offset collisions and resulting intrusion influences the number of fatalities and the restraint system interaction, a comparison of fatality ratios provides a means for reducing the influence of collision factors and isolating the effect of restraint and seating position. These data revealed that, regardless of occupant gender and belt use, the FP ratio is significantly lower ($p<0.001$, test for equality of ratios) when an air bag is present than when one is not. In other words, front passengers account for a lower proportion of all fatalities when an air bag is present than when one is not. Consider, for example, unbelted female occupants with an air bag. Approximately 22% of these fatalities involve occupants seated in the front passenger position. Similarly, front passengers account for about 21% of the fatalities sustained by belted females with an air bag. When an air bag is not present, however, the fatality ratio increases to approximately 35% for both belted and unbelted occupants. A similar trend is seen for male occupants, where the difference in the magnitude of the ratios between male and female is explained by the higher proportion of front passengers who are female. These data suggest that an air bag has more fatality-mitigating benefit for front passengers than for drivers, once the outcome is controlled for differences in belt use, gender, and collision factors. This finding is based on a limited number of cases, however, so the trend should be interpreted with caution.

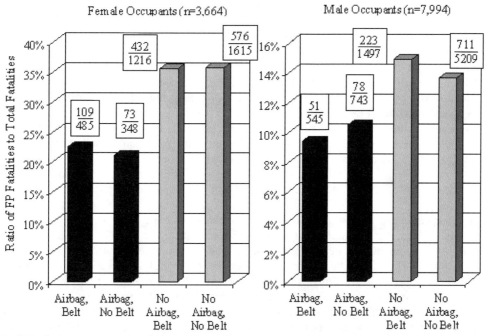

Figure 7. U.S. front passenger fatalities in frontal impacts as a percentage of total frontal impact fatalities – 1998 FARS (Kent et al. 2000). In cases with an airbag, front passengers account for a lower percentage of fatalities, indicating increased effectiveness for front passengers compared to drivers.

The size of the vehicle may influence air bag effectiveness, but this has not been quantified precisely. As discussed previously, a larger vehicle and its intrinsic better crash performance (e.g., Evans 2001) should exhibit decreased air bag effectiveness relative to a smaller vehicle, all other things being equal. All other things are not equal, however. For example, a smaller car is more likely to experience intrusion in a frontal crash due to the shorter distance available for frontal crush. This intrusion can re-orient the steering column, thereby changing the air bag deployment direction and degrading its performance (Zuby and O'Neill 2001). Furthermore, larger vehicles have different belt usage rates, different types of occupants, and other factors that confound air bag effectiveness estimates. In 1994, Werner and Sorenson found no strong relationship between vehicle size and the injury reductions attributable to an air bag (Table 3). In contrast to this finding, Edwards (1994) found a pronounced difference in air bag effectiveness by vehicle model in a study of 1,004 fatalities in 3.36 million Chrysler vehicles in the U.S. fleet. Normalized (to a young male) frontal impact fatality rate changes varied from -53.5% (i.e. a 53.5% reduction in air bag-equipped models) to +11.6% (i.e., an increase in fatality rate with the air bag-equipped model), depending on the particular Chrysler model considered. There are obvious influences other than vehicle size in the Edwards study, however, including significant demographic effects among vehicle models. The latest report by NHTSA (1999) does not identify a clear relationship between vehicle size and air bag effectiveness. They find that fatality reductions in purely frontal crashes are higher for the lightest and heaviest vehicles than for those in the medium category (Table 8).

3c. Crash Severity, Deployment Threshold, and Depowering

That air bags can cause injuries is universally accepted in the literature and seems to be beyond challenge. At question is whether the tradeoff between injury mitigation and injury causation results in an overall benefit. Several studies have attempted to evaluate various air bag design philosophies, particularly with respect to deployment threshold and depowering, within this context. This section describes these efforts. Tradeoff evaluation in terms of air bag costs and cost benefit will be discussed in the subsequent section.

Using the normalized repair costs as a surrogate for collision severity, Werner and Sorenson (1994) performed a comparison of air bag deployment threshold for different vehicle models insured by State Farm. They found significant differences in injury outcome for different models, and included an evaluation of one particular vehicle model that underwent depowering (which they quantified via a tank test and a detailed description of the design changes). They found that abrasion, contusion, and laceration rates were lower after depowering, and that there was no discernable degradation of protection at higher injury severity levels. In a follow-up to that study, Werner et al. (1996) compared injury claims in low-speed crashes (change in velocity (ΔV) " 24 km/h) for two similar vehicles (Dodge Neon, Honda Civic) with different air bag deployment thresholds. For the vehicle with the lower deployment threshold (the Neon), they found upper extremity injuries and facial injuries to be more common than in the Civic. Overall, about 20% of Neon drivers reported an injury in these low-speed crashes, compared to 14% of Civic drivers. Relative risk of injury in these crashes for Neon drivers compared to Civic drivers was found to be 1.53 (95% confidence limit: 1.00 to 2.30). Interestingly, when the injury performance of these two vehicles in higher severity crashes (upper 30% in terms of repair costs) was compared, they found that the car with the higher deployment rate and more aggressive deployment (the Neon) still resulted in higher injury risk (Werner et al. 1997). Fifty percent of Neon drivers reported an injury, while only 35% of Civic drivers did. The relative risk of injury for Neon drivers compared to Civic drivers was found to be 1.83 (95% confidence limit: 1.16 to 2.96). In particular, upper arm and wrist/hand injury rates were higher in the Neon.

Accompanying the depowering effort were concerns that crash protection would be compromised in some cases, particularly for large occupants in severe crashes. The NHTSA SCI group is currently investigating the performance of depowered air bags, and has compiled information on 472 cases as of January 1, 2001. These data are currently under internal NHTSA review and analysis, and no determinations have been made. Zuby et al. (2001) evaluated air bag deployments in NASS/CDS in an attempt to determine whether any fatalities could be attributed to insufficient restraint from a depowered air bag. They could not definitively identify a case and concluded that, in many cases of fatal injuries to an air bag-exposed occupant, gross deformation of the occupant compartment precluded the air bag from having much effect regardless of depowering. This finding was reiterated by Augenstein et al. (2002) based on a comparison of trauma patients restrained by a full-power or by a depowered bag and by O'Neill (2000) who, two years after the changes to FMVSS 208 allowed depowering, had been unable to identify a case of insuffient restraint from a depowered air bag.

Dalmotas et al. (1996) looked at injury and harm as a function of collision severity and gender for a Canadian population of 409 air bag-equipped vehicles and nearly universal belt use. In a comparison of non-deployment crashes in NASS/CDS, they showed that harm increases with air bag fitment by 77.8% in crashes with ΔV below 24 km/h and by 25.6% in crashes having ΔV from 24 km/h to 39 km/h. Only at ΔV values above 39 km/h did the harm begin to decrease in cars with an air bag. Despite the clear transition from air bag detriment to air bag benefit, 72.7% of the deployments studied by Dalmotas et al. were in crashes with ΔV below 26 km/h, which compares well with the findings of Lau et al. (1997) for crashes in the U.S (Figure 8). Lau et al. (1997) supported Dalmotas et al. with their finding that crashes with ΔV less than 32 km/h have a higher risk of minor injury if an air bag deploys. They concluded that nearly half of the minor injuries in their study were due to interaction with the deploying air bag. This finding, which is supported by evaluations of crashes in the U.S. (Stucki and Biss 2000), Germany (Otte 1995) and the United Kingdom (Parenteau et al. 1999), supports the notion that, for more than half of crashes where an air bag deploys, a case can be made that they cause more injury than they prevent (Graham 1997a) - a ratio that may be improved by a higher average deployment threshold.

The deployment threshold question has a large gender and belt use component. In a study of 51,031 paired drivers and passengers, Cummings et al. (2002) found a nonsignificant trend for the air bag benefit, in terms of fatality rate, to be slightly greater for females. In contrast, Dalmotas et al. (1996) showed that belted male drivers realize an 11.6% decrease in overall harm if they have an air bag, but that belted female drivers do not realize a safety benefit from air bags, regardless of the injury metric considered (Table 9). These studies are not necessarily contradictory, however, since differences in belt usage were present between the two populations considered. Werner et al. (1996) found that females appeared to be sustaining more injuries than males, particularly to the upper extremities. Segui-Gomez (2000) found that air bag deployment in frontal or near-frontal crashes decreases the probability of having a severe or fatal injury (AIS 4+), but that air bag deployment in low-severity crashes increases the probability that the driver (particularly a woman) will sustain AIS 1-3 injuries. The most significant finding of the study, however, is that air bag deployment exerts a net injurious effect in low-severity crashes and a net protective effect in high-severity crashes, but that the transition severity is dramatically lower for males than for females. When different metrics of injury are used, the transition speed changes, but is always lower for males than for females (Table 10). So, for example, if the maximum Functional Capacity Index (MFCI) is used as the injury metric, air bags become beneficial for males in crashes at or above 8.8 km/h

ΔV. For females, air bags do not become beneficial until crash ΔV reaches 41.1 km/h. The new FMVSS 208 includes a small female dummy as part of the performance assessment of air bags to address the concerns for balancing restraint loads for the adult male and female.

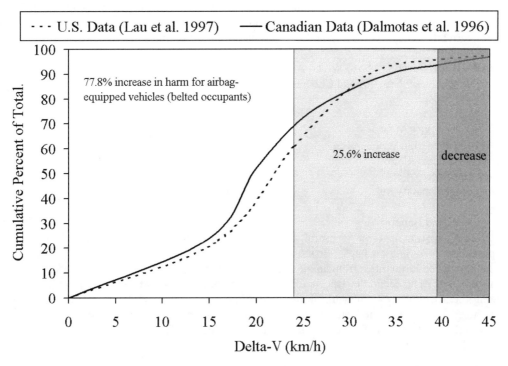

Figure 8. Cumulative distribution of ΔV frequency for airbag-deployment cases in the U.S. and in Canada (modified from Dalmotas et al. 1996 and Lau et al. 1997). Also shown is the change in harm for belted occupants associated with airbag fitment for three collision severity ranges (based on Dalmotas et al. 1996).

Table 9 – Overall Injury Risk and Harm Rates for Belted Male and Female Drivers as a Function of Airbag Availability (Dalmotas et al. 1996).

Performance Measure	Percent Change with an Airbag
Male	
AIS 1+ Injury Probability	5.5%
AIS 2+ Injury Probability	-12.9%
AIS 3+ Injury Probability	10.4%
Mean Number of Injuries	-7.8%
Mean Level of Harm	-11.6%
Female	
AIS 1+ Injury Probability	13.2%
AIS 2+ Injury Probability	8.5%
AIS 3+ Injury Probability	41.1%
Mean Number of Injuries	27.4%
Mean Level of Harm	9.2%

While none of the studies presented here discuss the technical consequences of increasing air bag deployment threshold (e.g., sensing delays in high severity crashes or pole crashes), the field performance studies are consistent in their finding that air bag effectiveness would improve if deployments could be avoided in lower severity crashes. Based on the findings of Segui-Gomez (2000), maximum benefit for all drivers would occur with an air bag deployment threshold ΔV as high as 36.5 km/h (Table 10). There are, however, clear technical tradeoffs associated with an increase in deployment threshold. In general, a higher threshold will tend to increase the risk of late deployment in a severe crash since the sensor will have to gather additional information before "deciding" whether to initiate deployment. Intelligent, "learning" algorithms (Schubert 2001, Singh and Song 2002) and the use of pre-crash sensing (Thiesen et al. 2002) are techniques currently being investigated and implemented to improve crash discrimination.

Table 10 – Crash Severity Above Which Airbag Deployment has a Net Protective Benefit in Frontal or Near-Frontal Crashes (modified from Segui-Gomez 2000)

Injury Metric	All Drivers (km/h)	Female Drivers (km/h)	Male Drivers (km/h)
3-level MAIS (none, minor, severe)	32.8	52.0	12.9
7-level MAIS (0,1,2,3,4,5,6)	36.5	62.2	9.7
Injury Severity Score	16.6	27.5	5.2
Max. Functional Capacity Index	25.3	41.1	8.8

4. Air bag Costs and Benefits

Benefit-cost analyses are a useful, albeit imperfect, method to assess and rank different public health measures, including air bags. Economic metrics are imperfect because the societal acceptance of certain outcomes (child fatalities from an air bag, for example) may not relate in any sense to the resulting economic tradeoff. The economic aspects of air bags cannot be ignored, however, particularly in terms of prioritizing them within the framework of alternative injury prevention strategies. This section discussed the societal costs that have been incurred as a result of air bags, and attempts to put these costs in perspective relative to the benefits derived from air bags. The benefit-cost of air bags is also compared to other public health strategies.

4a. Increased Costs Associated with Air bag Deployment

Peterson and Hoffer (1994) evaluated insurance claims for 87 automobile models manufactured between 1989 and 1991, including some vehicles equipped with an air bag. They found that injury losses and collision losses are never mitigated and usually worsen significantly for air bag-equipped cars relative to cars equipped with only a seatbelt. The increase in severity of collision claims was corroborated by Mirnazari and Henning (1999), who found an overall increase to the severity of collision claims by about 4.6% for air bag-equipped vehicles. They also found that the increase in cost is more pronounced for older cars than for newer cars since the cost of replacing the air bag is relatively independent of the car's age, while the overall cost of the car decreases as it ages. On other words, vehicles retain relatively high market value during the first few years of operation, and collision repair is justified even in the event of an air bag deployment. Werner and Sorenson (1994) estimated that virtually all crashes with air bag deployment result in total loss once a vehicle reaches seven years of age (Table 11). Using low-speed crash tests, Langweider and Hummel (1998) assessed the economic consequences of air bag activation 1) with no one in the passenger seat and 2) at low crash severities where the air bag may not have been needed. Repair costs were increased by 8%-31% when the driver air bag needed to be replaced and 47%-56% if the car had driver and passenger air bags. In field crashes, the authors stated that these costs are even higher, with the costs for the air bag and air bag-induced damage (windshield, dash, etc.) two to three times as high as the repairs due to the collision itself.

4b. Benefit and Cost Analyses of Air bags

In their analysis of proposed changes to MVSS 208, NHTSA (1974) performed one of the earliest economic assessments of air bags. They projected a benefit-cost ratio for the "air cushion-lap belt" system of 3.6 to 6.0. In other words, the societal benefits derived from an air bag would outweigh the costs. Patrick (1975) challenged these findings and performed his own assessment of several different restraint systems. He found a benefit/cost ratio greater than one, but his values were lower than those estimated by NHTSA. He projected that the benefit/cost ratio for an air bag plus 20% belt usage would be 2.31, dropping to 2.1 for the passive air bag system alone. In contrast, his estimate for mandatory three-point harness use was 90.5 benefit/cost the first year, decreasing to 10.1 the tenth and following years (non-constant because the author assumed a 10% turnover in the fleet, with the cost of new belt systems counting against the ratio). The primary advantage of the mandatory harness system was that essentially all of the fleet was equipped in 1975, while other systems would require approximately 10 years to permeate the fleet.

Table 11 –Additional Premium Cost Per Policy, $3,000 Airbag System Replacement Cost by Age of Vehicle (Werner and Sorenson 1994)

Vehicle age (years)	Average deployments per 1,000 insured vehicles in one year	Percent repaired	Vehicles repaired per 1,000 insured vehicles in one year	Marginal additional premium cost per policy ($ U.S.)
0-1	4.0	70	2.8	8.4
1-2	4.0	59	2.4	7.2
2-3	4.0	56	2.2	6.6
3-4	4.0	40	1.6	4.8
4-5	4.0	21	0.8	2.4
5-6	4.0	10	0.4	1.2
6-7	4.0	<10	<0.4	<1.2

In one of the early attempts to prioritize passive restraint systems, Warner (1979) listed the "top twenty" occupant protection countermeasure targets, which represented 71% of all societal losses from car occupants (Table 12). He found that the right-front passenger appeared in only two of the top thirteen targets and accounted for less than 18% of the total loss. Based on these findings, Warner recommended that air bags be installed only on the driver side and, in an impressive display of foresight, predicted that the side effects of large passenger air bags (danger to children among them) carry with them the risk of a consumer backlash regarding air bags.

Table 12 – Top Twenty Occupant Protection Countermeasure Targets for 1978 Model Year Vehicles as Determined by Warner (1979)

Rank	Vehicle Type	Occupant	Crash Mode	Crash Severity (mi/hr)	Percent of Loss
1	Large car	Driver	Front	0-30	13.0
2	Small car	Driver	Front	0-30	11.8
3	Large car	Driver	Side	0-20	8.4
4	Small car	Driver	Side	0-20	7.0
5	Small car	RFP	Front	0-30	5.5
6	Large car	RFP	Front	0-30	5.0
7	Small car	Driver	Front	30-40	4.4
8	Small car	Driver	Rollover	0-50	4.3
9	Large car	Driver	Rollover	0-50	4.3
10	Small car	Driver	Side	20-30	4.1
11	Large car	Driver	Front	30-40	3.6
12	Small car	Driver	Front	40-50	3.4
13	Small car	Driver	Rear	0-30	3.3
14	Large car	RFP	Side	0-20	3.3
15	Small car	RFP	Side	0-20	3.3
16	Small car	Driver	Front	50+	3.3
17	Large car	Driver	Front	0-30	3.1
18	Large car	Rear	Front	0-30	3.1
19	Large car	Driver	Side	20-30	2.9
20	Large car	Driver	Front	40-50	2.8

Table 13 – Cost-Effectiveness of Selected Crash Injury Prevention and

Other Public Health Investments (Graham et al. 1998, Lissy et al. 2000)

Crash Injury Intervention	Target Population	Net Cost Per QALY Saved
Lap/shoulder belts (assuming 50% use)	Front seat occupants	< $0
Compulsory helmet use compared to voluntary helmet use	Motorcyclists	< $0
Daytime running lights	All motor vehicles	< $0
Frontal airbags compared to lap/shoulder belt only (assuming 50% use)	Drivers	$24,000
	Drivers and Right-Front Passengers	$61,000
55 mph speed limit compared to 65 mph	Rural interstate travelers	$82,000
Add shoulder belts to lap belts in back seat (assuming 9% use)	Passengers in rear outboard seats	$160,000
	Passengers in rear center seats	$2,400,000
Cellular phone restrictions	All drivers	$700,000
Cancer Intervention		
Restriction of cigarette sales to minors compared to no restrictions	All children <18	$950
Pap smear every four years compared to no screening	All women 20-75	$16,000
Pap smear every year compared to every three years	All women >65	$62,000
Infectious disease intervention		
Immunization to prevent Hepatitis B virus	Perinatal, infants, adolescents	$3,300-$33,000
Voluntary screening and counseling to prevent HIV transmission	Patients 15-54 in an acute care setting and partners	$59,000
Coronary disease intervention		
Coronary angiography and revascularization if indicated, compared to medication only	Acute myocardial infarct (AMI) patients, 45-74, positive exercise test or angina, and prior AMI	$17,500-$45,000
	AMI patients, 45-74, negative exercise test	$63,000-$340,000

Lissy et al. (2000) compared the cost per quality-adjusted life year (QALY) saved for various highway safety investments, including driver and passenger air bags, which Graham et al. (1998a) compared with some common interventions for cancer prevention, infectious disease prevention, and coronary disease prevention (Table 13). Driver-side air bag systems cost about $24,000 per QALY saved, while dual (driver and passenger) air bag systems cost about $61,000 per QALY saved. The passenger air bag is less cost-effective than the driver air bag for two primary reasons: 1) there are 2.6 times more drivers than passengers and 2) the effectiveness of passenger air bags is negative for children (Graham 1997b). Negative effectiveness for children has a large effect on cost-benefit because children killed by an air bag lose 75 years on average, while a driver killed by an air bag loses 35 years (Graham 1997b). To put these air bag costs in perspective, consider that 50% lap-shoulder belt use is cost beneficial (in other words, the cost per QALY saved is less than zero), as are compulsory helmet use for motorcyclists and daytime running headlights. The cost-benefit of air bags, therefore, is not as good as that for some occupant protection measures. Air bags compare well with other well-accepted measures in preventative medicine, however (Graham 1997a). For example, the cost-benefit for dual air bag systems is similar to decreasing 65 mph speed limit to 55 mph on rural interstates, which costs approximately $82,000 per QALY saved. Increasing the frequency of pap smears for women over 65 from once every three years to once every year has about the same cost effectiveness for cancer prevention as driver-and-passenger frontal air bag systems ($62,000 per QALY saved) have for injury prevention. As another example, Hepatitis B vaccination for perinatals, infants, and adolescents has similar cost effectiveness for infectious disease prevention as driver-only air bag systems ($3,300-$33,000 per QALY saved) have for injury prevention.

Levitt and Porter (2001) estimated that air bags cost approximately $1.8 million per life saved, while seatbelts cost approximately $30,000. Morrall (1986) published an important study of the cost-effectiveness of federal health and safety regulations. Based on a cost per life saved, passive restraints were found to be one of the most effective interventions that the government could mandate. Morrall's estimate for the cost per life saved with passive restraints was $300,000 as compared to $132,000,000 for asbestos regulation and up to $72,000,000,000 for formaldehyde regulation to prevent deaths. On other words, air bags provide orders of magnitude more public benefit in lives saved than many health initiatives. This approach to managing risk sensibly was later picked up in the Reagan administration as a practical approach to setting regulation, and it has been discussed in the UK (BMA 1987) as well.

5. Special Topics
This section discusses performance aspects of air bags that do not fit conveniently within the previous sections. First, the behavioral aspects of air bag marketing and legislation are addressed. The limited information regarding the field performance of side air bags and facebags is then presented.

5a. Risky Behavior and Air bags
The Peltzman hypothesis (1975), also called the principle of offsetting behavior, states that a person's behavior will change in response to a perceived change in protection. Most assessments of air bag performance assume that this change in behavior is small, an assumption that Evans (1999) refers to as the "naïve assumption". The effect of offsetting behavior is the subject of extensive literature, particularly with regard to automobile safety, and an exhaustive overview is beyond the scope of the current discussion. The behavioral aspect of air bag legislation and marketing should be borne in mind, however, in the interpretation of field performance studies.

Peterson and Hoffer (1994) evaluated insurance claims for 87 automobile models manufactured between 1989 and 1991, including some vehicles equipped with an air bag. They found that injury losses and collision losses are never mitigated and usually worsen significantly for air bag-equipped cars relative to cars equipped with only a seatbelt. While there are likely many factors contributing to this result, the authors concluded that this finding supports the Peltzman hypothesis in that driver behavior becomes more risky to compensate for a perceived improvement in vehicle safety, and that at-risk consumers prefer air bag vehicles.

Some researchers have found that regulatory attempts to improve automotive safety through product design can be at least partially offset by driver behavioral changes (e.g., Adams 1983, Evans 1991, Evans 1999). Adams (1985), for example, claimed that seat belt use laws induced an increase in risky behavior by drivers and an accompanying increase in risk to other road users (e.g., pedestrians and pedalcyclists). One example of risky behavior that may result from the presence of an air bag is a decrease in seat belt use. A tendency to not wear the seat belt in air bag-equipped cars may result from two phenomena. First, a driver may recognize that the combination of a seatbelt and an air bag provide the greatest level of protection but, under the principle of "risk homeostasis" (Evans 1991), he may be willing to forego the additional benefit of a seatbelt if he believes that an air bag provides a level of protection comparable to the level of protection he had previously had with a seatbelt. Second, the driver may erroneously assume that an air bag alone provides an adequate level of protection. To assess this behavior, Williams et al. (1990) performed an observational study of seatbelt use rates in 1,628 cars with air bags and 34,223 cars without. They found a slightly higher usage rate (66% vs. 63%) in the vehicles with air bags, and concluded that there was no evidence for the speculation that drivers with air bags will reduce their belt use. This finding is supported by national seatbelt use studies, which show a steady increase in seatbelt use rate despite the rapid penetration of air bags into the fleet (Shelton and Lago 1996). Another risky behavior that has been attributed to the introduction of air bags is increased aggressiveness. Hoffer et al. (1994) analyzed two independent data sets to test whether differences in driver behavior exist between cars equipped with air bags and those not so equipped. They found that relative injury claims increase following adoption of an air bag system, and that the increase does not diminish over time. They concluded that this result appears to be attributable to offsetting behavior. They determined that air bag-equipped cars, as described in Virginia State Police accident reports, tend to be driven more aggressively and that this increased aggressiveness appears to offset the effect of the air bag for the driver and increases the risk of death to others. The claim that increased risk-taking may completely offset the benefit of an air bag is not supported by studies that find an overall reduction in injuries for air bag-equipped vehicles, but a search of the literature reveals no studies that definitively quantify the role of offsetting behavior in decreasing the safety benefit of air bags.

5b. Side Air bags and Facebags

Due to higher seatbelt use rates in Australia and Europe, smaller frontal-impact "face air bags" are sometimes installed instead of the larger air bags common in the U.S. Fildes et al. (1992a) performed an economic analysis of full-size air bags as an alternative to the smaller facebags common in the high belt-use (94% in the front seat) Australian environment. Despite this high belt use rate, they found that full-size air bags had greater trauma reduction and a higher benefit/cost ratio (1.17) than facebags (0.58-0.98 depending on the assumptions made). In an extension of this study, Fildes et al. (1992b) estimated harm reduction for several different occupant protection systems, including a driver full-size air bag, a driver facebag, and a front-passenger air bag. They found that the introduction of full-size driver air bags into the entire Australian fleet would save $476 million (1992 AUS) annually, while the fleet-wide introduction of the driver facebag resulted in an overall savings of only $212 million. One reason given for a full-sized driver air bag is the frequent involvement of unbelted occupants in fatal crashes. Even today, the rate of unbelted occupant fatalities approaches 50% in European countries like Sweden and Germany, so the restraint potential of the full air bag has practical benefits in these cases. This also points to the atypical circumstances of fatal crashes, as the observed belt wearing rates exceed 90% in these countries.

Very few field-based studies of side air bag performance have been published. Langweider and Hummel (1998) presented a case of side air bag deployment, but it was unclear whether the air bag exacerbated or mitigated the resulting injuries. Baur et al. (2000) compared 14 crashes with a side air bag system with 10 similar crashes without the system. They found a trend of increasing low severity injuries and decreasing severe injuries with the air bag system, but could not draw any statistically significant conclusions. Dalmotas et al. (2001) discussed 24 case studies of crashes involving side air bags in Canada. They found no significant air bag-related injuries, and noted that the air bag may have reduced the severity of some injuries, but could not draw firm conclusions regarding the efficacy of side air bags or their potential for causing injury. One occupant with a side air bag sustained fatal injuries (not necessarily related to the air bag). Chidester and Roston (2001) presented the NHTSA SCI findings for 48 crashes involving side air bag deployments. This study identified one case of an AIS 3 thoracic injury sustained as a result of loading from the side air bag cover, and one case of fatal thoracic injuries sustained by an occupant with an air bag (again, not necessarily related to the air bag). Despite the air bag-related injury and the fatality, these authors stated that "in all cases the air bag provided an increase in occupant protection", but did not quantify the magnitude of this increase.

6. Conclusions

Quantifying all of the consequences of widespread implementation of air bags into the vehicle fleet is extremely difficult, with thousands of factors potentially confounding interpretation of the field data. Some of these confounders cannot be addressed by any statistical method. For example, laboratory tests verify that the initial position of the occupant has a substantial influence on the potential for injury and on the effectiveness of a restraint system, yet crashes in the field rarely provide sufficient evidence to estimate the initial position of the occupants. Individual variations in injury tolerance also cannot be estimated reliably, and gross indicators such as age and gender must be used. As the number of crashes involving air bags increases, it will eventually become possible to remove the effect of all non-systematic biases in the data. Unfortunately, there will be no way to know whether some factors introduce systematic bias regardless of the size of the dataset, and many metrics must be considered in the assessment of air bag performance.

This chapter has therefore presented a broad but not necessarily deep overview of numerous metrics describing the air bag's field experience (Table 14). We have described characteristics of air bags that must be considered good. For example, fatalities for drivers and passengers in frontal crashes have been reduced due to air bags, as have severe head injuries. We have also described characteristics that must be considered bad: children can be killed by an air bag and air bags cause more harm than they mitigate in as many as half of the situations in which they deploy. We must therefore conclude that there is room for improvement. One area in which large performance gains can be made is in improved discrimination between those crashes that require deployment and those that do not. Since the harm from air bags and the benefit from air bags occur at opposite ends of the collision severity spectrum, improvements may be made by increasing the deployment threshold. Both improved discrimination and higher deployment thresholds will require advances in sensing technology so that performance is not lost in crashes where an air bag is needed.

We must also conclude from this chapter that the performance of air bags is improving. Researchers are gaining the field experience necessary to define better the types of crashes that benefit from air bag deployment. The cases of children killed by passenger air bags have almost disappeared, as have adult fatalities from the driver air bag, without a perceivable reduction in the performance of air bags in severe crashes. The federal legislation and market impetus are in place for unprecedented understanding and optimization of air bag system performance.

7. References

AAAM (1990) The Abbreviated Injury Scale – 1990 Revision. Association for the Advancement of Automotive Medicine, Des Plaines, IL.

Adams, J. (1983) Public safety legislation and the risk compensation hypothesis; the example of motorcycle helmet legislation. Environment and Planning C: Government and Policy 1:193-203.

Adams, J. (1985) Smeed's law, seat belts, and the emperor's new clothes. In: Evans, L. Schwing, R. (eds.) Human Behavior and Traffic Safety, pp. 193-248. Plenum Press, New York, NY.

Aldman B., Andersson A., Saxmark O. (1974) Possible effects of air bag inflation on a standing child. Proc. IRCOBI Conference, Lyon, France.

Atkinson, P., Mclean, M., Telehowski, P., Khan, I., Martin, S., Hariharan, P., Atkinson, T. (2002) An analysis of recent accidents involving upper extremity fractures associated with air bag deployment. Paper 2002-01-0022, Society of Automotive Engineers, Warrendale, PA.

Augenstein, J., Perkeck, E., Williamson, J., Stratton, J., Digges, K., Lombardo, L. (1997) Air bag induced injury mechanisms for infants in rear facing child restraints. Paper 973296, Society of Automotive Engineers, Inc., Warrendale, PA.

Augenstein, J., Perdeck, E., Stratton, J., Digges, K., Steps, J. (2002) Performance of Depowered Air Bags in Real World Crashes. Paper 2002-01-0186, Society of Automotive Engineers, Warrendale, PA.

BMA (British Medical Association) (1987) Living with Risk. Wiley Medical Publication, John Wiley & Sons.

Backaitis, S. and Roberts, V. (1987) Occupant injury patterns in crashes with air bag equipped government sponsored cars. Paper 872216, Society of Automotive Engineers, Warrendale, PA.

Baker, S., O'Neill, B., Haddon, W. (1974) The Injury Severity Score: a method for describing patients with multiple injuries and evaluating emergency care. J. Trauma 14(3):187-196.

Baker, R.S., Flowers, C.W. Jr, Singh, P., Smith, A., Casey, R. (1996) Corneoscleral laceration caused by air-bag trauma. Am J Ophthalmol. 121(6):709-11.

Bandstra, R.A., Carbone, L.S. (2001) Unusual basal skull fracture in a vehicle equipped with an air bag. Am J Forensic Med Pathol. 22(3):253-5.

Barry, S., Ginpil, S., O'Neill, T. (1999) The effectiveness of air bags. Accident Analysis and Prevention, 31:781-7.

Table 14 - Summary of Restraint Performance in the Field

Lap/Shoulder Belt	Airbag	Airbag + Belt	Analysis Metric	Reference
31%	18%		Fatality effectiveness	Wilson and Savage (1973)
37%	57%	57%	Driver fatality effectiveness in frontal crashes	NHTSA (1974)
	25%, 28%		Fatality and injury effectiveness	Patrick (1975)
32%	25%		Fatality effectiveness	Huelke et al. (1979)
40%-50%	20%-40%	45%-55%	Fatality effectiveness	NHTSA (1984)*
42%±4%	18%±4%	46%±4%	Fatality effectiveness	Evans (1991)
	22%		Fatality reduction in frontal crashes (fleetwide)	Edwards (1994)
	23%		Fatality effectiveness	Lund and Ferguson (1995)
45%	11%	50%	Fatality effectiveness in all crashes	NHTSA (1999)
	19%-31%		Fatality effectiveness in frontal crashes	NHTSA (1999)
	15,600		Projected annual lives saved	NHTSA (1974)
	842		Total lives saved in 1997	NHTSA (1999)
9100			Total lives saved in 1997	Martin et al. (2000)
3.3	2.7		Mean ISS in a population of similar crashes	Mohan et al. (1976)
	6%, 18%		AIS 2+ and 3+ injury effectiveness of combined driver/passenger airbags	Pursel et al. (1978)
	-34%		AIS 2+ injury effectiveness of passenger airbags	Pursel et al. (1978)
64%	42%	64%	Serious injury effectiveness in all crashes	NHTSA (1999)
	8%-56%		Increase in repair costs associated with deployment	Langweider and Hummel (1998)
$30,000	$1.8 million		Cost per life saved (as used)	Levitt and Porter (2001)
	$300,000		Cost of passive restraints per life saved (note that the corresponding costs for asbestos and formaldehyde regulation were $132 million and $72 billion)	Morrall (1986)
	$24,000 $61,000		Cost per QALY saved for driver (top) and dual (bottom) systems	Lissy et al. (2000)
<$0			Cost per QALY saved (assuming 50% usage)	Graham (1975b)
	71.1%		Increase in harm to the face and upper extremities when an airbag deploys	Dalmotas et al. (1996)
	33%, 67%, 31% (belted), 84% (unbelted)		Various estimates of the increased fatality risk to children associated with a passenger airbag	Ferguson et al. (1996), Graham (1998b), Glass et al. (2001)
	232		Total number of airbag fatalities investigated by the NHTSA SCI group as of May 2001 (confirmed + unconfirmed)	Chidester and Roston (2001)

*see Thompson et al. (2002) for a discussion of NHTSA's 1984 estimates.

Baur, P., Lange, W., Messner, G., Rauscher, S., Pieske, O. (2000) Comparison of real world side impact/rollover collisions with and without thorax air bag/head protection system: a first field experience study. Proceedings of the Association for the Advancement of Automotive Medicine, pp. 187-201, Des Plaines, IL.

Blower, D. and Campbell, K. (1994) Comparison of occupant restraints based on injury-producing contact rates. Paper 942219, Society of Automotive Engineers, Inc., Warrendale, PA.

Brown, C.R. (1998) Dental injuries as a result of air bag deployment. Pract Periodontics Aesthet Dent. 10(7):856, 859.

Buckley, G., Setchfield, N., Frampton, R. (1999) Two case reports of possible noise trauma after inflation of air bags in low speed car crashes. BMJ 318:499-500.

Chidester, A. and Roston, T. (2001) Air bag crash investigations. Paper 246, Conference on the Enhanced Safety of Vehicles (ESV), National Highway Traffic Safety Administration, Washington, DC.

Crandall, J., Kuhlmann, T., Martin, P., Pilkey, W., Neeman, T. (1994) Differing patterns of head and facial injury with air bag and/or belt restrained drivers in frontal collisions. Proc. Advances in Occupant Restraint Technologies: Joint AAAM-IRCOBI Special Session, Lyon, France.

Cullinan, M., Merriman, T. (2001) Oesophageal rupture resulting from air bag deployment during a motor vehicle accident. ANZ J Surg. 71(9):554-5.

Cummings, P., McKnight, B., Rivara, F., Grossman, D. (2002) Association of driver air bags with driver fatality: a match cohort study. BMJ 324:1119-22.

Dalmotas, D., Hurley, R., German, A. (1996) Supplemental restraint systems: friend or foe to belted occupants? Proceedings of the Association for the Advancement of Automotive Medicine, pp. 63-75, Des Plaines, IL.

Dalmotas, D., German, A., Tylko, S. (2001) The crash and field performance of side-mounted air bag systems. Paper 442, Conference on the Enhanced Safety of Vehicles (ESV), National Highway Traffic Safety Administration, Washington, DC.

Duma, S., Kress, T., Porta, D., Woods, C., Snider, J., Fuller, P., Simmons, R. (1996) Air bag-induced eye injuries: a report of 25 cases. J. Trauma 41(1):114-119.

Durbin, D., Kallan, M., Elliott, M., Arbogast, K., Cornejo, R., Winston, F. (2002) Risk of injury to restrained children from passenger airbags. Proc. Annual Scientific Conference of the Association for the Advancement of Automotive Medicine, AAAM, Des Plaines, IL.

Edwards, W. R. (1994) An effectiveness analysis of Chrysler driver air bags after five years exposure. Paper 94-S4-O-09, 14th Conference on the Enhanced Safety of Vehicles (ESV), National Highway Traffic Safety Administration, Washington, DC.

Evans, L. (1986) The effectiveness of safety belts in preventing fatalities. Accident Analysis and Prevention 18:229-241.

Evans, L. (1988) Restraint effectiveness, occupant ejection from cars, and fatality reductions. General Motors Research Publication GMR 6398, General Motors Corporation, Warren, MI.

Evans, L. (1989) Air bag effectiveness in preventing fatalities predicted according to type of crash, driver age, and blood alcohol concentration. Proc. 33rd Annual Scientific Conference of the Association for the Advancement of Automotive Medicine, pp. 307-322, Des Plaines, IL.

Evans, L. (1991) Traffic Safety and the Driver. Van Nostrand Reinhold. New York, NY

Evans, L. (1999) Transportation Safety. In Handbook of Transportation Science, R.W. Hall Editor, Kluwer Academic Publishers, Norwell, MA.

Evans, L. (2001) Causal influence of car mass and size on driver fatality risk. American Journal of Public Health 91:1076-81.

Federal Register (1977) U.S. Federal Register, Vol. 42, No. 128, Part 571 – Federal Motor Vehicle Standards: Occupant protection systems, Docket No. 75-14, Notice 10, pages 34289-34305.

Ferguson, S., Braver, E., Greene, M., Lund, A. (1996) Initial estimates of reductions in deaths in frontal crashes among right-front passengers in vehicles equipped with passenger air bags. The Chronicle of ADTSEA, 44(4).

Fildes, B., Cameron, M., Vulcan, A., Digges, K., Taylor, D. (1992a) Air bag and facebag benefits and costs. Proceedings of the International Research Council on the Biomechanics of Impact (IRCOBI) Conference, pp. 323-335.

Fildes, B., Cameron, M., Vulcan, A., Digges, K. (1992b) Injury mitigation for a range of vehicle safety measures. Proceedings of the Association for the Advancement of Automotive Medicine, pp. 93-108, Des Plaines, IL.

Foret-Bruno, J-Y., Trosseille, X., Page, Y., Huere, J-F, Le Coz, J-Y., Bendjellal, F., Diboine, A., Phalempin, T., Villeforceix, D., Baudrit, P., Guillemot, H., Coltat, J-C (2001) Comparison of thoracic injury risk in frontal car crashes for occupants restrained without belt load limiters and those restrained with 6 kN and 4 kN belt load limiters. Stapp Car Crash Journal 45:205-224.

Frampton, R., Sferco, R., Welsh, R., Kirk, A., Fay, P. (2000) Effectiveness of air bag restraints in frontal crashes – what European studies tell us. Proceedings of the International Research Council on the Biomechanics of Impact (IRCOBI) Conference, pp. 425-438, Montpellier, France.

Garcia, R. (1994) Air bag implicated in temporomandibular joint injury. Cranio 12:125-127.

Gault, J.A., Vichnin, M.C., Jaeger, E.A., Jeffers, J.B. (1995) Ocular injuries associated with eyeglass wear and air bag inflation. J Trauma 38(4):494-7.

Glass, R.J., Segui-Gomez, M., Graham, J. (2000) Child passenger safety: decisions about seating location, air bag exposure, and restraint use. Risk Anal. 20(4):521-7.

Goldberg, M.A., Valluri, S., Pepose, J.S. (1995) Air bag-related corneal rupture after radial keratotomy. Am J Ophthalmol. 120(6):800-2.

Gotschall, C., Eichelberger, M., Morrissey, J., Better, A., Reardon, J., Bents, F. (1997) Nonfatal air bag deployments involving child passengers. Paper 973297, Society of Automotive Engineers, Inc., Warrendale, PA.

Gottesman, M., Sanderov, B., Ortiz, O. (2002) Carotid artery dissection and stroke caused by air bag injury. Am J Emerg Med. 20(4):372-4.

Graham, J. (1997a) Statement before the National Transportation Safety Board, Supplemental Restraint Panel, Washington, DC.

Graham, J. (1997b) Statement before the National Transportation Safety Board, Effectiveness Panel, Washington, DC.

Graham J., Corso, P., Morris, J., Segui-Gomez, M., Weinstein, M. (1998a) Evaluating the cost-effectiveness of clinical and public health measures. Annu. Rev. Public Health, 19:125-152.

Graham J., Goldie, S., Segui-Gomez, M., Thompson, K., Nelson, T., Glass, R., Simpson, A., Woerner, L. (1998b) Reducing risks to children in vehicles with passenger air bags. Pediatrics 102(1).

Gross, K., Haidar, A., Basha, M., Chan, T., Gwizdala, C., Wooley, R., Popovich, J. (1994) Acute pulmonary response of asthmatics to aerosols and gases generated by air bag deployment. Am J Respir Crit Care Med 150:408-14.

Gross, K., Koets, M., D'Arcy, J., Chan, T., Wooley, R., Basha, M. (1995) Mechanism of induction of asthmatic attacks initiated by the particles generated by air bag system deployment. J Trauma 38:521-7.

Han, D.P. (1993) Retinal detachment caused by air bag injury. Arch Ophthalmol. 111(10):1317-8.

Highway Loss Data Institute (1991) Driver injury experience in 1990 models equipped with air bags or automatic belts. Highway Loss Data Institute, Arlington, VA.

Hoffer, G., Millner, E., Peterson, S. (1994) Are drivers of air bag-equipped cars more aggressive?: a test of the Peltzman hypothesis. The Journal of Law and Economics 38: 251-264.

Hollands, C.M., Winston, F.K., Stafford, P.W., Shochat, S.J. (1996) Severe head injury caused by air bag deployment. J Trauma. 41(5):920-2.

Horsch, J.D., Culver, C.C. (1979) A study of driver interactions with an inflating air cushion. Paper No. 791029, 23rd Stapp Car Crash Conference, P-82, Society of Automotive Engineers, Inc., Warrendale, Pennsylvania, USA.

Horsch, J.D., Lau, I.V., Andrzejak, D.A., Viano, D.C., et al. (1990) Assessment of air bag deployment loads. Paper No. 902324, 34th Stapp Car Crash Conference, P-236, Society of Automotive Engineers, Inc., Warrendale, Pennsylvania, USA.

Huelke, D., Sherman, H., Murphy, M. (1979) Effectiveness of current and future restraint systems in fatal and serious injury automobile crashes, Paper 790323, Society of Automotive Engineers, Warrendale, PA.

Huelke, D., Compton, C., Studer, R. (1985) Injury severity, ejection, and occupant contacts in passenger car rollover crashes. Paper 850336, Society of Automotive Engineers, Warrendale, PA.

Huelke, D. (1995) An overview of air bag deployments and related injuries. Case studies and a review of the literature. Paper 950866, Society of Automotive Engineers, Warrendale, PA.

Huelke, D., Gilbert, R., Schneider, L. (1997) Upper-extremity injuries from steering wheel air bag deployments. Paper 970493, Society of Automotive Engineers, Warrendale, PA.

Kent, R., Crandall, J., Bolton, J., Duma, S. (2000) Driver and right-front passenger restraint system interaction, injury potential, and thoracic injury prediction. Proceedings of the Association for the Advancement of Automotive Medicine, pp. 261-282, Des Plaines, IL.

Kullgren, A., Krafft, M., Malm, S., Ydenius, A., Tingvall, C. (2000) Influence of airbags and seatbelt pretensioners on AIS1 neck injuries for belted occupants in frontal impacts. Stapp Car Crash Journal 44:117-125.

Langweider, K. and Hummel, T. (1998) The effect of air bags on injuries and accident costs. Paper 98-S1-W-27, 16th Conference on the Enhanced Safety of Vehicles (ESV), National Highway Traffic Safety Administration, Washington, DC.

Lau, V., Horsch, J.D., Viano, D.C., Andrzejak, D.V. (1993) Mechanism of injury from air bag deployment loads. Accident Analysis & Prevention, 25(1):29-45.

Lau, E., Ray, R., Cheng, L. (1997) Deployment of air bags in traffic accidents: characteristics and consequences. Proceedings of the Association for the Advancement of Automotive Medicine, pp. 371-385, Des Plaines, IL.

Levitt, S. and Porter, J. (2001) Sample selection in the estimation of air bag and seat belt effectiveness. The Review of Economics and Statistics, 83(4):603-15.

Levy, Y., Hasson, O., Zeltser, R., Nahlieli, O. (1998) Temporomandibular joint derangement after air bag deployment: report of two cases. J Oral Maxillofac Surg. 56(8):1000-3.

Lissy, K., Cohen, J., Park, M., Graham, J. (2000) Cellular phones and driving: weighing the risks and benefits. Risk in Perspective 8(6), Harvard Center for Risk Analysis, Cambridge, MA.

Lund, A. and Ferguson, S. (1995) Driver fatalities in 1985-1993 cars with air bags. The Journal of Trauma: Injury, Infection, and Critical Care, 38(4):469-475.

Malliaris, A., Hitchcock, R., Hansen, M. (1985) Harm causation and ranking in car crashes. Paper 850090, Society of Automotive Engineers, Warrendale, PA.

Malliaris, A., Digges, K., DeBlois, J. (1995a) Evaluation of air bag field performance. Paper 950869, Society of Automotive Engineers, Warrendale, PA.

Malliaris, A., Digges, K., DeBlois, J. (1995b) Injury patterns of car occupants under air bag deployment. Paper 950867, Society of Automotive Engineers, Warrendale, PA.

Malliaris, A., Digges, K., DeBlois, J. (1996) Air bag field performance and injury patterns. Paper 960659, Society of Automotive Engineers, Warrendale, PA.

Malliaris, A. and DeBlois, J. (1998) Overstated safety belt use rates – evidence, consequences, and remedies. Paper 980351, Society of Automotive Engineers, Warrendale, PA.

Martin, P., Crandall, J., Pilkey, W. (2000) Injury trends of passenger car drivers in frontal crashes in the USA. Accident Analysis and Prevention 32:541-557.

Mcfeely, W.J. Jr, Bojrab, D.I., Davis, K.G., Hegyi, D.F. (1999) Otologic injuries caused by air bag deployment. Otolaryngol Head Neck Surg. 121(4):367-73.

McKay, M., Fitzharris, M., Fildes, B. (1999) Driver injury patterns in the United States and Australia: does belt wearing or air bag deployment make a difference? Proceedings of the Association for the Advancement of Automotive Medicine, pp. 409-424, Des Plaines, IL.

Mertz, H. and Weber, D. (1982) Interpretations of the impact responses of a three-year-old child dummy relative to child injury potential. Proc. 9th International Technical Conference on Experimental Safety Vehicles (ESV), Kyoto, Japan.

Mertz, H., Driscoll, G., Lenox, J., Nyquist, G., Weber, D. (1982) Responses of animals exposed to deployment of various passenger inflatable restraint system concepts for a variety of collision severities and animal positions. Proc. 9th International Technical Conference on Experimental Safety Vehicles (ESV), Kyoto, Japan.

Mertz, H. (1988) Restraint performance of the 1973-76 GM air cushion restraint system. Paper 880400, Society of Automotive Engineers, Warrendale, PA.

Miller, H. (1995) Injury reduction with smart restraint systems. 39th Annual Proceedings of the Association for the Advancement of Automotive Medicine, pp. 527-541, Des Plaines, IL.

Mirnazari, M. and Henning, N. (1999) Air bags and their impact on collision insurance losses. Paper 1999-01-1063, Society of Automotive Engineers, Warrendale, PA.

Mohan, D., Zador, P., O'Neill, B., Ginsburg, M. (1976) Air bags and lap/shoulder belts – a comparison of their effectiveness in real world, frontal crashes. Proceedings of the Association for the Advancement of Automotive Medicine, Des Plaines, IL.

Montalvo, F., Bryant, R.W., Mertz, H.J. (1982) Possible positions and postures of unrestrained front-seat children at instant of collision. Paper 826045, Proc. 9th International Technical Conference on Experimental Safety Vehicles (ESV), Kyoto, Japan.

Morrall J.F. (1986) A Review of the Record. Regulation 25-34, Noember/December.

Mouzakes J, Koltai PJ, Kuhar S, Bernstein DS, Wing P, Salsberg E. (2001) The impact of air bags and seat belts on the incidence and severity of maxillofacial injuries in automobile accidents in New York State. Arch Otolaryngol Head Neck Surg.: 127(10):1189-93.

NHTSA (1974) Analysis of effects of proposed changes to passenger car requirements of MVSS 208. National Highway Traffic Safety Administration, U.S. Department of Transportation, Washington DC

NHTSA (1984) FMVSS 208 regulatory impact analysis. National Highway Traffic Safety Administration, U.S. Department of Transportation, Washington DC

NHTSA (1999) Fourth report to Congress – effectiveness of occupant protection systems and their use. National Highway Traffic Safety Administration, U.S. Department of Transportation, Washington DC.

O'Halloran, H.S., Draud, K., Stevens, J.L. (1998) Primary enucleation as a consequence of air bag injury. J Trauma 44(6):1090.

O'Neill, B. (2000) Air bag test requirements under proposed new rule. Statement before the Transportation Subcommittee, U.S. House of Representatives Appropriations Committee, Washington DC.

Otte, D. (1995) Review of the air bag effectiveness in real life accident demands – for positioning and optimal deployment of air bag systems [sic]. Paper 952701, Society of Automotive Engineers, Warrendale, PA.

Parenteau, C., Shah, M., Desai, T., Frampton, R. (1999) US and UK belted driver injuries with and without air bag deployments – a field data analysis. Paper 1999-01-0633, Society of Automotive Engineers, Warrendale, PA.

Patrick, L.M., Nyquist, G.W. (1972) Air bag effects on the out-of-position child. Paper No. 720442, Society of Automotive Engineers, Warrendale, PA.

Patrick, L.M. (1975) Passive and active restraint systems – performance and benefit/cost comparison. Paper 750389, Society of Automotive Engineers, Warrendale, PA.

Pearlman, J.A., Au Eong, K.G., Kuhn, F., Pieramici, D.J. (2001) Air bags and eye injuries: epidemiology, spectrum of injury, and analysis of risk factors. Surv Ophthalmol 46(3):234-42.

Peltzman, S. (1975) The effects of automobile safety regulation. J. of Political Economy 83:677-725.

Perez, J., Palmatier, T. (1996) Air bag-related fatality in a short, forward-positioned driver. Ann Emerg Med. 28(6):722-4.

Peterson, S. and Hoffer,G. (1994) The impact of air bag adoption on relative personal injury and absolute collision insurance claims. J. of Consumer Research 20:657-662.

Phen, R., Dowdy, M., Ebbeler, D., Kim, E-H, Moore, N., VanZandt, T. (1998) Advanced Air Bag Technology Assessment, Final Report, prepared for NHTSA and NASA by the Jet Propulsion Laboratory, Jet Propulsion Laboratory, California Institute of Technology, Pasadena, California.

Polk, J. and Thomas, H. (1994) Automotive air bag-induced second-degree chemical burn resulting in *Staphylococcus aureus* infection. JAOA 94(9):741-743.

Powell, M. and Lund, A. (1995) Leading edge deployment speed of production air bags. Paper 950870, Society of Automotive Engineers, Inc., Warrendale, PA.

Pursel, H., Bryant, R., Scheel, J., Yanik, J. (1978) Matching case methodology for measuring restraint effectiveness. Paper 780415, Society of Automotive Engineers, Warrendale, PA.

Reed, M.P., Schneider, L.W., Burney, R.E. (1994) Laboratory investigations and mathematical modeling of air bag- induced skin burns. SAE 942217 38th Stapp Car Crash Conference, Society of Automotive Engineers, Inc., Warrendale, PA.

Richter, H., Stalnaker, R., Pugh, J. (1974) Otologic hazards of airbag restraint systems. Paper 741185, Society of Automotive Engineers, Inc., Warrendale, PA.

Roccia, F., Servadio, F., Gerbino, G. (1999) Maxillofacial fractures following air bag deployment. J Craniomaxillofac Surg. 27(6):335-8.

Rouhana, S.W., Webb, S.R., Wooley, R.G., McCleary, J.D., et al. (1994) Investigation into the noise associated with air bag deployment: Part I – Measurement technique and parameter study. 38th Stapp Car Crash Conference, Paper No. 942218, Society of Automotive Engineers, Inc., Warrendale, Pennsylvania, USA.

Rouhana, S.W., Webb, S.R., Dunn, V.C. (1998) Investigation into the noise associated with air bag deployment: Part II - Injury risk study using a mathematical model of the human ear. Paper No. 983162, 42nd Stapp Car Crash Conference, P-337, Society of Automotive Engineers, Inc., Warrendale, Pennsylvania, USA.

Ruiz-Moreno, J.M. (1998) Air bag-associated retinal tear. Eur J Ophthalmol. 8(1):52-3.

Schubert, P. (2001) Robust automated airbag module calibration. Paper 2001-01-0159, Society of Automotive Engineers, Inc., Warrendale, Pennsylvania, USA.

Segui-Gomez, M. (2000) Driver air bag effectiveness by severity of the crash. American Journal of Public Health 90:1575-1581.

Shelton, T. and Lago, J. (1996) National occupant protection use survey. Fifteenth International Technical Conference on the Enhanced Safety of Vehicles (ESV), National Highway Traffic Safety Administration, Washington, DC.

Singh, G. and Song, H. (2002) Intelligent algorithms for early detection of automotive crashes. Paper 2002-01-0190, Society of Automotive Engineers, Warrendale, PA.

Smally, A.J., Binzer, A., Dolin, S., Viano, D. (1992) Alkaline chemical keratitis: eye injury from air bags. Ann Emerg Med. 21(11):1400-2.

Smith, G., Hurite, S., Yanik, A. (1972) Human volunteer testing of GM air cushions. Paper No. 720443, Society of Automotive Engineers, Warrendale, PA.

Smith, G. (1973) Field testing of 1000 air cushion equipped vehicles. Proceedings of the Association for the Advancement of Automotive Medicine, pp. 443-464, Des Plaines, IL.

Smith, G. (1977) Air bag update – recent crash case histories. Paper 770155, Society of Automotive Engineers, Warrendale, PA.

Smith, R. and Moffatt, C. (1975) Accident experience in air bag-equipped cars. Proceedings of the Association for the Advancement of Automotive Medicine, pp. 60-79, Des Plaines, IL.

Stalnaker, R.L., Klusmeyer, L.F., Peel, H.H., White, C.D. et al. (1982) Unrestrained, front seat, child surrogate trajectories produced by hard braking. Proc. 26th Stapp Car Crash Conference, SAE paper 821165, Society of Automotive Engineers, Warrendale, PA.

Struble, D. (1998) Air bag technology: what it is and how it came to be. Paper 980648, Society of Automotive Engineers, Warrendale, PA.

Stucki, S. and Biss, D. (2000) A study of the NASS-CDS system for injury/fatality rates of occupants in various restraints and a discussion of alternative presentation methods. Proceedings of the Association for the Advancement of Automotive Medicine, pp. 93-114, Des Plaines, IL.

Theisen, M., Eisele, S., Rolleke, M. (2002) The computation of airbag deployment times with the help of precrash information. Paper 2002-01-0192, Society of Automotive Engineers, Warrendale, PA.

Thompson, K., Segui-Gomez, M., Graham, J. (2002) Validating benefit and cost estimates: the case of airbag regulation. Risk Analysis 22(4):803-11.

Van der Linden, W.J. (2002) Dislocated fracture of the mandibular condylar process after air bag deployment: report of a case. J Oral Maxillofac Surg. 60(1):113-5.

Viano, D. (1988) Limits and challenges of crash protection. Accident Analysis and Prevention 20(6):421-429.

Viano, D. (1991) Effectiveness of safety belts and air bags in preventing fatal injury. Paper 910901, Society of Automotive Engineers, Warrendale, PA.

Viano, D. (1995) Restraint effectiveness, availability and use in frontal crashes: implications to injury control. J. Trauma 38(4):538-546.

Walz, F.H., Mackay, M., Gloor, B. (1995) Air bag deployment and eye perforation by a tobacco pipe. J Trauma 38(4):498-501.

Warner, C. (1979) Passive protection: promulgation or politics. Proceedings of the Association for the Advancement of Automotive Medicine, pp. 135-148, Des Plaines, IL.

Werner, J. and Sorenson, W. (1994) Survey of air bag involved accidents an analysis of collision characteristics, system effectiveness and injuries. Paper 940802, Society of Automotive Engineers, Warrendale, PA.

Werner, J., Roberson, S., Ferguson, S., Digges, K. (1996) Air bag deployment frequency and injury risks. Paper 960664, Society of Automotive Engineers, Warrendale, PA.

Werner, J., Roberson, S., Ferguson, S., Digges, K. (1997) Injury risks in cars with different air bag deployment rates. Paper 970491, Society of Automotive Engineers, Warrendale, PA.

Williams, A., Wells, J., Lund, A. (1990) Seat belt use in cars with air bags. American Journal of Public Health 80(12):1514-1516.

Wilson, R. and Savage, C. (1973) Restraint system effectiveness – a study of fatal accidents. Proceedings of Automotive Safety Engineering Seminar, General Motors Corporation, Warren, MI.

Yaremchuk, K. and Dobie, R. (1994) Otologic injuries from air bag deployment. Otolaryngology – Head and Neck Surgery 125(3):130-134.

Zador, P. and Ciccone, M. (1991) Driver fatalities in frontal impacts: comparison between cars with air bags and manual lap belts. Insurance Institute for Highway Safety, Arlington, VA.

Zhou, Q., Rouhana, S., Melvin, J. (1996) Age effects on thoracic injury tolerance. Paper number 962421, Stapp Car Crash Conference, Society of Automotive Engineers, Warrendale, PA.

Zuby, D. and O'Neill, B. (2001) Steering column movement in severe frontal crashes and its potential effect on airbag performance. Paper 136, Proceedings of the International Conference on the Enhanced Safety of Vehicles (ESV), National Highway Traffic Safety Administration, U.S. Department of Transportation, Washington, DC.

Zuby, D., Ferguson, S., Cammisa, M. (2001) Analysis of driver fatalities in frontal crashes of air bag-equipped vehicles in 1990-98 NASS/CDS. Paper 2001-01-0156, Society of Automotive Engineers, Warrendale, PA.

II. FEDERAL AIR BAG LEGISLATION

Public Law 89-564
89th Congress, S. 3052
September 9, 1966

An Act

80 STAT. 731

To provide for a coordinated national highway safety program through financial assistance to the States to accelerate highway traffic safety programs, and for other purposes.

Be it enacted by the Senate and House of Representatives of the United States of America in Congress assembled,

TITLE I—HIGHWAY SAFETY

Highway Safety Act of 1966.

SEC. 101. Title 23, United States Code, is hereby amended by adding at the end thereof a new chapter: 72 Stat. 885.

"Chapter 4.—HIGHWAY SAFETY

"Sec.
"401. Authority of the Secretary.
"402. Highway safety programs.
"403. Highway safety research and development.
"404. National Highway Safety Advisory Committee.

"§ 401. Authority of the Secretary

"The Secretary is authorized and directed to assist and cooperate with other Federal departments and agencies, State and local governments, private industry, and other interested parties, to increase highway safety.

"§ 402. Highway safety programs

"(a) Each State shall have a highway safety program approved by the Secretary, designed to reduce traffic accidents and deaths, injuries, and property damage resulting therefrom. Such programs shall be in accordance with uniform standards promulgated by the Secretary. Such uniform standards shall be expressed in terms of performance criteria. Such uniform standards shall be promulgated by the Secretary so as to improve driver performance (including, but not limited to, driver education, driver testing to determine proficiency to operate motor vehicles, driver examinations (both physical and mental) and driver licensing) and to improve pedestrian performance. In addition such uniform standards shall include, but not be limited to, provisions for an effective record system of accidents (including injuries and deaths resulting therefrom), accident investigations to determine the probable causes of accidents, injuries, and deaths, vehicle registration, operation, and inspection, highway design and maintenance (including lighting, markings, and surface treatment), traffic control, vehicle codes and laws, surveillance of traffic for detection and correction of high or potentially high accident locations, and emergency services. Such standards as are applicable to State highway safety programs shall, to the extent determined appropriate by the Secretary, be applicable to federally administered areas where a Federal department or agency controls the highways or supervises traffic operations. The Secretary shall be authorized to amend or waive standards on a temporary basis for the purpose of evaluating new or different highway safety programs instituted on an experimental, pilot, or demonstration basis by one or more States, where the Secretary finds that the public interest would be served by such amendment or waiver.

Uniform standards.

"(b)(1) The Secretary shall not approve any State highway safety program under this section which does not—

"(A) provide that the Governor of the State shall be responsible for the administration of the program.

"(B) authorize political subdivisions of such State to carry out local highway safety programs within their jurisdictions as a

Pub. Law 89-564   - 2 -   **September 9, 1966**
80 STAT. 732

part of the State highway safety program if such local highway safety programs are approved by the Governor and are in accordance with the uniform standards of the Secretary promulgated under this section.

"(C) provide that at least 40 per centum of all Federal funds apportioned under this section to such State for any fiscal year will be expended by the political subdivisions of such State in carrying out local highway safety programs authorized in accordance with subparagraph (B) of this paragraph.

"(D) provide that the aggregate expenditure of funds of the State and political subdivisions thereof, exclusive of Federal funds, for highway safety programs will be maintained at a level which does not fall below the average level of such expenditures for its last two full fiscal years preceding the date of enactment of this section.

"(E) provide for comprehensive driver training programs, including (1) the initiation of a State program for driver education in the school systems or for a significant expansion and improvement of such a program already in existence, to be administered by appropriate school officials under the supervision of the Governor as set forth in subparagraph (A) of this paragraph; (2) the training of qualified school instructors and their certification; (3) appropriate regulation of other driver training schools, including licensing of the schools and certification of their instructors; (4) adult driver training programs, and programs for the retraining of selected drivers; and (5) adequate research, development and procurement of practice driving facilities, simulators, and other similar teaching aids for both school and other driver training use.

"(2) The Secretary is authorized to waive the requirement of subparagraph (C) of paragraph (1) of this subsection, in whole or in part, for a fiscal year for any State whenever he determines that there is an insufficient number of local highway safety programs to justify the expenditure in such State of such percentage of Federal funds during such fiscal year.

Use of funds.

"(c) Funds authorized to be appropriated to carry out this section shall be used to aid the States to conduct the highway safety programs approved in accordance with subsection (a), shall be subject to a deduction not to exceed 5 per centum for the necessary costs of administering the provisions of this section, and the remainder shall be apportioned among the several States. For the fiscal years ending June 30, 1967, June 30, 1968, and June 30, 1969, such funds shall be apportioned 75 per centum on the basis of population and 25 per centum as the Secretary in his administrative discretion may deem appropriate and thereafter such funds shall be apportioned as Congress, by law enacted hereafter, shall provide. On or before January 1, 1969, the Secretary shall report to Congress his recommendations with respect to a nondiscretionary formula for apportionment of funds authorized to carry out this section for the fiscal year ending June 30, 1970, and fiscal years thereafter. After December 31, 1968, the Secretary shall not apportion any funds under this subsection to any State which is not implementing a highway safety program approved by the Secretary in accordance with this section. Federal aid highway funds apportioned on or after January 1, 1969, to any State which is not implementing a highway safety program approved by the Secretary in accordance with this section shall be reduced by amounts equal to 10 per centum of the amounts which would otherwise be apportioned to such State under section 104 of this title, until

150

September 9, 1966   - 3 -   Pub. Law 89-564
80 STAT. 733

such time as such State is implementing an approved highway safety program. Whenever he determines it to be in the public interest, the Secretary may suspend, for such periods as he deems necessary, the application of the preceding sentence to a State. Any amount which is withheld from apportionment to any State under this section shall be reapportioned to the other States in accordance with the applicable provisions of law.

"(d) All provisions of chapter 1 of this title that are applicable to Federal-aid primary highway funds other than provisions relating to the apportionment formula and provisions limiting the expenditure of such funds to the Federal-aid systems, shall apply to the highway safety funds authorized to be appropriated to carry out this section, except as determined by the Secretary to be inconsistent with this section. In applying such provisions of chapter 1 in carrying out this section the term 'State highway department' as used in such provisions shall mean the Governor of a State for the purposes of this section.

"(e) Uniform standards promulgated by the Secretary to carry out this section shall be developed in cooperation with the States, their political subdivisions, appropriate Federal departments and agencies, and such other public and private organizations as the Secretary deems appropriate.

"(f) The Secretary may make arrangements with other Federal departments and agencies for assistance in the preparation of uniform standards for the highway safety programs contemplated by subsection (a) and in the administration of such programs. Such departments and agencies are directed to cooperate in such preparation and administration, on a reimbursable basis.

"(g) Nothing in this section authorizes the appropriation or expenditure of funds for (1) highway construction, maintenance, or design (other than design of safety features of highways to be incorporated into standards) or (2) any purpose for which funds are authorized by section 403 of this title.

"§ 403. Highway safety research and development

"The Secretary is authorized to use funds appropriated to carry out this section to carry out safety research which he is authorized to conduct by subsection (a) of section 307 of this title. In addition, the Secretary may use the funds appropriated to carry out this section, either independently or in cooperation with other Federal departments or agencies, for (1) grants to State or local agencies, institutions, and individuals for training or education of highway safety personnel, (2) research fellowships in highway safety, (3) development of improved accident investigation procedures, (4) emergency service plans, (5) demonstration projects, and (6) related activities which are deemed by the Secretary to be necessary to carry out the purposes of this section.

"§ 404. National Highway Safety Advisory Committee

"(a) (1) There is established in the Department of Commerce a National Highway Safety Advisory Committee, composed of the Secretary or an officer of the Department appointed by him, who shall be chairman, the Federal Highway Administrator, and twenty-nine members appointed by the President, no more than four of whom shall be Federal officers or employees. The appointed members, having due regard for the purposes of this chapter, shall be selected from among representatives of various State and local governments, including State legislatures, of public and private interests contributing to,

Pub. Law 89-564   **- 4 -**   **September 9, 1966**

80 STAT. 734

affected by, or concerned with highway safety, and of other public and private agencies, organizations, or groups demonstrating an active interest in highway safety, as well as research scientists and other individuals who are expert in this field.

"(2)(A) Each member appointed by the President shall hold office for a term of three years, except that (i) any member appointed to fill a vacancy occurring prior to the expiration of the term for which his predecessor was appointed shall be appointed for the remainder of such term, and (ii) the terms of office of members first taking office after the date of enactment of this section shall expire as follows: ten at the end of one year after such date, ten at the end of two years after such date, and nine at the end of three years after such date, as designated by the President at the time of appointment, and (iii) the term of any member shall be extended until the date on which the successor's appointment is effective. None of the members appointed by the President other than Federal officers or employees shall be eligible for reappointment within one year following the end of his preceding term.

Pay.

"(B) Members of the Committee who are not officers or employees of the United States shall, while attending meetings or conferences of such Committee or otherwise engaged in the business of such Committee, be entitled to receive compensation at a rate fixed by the Secretary, but not exceeding $100 per diem, including traveltime, and while away from their homes or regular places of business they may be allowed travel expenses, including per diem in lieu of subsistence, as authorized in section 5 of the Administrative Expenses Act of 1946 (5 U.S.C. 73b-2) for persons in the Government service employed intermittently. Payments under this section shall not render members of the Committee employees or officials of the United States for any purpose.

60 Stat. 808;
75 Stat. 339,
340.

"(b) The National Highway Safety Advisory Committee shall advise, consult with, and make recommendations to, the Secretary on matters relating to the activities and functions of the Department in the field of highway safety. The Committee is authorized (1) to review research projects or programs submitted to or recommended by it in the field of highway safety and recommend to the Secretary, for prosecution under this title, any such projects which it believes show promise of making valuable contributions to human knowledge with respect to the cause and prevention of highway accidents; and (2) to review, prior to issuance, standards proposed to be issued by order of the Secretary under the provisions of section 402(a) of this title and to make recommendations thereon. Such recommendations shall be published in connection with the Secretary's determination or order.

"(c) The National Highway Safety Advisory Committee shall meet from time to time as the Secretary shall direct, but at least once each year.

"(d) The Secretary shall provide to the National Highway Safety Committee from among the personnel and facilities of the Department of Commerce such staff and facilities as are necessary to carry out the functions of such Committee."

Repeal.

SEC. 102. (a) Sections 135 and 313 of title 23 of the United States Code are hereby repealed.

(b)(1) The analysis of chapter 1 of title 23, United States Code, is hereby amended by deleting:

"135. Highway safety programs."

152

September 9, 1966　　- 5 -　　Pub. Law 89-564

80 STAT. 735

(2) The analysis of chapter 3 of title 23, United States Code, is hereby amended by deleting:

"313. Highway safety conference."

(3) There is hereby added at the end of the table of chapters at the beginning of title 23, United States Code, the following:

"4. Highway safety_____ 401".

SEC. 103. Section 307 of title 23, United States Code, is amended (1) by inserting in subsection (a) thereof immediately after "section 104 of this title" the following: ", funds authorized to carry out section 403 of this title," and (2) by adding at the end of such section the following new subsection:

"(d) As used in this section the term 'safety' includes, but is not limited to, highway safety systems, research, and development relating to vehicle, highway, and driver characteristics, accident investigations, communications, emergency medical care, and transportation of the injured." "Safety."

SEC. 104. For the purpose of carrying out section 402 of title 23, United States Code, there is hereby authorized to be appropriated the sum of $67,000,000 for the fiscal year ending June 30, 1967; $100,000,000 for the fiscal year ending June 30, 1968; and $100,000,000 for the fiscal year ending June 30, 1969. Appropriations.

SEC. 105. For the purpose of carrying out sections 307(a) and 403 of title 23, United States Code, there is hereby authorized to be appropriated the additional sum of $10,000,000 for the fiscal year ending June 30, 1967; $20,000,000 for the fiscal year ending June 30, 1968; and $25,000,000 for the fiscal year ending June 30, 1969.

SEC. 106. All facts contained in any report of any Federal department or agency or any officer, employee, or agent thereof, relating to any highway traffic accident or the investigation thereof conducted pursuant to chapter 4 of title 23 of the United States Code shall be available for use in any civil, criminal, or other judicial proceeding arising out of such accident, and any such officer, employee, or agent may be required to testify in such proceedings as to the facts developed in such investigation. Any such report shall be made available to the public in a manner which does not identify individuals. All completed reports on research projects, demonstration projects, and other related activities conducted under sections 307 and 403 of title 23, United States Code, shall be made available to the public in a manner which does not identify individuals. Reports of highway traffic accidents.

TITLE II—ADMINISTRATION AND REPORTING

SEC. 201. The Secretary shall carry out the provisions of the Highway Safety Act of 1966 (including chapter 4 of title 23 of the United States Code) through a National Highway Safety Agency (hereinafter referred to as the "Agency"), which he shall establish in the Department of Commerce. The Agency shall be headed by an Administrator who shall be appointed by the President, by and with the advice and consent of the Senate, who shall be compensated at the rate prescribed for level V of the Federal Executive Salary Schedule established by the Federal Executive Salary Act of 1964. The Administrator shall be a citizen of the United States, and shall be appointed with due regard for his fitness to discharge efficiently the powers and the duties delegated to him. The Administrator shall have National Highway Safety Agency.

78 Stat. 419.
5 USC 2211.

153

Pub. Law 89-564   **- 6 -**   **September 9, 1966**

80 STAT. 736

no pecuniary interest in or own any stock in or bonds of any enterprise involved in (1) manufacturing motor vehicles or motor vehicle equipment, or (2) constructing highways, nor shall he engage in any other business, vocation, or employment. The Administrator shall perform such duties as are delegated to him by the Secretary. On highway matters the Administrator shall consult with the Federal Highway Administrator. The President is authorized to carry out the provisions of the National Traffic and Motor Vehicle Safety Act of 1966 through the Agency and Administrator authorized by this section.

Ante, p. 718.

Report to President and Congress.

SEC. 202. (a) The Secretary shall prepare and submit to the President for transmittal to the Congress on March 1 of each year a comprehensive report on the administration of the Highway Safety Act of 1966 (including chapter 4 of title 23 of the United States Code) for the preceding calendar year. Such report should include but not be restricted to (1) a thorough statistical compilation of the accidents and injuries occurring in such year; (2) a list of all safety standards issued or in effect in such year; (3) the scope of observance of applicable Federal standards; (4) a statement of enforcement actions including judicial decisions, settlements, or pending litigation during the year; (5) a summary of all current research grants and contracts together with a description of the problems to be considered by such grants and contracts; (6) an analysis and evaluation of completed research activities and technological progress achieved during such year together with the relevant policy recommendations flowing therefrom; (7) the effectiveness of State highway safety programs (including local highway safety programs) and (8) the extent to which technical information was being disseminated to the scientific community and consumer-oriented material was made available to the motoring public.

Recommendations for additional legislation.

(b) The annual report shall also contain such recommendations for additional legislation as the Secretary deems necessary to promote cooperation among the several States in the improvement of highway safety and to strengthen the national highway safety program.

Report to Congress.

SEC. 203. The Secretary of Commerce shall report to Congress, not later than July 1, 1967, all standards to be initially applied in carrying out section 402 of title 23 of the United States Code.

Effect of alcohol on highway safety and drivers, study.

SEC. 204. The Secretary of Commerce shall make a thorough and complete study of the relationship between the consumption of alcohol and its effect upon highway safety and drivers of motor vehicles, in consultation with such other government and private agencies as may be necessary. Such study shall cover review and evaluation of State and local laws and enforcement methods and procedures relating to driving under the influence of alcohol, State and local programs for the treatment of alcoholism, and such other aspects of this overall problem as may be useful. The results of this study shall be reported to the Congress by the Secretary on or before July 1, 1967, and shall include recommendations for legislation if warranted.

SEC. 205. The Federal Highway Administrator and any other officer who may subsequent to the date of enactment of this Act become the operating head of the Bureau of Public Roads shall receive compensation at the rate prescribed for level IV of the Federal Executive Salary Schedule established by the Federal Executive Salary Act of 1964.

78 Stat. 419.
5 USC 2211.

SEC. 206. Section 105 of title 23, United States Code, is hereby amended by adding the following subsection at the end thereof:

Priority projects.

"(e) In approving programs for projects on the Federal-aid systems pursuant to chapter 1 of this title, the Secretary shall give priority to those projects which incorporate improved standards and features with safety benefits."

154

September 9, 1966   - 7 -   Pub. Law 89-564
                                          80 STAT. 737

SEC. 207. In order to provide the basis for evaluating the continu- Estimate of
ing programs authorized by this Act, and to furnish the Congress with cost.
the information necessary for authorization of appropriations for
fiscal years beginning after June 30, 1969, the Secretary, in cooperation
with the Governors or the appropriate State highway safety agencies,
shall make a detailed estimate of the cost of carrying out the provisions
of this Act. The Secretary shall submit such detailed estimate and
recommendations for Federal, State, and local matching funds to the
Congress not later than January 10, 1968.

SEC. 208. This Act may be cited as the "Highway Safety Act of Short title.
1966".

Approved September 9, 1966, 1:11 p.m.

LEGISLATIVE HISTORY:

HOUSE REPORTS: No. 1700 accompanying H. R. 13290 (Comm. on
               Public Works) and No. 1920 (Comm. of Conference).
SENATE REPORT No. 1302 (Comm. on Public Works).
CONGRESSIONAL RECORD, Vol. 112 (1966):
    June 24: Considered and passed Senate.
    June 27: Reconsidered and passed Senate.
    Aug. 18: Considered and passed House, amended, in lieu of
             H. R. 13290.
    Aug. 31: House agreed to conference report.
    Sept. 1: Senate adopted conference report.

Analysis of Effects of Proposed Changes to
Passenger Car Requirements of MVSS 208

Prepared by:
Motor Vehicle Programs
National Highway Traffic Safety Administration
And Transportation Systems Center

U.S. Department of Transportation

August, 1974

Abstract

This report analyzes the impact of proposed changes to Motor Vehicle Safety Standard 208 "Occupant Crash Protection" as they apply to passenger cars (Figure 1). The proposal of March 19, 1974, would extend the present 3 options including interlock-belt systems to September 1, 1976, and thereafter would require passive protection supplemented by lap belts with sequential warning if needed for rollover protection in front seats. Lap belts would be required in rear seats.

Interlock-belt system effectiveness was compared with air cushion-lap belt system effectiveness in terms of reduced deaths and injuries (Figure 12 and 13). If the total passenger car population were equipped with the interlock-belt system we could expect 7,000 fewer fatalities and 340,000 fewer injuries annually (Figures 7 and 8). Comparable figures for the air cushion-lap belt system are 15,600 and 1,000,000, respectively (Figures 9 and 10). Using three different techniques for economic analysis, the benefit/cost ratios range from 2.9 to 5.2 fort he interlock-belt system and from 3.6 to 6.0 for the air cushion-lap belt system.

In conclusion, results of this study show that the proposed rulemaking, represented by the air cushion-lap belt system, is clearly superior to the interlock-belt system in the reduction of fatalities and injuries. Furthermore, the air cushion-lap belt system is fully justified even from an economic point of view since its benefit/cost ration and its incremental benefit/cost ratio (incremental benefits/incremental costs) are substantially in excess of 1.0.

U.S. Department of Transportation

National Highway Traffic Safety Administration

Final Regulatory Impact Analysis

Amendment to Federal Motor Vehicle Safety Standard 208 Passenger Car Front Seat Occupant Protection

FINAL REGULATORY IMPACT ANALYSIS

AMENDMENT TO FMVSS NO. 208

PASSENGER CAR FRONT SEAT
OCCUPANT PROTECTION

NATIONAL HIGHWAY TRAFFIC SAFETY ADMINISTRATION
PLANS AND PROGRAMS
OFFICE OF PLANNING AND ANALYSIS

JULY 11, 1984

SUMMARY

In October 1983, the Department of Transportation published a Notice of Proposed Rulemaking (NPRM) which proposed several alternative amendments to FMVSS No. 208, Occupant Crash Protection. The Preliminary Regulatory Impact Analysis (PRIA) accompanying the NPRM discussed the uncertainty involved in determining the effectiveness of restraint systems, safety benefits, insurance savings/costs, as well as consumer and other costs that could be anticipated under various alternatives and solicited comments on this subject. In response to the NPRM, over 7,800 commenters offered their views about various aspects of the proposed rulemaking, including the automobile manufacturers, insurance companies, consumer groups, and other interested parties. In May 1984, the Department published a Supplemental Notice of Proposed Rulemaking (SNPRM) asking for comments on four additional alternatives, as well as other issues. There were over 130 comments to the SNPRM. In preparation for this rulemaking, the Department of Transportation conducted comprehensive analyses of pertinent comments and of all accident data and other material available in its files. On the basis of these analyses, the agency sought to determine the effects on benefits and costs of the proposed alternatives to improve passenger car occupant protection.

While many of the uncertainties still remain, notably the uncertainty surrounding the precise level of potential usage of automatic belts, the summary data below are based on the best currently available estimates.

Effectiveness.

Effectiveness of an occupant restraint system is defined as the percentage
reduction in fatalities or injuries for restrained occupants as compared to
unrestrained occupants. In this analysis, the agency reviewed all pertinent
accident data in order to develop a range of estimates of the effectiveness
for air bags without belts, with lap belts, and with three point belts;
manual lap belts, manual lap and shoulder belts; and automatic belts. The
results of the effectiveness evaluation are as follows:

TABLE 1
PERCENT EFFECTIVENESS

	Manual Lap Belt	Manual Lap/ Shoulder Belt	Automatic Belt	Air Bag Alone	Air Bag With Lap Belt	Air Bag With Lap/ Shoulder Belt
Fatalities	30-40	40-50	35-50	20-40	40-50	45-55
AIS 2-5 Injuries	25-35	45-55	40-55	25-45	45-55	50-60
AIS 1 Injuries	10	10	10	10	10	10

According to these estimates, there is no single system more effective than
the manual lap/shoulder belt when used; but using this system with an air bag
as a supplement provides the most effective system for both fatalities and
AIS 2-5 injuries.

Throughout the analysis, the safety benefits and insurance premium changes
will be presented as a range of values. These ranges reflect the low and
high effectiveness estimates.

Safety Benefits

Based on projected fatalities and injuries and using the range of
effectiveness estimates and a range of automatic and manual seat belt usage,
estimates were made of the incremental reductions in fatalities, AIS 2-5
injuries, and AIS 1 injuries for all automatic restraint systems (air bags
without seat belts, air bags with lap belts, air bags with lap/shoulder belts
and automatic belts) and for mandatory use laws if they are effective in all
states. Estimates are provided across a broad range of usage (20-70 percent)
for automatic belts and a narrower range (40-70 percent) for mandatory use
laws because the precise level of future usage is uncertain. Below are the
results of this analysis:

TABLE 2
INCREMENTAL REDUCTION IN

	Fatalities	AIS 2-5 Injuries	AIS 1 Injury
Air Bags Only (No Lap Belt Usage)	3,780-8,630	73,660-147,560	255,770
Air Bags With Lap Belt (12.5% Usage)	4,410-8,960	83,480-152,550	255,770
Air Bags With Lap Shoulder Belt (12.5% Usage)	4,570-9,110	85,930-155,030	255,770
Automatic Belts			
20% Usage	520-980	8,740-15,650	22,760
30%	1,420-2,280	24,370-37,440	52,640
40%	2,320-3,590	39,990-59,220	82,510
50%	3,230-4,900	55,610-81,000	112,380
60%	4,130-6,200	71,240-102,790	142,250
70%	5,030-7,510	86,860-124,570	172,120
Mandatory Belt Use Laws (in all states)			
40% Usage	2,830-3,590	47,740-59,220	82,510
50%	3,860-4,900	65,300-81,000	112,380
60%	4,890-6,200	82,860-102,790	142,250
70%	5,920-7,510	100,430-124,570	172,120

Insurance Premium Changes

Based on the projected loss experience of the insurance industry resulting from an automatic occupant protection requirement, insurance premiums should change for various automobile insurance coverages, as well as for health insurance and life insurance. These results are summarized below:

TABLE 3
SUMMARY OF POTENTIAL EFFECTS
ON INSURANCE PREMIUMS FROM
AUTOMATIC RESTRAINT REQUIREMENTS

Air Bags	Per Vehicle Annual Savings ($)	Per Vehicle Lifetime Savings ($)	Total Annual Savings 1990 Fleet Equivalent ($M)
Automobile Insurance			
Savings-Safety	9-17	62-115	1,108-2,046
Loss-Deployment	(3)	(18)	(312)
Health Insurance	4-8	29-54	521-962
Life Insurance	0-1	3-7	62-136
Total	10-23	76-158	1,379-2,832
Automatic Belts (For 20 Percent Assumed Usage)			
Automobile Insurance	1-2	5-14	89-243
Health Insurance	0-1	2-7	42-114
Life Insurance	0	0-1	7-14
Total	1-3	7-22	138-371
Automatic Belts (For 70 Percent Assumed Usage)			
Automobile Insurance	10-14	65-94	1,146-1,676
Health Insurance	5-7	31-44	539-788
Life Insurance	1	4-6	71-106
Total	16-22	100-144	1,756-2,570

Consumer Cost

The following table presents current estimates of the consumer cost of different automatic restraints (air bags and automatic belts) as well as the incremental fuel cost over the lifetime of the vehicle resulting from the additional weight of such restraints.

TABLE 4
PER VEHICLE COST IMPACTS

	Incremental Cost	Lifetime Energy Costs	Total Incremental Cost Increase
Automatic Belt System (2-pt. or 3-pt. Non-Power, High Volume, Driver and Front Right)	$40	$11	$51
Air Bag - Driver Only (High Volume)	$220	$12	$232
Air Bag - Full Front (High Volume)	$320	$44	$364

Net Dollar Costs

The results of a lifetime net dollar cost analysis for air bags and automatic belts are shown in the following table. The analysis considers only the costs related to motor vehicle ownership; it does not include economic costs to society, or values for the pain and suffering experienced by the victims of motor vehicle accidents. Thus, lifetime dollar costs include retail price increases and fuel cost increases and lifetime dollar benefits include only insurance premium reductions. The range of lifetime net dollar costs is $206-$288 per car for air bags at 12.5 percent lap belt usage. For automatic

165

TABLE 5

SUMMARY OF SAFETY BENEFITS AND NET DOLLAR COSTS OR BENEFITS FOR AIR BAGS AND AUTOMATIC BELTS
(COSTS ON A PER CAR BASIS)

	----SAFETY BENEFITS----		INCREMENTAL LIFETIME COSTS	LIFETIME INSURANCE PREMIUM REDUCTIONS	LIFETIME NET DOLLAR COST OR (BENEFITS)
	FATALS	AIS 2-5 INJURIES			
Full Front Air Bag With Lap Belt					
No Usage of Lap Belt	3,780-8,630	73,660-147,560	$364	$66-154	$210-298
12.5% Usage of Lap Belt	4,410-8,960	83,480-152,550	364	76-158	206-288
Driver and Front Right Air Bag with Lap Belt (Center Seat Exempt)					
No Usage of Lap Belt	3,710-8,490	72,480-145,408	354	64-151	203-290
12.5% Usage of Lap Belt	4,340-8,810	82,260-150,370	354	74-155	199-280
Driver Air Bag with Lap Belt					
No Usage of Lap Belt	2,680-6,250	56,330-114,370	232	36-100	132-196
14.0% Usage of Lap Belt	3,200-6,520	64,820-118,680	232	44-104	128-188
Driver and Right Front Automatic Belt (Center Seat Exempt)					
20% Usage	520-980	8,740-15,650	51	7-22	29-44
70% Usage	5,030-7,510	86,860-124,570	51	100-144	(49)-(93)
Driver Automatic Belt					
20% Usage	270-580	5,260-10,370	26	0-8	18-26
70% Usage	3,610-5,440	67,160-96,770	26	65-99	(39)-(73)

Note: ( ) means dollar benefits (insurance premium reductions) exceed dollar costs.

belts, net dollar costs vary by belt usage rates because the insurance benefits vary by belt usage rates. At 20 percent usage, lifetime insurance benefits range between $7-$22 per car resulting in a lifetime net cost per car of $29-$44, while at 70 percent usage lifetime insurance benefits are $100-$144 per car, resulting in a net dollar savings of $49-$93 per car.

Breakeven Points

Several breakeven points were calculated throughout the analysis. The breakeven points indicate where benefits of one alternative equal another, or where costs equal benefits, etc.

Figure I shows the fatality reduction breakeven points between automatic belts and air bags for a variety of combinations within the ranges of usage and effectiveness as they apply to these two restraint systems.

For example, the combination of the high level of effectiveness for automatic belts (50 percent) and the low effectiveness for air bags (20 percent) result in a breakeven point at a usage level of 44 percent. That is, with 44 percent automatic belt usage, the safety benefits provided by these two systems are equal.

Figure 2 shows breakeven points for costs related to automatic belts using low and high effectiveness estimates. The breakeven point occurs when lifetime costs (retail price increases and additional fuel costs) equal lifetime insurance premium reductions. At the high effectiveness level, the breakeven point occurs at the 32 percent usage level. At the low effectiveness level, the breakeven point occurs at the 44 percent usage level.

Air bag systems do not attain similar breakeven points. The estimated lifetime cost of a full front air bag system is $364, while lifetime insurance premium reductions range from $76-$158 at 12.5 percent lap belt usage for low and high estimates of effectiveness respectively. Based on these estimates, there is no point at which air bag insurance savings would equal air bag costs. This is true for all air bag configurations--full front, driver only, and driver and front right seats (center seat exempt). It should be noted, however, that these are not "societal" breakeven points as they do not include lost productivity and other costs to society.

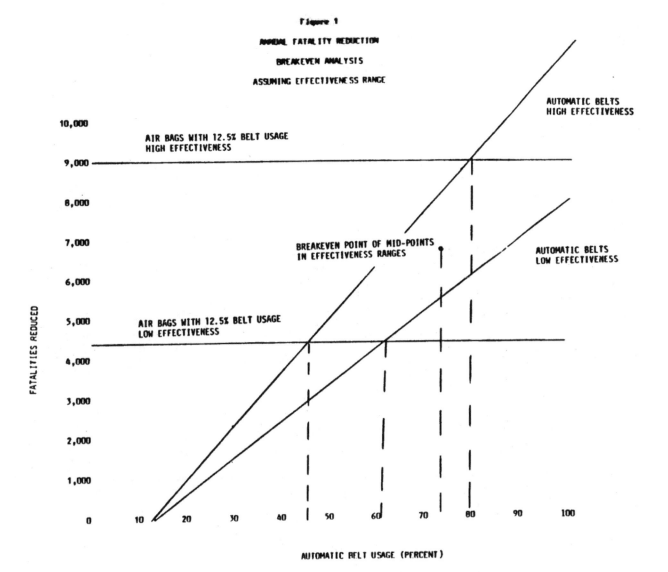

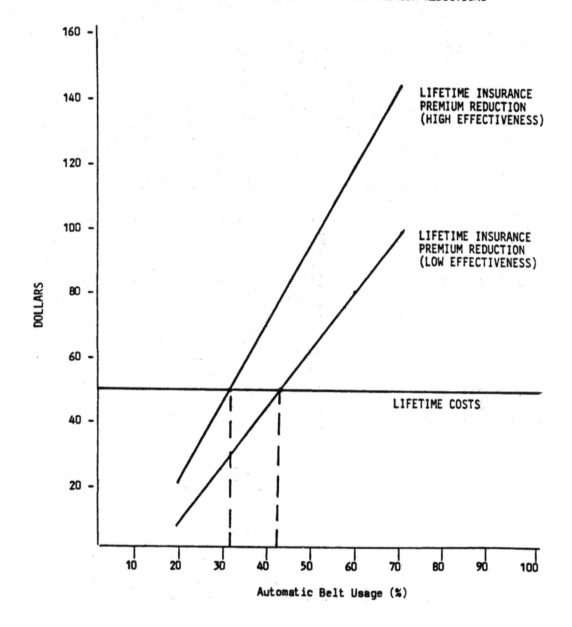

Figure 2

BREAKEVEN POINT ANALYSIS FOR AUTOMATIC BELTS
LIFETIME COSTS VERSUS LIFETIME INSURANCE PREMIUM REDUCTIONS

Benefits of the Final Rule

The Final Rule calls for a gradual introduction of automatic restraints during model years 1987-89 and a full implementation of the automatic occupant protection requirement of FMVSS 208 effective September 1, 1989, unless two-thirds of the U.S. population are covered by mandatory safety belt use laws. Tables 6 and 7 show the reductions in fatalities and AIS 2-5 injuries, respectively, over the life of cars sold during model years 1987-89. Reductions are shown for two possible scenarios that satisfy the Final Rule's implementation schedule: under the first scenario automatic belts would be used in 10, 25 and 40 percent of the fleet, respectively, for the first, second and third year; under the second scenario air bags would be used in 6.67, 16.67 and 26.67 percent of the fleet, respectively (the Final Rule allows an extra credit of 1.5 for each car that provides automatic protection with a system other than seat belts for the purpose of meeting the percentage requirements of the Final Rule). These benefits should be added to those that accrue under full implementation (see Table 2) which begins in model year 1990.

TABLE 6
INCREMENTAL REDUCTION IN FATALITES
OVER THE LIFETIME OF THE MODEL YEAR FLEET
CENTER SEAT EXEMPT
BASED ON LOW-HIGH EFFECTIVENESS ESTIMATES

	MY 1987 10% Automatic Belts, 6.67% Air Bags	MY 1988 25% Automatic Belts; 16.67% Air Bags	MY 1989 40% Automatic B 26.67% Air
Air Bags Only	250-570	620-1,420	990-2,260
Air Bags with Lap Belt (12.5% Usage)	290-590	720-1,470	1,160-2,350
Air Bags with Lap/ Shoulder Belts (12.5% Usage)	300-600	750-1,500	1,200-2,390
Automatic Belts (20% Usage to 70% Usage)	50-100 500-750	130-250 1,260-1,880	210-390 2,010-3,000

TABLE 7
INCREMENTAL REDUCTION IN AIS 2-5 INJURIES
OVER THE LIFETIME OF THE MODEL YEAR FLEET
CENTER SEAT EXEMPT
BASED ON LOW-HIGH EFFECTIVENESS ESTIMATES

	MY 1987 10% Automatic Belts, 6.67% Air Bags	MY 1988 25% Automatic Belts; 16.67% Air Bags	MY 1989 40% Automatic B 26.67% Air
Air Bags Only	4,830-9,700	12,080-24,240	19,330-38,780
Air Bags with Lap Belt (12.5% Usage)	5,490-10,030	13,710-25,070	21,940-40,100
Air Bags with Lap/ Shoulder Belts (12.5% Usage)	5,650-10,200	14,120-25,480	22,590-40,770
Automatic Belts (20% Usage to 70% Usage)	870-1,570 8,690-12,460	2,190-3,910 21,720-31,140	3,500-6,260 34,740-49,830

Table 8 shows the reductions of fatalities and AIS 2-5 injuries that would occur if states containing a total of 67 percent of the Nation's population enacted mandatory use laws, without the implementation of the automatic restraint requirements of Standard 208. Of course, benefits would be higher if additional states passed mandatory use laws.

TABLE 8
ANNUAL SAFETY BENEFITS OF
MANDATORY USE LAWS
AFFECTING 67% OF THE POPULATION

INCREMENTAL FATALITY REDUCTION

USAGE	EFFECTIVENESS		
	LOW (40%)	MID-POINT (45%)	HIGH (50%)
40%	1,900	2,160	2,410
70%	3,970	4,500	5,030

INCREMENTAL AIS 2-5 INJURY REDUCTION

	LOW (45%)	MID-POINT (50%)	HIGH (55%)
40%	31,990	35,800	39,680
70%	67,290	75,310	83,460

950865

Regulatory History of Automatic Crash Protection in FMVSS 208

Stephen R. Kratzke
National Highway Traffic Safety Administration

ABSTRACT

This paper summarizes the regulatory history of the automatic crash protection requirements in Federal Motor Vehicle Safety Standard 208. It is intended to give the reader an overview of the regulatory history involved in Standard 208, from its beginning in 1968 as a requirement that passenger cars be equipped with seat belts to its present requirement that, as of 1998, all passenger cars and light trucks must be equipped with air bags. It also discusses and summarizes the various court cases that have challenged different aspects of the automatic crash protection requirements.

INTRODUCTION

This paper traces the evolution of the occupant protection requirements in Federal Motor Vehicle Safety Standard 208 from its beginnings as a requirement for seat belts to be installed in passenger cars to its current requirements that each passenger car provide an air bag and a manual lap/shoulder belt for the driver and right front passenger position beginning in the 1998 model year (September 1, 1997) and that light trucks and vans provide an air bag and a manual lap/shoulder belt for the driver and right front passenger beginning in the 1999 model year (September 1, 1998). The purpose of this paper is to give the reader an understanding of how these requirements evolved into their present form and what purpose they are intended to serve.

THE FIRST OCCUPANT PROTECTION REQUIREMENTS IN STANDARD 208

ORIGINAL OCCUPANT PROTECTION REQUIREMENTS - Standard 208 was one of the 19 original Federal motor vehicle safety standards. It required that passenger cars provide a seat belt at every forward-facing designated seating position. This requirement took effect on January 1, 1968. There were no crash testing requirements to evaluate the protection afforded to vehicle occupants.

INITIAL AUTOMATIC PROTECTION REQUIREMENTS - It was not long, however, before the National Highway Traffic

Safety Administration (NHTSA) began to explore the possibility of requiring automatic crash protection in motor vehicles. Vehicles that provide automatic crash protection protect their occupants by means that require no action by the vehicle occupants. The two types of automatic crash protection that have been offered for sale on production vehicles are automatic seat belts and air bags. The effectiveness of a vehicle's automatic crash protection is assessed through crash testing. A vehicle must comply with specified injury criteria, as measured on a test dummy, when tested in a 30 mph crash test.

On July 2, 1969, NHTSA published an advance notice of proposed rulemaking (ANPRM) requesting comments on the merits of crash protection systems that protect vehicle occupants by means that require no action on the part of the occupants.[1] This notice specifically mentions the possibility of meeting such a requirement by means of air bags. After evaluating these comments, NHTSA published a notice proposing to require automatic crash protection for all passenger cars beginning January 1, 1972 and for all light trucks and vans beginning January 1, 1974.[2]

On November 3, 1970, NHTSA published a final rule that required automatic crash protection for all passenger cars as of July 1, 1973, and for most light trucks and vans as of July 1, 1974. Compliance would have been determined by a crash test with test dummies in the front outboard seats.[3] In the preamble to this rule, the agency made clear its position that automatic protection would supplant the need for vehicle occupants to use manual seat belts. NHTSA said: "Several comments recommended that the requirement for seat belts be retained, citing the benefits of keeping the driver in his seat during violent maneuvers and the possibility of failure of a passive system. It is the [NHTSA's] position that the possible benefits of required seat belts would not justify the costs to the manufacturers and to the public. Only a small percentage of the public uses the upper torso restraints that are presently furnished with passenger

---

[1] 34 FR 11148

[2] 35 FR 7187; May 7, 1970

[3] 35 FR 16927

cars."[4] In other words, passenger cars, light trucks, and vans would no longer be required to provide manual seat belts at each designated seating position once the automatic protection requirements took effect.

NHTSA received many petitions for reconsideration of this rule. In response to those petitions, NHTSA published a notice that postponed the effective date of the automatic protection requirements for passenger cars from July 1, 1973 until August 15, 1973, and granted a similar 45 day postponement for light trucks and vans, to correspond more precisely to the manufacturers' changeovers for a new model year's production.[5] This notice also repeated NHTSA's previous conclusion that the existing test dummy specifications published by the Society of Automotive Engineers were "the best available." The notice went on to state that, "NHTSA is sponsoring further research and examining all available data, however, with a view to issuance of further specifications for these devices."[6]

As indicated above, the automatic protection requirements did not require the use of any particular technology to achieve the desired result. Instead, manufacturers were free to use any means they chose so long as their vehicles provided the specified level of protection with no action by vehicle occupants. Up to this point, however, the regulatory notices dealing with automatic protection and the public comments responding to those notices had focused almost exclusively on air bag systems. In a July 8, 1971 final rule,[7] NHTSA added explicit language to Standard 208 to acknowledge that automatic belts could be used to meet the automatic protection requirements and to state the applicability of various requirements in the Standard to automatic belts (called "passive belts" at that time). On that same day, the agency published a proposal that automatic protection systems have a means of emergency release.[8] NHTSA suggested that automatic belts could use a spool-out mechanism and air bags would meet this requirement by deflating.

In response to petitions for reconsideration of the July 8, 1971 final rule on automatic belts, NHTSA excluded automatic belts from the assembly performance and webbing requirements. The agency explained that this change was made to allow manufacturers as much freedom in the design of automatic belt systems to fit the particular crash pulse of each car as they have in the design of other types of automatic protection systems.[9]

In addition to broadening the focus of its automatic protection rulemaking to recognize means other than air bags, the agency also introduced into its automatic protection rulemaking the concept of an ignition interlock system. Several vehicle manufacturers asked NHTSA to delay the date by which automatic protection had to be installed in passenger cars because of unresolved technical problems with automatic protection systems. On October 1, 1971,[10] NHTSA proposed to postpone the effective date for mandatory automatic protection from August 15, 1973 until August 15, 1975. However, if a car produced between those dates did not provide automatic protection, it had to be equipped with an interlock system that would prevent the engine from starting if any front seat occupants did not have their manual seat belts fastened. Front outboard seating positions would also be subject to a crash test with a test dummy in each such seating position and the manual belt system fastened around the test dummy. The agency explained its proposal as follows: "It is intended by this option to provide a high level of seat belt usage, and to increase the life- and injury-saving effectiveness of installed belt systems, in the interim period before [automatic] systems are required." The ignition interlock option was adopted as proposed in a rule published February 24, 1972.[11]

THE COURT DECISION IN CHRYSLER V. DOT

Shortly after the March 10, 1971 rule was published requiring automatic protection in new vehicles, Chrysler, Jeep, American Motors, Ford, and the Automobile Importers of America filed lawsuits in the U.S. Court of Appeals for the Sixth Circuit challenging the automatic protection requirements. These lawsuits raised three primary arguments:

1. The automatic protection requirements were not "practicable," as required by NHTSA's authorizing legislation, because the technology needed to comply with automatic protection was not sufficiently developed as of that time. The manufacturers argued that NHTSA had no authority to establish a safety standard that required the industry to improve upon the existing technology. Under this view, automatic protection should not be required until devices to meet the requirement were sufficiently developed as of the date of the rulemaking so as to permit ready installation.

2. The automatic protection requirements did not "meet the need for motor vehicle safety," as required by NHTSA's authorizing legislation, because seat belts offered better occupant protection than automatic protection.

3. The automatic protection requirements were not "objective," as required by NHTSA's authorizing legislation, because the existing SAE Recommended Practice did not adequately specify sufficient details for the construction of the test dummy.

The Sixth Circuit announced its decision on these lawsuits on Dec 5, 1972 in an opinion titled Chrysler v. DOT, 472 F.2d 659 (6th Cir. 1972). The court ruled in favor of NHTSA on the first argument, stating that "the agency is empowered to issue safety standards which require improvements in existing technology or which require the development of new technology, and it is not limited to issuing standards based solely

---

[4] 35 FR 16928

[5] 36 FR 4600; March 10, 1971

[6] 36 FR 4602

[7] 36 FR 12858

[8] 36 FR 12866; July 8, 1971

[9] 36 FR 23725; December 14, 1971

---

[10] 36 FR 19266

[11] 37 FR 3911

on devices already fully developed."[12] The court also upheld the agency on the second point raised by the manufacturers, ruling that the question of whether to require automatic protection was delegated to NHTSA, that there was substantial support in the record for the decision to mandate automatic protection, and so the court had no basis for substituting its judgment for that of the agency.[13] However, the court found in favor of the manufacturers on the third argument. The court concluded that the SAE J963 test dummy incorporated in Standard 208 was not defined with sufficient specificity to meet the statutory requirement for "objectivity." Because of this shortcoming, this issue was remanded to the agency with instructions to delay the automatic crash protection requirement until a reasonable time after objective dummy specifications had been issued.[14]

The Sixth Circuit was petitioned by Ford to clarify the effect of its Chrysler decision on the requirement in Standard 208 for crash testing of the manual belts at front outboard seating positions in cars to be equipped with an ignition interlock. The Sixth Circuit announced its decision on this petition on February 2, 1973 in an opinion titled Ford v. NHTSA, 473 F.2d 1241 (6th Cir. 1973). The court ruled that its conclusion that the test dummy was not objective was equally applicable to this crash testing, and ordered the agency to delay the crash testing requirements for manual belt systems at front outboard seating positions in cars equipped with an ignition interlock.

The Ford v. NHTSA opinion was particularly problematic, because it overturned on February 2 an option that was scheduled to take effect on August 15 of that same year. All car manufacturers except General Motors had intended to choose the ignition interlock option, but it had now been declared invalid. On April 20, 1973, NHTSA addressed this problem by proposing to delete the requirement for crash testing of manual belts at the front outboard seats in cars equipped with an ignition interlock system.[15] NHTSA adopted this proposal in a June 20, 1973 final rule.[16] That June 20 final rule also announced NHTSA's position that the decision in the Chrysler case invalidated the automatic protection requirements, regardless of whether the language mandating automatic protection remained in the text of the standard. The agency also announced in the June 20 rule that additional rulemaking would be needed to reestablish an effective date for automatic protection requirements.

NHTSA proposed to adopt much more detailed test dummy specifications in a notice published April 2, 1973.[17] In fact, these proposed specifications were the existing specifications General Motors used for its Hybrid II test dummy. The Hybrid II test dummy was adopted as the test dummy to be used in

Standard 208 compliance testing in a final rule published on August 1, 1973.[18] That August 1 rule also repeated the agency's previous announcement that it would give the public further notice and opportunity for comment before making a final decision on whether to reinstate the automatic protection requirements.

THE IGNITION INTERLOCK EXPERIENCE

As of August 15, 1973, all new cars had to be equipped either with automatic protection or an ignition interlock for both front outboard seating positions. General Motors sold a few thousand of its 1974 model year cars equipped with air bags that met the automatic protection requirement. Every other 1974 model year car sold in the United States came with an ignition interlock, which prevented the engine from operating if either the driver or front seat outboard passenger failed to fasten their manual seat belt.

In a March 19, 1974 notice, NHTSA described the public reaction to the ignition interlock as follows: "Public resistance to the belt-starter interlock system currently required (except on vehicles providing [automatic] protection) has been substantial, with current tallies of proper lap-shoulder belt usage on 1974 models running at or below the 60% level. Even that figure is probably optimistic as a measure of results to be achieved, in light of the likelihood that as time passes the awareness that the forcing systems can be disabled, and the means for doing so will become more widely disseminated, ..."[19] There were also speeches on the floor of both houses of Congress expressing the public's anger at the interlock requirement. On October 27, 1974, President Ford signed into law a bill that prohibited any Federal motor vehicle safety standard from requiring or permitting the use of any seat belt interlock system. In response to this change in the law, NHTSA published a final rule on October 31, 1974 that deleted the interlock option from Standard 208 effective immediately.[20]

While the interlock option was still in place, NHTSA also addressed the subject of automatic belts in response to a petition from Volkswagen. That company was going to introduce in its 1975 model year Rabbit models an automatic belt system that consisted of an upper torso restraint and knee bolsters, and asked for some changes and clarifications to the requirements of Standard 208 to ensure that this new belt system would comply with the standard. NHTSA proposed to require a manual single point emergency release mechanism on all automatic belts to allow vehicle occupants postcrash egress from the vehicle.[21] In the final rule adopting this proposal[22], NHTSA noted that one commenter had objected to the emergency release mechanism for automatic belts because the mechanism could be used to

---

[12] 472 F.2d 673

[13] 472 F.2d 674-675

[14] 472 F.2d 681

[15] 38 FR 9830

[16] 38 FR 16072

[17] 38 FR 8455

---

[18] 38 FR 20449

[19] 39 FR 10272

[20] 39 FR 38380

[21] 39 FR 3834; January 30, 1974

[22] 39 FR 14593; April 25, 1974

disconnect the belts in non-emergency situations. NHTSA responded that the advantages of having an emergency release mechanism outweighed the disadvantages of possible abuse.[23] Another commenter suggested that a lever or pushbutton that allowed the belt to spool out from the retractor instead of separating would be a more appropriate emergency release mechanism. NHTSA responded by stating that it believed the uniformity of having push-button action to release all motor vehicle seat belt assemblies was more compelling than the advantages suggested by the commenter of permitting a spool-out release.[24] The agency also provided more information about how it would determine whether an automatic belt system provided the necessary protection with "no action by vehicle occupants," as specified in Standard 208. Volkswagen did in fact introduce automatic belts as an optional piece of equipment on its 1975 Rabbit models.

THE COLEMAN DECISION

Standard 208 had never allowed the ignition interlock option to be anything more than an interim measure on the way to full automatic protection. The interlock was scheduled to expire as a permissible option on August 15, 1975 in any event. Thus, the change in the law eliminated the ignition interlock 10 months before it would have expired anyway. The more significant question at this point was whether rulemaking would be initiated to reinstate the automatic protection requirements and when any such requirements would take effect.

In a March 19, 1974 notice,[25] when the interlock option was still in place, NHTSA proposed to reinstate the automatic protection requirements for front outboard seating positions in passenger cars as of September 1, 1976. A little more than one year later, NHTSA again proposed to extend until September 1, 1976 the period during which manufacturers could comply with Standard 208 simply by installing manual belts at front seating positions.[26] However, this later notice did not propose any specific date for reinstating automatic protection requirements. In fact, this April 1975 notice announced that "A decision has not yet been made on the long-term requirements for occupant crash protection." The proposed extension of existing requirements until September 1, 1976 was made final in an August 13, 1975 rule.[27] The August 1975 rule stated: "While the NHTSA recognizes that the present crash protection options will in all likelihood be in effect for some period after August 31, 1976, the agency has not proposed more than the 1-year extension. ... The NHTSA intends to propose the long-term requirements for occupant crash protection, both for passenger cars and for light trucks and MPV's, as soon as possible."

The Secretary of Transportation, William T. Coleman, Jr., published a proposal on June 14, 1976 that had several noteworthy aspects.[28] This notice announced that Secretary Coleman was personally taking responsibility for the decision on automatic crash protection requirements in Standard 208. All previous decisions had been left to the NHTSA Administrator, subject to the review and approval of the Secretary. This notice proposed five options for dealing with occupant protection in frontal crashes. These were:

1. Continue existing manual belt requirements and continue further research to identify effective means of automatic protection.

2. Continue existing manual belt requirements and try to encourage States to pass mandatory belt use laws, which would substantially increase use of manual belts.

3. Continue existing manual belt requirements and conduct a Federally sponsored field test of automatic protection in vehicles used on the public roads.

4. Reinstate automatic protection requirements.

5. Require manufacturers to provide consumers with the option of ordering automatic protection in some or all of their models.

In accordance with this decision, NHTSA published a proposal[29] and final rule[30] extending the period during which manufacturers had the option of complying with Standard 208's occupant protection requirements by simply installing manual seat belts at all designated seating positions. That option was now extended until August 31, 1977, to allow time for the Secretary to announce his final decision on the automatic protection requirements.

On December 6, 1976, Secretary Coleman announced his decision on automatic protection. This decision was not published in the Federal Register, but written copies were placed in the public docket at the Department of Transportation. Secretary Coleman called on auto manufacturers to join with the Federal government in a demonstration program so that approximately 500,000 cars with passive restraint systems would be offered for sale at a reasonable cost to consumers in the 1979 and 1980 model years (i.e., the period beginning Sept. 1, 1978 and ending August 31, 1980). Secretary Coleman also announced that the Department of Transportation would make additional efforts to promote seat belt use during this period. At the end of this demonstration project, the Secretary concluded that the Department could make a more informed choice about the need for automatic protection requirements, based upon the real world experience gained from the 500,000 production vehicles that would be equipped with automatic protection and updated estimates of likely manual belt use in the future.

A notice appeared in the January 27, 1977 Federal Register[31] announcing Secretary Coleman's December 6, 1976 decision and "incorporating it by reference" in the Federal Register

---

[23] 39 FR 14593-14594

[24] 39 FR 14594

[25] 39 FR 10271

[26] 40 FR 16217; April 10, 1975

[27] 40 FR 33977

---

[28] 41 FR 24070

[29] 41 FR 29715; July 19, 1976

[30] 41 FR 36494; August 30, 1976

[31] 42 FR 5071; January 27, 1977

notice. The January 27 notice also extended indefinitely the option for manufacturers to comply with Standard 208 by installing belts at all designated seating positions in the vehicle.

THE ADAMS DECISION

Less than two months after the publication of the Federal Register notice announcing Secretary Coleman's final decision on the automatic protection requirements in Standard 208, a notice was published announcing that the new Secretary of Transportation was reexamining that decision. Thus, on March 24, 1977, a notice was published in the Federal Register[32] announcing that the new Secretary of Transportation, Brock Adams, was conducting a reexamination of Secretary Coleman's decision on automatic protection. This notice asked for public comments on the following three alternatives:

       1. Leave the Coleman decision in place;
       2. Reinstate the automatic protection requirements; or
       3. Seek to raise the usage level for manual belts by encouraging the States to pass belt use laws.

Before a final decision was announced on the automatic protection requirements for cars, NHTSA published a notice extending indefinitely the existing occupant protection requirements for light trucks and vans. This notice was published on June 2, 1977[33], and allowed manufacturers to either install automatic protection for the front outboard seating positions or to install manual seat belts at those seating positions. NHTSA explained in this notice that the agency had originally intended that manufacturers would have had the benefit of the experience with installing automatic protection in passenger cars before automatic protection systems were required in light trucks and vans. Since the Secretary of Transportation was in the process of deciding whether automatic protection systems should be required in passenger cars, it seemed premature to require those systems in light trucks and vans. Hence, NHTSA announced an indefinite extension of the option for light trucks and vans to meet the occupant protection requirements simply by installing manual seat belts at the front outboard seating positions. The agency noted that this indefinite extension did not preclude future rulemaking to modify the occupant protection requirements for light trucks and vans, but promised that notice and opportunity for comment would be provided prior to any modifications.[34]

On July 5, 1977, a final rule reinstating automatic protection requirements for passenger cars was published in the Federal Register.[35] The Department of Transportation concluded that automatic protection was necessary even though manual seat belts were "highly effective" at preventing injury and ejection,

because so few vehicle occupants used their manual belts.[36] The Department rejected the option of seeking mandatory seat belt use laws in the various States, because the prospects for success looked so poor.[37] The Department concluded that the demonstration programs called for in the Coleman decision were unnecessary, because they would have further delayed the mandatory introduction of occupant protection systems the Department had already found to be technologically feasible, practicable, and capable of offering substantial life-saving potential at reasonable costs.[38]

Accordingly, the July 5, 1977 rule required that all 1982 model year cars with a wheelbase over 114 inches be equipped with automatic protection. All 1983 model year cars with a wheelbase over 100 inches would be required to offer automatic protection, and all 1984 model year cars would be required to offer automatic protection. This gradual phased-in approach was intended to give vehicle manufacturers additional leadtime to overcome the greater difficulties of installing air bags in smaller cars and to give those efforts the benefit of the manufacturers' experience with installing air bags in larger cars.

Six petitions for reconsideration of this rule were filed. In addition, a group called the Pacific Legal Foundation filed a petition for a review of the rule in the U.S. Court of Appeals for the District of Columbia Circuit and asked the Department to stay the effectiveness of the automatic protection requirements for a period of time equal to the length of the judicial review. The vehicle manufacturers that filed petitions for reconsideration questioned the Department's analyses of the effectiveness of automatic protection systems. The Pacific Legal Foundation charged that the Department had failed to consider the public reaction to automatic protection and had ignored potential hazards posed by automatic protection systems. Ralph Nader and the Center for Auto Safety charged that the Department had improperly delayed implementation of the automatic protection requirements and that the Department had no authority to phase-in the automatic protection requirements gradually rather than requiring full compliance by the effective date. These petitions were denied in all significant respects by a December 5, 1977 notice in the Federal Register.[39]

THE COURT DECISION IN PACIFIC LEGAL FOUNDATION v. DOT

As indicated above, a group called Pacific Legal Foundation filed a lawsuit challenging the Adams decision to reinstate automatic protection requirements even before the Department had responded to the petitions for reconsideration of the Adams decision. After the Department denied their petitions, Ralph Nader and the Center for Auto Safety filed their own lawsuit challenging the Adams decision in the same court.

---

[32] 42 FR 15935

[33] 42 FR 28135

[34] 42 FR 28136; June 2, 1977

[35] 42 FR 34289

[36] 42 FR 34290; July 5, 1977

[37] 42 FR 34291-34292; July 5, 1977

[38] 42 FR 34291

[39] 42 FR 61466

The Pacific Legal Foundation argued that the Adams decision should be overturned because of three major shortcomings. First, the group argued that the data did not support the Secretary's findings on the effectiveness of air bags. Second, Pacific Legal Foundation argued that the Adams decision was unlawful because it failed to consider public reaction to the automatic protection requirements. Third, the group charged that the rule ignored collateral dangers to public safety posed by air bags. Ralph Nader and the Center for Auto Safety also challenged the rule, arguing that the Department had no authority to delay the effective date for the automatic protection requirements until the 1982 model year or to phase-in the requirements over three successive model years.

The D. C. Circuit announced its decision on this matter on Feb 1, 1979 in an opinion titled Pacific Legal Foundation v. DOT, 593 F.2d 1338 (D.C. Cir. 1979). The court upheld the Adams decision on all grounds. This court decision contains two especially significant findings. First, the court agreed with Pacific Legal Foundation that the Department must consider the likely public reaction to mandates in the safety standards in order to fulfill its responsibility to ensure that the standard is "practicable" and "meets the need for safety."[40] In this case, the court found that the Department had adequately considered the anticipated public reaction to the automatic protection requirements, notwithstanding the Department's claim that it was not required to consider the public reaction when promulgating safety standards. Second, the court expressly found that the Department could use a phase-in schedule when needed "to tailor safety standards to engineering reality."[41]

THE PECK DECISION

There had been only minor rulemaking notices from NHTSA dealing with the subject of automatic protection after the December 5, 1977 response to petitions for reconsideration of the Adams decision. Perhaps the most noteworthy rulemaking on automatic protection between 1978-1980 had to do with the emergency release mechanism on automatic belts. On May 22, 1978,[42] NHTSA published a notice proposing to allow automatic belts to use an emergency release mechanism other than the push-button, single-point release that had been required since the April 1974 final rule. This proposal was in response to a GM petition to allow a "spool release" design as the emergency release mechanism for automatic belts. The agency followed up this proposal with a November 13, 1978 final rule[43] that allowed automatic belts to use any emergency release mechanism that is a single-point release and that is accessible to a seated occupant. The agency explained its action thus: "This amendment will allow manufacturers to experiment with various [automatic] belt designs before the effective date of the [automatic protection] requirements and determine which

---

[40] 593 F.2d 1345-1346

[41] 593 F.2d 1348

[42] 43 FR 21912

[43] 43 FR 52493

designs are the most effective and at the same time acceptable to the public."[44]

However, NHTSA's actions in 1981 ended this period of relative stability regarding the regulatory requirements for automatic protection. On February 12, 1981,[45] a notice signed by Drew Lewis, the new Secretary of Transportation, was published in the Federal Register. This notice proposed to delay the start of the phase-in for cars to be equipped with automatic protection for one year (from model year 1982 to model year 1983). The proposal to delay the phase-in was based upon the economic difficulties then confronting the automobile industry.

On April 9, 1981, two notices signed by Secretary Lewis were published in the Federal Register. The first was a final rule delaying for one additional year the start of the phase-in of automatic protection.[46] The second was a notice proposing three alternative courses of action regarding the future status of automatic protection requirements.[47] The three alternatives on which the Department sought comment were:

1. Retain the new phase-in schedule, but reverse the sequence of vehicles. In other words, small cars would be required to provide automatic protection first (in model year 1983), then mid-size cars (in model year 1984), and finally large cars (in model year 1985).
2. Allow one additional year of leadtime, but eliminate the phase-in. In other words, all cars would have to provide automatic protection beginning in model year 1984.
3. Rescind the requirements for automatic protection.

A final rule signed by NHTSA Administrator Raymond Peck that rescinded the automatic protection requirements for cars was published on October 29, 1981.[48] This rule indicated that there was significant uncertainty about the public acceptability of automatic protection. Vehicle manufacturers had overwhelmingly indicated that they would comply with the automatic protection requirements by installing detachable automatic belts in their new cars. The agency found substantial uncertainty about the likely use rates of these detachable automatic belts and announced that it could not reliably predict that detachable automatic belts would produce even a 5 percentage point increase in belt use over the existing belt use rates for manual belts.[49] The uncertain benefits combined with the relatively substantial costs for automatic protection led NHTSA to conclude that the automatic restraint requirement was no longer reasonable or practicable.

LEGAL BATTLE CULMINATING WITH SUPREME COURT DECISION ON AUTOMATIC PROTECTION

---

[44] 43 FR 52494

[45] 46 FR 12033

[46] 46 FR 21172

[47] 46 FR 21205

[48] 46 FR 53419

[49] 46 FR 53423

Throughout the course of this standard, groups have looked to the Federal courts to overturn decisions on the automatic protection requirements with which the group is dissatisfied. This rescission of the automatic protection requirements proved no different. On November 23, 1981, State Farm Mutual Automobile Insurance Company, joined by several other petitioners, filed a lawsuit in the U.S. Court of Appeals for the District of Columbia Circuit challenging the rescission of the automatic protection requirements.

That court announced its decision on June 1, 1982 in an opinion titled State Farm v. DOT, 680 F.2d 206 (D.C. Cir. 1982). The court unanimously reversed NHTSA's rescission of the automatic protection requirements. The court found that NHTSA had failed to consider or analyze obvious alternatives to rescission and had offered no evidence to show that the 1977 conclusion that seat belt usage would increase with automatic belts was no longer true.[50]

On September 8, 1982, NHTSA filed a petition with the U.S. Supreme Court asking the Supreme Court to review the D.C. Circuit Court's decision. The Supreme Court granted that petition on November 8, 1982.[51] The Supreme Court announced its decision on this matter on June 24, 1983 in an opinion titled Motor Vehicle Manufacturers Association v. State Farm, 463 U.S. 29 (1983). The Supreme Court unanimously ruled that the rescission of the automatic protection requirements was unlawful, or "arbitrary and capricious," because the agency had failed to consider obvious alternatives to rescission and explain why alternatives short of rescission were not chosen.[52] The Court noted that air bags, nondetachable automatic belts, and detachable automatic belts were three existing technologies that could be used to comply with the automatic protection requirement. The Supreme Court said that, even if NHTSA were correct that detachable automatic belts would yield few benefits, that fact alone would not justify rescission. Instead, the Court said that fact would justify only the modification of the automatic protection requirements to prohibit compliance by means of detachable automatic belts.[53] The Court reasoned that the necessary next step to justify rescission was for NHTSA to adequately explain why it no longer believed that compliance

---

[50] 680 F.2d 230

[51] 459 U.S. 987 (1982)

[52] All nine Justices joined in the opinion finding the rescission of the automatic protection requirements was arbitrary and capricious because it failed to consider alternatives to rescission, such as permitting compliance only by means of air bags or nondetachable automatic belts, and explain why those alternatives were not adopted. However, in a separate opinion for four of the nine Justices, Justice Rehnquist expressed the view that the rescission of requirements permitting compliance with automatic protection by means of detachable automatic belts was satisfactorily justified and would not be considered "arbitrary and capricious." 463 U.S. 57-59 (1983).

[53] 463 U.S. 47 (1983)

---

with the automatic protection requirements by means of either air bags or nondetachable automatic belts would be an effective and cost-beneficial way of saving lives and preventing injuries.[54] Since NHTSA had not offered such an explanation, its rescission of the automatic protection requirements was unlawful. Accordingly, the Department was ordered to conduct further rulemaking on the status of the automatic protection requirements.

THE DOLE RULE AND THE RESULTING LEGAL CHALLENGE

Rulemaking on the automatic protection requirements was begun again on October 19, 1983, when the Department of Transportation published a notice seeking comments on a wide range of alternative actions the Department might take with respect to automatic protection in response to the Supreme Court's decision.[55] This notice sought comments on three broad possibilities - retain the automatic protection requirements in their existing form and establish a new compliance date, amend the automatic protection requirements, e.g., to preclude detachable automatic belts, or rescind the automatic protection requirements. In addition, this notice asked for comments on three other supplementary actions that could be taken in conjunction with any of these three broad possibilities. The supplementary actions were: (1) conduct a demonstration program for automatic protection, along the lines of the Coleman decision, (2) seek legislation to encourage States to pass mandatory seat belt use laws, and (3) seek legislation to require vehicle manufacturers to provide consumers with an option to select air bags or automatic belts, instead of manual belts, in their new cars.[56]

The Department received more than 6,000 comments on this notice. After reviewing the comments, the Department published a supplemental notice on May 14, 1984.[57] This notice asked for further comment on alternatives being considered regarding State laws mandating seat belt use, mandatory demonstration programs for automatic crash protection, and the possibility of requiring driver's side air bags on all small cars.

A final rule reinstituting automatic crash protection for cars was signed by Elizabeth Dole, the Secretary of Transportation and published on July 17, 1984.[58] This rule provides for a phase-in of automatic protection in cars beginning September 1, 1986 (the 1987 model year). All cars manufactured after September 1, 1989 (the 1990 model year) are required to provide automatic protection. During the phase-in of automatic protection, the rule encourages manufacturers to choose to install air bags instead of automatic belts by providing a 1.5 car credit for cars with driver air bags and any type of automatic protection

---

[54] 463 U.S. 49-51 (1983)

[55] 48 FR 48622

[56] 48 FR 48632

[57] 49 FR 20460

[58] 49 FR 28962

for the passenger. This rule also encouraged States to pass mandatory seat belt use laws. It did so by including a provision that the automatic protection requirements might be eliminated if the Secretary of Transportation made a determination by April 15, 1989 that enough States had enacted mandatory seat belt use laws that met criteria specified in Standard 208.

There were 16 petitions for reconsideration of this rule. NHTSA published a notice responding to those petitions on August 30, 1985.[59] This notice denied requests to change the phase-in schedule and to eliminate the possibility that automatic protection requirements might be rescinded if enough States enacted mandatory belt use laws. This notice also expanded NHTSA's efforts to encourage manufacturers to install air bags instead of automatic belts by adding a new one car credit. Under this new one car credit provision, manufacturers could comply with the automatic protection requirement by installing an air bag for the driver and manual lap/shoulder belt for the passenger. The one car credit provision was scheduled to remain in effect until the phase-in for automatic protection ended on August 31, 1989.

Shortly after the publication of this response to the petitions for reconsideration, a lawsuit challenging the July 1984 final rule was filed in the U.S. Court of Appeals for D.C. Circuit by State Farm Insurance Company and the State of New York. State Farm's primary argument was that the rule was unlawful because it would revoke automatic crash protection if enough States enacted mandatory belt use laws. New York argued that the rule was unlawful because it failed to require air bags or nondetachable automatic belts as the only permissible means of automatic protection.

The court announced its decision on September 18, 1986 in an opinion titled State Farm v. Dole, 802 F.2d 474 (D.C. Cir. 1986). The court upheld the Secretary's 1984 rule in all respects. The court indicated that State Farm's objection was based upon that insurer's speculation about what the Secretary might do if States passed laws and if those law were determined to comply with certain criteria. The court said that State Farm was not entitled to any legal relief until these potential events and determinations had actually occurred. In response to New York's argument that NHTSA should not have allowed detachable belts to be permitted as automatic protection, the court said: "The Safety Act does not require the Secretary to adopt the technological alternative providing the greatest degree of safety. The Act expressly permits the Secretary to consider such factors as reasonableness and practicality in addition to safety features. Both the Supreme Court and this court, moreover, have recognized the Secretary's authority to consider such factors as cost and public acceptance."[60]

THREE SIGNIFICANT RULES THAT GREW OUT OF THE DOLE RULE ON AUTOMATIC PROTECTION

USE OF A MORE ADVANCED TEST DUMMY - While the rulemaking that led to Secretary Dole's decision was underway, General Motors filed a petition for rulemaking asking

that a new test dummy it had developed, called the Hybrid III dummy, be permitted to be used in testing compliance with the provisions of Standard 208, including the automatic protection requirements. NHTSA granted this petition on July 20, 1984. On April 12, 1985,[61] NHTSA published a notice proposing to incorporate the Hybrid III test dummy as a permissible alternative for Standard 208 compliance testing. NHTSA explained its proposal as follows: "Based on its review of the available test data, the agency recognizes that the Hybrid III test dummy represents an appreciable advancement in the state-of-the-art of human simulation and that its use would be beneficial for continued improvement in vehicle safety."[62] This notice proposed to take advantage of the enhanced capabilities of the Hybrid III by adding new injury criteria for the neck, lower leg, facial laceration, and chest deflection. This notice also proposed to require the use of the Hybrid III test dummy for all Standard 208 compliance testing for cars manufactured on or after September 1, 1991.

NHTSA published a final rule in this area on July 25, 1986.[63] This rule adopted the Hybrid III test dummy for Standard 208 compliance testing and adopted the proposal that the Hybrid III be the exclusive dummy for Standard 208 compliance testing as of September 1, 1991. However, the July 1986 rule adopted only one additional injury criterion for use with the Hybrid III test dummy, the proposed one for chest deflection. In response to petitions for reconsideration of this July 1986 rule, NHTSA postponed the date for mandatory use of the Hybrid III test dummy to allow more time to examine technical issues that might arise from the use of this new test dummy.[64]

After completing its technical examination of these issues, NHTSA published a notice on December 10, 1992[65] proposing that the Hybrid III test dummy would be the only dummy used in Standard 208 compliance testing beginning September 1, 1996. After reviewing the comments on this proposal, NHTSA published a final rule giving one additional year of leadtime before requiring exclusive use of the Hybrid III test dummy.[66] The Hybrid III test dummy is now the only dummy that will be used for Standard 208 compliance testing of vehicles manufactured on or after September 1, 1997 (the 1998 model year).

CRASH TESTING OF MANUAL BELTS IN LIGHT TRUCKS AND VANS - On April 12, 1985, NHTSA published a notice that proposed, among other things, that the occupant protection afforded by manual seat belts installed in front outboard seating positions of light trucks and vans be evaluated according to the same crash test used to evaluate automatic

---

[59] 50 FR 35233

[60] 802 F.2d 474, at fn.23, 486-87 (1986)

---

[61] 50 FR 14602

[62] 50 FR 14603

[63] 51 FR 26688

[64] 53 FR 8755; March 17, 1988

[65] 57 FR 58437

[66] 58 FR 59189; November 8, 1993

protection.[67] NHTSA adopted this proposal in a final rule published November 23, 1987.[68] The November 1987 rule required light trucks and vans manufactured on or after September 1, 1991 (the 1992 model year) to be certified as complying with this crash testing requirement.

"ONE CAR CREDIT" FOR DRIVER AIR BAGS EXTENDED UNTIL AUGUST 31, 1993 - On June 11, 1986, Ford Motor Company filed a petition asking NHTSA to permit the production of cars with driver air bags and no automatic protection for the front seat passenger after September 1, 1989, the date on which this "one car credit" provision was scheduled to expire. In its petition, Ford said that, if this request were granted, Ford would "in all likelihood" install driver air bags in the majority of its North American-designed cars. NHTSA published a notice proposing to grant Ford's petition on November 25, 1986.[69] In that notice, NHTSA indicated that the proposed extension of the one car credit "would encourage the orderly development and production of passenger cars with full-front air bag systems."[70]

Comments supporting the proposed extension of the one car credit were submitted by air bag suppliers, insurance companies and their trade associations, vehicle manufacturers and their trade associations, and researchers and other organizations involved in highway safety issues. The only comments opposing this extension were submitted by the Center for Auto Safety and Robert Phelps, a private citizen. After considering these comments, NHTSA published a final rule extending the one car credit until August 31, 1993.[71] The agency explained that the extension of the one car credit would promote the widespread introduction of air bags. In addition, the agency concluded that there were a number of technical issues that still needed to be resolved before widespread introduction of passenger air bags would occur.

One petition for reconsideration of this rule was filed by Public Citizen, a group that had not previously participated in this rulemaking. NHTSA denied this petition in a notice published November 5, 1987.[72] In the denial, NHTSA again explained that it was seeking to encourage manufacturers to install air bags, instead of automatic belts, and that all available evidence indicated that the extension of the one car credit had increased the likelihood of widespread use of air bags.

Public Citizen filed a lawsuit in the U.S. Court of Appeals for the D.C. Circuit challenging NHTSA's extension of the one car credit until 1993. The court announced its decision on July 15, 1988 in an opinion titled Public Citizen v. Steed, 851 F.2d 444 (D.C. Cir. 1988). The court unanimously upheld the extension of the one car credit provision. The court specifically

found that NHTSA had extended the one car credit to encourage greater installation of air bags, because the agency believed that air bags would offer long-term overall safety gains for vehicle occupants. The court said that, even if it accepted Public Citizen's assertion that the one car credit extension would permit a near-term reduction in front seat occupant protection, "[i]t is within NHTSA's province to balance estimated long-term safety benefits against the possibility of a marginal short-term reduction in safety."[73]

AUTOMATIC PROTECTION IN LIGHT TRUCKS AND VANS

NHTSA had not taken up the question of automatic protection in light trucks and vans again since its June 1977 notice indefinitely suspending the requirements for automatic protection in those vehicles. However, the agency's rulemaking on crash testing of manual belt systems in light trucks and vans had raised the issue of occupant protection in those vehicles. On January 9, 1990, NHTSA published a notice proposing to require automatic protection to be phased in for light trucks and vans in a manner that closely paralleled the recently completed phase-in of automatic protection for passenger cars.[74] A final rule requiring automatic protection in light trucks and vans was published on March 26, 1991.[75] This rule requires that 20 percent of each manufacturer's model year 1995 production of light trucks and vans provide automatic protection, 50 percent of model year 1996 production provide automatic protection, 90 percent of model year 1997 production provide automatic protection, and all 1998 light trucks and vans must provide automatic protection.

AIR BAGS REQUIRED AS THE MEANS OF AUTOMATIC PROTECTION

On December 18, 1991, then President Bush signed into law the Intermodal Surface Transportation Efficiency Act (ISTEA). Among other things, ISTEA requires that all passenger cars manufactured on or after September 1, 1997 (the 1998 model year) and light trucks and vans manufactured on or after September 1, 1998 (the 1999 model year) provide air bags at the driver and right front passenger positions. In response to this mandate, NHTSA published a notice on December 14, 1992.[76] This notice proposed to require that passenger cars and light trucks and vans comply with the automatic protection requirements by installing air bags and manual lap/shoulder seat belts at the driver and right front passenger positions. This NPRM also proposed to require that these vehicles have a label on the sun visor providing occupants with important safety information about the air bags and advising occupants that they must always wear their safety belts for maximum safety

---

[67] 50 FR 14589; April 12, 1985

[68] 52 FR 44898

[69] 51 FR 42598

[70] 51 FR 42599

[71] 52 FR 10096; March 30, 1987

[72] 52 FR 42440

[73] 851 F.2d 444, at 449 (1988).

[74] 55 FR 747

[75] 56 FR 12472

[76] 57 FR 59043

protection in all types of crashes. This proposal was adopted as a final rule in a notice published September 2, 1993.[77]

SUMMARY

The regulatory history of the automatic protection requirements in Standard 208 is obviously a long and complicated one that begins in the late 1960's and will continue into the late 1990's. However, there are eight major events that have been especially significant in that history. These are:

1. The 1972 court decision in Chrysler v. DOT. In this case, the court overturned the automatic protection requirements, although the court found that NHTSA had authority to promulgate such requirements and that the automatic protection requirements that were then established met the need for motor vehicle safety. However, the requirements were invalid because the specifications for a test dummy were inadequate.

2. The 1973 ignition interlock option and the 1974 Congressional disapproval of such an option. NHTSA thought it had an alternative to automatic protection that would allow it to achieve roughly the same benefits as automatic protection without getting into the technical and cost issues associated with automatic protection. However, the ignition interlock was so unpopular that Congress amended Federal law to provide that NHTSA could not require or even permit manufacturers to comply with a safety standard by means of an interlock.

3. The 1976 decision by Secretary Coleman to implement a demonstration program. This program was intended to resolve any lingering concerns about the public acceptability and real world effectiveness of automatic protection.

4. The 1977 decision by Secretary Adams to reimpose the automatic protection requirements on a phased-in schedule. This would have required passenger cars to provide automatic protection beginning in the 1982 model year.

5. The 1981 decisions by Secretary Lewis and Administrator Peck to rescind the automatic protection requirements. This rescission was based on the changed economic circumstances and the likely insignificance of the safety benefits that would result if vehicles provided automatic protection by means of detachable automatic belts.

6. The 1983 decision by the Supreme Court declaring the 1981 rescission of the automatic protection requirements unlawful. This decision guided the Department's subsequent consideration of these requirements.

7. The 1984 decision by Secretary Dole to reinstate the automatic protection requirements on a phased in schedule. This decision became the first requirement for automatic protection that actually went into effect.

8. The 1991 Federal law requiring that air bags, supplemented by manual lap/shoulder seat belts, be the means of automatic protection offered in all new cars by the 1998 model year and in all new light trucks and vans by the 1999 model year.

Most of the comments on the NHTSA proposal to implement the 1991 Federal law mandating air bags were directed toward the agency's proposed language for labels to be required on sun visors and on the proposed exemption procedures, with almost nothing said about air bags. This probably reflects the fact that all the commenters knew that Federal law mandated an air bag requirement, regardless of the comments. It is nevertheless ironic that 24 years after an automatic protection requirement was first proposed, the requirement to provide air bags in all passenger cars and light trucks and vans was adopted with so little comment. One would not have predicted this after all the regulatory notices, lawsuits, and other high profile actions that have been associated with the automatic protection requirements.

---

[77] 58 FR 46551

Fourth Report to Congress

Effectiveness of Occupant Protection Systems and Their Use

National Highway Traffic Safety Administration
U. S. Department of Transportation
Washington, D. C. 20590

May 1999

EXECUTIVE SUMMARY

The Intermodal Surface Transportation Efficiency Act (ISTEA) of 1991, enacted by Congress on December 18, 1991, directed the Secretary of Transportation to report on the effectiveness of occupant protection systems based on their actual use, and on lap and shoulder belt use by the public and various groups at both the state and national levels (Section 2508 (e)). This is the fourth report on the effectiveness of occupant protection systems and safety belt use.

The major findings of this report are presented below.

System Effectiveness

o   Air bags provide *fatality protection* in potentially fatal crashes. Drivers protected by air bags experienced reduced fatality risk of 31 percent in purely frontal crashes (12:00 point of impact on the vehicle), 19 percent in all frontal crashes (10:00 to 2:00), and 11 percent in all crashes.

o   Based on 11 percent effectiveness in all crashes, it is estimated that air bags have saved 2,263 lives from 1987 through 1997, including 842 lives saved in 1997 alone.

o   Driver air bags appear to be about as effective in reducing fatality risk in purely frontal crashes for light trucks (36 percent) as they are in passenger cars (31 percent).

o   With the increase in available numbers of fatal crashes involving driver air bag-equipped cars over the past few years, it is possible to estimate, separately, the effects of driver air bags when the driver was belted and when the driver was unbelted. Air bags provide about a 9 percent reduction in fatality risk for the belted driver (relative to a belted driver without air bags), and 14 percent for the unbelted driver in all crashes.

o   The 9 percent effectiveness of air bags for belted drivers, coupled with the 45 percent effectiveness of lap-shoulder belts, yields an estimated 50 percent fatality-reducing effectiveness for the air bag plus lap-shoulder belt system when safety belts are used.

o   In purely frontal crashes, passenger air bags appear to be about as effective (32 percent) for right-front passengers age 13 and older as driver air bags (31 percent) are for drivers.

o   For right-front passengers less than 13 years old, analysis of frontal crashes shows a higher fatality risk in cars with dual air bags than for children in comparable cars without passenger air bags. Given the limited data, it is impossible to quantify the increase in risk accurately at this time.

o   As early as December 1991, the agency issued a consumer advisory warning against placing rear-facing child safety seats in front of passenger side air bags.

o   Concerning ***overall injury reduction for drivers***, for serious injury, the air bag plus lap-shoulder belt (when used) and manual lap-shoulder belts alone each provided about 64 percent reduction in injury risk, while automatic belts exhibited 49 percent effectiveness when they were used. The estimated effectiveness of the air bag alone was 42 percent (not statistically significant). The number of occupants in passenger air bag seating positions is still relatively small, and thus, no analyses were conducted.

o   The combination of an air bag plus use of the lap-shoulder belt provides the greatest moderate injury protection (66 percent) followed by manual lap-shoulder belts (53 percent), automatic belts (51 percent) and the air bag alone (10 percent, nonsignificant).

o   Exploratory analyses of these data indicate that current air bags involve a trade-off among certain types of injury. The addition of an air bag to the lap-shoulder belt user increases head injury protection at both the moderate and serious injury levels, as well as chest injury protection at the moderate injury level, while at the same time increasing the risk of moderate and serious arm injury. However, injuries to the head and chest pose much greater threat to life than do arm injuries.

o   Certain challenges regarding air bag deployment have materialized, which are discussed in this report. The first involves the increased risk of upper extremity injury associated with air bag deployment. The second, and more challenging issue, involves the child-passenger air bag interaction. On November 22, 1996, NHTSA announced a comprehensive approach to preserve the safety benefits of air bags while minimizing their danger to children and at-risk adults. Its approach centers on accelerating the development of advanced air bag technology for future vehicles with the intent of having systems available for 1999 models. More immediate measures include adopting enhanced warning labels, reducing the aggressivity of air bags, continuing to allow the use of manual on-off switches in vehicles without a rear safety to protect children, and allowing dealers to install a manual on-off switch for any eligible owner who requests it and receives approval from NHTSA.

o   As of September 1, 1998, NHTSA has confirmed 90 crashes where the deployment of the passenger-side air bag resulted in 24 serious injuries, one fatal abdomen injury, and 65 fatal head or neck injuries to infants or children. Twenty-four involved infants in rear-facing child seats, including fifteen deaths. Of the remaining 51 fatalities (children not in rear facing child safety seats), 2 were in forward facing child safety seats that were improperly secured to the vehicle. Forty six of the 51 were out position, unrestrained or improperly restrained at the time of the crashes, 42 of which involved pre-impact braking and/or out-of-position children, placing the child in proximity to the deploying air bag. Three of the 51 fatally injured children, not in a rear facing child safety seat, were determined to have been wearing the lap and shoulder belt. However, it is unknown whether the children were seated in a correct posture position and if the belts were snug. Pre-impact braking, coupled with improper or no safety belt use, generally results in the child moving forward into proximity with the passenger-side air bag prior to the actual crash and subsequent air bag deployment

Public Information and Rulemaking

o   In addition to pursuing technological advancements, NHTSA has launched a comprehensive public education program designed to: (1) alert the public to the dangers air bags pose to children and at-risk adults, (2) increase the correct use of safety belts, and (3) increase the proper positioning and use of child safety seats.

o   To ensure that infants and children ride safely, with or without a passenger-side air bag, NHTSA issued a strong warning in a press release dated October 27, 1995. This warning and advisory urged care givers to follow three "rules":

--   Make sure *all* infants and children are properly restrained in child safety seats or lap and shoulder belts for every trip.

--   The *back seat* is the safest place for children of any age.

--   Infants riding in rear-facing child safety seats should *never* be placed in the front seat of a vehicle with a passenger-side air bag.

o   On November 9, 1995, NHTSA published a request for comments to inform the public about its efforts to reduce the adverse effects of air bags and to invite the public to share information and views with the agency (60 FR 56554). The request for comments focused on possible technological changes to air bags to reduce their adverse effects, including possible regulatory changes.

o   In early 1996, the Department and NHTSA successfully led the effort to create a private-public partnership which would undertake and fund a national program to address air bag safety issues. The resulting coalition, now called the Air Bag and Seat Belt Safety Campaign (ABSBSC), and NHTSA worked closely together on a comprehensive set of program activities designed to preserve the lifesaving benefits of air bags, to alert the public to the proper use of those devices, and to increase the correct use of safety belts and child safety seats.

o   On August 6, 1996, NHTSA published a notice of proposed rulemaking, proposing amendments to NHTSA's occupant crash protection standard and child restraint standard to reduce the adverse effects of air bags, especially those on children. The agency proposed that vehicles without advanced passenger-side air bags would be required to have new, attention-getting warning labels and permitted deactivation of the passenger-side air bag. NHTSA also proposed to require rear-facing child seats to bear new, enhanced warning labels. Finally, this notice discussed the agency's research on other air bag issues, such as technology to reduce arm and other injuries to drivers.

o   On November 27, 1996, NHTSA published a final rule (61 FR 60206) requiring vehicles with air bags to bear three new warning labels. Two of the labels replace existing labels on the sun visor. The third is a temporary label on the dash. These labels would not be required on vehicles having an advanced passenger-side air bag. This rule also requires rear-facing child seats to bear a new, enhanced warning label. The domestic and import vehicle manufacturers are sending letters to the owners of passenger air bag-equipped vehicles apprising them of the adverse effects of air bags. The sun visor labels will be included with the letter.

o   Continuing the Department of Transportation's comprehensive effort to preserve the benefits of air bags and minimize their risk, a Notice of Proposed Rulemaking was published in January, 1997, offering proposals for deactivating air bags. In parallel with this effort, rulemaking deliberations proceeded concurrently on the issue of reducing the aggressivity of air bags. At the same time, a final rule was issued, extending until September 1, 2000, the time period during which vehicle manufacturers would be permitted to offer manual cut-off switches for the passenger-side air bag for vehicles without rear seats or with rear seats that are too small to accommodate rear-facing infant seats.

o   On March 19, 1997 (62 FR 12960) the final rule, allowing manufacturers to quickly implement redesigned air bags, was issued. The new rule paved the way for manufacturers to install air bags that were redesigned, with the goal of mitigating the injurious effects of air bags, while maintaining their proven benefits. Manufacturers responded by installing air bags that are 20 percent to 35 percent less powerful. The new rule also provides manufacturers and suppliers with additional time to develop a variety of advanced air bag technologies to tailor air bag deployment more appropriately to crash severity, occupant size and position, seat belt use and other vehicle factors. This rule provided changes that would affect new production vehicles. The issue of what to do about vehicles already on the road came next. In addition, this release repeated prior agency warnings of the dangers of placing a rear-facing infant seat in front of an air bag, and broadened the previous warnings to apply to older children and even adults who may ride unrestrained.

o   On November 21, 1997, the agency announced the final rule permitting deactivation. Auto dealers and service outlets could begin deactivating air bags on January 19, 1998.

o   To give manufacturers and suppliers the maximum benefits of currently unproven but promising advanced air bag technology, the agency has decided that the optimal path for future rulemaking should be performance-based requirements. This approach would permit varied choices and options in order to provide manufacturers with utmost flexibility to design new systems. Toward this end, the agency published a Notice of Proposed Rulemaking dealing with advanced air bags on September 18, 1998, with a Final Rule to be published one year later.

Safety Belt Use

In April, 1997, the Department and NHTSA presented to President Clinton a comprehensive set of program initiatives designed to increase national seat belt and child safety seat use: The Presidential Initiative to Increase Seat Belt Use Nationwide. Strategies supporting the initiative include: (1) the formation of public-private partnerships, (2) improved seat belt and child passenger safety legislation, (3) high visibility enforcement of those laws, and (4) intensive public information and education programs. This initiative was launched in October 1997 as the Buckle

Up America campaign. This campaign encourages the increased use of seat belts and child safety seats through more effective public education about and enforcement of existing seat belt and child safety seat use laws and the passage of standard (primary) enforcement provisions in state belt use laws, including other improvements to strengthen those laws.

As of December, 1997, state surveys indicated safety belt use rates ranging from 48 percent in Arkansas and Mississippi to 88 percent in California. An estimate of national safety belt use is derived through a population-weighted average of these state use rates. *The national safety belt use rate as of December 1997 was estimated to be 69 percent.*

Belt use information is not routinely collected for the military, government employees, or law enforcement personnel. However, based upon the existence of mandatory use policies, training programs, and promotional campaigns, use among these groups is expected to be higher than in the general population.

The overall observed safety belt use rate in NHTSA's 1996 National Occupant Protection Use Survey (NOPUS) moving traffic study was 61.3 percent, compared to the 58.0 percent observed in 1994. Shoulder belt use observed in the 1996 moving traffic study was 64.4 percent for passenger car occupants compared to 62.8 percent in 1994 and 56.4 percent for light truck occupants, compared to 50.2 percent in 1994.

Direct comparison of findings between the NOPUS and state surveys is difficult, primarily because of the differences in vehicle and occupant coverage. However, a rough comparison of overall use can be made between the state-based estimate for 1996 of 68 percent and the NOPUS estimate for passenger car drivers and passengers of 64.4 percent. In this comparison, the state based estimate falls within the 95 percent confidence interval of the NOPUS estimate.

The combination of surveys that has been used to measure safety belt use over the past several years also provides insight with regard to changes in use rates. Until 1990, the 19 cities survey was used as the index of national use. In 1990, that index for passenger car drivers was 49 percent. The NOPUS estimate of belt use among passenger car drivers in 1996 is 65.1 percent. The difference of 16 percentage points between the 19 cities index and the NOPUS estimate is consistent with the 15 percentage point change in use indicated by the aggregate of state surveys between 1990 and 1996 (i.e., 53 percent in 1990 and 68 percent in 1996).

State surveys provide an essential source of information for monitoring progress in the states. The NOPUS provides a probability-based sample of national use with the ability to estimate statistical error. In addition, the NOPUS provides a unique source of detailed information concerning restraint use by vehicle type, age, gender, race, geographic area, time of day, day of week, urbanization, etc.

U.S. Department of Transportation

National Highway Traffic Safety Administration

http://www.nhtsa.dot.gov

FINAL ECONOMIC ASSESSMENT

FMVSS NO. 208 ADVANCED AIR BAGS

Office of Regulatory Analysis & Evaluation
Plans and Policy
May, 2000

EXECUTIVE SUMMARY

This Final Economic Assessment analyzes the potential impact of new performance requirements and test procedures for advanced air bag systems. Consistent with the National Highway Traffic Safety Administration Re-authorization Act of 1998, which is part of the Transportation Equity Act for the 21st Century (TEA 21), the intent of this rulemaking is to minimize risks caused by air bags to out-of-position occupants, especially infants and children, and to improve occupant protection provided by air bags for belted and unbelted occupants of all sizes. To achieve these goals, NHTSA is requiring vehicles to meet test procedures that broaden the scope of the current standard to ensure that occupants are properly protected under a wider variety of crash circumstances.

Test Requirements

The risk of injury from air bags arises when occupants are too close to the air bag when it inflates. Generally, those most at risk from injury are infants, young children, and out-of-position drivers. To address these concerns, new tests employ crash dummies representing infants, 3-year olds, 6-year olds, and 5th percentile female drivers. There are a variety of tests to protect these at-risk occupants. These tests generally require either that the air bag be suppressed if certain risk conditions exist or that deployments occur at levels that produce a low probability of injury risk. For purposes of this analysis, it is assumed that manufacturers will choose the low-risk deployment option for drivers. On the passenger side, the costs and benefits of two options are examined. Option 1 assumes the automatic suppression test will be met by using a weight sensor for the infant, 3 and 6 year old dummies or a weight and presence sensor. Option 2 assumes a weight sensor for infants and a low-risk air bag for the 3 and 6 year old dummies for the out-of-position tests.

The assessment analyzes three alternative sets of high speed tests to preserve and enhance air bag protection. Each set of tests includes belted and unbelted full frontal perpendicular rigid barrier tests using 5th percentile female and 50th percentile male crash dummies, 30 degree oblique tests into a rigid barrier using unbelted 50th percentile male dummies, and 40 percent offset frontal deformable barrier tests using belted 5th percentile female dummies. While Alternatives 1 and 2 both require a 0 to 48 kmph (0-30 mph) belted test for the 5th percentile female and 50th percentile male dummy, the primary difference between Alternatives 1 and 2 is their treatment of unbelted occupants. Alternative 1 would require an unbelted 32 to 40 kmph (20 to 25 mph) frontal rigid barrier test, while Alternative 2 would require an unbelted 32 to 48 kmph (20 to 30 mph) frontal rigid barrier test. Alternative 3 is the final rule. It is the same as Alternative 1 (an unbelted 32 to 40 kmph [20 to 25 mph] frontal rigid barrier test), but increases the speed of the belted test for the 50th percentile male dummy test to 0-56 kmph (0-35 mph). Chapter I provides the detail of the three alternative sets of high speed tests.

NHTSA is also upgrading the injury criteria applicable to the existing 50th percentile male dummy, and applying appropriate injury criteria to each of the new dummies in this rule. These criteria are used to assess the risk of injury. The new criteria will change the way head injuries are measured, include a measure of neck injury, and reduce the allowable chest deflection during the tests.

Technical Feasibility

The agency has tested three vehicles to most of the proposed tests. These are the Dodge Intrepid, the Toyota Tacoma, and the Saturn SL1. The Saturn passed all of the 30 mph rigid barrier tests, the static low risk deployment tests on the driver side, and the 35 mph belted test with the 50th percentile dummy. It did not meet the static low risk deployment tests on the passenger side. However, with the addition of a weight sensor, the agency believes the 1999 Saturn could pass the passenger side suppression tests. The Saturn performed better in these tests overall than the Intrepid or Tacoma. The Saturn SL1 has a soft crash pulse and it has a different air bag design than most vehicles with an unusual tether design in the center of the air bag. The agency believes that, at a minimum, different designs, more advanced sensors, and multi-stage inflators would be required in many vehicles to pass all of the tests considered in the three alternatives.

The agency also tested 11 other vehicles to understand how they would perform in different test conditions, most notably, in the high speed unbelted tests. These tests show that model year (MY) 1998-99 air bags generally meet our new injury criteria for unbelted 50th percentile male dummies in a 48 kmph (30 mph) unbelted test and for 5th percentile female dummies in a 32 kmph (25 mph) unbelted test. Five of twelve vehicles tested met our new injury criteria for 5th percentile female dummies in 48 kmph (30 mph) unbelted tests on the driver side and five of eleven met the new criteria on the passenger side. The data suggests that, at a minimum, design changes, such as recessing the air bag, improving fold patterns, and installing internal baffles in the air bag to assure safer deployment would be required for 50th male and 5th female dummies to simultaneously meet our injury criteria in 48 kmph (30 mph) belted and unbelted tests. The body of tests suggests that meeting the injury criteria for both the 50th percentile male and 5th percentile female in unbelted 48 kmph (30 mph) rigid barrier tests, while at the same time meeting the out-of-position tests, is a complex job. Adding pretensioners to belt systems may be needed by some vehicles to meet the 35 mph belted test with the 50th percentile male dummy.

It should be noted that there is significant complexity in air bag testing and technology that will be required by this final rule. We are requiring the use of a new test dummy (the 5th percentile female dummy) in high speed tests, adding a new test (offset belted), adding new neck injury criteria, and making existing injury criteria more stringent (chest deflection). We are also adding an entire new series of low-speed tests, which will require manufacturers to install air bag suppression systems or low risk deployment systems, or both. Simultaneously meeting the performance requirements of the low speed tests and the unbelted test speed will require the introduction of risk reduction technologies and increase the technical complexity in system design.

Benefits

The assessment provides analyses of the safety benefits from tests that reduce the risk of injury from air bags in low-speed crashes, as well as from tests that improve the overall effectiveness of air bags in high speed crashes. The agency estimates that in a fleet fully equipped with pre-model year (MY) 1998 air bags, there would be 46 drivers, 18 infants, 105 children, and 18 adult passengers (187 occupants in total) at risk of being killed by air bags annually because they were out of position when the air bag deployed in low speed [< 40 kmph (25 mph) delta-v] crashes. A variety of technologies would be required to prevent these deaths, including weight or presence sensors to suppress the air bag, multi-stage inflators, and low risk deployment air bags. Of the

187 potential at-risk fatalities, NHTSA estimates that suppression technologies could prevent up

to 93 fatalities, low-risk air bags could prevent up to 154 fatalities, and multi-stage inflation

systems could prevent up to 179 fatalities when combined with weight sensors used to suppress

the air bag. Thus, more than 95 percent of the fatalities seen to date in low speed deployments

could be eliminated by technologies used to meet the test requirements.

NHTSA also estimates that a fully air bag equipped fleet would result in serious to critical severity

(MAIS 3-5) nonfatal injury caused by air bags to 38 drivers, 9 infants, 200 children, and 15 adult

passengers that would be out of position in low speed crashes. Of these 262 serious but nonfatal

injuries, suppression technologies could prevent 151 injuries, low-risk air bags could prevent 191

injuries, and multi-stage inflation systems could prevent up to 252 injuries when combined with a

weight sensor. Thus, more than 95 percent of the air bag caused injuries in low speed

deployments could be eliminated by technologies used to meet the test requirements.

There is some question about the reliability of suppression and low risk deployment

countermeasures and further development of these countermeasures is necessary. To the extent

that these systems are not as reliable as assumed, children and small adults would continue to be

at risk. Even if suppression and low risk deployment technologies are completely reliable, there

will remain some out-of-position individuals subject to the full force of the air bag under certain

circumstances. The risks to out-of-position individuals could be greater with an air bag designed to provide a 30 mph unbelted performance compared to an air bag designed to provide 25 mph unbelted performance.

In addition to minimizing the risk to out of position occupants, this rulemaking seeks to improve occupant protection provided by air bags for both belted and unbelted occupants of all sizes, with new tests and new injury criteria. Among the tests this analysis examines are three different high speed tests that would improve the performance of air bags. These include the 25 mph offset test for belted 5th percentile female dummies, the 30 mph rigid barrier test for both belted and unbelted 5th percentile female dummies, and the 35 mph rigid barrier test for belted 50th percentile male dummies. A variety of technologies could be used to comply with these tests including modified air bag fold patterns, improved inflators, added sensors, multi-stage inflators, and pretensioners. Air bag systems designed to comply with the 25 mph offset test would, over the lifetime of one model year's production, save 20-28 more lives and prevent 134-262 more nonfatal injuries than the pre-MY 1998 baseline vehicles. Systems designed to the 30 mph tests with the 5th percentile female dummy would save 23 more lives (4 belted and 19 unbelted) and prevent 184 more nonfatal injuries (43 belted and 141 unbelted). Systems that meet the 35 mph rigid barrier test with the belted 50th percentile male dummies would save from 0-4 more lives and prevent 256 to 486 more nonfatal injuries.

Table E-1 summarizes the estimated benefits from the low speed tests and from the high speed tests, excluding the difference for the unbelted high speed tests (25 mph or 30 mph).

Table E-1
Estimated Range of Benefits for Low Speed Tests,
Offset Tests, 5th Female and 50th Male Belted Tests, and New Injury Criteria

Alternative	Fatalities Reduced	Injuries Reduced
#1	117 - 211	328 - 557
#2	136 - 230	469 - 698
#3	117 - 215	584 - 1,043

The most contentious issue of this rulemaking is whether the unbelted tests should be set at 40 kmph (25 mph) or 48 kmph (30 mph). Estimates of the relative impact of the unbelted high speed tests are subject to a degree of uncertainty for several reasons, not the least of which is the fact that no vehicles were ever subject to a 25 mph unbelted standard. We cannot estimate the most likely difference between setting the unbelted tests at the two different levels, because it depends on how the manufacturers would meet the alternative performance requirements.

In the preamble to the final rule, we discuss in detail our reasons for believing that it is unlikely that vehicle manufacturers will significantly depower their air bags compared to the MY 1998-2000 fleet. Vehicle manufacturers have not depowered their air bags so much that they minimally comply with the sled test. Crash tests and field experience to date with vehicles certified to the sled test have indicated that there has not been a loss of frontal crash protection compared to pre-MY 1998 vehicles. If, as we expect, the manufacturers keep the same level of power as they currently have in MY 1998-2000, even with a 25 mph unbelted test requirement, then the difference in actual benefits between the two test speeds would be small or even eliminated.

At the same time, we cannot rule out the possibility that air bags will be significantly depowered. To account for this possibility, we calculated a "worst case" scenario comparing the benefits at the minimum performance requirements of each speed. We derived point estimates using two different methods and different sets of assumptions. We estimate that vehicles designed with 30 mph air bags could provide 229 or 394 more lives saved than vehicles designed with minimally compliant 25 mph air bags. However, we also estimate that 30 mph air bags could result in an additional 1,345 serious injuries[1] compared to vehicles designed with 25 mph air bags. These point estimates do not necessarily define the full range of possible outcomes due to uncertainty regarding both data and assumptions under each method.

The total benefits from tests that reduce the risk of injury and tests that improve occupant protection are combined in Table E-2 for the three alternatives. The range of benefits provided in Table E-2 assume the worst case difference between vehicles designed to meet the 25 mph unbelted test and vehicles designed to meet the 30 mph unbelted test at the low end of the range and assume there is no difference in benefits between the 25 mph unbelted test and the 30 mph unbelted test at the high end of the range. The high end of the range is based upon the assumption that manufacturers might make no changes in their current vehicles even with a 25 mph unbelted standard.

---

[1] The less aggressive single-stage air bag that can be designed to a 25 mph unbelted test can result in fewer air bag caused injuries at low speeds than an air bag designed to a 30 mph unbelted test. Thus, single-stage air bags designed to a 30 mph unbelted test can prevent more fatalities, while single-stage air bags designed to a 25 mph unbelted test can prevent more injuries. Multi-stage air bags are assumed to provide the same level of benefits during the first stage, whether the second stage is designed for a 25 mph unbelted test or a 30 mph unbelted test.

The agency estimates that the 30 mph generic sled test is roughly equivalent to a 22 mph rigid barrier perpendicular (0 degree) crash. During the depowering rulemaking, we looked at the relative safety consequences of an air bag designed to just meet the performance requirements associated with a 30 mph generic sled test. The agency estimated the fatality impacts of designing a vehicle to minimally meet the performance requirements imposed by the current 30 mph generic sled test and compared these to the fatality impacts of designing a vehicle to just meet the 25 mph unbelted rigid barrier test. Assuming there is no impact on air bag size, air bags designed to the 25 mph unbelted rigid barrier test could save 64 to 144 more lives than air bags designed to the generic sled test (assumed to be 22 mph). Assuming air bags designed to the generic sled test would be reduced in size and provide no benefit in partial frontal impacts, since the 25 mph unbelted rigid barrier test includes an up to 30 degree oblique test for the 50th percentile dummy while the generic sled test has no angular component, 282 to 308 more lives (this range includes the 64 to 144 estimates mentioned earlier) could be saved by air bags designed to the 25 mph unbelted rigid barrier test with the oblique test than lives saved by air bags designed to just comply with the generic sled test.

Costs

Potential compliance costs for the Final Rule vary considerably and are dependent upon the method chosen by manufacturers to comply. Methods such as modified fold patterns and inflator adjustments can be accomplished for little or no cost, given enough leadtime. More sophisticated solutions such as proximity sensors can increase costs significantly. Dynamic presence sensors (the technology assumed for the high end costs of Option 1) are not available at this point in time. They have not been refined to the point that they are in use in vehicles and are not required by tests in any Alternative. However, they have the potential to provide more benefits on the passenger side than weight sensors or low risk air bags. Dynamic presence sensors could be used by manufacturers to meet the test requirements in the future. As such, the cost and benefits of these systems have been estimated. The range of potential costs for the compliance scenarios examined in this analysis is $21-$128 per vehicle (1997 dollars). This amounts to a total potential annual cost of up to $2 billion, based on 15.5 million vehicle sales per year.

Property Damage Savings

Compliance methods that involve the use of suppression technology have the potential to produce significant property damage cost savings because they prevent air bags from deploying unnecessarily. This saves repair costs to replace the passenger side air bag, and frequently to replace windshields damaged by the air bag deployment. Property damage savings from these requirements could total up to $85 over the lifetime of an average vehicle. This amounts to a potential cost savings of $1.3 billion.

Net Cost Per Fatality Prevented

Estimates were made of the net costs per equivalent fatality prevented. The low end of the range for both Alternative1 and Alternative 3 Option 1 scenarios produced no positive net benefits. This reflects the conflicting impacts on fatalities and injuries that result from air bags designed to just meet an unbelted 25 mph test. Lives are not saved in high speed crashes, but nonfatal injuries are prevented in lower speed crashes. The positive impact on nonfatal injuries almost totally offsets the negative impact on fatalities. For the high end of the Option 2 scenarios, property damage savings have the potential to offset all, or nearly all of the cost of meeting this final rule.

In these cases, both net costs and safety impacts are positive so there is no cost per equivalent fatality, just cost savings and safety benefits.

Conclusions

Table E-2 summarizes the costs and benefits of the different Alternatives.

Table E-2
Summary of Costs and Benefits

	Cost Per Vehicle	Total Costs (Billions)	Lifetime Property Damage Savings Per-Vehicle	Total Property Damage Savings (Billions)	Net Consumer Costs (Savings) (Billions)
Alternative 1, Option 1	$21-$124	$0.32 - $1.93	$12-$85	$0.19-$1.31	$0.13-$0.61
Alternative 1, Option 2	$24-$65	$0.37-$1.01	$12-$85	$0.19-$1.31	$0.18-($0.30)
Alternative 2, Option 1	$21-$125	$0.32-$1.93	$12-$85	$0.19-$1.31	$0.13-$0.62
Alternative 2, Option 2	$24-$66	$0.37-$1.02	$12-$85	$0.19-$1.31	$0.18-($0.29)
Alternative 3, Option 1	$23-$128	$0.36-$1.98	$12-$85	$0.19-$1.31	$0.17-$0.67
Alternative 3, Option 2	$27-$68	$0.41-$1.06	$12-$85	$0.19-$1.31	$0.22-($0.25)

	Annual Fatalities Prevented	Annual Nonfatal Injuries Prevented	Total Equivalent Fatalities Prevented	Net Cost Per Fatality Saved (Millions)	
Alternative 1, Option 1	-233 to 211	1,710-1,902	-24 to 316	NS - $1.9M	
Alternative 1, Option 2	-202 to 209	1,756-1,891	6 to 313	$30.9M - NC	
Alternative 2, Option 1	162 to 230	498-2,059	168 to 342	$0.8M - $1.8M	
Alternative 2, Option 2	204 to 228	861-2,048	231 to 339	$0.8M - NC	
Alternative 3, Option 1	-233 to 215	1,966-2,388	-5 to 356	NS - $1.9M	
Alternative 3, Option 2	-202 to 213	2,012-2,377	25 to 353	$9.0M - NC	

NS = Negative safety benefits
NC = No cost, or a net cost savings

Alternative 1 includes: 20-25 mph unbelted test, 0-30 mph belted test, 0-25 mph offset belted test
Alternative 2 includes: 20-30 mph unbelted test, 0-30 mph belted test, 0-25 mph offset belted test
Alternative 3 includes: 20-25 mph unbelted test, 0-30 mph belted test for 5th female, 0-35 mph belted test for 50th male, 0-25 mph offset belted test
Option 1 includes, passengers up to 6 years old suppression, driver low risk air bag
Option 2 includes, infant suppression, passenger low risk air bag, driver low risk air bag

2001-01-0157

Advanced Air Bag Systems and Occupant Protection: Recent Modifications to FMVSS 208

John E. Hinger and Harold E. Clyde
Exponent® Failure Analysis Associates, Inc.

Copyright © 2001 Society of Automotive Engineers, Inc.

ABSTRACT

Because of a rising number of air bag related injuries and specific Congressional instructions, FMVSS 208 was revised in March 1997. At that time, the changes allowed manufacturers to quickly implement redesigned air bags that were less powerful with the goal of reducing air bag related injuries. The legislature has since mandated additional revisions to FMVSS 208 to ensure use of new technologies for the protection of occupants of varying stature. This paper presents an overview of some of the more significant modifications to FMVSS 208 and discusses some of the challenges for advanced air bag systems.

INTRODUCTION

Frontal crash protection for vehicle occupants has been extensively debated for many years. Frontal crashes are the leading cause of fatalities to front seat occupants, particularly to unrestrained occupants. On May 12, 2000, the National Highway Traffic Safety Administration (NHTSA) published an amended version of Federal Motor Vehicle Safety Standard (FMVSS) 208 for occupant safety in motor vehicles. Changes in FMVSS 208 were made pursuant to Congressional instruction to NHTSA through the Transportation Equity Act for the 21st Century which required NHTSA to "improve occupant protection for occupants of different sizes, belted and unbelted ... while minimizing the risk to infants, children, and other occupants from injuries and deaths caused by air bags, by means that include advanced air bags."[1] The new requirements will help ensure that advanced air bag technologies, meant to improve the injury reducing capabilities of air bag systems, will be installed in future motor vehicles. The key modifications, included in this standard, were designed to reduce the risk of injury associated with the passive occupant protection provided by an air bag, including that to small children and out-of-position occupants, and to increase new vehicle testing requirements that are to include the use of additional sizes of test dummies.

Previous SAE papers have presented the regulatory history of air bag systems through the 1990s. This paper presents recent regulatory changes and how these have modified FMVSS 208. These topics include the phased-in adaptation of advanced air bag technologies through 2010; details of the new testing requirements, injury criteria, and use of additional sizes of crash test dummies; and a discussion of the current issues regarding these technologies. Implementation of the new requirements will occur in two stages with a phase-in period for each stage. The first stage is intended to minimize the risk of air bags to vehicle occupants, especially women drivers and children in the front passenger seat. The second stage is designed to improve protection for belted occupants and represents a fundamental change in occupant protection philosophy since there will be higher test speed requirements for restrained dummies than for unrestrained dummies, causing designers to focus on each test condition independent of the other.

FRONTAL INJURY

Air bag systems have been effective in reducing injuries and death to front seat occupants during collisions. Air bags have reduced driver fatality risk by 31% in pure frontal crashes and by 11% in all crashes [2], and reduced fatalities for right-front occupants, age 13 and older, by 32% in pure frontal crashes.[2] It is estimated that the combination of seat belts and air bags are 75 percent effective in preventing serious head injuries and 66 percent effective in preventing serious chest injuries.[3] Since the incorporation of air bags into passenger vehicles in 1986 and through August 1, 2000, it is estimated that air bags have saved the lives of 5,899 front seat occupants.[4] NHTSA estimates that air bags may save up to 3,200 lives annually when all passenger cars and light trucks are equipped with air bags. The benefits of air bags have not been without some occupant safety cost. As of August 1, 2000, there have been 167 confirmed air bag-related fatalities.[4] Of these fatalities, 99 were children, 62 were drivers and 6 were passengers. The one common fact to each of these air bag fatalities is that the occupant was very close to the air bag when it started to deploy.[1] During May, 2000,

205

NHTSA issued a final rule amending FMVSS 208 to improve the frontal crash protection of air bags for all occupants and to reduce the risk of air bag induced injuries to occupants, particularly small women and children.

PREVIOUS RULEMAKING

Previous rulemaking has attempted to address some of the issues surrounding the risk of injuries to small women and children. This rulemaking has regulated the air bag warning labels provided in vehicles, the availability of on-off switches for air bags, the option of a sled test to certify vehicles with a "depowered" air bag design and the requirements for deactivation of air bags.[5]

TEA 21

The Transportation Equity Act for the 21st Century (TEA 21) was enacted in June 1998, by Congress requiring NHTSA to issue a rule amending FMVSS 208 "to improve occupant protection for occupants of different sizes, belted and unbelted, under Federal Motor Vehicle Safety Standard No. 208, while minimizing the risk to infants, children, and other occupants from injuries and deaths caused by air bags, by means that include advanced air bags."[1] To achieve these goals, NHTSA is requiring vehicles to meet broader test requirements with an assortment of new dummies to ensure that occupants are properly protected under a wider variety of crash conditions.

FMVSS 208

The purpose of FMVSS 208 is to reduce the number of deaths of vehicle occupants, and severity of injuries, by specifying vehicle crashworthiness requirements and specifying equipment requirements for active and passive restraint systems.[1] Originally, the standard specified the types of restraints required and was later amended to specify performance requirements for test dummies seated in the front outboard seating positions of passenger cars. Previously, FMVSS 208 required all passenger cars manufactured after September 1, 1997, and light trucks manufactured after September 1, 1998 to be equipped with driver and passenger air bags, along with manual lap and shoulder belts. Specific sun visor warning labels have also been required. Currently, for unbelted occupants, manufacturers have the option of certifying vehicles with a 48 km/h (30 mph) frontal barrier crash test or with an approximately 48 km/h delta-V generic pulse sled test. For belted occupants, the vehicles are to be certified with 48 km/h frontal and oblique barrier tests. To comply with FMVSS 208, instrumented test dummies are positioned in the front outboard seating positions for each test and must meet specific injury criteria established in the standard.[6]

MODIFICATIONS TO FMVSS 208

The recent modifications to FMVSS 208 will ensure that advanced air bag technologies will be adapted to make vehicles more effective in protecting occupants and reduce the risk of air bag induced injuries. These modifications specifically address small stature drivers, child occupants and the average male. The standard requires new dynamic and static testing, including changing the way head injury risk is measured, adding new neck injury criteria and reducing the allowable chest deflection. New warning labels have also been required for vehicles with advance air bag systems.

SMALL STATURE OCCUPANTS

TEA 21 specifically targeted the adaptation of advanced air bag technologies to improve the occupant protection of small stature drivers. Specific language in the law requires that the protection of small drivers be taken into account for all occupant protection requirements.

The 5th percentile female dummy has been added to most of the new crash test requirements. It will be used in both belted and unbelted vehicle crash tests and for out-of-position occupant testing.

CHILDREN

Manufacturers have the option of selecting one of two alternative test methods to reduce the risk of air bag-induced injuries to infants in child restraint systems. The manufactures' options include either suppressing the air bag deployment or deploying the passenger air bag in a low risk manner in the presence of a 12-month-old Child Restraint Air Bag Interaction (CRABI) dummy in a rear facing child safety seat (RFCSS) or a convertible child restraint in the rear-facing mode.

To reduce the risk of air bag-induced injuries to small children in the front seat, manufacturers will conduct tests using 3-year-old and 6-year-old child dummies. The manufacturer has the option of either suppressing the air bag deployment if a child is present, deploying it in a low risk manner if a 3-year-old and 6-year old child dummy is out-of-position or suppressing the air bag deployment when an occupant is out-of-position.

Additional injury parameters were developed with limits that vary between infant, child, and adult dummies. The required warning labels will still state that children are safest in the back seat.

PHASE-IN REQUIREMENTS

The new requirements of FMVSS 208 are scheduled to take effect over a seven-year period. This period is divided into two phases. The first phase occurs over

4 years and addresses the reduction of injury due to occupant air bag interactions. The second phase occurs over 3 years and addresses increased occupant protection of restrained occupants.

FIRST PHASE – AIR BAG INJURY REDUCTION

The first phase-in period adds the requirements of 5th percentile dummy barrier testing (frontal and offset), air bag suppression or multistage inflation in the presence of rear facing child seats, 3 and 6 year-old child dummies or out-of-position occupants, and 5th percentile out-of-position driver occupants.

The first phase-in requires that a percentage of each manufacturers light vehicle production meet the requirements of the standard by a specified date. If a manufacturer exceeds the percentage required early in the phase-in, they receive credit for future years, but must be 100% compliant by September 1, 2006. The first phase-in is follows:

35% of production beginning September 1, 2003

65% of production beginning September 1, 2004

100% of production beginning September 1, 2005

SECOND PHASE – RESTRAINED OCCUPANTS

The second phase-in requires that the speed of frontal barrier testing with belted 50th percentile male dummies be raised from 48 km/h to 56 km/h (30 mph to 35 mph). The second phase-in also allows manufacturers to receive future credit for meeting the requirements earlier. All vehicles must 100% compliant by September 1, 2010. The second phase-in is follows:

35% of production beginning September 1, 2007

65% of production beginning September 1, 2008

100% of production beginning September 1, 2009.

DYNAMIC TESTING REQUIREMENTS

The dynamic test matrix for FMVSS has been drastically modified by the new requirements. The previous crash test requirements of FMVSS 208 required only the use of a belted and unbelted 50th percentile male dummy in a 48 km/h rigid and oblique barrier test. This could be supplemented with the 48 km/h sled test option for air bag certification.

The new requirements include testing with belted and unbelted 5th percentile female and 50th percentile male dummies into a frontal rigid barrier. Additionally, the unbelted 50th percentile male is used in an oblique rigid barrier test and a belted 5th percentile female dummy is used in a deformable offset barrier test. Chart 1

summarizes the test dummy size, restraint use, barrier type, impact speed and angle for the new requirements.

Chart 1. FMVSS 208 required crash test matrix

ATD	Seat Belt	Barrier Type	Crash Type	Speed (mph)	Angle
50th	Yes	Rigid	Frontal	0-30*	90
5th	Yes	Rigid	Frontal	0-30	90
50th	No	Rigid	Frontal	20-25	90
5th	No	Rigid	Frontal	20-25	90
50th	No	Rigid	Frontal	20-25	30
5th	Yes	Deform able	Offset Frontal	0-25	90

* Raised to 35 mph starting in the 2008 model year

The 40% offset deformable barrier test, with a 5th percentile female dummy, is believed to simulate real world crashes and has been designed to test air bag systems for deployment and driver interaction with the air bag system.

NEW STATIC TEST REQUIREMENTS

Additional static tests of air bag systems are now required to determine the systems interaction with small adults, children, and infant restraint devices. These tests include static testing of the air bag system with out-of-position occupants. The system must either inflate in a low risk manner or suppress the deployment of the air bag if an out of position driver were detected.

Upon the detection of a small right front seat occupant, the system must either inflate at a low-speed impact inflation rate, or suppress the air bag deployment, or suppresses the air bag deployment if the child moves close to the air bag during an impact.

With the presence of a child safety seat in the right front position, the system must either inflate at a low-speed impact inflation rate, or suppress the inflation of the air bag.

INJURY CRITERIA

In addition to Head Injury Criteria (HIC), femur load, chest deflection and chest acceleration, new standards are established for neck injury and chest deflection.

The calculation formula and upper acceptable limit of HIC has been modified. The new formula calculates HIC based on maximum head accelerations over a 15 millisecond time frame, instead of the previous 36 millisecond duration. The maximum acceptable limit for the HIC_{15} calculation is not to exceed the value of 700 for the 50th percentile male and 5th percentile female dummies.

Chest acceleration is not to exceed 60 G's for the 50th percentile male and 5th percentile female dummies. While chest deflection shall not exceed 52 mm for the

5th percentile female and 63 mm for the 50th percentile male.

Femur loads shall not exceed 6805 N for the 5th percentile female and 10,000 N for the 50th percentile male.

The neck injury combines neck load and moments into an injury standard. This new value is called Nij. Rather than evaluating each neck load and moment separately, Nij combines axial neck load (Fz) and bending moment (Mocy) into a single cumulative score. The Nij is defined as:

$$Nij = \left(\frac{Fz}{Fzc}\right) + \left(\frac{Mocy}{Myc}\right) \qquad (1)$$

Where Fzc is the critical axial load of 2800 N in both tension and compression. Myc is the critical bending moment about the occipital condyle and is 93 Nm in flexion and 37 Nm in extension.

The Nij cannot exceed 1 at any measured load condition during the test event. Additionally the peak axial force (Fz) cannot exceed 1490 N in tension or 1820 N in compression.

OTHER ISSUES

The 1997 revisions to FMVSS 208 for certification allowed manufacturers to use depowered air bags with the use of the sled test option. The sled test option for certification will be eliminated because of the variability in the structural characteristics of vehicles within the US fleet and the direct consequences on test conditions and results due to these variations. Sled testing did not account for variations in crash pulse to individual vehicle designs since a single pulse was used for all vehicles regardless of vehicle weight, stiffness, or size class. Therefore, NHTSA has determined that only full-scale crash tests would be allowed for certification.

ADAPTATION OF ADVANCED AIR BAG TECHNOLOGY

Under the requirements of TEA 21, FMVSS 208 has been modified to both allow the use of advanced technologies, and provide a means by which a manufacturer can certify new technologies for use in production vehicles. The regulations within 208 are specifically vague in the requirements of which advanced technologies will be allowed. This vagueness in requirements was designed so that new technologies (currently under development or yet to be invented) would not be stifled in their development by the regulations.

One area under development is the use of multi-level inflation systems. These systems could provide differing levels of protection for low-speed and high-s collisions or be tailored to reduce the injury potential for out of positions occupants. Vehicles equipped with a multi-level inflation system will be required to meet the injury criteria with a low risk (level) deployment in a rigid barrier test with unbelted 5th percentile dummies in both outboard seating positions at 26 km/h (16 mph). This test, combined with the unbelted 32 km/h (20 mph) rigid barrier test requirements, provides the flexibility to develop a multi-level inflation system that can protect occupants in low speed crashes and reduce the injury to out-of-position occupants when a high-level deployment occurs. High-level deployments provide additional protection to occupants in severe crashes.

Another area under development is a weight sensor that can detect the presence of a child (infant to 6-year-old) and suppress the deployment of the passenger air bag. Currently, there is no suitable test dummy to test suppression systems for children ages 7 to 12.

AIR BAG WARNING LABELS

Since 1995 model year, NHSTA has specified the content of vehicle warning labels for supplemental restraint systems. These original warning label requirements were subsequently revised by NHTSA for 1997 model year vehicles. For vehicles equipped with advance air bag systems a new sun visor-warning label will be used. Also, if manufacturers deem it necessary to supplement the new required label with additional information on a separate label they will now permitted to do this. The new required label should be visible with the sun visor in the down position. The new required label is shown in Figure 1.

Additionally, a new removable label should be placed on the right side of the occupant compartment near the supplemental restraint system prior to its sale. This new label is shown in Figure 2.

Figure 1. Sun visor warning label

Figure 2. Removable dash label

Label Outline, Vertical and Horizontal Lines Black

Bottom Text Black with
White Background

Top Text Black with
Yellow Background

This Vehicle is Equipped with Advanced Air Bags

Even with Advanced Air Bags

Children can be killed or seriously injured by the air bag.

The back seat is the safest place for children.

Always use seat belts and child restraints.

See owner's manual for more information about air bags.

FUNDAMENTAL CHANGE IN OCCUPANT PROTECTION PHILOSOPHY

All previous standards only addressed injury criteria with a 50th percentile male dummy. No regulatory consideration was given to out of position occupants or occupants of varying statures. Additionally, the requirements for belted and unbelted occupants remained identical. By separating the belted occupant requirements from the unbelted occupant, NHSTA has made a philosophical change in the viewing of injuries for occupants.

This change allows for better protection of smaller stature occupants. The physics of slowing an unbelted occupant and a belted occupant vary greatly. A belted occupant will rely primarily on the belt system to provide ride-down time and limit deceleration forces and the unbelted occupant will rely on the air bag system to reduce contact forces with the vehicle interior. For the belted occupant, the air bag system is truly a supplemental system to the seat belt system, while the unbelted occupant relies on the air bag system as their primary protection.

To date, air bags have been designed for the benefit of belted and unbelted 50th percentile male occupant in a 30 mph barrier collision due to certification requirements. Under the recent modifications to FMVSS 208, air bag systems in the future will be designed to benefit occupants of various sizes, regardless of belt usage.

The law in 49 of the 50 states currently requires the use of seat belts. However, the national seat belt usage rate in the United States is only about 69%.[3] The recent modifications to FMVSS will allow air bag system designers the flexibility to provide protection to belted occupants, while reducing the injury potential to unbelted occupants, particularly to occupants that have historically had some increased risk of injuries from deploying air bags.

CURRENT ISSUES

Prior to modifying FMVSS 208, NHTSA and NASA agreed to cooperative effort that leveraged NASA's expertise in advance technologies to understand the parameters affecting air bag systems, to assess air bag technology state-of-the-art and to identify new concept for air bag systems. The Jet Propulsion Laboratories (JPL) was selected by NASA to undertake this investigation. Due the volume of information contained in JPL's report, only highlights of the findings will be presented and readers interested in more details should read the report.[7]

"Air bag systems are a significant engineering design challenge because they deploy rapidly and with great force toward an approaching occupant. Their deployments are based on predictions of crash severity early in the event, ..."[7]

Advanced restraint systems can improve the safety of air bag systems by providing more information about the type and severity of crash and tailoring the deployment characteristics to individual occupants. "Improving air bag safety is an incremental process, and implementation of advance technology will be evolutionary."[7]

It was anticipated that advanced technologies such as improved crash sensors, belt use sensors, seat position sensors, automatic suppression, two-stage inflators, compartmented air bags would be available by model year 2001, and by model year 2003, more sophisticated occupant sensing systems could be incorporated to suppress inflation of the air bag system when it has a high likelihood of injuring a front seat occupant. Along with these technologies, there comes a predicted risk of air-bag-induced injuries from the unreliability of the advanced systems. The development of advanced restraint systems is influenced by government regulatory requirements and industrial costs. Chart 2 identifies advanced air bag technologies that are currently under development or could be developed in the future and the expected readiness date of these technologies as determined by JPL.[7]

Advanced air bag technologies can make air bag systems safer, but only if the technologies are deemed reliable. Without reliable technologies, an air bag system may deploy when not needed, may not deploy when needed or may deploy in a manner not optimal for the occupant. It will remain the goal of air bag designer to only implement new technologies when they have been determined to be reliable.

Chart 2. Advance Air Bag Technologies

Technology	Description	Readiness Date
Pre-crash sensing	Remote sensing for early crash severity determination	Could be available by MY2001
Belt use sensor	Determines whether or not a seat belt is being used	Could be available by MY2000
Belt spool-out sensor	Aid in determining occupant size	Could be available by MY2001
Seat position sensor	Used to estimate driver size and proximity to air bag	Could be available by MY2001
Occupant classification sensor	Measure occupant weight and presence	Could be available by MY2000
Occupant proximity sensor	Provide range information between occupants and interior	Could be available by MY2000/2001
Inflatable seat belts	A portion of seatbelt is inflated to augment the belt function	Could be available by MY2001

ADOPTED ADVANCE AIR BAG TECHNOLOGIES

Currently, some manufacturers have already begun to incorporate advanced air bag technologies into their newer vehicles. As of MY2000, advanced air bag technologies that have been adopted in some vehicles include: dual-stage passenger air bags, advanced crash severity sensors, belt use sensors, dual-threshold deployments for driver and passenger air bags, and driver seat position sensors.

CONCLUSIONS

The modifications to FMVSS 208 will provide greater protection to small stature drivers, front seat child passengers and front seat infant passengers. New air bag designs that are developed to meet the new requirements will create less risk of serious air bag induced injuries to occupants than current air bag designs. It is expected that new air bag system designs will incorporate advanced air bag technologies.

In addition to using a larger family of test dummies to evaluate air bag system performance, new designs will be required to meet more stringent injury criteria.

The new requirements of FMVSS 208 will be phased-in over a seven-year period. The first phase, covering occupant air bag injury reduction, is to begin with model year 2003, and the second phase, covering increased protection for restrained occupants, will be completed by model year 2010.

Seat belts still offer the most effective protection to occupants and will remain the primary restraint system for all occupants. The continuous development and manufacture of improved occupant crash protection systems in motor vehicles remains the goal of FMVSS 208.

This paper provides a summary the most recent modifications to FMVSS 208. Although, the new requirements have been adopted, some aspects of the new requirements are still being addressed through the rulemaking process and could lead to further changes in the standard prior to its implementation in 2003.

REFERENCES

1. Department of Transportation, NHTSA, 49CFR Parts 552, 571 and 595, [Docket No. NHTSA 007013; Notice 1], May 2000.
2. Fourth Report to Congress, Effectiveness of Occupant Protection Systems and their Use, NHTSA May 1999
3. NHTSA Safety Fact Sheet, 11/2/99.
4. NHTSA Special Crash Investigation report, 8/1/00.
5. Recent Regulatory History of Air Bags, SAE 980650.
6. Code of Federal Regulations, Transportation, part 571.208, October 1, 1997.
7. Advance Air Bag Technology Assessment, JPL, April 1998.

CONTACT

John E. Hinger is a managing engineer with Exponent Failure Analysis Associates. He has previously worked in the automotive field for several automotive manufacturers. He has a B.S. in mechanical engineering from the University of Illinois. He may be contacted at jhinger@exponent.com.

Harold E. Clyde, P.E. is a senior engineer with Exponent Failure Analysis Associates. He has a M.S. in mechanical engineering from Brigham Young University. He may be contacted at hclyde@exponent.com. Either author may be contacted through Exponent at (650) 688-7282.

2001-01-0165

Theoretical Evaluation of the Requirements of the 1999 Advanced Airbag SNPRM -
Part One: Design Space Constraint Analysis

**Tony R. Laituri, N. Sriram, Brian P. Kachnowski,
Brion R. Scheidel and Priya Prasad**

Ford Motor Co.

Copyright © 2001 Society of Automotive Engineers, Inc.

ABSTRACT

In the 1999 Supplemental Notice for Proposed Rulemaking (SNPRM) for Advanced Airbags, the National Highway Traffic Safety Administration (NHTSA) sought comments on the maximum speed at which the high-speed, unbelted occupant test suite will be conducted, i.e., 48 kph vs. 40 kph. To help address this question, an analysis of constraints was performed via extensive mathematical modeling of a theoretical restraint system. First, math models (correlated with several existing physical tests) were used to predict the occupant responses associated with 336 different theoretical dual-stage driver airbag designs subjected to six specific Regulated and non-Regulated tests. Second, the pertinent, predicted occupant responses for all 336 designs were compared with a set of generic acceptance criteria for the six distinct performance constraints (where two of the six represented the aforementioned "high-speed" unbelted occupant test suite and where "high-speed" was set equal to either 48 or 40 kph). Finally, statistics were generated to help evaluate the stringency of the various performance constraints.

Results from the assessment for a modeled, prototype, mid-sized passenger car included the following: (1) None of the 336 theoretical dual-stage driver airbag designs satisfied the generic acceptance criteria set when the unbelted rigid fixed barrier testing constraints were run at 48 kph, (2) 21 of the 336 satisfied the generic acceptance criteria set when the unbelted rigid fixed barrier testing constraints were run at 40 kph, and (3) When considering the discarded designs, nearly all of them were predicted to not comply due to (at least one of) the unbelted occupant performance constraints of the generic acceptance criteria set.

1.0 BACKGROUND

The Federal Motor Vehicle Safety Standard No. 208 (FMVSS 208) was amended in 1997 to include an alter tive to the 48 kph, full-vehicle, rigid fixed barrier test inv

ing unbelted occupants. This amendment involved what came to be known as the "generic sled test," i.e., an unbelted, mid-sized male instrumented test dummy subjected to a 48 kph ΔV, 17.2 G, 125 ms half-sine wave, "generic" crash pulse on a hydraulically-controlled sled. This sled test alternative was made available by NHTSA in order to expedite the introduction of depowered airbags -- a means to reduce further the already low real-world risk of injury related to airbag inflation.

The dual-stage airbag inflator is one of the possible design elements of an advanced restraint system that is anticipated to help reduce even further the risk of airbag inflation-related injury. The function of the dual-stage inflator is to provide two separate injections of gas into the airbag. If dual-stage inflators are activated by new, advanced-technology crash-sensing systems, occupant restraint systems could potentially be more adaptable and yield airbag energies more commensurate with vehicle crash severities, i.e., lower-energy inflations for moderate-severity crashes and higher-energy inflations for the more severe crashes.

In the midst of significant manufacturer and supplier efforts to voluntarily develop and implement these 1st-generation advanced restraint systems, NHTSA issued a Notice for Proposed Rulemaking (NPRM) (NHTSA, 1998) and a Supplemental Notice for Proposed Rulemaking (NHTSA, 1999). Therein, NHTSA proposed broad-ranging new elements to Regulated testing -- four of present interest: (1) An end to the aforementioned 1997 amendment to FMVSS 208, thereby reverting to the high-speed rigid fixed barrier vehicle testing with an unbelted, mid-sized male Hybrid III test dummy (HIII50), (2) The introduction of high-speed rigid fixed barrier vehicle testing with an unbelted, small-sized female Hybrid III test dummy (HIII05) in addition to the aforementioned unbelted HIII50 high-speed rigid fixed barrier vehicle test, (3) The introduction of a set of static, out-of-position occupant (OOPO) tests, and (4) The intro-of new neck injury criteria for both instrumented

The present study provided a theoretical, mathematical model-based analysis of the predicted effects of Regulated and non-Regulated performance constraints on available design space, i.e., the designs that satisfied the related set of generic acceptance criteria.

2.0 MATHEMATICAL MODEL DEVELOPMENT

2.1 MODELS FOR DYNAMIC PERFORMANCE TESTS

The first step toward attempting to predict the effects of alternative frontal impact Regulations on airbag designs involved development of occupant response models of existing physical tests. The modeling software chosen was Madymo3D developed by TNO. The physical tests involved a non-production, mid-sized passenger car equipped with a prototype advanced restraint system. The airbag system included a dual-stage inflator. The seat belt system included a load-limited retractor and a pyrotechnic pretensioner at the belt buckle. The validation cases consisted of variations on crash type, speed, severity, restraint level, occupant size, and occupant seating position (see Figure 1 for one example, i.e., the USA New Car Assessment Program (NCAP) test).

Six validation cases were studied in order to attempt to correlate math model occupant responses with those of physical experiments while applying consistent modeling practices in all of the cases (see Appendix 1). During the validation process, special attention was given to six occupant responses: head acceleration, upper neck fore and aft shear force, upper neck tensile axial force, corrected upper neck extension moment, chest acceleration, and chest deflection. Additionally, pelvis displacement and femur load comparisons between test and simulation were made with intent to gain acceptable model correlation with the time histories of the lower-body responses observed in the tests. Other comparisons included column stroke and seat belt forces (when applicable).

The results of the validation effort and detailed comments on the applied modeling procedures are given in Appendix 1.

Given acknowledged test-to-test variability of occupant responses, the correlations between the simulated occupant responses and the actual, physical test case responses (for all of the validation cases shown in Appendix 1) were deemed acceptable.

2.2 MODELS FOR STATIC PERFORMANCE TESTS

In an effort to reduce even further the airbag inflation-related risks to occupants in contact with the airbag module at the time of airbag deployment, static, dummy chest-on-module testing is also a proposed Regulated test condition.

Previous research (Laituri, et al., 1999) was used to relate various inflator predictor variables to a pertinent occupant response for the static, HIII50 chest-on-module test condition (ISO, 1998). Regression analysis was used to show good agreement when the predictor variable was a tank-test derived parameter (designated as the "10ms-windowed inflator thrust variable") and the response variable was the peak viscous criterion, V^*C_{max} (for the HIII50 in the aforementioned test condition of Figure 2). However, sternal velocity has also been considered as a measure of the thoracic injury risk. Accordingly, after revisiting the data used to generate the aforementioned thrust variable-V^*C_{max} regression equation, the following linear regression equation (with a correlation coefficient, R^2, equal to 0.87 for 18 data points) was derived:

$$SV_{HIII50} = 0.67 + 0.0122\left(\dot{m}_{max}\big|_{10ms}\sqrt{\gamma RT_{inf}}\right) \quad (1)$$

where SV_{HIII50} is sternum velocity in m/s, $\dot{m}_{max}\big|_{10ms}$ is the peak 10-ms windowed inflator mass flow rate in kg/s, γ is the inflator gas specific heat ratio, and T_{inf} is the modeled inflator exit gas temperature in degrees Kelvin. The term in parentheses in Eq. (1) is the 10ms-windowed inflator thrust variable in N.

Moreover, the chest-on-module test condition with the small female instrumented test dummy needed to be considered (since it is one of the proposed Regulated tests). Physical tests in which nominally identical airbag systems interacted with both the HIII50 and HIII05 in the chest-on-module test condition were analyzed. For a small sample (N=3), the following approximation resulted:

$$SV_{HIII05} \approx 1.1(SV_{HIII50}) \quad (2)$$

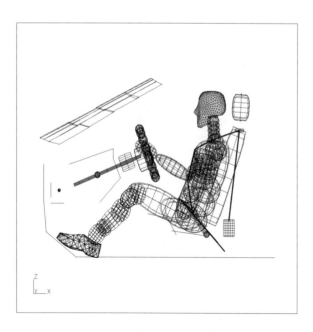

Figure 1: Example of Madymo3D Occupant Model

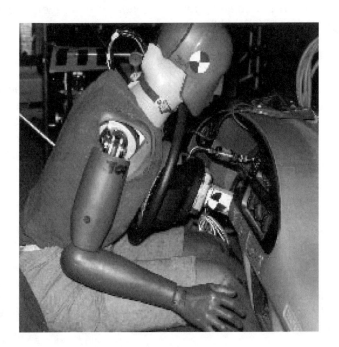

Figure 2: Chest-on-Module Test Condition (HIII50)

These mathematical models served as the foundation for an extensive parametric study to help better understand the effect of alternative frontal impact Regulations on potential airbag designs.

3.0 DESIGN SPACE SURVEY TECHNIQUE

3.1 OBJECTIVES

In order to attempt to predict the effects of the aforementioned performance constraints on available airbag design space, four major topics were addressed: (1) Specification of the sampled design space for seat belts and airbags, (2) The selection of performance tests to which the resulting studied designs were subjected, (3) The means by which the related occupant responses were estimated, and (4) The generic criteria by which suitable (acceptable) designs were identified.

3.2 DESIGN SPACE SAMPLING: SEAT BELTS

Recent research has indicated that retractors with a load limit set at approximately 4 kN can provide an effective balance between the required restraining belt load and rib fracture risk potential in real-world crash events (Foret-Bruno, et al., 1998). Additionally, occupants in real-world crashes do not always have their seat belts on as tightly as test dummies in controlled laboratory tests (Bauberger, et al., 1996). Therefore, when a belt was called for in this study, the modeled seat belt was assumed to be a design constant that consisted of both a 4 kN load-limited retractor and a pyrotechnic, buckle pretensioner.

3.3 DESIGN SPACE SAMPLING: AIRBAGS

The airbag design variables examined in this study included airbag venting, airbag size, and inflator characteristics.

Airbag venting is typically designated as the "number of holes" x the "vent hole diameter." A range of vent sizes typical of both full-powered and depowered airbag designs was considered in this study, viz., 2x10 mm, 2x15 mm, 2x20 mm, 2x25 mm, 2x30 mm, 2x35 mm, and 2x40 mm.

Driver airbag size is typically specified by its unfolded diameter. A range of airbag sizes typical of present depowered airbags was considered in this study, viz., 610 mm, 648 mm, and 673 mm.

An internal tether was assumed present, but primarily for purposes of design space sampling simplification, its length was assumed fixed at 254 mm.

Prior experimental research on the potential relationship between responses for in-position, HIII50s and out-of-position, HIII50s helped identify the related design challenge associated with inflator selection (Prasad, et al., 1996). Accordingly, a rather fine resolution of the design space associated with the dual-stage inflator was warranted for this study.

Nineteen theoretical inflators and one experimental, prototype dual-stage driver inflator were studied. The prototype inflator served as a baseline from which the other theoretical inflators were derived. The nineteen variations were derived by scaling the mass flow rates of the baseline Stage I and Stage II contributions of the inflator by /- 25% increments from the baseline (while keeping the gas constituents constant) as shown in Figure 3. More details concerning these theoretical inflators are found in Table 1.

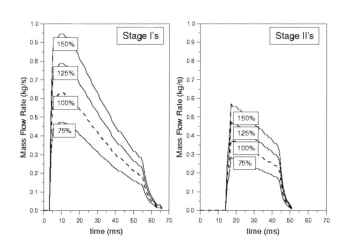

Figure 3: Theoretical Dual Stage Inflator Constructs

Table 1:
Tank Test Details on Studied, Theoretical, Dual-Stage Inflators (28.3 liter, Air-filled Tank)

Stage I Scale	Stage II Scale	Peak Tank Pressure	Peak 0 ms-based Tank Pressure Rise Rate	Peak 10 ms-based Tank Pressure Rise Rate	Generated Gas Mass
(% of Ref)	(% of Ref)	(kPa g)	(kPa/ms)	(kPa/ms)	(kg)
75	0	210.76	6.49	6.28	0.0167
100	0	277.20	8.62	8.30	0.0223
125	0	342.75	10.74	10.29	0.0279
150	0	407.68	12.85	12.26	0.0335
75	75	259.43	8.26	7.66	0.0238
100	75	325.58	10.17	9.48	0.0294
125	75	391.01	12.05	11.30	0.0349
150	75	455.91	13.90	13.18	0.0405
75	100	275.69	8.98	8.28	0.0261
100	100	341.76	10.88	10.08	0.0317
125	100	407.15	12.75	11.87	0.0373
150	100	472.04	14.60	13.65	0.0429
75	125	291.96	9.69	8.90	0.0285
100	125	357.95	11.58	10.69	0.0340
125	125	423.31	13.45	12.47	0.0396
150	125	488.19	15.29	14.23	0.0452
75	150	308.22	10.40	9.53	0.0308
100	150	374.15	12.29	11.30	0.0364
125	150	439.48	14.15	13.07	0.0420
150	150	504.34	15.99	14.02	0.0476

3.4 CONSIDERED PERFORMANCE TESTS

Tests used to help assess vehicle crashworthiness may be broken into two groups: Regulated and non-Regulated. For purposes of this study, six performance tests (and the related performance constraints) were considered:

Three "Regulated" tests:

• The high-speed, unbelted occupant SNPRM test suite consisting of rigid fixed barrier, full-vehicle testing with both HIII50 and HIII05 unbelted drivers. This two-test suite of tests was studied with "high-speed" set to equal either 48 or 40 kph (hereafter, designated as "RFB48" and "RFB40," respectively.)

• The static, out-of-position occupant test involving

HIII05 in the chest-on-module condition (hereafter, designated as "OOPO HIII05.")

Three "non-Regulated" tests:

• The USA New Car Assessment Program test for frontal impact involving rigid fixed barrier, full-vehicle testing at 56 kph with a belted HIII50 driver (hereafter designated "NCAP.").

• The lower-speed, airbag deployment threshold test involving deformable barrier, full-vehicle testing with a fully-forward belted HIII05 driver. The test chosen as a surrogate for this crash environment was the belted HIII05 subjected to the generic sled test pulse (hereafter, designated as "Canadian test surrogate" or, for short, "Canadian.")

• The static, out-of-position occupant test involving the HIII50 in the chest-on-module condition (ISO Technical Report 10982, 1998) (hereafter, designated as "OOPO HIII50.")

It should be noted that these tests, while numerous, should be considered non-comprehensive with respect to the actual, total number of actual performance constraints that will be placed upon future airbag designs. Further discussion of this topic is left for a later section.

3.5 OCCUPANT RESPONSE PREDICTION PROCEDURES

3.5.1 DYNAMIC PERFORMANCE TESTS

The occupant responses associated with performance tests involving dummies in motion when the airbags are deployed were estimated via the Madymo3D math modeling methods discussed in Section 2.1 and Appendix 1. See Table 2 for more details regarding the monitored, dynamic, in-position occupant responses.

3.5.2 STATIC OOPO PERFORMANCE TESTS

The occupant responses associated with performance tests involving dummies at rest when the airbags are deployed were estimated via the regression and body-size sternum velocity approximations discussed in Section 2.2. See Table 2 for more details regarding the monitored, static, out-of-position occupant responses.

Note that, since the studied inflators were theoretical, the variables on the right-hand side of Eq. (1) needed to be determined for each the inflators.

The $\dot{m}_{max}\big|_{10ms}$ term was derived from a Madymo3D mathematical model of a tank test involving a studied inflator. The T_{inf} term was modeled as a constant (Wang, et al., 1988) and estimated from the TNO Madymo Tank Analysis code (Madymo Utilities Manual, 1999). Both the gas constant, R, and the constant-pressure gas specific heat, c_p, were derived from Amagat's rule for mixtures (Van Wylen, et al., 1978). The ideal gas equation was then used to derive both the constant-volume specific heat, c_v, and γ for the gas mixture. The results of these calculations are shown in Table 3. With the 10ms-windowed inflator thrust variable calculated for each of the theoretical inflators, SV_{HIII50} and SV_{HIII05} were estimated via Eqs. (1) and (2), respectively.

3.6 GENERIC ACCEPTANCE CRITERIA

In order to assess compliance or non-compliance of the studied designs subjected to the aforementioned Regulated and non-Regulated performance tests, generic acceptance criteria were derived. Specifically, pertinent occupant responses associated with each of the performance tests were compared with both published injury assessment reference values (IARVs) and proposed injury criteria performance limits (ICPLs) with applied, generic safety factors, defined as margins to increase confidence in compliance. Further details follow.

3.6.1 PUBLISHED IARVS

The published occupant response IARVs of Table 4 were essential in the generation of the set of generic acceptance criteria (designated, "SGAC").

Table 2:
Summary of Performance Tests and
the Related Pertinent Occupant Responses
and Prediction Methods

Performance Test	Dummy	Pertinent Occupant Responses	Source of Predicted Occupant Response
High-Speed Unbelted Test	HIII50	HIC15, Upper Neck Axial Force, Upper Neck Shear Force, Upper Neck Moment, Upper Neck $N_{TE,}$ Chest G's (3ms cumdur), Chest Deflection	Madymo3D Model
High-Speed Unbelted Test	HIII05	HIC15, Upper Neck Axial Force, Upper Neck Shear Force, Upper Neck Moment, Upper Neck $N_{TE,}$ Chest G's (3ms cumdur), Chest Deflection	Madymo3D Model
NCAP	HIII50	CPI	Madymo3D Model
Canadian	HIII05	Upper Neck Moment	Madymo3D Model
OOPO HIII50	HIII50	Sternum Velocity	Regression Model
OOPO HIII05	HIII05	Sternum Velocity	Regression Model

3.6.2 PROPOSED INJURY CRITERIA PERFORMANCE LIMITS

NHTSA's NPRM called for a new Regulated response, viz., the upper neck combined loading criterion, N_{ij} . The ij indices represent various combinations of tension, compression, extension, and flexion. Since combined tension-extension loading is generally considered the most prevalent loading condition associated with airbags in frontal impacts, only N_{TE} was considered in this study. Additionally, only NPRM-related estimates of N_{TE} were made in this study since the SNPRM had not been published when the mathematical model-based simulations were being first

Table 3:
Inflator Characteristics for Theoretical Inflators derived
from Tank Test Models and Amagat's Rule of Mixtures

Stage I Scale	Stage II Scale	Peak 10ms-based Mass Flow Rate	Inflator Gas Temp	$\dot{m}_{max}\big|_{10ms}\sqrt{\gamma RT_{inf}}$
(% of Ref)	(% of Ref)	(kg/s)	(K)	(N)
75	0	0.462	672	264
100	0	0.616	671	351
125	0	0.771	671	439
150	0	0.925	670	526
75	75	0.705	594	378
100	75	0.853	608	462
125	75	1.001	619	548
150	75	1.150	626	633
75	100	0.794	578	420
100	100	0.940	594	504
125	100	1.088	605	588
150	100	1.236	614	673
75	125	0.882	563	460
100	125	1.029	582	546
125	125	1.175	594	630
150	125	1.322	604	715
75	150	0.971	552	502
100	150	1.117	570	587
125	150	1.264	583	671
150	150	1.410	593	755

215

Table 4:
AAMA-Proposed IARVs and NHTSA-Proposed ICPLs

Body Part	Response	HIII05	HIII50	Source
Head	HIC15	779	700	(AAMA, 1998)
Upper Neck	Ext. Moment	39 N-m	77 N-m	(AAMA, 1998)
Upper Neck	Neck Tensile Force	2070 N	3290 N	(AAMA, 1998)
Upper Neck	Neck Shear Force	2068 N	3100 N	(AAMA, 1998)
Chest	Chest Acceleration	73 G's (3ms cumdur)	60 G's (3ms cumdur)	(AAMA, 1998)
Chest	Chest Acceleration	60 G's (3ms cumdur)	60 G's (3ms cumdur)	(NHTSA, SNPRM, 1999)
Chest	Chest Deflection	53 mm	65 mm	(AAMA, 1998)
Chest	Sternum Velocity	8.2 m/s	8.2 m/s	(AAMA, 1998)
Upper Neck	Combined Tension-Extension Response	1.0	1.0	(NHTSA, SNPRM, 1999)

conducted. The equation for N_{TE} is

$$N_{TE} = \max\left(\frac{F_z(t)}{F_{z_1}} \quad \frac{M_y(t)}{M_{y_1}}\right) \qquad (3)$$

where F_z is the upper neck tensile force and M_y is the corrected upper neck extension moment. F_{z_1} was defined as 3200 N for HIII05 and 3600 N for HIII50, and M_{y_1} was defined as 60 N-m for HIII05 and 125 N-m for HIII50. It should be noted that all upper neck moments reported herein were "corrected," i.e., neck moments at the upper neck load cell were transferred to be about the occipital condyles. NHTSA's proposed ICPL for N_{TE} of 1.0 was used to generate the related generic acceptance criterion.

3.6.3 NON-REGULATED TEST METRICS

3.6.3.1 NCAP

The NCAP test score is based on the combined probability of injury, CPI, defined as

$$CPI = P_{head} + P_{chest} - (P_{head} \times P_{chest}) \qquad (4)$$

where P_i represents the probability of life-threatening injury attributed to either the head or chest.

The equations used to calculate the NCAP score were

$$P_{head} = \frac{1}{(1 + e^{(5.02 - 0.00351 \times HIC36)})} \qquad (5)$$

and

$$P_{chest} = \frac{1}{(1 + e^{(5.55 - 0.0693 \times ChestGs)})} \qquad (6)$$

subject to the following scoring system:

5 star: $0.0 < CPI \le 0.10$

4 star: $0.10 < CPI \le 0.20$ $\qquad$ (7)

3 star: $0.20 < CPI \le 0.35$

Since NCAP is not a Regulated crash test, the CPI had to be chosen in order to quantify the related generic acceptance criterion. Accordingly, a CPI score of 0.20 was chosen.

3.6.3.2 CANADIAN SURROGATE TEST

This test, involving a belted HIII05, was previously described in Section 3.4. The HIII05 upper neck extension moment is typically the occupant response of interest in this performance test. However, since this is not a regulated test, the magnitude of moment had to be chosen in order to quantify the related generic acceptance criterion. Accordingly, the published IARV, 39 N-m of Table 4, was chosen.

3.6.4 OOPO HIII50 AND OOPO HIII05

The OOPO HIII50 and OOPO HIII05 (chest-on-module) performance tests were previously discussed in Section 2.2. For this study, these tests were assessed by the AAMA IARVs of Table 4, i.e., sternum velocities of 8.2 m/s were used to generate the related generic acceptance criteria.

Note that the HIII05 is the targeted Regulation device for this mode of testing (NHTSA, 1998), but for the purposes of this study, the OOPO HIII50 test was considered as a non-Regulated performance test.

Moreover, whereas the chest-on-module performance tests primarily help assess <u>thoracic</u> injury risks, another static OOPO test was proposed to better help assess <u>neck</u> injury risks (NHTSA, 1998). By not accounting for this test, the studied OOPO performance constraints were deemed non-comprehensive.

3.6.5 SAFETY FACTORS

Safety factors, defined as margins to ensure confidence in compliance (to compensate for system variability), were in this study. The <u>generic</u> safety factors applied to ere

$$SF_{reg} \equiv \frac{IARV \ or \ ICPL}{response} = 1.25 \qquad (8)$$

and those applied to non-Regulated responses were

$$SF_{nonreg} \equiv \frac{IARV \ or \ Target}{response} = 1.11 \qquad (9)$$

Eqs. (8) and (9) could also be considered as generic compliance margins of 20 and 10%, respectively. It is accordingly noted that, for the purposes of this study, greater compliance margins was applied to Regulated responses than to non-Regulated responses cf., Table 5. Additionally, it should be noted that, since discussion continued as to which occupant response (chest acceleration, chest deflection, sternal velocity, or the viscous criterion) better predicts OOPO thoracic injury potential, less stringent weighting was applied to the estimated OOPO-related occupant responses. Again, it should be recalled that generic safety factors were applied in this theoretical study; Precise statistical compliance factors and specific corporate emphasis were beyond the scope of this study.

The collective results of Sections 3.4 through 3.6.5 were used to derive the set of generic acceptance criteria for all studied performance tests, e.g., for the NCAP performance test, the related generic acceptance criterion was equal to 0.18 (i.e., SF_{nonreg} x $Target_{NCAP}$ = 0.9 x 0.2).

With both a means to estimate the pertinent occupant occupant responses for the performance tests for a wide variety of designs and a means to assess the stringency of the accompanying performance constraints, the design space study was then conducted.

4.0 DESIGN SPACE SCREENING EXERCISE (DSSE)

In order to predict the effects that alternative frontal impact test Regulations would have on the number of candidate dual-stage driver airbag designs (when assessed subject to a set of generic acceptance criteria established for the six considered performance tests), the following computational exercise was executed:

Step 1:
All of the studied airbag design variable combinations were subjected to mathematical simulations of each of the potential, unbelted FMVSS 208 tests.

Step 2:
All of the studied airbag design variable combinations were subjected to mathematical simulations of both the NCAP and Canadian tests.

Step 3:
All of the studied airbag design variable combinations were subjected to the mathematical model representations of HIII50 and HIII05 chest-on-module tests.

Table 5:
Safety Factor Schedule for Generic Acceptance Criteria

Performance Test	Dummy	Pertinent Occupant Responses	Safety Factor for Constraint-Related Occupant Responses
High-Speed Unbelted Test	HIII50	HIC15, Upper Neck Axial Force, Upper Neck Shear Force, Upper Neck Moment, Upper Neck $N_{TE,}$ Chest G's (3ms cumdur), Chest Deflection	1.25
High-Speed Unbelted Test	HIII05	HIC15, Upper Neck Axial Force, Upper Neck Shear Force, Upper Neck Moment, Upper Neck $N_{TE,}$ Chest G's (3ms cumdur), Chest Deflection	1.25
NCAP	HIII50	CPI	1.11
Canadian	HIII05	Upper Neck Moment	1.11
OOPO HIII50	HIII50	Sternum Velocity	1.11
OOPO HIII05	HIII05	Sternum Velocity	1.11

Step 4:
Each of the predicted occupant responses from Steps 1-3 were compared with the set of generic acceptance criteria (with SF_{reg} = 1.25 and SF_{nonreg} = 1.11).

For the purposes of this study, designs predicted to satisfy the set of generic acceptance criteria were binned as "acceptable" while designs that did not satisfy were binned as "unacceptable."

Step 5:
Statistics such as "Evaluated Designs Deemed Unacceptable (%)" were calculated for each performance constraint, e.g., Evaluated Designs Deemed Unacceptable (%) = 92% when considering the HIII50 occupant Chest G's in the rigid fixed barrier case at 48 kph (i.e., 310 designs of the studied 336 did not have the predicted 3ms cumdur Chest G's to be less than 48 G's (= 60/1.25)).

This 5-step exercise, designated the "Design Space Screening Exercise" (or "DSSE"), can be expressed in mathematical notation as follows:

$$\begin{Bmatrix} vent \\ bagsize \\ stage1scale \\ stage2scale \end{Bmatrix} = \begin{Bmatrix} 2x10, 2x15, 2x20, 2x25, 2x30, 2x35, 2x40mm \\ 610, 648, 673mm \\ 0.75, 1.00, 1.25, 1.50 \\ 0.0, 0.75, 1.00, 1.25, 1.50 \end{Bmatrix} \quad vs \quad \{constraints\} \quad (10)$$

where the studied performance constraints were

and recall

$$FMVSS208|_{unbelted}Options = \begin{Bmatrix} RFB48-HIII50 & RFB48-HIII05 \\ RFB40-HIII50 & RFB40-HIII05 \end{Bmatrix} \quad (12)$$

The related crash pulses associated with this specific study of a prototype mid-sized, passenger car are shown in Figure 4. Note that, due to rebound effects in the barrier tests, integration of the crash pulse deceleration does not exactly yield the speeds in the aforementioned case designations.

5.0 RESULTS OF DESIGN SPACE SAMPLING EXERCISE

One theoretical airbag deployment strategy was assessed via the DSSE, specifically, a "Non-Uniform Deployment Schedule" consisting of:

• Stage I-only deployments for all studied performance tests involving the HIII05 (irrespective of restraint level and crash speed/severity), and

• Stage I II deployments for the studied rigid fixed barrier crash tests, i.e., RFB40, RFB48, and NCAP, and

• Stage I II deployments possible in the OOPO HIII50 performance test.

5.1 RESULTS OF DESIGN SPACE SCREENING EXERCISE - HIII50

Given the application of the set of generic acceptance criteria, Figure 5 shows the screening (thrifting) effects on design space observed for the HIII50 (where the studied airbag deployment schedule called for deployment of Stage I II airbags). Note the following:

For the dynamic performance constraints:

- For the constraint set that included the RFB48 test suite, 92% of the 336 studied designs were discarded due to Chest G considerations, and

- For the constraint set that included the RFB40 test suite, 0% of the 336 studied designs were discarded due to Chest G considerations.

For the static, OOPO performance constraints:

- For the constraint set that included the RFB48 test suite, 50% of the 336 studied designs were discarded due to sternal velocity considerations, and

- For the constraint set that included the RFB40 test suite, 50% of the 336 studied designs were discarded due to sternal velocity considerations.

5.2 RESULTS OF DESIGN SPACE SCREENING EXERCISE - HIII05

Similarly, given the application of the set of generic acceptance criteria, Figure 6 shows the screening (thrifting) effects on design space observed for the HIII05 (where the studied airbag deployment schedule called for deployment of only Stage I airbags). Note the following:

For the dynamic performance constraints:

- For the constraint set that included the RFB48 test suite, 76% of the 84 studied designs were discarded due to N_{TE} considerations, and

- For the constraint set that included the RFB40 test suite, of the 84 studied designs were discarded due to N_{TE} ations, and

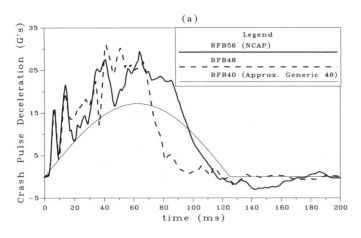

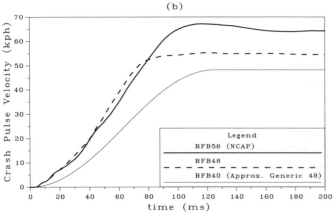

Figure 4: Crash Pulse Information

- For the constraint set that included the RFB48 test suite, 58% of the 84 studied designs were discarded due to chest deflection considerations, and

- For the constraint set that included the RFB40 test suite, 32% of the 84 studied designs were discarded due to chest deflection considerations.

For the static, OOPO performance constraints:

- For the constraint set that included the RFB48 test suite, 25% of the 84 studied designs were discarded due to sternum velocity considerations, and

- For the constraint set that included the RFB40 test suite, 25% of the 84 studied designs were discarded due to sternum velocity considerations.

5.3 RESULTS OF DESIGN SPACE SCREENING EXERCISE - RESTRAINT LEVEL CONSIDERATION

Another way to analyze the results of this study was to consider non-acceptance from the standpoint of restraint level. Table 6 illustrates that nearly all of the non-acceptable studied designs that were predicted to not satisfy the set of generic acceptance criteria were discarded due to a lack of compliance with one (or more) of the performance constraints associated with the unbelted occupant tests.

5.4 RESULTS OF DESIGN SPACE SCREENING EXERCISE - AGGREGATE

When the results of Sections 5.2 through 5.3 were considered collectively, the screening exercise yielded the results shown in Table 7.

Observations from Figures 5 and 6 and Tables 6 and 7 indicated a significant reduction in the available design space when contrasting the 40 vs. 48 kph high-speed unbelted occupant-influenced performance constraints of the SGAC. Specifically, when the generic acceptance criteria set contained the unbelted rigid fixed barrier testing constraints run at 48 kph, not a single design (in the design space studied) satisfied the considered set of generic acceptance criteria. This reduction was almost entirely attributable to the <u>unbelted</u> occupant portion of the performance constraints.

6.0 DESIGN SELECTION PROCEDURE

A procedure was developed to determine the extent to which a design satisfied the constraint set. Accordingly, this procedure was used to help select designs that best attempted to satisfy the chosen constraints.

The "Design Selection Procedure" was a natural extension of the "Design Space Screening Exercise."

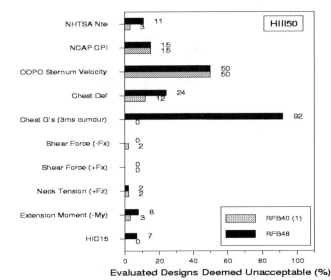

Figure 5: Results of Design Space Screening Exercise for Two High-Speed Unbelted Occupant Test Suites: HIII50 Test Dummy Considerations

(1) Assumption, from NHTSA 1999 SNPRM, that RFB40 approx. same severity as Generic48

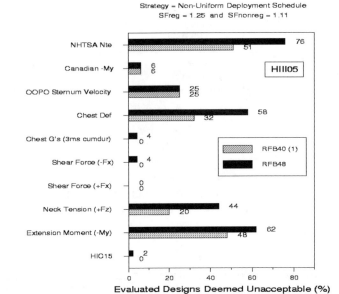

Figure 6: Results of Design Space Screening Exercise for Two High-Speed Unbelted Occupant Test Suites: HIII05 Test Dummy Considerations

(1) Assumption, from NHTSA 1999 SNPRM, that RFB40 approx. same severity as Generic48

This procedure to identify the design that best satisfied the constraints was established by defining a "Selected Airbag Design Acceptance Factor," i.e.,

"Selected Airbag Design Acceptance Factor" = minimum [maximum (all normalized Regulated and non-Regulated occupant responses)]

over the domain of 336 studied designs, where

normalized occupant response =
predicted occupant response / related component of the SGAC.

The results of the design selection procedure (as before, applied with intent to contrast the effect of varied high-speed, unbelted constraints) are shown in Table 8.

Note that, for continuity purposes, the results of Table 7 are presented in the second column of Table 8.

The third column of Table 8 contains the design specifications for the airbag that was chosen (out of the 336 studied designs) subject to the aforementioned procedure.

The fourth column of Table 8 contains that design's "Selected Airbag Design Acceptance Factor" -- a derived number which can be interpreted as the level of compliance for the design that best satisfied the most-restricting performance constraints. Accordingly, from inspection of Table 8, no design was predicted to have a normalized occupant response lower than 0.85. Moreover, the 0.85 value implies both SGAC compliance and that approximately 15% of the design space remained available for purposes of design selection after the design space screening exercise was completed. This available design space could be possibly used to help address other constraints not considered in this study, e.g., the aforementioned OOPO performance tests for the neck, non-Regulated frontal impact performance tests such as the high-speed, deformable offset barrier test conducted by the Insurance Institute for Highway Safety, 5-star USA NCAP performance, etc.

Table 6:
Results of the Design Space Screening Exercise
with Respect to the Level of Restraint
(Non-Uniform Deployment Schedule)
with SF_{reg} = 1.25 and SF_{nonreg} = 1.11

Scenario	Studied Designs	% Non-Acceptable (present IARVs proposed ICPL) with	
		all constraints	unbelted only constraints
"High-Speed" Unbelted Occupant Component of SGAC			
RFB40[1]	336 (HIII50), 84 (HIII05)	94	93
RFB48	336 (HIII50), 84 (HIII05)	100	100

Table 7:
Results of the Design Space Screening Exercise
with Aggregate Considerations
(Non-Uniform Deployment Schedule)
with SF_{reg} = 1.25 and SF_{nonreg} = 1.11

"High-Speed" Unbelted Occupant Component of SGAC	Studied Designs	Accepted Designs (present IARVs proposed ICPL)
RFB40[1]	336 (HIII50), 84 (HIII05)	21
RFB48	336 (HIII50), 84 (HIII05)	0

Table 8:
Results of the
Design Space Screening Exercise
and the
Design Selection Procedure

"High-Speed" Unbelted Occupant Component of SGAC	Number of Accepted Designs [2] (out of 336)	Selected Airbag Details [3]	Selected Airbag Design Acceptance Factor	Limiting Factor
RFB40 (approx by Generic48 [1])	n = 21 (6%)	Inflator Stage I = 100% Inflator Stage II = 75% Airbag=648mm 2x20 mm vents	0.85	Sternum Vel (OOPO-HIII50)
RFB48	n = 0 (0%)	Inflator Stage I = 125% Inflator Stage II = 100% Airbag=610mm 2x25 mm vents	1.07	Chest G's (in position - unbelted HIII50)

(1) Assumption, from NHTSA 1999 SNPRM, that RFB40 appro
same severity as Generic48

(1) NHTSA 1999 SNPRM

ercise"

The RFB48 case, in contradistinction to the RFB40 case, was estimated to have a 1.07 value. This prediction indicated that, of the domain of 336 studied designs, none were found to satisfy the SGAC, i.e., an absence of available design space where the design that best satisfied was 7% higher than dictated by the SGAC.

Finally, it should be noted from inspection of Table 8 and Table 3 that for the designs selected to best satisfy the studied performance constraints, the RFB48-based design was predicted to require higher mass flow rates for both Stage I and Stage II than those required for the RFB40-based design.

7.0 CONCLUSIONS

Advanced airbags with dual-stage inflators have been identified as a potential means to further enhance occupant protection when used in conjunction with advanced-technology, crash-sensing systems. With intent to comment on the maximum speed at which the high-speed, unbelted occupant test suite of the 1999 SNPRM for Advanced Airbags will be conducted (48 vs. 40 kph), an extensive parametric study involving 336 different theoretical dual-stage driver airbag systems in a prototype mid-sized passenger car was conducted.

The two candidates for the high-speed, unbelted occupant performance test suite were considered in conjunction with four other performance tests, viz., NCAP, Canadian, OOPO HIII50, and OOPO HIII05 via mathematical modeling. By comparing the pertinent, predicted occupant responses with a set of generic occupant response acceptance criteria derived for the studied performance tests (which involved safety factor-weighted IARVs/ICPLs and chosen non-Regulated targets), available design space was estimated, i.e., the number of theoretical dual-stage driver airbag designs that satisfied the set of generic occupant response acceptance criteria.

The conclusions drawn from this study included:

• The introduction of both the HIII05 and the HIII50 into the proposed, future Regulation was predicted to significantly reduce the available design space (considered in this study).

• None of the 336 studied theoretical dual-stage driver airbag designs satisfied the generic acceptance criteria set when the unbelted rigid fixed barrier testing constraints were run at 48 kph.

• 21 out of the 336 satisfied the generic acceptance criteria set when the unbelted rigid fixed barrier testing constraints were run at 40 kph.

• When considering the discarded designs, nearly all of them were predicted to not comply due to (at least one of) the unbelted occupant performance constraints of the generic acceptance criteria set.

• For the designs selected to best satisfy the studied, non-comprehensive performance constraints, the RFB48-based design was predicted to require higher mass flow rates for both Stage I and Stage II than those required for the RFB40-based design. Accordingly, the anticipated positive real-world benefits of dual-stage driver inflators may be reduced. The estimated, projected significance of this conclusion is beyond the scope of the present study and was left for the second part of this research.

• Given the above conclusions (and the attendant assumptions), the 40 kph high-speed unbelted occupant test suite was considered to be significantly more amenable to the application of dual-stage driver airbags than its 48 kph counterpart.

It should be noted that, implicit to these results, the static OOPO requirement was only partially addressed; No HIII05 neck tests were considered. Also, other non-Regulated frontal impact performance tests such as the high-speed, deformable offset barrier test conducted by the Insurance Institute for Highway Safety were not considered. Accordingly, by the addition of those performance constraints, the available design space may be even more limited.

ACKNOWLEDGMENTS

The authors would like to thank the following contributors to this study: Kris Warmann for providing much of the information regarding the prototype hardware; Linda Rink for providing baseline inflator characteristics predictions; T.C. Weng for discussions regarding many of the occupant model inputs; Scott Schmidt and Dave Clark for providing the AAMA reference material; Jeff Nadeau, Stacy Nadeau, and Phil Przybylo for reviewing this document.

REFERENCES

American Automobile Manufacturers Association (AAMA), "Proposal for Dummy Response Limits for FMVSS 208 Compliance Testing," Docket No. NHTSA 98-4405, Notice 1, AAMA S98-13, Attachment C, December 1998.

Bauberger, Alfred and Dieter Schaper, "Belt Pretensioning and Standardized 'Slack' Dummy," 96-S10-W28, 15th International Technical Conference on Enhanced Safety of Vehicles, Melbourne, Australia, May 1996.

Foret-Bruno, J-Y, et al., "Thoracic Injury Risk in Frontal Car Crashes with Occupant Restrained with Belt Load Limiter, 42nd Stapp Car Crash Conference Proceedings, Paper 983166, November 1998.

ISO Technical Report 10982, "Road Vehicles-Test Procedures for Evaluating Out-of-Position Vehicle Occupant Interactions with Deploying Air Bags, March 15, 1998.

Laituri, Tony R. and Priya Prasad, "Correlation of Driver Invariables with the Viscous Criterion for the

Mid-Sized Male, Instrumented Test Dummy in the Chest-on-Module Condition, SAE Paper 1999-01-0763.

Madymo Database Manual, Version 5.3, TNO Road-Vehicles Research Institute, 1997.

Madymo Utilities Manual, Version 5.4, TNO Road-Vehicles Research Institute, 1999.
National Highway Traffic Safety Administration, Preliminary Economic Assessment, FMVSS No. 208, Advanced Airbags, Office of Regulatory Analysis & Evaluation Plans and Policy, August 1998.

National Highway Traffic Safety Administration, Supplemental Notice for Proposed Rulemaking, FMVSS No. 208, Advanced Airbags, September 1999.

Prasad, Priya, and Tony R. Laituri, "Consideration of Belted FMVSS 208 Testing," 96-S3-O-03, 15th International Technical Conference on Enhanced Safety of Vehicles, Melbourne, Australia, May 1996.

SAE J211, Instrumentation for Impact Test - SAE J211 SAE Recommended Practice, October 1988.

Van Wylen, Gordon J., and Richard E. Sontag, Fundamentals of Classical Thermodynamics, John Wiley & Sons, 1978.

Wang, J.T. and Donald J. Nefske, "A New CAL3D Airbag Inflation Model," SAE Paper 880654, 1988.

DEFINITIONS, ACRONYMS, ABBREVIATIONS

ACRONYMS

AAMA	American Automobile Manufacturers Association
CPI	Combined Probability of Injury
DSSE	Design Space Screening Exercise
FMVSS	Federal Motor Vehicle Safety Standard
HIC	Head Injury Criteria
HIII05	Hybrid III, Small-female test dummy
HIII50	Hybrid III, Mid-sized male test dummy
IARV	Injury Assessment Reference Value
ICPL	Injury Criteria Performance Limit
ISO	International Standards Organization
NCAP	New Car Assessment Program
NHTSA	National Highway Traffic Safety Administration
NPRM	Notice for Proposed Rulemaking
OOPO	Out-of-Position Occupant (chest-on-module)
SF	Generic Safety Factor
SGAC	Set of Generic Acceptance Criteria
SNPRM	Supplementary Notice for Proposed Rulemaking
SV	Sternum Velocity

APPENDIX 1

The validation cases (and their accompanying hardware) are outlined in Table A1. Note that "Generic" refers to "generic pulse" and "Rigid" refers to "rigid fixed barrier." Hereafter, the case numbers of Table A1 will serve as designations for the various test type/speed/severity/restraint level/occupant size/seating position variables.

A comparison of the aforementioned six modeled and physical occupant responses for the "NCAP" case, i.e., Case 1, is shown in Figure A1 where all occupant responses (for both model and test) were filtered subject to SAE J211 protocols (SAE J211, 1988). Unless otherwise noted on the model/test comparison plots, the ordinate range for each of the occupant responses in Figure A1 (and the validation plots thereafter) was determined from recently-published response limits known as injury criteria performance limits (ICPLs) (NHTSA, 1998). It should be noted that, in general, the response limits are dummy-size dependent. Displaying the results in this manner helped identify responses near proposed ICPLs. The model-test comparisons for Cases 1 through 6 are found in Figures A1 through A6, respectively.

Table A1:
Validation Cases for Madymo3D Modeling (Driver)

Case Number	Type	Speed (kph)	Test Count	Severity	Restraint Level	Occ	Seat Position	Inflator
NCAP (Case 1)	Vehicle	56	1	Rigid	Belt Bag	HIII50	Mid	Stages I II
Canadian (Case 2)	Sled	48	1	Generic	Belt Bag	HIII05	Full-Forward	Stage I
FMVSS 208 (Case 3)	Sled	48	1	Generic	Bag	HIII50	Mid	Stage I
FMVSS 208 (Case 4)	Vehicle	48	2	Rigid	Bag	HIII50	Mid	Stages I II
FMVSS 208 (Case 5)	Vehicle	48	1	Rigid	Bag	HIII05	Full-Forward	Stage I
FMVSS 208 (Case 6)	Vehicle	48	1	Rigid	Bag	HIII05	Full-Forward	Stages I II

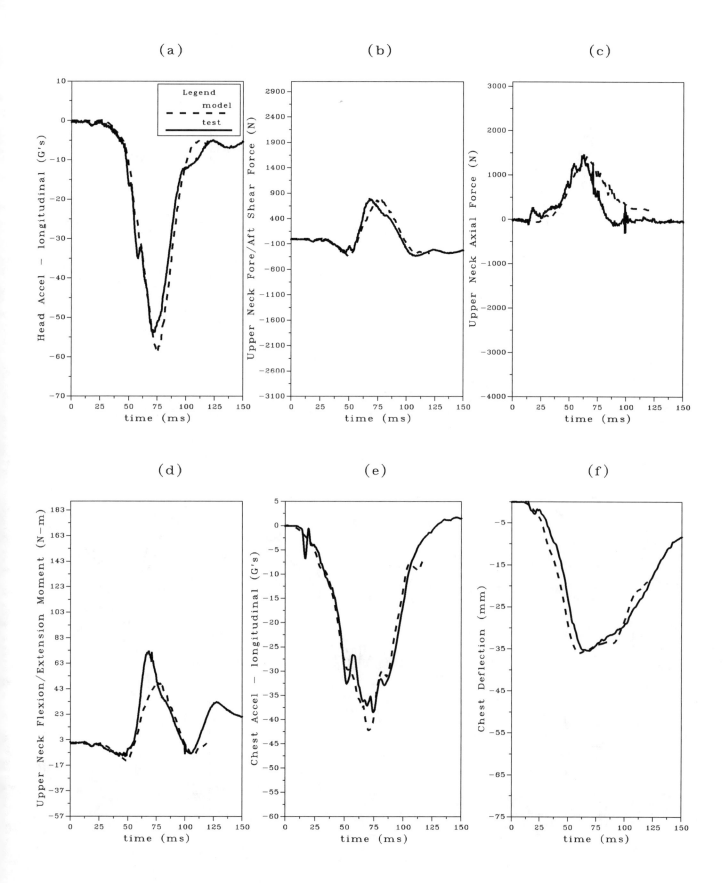

Figure A1: Mathematical Model Validation (Occupant Responses for Case 1)

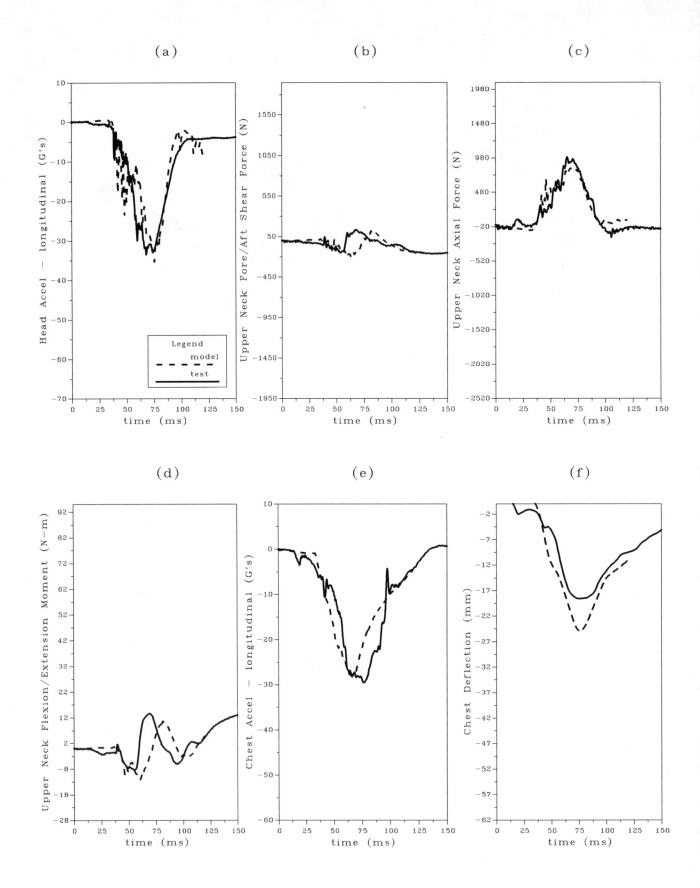

Figure A2: Mathematical Model Validation (Occupant Responses for Case 2)

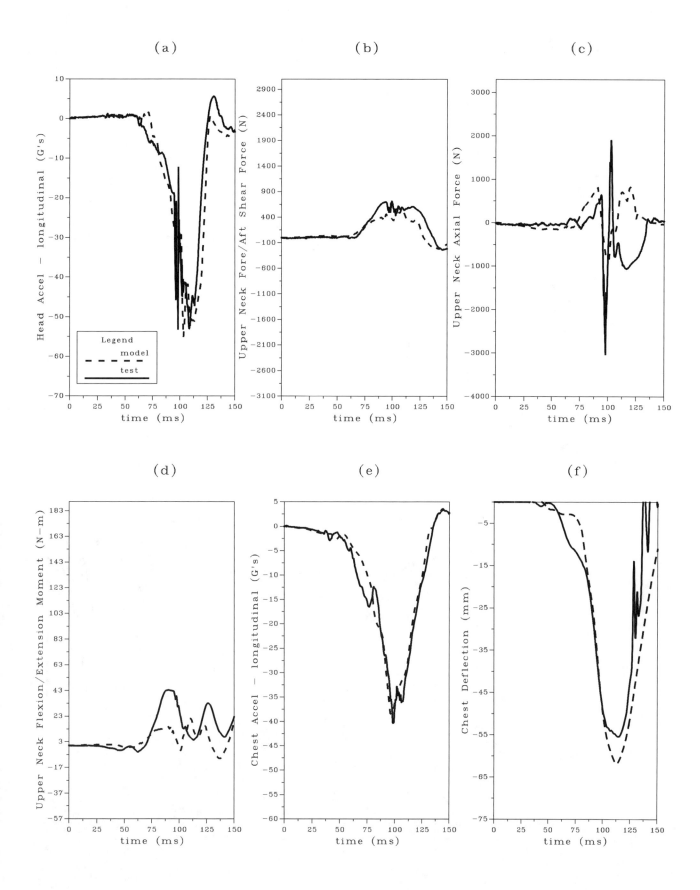

Figure A3: Mathematical Model Validation (Occupant Responses for Case 3)

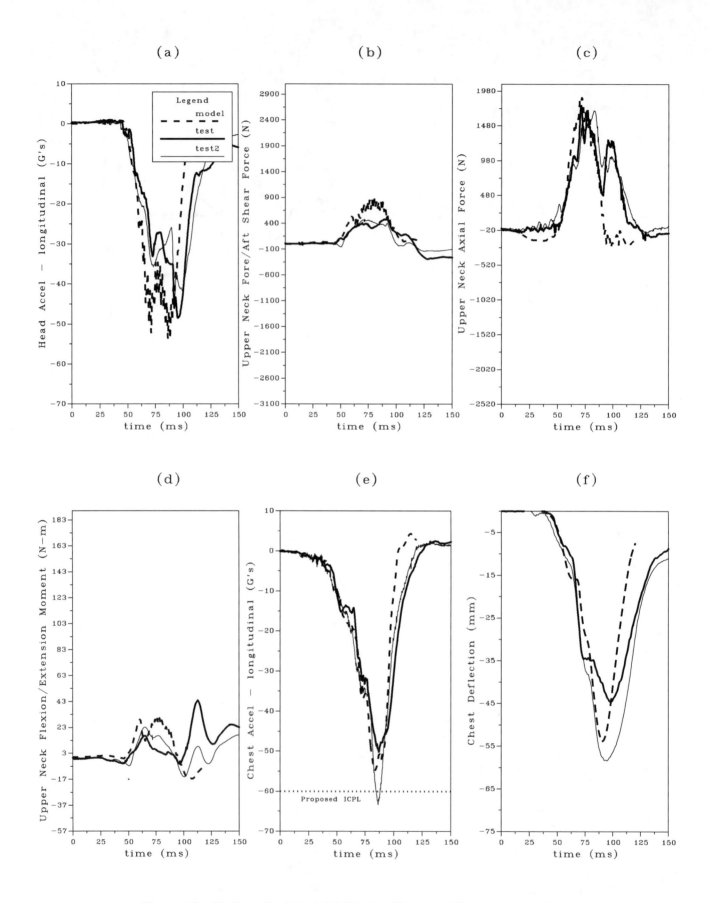

Figure A4: Mathematical Model Validation (Occupant Responses for Case 4)

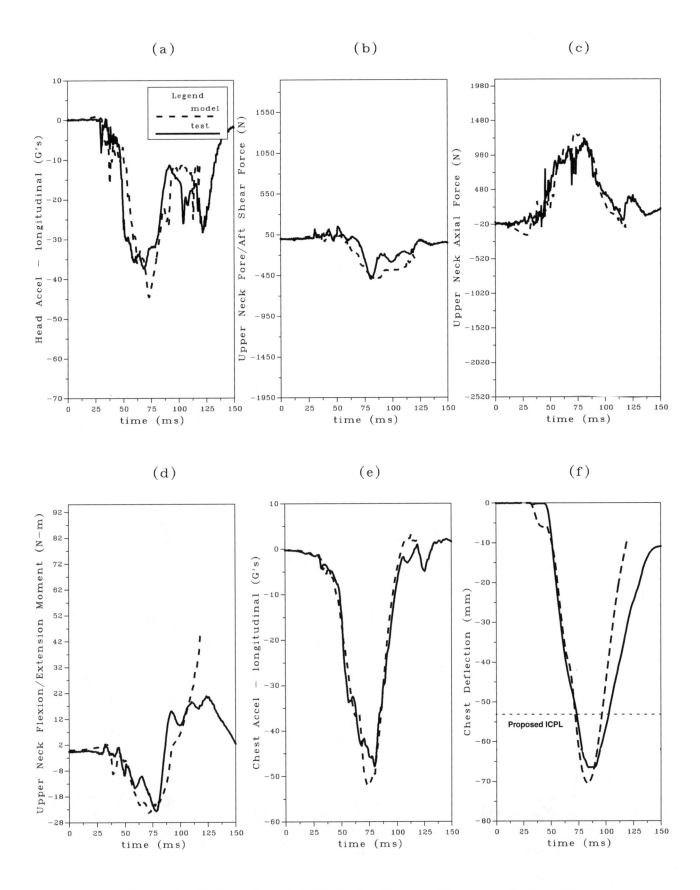

Figure A5: Mathematical Model Validation (Occupant Responses for Case 5)

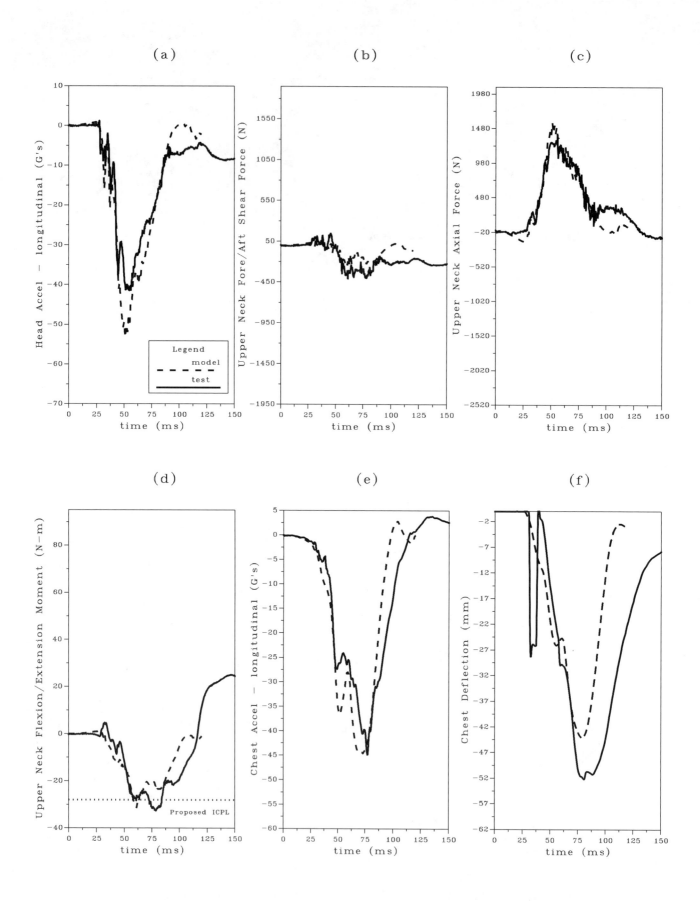

Figure A6: Mathematical Model Validation (Occupant Responses for Case 6)

Pertinent details of the model(s) included:

Simulation code: Madymo3D version 5.3

Dummies: Rigid-link representations from TNO database v. 5.3.1 with later-added facetted head(s)

Airbags: Finite element representations

Airbag unfolding: Actual initial size via initial metric approach (with no self contact)

Airbag inflator characteristics: Time-dependent mass flow rate and temperature from supplier

Airbag inflation: 6-jet array to better represent observed airbag trajectory

Airbag tethers: 9-strap array to represent a circular ring

Airbag material: Density, thickness, and isotropic modulus of elasticity estimated from testing

Airbag leakage: Discrete vent and seam losses modeled via discharge coefficient (Note: A seam leakage estimate was based on airbag diameter and was introduced via a modified vent discharge coefficient)

Airbag permeability: Estimated from free inflation testing

Airbag trigger times: In accordance with specific events or estimated from thumb rules

Airbag energy losses due to module: Modeled as "rigid" mesh with a time lag from the airbag trigger time that is directly proportional to the inflator's peak tank pressure rise rate

Seat belt retractor: Estimated force-deflection characteristic from supplier testing

Seat belt pyrotechnic pretensioner: Modeled as an initially-locked, pretensioned spring

Seat belt pyrotechnic pretensioner trigger times: Set equal to airbag trigger times

Seat belt slack: Slight amounts present (approximated)

Dummy position: In accordance with specific event (and instrumented dummy size)

Crash pulse: Estimated from average of B-pillar left and right rocker decelerations

Intrusions: Time-dependent functions from vehicle structural models and post-crash inspections

Column stroke: Predicted via a combination of the force-deflection characteristic and interaction with intr

components which have prescribed motion

Steering Wheel: Compliant

Occupant chest/steering wheel contacts: Conducted with "Evaluation" feature of Madymo3D

III. AIR BAG DEVELOPMENT

Development of Advanced Restraint Systems for Minicars RSV _____

CHARLES STROTHER, MICHAEL U.
FITZPATRICK, and TIMOTHY P. EGBERT
Minicars, Inc. EngineeringStaff

ABSTRACT

Design and development of the Research Safety Vehicle (RSV) restraint systems has provided a unique opportunity to integrate advanced restraint system concepts into a vehicle whose structure has been designed to be structurally superior in a crash environment. This combination has enabled Minicars to "tune" both structure and restraints to perform in a mutually complementary manner. Because of this integrated approach to system design, the restraint systems have shown in actual vehicle crashes to be capable of protecting vehicle occupants to velocities in excess of 50 mi/h. Although the restraints that perform this task are advanced systems incorporating the latest techniques in restraint design, such as dual bags, stroking airbag mounting surfaces, and so forth, the restraint components are simple and producible. This paper describes occupant restraint systems designed for the RSV.

INTRODUCTION

During Phase I of the RSV program, Minicars performed an analysis to determine the necessary characteristics of a safe, efficient, economical vehicle for the 1980's time frame. As part of the study, a benefit/cost analysis was performed to identify that combination of safety features that would result in the highest safety payoff (benefits minus costs). Since the most important occupant protection is afforded by restraint systems, extensive analyses were performed to determine potential restraint systems which could be considered for the Research Safety Vehicle. Several candidate restraints were selected: standard 3-point belts; force-limited 3-point belts; preloaded, force-limited 3-point belts; force-limited airbelt systems; airbag restraints; and lap belts (for rear seat positions only). The results of these analyses (table 1) indicated the airbag restraint would provide the highest safety payoff for both the driver and right front seat passenger. Although lap belt restraints would provide the highest safety payoff for rear seat passengers, force-limited 3-point belts were nevertheless selected because it was felt that, in the context of a research program, a more advanced system was warranted, especially in view of the fact that its impact on the total restraint safety payoff was so slight.

Critical to the performance of any restraint system is the deceleration produced on the occupant compartment by the vehicle fore-structure. The relatively long RSV front end is constructed from foam-filled sheet metal. This configuration produces mild compartment deceleration profiles (fig. 1) without creating significant compartment intrusion, even in severe frontal crashes.

Table 1. Ranking of restraint systems[a]

Driver	Right front passenger	Rear passenger (2)	B/C ratio	Safety payoff (billion $)
AB	AB	LAP	2.99	4.33
AB	AB	FL-3 PT	2.85	[b]4.29
AB	AB	3 PT	2.88	4.28
AB	AB	FL-ABLT[c]	2.81	4.27
AB	FL-ABLT	3 PT	2.96	4.19
AB	FL-ABLT	FL-ABLT	2.88	4.17

[a]Assumes radar accident avoidance system with 1.2 g brakes.
[b]RSV choice.
[c]Force-limited air belt.

DRIVER RESTRAINT SYSTEM

The RSV driver restraint is comprised of an energy-absorbing steering column and mounting system, a steering wheel assembly that houses the airbag, and a knee restraint.

The steering column energy-absorbing (EA) unit (fig. 2) is a telescoping absorber in which the aft section, a 1 3/4-inch square tube, connects with the steering wheel assembly and the forward section, a 3-inch square tube, connects to the vehicle cowl.

Energy is absorbed during telescoping by a rollerless tape mechanism. A continuous length of steel tape is attached at its ends to the right and left side interior walls of the forward tube and at its middle to the forward end of the aft section of the column (fig. 3a). The tape forms two loops, each having a diameter approximately equal to one-half the difference of the forward and aft column section widths. As the column telescopes (fig. 3b), the tape loops roll off the sides of the forward section and onto the sides of the smaller aft section, producing a column collapse force given by

$$F_c(x) = \frac{2\sigma_y t^2 W(x)}{D}$$

where x = column stroke (in)
 σ_y = yield strength of the tape material (psi)
 t = tape thickness (in)
 W(x) = tape width at the loop as a function of column stroke
 D = loop diameter (in)

The parameters above have been set to result in a theoretical stroking level of 1 750 pounds over the first 5 inches of its length. Thereafter the width of the bands increases to accommodate large adult drivers at the highest impact severity.

As with other steering columns, the RSV column is subjected to severe bending moments. The most severe of these is the counterclockwise[1] bending moment associated with the upward force at the steering wheel rim end of the column. To counteract this moment without introducing excessive friction, a roller is provided at the aft upper end of the forward section of the column to bear on the top flat of the aft column section. A nylon slide pad is provided on the bottom forward end of the aft section to bear on the

---

[1] As viewed from the driver side of the vehicle.

Figure 1. RSV compartment response in 50-mi/h barrier impacts.

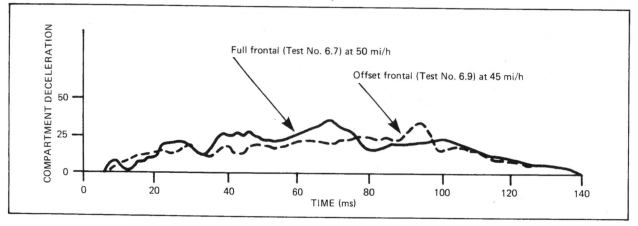

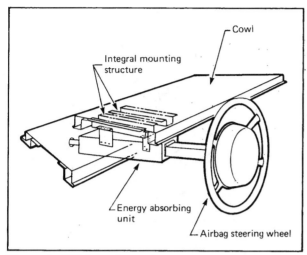

Figure 2. RSV steering column system.

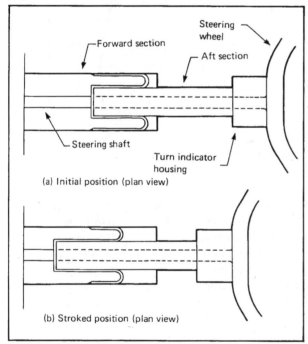

Figure 3. RSV column EA mechanism.

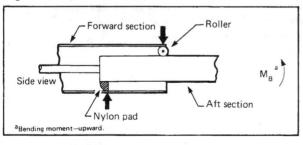

Figure 4. Nylon slide pad.
[a] Bending moment—upward.

lower interior surface of the forward column section, as shown in figure 4.

Significant lateral bending moments arise when the vehicle is involved in an angular or offset frontal collision and/or when the driver is not directly behind the column at impact. To counter these lateral bending moments, nylon pads are provided on the forward section of the column at its aft end on the interior vertical walls of the tube. These pads bear against the right and left sidewalls of the column aft section. Two shoulder bolts, inserted vertically through the forward section of the column, contain nylon rollers which bear on the side walls of the aft column section as shown in figure 5.

The steering shaft is a conventional (Ford) telescoping unit with a splined aft end. Bearings for this shaft are provided at the firewall at the point it enters the aft section of the column and where it connects to the steering wheel. Also, attached to the aft end of the column is a Ford turn indicator/ignition/column lock assembly.

The steering wheel assembly is comprised of a wheel rim, inflator, reaction plate/retaining ring assembly, airbag configuration, and a bag cover as shown in the exploded view of figure 6. The wheel rim is a General Motors Air Cushion Restraint System (GM ACRS) rim, a significantly more rugged rim than any other commercially available model. The wheel rim has four spokes, a diameter of 15 1/2 inches, and a depth of 4 inches. The inflator is a modification of a unit manufactured by Thiokol Corporation to meet GM ACRS specifications, the inflator propellant charge being increased by about 10 percent. The bag system is clamped to a large (9-3/4-inch diameter) reaction plate/retaining ring assembly.

The RSV driver bag system is a dual-bag configuration. Both the inner and outer bags are cylindrical in shape (fig. 7). The inner bag column is 1.0 cubic foot; the larger, outer bag has a volume of 1.7 cubic feet. The bags are constructed of low permeability, plain weave 840 denier nylon No. 6. Venting in both bags is accomplished solely by bag porosity. The airbag cover is vacuum-formed of 1/16-inch thick, low density polyethylene. Although this material is flexible, it holds its shape and is environmentally rugged. It is impervious to

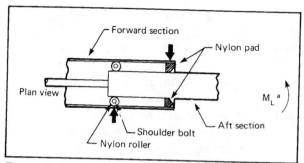

Figure 5. Insertion of shoulder bolts.

all common solvents. The tear pattern, preslit in the aft face, is covered by decorative tape.

The lower body energy of the driver is absorbed by the foam knee restraint, attached to a sheet metal backing plate on the lower dashboard (fig. 8). As can be seen, the backing pan is attached to the cowl and the firewall, and is oriented at about a 45-degree angle. The knee restraint consists of about 10 inches of rigid foam faced with a protective and decorative elastic foam-backed cover material.

System Operation

Operation of the RSV driver restraint system is best described by relating the sequence of events which occur during the most severe frontal impact condition for which the system is designed a 50-mi/h barrier impact. A sequence of five sketches depicting critical moments in the frontal crash event is shown in figure 9. Sketch (a) shows the occupant, restraint, and vehicle at the moment of bumper contact with the barrier.

Sketch (b) shows the time of sensor switch closing that occurs in the bumper switches at about 9 milliseconds into the event. At this point, the driver has moved forward only a fraction of an inch.

About 30 milliseconds into the event the occupant begins to experience deceleration from his restraint system, as shown in (c). The small inner bag, which receives gas directly from the inflator, can respond very quickly.

At about 80 milliseconds into the 50-mi/h barrier impact (d), the force exerted by the airbag system on the driver's chest exceeds the stroking force (about 1 750 pounds) of the steering column. Since the effective

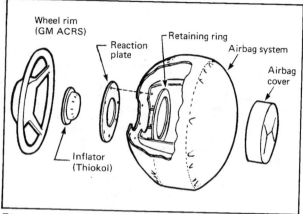

Figure 6. Steering wheel assembly—exploded view.

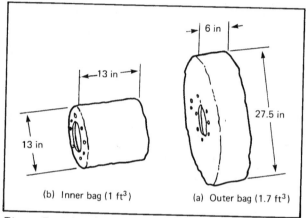

Figure 7. RSV airbag configuration.

weight of the 50th-percentile upper body is about 60-65 pounds in this situation, the force-limiting nature of the collapsible steering column produces chest deceleration levels of about 30 g during column stroking. The larger cubic foot outer bag, inflated by the gas vented from the inner bag, captures and prevents the head from whipping forward.

Sketch (e) shows the end of column stroking. At about 100 milliseconds, the driver reaches maximum forward translation in the compartment. The steering column has stroked about 4 1/2 inches and the knees have penetrated the lower dash about 6 inches (e). At this instant the occupant begins a very mild rebound into the seat.

Driver Restraint System Performance

To date, 36 sled and three vehicle crash tests have been performed in Phase II of the

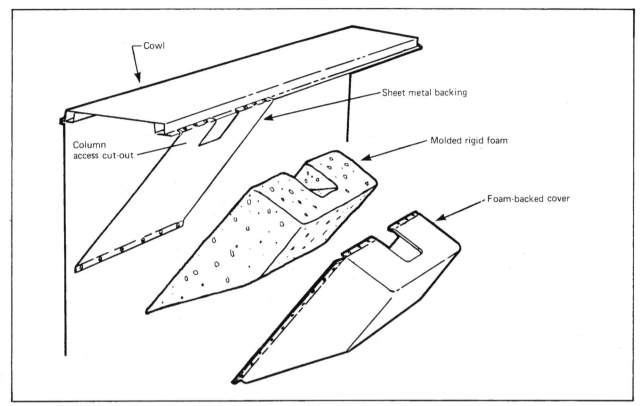

Figure 8. RSV driver knee restraint.

RSV effort to investigate various driver restraint system configurations. Because most of these tests were conducted to develop the system rather than evaluate system performance, it is not possible at this time to present a data package that completely defines the RSV driver restraint performance. Only developmental tests that demonstrate the performance of the final system configuration are included.

Sled Test Nos. 57, 58, 61, and 63 with vehicle crash Test Nos. 6.7 and 6.9 show the high-speed performance of the driver system with a range of driver somatotypes. In sled Test No. 57, the 50th-percentile male driver dummy was subjected to a simulated 45-mi/h perpendicular flat barrier impact. During this test, the column stroked about 3 1/2 inches. Because peak femur loads were about 1 800 pounds — somewhat higher than desired — the knee restraint was softened on subsequent tests. The results of this test are presented in figure 10.

In vehicle Test No. 6.7, an RSV structure with a 50th-percentile male dummy in the driver position was impacted into the flat barrier at 51.8 mi/h. Because at the time this test was run a somewhat stiffer vehicle front end was anticipated, the column collapse force was about 250 pounds higher than that used later in Test No. 57. The results of this test are presented in figure 11. Had the final configuration column been employed in this test, chest and head injury measures would have been somewhat lower because of the lower collapse force of the column. The recorded injury measures, however, are significantly below accepted injury criteria.

Essentially the same combination of vehicle structure and column design were tested in Test No. 6.9, in which the vehicle was tested in an offset barrier (right side engagement) at an impact speed of 45 mi/h. No column stroke resulted because of extremely low compartment deceleration levels experienced. Injury measures were below the accepted criteria, despite the lack of column stroking.

In sled Test No. 58, a 95th-percentile male driver dummy was subjected to a simulated 45-mi/h perpendicular, flat-barrier impact. In this test, a column stroke characteristic was made constant at 1 750 pounds throughout 8

EXPERIMENTAL SAFETY VEHICLES

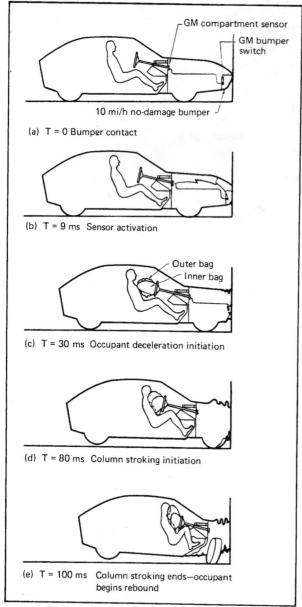

Figure 9. Driver restraint operation during critical moments in a 50-mi/h barrier impact.

inches of column design collapse. The results are given in figure 12. A peak chest deceleration of only 50 g's occurred at the end of column stroke.

In sled Test No. 61, the column stroking characteristic of Test No. 58 was modified to increase the stroking force to about 2 300 pounds during the last 3 inches of column collapse. The results of this 50-mi/h test are presented in figure 13. As shown in the chest resultant acceleration plot, the increase in column stroking force level was too abrupt; hence, a 68-g peak acceleration resulted. The column stroked about 7 3/4 inches. The increase in stroking level was subsequently made more gradual. On the basis of these two sled tests, it is estimated that the 95th-percentile male driver will be protected to barrier equivalent impact speeds between 45 and 50 mi/h.

The RSV driver restraint system was tested to determine its effectiveness in protecting lightweight drivers. A representative plot of system performance in the perpendicular barrier impact mode when used by drivers represented by the 5th-percentile female dummy is shown in figure 14. The results are from Test No. 63, a simulated 45-mi/h impact. With such a lightweight driver, the column stroke was very small — only a fraction of an inch. Minimal stroking with lightweight drivers results because most of the occupant kinetic energy absorbed in the compartment is by virtue of bag penetration and lower dash crush.

Protection Capability — Summary

On the basis of the sled and vehicle crush tests conducted to date, it is estimated that the majority of RSV driver restraint system users will be protected in frontal impacts of a severity comparable to a 50-mi/h barrier impact. Protection levels for large drivers, represented by the 95th-percentile male dummy, are estimated to be slightly below this level — in the 45-50 mi/h range.

RIGHT FRONT PASSENGER RESTRAINT SYSTEM

Description

The restraint system selected for the RSV front passenger position, like the driver restraint, uses the dual airbag concept. Unlike the driver system, however, it was designed in accordance with two constraints not imposed on the design of the driver system. For the front seat passenger, it was necessary to develop a passive passenger restraint system that would not only meet the requirements

SECTION 4: TECHNICAL SEMINARS

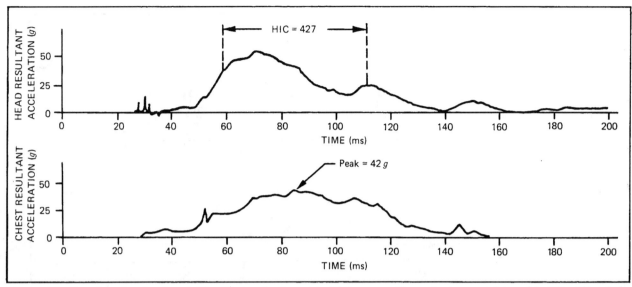

Figure 10. Results of a simulated 45-mi/h barrier impact—50th-percentile male dummy driver (sled Test No. 57).

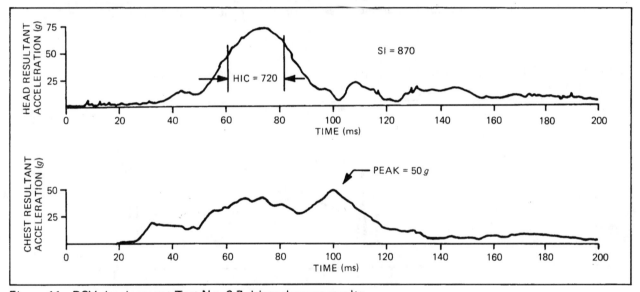

Figure 11. RSV development Test No. 6.7 driver dummy results.

for occupant protection, but would also maintain standard chest-to-dash distances so that ingress and egress and normal passenger movement in the compartment would not be hindered. It was also desirable that the system afford maximum protection for the out-of-position child in the front passenger seat and that the system be entirely mass-producible. Comprised of dual-chamber airbags, a Thiokol solid propellant gas generator, a stroking dash unit that houses the gas generator, and a knee pad, the system is linked to the vehicle structure via an energy-absorbing aluminum honeycomb that crushes at approximately 3 700 pounds (fig. 15).

System Operation — Normally Seated Passengers

System operation commences approximately 9 milliseconds after bumper contact. The bumper sensor signals the gas generator to initiate gas flow into the relatively small (3.8 cubic foot) torso bag (see fig. 16 for inflation sequence). Because of the small bag size, the bag fills very quickly. Chest g's begin to increase approximately 25-30 milliseconds after bumper contact.

Airbag. As the torso penetrates the lower torso bag, torso g's and torso bag pressure begin to increase. This increased pressure

EXPERIMENTAL SAFETY VEHICLES

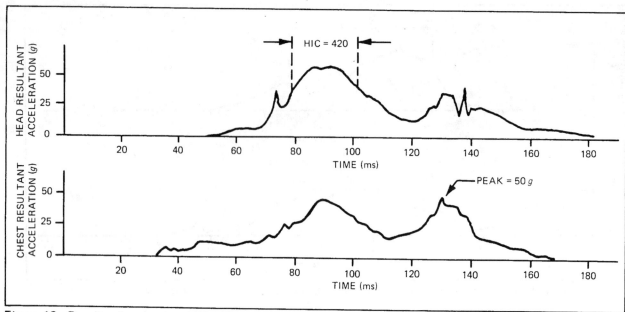

Figure 12. Results of a simulated 45-mi/h barrier impact—95th-percentile male dummy driver (sled Test No. 58).

Figure 13. Results of a simulated 50-mi/h barrier impact (tapered band)—95th-percentile male driver dummy (sled Test No. 61).

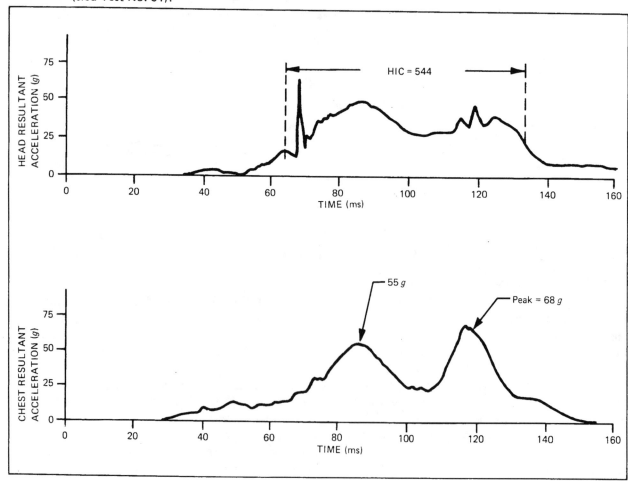

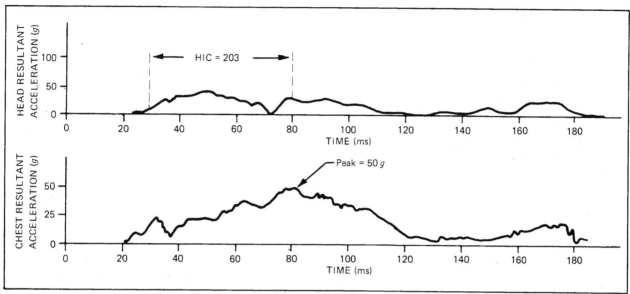

Figure 14. Results of a simulated 45-mi/h barrier impact—5th-percentile female driver dummy (sled Test No. 63).

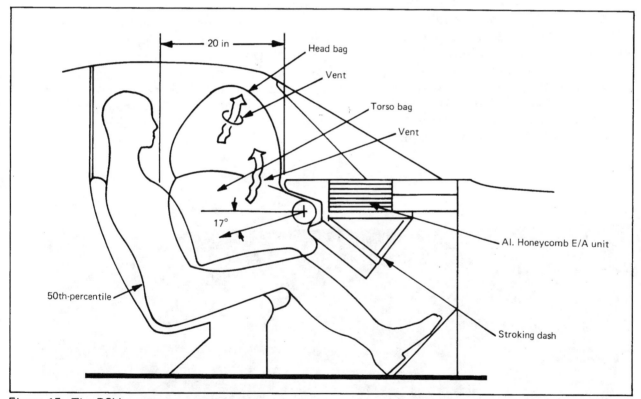

Figure 15. The RSV passenger restraint.

diverts a larger portion of the gas flow to a vent between the torso and head bags, causing the head bag to inflate. Approximately 50 milliseconds after bumper contact, the head bag is completely inflated. Since the head does not require support until after the torso has been somewhat retarded, the head bag need not deploy as rapidly as the torso bag. Thus, the dual-bag feature of the system enables the gas to be used twice — first to inflate the torso bag to slow the torso, and second to inflate the head bag to retard the head. Other advantages are inherent in the dual-bag system. For example, because the chest has a higher mass-to-area ratio than the head, it requires a higher bag pressure. This

EXPERIMENTAL SAFETY VEHICLES

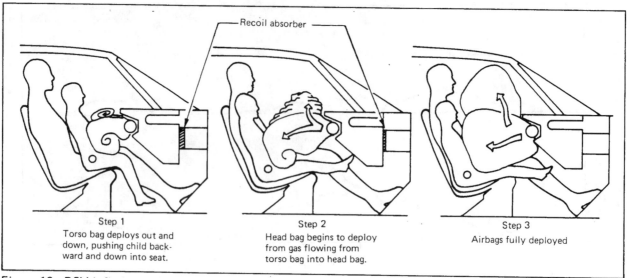

Figure 16. RSV inflation sequence.

requirement is ideally satisfied by the dual bag system since the volume and venting features of the two bags can be individually tailored to satisfy these differing head and chest requirements. Another very important advantage is that the inherently quick response time of the small torso bag and the tailoring characteristic cited combine to provide a very stroke-efficient airbag system.

The inflated RSV passenger bag is shown in figure 17. As can be seen, the membrane separating the two bags acts as a tension member to prevent the restraint bag from presenting a spherically-shaped bag front to the passenger at initial contact. With a membrane tensile force at the bag center, the bag front is nearly flat. This causes chest g's to increase more rapidly early in the event (since the area of chest contact is increased) with a correspondingly higher percentage of passenger energy absorbed in the efficient "ride down" mode. The blunt profile also provides

Figure 17. Inflated RSV passenger bag.

SECTION 4: TECHNICAL SEMINARS

more passenger stability — especially in oblique impacts — by preventing him from sliding off the bag to either side.

Dash Assembly. At approximately the same time the head bag becomes fully inflated, the knees contact the knee pad and apply force to the dash. When the combined force of the knees and the airbag become greater than the resisting force of the energy-absorbing honeycomb located in series with the dash, the dash begins to stroke. For 50-mi/h sled impacts, about 4 inches of dash stroke has been found to be ideal for bringing a 50th-percentile male dummy to rest in the compartment with minimum injury levels.

The stroking dash is preferred over a conventional airbag system for three reasons:

1. Very low chest amplification factors (ratio of torso *g*'s to crash pulse *g*'s) are obtained because the effective spring constant of the restraint can be reduced to near zero. Since the amplification factor can be shown to be proportional to this effective spring constant (which is quite high for a conventional airbag behaving like a pneumatic spring), it is important to reduce this spring constant to the lowest possible value. Because the honeycomb material is a constant force device, its spring rate is zero.

2. Increased distance with which to bring the passenger to rest is possible with the stroking dash because it moves forward from its conventional dash location (20 inches chest-to-dash with dash unstroked) during the crash. The additional stroke provided by the stroking dash further increases occupant stopping distance.

3. Lower rebound velocities are possible since a higher ratio of absorbed-to-stored energy is made possible by the crush efficient honeycomb energy absorber.

System Operation — Out-of-Position Child

A restraint system designed to protect the out-of-position child must incorporate features that prevent injuries due to inadvertent airbag deployment in noncrash and minor accidents. It must also ensure that design considerations for the out-of-position occupant will compromise or reduce the protection afforded by the system to the normally seated occupant.

These imposed constraints have been satisfied by incorporating several features into the RSV passenger restraint system, including a special bag-folding technique to reduce effective bag mass, low inflator mount and proper adjustment of inflator "down angles" for bag impact at the child's center-of-gravity, and a recoil absorber.

Performance and Test Results

The results of recent sled and car crash tests are summarized in table 2. The results show good system performance. The full size range of vehicle passengers from 6-year-old through 95th-percentile male have been protected to injury levels below the allowable limits at 45 mi/h. In addition, in all vehicle crash tests conducted to date (Test Nos. 6.5, 6.7, and 6.9), the system has performed extremely well.

Crash Test 6.7. The first 50-mi/h barrier test (No. 6.7) was conducted on May 12, 1976, to verify the vehicle structure and restraint system performance at this crash severity. These systems had been successfully tested previously (Test 6.5) at 40 mi/h Banier Equivalent Velocity (BEV). A 50th-percentile male dummy was positioned in the passenger side of the compartment. Actual crash velocity was 51.1 mi/h. Test summary results are presented in table 3.

The overall passenger trajectory, based upon high-speed movies, was quite good. The only anomaly was the high femur loads, which resulted because the knees impacted the inflator mount. To prevent this, the bag was aimed a few degrees lower in Test 6.9. The airbag was then able to support a greater percentage of the total body load as well as provide cushioning between the knees and the inflator mount.

Crash Test 6.9. The first offset barrier crash was conducted on July 11, 1976, to evaluate restraint performance in an RSV structure in which intrusion was apt to be excessive (that

243

EXPERIMENTAL SAFETY VEHICLES

Table 2. Test results—right front passenger

RSV test	Dummy	Velocity (mi/h)	Peak resistant chest (g)	HIC	Femur loads (lb)		Remarks
					Left	Right	
21	50th	49	27	329	1 300	900	
22	50th	43	44	374	1 050	1 050	
23	50th	46	44	355	1 100	1 170	
24	6-yr-old	46	38	441	[a]n/a	[a]n/a	
25	95th	45	57	628	2 600	2 100	New knee bolster
26	5th	45	60	264	800	1 000	Hard pulse
6.5[b]	50th	41	40	311	900	450	Frontal barrier test
45	50th	40	45	668	1 150	1 000	New lightweight stroking dash
46	50th	51	43	952	1 800	1 700	Knees hit inflator mount
6.7[b]	50th	51	46	722	3 200	1 800	Frontal barrier test—knees hit inflator mount
6.9[b]	50th	45	30	187	980	690	Offset barrier crash—good test

[a]Not applicable.
[b]Vehicle crash.

Table 3. Right front passenger restraint system—
Test 6.7

Item		Test result
HIC		722
CSI		553
Peak resultant chest g's	Peak chest g	46
Chest amplification factor (same as Test 6.5)	Peak crash g	1.2
Femur loads (lb):		
Left		1 800
Right		3 200

is, only approximately half the frontal area of the car would be reacting the load). The barrier was offset toward the right front passenger side of the vehicle. A 50th-percentile male dummy was positioned in the passenger seat. Test speed was 45 mi/h. Results of the test are summarized in table 4. As can be seen, the passenger injury measures were extremely low for a vehicle of the subcompact class impacting the barrier at 45 mi/h.

Static Out-of-Position Child Tests. Combined sled test data and car crash results have shown the right front passenger system to function well in frontal crashes at impact speeds in excess of 45 mi/h for the anthropometric size range of passengers from 6-year-old child to 95th-percentile male. The

restraint system has performed well in all recent tests. Injury measures have been substantially below the NHTSA injury criteria limits.

Specifically, the 6-year-old child experienced low injury measures for a 46-mi/h frontal sled test (see table 2—sled run No. 24).

Another series of tests is being conducted to demonstrate restraint performance for the 6-year-old seated in an out-of-position configuration. Previous programs conducted for NHTSA seem to show that it is conservative (higher injury levels are experienced) to statically conduct out-of-position tests where the airbag impacts the seated child dummy leaning against the undeployed airbag within a nonmoving vehicle.

To date fourteen such tests have been conducted. Results indicate that although the system works well — dummy torso levels and

Table 4. Right front passenger restraint system—
Test 6.9

Item	Test result
HIC	189
CSI	261
Peak resultant chest g	30
Chest amplification factor	1.1
Femur loads (lb):	
Left	980
Right	690

velocities are well within criteria limits — head accelerations and consequently HIC values are very high when the dummy head contacts the seat back. Attempts have been made to reduce head accelerations and HIC values by increasing the seat back padding thickness, softening the comfort contour seat force limiters,[2] and even lowering the inflator charge; however, HIC values at seat back impact remain in excess of 1 000.

Testing in this mode is continuing. In forthcoming tests it is planned to adjust the down angle of the inflator. This should reduce the velocity of the bag front in the horizontal rearward direction, thereby lowering the dummy velocity at the instant of seat back impact.

RSV REAR PASSENGER RESTRAINT SYSTEM

System Description

The rear passenger restraint system used in the Minicars RSV is a single-retractor, force-limited, 3-point lap and shoulder belt harness (fig. 18). The basic system consists of a reconfigured and slightly modified 1976 Chevette seatbelt system with force-limiters located at the anchor points. Although it looks and operates like the Chevette front seatbelt system, the RSV system more efficiently protects the occupant during a crash. Increased efficiency is attributed to three modifications made to the basic Chevette system: 1) the upper anchor point was relocated to provide a more advantageous angle through which to transfer forces from the vehicle to the occupant; 2) the standard nylon webbing was replaced with low-stretch polyester webbing; and 3) force-limiting devices were inserted at the three anchor points. The upper anchor was relocated and the webbing material replaced so that a significant restraining force could be applied to the occupant early in the crash event. Force-limiting softens the peak forces applied to the occupant and allows efficient use of the stroking space within the passenger compartment.

---

[2] The RSV front seat backs were originally attached to the interior roof via force limiters.

Implementation and Packaging

The RSV force-limited belt system is extremely simple to implement. Polyester seatbelt webbing, which looks and feels like the more common nylon, is presently being used in some production belt systems. No production tooling changes are introduced by changing to polyester belts.

Force-limiting is likewise easy to implement. The Minicars RSV system incorporates a small, lightweight, add-on mechanism at each anchor location of the production seatbelt hardware. The mechanism itself is a stamped metal piece with two ears pierced by a 0.25-inch diameter pin. A mild steel tape (0.90-inch thick) passes around the pin. Figure 19 shows the torso belt and retractor configuration with the force-limiter mechanism and figure 20 illustrates the lap belt force-limiting assembly.

Dynamic Testing

The RSV rear seat restraint was sled-tested using 50th-percentile male, 5th-percentile

Figure 18. Rear seat restraint configuration.

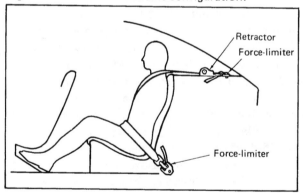

Figure 19. Torso belt force-limiter assembly.

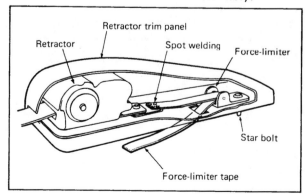

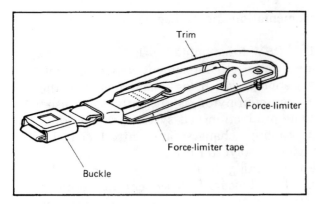

Figure 20. Lap belt force-limiter assembly.

female, and 6-year-old child dummies in the frontal impact mode.

Test results indicate improved performance over those of conventional nonforce-limited systems using nylon webbing. Tables 5 through 7 summarize test results for the system using the head injury criterion measure. Chest accelerations for all dummy sizes were less than 50 g's, indicating that the chest was well protected.

When a seatbelt restraint system is used, HIC is influenced strongly by the magnitude of the centrifugal acceleration component caused by the rapid forward head rotation. Force-limiting reduces this component but does not decrease it as dramatically as it decreases chest decelerations. At the present time, the effect of head rotation in relation to actual injury is not well known. The HIC was originally developed to determine the severity of head strike on hard objects, with an HIC of 1 000 indicating skull fracture. Although using HIC as an injury measure for seat-belted occupants when no head strike occurs is open to question, in the absence of a better injury measure, the HIC measurement was used in the RSV program.

The following tabulations are the estimated velocities to which the restraint systems are effective.

	RSV rear seat restraint
50th-percentile male	44
5th-percentile female	37
6-year-old child	35

It will be noted that the efficiency of the force-limited seatbelt was greater for the largest dummy than for the two smaller dummies. The restraint forces applied are the same for large and small occupants, because the system cannot distinguish occupant size. Higher accelerations with correspondingly higher injury measures are created for the smaller occupant mass.

Table 5. 50th-percentile male dummy

Run No.	Incoming velocity (mi/h)	HIC
28	37.2	466
29	42.6	875
30	45.3	1 225
31	45.0	892
32	45.0	924
40	40.3	684

Table 6. 5th-percentile female dummy

Run No.	Incoming velocity (mi/h)	HIC
41	39.5	1 008
42	34.3	806
43	35.5	1 179

Table 7. 6-year-old child dummy retractor system

Run No.	Incoming velocity (mi/h)	HIC
41	39.5	1 588
42	34.3	823
43	35.5	1 003

One thing that seems evident based upon the development of the RSV rear seat restraint system is the ease with which present-day belt systems can be modified to give very much improved performance. Hardware is readily available that is excellent for this purpose, and only two minor changes — the use of polyester webbing and force-limiting anchors — are necessary to produce signif-

SECTION 4: TECHNICAL SEMINARS

icantly more effective seatbelt restraint systems with greatly reduced injury levels.

SIDE IMPACT AND ROLLOVER PROTECTION

All RSV occupants are passively protected in lateral collisions and rollovers by the well-designed vehicle interior. Additionally, rear seat passengers are provided with an advanced seatbelt restraint system which when worn offers unquestioned benefit.

RSV side impact and rollover protection is afforded by the specially constructed gullwing door, which is the most significant feature of side impact and rollover restraint. The exterior structure of the RSV door works in conjunction with the interior padding to attenuate occupant lateral impacts. Crashworthy door latches prevent occupant ejection during lateral and rollover collisions.

The RSV door interior is contoured as illustrated in figure 19. As can be seen, door padding is extensive in the shoulder and hip levels. The most critical area for padding is immediately adjacent to and just forward of the driver and front passenger seating positions. Substantial padding is required in this area of the door to provide adequate protection in side impacts. In this mode the occupant translates laterally with perhaps a slight forward component due to his vehicle's initial forward velocity and/or the less than perpendicular orientation of the vehicles at impact. These considerations result in contouring the center portion of the door as seen in figure 19. The center portion, about 15-20 inches wide, is angled to parallel the 22-degree seat back angle. A cross-sectional view of this portion of the door is presented in figure 20. As can be seen, the shoulder and hip impact areas are provided with foam pads. The vinyl-covered ensolite vehicle skin provides damage protection during day-to-day use. Two vehicle-to-vehicle side impact tests (7.2 and 6.8) have been conducted to date using essentially this door padding configuration. Although the impact configuration and resulting dummy injury measures were comparable in the two tests, Test 6.8 is more significant and more interesting for several reasons:

1. The RSV side structure employed in this test is more representative of the final design,
2. Unbelted near and far side front seat occupants were positioned in the struck vehicle, and
3. The vehicle interior was extensively simulated to observe occupant contacts during the event.

In Test 6.8, the RSV driver side door was impacted by a Ford Pinto traveling at 34.7 mi/h. The Pinto, oriented 300 degrees to the RSV, contacted it at the A-pillar/door conjuncture. The two unbelted dummy occupants were 50th-percentile male surrogates seated in the front seat locations. The front seat area was provided with mock-ups of every conceivable contact surface in anticipation of impact from the far side (passenger) dummy. Included as mock-ups were a steering column assembly, right side dashboard, gear shift lever, and an item called a "1-G hip pad," suggested as a feature for the driver that would keep his lower body behind the wheel during severe vehicle maneuvers in the absence of lap belts.

The results of this test, tabulated in table 8, were viewed as very satisfactory. The far-side occupant certainly would not have sustained significant injury despite the absence of belt restraint. The near-side door pad produced very low driver accelerations and was penetrated only nominally (about 1 1/2 inches).

Side Glazing. Head impact energy is absorbed and ejection is prevented by using fixed side glazing. Side glazing constituents, dimensions, arrangement, and attachment to the door frame are presented in figure 21. The glazing consists of a membrane to which 100

Table 8. Dummy results from RSV crash test 6.8

Item	Driver	Front passenger
Head HIC	270	465
Chest acceleration peaks (g):		
A-P	20	27
L-R	25	[a]47
I-S	10	13
Result	32	[b]53
Pelvic left-right peaks (g)	22	18

[a]3 ms duration level = 40 g.
[b]3 ms duration level = 44 g.

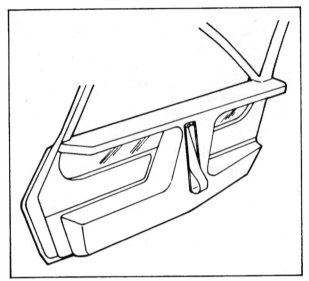

Figure 21. RSV door padding contour.

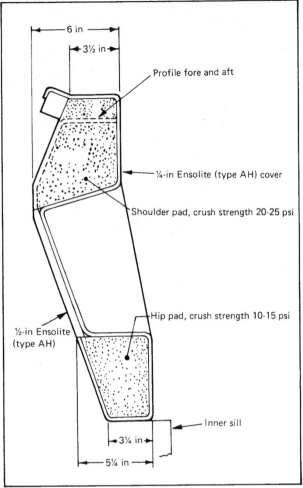

Figure 22. Cross-section of center portion of the door.

mil glass is autoclave-bonded. The glass functions only during normal vehicle operation. At impact, the glass breaks away from the membrane.

The membrane consists of Mylar (5 mils thick) and architectural vinyl (15 mils thick), the thicknesses of which were determined from the results of twenty-four head impact tests conducted on the Minicars VEAC I sled. In these tests, flat samples of side glazing were fixed in place and impacted at 20 mi/h by a 12.5-pound head form. Different combinations of Mylar and PVB thicknesses were evaluated. The results of some of these tests were presented in table 9. Note that the HIC measurements are about one-half that allowed in the FMVSS 208 injury criteria and that rebound velocities are about 40 percent of the impact velocity. The Mylar/PVB membrane is initially formed to have tabs about 3 1/2 inches long extending beyond the glass boundaries at its four sides. These tabs are folded back on themselves around a length of wire and heat-bonded to form a loop. These membrane loops fit inside the window frame channel as shown in figure 21. Spring steel clips are then inserted by special pliers into the frame channel to fix the glazing. Molding is then placed around the window to conceal the clips and provide a weather seal. To remove the window, the molding is stripped away and a screwdriver is used to extract the clips.

Side glazing can be removed in an emergency by pulling on a finger ring attached to a hidden wire imbedded in the membrane (not shown). The wire will cut the membrane on three sides, resulting in an exit opening.

SUMMARY

Protective features of the RSV compartment interior, and present performance results in simulated and/or actual vehicle impacts have been described. Impressive levels of frontal impact protection are provided front seat occupants by the use of advanced passive airbag restraints. Incorporated in both the driver and front passenger systems are two significant features which, with the specially configured vehicle front end, account for

SECTION 4: TECHNICAL SEMINARS

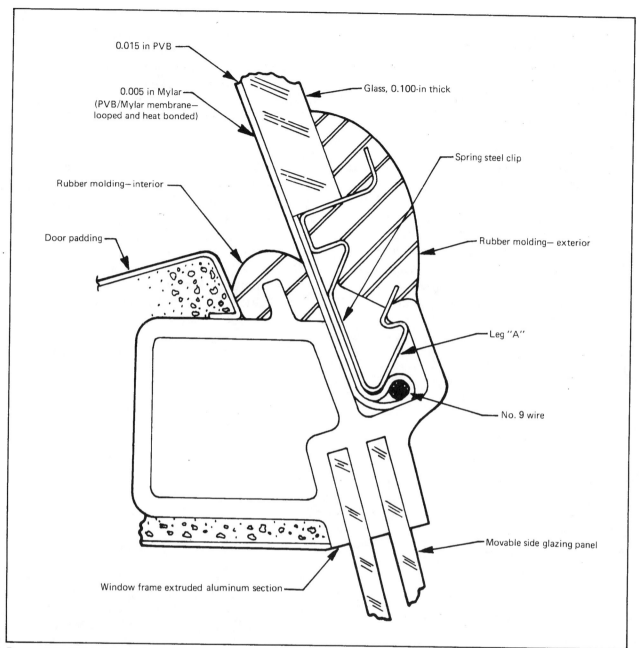

Figure 23. RSV side glazing—sectional view of the lower attachment area.

Table 9. Glazing head impact results—20-mi/h impacts with 12.5-lb headform

Test No.	Impact speed (mi/h)	Recorded HIC	Head penetration (in)	Rebound velocity (mi/h)
19	20.4	458	5.0	8.5
22	20.1	540	4.1	7.8
23	18.8	491	3.3	8.3

protection levels of approximately 50 mi/h BEV. These features are (1) the dual-bag concept that achieves the fast response times required of systems capable of 50-mi/h protection, and (2) the force-limiting capability that enables occupant kinetic energy to be efficiently absorbed within the vehicle compartment, thus minimizing injury indices.

The dual bag concept is implemented in the driver restraint by two concentric (inner and outer) bags. The inner (torso) bag receives gas directly from the inflator and "vents" to the outer head bag. In the passenger restraint, a large partitioned bag is divided into an upper and a lower chamber. The lower chamber torso receives air directly from the gas inflator and "vents" to the upper head chamber.

Force-limiting in the driver system occurs in the telescoping of the steering column. In the front passenger restraint it occurs in the stroking of the right side dashboard. Force-limiting was also incorporated into the RSV rear seat 3-point restraint system to extend occupant protection capability approaching 45 mi/h with retractors. This capability exceeds that provided by conventional belt systems.

RSV occupants are protected in side impact and rollover accidents by the strong shut faces and secure door latches of the gullwing doors as well as the well-padded, roomy interior and the fixed side glazing. Vehicle-to-vehicle side impact test results indicate that RSV near- and far-side occupants would receive minimal injury in a 35-mi/h impact with a vehicle of the Pinto weight class.

The Daimler-Benz Development of a Final Production Air Bag System for the U.S.A.

HANS JUERGEN SCHOLZ
Research and Development,
Passenger Car Division
Daimler-Benz **AG**
Sindelfingen, Germany

ABSTRACT

After more than fourteen years of research and development, Daimler-Benz has arrived at the point where we will place into production the air bag system, one of the most debated and controversial technologies in the entire history of the automobile.

This passive occupant restraint system, which will be automatically triggered in an emergency, has to meet not only the most stringent technical requirements, but must also guarantee, especially in view of the present legal situation, the highest degree of reliability.

In this context it should be mentioned that the changes in U.S. Federal Motor Vehicle Safety Standard 208 (FMVSS 208) have also increased the challenges in complying with this regulation. It definitely was no easy task for a vehicle manufacturer to interpret the future of this Standard correctly and make meaningful decisions accordingly.

THE DAIMLER-BENZ AIR BAG SYSTEM

The Daimler-Benz air bag system, scheduled for production starting with model year 1982 in the U.S.A. and Canada, consists of four (4) basic components:
- The sensor
- The gas generator
- The air bag
- The knee bolster.

THE SENSOR

The brain of the air bag system, without any doubt, is the triggering system, i.e. the sensor. The sensor has four main functions:

1. To detect an accident
2. To react correctly accordingly
3. To monitor the readiness of the air bag system
4. Storing of information.

Proper functioning includes timely triggering and avoidance of a false signal leading to an inadvertent deployment.

To achieve these somewhat simplified foregoing requirements, electronic components are being used which already have found widespread applications in the automobile industry.

The electronic sensor shown in illustration 1 has been developed together with Robert Bosch Company for ready installation on the assembly line.

Recognition of Impact Severity and Triggering

The data input of the vehicle deceleration occurring during an impact given by an accelerometer serves as a medium to recognize impact severity. The receiver installed in the sensor is a viscous damped mass consisting of a spring bar with inertial mass (illustration 2). A strain gauge mounted on the bar changes resistance resulting in an electronic signal occurring during any deceleration along the center line of the vehicle.

The directional sensitivity of the receiver decreases as a cosine function, i.e. as the direction of impact deviates from the center line of the vehicle [Diagram 1]. The acceleration signal will be processed by an integrated circuit. The signal finally arrives at an integrator after amplifying and limiting the amplitude of the signal minus a constant a_s (illustration 3). The threshold a_s has been set in such a way that vehicle accelerations/decelerations under normal vehicle operations will not be registered and cannot result in an inadvertent deployment. The indicator computes from the acceleration signals received as a change in velocity. Should this value exceed the set values ΔV_1, and ΔV_2, then the final steps Z_1, Z_2, and Z_3 will be sought. The triggering signal, however, will only be released when the arming switch (mercury switch in the schematic) is in a closed position. This switch is velocity sensitive and installed to protect against inadvertent deployment and to increase reliability. The mercury switch will be closed during vehicle deceleration on the center line axis of approximately 1.7g.

When the first threshold ΔV_1 is attained, the final stage Z_1 will give a triggering signal to one of the two gas generators of the front passenger air bag system (front seat passenger 1, diagram 2). If ΔV_2 is also reached, Z_2 will give a triggering signal to the gas generator installed in the hub of the steering wheel as the second of a bi-level system. The second gas generator of the front seat passenger system can only be triggered after a constant time lapse after the receipt of ΔV_1. A pre-condition certainly is that also ΔV

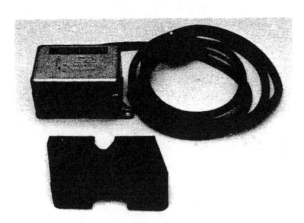

Illustration 1. Electronic sensor with connecting cable, 10-pin connector and mounting bracket.

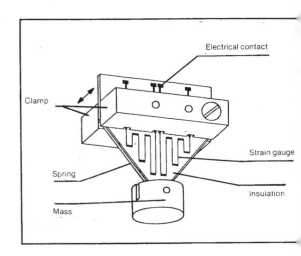

Illustration 2. Accelerometer.

SECTION 5: TECHNICAL SESSIONS

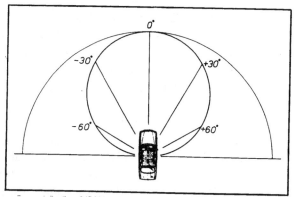

Diagram 1. Schematic of directional sensitivity.

has been reached, otherwise Z_3 will be triggered simultaneously with Z_2 (example 2 of the signal

schematic). With the limiting values of ΔV_1 and ΔV_2, the triggering function is tailored exactly to the severity of impact: at frontal impact speeds against a solid plane object with a speed of less than 12 km/h, neither system will be activated. At an impact speed between 12 and 18 km/h, ΔV_1 will be activated; within the approximate impact range between 18 km/h and 25 km/h, ΔV_2 will also become operational.

Monitoring and Storing of Data

Each single ignition circuit, as well as the sensor, are continuously monitored for proper function *at the highest degree possible*. The ignition circuits are, while the vehicle is in operation,

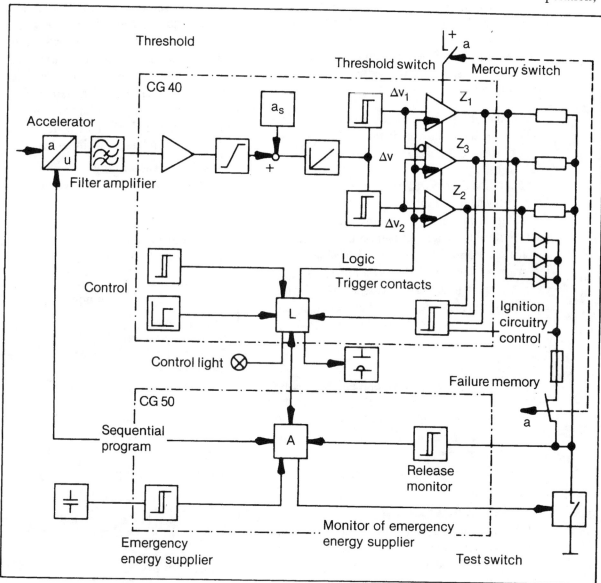

Illustration 3. Schematic of sensor.

checked for circuit continuity, while the electrical functions of the air bag components and triggering device are monitored during each starting procedure. Interruptions or breaks in the ignition circuitry, malfunctions of the triggering components, or failure of the capacitor will be indicated by a warning light on the dashboard. Proper function of the warning lamp bulb is observed by being lit for approximately 10 seconds during the first turn of the ignition key. If the system is functional, the control lamp will extinguish after 10 seconds. Any detection of a malfunction that would lead to an inadvertent deployment will cause the safety switch to remain in the open position. Should there be a malfunction of any electronic part, the safety switch will prevent activation and consequent deployment of the system.

The sensor is designed in such a way that after deployment, a specially trained person can immediately recognize

- Whether the system was operational, or
- Whether the deployment took place prior to or during the impact.

The operational time will be registered by a Coulomb-cell which will show the time lapse between reporting an inoperative system and actual impact. With this system, it can be observed how much time elapsed between the illumination of the warning lamp and the accident. A simple overload fuse will also give evidence whether the air bag system was triggered by an accident and an inadvertent deployment could not be claimed.

The Gas Generator

Daimler-Benz began development of a solid propellant gas generator in 1970. Meanwhile, this particular mode of generating gas has become the state-of-the-art. The advantages of this system in comparison to common pressure generators or hybrid systems are mainly

- Less weight
- Smaller envelope
- Less influence by temperature and
- Ease of producing the exact amount of gas in the passenger air bag as dictated by the severity of impact. (The bag will be filled to either 50% or 100% of capacity.)

The solid propellant gas generator is a by-product of the space motor development. The product of ignition is NOT an explosion, but an exact consumption of the solid propellant with a concurrent pre-determined expansion of generated gases.

This discrimination between explosion and programmed expansion of the gases is of utmost importance especially with the first application of pyrotechnics in the civilian automotive field. This will avoid unjustified prejudice in the future.

To make it clear, the air bag gas generator has absolutely no relation to explosives and can never be misused as such due to the burning rate of the material.

For this reason, in the Federal Republic of Germany the propellant has the same classification as pyrotechnical fireworks. In the United States of America, this material has been classified by the Department of Transport, Bureau of Hazardous Materials, as a flammable solid. In Canada, the solid propellant will not be subject to the Canadian Explosives Act.

The two vendors of Daimler-Benz are

- Bayern-Chemie, GmbH, Munich, and
- Thiokol Corporation, Brigham City, Utah, U.S.A.

The Bayern-Chemie gas generator (illustration 4) was developed jointly with Daimler-Benz starting in 1970 and will be offered beginning

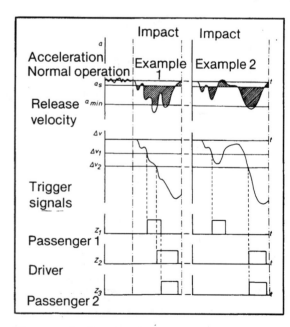

Diagram 2. Sensor signal evaluation.

SECTION 5: TECHNICAL SESSIONS

Illustration 4. Bayern-Chemie Solid propellant gas generator.

December 1980 in Germany as an option in the 'S' class passenger cars. This is for the driver position only, the front seat passenger will receive a Type 2 ETR (Emergency Tensioning Retractor). The Thiokol gas generator (illustration 5) has been adapted during the last five years to our passenger cars, and it is intended to install the air bag system in all passenger cars marketed in the United States and Canada beginning with model year 1982.

The Thiokol gas generator contains approximately 100 grams of pelletized solid propellant. An ignitor with a single bridge wire—called a squib—is located in the center of the housing. Metal filters and neutralizers are placed outside the combustion chamber which limit the exhaust of particles and cool down the gases.

The bridge wire will be heated electrically after a sensor signal is received. Consequently, the highly sensitive initial charge is triggered whose pressure and high temperature in turn escapes through ports and ignites the slow-burning pelletized solid propellant. This burning process takes approximately 50 ms. The maximum pressure in the combustion chamber is close to 180 bar. The gas filling the air bag consists of roughly 95% of non-toxic nitrogen.

In addition to the governmental tests and certification, the complete gas generator has been subjected to tests under most stringent accelerated test conditions and exposed to shock, drop, vibration, humidity, and extreme temperature loads. The generator has proven to withstand the

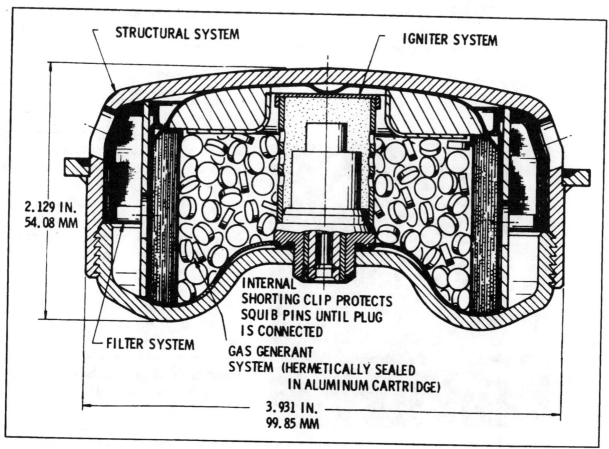

Illustration 5. Thiokol solid propellant gas generator.

255

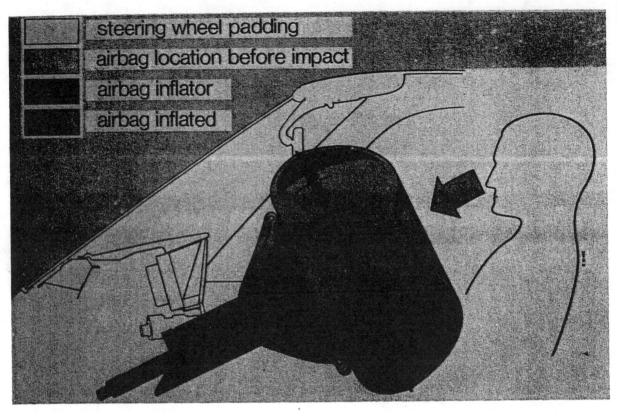

Illustration 6. Driver air bag.

aging process for the expected lifetime of an automobile.

Furthermore, studies of the endurance of the gas generator under extreme conditions have been conducted while installed in an automobile.

Illustration 7. Passenger air bag with 2 chambers and trapezoidal shape.

The representative of Thiokol Corporation, Mr. George Kirchoff, will report following immediately after my presentation.

The Air Bag

The actual protective element is the air bag, sometimes also referred to as an air cushion. Depending on the ambient temperature the bag will be deployed within 25–35 ms on the driver side, and 40–50 ms on the passenger side and completely fill, in a deployed condition, the available space between occupant and the steering wheel or the dashboard. The bag consists of Neopren-coated Nylon and has, depending on the selected energy absorption requirements, correspondingly tailored venting holes.

The bag in the driver position is oval shaped, has a volume of 60 liters and actually are two circular pieces vulcanized together at the circumference (illustration 6). The side facing the driver is constrained in its outward expansion by straps which limit the motion of the bag. This guarantees that under normal conditions the driver's face will never contact the bag during deployment.

The passenger bag is trapezoidal (illustration 7), has a volume of 145 liters and also consists of two vulcanized pieces.

A seam divides the passenger bag volume into two equal parts. Chamber 1 is connected with the gas generators directly over the inlet ports and will be filled in the first stage of deployment. Only after activation of the second generator, the overpressure rips the seam open and the total bag will be filled and deployed. Facing the dashboard are four venting holes which guarantee a relatively smooth "ride-down" of the front seat passenger.

The extended deployment time of the passenger air bag, in contrast to the driver bag, is attributed to the fact that a pre-determined 15 ms time delay signal is given by the sensor between triggering the first gas generator (first level) and the second gas generator (second level). The trapezoidal shape is designed to allow for a slanted contact area that gives more cushioning in the head impact area.

The driver, as well as passenger air bag, has in addition to the vulcanized seam, stitching with high strength thread. This guarantees a tight seam under extreme conditions, for example, extremely low temperatures.

Knee Bolster

To brace occupants during a frontal impact, only three locations on the human body can basically be used:

- The chest
- The pelvis
- The thigh

Inasmuch as a passive restraint system like the air bag cannot restrain the pelvis, the retaining

Illustration 9. Knee bar with steering column.

of the lower body and extremities in an accident can only be accomplished by restraining knee and thigh movement. In order to keep the load within a tolerable limit, the restraint must have the ability to deform according to design.

The requirements for the knee bolster are:

- No limitation in comfort while operating the passenger car
- Prevent submarining
- Control body movement
- Limit restraining forces to a tolerable level.

Illustration 8 shows the arrangement of a knee bolster with the steering elements of the driver position.

The knee bolster will have absorbed, in a 50 km/h frontal impact into a solid barrier, an energy of approximately 450 joules per femur (50% dummy). Due to the envelope given while still offering an optimum in seating comfort, the deformation distance is limited to 140 mm.

The knee bolster shown consists of a corrugated metal tube with a diameter of 145 mm and a thickness of 0.4 mm (illustration 9). By using a corrugated tube in comparison to a simple tube

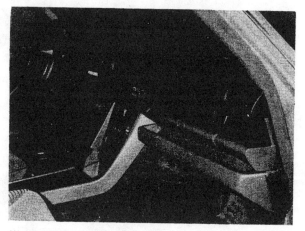

Illustration 8. Mercedes-Benz S-Class interior for model year 82.

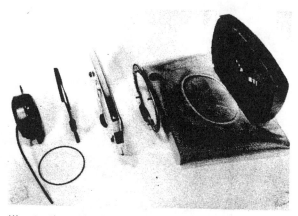

Illustration 10. Driver air bag components.

EXPERIMENTAL SAFETY VEHICLES

Illustration 11. Passenger air bag components.

which would need a metal thickness of 1.5 mm to have the same deformation qualities, much weight could be saved.

In addition to the four most important components of the air bag system, many other parts are being used which will be shown in order:

- Illustration 10—driver air bag module
- Illustration 11—front passenger air bag module
- Illustration 12—harness with sensor
- Illustration 13—total picture driver position
- Illustration 14—total picture front passenger position
- Illustration 15—air bag control light on the dashboard

PROBLEMS ASSOCIATED WITH THE SYSTEM

The Daimler-Benz air bag development is based on an all-encompassing research program. Until freezing the total concept, during the testing and improvement phase of each single component, in excess of 2,000 air bag tests have been conducted at Daimler-Benz alone, not withstanding the numerous tests by our suppliers.

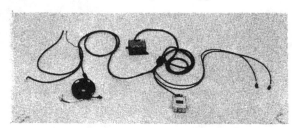

Illustration 12. Sensor with cables, connectors and emergency energy supply.

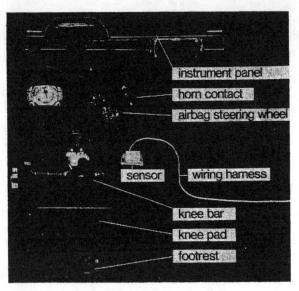

Illustration 13. Total system driver.

These tests were:

- Noise pressure-level tests while inflating the bag
- Simulated crashes using white bodies on the Hy-GE sled with differing speeds and varying occupant sizes
- Full scale crash tests with different vehicle models, velocities, and impact angles
- Environmental test programs for gas generators and sensors
- EMI-tests
- Tests using volunteers to study the reaction of human beings.

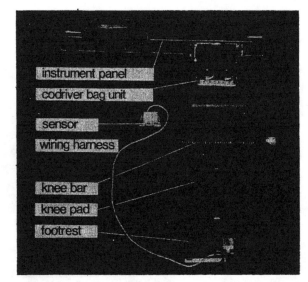

Illustration 14. Total system front seat passenger.

Illustration 15. Control lamp in dashboard.

During the course of the air bag development, problem areas could be identified and technical solutions developed; others, however, are areas over which the manufacturer has no control. Two typical samples are listed below.

Electro-Magnetic Interference (EMI)

Automobiles may be exposed during operation to high frequency transmitting sources, like radio towers, for instance. In such a case, the high frequency beam may be induced in the electrical harness of this system. The EMI compatibility of the air bag electrical system has been tested first by exposing parts of the vehicle body to radiation and subsequently on the total body was exposed using radar and radio stations.

The test data has shown the urgent need of shielding to reduce the electrical flux in the harness and avoid possible inadvertent deployment. For this reason special efforts have been made to shield the sensor by means of filters. The connecting plug and the gas generator are equipped with a suppression coil. By these measures we could solve within the state-of-the-art the problems.

Disposal of Gas Generators

In closing, I want to draw your attention to another problem area which will increase with the air bag system installations in an increasing number of passenger cars. This is the proper disposal of gas generators. Up to now there is no procedure available for disposing of gas generators, especially addressing the abandoned vehicles. If this problem is not solved, it is unavoidable that gas generators will be removed, subject to misuse, and damages or injuries may occur.

This problem cannot be solved by the vehicle manufacturer but should be addressed by the whole rulemaking body.

Road Test Programs

Along with lab testing programs is a real-life testing on the road.

Presently we have approximately 500 passenger cars equipped with air bags on the road and expect by the end of January 1981 to increase this number to 750. The majority—600 passenger cars—are already produced on the assembly line, no longer at the R & D shop. This measure should enable everybody concerned to acquaint themselves with the system and recognize problem areas in time prior to fully phasing into the production process. The proper training of field and service personnel has also begun.

In spite of these difficulties, Daimler-Benz has opted for the air bag system instead of the passive belt.

When a rulemaking body requires a passive occupant restraint system, it cannot simultaneously deny the comfort or convenience of the passenger—priority concerns in a sedan, for example—by the design or function of such a system.

In comparison to this system, the passive belt is, according to the present state-of-the-art, no alternative in the areas of comfort and convenience, even though almost identical in its protective value.

The Development of an Advanced Airbag Concept

**Lennart Johansson, Jan Billig,
Hugo Mellander,**
Volvo Car Corporation
Bernd Werner, Peter Hora,
Bayern Chemie GmbH

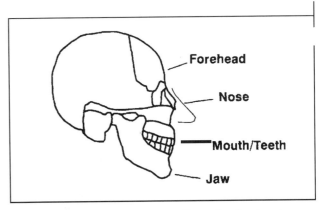

Figure 2. Facial injuries are mostly to the forehead, nose, mouth/teeth and jaw.

Abstract

Interaction between the head of the driver and the steering wheel may cause facial injuries, especially in high speed accidents.

Air cushion technology provides the means of distributing and reducing the inertia forces acting on the head and the face.

This paper describes the development of an airbag system, including the electrical sensor which is integrated in the steering wheel, and it focuses on the problems of positioning the sensor in the steering wheel.

Information from necessary sensor testing, both on rough roads and in crash conditions, is presented.

The improvements which are achieved, in terms of reduced violence to the face, have also been assessed experimentally by using a test dummy with a load sensing face.

Introduction

In connection with the increased usage of seat belts in Europe, the number of serious injuries in collisions has decreased.

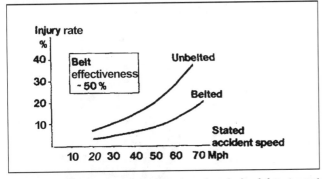

Figure 1. Injury rates for belted and unbelted front seat occupants at different stated accident speeds. (Ref. 1)

However, despite the belt, there is still the problem of skull and facial injuries sustained by the driver when his head hits the steering wheel in high speed accidents. Volvo's accident statistics show that 10% of drivers in head-on collisions receive skull and facial injuries. See figure 2.

Since the total pattern of injuries decreases with the usage of belts, the remaining facial injuries constitute a greater proportion of the injuries suffered than previously. Normally, the injuries are relatively slight, but this type of cosmetic injury often creates psychological problems for the victims, and rehabilitation can, in some cases, take years.

Our goal was to optimise an airbag system for the belted driver and, by using a high level of integration, to produce a cost-effective and compact system. A bag in the steering wheel would protect against skull and facial injuries in head-on collisions and increase the chances of survival in high speed collisions. Trimming a bag for a restrained occupant meant that the bag could be made smaller than the US-system (for unrestrained occupants).

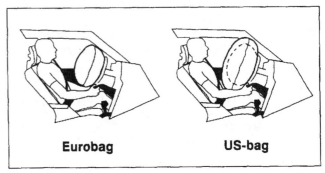

Figure 3. The pictures show the difference in size between a bag for a belted driver (Eurobag) and a bag complying with FMVSS 208.

The results after computer simulation and sled tests showed that a bag with a volume of approximately 35 litres (diameter 550 mm) without internal straps gave the best results.

A small bag has advantages with regard to out-of-position-exposure. Omission of the straps meant that the bag could be made relatively simple in design. If a bag material with a coating which can be welded is used, a cheap method of manufacture can be developed whereby two discs (one of which has a pre-welded disc with the necessary attachment holes for the inflator) are welded together. (See figure 4).

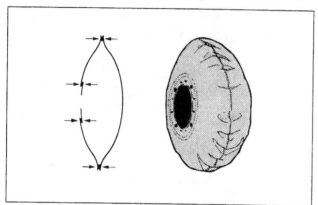

Figure 4. Three parts welded together to a bag.

Since the inflator only needed to generate gas for a 35 litre bag, its dimensions could be reduced as shown in figure 5.

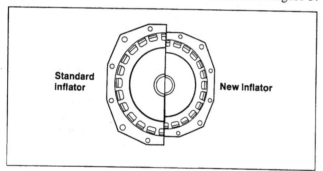

Figure 5. A Eurobag inflator (~15% smaller in diameter and ~20% lower in height) compared with the conventional inflator.

This gives not only a reduction in weight, but the new inflator also occupies less space in the steering wheel.

System layout

The whole system is located in the steering wheel, i.e. the cover, the bag, the inflator, the sensor, including diagnostic and reverse energy units, the connector coil and the warning lamp. (Figure 6)

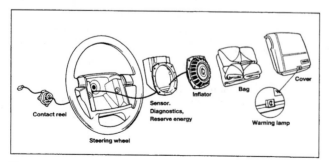

Figure 6. Eurobag concept.

The advantages of this are, among other things, that the sensor can be fitted directly onto the inflator and, because of the fact that all system components are in the steering wheel, the system could be installed as an aftermarket item.

Sensor development

The Eurobag sensor, fitted in the steering wheel, is basically an electronic software controlled microprocessor system. The trigger algorithm follows "standard" acceleration signals within electronic airbag systems, i.e. above a certain acceleration threshold and when a certain velocity change has occurred, the power switch is triggered to ignite the squib. The threshold depends on the crash behaviour of the actual car.

The development of the sensor was directed towards Volvo 740/760. The 760 model also contains a tiltable steering wheel (tiltable ±7°), and this must be considered in the development of the sensor. (figure 7)

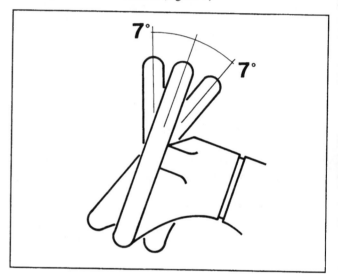

Figure 7. The 760 steering wheel can be tilted to three positions (±7° from the normal position).

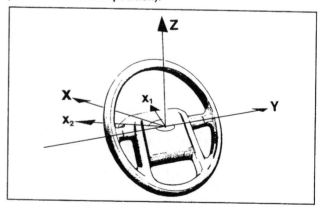

Figure 8. Steering wheel with 5 accelerometers in the different directions.

Development of software

The sensor development began with the accumulation of measuring data. With the aid of a measuring unit with five accelerometers fitted in the steering wheel, signals from a number of different collisions were registered, to be analysed later.

The accelerometers were placed in the X, Y and Z axes and ±45° from X. (figure 8)

Tests were carried out at different speeds from 4 mph to 35 mph and were of types 0°, 30°, 90°, 180°, pole and underride. (figure 9)

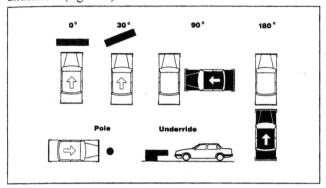

Figure 9. Different types of collision situations, where measurements of acceleration in the steering wheel were registered.

Since the sensor should not trigger during different types of rough-road driving, measurements were carried out on that type of surface as a reference. Examples of "non-triggering situations" are curbs, road depressions and different types of bumps. (figure 10)

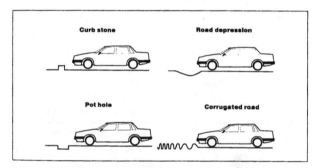

Figure 10. Examples of rough-road situations.

- Curb-stone testing was performed against a 100 mm high curb at 35 mph and a 150 mm high curb at 30 mph. After such curb-stone driving, the car is damaged, but the driver unharmed.
- Road depressions at speeds up to 55 mph and with maximum load in the car.
- Driving on pot-holes and Belgian pavé with wheels both locked and rolling.
- Test driving was also performed with unbalanced wheels at different speeds.

The accumulation of signals was also supplemented by measurements made when hitting the steering wheel with different objects, e.g. a hammer and a fist. All tests were done with the steering wheel tilted at the three different angles. The measured acceleration levels from the different tests were used as input data to a computer simulation program, which simulates the sensor's signal processing. The calculated times were compared with the requirement according to the 125 mm criteria.

The trigger time that the total system must meet was determined by the conventional 125 mm requirement, i.e. the time required for an unbraked mass to move 125 mm in a crash. (figure 11). This is to avoid interaction between a deploying bag and the occupant.

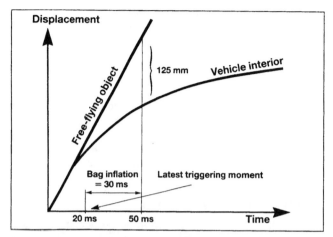

Figure 11. Unbraked mass (the "worst case"" of an unrestrained driver) may move, due to inertia, a maximum of 125 mm before the bag is fully inflated.

Further analysis work has shown that when belts are used, a longer time delay can be accepted. But for the development of this sensor, the 125 mm requirement was adhered to.

After a number of development stages, a signal processing algorithm was obtained which fulfilled the set triggering criteria, without inadvertent triggering during rough-road simulation.

The algorithm is subdivided into four paths. Path 1 integrates the acceleration signal once to a delta velocity proportional voltage. Path 2 integrates the acceleration signal twice to a voltage that is proportional to the displacement of a free mass. Path 3 integrates the negative acceleration once to a velocity proportional signal and is used to recognize the direction of the crash. Path 4 compares the acceleration with a certain threshold and controls the connection of the signals evaluated in Path 1 to Path 3. In the logic block, the signals are connected together depending on certain thresholds and timing functions.

In general it can be said that in a normal crash, the trigger signal for the squib is generated by Path 1, if a certain threshold is exceeded. Path 2 to Path 4 are used to distinguish between a crash situation when a trigger signal should be generated and, for example, an acceleration signal generated in rough road conditions where no triggering is required.

This algorithm was implemented in the hardware solution which was developed simultaneously. (figure 12)

The programme also includes a diagnostic section which continuously controls the main components of the system. Also included is a part of the programme which calibrates the acceleration transducer during the final test in the fac-

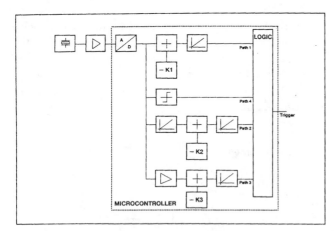

Figure 12. Schematic block diagram showing the signal conditioning parts of the sensor.

tory via software and compensates the acceleration transducer sensitivity over various temperatures. If a failure is detected a failure lamp lights up.

Development of hardware

With the help of simulation and practical tests it was found that an accelerometer in the direction of the steering shaft was sufficient to give acceleration signals which could be used. The hardware was built up around a microprocessor with the necesssary protective circuits and watchdog. The unit was fitted with a capacitor for a reserve energy supply in case the ordinary voltage feed is cut off in a collision. The voltage supply to the sensor is fed through a contact reel of clockspring type since the slip rings for the horn were found unreliable. The tests performed with slip rings showed that in cold conditions condensation could freeze on the tracks and cause a voltage cut for up to 10 minutes. Consequently the voltage feed via the slip rings is used for redundancy.

The hardware concept, shown in the attached block diagram, is structured in three major parts: acceleration sensor with amplifier; microcontroller and memory; power switch and power supply.

Acceleration sensor with amplifier

Piezoelectric acceleration sensor with high shock resistance; amplifier as impedance converter and filter; diagnostic unit for acceleration sensor.

Microcontroller and memory

Triggering of power switch for squib ignition depending on implemented software controlled trigger algorithm; controlling of all diagnostic functions; software stored in the memory.

Power switch and power supply unit

Low impedance power switch for squib ignition and failure lamp control with integrated diagnostic functions; voltage regulator for conditioning of the car supply voltage; overvoltage protection and switch for reserve energy supply on board. If faults occur in any of the components, the diagnostic unit gives a signal to a lamp located in the steering wheel cover.

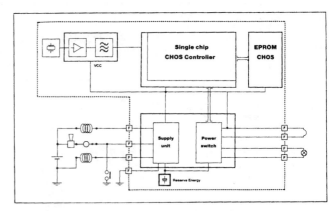

Figure 13. Block diagram showing the hardware concept.

The final prototypes were verified in a large number of sled tests and full-scale collision tests. The system fulfilled all criteria. Furthermore, a number of different rough-road tests were performed, and these proved positive, i.e. the sensor did not trigger.

Crash test results

Full-scale collisions have been performed from 0° and 30°, pole and under-ride at speeds of 30, 35 and 40 mph. The Eurobag system was found to function well as a complement to the seat belt in all the above situations. The sensor located in the steering wheel fulfils the specified requirements and functions with the steering wheel tilted in different positions.

The tests showed that the Eurobag concept reduced Hic by approximately 25–30%, see figure 14.

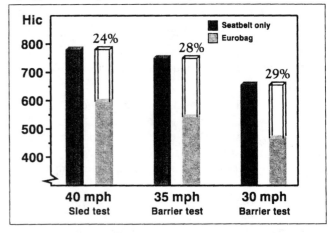

Figure 14. Typical values from the evaluation tests.

Another advantage with the Eurobag was that it reduced the forward movement of the head by approximately 20%. The forces on the torso belt were reduced by approximately 10%, see figure 15.

Without a Eurobag there is a variation in measured Hic values due to different head impact location in the tests with

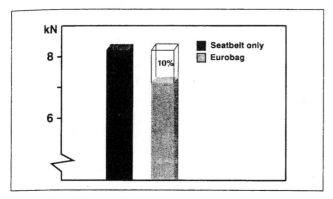

Figure 15. 10% reduction of torso belt forces with Eurobag.

the steering wheel tilted at different angles. However, with the Eurobag concept, a much more consistent behaviour was observed, with lower Hic values.

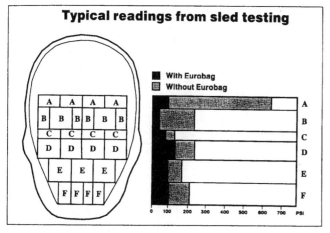

Figure 16. Comparison of facial pressure with and without Eurobag. The bar diagram shows the highest achieved pressure in any of the measuring plates with the same designation.

In order to confirm the reduction of facial injuries, a load-sensitive face dummy was used in the different collision situations. With the Eurobag concept, the facial pressure is reduced by an average of 80%, (figure 16).

Conclusions

This project has shown that a small bag, optimized for belted drivers is an excellent complement to the seat belt. It protects the head against impact against the steering wheel and thereby reduces the risk of head and facial injuries in head-on collisions.

The possibility to position the sensor in the steering wheel was successfully demonstrated. By using an electronic sensor, a sufficiently complex conditioning of the signal could be made to meet the non-triggering and triggering requirements and also to diagnose and indicate faults in the system. There is now the possibility to choose between a body-mounted crash sensor or a sensor located directly on the safety system.

The advantages and disadvantages of these two approaches have to be further investigated and they will probably vary for different categories of cars.

References

(1) H Norin, G Karlsson, J Korner: "Seat Belt Usage in Sweden and its Injury Reducing Effect". SAE 840194.

(2) S Koyabashi, K Honda, K Shitanoki: "Reliability Considerations in the Design of an Airbag System". ESV 1987.

(3) G F Kirchoff et al: "Advanced Concepts for Driver Air Cushion Systems". ESV 1985.

(4) D S Breed, V Castelli: "Problems in Design and Engineering of Airbag Systems". SAE 880724.

950347

Investigation of Sensor Requirements and Expected Benefits of Predictive Crash Sensing

William R. Swihart
TRW Transportation Electronics Div.

Albert F. Lawrence IV
TRW Electronic Systems and Technology Div.

ABSTRACT

Increased motor vehicle safety standards for frontal and side impact crashes are requiring quicker sensor response to deploy air bags and safety related components. A reliable method of pre-arming crash safety systems may provide the potential for enhanced occupant protection by decreasing the time required to discriminate between real crash events and non-events. The reduced discrimination times will allow safety engineers more design flexibility in tailoring air bag inflation rates, pressures and belt pre tensioning systems to further improve their effectiveness in severe crash events and to prevent air bag induced injury due to an out of position occupant.

The paper investigates the initial system engineering requirements for a pre-arming sensor and establishes a baseline approach for further study. Prototype Forward Looking Automotive Radar sensors are used to investigate and test performance of existing radar based sensors to determine the ability of current devices to provide pre-arming capability and provides direction for further sensor development.

INTRODUCTION

Radar applications for automobiles have been widely researched for many years to provide collision warning, blind spot detection and autonomous or semi-autonomous cruise control. Deployment of radar systems with many of these features appears imminent within this decade given the development of low cost signal processing and radar transceiver functions. Cost competitiveness with existing systems, usefulness and benefit of this advanced technology will determine their success in the marketplace. Therefore, system designers and engineers have been looking at other potential functions that can enhance the usefulness and desirability of proposed systems. One approach serves to increase the utility of a forward looking radar system by providing the range and velocity information already generated by an intelligent cruise control radar to a single point electronic crash sensor module. This information can be incorporated into the air bag firing decision algorithm to decrease the time required to fire the air bag. The faster time to fire signal can significantly enhance occupant protection and reduce air bag induced injuries to occupants by tailoring the bag deployment. Tailoring of the inflation pressures [1] and energy management of inflating air bags can significantly effect the loading severity and ultimately the protection of the occupant. Loading severity is increased for those cases where occupants may be standing or may be thrown forward against the deployment door due to pre-event braking or an avoidance maneuver. Research has shown that small adults and children are more susceptible to this out-of-position condition. Faster crash sensing and optimized energy management in the occupant restraint system may improve the protection in these cases and in overall occupant protection.

FASTER TIME TO FIRE DECISIONS

Currently, single point electronic crash sensors employ solid state accelerometers and in some cases safing or arming sensors to detect events. The microprocessor in the electronic module performs calculations to discriminate between must fire and no fire events.

For frontal crash sensing, the required time to fire depends on many factors but typically extends into 10 to 20 ms range for 48.3 KPH barrier events. Most crash sensing algorithms convert the event accelerations into physical parameters and must use a significant portion of the required time to fire in order to arrive at a correct and reliable decision. Rather than improve the quality of the event accelerations to decrease the time to fire, possibly through costly structural changes to the vehicle, pre-event sensing uses information about the impending event. Current state of the art electronic crash sensors reliably provide the firing decision before the required time to fire but could be substantially improved by using range and relative velocity data from a forward looking radar sensor.

The improvement in time to discriminate between real crash events and non events was investigated by using two series of crash events for modern automobiles. Two different types of vehicles were selected for this study. The first crash series is from a typical mid-sized platform with unibody construction while the second crash series is from a typical minivan platform. The selected crash types are listed in Table 1 along with their respective required time to fire and expected time to fire using current state of the art electronic sensors. The algorithms and thresholds were modified to make appropriate use of the range and relative velocity data from a radar sensor. As can be seen from the results in Table 1, a significant improvement in time to fire can be realized with a priori knowledge of an impending event. The smallest improvement still showed a respectable ten (10) per cent improvement. This data also provides some clues as to the vehicle types and crash events that may benefit most from predictive crash sensing.

TABLE 1

Predictive Crash Sensing Performance

RTTF - Required Time to Fire, ms

TTF1 - Time to Fire With Out Predictive Crash Sensing

TTF2 - Time to Fire With Predictive Crash Sensing, ms

NF - No Fire

Mid Sized Platform, Sedan

Event	RTTF ms	TTF1 ms	TTF2 ms	Delta ms
12.9 KPH Head On Barrier	NF	NF	NF	NF
22.5 KPH Pole	95.0	88.2	52.2	36.0
25.8 KPH 30 Degree Barrier	80.0	61.5	12.6	48.9
48.3 KPH 30 Degree Barrier	55.0	53.6	13.9	39.7
48.3 KPH Head On Barrier	20.0	14.8	10.6	4.2

Minivan Platform

Event	RTTF ms	TTF1 ms	TTF2 ms	Delta ms
11.3 KPH Head On Barrier	NF	NF	NF	NF
29.0 KPH Pole	86.0	74.6	63.0	11.6
48.3 KPH 30 Degree Barrier	50.0	15.0	11.6	3.4
48.3 KPH Head On Barrier	15.0	10.8	9.5	1.3
56.4 KPH Head On Barrier	13.0	10.1	8.8	1.3

Consistently greater reduction in time to fire was demonstrated for the sedan than for the minivan. Although a detailed analysis of the individual structures has not been performed for this paper, the crash signatures of the two vehicles suggest that the stiffer and shorter structure of the minivan provides much greater energy transfer earlier in the event in which to base a firing decision. Therefore, the improvements in time to fire may not be as great for a structure of this type. Greater reductions in time to fire were observed for slower developing crashes and obliques. High speed perpendicular barrier crash events typically exhibit more energy transfer earlier in the event than that of the other crashes. Generally, the more energy available in the early stages of the event the smaller the expected improvement in time to fire. Additionally, the improvement in time to fire due to predictive crash sensing may be limited in extremely short, four (4) to eight (8) ms, time to fire requirements such as those found in short stiff vehicle structures or side impact crash events.

This data does not however indicate whether either vehicle structure or which specific crash types are the most severe in terms of occupant injury. It is also not a straightforward conclusion that reductions in time to fire will translate directly to significant improvements in occupant protection. Additional study is required to correlate the improvement in time to fire and occupant protection for specific crash types and vehicle structure. It is hypothesized by the authors that reductions in injury due to inflating airbags can be reduced for those cases where a faster time to fire results in deploying the air bag prior to the occupant coming into contact with the air bag module.

The authors have not extended the predictive crash sensing approach to consider making a firing decision only on the a priori knowledge of an impending event. The benefit of such an approach may allow initiation of the air bag prior to impact. However, this extension would require extensive research on target acquisition, tracking and destructive power of the object on the collision path.

REDUCING INJURY

Extensive research and testing have been performed to assess the injury to occupants during severe crash events. With the widespread use of airbags, much of this research has focused on the deployment loads of airbags and the resulting potential of injury to occupants that are out of position or in close proximity to the deploying air bag. Some studies have shown that loading of the airbag can influence injury [3] in real world crashes.

The design of an effective restraint system involves many factors beyond the control of the safety restraints designer relating to crash dynamics, vehicle crush characteristics, occupant size, weight, occupant position, seat belt use and many others. Some design variables available to the air bag designer are air bag inflation rate, ultimate or maximum pressure and venting. The deployment and pressurization rate of the air bag represents a compromise between not having sufficient pressure in the bag to restrain the occupant as he begins to move relative to the seat and that of having so much impulsive energy as to induce injury to an occupant in close proximity to the deploying bag. An earlier firing signal permits more time to inflate the bag and hence reduce or lower deployment loads. High and low levels of inflation aggressiveness have been considered in the referenced research. Most testing has been performed with instrumented test dummies or surrogate subjects, [4,5] that were positioned over air bag modules prior to deployment. These studies present injury values for position of subjects in relation to a deploying airbag. The injury values are based on the peak loading obtained for chest compression, chest velocity and head injury criteria (HIC). Research and test data derived from using swine as test subjects to determine the mechanism and timing of blunt injury [6] have established these injury criterion as useful in subsystem design to mitigate injury.

Other studies have developed models for the timing and mechanisms of injuries and have validated the injury criteria used in [4,5].

The reduction in injury to an occupant as a result of predictive sensing was assessed using data from [4,5]. Data is presented for chest velocity, chest compression and HIC for surrogates positioned at various longitudinal positions and at three distances from the deploying air bag. A relative reduction in injury parameter was compiled based on this data within a group of tests that contained data at different spacing. This data was then averaged to eliminate any one set of data dominating the results. As expected, the further the surrogate was from the deploying bag, the greater reduction in injury parameter thus indicating a potential reduction in injury. The results from this analysis are shown in Figure 1. These results only consider the effect of increased spacing between the subject and the deploying air bag. Although this is an important component in reducing injury, an out-of-position occupant could still be subjected to large deployment forces if positioned near the deploying bag. Additional benefit may be achieved by reducing the inflator mass flow rate and increasing inflation time to both the in- and out-of-position occupants.

This data indicates a general trend in reducing injury and should not be taken as an absolute. The differences in surrogate position in relation to the center of the wheel caused large variations in the injury parameters. In some cases, only small reductions were noted in one or more of the parameters.

The improvement in injury values is dramatic for slight separations from the air bag module. Translating this separation to a requirement for faster air bag firing time was determined by reviewing the head movement found in a typical sedan during barrier events. Figure 2 presents the required improvement in time to fire assuming free body displacement by 25 and 50 mm at various speeds. Superimposed on this figure are film analysis data showing the time required for the head to move 25 mm at maximum head velocity. Test data is for an unbelted occupant. Test data is for barrier events for an unbelted occupant. Therefore, it is expected that improvements in time to fire due to predictive sensing in the order of 2 to 8 ms may provide a means of reducing injury.

SENSOR REQUIREMENTS

Various technologies have been investigated to provide range and relative velocity data for automotive applications such as autonomous cruise control, collision warning, collision avoidance and object warning. Most researchers have concluded that radar provides the best compromise for automotive applications due to its all weather performance [7], ability to be integrated into a vehicle and potential for low cost. Safety issues are obviated given the use of low power millimeter wave frequencies. The potential exists that the function of predictive crash sensing can be incorporated into forward looking radar sensors that are currently being developed for autonomous cruise and collision warning. A brief description of sensor operation is presented followed by a discussion of the key sensor requirements and their impact on overall effectiveness. A review of the current capabilities is discussed.

In a generic sense, any remote sensor will undergo search, target identification, trajectory computation, and threat/non threat decision as part of their operational flow. Given the dynamic nature of real world driving scenarios, the sensor needs to execute the described flow for every target representing a potential threat and needs to do so concurrently to meet response time considerations. A challenge for the system designer is to resolve the details of each step to fit general driving scenarios to maximize efficiency and/or optimize performance for given cost without generating false alarms. In this case a false alarm is defined as incorrectly identifying a target as a real threat and thus leaving the system susceptible to low speed deployment.

In the search mode, the sensor monitors its coverage space for potential threats. In a multiuse sensor, this mode is sometimes referred to as surveillance mode. This function needs to be carried out concurrently as a specific target is being tracked to insure that the entire coverage space is being monitored for safety reasons. As return energy is detected, the sensor must process this data to determine whether there is a target constituting a potential threat, whether the returns are noise, or are objects which are not a threat. This process can be challenging because real threats are immersed in clutter and because simple radar signal processing can not determine where the road is going. In some instances, tracking the target is necessary in order to estimate its trajectory to determine if it is on a collision path. Additional parameters such as speed, throttle angle, braking level, wheel slip, and steering angle could be incorporated with threat trajectory to increase confidence in the prearming decision. This additional platform information may be available on the vehicle's information network as a result of the other functions of the radar sensor.

One of the most critical requirements for the sensor is its response time, specifically the time from when a threat enters the field of view of the sensor to a determination that the threat has been confirmed and finally communicated to crash sensor processor. Required response time is a function of the worse case dynamics, distance, velocity, and acceleration of both the platform and the threat object. Figure 3 illustrates the response time needed as a function of distance and speed assuming zero deceleration. For typical range coverage and vehicle speeds a minimum time to detect a target and determine if it is on a collision path is about 1 second. In most situations when the driver becomes aware of a threat, defensive actions may increase the allowable response time. One second should be adequate for most situations. Current devices being considered are able to identify and track targets in a few milliseconds.

Another critical parameter is the spatial coverage of the sensor; this coverage directly affects how much time is available to respond taking into account the direction the threat is incident from. In order to evaluate where a target is first detected, probability of detection contours were computed assuming typical target radar cross-sections and using the sensors range, horizontal and vertical beam widths, and dynamic range. An analysis was performed to calculate the time in which the sensor had to respond assuming a target such as another vehicle was first detected when 50% of its frontal edge was within the probability of detection contour. Figure 4 and 5 illustrate the results of this analysis for a sensor with 8° and 60° total horizontal coverage respectively. In the case of an 8° coverage sensor which is typical for an autonomous cruise control or forward collision warning sensor, timely identifications and response are achievable for angles of incidence up to 15 to 20°. In future products, as the horizontal coverage of the sensor is expanded to 60°, predictive sensing warnings to angles of 35° incidence are achieved.

The existence of many pathological cases in real driving conditions complicates the signal processing and results in the majority of work in designing systems that are reliable and behave as expected. One of these cases is discriminating between a collision and a near miss. Specification of range and velocity accuracy, both systematic bias and random components, and latency are straight forward to derive given the desired response time and trajectory accuracy. While range and velocity information by themselves enable the sensor to determine that a collision may occur, inherent in the calculation is the assumption that the paths of the threat and the platform will cross. In many real driving scenarios the paths of the threat and the platform come very close but do not cross; an example of this is opposing traffic on a two lane undivided highway. By adding the additional requirement of sensing the horizontal direction of the threat from the direction of travel of the platform, distinguishing between a collision and a near miss is possible. Figure 6 is a comparison of the angular signature of two scenarios; the first is a vehicle in the path of the platform on a slight curve, and the second is a vehicle in an adjacent lane. In the first case, a collision warning would be issued based on the target angle tending toward zero. For the second example, target angle begins to diverge; this characteristic can be used to discriminate a near miss. Note that for a sensor with wide horizontal angular coverage, angular data allows early recognition of a near miss, but for typical forward looking sensors with angular coverage on the order of 8°, the object leaves the field of view approximately the same distance as when angular signature departs from that of a target which will collide.

SUMMARY

It is clear that earlier time to fire decisions can result in reduced injury to occupants and that radar sensors can be used to provide information about an impending event that can be used in the decision making process to fire the air bag. More study is required to refine the decision algorithms to provide variable thresholds as a function of threat and platform information. A first analysis of the generic sensor requirements reveals that current radar sensors envisioned for autonomous cruise control have adequate spatial coverage and range to provide this dual use function for small angle of incidence threats and for typical driving scenarios and speeds. It is further envisioned that extensive work may be required to characterize the many pathological cases encountered in real world driving conditions and that signal processing advances are required to assess the threat of specific targets. Wider field of view may also be required to extend angle of incidence of threats, although it is not clear that all of these scenarios will result in the need to fire a frontal air bag.

ACKNOWLEDGEMENTS

The authors wish to thank Thomas H. Vos, Ronald M. Muckley, David J. Bauch and Mark Carlin for their help in preparation of this paper.

[1] W. R. Carey, T. J. Wissing, R. G. Gehrig, G. W. Goetz, and D. A. Larson, "Energy Management in IORS", SAE Paper No. 720418, Second International Conference on Passive Restraints, May 1972.

[2] L. Patrick and G. Nyquist, "Airbag effects on the out of position Child." SAE Paper No. 720442, Second International Conference on Passive Restrainsts, Detroit, May 1972.

[3] H. Mertz, "Restraint System Performance of the 1973-76 GM Air Cushion Restraint System", SAE Paper 880400, February 1988.

[4] J. W. Melvin, J. D. Horsch, J. D. McCLeary, L. C. Wideman, J. L. Jensen and M. J. Wolanin, "Assessment of Air Bag Deployment Loads with the Small Female Hybrid III Dummy", SAE Paper 933119.

[5] John Horsch, Ian Lau, Dennis Andrzejak, David Viano, John Melvin, Jeff Pearson, David Cook, and Greg Miller, "Assessment of Air Bag Deployment Loads", SAE Paper 902324, Thirty-fourth Stapp Car Crash Conference, November, 1990.

[6] Ian V. Lau and David C. Viano, "How and When Blunt Injury Occurs - Implications to Frontal and Side Impact Protection," SAE Paper 881714, October 1989.

[7] K.W. Chang, H. Wang, G. Shreve, J. Harrison, M. Core, J. Yonaki, A. Paxton, M. Yu, C. H. Chen, G. S. Dow, B. Allen, K. Tan and P. Moffa, "Forward Looking Automotive Radar Using W-band Single-Chip Transceiver," Transactions of the IEEE.

FIGURE 1

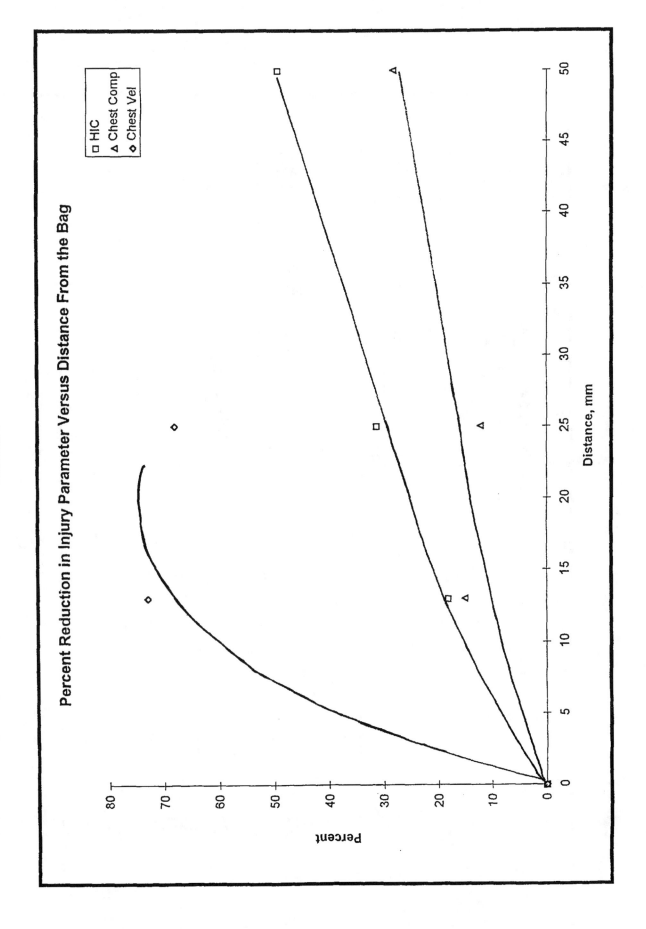

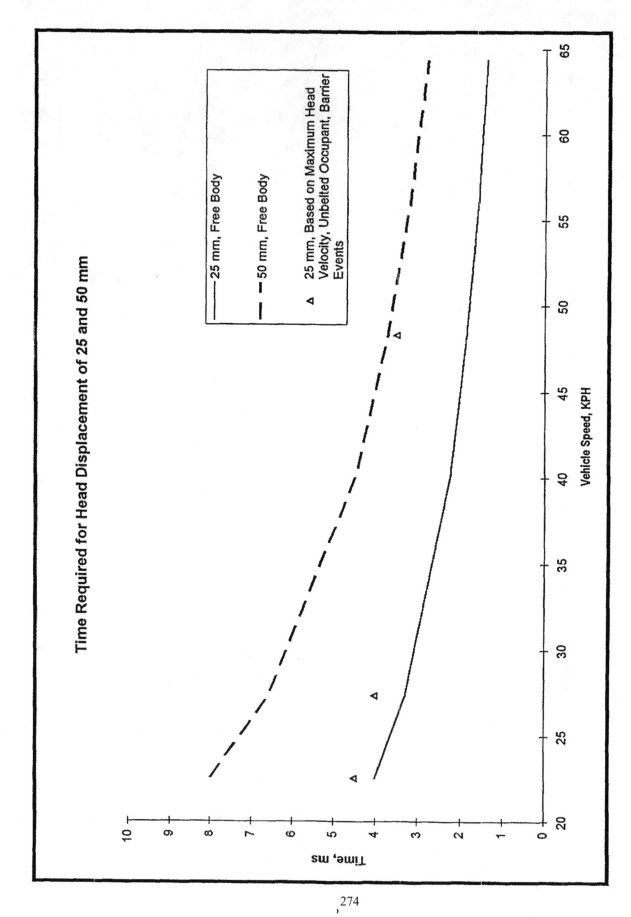

FIGURE 3

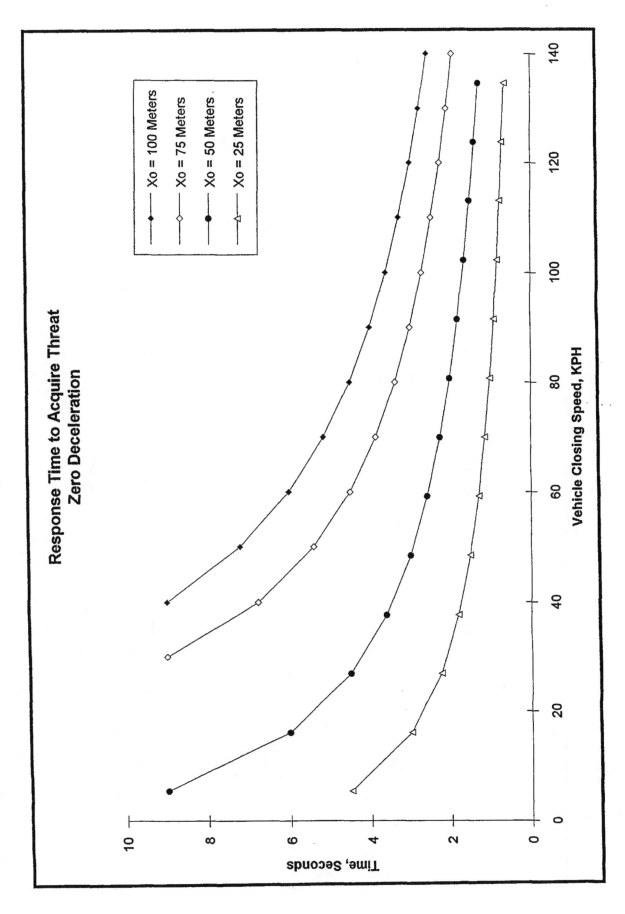

Figure 4

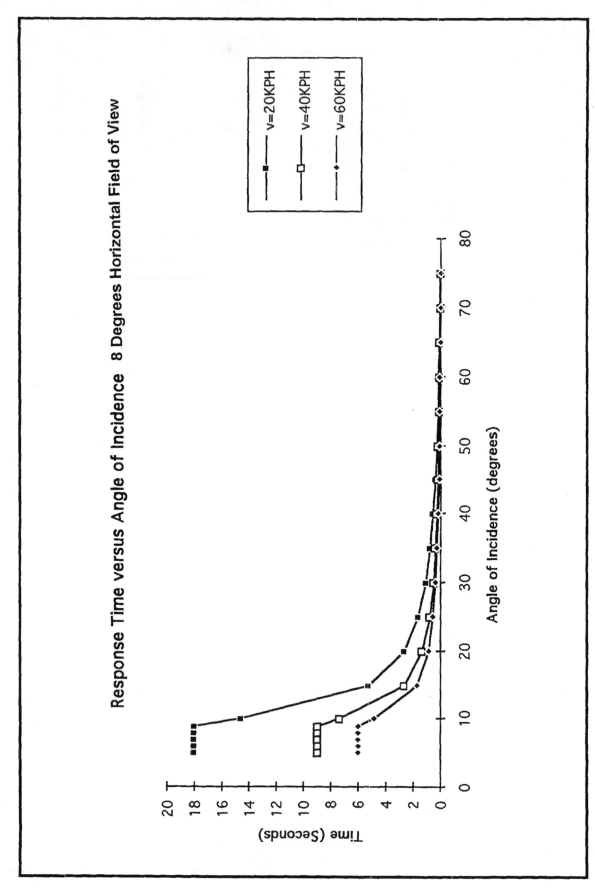

Figure 5

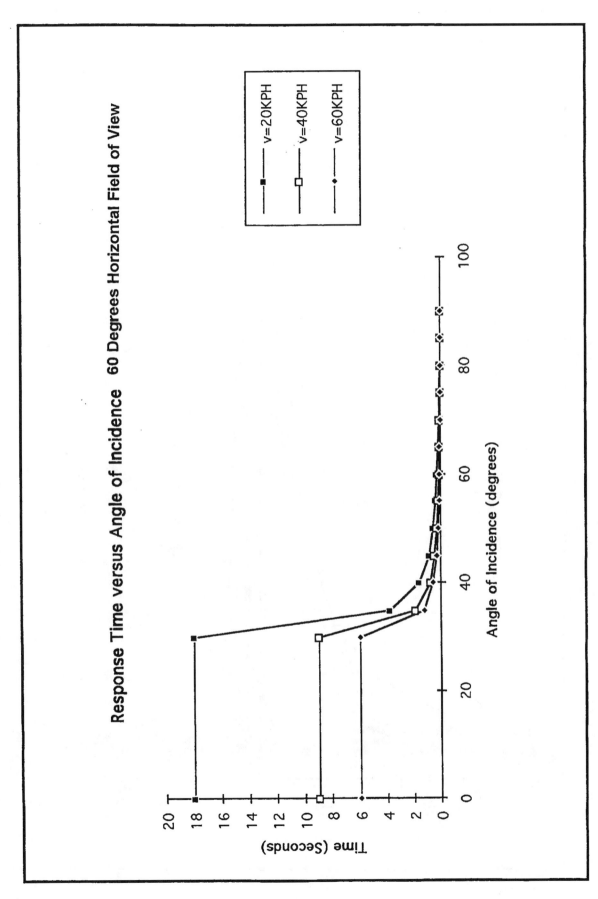

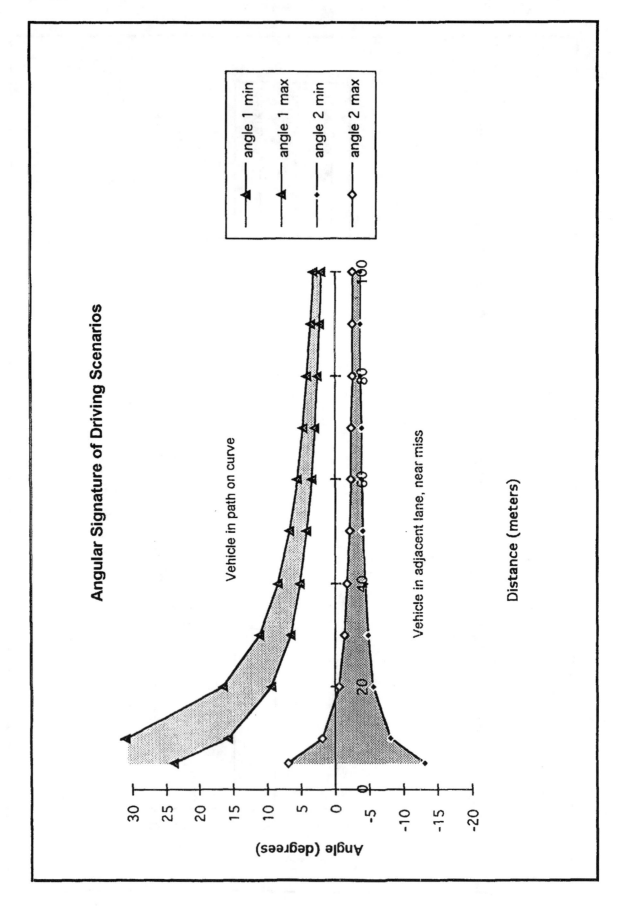
Figure 6

960226

The BMW Seat Occupancy Monitoring System: A Step Towards "Situation Appropriate Airbag Deployment"

Copyright 1996 Society of Automotive Engineers Inc

Klaus Kompaß
Bayerische Motoren Werke AG

Michel Witte
Interlink Electronics Europe S A R L

OVERVIEW

Future developments of airbag systems, which are now considered as standard equipment in cars will focus on three main topics

- new developments in airbag application fields which have not previously been fully investigated ie side airbags,

- cost reduction and cost effectiveness of existing systems via the introduction of uniformity in design thus maintaining and guaranteeing quality levels and product availability

- automatic adaptation of both existing and future airbag systems to the requirements of individual vehicles

The aim of this presentation is to illustrate this final point We hereby present a system which avoids unnecessary airbag deployment when the passenger seat is unoccupied

Due to the increasing amount of cars equipped with passenger airbags, there are now a larger number of accidents during which passenger airbags are activated although nobody is actually sitting in the passenger seat Even when one discounts those accidents which result in car "write-offs", the unnecessary deployment of the passenger airbag increases repair costs in an unjustified way In order to avoid this type of situation, all BMW vehicles as of June 1994are equipped with a seat occupancy monitoring system (SOMS) installed in the passenger seat

The SOMS consists of a polymer sandwich with integrated sensors whose resistance values change proportionally with applied force The sensor mat layout is tailored to each individual seat design Due to this flexibility of design the sensor matrix may be

mounted in three different ways

- between the seat foam and the seat cover,

- between two layers of foam,

- between the shell of the seat and the foam

Sensor resistance is measured electronically The signal "passenger seat unoccupied" is obtained and the airbag-ignition-circuit is switched off only when the sensor resistance lies between two pre-defined values Outside of these two values, and this does include if the system is damaged the seat is always considered occupied and the airbag-ignition-circuit remains "on standby"

This presentation sets out to explain the technology on which the system is based It also details the history of the system's conception and development as well as the different criteria it must satisfy

INTRODUCTION

PREFACE

"Do you work in airbag development? There is nothing left to develop on an airbag, you find them in most cars nowadays!"

Often one hears this or similar responses when explaining to friends or neighbours what you do for a living One would like at this point to inform them of new generator fuels improved coatings optimum inflation times but one also knows that the explanation would bring only a polite smile from your "audience

Is your neighbours reaction justified? Certainly not! It is considered by a lot of people that we are now in a phase of airbag development which is comparable to safety belt development at times when all cars were fitted with static safety belts Similar progress as that made in the safety belt arena (inertia belts, pre-tensioners, force limiters etc) will surely follow in the ongoing development of airbag systems Today's buzzword is "SITUATION APPROPRIATE AIRBAG DEPLOYMENT"

SITUATION APPROPRIATE AIRBAG DEPLOYMENT

This phrase encompasses a vision of a fully automated airbag system whose activation is specifically determined by the safety needs at a precise moment

- if a passenger is seated on the front edge of the seat the force of deployment must be reduced

- reboard (rear-facing) baby seats must not be hit by the airbag

- passengers who are already protected by seat belts only need the additional protection offered by an airbag in cases of higher collision speeds,

- airbag is not required to deploy if no passenger is present in the seat,

- the automatic control system reduces the effect of temperature on the combustion of the airbag fuel

STATISTICS

In order to underline the necessity of these developments we would like to present the following facts and figures

BMW's Accident Research Centre states that a BMW vehicle is in average occupied by only 1 4 passengers This means that a passenger is present in only 41% of BMW vehicles This figure is indicative of the general situation in Germany In the USA this figure is far lower with only 32 out of every 100 vehicles carrying a passenger

If one supposes theoretically that all vehicles are equipped with airbags for both driver and passenger as well as pyrotechnical safety belt tensioners and that the total number of vehicles is 30 million one can safely assume there will be in the approximately 100 000

accidents each year where the collision speed lies between 25 and 40 km/h Between these speeds the airbag must be activated and car damages can still be repaired For your information the safety belt tensioners would be activated at lower collision speeds, but it is not our aim to fully explain the statistics of such events As a passenger is present in only 41% of the 100 000 collisions in which the airbag must be deployed, there are 59,000 unnecessary actuations of the passenger airbag Assuming the not unreasonable amount of DM 3000 - (£1250, US $1900) per vehicle to repair the damage caused by the passenger airbag deployment alone, one can quite easily see the dramatic annual economical effect of some DM 177 million (£71m, US $111m) per year unnecessary airbag activation can have

These figures illustrate the need for measures to avoid unnecessary airbag deployment

DEVELOPMENT OF AUTOMOTIVE SEAT OCCUPANCY MONITORING SYSTEM (SOMS)

AIRBAG TASK FORCE OF GERMAN CAR MANUFACTURERS

At the end of the 1980's BMW had already gathered initial ideas and explored various concepts for SOM systems At the same time BMW began with a one development

As the basic statistical data is not influenced by vehicle make or model it was decided, in 1990 by five leading German automotive manufacturers (Audi, Mercedes-Benz Porsche Volkswagen and BMW) to form a task force and to co-operate to find a solution to the problem and to develop an Seat Occupancy Monitoring System for passenger seats

Their work concentrated on specific areas

- Definition of the specification for such a system,

- collection of ideas for proposed solutions (it was decided to ask for solutions from as many sources as possible),

- comparison and evaluation of the proposed solutions

REQUIREMENTS OF THE AIRBAG TASK FORCE

The primary requirement of the SOM system is to offer a fail safe system, in the event of damage or inconsistent signals, the system must act as if the passenger seat is occupied This is in order to ensure airbag activation during a collision even if the SOM system is restricted in any way It is, under such circumstances preferable to activate the airbag and to save say a briefcase than to risk non-deployment of the airbag a passenger being present

The SOM system

- must recognise the presence of a passenger only (ie not a briefcase, box of groceries, etc), but not necessarily the position or posture of the person

- is not required to differentiate between a baby seat and a passenger

- must give a "passenger seat occupied" signal when a passenger is sitting in any position in the seat,

- should not influence the design of the car interior,

- should have no effect on the seat comfort

- should guarantee a safe function throughout the lifetime of the vehicle (for this point one must also consider high mechanical and environmental constraints)

- should detect a safe weight of

12kg or more,

- should offer a "seat unoccupied" signal for weights below 12kg,

- must guarantee that the "seat unoccupied" signal is not influenced by any interference,

- must have the capability to be monitored or controlled via two data lines connected to the central "airbag release control unit",

- must be easily adapted to different car seats, interiors, etc,

- must be compatible with seat heating systems where necessary

CHOOSING A SUITABLE SYSTEM

All offered SOM systems where evaluated based on the task force's initial guidelines. In order to decide for the most appropriated solution a cataolgue for evaluation was created

Only the major points of the original specification have been highlighted in the following descriptions of the various systems (and the associated arguments for and against each option) We must stress that the following technologies are only those which were presented at the time of the task force Due to increasing interest in such systems, in the future much more research time shall be spent to improve these technologies, so from time to time evaluation results may change This report however, can only discuss the situation as it is now

The proposed technologies were

Capacitive measurement systems

Capacitive systems use the car's seat and body as capacitor's electrodes The dielectric separating the electrodes is made up of either air (in the case of an unoccupied seat) or a human body The change in capacitance caused by the change in dielectric can be measured If the capacitance falls between two pre-defined limits the seat is considered to be occupied

This option is attractive because of

- easy integration into the seat,

- robustness,

- cost effectiveness,

- availability of components

The problem with capacitive systems is that with a total capacitance of 150pF, the difference between the detection of occupied and unoccupied seats allows only 15 to 20pF This option was therefore evaluated as being too sensitive to interferences

UHF-Sensors

Seat occupancy detection can be performed in various ways using UHF sensors

- Microwave rays can measure distances between obstacles, however, additional supplementary information is required, such as the position of the seat in order to differentiate between a passenger sitting in the rear of the vehicle and an unoccupied front passenger seat

- It is possible to measure dynamic movements of an obstacle located between transmitting and receiving antennae One problem however, remains it is impossible to differentiate between an unoccupied seat and an inactive (perhaps sleeping) passenger

- A third possibility is to detect reductions in radiation damping between the transmitter and the receiver

The criteria against UHF systems were the high costs involved for example the fitting of antennae into the dashboard and the back of the seat Furthermore there are psychological barriers to overcome due to continuous exposure of the passenger to low level radiation

Infra-Red-Sensors

Reflected body temperature is considered to be a measurable variant Sensors are used to measure the reflection of or the change in temperature over a pre-determined period of time

This type of system was not successful as no accurate (and regular) signal could be obtained due to differences in environmental temperature, clothing, seat heating, etc)

Optical detection systems

Again there was more than one way to use optical systems

- Light barriers supervise a defined zone within the passenger seat area If the light beam is interrupted the seat is deemed to be occupied The problem with this option was the difficulty to define accurately enough the required zone to be supervised It was not possible to obtain a reliable signal As a result this type of system was judged unfeasible

- Optical displacement detectors such as those used in security systems, were evaluated The idea was to detect the movement of the passenger, again, within a pre-determined area The system proved to be too insensitive In addition to this it was impossible to exclude the possibility that an inactive (sleeping) passenger could be detected as an unoccupied seat

- Pattern recognition CCD-Chip mounted in front of the passenger seat area registers the image it "sees" This image is electronically compared to the image of an unoccupied seat This system is very innovative, but at the time of it's presentation many negative aspects were noted cost complexity of the system assembly difficulty expensive adaptation Furthermore, a safe detection would also be made more difficult in varying light (floodlights, etc)

Radar / Ultrasonic-Sensors

Similar to the UHF-Sensors waves from a transmission antenna are influenced by the obstacle in their path (ie human body) when the seat is occupied The problems which arose were almost identical to those experienced with the UHF-Sensors expensive difficult to integrate plus this system does not give information when the passenger is sitting out of position

Fibre-Optics

When the seat is occupied fibre optic strands embedded in the seat are distorted This distortion or bend, alters the intensity of the light travelling through the strands which is translated into a "seat occupied" signal The major problems with this system is it's high fitting cost plus the results it can achieve can be affected by the change, through time, of the tension of the seat cover

Piezoelectric-Cable

The principle of this system is based on continuous changes of the voltage of the piezo-elements when the seat is occupied due to the passenger's plus the car's own vibrations during driving This change in voltage is evaluated and indicates whether or not the seat is occupied The trials by BMW failed due to the driving comfort experienced in modern vehicles On a smooth road the test sample indicated an "unoccupied seat" when in fact the passenger was sitting still (eg sleeping) because the vibrations of the passenger could not be identified

Pressure Measuring System based on Force Sensing Resistors (FSR[TM])

A system was offered by INTERLINK ELECTRONICS EUROPE a Luxembourg based company Their system could measure static forces thanks to an array of FSR[TM] sensors All requirements listed by the German Airbag Task Force seemed to be satisfied by the Interlink system As an added bonus the costs involved corresponded favourably with the original price targets of the automotive industry

Functional properties of the FSR

The **F**orce **S**ensing **R**esistor (FSR) is based on a reproducible surface-effect The technology had been patented worldwide by INTERLINK ELECTRONICS
The typical construction of a FSR[TM] as it is shown in figure 1 is based on a sandwich of two polymer films or sheets and a spacer On one sheet a conducting pattern is deposited (screen-printing) in the form of interdigitating electrodes On the other sheet a proprietary semiconductive polymer is deposited Finally the two sheets are faced together so that the conducting fingers are shunted by the conducting polymer

Structure of the force-sensing resistor

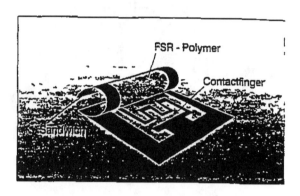

Figure 1 **BASIC FUNCTION OF THE INTERLINK FSR ARRAY**

If a force is applied upon the FSR the electric resistance of the sensor changes according to the amount of the force Depending on the size of the used force actuator and the amount of the force a more or less greater number of interdigitating electrodes are short circuited As all these resistance bridges are shunted by the semiconductive polymer the device follows the law of Ohm

$$(1/R_{tot}) = (1/R_1) + \quad +(1/R_N)$$

If the force with the corresponding actuator-surface and the resistance are marked up in a diagram with log-log coordinates as it is shown in figure 2 you can see an inversely proportianal relation between the force and the resistance The course of the shape can be adapted by INTERLINK to the specific application

cycles are guaranteed Extented tests proved that the used materials do not react with humidity

Resistance/pressure curve of the FSR - element

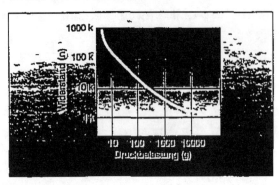

Figure 2

The FSR can be used as an intelligent switch By defining different thresholds you can get another degree of freedom for the optimal adoption to the application

In the case of the SOM the sensor is designed in such a way that a child with a weight of 12 kg is detected in any position in the seat If the resistance is above a fixed threshold the seat is "unoccupied" If on the other hand the measured resistance is below a fixed threshold the system gives the message "seat occupied

The sitting-surface had been fixed to a circle area diameter of 120 mm By interrelating the force to the surface it is possible to avoid that a briefcase or a bag etc causes a decrease of the resistance below a fixed threshold

Mechanical characteristics of the FSR

The laminated thermoplastic supporting foils can be used at a temperature of -40 to 170 C The spacer is resistant to a temperature of -40 to 150 C At temperatures above 120 C we recommend a temperature-compensation because of the hardening of the spacer-foil
The FSR can be designed in any possible form that can be shaped on a two-dimensional CAD The thickness of the FSR is of about 0 5mm To facilitate the application of the FSR the sensor can be furnished with an adhesive on the backside One million of

Figure 3 FSR-SENSORELEMENT

One of the most important advantages of the FSR™ in comparison with membrane switches is the fact that you need only a travel of about 70 μm to release a defined signal

ADAPTATION OF THE FSR™ TO THE SEAT

Each car manufacturer has its own conception of optimal seat-ergonomics and -comfort In addition the costs are an important factor in the design of car-seats So actually you can find three different types of seats

- full seatfoam on a spring-wire- or on a spring-mat-support
- full seatfoam on a seat-shell
- rubber-hair-mat with cushion-foam on a spring-wire- or on a spring-mat-support

In order not to modify the construction of a seat the FSR™-sensor-mat has to be adopted to the seat Beside the specifications of the task force as mentioned above the large tolerances for the seat design and mounting must be considered
For each seat-type Interlink developed one

specific sensor mat The sensor-mat is designed in such a way that different kinds of seat cover can be used For the future it would be easier and less expensive if the sensor mat could be integerated into the seat foam

DESIGN AND LAYOUT

The criterias for the design of the sensor mat are the following

- specifications of the task force,
- place of installation of the sensor mat
- optical tracing of the sensor mat on the seat cover,
- stress in vertical direction and torsional stress
- full redundancy and
- costs

The different places of installation lead to divergent designs of the sensor mat

Assembly between the seat foam and the seat cover

This alternative is selected if there is not enough support space in the area of the spring-wire- or of the spring-mat-support In this case the main design criterias are

- the seat ventilation
- the seat ergonomics
- the mechanical stress and
- the mounting tolerances

Figure 4 BMW SEAT WITH SOMS

For this alternative the sensor mat is designed in the shape of fishbones This design allows an optimal adoption to the torsional stress To ensure an optimal ventilation of the seat and to avoid optical tracing on the seat cover the sensor mat is designed in such a way that the support foil is placed near the groove of the seat To facilate the mounting process the mat is placed on a felt support By using felt the sensor is uncoupled with the seat cover The different sensor elements are designed in such a way that there are no influences from different seat cover tensions or from initial loadings that are caused trough the mounting of the seat

Assembly between the seat foam and the closed seat shell

This assembly alternative allows the design of a smaller sensor-mat The tolerances of the seat foam and the seat shell are important criterias for the design of the sensor By mounting the sensor mat directly on the seat shell there is only little mechanical stress The problem in this case however is the curve of the shell Therefore the sensor must be designed in such a way that there are no initial loadings on the sensor

Redundancy - fail safe

The FSR™-sensor-mat is designed so that each sensor element is part of the conductor The conductors are traced all over the sensor mat and form two of the four connectors of the sensor A diode is placed between the other two connectors So it is possible to check the circuit and the connectors by a simple short circuit test in low-resistance direction of the diode The failure of one or few elements of the sensor mat will have little influence on the function of the sensor mat A failure of the circuit or a short circuit however can have fatal consequences
For the measurement of the applied force the circuit is switched in high-resistance direction of the diode The electronics is designed so that all failures cause low resistance and thus the signal "seat occupied"

Low-Pass-Filter

The looped conductors of the sensor mat are printed in a constant distance The sensor mat has a defined natural capacitance The FSR™ sensor-mat is a very primitive low pass filter Anyway one will include on the interface connection an additional low pass filter This will be more than enough to protect the system against EMI influences

The mechanical traction relief

To ensure an unproblematic handling during the assembly, the conectors and the crimping of the sensor mat were molded in hotmelt The device can support tensions up to 100 N

ELECTRONICS

SOMS Control Unit

The raw data provided by the sensor array is not by itself sufficient to avoid unnecessary passenger airbag deployment

In order to

- detect any situation accurately and safely,
- ensure the system is fail safe,
- switch the occupied seat to unoccupied status,
- design and integrate the diagnostic system, etc

a supplementary SOMS-control unit is required This unit transmits via a serial interface the status of detection as made by the sensor array to the airbag control unit The SOMS control unit no larger than a matchbox, is fitted to the passenger seat and measures, in cycles the resistance of the array The resistance is determined in two partial measurements in both the normal and also in the inverse direction of the diode which is integrated into the sensor array The measurements are evaluated by the software and transmitted to the airbag control unit as one of four possible "messages"
> »seat occupied
> »seat unoccupied
> »system failure due to a short circuit
> »system failure due to an open circuit

If either of the latter two situations are transmitted or data is interrupted or stopped the airbag control unit detects an error Any error leads to a 'seat occupied" status

In the event of a change in status from "occupied to "unoccupied" the control unit maintains an 'occupied safety timeframe of two minutes This is a built-in safety mechanism to ensure no false signals are transmitted if the passenger is simply changing position in the seat The switching

from "unoccupied" to "occupied" is of course, immediate

Integration of the Airbag Control Unit

The introduction of central airbag control units, ZAE and BAE, which are also developments of the task force, made it possible to process the information transmitted by the SOM system The airbag control unit is able to compare different inputs and avoid incorrect reactions In the past, based on older airbag control units, the decision to deploy the airbags was made when the seat belt was not attached It was, therefore possible that in an accident where the driver was wearing his seat belt but no passenger was present (and therefore the passenger seat belt was not fastened), the driver's safety belt tensioners were activated (and the driver's airbag not deployed) but the passenger airbag was uselessly activated This could easily lead to the confusing situation for the driver who may find it difficult to understand why his airbag did not deploy but his not present passenger's did The new generation of airbag control units decides if in addition to the seat belt tensioner an airbag is needed and espacialy based on the SOMS information if the passenger airbag is needed

Tests

The FSR^{TM} sensor array the SOMS Control Unit, as well as the complete system, have been exposed to environmental functional and "abuse" testing to evaluate the system's suitability for daily use The tests were designed to ensure
> »seat comfort is not compromised for the sake of the SOMS system

> »seat heating must not influence the function or durability of the sensor and vice versa

> »the system does not accidentally switch from "occupied" to "unoccupied" even under extremely violent driving conditions

Crash tests have shown that in an accident an "unoccupied" seat is correctly detected

Safety Philosophy

with regard to Baby Seats

A question which arises regularly in this context, and we have not, so far, addressed during this presentation, is

»Can a SOMS detect a reboard baby seat in order to avoid airbag deployment which could be harmful for the child?

The answer is an explicit **NO !** We have continuously stressed that the SOM system is required to give an "occupied" signal whenever a dubious result is transmitted and, therefore, there may be occasions when an airbag is unnecessarily deployed A system which deactivates the airbag when a reboard baby seat is fitted to the passenger seat, must do the opposite of the SOM system This leads to a conflict of interests when an SOM system is "asked" to detect baby seats The development of a baby seat recognition system, is therefore, considered as high priority by most automotive manufacturers

It is possible that within the next two years myself, or a colleague of mine taking part in this conference will make a presentation on this development

In the meantime, as long as no automatic baby seat recognition system is available, the passenger airbag must either be switched off or the fitting of reboard baby seats to passenger seats forbidden

FINANCIAL CONSIDERATIONS

A rebate on fully comprehensive insurance comparable to that offered to drivers of cars fitted with ABS is not expected from insurance companies A financial advantage for the owners of cars equipped with SOMS results from favourable insurance classification which is regularly adapted in respect to average damage repair costs As the average repair costs will decrease for cars equipped with SOMS one can count on an indirect rebate as a result of lower classification

Furthermore car owners who do not have fully comprehensive cover and are, therefore liable to pay any damage costs by themselves have a direct advantage with an SOMS fitted car as repair costs are lower

SUMMARY

In parallel with new developments in airbag technology, Situation Appropriate Airbag Deployment will remain high on the priority list of Airbag Development groups This aim will result in an airbag system which can adapt itself to the requirements of any specific situation

The first step in this direction is represented by the BMW Automotive Seat Occupancy Detection system shown in this presentation

BMW is now manufacturing many of it's vehicles with a supplementary restraint system the passenger airbag The unpleasant side-effect of customer dissatisfaction caused by unnecessary airbag deployment represented for BMW a challenge which they have answered with the help of this innovative technology

980646

An Innovative Approach to Adaptive Airbag Modules

Shawn Ryan
Advanced Development Group, Delphi Interior & Lighting Systems

Copyright © 1998 Society of Automotive Engineers, Inc.

ABSTRACT

An airbag module with adaptive capability can be achieved by several methods. Separate inflators, a single hybrid inflator with multiple heaters, or a dual-stage pyrotechnic inflator can all be used to control the inflation level of an airbag. In addition, Pyrotechnically Actuated Venting (PAV) offers an innovative method for regulating airbag inflation energy. PAV allows a controlled amount of gas to be vented out of the module before it enters the cushion. This paper discusses several methods for achieving a variable output airbag module. Static and sled tests were conducted to evaluate PAV in comparison to other adaptive restraint modules.

IINTRODUCTION

The airbag is an important part of a vehicle's occupant restraint system. In combination with the knee bolsters, seat belts, and a steering column, the airbag manages the energy of an occupant during a frontal collision. According to the Special Crash Investigation (SCI) database, airbags have saved over 2500 lives. Through November 1997, SCI has attributed 87 deaths to airbags. Forty-nine of these deaths have been infants and children who were riding in the front passenger seat. Most of the 35 drivers whose deaths are associated with airbags have been smaller adults. In many of these instances the occupant was unbelted, or improperly belted.

There are two primary modes by which an airbag can cause injury to the occupant. The first is punch-out force. Punch-out injuries occur when the occupant is very close to the module. During punch-out, the cushion volume is small, the pressure is high, and the occupant inhibits the cushion from unfolding. Punch-out loading can cause injuries to the occupant's chest. The second mode is membrane force. Membrane force occurs as the cushion is expanding and interacts with the occupant before it has achieved its designed shape. Membrane loading can cause injuries to the occupant's head and neck.

One method of reducing the risk of injury is to reduce the power at which the airbag deploys. Recent legislative changes have been implemented to allow less aggressive airbags. The risk of injury from punch-out and mem-

brane loading is reduced with less aggressive airbags. Less aggressive airbags may provide less restraint to normally seated, unbelted occupants in high severity crash scenarios.

The current amendment to the Federal Motor Vehicle Safety Standard 208 which allows the use of less aggressive airbags expires on September 1, 2001. The National Highway and Traffic Safety Administration has indicated future regulations will restore high severity crash testing and add lower severity testing with smaller occupants. Once high severity, unbelted test requirements are restored, adaptive airbag modules will maintain the reduced risk of injury initially achieved through depowering.

DISCUSSION

While there are many possible approaches to achieve variable output airbag modules, only a few are considered in this paper. This paper discusses separate inflators; dual stage pyrotechnic inflators with separate combustion chambers; dual heater hybrid inflators; and PAV. All of these systems require the development of additional sensing technology not included in this discussion.

SEPARATE INFLATORS – One method of achieving a variable output airbag module is by using a primary and a secondary inflator in a module. The inflators may use sodium azide, hybrid, or non-azide technology. The primary inflator would provide restraint for low level deployments. The secondary inflator would provide additional gas for higher severity deployments. For a high level deployment, both inflators would deploy simultaneously. For a low level deployment, only the primary inflator would deploy. An offset time between the two inflators may be introduced to achieve additional levels of restraint. The total energy and mass input to the cushion is the same with an offset deployment as a high level deployment, however more of the energy is input to the cushion later in the event.

Dual inflator systems offer some unique advantages. Different types of inflators using different technologies can be used in the same module providing flexibility in the design. The primary and secondary inflators may be tuned independently for greater control of the high and low level tank curves. In many cases, a validated design can be used for the primary inflator. Communization between the primary and secondary inflator can reduce manufacturing costs.

Dual inflator systems also have some drawbacks. Dual inflator systems are larger, have more mass, and cost more than a single inflator using the same technology. When only the primary inflator is required for restraint, disposal of the live secondary inflator becomes an issue. Finally, since there are two separate combustion processes, the variability doubles over a single inflator for a given output.

DUAL CHAMBERED PYROTECHNIC INFLATORS – The simplest embodiment dual stage pyrotechnic inflator has a single body with separate combustion chambers. Pyrotechnic inflators produce gas and heat from the combustion of a gas generant. For a high level deployment, both chambers are ignited simultaneously. For a low level deployment, only the primary chamber is ignited. The primary and secondary chambers produce energy independently. Because the primary and secondary inflators in a separate inflator system also produce energy independently, it is expected that a dual chambered pyrotechnic inflator could provide similar performance to a separate inflator system using the same technology. The total energy and mass input to the cushion would be the same with an offset deployment as a high level deployment, however, more of the energy would be input to the cushion later in the event.

Dual level pyrotechnic inflators are less expensive, smaller, and lower mass than separate primary and secondary inflators with the same technology. Current designs indicate dual level pyrotechnic inflators using advanced non-azide technology can be made smaller than other dual level systems.

There are also some drawbacks to dual chambered pyrotechnic inflators. When only the primary stage is required for restraint, disposal of the live secondary stage becomes an issue. Since there are two separate combustion processes, the variability doubles over a single output inflator for a given output.

DUAL HEATER HYBRIDS – Another method of achieving a variable output airbag module is by using a primary and a secondary heater in a hybrid inflator. Hybrid inflators consist of stored gas and a pyrotechnic material. The pyrotechnic material is ignited to produce heat and some additional gas. The heat expands the stored gas to fill the cushion. For high level deployment, both heaters are ignited simultaneously. For low-level deployment, only the primary heater is ignited. The dual heater hybrid systems differ from the separate inflator systems and dual chamber pyrotechnic inflators in that the total mass produced by the dual heater hybrid inflator remains relatively constant.

To achieve intermediate restraint levels, an offset time between the ignition of the two heaters could be introduced. Although the total energy produced by the two heaters would be the same as a high level deployment, the time interval that the heaters produce the energy would lengthen. Because the mass would flow out of the inflator, the heat transferred to the gas would change. In general, the heat transferred to the gas would decrease as the offset time increased. Initial testing has shown the longer the offset time, the lower the peak tank pressure. The exact relationship between peak tank pressure and offset time will depend on the inflator design.

As with dual chamber pyrotechnic inflators, dual heater hybrids can be made smaller, less expensive, and with less mass than separate primary and secondary inflators using the same technology. By varying the offset time, the total heat energy input to the cushion could be adjusted between the high and low levels.

A dual heater hybrid inflator presents some challenges. Dual heater hybrids must perform under a variety of internal operating pressures. This increases design considerations and makes it more difficult to control variability. In particular, the effect of offset time on restraint depends heavily on the internal design of the inflator. The ability to independently tailor the high and low level tank curves is also limited by the inflator design geometry.

PYROTECHNICALLY ACTUATED VENTING (PAV) – PAV controls the inflation by allowing a portion of the gas produced by the inflator to be vented out of the module. The maximum amount of gas vented out of the module is directly related to the portion of inflator ports aligned with a vent slot in the module. This amount can be tailored by design geometry to meet module performance requirements. It is important to understand that the average pressure in the cushion does not drive the mass flow out of the vent slot. Instead, the mass flow out of the module is driven by localized pressure from the inflator ports. The localized pressure forces a large percentage of gas from the inflator out of a relatively small opening. For a normally open system, a mechanism is incorporated into the module such that when actuated it closes the vent slot. A normally closed system would open the vent slot when the mechanism is actuated. A normally open PAV system was used in this study.

In order to respond to information from a crash severity sensor, the mechanism must move within milliseconds. Also, the mechanism must have sufficient power to overcome resistance due to impingement of gas from the inflator. An initiator can actuate the mechanism. An initiator is a device that converts an electrical signal into a pyrotechnic event. For a high output, the inflator would be deployed and the mechanism would be actuated at the same time. The mechanism can close the vent in less than two milliseconds, causing all of the gas produced by the inflator to enter the cushion. When only the inflator is deployed, a portion of the gas escapes through the vent slot and the result is low level deployment.

By altering the offset time between the inflator and initiator deployment, the inflation level for a PAV module can be controlled. Figure [1] shows the average pressure curves in a 60 L closed tank for a passenger PAV system with different offset times. Figure [2] shows the average pressure curves in a 60 L closed tank for a driver PAV system with different offset times. As the offset time increases, the tank peak pressure decreases from the high level to the low level. In offset deployments, the tank pressure follows the low level pressure curve until the mechanism is actuated.

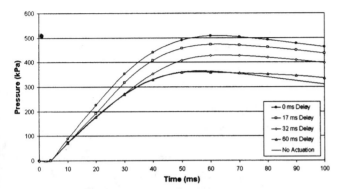

Figure 1. Pyrotechnically Actuated Venting Average 60 L Tank Pressure Passenger Hybrid Inflator

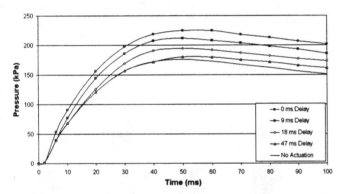

Figure 2. Pyrotechnically Actuated Venting Average 60 L Tank Pressure Driver Non-Azide Inflator

PAV modules have several advantages. In a PAV system, the offset time directly affects the total energy input to the cushion. Since PAV works on a module level, it can be applied to single output inflators that are already validated. From Figure [1] and Figure [2], there is little difference between the low level and a long offset time. This allows the initiator to always be fired to eliminate disposal concerns. The tank testing indicated PAV did not increase the variability of the system.

Although inflator design is simplified with PAV, the module design considerations increase. The design must ensure that the mechanism operates with the same reliability as other adaptive modules. It is difficult to apply PAV to designs where the ports are not located on the circumference of the inflator, and/or the inflator ports are not localized. The ability to tune the low level is limited to scaling the magnitude of the high level tank pressure curve.

TESTING – In order to compare the performance of variable output airbag modules, several tests were conducted. For these tests, a generic mid-sized car environment was simulated. The passenger adaptive modules utilized separate hybrid inflators, a dual heater hybrid inflator, and a single output hybrid inflator with a PAV mechanism. The passenger adaptive modules were compared to a baseline module, which used a single output sodium azide inflator. The driver adaptive modules utilized separate hybrid inflators, a dual heater hybrid inflator, and a single output non-azide inflator with a PAV mechanism. The driver adaptive modules were also compared to a baseline module, which used a single output sodium azide inflator.

One sled test series focused on evaluating the high level performance of the adaptive modules. For the purposes of this evaluation it was assumed that the high level performance should be designed to restrain an unbelted 50th percentile male Anthropomorphic Test Device (ATD) in a high severity crash. The sled pulse approximated a 30 mph barrier impact for a mid sized vehicle. This test simulated the government requirement for U.S. vehicles before changes were made to allow depowering. For the separate inflators, both inflators were deployed at the same time; for the dual heater hybrid, both heaters were ignited at the same time; and for PAV, the inflator and the mechanism were actuated at the same time.

In each test series, the ATD was instrumented to provide the following data: Head Injury Criteria (HIC), neck flexion, neck extension, neck shear forward, neck shear rearward, neck axial tension, neck axial compression, chest acceleration, chest deflection, chest viscous criteria, and femur loads. This data, along with film and kinematics analysis was reviewed to evaluate each adaptive module. In this test series, the neck axial compression data subset best illustrated the overall differences between each module. Other data subsets when considered individually may not reflect the same trend. In this series, the modules that provided less restraint allowed the passenger ATD to penetrate far enough into the cushion to make contact with the windshield. The cushions that allowed more contact with the windshield resulted in higher neck axial compression. Film and kinematics analysis showed all of the passenger adaptive modules were allowed more penetration than the baseline. This could be adjusted by using higher output inflators in each adaptive module and/or tailoring the cushion venting. The passenger ATD neck axial compression numbers for this test series are shown in Figure [3]. The data is graphed in terms of percentage of Injury Assessment Reference Value (IARV). A value below 100% IARV is desirable.

The driver ATD data indicates the separate inflator system had the best correlation to the baseline performance. Film and kinematics analysis showed dual heater hybrid and PAV driver modules allowed more penetration into the cushion by the ATD than the baseline. This can be adjusted by tuning the inflator outputs or module characteristics such as cushion venting. As with the passenger side, the systems that provided less restraint allowed the driver ATD to make contact with the windshield, thus increasing neck axial compression. The driver ATD neck axial compression numbers for this test series are shown in Figure [4].

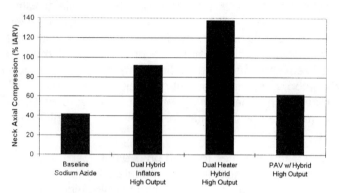

Figure 3. Generic Mid Size Car Hyge Sled Test Data: FMVSS 208 Pulse Unbelted 50th Percentile Male Passenger -Mid Seating Position

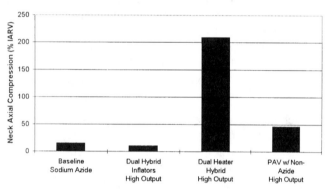

Figure 4. Generic Mid Size Car Hyge Sled Test Data: FMVSS 208 Pulse Unbelted 50th Percentile Male Driver-Mid Seating Position

Another sled test series evaluated the low level restraint of the adaptive modules. For this evaluation, the generic AAMA pulse was used with unbelted 5th percentile female ATDs seated full forward. For the separate inflators, only the primary inflator was deployed; for the dual heater hybrid, only the primary heater was ignited; and for PAV, only the inflator was deployed.

The passenger ATD data indicates good correlation between all of the low output adaptive modules. In this test series, the baseline modules provided a stiffer cushion. This resulted in higher chest and neck numbers for the baseline. In this series, the neck extension data subset best illustrates the difference in performance for each system. The passenger ATD neck extension numbers are shown in Figure [5]. The higher neck extension from the baseline system is due to higher membrane loading from the cushion under the ATD's chin.

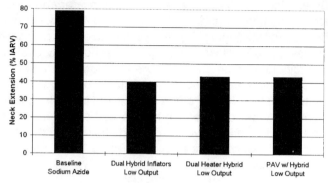

Figure 5. Generic Mid Size Car Hyge Sled Test Data: AAMA Pulse Unbelted 5th Percentile Female Passenger - Seated Full Forward

The neck axial tension data subset provided the best illustration of overall driver module performance in this test series. The driver neck axial tension numbers are shown in Figure [6]. As with neck extension, axial tension is often related to membrane loading from the cushion under the ATD's chin. The stiffer baseline cushion resulted in greater membrane force than the PAV and the separate inflator driver modules. Film and kinematics analysis showed the driver dual heater hybrid modules allowed the driver ATD to penetrate far enough into the cushion to trap the cushion under the ATD's chin. This resulted in a higher neck axial tension than the baseline. The low level performance of the driver dual heater hybrid system can be adjusted by tuning the inflator output and/or modifying the module.

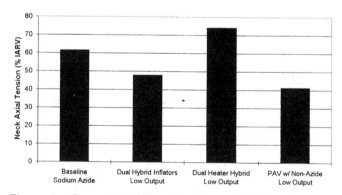

Figure 6. Generic Mid Size Car Hyge Sled Test Data: AAMA Pulse Unbelted 5th Percentile Female Driver - Seated Full Forward

A test series was conducted to evaluate Out Of Position (OOP) occupant performance for each module. The tests were conducted on static fixtures that simulated an OOP 5th percentile female. The passenger ATD was positioned for maximum neck interaction. The driver ATD was positioned for high neck interaction with some chest loading.

The passenger ATD data demonstrates decreased membrane loading for all low outputs over the baseline. All adaptive passenger modules showed decreased membrane loading at low level than at high level. Again, the neck extension data subset most clearly demonstrates the overall performance for each system. Passenger ATD neck extension data is shown in Figure [7].

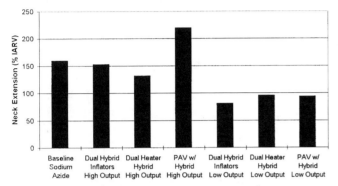

Figure 7. Generic Mid Size Car Out of Position Test Data: Static Fixture 5th Percentile Female Passenger

All adaptive driver modules showed decreased membrane loading at low level than at high level. The low level separate inflator system did not demonstrate and improvement over the baseline system. The neck axial tension data subset best illustrates the performance of each system. Driver ATD neck axial tension data is shown in Figure [8].

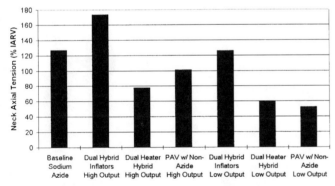

Figure 8. Generic Mid Size Car Out of Position Test Data: Static Fixture 5th Percentile Female Driver

CONCLUSION

Several methods for creating a variable output airbag module were compared. Each method has potential advantages and trade-offs over other systems. With an offset deployment, separate inflator and dual chambered pyrotechnic systems would provide a maximum amount of energy to the cushion late in the event. Normally open PAV systems provide energy to the cushion that decreases from the full level to the low level as offset time increases. The effect of offset deployments in a dual heater hybrid depends on the inflator design.

Tank testing proved Pyrotechnically Actuated Venting can alter the energy delivered to the cushion. The sled test series with the unbelted 50th percentile male ATD indicated that most of the adaptive modules required additional tuning to provide restraint equivalent to the baseline in a high severity scenario. The sled test series with the unbelted 5th percentile female ATD demonstrated a properly tuned low level is capable of reducing the risk of membrane loading to a small occupant in a low severity scenario. The OOP testing indicated low levels also decrease the risk of injury over high levels when the occupant is in close proximity to the airbag. The OOP testing confirmed that by altering the energy delivered to the cushion, PAV is capable of altering occupant performance.

ACKNOWLEDGMENTS

Many people contributed their insight and assisted in gathering data. In particular, Allen Starner, Bob Neiderman, James Webber, and Alex Damman were instrumental in the completion of this paper.

ABBREVIATIONS

AAMA: American Automotive Manufacturers Association
ATD: Anthropomorphic Test Device
IARV: Injury Assessment Reference Value
OOP: Out Of Position
PAV: Pyrotechnically Actuated Venting
SCI: Special Crash Investigations database

980648

Airbag Technology: What it is and How it Came to Be

Donald E. Struble
Collision Safety Engineering

Copyright © 1998 Society of Automotive Engineers, Inc.

ABSTRACT

Since air bags emerged as an occupant protection concept in the early '70s, their development into a widely-available product has been lengthy, arduous, and the subject of an intense national debate. That debate is well documented and will not be repeated here. Rather, operating principles and design considerations are discussed, using systems and components from the developmental history of airbags as examples.

Design alternatives, crash test requirements, and performance limits are discussed. Sources of restraint system forces, and their connection with occupant size and position, are identified. Various types of inflators, and some of the considerations involved in "smart" systems, are presented. Sensor designs, and issues that influence the architecture of the sensor system, are discussed.

INTRODUCTION

When a vehicle crashes, it is acted upon by collision forces that tend to change its velocity. In a direction opposite to these forces, everything in the vehicle that can move does so. This includes the occupants, with their various articulated segments. A restraint system has the purpose of intercepting these motions and managing or eliminating the "second collisions" of an occupant's parts with potential contact surfaces, so that injuries can be mitigated to the extent possible.

An occupant restraint is a system, and works in harmony with the vehicle structure. The restraint system may include an air bag, which itself is a collection of components designed to work with each other and in cooperation with other parts of the vehicle.

RESTRAINT SYSTEMS IN FRONTAL CRASHES

In a crash, potential contact surfaces in the vehicle experience changes in their velocities, and to some extent, their directions of travel. Figure 1 shows the velocity-time history, in an actual crash, for the instrument panel. In contrast, a free particle would keep moving as before the crash. This behavior would be represented in Figure 1 as a horizontal line at 35 mph, in comparison to the descending lines associated with a belted occupant. As time proceeds, the velocity differential between the occupant and the vehicle builds up. One might think of this as an accrual of a "velocity debt," which has to be paid back in order for the vehicle and its occupants to achieve a final common velocity (perhaps zero). In Figure 1, this debt would be the vertical distance between lines representing a potential contact surface, such as the instrument panel, and those representing the occupant. The debt is due and payable when contacts are made between the occupant and one or more contact surfaces. Figure 1 shows that debt management starts earlier for a belted occupant, and that the debt itself is much reduced.

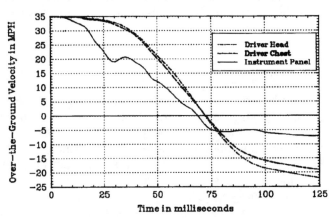

Figure 1. Velocity Histories in 35 mph Barrier Crash.

In this particular vehicle, there was 10.71 inches of space between the sternum and the steering wheel, and 23.18 inches between the head and the windshield, at the beginning of the crash. Figure 2 shows that a free particle would use up these distances in 54 and 75.6 milliseconds, respectively. If an unrestrained occupant behaves similarly, we would expect debt repayment to occur in earnest at about 76 milliseconds, as the head reaches the windshield, and the debt amounts to about the final velocity change, or delta-V (ΔV), of the vehicle.

Of course, these numbers will vary, depending on the crash, the initial position of the occupant, etc. Figures 1 and 2 show estimated occupant velocities and displacements, based on accelerometer data. These indicate that debt management began at about 25 milliseconds for a belted occupant. For an unbelted occupant, the restraint process would have to begin by 50 milliseconds at the latest.

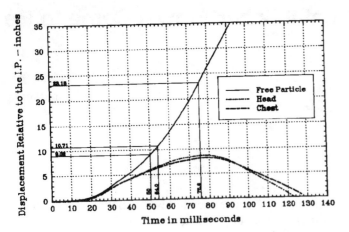

Figure 2. Relative Displacements in a 35 mph Barrier Crash.

The purpose of the restraint system is to intervene early, before the velocity debt becomes unmanageable, and to initiate a repayment program in the form of occupant accelerations - or decelerations (negative accelerations), as the case may be. An early start reduces the debt that has to be paid back, and at the same time, a good restraint system will tend to lengthen the payback period (to perhaps 150 milliseconds or more). If the crash is not excessively severe, the restraint system can keep the debt payments down to a level the human body can tolerate. The action of the restraint system in meting out accelerations to the occupant while the vehicle itself is still accelerating is known as "ride down."

The earliest restraints tended to be lap belts, which were thought to promote occupant retention within the vehicle, but which obviously could not provide much restraint for the upper body during severe frontal crashes. In the U.S., the usage of such systems was disappointingly low, and the prospects for shoulder belt usage were even more discouraging [DOT 78]. Thus, in the late '60s, air bags were seen as forcible intervention (debt management) for the vast majority occupants who would otherwise be unrestrained. This forcible intervention became known as "passive protection," and airbag requirements were promulgated by the NHTSA with unrestrained occupants in mind. The U.S. thus embarked on a technological odyssey in which the occupant protection standard, FMVSS 208, did not require dynamic (crash) testing of the most effective restraint system then and now available - seat belts. Meanwhile, the airbag designer was saddled with the task of protecting those who failed to use their available seat belts, and who would presently be in violation of the law in 49 states.

DEVELOPMENT OF THE AIRBAG CONCEPT

In this historical context, it is easy to see how the concept of air bags came into being: the forcible intervention would come in the form of an inflatable bag which would leap into the gap between the occupant and the vehicle interior. Once there, the bag's internal pressure would be uniformly applied to the occupant over a large area, not only meting out accelerations in controlled doses, but minimizing the stress on the occupant by distributing restraint the forces over as much area as possible. The result would be a more gradual pay down of the velocity debt, ideally with a minimum interest charge (in other words, rebound velocity). Unfortunately, the question of how to accomplish all this would not be a trivial one.

Of course, bags could not be pre-inflated because of the possibility of actually causing an accident. Since inflation had to begin after the crash started, the first step was to determine whether a bag could be inflated quickly enough. Figure 3 shows the time budget used by General Motors in their airbag vehicles of the mid-70s. These were very large sedans, so the time budget for today's vehicles would tend to be shorter, depending on such variables as the vehicle size, object struck, structural engagement, etc. In actual fact, the air bag must be inflated in a considerably shorter time, typically 25 to 30 milliseconds, because at the beginning of the event, the crash detection system requires a certain amount of time to do its work.

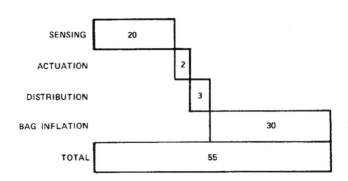

Figure 3. Typical Time Budget for Large Cars -GM

In addition to filling the air bag quickly, the inflation gas itself should not present hazards to the occupants or any one else involved with the vehicle from production to salvage. High-pressure air, stored in a tank and released with a quick-action valve, probably seemed an obvious choice to early airbag developers.

DRIVER-SIDE AIRBAG SYSTEMS

Design of a driver-side airbag system is heavily influenced by the available volume in the steering wheel hub. This volume is entirely insufficient to house a high-pressure tank, stowed air bag, and other hardware. Consequently, the only way to implement a stored-gas system on the driver side was to locate the tank elsewhere, and pipe the gas up the steering column. A prototype system, in which a single bottle inflated both driver and passenger systems, was developed by Volkswagen for its Type I Beetle, but without satisfactory results [Seiffert 72]. At Ford, after a long development effort to achieve adequate inflation time, a cooperative program was started with the automotive supplier Eaton, Yale, and Towne to develop a system for the right front passenger only [Frey 70]. The justification for this approach was that the driver was already protected by the combination of compressible steering column, padded hub, and yielding rim; it was the right front passenger who had the greatest need of an air bag. (On the other hand, there is almost always a driver present, whereas this seat is occupied only about 40 percent of the time, reducing the available benefit there.) In any case, a fleet of 831 1972 Mercury Montereys was fitted with passenger-side air bags for field testing, of which 126 were delivered to the U.S. Government.

SOLID PYROTECHNIC INFLATORS – In the meantime, General Motors and Chrysler were trying to find a gas source that could be fitted in the steering wheel hub. Chrysler's efforts utilized smokeless powder, but did not result in hardware being integrated into vehicles. GM, on the other hand, developed a driver system using the combustion of sodium azide to produce inflation gas. Sodium azide is a highly toxic and unstable compound and thus requires special care in production and disposal, but it has chemical reaction characteristics suitable for inflating air bags, and its primary combustion product is harmless nitrogen gas. Sodium azide has thus been the main ingredient of gas generants for driver's systems from the earliest times until the present, although it seems destined to be replaced by more environment-friendly materials.

Gas generant comes in solid form and is like a solid rocket propellant, in that the chemical reaction rate at any instant depends on the exposed surface area and the temperatures created by the reaction itself. These are controlled by the size, shape, and sodium azide content of the "grain" -- a rocket propellant term. Typically, the grain is about 60 per cent sodium azide, it is packaged in pellet form, and there is 75 to 100 grams of it in a driver's system.

The sodium azide pellets are part of a unit known as the inflator. The inflator housing is hermetically sealed to keep the pellets isolated and protected from vandalism until they are needed. Holes around the circumference of the housing allow the nitrogen gas to escape and provide axisymmetric bag filling, without creating any net thrust (for safety during handling). The inflator also contains a labyrinth of screens and perhaps baffles, to filter out particulates and to cool the hot nitrogen gas. Finally, the inflator contains a squib with electrical connections, so that the ignition of the pellets may be started. Figure 4 shows an inflator produced by Morton Thiokol.

AIRBAG MODULE – The inflator is part of a larger assembly called the airbag module, or simply module, which is the unit that actually gets installed in the vehicle. In addition to the inflator, the module contains the stowed air bag, which is securely fastened to the side of the module away from the occupant, so as to contain the pressure of the nitrogen gas when inflated. The exterior surfaces of the module include the bag cover, which is typically plastic. The cover has molded-in lines where the material is weaker, allowing it to be split by the pressure of the inflating bag. The cover opens up, typically like petals of a flower, allowing the air bag to unfurl. A typical airbag module, installed in an airbag steering wheel, appears in Figure 5.

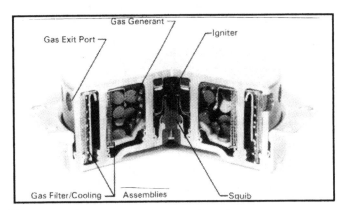

Figure 4. Typical Driver System Gas Generator - Morton Thiokol

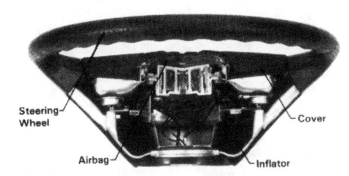

Figure 5. Driver Airbag Inflator and Module Assembly.

Clearly, the airbag module occupies precious real estate in the steering wheel hub, and significantly increases its mass. The volume must be kept to a minimum so as to avoid blocking the driver's view of the instrument cluster, or being in the way during emergency steering maneuvers. The added mass adds to the rotational inertia of the steering wheel; since this could adversely affect steering system returnability, the module mass must also be minimized. These factors have fostered the development of lighter and smaller inflators, and the use of thinner airbag material (so that the stowed air bag can be smaller and lighter).

The airbag cover must be durable, since it is often contacted by the driver. Since it is one of the most prominent features of the driver's station, it is important to the visual appeal of the vehicle, so the bag cover must also be attractive. It must protect the air bag from moisture, spilled liquids, etc. Most importantly, it must not impede the unfurling of the air bag, nor pose a hazard to vehicle occupants during deployment.

ELECTRICAL CONNECTIONS – Where the steering wheel meets the column, there is relative rotational motion between the airbag module electrical leads and the stationary wiring harness in the column. In the first-generation GM system, electrical contact throughout the steering wheel motion was maintained via a special slip ring assembly. Redundant electrical contacts were used to handle threats to reliability posed by electrical noise due to friction. In the 1985 Ford Tempo/Topaz, and in many later designs, a spiral wire (like the main spring in a clock), having enough travel to accommodate steering wheel turns from lock to lock, was used. A down side of the clock spring design is that repair personnel must be sure it is properly "rewound" when a steering wheel is replaced.

STEERING WHEEL – The steering wheel is not only the "launch pad" for the air bag; it (and to a lesser extent, the windshield) also serves as the reaction surface. This terminology means that restraint forces developed in the air bag itself have to go somewhere, and the primary load path is through the steering wheel and into the column. Consequently, the design of an airbag wheel goes beyond merely accommodating the volume, mass, and electrical connections of the module. The strength and stiffness of the spokes and rim must be sufficient to provide a stable reaction surface, but as yielding as possible in non-deployment accidents involving unbelted occupants. The General Motors first-generation airbag wheel was the result of a considerable development effort, and seems to have served as the point of departure for subsequent airbag steering wheel designs.

STEERING COLUMN – In general, steering columns are designed to limit rearward displacement (relative to the compartment) in frontal crashes, as regulated by FMVSS 204, and they are also designed to limit the forces due to occupant contact, by being able to stroke forward in the event of such a contact. In this sense, the steering column is part of the restraint system, even in a non-airbag vehicle.

In an airbag vehicle, however, contact is deliberately made with the driver. If the driver is unbelted (as was assumed to be the case in the early 18

70s), virtually all of the upper body restraint forces pass through the column. A conventional column might stroke forward under these conditions, but at such a low force that the stroking element reaches the end of its available travel and abruptly "bottoms out" against some mechanical limit. An airbag column, on the other hand, is designed to move forward in a controlled fashion, which means that its inertia, and static friction in the column, must be dealt with carefully. Typically, this is done by starting the static stroke vs. force characteristic at a reduced level (perhaps by "pre-stroking" the energy-absorption unit), and then increasing the static force as stroke progresses. This generally means a redesign of the stroking element, and a strengthening of the cowl structure and bracketry through which the column passes

the loads on to the rest of the vehicle. GM's structural modifications are shown in Figure 6.

The legacy of research done at GM in the mid- to late-'60s on energy-absorbing steering columns is clearly seen in the design philosophy of their first-generation airbag systems. The steering column was the primary energy-absorbing element; the air bag's function was to leap into the gap between the driver and the steering wheel, couple the driver to the mechanical energy-absorbing elements in the column, and provide a more uniform application of restraint loads on the occupant's upper body.

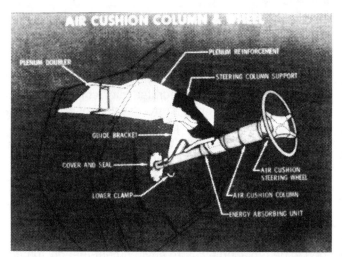

Figure 6. Structural Modifications Required for Driver System - GM

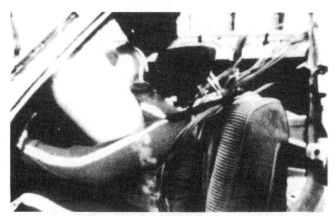

Figure 7. Belted Occupant Kinematics - Mercedes-Benz System

The column is aided and abetted in performing this duty if the torso is perpendicular to it. If a driver has a lap belt, the lap belt will limit the forward stroke of the pelvis and cause the upper body to pivot forward, thus helping to achieve the desired alignment. See Figure 7. For unbelted drivers, however, this does not occur; the whole body tends to translate forward with little articulation until interior contacts are made (typically, with the knees first). In this situation, we may well have a mostly erect torso moving straight forward toward an inclined steering wheel. As seen in Figure 8, a GM illustration of its first-generation airbag system, this geometry can cause the air bag to develop a wedge shape in side view. More importantly, it can cause a significant upward force component to be applied to the end of the column. See Figure 9.

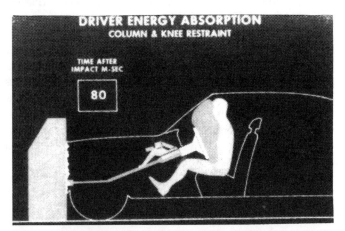

Figure 8. Unbelted Driver Kinematics - GM System

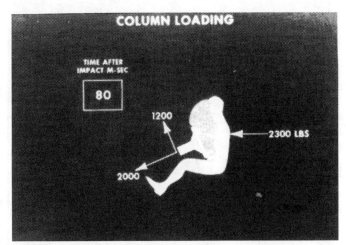

Figure 9. Column Loading - First-Generation GM System

These (off-axis) up-loads considerably increase the force levels required to stroke the column. The situation is rather like trying to close a chest of drawers by pushing on one side of the drawer. It may "jam," and resist closing altogether. To overcome such difficulties, one tends to design the stroking element for greater lateral stiffness, and less sensitivity to the minor misalignments that result from off-axis loads. An example of such a design is the steering column found in the Minicars Research Safety Vehicle (RSV), shown in Figure 10 [Struble 79]. Despite the lack of front-seat belt restraints, this system achieved dummy injury measures well below the FMVSS 208 criteria set for 30 mph, at delta-Vs in excess of 50 mph. At these speeds, one needed to use all the interior occupant stroking space that one could get. Reducing the articulation of the body segments (recall the lack of a lap belt) tended to keep the head away from the windshield and header, and was helpful in increasing the available stroking distance.

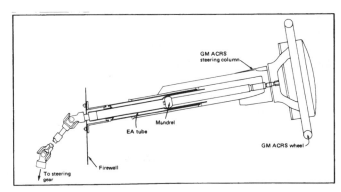

Figure 10. Minicars RSV Wheel and Column

Another approach to dealing with column up-loads is to reduce the angle between the torso and the steering wheel. For belted occupants, this is achieved by the lap belt, as described above. For unbelted occupants, particularly in the extreme crash conditions addressed in the RSV program, the need to reduce the angle resulted in a steering column rather more horizontal than most. Of course, such architecture causes the steering shaft to penetrate the dash panel at a water line much higher than the steering gear, so an intermediate shaft with double U-joints was required to make the connection. It is unknown how most drivers would have liked the more vertical steering wheel, or the closer positioning to the sternum.

<u>Air bag</u> – A third approach to up-load difficulties is to reduce the dependence on column stroking, which brings us to the design of the air bag itself. The first-generation GM air bag was about 22 inches in diameter when deployed, and extended about 10 inches rearward of the wheel rim [Campbell 72]. It was pressurized to approximately 3 psi, absent occupant loading, and was about 2.8 cubic feet in volume. This was larger than most current designs, such as the Mercedes-Benz bag shown in Figure 11, which tend to run about two cubic feet, or 60 liters, in volume.

Figure 11. Inflated Mercedes-Benz Air Bag

The GM bag material itself was neoprene-coated nylon. The neoprene coating served to reduce the porosity of the bag, thus extending the duration of bag inflation. It was discovered, however, that the bag tended to act as a pneumatic spring, storing energy as the occupant stroked forward into it, but then returning the energy to the occupant later [Klove 72]. This resulted in undesirable amounts of occupant rebound into the seat back and head restraint. To counteract this tendency, the neoprene coating was subsequently removed; forcing gas through the fabric pores increased the energy dissipated (as opposed to energy being stored and subsequently returned).

Subsequent bag designs have included actual holes, or vents, which took on a larger share of the task of absorbing occupant energy. The vents are generally placed on the back side of the bag, away from the occupant, and are so located as to avoid blockage (by the steering wheel spokes, instrument cluster brow, windshield, etc.). Figure 12 is typical. The total vent area is determined during the development process so as to provide the best compromise among the disparate demands being made on the restraint system.

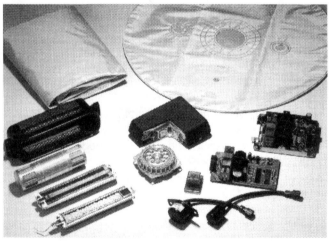

Figure 12. Typical Components, Including Vented Air Bag - Takata

As production volumes have increased, the bag has become perhaps the most labor-intensive component to manufacture and inspect. Typically, it has at least two major sections, which have to be sewn together in the traditional manner, by a worker at a sewing machine. In addition, reinforcements have to be added at the vent holes, the edge or edges where it is attached to the module, and at the tethers (if any). More recent developments indicate that bags may be woven, by machine, in a single piece.

Tethers are straps that connect the front and back surfaces of the bag. They have become more widespread in recent years, and their purpose is to limit the travel of the bag front during deployment. For small occupants seated close to the steering wheel, tethers can reduce the occupant accelerations generated when the bag front contacts the sternum. Tethers also increase the bag's aspect ratio (diameter divided by depth), improving the ability of the bag to protect occupants who load one side of the bag more than the other (due to angularity of impact or being out of position, for example).

The air bag used in the Minicars RSV had two chambers, as indicated in Figure 13 [Struble 79]. The inner chamber was connected directly to the inflator and filled first. Its relatively small size (28 l, or 1.0 cubic feet) provided a quick fill time, so it could jump into the gap between the steering wheel and the chest as soon as possible. It was aimed directly at the sternum, and this geometry was instrumental in establishing the column angle. The chest bag was vented to the outer bag (48 l, or 1.7 cubic feet), which filled more slowly and provided restraint to the head. The timing of the chest and head bag inflation could be tuned somewhat by the size of the vent in the chest bag. Once the gas had been re-used in the head bag, it was vented to the atmosphere.

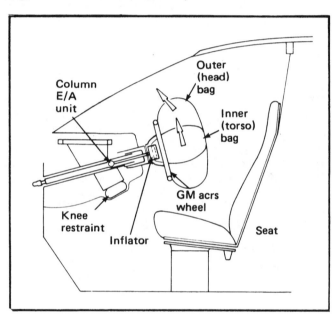

Figure 13. Minicars RSV Driver Airbag System

While the air bag inflates faster than the blink of an eye, a look at high-speed films will reveal that it is hardly instantaneous. Indeed, there is an inflation sequence, and a review of a bag pressure time history, as seen in Figure 14, will reveal some distinct phases. First, there is a relatively high pressure spike of relatively short duration, associated with inflation gas being pumped into an extremely confined space behind the folded air bag. The bag is pressed against the bag cover with sufficient force to split the cover seams, and the cover opens. In very short order, there is a volume increase behind the air bag - a high percentage increase because the volume was so low to start with. Consequently, the bag pressure drops, typically to zero. This is known as the punch-out phase, because the bag is punching out through the airbag cover [Lau 93]. In the test from which Figure 14 was derived, the occupant was not close enough to interfere with bag deployment, and the punch-out phase lasted about nine msec. If the bag has to move the occupant

300

out of the way to achieve fill, both the peak pressure and the duration of the punch-out phase can increase.

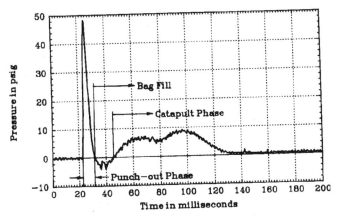

Figure 14. Airbag Pressure-Time Curve

The end of the punch-out phase marks the beginning of the bag fill phase. Now the airbag material has some velocity, and hence some momentum, as the cover doors swing open. Thus the fabric keeps moving, and as a result a negative gage pressure is created in the bag. During this time, the bag is seen in high speed films to be unfolding rapidly, with numerous sharp creases in the fabric. Bag unfurling motions are highly complex and three-dimensional, and the pattern depends on the folding process. The lateral portions of the bag may even appear to be sucked in. Of course, the inflator is continuing to produce gas throughout this time.

Finally, the bag material reaches the geometric limits of its travel, and the pressure climbs back up through zero. This is the earliest point at which bag pressure is available as an occupant restraint mechanism (just over 45 msec in Figure 14). In very short order, the creases come out of the material, and the bag assumes its inflated shape. The actual time budget used in the design for the system of Figure 14 is unknown, but it appears that 50 msec would have been an attainable goal.

AIRBAG PERFORMANCE – Airbag design is driven by a number of considerations, among which are the usual villains of cost, weight, and size. Occupant protection performance is dictated by the requirements of FMVSS 208, which has specified that dummy injury measures for the head, chest, and femurs be within certain limits for 30 mph frontal barrier crashes at any horizontal angle up to plus or minus 30 degrees from vehicle center line. FMVSS 208 has required that tests be run with the dummy occupants, representing 50th percentile males, unbelted. Generally, the angled barrier crashes involve longer stopping distances and softer crash pulses (vehicle acceleration versus time) and in that sense are less severe, but they do pose potential difficulties associated with the occupant moving into the bag at an angle. Another crash condition is the 35 mph frontal barrier test used in NHTSA's New Car Assessment Program (NCAP), in which the dummy occupants are belted. This test is not required by the safety standards, but it has nevertheless become a de facto design requirement. These dichotomous test conditions mean that a single bullet (i.e., one system design) has to be fired at two targets. It is thus not surprising to find, in the earlier airbag designs NCAP-tested at 35 mph, that the FMVSS 208 injury criteria were exceeded. In fact, while some vehicles equipped with belts only have been meeting all the 208 criteria since the beginning of the 35-mph NCAP tests, it was not until 1988 that an airbag-equipped vehicle did so.

LOWER BODY RESTRAINT – The lack of a lap belt in the 30 mph test means that lower body restraint has to be provided by other means, typically by resisting the forward movement of the knees. The hardware involved is variously known as a knee bolster, knee restraint, or knee blocker. In any case, the design concept involves the knees engaging a deformable structure which limits knee movement to some extent, while maintaining femur loads within acceptable limits. These femur loads are transmitted to the pelvis, providing pelvic restraint. Attention has to be paid to the possibility of knee contact with the steering column, particularly during angular impacts.

Needless to say, the action of the knee restraint, and its effect on occupant kinematics, depends on the initial spacing between the knees and the restraint. Generally, the designer would like the spacing to be small, but care must be taken not to interfere with the operation of the foot pedals -- particularly the brake pedal. On the other hand, the need to reach the pedals tends to cause the driver to position the seat so that the knees are placed at a fairly uniform distance from the knee bolster, regardless of occupant size.

In a 35 mph NCAP test, pelvic restraint provided by the knee bolster, when combined with a lap belt, could be excessive. One option for dealing with this is to sew in one or more loops in the lap belt, which can pull out at a force level sufficient to provide occupant retention in non-frontal accidents, but allow enough pelvic motion to avoid unacceptable occupant kinematics. Webbing material with varying stretch characteristics can also be chosen. It's a bit of a balancing act.

UPPER BODY RESTRAINT – Similarly, the non-use of the torso belt in the FMVSS 208 test results in all the upper-body restraint being provided, in that test, by the air bag. The safety standard therefore effectively establishes the force and stroke requirements for the air bag, which may not be optimal for smaller occupants seated closer to the bag (see below). Even for a normally-seated 50th percentile male, the addition of a shoulder belt in the 35 mph NCAP test may mean an excessive amount of restraint force in that test, and indeed this could be the cause of the chest accelerations exceeding 60 Gs in NCAP tests of the earlier airbag cars. One approach to dealing with this situation is to adjust the timing of restraint forces from the air bag and the belt so that the peaks do not coincide.

An insight into this timing is provided in Figure 15, which shows the timing of chest accelerations and torso belt loads. We see that the belt loads reach their peak at about 50 msec, which closely corresponds to time budgeted for airbag inflation that was mentioned earlier. After 50 msec, we see the torso belt loads falling off while the chest accelerations continue more or less level until about 80 msec. The likely explanation is that the timing of the belts and airbag inflation has been adjusted so that the air bag picks up where the torso belt leaves off, in terms of providing upper torso restraint.

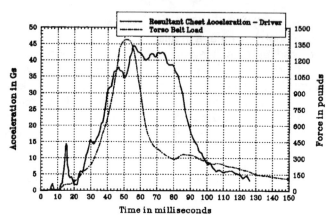

Figure 15. Belt Load and Chest Acceleration in 35 mph Barrier Test - 1994 Volvo 850

VARIABLES AFFECTING PERFORMANCE LIMITS – There are other variables not addressed by government standards or tests. Factors considered by the manufacturers probably vary, but some that come to mind are occupant size (5th percentile female through 95th percentile male), distance between the airbag cover and the sternum, and object struck (e.g., pole, offset barrier, etc.). Generally, protection of larger occupants is stroke-limited, which is to say that as crash severity is increased, some specified injury criteria limit is reached when the occupant stroke becomes excessive, resulting in bottoming out of the air bag or steering column, or contact with the interior. Protection of smaller occupants tends to be acceleration-limited. This is because restraint systems have to generate enough force to protect the many larger occupants in the population. In the same crash, these same force levels applied to smaller occupants will result in larger occupant accelerations. Smaller occupants may thus reach acceleration limit values at lower crash severities than larger occupants will.

When the object struck is not a barrier, the crash pulse may be softer, but compartment intrusion may also be greater than in a barrier crash at the same speed, due to the concentration of crash forces on only a part of the structure. This could be reflected in displacement or rotation of the steering column, which could affect the "aim" of the airbag restraint forces, and the resulting occupant kinematics. It could also affect the sensing time, possibly producing a later deployment. This is a function of the sensor system design, to be discussed later.

If the driver is sitting closer to the air bag than in the design condition, contact between the deploying bag front and the occupant can occur earlier, and at a higher contact velocity. This could increase the accelerations experienced by the occupant, particularly if he or she is small.

DESIGN PARAMETERS – To deal with all these requirements, some of them in conflict with one another, the designer and developer of airbag systems has a number of parameters to work with. The crash pulse is an important one, albeit one that the airbag engineer may have little control of. Another one is the inflator characteristic, which is generally expressed in terms of the time histories of pressure when the inflator is discharged into a fixed, closed volume (the so-called "tank test"). Typical tank test results are shown in Figure 16. Other variables include the column stroke characteristic (force vs. distance), steering wheel location, airbag vent area, bag volume, bag diameter, tether length, seat belt anchorage points, webbing stretch characteristics, and seat cushion stiffness.

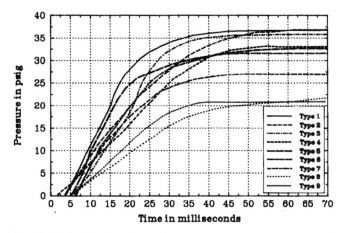

Figure 16. Tank Test Curves for Morton-Thiokol Inflators - 60 l, 22°C

PASSENGER-SIDE AIRBAG SYSTEMS

Perhaps the feature that most consistently distinguishes passenger-side systems from their driver-side counterparts is the location of the launch pad: driver systems are mounted on the steering wheel, and passenger systems are mounted on, and part of, the instrument panel. This distinction causes a significant difference in the longitudinal distance between the airbag mount and the occupant's sternum. Moreover, the passenger side lacks a stroking element, like the steering column, that could be used for absorbing energy.

One approach to transferring driver's side airbag technology to the passenger's side would be to blur these distinctions. In other words, move the launch pad aft and mount it on a stroking element. This concept led directly a design using a "bag-bolster" positioned rearward about even with the steering wheel. A bag-bolster system was sled-tested at Calspan [Romeo 75] at 75 kph (47 mph), but concerns remained regarding public acceptance of the appearance and ease of ingress and egress of such a design. Subsequent concepts for a passenger-side air bag have avoided significant changes in the instrument panel location.

In the Minicars RSV program, the objectives for the right front passenger system included the ability to undergo a crash at 50 mph delta-V, and still provide occupant protection within the 30 mph FMVSS 208 criteria. Originally, it was thought that a stroking dash (albeit conventionally-positioned) would be required to meet such ambitious goals. However, it turned out that the performance goals could be achieved, with room to spare, using venting alone. All other passenger-side airbag systems, as far as is known, have similarly relied upon venting for energy absorption.

BAG GEOMETRY – In the development process, it may be tempting to extrapolate from a driver side system by starting with a deeper (in the longitudinal direction) version of a driver bag, mounted on the instrument panel. This approach would not be valid because the aspect ratio of such a design would not be nearly high enough; the bag would buckle or be pushed aside by the occupant's motion. The bag needs to be wider (laterally) and taller (vertically) for two reasons: (1) to avoid instability (buckling) when loaded in compression, and (2) to handle the greater lateral deviation in an occupant's path during an angular collision, due to the occupant being farther from the bag when the crash event starts. This is true even when there is not a designated center seated position in the front seat. If there is a center position, and airbag protection is provided for that occupant, the bag must be wider still.

These considerations lead to a considerably larger air bag. The first-generation GM bag, designed to protect both center and right front occupants, had a volume of about 14 cubic feet. Most present bags are much smaller. The inflated shape is typically a lateral cylinder with vertical ends, but the cross-section is not necessarily circular. See Figure 17.

The Minicars RSV passenger-side air bag was dual-chambered, for reasons that can be understood by comparing it to its counterpart on the driver's side. Refer to Figure 18. In this design, the inflator emptied directly into the lower, or torso bag, where earlier application of restraint forces was needed. This bag had a volume of about 2.75 cubic feet (78 l). The torso bag was vented to the upper, or head bag, to pick up the head somewhat later. The volume of the head bag was about 3.0 cubic feet (85 l). Gas from the head bag was then vented to the atmosphere.

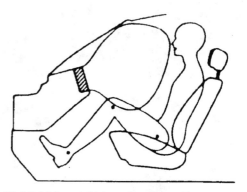

Figure 17. Mid-Mount Airbag Configuration

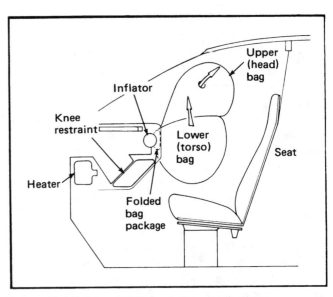

Figure 18. Minicars RSV Dual-Chambered Passenger Airbag System

The first-generation GM system also had a dual-chambered bag, but for entirely different reasons. One chamber was intended specifically for knee restraint. Since knee restraint loads are more concentrated than, say, the loads applied to the chest or the head, the knee bag operated at a much higher pressure. Due to its smaller size, it filled first. The other chamber, being larger, filled relatively more slowly and provided restraint to the torso and head. See Figure 19.

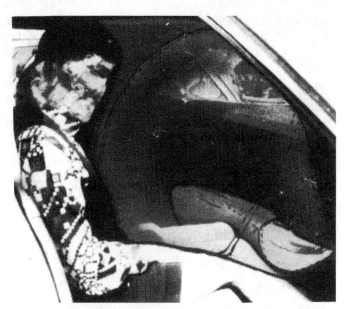

Figure 19. First-Generation GM Low-Mount Airbag with Knee Bag

TYPES OF AIRBAG MOUNTS – Compared to driver systems, passenger air bags have many more design options regarding their integration with the vehicle. The location and orientation of the module gives rise to some terminology regarding the type of mount. At the time it was designed, the Minicars RSV air bag was called a high-mount system, but as indicated in Figure 17, it would be called a mid-mount configuration in today's parlance. The airbag module is located on the aft face of the instrument panel, about where a glove box might traditionally be located. In this position, the air bag cannot, and is not intended to, provide lower-body restraint. The Minicars system was typical of many in that it was designed to be tested without the use of lap belts. Thus the lower-body restraint had to be provided by other means, such as a knee bolster. The module and the knee bolster make it very difficult to provide a traditional glove box in the instrument panel. Obviously, there is also a significant impact on the location and routing of heating, ventilation, air conditioning, and other components in the instrument panel. The bolster itself is similar in concept to those used on the driver's side, except that the designer does not have to deal with potentially hard contact surfaces presented by the steering column.

In another contrast to the driver's side, there is no requirement for the passenger to reach the pedals. For bench-seat vehicles, the passenger seat position may be determined by the driver. With bucket seats, the position may be a function of who is seated in the right rear seat, and how big they are. Variability in the size of the right front passenger provides (literally) another dimension. In any event, one can expect considerably more variation in the knee-to-knee-bolster distance at the beginning of the crash. A small occupant, seated far from the bag, could tend to submarine under the bag.

General Motors, in the design of their mid-1970s system, used an inflatable knee restraint to provide added tolerance to such variations. (Recall the very large size of the vehicles involved.) To effect the design, the airbag module was located lower on the instrument panel in what has come to be known as a low-mount configuration, illustrated in Figure 20.

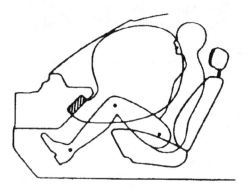

Figure 20. Low-Mount Airbag Configuration

A third design is known as a top-mount (or dash-top) configuration. In this system, the deploying bag is not directed aft at the chest or down at the knees, but rather upward toward the windshield. The airbag cover is typically on the top of the instrument panel, as seen in Figure 21. Top-mount designs were proposed in the early '70s, but were not used in either the GM or Minicars systems. More recently, this configuration has become widely used, for reasons to be discussed below.

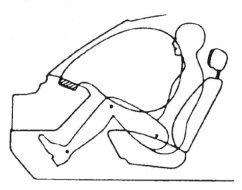

Figure 21. Top-Mount or Top-Dash Configuration

HOW RESTRAINT FORCES ARE DEVELOPED – When the occupant presses against an air bag, the pressure in the bag is transmitted directly across the layer of fabric, onto the occupant's body. Restraint forces are thus generated, but it would be a mistake to attribute all such forces to this mechanism (gas pressure).

Consider, for example, that with a 50th percentile male dummy and the seat in its middle position, there may be 560 mm (22 in) between the chest and the instrument panel. If the air bag is to fill this gap in 25 msec (say), the deploying air bag surface must move rearward at an average speed of at least 50 mph. Of course, the instantaneous speed varies, so one would expect the peak to be much higher. Indeed, film analysis of the deployment of various driver-side systems showed peak speeds ranging from about 100 mph to over 200 mph [Kossar 92]. Speeds at the time of facial contact were lower, of course, and would depend on where the seat is positioned.

When the air bag contacts the occupant, there is a momentum transfer between the two that depends on the portion of the air bag's mass brought to rest against the occupant, its velocity at contact, and the portion of the occupant's mass involved in the contact. This phenomenon is called bag slap, and can generate restraint forces when the bag pressure is low or even negative. See Figure 22. If multiple layers of fabric are involved (as when the bag is still partly folded, for example) and the brunt of the impact is taken by just the sternum or the head, the increased effective mass of the bag and the reduced effective mass of the occupant will combine to produce higher occupant accelerations. Obviously, if the occupant is initially positioned closer to the air bag, contact occurs sooner, earlier in the bag unfolding sequence, and possibly at a higher contact velocity.

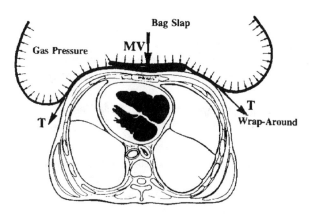

Figure 22. Sources of Restraint Forces

Clearly, bag slap is a greater concern for occupants having smaller mass, seated closer to the bag, and so positioned that the deployment forces are directed higher on the body.

Once momentum has been transferred to the occupant, the bag slap phase of the restraint process gives way to the catapult phase, wherein the occupant and aft bag material move together. As the occupant strokes into the bag, the pressure builds and generates restraint forces, as mentioned before. At the same time, membrane tension is created in the bag fabric, which is partially wrapped around the occupant. Because of occupant penetration, the bag tension has a rearward component at the locations where the fabric and the occupant cease to be in contact. These rearward components, or wrap-around forces, contribute significantly to restraint action during the catapult phase. This is particularly true on the passenger side, because the bag tends to be wider and deeper.

EFFECTS OF OCCUPANT SIZE AND POSITION – Since passengers don't operate foot pedals, they can choose to sit closer to the instrument panel (or farther away, of course) than they might on the driver side. In fact, the position of a bench seat may be controlled by the driver; with bucket seats, the position might depend more on the presence and size of a rear occupant than anything else. At the same time, the front passenger doesn't have to be of driving age. One finds, therefore, a much wider range of occupant sizes, weights, ages, and seat positions on the passenger side.

If the occupant is short in stature, the bag may be directed more at the head and neck, and less at the torso. In this way, the relatively small effective mass of a child, for example, could be reduced still further. If the membrane tension is sufficiently high, coupled with a small effective mass, the bag fabric may actually snap taut, propelling the head rearward.

If restraint performance is optimized for a normally-seated 50th percentile male with the seat in mid-position, one must expect some degradation of performance when any, or some combination of, these conditions is varied. Clearly, a system acting as if an unbelted 160-pound occupant is present is going to be much too powerful to be optimal for a child weighing a fifth as much. While the Minicars RSV system may have presented the most potential for occupant protection at the highest severities, at the same time it may also have posed a higher risk to small and out-of-position occupants. Short of tailoring the restraint system for the conditions (see below), the best one can do is make the greatest accommodation for the occupants most at risk (e.g., children, small adults, the elderly), while keeping performance for the "nominal" occupant and seat position within the limits imposed by Government testing.

Within a given airbag configuration, one can vary such parameters as the inflator charge, the angle and position at which the inflator is mounted, the fabric weight and vent area, and the folding pattern. Obviously, lighter fabric weight tends to reduce bag slap and stowage volume. Typically, the folding pattern has been found to be very important in addressing the needs of small or out-of-position occupants.

Another approach tried in the '70s was the so-called aspirated system [Katter 75]. This design concept stemmed, at least in part, from early concerns regarding overpressure in the compartment due to bag deployment. (Consider, for example, that the U.S. ESV family sedan designs of the early '70s had air bags in both front and rear seats [Alexander 74].) It was thought that overpressure might be alleviated if a supersonic ejector concept could be adapted to pump compartment air into the air bag. Such devices worked through viscous mixing, in a diffuser, of the air streams from the primary source (the inflator) and the secondary source (the compartment). The secondary flow continued until the bag pressure became high enough to stall the diffuser, at which time a check valve in the secondary air stream must close. Since the occurrence of stall depended on what the bag contacted, the system would naturally adjust the degree and rapidity of fill if the occupant were closer to the bag. Another advantage was the reduced inflation requirements for the inflator.

As it turned out, bag folding technique had a much larger influence than aspiration on the results for out-of-position children [Romeo 78]. As of this writing, there has been no known further development of aspirated systems.

TYPES OF INFLATORS – The volume of a passenger-side air bag is much larger than that of the driver-side bag, so one might expect a passenger-side inflator to need proportionately more gas generant. Of course, the packaging constraints are altogether different, meaning that there is little incentive to make a passenger-side inflator look much its driver-side counterpart. Rather, it is typically a circular cylinder, mounted so that its long axis is lateral to the vehicle. As with a driver-side inflator, it typically has a hermetically-sealed metal housing, for the same reasons. Generally, nitrogen gas passes through a complex of screens and perhaps baffles, and exits through a series of holes or slots. These openings are evenly distributed from one end of the cylinder to the other, so as to provide an even fill across the width of the air bag.

Of course, other inflator configurations are possible. In the late '70s, when the inflator business was at its nadir, Calspan Corporation developed a design using two ganged driver inflators to fill a passenger bag [Romeo 78]. The primary motivation was the lack of a production passenger inflator, but utilizing two driver units with a common manifold has certain other advantages, discussed below.

Finally, we return to the type of inflator thought of first -- stored gas. Packaging constraints permit the storage bottle to be sufficiently close to the air bag to allow rapid inflation, but the same constraints cause the pressure in the bottle to be on the order of 3000 pounds per square inch. (The smaller the bottle, the higher the pressure required to store an amount of gas sufficient to inflate the bag.) Such pressures raise concerns about leakage (and thus reliability), when one considers that the system must remain absolutely leak-proof for perhaps 20 years. Nevertheless, the mid-70s passenger side systems by General Motors utilized stored-gas inflators, and successful deployments have occurred in cars of relatively advanced age.

The passenger-side module, shown in Figure 23, employed a membrane that was pierced to start the inflation process, and a manifold to distribute the inflation gas to the air bag. A characteristic of stored-gas systems is that the pressure in the tank is at its highest at time zero; as gas escapes from the tank (at the speed of sound), the remaining volume expands adiabatically (without heat loss), and as it does, the temperature drops. At the same time, the tank pressure drops.

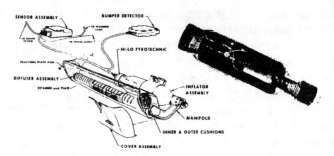

Figure 23. First-Generation GM Passenger Air Bag, with Inflator

The higher the storage pressure and the smaller the tank volume, the more steeply the pressure declines. At the same time, the peak noise level increases [Jones 71]. Questions have been raised regarding hearing loss, but the concerns seem to have subsided with the advent of pyrotechnic inflators on the passenger side. This is because pyrotechnic systems provide a more even gas flow, which reduces to some extent the concern about inflation noise. Because the gas generant grain can be modified to some extent without changing the inflator housing, the gas flow is more readily tailored or adjusted than with a stored-gas system.

At the same time, the combustion temperature in a pyrotechnic inflator is much higher than ambient, such that ambient temperature variations do not have much effect on gas generation. Stored-gas systems are thought to be more sensitive in this regard.

Since the inflation gas expands as it passes through the opening and into the air bag, it cools (and in fact gets very cold). If the gas pressure within the bag were insufficient, it could be increased by heating the gas. This thought gives rise to the concept of an "augmented" or "hybrid" inflator, in which heat is applied to the gas on its way to the air bag.

The GM passenger-side systems of the mid-1970 utilized this concept. A small charge of pyrotechnic materials could be ignited, not for the purpose of materially increasing the amount of inflation gas, but rather to add energy to the inflation gas by raising its temperature, and causing it to expand more. The increased temperature reduced the sensitivity to variations in ambient temperature [Seiffert 72].

DUAL-LEVEL INFLATORS – The engineers at Calspan were not the only ones to design a passenger system using two driver inflators. A 1979 paper described a Mercedes-Benz design in which "incremental deployment" of one or both inflators could advantageous in low-speed impacts, or for out-of-position occupants or children [Reidelbach 79]. The wording of the paper suggests that such a feature was not implemented, nor was it at Calspan. With GM's design, however, the choice could be made between a "low-level" deployment in which just the stored gas was released, or a "high-level" deployment which also involved the ignition of the augmented inflator charge.

Of course, to make a choice there must be some logic employed, and there must be sensors to provide the inputs. With the GM system, the choice involved the nature of the crash pulse. GM's design employed two impulse detectors on the bumper near the frame attachments, plus sensors on the dash panel. Generally, the trigger level was lower for the bumper units than for the dash panel sensors. Lower-speed crashes, underrides, center pole impacts, etc. might cause the lower-level units to trigger, but not necessitate a maximum-level inflation (as would be achieved by igniting the augmented pyrotechnic charge). Therefore, the logic was this: to release the stored gas if the lower-level sensors triggered, but to ignite the augmented charge only if the higher-level sensor triggered as well. It is noteworthy that the logic involved only the nature of the crash as experienced by the vehicle; no decisions were made on the basis of conditions inside the compartment.

FROM DUAL-LEVEL INFLATORS TO "SMART" SYSTEMS – The decision whether to have a high- or a low-level deployment could be made on the basis of factors other than the crash characteristics seen by the vehicle. For example, Is the seat occupied? If the seat is occupied, is the belt being worn? In this electronic age, we could envision a so-called "smart" system, in which the deployment logic could be based on detecting the presence of a child seat (particularly a rear-facing one), discriminating between humans and various objects in the seat, and/or detecting the size of the occupant, his or her proximity to the instrument panel, etc. Again, it is noteworthy that all considerations involve the occupant and conditions inside the passenger compartment, in contrast to the logic utilized in the first-generation General Motors air cushion system of the mid-70s.

The term "smart" has more recently acquired an official definition, thanks to the NHTSA. In a rule issued in late 1996 [NHTSA 96], an airbag system is considered "smart" if:

- It does not deploy if the mass on the seat is 30 kg or less.
- It does not deploy if a rear facing child seat or out-of-position occupant is present.
- It does deploy if a properly belted child is present and there is no risk of injury.

Obviously, the ability to adjust the air bag's deployment to these and other factors is highly dependent on sensor technology, which is still in a state of intensive development at this writing.

Among the competing sensor technologies are the following:

- Sensing occupant weight by detecting the seat cushion deflection. Such a device could be fooled by heavy objects, would not be sensitive to occupant position, and would require extensive cushion redesign to incorporate the sensor.
- Infrared sensors to measure body heat. Here, the challenge is to distinguish the body heat "signal" (about 37°C) from the high "noise" levels due to variables such as compartment temperature (which can vary from -20°C to 70°C), heated seats, and heat-absorbing clothing.
- Detecting changes in capacitance due to an occupant. Such technology could detect occupant position, and sensors could be located in a variety of places, but at the same time they could be fooled by conductive materials, including moisture.
- Ultrasound could be used, as during pregnancy, and sensors could be located in a variety of places, but would be sensitive to temperature and humidity. Large objects could block sound waves.
- A semiconductor-based "seeing eye" could discriminate between the visual appearance of an occupied seat and an empty one, and thus could detect occupant presence and position. However, high resolution plus pattern recognition equals high cost.

All of these systems are "passive" in the sense that occupant does not have to (and indeed, cannot be expected to) wear a reflector or transponder to "talk back" to the sensor or sensors. An "active" system, by way of contrast, includes such a device, which makes the sensor's task easier and improves the quality of communication between sensor and object. This concept would be practical for specific hardware designed to fit on an automobile seat, such as a child seat. A rear-facing child seat could have a reflector or transponder which could cause the air bag to be depowered or deactivated if the child seat is placed in the front seat, despite warnings to the contrary. Alternatively, a reflector or transponder on a forward-facing child seat could provide quality information on the child seat's presence and distance from the airbag module. The problem, of course, is the installed base of millions of child seats not so equipped.

CRASH SENSORS

At the instant of contact, the vehicle will not yet "know" that it is in a crash. The best that can be done is to have sensors and associated electronics continually on "sentry duty," keeping track of velocities and accelerations, and looking for the telltale signs that a crash, and not just a hard bump, has begun. The sensor system will also have to distinguish the direction of the crash, and whether the crash severity warrants a deployment. This is a tall order and takes some time, called sensing time -- perhaps on the order of 20 to 25 milliseconds. What is left of the first 50 milliseconds can be devoted to filling the air bag.

DEPLOYMENT THRESHOLD – Clearly, there are circumstances in which airbag deployment is undesirable. In non-frontal impacts (e.g., side, rear, and rollovers), the injury hazards may not be amenable to reduction by an air bag. In any case, airbag deployment necessitates replacement, which may add considerably to the repair bill. In property-damage-only accidents, airbag replacement cost could represent a significant part of the total. If airbag deployment were perceived as unnecessary, and if the replacement cost were high, negative reaction could hinder the public acceptance of air bags.

Similarly, the injury potential may be low in low speed non-deployment accidents, particularly if the belts are being worn; at the same time, the energy of airbag deployment could actually increase the probability of injury. Therefore, for frontal impacts a deployment threshold is created. Below the threshold the air bag should never deploy, and above the threshold it always should. Of course, every crash has a different signature, so for design purposes the threshold is described in terms of barrier impacts. At one time, the Government was proposing a 15 mph barrier impact as the minimum to initiate airbag deployment [NHSB 70], probably with an eye to the 15 mph interior impact requirements of FMVSS 201 [FMVSS 201]. In any event, a requirement for deployment threshold was never promulgated, but similar considerations for protecting unrestrained occupants has tended to cause thresholds to be set at about 12 mph in frontal barrier impacts.

Of course, variability in crashes means that the threshold has to have some thickness. In other words, there has to be a "gray area" within which deployment may or may not occur. The specification for this threshold can, and probably does, vary with vehicle design, but generally the specification does reflect a gray area. The low end of the gray area, below which the bag should never deploy, may be set to 8 mph in a barrier crash, and the upper end, above which it should always deploy, may be 14 mph, again in a barrier crash. The sensor system is designed to respond only to the frontal component of an accident, so deployment occurs only if the frontal component exceeds the threshold.

The setting of the deployment threshold can depend on many factors, including the usage of seat belts. Some systems have the ability to adjust the threshold up or down automatically, based on whether the belts are being worn, since for seat belt users the potential benefits of air bags are reduced at low speeds. Such decision-making capability is one aspect of a "smart" restraint design.

RELIABILITY CONSIDERATIONS – Deployment threshold is related to another topic - reliability. Air bags are unique among automotive systems. They are different from brakes, for example, which can be disassembled for inspection or maintenance, and which can give can give cues regarding their condition whenever they are used. On the other hand, air bags may remain unused for long periods -- perhaps 20 years -- but they must remain fully ready to perform when needed, and not deploy when not needed. Therefore, an airbag system must have a built-in readiness tester that can immediately inform the operator or the repair technician when there is a problem, and a diagnostic system that will indicate where the problem is. Even so, the consequences of an error (especially a failure to deploy when needed) can be so serious that reliability targets have been set at levels comparable to man-rated space missions [Jones 70]. These levels are achieved by quality assurance inspection and testing at a 100 percent level, as opposed to a statistical sampling process. In other words, every step in the manufacture and assembly of every airbag system entails inspection, testing if appropriate, and documentation, and every vehicle is subjected to a complete diagnostic test procedure before it is offered for sale. Thereafter, every system is electronically tested every time the engine is started.

ELECTRICAL CONNECTIONS – If the airbag system wires were manipulated by service personnel, there would be increased odds of a wire being cut, incompletely re-connected, or not re-connected at all. Therefore, the airbag wiring is contained in a separate and independent harness, wrapped with a material of a distinctive color (typically yellow), and routed inconspicuously. Harness routing is also chosen so as to minimize the likelihood of being pinched or crushed during the crash.

The connections to this harness are crucial to system reliability. Typically, each connector employs dual contacts, gold plated to prevent any compromise of electrical continuity due to corrosion. As shown in Figure 24, part of a bar code may be printed on each of the mating connector housings, so a bar-code reader can verify and record that connector is correctly assembled [Kobayashi 87].

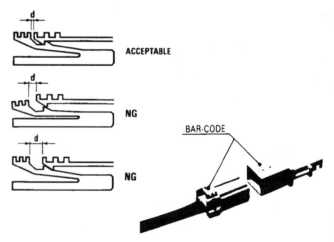

Figure 24. Machine-Checkable Airbag Harness Connector - Honda

THE NEED FOR FIELD TESTING – Because of the extremely stringent reliability requirements, and the relative scarcity of accidents warranting an airbag deployment, it is virtually impossible to design a test protocol, prior to vehicle introduction, which would give adequate insights regarding reliability issues [Jones 71]. Thus we find manufacturers being very cautious, with the early airbag designs being introduced in limited quantities, in fleet settings. A notable example is GM's 1000-car "green fleet" in 1973 [Smith 73].

Another example is the NHTSA-sponsored airbag retrofit program for 539 police cars in 1983-85 [Romeo 84].The intent of this program was to design, test, and evaluate a driver airbag retrofit system, and using such production hardware as existed at the time, to manufacture and install retrofit kits into Highway Patrol vehicles in various states. Sensors and diagnostic systems were supplied by the Technar Corporation, the gas generator came from Bayern Chemie, and the steering wheel and airbag module were made by Takata Corporation. Important lessons were learned, but plans to offer a retrofit kit on a wider scale were thwarted by the inability to obtain adequate products liability insurance [DeLorenzo 86].

Other important field test fleets in the U.S. were the 5300 Ford Tempos for the U.S. Government in 1985 [Maugh 85], further described below, and Chrysler's introductory fleet in 1988 [Edwards 91]. Deployment accidents in all these fleets were investigated in detail [Mertz 88], and valuable insights were obtained from this field experience.

HOW RELIABILITY REQUIREMENTS AFFECT SENSOR SYSTEM ARCHITECTURE – Basically, reliability entails the avoidance of two kinds of errors: false negatives and false positives. In other words, a sensor should avoid a failure to trigger when it is supposed to, and avoid triggering when it is not supposed to. The probability of a false negative can be reduced by placing more sentries on duty, and empowering any one of them to sound the alarm. In engineering terms, one would have multiple sensors connected in parallel. However, the probability of a false positive (i.e., an unwanted deployment) increases with the number of sensors in parallel, since the probability of a false positive for the system as a whole is a combination of the individual false positive probabilities.

On the other hand, the system-wide probability of a false positive can be reduced by placing two or more sensors in series. In this system, any one sentry can squelch the alarm. Consequently, the probability of a false negative increases with the number of sensors in series.

If several sensors are employed, many combinations of series and parallel connections are possible. In the General Motors system of the 1970s, the driver and low-level passenger circuits were connected through sensors on both the positive and the negative (ground) side; i.e., in series. On each side, there was a low-level G switch and a bumper switch in parallel. The same parallel combination was used on the negative side of the high-level passenger circuit, but the positive side was connected through a separate high-level G switch [Louckes 73].

See Figure 25. The G switches, actuation circuits, backup energy source, diagnostics, and crash monitoring and recording devices were included in a single component mounted in the passenger compartment.

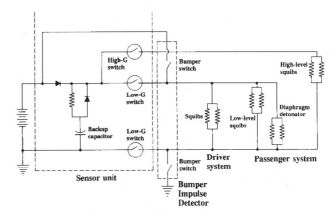

Figure 25. Sensor Circuit, First-Generation GM System

The result of this logic was that a low-level crash would trigger a driver airbag inflation and a low-level passenger system inflation. A high-level passenger system inflation would occur only if the low-level circuit were completed on the negative side, and the high-level sensor closed (on the positive side).

The switch used in GM's Bumper Impulse Detector was a prototype of sensor designs to come, in that it consisted of an inertial mass and a spring; for the switch to close, the acceleration would have to be strong enough to overcome the resistance of the spring and would have to last long enough for the mass to travel the requisite distance to reach an electrical contact. See Figure 26. Consequently, this device was a mechanical integrator of accelerations, in which switch closure depended, for the most part, on the velocity change. It was calibrated to close in an 11 mph barrier crash, and its purpose was to achieve early detection of the crash. (Recall that the G switches, in contrast, were well aft, in the passenger compartment.) The Bumper Impulse Detector employed dual contacts and springs in series.

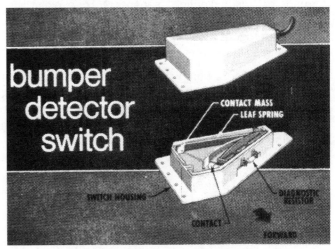

Figure 26. Bumper Impulse Detector, First-Generation GM System

The G switches consisted of a mass on a wire, which formed a pendulum, as shown in Figure 27. The mass was held in its aft most position by a magnet; if the vehicle acceleration were strong enough, the mass could break free of the magnet and move forward in a wedge-shaped groove which extended to 37 degrees either side of straight ahead. The low-level G switch was calibrated (by means of the magnet strength) to close at an acceleration level corresponding to an 11 mph barrier crash. Obviously, the acceleration level would vary with the vehicle and the location of the switch, but in the GM design the level was 18 Gs. The high-level G switch was reportedly calibrated to close at 30 Gs [Louckes 73].

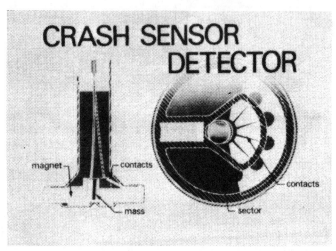

Figure 27. G-Switch Sensor, First-Generation GM

The bumper presented a very harsh environment for any electrical device, including the Bumper Impulse Detector. Designers of subsequent systems have refrained from placing sensors in such a location.

The 1985 Ford Tempo/Topaz system (driver side only) employed five sensors, which were functionally divided into two groups. One group, the secondary or "safing" sensors, were intended to minimize the probability of false positives (inadvertent deployments), and were therefore, as a group, wired in series with the group of primary or "discriminating" sensors. To avoid raising false negatives (failure to deploy), the safing sensors were set to trigger at lower crash severities than the discriminating sensors were.

The safing sensors were wired in parallel with each other and were located on vehicle center line at two locations: the top of the radiator support and on the dash panel. The discriminating sensors were connected in series with the safing sensors, which could thus switch Aon@ frequently without necessarily resulting in a deployment. See Figure 28. The discriminating sensors, wired in parallel with each other, were located on the left and right front fender aprons, and at the top center of the radiator support, as indicated conceptually in Figure 29.

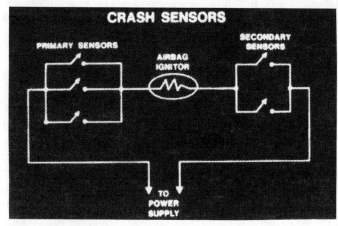

Figure 28. 1985 Ford Tempo/Topaz Sensor Circuit

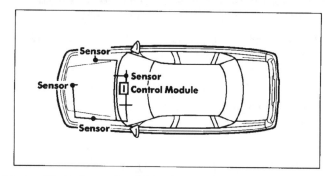

Figure 29. Schematic of Distributed-Sensor System

The sensors were manufactured by the Breed Corporation, and employed a ball that, at a specified deceleration, would pull away from a bias magnet and move through a tube toward a set of electrical contacts, which would have to be bridged to activate the inflator. A small clearance gap between the ball and the tube damped the response to higher-frequency accelerations. This concept is shown in Figure 30. It was simple and reliable, but sensitivity to cross-axis accelerations and temperature variations made its performance more difficult to predict in low speed impacts.

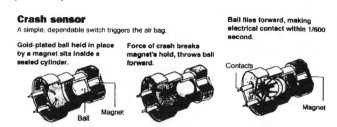

Figure 30. Ball-in-Tube Sensor Function

SENSOR CALIBRATION – Beyond meeting the specifications for deployment threshold, there is a second requirement that influences how sensors are calibrated: sensing time, or time-to-fire. Recall that the sensing time budget is typically 25 or 30 milliseconds. In this time frame, the discriminating sensors must make their decision. Figure 31 shows the time history of ΔV accumulation for four hypothetical situations involving different speeds and objects struck. The upper curve describes an obvious deployment accident, and it has distinguished itself from the others by 15 msec or so. The remaining three curves include one other deployment event and two for which deployment is probably not warranted; yet in the first 30 msec of the crash it is difficult to tell which is which.

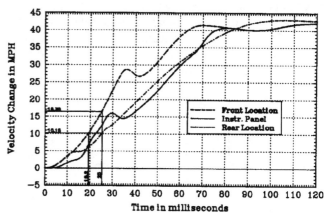

Figure 32. Typical Structural Responses for 35 mph Barrier Crash

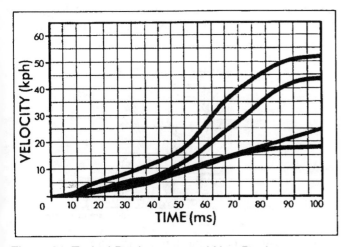

Figure 31. Typical Deployment and Non-Deployment Events - NEC

Recognizing that the crash pulse, and hence the ΔV accumulation, depend on where the sensor is placed, one can attempt to sort out the curves of Figure 31 in a shorter time by placing a the discriminating sensors at a variety of locations on the vehicle. Note, however, that the ΔV required to trigger the sensors will be well below the ΔV ultimately experienced in the crash.

We can illustrate the considerations involved by assuming that a sensor is a ΔV switch; i.e., that its trigger decision is made solely on the basis of the ΔV seen at the sensor location. (As indicated previously, this is not a bad assumption for traditional mechanical sensors.) Figure 32 shows the ΔV accumulation in a typical 35 mph barrier crash. If one needed a 25 msec sensing time for such conditions, and if one had a sensor at a rear location (such as near the B-pillar), one would need about a 10 mph trigger level. If that level proved to be too low (for deployment threshold considerations, say), then one could choose to move the sensor to a forward location; there, it could achieve a 25 msec sensing time with a trigger level of just over 16 mph. Alternatively, the same trigger level, in the forward location, would reduce the sensing time to just over 19 msec.

The results of these considerations are seen in the design of the Ford Tempo/Topaz system. Calibration variables for the sensors included the distance between the ball's rest position and the contacts, its clearance between the guide tube inner walls, and the strength of the bias magnet. These variables were tuned by subjecting them to haversine acceleration pulses of various amplitudes and durations, worked out by Breed to meet Ford's requirements regarding deployment threshold and sensing time. For the discriminating sensors, the "must deploy" threshold in 14 mph barrier crashes was tested with "soft," "medium," and "hard" haversine pulses at 10 mph, and the "no deploy" threshold in 8 mph barrier crashes was tested with "soft," "medium," and "hard" pulses at 6.5 mph. Safing sensor calibration pulses were at 3.5 mph for the forward sensor and 1.1 mph for the aft one. At Ford, each and every sensor was tested twice: once during the manufacturing process and again after final sensor assembly.

This distinction between safing and discriminating sensors, and the use of haversine pulses to test their calibration, has become typical of sensor system design to the present.

TYPES OF SENSOR DESIGNS – Airbag sensor concepts since the '70s have shown their origins in the GM Bumper Impulse Detector: an inertial mass being required to overcome a bias force (from a magnet, for example) and travel a certain distance before electrical contact is made. One implementation of this concept was the ball-in-tube device described previously. Another was the widely used "rolamite" sensor developed by the Technar Corporation (now a part of TRW). See Figure 33. A roller was wrapped with a spring band, which tended to keep the roller pushed against a stop. In a crash, the roller moved forward without slipping (or damping), unrolling the spring band in so doing, and covering the distance to the electrical contacts (again, gold-plated) if the crash was severe enough. The rolling action reduced friction to a minimum; the surface on which the rolling occurred was slightly arched. The calibration level was determined by the spring band and the mass of the roller. For a particular application, each sensor would be calibrated via a set screw, adjusting the position of the roller at rest.

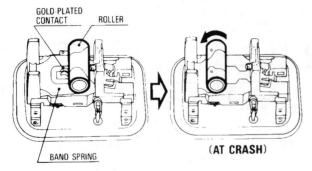

Figure 33. Rolamite Sensor Action - Honda

Another electro-mechanical sensor is the gas-damped diaphragm type, shown in Figure 34. A diaphragm, to which the inertial mass is mounted, provides both the bias force and segregation of the gas volumes. Gas is permitted to pass from one chamber to another through one or more orifices, creating a damped response.

As noted previously, such systems are electrical switches that respond to velocity change, by perform a mechanical integration of acceleration. Another approach would be to measure the strain in the bias spring, by making that spring a piezoresistive or piezoelectric element in a solid state electrical circuit. This results in a very small device in which the only motion is beam bending. The acceleration can be integrated digitally, or processed in other ways. For example, acceleration and velocity could be used together in some mathematical formula or algorithm [NEC 95].

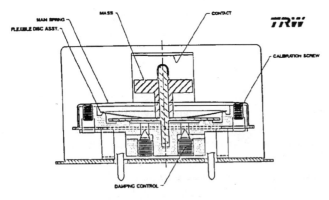

Figure 34. Gas-Damped Sensor - TRW

Electronic sensor systems could be tuned to a particular vehicle by adjusting the parameters or code stored in firmware. This approach opens the door to a great deal of design flexibility, and does so with reduced weight, cost, and complexity (and thus increased reliability). At this writing, such systems are common, and there is every reason to expect them to predominate in the future.

FROM MULTI-POINT TO SINGLE-POINT SYSTEMS – Both the early GM and Ford systems involved multiple sensors at multiple locations, and we have seen how such a strategy results in different calibrations for different sensors. It also results in a considerable wiring harness, which, as we have seen before, entails special care to maintain reliability.

It is desirable, from both cost and reliability perspectives, to decrease the amount of wiring in the air bag system. A Breed Corporation concept carried this objective to its logical conclusion: a purely mechanical system, located at the inflator, that both sensed the crash and initiated the bag inflation [Breed 85]. In this system, a spring-loaded firing pin was held in place by a lever, which itself was held in place by a bias spring. See Figure 35. The lever could be moved by a sensing mass, and if the motion were sufficient, the firing pin would be released, and propelled into a stab primer.

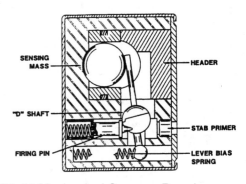

Figure 35. All Mechanical Sensor - Breed

Having no sensors any closer to the crush zone than the steering wheel hub caused some concern regarding sensing time. However, the reduction of axial play in the column produced sensing times on the order of 30 milliseconds in 30 mph frontal barrier tests. The system was intended to be retro-fitted into Chevrolet Impala police vehicles.

The general trend, however, has been to stay with electrical sensing elements, so that the increasing computational power, cost-effectiveness, space efficiency, and reduced weight of digital electronics can be used to fullest advantage. This approach allows the integration of sensors, diagnostics, backup power supply, and the logic elements into a single sealed unit. These are known as single-point systems, although separate safing and discriminating sensors are generally employed. Since the electronic environment is much less hostile in the occupant compartment than in the engine compartment, the air bag electronic module is often placed on the tunnel, near or at the dash panel. Because no information comes directly from the crush zone, increased emphasis is placed on signal processing and digital logic.

CONCLUSION

Among all safety system concepts that have ever found their way into production automobiles, the air bag arguably represents the most dramatic departure ever taken from traditional automotive technology. It started in a time of technological optimism when Americans were heading to the Moon, yet refusing to fasten their seat belts. At that time, the air bag seemed like a way to solve a behavioral problem by technological means.

Since then, many nations have shown the way in achieving widespread belt use, and American belt use, while still far behind, has risen to levels not imagined in the '70s or early '80s. Belt systems have thus garnered much of the safety benefit that was assumed to be available to air bags. At the same time, the operating airbag concept has changed from being the only restraint widely used, to being strictly supplementary to the belts. We find that air bags are playing a more limited role, and are contending for a smaller portion of the available safety benefit.

Notwithstanding these developments, air bags were greeted enthusiastically by the public (with perhaps some help from the various passive belt designs). They became almost a litmus test of automotive safety, The development of these devices, and their availability in large numbers, represent astounding technological achievements, and are tributes to the many thousands of individuals, beyond the few cited in this paper, who made it all happen. Air bags are now busily saving lives and reducing injuries, though not necessarily in the numbers originally envisioned.

However, there have also been continued warnings of technological problems and potential adverse side effects, and some of these have come to pass as vast numbers of airbag-equipped vehicles have taken to the road. We are now engaged in redoubled efforts to address these, and the near future will contain many new developments. The work is not yet finished.

REFERENCES

1. [Alexander 74] *An Evaluation of the U.S. Family Sedan Experimental Safety Vehicle (ESV) Project*, GH Alexander, RD Vergara, JT Herridge, W Millicovsky, and MR Neale, Final Report, Contract DOT-HS-322-3-621-1, October 1974.
2. [Breed 85] "The Breed All-Mechanical Airbag Module," A Breed, SAE Paper 856014, *Proceedings, Tenth International Technical Conference on Experimental Safety Vehicles*, 1985.
3. [Campbell 72] "Air Cushion Restraint Systems Development and Vehicle Application," DD Campbell, SAE Paper 720407, May 1972.
4. [DeLorenzo 86] "Supplier cuts off air bag retrofitter - Lack of adequate liability insurance cited," M DeLorenzo, Automotive News, 10 October 1986.
5. [DOT 78] "New DOT Study Finds Only 14% of Drivers Use Auto Safety Belts," Press Release, U.S. Department of Transportation, 15 December 1978.
6. [Edwards 91] "A Preliminary Field Analysis of Chrysler Airbag Effectiveness," WR Edwards, *Proceedings, Thirteenth International Technical Conference on Experimental Safety Vehicles*, November 1991.
7. [FMVSS 201] National Highway Traffic Safety Administration, FMVSS 201, *Interior Impact Protection*.
8. [Frey 70] "History of Air Bag Development," SM Frey, *Proceedings, International Conference on Passive Restraints*, May 1970.
9. [Jones 70] "Inflatable Passive Air Restraint System Crash Sensors," TO Jones, *Proceedings, International Conference on Passive Restraints*, May 1970.
10. [Jones 71] "Crash Sensor Development," TO Jones and OT McCarter, SAE Paper 710016, January 1971.
11. [Katter 75] *Development of Improved Inflation Techniques*, LB Katter, Final Report, Contract DOT-HS-344-3-690, September 1975.
12. [Klove 72] "Special Problems and Considerations in the Development of Air Cushion Restraint Systems," EH Klove, Jr. and RN Oglesby, SAE Paper 720411, May 1972.
13. [Kobayashi 87] "Reliability Considerations in the Design of an Air Bag System," S Kobayashi, K Honda, and K Shitanoki, *Proceedings, Eleventh International Technical Conference on Experimental Safety Vehicles*, May 1987.
14. [Kossar 92] "Air Bag Deployment Characteristics," LK Sullivan and JM Kossar, Final Report No. DOT HS 807 869, February 1992.
15. [Lau 93] "Mechanism of Injury from Air Bag Deployment Loads," IV Lau, JD Horsch, DC Viano, and DV Andrzejak, *Accident Analysis and Prevention*, Pergamon Press, Vol. 25, No. 1, February 1993.
16. [Loukes 73] "General Motors Driver Air Cushion Restraint System," TN Loukes, RJ Slifka, TC Powell, and SG Dunford, SAE Paper 730605, May 1973.
17. [Maugh 85] "Supplemental Driver Airbag System - Ford Motor Company Tempo and Topaz Vehicles," RE Maugh, SAE Paper 856015, July 1985.
18. [Mertz 88] "Restraint Performance of the 1973-76 GM Air Cushion Restraint System," HJ Mertz, SAE Paper 880400, February 1988.
19. [NEC 95] Advertising material from NEC Technologies, Inc., 1995.
20. [NHTSA 96] *Final Rule on Labels*, National Highway Traffic Safety Administration, 27 November 1996.
21. [NHSB 70] National Highway Safety Bureau: Proposed Amendment to Motor Vehicle Safety Standard 208, *Occupant Crash Protection*, 3 November 1970.
22. [Reidelbach 79] "Advanced Restraint System Concepts," W Reidelbach and H Scholz, SAE Paper 790321, February 1979.
23. [Romeo 75] "Front Passenger Passive Restraint for Small Car, High Speed, Frontal Impacts," DJ Romeo, SAE Paper 751170, November 1975.
24. [Romeo 78] *Front Passenger Aspirator Air Bag System for Small Cars*, DJ Romeo, Final Technical Report, Phase II, Contract DOT-HS-5-01254, March 1978.
25. [Romeo 84] "Driver Air Bag Police Fleet Demonstration Program - A 15-Month Progress Report," DJ Romeo and JB Morris, SAE Paper 841216, October 1984.
26. [Seiffert 72] "Development Problems with Inflatable Restraints in Small Passenger Vehicles," UW Seiffert and GH Borenius, SAE Paper 720409, May 1972.

27. [Smith 73] "The 1,000 Car Air Cushion Field Trial Program," GR Smith and MR Bennett, *Proceedings, Automotive Safety Engineering Seminar*, June 1973.
28. [Struble 79] "Status Report of Minicars= Research Safety Vehicle," DE Struble, *Proceedings, Seventh International Technical Conference on Experimental Safety Vehicles*, June 1979.

CONTACT

Donald E. Struble, Ph.D.
Collision Safety Engineering, Inc.
2320 West Peoria Avenue, Suite B-145
Phoenix, AZ 85023   Phone: (602) 395-1011

982293

Direct Thermal Detection for Front Passenger Seat Airbag Suppression

David K. Lambert
General Motors Research and Development Center

Copyright © 1998 Society of Automotive Engineers, Inc.

ABSTRACT

Direct thermal detection of the passenger is considered for use in a system that would suppress deployment of the passenger-side airbag for a rear-facing infant seat. The temperature at the surface of the front passenger's seat is compared with the temperature at the surface of the driver's seat. If the two temperatures differ by more than a preset amount the airbag would be suppressed. It is shown that when the ambient temperature equals the passenger's skin temperature (at about 35 °C) seat surface temperature does not distinguish a normally seated adult from a rear facing infant seat. Attempts to circumvent this problem by adding a heater were either too slow or too insensitive for an airbag suppression system. However, direct thermal detection may add reliability to some other type of airbag suppression system, such as one based on passenger weight.

INTRODUCTION

Drivers are advised to never let children age 12 or under ride in the front-passenger seat, especially in a rear-facing infant seat (RFIS). However, it has been suggested that children that are in the front seat might be protected by "smart airbag" systems that would automatically decide whether the airbag should deploy. This paper examines the possibility of using direct thermal detection of the front-seat passenger to help decide whether to allow the passenger-side airbag to deploy. It is assumed that if a normally seated adult occupies the front-passenger seat, the seat surface is at approximately the same temperature as the driver's seat surface. If an infant in a RFIS occupies the seat, the temperature measured at the passenger's seat surface would typically not be close to that of the driver's seat surface. A simple logic diagram that illustrates this is shown in Fig. 1. This paper is concerned with the performance of direct thermal detection.

The ideal suppression system would always make correct decisions. While direct thermal detection cannot do this by itself, it may be useful as a way to increase the reliability of decisions based on another approach. Other sensors that have been suggested include those that measure the passenger's weight [1-6], measure the electrical capacitance between suitably positioned electrodes [1-7], measure mechanical vibrations [8], use ultrasound [9,10], use microwaves [10], use geometric optics [10-14], and that detect thermal infrared emissions [9,15]. Ways have been suggested for a suppression system to combine multiple inputs and thereby reach a more reliable decision than would be possible from a single input [3,4,9,10,14,15]. Suppression systems are assessed in a recent report that was commissioned by the National Highway Traffic Safety Administration [16].

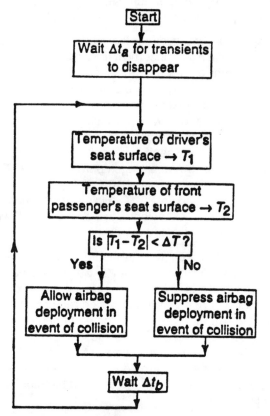

Figure 1. Logic diagram of a system that uses direct thermal detection to decide whether to suppress deployment of the passenger-side airbag.

315

IMPLEMENTATION

The use of direct thermal detection rests on the premise that the skin temperature under a normally seated adult passenger is controlled by their core body temperature, 36 to 38 °C for a person in good health. Normally, vehicle occupants try to be as comfortable as possible by adjusting the vehicle's heating, ventilation and air conditioning system appropriately, and by wearing appropriate clothing for the weather. Consequently, the equilibrium temperature measured at the surface of the driver's seat would typically be higher than the equilibrium temperature measured under a RFIS. If ambient temperature is above skin temperature, however, the driver's seat surface is cooled and this relationship is reversed.

The body is a source of heat. At low ambient temperature, heat loss occurs primarily through the skin by conduction, convection, and radiation. At high ambient temperature in a dry climate, evaporation of sweat predominates. As the ambient temperature continuously increases from 20 °C up to 50 °C, there is an intermediate temperature (near 35 °C) at which the average skin temperature equals the ambient temperature. At this temperature, all of the net heat transfer from the body is due to evaporation (mostly from the skin but also from the lining of the respiratory tract). Evidently, in this situation, it is impossible for a temperature sensor alone to distinguish a normally seated adult passenger from a RFIS.

The speed at which direct thermal detection is able to respond is also a concern. It is clearly impossible for direct thermal detection to detect that an occupant has been thrown into an injury prone position during a collision. Preliminary tests showed that direct thermal detection may be able to provide an initial response within 5 s. For example, after a person wearing normal indoor clothing sat on a cloth seat, the seat surface temperature increased from the ambient temperature of 22 °C to 28 °C in 6 s. After an additional 10 s, the seat-top temperature had increased to 30 °C. However, a much longer response time is expected if the passenger is wearing heavy winter clothing.

The rapid approach to equilibrium seen in the above example is explained as follows. Before the passenger sits down there is a boundary layer of air, next to their skin, that has been warmed to skin temperature. When they sit down, this warm air is forced through their clothing into the seat. Consequently, the initial time constant is set by the thermal response time of the sensor. The effect of a vehicle's seat cover material on skin temperature and thermal comfort has been investigated by Fung and Parsons [17].

One way to implement direct thermal detection is to compare the temperature of the driver's seat surface with the temperature of the passenger's seat surface, as shown in Fig. 1. One or more thermistors at the seat surface can be used to perform the temperature measurement. A comparison of the signals from one thermistor at each of these two locations is shown in Fig. 2. One thermistor was under the driver in a 1998 Pontiac Grand Am GT. The other thermistor was under the passenger. Both thermistors were Fenwal Electronics type GA52J16 mounted in identical BNC-to-banana plug adapters that had been fastened to the center of both the passenger's seat and the driver's seat. The thermistor's thermal response time is specified to be 4 s in still air. The driver and passenger sat down on the two sensors at the same time. The temperature in the vehicle was about 58 °C and the slope of thermistor resistance as a function of temperature was about 3.5 KΩ/°C. After the two people sat down, since the ambient temperature was above skin temperature, the seat surface temperature decreased. The difference between the resistances of the two thermistors rose to a maximum value and then returned to zero. The difference did not decrease to half its maximum value until 48 s after the people sat down.

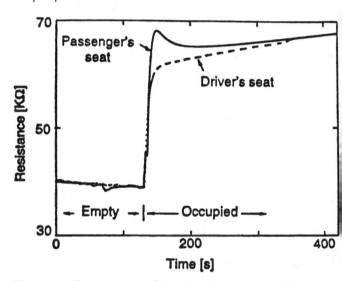

Figure 2. Comparison of the signals measured by thermistors at the surface of the driver's seat and the passenger's seat. Both seats were occupied simultaneously.

To investigate how the sensor responds with the vehicle interior temperature at about body temperature, the vehicle was parked with the engine running and its own heater was used to maintain the interior temperature at about 35 °C. As shown in Fig. 3, in this situation it is impossible to tell whether or not there is a normally seated adult on the passenger seat, simply from the measured temperature at the seat surface.

The ambiguity when skin temperature equals ambient temperature is a fundamental problem. In an attempt to overcome it, two approaches were tried. Both involved supplying heat to the seat surface. Both observed the resultant temperature in an attempt to distinguish between a RFIS and a normally seated adult. Neither proved successful.

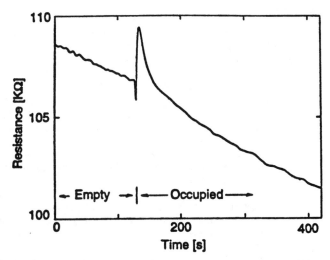

Figure 3. Measured effect of an adult passenger sitting down on the seat when the ambient temperature is approximately equal to the passenger's skin temperature.

The first approach used a flat-plate heater. The objective was to distinguish a person from an inanimate object (such as the bottom of a RFIS) by observing the person's ability to regulate their skin temperature.

The flat-plate heater consisted of two aluminum plates, 0.1 × 3.6 × 10 cm, each of which contained a thermistor (Fenwal Electronics type GA52J16). A glass microscope slide was sandwiched between the two plates to provide a known thermal resistance. An electrical heater (1.3 W) that could be switched on and off was attached to the bottom side of the bottom plate. The flat-plate heater used in the present experiment was conceptually similar to a heat-flux meter developed by Xizhong, Zizhu, and Genhong for physiological studies [18].

To understand how the device was expected to behave, consider the electrical analog, shown in Fig. 4, which can be used to derive the temperature distribution that occurs after reaching thermal equilibrium. Temperature is measured at two positions, A and B, between the heat source and the body core, both with the heater on and with the heater off. Let these measured values at A and B with the heater off be T_A^* and T_B^*, respectively. Let those with the heater on be T_A^* and T_B^*, respectively. The (unknown) core body temperature is assumed constant at T_0. In principle, T_0 can be obtained from the measured values:

$$T_0 = (T_B T_A^* - T_A T_B^*)/[(T_A^* - T_A) + (T_B - T_B^*)]. \quad \text{(Eq. 1)}$$

It should also be possible to use the measured temperatures to determine the ratio of the (unknown) thermal resistance from B to the body core, to the (known) thermal resistance from A to B. The initial hope was that by measuring two independent quantities it would be possible to distinguish between a RFIS and a normally seated adult even when the ambient temperature was about 35 °C.

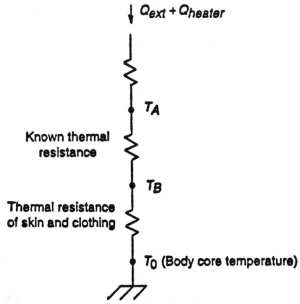

Figure 4. Electrical analog of the flat-plate sensor in contact with a person on the seat, after reaching thermal equilibrium. Here T_A and T_B are the measured temperatures, T_0 is the unknown but regulated body core temperature, and Q_{ext} and Q_{heater} are the heat fluxes supplied by the environment and the heater, respectively.

In practice, it took minutes for the measured temperatures to reach equilibrium. This is too long to be useful for airbag suppression. The response time was dominated by the thermal response of the body. The device itself had a thermal response time of about 20 s. In physiological studies with their heat flux meter pressed directly against the patient's skin, Xizhong, Zizhu, and Genhong waited 10 min between measurements.

A second approach used a self-heated thermistor that was embedded in a plastic button. By cycling the thermistor current on and off if was possible to determine both the temperature at the seat surface and the ability of material near the sensor to remove heat. Response was much faster than the flat-plate heater since there was no need to wait for thermal equilibrium.

The device used for this approach was similar to the thermal diffusion probe developed by Arnaud et al. [19]. To test the concept, the thermistor used was Fenwal Electronics type GB31J1. It was epoxied into a plastic button 9 mm in diameter and 2 mm thick. The "on" current through the thermistor was 2.8 mA. The current was switched on and off with a 50% duty cycle. The thermistor voltage was measured twice each cycle, once just after the current was switched on, and once just before it was switched off. The two corresponding temperatures provided information about the temperature at the seat surface and about the ability of material near the sensor to remove heat from it while the current is on. However, as shown in Fig. 5, with a 5 s cycle time and with the vehicle's interior at about 35 °C the effect on the signal caused by a passenger sitting on the sensor was not significant.

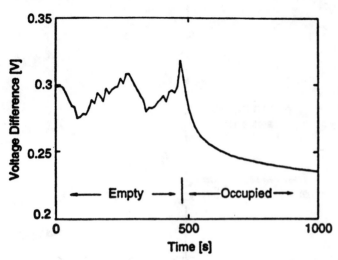

Figure 5. Measured effect of a passenger sitting down on the pulsed thermistor sensor when the ambient temperature was approximately equal to the skin temperature. The quantity plotted is the difference in thermistor voltage between the beginning and end of the current pulse.

CONCLUSION

The most serious problem with the use of direct thermal detection for automatic suppression of airbag deployment is that it cannot distinguish between a RFIS and a normally seated adult if the ambient temperature is about 35 °C. Two different attempts were made to use a heater to overcome direct thermal detection's lack of performance at about 35 °C. Neither proved successful

Direct thermal detection is also too slow to determine whether an occupant has been thrown into an injury prone position by precollision braking. The data reported here suggests that if the driver and front-seat passenger are wearing light clothing, the response time of direct thermal detection could be as fast as 10 s. The response time would be longer if either the driver or front-seat passenger were in heavy winter clothing.

For these reasons, direct thermal detection by itself is not suitable as the sensor for passenger-side airbag suppression. It may be possible to use direct thermal detection in conjunction with another type of sensor, such as a weight sensor, to increase the reliability of the deployment decision. Many of the issues involved in developing a reliable suppression system are discussed in Ref. 16.

ACKNOWLEDGMENTS

The author would like to thank Rex L. Harris and Galen E. Ressler for helpful conversations.

REFERENCES

1. E. Vollmer, "Inflatable air bag safety device for motor vehicles," U.S. Patent 5,161,820 (10 Nov 1992).
2. B. K. Blackburn, J. F. Mazur, and S. B. Gentry, "Occupant sensing apparatus," U.S. Patent 5,232,243 (3 Aug 1993).
3. C. E. Steffens, Jr., T. H. Vos, S. B. Gentry, J. F. Mazur, and B. K. Blackburn, "Method and apparatus for controlling an actuatable restraining device in response to discrete control zones," U.S. Patent 5,413,378 (9 May 1995).
4. J. F. Mazur, B. K. Blackburn, and S. B. Gentry, "Method and apparatus for sensing a rearward facing child restraining seat," U.S. Patent 5,454,591 (3 Oct 1995).
5. B. K. Blackburn, J. F. Mazur, and S. B. Gentry, "Occupant sensing apparatus," U.S. Patent 5,494,311 (27 Feb 1996).
6. F. Zeidler, V. Petri, R. Mickeler, and M. Meyer, "Device for detecting seat occupancy in a motor vehicle," U.S. Patent 5,612,876 (18 Mar 1997).
7. T. Yasuda, J. Sato, and M. Ohashi, "Capacitive occupancy detector apparatus," U.S. Patent 4,796,013 (3 Jun 1989).
8. A. L. Browne, D. A. Stephenson, and W. Kiel, "Using seat mounted accelerometers to differentiate between normally seated passengers and infants in infant seats," S.A.E. Paper 933092.
9. A. P. Corrado, S. W. Decker, and P. K. Benbow, "Automotive occupant sensor system and method of operation by sensor fusion," U.S. Patent 5,482,314 (9 Jan 1996).
10. D. S. Breed, V. Castelli, W. C. Johnson, W. E. DuVall, and R. M. Patel, "Vehicle occupant position and velocity sensor," U.S. Patent 5,653,462 (5 Aug 1997).
11. D. J. Tupman, "Optical seat switch," U.S. Patent 4,683,373 (28 Jul 1987).
12. J. H. Kamei, C. L. Boon, and P. F. Stevens, "Automotive occupant sensing device," U.S. Patent 5,528,698 (18 Jun 1996).
13. J. H. Semchena, E. M. Faigle, R. J. Thompson, J. F. Mazur, and C. E. Steffens, Jr., "Apparatus and method for controlling an occupant restraint system," U.S. Patent 5,531,472 (2 Jul 1996).
14. H. Takahashi, M. Hamada, H. Fujishima, M. Naito, K. Sasaki, J. Tsuchiya, and T. Maeda, "Air bag apparatus for passenger seat," U.S. Patent 5,702,123 (30 Dec 1997).
15. S. B. Gentry, J. F. Mazur, and B. K. Blackburn, "Method and apparatus for detecting out of position occupants," U.S. Patent 5, 330,226 (19 Jul 1994).
16. R. L. Phen, M. W. Dowdy, D. H. Ebbeler, E-H. Kim, N. R. Moore, and T. R. VanZandt, "Advanced air bag technology assessment," Jet Propulsion Laboratory Report No. 98-3 (April 1998).
17. W. Fung and K. C. Parsons, "Some investigations into the relationship between car seat cover materials and thermal comfort using human subjects," J. Coated Fabrics 26 (1996) 147.
18. Z. Xizhong, D. Zizhu, and Z. Genhong, "Application of the heat flux meter in physiological studies," J. Therm. Biol. 18 (1993) 473.
19. F. Arnaud, G. Delhomme, A. Dittmar, P. Girard, L. Metchiporouk, C. Martelet, R. Cespuglio, and W. H. Newman, "A micro thermal diffusion sensor for non-invasive skin characterization," Sensors and Actuators A, 41-42 (1994) 240.

1999-01-0761

Occupant Classification System for Smart Restraint Systems

K. Billen, L. Federspiel, P. Schockmel and B. Serban
I.E.E. International Electronics & Engrg.

W. Sherrill
I.E.E. Automotive USA, Inc.

Copyright © 1999 Society of Automotive Engineers, Inc.

ABSTRACT

The primary function of the Occupant Classification System is to provide reliable passenger seat occupancy information to the automobile's central processing unit to control airbag deployment.

Our Occupant Classification (OC) sensor system is based on analysis of the seat occupancy pressure profile, which discerns human like from human unlike profiles. If the occupancy is identified as a person, an allocation into one of four morphologic ranges is made. Accordingly, children, light adults, heavy children, medium adults, heavy adults, etc. can be discerned (Figure 1).

INTRODUCTION

In reply to: contrast to actual one-stage airbag systems, future restraint systems will offer a multitude of triggering possibilities. The goal of such systems is to reduce the risk and magnitude of injuries by automatically adapting the airbag and seat belt pretensioner to the driving status of the vehicle, its occupants, and the crash severity.

Several systems are in development. They are based on different measurement principles, such as strain gauges, liquid pressure, optical fibers, capacitive, ultrasonic, infrared, etc. IEE's sensor is based on FSR, (Force Sensing Resistor) technology. The sensor is assembled in the seat and captures a pressure related seating profile.

Literature and tests performed with hundreds of persons indicate a correlation between the anthropometric characteristics of the person and its corpulence. The analysis of characteristic parameters extracted from the pressure profile allows classification of seat occupancy (Figure 2).

Child Seat

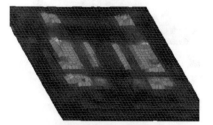

5th Percentile Female

50th Percentile Male

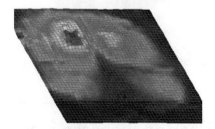

9 Year Old Child

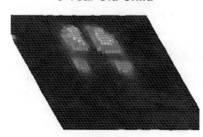

Figure 1. Pressure Profiles Of A Child Seat (Human Unlike) And 3 Persons Belonging To Different Weight Ranges

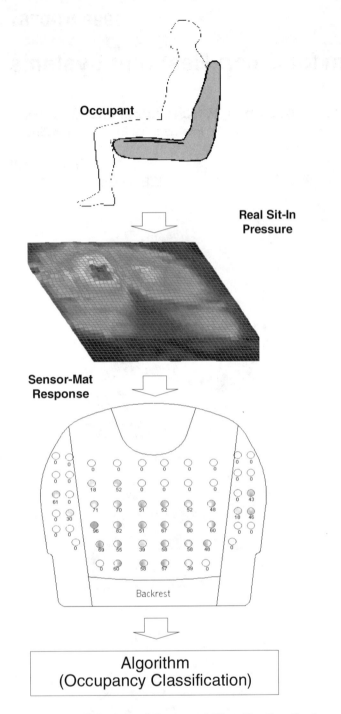

Figure 2. Principle of Occupant Classification System

DESCRIPTION OF PROPOSED SYSTEM

SYSTEM OVERVIEW – The total system consists of the following sub systems (Figure 3):

- OC Sensor Mat (FSR® technology)

  The sensor mat consists of two polymer films screen printed with conductive patterns and connecting leads, then sandwiched together with a polymer spacer sheet in between (Figure 4). Up to 100 single cells can be integrated in one sensor mat. A special circuit design provides access to each cell.

- Measurement Unit (Microcontroller, ASIC)

  With an ASIC controlled by a microcontroller all FSR®-cells are measured and the resistance values are digitized. A specific algorithm is used handling the digital resistance matrix to provide an object classification that is used by the airbag control unit.

- Airbag Control Unit

  This unit controls the deployment of the airbag using all available crash information.

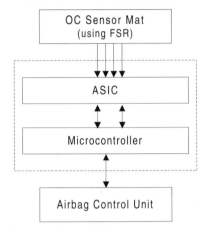

Figure 3. System Overview

FSR TECHNOLOGY – The **F**orce **S**ensing **R**esistor® sensor consists of two facing substrates on which proprietary pressure-sensitive films and electrodes are printed (Figure 4).

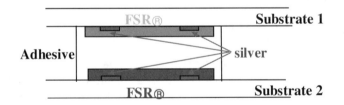

Figure 4. FSR® technology

When a force is applied on a sensor cell its electrical resistance decreases. This decrease is a monotonous function of the applied force.

SENSOR MAT – The occupant classification sensor consists of a mat with multiple sensor cells whose resistance values are acquired and converted into a set of digital values which provide a discrete pressure profile (Figure 5).

The sensor is assembled between the seat trim and cushion. Each unique seat design requires specific sensor adaptation. Since 1995, starting with the **P**assenger **P**resence **D**etection System, IEE has adapted its sensors to more than 50 seat types. Figure 6 shows an equivalent circuit of the OC mat. The FSR® cells (R) are connected by printed silver leads (R_l) in a matrix configuration. Fixed resistors (R_f) at the end of each row and column connection allow for several hardware self-tests, e.g., short circuits, wire breaks, leakage current, etc. to check the sensor mat integrity.

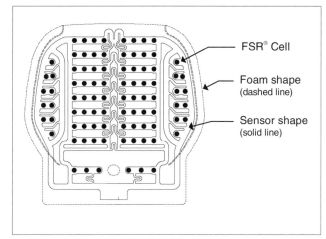

Figure 5. OC Sensor Layout

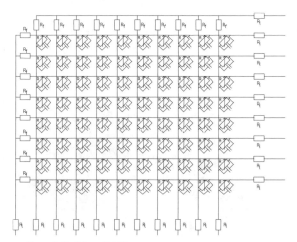

Figure 6. OC Circuit Diagram

MEASUREMENT UNIT – This electronic unit consists of four main components: the microprocessor, the ASIC, the EEPROM and the communication interface. The microprocessor controls all operating sequences and runs the classification algorithm. The ASIC is the unit that interfaces the FSR® cells with the microprocessor. All the measurement steps performed by the ASIC are controlled by the microprocessor via serial synchronous interface. The EEPROM stores all of the configuration data and relevant system data in case of a crash. The communication interface allows the right data flow between the OC electronics unit and the airbag control unit according to the OEM protocols. Therefore, every OC electronics unit is customer dedicated.

DATA PROCESSING – The processing of the data from the acquisition to the classification is shown in Figure 7. The discrete pressure profile is analyzed by an algorithm, which evaluates several parameters and calculates the digital pressure data. Different filters and memory functions make the system insensitive to the dynamic environment and movements of the seat occupancy. The mathematical and logical evaluation of all the calculated parameters determines the classification (airbag control data).

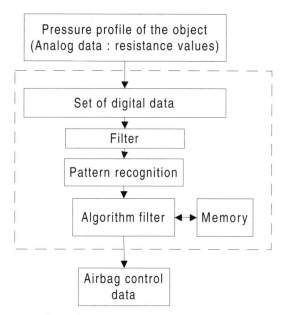

Figure 7. Data Processing

The primary function of the OC-Algorithm is the classification of objects. First, we make a distinction between human beings and non-human beings, then further distinguish according to human anthropometric characteristics. Typical objects of the non-human class are child seats, cases or bags.

MAIN PROFILE PARAMETERS – The complete occupant identification and classification is made through the evaluation of specific functions, called *Profile Parameters*. The most important ones are described below:

The **Object Parameter** is used to discern human like and human unlike occupancies. This parameter relies on the fact that the interaction between the human buttock and the seat surface leads always to a compact pressure profile while all other types of occupancies lead to non-homogenous profiles.

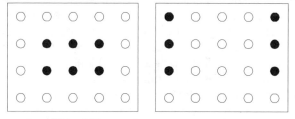

Figure 8. Object Parameter

Figure 8 depicts two distinct types of pressure profiles, both cases have only six activated sensor cells. A very compact activation pattern (left) is typical for human beings, while child seats usually generate a dissipated, fuzzy pattern (right). The *Object Parameter* is able to make a clear distinction between these two patterns by analyzing their topological distribution despite having the same amount of activated cells in both patterns. Figure 9 shows the calculation of this parameter for different people and child seats. It can be seen that at the same weight (e.g. at 40kg in the graph) the *Object Parameter* may have different values if the occupant is a child seat or a human being.

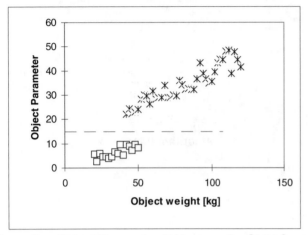

Figure 9. Object Parameter (* Humans & - Child Seats)

The **Coherence Parameter** is an additional parameter used for the profile identification. It is used to evaluate the size and the shape of an activation pattern. An additional information concerning profile discontinuities is also provided.

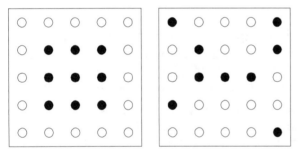

Figure 10. Coherence Parameter

Figure 10 shows the role of the *Coherence Parameter* in the occupant classification. Again, an equal number of cells are activated in both cases. In the left profile the cells are forming a coherent area, while they are distributed in a discontinuous manner in the right (non-human) profile. The profiles are easily distinguishable from the other by the absolute value of this parameter. The *Coherence Parameter* has a similar behavior as previous but depends more significantly on the weight of a human being as it is shown in Figure 11.

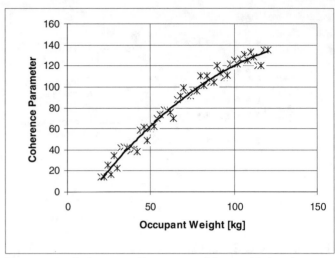

Figure 11. Coherence Parameter (Human Beings)

The third parameter, **Width Parameter**, allows classification of the human passenger by major anthropometric class.

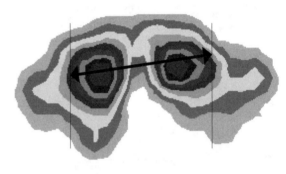

Figure 12. Width Parameter

As shown in Figure 12, the Width Parameter contains the information concerning the geometry of the buttock pressure profile.

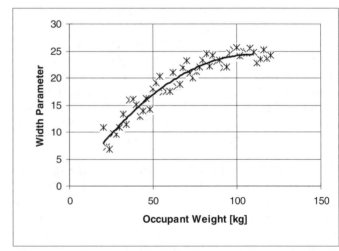

Figure 13. Width Parameter (Humans)

Figure 13 shows the dependency of the *Width Parameter* with weight for human beings. According to the actual implementation of the algorithm the Width Parameter will not be calculated if the occupant is not a human being.

The **Profile Quality** function is the parameter, which allows a stable classification during the seat occupancy. It quantifies the overall profile quality or *goodness* by combining the previously defined parameters and seat specific parameters. Its evaluation provides critical information to the filtering mechanism. Thus, by using the information about the object position, the activation pattern, the profile orientation and by following a pre-programmed decision tree, the filtering procedure is able to provide at any moment the correct passenger classification with no regard to the instantaneous passenger position on the seat.

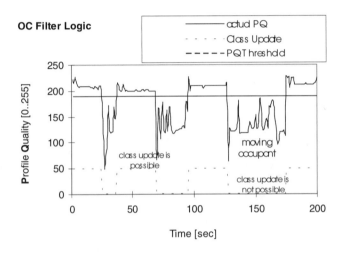

Figure 14. OC Filter Logic using a *Profile Quality Parameter*

Figure 14 shows how the classification update is made when the Profile Quality has values higher then a pre-determined threshold. This Threshold is itself a function of history and occupant position.

CONCLUSION

The morphology-based occupant classification system is able to provide an instantaneous classification. The system efficiently discerns child seats and other non-human seat occupants from human ones. The human occupants (passengers) are classified with regard to their corpulence and to other anthropometric properties. A precise position of their center of gravity on the seat is also provided.

ACKNOWLEDGMENTS

The authors would like to express their gratitude to the OEM's and organizations that are supporting IEE in the development of the Occupant Classification system.

REFERENCES

1. The BMW Seat Occupancy Monitoring System: A Step Towards "Situation Appropriate Airbag Deployment*, Klaus Kompass, Michel Witte, SAE Paper

2. Automatic Passenger Presence Detection and Child Seat Orientation Detection, Andreas Hirl, Peter Popp, Joachim Uhde, Paul Schockmel, SAE Paper

3. Situation Appropriate Airbag Deployment: Child Seat Presence and Orientation Detection (CPOD), Thierry Goniva, I.E.E. S.àrl

4. Real World Development of Adaptive Restraint Systems through the Use of Anthropometrically scaled Occupant Simulation Models, L. Michaelson, R. Hoffmann, Alzenau

5. Anthropologischer ATLAS, Alters- und Geschlechtsvariabilität des Menschen, Dr. B. Flügel, Dr. H. Greil/ Prof. K. Sommer

6. 1998 Anthropometric Survey of U.S. Army Personnel: Methods and Summary Statistics, United States Army Natick Research, Development and Engineering Center, Natick, Massachusetts 01760-5000, C. Gordon, Th..Churchill, Ch. E. Clauser, B. Bradtmiller, J.T. McConnville, I. Tebbetts, R. A. Walker

7. Occupational Ergonomics, Theory and Applications, James F. Annis and John T. McConville edited by Amit Bhattacharya, James D. McGlothlin

8. Body Dimensions of People; Terms and Definitions, measuring procedures, DIN 33402, Part 1

CONTACT

William F. Sherrill- U.S. Operations Manager
IEE Automotive USA, Inc.- Livonia, MI 48150
Phone 734.458.1700 Fax 734.458.7330
E-mail: IEE_Automotive@Compuserve.com
Website: http://www.iee.lu

Influence of Air Bag Folding Pattern on OOP-injury Potential

Yiqin Mao and Hermann Appel
Institute of Automotive Technology, TU Berlin, Germany

Copyright © 2001 Society of Automotive Engineers, Inc.

ABSTRACT

Four different air bag folding patterns are discussed in this paper. The influence of air bag folding pattern on inflation characteristics and occupant OOP-injuries is simulated and evaluated. The results indicate that different folding patterns strongly influence their aggressiveness. Compared with the other foldings, the leporello-folding, corresponding to the high opening pressure and high "punch out"-, "bag slap"-, "membrane loading"-effect is proved to be a critical folding pattern for the OOP-Injuries.

This study also shows that the combination of design, test and simulation provides a comprehensive and effective tool to investigate the influence of air bag design parameters on the OOP-injury and to develop the "Advanced Air Bags".

INTRODUCTION

Although the air bag has been proved to be an effective device to reduce injury risk in automobile collision, there is evidence of risk of fatal injuries especially to out-of-position small female and child occupants who interact with the air bag during its deployment. In order to reduce the injury potential of air bags NHTSA has proposed a change in FMVSS208, so called SNPRM (Supplemental Notice of Proposed Rulemaking), to require "Advanced Air Bags".

The folded air bag is an important component of the air bag module. Tests have already indicated that the folding pattern has a significant influence on occupant injury under OOP-conditions[1]. In this paper the air bag folding is of special interest. Four air bag folding patterns - leporello-folding (conventional), raff-folding (also know as petri-folding), stochastical folding and Z-folding, abbreviated to L-, R-, S- and Z-folding respectively - are studied. Their influences on OOP-injury potential are evaluated and compared with each other.

Finite Element Technique is widely used in numerical simulation for occupant safety. However, exactly modeling the folded air bags and simulating the interaction between air bag and dummy during deployment are very complex problems. Especially for R-, S- and Z-folding with crumpling, distorting or crushing, generating folded air bag meshes is very difficult. Therefore a pre-processor named "Fold-NEGE" is developed to generate the nods and elements of folded air bags and is combined with PAM-SAFE program during the simulation.

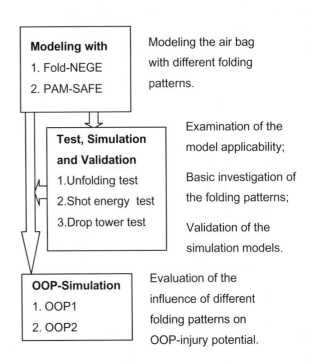

Figure 1: Methodology for the investigation of the influence of the folding pattern on the OOP-injury potential of the air bag.

It is well known that OOP test results are very sensitive to system parameters and it is very difficult to achieve the identical results through repeated laboratory test [2]. So three kinds of tests, unfolding test, shot energy test and drop tower test are conducted in this work. Compared with OOP test they are simple and economical to point out the tendencies of the characteristic of the air bag module. Due to eliminated system parameters high reproducibility can be achieved. Additionally conclusions for OOP situations are possible because of the small distance of the impactor / drop rig to the air bag system in the initial state, that is similar with OOP-situation. Therefore they are very suitable for the basic investigation regarding the OOP-injury potential of an innovative measure on the air bag system and for validating the simulation models.

The final OOP simulation will be done only after successful validation of the simulation models and plausible evaluation of the influence of the folding pattern on the injury potential of the air bag system with these three tests. With the verified simulation models the expected occupant loads in OOP situations are calculated regarding differently folded air bag systems, then an evaluation of the OOP-injury potential of the different folding patterns can be achieved.

Under this consideration the strategic methodology is represented in figure 1.

FOLDING PATTERNS AND MODELING

Four folding patterns are discussed in this paper. They are L-, R-, S- and Z-folding.

L-folding is the conventional folding scheme for air bags. Its folding lines are straight and the air bag is folded in accordion-type layers to a package which is generally located directly above the inflator (figure 2). The individual folds are situated above and into one another and form a package with self blocking. This type of folding is implemented in the PAM-SAFE program and can be modeled easily. However the manufacturing is complex.

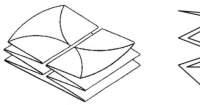 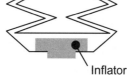

Perspective view (extract from[1])   Cross-section view

Figure 2: Geometry of the L-folding

With the **R-folding** the air bag is formed to a concentric 3D wave structure around the inflator and then pressed radially to a package in air bag module (figure 3). The main folding lines run along closed circles around a center and radiate from it.

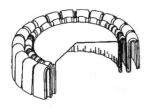 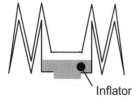

Perspective view (extract from[1])   Cross-section view

Figure 3: Geometry of the R-folding

The **S-folding** represents the type of folding, whose folding lines are not regular, but stochastical (figure 4). The flat air bag is simply pressed (shifted) from the outside to the center and packed into the air bag module.

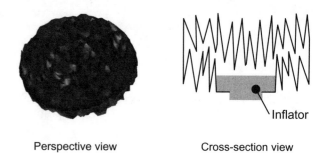

Perspective view   Cross-section view

Figure 4: Geometry of the S-folding

The **Z-folding** comes from the "peter patent" [3]. The fabric is folded like accordion vertically around the scope of the gas inflator and above it (figure 5). Regarding the position of the upper and lower surface, especially to the folding lines, the Z-folding is similar to the R-folding. The folding process is not easy, that leads to high production cost.

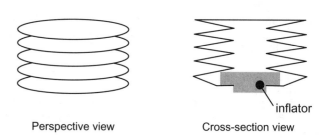

Perspective view   Cross-section view

Figure 5: Geometry of the Z-folding

Considering the folding technique and economy the characteristics of the folding patterns are summarized in table 1.

	Location of the fabric package to inflator	Location of the upper to lower surface	Main folding lines	Economy
L-folding	above	In / above	straight-lined	traditionally good
R-folding	around / above	above	circularly	good
S-folding	around / above	above	circularly	very good
Z-folding	around / above	above	circularly	not good

Table 1: Characteristics of the folding patterns

In order to examine the influence of folding patterns directly the folding pattern is assumed as the single design parameter to be researched on the air bag system. All the tests and simulations are carried out with a driver-side air bag module that has the same flat air bag configuration and the same inflator. The folding patterns vary among the L, R- S- and Z-folding. It is generally time consuming to generate a mesh which can demonstrate actual folded bag configuration and can be packed in the limited space of the house. In order to describe the R-, S- and Z-folding realistically and to generate the air bag mesh, a pre-processor "Fold-NEGE" is developed. With this pre-processor **the finite element model** of the air bag in the folded configuration is obtained.

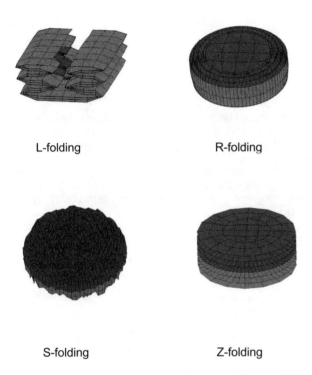

Figure 6: Modeling the folded air bag with "Fold-NEGE"

TEST, SIMULATION AND VALIDATION OF FUNDAMENTAL TESTS

As indicated in the introduction three kinds of tests* - unfolding test, shot energy test and drop tower test - are carried out in this work. They serve on the one hand as the fundamental research about the influence of the folding pattern on the unfolding process and injury potential on the other hand as the validation of air bag simulation model.

In the **unfolding test** it is of interest to show which influences the folding patterns have on air bag kinematics and physical parameters (e.g. pressure, volume, etc.) during the deployment. The pressure and volume curves for the four fold schemes are compared in figure 7.

*: The tests are carried out on the support of an air bag supplier.

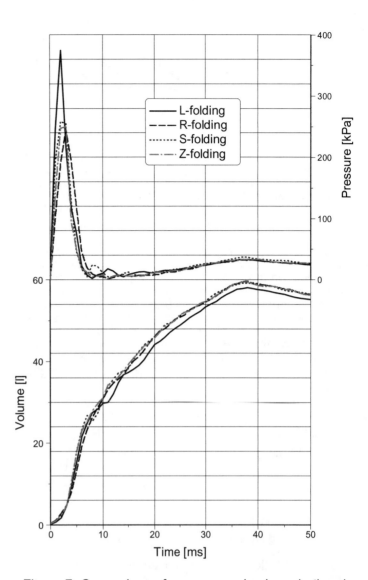

Figure 7: Comparison of pressure and volume in the air bag during the unfolding test

The air bag characteristics during the unfolding test in each case with L-, R-, S- and Z-folding are summarized in table 2.

	max. opening pressure [kPa]	max. work pressure [kPa]	max. expansion [mm]
L-folding	374,5	33,5	358
R-folding	244,6	32,5	326
S-folding	258,9	34,0	330
Z-folding	249,7	33,8	336

Table 2: Comparison of the influence of the folding pattern on the unfolding test

It is important to note that up to initial 10 ms the pressures are strongly different during the opening phase. This is mainly caused by the damping effect (self blocking effect) of the air bag [4, 5]. In the procedure of the air bag unfolding with the L-folding the volume increases a little bit slower because of self blocking and fabric friction, through which a higher opening pressure is generated. The opening pressure of the L-folding is 374,5 kPa, about 50% higher than that of the other foldings, however the work pressures show no large difference at approx. 34 kPa. Additionally the max. air bag expansion with L-folding is 358 mm, approximately 20 to 30 mm longer than with the other types of folding.

A high stored energy results from the high opening pressure. This stored energy affects through the "punch out"-effect on the occupant in OOP situation, so that the air bag with L-folding with higher opening pressure leads to a hard contact with the occupant, which causes higher load and injury potential. For explaining this stored energy a highly simplified mechanical model is constructed (figure 8). The energy stored by the gas pressure in the air bag is represented by a pre-strained spring and the thicker spring corresponds to the higher pressure.

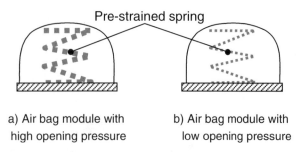

Figure 8: Model of the stored energy in the opening phase

In the **shot energy test** a half sphere shaped, hanging impactor which simulates the shape of the head was hit by a deploying air bag (figure 9). The shot energy is calculated with the formula E=mgh, where m is the mass of impactor and h is the max. height that the impactor travels. This gives a statement of unfolding energy.

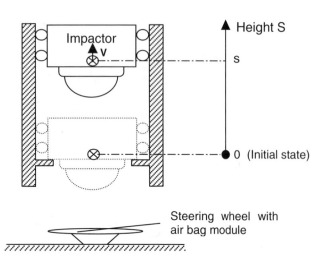

Figure 9: The shot energy test

The simulation results with an initial distance of 25cm between the impactor and the air bag module are shown in table 3. With L-folding the impactor achieved a max. height of 64.9 mm, almost two times higher than that of the other three foldings. Proportionally the air bag with L-folding shows the highest energy of 6.87J. Generally the R-, S- and Z-folding have similar results.

	L-folding	R-folding	S-folding	Z-folding
max. height [mm]	64,9	24,9	28,8	27,8
shot energy [J]	6,87	2,64	3,05	2,94

Table 3: Simulation results of the shot energy test

The characteristic for the **drop tower test** is the 3 millisecond acceleration (a3ms) and the speed change (Δv) of the drop rig body which simulates the occupant thorax (figure 10). The simulation results are shown in table 4. Similar to the shot energy test it can be observed that the air bag with the L-folding indicates a higher injury potential than the remaining types of folding.

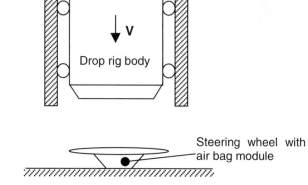

Figure 10: The drop tower test

	L-folding	R-folding	S-folding	Z-folding
a_{3ms} [g]	73,2	51,7	44,6	50,3
Δv [m/s]	14,78	11,99	12,25	12,24

Table 4: Simulation results of the drop tower test

The evaluation and the improvement of the model validity proceed usually through the comparison of the model behaviour with the experimentally determined system characteristic. The model **validation** serves the examination and the adjustment of the model parameters and forms the basis for the execution of the simulation. Experiences show that only partial model parameters have sufficient accuracy, since not all processes can be determined exactly, for example, friction and damping due to non-linearity and unsteadiness. However by limiting the appropriate model conditions the uncertain parameters can be determined through sensitivity analysis and be adapted to real conditions with the model validating [6].

OOP-SIMULATION

As pointed out in methodology, the OOP-simulation as ultimate simulation is conducted after successful investigation of the air bags with different folding patterns in simple tests. The goal is above all an estimation of the influence of the folding pattern on the OOP-injury potential and the possibility to achieve advanced air bag fulfilling the SNPRM by favourable folding. The biomechanic occupant loading is evaluated according to the injury criteria of SNPRM.

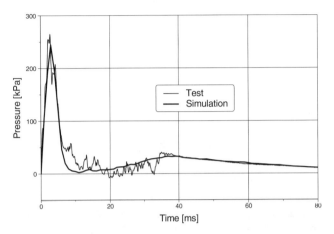

Figure 11: Comparison of pressure between test and simulation in the unfolding test with the R-folding

For effective study of fundamental effects of the folding patterns on the interaction with occupant in OOP situation, it is decided to conduct static OOP-tests. There is some ambiguity in the interpretation of ISO OOP-positions [ISO DTR 10982, 1996] [7]. In an effort to simulate the worst case of occupant loading the simulation comprised two main configurations with the 5% female HYBRID III dummy under the consideration of ISO OOP-Positions: 1. head centered on module (OOP1) (figure 13) and 2. chest centered on module (OOP2) (figure 14). The head or the thorax was positioned in direct contact with the air bag module with the spine parallel to the plane of the steering wheel. At the same time the middle axis of the air bag module passes through the center of gravity of the dummy's head or chest.

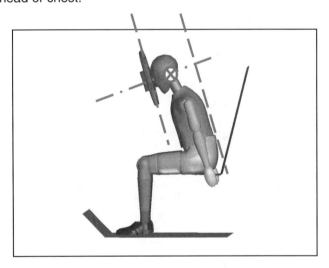

Figure 13: OOP1-position: head centered on module

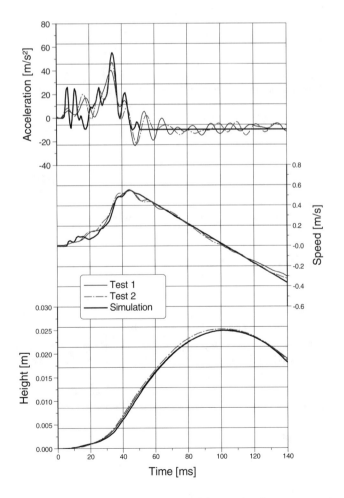

Figure 12: Comparison of impactor kinematics between test and simulation in the shot energy test with the R-folding

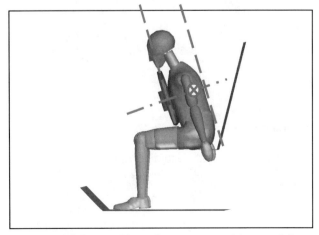

Figure 14: OOP2-position: chest centered on module

The simulation results in **OOP1** position indicate that the head acceleration with L-folding is clearly much higher than that of the other foldings in the first contact phase with the deploying air bag. The max. head and neck loads are shown in table 5 and figure 15. The HIC value of L-folding is 558, about 55% to 62% higher than that of R-, S- and Z-folding. Although the HIC values with all four foldings are under the HIC limit value, the neck loading dominates as the most dangerous injury mechanism in this OOP1 position. In this position the NIJ values with all the folding patterns exceed the injury criterion of NIJ=1. With 1,55 the NIJ value of L-folding is far higher than that of the other foldings.

other on the left side in the view of occupant. Analysis of the simulation shows that a gap forms in the middle, while the air bag deploys outward, and the bag slides from the chin. However the air bags with R-, S- and Z-folding spread out from the center. So the chin was hit not heavily by the air bag with L-folding and it causes a smaller flexion moment of the neck. A schematic representation is shown in figure 16.

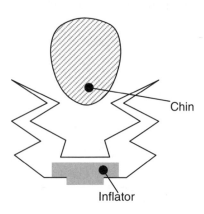

Figure 16: Schematic representation for the explanation of the neck flexion moment with L-folding

	Head loading	Neck loading			
	HIC_{15}	Fz [kN]	My(flex.) [Nm]	My(ext.) [Nm]	NIJ
L-folding	558	3,02	51,05	40,82	1,55
R-folding	307	1,61	105,00	19,88	1,16
S-folding	328	1,55	101,04	28,09	1,11
Z-folding	344	1,95	95,21	30,11	1,19
limit value of SNPRM	700				1,0
Intercept value		(3,37)	(155)	(62)	

Table 5: Occupant loading in OOP1-position

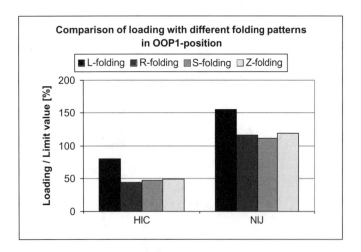

Figure 15: Comparison of the HIC and NIJ value regarding their limit values in OOP1-position

Almost all loads with L-folding are clearly much higher than those with the other folding patterns except a lower flexion moment of 51,05 Nm, which corresponds about half of the values of the remaining types of folding. This reduction for the L-folding can be traced to the unfolding process of air bag with L-folding. With L-folding the air bag is situated mainly above the inflator and is concentrated into two sections, one on the right side the

Positioning the dummy with chest centered on the air bag module in **OOP2** position the loads applied to the chest and neck are significant. Table 6 and figure 17 show the simulation results. The chest injury criterion CTI with L-folding exceeds the critical limit value, while this value with other foldings remains under the limit. The neck loads with all folding patterns are uncritical, however the NIJ value with L-folding is higher than that with the remaining foldings.

	Chest loading				neck loading
	a_{3ms} [g]	a_{max} [g]	deflect. [mm]	CTI	NIJ
L-folding	59,36	64,14	42	1,21	0,79
R-folding	39,75	42,85	33	0.87	0,62
S-folding	40,45	44,44	37	0.93	0,68
Z-folding	38,53	41,79	37	0.90	0,61
limit value of SNPRM	60		52	1,0	1,0
intercept value		(90)	(84)		

Table 6: Occupant loading in OOP1-position

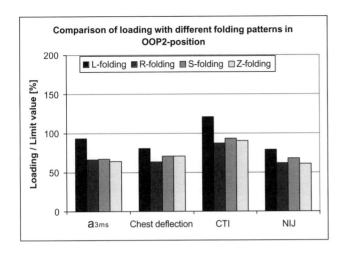

Figure 17: Comparison of loading regarding their limit values in OOP2-position

The different system responses, i.e. the different loads of the dummy, can be traced again to the folding pattern. Since with the L-folding the air bag fabric is in and above one another and forms a self blocking, a rapid pressure increase with substantially higher opening pressure is obtained in the opening phase, from which a higher stored energy results. This stored energy affects the dummy through the "punch out"-effect. So the air bag with the L-folding has a hard contact with the dummy's head (OOP1)/chest (OOP2), which leads to clearly higher head and neck(OOP1)/chest (OOP2) loading.

Because of whip-strike-like unfolding from the inside toward the occupant the air bag with L-folding has considerable "bag slap"- and "membrane loading"- effects, which also lead to higher occupant loading.

In comparison with the L-folding the upper layer of an air bag with R-, S- and Z-folding is always above the fabric lower layer. Thus, no self blocking develops during the unfolding process and they have similar unfolding behaviour. The air bag spreads out radially. Due to this characteristic the R-, S- and Z-folding result in reduction of "punch out"-, "bag slap"- and "membrane loading"- effects and have a much more "gentle " unfolding process, which leads to the lower occupant loading.

CONCLUSION

In this paper the influence of folding patterns on the injury potential is examined and the folding patterns are proved to be an effective constructional parameter to achieve the air bag with „low risk". The folding patterns have strong influence on the OOP-injury potential. The L-folding indicates a substantially higher injury potential. The R-, S- and Z-folding have no significant difference and show gentler deployment with less aggressiveness. The similarities as well as the differences are essentially due to the location of the upper to lower bag surface and the location of the fabric package to the inflator.

The main performances concerning the different folding patterns are summarized as follows:

- The air bag with L-folding is folded in accordion-type layers and packaged directly above the inflator. The layers of fabric are situated above and inside one another. However the air bag with the R-, S- or Z-folding is mainly folded around the inflator. The air bag upper layer always lies over lower layer and only one layer of air bag fabric is situated directly over the inflator. Just because of this similarity of R-, S- and Z-folding their behaviour is not considerably different.

- During air bag deployment with L-folding the pressure affects at first only up to the respective folding edges. Only after expansion of the folding edge the air bag continues to spread. Because of the location of the upper to lower bag layer with R-, S- and Z-folding there is no such self blocking during the air bag unfolding.

- The self blocking leads to high opening pressure with high stored energy. So the air bag with L-folding has a substantially higher opening pressure which causes a stronger "punch out"-effect on the occupant than that of the R-, S- and Z-folding.

- Because of the location of the fabric package to the inflator the entire bag package with L-folding must be whip-strike-like ejected from the module. The fabric acceleration in direction towards the occupant is higher. At the meantime the air bag fabric with L-folding deploys also through its "local unfolding" with relative motion between bag and occupant. So the L-folding leads to considerable "bag slap"- and "membrane loading"-effect.

- However the air bag with R-, S- or Z-folding begins to deploy with the center portion over the inflator and is pulled out of the module. It inflates immediately in a radial direction and does not travel so far toward the occupant during the initial deployment phase. The fabric acceleration in direction towards the occupant is lower [8].

- Because of the special features of the folding pattern the air bag with L-folding has a considerably higher injury potential. The R-, S- and Z-folding show comparable gentler development processes and reduced injury potential. With the consideration of economy the stochastic folding is recommendable.

- Independent of sensor technology and inflator characteristic the air bag deployment behaviour can be controlled through folding pattern. For example, with stochastic folding a less aggressive air bag can be obtained.

- It can be supposed that air bag with favourable folding, supported by improved sensor technology and stage inflator, can fulfil the requirement of the SNPRM

The simulation results which correspond to the test results in unfolding test, shot energy test and drop tower test indicate that the effects of the folding patterns on the injury potential are successfully simulated. Based on the air bag models with different folding patterns which are validated with simplified test configurations it is possible to simulate OOP problems and to reveal the influence of the different folding patterns on the OOP-injury potential qualitatively correctly. The methodology used in this work is proved to be successful.

This study also shows that the combination of design, test and simulation provides a comprehensive and effective tool to investigate the influence of air bag design parameters on the OOP-injury and to develop the "Advanced Air Bags".

CONTACT

Yiqin Mao and Prof. Hermann Appel can be reached through the Institute of Automotive Technology, TU Berlin, Germany or email ymao@kfz.tu-berlin.de and appel@kfz.tu-berlin.de.

REFERENCES

1. Malczyk, A. and Adomeit, H-D.: "The Airbag Folding Pattern as a Means for Injury Reduction of Out of Position Occupants", SAE 952704
2. Sheng,J.; Mu,W.; Chen,C.; Bayley,G.: The Investigations of Effects of Air Bag System Parameters on Out-of-Position Occupant Response. VDI-Tagung "Innovativer Kfz-Insassen- und Partnerschutz", Berlin, VDI-Berichte Nr.1471, 1999
3. Wils, Oliver.: Untersuchung des Schutz- und Gefährdungspotentials von Airbagsystemen und hierzu alternativen Schutzeinrichtungen unter Zuhilfenahme der rechnerischen Simulation. F84 D13/95, Diplomarbeit an der TU Berlin
4. Adomeit, H.-D.; Meißner, Dirk: Ergänzung im Rückhaltesystem oder Basis für neue Konzepte des Insassenschutzes. Bag & Belt'94, Köln, 1994
5. Adomeit, H.-D.; Malczyk, Axel: Verminderte Out-of-Position-Risiken durch PETRI-gefaltete Airbags. Bag & Belt'96, Bonn, 1996
6. Kramer, Florian: Passive Sicherheit von Kraftfahrzeugen. ATZ-MTZ-Fachbuch. Vieweg, ISBN 3-528-06915-5, 1998
7. Bass, Cameron R.; Crandall, Jeff R.; Pilkey, Walter D.: Out-of-Position Occupant Testing, Report OOPS3, University of Virginia, Automobile Safety Laboratory, 1998
8. Mao, Yiqin; Appel, Hermann: Numerical Investigation of Air Bag Fold Scheme. International Federation of Automotive Engineering Societies (FISITA)'1998, Paris

IV. LABORATORY PERFORMANCE OF AIR BAGS – OCCUPANT RESTRAINT

740578

Human Volunteer and Anthropomorphic Dummy Tests of General Motors Driver Air Cushion System

COPYRIGHT 1974
SOCIETY OF AUTOMOTIVE
ENGINEERS, INC.

George R. Smith, Emery C. Gulash, and Roy G. Baker
Environmental Activities Staff, General Motors Corp.

VEHICLE RESTRAINT SYSTEMS have generally been evaluated solely with anthropomorphic dummies and defined measures of injury criteria. With a restraint as different as the air cushion, General Motors (GM) has taken an additional step in an attempt to understand the interface between the occupant and restraint system. A program was established to test the driver air cushion system in direct frontal impacts on an impact sled using human volunteers. This program, conducted by Southwest Research Institute, San Antonio, Texas, under contract to General Motors Corporation, was similar to tests on the passenger side air cushion which were done at Holloman Air Force Base for the National Highway Traffic Safety Administration. In both programs, the tests were conducted under the most safe and controlled conditions possible, and typify only what the air cushions might accomplish in direct, head-on impacts.

The primary concern for both GM and Southwest Research Institute (SwRI) was to safeguard the rights and welfare of the human volunteers. The test program was designed to accomplish this.

This paper describes the test procedures and presents observations on the human kinematic and physical response to the impacts. Data is provided on the dummy (Hybrid II 50th percentile and 95th percentile) and human tests.

APPROACH

The general approach to this test program involved the stepped-severity concept which was successfully utilized at Holloman Air Force Base on the passenger system tests. The human volunteers were exposed to increasingly severe impact environments beginning at a level approximately equivalent to a 13 mph (21 kph) barrier crash and culminated as planned at a 30 mph (48 kph) barrier crash condition. Tests were conducted at eight different impact severity levels, with each level used as a step toward the final 30 mph barrier equivalent. Each level contained four dummy tests (three 50th percentile and one 95th percentile) and five human tests. The 50th percentile dummy was a Hybrid II and the 95th percentile was a modified Sierra 292-895. The decision to proceed from one severity level to the next was based on the volunteer's comments and information obtained from the transducers and high-speed motion pictures. A physician was present to perform a short pre-test physical

--------- ABSTRACT ---------

Dynamic tests of the General Motors driver air cushion system using human volunteers were conducted at Southwest Research Institute. Forty human tests were conducted at eight different impact severities. Thirty-two anthropomorphic dummy tests were made under similar conditions. The test work proceeded as planned through impacts equivalent to a 30 mph (48 kph) barrier crash of a full size vehicle. No significant injuries were experienced by the volunteers. The extent of trauma was generally limited to minor abrasions, ecchymosis, and erythema. In comparable tests, the anthropomorphic dummies' response to impact was more exaggerated than the humans'.

335

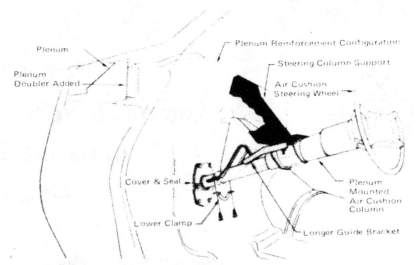

Fig. 1 - Main elements of GM driver air cushion system

exam and a post-test exam. As a part of these exams, he reviewed the recorded physiological data.

The tests were conducted by a team comprised of GM and SwRI personnel. GM had basic responsibility for the test buck and hardware while SwRI was responsible for the humans and impact sled facility. All test protocol, from schedules to checklists, revolved around one basic creed — human volunteer safety is of the highest priority. Care was taken in the planning and design of the experiments. Emphasis was placed on establishing test procedures, defining assignments, preparing and following checklists and reviewing emergency procedures.

Impact simulation at SwRI was based on an average velocity profile and corresponding acceleration waveshape which was supplied by GM. These average traces were compiled from full size vehicle barrier tests conducted by GM. SwRI was asked to match these traces as closely as possible with emphasis placed on matching the velocity profile. The severity of the test from the volunteer's standpoint was considered more closely related to the vehicle velocity than the instantaneous vehicle acceleration.

Extensive system testing preceded the first exposure of humans to an air cushion. Validation of the system (test buck, hardware and ACRS) was conducted at the GM Safety Research and Development Laboratory. Initial SwRI tests included sled correlation (dummy), animal (baboon) and harness indoctrination (human).

GENERAL MOTORS DRIVER ACRS DESCRIPTION

The restraint system evaluated in the test program was similar to production equipment installed in the 1000 Car GM Field Trial Program. The main elements of the driver air cushion system are an energy absorbing steering column, a steering wheel which houses a chemical gas generator and air cushion, and a knee restraint to absorb lower torso energy (Figure 1).

The steering column's energy absorption is accomplished by steel balls extruding a track between two telescoping jackets during column compression. The column has a dual stage energy absorber to obtain an improved energy absorption rate for the air cushion system. Initial low compression load is created by pretracking the balls through the first inch of travel to reduce the initial inertia spike on the driver. The columns differed slightly from regular production air cushion equipment because they were installed without a shift lever, ignition lock, hazard warning switch, turn signal lever and were locked to prevent rotation.

The steering wheel hub supports the air cushion module and the inflated cushion is supported by the spokes and wheel rim. The air cushion module is composed of a chemical gas generator, a neoprene coated nylon air cushion, an inner container and an outer vinyl cover.

The knee restraint, mounted below the instrument panel, consists of a sheet metal retainer, flexible foam and vinyl skin. Its purpose is twofold, to control the driver kinematics during a frontal impact by restraining the lower torso and directing the upper torso and head into the steering wheel air cushion, and to mitigate or prevent injury to the legs.

A complete description of the system's overall operation may be found in SAE Paper No. 730605, "General Motors Driver Air Cushion Restraint System," by T. N. Louckes et al.

TEST ENVIRONMENT

The GM designed body buck (Figure 2) resembled the buck used at Holloman and simulated the interior dimensions of a 1973 Oldsmobile Delta 88. An Oldsmobile production cloth bench seat was employed along with a special head

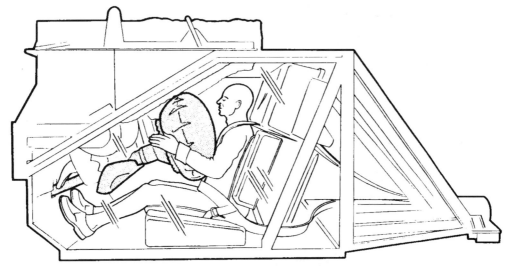

Fig. 2 - Primary restraint - driver air cushion system

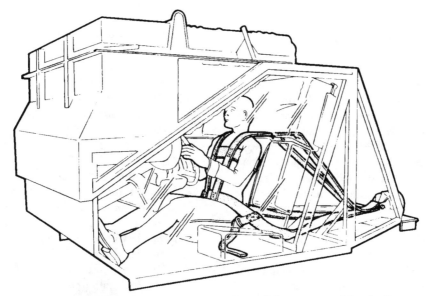

Fig. 3 - Secondary restraint - aircraft harness and lap belt

restraint which covered the entire width of the buck. The bench seat could be adjusted fore and aft to accommodate different size volunteers.

The simulated windshield was fabricated from styrofoam and shaped to the contours of the conventional windshield so the air cushion would interact against it in the same manner as it would in an automobile. The instrument panel above the knee restraint pad was a sheet metal mock-up covered by foam. A piece of Ensolite was inserted between the top of the knee restraint and the bottom of the instrument panel to prevent any volunteer hand or arm involvement with the bracketry.

In addition to the air cushion, which was the primary restraint, two backup restraint systems were provided as additional safety precautions. The air cushion system activated as intended in every test and the two backup systems were never needed. However, they did provide additional confidence for the volunteers in the test program.

The secondary restraint was a belt harness (Figure 3) that effectively restrained the shoulders, chest, lap, and legs. It was attached to a single bolt and separation nut assembly mounted behind the seat on the body buck.

The system was controlled by proper inflation of the air cushion. As the cushion blossomed, it separated the two flaps on the inner cover of the module. Two interrupt wires were attached across these flaps. As the flaps separated, the interrupt wires were broken which caused an electric circuit to feed current to the separation nut and released the belt harness. This allowed the volunteer to interact freely with the air cushion restraint system. If the wires were not broken because the air cushion did not fire upon signal, the interrupt wires prevented any current flow to the release system and

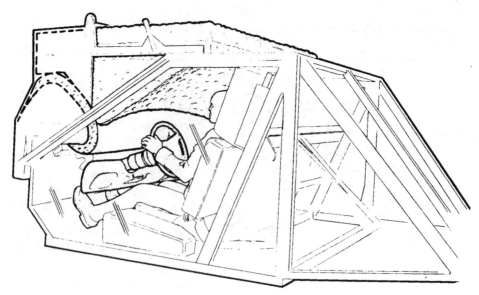

Fig. 4 - Tertiary restraint - energy absorbing interior

the belt harness then functioned as the primary restraint.

The belt harness system was also used in low speed volunteer indoctrination tests. Each of the volunteers was subjected to a simulated crash roughly equivalent to a 10 mph (16 kph) barrier crash while restrained by the belt harness system (no air cushion deployed). The purpose of this phase of tests was to acquaint the volunteers with the test procedures and instrumentation as well as building confidence in the secondary restraint. Some volunteers expressed that they received slight discomfort from the belts due to their tightness and some soreness to the neck caused from flexion during impact.

The tertiary restraint system involved the interior of the body buck (Figure 4). The windshield and instrument panel were specially designed to distribute energy in the event of failure of both the air cushion and belt restraints. The steering column's compression provides for energy absorption. Additional protection was offered, primarily against ejection by a clear plastic panel added across the left-hand body buck opening, and a special looping of the harness system around steel beams on the buck, which limits excessive excursion. Rebound protection was provided by the slab foam head restraint, which extends across the entire width of the buck.

The SwRI impact simulator is an "impact-with-rebound" type and is basically an MTS Model 858.05. The sled is accelerated up to the required speed by the use of natural-rubber bungee cords. The sled impacts an SwRI designed rebound deceleration programmer. The impact of the sled forces the programmer's piston back until the sled is brought to zero velocity. The piston then returns its stroke accelerating the sled in the opposite (rebound) direction. After the sled rebounds from the programmer, it is slowed by stretching the bungee cords and is brought to a complete stop by the rebound brakes.

A 30 mph (48 kph) barrier crash pulse of a full size vehicle was simulated on the SwRI impact sled. The lower severity levels were scaled from this pulse.

INSTRUMENTATION

The data collected during the course of this experiment included both mechanical and physiological parameters. During the test program, the data served two purposes, to assure continuous subject safety and to help evaluate the effectiveness of the driver ACRS.

Head and thorax accelerations were measured on both the volunteers and anthropomorphic dummies. The mounting techniques for the external accelerometers were the same for both humans and dummies. The triaxial head accelerometer was mounted on the left side of an adjustable plastic headband which was removed from the inside of a welder's helmet. The adjustment knob was removed from the back side of the headband and replaced by a strip of Velcro to provide infinite adjustment for head size. The accelerometer was located on the skull over the left temporoparietal region 1.5 to 2.0 inches (3.8 to 5.0 centimeters) above the Frankfort Plane and superior to the tragus. An Ensolite cup was placed over the accelerometer for the subject's protection.

The triaxial thorax accelerometer was mounted on a rubber-backed aluminum plate and covered by a 4 x 8 x 1 inch piece of Ensolite. The rubber backing was placed toward the subject and was centered over the left erector spinal muscle of the back at the level of the T6 vertebra. It was positioned off the mid sagittal plane to prevent

Fig. 5 - Volunteer accelerometer instrumentation

injury to any protruding spinous process. A strip of 2-inch wide Velcro was fastened tightly around the subject's thorax as he expelled air, fixing the accelerometer to him. The chest strap portion of the belt harness was then tightened around the Velcro strap. Figure 5 shows a volunteer in the body buck with the instrumentation attached.

The output from the standard internal head, chest, and femur transducers in the anthropomorphic dummies were also recorded. The physiological data monitored included electrocardiogram, pulse rate, respirations, and blood pressure.

VOLUNTEER PREPARATIONS

Healthy young males were recruited as test subjects by SwRI. All were from the San Antonio area and had diversified backgrounds, from students to firemen. All subjects were required to take and pass a rigorous physical examination. The exam was administered at SwRI's medical facility with the individual being sent to a radiology laboratory for X-ray studies. The physical investigation included an examination of the heart, lungs, ears, nose and throat, eyes, hearing, reflexes, muscle and joint motion, pulse rate and blood pressure. The individual was also subjected to a "double masters" test to determine ECG and cardiac status during and immediately post exercise. Baseline enzyme studies including creative phosphokinase (CPK), lactic dehydroginase (LDH) and serum glutamic transaminase (SGOT) were performed on the volunteers. All individuals accepted in the test program were required to have an X-ray evaluation of the entire spine and all major bones and joints by a series of 22 views to rule out significant arthritis with spurs or lipping, marked cortical thinning, joint abnormalities with calcification or foreign bodies, ununited or poorly united fractures. As a result of the intensive examinations, nine of the candidates who applied were disqualified for various reasons.

In addition, certain liberal anthropometric size ranges were adhered to. The subjects selected were able to place their feet flat on the toe pan while harnessed in the number two seat position (one notch rear of full forward), their seated height measurement fell in the range of 34.2 - 37.5 inches (87 - 95 centimeters) and their weights did not exceed 220 pounds (100 kilograms). These requirements were established because they represent the range of the 50th and 95th percentile dummies' sizes which were utilized in the tests conducted for this program at the GM Safety Research and Development Laboratory.

A complete set of anthropometric dimensions on each volunteer was also completed and is given in Table I. The dimensions included weight, stature, sitting height, acromion-radiale length, forearm grip distance, buttock-knee length, and sitting knee height.

Thirty-two individuals qualified for the tests. Each of these underwent an indoctrination test with the harness restraint system (no air cushion) and a full complement of instrumentation.

Twenty-six individuals participated in the air cushion tests. They ranged in age from 19 to 30 years. Their statures ranged from 67 to 74 inches (171 to 188 cm) and their weights from 138 to 208 pounds (62 to 94 kg).

Immediately before the scheduled test, a short physical exam was performed on the volunteer by the physician in attendance. The subject was then instrumented for ECG (Ag/Ag Cl electrodes) and blood pressure (sphygmomanometer cuff). Clothing consisted of tight fitting cotton long sleeve "ski pajama" tops and bottoms. A cotton "thermal" undershirt was worn beneath the pajama top. Rubber soled, low top gym shoes were worn in every test.

Pliable plastic goggles, earplugs, mouthpiece, and a rubber bathing cap were worn by the volunteers. The bathing cap was worn to provide a grip for the accelerometer mount. All jewelry was removed for the test. Cold cream was applied to the volunteer's nose, chin, thumb and wrist area to minimize possible abrasions from the deploying air cushion.

In all tests the volunteer was placed inside the body buck in a forward facing position and centered laterally behind the steering wheel. The seat was adjusted to allow the volunteer to place his feet firmly on the toe pan. His hands were placed in the 9 and 3 o'clock positions on the steering wheel with the thumbs locked around the rim. His head was upright in a normal driving position. The belt harness was tightened to the point where it became painful to the volunteer. As described previously, the belt harness always released at the start of an air cushion deployment because the cushion deployed properly. Each

Table I - Volunteer Anthropometrics

VOLUNTEER NUMBER	AGE Years	WEIGHT Lbs.	STATURE cm	SITTING HEIGHT cm	ACROMION-RADIALE cm	FOREARM-GRIP cm	BUTTOCK-KNEE cm	SITTING KNEE HEIGHT cm
1	22	173	183.5	93.5	34.7	39.0	63.0	57.0
5	25	191	178.5	92.3	38.0	38.7	59.3	54.4
9	23	160	177.0	88.0	39.3	41.0	59.5	55.5
13	20	146	178.6	90.6	38.8	37.8	60.3	56.8
14	25	186	176.0	90.0	37.0	38.4	59.5	55.5
15	25	156	185.5	95.0	38.4	39.1	60.2	57.2
18	30	200	185.5	95.0	40.0	40.0	62.0	57.5
20	24	171	188.0	95.0	40.0	40.5	57.5	58.5
21	24	184	179.0	91.3	37.5	39.5	59.6	56.4
22	25	168	174.0	92.0	37.4	37.7	56.0	55.0
23	21	199	180.5	94.0	41.0	39.0	57.5	57.5
24	22	176	181.5	93.5	38.5	37.5	59.4	56.5
27	25	170	177.0	89.5	36.0	35.5	60.0	52.0
28	26	165	177.0	93.5	35.0	38.0	57.5	53.8
29	25	208	183.0	93.0	37.5	38.8	65.0	57.0
30	24	150	176.0	88.5	36.0	35.5	59.0	54.5
31	21	179	181.0	93.5	38.6	39.7	61.5	58.0
33	25	202	180.0	94.5	37.5	38.0	61.5	58.0
34	19	138	176.0	88.5	36.0	38.0	59.0	55.3
35	21	162	178.0	92.5	40.0	39.5	59.0	55.4
36	20	160	184.0	89.5	40.0	40.5	65.5	57.3
37	27	150	176.0	91.0	36.0	36.4	56.5	52.5
38	28	175	171.0	87.5	35.5	36.0	59.0	53.0
40	20	181	176.5	91.0	37.0	38.0	57.5	54.0
41	21	182	176.0	90.5	38.0	37.5	58.0	53.5
42	19	174	178.4	91.3	37.8	38.2	61.2	56.2
50th Dummy	-	164	-	90.7	34.8	39.1	58.7	55.9
95th Dummy	-	217	185.7	96.5	39.4	41.2	64.0	59.4

volunteer was instructed to brace his whole body for an impact and to try and keep his hands on the steering wheel throughout the test.

Following the impact the volunteer was asked to relax and remain seated until the investigators had discussed the ride with him. The physician assessed and recorded the trauma, if any existed. The volunteer was then removed from the body buck and asked to fill out several subjective reports.

Many measures were taken to insure a safe ride for the volunteers and no significant trauma was experienced in any of the tests. However, as an added precaution, the physician provided certain emergency and resuscitative measures and devices for immediate employment in the impact laboratory. These included a cardiac defibrillator, tracheotomy set, tourniquets, splints, suture set, I.V. fluids, drugs for I.V. use and morphine.

The SwRI physician involved in the test program is board certified in surgery with a specialty in the diagnosis, treatment and surgery of trauma. If a subject had been injured severely to require movement to a hospital for more definitive treatment than could be administered at SwRI, arrangements for rapid transport to a nearby hospital, as well as emergency care, had been made. Fortunately, none of the backup precautions were used.

DYNAMIC TESTS

The air cushion deployed as intended in all of the tests. The single point release system of the secondary restraint harness system also operated as intended during the course of the experiment and the dummies and humans were free to interact with the driver air cushion restraint system.

The deploying cushion always began to blossom before the subject began to translate forward. During the deployment process, as the cushion mushroomed, it wiped across the face and forearms on the majority of subjects irregardless of test severity. The knees made contact with the knee restraint pad before the upper torso and head impacted the inflated air cushion. The toe pan and knee restraint loads decreased when the bag pressure began to increase from upper torso impact.

After initial head contact, there was little additional rotation of the head into or over the cushion, minimizing the amount of neck flexion. If the volunteers' hand came off the wheel, it generally traveled forward striking the simulated instrument panel with the back of the wrist.

The knee restraint effectively controlled the human kinematics by absorbing some lower torso impact energy through deformation and by directing the upper torso and head into the steering wheel air cushion. Additional energy was absorbed through the lower legs into the toe pan.

The steering column absorbed part of the upper torso energy. At the higher velocities, this was

accomplished in two stages. The first stage occurred when the column compressed due to the force exerted through the arms of the subject. The second stage of collapse occurred when the subject's upper torso and head loaded the steering column through the air cushion. The air cushion was effective in distributing the forces on the upper torso and head. There was, however, one case (Test 636 at the 27.5 mph level) in which the subject's upper abdomen contacted the steering wheel lower rim with sufficient force to permanently deform it 0.3 inches. There was no apparent internal trauma to the subject.

The volunteers traversed essentially in the same longitudinal plane of the vehicle on rebounding from the air cushion restraint system as they did during impact. The head rebound velocities of the volunteers was considerably less than the dummies tested at SwRI.

Only minor trauma was experienced by the volunteers during the test series. It would fit severity codes of 0 (no injury) and 1 (minor) of the abbreviated injury scale of the American Medical Association [1], where scale limits range from 0 (no injury) to 6 (fatal). The trauma in these tests was limited to erythema, abrasions, and ecchymosis with exception to three minor lacerations which were due to the construction method of the safety goggles used. A full listing of trauma, as compiled by the SwRI physician, may be found in Table II.

The areas where trauma was most frequently observed were the arms, face (mouth, nose, and chin), and the chest. Observed trauma to the legs occurred in only four tests. This minimal amount of trauma is particularly notable because no protective gear was worn on the legs, only cotton leotards. The majority of all trauma was apparently deployment induced. Patterns of trauma did not change over the range of impact severity levels. Superficial trauma was essentially the same at the 30 mph (48 kph) level as it was at 13 mph (21 kph).

A listing of the symptoms recorded on the physical symptom survey sheet by each of the volunteers may be found in Table III. No subjects recorded experiencing severe pain and only in two tests was moderate pain recorded.

According to the SwRI physician, no pathologically significant electrocardiogram (ECG) abnormalities could be demonstrated in any volunteer during the course of these tests. He felt the most severe adverse findings encountered in the test

---

Footnote [1] - Developed by the American Medical Association Committee on Medical Aspects of Automotive Safety, in cooperation with physicians representing medical specialties most involved in the diagnosis, care and treatment of crash injuries, and General Motors Corporation.

Table II – Observed Trauma

Approximate Barrier Equiv.	Run No.	Subject	Findings
13.0 MPH (21 kph)	508	33	Erythema and abrasions of forearm.
	509	18	Abrasions, contusions and erythema of chin, lower lip and both forearms. Sensitive area in rt. elbow for 72 hours.
	511	40	Ecchymosis and erythema lower lip and chin. Erythema both thenar areas.
	512	24	Ecchymosis both antecubital spaces and both thenar areas. Erythema mid-chest (minimal).
	513	34	Ecchymosis and erythema left antecubital space. Erythema right epigastrium.
18.0 MPH (29 kph)	543	23	Abrasions and contusions lt. antecubital space.
	544	9	Abrasions and contusions both forearms. Erythema periumbilical region.
	545	28	Abrasions chin and tip of nose. Erythema chest and epigastrium.
	559	30	Abrasions chin, lower lip and left upper arm. Contusion right thenar area. Also had mild confusion (no sequellae).
	560	5	Erythema right cheek and lower lip. Abrasions tip of nose and right abdomen.
20.0 MPH (32 kph)	577	15	Erythema both wrists. B/P dropped to 74/48, pulse became weak and thready because of hypoglycemia — recovered with coke and candy bar.
	583	21	Abrasions both forearms. Contusion left pretibial region from bolt head on dash.
	584	22	Erythema both thenar areas. Abrasion rt. wrist, nose, lower lip. Wheel rim contact was present.
	585	13	Erythema both antecubital spaces. Abrasion left wrist. Numbness both thumbs followed in 24 hrs. by swelling and tenderness left thumb. Disappeared in 36 hrs.
	588	35	Erythema epigastrium (minimal). Abrasions both wrists and forearms.
22.0 MPH (35 kph)	609	38	Erythema both wrists and forearms. Abrasions left forearm, antecubital space, chin and upper lip. Headache immediately post run. Had wheel rim contact.
	610	41	Erythema both antecubital spaces and in umbilical area. Had wheel rim contact.
	611	27	Erythema both antecubital spaces, abrasion right wrist and left knee. Steering wheel rim contact with breath knocked out.
	612	29	Erythema both forearms, both antecubital spaces. Abrasion right wrist.
	613	36	Erythema both wrists, both antecubital spaces and right epigastrium. Had wheel rim contact and his breath knocked out.

Observed Trauma (cont.)

24.0 MPH (39 kph)	635	28	Erythema chest and epigastrium abrasion thenar area. Had wheel rim impact.
	636	14	Erythema epigastrium and left chest. Both tibial regions. Abrasions both wrists and forearms. Had severe rim impact and breath knocked out.
	637	18	Abrasions both anticubital spaces and left patellar area. Laceration bridge of nose from goggles. Limitation of motion by pain left wrist.
	638	9	Erythema chest - minimal, abrasions both wrists.
	639	1	Erythema both antecubital spaces. Abrasions both forearms.
26.0 MPH (42 kph)	654	37	Erythema both wrists, both antecubital spaces and epigastrium.
	655	20	Erythema both antecubital spaces. Laceration bridge of nose from goggles.
	656	23	Erythema both wrists and both antecubital spaces. Saw stars due to blow of bag on nose.
	657	31	Ecchymosis both thenar areas and both antecubital spaces.
	658	42	Erythema and abrasions both wrists, forearms right hand. Hypoglycemia with hypotension post impact with recovery after coke and candy. Tremulousness, perspiration and nervousness disappeared with candy.
28.0 MPH (45 kph)	674	21	Erythema both antecubital spaces. Abrasion left wrist. Laceration 4th web space on right hand.
	675	1	Erythema both antecubital spaces. Abrasions both wrists. Ecchymosis both antecubital spaces.
	676	15	Erythema both antecubital spaces.
	677	28	Erythema costoxiphoid region and both antecubital spaces. Had minimal wheel rim impact.
	678	24	Erythema left knee, both wrists, both antecubital spaces. Pain in left elbow.
30.0 MPH (48 kph)	687	13	Erythema both antecubital spaces. Ecchymosis right wrist. Felt momentarily dazed.
	688	37	Erythema both antecubital spaces.
	689	31	Erythema both antecubital spaces. Contusions both wrists, lacerations bridge of nose from goggles.
	690	35	Erythema both antecubital spaces and in epigastrium. Ecchymosis both wrists and left forearm. Minor laceration left forefinger.
	691	9	Erythema both wrists and left elbow. Abrasion left knee with slight bleeding.

Table III – Physical Symphony Survey Response

FRONT BACK

FRONT BACK

RIGHT LEFT RIGHT

SEVERITY CODE

1. No Symptoms
2. Mild Pressure
3. Moderate Pressure
4. Slight Discomfort
5. Definite Discomfort
6. Mild Pain
7. Moderate Pain
8. Severe Pain

Approximate Barrier Equiv.	Test No.	Subject	Symptom Area and Severity Response
13.0 MPH (21 kph)	508	33	B4, E3, G3
	509	18	B4, H4, J4
	511	40	B6
	512	24	H6, J6
	513	34	H4, J4
18.0 MPH (29 kph)	543	23	B5, H6, J6
	544	9	B2, H2, I2, J2
	545	28	B4, G4
	559	30	B6, D4, E4, F4, G4, H3, I3, J3, K3
	560	5	B4
20.0 MPH (32 kph)	577	15	D2, E2, F2, J2
	583	21	O7, J7, H7
	584	22	B6, G4, H4
	585	13	H2, J2, O2
	588	35	–
22.0 MPH (35 kph)	609	38	B7, N2, O2
	610	41	G6, B4, H4, J4, N6, O6
	611	27	D5, E5, F5, H4, J4
	612	29	B7, E7, O7, H7, J7, F6, N6
	613	36	E3, D2, F2, G2
24.0 MPH (39 kph)	635	28	H6
	636	14	G4, N2, O2
	637	18	B2, H2, J2, N2, O2
	638	9	B2, K2
	639	1	H2, J2, T2, V2, EE2, FF2
26.0 MPH (42 kph)	654	37	–
	655	20	B4, H4
	656	23	A2, B2, C2, H3, I4, K5
	657	31	B6, H6, J6
	658	42	H2, J2, N2, O2
28.0 MPH (45 kph)	674	21	J6, N3, O3
	675	1	A2
	676	15	–
	677	28	K6
	678	24	H3, J3, A2, B2, G2, N2, O2
30.0 MPH (48 kph)	687	13	–
	688	37	–
	689	31	B6, K5, H4, J4, W4
	690	35	–
	691	9	B2, H2, J2, K2

Table IV – Driver Air Cushion Physical Test Data — Dummies and Humans

Approx. Barrier Equivalent	Test Number	Subject Number	Height Inches	Weight Pounds	Head Max Resultant Accel Impact-g Ext (Int)	Rebound-g Ext (Int)	Head Injury SI Ext (Int)	HIC Ext (Int)	Thorax Max Resultant Accel Impact-g Ext (Int)	Rebound-g Ext (Int)	Thorax SI Ext (Int)	Femur Loads lb (Dummy) Left	Right
13	499	50D		164	ND[2] (22)	36 (33)	ND[2] (180)	ND[2] (110)	16 (17)	12 (17)	10 (80)	300	580
	500	50D			23 (24)	40 (35)	ND (220)	160[2] (130)	20 (21)	14 (16)	60 (90)	810	550
	501	95D		217	24 (26)	24 (29)	200 (200)	140[2] (100)	21 (26)	12 (16)	80 (110)	1150	570
	502	50D			23 (22)	36 (35)	210[1] (200)	120[1] (110)	17 (18)	10 (18)	80 (90)	660	610
	508	33	70.9	202	22	34	210	100	13	11	40		
	509	18	73.0	200	15	12	160[2]	90	14	8	30		
	511	40	69.5	181	17	25	210[2]	120	17	12	40		
	512	24	71.5	176	ND[2]	ND[2]	ND[2]	ND[2]	10[1]	5[1]	20[1]		
	513	34	69.3	138	16	20	210[2]	110	14	6	30		
18	533	95D			ND	ND	ND	ND	ND	ND	ND	ND	ND
	540	50D			ND[2] (33)	57 (46)	ND[2] (450)	ND[2] (220)	37 (34)	11 (20)	220 (210)	1150	1110
	541	50D			ND[2] (37)	ND[2] (44)	ND[2] (380)	ND[2] (180)	32 (33)	11 (16)	170 (180)	1160	910
	542	95D			ND[2] (42)	47 (44)	ND[2] (560)	ND[2] (260)	41 (44)	22 (22)	310 (260)	1590	680
	543	23	71.1	199	28	45	290	140	23	13	90		
	544	9	69.7	160	42	45	370[1]	260	25	12	120		
	545	28	69.7	165	34	52	490[1]	290	29	17	150		
	559	30	69.3	150	ND[2]	56	ND[2]	ND[2]	23	13	110		
	560	5	70.3	191	39	56	400[1]	200	26	12	110		
20	565	50D			33 (37)	45 (42)	350 (390)	160 (190)	40 (38)	14 (21)	220 (230)	1420	800
	566	50D			32 (35)	68 (50)	460[1] (390)	270 (210)	47 (37)	15 (22)	290 (250)	1220	760
	567	50D			38 (38)	55 (49)	440 (420)	220 (210)	34 (34)	14 (20)	220 (210)	1140	1100
	568	95D			36 (41)	42 (41)	340 (450)	160 (220)	36 (42)	17 (20)	260 (240)	1540	600
	577	15	73.0	156	ND	ND	ND	ND	ND	ND	ND		
	583	21	70.5	184	38	38	390	210	21	13	120		
	584	22	68.5	168	30	52	500	290	28	28	200		
	585	13	70.3	146	34	49	380	170	30	17	140		
	588	35	70.1	162	ND	38	ND	ND	28	26	180		
22	605	50D			36 (40)	56 (45)	450 (420)	210 (220)	35 (36)	8 (17)	180 (230)	1340	890
	606	50D			38 (42)	51 (41)	450 (390)	200 (190)	34 (36)	11 (16)	190 (210)	1280	900
	607	50D			38 (43)	50 (45)	420 (410)	190 (190)	34 (35)	8 (15)	150 (200)	1380	880
	608	95D			44 (37)	34 (33)	290 (290)	130 (150)	36 (42)	13 (13)	140 (190)	1130	920
	609	38	67.3	175	32	27	220	120	25	6	110		
	610	41	69.3	182	40	15	230	160	29	6	120		
	611	27	69.7	170	46	41	360	160	46	17	160		
	612	29	72.0	208	41	42	410	250	25	10	110		
	613	36	72.4	160	49	47	430	260	34	9	150		
24	626	50D			40 (43)	54 (48)	560 (530)	240 (250)	39 (43)	11 (16)	240 (300)	1490	1170
	627	50D			55 (45)	58 (48)	550 (460)	250 (220)	44 (39)	11 (19)	230 (250)	1500	1150
	633	50D			57 (49)	64 (51)	670 (540)	290 (270)	42 (44)	9 (16)	250 (280)	ND	1170
	634	95D			46 (36)	43 (36)	400 (380)	200 (200)	70 (66)	12 (19)	290 (370)	1690	810
	635	28	69.7	165	37	42	430[1]	250	38	13	160		
	636	14	69.3	186	48	25	390	270	46	6	230		
	637	18	73.3	200	32	30	300[1]	220	20	11	100		
	638	9	69.7	160	45	48	400	180	30	11	130		
	639	1	72.2	173	42	22	310	240	25	8	110		
26	650	50D			45 (47)	53 (46)	560 (530)	270 (290)	49 (48)	14 (19)	310 (360)	1440	1090
	651	50D			50 (51)	68 (58)	730 (720)	310 (300)	38 (44)	17 (20)	290 (360)	1300	1080
	652	50D			46 (54)	48 (45)	470 (510)	210 (280)	48 (43)	12 (17)	280 (310)	1440	1140
	653	95D			42 (46)	40 (46)	500 (620)	260 (340)	51 (51)	12 (19)	320 (390)	1870	910
	654	37	69.3	150	35	43	320	160	29	13	160		
	655	20	74.0	171	47	41	610[2]	290	25	13	110		
	656	23	71.1	199	30	47	310	150	17	10	90		
	657	31	71.3	179	35	10	250	170	19	13	110		
	658	42	70.2	174	38	39	470	240	35	14	200		
28	670	50D			45 (44)	68 (52)	630 (650)	290 (330)	41 (41)	14 (20)	290 (360)	1420	1160
	671	50D			45 (48)	72 (58)	790 (770)	360 (390)	43 (46)	14 (22)	320 (400)	1600	1180
	672	50D			50 (52)	58 (50)	620 (630)	260 (290)	48 (46)	15 (19)	350 (380)	1510	1100
	673	95D			40 (41)	49 (50)	600 (670)	300 (360)	49 (66)	20 (19)	340 (410)	1810	990
	674	21	70.5	184	ND	ND	ND	ND	33	11	190		
	675	1	72.2	173	65	30	550	340	37	10	200		
	676	15	73.0	156	45	37	470	260	37	9	200		
	677	28	69.7	165	39	39	450	260	37	17	240		
	678	24	71.5	176	38	57	500	240	33	9	170		
30	683	50D			53 (56)	55 (46)	630 (750)	280 (380)	53 (53)	12 (18)	390 (460)	1620	1170
	684	50D			48 (50)	74 (61)	960 (860)	490 (420)	45 (45)	13 (21)	380 (440)	1570	1180
	685	50D			43 (48)	67 (63)	730 (920)	360 (450)	41 (45)	17 (23)	360 (430)	1610	1140
	686	95D			45 (42)	46 (52)	610 (800)	320 (410)	48 (71)	22 (22)	410 (510)	2040	890
	687	13	70.3	146	49	61	680	280	38	14	250		
	688	37	69.3	150	44	44	670	260	33	15	210		
	689	31	71.3	179	71	44	700	460	38	16	260		
	690	35	70.1	162	48	46	520	310	36	16	220		
	691	9	69.7	160	58	67	860	380	38	12	220		

ND - No Data   [2] Excessive "Ringing"   [1] Corrected for noise in trace   [3] No G_z

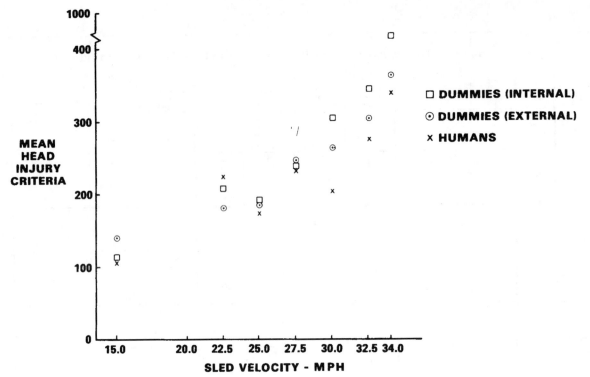

Fig. 6 - Mean HIC vs. Impact Severity (sled velocity in a full size vehicle 30 mph barrier crash simulation is approximately 34 mph) — dummies and humans

program were two volunteers (Tests 577 and 658) who demonstrated hypoglycemic symptoms which were secondary to the impact. Neither volunteer had eaten for approximately 18 hours pretest due to misunderstanding of instructions in regard to food intake. Both, however, responded to ingestion of a candy bar and soda pop without any sequellae.

Almost all of the volunteers immediately agreed to additional tests at higher severity levels. Four of the five individuals in the 34 mph group stated they would proceed to higher level impacts.

In describing the test experience, many volunteers expressed that they felt a "stinging" sensation on their face around the mouth. This was apparently caused by the wiping of bag across the face as it deployed. This same feeling was also felt on the hands and forearms. The volunteers could also feel the impact to the knee restraint, however, most did not record experiencing any discomfort. There were two individuals who recorded having moderate pain in their lower leg which quickly subsided (at 20 and 22 mph barrier equivalent, 32 and 35 kph).

DUMMY AND HUMAN DATA

Comparison of the dynamic response of dummies and humans was not the main objective of this program. However, since the dummies and humans were run under identical conditions at SwRI, the photographic and transducer data obtained lends itself to some comparison. A general overview of this test series reveals that the anthropomorphic dummies' dynamic response was conservative compared to the human response.

Evidence of these points may be seen in Table IV, Driver Air Cushion Physical Test Data — Dummies and Humans. This chart is a conglomerate of data gathered from the transducers. The Subject information was obtained from the physician. The 50D and 95D designate 50th percentile and 95th percentile dummies, respectively. The sled parameters are numbers calculated from Visicorder traces. Head, chest, and upper leg acceleration, force, and injury number data was tabulated from the information recorded on magnetic tape.

Some variation exists in the test results produced by the different accelerometers. Obviously, with the humans the transducers were not rigidly attached to the different body segments. A great effort was made in designing the mounts to reduce the transient differences in the relative motion between the body segment and transducer. The variation in the injury numbers for a single dummy test was due to the differences between the accelerations recorded on the internal and external transducers. From test to test, variation in injury numbers was also due to changes in the true accelerations. Figures 6 and 7 depict the mean injury numbers (head injury criteria and thorax severity index) obtained from the dummy internal and external, and human accelerometers. The mean numbers for dummies generally came from three 50th and one 95th percentile tests. The human sizes varied between these two limits. The

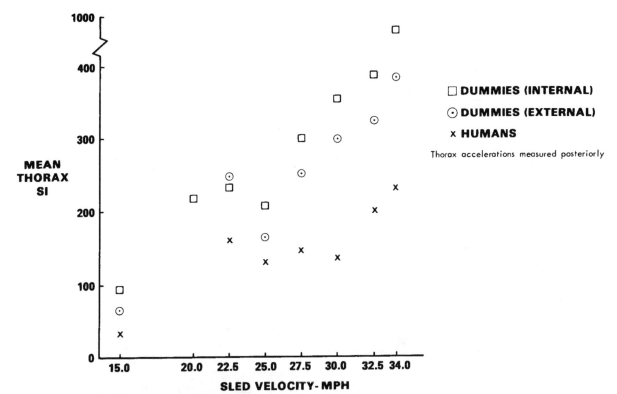

Fig. 7 - Mean Thorax SI vs. Impact Severity (sled velocity in a full size vehicle 30 mph barrier crash simulation is approximately 34 mph (48 kph) — dummies and humans

figures show that at all but one severity level, the mean HIC and mean thorax SI for humans was less than that recorded by the dummies. This difference could in part be attributed to the difference in response of human and dummy skin surfaces. It is also due to the difference in dummy and human kinematics. A review of the high-speed movie film reveals that the dummy's response is usually more exaggerated than a human's. Again, muscle tone plays a key role in this difference.

In addition to the variance in injury numbers, dummy interaction with the air cushion system differed from humans. Steering wheel lower rim deformation was common in dummy tests and appeared only once in the human tests (Test 636).

The amount of column compression differed greatly from dummies to humans. Dummies compressed the column more, sometimes exceeding six inches, while the humans rarely exceeded four inches.

At the higher sled severities, the humans often compressed the column in two stages while in the dummy tests, once the column compression began, it continued without a pause. The difference can be accounted for in the strength of the humans' arms. Because the volunteers were in a braced position, their arms were fairly rigid and could transmit a force to the steering system. Even though this force was not axial down the column, the resultant was sufficient to begin the compression.

The dummies, on the other hand, did not have any muscle tone in their arms (beyond the 1g joint setting) and column compression did not begin until their upper torso impacted the inflated air cushion. The second stage of compression in the human tests occurred when their upper torsos contacted the inflated air cushion. The time of peak impact pressures in the inflated air cushion was later in humans than the dummies (at 30 mph, peak impact pressure occurred between 75 and 100 msec after onset of sled deceleration in dummy tests and between 100 and 125 msec with humans). The humans were able to "hold themselves off" the air cushion longer. This was the reason humans did not experience severe impacts with the steering wheel lower rim.

Along the same line, the dummies usually made contact with the inflated air cushion (or steering wheel) about the time they contacted the knee restraint, while the human knees hit before their upper torso. Again, arm and leg muscle tone was a key factor.

A time comparison of events during four 30 mph (48 kph) barrier simulations is presented in Figure 8. This graphically portrays the difference in the sequence in which the human and dummy respond to impact. In two of these tests, a 50th percentile Hybrid II dummy was used and in the other two, different volunteers. Certain anthropometrics of the dummy and volunteers are charted in Table V for comparison purposes. Volunteer

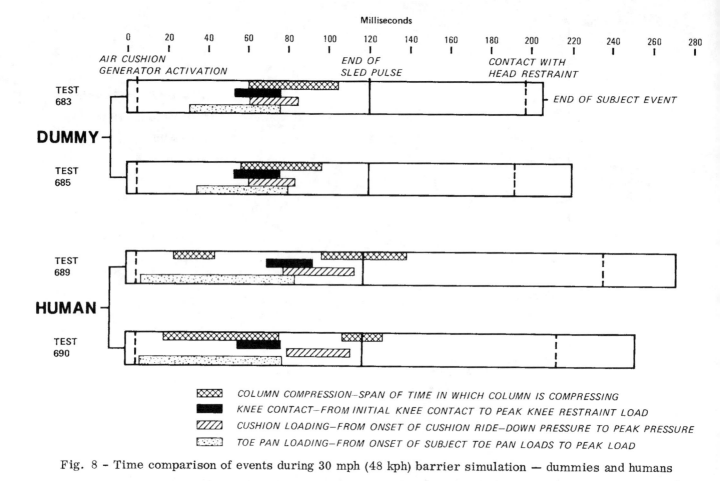

Fig. 8 - Time comparison of events during 30 mph (48 kph) barrier simulation — dummies and humans

Table V - Anthropomorphic Comparison of Hybrid II and Volunteers 31 and 35

Subject	Height in.(cm)	Weight lb.(kg)	Sitting Height in. (cm)	Buttock Knee in. (cm)
50th Dummy	—*	164 (74.4)	35.7 (90.7)	23.1 (58.7)
Volunteer #31 Test 689	71.3 (181)	179 (81.3)	36.8 (93.5)	24.2 (61.5)
Volunteer #35 Test 690	70.1 (178)	162 (73.5)	36.4 (92.5)	23.2 (59.0)

* Hybrid II Dummy does not stand erect

Number 35 in Test 690 was very close in size and weight to the Hybrid II dummy. The volunteer for Test 689 (Number 31) was close in size but greater in weight.

The time comparison of events displays the fundamental differences between a dummy and human volunteer with muscle tone. The events graphed include column compression, knee contact, cushion loading, and toe pan loadings. For each event, time is the only variable. Column compression involves the periods of time during which the column compresses. Knee contact is the time from which initial knee contact is made until the loads applied to the knee restraint peaked. These forces were measured by uniaxial load cells mounted to the knee restraint attachment points. The cushion loading bar represents the time of the onset of cushion pressure increase due to loading from the subject's upper torso and head until peak pressure. The pressure was measured by a transducer attached to the back of the air cushion module. The toe pan loading bar depicts the time span of the onset of toe pan loads until the peak before which the loading decreases. Load cells were mounted beneath the toe pan plate to measure the forces.

The sequence of these four events in the two dummy and two human tests is the same with the exception of column compression. In both dummies and humans the toe pan loads increased before the knees contacted the knee restraint. After knee contact, air cushion pressure began to increase. This, however, is where the similarity ends. The humans rode down the impacts longer with their legs than the dummies did. In the human tests, significant toe pan loading began about 10 milliseconds after the sled deceleration began (all times quoted are in reference to the start of sled deceleration, time zero). The dummies did not impart significant loads until about 30 milliseconds. The humans continued applying loads for at least 70 milliseconds before they began to drop off, while in the dummies, this time span was about 45 milliseconds. In these four tests, the time of knee contact with the knee restraint was about the same. The time at which air cushion pressure began to increase from upper torso loading was later in humans than the dummies. The humans also loaded the cushion for a longer period of time.

The most noticeable difference occurs in the column compression. In the human tests the first stage of compression occurs before contact was made with the knee restraint. A second stage of compression occurs after the knee restraint loads have peaked. In the dummy tests, on the other hand, total column compression occurred during one period of time and it was after contact was made with the knee restraint.

The volunteers completed the total event in a longer period of time as marked by the end of significant head accelerations. They were able to dissipate kinetic energy over a longer time span thus reducing the overall severity.

A review of the data obtained in this test program, including calculated injury criteria, motion pictures, and restraint system hardware indicates that the anthropomorphic dummies' response to the impact established them as a conservative measure of the human volunteers.

Injury numbers for the dummies were generally greater than those of the humans. Dummy femur loads approached and sometimes exceeded biomechanical estimates of human tolerance yet in corresponding human tests the volunteers did not even experience mild pain. The time duration of events was longer for the humans than the dummies. Steering wheel rim deformation was evident in many dummy tests while it only occurred in one volunteer test. Subjective response by the volunteers indicated that at times, the trauma experienced was so minimal they did not know it existed.

SUMMARY

The dynamic impact tests at Southwest Research Institute for the first time exposed human volunteers to production-like driver air cushion system deployments at impact levels equivalent to a 30 mph barrier crash (48 kph). No significant injury was produced. At no time was it necessary for the secondary restraint systems to be utilized and the entire program schedule was carried out as planned.

It must be noted, however, that all tests were conducted under the most safe and controlled conditions possible, and typify only what the air cushions might accomplish in direct, head-on impacts up to 30 mph (48 kph).

In comparing the test results, the anthropomorphic dummies' response to impact was conservative compared to the human volunteers.

933121

Thoracic Biomechanics with Air Bag Restraint

**Narayan Yoganandan, Frank A. Pintar, David Skrade,
Wayne Chmiel, John M. Reinartz, and Anthony Sances, Jr.**
Medical College of Wisconsin and V. A. Medical Center

ABSTRACT

The objective of the present study was to determine the biomechanics of the human thorax in a simulated frontal impact. Fourteen unembalmed human cadavers were subjected to deceleration sled tests at velocities of nine or 13 m/s. Air bag - knee bolster, air bag - lap belt, and air bag - three-point belt restraint systems were used with the specimen positioned in the driver's seat. Two chest bands were used to derive the deformation patterns at the upper and lower thoracic levels. Lap and shoulder belt forces were recorded with seatbelt transducers. After the test, specimens were evaluated using palpation, radiography, and a detailed autopsy. Thoracic trauma was graded according to the Abbreviated Injury Scale based on autopsy findings. Peak thoracic deformations were normalized with respect to the initial chest depth to facilitate comparison between the specimens.

Results indicated that under any restraint combination, regional differences exist in the deformation response between the upper and lower thoracic levels. The air bag - knee bolster tests indicated more uniform compressions of the thorax (based on chest band contours), demonstrated greater maximum lower chest deflections, produced fractures in the lower region of the thorax due to steering wheel contact, allowed greater hip and torso excursion, and produced significant steering wheel and column loading with permanent deformations. The air bag - lap belt experiments indicated uniform compressions of the thorax (chest band contours), produced minimal fractures, allowed greater torso excursion but less hip excursion, and produced significant steering wheel and column loading with residual deformations. The air bag - three-point belt system tests indicated high localized compressions of the thorax (chest band contours), produced multiple rib fractures consistent with shoulder belt loading, allowed less hip and torso excursion, and produced virtually no steering wheel and column loading. Based on the contours of thoracic deformation, kinematics and injury patterns, the biomechanical response of the human thorax is different between air bag - three point belt loading compared to the air bag - knee bolster/lap belt restraint combination.

INTRODUCTION

Frontal crashes in a vehicular environment are responsible for significant societal costs. Many recent surveys indicate that frontal impacts constitute over one-half of all passenger vehicle collisions. Increased use of seatbelts due to legislation in 42 states and the District of Columbia have significantly reduced trauma in these circumstances. Further, introduction of a passive restraint system (air bag) has been reported to be effective in reducing fatalities to the driver (1,3). While the improvements in mitigating trauma due to the presence of restraint systems are reported in the literature, studies have been retrospective using field accidents as the primary input. The human thorax is one of the most frequently injured regions in a frontal impact, and biomechanical evaluation should be based not only on epidemiological studies but also on controlled laboratory experimentation. Epidemiological studies reveal the type of trauma but fail to characterize the entire biomechanical response during impact. Controlled laboratory simulations of frontal impact using human cadavers and sled equipment under real-world restraint conditions will assist in the delineation of human tolerance, identification of injuries to the hard and soft tissues, advance mathematical models, develop anthropomorphic test devices with improved biofidelity, and evaluate the injury mitigating characteristics of vehicular interiors.

Considerable research has been expended in the last three decades to study the impact biomechanics of the human chest (6). Studies have primarily used blunt impact loading to the sternum to define human tolerance. In principle, these data are applicable to an unrestrained occupant but do not simulate the effect of restraint loading in a frontal impact situation. While some investigations have been conducted using belted human cadavers, to the best of our knowledge, very few studies describe the response of the human thorax under supplemental air bag (with and without belt) restraint systems. Early air bag and lap belt studies with living experimental animals (female baboons) were conducted at the Holoman Air Force Base in the late sixties by Snyder and co-workers (8,9). These studies indicated that the baboon tolerance to frontal impacts of 48-64 kph is above 57G. Walsh et al used one human cadaver specimen in two car to car tests (10) and Kallieris et al reported autopsy data from sled tests with ten unembalmed human cadavers positioned in the front passenger seat restrained by an air bag and a knee bar (5). Understanding of the thoracic biomechanics with air bag loading is of importance because an increased number of vehicles are being equipped with this restraint system. Consequently, this study was developed to investigate the biodynamics of the human thorax in simulated frontal impacts under air bag with knee bolster, air bag with lap belt, and air bag with three-point belt systems.

MATERIALS AND METHODS

A. SPECIMEN SELECTION AND PREPARATION: Fourteen unembalmed human cadavers were used in the study. The subjects ranged in age from 29 to 81 years (mean: 64), height from 150 to 182 cm (mean: 169), and weight from 41 to 85 kg (mean: 64). There were six females and eight males. The seated height ranged from 72 to 93 cm (mean: 87). Table 1 includes a summary of specimen data. The subjects were screened for Hepatitis A, B, and C, as well as the human immunodeficiency virus. In addition, they were selected based on evaluation of medical records and radiography to exclude specimens with severe degenerative disease and/or trauma. All studies were performed in a hospital environment and even though the specimens were negative for infection, precautions similar to that taken in an operating room were followed in the present study.

Specimens were prepared and geometrical data were obtained according to standard procedures. Briefly, anthropomorphic information such as seated height, chest circumference and chest depth were obtained according to the National Highway Traffic Safety Administration (NHTSA) guidelines. The specimens were pressurized to approximate the *in vivo* pulmonary and vascular characteristics (7).

B. SPECIMEN INSTRUMENTATION: Two NHTSA chest bands were mounted on each preparation (2,4). A chest band consists of a series of strain gauges mounted on a metal strip. The signals from the strain gauges can be used to determine the cross-sectional shape of the thorax. The first (upper level) chest band covered the midsternum region approximately at the anterior level of the fourth rib, while the second (lower level) chest band covered the xyphoid process approximately at the sixth rib anteriorly (12). All preparations were instrumented with 40 gauge chest bands except specimens 2F (upper and lower), and C3, 14B, and 6C (lower), in which 24 gauge chest bands were used. The lap or lap and shoulder belts were instrumented with load cells in tests using belt restraints. The details of the restraint systems used in the present study are given in the following Section "D. Restraint Combinations". A velocity transducer and an accelerometer placed on the sled buck recorded the velocity and acceleration profiles of the sled equipment.

C. TESTING PROCEDURE: Eight of the 14 specimens were tested at 13 m/s (high velocity), and the remaining six specimens were tested at nine meters per second (low velocity). A horizontal deceleration sled (MTS Systems Corporation, Minneapolis, MN) was used in the study. The test was filmed at 1000 frames/second with an onboard camera (Model 160-1B, Instrumentation Marketing Corporation, Burbank, CA) placed on the driver side . Kinematics of the specimen during impact were derived by placing photo targets at the head, first thoracic vertebral level, hip, knee, and shoulder regions. All data were

gathered using an onboard digital data acquisition system (DSP Technology, Inc., Fremont, CA) at a sampling rate of 12,500 Hz in accordance with the Society of Automotive Engineers SAE J211b specifications. The chest band and seatbelt force signals were filtered at SAE Class 1000. After the test, the specimens were palpated, radiographs were taken, and a detailed autopsy was conducted. Thoracic trauma was graded according to the Abbreviated Injury Scale (AIS, 1990 revision) using autopsy findings.

D. RESTRAINT COMBINATION: The restraints included air bag - knee bolster (AK), air bag - lap belt (AL), and air bag - three-point belt (A3) combinations. The air bag - lap belt combination was chosen to load the thorax anti-symmetrically. For the high velocity tests there were a total of three specimens in the AK, three in the AL, and two in the A3 categories. For the low velocity, there were three specimens in the AK, one in the AL, and two in the A3 categories (Table 2). Under each of these restraint variables one test was run with Hybrid III mannikin prior to the specimen experiment. The standard and commercially available air bag for the Ford Tempo was used in all tests. A standard lap and shoulder belt combination (six percent elongation) was used for the three-point belt, and an extruded polystyrene foam (Dow Chemicals Co., Midland, MI) was used for the knee-bolster test. The polystyrene foam material was cut into the shape of a knee bolster (but with increased thickness) and attached to the sled buck. Initial sled tests were conducted with the Hybrid III mannikin at a velocity of 9-13 m/s to determine the amount of knee excursion and the indentation into the knee bolster to evaluate the suitability for specimen tests.

E. THORACIC DEFORMATIONS AND INJURY EVALUATION: Data processing of the chest band output included a transformation of the individual curvature signals to obtain the deformation histories. The thoracic deformation contours were computed using the RBAND_PC Software from NHTSA (2,4). From these histories, the pattern, the peak magnitude, and the time of occurrence of the peak chest deflections were extracted for the upper and the lower chest bands. Peak chest deflections were obtained from these contours according to accepted techniques (12). Normalized chest deflections were computed by dividing the peak deformation by the initial undeformed chest depth. Statistical analysis of these parameters were conducted using a two-factor analysis of variance (ANOVA).

RESULTS

In all experiments the air bag began to deploy around 6-8 ms after impact initiation. Approximately 40 ms into the impact event the air bag was fully deployed. Kinematics of the specimen varied among the three restraint systems (Figure 1-3). In the air bag-knee bolster restraint system experiments (Figure 1), the specimen kinematics indicated a forward motion and contact of the knee with the knee bolster approximately at the time of full deployment of the air bag. With the knees further loading the knee bolster, the lower torso initially contacted the air bag in the region of the lower rim of the steering wheel. The upper torso loaded the air bag and the lower regions of the thorax continued to load the air bag and the lower wheel rim up to approximately 100 ms for high velocity tests and 70 ms for the low velocity sled experiments. The rebounding of the specimen initiated following these

Table 1: Specimen Data

Specimen ID	Age	Sex	Height (cm)	Weight (kg)	Seated height (cm)
2F	67	F	164	50	82
C3	64	M	166	70	84
14B	58	M	184	73	87
5H	67	F	150	57	72
6C	68	M	174	59	93
U8	29	F	170	41	93
M9	71	M	169	81	87
24G	76	M	168	81	88
5Z	75	F	180	85	91
126	64	F	168	54	88
7E	81	M	182	62	94
L28	67	F	154	46	78
2Y9	59	M	175	78	91
3P	56	M	168	63	89

Table 2: Test Matrix

Restraint System	Velocity	Sample Size
Air bag-Knee bolster (AK)	High	3
	Low	3
Air bag-Lap belt(AL)	High	3
	Low	1
Air bag-3 point belt (A3)	High	2
	Low	2
	TOTAL	14

kinematics with continuing deflation of the air bag and unloading of the steering wheel and the column. Residual deformations occurred to the steering wheel in these tests.

In the AL tests (Figure 2), the pelvis was restrained by the lap belt during the period of full deployment of the air bag. The upper torso pivoted around the restraint and contacted the air bag following full deployment. Loading of the steering wheel and column occurred producing permanent deformations. The upper thoracic region contacted the air bag due to the pivoting action by the lap belt restraint.

With the A3 restraint system combination (Figure 3), the belt loaded the specimen prior to air bag contact. The shoulder belt force reached its maximum value at approximately 60-70 ms and during this period the specimen was in contact with the air bag for approximately 20 ms. Little or no steering wheel and column deformations occurred in these experiments. The specimen began to rebound after approximately 100 ms for the high velocity and 80 ms for the low velocity tests without permanently deforming the wheel and column.

Table 3 includes a brief summary of the biomechanical variables for each specimen. Chest deformations were dependent on the velocity and the thoracic level. In the high velocity tests, the mean normalized peak chest deflections at the upper thoracic level were 0.27, 0.25, and 0.34 for the AK, AL, and A3 restraint combinations, respectively (Figure 4). For the low velocity tests, the mean normalized peak deflections were 0.17, 0.17, and 0.21 for the AK, AL, and A3 restraint combinations, respectively (Table 4). The magnitude of these peak deformation variables at the lower thoracic level

were 0.35, 0.14, and 0.24 for the high velocity experiments, and 0.29, 0.17, and 0.24 for the low velocity experiments, respectively.

For all the restraint combinations, increasing velocity resulted in increasing upper thoracic level deflections ($p < 0.05$). The type of restraint system had a significant effect ($p < 0.05$) on lower thoracic deflections. Furthermore, deflections at the lower level were less with belt restraint combinations (AL or A3) compared to the air bag - knee bolster (AK) system (Table 4). At the upper level however, peak chest compressions were greater with the air bag three-point belt combination. While no differences were apparent in the time of occurrence of the upper and lower peak chest deflections between the knee bolster (AK) and lap belt (AL) systems, the introduction of the three-point belt system produced early peaks ($p < 0.01$ at the upper and lower levels).

Figures 5, 6, and 7 illustrate the representative thoracic deformation contour patterns for all the restraint combinations at high velocity. Similar patterns were found in the low velocity tests. The shape of the contours at the upper and lower thoracic levels were similar for the AK and AL tests; they indicated nearly uniform compression of the torso in the anterior region with air bag contact. In contrast, tests with the three-point belt combination demonstrated a significant localized compression in the region of the shoulder harness (Figure 7). The shape of these deformation contours were strikingly similar to those obtained using three-point belt (no air bag) restraint in frontal impact tests conducted in our laboratory under a similar environment (12). As stated earlier

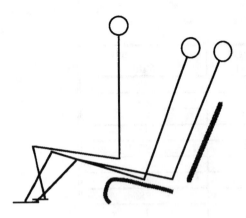

Figure 1: Stick figure representation of the specimen kinematics in the AK tests. The three positions (from right to left) indicate zero time, time of air bag inflation, and time of maximum body excursion.

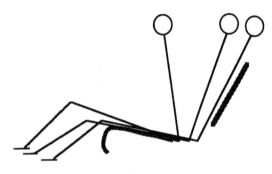

Figure 2: Stick figure representation of the specimen kinematics in the AL tests. The three positions (from right to left) indicate zero time, time of air bag inflation, and time of maximum body excursion.

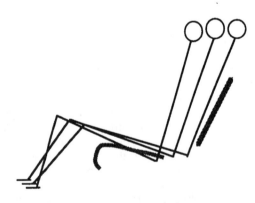

Figure 3: Stick figure representation of the specimen kinematics in the A3 tests. The three positions (from right to left) indicate zero time, time of air bag inflation, and time of maximum body excursion.

the magnitudes of peak deflections were different among the three restraint systems.

The number of rib fractures varied from zero to 13 in all specimens. At the high velocity, for the AK, AL, and A3 restraint combinations, there were a mean of 7.0 (range: 0-11), 2.0 (range: 0-3), and 11.5 (range: 10-13) rib fractures with the corresponding mean AIS ratings of 2.3, 1.3 and 3.5, respectively. More than one fracture occurred to an individual rib for both the A3 tests, while two out of three tests in the AK category produced two fractures of the same rib. In contrast, experiments with the AL restraint system did not produce multiple fractures of the same rib. The multiple rib fractures in the A3 tests illustrated in Figure 8 were primarily concentrated in the region where the shoulder belt loaded the chest. In the AK tests, however, the rib fractures occurred in the region where the steering wheel loaded the lower torso of the specimen (Figure 9). Fractures in the air bag - lap belt experiments occurred in the anterior middle region of the torso due to air bag and steering wheel/column loading (Figure 10).

At the low velocity, for the AK, AL, and A3 restraint system combinations, there were a mean of 3.0 (range: 0-6), 8.0, and 3.5 (range: 3-4) rib fractures with the associated AIS rating of 1.7, 3.0, and 2.0, respectively. These tests did not result in more than one fracture to the damaged rib. In addition, similar severity of trauma was not apparent on posttraumatic radiographs.

DISCUSSION

In a previous paper (12), we reported the response of the human surrogate thorax under simulated frontal impacts at a deceleration velocity of 13 m/s. Both unembalmed adult human cadavers and 50th percentile Hybrid III mannikin positioned in the driver's seat were evaluated for thoracic deformation contours with a three-point belt restraint system. The present study is an extension of this research to investigate the biodynamics of the human thorax under varying restraint combinations and velocities. As stated in the introduction, because of the increased availability of passive restraint systems (air bags and automatic belts), all tests were conducted with an air bag as one of the restraint components.

Investigations into the biodynamic response of the thorax have been an area of research because of its effect on injury assessment, treatment, and mitigation. Considerable efforts have been advanced to study the mechanisms of thoracic trauma and delineate the associated biomechanical parameters to define tolerance criteria (6,8,9). These studies have included in vivo animals (female baboons) with air bag and lap belt restraints, human cadavers with and without belt restraints, and anthropomorphic test devices (ATD) in sled simulations. Response of the ATD with air bag supplemental restraint system has also been studied using sled tests and mathematical analogues. In contrast, experimentation with human cadavers, a vital part of impact biomechanics research, under air bag restraint combinations has been very limited.

In earlier studies, Walsh and co-workers (10,11) conducted car to car offset and frontal impact tests at a closing velocity of 27 m/s. One human cadaver was tested twice (front seat passenger followed by driver) with an air cushion restraint system. In the first impact no injuries were observed on radiography. In the subsequent test, two nondisplaced rib fractures were identified at autopsy (AIS=2). A "caliper type" device, using a hinged front arm turning a calibrated rotary potentiometer, was used to measure the anteroposterior deflection at one location of the chest. The deflectometer data was lost in the first test. In the second test, a deflection of 5 cm was reported. This is in general agreement with the results of our study. However, since neither the location of the deflectometer nor the temporal analysis of the deflection was reported, further comparisons with the present study are n

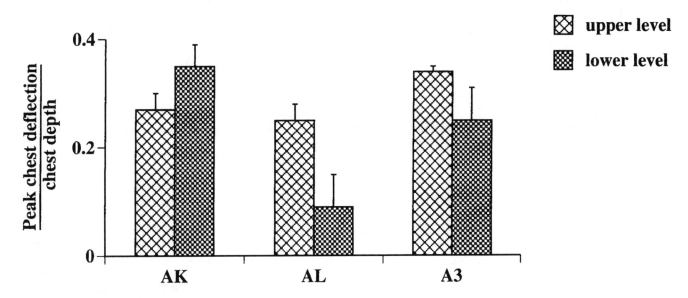

Figure 4: Bar chart representation of the mean normalized peak deflections at the upper and lower thoracic levels as a function of restraint system for the high velocity tests. AK = air bag - knee bolster, AL = air bag - lap belt, A3 = air bag - three-point belt. Error bars indicate standard error of the mean.

Table 3: Summary of Biomechanical Data

ID	Restraint type	Vel. (m/s)	C_{max} Up	C_{max} Low	*Time (ms) Up	*Time (ms) Low	Belt load (N) Shoulder	Belt load (N) Lap	Rib Fractures	Thoracic AIS@
2F	AB/lap	13	0.21	0.00	85	-	NA	4057	3	2
C3	AB/lap	13	0.31	0.07	106	71	NA	4573	3	2
14B	AB/lap	13	0.22	0.20	108	108	NA	4439	0	0
5H	AB/3pt	13	0.33	0.29	58	58	5747	3941	13	3
6C	AB/3pt	13	0.35	0.18	73	63	5791	3799	10	4
U8	AB/kb	13	0.22	0.27	112	97	NA	NA	0	0
M9	AB/kb	13	0.33	0.40	92	112	NA	NA	11	4
24G	AB/kb	9	0.20	0.31	100	100	NA	NA	0	0
5Z	AB/kb	13	0.27	0.37	110	110	NA	NA	10	3
126	AB/kb	9	0.19	0.37	90	90	NA	NA	6	3
7E	AB/kb	9	0.12	0.20	90	90	NA	NA	3	2
L28	AB/3pt	9	0.22	0.26	80	70	2713	1214	3	2
2Y9	AB/lap	9	0.17	0.17	110	110	NA	2224	8	3
3P	AB/3pt	9	0.19	0.21	74	71	2739	1176	4	2

*: Measured from the time of impact
@: AIS rating refers to thoracolumbar spine, pelvis, thoracic contents and abdomen
AB: air bag, kb: knee bolster, lap: lap belt only, 3 pt: three-point belt
NA: Not applicable

Table 4: <u>Summary of Peak Chest Deformation Histories</u>

Restraint System	Velocity	Upper chest band		Lower chest band		Overall mean maxima
		deflection	t(ms)	deflection	t(ms)	
Air bag-Knee-bolster (AK)	High	0.27	105	0.35	106	0.35
	Low	0.17	93	0.29	93	0.29
Air bag-Lap belt(AL)	High	0.25	100	0.09	90	0.25
	Low	0.17	110	0.17	110	0.17
Air bag-3 point belt (A3)	High	0.34	66	0.24	61	0.34
	Low	0.21	77	0.24	71	0.24

possible. The study concluded that human cadavers are vital to the design, development and evaluation of restraint systems.

In more recent research, Kallieris et al (5), reported a series of ten human cadaver tests (ages from 18 to 66 years) with air bag - knee bar restraint in the right front passenger position. The tests were conducted at a velocity of 14 m/s. While zero to 20 rib fractures were found at autopsy (mean: 4), the present study identified zero to 11 rib fractures (mean: 7) in the AK series. In the Kallieris et al study, the thoracic AIS and MAIS ranged from zero to 4, (mean AIS=1.1, mean MAIS=2.6). For the MAIS, head, thorax, vertebral column, and extremity injuries were used. Internal injuries to the thorax/abdomen did not occur in any specimen (5). These results are in good agreement with our study which also did not produce any vital organ trauma under all three air bag restraint combinations. However, larger differences in the thoracic injury severity were observed (mean AIS rating for AK tests at 13 m/s=2.3 in the present study versus a mean AIS rating of 1.1 in the Kallieris et al study). Kinematics of the specimen, including interaction with the deformable knee bolster in contrast to the relatively rigid knee bar, and the variations in the demographics of the subjects may contribute to these differences in the injury severity.

The kinematics of the specimen were such that in the A3 series, the belt loaded the upper torso prior to air bag contact with approximately 40% of the peak shoulder belt force, thereby subjecting the thorax to compression. Approximately 60 to 70 ms into the impact, shoulder belt forces reached peak values. Similar observations were made in previous studies using three-point belt loading without an air bag (11). Because of the restraint and subsequent compressions of the thorax by the belt, little or no steering wheel and column deformations occurred secondary to air bag deployment and contact. This was true for both high and low velocity tests. Consequently, in these experiments, the belt loads the chest to a considerable extent.

In contrast, the AK series produced steering wheel and column deformations in all tests secondary to direct impact of the wheel by the lower torso. This was determined by residual deformations of the lower region of the steering wheel and confirmed by kinematic analysis. Multiple fractures occurred in these tests with the exception of one specimen (29 year old small female) at the high velocity. These injuries may have occurred secondary to the loading and residual displacements of the wheel and column. Inherent bias in the sample may also be a contributing factor.

For the AL series, the kinematics indicated an initial pelvic restraint by the lap belt which allowed the upper torso to pivot and contact the air bag following deployment. There was no direct impact of the specimen torso on the steering wheel.

This phenomenon is directly in contrast to the AK series where the restraining effect of the pelvis occurred following knee to knee bolster contact, reducing the pivoting action.

The pattern of the thoracic deformation contours for the air bag three-point belt restraint indicated considerable local compressions of the thorax due to asymmetrical loading by the shoulder belt (Figure 7). In the other two restraint cases however, the thorax was loaded more uniformly resulting in symmetrical deformations (Figures 5,6). This represents the loading of the chest by the air bag restraint in contrast to the asymmetrical loading of the thorax by the shoulder belt in the air bag - three-point belt case. The shape of the deformation contours in the air bag - three-point belt tests were remarkably similar to our earlier studies (12) wherein frontal impact simulations were conducted with three-point belt (no air bag) restraint. Furthermore, autopsy data in these two varying restraint combinations indicated multiple rib fractures with similar injury pattern. There was virtually no steering wheel/ column deformation in any of the present three-point belt restraint experiments. Therefore, it may be appropriate to conclude that, the thoracic loading, the resulting deformation patterns, and the ensuing pathology in three-point belt restrained cadavers (with and without the activation of the air bag) primarily stems from the asymmetrical loading of the shoulder belt with lesser contribution from the air bag or column loading. Peak upper and lower chest deflections which occurred significantly earlier ($p<0.01$) in the A3 case compared to the other two restraint combinations (AK,AL) may further reinforce this hypothesis.

Although the chest bands revealed uniform compressions at a given thoracic level in both AK and AL series, the differences in the rib fractures may be explained by the direct impact of the torso onto the steering wheel in the AK series. In contrast, in the A3 series, initial shoulder belt loading may account for the multiple rib fractures. The pivoting action in the AL tests and the initial restraining action in the A3 tests manifest as greater deflections at the upper chest level than its lower counterpart. However, in the AK series, the lower chest deflections were greater compared to the upper chest deflections presumably because of direct contact of the lower torso with the steering wheel and lack of pivoting due to the absence of a belt restraint. The orientation of the specimen and the wheel-bag attitude results in the primary contact at the lower torso level.

To the best of our knowledge, the present series represent the largest body of biomechanical data on human cadaver tests with air bag restraint combinations. Although a total of 14 specimens were used in the present study, the statistical analysis has limitations. This is due to the complexity of the test matrix when multiple factors are included. The trends in

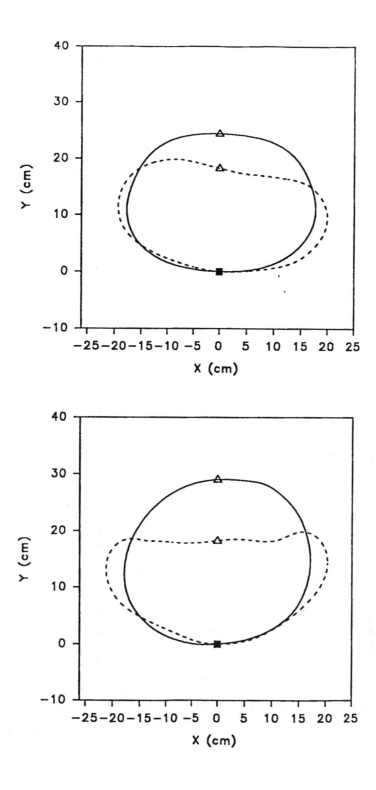

Figure 5: Thoracic deformation contours at the upper chest level (top) and at the lower chest level (bottom) as viewed from head to toe (superior to inferior direction). Solid lines represent the initial unloaded state and dotted lines represent the deformed state. This data was obtained for the specimen restrained by the air bag - knee bolster combination. Notice the uniform loading of the thorax by the restraint system. Sternum is shown by a triangle and spine is shown by a solid square.

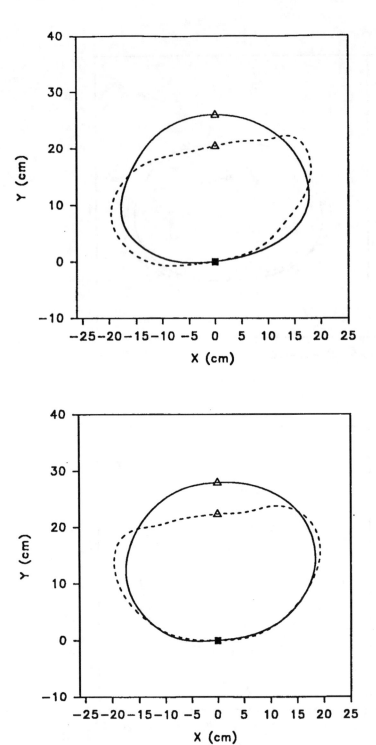

Figure 6: Thoracic deformation contours at the upper chest level (top) and at the lower chest level (bottom) as viewed from head to toe (superior to inferior direction). Solid lines represent the initial unloaded state and dotted lines represent the deformed state. This data was obtained for the specimen restrained by the air bag - lap belt combination. Notice the uniform loading of the thorax by the restraint system. Sternum is shown by a triangle and spine is shown by a solid square.

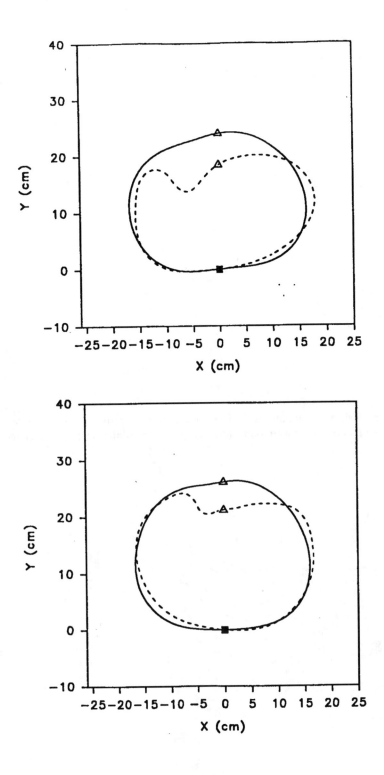

Figure 7: Thoracic deformation contours at the upper chest level (top) and at the lower chest level (bottom) as viewed from head to toe (superior to inferior direction). Solid lines represent the initial unloaded state and dotted lines represent the deformed state. This data was obtained for the specimen restrained by the air bag -- three-point belt combination. Notice the focal loading of the thorax by the restraint system. Sternum is shown by a triangle and spine is shown by a solid square.

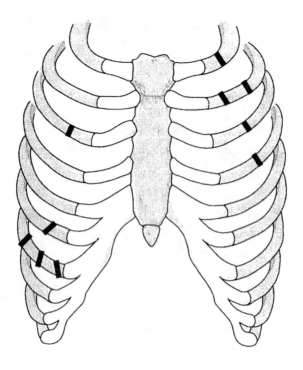

Figure 8: Injury pattern in the air bag -- three-point belt tests. The rib fractures are primarily concentrated in the region where the shoulder belt loaded the thorax. There were no residual deformations of the steering wheel and column.

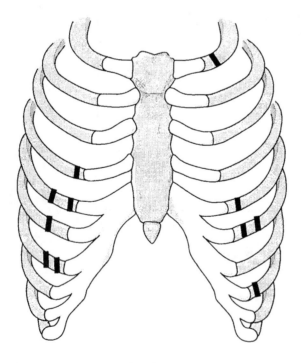

Figure 9: Injury pattern in the air bag - knee bolster tests. The specimen impacted the lower region of the wheel resulting in rib fractures at the lower rib cage. Maximum lower thoracic deformations were higher for these tests. Steering wheel and column indicated permanent deformations.

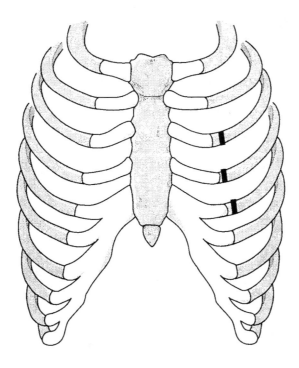

Figure 10: Schematic representation of the injury pattern in the air bag - lap belt tests. The three rib fractures (shown by solid lines) occurred due to contact of the torso with the air bag and deformations of the steering wheel and column.

the significance of the biomechanical data however, which were confirmed by kinematics, can be further reinforced by additional experimentation. Similar precautions also apply while interpreting thoracic trauma based on autopsy and using the AIS rating. It should be recognized that the AIS rating is not directly applicable to autopsy data; in contrast it is primarily based on clinical assessment and radiography. It is well known that in real-world situations rib fractures on clinical radiographs are routinely missed due to the complex anatomy of the rib cage and patient conditions which may preclude detailed radiography. In contrast, autopsy has an added flexibility to identify trauma in greater detail. However, clinical trauma assessment such as pneumothorax is hard to resolve in pathological tissue. The rib fractures which formed the principal basis for the thoracic trauma rating may be somewhat overestimated from a clinical standpoint. The differences can be resolved if a critical evaluation is made of the posttraumatic radiographs using double blind techniques with pathological findings. This topic deserves attention and will be the subject of another study.

ACKNOWLEDGMENT We wish to acknowldege the support and guidance of Richard M. Morgan and Rolf H. Eppinger of NHTSA in the conduct of this research. This study was supported in part by DOT NHTSA Grant DTNH22-89-Z-07305 and the Department of Veterans Affairs Medical Research Service. All findings and views reported in this manuscript are based on the opinions of the authors and do not necessarily represent the consensus or views of the funding organizations. We also wish to acknowledge the assistance of Karl Pintar, M.D., Chief Pathologist (Retired) who conducted the pathological evaluations and provided guidance during the sled tests.

REFERENCES

1. "Driver Deaths Down Substantially in Cars Equipped with Air Bags," Status Report, Insurance Institute for Highway Safety, Vol. 27, No. 12, October 3, 1992.
2. Eppinger RH: On the development of a deformation measurement system and its application toward developing mechanically based indices. Paper No. 892426, Proc 33rd Stapp Car Crash Conference, Society of Automotive Engineers, Washington, D.C., October 4-6, 1989, , pp 21-28.
3. "Evaluation of the Effectiveness of Occupant Protection," Interim Report, U.S. Department of Transportation, National Highway Traffic Safety Administration, June 1992.
4. Hagedorn AV, Eppinger RH, Morgan RM, Pritz HB, Khaewpong N: Application of a deformation measurement system to mechanical systems. Intl Research Council on Biokinetics of Impact, Berlin, Germany, September, 1991.
5. Kallieris D, Mattern R, Schmidt G, Klaus G: Comparison on three-point belt and air bag-knee bolster systems. Injury criteria and injury severity at simulated frontal collisions. Intl Research Council on Biokinetics of Impact, Cologne, Germany, Sept., 1982, pp 166-183.
6. Melvin JW, Weber K, eds: Review of biomechanical impact response and injury in the automotive

environment, Report No. DOT HS 807 042, U.S. Department of Transportation, National Highway Traffic Safety Administration, Washington, D.C., 1985, 201 pp.

7. Robbins D H , Lehman R J, Nusholtz G S , Melvin J W, Benson J B , Stalnaker R L, Culver R, "Quantification of Thoracic Response and Injury: The Gathering of Data," Report No. UM-HSRI-82-33-1, The University of Michigan, Highway Safety Research Institute, August 31, 1982.

8. Snyder RG, Snow CC, Young JW, Crosby WM, Price GT: Pathology of trauma attributed to restraint systems in crash impacts. Aerospace Med, 39(8): 812-829, 1968.

9. Snyder RG, Young JW, Snow CC: Experimental impact protection with advanced automotive restraint systems: Preliminary primate tests with air bag and inertia reel/inverted-Y yoke torso harness: Paper No. 670922, Proc. 11th Stapp Car Crash Conference, Society of Automotive Engineers, Anaheim, California, October 10-11, 1967, pp 406-431.

10. Walsh MJ, Kelleher BJ: Evaluation of air cushion and belt restraint systems in identical crash situations using dummies and cadavera. Paper No. 780893, Proc. 22nd Stapp Car Crash Conference, Society of Automotive Engineers, Ann Arbor, Michigan, October 24-26, 1978, pp. 295-339.

11. Walsh MJ, Romeo DJ: Results of cadaver and anthropomorphic dummy tests in identical crash situations. Paper No. 760803, Proc. 20th Stapp Car Crash Conference, Society of Automotive Engineers, Dearborn, Michigan, October 18-20, 1976, pp. 107-131.

12. Yoganandan N, Skrade D, Pintar FA, Reinartz J, Sances A: Thoracic deformation contours in a frontal impact. Paper No. 912891, Proc. 35th Stapp Car Crash Conference, Society of Automotive Engineers, San Diego, California, November 18-20, 1991, pp 47-63.

942216

The Performance of Active and Passive Driver Restraint Systems in Simulated Frontal Collisions

Dimitrios Kallieris, Kirsten Marion Stein, and Rainer Mattern
University Heidelberg

Richard Morgan and Rolf Eppinger
NHTSA

ABSTRACT

The study reports on the results of frontal collisions with 16 cadavers and two Hybrid III dummies with impact velocities of 48 km/h to 55 km/h and a mean sled deceleration of 17 g; mounted to the sled was the front part of a passenger compartment. The cadavers were restrained in the driver position with either 3-point belts (6% and 16 % elongation) and/or air bag with knee bolster and one case was unrestrained. In most cases, both a 12-accelerometer thoracic array and 2 chest bands were employed. In some cases the acceleration at Th6 was measured. The cadavers were autopsied and the injury severity was rated according to the AIS 90. Maximum resultant Th1, Th6, and Th12 accelerations or sternum accelerations in x-direction ranged from 35g to 78g when using 3-point belts and produced injuries ranging from a few rib fractures to unstable chest wall (flail chest). The same range of acceleration values were observed when a driver air bag was used, but three of the four the subjects remained uninjured. Chest deflections of 3 cm to 7 cm were observed by using 3-point belts and/or air bags with knee bolster. The sternum fracture was the typical injury by using 3-point belt systems. The chest band contours showed the torso belt produced a high local compression with resulting higher injury risks, while the air bag appeared to load a larger area of the chest front with less injury risk. Cervical spine injuries of AIS 1 and AIS 2 were observed, independent of using 3-point belt and/or air bag systems. The 3-point belt combined with driver air bag appears to be the best solution for the protection of the head and thorax during frontal collisions. However, further modification of the torso belt's yield characteristic would appear to offer further safety improvements.

INTRODUCTION

The frontal collision is the most frequent occurring accident type and is the accident type toward which the ma-

jority of safety measures have been directed. Due to extensive developments in passive safety and the effective combination of active and passive safety systems, a high standard of injury reduction for frontal collisions has been achieved. The aim of this study is to investigate the behaviour the standard 3-point belt, an air bag only, and belt plus air bag systems in frontal collisions. The collision characteristics of 50 km/h to 55 km/h with a mean sled deceleration of 17g were chosen to represent common accident types.

METHOD

TEST SUBJECTS - The test subjects were 16 unembalmed cadavers in the age range 20 to 63 years. The research content and the procedures governing the procurement, treatment, and disposition of human surrogates used in this program conform to all requirements of NHTSA Order 700-4, Ethical Use of Human Surrogates in NHTSA-Sponsored Research. Furthermore, two tests with Hybrid III dummy were also performed.

TEST EQUIPMENT - The tests were performed on the University of Heidelberg's deceleration sled. Mounted to the sled was the front part of a passenger compartment of a mid-sized car. Test subjects were positioned in the driver's seat and restrained by either a 3-point belt (6% or 16% elongation), a driver side air bag-knee bolster, or a 3-point (16% elongation) belt with supplemental driver side air bag combination; in one case the cadaver was unrestrained. Figure 1 illustrates the experimental configuration. Frontal collisions were simulated with impact velocities of about 48 to 55 km/h and a trapezoidal deceleration pulse with a mean value of 17 g. Table 1 shows the test matrix.

INSTRUMENTATION - In all tests, the subject's thorax was instrumented with a twelve-accelerometer array (Robbins et al. 1976, Eppinger, et al, 1978). In some of the

Table 1
Test matrix

Test No	Restraint system	velocity [km/h]	sled decel. [g]	age	sex	chest circumf. [cm]
T7	unrestrained	47	17	57	m	93
T3	3 point belt	50	17	63	m	98
T4	3 point belt	48	16	34	m	92
T6	3 point belt	49	18	34	m	94
T9	3 point belt	50	18	36	m	96
T11	3 point belt	55	17	20	m	95
T12	3 point belt	50	20	37	m	93
T13	3 point belt	49	15	29	m	78
T14	3 point belt	48	14	52	f	78
T2	driv. air bag - knee bolster	48	17	41	m	98
T5	driv. air bag - knee bolster	47	16	31	m	-
T8	driv. air bag - knee bolster	49	17	25	m	88
T10	driv. air bag - knee bolster	47	17	38	m	95
T1	driv. air bag - 3 point belt	53	18	62	m	102
T15	driv. air bag - 3 point belt	48	14	47	f	100
T16	driv. air bag - 3 point belt	48	14	32	m	102

tests, a triaxial accelerometer array was attached to Th 6. The shoulder belt force was also measured in some tests.

To measure thoracic contours during dynamic loading without disturbing the thorax, the External Peripheral Instrument for Deformation or Chest Band, developed by the National Highway Traffic Safety Administration (Eppinger 1989) was used. The chest band consists of a high strength steel alloy strip, 140 cm x 1.25 cm x 0.025 cm with between 16 & 40 strain gauges bonded at predetermined locations along its length. The steel strip with attached gauges is encased with Flexane 80; a two component urethane rubber for the casting of durable, resilient, medium to hard devices. Each set of gauges, in a four-arm Wheatstone bridge configuration, constitutes one data channel. The bridge is configured such that it is sensitive to the longitudinal bending and therefore each gauge/channel provides an output proportional to local curvature at its location. The chest bands were mounted at the level of the 4th and 8th ribs; Figure 2 shows the location of the installed chest bands at the cadaver thorax.

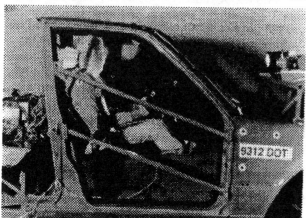

Figure 1: Test configuration

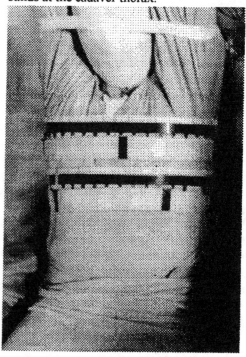

Figure 2: Location of the chest bands

Prior to conducting a sled test, the initial contour of the subject was checked against eight seperate actual physical dimensions by measuring with an anthropometer and all were accurate to within 8 percent.

KINEMATICS - Three highspeed cameras (frame rate 1000/s), two onboard and one stationary, were used to document impact dynamics.

AUTOPSY- INJURY SEVERITY - For each cadaver, a full autopsy was performed. The injuries were coded according to the AIS 1990.

DATA ANALYSIS - The data were recorded in analog format and were digitized at 1600 samples per second and subsequently filtered with a digital Butterworth filter channel class 180. The thoracic deformation contours were computed using the RBAND-PC Software from NHTSA. A program was equipped to evaluate the deformation of the front chest wall in relation to the fixed vertebral column.

To evaluate the Viscous Criterion (VC) at the level of the 4th and 8th rib, the deformation velocity was defined through differentiation of the deformation, the deformation-time-histories and the deformation velocity-time-histories were smoothed. For the evaluation of the compression, the chest depth was used (instantaneous deformation/chest depth, Lau and Viano 1986).

RESULTS

KINEMATIC BEHAVIOUR - In all the experiments using driver air bag, the bag began to deploy 8 ms after impact initiation and 30 ms into the impact event, the air bag was fully deployed. The bag volume was 72 liters. The kinematics of the test subjects varied among the three different restraint systems used.

In the air bag - knee bolster tests, the subjects translated forward during the first 40 ms at which time the knees make contact with the knee bolster. The head's forward and upward displacement is stopped by the head contacting the windscreen at 60 ms after the crash. Extension of the cervical spine follows with subsequent rebound.

The trunk of the cadaver was not always restrained symmetrically by the driver air bag and a rotation of both the head and torso was sometimes observed.

When using 3-point belt, the specimen kinematics also showed a forward movement for the first 40 ms after the crash initiation. The knees then contacted with the knee bolster. Head rotation began 55 ms after the crash began with the maximum head-neck bending observed 120 ms after the crash. Rebound followed. Usually, a rotation of the right shoulder around the diagonal belt was observed. A tangential head impact resulted against the steering wheel (dependent on driver's seated height) with the specimen

then rebounding; the rebound phase was finished at approximately 200 ms after the crash.

With the 3-point belt/driver air bag restraint system a combined series of motions were seen. The specimen first moves forwards, 40 ms after the crash the knee contact with the knee bolster and full deployment of the air bag was reached. 90 ms after the crash the front of the head dipped into the air bag.

Comparable tests with belted cadavers and dummies generally show a similar kinematic behaviour. However, the cadaver appeared more flexible than the dummy. A rotation of the head and torso-of the cadaver around the shoulder belt in counter-clockwise direction was observed test during the forward movement, whereas, with the dummy only, a small rotation of the head was observed during the forward translation phase.

MECHANICAL RESPONSE - Unique chest deformations were seen for each of the restraint systems used. Chest contours of a 49 km/h frontal collision by using a 3-point belt (16% elongation) is shown in Figure 3. At the top of the figure the contour at the level of the 4th rib is illustrated, at the bottom, the one at the level of the 8th rib. Additionally, the evaluated chest deflection over time is shown for both levels. Typically, a 3-point belt causes high local compression and deformation of the chest along the shoulder belt path. The maximum deflection at the upper level was 5,6 cm, 66 ms after the crash. At the lower level, a deflection of 4,7 cm at 77 ms after the crash was observed. Fractures of the sternum and the ribs were found along the shoulder belt path for this cadaver test (Fig. 6, T14).

Figure 4 also illustrates chest contours at two levels during the loading phase. When using a driver air bag - knee bolster system; the impact velocity was 49 km/h (T 8). Chest deflection of 2 cm and 4 cm at the upper and lower chest level were observed for the times of 95 ms and 87 ms. The compression levels and deformation patterns at the front of the chest were completely different than those induced by the 3-point belt. The 25 year old cadaver used in this test remained uninjured.

Chest contours seen when a combination 3-point belt (16 % elongation) and driver air bag was used are illustrated in Figure 5. The impact velocity was 48 km/h (T16). Maximum chest deflections of 5.44 cm at the upper level and 5.85 cm at the lower level were observed and occurred at about 90 ms after the crash. The chest contours show that with both a belt and air bag system were present, that the belt system appeared to dominate. The location and pattern of fractures for this test (Fig 6) resembled those of air bag only tests while a second identical test (T15) had a fracture pattern typical of a 3-point belt system.

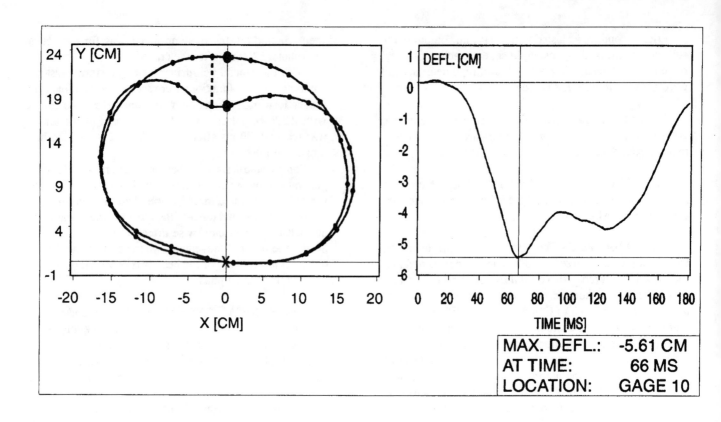

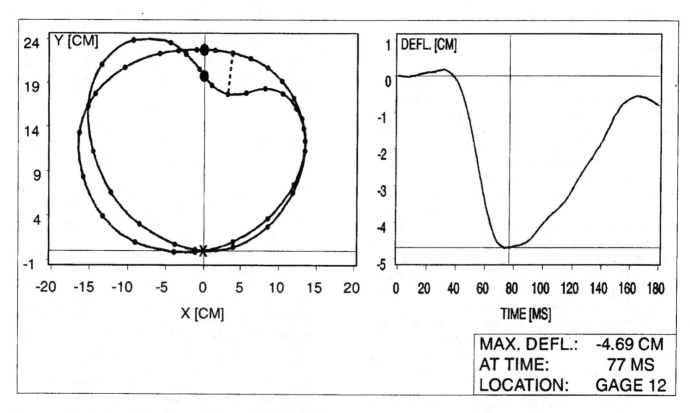

Figure 3: Thoracic deformation contours and deflection time histories at the level of the 4th rib (top) and at the level of the 8th rib (bottom): initial unloaded state and the maximum deformed state. The cadaver (T14) was restrained by a 3-point belt. X: Spine, O: Sternum

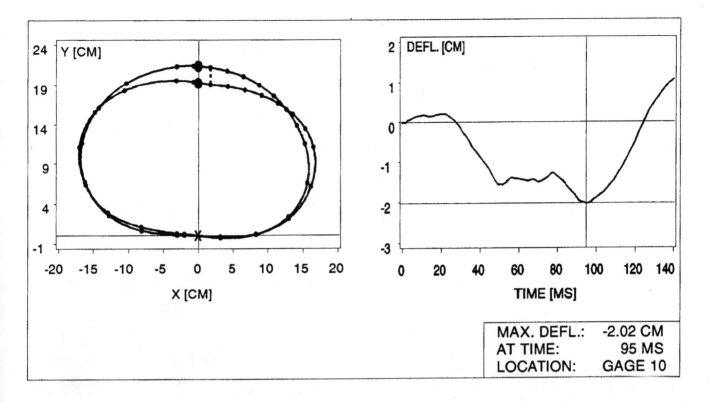

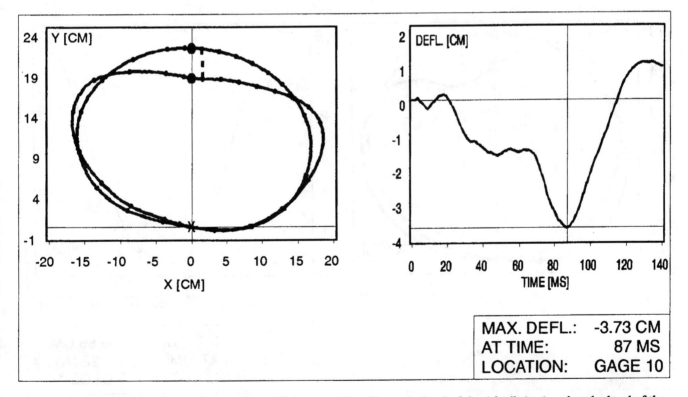

Figure 4: Thoracic deformation contours and deflection time histories at the level of the 4th rib (top) and at the level of the 8th rib (bottom): initial unloaded state and the maximum deformed state. The cadaver (T8) was restrained by the air bag - knee bolster combination. X: Spine, O: Sternum

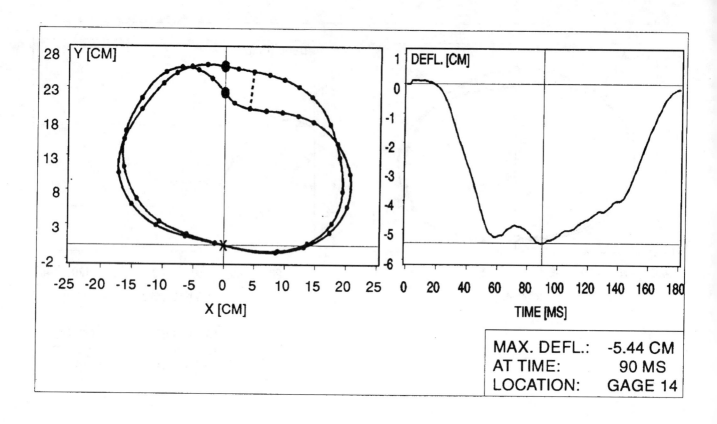

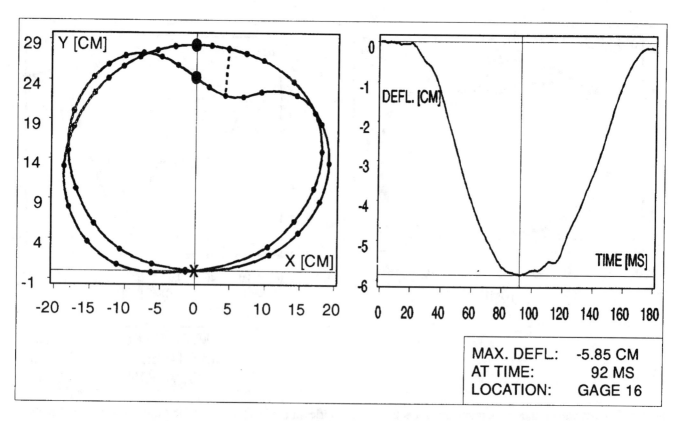

Figure 5: Thoracic deformation contours and deflection time histories at the level of the 4th rib (top) and at the level of the 8th rib (bottom): initial unloaded state and the maximum deformed state. The cadaver (T16) was restrained by the 3-point belt / driver air bag combination. X: Spine, O: Sternum

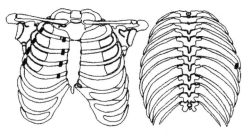

T 1; 53 km/h; 18 g;
6% 3 Pt. belt, driver-airbag
m; 62 y.; 82 kg; 181 cm;
TOAIS 4; NRRF 9

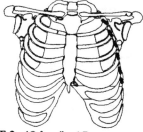

T 2; 48 km/h; 17g;
driver-airbag
m; 41 y.; 90 kg; 183 cm
TOAIS 3; NRRF 6

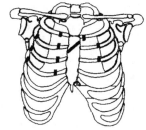

T 3; 50 km/h; 17 g;
6 % 3-Pt.-belt
m; 63 y.; 69 kg; 168 cm;
TOAIS 4; NRRF 10

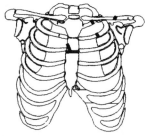

T 6; 49 km/h; 18 g;
8 % 3-Pt.-belt
m; 34 y.; 81 kg; 167 cm;
TOAIS 3; NRRF2

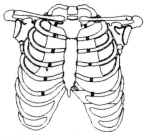

T 7; 47 km/h; 17 g;
unrestr
m; 57 y.; 67 kg; 170 cm;
TOAIS 4; NRRF 12

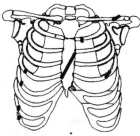

T 9; 50 km/h; 18 g;
8 % 3-Pt.-belt
m; 36 y.; 76 kg; 180 cm
TOAIS 4; NRRF 6

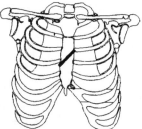

T 11 ; 55 km/h; 17 g;
6 % 3 Pt. belt;
m; 20 y.; 86 kg; 177 cm;
TOAIS 2; NRRF 0

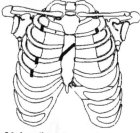

T 12; 50 km/h; 20 g;
8 % 3 Pt. belt;
m; 37 y.; 68 kg; 174 cm;
TOAIS 3; NRRF 5

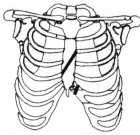

T 13; 49 km/h; 15 g;
16 % 3 Pt. belt;
m; 29 y.; 68 kg; 184 cm
TOAIS 2; NRRF 2

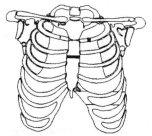

T 14; 48 km/h; 14 g;
16 % 3 Pt. belt;
w; 52 y.; 68 kg; 168 cm;
TOAIS 2; NRRF 1;

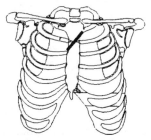

T 15; 48 km/h; 14 g;
16% 3Pt.belt,driver airbag;
w; 47 y.; 76 kg; 169 cm;
TOAIS 2; NRRF 0;

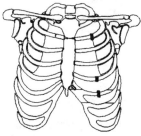

T 16; 48 km/h; 14 g;
16% 3Pt.belt,driver airbag;
m; 32 y.; 85 kg; 185 cm;
TOAIS 2; NRRF 3;

Figure 6: Injury pattern in frontal collision tests by using 3-point belt - driver air bag -knee bolster and 3-point belt - driver air bag combined restraint systems (NRRF: Number of Rib Fractures).

MEDICAL FINDINGS - Twelve of the sixteen tests conducted show thoracic injuries. The most frequent injury was rib fracture. The number of rib fractures and the injury pattern seem to be influenced by both the age of the cadaver and the type of restraint system used. This statement is a heuristic argument based on observations of the 16 human cadaver tests reported in this paper. There are not enough tests to conduct extensive statistical analysis.

Clavicular, sternal, and rib fractures occurred along the shoulder belt path. Furthermore, one liver rupture was observed when a 3-point belt was used. The locations of the observed fractures for all cases where the thorax was injured are illustrated in Figure 6 along with the maximum assessed torso AIS.

Fifty percent of the cases investigated also showed vertebral column injuries. They were located from the middle cervical spine to the upper thoracic vertebral column. The most frequent injuries were haemorrhages of the vertebral discs and lacerations of the ligaments. However, because of the small number of tests and the apparent strong influence of age, performance differences between the various restraint modes was not discernable. All the vertebral injuries observed are illustrated in Figure 7. The injury severity was scaled according to the AIS 90 for the vertebral column and ranged from AIS 0 to AIS 5.

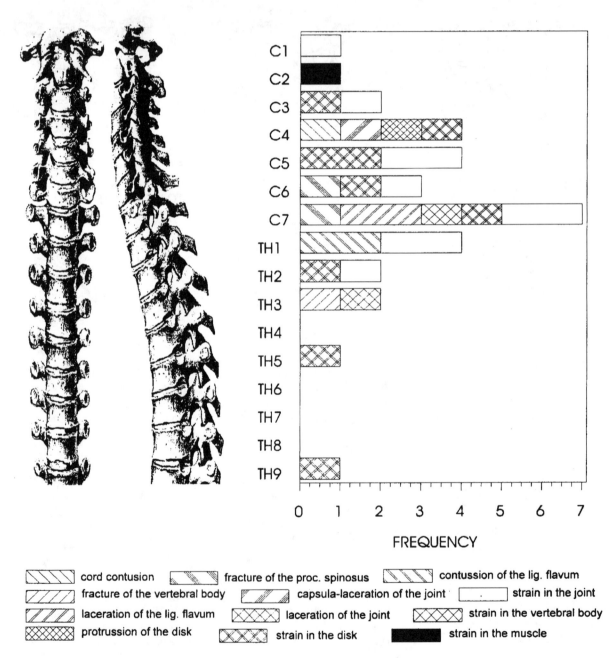

Figure 7: Location and kind of vertebral-injuries in frontal collision tests by using 3-point belt / driver air bag - knee bolster and 3-point belt / driver air bag combined restraint systems.

Table 2
Mechanical responses, subject's age, restr. systems and thoracic injuries

| Run | Deflection [cm] | | max. Acceleration [g] | | | | | AGE | Restr. | Injuries NRF |
	upper	lower	u. St.	l. St.	Resultant Th1	Th6	Th12	Y [g]	System	Cl., St., liver
T7	-	-	136	111	93	-	76	57	unrestr.	12/liver
T3	-	-	-	-	-	44	-	63	3-Pt.-belt 6 %	10
T4	-5,4	-2,7	31	53	31	-	50	34	3-Pt.-belt 6 %	0
T6	-	-	-	-	-	37	-	34	3-Pt.-belt 8 %	Cl./2/St.
T9	-	-	-	-	-	43	-	36	3-Pt.-belt 8 %	Cl./6/St.
T11	-5	-5	-	-	58	-	-	20	3-Pt.-belt 6 %	0/St.
T12	-	-4	-	-	-	47	-	37	3-Pt.-belt 8 %	5/St./liver
T13	(-1,1)	-3	49	39	48	-	32	29	3-Pt.-belt 16 %	2/St.
T14	-5,6	-4,7	40	35	35	-	49	52	3-Pt.-belt 16 %	1/St.
T2	-	-	-	-	-	78	-	41	dr. air bag	6
T5	-	-3,2	33	52	38	-	58	31	dr. air bag	0
T8	-2	-4	54	44	56	-	49	25	dr. air bag	0
T10	-7	-7	53	74	48	-	52	38	dr. air bag	0
T1	-	-	-	-	-	49	-	62	3Pt.b.+dr.air bag 6%	9
T15	-5,4	-7,3	45	50	34	-	45	47	3Ptb.+dr.air bag 16%	St.
T16	-5,4	-5,9	30	23	35	-	31	32	3Ptb.+dr.air bag 16%	3 (Infrac)

RESPONSES - INJURIES - Table 2 summarizes the most important mechanical responses and the observed thoracic injuries for the 16 tests conducted. The highest accelerations and the most severe trunk injuries were observed by the unrestrained test subject (T 7). The acceleration maxima of the thorax ranged from 30 g to 50 g with sternal and rib fractures of a thoracic injury severity of AIS 2 or AIS 3 observed. Several cases were uninjured. Chest deflections between 2 cm and 7 cm at the level of the 4th and 8th rib were found. Sample evaluations of VC for two test conditions were performed. A test by using 3-point belt (T14) resulted in a VC value of 0.3 to 0.4 m/s, while a further test using 3-point belt / driver air bag (T16) yielded a value of 0.2 to 0.32 m/s.

Differences between a 6% and 16% yield limit torso belt characteristic were best observed with tests using the Hybrid III. Figure 8 illustrates deformation differences observed by the upper chest band while Figure 9 shows differences seen by the lower band when restrained by the two different 3-point belt systems. The higher yield belt produced less maximum deflection, but still produced the typical thoracic deformation pattern. When the 16% belt was used in a specimen test and in conjunction with an air bag, (T15 & T16) the deformation pattern still resembled a belt only test. However, the fracture pattern of one test looked more like a bag only test while the other looked like a belt only test.

DISCUSSION

The driver's kinematic behaviour appears specific for each of the restraint systems used. The use of a driver air bag - knee bolster system resulted in a displacement of the head in the x - z - plane with an impact of the front of the head against the windscreen or the front part of the roof; an extension of the cervical spine followed. In order to prevent this, it would appear that the joint use of a lap belt is necessary. When using only a 3-point belt, considerable hyperflexion of the neck occurs. However, the only translational displacement of the head theoretically expected when using an air bag could not be confirmed by the medical findings of the cervical spine. Except for the unrestrained test (T 7) with rib fractures and an liver injury, the thoracic injury severity was determined by the number of rib fractures or the sternum fracture. By using the combination 3-point belt / driver air bag, the injury pattern of the 3-point belt predominated; this is thought to be a result of both the stiffness and path of the shoulder belt. The rib fractures resulting from air bag use are located at the front axillar or medio-clavicular line; this is in agreement with findings of Yoganandan et al. (1993). The most fractures that were observed were infractions, a fracture type which is not visible at the chest front when using the conventionally x-ray examination; these findings are now possible through autopsy and the touching of the ribs.

In 3 of 4 cases where a driver air bag only system was used; a thorax AIS 0 resulted. Therefore, as also seen in an earlier study with passenger air bag - knee bolster (Kallieris et al. 1982), a tendency toward a lesser vulnerability is provided by the air bag in comparison to the 3- point belt.

By using the chest band, a temporal examination of the chest deformation was possible at the level of the 4th and 8th rib. By this means, the differences in dynamic behaviour of the thorax when using different restraint systems was investigated and typical contours for each type of restraint system were observed. By using an additional computer program, deflection vs time evaluations from the con-

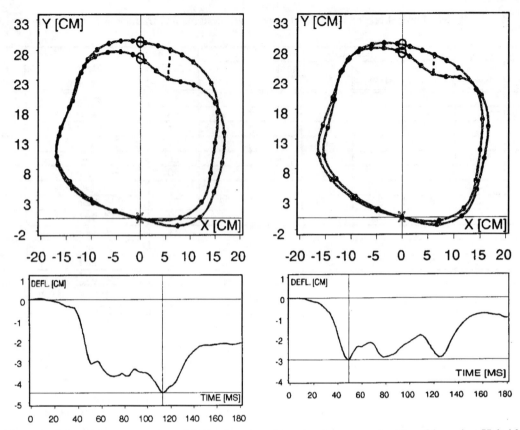

Figure. 8: Thoracic deformation contours and deflection time histories for upper chest band by using Hybrid III.
a) 6% belt. b) 16% belt.

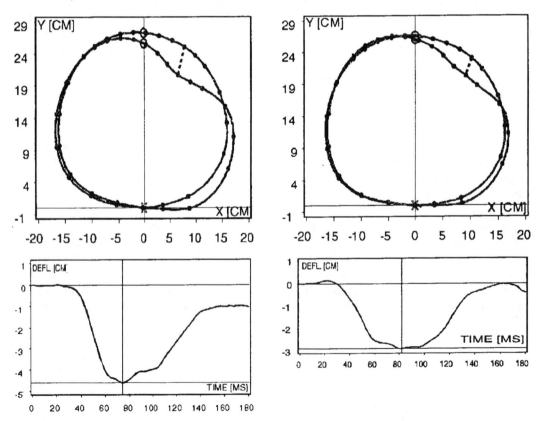

Figure 9: Thoracic deformation contours and deflection time histories for lower chest band by using Hybrid III.
a) 6% belt. b) 16% belt.

tour plots at the corresponding chest level were made and evaluated.

Chest deflections between 2 cm and 7 cm were observed for the different restraint systems. However, no specific deflection values concerning a restraint system could be observed. The investigated deflection values are in general agreement with the results published by Walsh et al. (1978), who measured with a different method and the results published by Yoganandan et al. (1993) who used the same method as the authors.

The chest band offers the advantages of being able to measure deformation anywhere underneath the band. Chest contours are produced which are particularly useful for uneven loading. However, it must operate on a system with a very large data acquisition capacity. Some of the operational problems are that generation of the contours is very sensitive to strain gauge failure, distance between gauges and the saturation of signals. Because of the complex kinematics the occupant experiences during a frontal collision, it is very difficult to optically evaluate the thorax deflection by the high speed films. The evaluation of the deflection according to the twofold integration of the acceleration difference spine-sternum is often inaccurate due to the rotation of the sternum-accelerometer. Therefore, the chest band method is best suited to measure the thorax deflection. However, more tests are required to optimise the settings for the use of this instrument.

When examining the response of the thorax when a belt/bag restraint system was used, both the 6% and 16% elongation belt produced belt-like deformations. This suggests that to obtain the thoracic injury mitigating benefits of a bag only restraint from a combination restraint system, that either a torso belt with greater elongation characteristic or one with a force limiter would be required.

CONCLUSIONS

For the impact conditions investigated in this study, a combination 3-point belt/driver air bag would appear to offer the best injury protection. However, because the best available torso belt system still produced belt-like deformations, it is anticipated that further performance improvements can be achieved with further modification of the torso belt yield characteristics.

The chest band is well suited to measure thoracic deformations during dynamic loading of the chest and allowed this detailed investigation of the differences in performance among the various restraint systems studied.

ACKNOWLEDGEMENTS

The sled buck, the steering wheel assembly, the seats, the air bags and the 3-point belts have been put at our disposal by the BMW AG, Munich. Additionally, the authors also thank BMW for the technical support.

REFERENCES

Eppinger RH, Augustyn K, Robbins DH (1978) Development of a Promising Universal Thoracic Trauma Prediction Methodology, SAE Paper No. 780891, Proc. 22nd Stapp Car Crash Conf. p. 209-268

Eppinger RH, (1989) On the Development of a Deformation Measurement System and Its Application Toward Developing Mechanically Based Injury Indices, SAE Paper No. 892426, Proc. 33d Stapp Car Crash Conf. , p. 21-28

Kallieris D, Mattern R, Schmidt G, Klaus G. (1982) Comparison on 3-Point Belt and Air bag - Knee Bolster Systems. Injury Criteria and Injury Severity at Simulated Frontal Collisions. Intern. Research Council on Biokinetics of Impact, Cologne, Germany, p. 166-183

Lau I, Viano D, (1986) The Viscous Criterion - Bases and Application of an Injury Severity Index for Soft Tissue, SAE Paper No. 861882, Proc. 30th Stapp Car Crash Conf. p. 123-142

Robbins DH, Melvin JW, Stalnaker RL, (1976) The Prediction of Thoracic Impact Injuries, SAE Paper No. 760822, Proc. 20th Stapp Car Crash Conf. p. 699-729

Walsh MJ, Kelleher BJ (1978) Evaluation of Air Cushion and Belt Restraint Systems in Identical Crash Situations Using Dummies and Cadavera, SAE Paper No. 760803, Proc. 20th Stapp Car Crash Conf. p. 107-131

Yoganandan N, Pintar FA, Skrade D, Chmiel W, Reinartz JM, Sances A, (1993) Thoracic Biomechanics With Air bag Restraint, SAE Paper No. 933121, Proc. 37th Stapp Car Crash Conf. p.133-144

950886

The Effect of Limiting Shoulder Belt Load with Air Bag Restraint

Harold J. Mertz, James E. Williamson, and Donald A. Vander Lugt
General Motors Corp.

ABSTRACT

The dilemma of using a shoulder belt force limiter with a 3-point belt system is selecting a limit load that will balance the reduced risk of significant thoracic injury due to the shoulder belt loading of the chest against the increased risk of significant head injury due to the greater upper torso motion allowed by the shoulder belt load limiter. However, with the use of air bags, this dilemma is more manageable since it only occurs for non-deploy accidents where the risk of significant head injury is low even for the unbelted occupant. A study was done using a validated occupant dynamics model of the Hybrid III dummy to investigate the effects that a prescribed set of shoulder belt force limits had on head and thoracic responses for 48 and 56 km/h barrier simulations with driver air bag deployment and for threshold crash severity simulations with no air bag deployment. For the belt/bag/occupant configuration that was evaluated, the analysis gives an optimum shoulder belt load limit that minimizes the dummy responses in the 56 km/h simulation without significantly increasing the risk of significant head injury in the non-deploy simulations.

INTRODUCTION

During the period from 1970 to 1977, Peugeot and Renault vehicles sold in France were equipped with 3-point lap/shoulder belt systems for the front outboard seating positions that incorporated a load limiter located in the shoulder belt webbing between the upper anchorage point and the occupant's shoulder (1,2). Three types of load limiting systems (A, B and C) were used. For each system the load was controlled by tearing of the stitching that was used to sew loops in the webbing. Thus, as the stitching tore, the added webbing from the loop allowed the torso to move forward relative to the car at a controlled force load. The Type A limiter had five loops with stitching tear loads ranging from 2.1 kN to 4.4 kN. The Type B limiter had five loops with the stitching tear loads

ranging from 7.2 kN to 8.0 kN. The Type C limiter was a single large loop with a tear initiation load of 5.5 kN and a mean tear load of 4.1 kN. The Association Peugeot-Renault (APR) collected accident data for 427 occupants who were involved in frontal accidents while wearing a 3-point lap/shoulder belt system with one of the load limiters. Mertz et al (3) analyzed the APR data and noted that the magnitude of the shoulder belt force could be estimated from the recorded amount of stitching tearing for 342 of these occupants. They conducted sled tests using a Hybrid III dummy and reproduced the various degrees of stitching tearing. The Hybrid III sternal deflections were correlated to the frequencies of AIS $\geq$ 3 thoracic injury calculated from the field accident data for similar stitch tearing. The resulting curve giving the risk of AIS $\geq$ 3 thoracic injury as a function of Hybrid III sternal deflection for shoulder belt loading is given in Figure 1. According to the analysis of Mertz et al (3), this curve can be converted to a risk curve for shoulder belt load for the Peugeot-Renault belt geometry without a load limiter by multiplying the Hybrid III chest deflection by 200 N/mm. The resulting thoracic injury risk curve is shown in Figure 2. Also shown on the figure are the load ranges for the Type A, B and C load limiters. Clearly, the lower the limit load, the lower the risk of AIS $\geq$ 3 of thoracic injury produced by the shoulder webbing. However, lower limit loads result in greater forward displacements of the torso with increased risk of head contact with the interior. The dilemma faced by Peugeot and Renault was to select a limit load that balanced the reduced risk of thoracic injury against the increased risk of head injury. With the use of air bags, this dilemma becomes more manageable since it only occurs for accidents below the threshold for air bag deployment. These accidents are minor and the risk of AIS $\geq$ 3 head injury is low even for the unbelted occupant. For the more severe deployment accidents the air bag will provide head as well as chest protection for the occupant restrained by a 3-point lap/shoulder belt system with a shoulder belt load limiter. This paper summarizes the results of a study done to

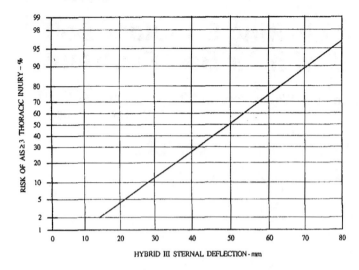

Figure 1: Risk of AIS ≥ 3 Thoracic Injury Due to Shoulder Belt Loading as a Function of Hybrid III Sternal Deflection (3).

investigate the effects that a set of shoulder belt force limits have on head and thoracic responses for 48 and 56 km/h barrier deployment simulations and for threshold crash severity simulations with no air bag deployment. The simulations were done using a validated occupant dynamics model of the Hybrid III dummy restrained by a generic lap/shoulder belt system and a generic driver air bag/steering wheel and column/knee restraint system.

MODEL DESCRIPTION AND RESULTS

The CAL-3D occupant dynamics model of the Hybrid III dummy was used for the analysis (4,5,6). The model included the following generic features: a 3-point belt system, a driver air bag and steering wheel and column, and a knee restraint. These features were validated by comparing model results to sled test results. Descriptions of the belt system along with

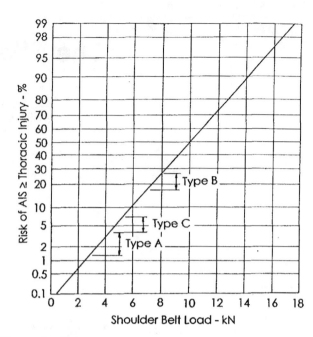

Figure 2: Risk of AIS ≥ 3 Thoracic Injury as a Function of Shoulder Belt Load for Peugeot-Renault Belt Geometry. Load Ranges for Type A, B, C Load Limiters are Shown (1,2).

pertinent occupant response data are given in Table 1 for the 48 and 56 km/h simulations and in Table 2 for the non-deploy simulations.

For the 56 km/h simulations, eight different shoulder belt load limiting conditions were evaluated in combination with the air bag/knee restraint system. These were the baseline 3-point belt system without the shoulder belt load limiting feature and seven belt configurations where the shoulder belt limit load was set at 0, 1, 2, 3, 4, 5 and 6 kN. Figure 3 shows the general characteristics of the load limiter. Note that in all simulations the shoulder belt load was not allowed to exceed its prescribed limit value. Figure 4 shows the generic velocity versus time pulse of the vehicle used for the simulations. With an 80 ms period to

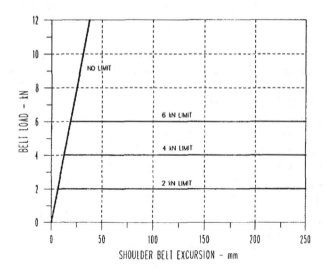

Figure 3: Characteristics of Shoulder Belts

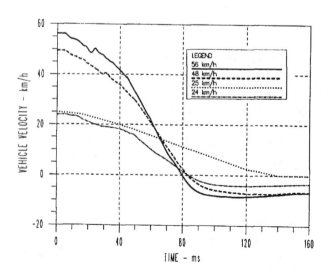

Figure 4: Vehicle Velocity - Time Pulses

Summary of Simulations with Hybrid III 50th Driver Occupant, Driver Air Bag Deployed

Barrier Speed (km/h)	Belt System		Head			Thorax		
	Load Limit (kN)	Excursion (mm)	36 ms HIC	15 ms HIC	AIS>/=4 Risk (%)	Max. Acc. (G)	Comp. (mm)	AIS>/=3 Risk (%)
56	None	24	1254	984	14.5	66	75	94
	6	25	1025	758	5.5	63	66	84
	5	51	987	641	3.1	61	54	61
	4	85	735	470	1.2	57	45	40
	3	123	630	371	0.6	54	39	27
	2	169	451	306	0.4	48	25	19
	1	221	708	419	0.8	53	38	25
	0	-	756	457	1.1	55	51	53
48	None	20	1105	732	5.0	63	67	85
	2	150	391	256	0.3	44	22	5

Table 2

Summary of Non-Deploy Simulations for 24 km/h Rigid Barrier and 25 km/h Mild Severity Pulses

Crash Pulse (km/h)	Belt System			Head					Thorax		
	Driver Size	Load Limit (kN)	Excursion (mm)	Forward Excursion (mm)	Max. Velocity (m/s)	Wheel Contact Velocity (m/s)	15 ms HIC	AIS>/=4 Risk (%)	Max. Acc. (G)	Comp. (mm)	AIS>/=3 Risk (%)
24	50th	None	8	439	5.7	None	25	<<0.1	23	23	5.5
	50th	2	151	565	6.1	None	10	<<0.1	16	13	1.5
	95th	None	11	493	6.2	None	25	<<0.1	22	30	12.1
	95th	2	227	682	6.8	4.8	*	-	15	13	1.5
25	50th	None	6	369	3.7	None	10	<<0.1	13	16	2.2
	50th	2	72	422	3.7	None	6	<<0.1	11	11	1.1
	95th	None	7	419	4.1	None	10	<<0.1	14	21	4.5
	95th	2	137	533	4.4	None	4	<<0.1	10	12	1.3

* Steering wheel impact not simulated.

reach zero velocity, the pulse is quite representative of the 56 km/h NCAP rigid barrier pulses of contemporary vehicles. The results shown in Table 1 are the 36 ms HIC values specified by FMVSS 208, the biomechanical based 15 ms HIC values (7), the peak chest acceleration and compression, and the maximum excursion of the shoulder belt webbing. Also shown are the risks of AIS $\geq$ 4 head injury based on the 15 ms HIC value and the injury risk curve of Prasad and Mertz (7) shown in Figure 5, and the risks of AIS $\geq$ 3 thoracic injury due to chest compression based on the injury risk curve of Mertz et al (3) shown in Figure 1. Figure 6 gives a comparison of the shoulder belt loads applied to the occupant for the baseline belt system without a load limiter and the 2 kN load limiter system. Note that while load limiter never allowed the load to exceed the prescribed limit value, the applied shoulder belt load was slightly greater due to D-ring friction.

Only two 48 km/h simulations were conducted, the air bag with the baseline belt system without a load limiter and the air bag with the belt system incorporating the 2 kN shoulder belt load limiter. The velocity-time pulse for the vehicle is shown in Figure 4. Again, it is representative of the 48 km/h rigid barrier pulse of contemporary vehicles.

For the non-deploy simulations, the baseline belt system without a load limiter and the belt system with the 2 kN shoulder belt limiter were evaluated using the Hybrid III dummy model and a 95 percentile adult male dummy model. The simulations were conducted with two crash pulses, a generic 24 km/h rigid barrier pulse and a more field representative, mild severity pulse (6.7 G, 25 km/h, 140 ms) that is used to evaluate out-of-position vehicle occupant interactions with deploying air bags (8). The velocity-time curve for each pulse is shown in Figure 4. For these simulations the steering wheel system was removed since the model is not validated for steering wheel head impact. Figures 7-10 depict the trajectories of a point on top of the head (122 mm above its center of gravity) relative to a generic steering wheel location. The maximum forward head excursion and the maximum head velocity relative to the vehicle were determined for each simulation and recorded in Table 2 along with other pertinent peak values. The occupant head trajectory was analyzed to determine if head to steering wheel contact could occur. If so, the contact velocity was calculated and given in Table 2. The 36 ms HIC values are not listed in Table 2 since they have no significance for non-head contact situations.

DISCUSSION OF RESULTS

Figure 11 is a plot of the maximum 36 ms HIC, 15 ms HIC, chest acceleration and chest compression as a function of the shoulder belt limit load for the 56 km/h simulation. Figure 12 shows the risks of AIS $\geq$ 4 head injury and the risks of AIS $\geq$ 3 thoracic injury as functions of the shoulder belt limit load for the 56 km/h simulation. All values were the lowest for the belt system with the 2 kN shoulder belt load limiter. Comparing the responses for the 2 kN shoulder belt limiter to the baseline belt system, the 36 ms HIC was reduced by 64 percent (1254 to 451), the 15 ms HIC was reduced by 69 percent (984 to 306), the chest acceleration was reduced by 27 percent (66 G to 48 G) and the chest compression was reduced by 67 percent (75 mm to 25 mm). The risk of AIS $\geq$ 4 head injury

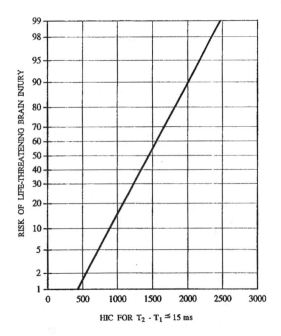

Figure 5: Risk of AIS $\geq$ 4 Brain Injury as a Function of 15 ms HIC (7).

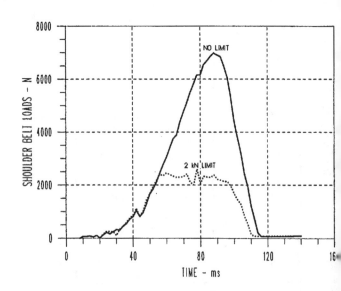

Figure 6: Comparison of Applied Shoulder Belt Load for Baseline System without Load Limiter and 2 kN Load Limiter, 50th Percentile Occupant, 56 km/h Rigid Barrier Simulation with Driver Air Bag Deployed.

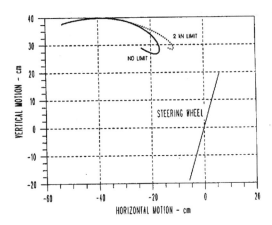

Figure 7: Head Trajectories of 50th Percentile Occupant Relative to Generic Steering Wheel for Baseline System without Load Limiter and with 2 kN Load Limiter, 25 km/h Mild Severity Pulse, No Air Bag Deployment, Seat in Mid Position.

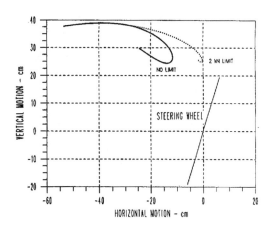

Figure 8: Head Trajectories of 95th Percentile Occupant Relative to Generic Steering Wheel for Baseline System without Load Limiter and with 2 kN Load Limiter, 25 km/h Mild Severity Pulse, No Air Bag Deployment, Seat in Mid Position.

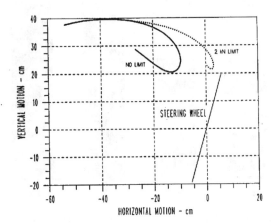

Figure 9: Head Trajectories of 50th Percentile Occupant Relative to Generic Steering Wheel for Baseline System without Load Limiter and with 2 kN Load Limiter, 24 km/h Rigid Barrier Pulse, No Air Bag Deployment, Seat in Mid Position.

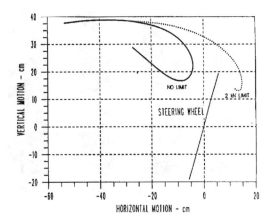

Figure 10: Head Trajectories of 95th Percentile Occupant Relative to Generic Steering Wheel for Baseline System without Load Limiter and with 2 kN Load Limiter, 24 km/h Rigid Barrier Pulse, No Air Bag Deployment, Seat in Mid Position.

was reduced from 14.5 percent to 0.4 percent, and the risk of AIS ≥ 3 thoracic injury was reduced from 94 percent to 19 percent. There was a corresponding reduction in the 48 km/h simulations with the 36 ms HIC being reduced by 65 percent (1105 to 391), the 15 ms HIC by 65 percent (732 to 256), the chest acceleration by 31 percent (63 G to 44 G) and the chest compression by 67 percent (67 mm to 22 mm). The risk of AIS ≥ 4 head injury was reduced from 5 percent to 0.3 percent, and the risk of AIS ≥ 3 thoracic injury reduced from 85 percent to 5 percent. Clearly, limiting the shoulder belt load in high severity deployment accidents will reduce the risk of significant head injuries and will have a dramatic effect on reducing the risk of significant thoracic injuries.

Figures 13 and 14 show the maximum forward head excursions and the maximum head velocities relative to the vehicle for the non-deployment simulations, respectively. The excursions of the shoulder belt webbing due to the 2 kN load limiter results in greater forward head excursions. For the 25 km/h mild severity simulations, the forward head excursion increased by 53 mm for the 50th percentile occupant and by 114 mm for the 95th percentile occupant. For the more severe 24 km/h rigid barrier simulation, the increases were greater, being 126 mm for the 50th percentile occupant and 189 mm for the 95th percentile occupant. There were small increases in the maximum head velocities relative to the vehicle with the 2 kN shoulder belt load limiter. For the 25 km/h mild severity simulation, there was no increase in the maximum head velocity for the 50th percentile

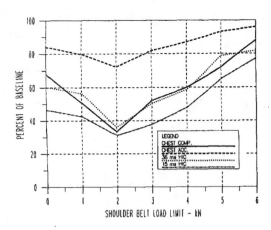

Figure 11: Peak Driver Responses as a Percent of Baseline Responses for Various Shoulder Belt Load Limits, 50th Percentile Driver, 56 km/h Simulations, Air Bag Deployed.

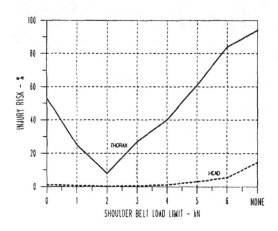

Figure 12: Risks of AIS ≥ 4 Head and AIS ≥ 3 Thorax Injuries for Various Shoulder Belt Load Limits, 50th Percentile Driver, 56 km/h Simulations, Air Bag Deployed.

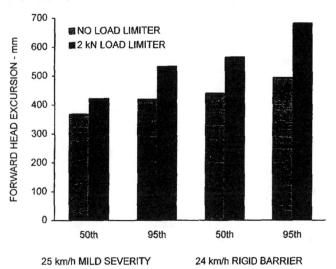

Figure 13: Peak Forward Head Excursions for Non-Deploy Simulations.

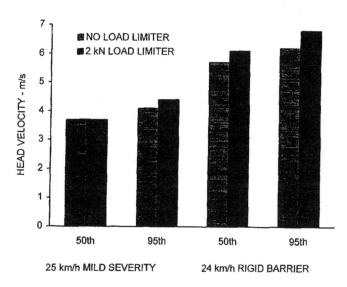

Figure 14: Peak Head Velocity for Non-Deploy Simulations.

occupant and only a 0.3 m/s increase for the 95th percentile occupant. For the more severe 24 km/h rigid barrier simulation, the increases in head velocity were 0.4 m/s for the 50th percentile occupant and 0.6 m/s for the 95th percentile occupant.

To determine if head-to-steering wheel contact could have occurred in these non-deploy simulations, a nominally positioned, generic steering wheel was superimposed on the head trajectories of the occupants (Figures 7-10). None of the occupants would have contacted the steering wheel with his head except the 95th male seated in the mid-seat position with the 2 kN load limiter in the more severe 24 km/h barrier simulation (Figure 10). For this simulation, the head velocity relative to the wheel at the time of contact would have been 4.8 m/s. Note that this simulation was conducted with the 95th occupant in the mid-seat position. If the seat was in the more customary rear position, then head-to-wheel contact would not occur

for this crash pulse.

For contemporary steering wheel designs, a head-to-wheel velocity of 5.4 m/s (12 mph) has been used as the threshold velocity for facial bone fracture and, consequently, the deployment threshold velocity for driver air bags (9). This threshold deployment velocity can be increased by reducing steering rim stiffness and mass. The 4.8 m/s head contact velocity predicted for the 95th percentile driver restrained by the belt system with the 2 kN load limiter in the 24 km/h rigid barrier non-deploy simulation is less than facial bone fracture threshold velocity of 5.4 m/s. In addition, the driver's ability to reduce his forward head excursion and velocity by bracing himself with his arms against the steering wheel rim was not simulated in the model. Also, both the 24 km/h rigid barrier and 25 km/h mild severity collision simulations are above the deployment threshold level required for the unrestrained driver. For these reasons, the risk of facial bone fracture if head-to-wheel contact had occurred in this simulation is considered small. Deployment of the driver's air bag would not be required due to this contact.

Simulations of the force limiting belt systems with a generic passenger air bag system were not conducted. However, one can deduce that similar reductions in head and thoracic injury risk would occur for the 2 kN load limiter although this load level may not be the optimum for the passenger air bag system. For the non-deploy conditions, the results would be identical to those given in Table 2 and Figures 8-10, 13 and 14 with no head-to-instrument panel contact for any of the occupants.

METHODS TO LIMIT SHOULDER BELT LOADS

To realize the effectiveness of shoulder belt load limiting indicated by the model analysis, the device must be capable of minimizing belt slack and loose webbing wrap while allowing the required upper torso excursion at the prescribed webbing load. Several commercial devices currently exist for taking up webbing slack and tightening belt wrap early in the collision. These are called belt pretensioners. The challenge is to develop a device which will allow the large amount of upper torso excursion at the prescribed belt load for the 56 km/h rigid barrier simulation with the air bag deployed. For the 50th percentile driver, the shoulder belt webbing excursion was 169 mm. Greater excursion would be needed for a 95th percentile driver for the same collision simulation. The "stitch tearing" concept used by Peugeot-Renault could provide the necessary webbing excursion, but the folded loops may not result in a "pleasing" packaging appearance. Deformable belt anchors have been utilized, but these devices have excursion restrictions due to packaging constraints. An attractive concept is to incorporate an energy dissipative mechanism on the webbing retractor reel. A clutch mechanism or a deforming element could be used to resist rotation of the webbing reel by friction forces or by the forces required to deform the element. The level of resistance could be designed to give the desired webbing load limit for the specific vehicle application. Sufficient webbing could be stored on the reel to allow for the needed torso excursion at the prescribed shoulder belt load level. Projects have been initiated to develop such hardware. Descriptions of the hardware and test results are planned for future publications.

SUMMARY

During the period from 1970 to 1977, Peugeot-Renault sold vehicles in France equipped with 3-point belt systems with three different types of load limiters located in the shoulder belt webbing. The dilemma faced by Peugeot-Renault was the selection of a limit force that would balance the reduced risk of significant thoracic injury due to belt webbing loading against the increased risk of significant head injury due to the greater upper torso motion allowed by the shoulder belt load limiter. With the use of air bags, this dilemma is quite manageable since a force level can be chosen that will minimize the risks of both head and thoracic injury. For the 48 km/h rigid barrier air bag deployment simulations that were conducted, the risk of AIS ≥ 3 thoracic injury due to shoulder belt loading of the 50th percentile driver's chest was reduced from 85 percent for the baseline belt system without a force limiter to 5 percent for the belt system with a 2 kN shoulder belt force limiter. The risk of AIS ≥ 4 head injury was reduced from 5 percent to 0.3 percent. For the 56 km/h simulations, the thoracic injury risk was reduced from 94 percent to 19 percent and the head injury risk was reduced from 14.5 percent to 0.4 percent. For low severity accidents where the air bag system is not deployed, the belt system with the 2 kN shoulder belt force limiter provided sufficient upper torso restraint so that either head-to-steering wheel contact will not occur or if it does occur, the risk of facial bone fracture due to the contact is small. Similar results and conclusions are expected when using the 2 kN shoulder belt force limiter system with a generic passenger air bag system even though this force level was chosen to minimize the head and thoracic responses of the 50th percentile male driver for the generic driver air bag system.

REFERENCES

1. Foret-Bruno, J. Y., Hartman, F., Thomas, C., Fayon, A., Tarriere, C. and Patel, A., "Correlation Between Thoracic Lesions and Force Values Measured at the Shoulder of 92 Belted Occupants Involved in Real Accidents", Twenty-Second Stapp Car Crash Conference, SAE 780892, October 1978.

2. Foret-Bruno, J. Y., Brun-Cassan, F., Brigout, C. and Tarriere, C., "Thoracic Deflection of Hybrid III Dummy Response for Simulations of Real Accidents", 12th Experimental Safety Vehicle Conference, Gothenburg, Sweden, 1989.

3. Mertz, H. J., Horsch, J. D., Horn, G. and Lowne, R. W., "Hybrid III Sternal Deflection Associated with Thoracic Injury Severities of Occupants Restrained with Force-Limiting Shoulder Belts", SAE 910812, SP-852, February 1991.

4. Fleck, J. T. and Butler, F. E., "Validation of the Crash Victim Simulation", DOT HS-806 279-282, 1981.

5. Wang, J. T. and Nefske, D. J., "A New CAL 3D Air Bag Inflation Model", SAE880654, 1988.

6. Deng, Y. C., "Development of a Mathematical Model for Predicting a Belt-Restrained Occupant Response in Automotive Crash", Crashworthiness and Occupant Protection in Transportation Systems, ASME H00854, Vol. 169, pp 223-234, 1993.

7. Prasad, P. and Mertz, H. J., "The Position of the United States Delegation to the ISO Working Group 6 on the Use of HIC in the Automotive Environment", SAE 851246, May 1985.

8. Guidelines for Evaluating Out-of-Position Vehicle Occupant Interactions with Deploying Air Bags, SAE Information Report J1980, November 1990.

9. Klove, E. H. and Oglesby, R. N., "Special Problems and Considerations in the Development of Air Cushion Restraint Systems", SAE 720411, May 1972.

ASSESSMENTS OF AIR BAG PERFORMANCE BASED ON THE 5TH PERCENTILE FEMALE HYBRID III CRASH TEST DUMMY

Dainius J. Dalmotas
Transport Canada
Canada
Paper Number 98-S5-O-07

ABSTRACT

Historically, assessments of frontal crash safety have been based primarily on the measured responses of 50th percentile male dummies in relatively high speed vehicle crashes against a rigid flat barrier. Under such test conditions, the ability of supplementary airbag systems to greatly reduce head injury potential is clearly evident in crash tests performed by Transport Canada and others. However, significant segments of the driving population travel routinely with their seats positioned ahead of the nominal mid-position used in 50th percentile male dummy tests. Moreover, most frontal impacts can be expected to produce softer vehicle deceleration signatures than those produced in flat rigid wall tests. The necessity of broadening the range of regulated crash conditions to which vehicles fitted with airbag systems are subjected is highlighted in crash tests performed by Transport Canada using 5th percentile female Hybrid III tests, with seats placed in their most forward positions. The neck loads observed in these tests far exceeded commonly referenced injury assessment values. The magnitudes of the neck loads were influenced not only by the aggressiveness of the airbag system, but also by the timing of the deployment of the airbag. The neck loads observed in low speed offset frontal crash tests often exceeded those observed in high-speed, rigid-wall tests, as a result of the timing of airbag deployment.

INTRODUCTION

The fitment of supplementary airbag systems is not mandatory in Canada. In the formulation of occupant protection standards governing occupant protection in frontal crashes, emphasis in Canada continues to be placed on regulating total system performance, rather than the specification of hardware. The technical requirements of Canada Motor Safety Vehicle Standard (CMVSS) 208 have been revised recently to reflect performance levels achievable with current technology. The revised performance requirements have only been satisfied consistently by vehicles fitted with supplementary airbag systems [1,2]. Given the highly integrated nature of the automobile industry in North America, it is anticipated

that most, if not all, new passenger-carrying vehicles sold in Canada will be fitted with supplementary airbag systems. Though no test with an unbelted dummy is specified in Canada, it is reasonable to expect that the design of most airbags fitted in Canada will continue to be strongly influenced by US regulatory requirements, which continue to emphasize the protection of unbelted occupants.

One major shortcoming of both Canadian and US regulatory requirements is that each front outboard seating position is tested with a dummy of 50th percentile male dimensions in one well-defined seating posture. Consequently, the performance levels achieved in the test may not be indicative of the levels of protection likely to be afforded to occupants of different stature. Of particular concern are possible adverse airbag-occupant interactions if the seat is located forward of the mid seat position. There is evidence from laboratory testing that the proximity of an occupant to the airbag module has a strong influence on the response of the neck and the chest [3,4].

FIELD PERFORMANCE

In order to gain an understanding of the field performance of supplementary airbag systems in Canada, Transport Canada, in the fall of 1993, initiated a directed study devoted to documenting the injury experience of occupants involved in crashes resulting in the deployment of an airbag system. The data collection methodology adopted for this study is similar to that used in the Fully Restrained Occupant Study (FROS) where the emphasis was on evaluating the collision performance of three-point seat belt systems [5]. The Air Cushion Restraint Study (ACRS) utilizes the resources of university-based collision investigation teams located across Canada. Each participating team is assigned a defined area of operation and case selection criteria. The study and findings are described in detail in previous publications [6,7,8].

Available Canadian evidence suggests that, as expected, airbags are highly effective in preventing serious or fatal head injury and facial fracture in high severity crashes, but that these gains are offset by bag-induced injuries in low severity crashes, when deployment is unwarranted if the belt system is being used. Female drivers are the most adversely affected in low-severity crashes.

The Canadian experience with airbags is consistent with the findings of a number of US studies. The introduction of the airbag has produced a variety of new injury mechanisms, such as facial injuries from "bag slap", upper extremity fractures, either directly from the deploying airbag module or from arm flailing, and thermal burns to the face and arms [9, 10, 11]. Among adults, most of the bag-induced injuries are minor in severity (AIS 1) as measured by the Abbreviated Injury Scale (AIS) [12]. However, upper extremity fractures rated AIS 2 or AIS 3 are not uncommon [13]. In the 1996 Report to Congress, NHTSA noted that the risk of serious (AIS 3) upper extremity injury to a belted driver may increase by some 40 percent with airbags [14]. Others have estimated that the risk of upper extremity injury among belted drivers may be increased by as much as a factor of 4 given airbag deployment [15]. Several studies have noted that the incidence of bag-induced upper extremity injury, particularly of upper extremity fracture, is far higher among female drivers than male drivers [8, 16,17]. The majority of the bag-induced arm fractures among belted female drivers occur in relatively low speed impacts [8].

In terms of overall fatality risk, the initial findings, at least for adults, are encouraging. Without exception, the effectiveness studies completed to date have shown that airbags reduce the risk of fatal injury among both drivers and adult passengers by some 11-14 percent, with the prevailing rates of seat belt usage in the US [18, 19, 20, 21].

Available evidence also suggests that airbags increase the overall risk of fatal injury among children under the age of 10 by some 21 percent [21]. In the US, NHTSA is investigating collisions involving airbag-related fatal or seriously injured occupants under its Special Crash Investigations (SCI) programme. Over 55 child deaths, directly attributable to airbag deployments, have been recorded to date under this programme. The vast majority of these deaths occurred in crashes of relatively minor severity. This death toll prompted NHTSA to relax the unbelted test requirements associated with FMVSS 208 in order to facilitate the rapid introduction of "depowered" airbag systems into the US.

At the time of writing, the SCI database also contained a total of 43 airbag-related adult fatalities. Of the 13 belted drivers represented in the database, 10 (77%) were females. All ten female victims were under 165 cm in height. The majority sustained fatal neck and/or head trauma. All three belted male drivers

sustained fatal chest trauma. Of the 21 unbelted drivers represented in the database, 16 (76%) were females. The majority of unbelted drivers, both males and females, sustained fatal chest trauma.

A monitoring programme, similar to the SCI, has also been implemented in Canada. To date, only one child death directly attributable to an airbag deployment is known to have occurred in Canada. At least four adult deaths directly attributable to an airbag deployment in a relatively low speed impact are known to have occurred in Canada. Three of the cases involved belted female drivers. The remaining case involved an unbelted male driver.

While most case studies of airbag-related deaths involve low to moderate speed collisions, it is important to recognize that the energy released by an airbag is independent of collision severity. As such, fatal bag-related injury can occur at all collision severities. With increasing collision severity, however, the injury outcome, in the absence of airbag deployment, becomes increasingly uncertain. Consequently, counts of airbag fatalities are limited to lower speed crashes where, in the absence of deployment, the occupant would have been expected to survive the crash.

JOINT TC/NHTSA CRASH TEST PROGRAMME

Based on an examination of the available data on the field performance of airbag systems in Canada, in 1996 Transport Canada implemented a major research programme to evaluate testing protocols which could be incorporated in Canada Motor Vehicle Safety Standard (CMVSS 208) to minimize the risk of bag-induced injury to belted occupants of short stature in frontal collisions. The crash test dummy selected for the programme was the 5th percentile Hybrid III female. In addition to representing a small adult, it has the advantage of representing, in size, a 12- to 13-year old child. Given the current recommendation in Canada, that all children aged 12 years or less, travel in a rear seat whenever possible, the 5th percentile female Hybrid III is an ideal dummy for the purposes of regulating front seat passenger-side protection.

Two series of full-scale vehicle crash tests were conducted as part of the programme. The first series involved 48 km/h rigid barrier crash tests with the seats in the full forward position. The second series of tests involved low-speed, offset-frontal crashes, utilizing the deformable barrier face and vehicle alignment protocols defined in Europe under Directive 96/79/EC. As in the rigid barrier tests, the 5th percentile Hybrid III was tested with the seat in the fully forward position.

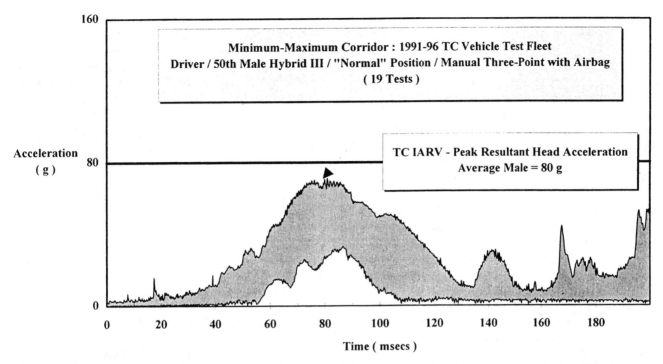

Figure 1. Range of Resultant Head Acceleration Responses Measured in 48 km/h Rigid Barrier Tests of First Generation Airbag with 50th Percentile Male Hybrid III (Driver Side).

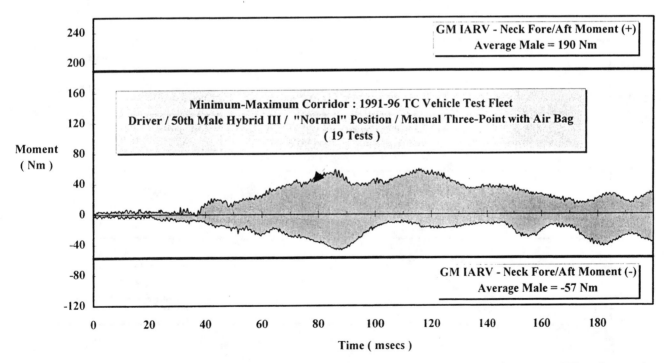

Figure 2. Range of Neck Extension Moment Responses Measured in 48 km/h Rigid Barrier Tests of First Generation Airbag Systems with 50th Percentile Male Hybrid III (Driver Side).

As part of a joint research agreement between Transport Canada and the NHTSA, the programme was expanded to include a representative sample of both first- and second-generation airbag systems and vehicles of different size classes. A total of 72 full-scale vehicle crash tests, utilizing one or two 5th percentile Hybrid III dummies, have been performed to date, generating a database of 124 individual 5th percentile Hybrid III dummy tests.

Baseline Responses - Mid-Size Male Hybrid III

In interpreting the results obtained in the tests with the 5th percentile female Hybrid III, it is informative to first consider the dummy responses typically measured in 48 km/h rigid barrier tests using the 50th percentile male Hybrid III dummy. The resultant acceleration-time histories of the head measured on the driver side in airbag tests with the dummy belted in 19 tests conducted by Transport Canada are presented in Figure 1. The fore/aft neck moment-time histories associated with the same tests are presented in Figure 2.

In a rigid barrier crash, the vehicle deceleration pulse generally produces deployment of the airbag early in the crash, typically within 15 to 25 milliseconds of the first contact with the barrier. This, in combination with the clearance between the steering wheel module and dummy, normally provided when the seat is in the mid-position, allows the airbag to inflate fully, prior to dummy contact. Under such circumstances, head and neck kinematics are well controlled and excessive forward flexion or rearward extension of the neck is avoided. In all 19 tests, the peak resultant head acceleration values were less than Transport Canada's Injury Assessment Reference Value (IARV) of 80 g [2]. Similarly, the peak fore/aft neck moments were all well below the IARV values of 190 Nm in flexion and 57 Nm in extension, derived by General Motors [22]. Although not presented, all peak neck shear forces and peak axial forces measured in this series of tests were also well below GM IARV values. Consequently, the tests would predict negligible risk of injury of the head or neck under the conditions represented. The near absence of bag-related fatalities among belted male drivers from head or neck trauma would support this conclusion.

5th Percentile Female Hybrid III Results

<u>Rigid Frontal Barrier Tests</u> - Driver-side response data generated with the 5th percentile female dummy are available for a total of 34 48 km/h rigid frontal barrier crash tests, in which the vehicle was equipped with a driver-side airbag and the bag deployed. The peak dummy response values and calculated injury indices

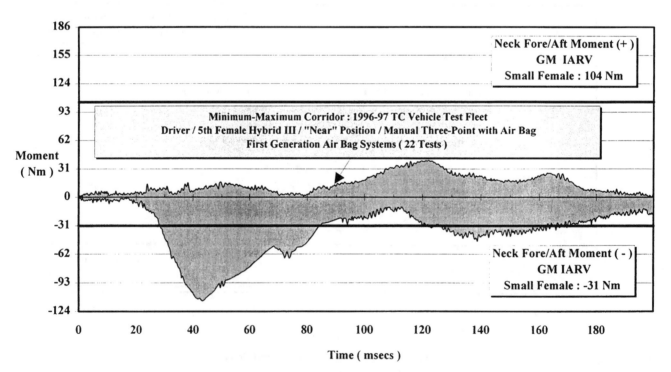

Figure 3. Range of Neck Extension Moment Responses Measured in 48 km/h Rigid Barrier Tests of First Generation Airbag Systems with 5th Percentile Female Hybrid III (Driver Side).

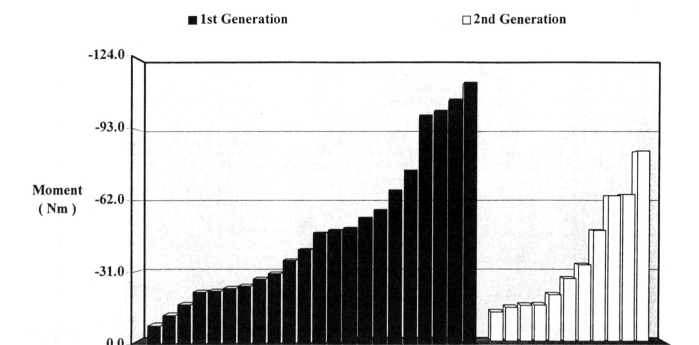

Figure 4. Peak Driver Neck Extension Moments Measured in 48 km/h Rigid Barrier Tests with 5th Percentile Female Hybrid III as a Function Air Bag Grouping.

obtained in this series of airbag tests are summarized in Appendix A1 for drivers, and Appendix A2 for front right passengers.

Given the imprecise nature of the term "depowered", the term "second-generation" is used in the present paper to denote vehicle models redesigned for model year 1998 to take advantage of the amendments to FMVSS 208 introduced to facilitate "depowering" of airbag systems in the US. The term "first-generation" is used to describe all pre-1998 airbag systems and 1998 airbag systems not yet redesigned at the time of vehicle purchase. It should be noted that the changes made to many 1998 vehicle models were not necessarily limited to reductions in the power output of the airbag module. Other components of the airbag system were frequently changed as well and, in some cases, the seat belt systems were redesigned. It should be also noted that six pre-1998 vehicles were modified by Transport Canada to reflect 1998 design changes to the airbag system and seat belt assemblies (if applicable). These vehicles are included in the second-generation airbag totals.

As most bag-related deaths in the case of belted female drivers are associated with neck trauma, the discussions below focus primarily on the fore/aft neck extension moments measured on the dummy. The range of neck responses observed in the first generation test series in rigid barrier tests with the 5th percentile female dummy in the driver's position is depicted in Figure 3.

The close proximity of the small dummy to the steering wheel results in the dummy interacting with the airbag while it is still expanding. This typically results in the head being forced upwards and rearwards as the bag continues to expand under the chin producing an extension-tension neck response. Maximum extension of the neck is generally observed some 40 to 50 milliseconds into the crash.

Complete driver neck response data are available for 22 of the 23 tests with first generation systems and 11 tests with second generation systems. A comparison of the peak neck extension moments observed in these tests is presented in Figure 4. As can be seen, both series of tests generated a wide range of peak values. In contrast to the results obtained using a mid-size male dummy, exceedances of the GM neck extension IARV for a small female (31 Nm) were common in this series of tests. The IARV was exceeded in 13 of 22 (59%) of the first-generation tests and in 5 of 11 (45%) of the second-generation tests. Peak values exceeding three times the IARV were observed in 4 (18%) of the first generation tests, the highest neck extension moment value being

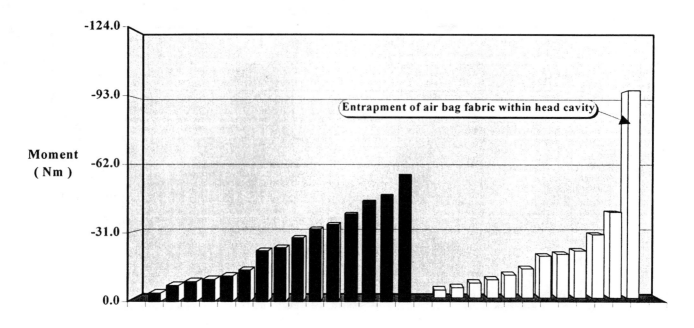

Figure 5. Peak Front Right Passenger Neck Extension Moments Measured in 48 km/h Rigid Barrier Tests with 5th Percentile Female Hybrid III as a Function Airbag Grouping.

113 Nm. The highest neck extension moment observed in the second generation tests was 84 Nm. The average peak neck extension moment observed in the second-generation test series was some 26% lower than the mean value observed in the first generation series of tests (36.6 Nm vs. 49.4 Nm).

The corresponding data for the passenger tests are presented in Figure 5. Passenger head and neck kinematics were far more complex than for the driver. Depending on the vehicle and design of the airbag system, the neck experienced either axial tension or compression accompanied by either forward flexion or rearward extension, with all possible combinations represented. In tests involving first-generation systems, exceedance of any neck IARV was observed only when the loading conditions produced a tension-extension response. The extension IARV was exceed in 6 out of 15 (40%) of the tests. However the maximum extension moment was only 58 Nm, less than half the maximum value recorded on a 5th-percentile driver. The extension IARV was exceed in 2 of 12 of the second-generation tests. In one of these, however, the airbag fabric very clearly penetrated the head cavity, despite the use of a protective neck shield. The neck response data for this test are therefore highly suspect. Excluding this test, the mean neck extension moment for the second generation test series was 16.3 Nm, or 38% less than the mean value of 26.1 Nm observed in the first-generation test series.

Offset Frontal Deformable Barrier Tests - The vast majority of tests conducted with the European offset deformable barrier face were conducted with a nominal impact speed of 40 km/h. This speed was selected since early testing indicated that the associated impact severity was sufficient to trigger the deployment of most, if not all, current airbag systems, while still representing a collision environment which is relatively innocuous to a belted individual, including belted occupants who travel with the seat fully forward. All tests were performed with a 40% vehicle offset to the barrier face as defined in Directive 96/79/EC. The driver- and passenger-side data generated by this series of 40 km/h tests are summarized in Appendices A.3 and A.4, respectively.

Complete neck response data for the driver's position in this series of 40 km/h impacts are available for 12 first-generation and 12 second-generation tests. The peak neck extension moments are presented in Figure 6. It is interesting to note that, despite the fact that the 40 km/h offset deformable barrier test condition is far less severe than the 48 km/h rigid barrier test condition, the offset tests produced higher peak neck response values.

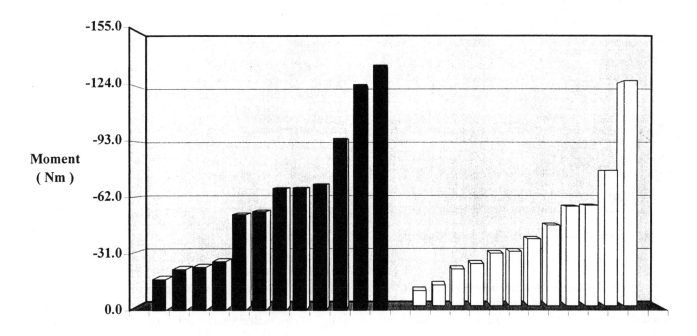

Figure 6. Peak Driver Neck Extension Moments Measured in 40 km/h Offset Frontal Deformable Barrier Tests with 5th Percentile Female Hybrid III as a Function Airbag Grouping.

The highest peak neck extension value observed in the first-generation test series was 134 Nm, while the corresponding highest peak value observed in the second generation test series was 127 Nm. Notwithstanding the similarity in maximum values, the mean peak neck extension moment observed in the second-generation test series was 36.3 Nm, a value approximately 42% lower that of the mean value of 62.7 Nm observed in the first generation test series.

The elevated neck moment values observed in the offset tests can be attributed to the timing of the airbag deployments. These occurred as late as 110 milliseconds into the crash. In a number of instances the initial clearance between the dummy and the delay in firing of the bag resulted in the dummies head being in contact with the airbag module at time of deployment (Figure 7).

The neck extension IARV was exceeded by the driver in 8 of 12 (67%) of the first-generation tests and in 6 of 12 (50%) of the second-generation tests. However, while peak neck extension values exceeding twice the IARV value were observed in 6 of 12 first-generation tests (50%), this was the case for only 2 of 12 (17%) of the second generation tests. That difference accounts for the much lower mean neck extension value noted above for the latter series of tests.

In the second generation test series, the influence of late bag deployment on neck response was far less pronounced than in the first generation test series. Indeed, the second lowest peak neck extension moment was recorded in the test which produced the latest airbag

Figure 7. Delayed Deployment (1st Generation Airbag)

389

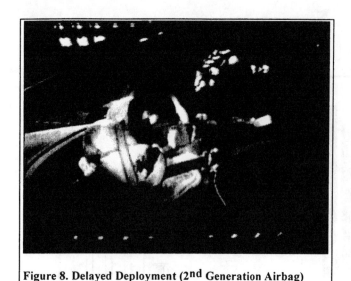

Figure 8. Delayed Deployment (2nd Generation Airbag)

deployment in the second-generation test series. At the time of deployment, the head was already in contact with the module. The tear pattern of the module cover and steering wheel design, in combination with the reduced power level of the airbag module, resulted in the airbag deploying laterally and sufficiently behind the steering wheel rim so that very little impact energy was transferred to the head (Figure 8). The peak driver neck extension moment observed in this test was 12 Nm.

Neck response data for the passengers in this series of 40 km/h impacts are available for 7 first-generation tests and 13 second-generation tests. The neck extension IARV was exceeded by the passenger in 4 of 7 (57%) first-generation tests and in 2 of 13 (15%) second-generation tests. The mean neck extension moment for the second-generation test series was 13.9 Nm, approximately 57% less than the mean value of 32.6 Nm observed in the first-generation test series.

The magnitude of the passenger neck moments was strongly influenced by the timing of the airbag deployment. This was true for both first and second generation vehicles. The highest neck moment observed in the second-generation test series was 58 Nm and was produced by the test associated with latest deployment (107 ms). The same vehicle model was also represented in the first-generation test fleet. The 1997 version of the same vehicle model produced a peak neck extension value of only 22 Nm. The much lower value likely reflects the earlier time of airbag deployment (34 ms).

<u>**Specialty Tests**</u> - As part of the above offset test series, a number of selected vehicle models were also tested at different impact severities. These tests were

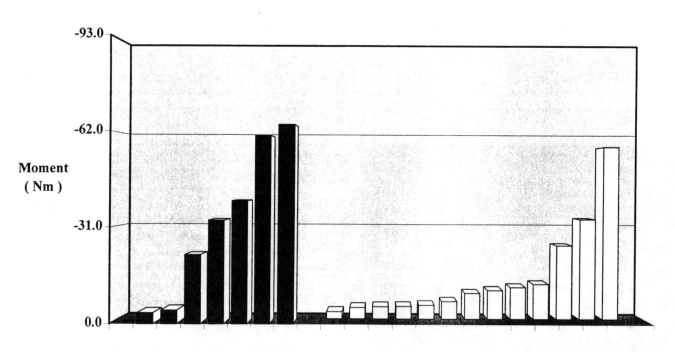

Figure 9. Peak Front Right Passenger Neck Extension Moments Measured in 40 km/h Offset Frontal Deformable Barrier Tests with 5th Percentile Female Hybrid III as a Function Airbag Grouping.

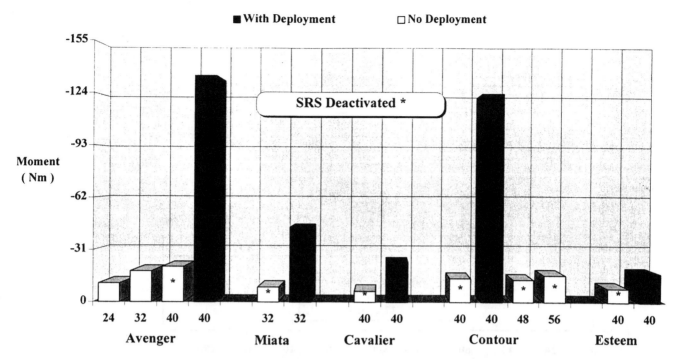

Figure 10. Peak Driver Neck Extension Moments Measured in Offset Frontal Deformable Barrier Tests with 5th Percentile Female Hybrid III as a Function Impact Speed and Airbag Deployment.

performed initially to establish the collision severity at which the airbag system would deploy in an offset deformable barrier test. In addition, it was intended to quantify the effects of deactivation, on the responses measured on a belted 5th percentile female Hybrid III, with the seat in the fully forward position, at collision severities at or just above the deployment threshold. Seven tests were performed with deactivated airbags. For one vehicle model, two additional tests, one at 48 km/h and one at 56 km/h, were performed with the airbag system deactivated. A detailed breakdown of the dummy responses measured in this series of tests is provided in Appendix A.5. The peak driver neck extension moments observed in this series of tests, are presented in Figure 10.

From the data presented, it can be observed that the peak neck moments obtained with airbag deployment always exceeded those which were obtained when the airbag system was deactivated. Indeed, in none of the tests performed with the airbag system deactivated was any commonly referenced IARV or regulated injury index exceeded. These results suggest that current airbag deployment thresholds are set too low, at least for belted drivers.

The above results also highlight the requirement for a low-speed test procedure to ensure that airbag systems are not only optimized for belted occupants but also that their performance is assessed over a range of different collision severities. In Figure 11, curves of vehicle deceleration versus time, typically observed in high-speed tests against rigid barriers, are compared with those observed in 40 km/h offset frontal deformable barrier tests with a 40% vehicle offset. Whereas the rigid wall test can be seen to produce very high vehicle decelerations very early in the crash sequence, the offset condition produces a "soft" deceleration pulse with peak decelerations relatively late in the collision. As can also be seen, the profile of the crash pulse in the offset test shows good agreement with generic sled pulses used to represent a typical collision.

The late deceleration peaks produced in the offset test often trigger airbag deployment. Under such situations, very high neck loads can be produced by the bag, whereas, in the absence of airbag deployment, the same occupant would be riding down the collision safely. With the advent of airbag systems, it can be seen that the relevance of the high speed rigid wall test has been greatly reduced.

Paired-Vehicle Comparisons - Many vehicle models represented in the first-generation test series differed from those in the second-generation series. The subset of vehicle models that was represented in both series of tests was examined separately, to see if these tests of paired vehicles showed any trends which differed from those observed in the main programme. The results for the paired vehicles are presented in Figures 12 and 13 for rigid- and offset-barrier tests, respectively.

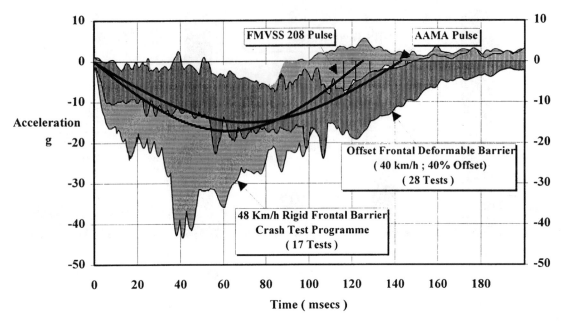

Figure 11. Comparison of Crash Pulses

From the results presented, it can be seen that the tests of paired vehicles produced trends similar to those noted in earlier discussions. The second-generation tests produced lower peak driver neck extensions, with the largest differences being observed in the offset tests. Given that the offset test is more representative of real-world crashes, this suggests that the magnitude of the benefits likely to be achieved with "depowering" could be greater than predicted on the basis of rigid barrier test data. Further support for this observation can be found in a comparison of the peak neck extension values, observed in static tests of one of the vehicle models represented in the paired-vehicle subset. Those results are presented in Figure 14.

In that series of static airbag tests, a 5th percentile female Hybrid III driver was subjected to a series of four separate airbag deployments. The baseline test was done with the seat in the fully forward position and the seat back in the most upright position. The dummy was then pivoted forward until the head was in contact with the module and retested. Additional tests were performed at two intermediate positions. The fifth static test took the form of an ISO-type "chin on hub" out-of-position test. As would be expected, the maximum neck extension moments increased with increasing proximity of the dummy to the airbag module. In tests where the dummy is in close proximity to the module, the reductions in peak neck loads achieved with second-generation airbag modules show much closer agreement with those predicted by the offset tests than with those predicted by the rigid barrier tests. It is also interesting to note that, while static out-of-position tests are frequently regarded to represent a "worst-case" scenario, even the "chin on hub" test produced a peak neck extension value that was lower than that observed in the full-scale vehicle offset test.

DISCUSSION

Low speed offset frontal crash testing, using belted 5th-percentile dummies in the fully forward seat position, overcomes two serious deficiencies which exist in current regulatory practices. The first deficiency is the absence of any requirements explicitly addressing the frontal protection requirements of drivers of short stature who, by necessity, often sit close to the steering assembly. In addition, current regulatory practices fail to ensure that optimum benefits are achieved over the range of collision severities represented in the field. Rigid wall tests, in themselves, provide little assurance that timely deployment of the airbag will be achieved in the "softer" collisions which account for the majority of real frontal crashes. The low-speed offset test should not be viewed as a substitute for the high speed barrier test. Rather, it

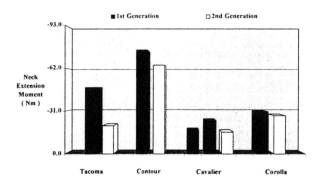

Figure 12. Peak Driver Neck Extension Moment :
1st Generation vs. 2nd Generation Systems
(48 km/h Rigid Barrier).

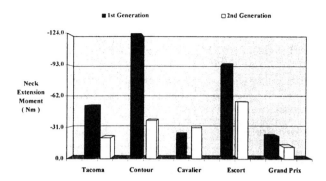

Figure 13. Peak Driver Neck Extension Moment :
1st Generation vs. 2nd Generation Systems
(40 km/h Offset Frontal).

should be viewed as a means of broadening the relevance of frontal protection standards to encompass a wider range of occupants sizes and collision severities. An added advantage of the low speed offset frontal test, as described in the present paper, is that it makes use of testing hardware already in widespread use around the world.

The findings of the present study suggest that changes in airbag design introduced in most 1998 models should help to reduce the incidence of serious or fatal bag-related injury among both drivers and right-front passengers. Further improvements in sensor technology are required, however, with respect both to the discrimination of collision severity and the assurance of timely airbag deployment. The frequency of late or delayed deployments observed in the present test programme suggests the need for additional, satellite crash sensors in the forward portions of the vehicle.

Not all aspects of the testing hardware or procedures developed or employed in the offset testing protocol have been finalized. Issues yet to be resolved completely include the design of the neck shield, and finalization of the dummy positioning procedure. Once these two issues are resolved, repeatability trials will be performed.

DISCLAIMER

The conclusions reached and opinions expressed in this paper are solely the responsibility of the author. Unless otherwise stated, they do not necessarily represent the official policy of Transport Canada.

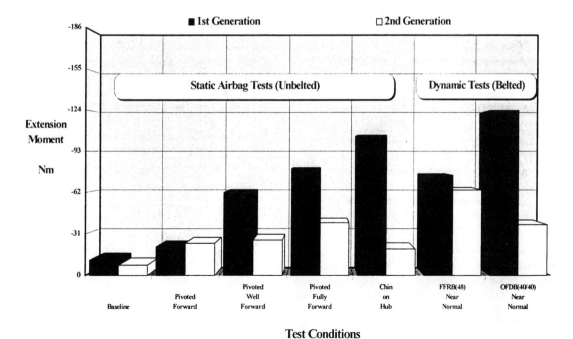

Figure 14. Peak Driver Neck Extension Moment as a Function of Test Condition.

REFERENCES

1 Dalmotas DJ. Welbourne ER: Improving The Protection Of Restrained Front Seat Occupants In Frontal Crashes. Proceedings of the 13th International Technical Conference on Experimental Safety Vehicles, Paris (France), November 4-7, 1991.

2 Welbourne ER: Specifying Performance Requirements To Reduce The Risk Of Closed Head Injury. Proceedings of the International Conference on Air Bags and Seat Belts: Evaluation and Implications for Public Policy, Montreal (Canada), October 18-20, 1992.

3 Horsch J. Lau I. Andrzejak D. Viano D. Melvin J. Pearson J. Cok D. Miller G: Assessment Of Air Bag Deployment Loads. Proceedings of the 34th Stapp Car Crash Conference, SAE Paper 902324, Society of Automotive Engineers, Warrendale, PA, 1990.

4. Melvin J. Horsch J. McCleary J. Wideman L. Jensen J. Wolanin M: Assessment Of Air Bag Deployment Loads With The Small Hybrid III Dummy. Proceedings of the 37th Stapp Car Crash Conference, SAE Paper 933119, Society of Automotive Engineers, Warrendale, PA, 1993.

5 Dalmotas DJ: Mechanisms Of Injury To Vehicle Occupants Restrained By Three-Point Seat Belts. SAE Technical Paper 801311, Society of Automotive Engineers, Warrendale, PA, 1980.

6 Dalmotas DJ. German A. Hendrick BE. Hurley RM: Airbag Deployments: The Canadian Experience. Journal of Trauma: Injury, Infection, and Critical Care, Vol. 38, No. 4, April, 1995.

7 Dalmotas DJ. Hurley RM. German A: Airbag Deployments Involving Restrained Occupants. SAE International Congress and Exposition, SAE Paper 950868, February 27-March 2, 1995.

8 Dalmotas DJ. German A. Hurley RM. Digges K: Air Bag Deployment Crashes in Canada, Paper No. 96-S1-O-05, Enhanced Safety of Vehicles Conference, Melbourne, Australia, May, 1996.

9 Marsh JC: Supplemental Air Bag Restraint Systems: Consumer Education And Experience. SAE Paper 930646, Society of Automotive Engineers, Warrendale, PA, 1993.

10 Huelke DF. Moore JL. Compton TW. Samuels J. Levine RS: Upper Extremity Injuries Related To Air Bag Deployments. Advances in Occupant Restraint Technologies: Proceedings of Joint AAAM-IRCOBI Special Session, Lyon (France), September 22, 1994.

11 Crandall JR. Kuhlmann TP. Martin PG. Pilkey WD. Neeman T: Differing Patterns Of Head And Facial Injury With Air Bag And/Or Belt Restrained Drivers In Frontal Collisions. Advances in Occupant Restraint Technologies: Proceedings of Joint AAAM-IRCOBI Special Session, Lyon (France), September 22, 1994.

12 American Association for Automotive Medicine: The Abbreviated Injury Scale, 1990.

13 Huelke DF. Moore JL. Compton TW. Samuels J: Upper Extremity Injuries Related to Airbag Deployments. Journal of Trauma, 38:482, 1995.

14 National Highway Traffic Safety Administration: Third Report to Congress: Effectiveness of Occupant Protection Systems and Their Use. U.S. Department of Transportation, December, 1996.

15 Kuppa SM. Yeiser, CW. Olson MB. Taylor L. Morgan R. Eppinger, R: RAID - An Investigation Tool to Study Air Bag/Upper Extremity Interventions. SAE International Congress and Exposition, Detroit, MI, 1997.

16 Werner JV. Roberson SF. Ferguson SA. Digges KH: Air Bag Deployment Frequency And Injury Risk. SAE International Congress and Exposition, SAE Paper 960664, February 26-29, 1996.

17 Bass CR. Duma SM. Crandall JR. Morris R. Martin P. Pilkey WD. Hurwitz S. Khaewpong N. Eppinger R. Sun E: The Interaction of Air bags with Upper Extremities. Poceedings of the 41st Stapp Car Crash Conference, SAE Publication P-315, 1997.

18 Kahane CJ: Fatality Reduction by Air Bags: Analysis of Accident Data Through Early 1996. Department of Transport, 1996.

19 Ferguson SA. Lund AK. Greene MA: Driver Fatalities in 1985-1994 Air Bag Cars. Insurance Institute for Highway Safety, 1996.

20 Ferguson SA: Update on Airbag Performance in the United States: Benefits and Problems. Insurance Institute for Highway Safety, 1996.

21 Ferguson SA. Braver ER. Greene MA. Lund AK: Preliminary Report: Initial Estimates of Reductions in Deaths in Frontal Crashes Among Right Front Passengers in Vehicles Equipped with Passenger Airbags. Insurance Institute for Highway Safety, 1996.

22 Mertz HJ. Anthropomorphic Test Devices. Accidental Injury: Biomechanics and Prevention, Springer-Verlag, New York, NY, 1993.

Appendix A.1 - 48 km/h Full Frontal Rigid Barrier Crash Test Series / 5th Percentile Female Hybrid III ATD / Driver Side Results

TC Test Number / Test Vehicle	Head Response — Resultant Head No Clip Acc. (g) SAE 1000	HIC (15 ms)	Neck — Axial Force + (N) SAE 1000	Axial Force −	Calc. Occipital Moment + (Nm) SAE 600	Occipital Moment −	Chest — Mid-Sternum Deflection (mm) SAE 600	Deflection SAE 180	Mid-Sternum VC (m/s) SAE 600	VC SAE 180	VC SAE 60	Resultant Chest No Clip Acc. (g) SAE 600	Chest Acc. SAE 180
First-Generation Test / Air Bag Deployment													
TC96-101D 1996 Toyota Tacoma	97.6	700	2802	−1768	8	−48	#N/A	−41.4	#N/A	#N/A	0.000	81.0	#N/A
TC96-102D 1996 Dodge Avenger	58.4	283	2710	−454	7	−101	#N/A	−29.7	#N/A	#N/A	0.265	38.3	#N/A
TC96-103D 1996 Mazda MPV	104.0	268	2612	−327	11	−99	#N/A	−22.7	#N/A	#N/A	0.164	40.2	#N/A
TC96-112D 1996 Merc Mystique	67.1	307	2593	−387	11	−75	#N/A	−24.4	#N/A	#N/A	0.210	48.0	#N/A
TC96-114D 1996 Chev Cavalier	52.7	181	1857	−603	13	−17	#N/A	−21.2	#N/A	#N/A	0.090	46.4	#N/A
TC96-115D 1996 Suzuki Esteem	61.9	309	1980	−1006	14	−36	#N/A	−30.6	#N/A	#N/A	0.269	49.5	#N/A
TC96-122D 1996 Mazda Miata	39.8	54	1793	−549	32	−49	#N/A	−42.3	#N/A	#N/A	0.337	49.3	#N/A
TC96-151D 1996 Toyota Corolla	50.1	154	1638	−432	10	−31	#N/A	−21.7	#N/A	#N/A	0.075	39.8	#N/A
TC97-101D 1997 GM Venture	65.6	334	2120	−276	#N/A	#N/A ND	#N/A	−34.7	#N/A	#N/A	0.363	50.4	#N/A
TC97-102D 1997 Jeep TJ	48.6	209	1669	−333	16	−25	#N/A	−48.2	#N/A	#N/A	0.518	41.2	#N/A
TC97-103D 1997 Hyundai Tiburon	41.3	75	1844	−521	40	−105	#N/A	−33.8	#N/A	#N/A	0.256	48.6	#N/A
TC97-104D 1997 Ford F150 PU	129.7	313	1812	−803	26	−28	#N/A	−35.1	#N/A	#N/A	0.267	80.5	#N/A
TC97-105D 1997 Saturn SL	42.7	149	1593	−403	17	−12	#N/A	−34.7	#N/A	#N/A	0.367	36.8	#N/A
TC97-106D 1997 Suzuki X90	58.6	290	2514	−1051	11	−66	#N/A	−35.2	#N/A	#N/A	0.226	59.6	#N/A
TC97-107D 1997 Dodge Dakota	47.5	106	1736	−720	8	−41	#N/A	−40.4	#N/A	#N/A	0.323	48.2	#N/A
TC97-110D 1997 Chev Cavalier	50.7	138	1469	−521	19	−24	#N/A	−29.4	#N/A	#N/A	0.311	38.7	#N/A
TC97-134D 1997 Toyota Rav4	65.8	403	2491	−800	31	−58	#N/A	−39.4	#N/A	#N/A	0.367	61.5	#N/A
TC97-153D 1997 Cherolet Malibu	48.8	194	2047	−736	6	−50	#N/A	−23.2	#N/A	#N/A	0.123	115.7	#N/A
TC97-161D 1997 Pontiac Grand Prix	47.7	165	1262	−86	11	−23	−17.0	#N/A	0.074	0.054	0.051	#N/A	32.7
TC97-162D 1997 Toyota Camry CE	53.9	208	1895	−233	10	−113	−30.7	#N/A	0.393	0.333	0.318	#N/A	43.3
TC97-164D 1997 Volkswagen Jetta GL	37.2	93	1250	−35	28	−23	−21.0	#N/A	0.200	0.182	0.146	#N/A	52.9
TC97-165D 1997 Ford Escort LX	52.0	178	1902	−255	2	−55	−23.7	#N/A	0.171	0.132	0.121	#N/A	59.2
TC98-105D 1998 Plymouth Voyager	52.9	255	1567	−368	12	−8	−43.0	#N/A	0.659	0.509	0.509	#N/A	46.7
Second-Generation Test / Air Bag Deployment													
TC97-201D 1996 Merc Mystique [M]	69.9	366	2385	−178	16	−64	−31.5	#N/A	0.507	0.421	0.411	#N/A	46.8
TC97-203D 1997 Chev Cavalier [M]	44.1	113	1446	−181	14	−16	−18.6	#N/A	0.218	0.190	0.162	#N/A	42.1
TC98-102D 1998 Nissan Altima	45.4	141	1481	−169	13	−16	−21.5	#N/A	0.155	0.140	0.127	#N/A	42.9
TC98-103D 1998 Honda Accord	50.4	225	1647	−329	2	−49	−32.3	#N/A	0.351	0.305	0.299	#N/A	48.4
TC98-106D 1998 Ford Explorer 2WD	48.2	154	2179	−276	13	−65	−39.8	#N/A	0.850	0.727	0.701	#N/A	61.4
TC98-107D 1998 Nissan Sentra	47.5	199	1363	−7	4	−15	−20.3	#N/A	0.133	0.117	0.099	#N/A	38.1
TC98-108D 1998 Dodge Neon	67.8	354	1996	−339	8	−13	−29.0	#N/A	0.358	0.295	0.288	#N/A	50.4
TC98-111D 1998 Mazda 626	52.4	220	2150	−663	4	−84	−23.9	#N/A	0.185	0.132	0.126	#N/A	49.3
TC98-112D 1998 Nissan Frontier	65.9	436	1626	−435	12	−34	−40.7	#N/A	0.509	0.413	0.408	#N/A	48.2
TC98-201D 1998 Toyota Corolla VE	60.5	324	1957	−355	5	−28	−18.3	#N/A	0.189	0.110	0.097	#N/A	37.9
TC98-205D 1998 Toyota Tacoma PU	87.7	545	2730	−436	10	−20	−42.5	#N/A	0.700	0.488	0.448	#N/A	62.1

Notes :

M — Vehicle modified to reflect 1998 design changes.

ND — No data. Transducer or data acquisition failure/malfunction.

Appendix A.2 - 48 km/h Full Frontal Rigid Barrier Crash Test Series / 5th Percentile Female Hybrid III ATD / Front Right Passenger Side Results

Full Frontal Rigid Barrier Crash Test Series	Head Response		Neck Response		Chest Response		
TC Test Number / Test Vehicle	Resultant Head No Clip Acc. (g) [SAE 1000]	HIC (15 ms)	Axial Force Positive/Negative (N) [SAE 1000]	Calculated Occipital Moment Positive/Negative (Nm) [SAE 600]	Mid-Sternum Deflection (mm) [SAE 600 / SAE 180]	Mid-Sternum VC (m/s) [SAE 600 / SAE 180 / SAE 60]	Resultant Chest No Clip Acc. (g) [SAE 600 / SAE 180]
First-Generation Test / Air Bag Deployment							
TC97-165P 1997 Ford Escort LX	48.5	197	2015 / -83	31 / -58	-32.5 / #N/A	0.274 / 0.209 / 0.203	#N/A / 45.7
TC97-164P 1997 Volkswagen Jetta GL	49.1	160	1849 / -293	17 / -40	-37.4 / #N/A	0.450 / 0.381 / 0.376	#N/A / 49.7
TC97-162P 1997 Toyota Camry CE	89.6	108	273 / -1694	78 / -7	-19.3 / #N/A	0.105 / 0.072 / 0.067	#N/A / 33.5
TC97-161P 1997 Pontiac Grand Prix	58.1	204	1638 / -88	8 / -45	-21.0 / #N/A	0.169 / 0.125 / 0.119	#N/A / 44.5
TC97-153P 1997 Cherolet Malibu	54.9	236	587 / -1190	57 / -23	#N/A / -15.1	#N/A / #N/A / 0.083	47.2 / #N/A
TC97-134P 1997 Toyota Rav4	152.8	564	2961 / -1536	4 / -33	#N/A / -33.1	#N/A / #N/A / 0.312	80.1 / #N/A
TC97-110P 1997 Chev Cavalier	62.2	343	1263 / -708	26 / -14	#N/A / -21.8	#N/A / #N/A / 0.128	50.9 / #N/A
TC97-107P 1997 Dodge Dakota	34.2	86	1596 / -432	33 / -48	#N/A / -33.0	#N/A / #N/A / 0.313	42.1 / #N/A
TC97-106P 1997 Suzuki X90	65.3	133	2019 / -362	43 / -35	#N/A / -36.4	#N/A / #N/A / 0.256	56.9 / #N/A
TC97-105P 1997 Saturn SL	49.9	205	1815 / -397	24 / -11	#N/A / -30.1	#N/A / #N/A / 0.140	47.6 / #N/A
TC97-104P 1997 Ford F150 PU	46.7	180	1124 / -406	30 / -4	#N/A / -34.4	#N/A / #N/A / 0.304	59.7 / #N/A
TC97-103P 1997 Hyundai Tiburon	53.4	267	990 / -495	25 / -10	#N/A / -26.6	#N/A / #N/A / 0.204	63.5 / #N/A
TC97-102P 1997 Jeep TJ	43.0	166	1552 / -273	13 / -29	#N/A / -38.1	#N/A / #N/A / 0.345	41.0 / #N/A
TC97-101P 1997 GM Venture	53.6	193	2003 / -394	35 / -24	#N/A / -26.5	#N/A / #N/A / 0.179	95.5 / #N/A
TC96-124P 1996 Dodge Caravan	72.6	87	812 / -793	98 / -9	#N/A / -26.0	#N/A / #N/A / 0.152	46.5 / #N/A
Second-Generation Test / Air Bag Deployment							
TC97-201P 1996 Merc Mystique [M]	62.0	384	757 / -725	47 / -4	-16.2 / #N/A	0.076 / 0.066 / 0.063	#N/A / 52.9
TC97-203P 1997 Chev Cavalier [M]	59.6	285	1764 / -416	27 / -8	-14.4 / #N/A	0.080 / 0.061 / 0.059	#N/A / 58.8
TC98-102P 1998 Nissan Altima	65.0	296	197 / -1342	74 / -11	-11.9 / #N/A	0.063 / 0.047 / 0.046	#N/A / 39.8
TC98-103P 1998 Honda Accord	56.3	269	1057 / -241	14 / -22	-23.1 / #N/A	0.190 / 0.174 / 0.156	#N/A / 44.6
TC98-105P 1998 Plymouth Voyager	63.6	318	1477 / -459	37 / -19	-30.6 / #N/A	0.340 / 0.270 / 0.254	#N/A / 51.9
TC98-106P 1998 Ford Explorer 2WD	43.8	155	1249 / -585	22 / -14	-21.2 / #N/A	0.156 / 0.113 / 0.103	#N/A / 46.7
TC98-107P 1998 Nissan Sentra	51.7	244	1066 / -292	29 / -20	-27.1 / #N/A	0.242 / 0.208 / 0.193	#N/A / 45.3
TC98-108P 1998 Dodge Neon	59.0	303	519 / -729	47 / -7	-19.9 / #N/A	0.132 / 0.087 / 0.078	#N/A / 48.7
TC98-111P 1998 Mazda 626	61.7	262	2783 / -222 S2	18 / -96 S2	-27.9 / #N/A	0.294 / 0.208 / #N/A	#N/A / 48.8
TC98-112P 1998 Nissan Frontier	70.7	356	1050 / -2126	45 / -30	-43.9 / #N/A	0.665 / 0.602 / #N/A	#N/A / 56.2
TC98-201P 1998 Toyota Corolla VE	88.0	559	578 / -1904	29 / -5	-19.0 / #N/A	0.062 / 0.050 / 0.048	#N/A / 47.4
TC98-205P 1998 Toyota Tacoma PU	59.0	300	1447 / -374	37 / -40	-35.8 / #N/A	0.417 / 0.344 / 0.342	#N/A / 66.9

Notes :

M   - Vehicle modified to reflect 1998 design changes.

ND - No data. Transducer or data acquisition failure/malfunction.

S2 - Peak value suspect. Penetration of airbag fabric into head cavity.

Appendix A.3 - Offset Frontal Deformable Barrier Crash Test Series / 5th Percentile Female Hybrid III ATD / Driver Side Results

TC Test Number / Test Vehicle	Head Response: Resultant Head No Clip Acc. (g) SAE 1000	HIC (15 ms)	Neck: Axial Force +/- (N) SAE 1000		Calculated Occipital Moment +/- (Nm) SAE 600		Chest: Mid-Sternum Deflection (mm) SAE 600	SAE 180	Mid-Sternum VC (m/s) SAE 600	SAE 180	SAE 60	Resultant Chest No Clip Acc. (g) SAE 600	SAE 180
40 km/h ; 40% Offset Frontal Test - First-Generation Test / Air Bag Deployment													
TC94-022D 1994 Dodge Caravan	53.9	226	1009	-32	19	-67	-25.0	#N/A	0.200	#N/A	0.158	#N/A	21.3
TC95-206D 1995 Ford Contour	74.5	367	2752	-505	17	-124	#N/A	-22.4	#N/A	#N/A	0.170	#N/A	42.4
TC96-002D 1996 Suzuki Esteem	42.5	85	1225	-229	19	-17	#N/A	-23.1	#N/A	#N/A	0.174	#N/A	28.2
TC96-021D 1996 Toyota Tacoma	93.2	648	3044	-705	20	-53	#N/A	-23.1	#N/A	#N/A	0.072	#N/A	29.7
TC96-024D 1996 Chev Lumina LS	71.7	240	2676	-308	5	-67	#N/A	-33.9	#N/A	#N/A	0.569	#N/A	58.6
TC96-025D 1996 Chev Cavalier	47.6	112	1330	-270	23	-24	#N/A	-21.9	#N/A	#N/A	0.169	#N/A	20.6
TC96-211D 1996 Dodge Avenger	82.9	338	4583	-644 S	11	-134 S1	#N/A	-37.6	#N/A	0.057	0.252	#N/A	77.8
TC97-205D 1997 Pontiac Grand Prix	27.5	52	693	-115	11	-23	-9.7	#N/A	0.035	0.062	0.021	#N/A	17.9
TC97-206D 1997 Toyota Camry	90.5	293	1763	-4	2	-54	-30.3	#N/A	0.266	0.076	0.217	#N/A	32.2
TC97-208D 1997 VW Jetta	38.0	19	832	-230	22	-27	-14.7	#N/A	0.081	0.188	0.045	#N/A	22.4
TC97-209D 1997 Ford Escort	45.7	138	755	-25	19	-94	-16.6	#N/A	0.113	0.118	0.091	#N/A	22.0
TC98-207D 1998 Dodge Caravan	31.6	66	1184	-37	1	-69	-21.8	#N/A	0.136	#N/A	0.104	#N/A	24.0
40 km/h ; 40% Offset Frontal Test - Second-Generation Test / Air Bag Deployment													
TC97-200D 1997 Merc Mystique [M]	50.9	187	896	-81	18	-38	-24.3	#N/A	0.265	0.127	0.210	#N/A	25.7
TC97-204D 1997 Chev Cavalier [M]	48.4	193	825	-41	8	-31	-12.5	#N/A	0.056	0.106	0.048	#N/A	21.4
TC98-101D 1998 Toyota Corolla	44.2	145	1370	-70	21	-24	-13.6	#N/A	0.103	0.220	0.046	#N/A	26.6
TC98-109D 1998 Toyota Tacoma	62.5	311	1535	-412	15	-21	-20.4	#N/A	0.173	0.491	0.103	#N/A	28.6
TC98-202D 1998 Nissan Altima	45.0	96	1499	-68	16	-57	-18.6	#N/A	0.120	0.181	0.098	#N/A	16.6
TC98-203D 1998 Ford Escort	41.0	87	1478	-6	35	-56	-14.1	#N/A	0.079	0.383	0.060	#N/A	18.1
TC98-204D 1998 Ford F150	29.4	27	646	-39	15	-9	-17.1	#N/A	0.094	#N/A	0.067	#N/A	21.0
TC98-206D 1998 Ford Explorer 2WD	62.3	183	2573	-57	1	-30	-24.3	#N/A	0.249	#N/A	0.167	#N/A	27.7
TC98-208D 1998 Dodge Neon	77.1	486	2829	-117	0	-127	-26.1	#N/A	0.334	#N/A	0.169	#N/A	38.7
TC98-209D 1998 Honda Accord	81.7	402	3495	-603	2	-77	-27.6	#N/A	0.764	#N/A	0.374	#N/A	41.1
TC98-210D 1998 Nissan Sentra	64.0	325	2246	-589	39	-46	-17.5	#N/A	0.235	#N/A	0.151	#N/A	29.8
TC98-211D 1998 Pontiac Grand Prix SE	57.1	131	2090	-22	22	-12	-27.3	#N/A	0.466	#N/A	0.306	#N/A	33.9

Notes :

M   - Vehicle modified to reflect 1998 design changes.

ND - No data. Transducer or data acquisition failure/malfunction.

S   - Full-scale setting of transducer exceeded.

S1 - Peak value suspect. Full-scale setting for x-axis neck shear force exceeded.

Appendix A.4 - Offset Frontal Deformable Barrier Crash Test Series / 5th Percentile Female Hybrid III ATD / Front Right Passenger Side Results

TC Test Number / Test Vehicle	Head Response — Resultant Head No Clip Acc. (g) SAE 1000	HIC (15 ms)	Neck Response — Axial Force Positive/Negative (N) SAE 1000	Calculated Occipital Moment Positive/Negative (Nm) SAE 600	Chest Response — Mid-Sternum Deflection (mm) SAE 600 / SAE 180	Mid-Sternum VC (m/s) SAE 600 / SAE 180 / SAE 60	Resultant Chest No Clip Acc. (g) SAE 600 / SAE 180
First-Generation Test / Air Bag Deployment / "Near" Position							
TC94-022P 1994 Dodge Caravan	82.2	202	482 / -527	84 / -4	-9.6 / #N/A	0.048 / 0.033 / 0.031	#N/A / 30.8
TC96-024P 1996 Chev Lumina LS	128.9	378	3507 / -289	23 / -61	#N/A / -14.7	#N/A / #N/A / 0.060	33.8 / #N/A
TC96-025P 1996 Chev Cavalier	74.8	22	435 / -235	43 / -4	#N/A / -13.3	#N/A / #N/A / 0.045	21.4 / #N/A
TC97-205P 1997 Pontiac Grand Prix	44.7	98	1125 / -208	31 / -22	-8.0 / #N/A	0.031 / 0.024 / 0.020	#N/A / 35.2
TC97-206P 1997 Toyota Camry	210.7	1640	2950 / -4050	58 / -64	-9.4 / #N/A	0.087 / 0.061 / 0.047	#N/A / 35.1
TC97-208P 1997 VW Jetta	25.7	36	1104 / -105	14 / -33	-11.7 / #N/A	0.038 / 0.028 / 0.026	#N/A / 24.4
TC97-209P 1997 Ford Escort	27.6	45	1231 / -29	19 / -40	-8.5 / #N/A	0.028 / 0.026 / 0.022	#N/A / 22.7
Second-Generation Test / "Near" Position							
TC97-200P 1997 Merc Mystique [M]	29.8	63	489 / -241	32 / -4	-8.6 / #N/A	0.020 / 0.016 / 0.015	#N/A / 23.0
TC97-204P 1997 Chev Cavalier [M]	44.9	63	599 / -527	18 / -4	-8.2 / #N/A	0.020 / 0.015 / 0.013	#N/A / 19.1
TC98-101P 1998 Toyota Corolla	67.3	373	1901 / -1872	66 / -9	-18.0 / #N/A	0.057 / 0.046 / 0.042	#N/A / 32.4
TC98-109P 1998 Toyota Tacoma	45.8	101	1650 / -108	4 / -25	-23.3 / #N/A	0.309 / 0.256 / 0.214	#N/A / 40.8
TC98-202P 1998 Nissan Altima	46.6	124	35 / -1948	91 / -4	-5.7 / #N/A	0.020 / 0.016 / 0.015	#N/A / 24.4
TC98-203P 1998 Ford Escort	29.6	12	276 / -565	41 / -2	-12.3 / #N/A	0.022 / 0.018 / 0.015	#N/A / 15.7
TC98-204P 1998 Ford F150	24.3	19	631 / -83	13 / -5	-16.9 / #N/A	0.079 / 0.066 / 0.061	#N/A / 23.5
TC98-206P 1998 Ford Explorer 2WD	100.3	83	1028 / -1300	67 / -12	-9.6 / #N/A	0.057 / 0.049 / 0.039	#N/A / 28.4
TC98-207P 1998 Dodge Caravan	38.9	117	961 / -41	19 / -6	-18.6 / #N/A	0.098 / 0.075 / 0.061	#N/A / 22.0
TC98-208P 1998 Dodge Neon	184.7	200	1142 / -83	30 / -11	-15.0 / #N/A	0.074 / 0.057 / 0.056	#N/A / 37.1
TC98-209P 1998 Honda Accord	61.4	297	546 / -1311	32 / -33	-13.4 / #N/A	0.043 / 0.034 / 0.031	#N/A / 21.4
TC98-210P 1998 Nissan Sentra	53.2	119	791 / -8	24 / -10	-15.0 / #N/A	0.051 / 0.035 / 0.029	#N/A / 18.4
TC98-211P 1998 Pontiac Grand Prix SE	112.1	365	2315 / -18	36 / -58	-16.8 / #N/A	0.041 / 0.031 / 0.028	#N/A / 21.5

Notes :

M   - Vehicle modified to reflect 1998 design changes.

Appendix A.5 - Other (Special) Tests / 5th Percentile Female Hybrid III ATD

Special Test Series / TC Test Number / Test Vehicle	Head Response		Neck Response		Mid-Sternum Deflection (mm)		Chest Response				
	Resultant Head No Clip Acc. (g) SAE 1000	HIC (15 ms)	Axial Force Positive/Negative (N) SAE 1000	Calculated Occipital Moment Positive/Negative (Nm) SAE 600	SAE 600	SAE 180	Mid-Sternum VC (m/s) SAE 600	SAE 180	SAE 60	Resultant Chest No Clip Acc. (g) SAE 600	SAE 180
48 Km/h Frontal Barrier Crash Test Series											
Driver Side : No Air Bag System Fitted / Air Bag Fitted - No Air Bag Deployment (Not triggered or suppressed)											
TC96-125D 1996 Ford Taurus [N1]	111.3	698	2447 / -816	14 / -30	#N/A	-39.6	#N/A	#N/A	0.264	51.9	#N/A
TC97-108D 1997 Hyundai Elantra	109.9	384	2399 / -308	28 / -40	#N/A	-52.7	#N/A	#N/A	0.643	64.0	#N/A
Passenger Side : No Air Bag System Fitted / Air Bag Fitted - No Air Bag Deployment (Not triggered or suppressed)											
TC97-108P 1997 Hyundai Elantra	59.1	338	2047 / -162	70 / -28	#N/A	-31.0	#N/A	#N/A	0.192	66.2	#N/A
Offset Frontal Deformable Barrier Crash Test Series											
Driver Side : 24 km/h ; 40% Offset Test - No Air Bag System Fitted / Air Bag Fitted - No Air Bag Deployment (Not triggered or suppressed)											
TC96-204D 1996 Dodge Avenger	62.7	206	644 / -597	18 / -11	#N/A	-7.9	#N/A	0.228	0.010	20.2	#N/A
Driver Side : 32 km/h ; 40% Offset Frontal Test - First-Generation Test / Air Bag Deployment											
TC95-021D 1995 Mazda Miata	116.3	490	4170 / -425	15 / -45	#N/A	-25.9	#N/A	#N/A	0.204	97.7	#N/A
Driver Side : 32 km/h ; 40% Offset Test - No Air Bag System Fitted / Air Bag Fitted - No Air Bag Deployment (Not triggered or suppressed)											
TC95-127D 1996 Mazda Miata	46.2	115	809 / -501	24 / -9	#N/A	-12.2	#N/A	#N/A	0.018	17.1	#N/A
TC96-202D 1996 Dodge Avenger	55.5	179	790 / -470	10 / -18	#N/A	-9.9	#N/A	0.500	0.014	19.7	#N/A
Driver Side 40 km/h ; 40% Offset Test - No Air Bag System Fitted / Air Bag Fitted - No Air Bag Deployment (Not triggered or suppressed)											
TC96-205D 1996 Suzuki Esteem	49.2	191	892 / -279	14 / -8	#N/A	-22.9	#N/A	0.223	0.055	23.6	#N/A
TC96-207D 1996 Chev Cavalier	52.6	131	978 / -235	33 / -7	#N/A	-20.0	#N/A	0.053	0.033	21.5	#N/A
TC96-209D 1996 Merc Mystique	45.1	135	978 / -355	16 / -14	#N/A	-20.6	#N/A	0.028	0.091	28.8	#N/A
TC96-210D 1996 Dodge Avenger	54.1	235	527 / -321	16 / -21	#N/A	-13.1	#N/A	0.227	0.018	22.1	#N/A
Driver Side : 48 km/h ; 40% Offset Test - No Air Bag System Fitted / Air Bag Fitted - No Air Bag Deployment (Not triggered or suppressed)											
TC95-002D 1995 Merc Mystique	48.5	205	488 / -29	15 / -13	-29.1	#N/A	0.215	#N/A	0.164	#N/A	32.2
Driver Side : 56 km/h ; 40% Offset Test - No Air Bag System Fitted / Air Bag Fitted - No Air Bag Deployment (Not triggered or suppressed)											
TC97-163D 1997 Merc Mystique	42.8	141	593 / -36	24 / -16	-34.1	#N/A	0.255	0.057	0.215	#N/A	31.2
Driver Side : 40% Offset Test - First-Generation Test / Specialty Test : Simulated Bracing Posture											
TC96-212D 1996 Dodge Avenger	68.8	258	3821 / -29	15 / -92	-37.7	#N/A	0.547	0.103	0.409	#N/A	45.5
Front Right Passenger Side : 40 km/h ; 40% Offset Test - No Air Bag System Fitted / Air Bag Fitted - No Air Bag Deployment (Not triggered or suppressed)											
TC96-021P 1996 Toyota Tacoma	26.8	45	816 / -317	32 / -6	#N/A	-24.1	#N/A	#N/A	0.064	20.1	#N/A
TC96-207P 1996 Chev Cavalier	24.1	35	866 / -229	33 / -8	#N/A	-22.6	#N/A	#N/A	0.041	19.3	#N/A
Front Right Passenger Side : 48 km/h ; 40% Offset Test - No Air Bag System Fitted / Air Bag Fitted - No Air Bag Deployment (Not triggered or suppressed)											
TC95-002P 1995 Merc Mystique	28.4	57	939 / -41	30 / -8	-11.9	#N/A	0.038	0.024	0.022	#N/A	26.2
Front Right Passenger Side : 56 km/h ; 40% Offset Test - No Air Bag System Fitted / Air Bag Fitted - No Air Bag Deployment (Not triggered or suppressed)											
TC97-163P 1997 Merc Mystique	91.0	353	1114 / -270	48 / -10	-14.1	#N/A	0.044	0.030	0.029	#N/A	31.9

Notes :

N1 - No deployment of driver-side airbag. Fault attributed to lack of adequate power in power supply substituted for original vehicle battery.

2001-22-0008

Stapp Car Crash Journal, Vol. 45 (November 2001), pp.
Copyright © 2001 The Stapp Association

The Influence of Superficial Soft Tissues and Restraint Condition on Thoracic Skeletal Injury Prediction

Richard W. Kent, Jeff R. Crandall, James Bolton
University of Virginia Automobile Safety Laboratory

Priya Prasad, Guy Nusholtz, Harold Mertz
Alliance of Automobile Manufacturers

ABSTRACT – The purpose of this study is to evaluate the hard tissue injury-predictive value of various thoracic injury criteria when the restraint conditions are varied. Ten right-front passenger human cadaver sled tests are presented, all of which were performed at 48 km/h with nominally identical sled deceleration pulses. Restraint conditions evaluated are 1) force-limiting belt and depowered airbag (4 tests), 2) non-depowered airbag with no torso belt (3 tests), and 3) standard belt and depowered airbag (3 tests). Externally measured chest compression is shown to correspond well with the presence of hard tissue injury, regardless of restraint condition, and rib fracture onset is found to occur at approximately 25% chest compression. Peak acceleration and the average spinal acceleration measured at the first and eighth or ninth thoracic vertebrae are shown to be unrelated to the presence of injury, though clear variations in peaks and time histories among restraint conditions can be seen. The maximum viscous criterion is found to correspond with injury, but only because it increases with the maximum chest compression. A simple analytical study is presented to elucidate the observed restraint condition dependence of rib fracture location and the restraint insensitivity of injurious maximum chest compression. Computed tomography images of a loaded torso are presented to show the load-distributing effect of the soft tissues superficial to the rib cage.

KEYWORDS – Thorax, Thoracic Injury, Restraint Systems, Injury Criteria, Cadaver Tests

---

INTRODUCTION

Thoracic trauma is a contributing factor in nearly 70% of all vehicular collision deaths and disabling injuries and is the principle causative factor in approximately 30% (Mulligan et al. 1994). In passenger vehicle collisions, chest injuries rank second only to head injuries in overall number of fatalities and serious injuries (Cavanaugh in Nahum and Melvin 1993); they also rank second to head injuries in overall societal harm (Malliaris 1985). The design and optimization of occupant restraint systems is driven to a large extent, therefore, by the requirement to reduce thoracic injury risk in a frontal collision.

The efficacy of restraint systems is assessed using a variety of tools. Human volunteer, human cadaver, or animal tests may be performed to evaluate and to compare systems. These types of tests, however, are expensive, laborious, and have limited repeatability, so anthropomorphic test devices (ATDs), or dummies, are often used as surrogates for humans.

ATDs are instrumented to measure various mechanical parameters, including sternal displacement relative to the spine (chest deflection), acceleration measured at the chest center of gravity (cg), and others. These mechanical parameters, or combinations thereof, are often correlated with the presence of injury in similar cadaver tests or field collisions, and used as predictors of injury risk. Used in this manner, these mechanical parameters are referred to as injury criteria.

In order to be useful for injury prediction, an injury criterion or a corresponding injury risk function must satisfy two requirements: it must be able to differentiate injurious loading conditions from non-injurious loading conditions, and it must be able to do this for the set of loading conditions that span the range of interest (including future advances in restraint system technology). When comparing the occupant protection benefits of different restraint conditions, it is necessary 1) that an injury criterion be equally effective as an injury predictor for all restraint conditions being evaluated and 2) that critical values of the criterion (values at which injury occurs) do not vary as a function of the restraint condition (or that this variation is quantifiable). If the effectiveness of the injury criterion depends on the specifics of the restraint, then any comparative studies of restraint conditions may be biased. In terms of contemporary restraint designs, therefore, a criterion must be applicable for blunt loading from the steering wheel hub and rim, belt loading, airbag loading, and combined belt-and-airbag loading. The purpose of this study is to evaluate the hard tissue injury-predictive value of various thoracic injury criteria when the restraint conditions are varied.

BACKGROUND

Thoracic injury criteria have been evaluated using a variety of loading conditions. Due to the low rates of seatbelt use in the 1960s and 1970s, early studies of thoracic injury and impact response focused on loading experienced by an unbelted subject (see Kroell 1994 for a review of thoracic impact studies performed prior to 1980). Force, acceleration, and deflection characteristics of the thorax were developed using blunt objects representing a steering wheel hub or instrument panel, and loading rates were representative of those experienced by unrestrained occupants in severe collisions (e.g. Nahum et al. 1970, Kroell et al. 1971, Mertz and Gadd 1971, Kroell et al. 1974). The chest deflection in response to the applied force was found to be a good predictor of thoracic injury risk, as defined by rib fractures, for this loading condition. When soft tissue injuries were considered, the maximum viscous criterion, VC_{max}, was found to be an effective predictor of injury for chest deflection rates between 3 m/s and 30 m/s (Lau and Viano 1986). As seatbelt use rates increased, it became necessary to evaluate thoracic response and injury criteria under localized belt impingement and at lower loading rates. Researchers have used the early blunt loading data, belted human cadaver and animal tests, dummy tests, and field data to evaluate thoracic injury criteria under localized belt loading (e.g., Cromack and Ziperman 1975, Fayon et al. 1975, Cesari and Bouquet 1990, Mertz et al. 1991, Horsch et al. 1991, Crandall et al. 1994). In these studies of belted occupants, the peak belt force was often found to correlate well with the presence and magnitude of injury.

As airbag restraints became more common in the vehicle fleet, researchers recognized the need for a thoracic injury criterion appropriate for airbag loading and for combined belt-and-airbag loading.

Cheng et al. (1982) was among the first to present thoracic response to airbag loading. Schneider et al. (1989) performed impactor tests using a larger surface area (500 cm^2 vs. 340 cm^2) than was used in the Kroell tests. Morgan et al. (1994) evaluated a series of 63 driver-side cadaver sled tests performed with subjects restrained by a torso belt only, an airbag only, or a combined belt-and-airbag restraint. They used these tests to develop a combined model (CM) for the prediction of thoracic injury based upon their finding that chest deflection alone was not the best indicator of injury risk for the entire dataset. This CM includes the peak chest deflection, the peak resultant acceleration measured at the first thoracic vertebra (T1), and the age of the cadaver at death. Those authors found that specific characteristics of the restraint system influenced the probability of injury for a given magnitude of CM: there is a greater probability of injury with a "belt-like" restraint than with an "airbag-like" restraint for the same measured value of CM. In other words, the critical value of CM changes as a function of restraint type. Based on this finding, these authors recommended separate injury criteria for "belt-like" restraints and "airbag-like" restraints. The recommended "belt-like" criterion includes measured deflection terms only (no acceleration or viscous terms), while the "airbag-like" criterion includes both deflection and acceleration terms.

Recently, the National Highway Traffic Safety Administration (NHTSA) proposed the combined thoracic criterion CTI as a desirable criterion for the evaluation of diverse restraint conditions (Eppinger et al. 1999). Similar to the CM of Morgan et al., and based on much of the same data, CTI results from the finding that a two-parameter logistic regression model using the T1 peak acceleration and the maximum chest deflection as covariates correlated well with the probability of an AIS 3 or higher injury (again, defined by the number and distribution of rib fractures sustained by the cadaver) (see also Kuppa and Eppinger 1998).

In 2000, Hassan and Nusholtz revisited the development of CTI. They identified several confounding issues associated with the tests, data collection, and statistical model used to develop CTI. Cases of massive rib fractures (greater than 10 fractures) with relatively little measured chest deflection (8 percent to 16 percent) were identified as deleterious to the dataset. Examination of these tests reveals that, depending on the specific anthropometry of the cadaver subject, interaction with the lower steering wheel rim and airbag may create a complex loading environment on the ribs, and cadaver chest deformation measured at discrete locations in the anatomically transverse plane may not be a good indicator of injury risk - particularly when injury risk is defined almost exclusively by the number and location of rib fractures (Kent et al. 2000). Prasad (1999) recommended that several of the tests from the NHTSA database be excluded from consideration and found that the cadaver-based maximum chest deflection was a good predictor of chest injuries for the remaining tests, regardless of differences in restraint condition and the resulting load distribution on the anterior thorax. In 2001, Kent et al. evaluated the utility of the existing cadaver sled test database for injury criteria evaluation. They concluded that the cadaver-based instrumentation introduced systematic bias into the assessment of injury criteria. Specifically, chestbands, especially older models, are subject to errors under localized belt loading, particularly at higher-severity impacts (over 48 km/h), but are accurate for distributed airbag loading. Further, T1 acceleration has been shown to be an acceptable predictor of dummy chest cg acceleration for some restraint conditions, but not for others (Shaw et al. 2001). Using dummy-based measurements to predict cadaver injury in paired tests, Kent et al. (2001) found that the maximum chest deflection was a better predictor of hard tissue injury than CTI since the inclusion of the acceleration component decreased both the specificity and the selectivity of the injury prediction model.

It is in response to the limitations of the existing cadaver sled test database that the current series of cadaver sled tests is presented. The goal with these tests was to hold constant as many factors as possible except for the restraint condition. Contemporary restraint designs were utilized, including force-limiting and pretensioned belt systems and de-powered airbags. The tests were performed with occupants seated in the right-front passenger (RFP) position so that the confounding effect of steering wheel loading would not be present. Tests were run at a lower test speed than many of the existing tests so that the chestband contours would be more reliable for belt-dominated loading. Acceleration was measured at three locations on the spine so that the effect of measurement location could be assessed. While the statistical power of the entire existing database is not present in the series presented here, it is hoped that the prospective experimental design of this series will provide insights not possible in a retrospective evaluation of the existing cadaver sled tests.

METHODS

Ten human cadavers tests were performed (Table 1). The subjects were seated in the RFP position of a reinforced contemporary mid-size vehicle buck mounted on a deceleration sled, as shown in Figure 1. A hydraulic decelerator was used to shape the sled deceleration pulse. A nominal impact velocity of 48 km/h and an approximately trapezoidal vehicle deceleration pulse were used for all tests. Figure 2 shows a composite plot of the sled deceleration pulses for all tests. This pulse was based on a frontal barrier test of a popular mid-size passenger sedan. Subjects were restrained with one of three restraint conditions:

FLB/AB - a three-point belt system with a buckle pretensioner and nominal 3.5-kN force-limiting retractor, the original-equipment instrument panel (IP), and a depowered airbag;

AB - a non-depowered airbag, lap belt, and the IP; and

SB/AB - a standard three-point belt system (non-pretensioned, non-force-limiting), a depowered airbag, and the IP.

Occupant positioning guidelines from the vehicle buck manufacturer (Ford Motor Company 1999) and the U.S. New Car Assessment Program (NCAP) were used, though the final seat and subject positions were dictated by several factors, including the cadaver's knee-to-IP distance, chest-to-IP distance, and clearance between the head and windshield roof header. The dominating factor was the head clearance, since head strikes against the windshield header have been found to generate artifactual thoracic acceleration peaks. These head strikes were avoided in all of the tests. In tests with a shoulder belt, head strikes could be avoided with occupants seated in or near the middle fore-aft seat adjustment position (Table 2). When a shoulder belt was used, the airbag deployment time was set to 13 ms after impact and, where applicable, the pretensioner was deployed 10 ms after impact. In the tests without a shoulder belt, however, it was necessary to position the subjects further rearward initially since the shoulder belt was not available to induce early neck flexion. For the tests without a shoulder belt, the airbag timing was dictated by the initial chest-to-IP distance. In all non-shoulder-belt tests, the airbag timing was such that the occupant contacted the airbag at approximately the time of maximum inflation (this was verified retrospectively using video analysis). This resulted in the level of airbag

Table 1 – Description of Cadaver Subjects (All tested at 48 km/hr)

Test No.	Age/ Gender	Mass (kg)	Ht. (cm)	Cause of Death	Preservation Method	Restraint Condition*
577	57/M	70	174	Pneumonia	Refrigerated	FLB/AB
578	69/F	53	155	Cerebrovascular accident (stroke) –advanced nodular diabetic glomerulosclerosis noted at autopsy	Frozen	FLB/AB
579	72/F	59	156	Myocardial infarction	Frozen	FLB/AB
580	57/M	57	177	Acute coronary insufficiency	Frozen	FLB/AB
650	40/M	47	150	Respiratory failure (pulmonary fibrosis consistent with scleroderma noted at autopsy)	Frozen	AB
651	70/M	70	176	Hypovolemic shock	Frozen	AB
652	46/M	74	175	Acute coronary insufficiency	Frozen	AB
665	55/M	85	176	Cardiopulmonary arrest	Frozen	SB/AB
666	69/M	84	176	Unspecified (failed to recover from abdominal surgery) – coronary insufficiency or an acute myocardial infarction were identified at autopsy.	Frozen	SB/AB
667	59/F	79	161	Myocardial infarction	Frozen	SB/AB

*FLB/AB - a three-point belt system with a buckle pretensioner and nominal 3.5-kN force-limiting retractor, the original-equipment instrument panel (IP), and a depowered airbag
AB - a non-depowered airbag, lap belt, and the IP
SB/AB - a standard three-point belt system (non-pretensioned, non-force-limiting), depowered airbag, and the IP

Figure 1. RFP test environment, lateral and oblique views.

inflation at occupant contact being visually similar in all ten tests.

Cadavers were preserved until the time of testing by refrigeration or freezing. Age, weight, height, and cause-of-death criteria were used to screen the cadavers. Subjects that were non-ambulant for an extended period prior to death were excluded from the study. Pre-test radiographs were taken to identify existing pathology. For some subjects, dual x-ray absorptiometry (DEXA) scans were performed to verify an acceptable level of bone mineral density. All test procedures were approved by the University of Virginia institutional review board.

Pulmonary and cardiovascular systems were pressurized to nominal *in vivo* levels (approximately

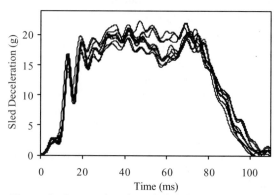

Figure 2. Composite plot of sled deceleration pulses for all tests.

Table 2 – Initial Seating Position and Deployment Timing
(all lengths in mm)

Test No.	Seat distance forward of mid-position	Knee to IP	Chest to IP	Head to windshield header	Airbag deployment (ms)	Pretensioner deployment (ms)
577	-25	129	510	not meas.	13	10
578	51	107	479	254	13	10
579	51	116	470	246	13	10
580	-25	128	522	310	13	10
650	-76	217	574	752	25	NA
651	*	267*	606*	728*	29*	NA
652	*	113*	527*	666*	18*	NA
665	0	101	401	227	13	NA
666	0	101	468	231	13	NA
667	0	101	470	250	13	NA

* Seat lowered approximately 70 mm and moved rearward approximately 121 mm from mid-position to preclude head strikes. Airbag inflation timing adjusted as described in the text.

10 kPa) immediately before testing. Cadaver instrumentation included tri-axial accelerometer arrays mounted on the posterior pelvis and at the levels of the twelfth and the eighth or ninth thoracic vertebrae (T12 and T8/9). Arrays of three orthogonal accelerometers and magnetohydrodynamic (MHD) angular rate sensors were mounted on the posterior surface of the head and posteriorly on T1, both approximately in the midsagittal plane. A uniaxial accelerometer was mounted on the body of the sternum, immediately below the manubrium. Chest deformation was measured using two 40-gage chest bands installed externally at the levels of the fourth and eighth ribs, measured laterally.

All instruments were sampled at 10 kHz. The maxima presented in the results sections were obtained from data filtered to 1000 Hz. To remove high-frequency peaks, 3-ms clip acceleration maxima, A_{clip}, are also presented for T8/9 and for T1, as is the average spinal acceleration (ASA). ASA is the mean acceleration over the time from 20% to 80% of the total change in velocity (ΔV) experienced by that instrument. The ASA calculation method is shown in Figure 3. Maximum chest deflection, regardless of location on the anterior aspect of the thorax, is presented in addition to the maximum chest deflection measured at the midsternal location. Chest deflection velocity and VC_{max} were determined by differentiating the data from the chestband profiles.

RESULTS

In all tests, the instrumentation functioned properly and well-controlled, repeatable occupant responses were achieved. The airbag and buckle pretensioner (where applicable) deployed properly in all tests except test 579. In this test, a broken circuit between the capacitive discharge squib ignitor and the pretensioner prevented pretensioner activation.

Thoracic Acceleration Responses

The cadaver acceleration responses, including ASA, were normalized to the standard 50th percentile male mass (75 kg) using the scaling procedure described by Eppinger, et al. (1984) (Table 3). Large specimen-to-specimen variability in maximum accelerations was observed, but trends among restraint conditions can be seen. Figure 4 shows the resultant acceleration-time histories measured at T1 and at T8/9 for all tests. These restraint-dependent trends can be seen in Figure 5, which shows the mean 3-ms clip maxima and ASA for each restraint condition. The FLB/AB condition resulted in the most repeatable responses, as indicated by the low standard deviation compared to the other two restraint conditions.

Compared to the FLB/AB restraint, the SB/AB and AB restraints resulted in significantly ($p < 0.05$) higher mean values of 3-ms clip acceleration and of ASA as measured at T1 (single-tailed heteroscedastic t-tests used for all tests of significance). Mean T8/9 ASA exhibited a similar trend: it was significantly lower for the FLB/AB restraint than for either the AB

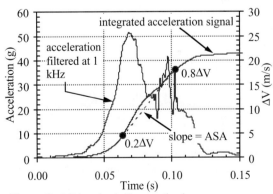

Figure 3. ASA calculation method.

Table 3 – Peak Kinematic Results from RFP Cadaver Tests (All Acceleration Values Scaled)

Test	577	578	579	580	650	651	652	665	666	667
Restraint	FLB/AB				AB			SB/AB		
T1 acceleration, g (time, ms)	33 (87)	27 (72)	31 (96)	32 (91)	48 (89)	68 (96)	70 (85)	76 (90)	71 (93)	67 (64)
T8/9 acceleration, g (time, ms)	41 (57)	43 (60)	40 (62)	42 (62)	35 (96)	44 (105)	58 (88)	51 (68)	44 (65)	56 (63)
T12 acceleration, g (time, ms)	39 (53)	48 (56)	47 (62)	45 (61)	48 (86)	58 (98)	63 (81)	79 (58)	46 (65)	69 (63)
Pelvis acceleration, g (time, ms)	33 (87)	27 (72)	56 (64)	55 (69)	80 (68)	51 (68)	55 (62)	61 (68)	68 (61)	113 (64)
T1 3-ms clip, g	30	26	28	31	47	57	53	64	66	45
T8/9 3-ms clip, g	39	39	38	42	32	42	43	50	40	44
T1 ASA, g	22	21	22	21	26	33	33	35	27	30
T8/9 ASA, g	21	23	23	22	22	29	27	35	28	32
Belt tension at D-ring, N	4478	4222	4896	4181	NA	NA	NA	7655	7105	6884
Belt tension at buckle, N[‡]	3400	3688	3559	3418	4084	3850	6424	4445	4232	4214
4th rib deflection, mm [%*] (time, ms)	50 [24%] (113)	52 [25%] (101)	88 [37%] (104)	67 [29%] (104)	0.2[†] [0%] (114)	25 [11%] (93)	31 [13%] (100)	99 [37%] (83)	89 [34%] (107)	95 [37%] (101)
8th rib deflection, mm [%*] (time, ms)	19 [8%] (64)	32 [15%] (60)	43 [17%] (98)	40 [17%] (60)	0.5 [0%] (94)	0.4 [0%] (33)	0.2 [0%] (27)	48 [16%] (71)	56 [21%] (101)	18 [7%] (99)
4th rib chestband sternal deflection, mm [%*] (time, ms)	48 [23%] (112)	52 [25%] (101)	81 [34%] (105)	66 [28%] (104)	0.0[†] [0%] (--)	24 [11%] (94)	28 [12%] (100)	75 [28%] (105)	83 [32%] (108)	93 [36%] (101)
4th rib sternal deflection velocity, m/s (time, ms)	2.7 (24)	2.6 (14)	3.7 (75)	2.7 (12)	0.0[†] (--)	2.0 (71)	2.2 (98)	5.3 (80)	2.2 (55)	6.9 (76)
4th rib sternal VC_{max}, m/s (time, ms)	0.2 (66)	0.3 (93)	1.0 (75)	0.3 (52)	0.0[†] (--)	0.1 (88)	0.2 (98)	1.3 (81)	0.4 (95)	2.2 (76)

*Values are the percentage of the initial chest depth measured by that chestband.

[†]Due to this subject's small stature, airbag loading was concentrated superior to the upper chestband.

[‡]Measurement taken on torso belt where applicable – in AB tests (650-652) it was taken on the lap belt.

restraint or the SB/AB restraint. The mean 3-ms clip acceleration at T8/9, however, was not significantly different for any restraint condition, though the SB/AB condition resulted in a slightly greater mean value than the AB condition or the FLB/AB condition.

The peak T1 and T8/9 acceleration occurred approximately 40 ms earlier in tests with the FLB/AB and the SB/AB restraints than in the tests with the AB restraint. This is a result of the shoulder belt loading the torso and generating spinal acceleration. Regardless of the restraint condition, two distinct regimes can be seen in the T8/9 acceleration-time histories shown in Figure 4. This phenomenon, which is most pronounced with the FLB/AB restraint, is a result of a timing lag between belt loading and airbag loading. These distinct loading phases were verified using high-speed video and are discussed in greater detail in an earlier paper (Kent et al. 2000).

Thoracic Deformation Responses

In these tests, the lower chestband did not measure significant deformations for any restraint condition. Deformation-based criteria are therefore discussed for the upper chestband only. To account for variations in anthropometry, chest deformations are presented in two forms: 1) the absolute deflection and 2) the deflection divided by the initial chest depth (Table 3). This normalized deflection will be referred to as chest compression (C_{max}).

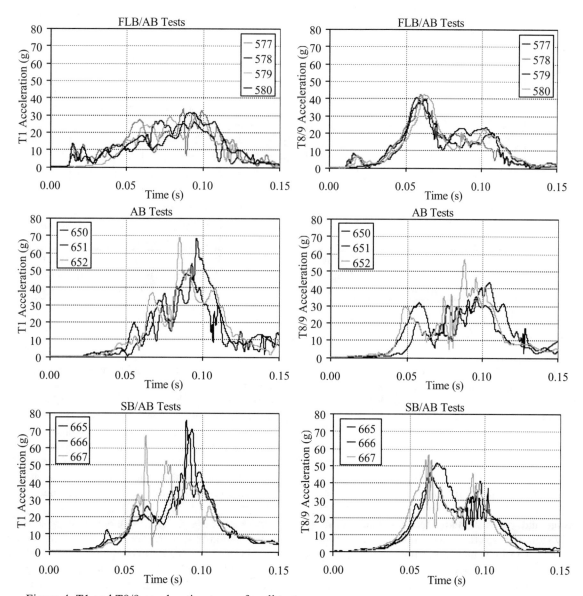

Figure 4. T1 and T8/9 acceleration traces for all tests

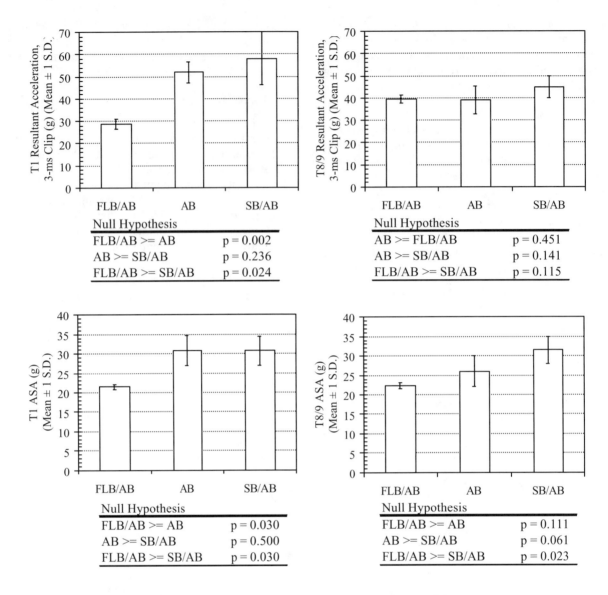

Figure 5. Comparison of acceleration-based parameters across restraint conditions.

In all tests but one, all chestband gages were below their full-scale values. In test 665, the gage located at the point of maximum curvature (i.e. under the belt) registered a constant value from approximately 95 ms to 105 ms. There was no evidence of gage failure, however, and it was concluded this was tenable output that simply exceed the gage full-scale. Since the signal clipped after the time of maximum chest deflection (83 ms), the profile is assumed to be valid up to that point. As expected, the maximum deflection and the maximum sternal deflection are strongly related, so they will not be discussed separately. The maximum deflection velocity, V_{max}, and VC_{max} will be presented for the mid-sternal location since this corresponds to the measurement location for the Hybrid III dummy.

The effect of seatbelt loading is seen clearly in the upper chestband profiles shown in Figure 6. In agreement with previous researchers (e.g. Yoganandan et al. 1994, Crandall et al. 1994), a region of localized deformation was observed in the area of belt loading – even with the force-limiting belt. The force-limiting belt did, however, result in generally less localized deformation (i.e. greater radii of curvature) compared to the standard belt, though the presence of injury also influences the degree of deformation (e.g. test 579). Figure 7 shows the deformation-based responses for each restraint condition. Mean maximum chest deflection exhibited a significant trend across restraint conditions: SB/AB generated the greatest maximum chest deflection – a mean of approximately 36%,

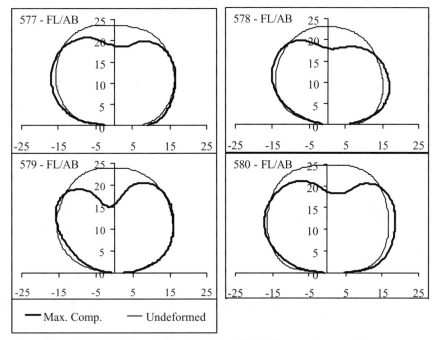

a. Force-limiting belt and airbag (FL/AB) restraint condition.

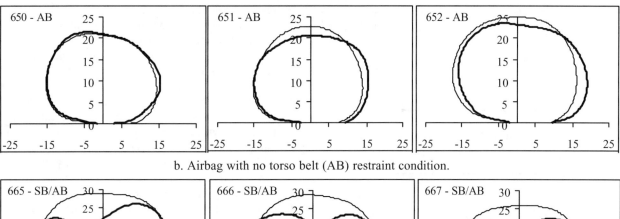

b. Airbag with no torso belt (AB) restraint condition.

c. Standard belt and airbag (SB/AB) restraint condition.

Figure 6. Comparison of chestband profiles at time of maximum chest compression.

followed by the FLB/AB condition (29%). The AB condition resulted in much lower mean maximum deflection – approximately 8%. In contrast to the acceleration results, the SB/AB condition had the most repeatable responses, while the AB and FLB/AB conditions resulted in much larger inter-test variability.

The maximum deflection velocity trend was identical to the trend for maximum deflection, but was not significant to the $p = 0.05$ level due to large inter-test variability. The SB/AB restraint resulted in the greatest mean (4.8 m/s), followed by the FLB/AB restraint (2.9 m/s), and the AB restraint (1.4 m/s).

p values were greater than 0.05 due to large inter-test variability. The SB/AB restraint resulted in a mean VC_{max} of 1.3 m/s, followed by the FLB/AB restraint (0.45 m/s), and the AB restraint (0.10 m/s).

Injury Results

Rib fractures were the most common injury observed at autopsy. Other injuries included a sternal fracture, a superficial liver laceration, a pelvic fracture, a clavicular fracture, and a separation of the sterno-clavicular joint (Table 4).

The FLB/AB restraint resulted in two cadavers sustaining injury, while two did not. The SB/AB restraint resulted in all three cadavers sustaining some form of injury. The AB restraint condition resulted in one cadaver sustaining bilateral simple fractures on the second and third ribs, while the other two were uninjured. This injured subject also sustained a large oblique iliac wing fracture, apparently a result of lap belt loading. As mentioned above, a primary goal of this test series was to avoid head strikes against the windshield header. The AB restraint condition, which results in increased forward excursion of the occupant within the vehicle, provided the greatest potential for these head strikes. Test 650 was the first performed at this restraint condition and, as a result, it was not known how much forward excursion the occupant would experience. The occupant was therefore placed in a sufficiently rearward position to avoid the risk of head contact. This positioning, combined with the lap belt loading, which tended to maintain the pelvis within the seat, resulted in the relatively diminutive subject sustaining airbag loading concentrated on the superior aspect of the anterior thorax. In contrast, the other subjects, which were taller and were also positioned farther forward relative to their torso lengths, sustained airbag loading more evenly distributed over the entire anterior thorax. This difference in kinematics and restraint interaction is probably a contributive factor for the difference in injury outcome experienced by the subjects restrained by the AB condition.

Predictors of Thoracic Injury

Several kinematic parameters were evaluated as thoracic injury predictors: C_{max}, ASA (T1 and T8/9), 3-ms clip maximum resultant acceleration (T1 and T8/9), V_{max}, and VC_{max}. Cross-plots are presented as a qualitative and visual means of comparing the injury-correlative value of each criterion. C_{max} is presented on the ordinate of six plots, while the abscissa shows A_{clip} (Figure 8), ASA (Figure 9), VC_{max} (Figure 10, top plot), or V_{max} (Figure 10, lower

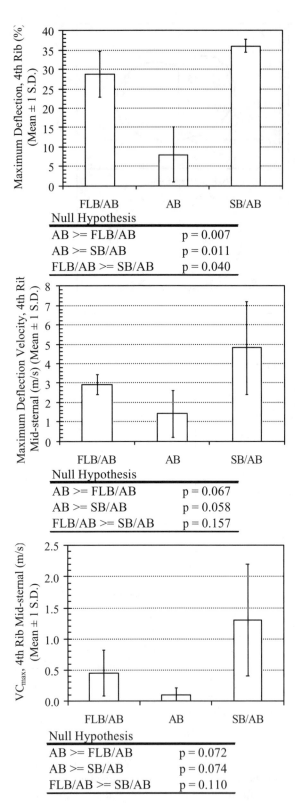

Figure 7. Comparison of deflection-based parameters across restraint conditions.

The maximum viscous criterion trend was also the same as the trend for maximum deflection. Again, all

Table 4 – Summary of Injuries to Trunk

Test No.		Injury Assessment	Number of Rib Fractures	AIS	MAIS
577	Force-limiting 3-point belt, depowered airbag (FL/AB)	No fractures, dislocations, or soft tissue injuries	0	0	0
578		Costochondral fractures on left ribs 4, 5, 6, 8; superficial liver laceration (right lobe)	4	3,2	3
579		Simple fractures of left ribs 2-8; Simple fractures of right ribs 2-5; sternal fracture at approximate location of belt path; separation of right sterno-clavicular joint	11	4, 2, 2	4
580		No fractures, dislocations, or soft tissue injuries	0	0	0
650	Lap belt, non-depowered airbag (AB)	Simple fractures of ribs 2 and 3 bilaterally; large oblique iliac wing fracture extending into the right sacroiliac joint	4	2,2	2
651		No fractures, dislocations, or soft tissue injuries	0	0	0
652		No fractures, dislocations, or soft tissue injuries	0	0	0
665	Standard 3-point belt, depowered airbag (SB/AB)	Simple fractures on right rib 2 and left ribs 3 and 4.	3	2	2
666		Simple fractures on right ribs 3 and 4 and left rib 6	3	2	2
667		Simple fractures on the left ribs 2-8 with 2 fractures on rib 5; simple fractures on the right ribs 2-6; midshaft clavicle fracture on right.	13	4	4

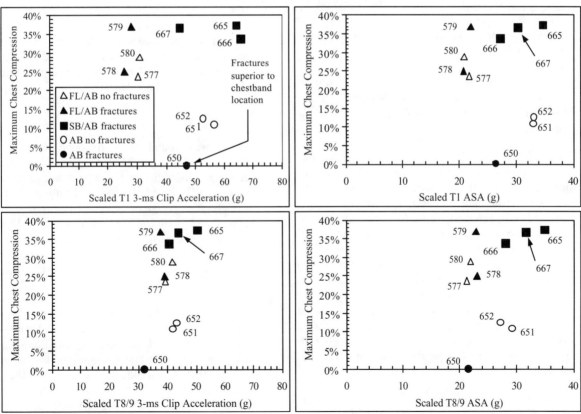

Figure 8. Maximum chest compression and 3-ms clip acceleration as injury predictors. Test numbers are listed for reference.

Figure 9. Maximum chest compression and ASA as injury predictors. Legend for Figure 8 applies.

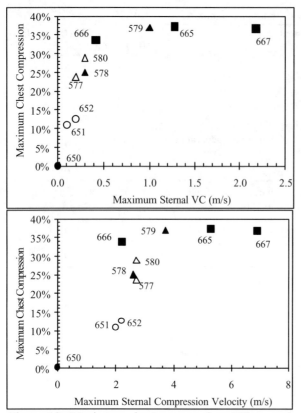

Figure 10. Maximum chest compression, VC_{max}, and maximum compression velocity as injury predictors. Legend for Figure 8 applies.

plot). In these figures, the different symbols (square, triangle, circle) correspond to the different restraint conditions and a filled symbol corresponds to an injurious test.

Figure 8, Figure 9, and Figure 10 show that maximum chest compression corresponds well with the presence of rib fractures. All cadavers that exhibited chest compression greater than 30% sustained rib fractures. Furthermore, only one cadaver with chest compression less than 25% sustained rib fractures. This cadaver, from test 250, is the subject discussed above that sustained airbag loading on the superior portion of the thorax and the rib fractures were located above the level of the chestbands. If this subject is not considered, then the lowest chest compression at which rib fractures were sustained was 25% (test 578). One subject (test 580) tolerated 29% chest compression without sustaining any injury.

Figure 8 shows that the 3-ms clip resultant acceleration measured at T1 and at T8/9 did not correspond with the presence of injury. Test 666 (66g) and test 667 (64g) had the greatest measured T1 A_{clip}, and both of those subjects sustained injury, but test 578 (26g) and test 579 (28g) also involved injured subjects and they had the lowest measured T1 A_{clip}. The T8/9 A_{clip} likewise did not discriminate injury tests from non-injury tests: all tests were clustered between 31g and 50 g, with no clear trend related to the presence of injury.

In Figure 9, the 3-ms clip shown in Figure 8 has been replaced with ASA, while C_{max} is maintained on the ordinate. This plot shows that ASA also did not contribute injury-predictive information. Test 652 (33g) and test 651 (33g), which were both non-injury tests, resulted in two of the three highest values of T1 ASA, while test 578 (21g) and test 579 (22g), both injurious, resulted in two of the lowest values of T1 ASA. T8/9 ASA was similarly non-predictive.

Figure 10 shows the maximum sternal compression velocity and sternal VC_{max} on the abscissa, again with C_{max} shown on the ordinate. This figure illustrates that V_{max} and VC_{max} were generally greater in tests with injury than in tests without. In fact, VC_{max} exhibited a reasonably good correspondence with injury. All subjects with VC_{max} greater than 0.4 m/s sustained injury and only one subject (test 578) sustained injury with VC_{max} below this level (again, excluding test 650). Compression velocity also exhibited correspondence with injury, though it is not as good as that for VC_{max}. All three subjects with V_{max} greater than 2.7 m/s sustained injury, but below that value there is no clear trend.

DISCUSSION

VC_{max} was found to correspond well with the presence of injury. This does not imply that this criterion should be used as a predictor of hard tissue injury. VC_{max} was developed using *in vivo* animal models that sustained soft tissue injuries. VC_{max} corresponds with the hard tissue injuries obtained in this study because it is correlated with maximum compression. The cross-plots presented in Figure 10 show that VC_{max} and V_{max} do not add hard tissue injury-predictive value to C_{max}.

In these tests, acceleration-based criteria also did not have utility as predictors of hard-tissue injury. Acceleration-based parameters may be useful as indicators of the restraint condition, as evidenced by the fact that clear trends were observed in these tests: the 3-ms clip and ASA were highest for the SB/AB condition and lowest for the FLB/AB condition. This trend was not significant, however, for any acceleration measure and additional testing or analysis is necessary before it is clear that acceleration is useful in this regard. The acceleration measured at the spine depends on the loading on the anterior thorax, as well as on the inertial, viscous, and elastic properties of the torso. As illustrated with these tests, the distributed force from an airbag-only restraint can generate high thoracic acceleration with very little chest deformation. For the purpose of evaluating hard-tissue injury potential when the restraint condition and therefore the magnitude of the applied force is changed, acceleration-based criteria were not found to be useful.

In contrast, the maximum chest compression appears to be causally related to the presence of rib fractures. This is a reasonable finding considering bone's elastic nature and insensitivity to loading rate. While many different specific failure criteria have been proposed for bone tissue (e.g., Keyak and Rossi 2000), it is clear that bone failure is strongly strain-dependent. The strain rate has to change by orders of magnitude before the failure strain of bone changes appreciably (e.g., McElhaney 1966). It is still unclear, however, whether the restraint condition affects the critical value of chest compression. For example, is more or less chest compression required to fracture ribs under belt loading as compared to blunt impactor loading or airbag loading? From the tests presented here, it is possible to conclude that the critical value of C_{max} for combined belt-and-airbag loading is approximately 25%-30% - a value that is consistent with previous researchers who have used different loading conditions. Kroell et al. (1971, 1974), for example, using a circular rigid impactor, concluded that rib fracture onset is generally above 20% compression and Viano et al. (1978) concluded that rib cage stability is compromised due to numerous rib fractures at 32% compression. Neither of these researchers drew these conclusions based on belt loading or combined belt-and-airbag loading, which is an indication that the critical value of chest compression may be relatively insensitive to the loading condition. A simplified analytical model can be used to illustrate this insensitivity relative to other mechanical parameters such as applied force or work, while also illustrating the restraint dependency of rib fracture location.

Analytical Study of Elastic Ring

Cadaver sled tests and finite element studies have shown that the distribution and the number of rib fractures are functions of the restraint configuration. Belt restraints consistently produce fractures along a path near the regions of loading under the belt (e.g. Patrick and Anderson 1974, Schmidt et al. 1974, Kallieris et al. 1982, Christian 1976, Newman and Jones 1984, Arajavi and Sanavirta 1989, Crandall et

al. 2000), while airbag restraints without belts produce fractures (Kallieris et al. 1998, Yoganandan and Pintar 1998) and stress concentrations (Plank et al. 1994) at anterolateral rib locations. Figure 11 shows the circumferential location and rib number for each fracture sustained by the 10 RFP test subjects presented here (the circumferential location is based on an elliptical simplification of the torso cross-section and is presented so that results for cadavers of different size may be compared – 0% corresponds to the middle of the sternum, 25% corresponds to the lateral-most aspect). For the tests with a torso belt, many of the fractures are located near the area of belt loading. Figure 12 presents additional cadaver data compiled for a study published by Crandall et al. (2000). These plots include more subjects (all tested on the driver side) and no combined loading tests. In these plots, the restraint dependence of fracture location is clear.

This diversity of rib fracture location indicates a different stress/strain distribution within the rib cage for each restraint condition. It is not necessarily true, therefore, that the same critical value of a mechanical parameter is appropriate for each of these loading conditions. An analysis of the shear and moment in the rib cage may be used to evaluate the correlations between peak applied force, peak compression, and the peak moment or shear in the rib cage model (proxies for stress or strain in the rib and therefore for injury). For this analysis, the rib cage is simplified as a circular ring of radius R, as shown in Figure 13. Belt loading is simplified as a point force, F, and airbag loading is simplified as a constantly distributed force of magnitude $F/\pi R$, which is applied from $-\pi/2 \leq \theta \leq \pi/2$. For this analysis, the viscoelastic contents of the thorax are neglected, as are the inertial effects and changes in moment and shear distribution due to deformation. These omissions are of minor concern, however, since the purpose of this simplified analysis is a comparison of the location of the peak moment and peak shear for two loading conditions and an analysis of the sensitivity of chest compression to loading conditions. The results of this analytical exercise should not be applied outside of this specific purpose.

Internal Moment, $M_p(\theta)$, and Shear, $V_p(\theta)$, for a Point Force Loading Condition - Consider the ring depicted in Figure 13a. This loading condition represents, in a highly simplified sense, a rib cage subjected to belt loading on the anterior aspect and a distributed reaction force on the posterior aspect. Due to symmetry, the problem can be reduced to the free-body diagram shown where $M_p(\theta)$, $V_p(\theta)$, and $P_p(\theta)$, are internal reactions to the applied load. Also from symmetry, $V_p(\theta) = 0$ at $\theta = \pi/2$ and the redundant force H_p is zero. The internal moment is therefore given by

$$M_p(\theta) = \frac{F}{2} R \sin\theta - M_p^* \qquad [1].$$

M_p^* is a redundant internal moment required to maintain the condition

$$\phi(0) = 0 \qquad [2]$$

where $\phi(\theta)$ is the rotation of the ring. The complimentary strain energy in the ring, U_p^*, can be described in terms of the cross-sectional moment of inertia, I, the modulus of elasticity, E, the ring radius, and the internal moment:

$$U_p^* = \frac{1}{2EI} \int_0^{\frac{\pi}{2}} M_p(\theta) R d\theta \qquad [3].$$

Applying the boundary condition of [2] and Castigliano's second theorem gives an expression for the redundant moment, M_p^*:

$$\phi(0) = 0 = \frac{\partial U_p^*}{\partial M_p^*} = \frac{R}{EI} \int_0^{\frac{\pi}{2}} M_p(\theta) \frac{\partial M_p(\theta)}{\partial M_p^*} d\theta \qquad [4].$$

The internal moment and shear can then be described in terms of the applied force and the radius:

$$M_p(\theta) = FR\left(\frac{\sin\theta}{2} - \frac{1}{\pi}\right) \qquad [5]$$

and

$$V_p(\theta) = \frac{1}{R} \frac{\partial M_p(\theta)}{\partial \theta} \qquad [6].$$

Internal Moment, $M_d(\theta)$, and Shear, $V_d(\theta)$, for the Distributed Loading Condition - Consider the ring depicted in Figure 13b. This loading condition represents, in a highly simplified sense, a rib cage subjected to airbag loading on the anterior aspect and a distributed reaction force on the posterior aspect.

The moments against which the internal moment must react are 1) the moment, M_w', due to the distributed load removed using the symmetric simplification, 2) the moment, $M_w(\theta)$, due to the

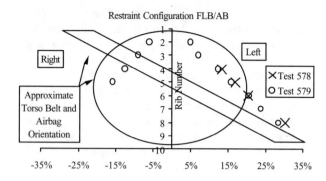

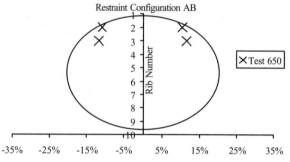

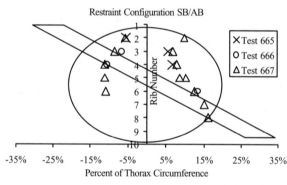

Figure 11. Location of all rib fractures sustained by subjects in the 10 RFP tests.

distributed load, $w(\theta)$, acting from $\theta = 0$ to $\theta = \theta$, and 3) the redundant moment, M_d^*, required to maintain the condition of equation [2]. Similarly to the point load case, the redundant moment M_d^* is found using the condition of equation [2] and the relationship shown in equation [4], which give an equation for the internal moment in terms of the applied force function and the radius:

$$M_d(\theta) = \frac{FR}{4}\left(\sin^2\theta - \frac{1}{2}\right) \qquad [7]$$

where

$$F = F_{tot} = \int_{-\frac{\pi}{2}}^{\frac{\pi}{2}} w(\theta) R d\theta \qquad [8]$$

and the internal shear is given by equation [6].

Ring Deflection for Each Loading Condition - The ring deflection for the point load condition can be determined using symmetry, complimentary energy, and Castigliano's second theorem:

$$\delta_p = \frac{\partial U^*}{\partial F} = \frac{2R}{EI} \int_0^{\frac{\pi}{2}} M_p(\theta) \frac{\partial M_p(\theta)}{\partial F} d\theta \qquad [9]$$

where δ_p is the displacement at $\theta = 0$ for the upper hemi-ring. For the distributed loading condition, the displacement at $\theta = 0$ for the upper hemi-ring can be adequately described by the relationship

$$\delta_d = \frac{k_z F R^3}{4EI} \qquad [10]$$

where k_z is an empirical constant defined by the distributed loading condition (e.g., Pilkey 1994). For the constantly distributed force considered here, $k_z \approx -0.17$.

Results of Analytical Analysis

Figure 14 shows the shear and moment distributions on the anterior thorax resulting from the same total force applied with each of these loading conditions. In the case of the point force, the peak shear, V_{pmax}, and peak moment, M_{pmax}, both occur at the point of force application ($\theta = 0°$). In contrast, when the force is distributed over the thorax, the peak shear, V_{dmax}, occurs at $\theta = \pm 45°$ and the peak moment, M_{dmax}, occurs at both $\theta = 0°$ and at $\theta = \pm 90°$. This is consistent with the experimental findings that rib fractures from belt loading often occur near the path of the belt, while rib fractures from airbag loading are often located anterolaterally.

Also of interest is the amount of loading condition-dependent variation in the mechanical parameters (applied force, work into the ring, and peak displacement). This can be evaluated by solving for the ratio (distributed/concentrated) for the displacement, force, and work values that correspond to a value of peak moment or shear (regardless of the location on the ring at which the peak occurs). A closed form solution for the ratio δ_d/δ_p can be found, as can ratios for the total force and the work into the

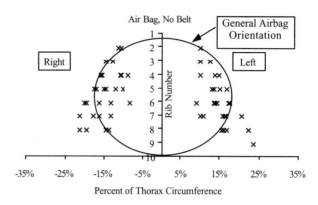

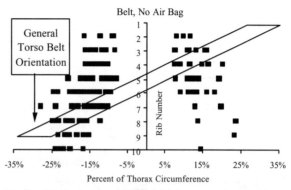

Figure 12. Location of bony rib fractures in 18 driver-side tests with belt-only restraint condition and airbag-only restraint condition (Crandall et al. 2000).

ring. The parameter that has the smallest ratio is the parameter least sensitive to changes in loading condition. Solving for these ratios, one obtains 1.86 for the deflection ratio, 2.55 for the force ratio, and 4.73 for the work ratio. The physical interpretation of these ratios is that, for example, 1.86 times more deflection is required to obtain a given internal moment if the force is distributed compared to a concentrated force. Similarly, in order to obtain a given internal moment, the rib cage must be subjected to 2.55 times more total force, and can absorb 4.73 times more work, for the case of a distributed force compared to a point force. While the simplicity of the model limits the conclusions that can be drawn, comparing ratios for the three mechanical parameters provides a basis for understanding the fundamental relationship between force, work, displacement and injury to the rib cage when the loading condition is varied. The injurious level of chest deflection is less sensitive to loading condition than is the injurious level of applied force or work.

Since most cadavers, regardless of loading condition, have been found to exhibit a rib fracture threshold of 20%-25% chest compression, it seems apparent that physically realizable changes in restraint condition do

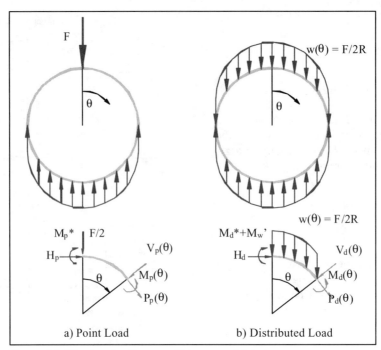

Figure 13. The two loading conditions considered.

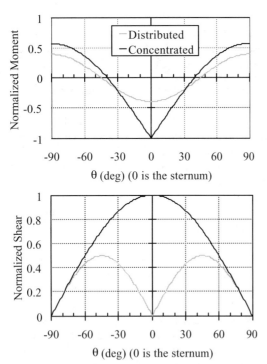

Figure 14. Shear and moment plots for each loading condition (total applied force is equal).

not change the critical value of chest deflection by 1.86 times. The analytical analysis presented above is a highly simplified representation of the rib cage only, which does not include any soft tissues. The point load case analyzed, therefore, does not correspond with any loading condition possible in an automotive environment – the analytical study is intended to bound the range of realistic conditions. In reality, the restraining force is always distributed, though the degree of distribution depends on the restraint condition.

CT Images of a Loaded Torso

In addition to the restraint itself, the thoracic soft tissues between the belt and rib cage may have a load-distributing effect. The degree of force distribution on the exterior thorax may not be indicative of the distribution at the level of the rib cage. The chestband profiles obtained during the SB/AB tests indicate a relatively concentrated force exerted on the anterior torso (see Figure 6). To date, however, we are aware of no studies that have related the shape of the exteriorly measured chestband profiles to the deformation sustained by the rib cage itself. It is possible that what appears to be highly concentrated force from the belt may actually be fairly well distributed at the depth of the rib cage, particularly for obese subjects or for females with breast tissue surrounding much of the rib cage. In fact, the load-distributing effect of the soft tissues may limit the loading condition dependency of the chest compression critical value. In an effort to understand better the role of soft tissue in load distribution, a non-obese human cadaver (168-cm tall, 75-year-old female) was subjected to concentrated, belt-like loading on the anterior torso.

Computed tomography (CT) images were obtained to quantify the shape of the exterior torso and of the rib cage at four levels of chest compression, including the limiting case where the posterior sternum contacted the anterior spine. These images, reproduced in Figure 15 next to outlines of the relevant structures, show that the soft tissue does act as a force distributor. The external profiles exhibit a more concentrated deformation than the internal rib cage profile (e.g. Figure 15d). It is also interesting to note that the depth of soft tissue anterior to the rib cage did not change significantly up to approximately 1.2 kN of applied force. While this single subject does not provide enough information to quantify all the possible differences between exterior (i.e. chestband) profiles and rib cage profiles, it does illustrate that these differences may be significant, particularly as they affect injury prediction for different restraint conditions. The force-distributing effect of the soft tissues, while it does confound cadaver-based deflection measurements, may simplify injury prediction by limiting the sensitivity of rib cage deformation to changing restraint conditions.

Study Limitations

As mentioned earlier, the use of chestbands to measure thoracic deformation is a potential source of error, and the error potential is restraint system-dependent. As discussed by Bass et al. (2000), chestbands are subject to errors under the localized loading caused by seatbelts. Similar findings have been presented in unpublished studies that are available publicly (Shaw et al. 1999, Hagedorn and Burton 1993). This potential error source may introduce systematic bias into the findings of this study. Nevertheless, the chestbands used in this study (40-gage) are the state of the art and chestbands are currently the only viable method for measuring dynamic chest deformation in a sled test. To maximize chestband reliability, all gage signals were inspected for evidence of clipping or other anomalies that could affect the measured profile. Further, static pretest profiles were verified against measured chest dimensions. The profiles shown in Figure 6 do appear to be reasonable responses for each restraint condition.

The use of cadaver injury as a model for *in vivo* injury is a limitation of this analysis. In this, as in any cadaver-based dataset, the dominant thoracic injury type is rib fractures. The presence of multiple rib fractures, while often associated with significant intrathoracic, intra-abdominal, and extra-cavity trauma (e.g. Pattimore et al. 1992) does not necessarily correlate with all types of thoracic

injuries sustained by the living. For example, Lee et al. (1997) found no clinically relevant correlation between thoracic skeletal injuries and acute traumatic aortic tear. For the purpose of evaluating injury criteria efficacy under diverse types of loading, however, cadaver testing is the best method currently available. In addition, while cadavers do not sustain soft tissue injuries representative of *in vivo* subjects, cadaver hard tissue injury is reasonably representative of the hard tissue injuries sustained by a frail population of living humans (Walfisch et al. 1985). Since approximately 61% of all AIS 2+ chest injuries in NASS are hard tissue injuries (rib fractures), and the maximum AIS is defined by fractures for approximately 72% of the occupants receiving MAIS 2+ chest injuries (Crandall et al. 2000), cadavers may be a reasonable representation of the living human population once differences in age and general physical condition are taken into account.

The finding that chest acceleration does not provide hard tissue injury-predictive value should not be interpreted to mean that the occupant's acceleration should not be limited in a crash. The human body does have an acceleration tolerance limit. The types of injuries sustained at accelerations above this limit are generally not, however, hard tissue injuries. Loss of consciousness, hemorrhaging, and other *in vivo* injuries, which have been associated with acceleration, cannot be evaluated using a cadaver model.

CONCLUSIONS

The right-front passenger position is a useful position to evaluate restraints and thoracic injury criteria because the confounding effect of the steering wheel is not present and, as a result, the thoracic response is dominated by loading from the restraint system.

The tests and analyses presented here indicate that the injury criterion best related to the presence of rib fractures is chest compression. Peak acceleration measured at any of three locations on the spine is not found to be a predictor of hard tissue injury, nor is it found to contribute hard tissue injury-predictive information to that provided by chest compression alone – even when the restraint condition is varied. Similarly, the average spine acceleration is not found to correspond with injury in these tests. Based on non-significant trends observed in these tests, acceleration-based criteria may be useful for differentiating tests with and without an airbag, though this utility has been merely observed and not proven here. The maximum viscous criterion and the

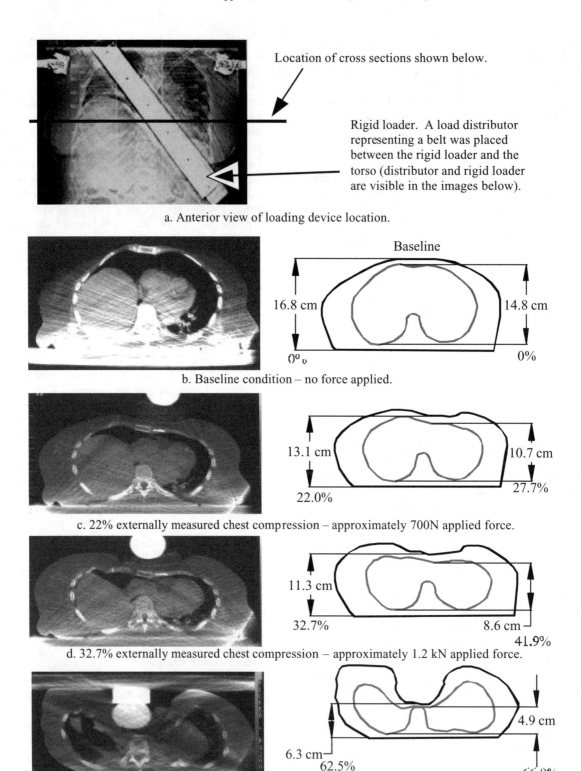

Figure 13. Comparison of externally measured deformation and rib cage deformation. CT images are shown on the left, line drawings of external profile and rib cage cross-sectional shape are shown on the right. Images are at the approximate location of the upper chestband used in the sled tests.

maximum compression velocity are found to correspond with injury, but this is hypothesized to be due to the fact that these two criteria correlate with maximum compression.

Compression is less sensitive to changes in restraint condition than is applied force or work. In the ten tests presented here, the compression level beyond which rib fractures occur is approximately 25% for combined belt-and-airbag loading, though the age and other characteristics of the individual subjects affect this level. This is consistent with the critical compression level found by other researchers using different loading conditions. The force-distributing effect of the superficial soft tissues contributes to the restraint condition insensitivity of injurious maximum compression.

ACKNOWLEDGMENTS

Funding for this study was provided by the Alliance of Automobile Manufacturers, though the findings do not necessarily represent the consensus views of the AAM. The authors gratefully acknowledge Ricky Bryant and Bryan Bush of the University of Virginia Automobile Safety Laboratory, who assisted with the sled testing.

REFERENCES

Arajavi, E. and Sanavirta, S. (1989) Chest injuries sustained in severe traffic accidents by seatbelt wearers. Journal of Trauma 29(1): 37-41.

Bass, C., Wang, C., Crandall, J. (2000) Error analysis of curvature-based contour measurement devices. Paper number 2000-01-0054, Society of Automotive Engineers, Warrendale, Pennsylvania.

Cavanaugh, J. (1993) The biomechanics of thoracic trauma. In Accidental Injury Biomechanics and Prevention, ed. A. Nahum and J. Melvin, pp. 362-390. Springer-Verlag, New York.

Cesari, D. and Bouquet, R. (1990) Behavior of human surrogates under belt loading. Proc. 34th Stapp Car Crash Conference, pp. 73-82, Society of Automotive Engineers, Warrendale, Pennsylvania.

Cheng, R., Yang, K., Levine, R., King, A., Morgan, R. (1982) Injuries to the cervical spine caused by a distributed frontal load to the chest. Proc. 26th Stapp Car Crash Conference, pp. 1-40, Society of Automotive Engineers, Warrendale, Pennsylvania.

Christian, M. S. (1976) Non-fatal injuries sustained by seat belt wearers, a comparative study. British Medical Journal, 2:1310-1311.

Crandall, J., et al. (1994) A Comparison of two and three point restraint systems. Proc. Joint Session of AAAM and International Research Council on the Biomechanics of Impact Conference, Lyon, France.

Crandall, J., Kent, R., Fertile, J., Martin, P., Mastrangelo, C. (2000) Rib fracture patterns and radiologic detection – a restraint-based comparison. Proc. 44th Annual Scientific Conference of the Association for the Advancement of Automotive Medicine. Chicago, Illinois.

Cromack, J and Ziperman, H. (1975) Three-point belt induced injuries: a comparison between laboratory surrogates and real world accident victims. Paper number 751141, Society of Automotive Engineers, Warrendale, Pennsylvania.

Eppinger, R. (1984) Development of dummy and injury index for NHTSA's thoracic side impact protection research program. Paper number 840885, Society of Automotive Engineers, Warrendale, Pennsylvania.

Eppinger, R., et al., (1999) Development of improved injury criteria for the assessment of advanced automotive restraint systems – II. National Highway Traffic Safety Administration, U.S. Department of Transportation, Washington, D.C.

Fayon, A., Tarriere, C., Walfisch, G., Got, C., Patel, A. (1975) Thorax of 3-point belt wearers during a crash (experiments with cadavers). Paper number 751148, Society of Automotive Engineers, Warrendale, Pennsylvania.

Hagedorn, A. and Burton, R. (1993) Evaluation of chestband responses in a dynamic test environment. Proceedings 22nd Annual Workshop on Human Subjects for Biomechanical Research, pp. 83-101.

Hassan, J. and Nuscholtz, G. (2000) Development of a combined thoracic injury criterion – a revisit. Paper number 2000-01-0158, Society of Automotive Engineers, Warrendale, Pennsylvania.

Horsch, J. D., Melvin, J., Viano, D., Mertz, H. (1991) Thoracic injury assessment of belt restraint systems based on Hybrid III chest compression. Proc. 35th Stapp Car Crash Conference, pp. 85-108, Society of Automotive Engineers, Warrendale, Pennsylvania.

Ford Safety Laboratories (1999) Hyge dummy positioning and F/A targeting sheet – 208 and seat back upright tests. Dearborn, Michigan, 1999.

Kallieris, D., Mellander, H., Schmidt, G., Barz, J., Mattern, R. (1982) Comparison between frontal impact tests with cadavers and dummies in a simulated true car restrained environment. Proc. 26th Stapp Car Crash Conference, pp. 353-367, Society of Automotive Engineers, Warrendale, Pennsylvania.

Kallieris, D., Conte-Zerial, P., Rizzetti, A., Mattern, R. (1998) Prediction of thoracic injuries in frontal crashes. Paper 98-S7-O-04 Proc. 16th ESV Conference.

Kent, R., Crandall, J., Bolton, J., Duma, S. (2000) Driver and right-front passenger restraint system interaction, injury potential, and thoracic injury prediction. Proc. 44th Annual Scientific Conference of the Association for the Advancement of Automotive Medicine, Chicago, Illinois.

Kent, R., Bolton, J., Crandall, J., Prasad, P., Nusholtz, G., Mertz, H., Kallieris, D. (2001) Restrained Hybrid III dummy-based criteria for thoracic hard tissue injury prediction. accepted for publication at the 2001 Conference of the International Research Council on the Biomechanics of Impact (IRCOBI), Isle of Man.

Keyak, J. and Rossi, S. (2000) Prediction of femoral fracture load using finite element models: an examination of stress- and strain-based failure theories. Journal of Biomechanics 33:209-214.

Kroell, C., Schneider, D., Nahum, A. (1971) Impact tolerance and response of the human thorax. Paper number 710851, Society of Automotive Engineers, Warrendale, Pennsylvania.

Kroell, C., Schneider, D., Nahum, A. (1974) Impact tolerance and response of the human thorax II. Paper number 741187, Society of Automotive Engineers, Warrendale, Pennsylvania.

Kroell, C. (1994) Thoracic response to blunt frontal loading. In Biomechanics of Impact Injury and Injury Tolerances of the Thorax-Shoulder

Complex, ed. Backaitis, S., pp. 51-79. Society of Automotive Engineers publication PT-45.

Kuppa, S. and Eppinger, R. (1998) Development of an improved thoracic injury criterion. Paper 983153, Proc. 42nd Stapp Car Crash Conference, Society of Automotive Engineers, Warrendale, Pennsylvania.

Lau, I.V. and Viano, D.C. (1986) The viscous criterion - bases and applications of an injury severity index for soft tissues. Proc. 30th Stapp Car Crash Conference, pp. 123-142, Society of Automotive Engineers, Warrendale, Pennsylvania..

Lee, J., Harris, J., Duke, J., Williams, S. (1997) Noncorrelation between thoracic skeletal injuries and acute traumatic aortic tear. Journal of Trauma 43(3):400-404.

Malliaris, A. C., et al. (1985) Harm causation and ranking in car crashes. Paper 850090, Society of Automotive Engineers, Warrendale, PA.

MeElhaney, J. (1966) Dynamic response of bone and muscle tissue. Journal of Applied Physiology 21:1231-1236.

Mertz, H., Gadd, C. (1971) Thoracic tolerance to whole-body deceleration. Paper number 710852, Society of Automotive Engineers, Warrendale, Pennsylvania.

Mertz, H., et al. (1991) Hybrid III sternal deflection associated with thoracic injury severities of occupants restrained with force-limiting shoulder belts," Paper 910812, Society of Automotive Engineers, Warrendale, Pennsylvania.

Morgan, R., Eppinger, R., Haffner, M., Yoganandan, N., Pintar, F., Sances, A., Crandall, J., Pilkey, W., Klopp, G., Dallieris, D., Miltner, E., Mattern, R., Kuppa, S., Sharpless, C. (1994) Thoracic trauma assessment formulations for restrained drivers in simulated frontal impacts. Paper number 942206, Society of Automotive Engineers, Warrendale, Pennsylvania.

Mulligan, G.W.N., Pizey, G., Lane, D., Andersson, L., English, C., Kohut, C. (1994) An introduction to the understanding of blunt chest trauma. In Biomechanics of Impact Injury and Injury Tolerances of the Thorax-Shoulder Complex, ed. Backaitis, S., pp. 11-36, Society of Automotive Engineers publication PT-45.

Nahum, A, Gadd, C., Schneider, D., Kroell, D. (1970) Deflections of the human thorax under sternal impact. Paper number 700400, Society of Automotive Engineers, Warrendale, Pennsylvania.

Newman J. and Jones I. (1984) A prospective study of 413 consecutive car occupants with chest injuries. Journal of Trauma, 24: 129-135.

Patrick, L.M. and Anderson, A. (1974) Three-point harness accident and laboratory data comparison. Paper 741181, Proc. 18th Stapp Car Crash Conference, Society of Automotive Engineers, Warrendale, Pennsylvania.

Pattimore, D., Thomas, P., Dave, S. (1992) Torso injury patterns and mechanisms in car crashes: an additional diagnostic tool. Injury: the British Journal of Accident Surgery 23(2): 123-126.

Pilkey, W. (1994) Formulas for stress, strain, and structural matrices. John Wiley & Sons, Inc. New York.

Plank, G. R., Kleinberger, M., and Eppinger, R H. (1994) Finite element modeling and analysis of thorax/restraint system interaction. Proc. 14th ESV Conference.

Prasad, P. (1999) Biomechanical basis for injury criteria used in crashworthiness regulations. Proc. International Research Conference on the Biomechanics of Impact (IRCOBI), Sitges, Spain.

Schmidt, G., Kallieris, D., Barz, J., and Mattern, R. (1974) Results of 49 cadaver tests simulating frontal collision of front seat passengers. Paper 741182 Proc. 18th Stapp Car Crash conference Society of Automotive Engineers, Warrendale, Pennsylvania.

Schneider, L., King, A., Beebe, M. (1989) Design requirements and specifications: thorax-abdomen development task, interim report, trauma assessment device development program. Report DOT HS 807 511, U.S. Department of Transportation, Washington, D.C.

Shaw, C., Wang, C., Bolton, J., Bass, C., Crandall, J., Butcher, J., Khaewpong, N., Sun, E., Nguyen, T. (1999) Chestband performance assessment using quasistatic tests. Injury Biomechanics Research: Proceedings of the 27th International Workshop, San Diego, California.

Shaw, C., Kent, R., Sieveka, E., Crandall, J. (2001) Spinal kinematics of restrained occupants in frontal impacts. accepted for publication at the 2001 Conference of the International Research Council on the Biomechanics of Impact (IRCOBI), Isle of Man.

Sutyak, J., Passi, V., and Hammond, J. (1997) Airbags alone compared with the combination of mechanical restraints and airbags: implications for the emergency evaluation of crash victims. Southern Medical Journal 90(9): 915-919.

Viano, D.C. (1978) Thoracic Injury Potential. Proc. 3rd International Meeting on Simulation and Reconstruction of Impacts in Collisions, pp. 142-156, IRCOBI. Bron, France.

Viano, D.C. and Lau, I.V. (1988) A viscous tolerance criterion for soft tissue injury assessment. Journal of Biomechanics 21(5): 387-399.

Walfisch, G., Chamouard, F., Lestrelin, D., et al. (1985) Predictive functions for thoracic injuries to belt wearers in frontal collisions and their conversion into protection criteria. Paper 851722, Proc. 29th Stapp Car Crash Conference, Society of Automotive Engineers, Warrendale, Pennsylvania.

Yoganandan, N., Skrade, D., Pintar, F., Reinartz, J., Sances, A. (1994) Thoracic deformation contours in a frontal impact. In Biomechanics of Impact Injury and Injury Tolerances of the Thorax-Shoulder Complex, ed. Backaitis, S., pp. 765-781. Society of Automotive Engineers publication PT-45.

Yoganandan N., Pintar, F. (1998) Biomechanics of human thoracic ribs. Journal of Biomechanical Engineering. 120(1):100-104.

Laboratory Reconstructions of Real World Frontal Crash Configurations using the Hybrid III and THOR Dummies and PMHS

Audrey Petitjean, Matthieu Lebarbe, and Pascal Potier
Centre Européen d'Etudes de Sécurité et d'Analyse des Risques, Nanterre, France

Xavier Trosseille
LAB PSA-RENAULT, Nanterre, France

Jean-Pierre Lassau
Service du don du corps, Institut d'Anatomie de l'UFR Biomédicale des Saints Pères,
Université René Descartes, ParisV, France

ABSTRACT – Load-limiting belt restraints have been present in French cars since 1995. An accident study showed the greater effectiveness in thorax injury prevention using a 4 kN load limiter belt with an airbag than using a 6 kN load limiter belt without airbag.

The criteria for thoracic tolerance used in regulatory testing is the sternal deflection for all restraint types, belt and/or airbag restraint. This criterion does not assess the effectiveness of the restraint 4 kN load limiter belt with airbag observed in accidentology.

To improve the understanding of thoracic tolerance, frontal sled crashes were performed using the Hybrid III and THOR dummies and PMHS. The sled configuration and the deceleration law correspond to those observed in the accident study. Restraint conditions evaluated are the 6 kN load-limiting belt and the 4 kN load-limiting belt with an airbag. Loads between the occupant and the sled environment were recorded. Various measurements (including thoracic deflections and head, thorax and pelvis accelerations and angular velocities on the dummies) characterize the dummy and PMHS behavior. PMHS anthropometry and injuries were noted.

This study presents the test methodology and the results used to evaluate dummy ability to discriminate both restraint types and dummy measurement ability to be representative of thoracic injury risk for all restraint types.

The injury results of the PMHS tests showed the same tendency as the accident study. Some of the criteria proposed in the literature did not show a better protection of the 4 kN load limiter belt with airbag restraint, in particular thoracic deflection maxima for both dummies. The four thoracic deflections measured on the THOR and Hybrid III dummies may allow more accurate analysis of the loading pattern and therefore of injury risk.

KEYWORDS – thorax tolerance, injury criteria, frontal crash, Hybrid III, THOR, PMHS, sled test.

INTRODUCTION

Studies of real-world car crashes show that a large proportion of severe injuries occur in frontal impacts. In particular, many thoracic injuries are sustained by elderly occupants.

Several proposals have been made for the evaluation of thoracic trauma using the measurement obtained from dummies. The first indicators of thoracic trauma were chest deflection or a linear combination of chest compression and age (Kroell et al. (1971), Neathery et al. (1975)) for a steering wheel-like loading. For shoulder belt loading, chest deflection remained an indicator of thoracic trauma but another chest compression/risk relationship was determined by Mertz et al. (1991). The emergence of the airbag restraint system generated other studies which assessed the current thoracic injury criteria, particularly in the case of a combined shoulder belt and air bag restraint system. For example, Horsch et al. (1991) and Mertz et al. (1997) found that injury thresholds differed according to the loading type. Morgan et al. (1994) discussed a solution for the determination of the restraint type before the application of the relevant thoracic criterion. Thoracic risk was also characterized by a criterion with a single limit applicable to all restraint types like the CTI proposed by Kuppa et al. (1998). In the same way, Kent et al. (2001) showed that one of the best predictors of thoracic risk for all restraint types (belt and/or air bag) was the maximum central thorax compression. Lastly, chest acceleration (Mertz et al. (1971)) and VC (Lau et al. (1986)) are still used as regulatory criteria.

The aim of this study was to assess the appropriateness of the various thoracic criteria issued from dummy measurements. Based on accident data, Foret-Bruno et al. (2001) showed the reduced probability of thoracic injury with a 4 kN load limiting belt and air bag compared with a 6 kN load limiting belt without air bag. This observation is used as the basis for evaluating the appropriateness of dummy thoracic criteria. The risks associated to both restraint types must be differentiated by the criteria, as they are on the road. In order to achieve this, sled tests were performed with the two restraint systems and a crash pulse representative of the real-world crashes studied. These tests were performed with PMHS (Post Mortem Human Subject) and Hybrid III and THOR 50th dummies. The appropriateness of the laboratory reconstruction was evaluated by comparing PMHS injuries to those observed in accident studies. Once the configuration was validated, the dummy criteria were evaluated by comparing their values with the probabilities of injury seen in the accident data.

In order to better understand the biomechanical phenomena associated with the two restraint types, chest deflections were measured at several locations on both dummies.

A large number of the dummy and environment responses (including thoracic deflections and head, thorax and pelvis accelerations and angular velocities on the dummies) were measured in order to validate a numerical model. The model was used for further analysis which was in progress at the time of writing.

METHODS

Sled test

In order to define a test configuration which would be representative of a real-world car crash, the deceleration pulse, the type of restraint, the belt anchorage coordinates and the environment geometry and mechanical properties were carefully reproduced for the PMHS and dummy tests.

Geometry: The global geometry adopted was taken from a current vehicle involved in accidents from the above study. All the elements were positioned according to the vehicle's geometry. The seat position and the belt anchorage were appropriate for a 50th percentile adult male. The coordinates of the most important points are indicated in Table A1 of the appendix. A side view of the sled is shown in Figure 1.

Steering wheel- Since the study investigated injuries caused by the restraint system only, the PMHS tests with the 6 kN load limiting belt were performed without the steering wheel to avoid any thoracic contact. Indeed, a preliminary cadaver test at 6 kN belt load limitation showed a large thoracic interaction with the steering wheel whereas Hybrid III or THOR tests only demonstrate a slight thoracic contact. Thus, the steering wheel was kept for the dummy test with 6 kN load limiting belt.

For PMHS, a head bolster was fitted forward of the steering wheel to avoid hard contact between his head and the vehicle environment in case the PMHS moved forward to this point.

Figure 1: General view of the sled.

Knee bolsters- The configuration included knee bolsters because accident studies showed knee contacts with the dashboard. Their position and angle were determined by a section of the dashboard of the reference vehicle. The bolsters used were 0.1m thick Expanded Polypropylene (EPP) blocks with a 0.2m square contact face and a density ranging from 45 g/L to 49 g/L.

Seat- The seat subframe was removed in order to allow load cells to be fitted under the seat to measure loads imposed by the dummy on the seat.

PMHS and Dummy- The feet of the different surrogates were fixed to the inclined foot rest to prevent them from moving during the test. This was achieved by holding the shoes onto the foot rest with metal bars. Thus, the rotation about the ankle during the crash is simplified, ensuring better repeatability. This, of course, was not representative of a real-world car environment but it was assumed that it did not influence the thorax tolerance.

PMHS specific equipment

Revascularization- The PMHS were revascularized to keep a vascular pressure close to physiological conditions (50 to 130 mmHg). An injection device was then developed and fitted on the sled to ensure the continuous injection of fluid during the crash. The injection system was designed to avoid an air/liquid mix and therefore to ensure correct injection pressure during the crash. The injection began 30 seconds prior to the crash and finished after the crash test. The liquid used was composed of four parts alcohol and one part Indian ink. This mix permanently marks damaged tissues. The aortic pressure was verified just before and during the test.

Re-inflation- The PMHS respiratory system was also re-inflated with 2.5 liters of air. The tracheal probe was closed using a clamp after the inflation and before definitive positioning of the subject.

Interfaces- Sensors were fixed on the head, vertebrae, sternum and at the sacrum using metal plates.

Protection- In order to protect the sled area from biological hazard, transparent Perspex panels were used to enclose the driver's compartment. The Perspex side panels were removable to allow access.

Sled test pulse: A EURONCAP pulse representing a 64 km/h impact against a deformable barrier with 40% of overlap was used with an inverse catapult (Figure 2). The use of this type of catapult allowed the initial surrogate position to be better maintained prior to impact. The pulse was highly reproducible.

Sled procedure: Each vehicle component that could have been deformed during the test (seat, steering wheel, belt, knee bolster, pyrotechnic pretensioner) was systematically replaced after each test.

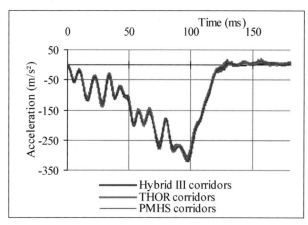

Figure 2: sled deceleration pulse corridors.

Restraint systems

A 4 kN load-limiting belt with airbag and a 6 kN load-limiting belt without airbag were evaluated. Time zero was the beginning of the pulse. The time to fire (TTF) of the pyrotechnic pretensioner was 15 ms. The airbag TTF was 20 ms. The airbag's volume was 65 liters.

Henceforth, in this article, the term "6 kN restraint" will be used to denote the 6 kN load-limiting belt restraint and the term "4 kN+AB restraint " will denote the 4 kN load-limiting belt with airbag restraint.

Test subjects

Table 1 shows the number of tests performed with each surrogate in each configuration.

PMHS were unembalmed. The last three were frozen a few weeks before being instrumented. They were at ambient temperature at the time of the test. Their sex, age, height and weight are noted in Table 2.

Table 1: Dummies and PMHS.

	Hybrid III	THOR	PMHS
4 kN + air bag	5	2	2 (C05, C22)
6 kN	4	2	2 (C17, C23)

Table 2: PMHS anthropometry.

	Sex	Age (years)	Mass (kg)	Height (cm)	Chest height (cm)	Thigh length (cm)
C05	F	78	70	169	87	50
C17	M	76	67	172	92	45
C22	M	81	60	174	86	48
C23	M	75	70	169	69	43.5

The PMHS were obtained through the Body Donations to Science office of the Anatomy Laboratory at Saints Pères University of Medicine in Paris V after approval of the experimental procedure by the ethical committee of the university.

The PMHS were tested for HCV, HBV, HIV, CMV, HTLV and a medical survey was documented. Cadavers suspected of bone fragility (subject confined to bed for a long period, subjects with bone cancer, metastasis, etc.) were excluded.

After the test, their mineralization rate was determined, and a three-point bending test was carried out on a section of the right and left fifth ribs.

Positioning procedure

Electromagnets were used to maintain the PMHS upper body in a realistic driver position. These electromagnets were fixed near the upper Perspex panel. Polar plates, held by the electromagnets, were linked by cables to the top of the head and to each sternal end of the clavicle (using a ring around the clavicle) (Figure 3). However, PMHS-C05 and C17 did not have their clavicles maintained and therefore were positioned more forward. The power supplied to the electromagnets was adjusted to allow the disconnection of the polar plate by the acceleration without perturbing the kinematics of the PMHS.

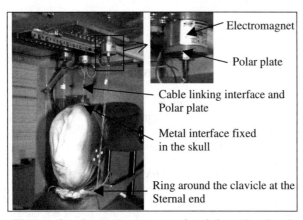

Figure 3: electromagnets maintaining the head and the thorax of the PMHS.

The initial positions of all surrogates were recorded using a 3D articulated arm recorder. The positions of selected metal dummy components or PMHS bone tuberosities were recorded before the test (Figure 4). As far as possible, the same surrogate position was maintained for each test. Figure 5, Figure 6 and Figure 7 show the surrogate pre-test positions in the XZ plane and in the laboratory coordinate system. These positions correspond with the side view shown in Figure 4.

Figure 5 shows that the repeatability of dummy positioning was good. Head CG, acromion and spine accelerometers on Figure 6 and Figure 7 shows that PMHS C05 and C17 are more forward compared to PMHS C22 and C23 and Hybrid III averaged position. Due to variation in anthropometry and in the number of electro-magnet used, the different PMHS were not positioned with the same repeatability as the dummies. The differences in the pre-test positioning may have had some effect on the test results.

Instrumentation

Environment: Vehicle environment measurements are indicated in Table A2 in the appendix.

Belt- Two load cells were placed on the lower and upper shoulder belt; a third one was positioned close to the anchorage of the lap belt on the outboard side of the sled. The displacement of the belt was determined by analysis of the high speed films.

Steering wheel- The steering wheel was mounted on a two-axis load cell allowing the loads on the steering wheel to be measured. An accelerometer was placed on the same part of the buck structure as the load cell in order to allow inertia compensation of the load cell measurements.

Knee bolster- Loads behind each knee bolster were measured by two-axis load cells. These loads were inertially compensated.

Seat- Loads under the seat were measured by four three-axis cells. These loads were inertially compensated. The pretensioner was isolated from the seat in order to measure only loads applied to the seat by the surrogate. The deformation of the anti-submarining ramp was also measured with a string potentiometer.

Two initial tests were performed without the dummy to check the validity of the vehicle environment measurements. Additional details are presented in the *Data Processing* section.

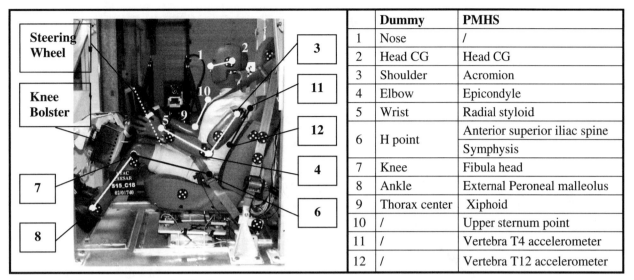

Figure 4: Main points recorded before the test.

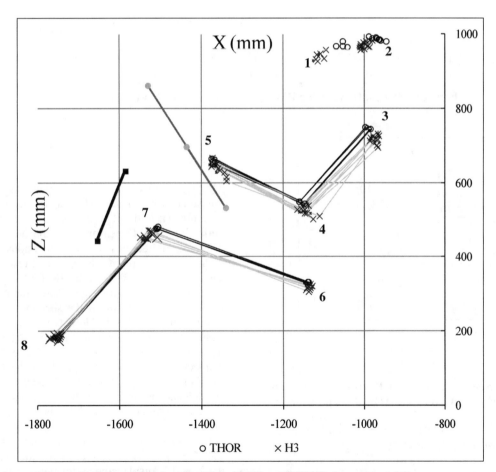

Figure 5: Schematic view for Hybrid III and THOR dummy position.

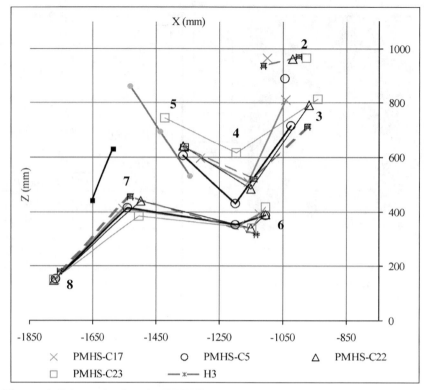

Figure 6: Schematic view for PMHS and averaged Hybrid III position.

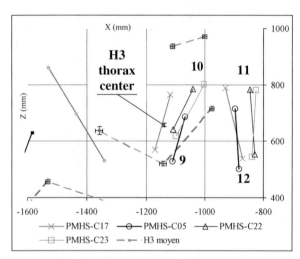

Figure 7 : Schematic view for PMHS sternum and spine position and averaged Hybrid III position.

Dummies and PMHS: The instrumentation is listed in Table A2. Specific details of dummy sensor positions are provided below (Figure 8).

Head- Tri-axial accelerations and angular velocities were measured at the center of gravity for the dummy and at the top of the head for the PMHS. The position of the PMHS head accelerometer relative to the Frankfort plane was recorded, allowing the calculation of the acceleration at the head center of gravity.

Thoracic deflections- The sternal deflection was measured for all the Hybrid III tests using the standard Hybrid III rod pot. The central sternal deflection is not measured on the THOR dummy. Thoracic deflections at four locations were also measured for some of the Hybrid III tests (repartition shown in Table 3) by four IR-TRACC (Infra-Red Telescoping Rod for the Assessment of Chest Compression) sensors (at the second and fifth ribs). The principle of the IR-TRACC is that the irradiance from an emitter infrared LED is detected by an infrared phototransistor in a telescopic tube connecting the spine to a point on the thorax. The irradiance is related to the distance between the emitter and the detector. These sensors measure the chest compression in the XY plane in the local co-ordinate system associated to the spine. A potentiometer measures the rotation of each IR-TRACC in the XY plane allowing the calculation of the X and Y displacements for the four points on the thorax. The Hybrid III was equipped with new ribs and a new lumbar spine during these tests.

Thoracic deflections were measured on the THOR by four CRUX units (at the third and fifth ribs). The

CRUX is a two bar linkage system that measures the displacement in three dimensions using three potentiometers to measure the rotations of the articulations of the linkage.

Table 3: instrumentation of the nine Hybrid III tests.

Hybrid III equipment	Thorax instrumentation	Restraint type	N° of test
Original ribs + original lumbar spine	Rod pot	4kN+AB	3
		6kN	2
new ribs + new lumbar spine	Rod pot, IR-TRACC	4kN+AB	2*
		6kN	2

* only one of the two IR-TRACC measurements is available

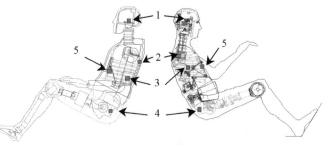

1: Head accelerometer
2: Upper spine accelerometer
3: Lower spine accelerometer
4: Pelvis accelerometer
5: Sternum accelerometer

Figure 8: Comparison of Hybrid III and THOR sensor locations.

Spine- Two accelerometers were mounted on the thoracic spine of each dummy. The locations of these sensors, which differ for the two dummies due to differences in their structures, are shown in Figure 8.

Pressure- PMHS internal pressures were measured in the right and left pleura, in the aorta, in the stomach and in the trachea.

Measurements

Coordinate systems: All the vehicle environment results were presented in a laboratory coordinate system (+X-axis rearward, +Z-axis upwards).

All the dummy results were presented in local coordinate systems in accordance with the SAE J211 convention regarding the signs of the measurements (+X-axis forward, +Z-axis downwards). This standard was also applied for PMHS results. Thus, the PMHS and dummy had identical axis orientations in their local coordinate systems.

Data processing: The steering wheel and the knee bolster loads applied by the surrogate were calculated by subtracting the inertia from measured loads, using the steering wheel accelerometer fixed on a interface common to the steering wheel and knee bolster sensors.

The seat loads were calculated by summing the loads of the four seat load cells and subtracting the inertia in the X-axis using the sled acceleration.

The CRUX measurements were processed using the software provided by GESAC, Inc (supplier of the THOR). The IR-TRACC measurements were processed using instructions from FTSS (the IR-TRACC supplier).

The displacement of the belt was measured using high-speed digital video (1000 frames/second) and Falcon software. The camera was situated 10 meters from the sled.

Filtering: All the filters used are documented in Table A2 in the appendix.

Scaling: As indicated in Table 2, the PMHS were not 50th percentile concerning their mass. In order to compare their responses with those of the dummies (Hybrid III and THOR: 78 kg), they were normalized using the procedures of Eppinger et al. (1984) except for the upper shoulder belt load (the maximum load remaining the same, whatever the surrogate mass, due to the load limiter). The scaling factor uses the mass ratio:

$$\lambda = \frac{78}{M}^{\frac{1}{3}}$$, where M is the PMHS mass.

The original responses were scaled as indicated in Table 4.

Table 4: Scaling factor.

Scaled measurement		Factor	Original measurement
Time	=	λ	Time
Length	=	λ	Length
Force	=	λ^2	Force
Velocity	=	1	Velocity
Acceleration	=	λ^1	Acceleration

RESULTS

Kinematic behavior

Figure 9 shows the times of the principal events observed during the test. Figure 10 shows photographs representing those principal events for the Hybrid III. A symbol is associated to each event represented on the photographs. These symbols correspond with those used in Figure 9. The chronology of the events was determined using films from a camera mounted on the right-hand side of the sled. The knee to knee bolster contact was determined using the knee bolster load curves. The air bag TTF was 20 ms but the deployment is only visible a few milliseconds later on the film. The event times presented for the Hybrid III are averaged for all the Hybrid III tests for each restraint type, as for the THOR tests.

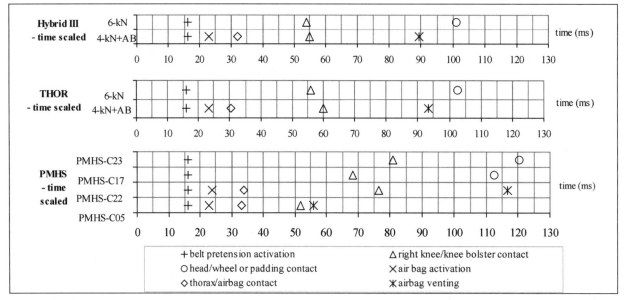

Figure 9: Film events

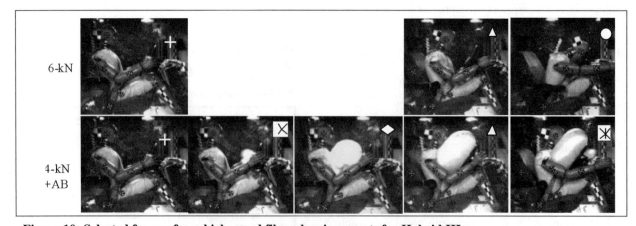

Figure 10: Selected frames from high speed films showing events for Hybrid III.

The pretension system was activated at 15 ms with both restraint types. In the case of the 4 kN+AB restraint, the airbag was activated at 20 ms. The airbag first made contact with the thorax between 30 and 35 ms, and then with the head for all surrogates. The dummy knee contacted the bolsters between 50 and 60 ms for both restraint types. This contact was slightly later with the 4 kN+AB restraint than with the 6 kN. PMHS knee contact with the bolsters occurred between 52 and 81 ms. This dispersion is partly due to anthropometry dispersion and initial positioning differences among PMHS. For the 6 kN restraint, as indicated in Figure 6, the symphysis and fibula head initial positions show that PMHS-C17 was positioned more forward than PMHS-C23 (40 mm). For the 4 kN+AB restraint, as indicated in Figure 6, the same points show that PMHS-C05 was positioned more forward than PMHS-C22 (45 mm). That is partly why the PMHS-C05 knee contact occurred before that of PMHS-C22, and the PMHS-C17 one before that of PMHS-C23. Moreover, PMHS-C23 had shorter thighs than the other PMHS. This may have slightly delayed the knee contact with the bolster.

For the 4 kN+AB restraint, airbag venting occurred between 89 and 94 ms for dummy. It occurred between 56 ms and 117 ms for PMHS. This dispersion is partly due to differences in the initial positioning and in anthropometry. As indicated in Figure 7, the T4 and T12 vertebrae position show that the thorax of PMHS-C05 was positioned further than that of PMHS-C22 (50 mm). PMHS-C05 highly loaded the airbag early in the crash which involved early airbag venting.

For the 6 kN restraint, the dummy head contacted the wheel between 101 and 103 ms. The THOR's head remained in contact with the steering wheel until 144 ms and the Hybrid III's head until 139 ms. For the 6 kN restraint, the PMHS head contacted the head bolster between 113 and 120 ms. PMHS head contact occurred later than dummy head contact partly because the head bolster was further from the PMHS than the steering wheel was from the dummy. In the case of the Hybrid III test with the 6 kN restraint, the thorax sometimes contacted the steering wheel lower rim.

Response of the surrogates

For the Hybrid III (H3) and THOR, all the following curves are response corridors based on all the response curves of test with a similar restraint type for each dummy. For example, for the Hybrid III with 6 kN restraint tests, external lap belt load corridor was built with the four external lap belt load measurements. Superior (resp. inferior) curve of the corridor corresponded with the maximum (resp. minimum) values of the four measurement curves at each time. The results shown in the tables in the appendix are maxima of both inferior and superior curves that delimit each corridor, with the time of occurrence. Therefore, there are two values given for each measurement. For the PMHS, no response corridors were constructed. Response curves are given for each test. Therefore, there is only one value given for each measurement.

Table A3 in the appendix presents the vehicle environment response data. Table A4 and Table A5 in the appendix present the surrogate response data.

Hybrid III dummy

Lower body: According to Figure 11, Figure 12 and Figure 13 there were no significant differences in Hybrid III lower body behavior for the two restraint types. Left knee bolster and seat loads are not presented but did not show any significant differences.

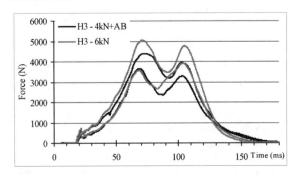

Figure 11: External lap belt load corridors for Hybrid III.

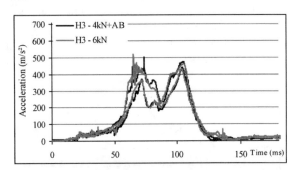

Figure 12: Sacrum resultant acceleration corridors for Hybrid III.

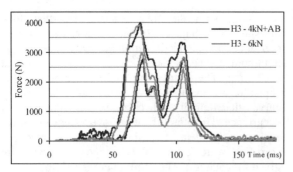

Figure 13: Right knee bolster resultant load corridors for Hybrid III.

Upper body: The different load limit levels of the two restraint systems are clearly reflected in the external shoulder belt load corridors (Figure 14).

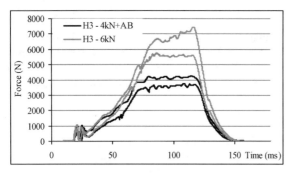

Figure 14: External shoulder belt load corridors for Hybrid III (not scaled).

Two different shapes of steering wheel resultant load corridor are shown in Figure 15. The 4 kN+AB restraint corridor exhibits a peak at 50 ms which is due to the loading of the airbag by the dummy's head and thorax (the first contact occurred around 31 ms)- and carries on until 145 ms. The 6 kN restraint corridor is close to zero until 100 ms when it rises because of the impact of the head on the steering wheel. This loading of the head lasted until 130 ms.

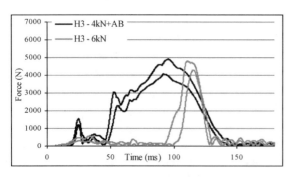

Figure 15: Steering wheel resultant load corridors for Hybrid III.

The shapes of both upper spine acceleration corridors are almost identical up to around 100 ms (Figure 16). The 4 kN+AB restraint corridor is slightly higher between 50 and 100 ms. This may be due to loading of the airbag on the dummy's upper body. The 6 kN restraint corridor rises at 100 ms, the time when the dummy's head hit the steering wheel.

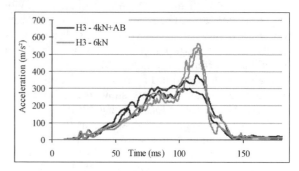

Figure 16: Upper spine resultant acceleration corridors for Hybrid III.

THOR dummy

Lower body: As with the Hybrid III, the THOR lower body behavior did not show any major differences between the two restraint types, as shown in Figure 17, Figure 18, and Figure 19.

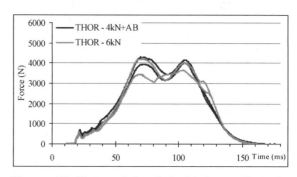

Figure 17: External lap belt load corridors for THOR.

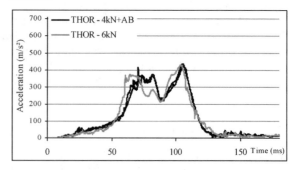

Figure 18: Sacrum resultant acceleration corridors for THOR.

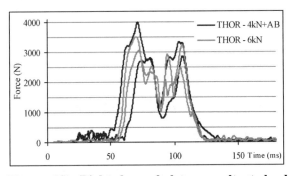

Figure 19: Right knee bolster resultant load corridors for THOR.

On Figure 18, only one sacrum measurement was retrieved from the 3 tests with the 6 kN restraint. Therefore, one curve is shown instead of a corridor. Left knee bolster and seat loads are not presented but did not show any significant differences.

<u>Upper body</u>: the observations for Hybrid III are also valid for THOR upper body corridors: there is a distinct difference in the external shoulder belt loads measured with the two restraint systems, which results from the different load limiters (Figure 20).

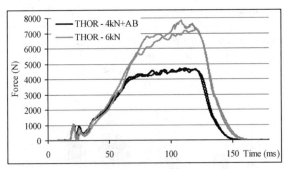

Figure 20: External shoulder belt load corridors for THOR (not scaled).

Figure 21 shows two different shapes of the steering wheel resultant load corridors. The 4 kN+AB restraint corridor rises at 50 ms - when the head and thorax loaded the airbag (the first contact occurred around 32 ms). This loading lasted until 150 ms. The 6 kN restraint corridor remains close to zero until 100 ms at which point the head impacted the steering wheel. This loading of the head lasted until 130 ms.

Both upper spine acceleration corridors are fairly similar in shape (Figure 22) although the 4 kN+AB restraint corridor is slightly higher between 50 and 100 ms. This may be due to loading of the airbag on the dummy upper body. With the 6 kN restraint the onset of loading occurred at 100 ms, the time when the dummy's head hit the steering wheel.

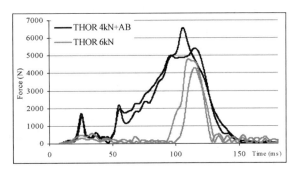

Figure 21: Steering wheel resultant load corridors for THOR.

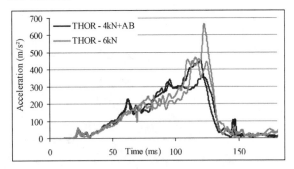

Figure 22: Upper spine resultant acceleration corridors for THOR.

PMHS: The dispersion in PMHS positioning and morphological characteristics generated mainly response corridors that are relatively wide. Thus, PMHS results were presented individually in order to better appreciate the behavior of each subject.

The dummy responses were of the same order of magnitude as those of the PMHS (Figure 23, Figure 24, Figure 25, Figure 26, Figure 27 and Figure 28). Therefore, this overall consistency concerning the measurements of all surrogates confirms there is no discrepancy of great importance between the crash conditions for dummy and PMHS tests.

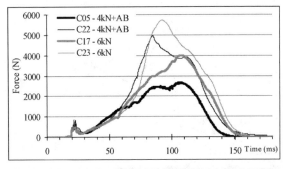

Figure 23: External lap belt load corridors for PMHS.

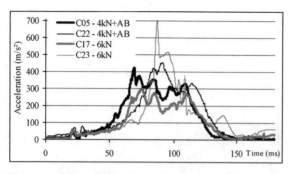

Figure 24: Sacrum resultant acceleration corridors for PMHS.

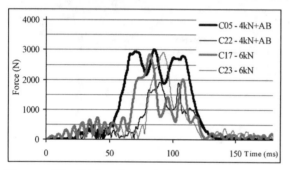

Figure 25: Right knee bolster resultant load corridors for PMHS.

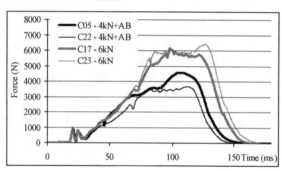

Figure 26: External shoulder belt load corridors for PMHS (not scaled).

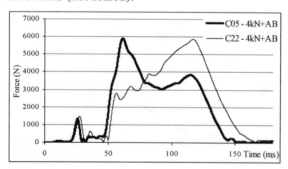

Figure 27: Steering wheel resultant load corridors for PMHS.

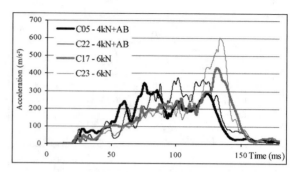

Figure 28: Upper spine resultant acceleration corridors for PMHS.

PMHS injury assessment: The detailed account of fractures shown in Table A6 in the appendix was obtained from the autopsy of the subjects. The injuries observed were coded using the AIS (which also takes into account soft tissue injuries). Standard characterization of the bones was performed, i. e. mineralization and a three-point bending test on the ribs (quasi-static test). C/L values and maximum bending load are shown below in Figure 29 and also in Table A6 in the appendix.

The mineralization values were consistent with common values (0.23 g/cm averaged with 30 PMHS by Fayon et al (1975)). Concerning the three-point bending test results, PMHS-C05 presented a low value compared to common values (194 N averaged with 30 PMHS by Fayon et al (1975)).

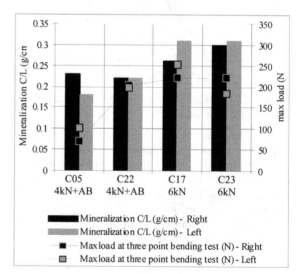

Figure 29: PMHS data - mineralization and maximum load at three-point bending test.

Dummy thoracic deflections

The initial belt path on the thorax was identical for all the Hybrid III tests, as shown in Figure 30, and for all THOR test, as shown in Figure 31. Pictures of initial belt path on the surrogates are shown in Figure A1 in the appendix.

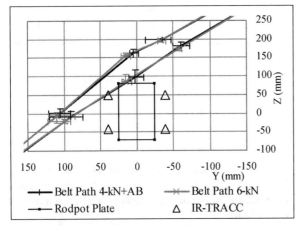

Figure 30: Belt path, IR-TRACC and rod pot plate position in the sled coordinate system with rod pot initial position as origin.

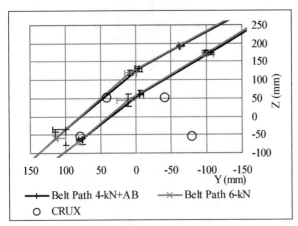

Figure 31: Belt path, CRUX position in the sled coordinate system with CRUX center position as origin.

The dummy thoracic deflection responses are presented for each test (no corridors were constructed).

Hybrid III: During the first part of testing, the Hybrid III dummy was equipped with ribs which had already been used. During the testing with the IR-TRACC, the Hybrid III dummy was equipped with new ribs. The results must be analyzed separately for both rib sets.

As shown in Figure 32, for both rib sets, the sternal deflections were higher with the 4 kN+AB restraint than with the 6 kN restraint although the deflections were higher with the original ribs (27/28 mm compared to 21/24 mm) than with those used with the IR-TRACC (21/24 compared to 18/19 mm). The new ribs may have been stiffer or the IR-TRACC may have influenced the rib mechanical behavior. With the 6 kN restraint, the peak in deflection that occurred at 107 ms may be due to slight contact of the steering wheel with the thorax. However, the maximum rod pot deflection with the 6 kN restraint remained lower than that of the 4 kN+AB restraint.

The rate of deflection was higher after 50 ms for the 4 kN+AB restraint which corresponded to the time when the airbag was loaded by the thorax.

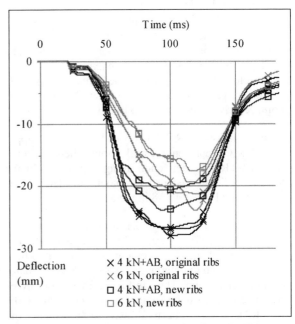

Figure 32: Rod pot deflections for Hybrid III.

The Hybrid III dummy was instrumented with IR-TRACC sensors for two tests with the 4 kN+AB restraint (the data was not available for one of the two) and for two tests with the 6kN restraint. The displacements were measured in the X and Y direction in the local coordinate system associated with the dummy spine.

Figure 33 shows the local coordinate system used to present the IR-TRACC deflection results with the IR-TRACC's in their initial position. The initial distance between the right and left IR-TRACC's was the same for the upper and lower IR-TRACC's. In Figure 34 and Figure 35, the lines that link the left and right IR-TRACC do not represent the real sternum shape but simply connect the pairs of IR-TRACC's.

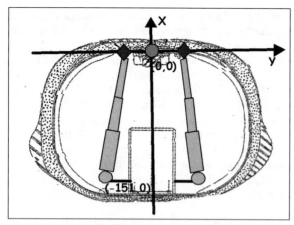

Figure 33: Schematic view in XY plane of Hybrid III thorax instrumented with IR-TRACC in initial position.

As indicated in Table A7 in the appendix, the IR-TRACC and the rod pot measurements did not reach their maximum values at the same moment. The lower right IR-TRACC X-axis maximum deflection is the closest to the rod pot one. Thus, IR-TRACC results are presented at the time when the lower right point X-axis deflection is maximum.

Figure 34 (lower IR-TRACC) and Figure 35 (upper IR-TRACC) represent the deflections when the lower right IR-TRACC deflection along the X-axis in the local coordinate system was maximum. The values are presented in Table A8 in the appendix.

The measurement of the upper left potentiometer was not available. Therefore, only the upper left IR-TRACC resultant displacement was known which was not enough to determine the X and Y coordinates of the sternal point. Thus, in Figure 35, arc of circles with the center at the spine attachment point of the IR-TRACC have been drawn to represent the possible positions of the upper left sternal point in the local coordinate system.

At the moment when the lower right deflection was at its maximum:
- Right side X-axis deflection was higher than on the left for both lower and upper points. The right/left difference was less significant with the 4 kN+AB restraint because the left deflections increased while the right deflections remained nearly the same as with the 6 kN restraint. The right/left lower deflection difference along the X-axis was 15 mm with the 6 kN restraint and 8 mm with the 4 kN+AB restraint.
- The upper right resultant deflection was the highest and the lower left was the smallest for both restraint systems.

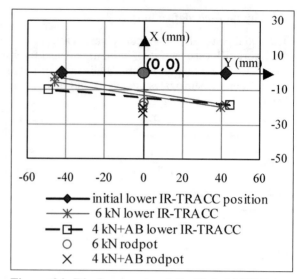

Figure 34: Displacements of the Hybrid III lower IR-TRACC in the X and Y direction in the local coordinate system at lower right maximum deflection.

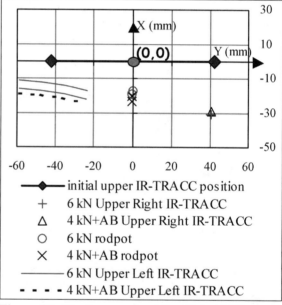

Figure 35: Displacements of the Hybrid III upper IR-TRACC in the X and Y direction in the local coordinate system at lower right maximum deflection.

THOR: The displacements in the X, Y and Z directions were measured in the local coordinate system associated with the dummy spine. As shown in Figure 31, the pre-test distance between the right and left lower CRUX points of measurement on the thorax is greater than for the upper CRUX units. In Figure 36 and Figure 37, the lines that links the left and right CRUX do not represent the real sternum shape but only connect the right pairs of CRUX units.

There was no central sternal deflection measurement such as the rod pot in the Hybrid III. In order to allow comparison with the Hybrid III IR-TRACC deflections, the CRUX deflections are also presented at the instant of maximum lower right CRUX deflection along the X-axis in the local coordinate system. The time of maximum lower right CRUX deflection was determined for each THOR test. The results are presented in Figure 36 and Figure 37.

It is interesting to note that the lower left CRUX indicates an outward deflection whereas the other points were deflecting inwards. This tendency was more significant for the 6 kN restraint (20 mm) that for the 4 kN+AB restraint (10 mm).

The greatest X and Y-axis resulting deflection was at the upper right point for both restraint types, as with the Hybrid III dummy. The upper right deflection amplitude remained the same for the 6 kN restraint and the 4 kN+AB restraint.

Given that the right side deflections remained nearly the same for the 6 kN and 4 kN+AB restraints and that the left side deflections were greater with the 4 kN+AB restraint, the left/right differences decreased with the 4 kN+AB restraint compared with the 6 kN restraint. For the upper deflections, the right/left difference was decreased from 35 (6 kN) to 23 mm (4 kN+AB) along the X-axis and from 9 (6 kN) to 2 mm (4 kN+AB) along the Y-axis. For the lower deflections, the right/left difference was decreased from 47 (6 kN) to 37 mm (4 kN+AB) along the X-axis and from 18 (6 kN) to 4 mm (4 kN+AB) along the Y-axis.

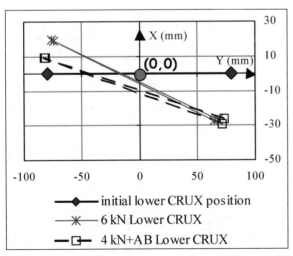

Figure 36: Displacements of the THOR lower CRUX in the X and Y direction in the local coordinate system at lower right maximum deflection.

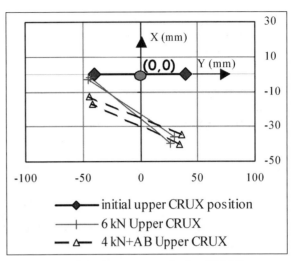

Figure 37: Displacements of the THOR upper CRUX in the X and Y direction in the local coordinate system at lower right maximum deflection.

For indication, the IR-TRACC and CRUX time history curves are shown in the appendix (Figure A2 to Figure A9).

Thoracic tolerance criteria

Thoracic Deflection: The analyses showed the rod pot deflections were always under the regulatory test limit of 50 mm. For both rib sets, the rod pot deflections were 18% lower for the 6 kN belt restraint than for the 4 kN+AB restraint.

Viscous Criterion (VC): The VC was calculated for the Hybrid III dummy only based on the rod pot deflection as indicated by Lau et al. (1986). The deflection velocity did not exceed 3 m/s. As recommended by Lau et al. (1986), the best thoracic criterion for this range of loading velocity is thoracic compression. The VC values are given for indication in Table 5.

Table 5: Hybrid III VC.

	4 kN+AB	6 kN
VC (m/s) - *original ribs*	0.12	0.05
	0.13	0.08
VC (m/s) - *new ribs*	0.08	0.05
	0.09	0.07

Belt or air bag-like loading and associated criteria: Thoracic criteria could be different for belt and air bag restraint. For example, Horsch et al. (1991) and Mertz et al. (1997) found that different chest deflection limits were required according to the loading type.

A solution for evaluating the thoracic risk would be first to determine the restraint type and then to apply the appropriate criteria.

Morgan et al. (1994) proposed to discriminate between restraint types by calculating a combination of thoracic compressions at five locations on the thorax (Morgan et al. (1994), equation 3). The criterion is calculated for the tests with Hybrid III instrumented with IR-TRACC (Table 6). The deflections used by Morgan et al. to calculate the criterion were PMHS external deflections whereas the deflections measured during these tests and used to calculate the criterion were Hybrid III internal deflections. It is not calculated for the THOR because it was not instrumented with a central deflection sensor.

Table 6: Analytical sorter to separate belt and air bag restraint tests for each Hybrid III tests.

$2.7 * X_{lower\_sym} + 2.5 * X_{upper\_sym} - 7.8 * X_{upper\_norm}$	
Hybrid III 4 kN+AB	Hybrid III 6 kN
9.3*	7.3
	8.2
Air bag like restraint	Belt like restraint

* only one of the two IR-TRACC measurements is available.

The tests with values greater than 9 are classified as air bag like and values less than 9 are classified as belt like restraints. All Hybrid III test analytical sorters are near the limit value. The 6 kN restraint tests were classified as belt like tests and the 4 kN + AB test was classified as an air bag like test.

One criterion is applied for air bag like restraint tests, combining chest compression, chest acceleration and age. Another is applied for belt like restraint tests combining chest compression and chest velocity (Table 7) (Morgan et al. (1994), equation 4 and 5). The critical value to separate AIS<3 and AIS>3 injury is –6 in the case of belt like restraint and 6.5 in the case of air bag like restraint.

Table 7: Predictor for thoracic injury according to air bag or belt like restraint.

Air bag like restraint criteria	Belt like restraint Criteria
AIS3 risk limit = 6.5	AIS3 risk limit = -6
4.1*	-9.6
	-9.7

* only one of the two IR-TRACC measurements is available.

Criteria for both restraints predict AIS<3. They are inferior to the PMHS lower values calculated by Morgan et al. (1994). The criteria were calculated from PMHS tests and thus may not be relevant for predicting injury from Hybrid III results using the same criteria values.

Two loading types are discriminated, which would allow to apply relevant criteria. However, it is not possible to compare the risk predicted for the different restraint types for Hybrid III using criteria determined by Morgan. Indeed Morgan et al. do not give injury risk curves, only a limit for AIS-3 injury. Consequently, the appropriateness of the criteria proposed by Morgan et al. (1994) can not be assessed.

However, risk curves for AIS3+ thoracic injury proposed by Mertz et al. (1997) were determined for shoulder belt and distributed loading for Hybrid III mid-male chest pot measurements. They can be used to estimate the risk for belt like and bag like restraint once the loading type is discriminated. The AIS3+ risk predicted using Mertz et al. risk curves is less than 0.1 % for the bag like restraint (Table A4 in the appendix, rodpot deflection under 35 mm) and from 3 to 5% for the belt like restraint (Table A5 in the appendix, rodpot deflection from 18 to 24 mm). These results show the same tendency as accident data, that is the thoracic risk associated with a 4 kN+AB restraint is lower than the risk associated with a 6 kN restraint.

3 ms clip value of resultant upper spine acceleration (3 ms spine acceleration): The 3 ms spine acceleration was higher for the 6 kN restraint for all surrogates, as shown in Table 8 and Table 9.

Table 8 : PMHS 3 ms spine acceleration m/s² (g).

	4 kN+AB		6 kN	
PMHS	C05	C22	C17	C23
	322 (32.8)	389 (39.7)	437 (44.5)	594 (60.6)

Table 9: Dummy 3 ms spine acceleration m/s² (g).

	Hybrid III		THOR	
Dummy	4 kN+AB	6 kN	4 kN+AB	6 kN
	300 (30.6)	516 (52.6)	350 (35.7)	454 (46.3)
	357 (36.4)	534 (54.4)	435 (44.3)	607 (61.9)

For the Hybrid III and PMHS, the 60 G limit was not exceeded whereas it was exceeded for the THOR.

Combined Thoracic Injury (CTI) Criterion: The CTI criterion was calculated for the last Hybrid III test series and for THOR tests as indicated by Kuppa et al. (1998):

$$CTI = 6.1832 + 0.0755As + 13.3052d \max$$

where :
- *"As"* is the 3 ms clip value of resultant upper spine acceleration,
- *"d max"* is the maximum compression for the different deflections on the thorax.

The deflections used by Kuppa et al. to calculate the CTI criterion were external whereas the deflections measured during these tests and used to calculate the CTI criterion were internal. However, the comparison of the CTI values between both restraint types is possible.

In the case of the Hybrid III, *"d max"* was the maximum of the five deflections measured (one measured by the rod pot and four measured by the IR-TRACC's). In the case of the THOR, *"d max"* was the maximum compression for the four CRUX deflections, given that the THOR was not instrumented with a central deflection sensor.

As shown in Table 10, the CTI criterion shows the restraint system 4 kN+AB generated a lower CTI value and so a lower probability of AIS>3 for both dummies.

Table 10 : CTI values.

Hybrid III		THOR	
4 kN+AB	6 kN	4 kN+AB	6 kN
-1.7	-0.3	-0.3	0.1
	-0.4	-0.8	1.2

The maximum of the upper spine and head resultant accelerations occurred at the same moment for Hybrid III and THOR with the 6 kN restraint. This phenomenon was also observed for PMHS-C17. Indeed, with the 6 kN restraint, the dummy head impact on the steering wheel and the PMHS head impact on the head bolster influenced the upper spine acceleration. The 3 ms clip value of resultant upper spine acceleration, and therefore the CTI criterion, was increased by this contact. The THOR had a higher CTI value than the Hybrid III.

DISCUSSION

Validity of the real world crash reconstruction

PMHS tests were performed in order to compare injuries sustained to those observed in real-world accidents.

Accident research allows one to calculate the risk of AIS3+ using the results of a logistic regression. J-Y. Foret-Bruno et al. (2001) (equation 1, Table 7) showed that, at 64 km/h (Equivalent Energy Speed 58 km/h), the probability of sustaining an AIS3+ thorax injury for an 80-year-old driver is:
16% with a 4 kN+AB restraint,
55% with a 6 kN restraint.

The four PMHS injury assessments of the present study were performed during autopsies. In the case of accident analysis, the injuries are counted using X-rays. According to Kent et al. (2002), some of the rib fractures sustained may not be detected with this technique. This may result in a higher number of fractures being recorded for the sled tests than for equivalent accident cases. Kent et al (2002) showed that numerous influencing parameters, including restraint type, caused underestimation of the number of fractures identified. However, there is no correction factor specific to the restraint system used during the sled tests of this study. Given the different techniques for rib fracture counting and given the small number of PMHS sled tests, there is not sufficient data to give any significant statistical results.

However, the tendency obtained with the PMHS AIS values was that the 4 kN+AB restraint offered better protection in the considered configuration. More, mineralization of the ribs showed the 4 kN+AB restrained PMHS had weaker bones than the 6 kN restrained PMHS. Three-point bending test results showed that PMHS-C05 had weaker bones whereas the other PMHS had equivalent bone strength. Further evidence of the greater effectiveness of the 4 kN+AB restraint is provided by these mineralization results and partially by the three-point bending test results.

Consequently, the tendency obtained with the PMHS AIS values is considered as representative of studied accident cases and validates the laboratory reconstruction. Therefore, a relevant dummy thoracic criteria should confirm the accident data, that is the 4 kN+AB restraint offered a better thoracic protection.

Discrimination of restraint type with dummy, study of the different thoracic criteria.

Deflections: According to Kent et al. (2001), the maximum central thorax compression was a good predictor of thoracic risk for all restraint types. However, the air bags used in Europe are smaller and less powerful than those used in the above study. These differences may result in different load distributions on the dummy upper body and may influence the appropriateness of this predictor. In the present study, the rod pot deflection was slightly higher with the 4 kN+AB restraint than with the 6 kN restraint. However, with the 4 kN+AB restraint, according to the accident data, fewer injuries were sustained. Foret-Bruno et al. (2001) showed the thoracic risk was reduced from 75 to 85% with a 4kN shoulder belt restraint compared to a 6 kN shoulder belt restraint. Consequently, the rod pot deflection which does not reflect such a large difference on the road can not be pertinent. It would at least be necessary to identify restraint types before applying it.

Both IR-TRACC and CRUX systems showed that the maximum deflection of the thorax was not located at the center of the sternum but at the upper right point. This is quite logical given that the belt passed near the upper right point. The deflection may be highest at the sternum when loads are applied directly in the area of the sternum or when they are more distributed as with airbag only restraint. Thus, the representativeness of the rod pot measurement as the maximum thorax deflection depends on the loading pattern, particularly the dissymmetry. Therefore, using the same central deflection limit may not be pertinent in determining the tolerance for all loading patterns and therefore all restraint types. The appropriateness of the X-axis maximum thoracic deflection for the whole thorax as a criterion is considered. The maximum deflection is located at the upper right point and was the same value for the 6 kN and 4 kN+AB restraints, for both Hybrid III and THOR. Consequently, the maximum thoracic deflection for the whole thorax does not reflect differences observed on the road, and thus should not be used as a criterion for injury risk.

The lower left point of measurement on the thorax tended to deflect relatively little (Hybrid III) or to move outward (THOR). Shaw et al. (2001) found the same results using the chest-band. The deflection was lower or the outward movement was higher with the 6 kN restraint. The 4 kN+AB restraint system loading pattern produced a reduction in the displacement differences between the four measurement points on the thorax, particularly left/right differences, along X

and Y-axis of the local coordinate system for THOR and along the X-axis of the local coordinate system for Hybrid III dummy. Lower left/right deflection differences and fewer injuries were linked. This could be used for the elaboration of a new criteria.

Belt or air bag-like loading and associated criteria: Using the procedure proposed by Morgan et al. (1994), the analytical sorter allows the separation between 4 kN + AB and 6 kN restraints as air bag and belt like restraints. The different criteria predict AIS<3. It is not possible to evaluate whether injury risk predicted for air bag like restraint is lower than the one predicted for belt like restraint using Morgan et al. criteria because they do not give injury risk curves. This method is determined using PMHS results. It would be interesting to determine an analytical sorter and criteria issued from dummy data Moreover, the Morgan et al. criteria associated with air bag like restraint take into account the spine acceleration. The latter is easily influenced, for example, in case of contact between the head and a hard vehicle component.

The risk predicted for air bag like and belt like restraint was estimated using the risk curves proposed by Mertz et al. (1997) once performed the separation between air bag like and belt like restraint according to Morgan et al. (1994) procedure. The Hybrid III rod pot measurements gave estimates of AIS3+ risk of less than 0.1% for 4 kN+AB and between 3 and 5% for 6 kN restraint. Consequently, the tendency obtained is considered to be representative of studied accident cases, even if the injury risk difference is small. The assessment of air bag like or belt like loading cannot be made from the central rod pot measurement of the Hybrid III. The additional measurements of several deflection points such as provided by the IR-TRACC are needed.

3 ms spine acceleration: The 3ms spine acceleration was higher for the 6 kN restraint. However, the head contact influenced the spine accelerations. Only the THOR values with the 6 kN restraint exceeded 60 G.

CTI: The CTI criterion was higher for the 6 kN restraint and higher for the THOR for each restraint system. It is important to notice that the CTI criterion utilizes the 3 ms clip value of resultant upper spine acceleration which was influenced by the head contact with the steering wheel.

The two latter criteria are in accordance with Foret-Bruno et al. (2001) accident studies. However, as they are highly influenced by head contact, it is not sure whether they distinguish between the type of thoracic restraint or between head impact and no head impact. As a consequence, they may not be appropriate to evaluate thoracic risk independently from head impact.

Comparison of THOR and Hybrid III responses

Tight control of the initial positioning helped with the comparison between dummies. This resulted in repeatable dummy responses.

Several measurements showed greater loading (exterior shoulder belt, steering wheel/ air bag) of the THOR thorax compared to that of the Hybrid III.

The Hybrid III and THOR dummies showed similar X-axis thorax deflection tendencies. The discrimination between restraint types was more marked with the THOR than with the Hybrid III concerning the thorax deflections. This may be due to the differences in thorax design or the loading of the thorax.

CONCLUSION

Existing accident data shows that the 4 kN load limiting belt + air bag restraint considered in the above study is far more effective for protecting the thorax in frontal crashes than the 6 kN system also considered. The PMHS injury results from the tests of the present study showed the same tendency. Consequently, the laboratory reconstruction was considered as valid. Thus, a relevant dummy criteria should demonstrate that thoracic protection was superior with the 4 kN+AB restraint in the considered configuration, as observed in accident data.

The sternal deflection criterion, indicated that the 4 kN load limiting belt + air bag restraint was less efficient than the 6 kN load limiting belt. This criterion was not appropriate to assess the effectiveness of the restraints used in this study without determining the thoracic loading type.

The procedure proposed by Morgan et al. (1994) allows the discrimination between restraint systems (4 kN+AB and 6 kN) and, thus, the application of the different relevant thoracic criteria. The assessment of the criteria proposed by Morgan et al. (1994) would require the development of injury risk curves. This would allow the comparison of the risk levels calculated for air bag like and belt like restraint. However, the discrimination between restraint systems (belt like or bag like) allows the estimation of injury risk using risk curves based on sternal deflection proposed by Mertz et al. (1997). The injury risk difference observed with this method shows the same tendency than the accident study observation, that is the thoracic protection is superior

with a 4 kN+AB restraint than with a 6 kN, even if it does not reflect the large difference observed with accident data.

The 3 ms upper spine acceleration and the CTI criteria agreed with the accidentology data but were highly influenced by head contact with the steering wheel. It is not possible to determine the appropriateness of these two criteria without head contact. Further tests would be necessary.

The thorax deflections showed asymmetric loading. Thorax deflections were higher for the THOR than for the Hybrid III. The greatest deflections were observed along the path of the belt (right-hand side measurement points of the thorax). This asymmetry is reduced for both Hybrid III and THOR with the 4 kN load limiting belt + air bag, the right side deflections remaining nearly the same as with the 6 kN load limiting and the left side deflections increasing. The reduction of the asymmetric dummy deflections corresponded with fewer injuries. The dummies identified different loading pattern on the thorax using four point deflection measurement systems. Concerning this point, THOR is more sensitive than Hybrid III. However, more data would be necessary to determine whether a new criterion and injury risk curves could be established based on such a system.

FUTURE WORK

The sled tests will be numerically modeled using the sled geometry and the mechanical properties of the vehicle environment elements. The Hybrid III sled test measurements will be used to validate the whole vehicle environment response. A human body model will be placed in this environment and tested numerically with the same crash pulse. This will allow us to study measurements not available during the sled tests and thus to better understand the thoracic tolerance. Moreover, the use of numerical models will allow each particular PMHS anthropometry and pre-test position to be reproduced. Therefore, the differences due to anthropometry and pre-test positioning will be taken into account allowing better understanding of the effects of the different restraints on PMHS biomechanical responses. Furthermore, existing thoracic criteria will be tested in other configurations close to that of tests of the present study. For example, without a steering wheel, one can evaluate the influence of the head impact on criteria using the 3 ms upper spine acceleration. Other numerical tests could be performed with an airbag restraint only to study thoracic deflections at several locations with a different loading pattern.

REFERENCES

Eppinger, R.H., Marcus, J.H., Morgan, R.M. (1984) Development of Dummy and Injury Index for NHTSA's Thoracic Side Impact Protection Research Program. SAE Technical Paper 840885. Society of Automotive Engineers, Warrendale, PA.

Fayon, A., Tarriere, C., Walfisch, G., GOT, C., Patel, A., (1975) Thorax of 3 Points Belt Wearers During Crash. Proc. 19th Stapp Car Crash Conference. Society of Automotive Engineers, Inc.

Foret-Bruno, J-Y., Trosseille, X., Page, Y., Huère, J-F., Le Coz, J-Y., Bendjellal, F., Diboine A., Phalempin, T., Villeforceix, D., Baudrit, P., Guillemot, H., Coltat, J-C., (2001) Comparison of Thoracic Injury Risk in Frontal Car Crashes for Occupant Restrained without Belt Load Limiters and Those Restrained with 6 kN and 4 kN Belt Load Limiters. Stapp Car Crash Journal 44:205-224.

Horsch, J., Melvin, J.W., Viano, D.C., Mertz, H.J. (1991) Thoracic Injury Assessment of Belt Restraint Systems Based on Hybrid III Chest Compression. Proc. 35th Stapp Car Crash Conference, pp85-108. Society of Automotive Engineers, Warrendale, PA.

Kallieris, D., Rizzetti, A., Mattern, R., Morgan, R., Eppinger, .R, Keenan, L. (1995) On the Synergism of the Driver Air Bag and the 3-Point Belt in Frontal Collisions. Proc. 39th Stapp Car Crash Conference, pp389-401. Society of Automotive Engineers, Warrendale, PA.

Kallieris, D., Otte, D., Mattern, R., Wiedmann, P. (1995) Comparison of Sled Tests with Real Traffic Accidents. Proc. 39th Stapp Car Crash Conference, pp51-58. Society of Automotive Engineers, Warrendale, PA.

Kent, R., Bolton, J., Crandall, J., Prasad, P., Nusholtz, G., Mertz, H., Kallieris, D. (2001) Restrained Hybrid III Dummy-Based Criteria For Thoracic Hard-Tissue Injury Prediction, Proc. 2001 International IRCOBI Conference on the Biomechanics of Impact, pp215-232. IRCOBI, Bron, France.

Kent, R.W., Crandall J., Patrie J., Fertile J. (2002) Radiographic Detection of Rib Fractures : a Restraint-Based Study of Occupants in Car Crashes. Traffic Injury Prevention, pp3:49-57.

Kroell, C.K., Schneider, D.C, Nahum, A.M. (1971) Impact Tolerance and Response of the Human Thorax. Proc. 15th Stapp Car Crash Conference, pp84-134. Society of Automotive Engineers, Warrendale, PA.

Kuppa, S., Eppinger, R.H. (1998) Development of an Improved Thoracic Injury Criterion. Proc. 42th Stapp Car Crash Conference, pp139-154. Society of Automotive Engineers, Warrendale, PA.

Lau, I., Viano, D.C. (1986) The Viscous Criterion-Bases and Applications of an Injury Severity Index for Soft Tissue, Proc. 30th Stapp Car Crash Conference, p123-142. Society of Automotive Engineers, Warrendale, PA.

Mertz. H., Gadd, C.W. (1971) Thoracic Tolerance to Whole-Body Decelration. Proc. 15th Stapp Car Crash Conference, pp135-157. Society of Automotive Engineers, Warrendale, PA.

Mertz. H., Horsch, J., Horn, G., Lowne, R. (1991) Hybrid III Sternal Deflection Associated with Thoracic Injury Severities of Occupants Restrained with Force-Limiting Shoulder Belts. SAE N°910812, Society of Automotive Engineers International Congress and Exposition, Warrendale, PA.

Mertz. H., Prasad, P., Irwin, A.L. (1997) Injury Risk Curves for Children and Adults in Frontal and Rear Collisions. Proc. 41st Stapp Car Crash Conference, pp13-30. Society of Automotive Engineers, Warrendale, PA.

Morgan, R.M., Eppinger, R.H., Haffner, M.P., Yoganadan, N., Pintar, F.A., Sances Jr, A., Crandall, J.R., Pilkey, W.D., Klopp, G.S., Kallieris, D., Miltner, E., Mattern, R., Kuppa, S.M., Sharpless, C.L. (1994) Thoracic Trauma Assessment Formulations for Restrained Drivers in Simulated Frontal Impacts. Proc. 38th Stapp Car Crash Conference, pp15-34. Society of Automotive Engineers, Warrendale, PA.

Neathery, R.F., Kroell, C.K., Mertz, H.J. (1975) Prediction of Thoracic Injury from Dummy Responses. Proc. 19th Stapp Car Crash Conference, pp295-316. Society of Automotive Engineers, Warrendale, PA.

Shaw, G., Crandall, J., Butcher, J. (2001) Biofidelity Evaluation of the THOR Advanced Frontal Crash Test Dummy. Proc. 2000 International IRCOBI Conference on the Biomechanics of Impact, pp11-29. IRCOBI, Bron, FRANCE.

Yoganadan, N., Skrade, D., Pintar, F.A., Reinartz, J., Sances, A. (1991) Thoracic Deformation Contours in a Frontal Impact. Proc 35th Stapp Car Crash Conference, pp. 47-63. Society of Automotive Engineers, Warrendale, PA.

ACKNOWLEDGMENTS

The authors would like to acknowledge the contribution of those who, in donating their bodies to science, made this study possible. We also acknowledge UTAC (Technical Union for the Automobile, Motor Cycle and Cycle) for its participation in this project. We also wish to acknowledge FTSS for having provided the IR-TRACC devices. Finally, we acknowledge Martin Page for his English assistance in the writing of this article.

APPENDIX

Table A1: Coordinates of the Important Geometric Points in the Sled Local Coordinate System.

Coordinates	X (mm)	Y (mm)	Z (mm)
Dummy Hx	-1117	750	310
External shoulder belt anchorage	-763	519	987
External lap belt anchorage	-927	450	97
Pretensioner anchorage	-1038	1009	172
Center of the steering wheel	-1436	762	696
Right knee bolsters (the 4 corners of the impact surface)	-1653	960	444
	-1653	760	444
	-1585	960	632
	-1585	760	632
Left knee bolsters (the 4 corners of the impact surface)	-1653	540	444
	-1653	740	444
	-1585	540	632
	-1585	740	632

Angle	Degree °
Angle of the steering wheel to the vertical	29.7°
Angle between the feet support and the horizontal	55°
Angle of the knee bolsters to the horizontal	70°

Figure A1: Initial belt path on THOR, Hybrid III and PMHS C22

Table A2: Vehicle environment and Dummy Measurements

Vehicle environment measurements			Filtering
Sled	✓	Acceleration X,Z	CFC60
Seat	✓ ✓ ✓	Back: Acceleration X,Z Seating : Load X,Z and Moment Y, and Acceleration X,Z Anti submarining sheet metal depression	CFC60 CFC60 CFC1000
Belt	✓ ✓ ✓ ✓	Upper Shoulder Load Lower Shoulder Load External Lap Load Unrolling (Tracking)	CFC1000 CFC1000 CFC1000 /
Steering wheel	✓ ✓	Load X,Z Acceleration X,Z	CFC60 CFC60
Knee bolsters	✓	Load X,Z (Right and Left)	CFC180

Dummy and PMHS measurements				Filtering
Head		✓ ✓	Acceleration X,Y,Z Angular Velocity X,Y,Z	CFC180 CFC60
Neck	Hybrid III	✓	Superior Neck: Load X,Z, Moment Y	CFC600
	THOR	✓ ✓	Superior Neck : Load X,Z, Moment Y Head angular displacement Y	CFC600 CFC600
Thorax	Hybrid III	✓ ✓ ✓ ✓ ✓	Thoracic deflection (X), thoracic deflection (resultant, angle) at four points Sternum : Acceleration X Spine Acceleration X,Y,Z (2 points) Spine angular Velocity Y Extensometers on ribs	CFC180 CFC1000 CFC180 CFC60 /
	THOR	✓ ✓ ✓ ✓	Thoracic deflection (X,Y,Z) at 4 points Sternum : Acceleration X Spine Acceleration X,Y,Z (2 points) Spine angular Velocity Y	CFC1000 CFC1000 CFC180 CFC60
	PMHS	✓ ✓	T4 and T12: Acceleration X,Y,Z and Angular Velocity Y internal pressures and extensometers on ribs	CFC180 CFC60 /
Pelvis	Hybrid III	✓ ✓ ✓	Inferior Lumbar: Load X,Z, Moment Y Sacrum Angular Velocity Y Sacrum : Acceleration X,Z	CFC600 CFC100 CFC180 CFC1000
	THOR	✓ ✓ ✓	Superior Lumbar: Load X, Z, Moment Y Sacrum Angular Velocity Y Sacrum : Acceleration X,Z	CFC600 CFC1000 CFC180 CFC1000
	PMHS	✓	Sacrum : Acceleration X,Z	CFC1000
Leg	Hybrid III THOR	✓ ✓	Femur F Z (Right and Left) Tibia F Z (Right and Left)	CFC600 CFC600

Table A3: Vehicle Environment Maxima Measurements Corridors.

In this table are presented the maxima of the two curves (inferior and superior) which delimit the response corridor.

4 kN + AB	Hybrid III		THOR		PMHS C05		PMHS C22	
	Max	Time (ms)	Max	Time (ms)	Max	Time (ms)	Max	Time (ms)
Upper shoulder belt (N) (not scaled)	3740	107	4645	112	4571	105	3694	112
	4247	114	4717	112				
Lower shoulder belt (N)	3093	111	4125	112	3732	107	3976	123
	4070	110	4300	113				
External lap belt (N)	3650	67	3943	71	2679	106	4964	85
	4400	72	4308	70				
Belt unrolling (mm)	125	123	201	127	91	144	130	126
	152	118	223	125				
Seat resultant loads (N)	9938	105	12814	102	7592	81	9301	105
	12706	101	13190	101				
Anti submarining ramp depression (mm)	24.4	104	23.1	105	23.3	108	28.3	115
	30.5	106	27.4	103				
Right knee bolsters resultant load (N)	2755	73	2803	80	3015	85	1925	89
	3996	71	3264	72				
Left knee bolsters resultant load (N)	2945	72	2809	73	3787	85	1863	89
	4343	71	3558	71				
Steering wheel resultant load (N)	4076	91	5184	112	5896	61	5861	116
	4902	95	6576	106				
Steering wheel resultant load on thorax (N)	2999	95	4031	115	/	/	/	/
	3513	86	4787	106				

6 kN	Hybrid III		THOR		PMHS C17		PMHS C23	
	Max	Time (ms)	Max	Time (ms)	Max	Time (ms)	Max	Time (ms)
Upper shoulder belt (N) (not scaled)	5767	85	6977	109	6156	97	6420	126
	7450	116	7844	108				
Lower shoulder belt (N)	4959	108	3940	67	5655	104	6194	107
	6140	111	5765	105				
External lap belt (N)	3580	68	3451	70	3996	106	5742	92
	5050	70	4217	72				
Belt unrolling (mm)	143	115	189	116	141	120	265	127
	180	115	213	118				
Seat resultant loads (N)	10476	103	11830	93	9484	95	13465	101
	14094	103	12933	95				
Anti sub-marining ramp depression (mm)	26.4	104	25.5	105	27.9	104	36.4	103
	18.4	103	27.4	104				
Right knee bolsters resultant load (N)	2975	72	3047	72	2851	81	2916	92
	3905	70	3509	70				
Left knee bolsters resultant load (N)	2995	72	3294	71	3302	82	3853	93
	4348	69	3878	69				
Steering wheel resultant load (N)	4295	115	5193	117	/	/	/	/
	4804	111	6432	114				
Steering wheel resultant load on thorax (N)	1746	109	/	/	/	/	/	/
	2557	109	/	/				

Table A4: Dummies Maxima Measurement Corridors for 4 kN Load Limiting Belt with Air Bag.

In this table are presented the maxima of the two curves (inferior and superior) which delimit the response corridor.

4 kN + AB	Hybrid III		THOR		PMHS C05		PMHS C22	
	Maximum	Time (ms)	Maximum	Time (ms)	Maximum	Time (ms)	Maximum	Time (ms)
Head resultant acceleration (m/s^2)	304	89	425	103	535	75	NA	NA
	419	95	447	102				
Head resultant angular rate (rad/s)	7.4	69	20.9	115	31.2	73	NA	NA
	14.5	71	23.9	121				
Upper neck X-axis load (N)	-437	64	-120	85	/	/	/	/
	-564	60	-161	70				
Upper neck Z-axis load (N)	1073	79	1401	88	/	/	/	/
	1351	61	1452	88				
Upper neck Y-axis moment (N/m)	10.8	105	-13.9	99	/	/	/	/
	29.5	89	-14.1	99				
Upper spine resultant acceleration* (m/s^2)	301	103	356	120	344	75	324	84
	378	113	454	119				
Lower spine resultant acceleration* (m/s^2)	333	104	311	89	389	89	376	101
	377	105	334	92				
Sternum X-axis acceleration* (m/s^2)	-289	108	-265	74	NA	NA	-262	120
	-470	81	-288	74				
central X-axis deflection (rodpot) (mm) -original ribs	-27	93	/	/	/	/	/	/
	-28	100	/	/				
central X-axis deflection (rodpot) (mm) - new ribs	-21	90	/	/	/	/	/	/
	-24	93	/	/				
Upper Right XY resultant deflection** (mm)	29	105	42	111	/	/	/	/
			43	114				
Upper Left XY resultant deflection** (mm)	20	89	15	64	/	/	/	/
			17	91				
Lower Right XY resultant deflection** (mm)	19	101	27	73	/	/	/	/
			31	113				
Lower Left XY resultant deflection** (mm)	12	100	16	116	/	/	/	/
			21	129				
Lumbar X-axis load*** (N)	-1035	67	2931	58	/	/	/	/
	-1400	77	8126 (peak)	58				
Lumbar Z-axis load*** (N)	-2715	104	-1011	102	/	/	/	/
	-3316	103	-1090	95				
Lumbar Y-axis moment*** (N/m)	245	110	47	114	/	/	/	/
	318	111	57	119				
Sacrum resultant acceleration (m/s^2)	442	104	423	106	427	69	455	90
	474	104	437	105				
Femur Right Z-axis load (N)	-1315	74	-1461	81	/	/	/	/
	-2312	64	-1717	82				
Femur Left Z-axis load (N)	-1595	72	-1471	73	/	/	/	/
	-2786	71	-2013	72				

*the sensor is not located at the same place; **IR-TRACC sensor for Hybrid III, CRUX sensor for THOR; ***Inferior lumbar sensor for Hybrid III, superior for THOR.

Table A5: Dummies Maxima Measurement Corridors for 6 kN Load Limiting Belt.

In this table are presented the maxima of the two curves (inferior and superior) which delimit the response corridor.

6 kN	Hybrid III		THOR		PMHS C17		PMHS C23	
	Maximum	Time (ms)	Maximum	Time (ms)	Maximum	Time (ms)	Maximum	Time (ms)
Head resultant acceleration (m/s^2)	886	110	NA	NA	634	130	NA	NA
	1034	110	NA	NA				
Head resultant angular rate (rad/s)	26.6	111	34.3	119	34	85	NA	NA
	36.7	108	43.4	118				
Upper neck X-axis load (N)	-690	103	-496	136	/	/	/	/
	-885	106	-558	139				
Upper neck Z-axis load (N)	1995	110	1620	119	/	/	/	/
	2400	109	2560	120				
Upper neck Y-axis moment (N/m)	70.5	119	-2.3	122	/	/	/	/
	94.7	117	20	121				
Upper spine resultant acceleration* (m/s^2)	378	113	447	117	434	131	603	133
	565	114	668	122				
Lower spine resultant acceleration* (m/s^2)	367	112	330	115	348	95	494	133
	406	114	465	123				
Sternum X-axis acceleration* (m/s^2)	-320	108	-268	112	-316	64	-626	69
	-711	107	-632	109				
central X-axis deflection (rot pot) (mm) - *original ribs*	-21	121	/	/	/	/	/	/
	-24	119	/	/				
central X-axis deflection (rot pot) (mm)- *new ribs*	-18	115	/	/	/	/	/	/
	-19	113	/	/				
Upper Right XY resultant deflection** (mm)	31	115	43	116	/	/	/	/
	31	113	45	108				
Upper Left XY resultant deflection** (mm)	16	122	8	126	/	/	/	/
	18	120	21	120				
Lower Right XY resultant deflection** (mm)	19	124	29	83	/	/	/	/
	20	115	31	82				
Lower Left XY resultant deflection** (mm)	8	90	23	132	/	/	/	/
	11	108	27	87				
Lumbar X-axis load*** (N)	-1371	77	-1067	111	/	/	/	/
	-1661	77	-1239	124				
Lumbar Z-axis load*** (N)	-2474	98	1090	121	/	/	/	/
	-3291	103	1933	120				
Lumbar Y-axis moment*** (N/m)	347	111	99	119	/	/	/	/
	388	106	131	124				
Sacrum resultant acceleration (m/s^2)	432	103	306	102	300	74	706	86
	463	105	427	103				
Femur Right Z-axis load (N)	-1450	73	-1789	71	/	/	/	/
	-2420	65	-2040	69				
Femur Left Z-axis load (N)	-1580	72	-1907	71	/	/	/	/
	-2933	69	-2530	69				

*the sensor is not located at the same place; **IR-TRACC sensor for Hybrid III, CRUX sensor for THOR; ***Inferior lumbar sensor for Hybrid III, superior for THOR

Table A6: PMHS Injury.

Test Number	Mineralization (g/cm) and max load (N) of a three point bending test of the ribs	AIS	Fractures			Soft tissues injuries
			Number	Location		
C05 4 kN + AB	Mineralization Right: 0.23 Left: 0.18 Max load Right: 70 Left: 101	AIS 2	**Total: 6** Sternum: 0 Right: 6 Left:0			/
C22 4 kN + AB	Mineralization Right: 0.22 Left: 0.22 Max load Right: 204 Left: 198	AIS 4	**Total: 19** Sternum: 1 Right: 14 Left: 4			/
C17 6 kN	Mineralization Right: 0.26 Left: 0.31 Max load Right: 221 Left: 254	AIS 5	**Total: 25** Sternum: 4 Right: 13 Left: 8			Right lung crushed Abrasion of the splenic capsule along the posterior side
C23 6 kN	Mineralization Right: 0.30 Left: 0.31 Max load Right: 220 Left: 183	AIS 4	**Total: 18** Sternum: 2 Right: 9 Left: 7			/

| rib fracture

|| no separated fracture at the joint between the bone and the cartilage

Table A7: X, Y and Z Thorax Deflections for Hybrid III and THOR.

In this table are presented the maxima of the two curves (inferior and superior) which delimit the response corridors.

	4 kN+AB restraint				6 kN restraint			
	Hybrid III		THOR		Hybrid III		THOR	
	Maximum	Time	Maximum	Time	Maximum	Time	Maximum	Time
Upper Right X deflection (mm)	-29	105	-41	110	-31	115	-42	123
			-43	114	-31	113	-43	110
Upper Left X deflection (mm)	NA	NA	-14	64	NA	NA	-8	126
			-17	91	NA	NA	-19	125
Lower Right X deflection (mm)	-18	101	-26	72	-19	116	-27	86
			-30	89	-20	115	-27	82
Lower Left X deflection (mm)	-10	94	+16	116	-4	129	+24	137
			+20	129	-6	129	+23	133
Upper Right Y deflection (mm)	-2	106	-7	169	-4	97	-12	95
			-9	180	-5	103	-14	97
Upper Left Y deflection (mm)	NA	NA	+5	130	NA	NA	-3	85
			+7	128	NA	NA	-9	126
Lower Right Y deflection (mm)	+4	35	-10	123	-4	124	-14	120
			-10	128	-7	132	-17	126
Lower Left Y deflection (mm)	-8	112	+2	131	-8	88	+6	111
			+5	125	-10	108	+6	132
Upper Right Z deflection (mm)	/	/	+5	108	/	/	+4	106
			+5	97	/	/	+5	103
Upper Left Z deflection (mm)	/	/	+3	68	/	/	+1	108
			+4	70	/	/	+4	108
Lower Right Z deflection (mm)	/	/	+3	71	/	/	+2	157
			+4	142	/	/	+4	159
Lower Left Z deflection (mm)	/	/	-6	128	/	/	-8	133
			-6	131	/	/	-9	141
Rod pot deflection (mm)- *new ribs*	-21	90	/	/	-18	115	/	/
	-24	93	/	/	-19	113	/	/

Table A8: IR-TRACC and CRUX Displacements at the time when the lower right deflection is maximum along the X-axis in the local coordinate system.

	6 kN								4 kN + AB							
	Hybrid III				THOR				Hybrid III				THOR			
	X	Y	Res.	T.	X	Y	Res.	T.	X	Y	Res.	T.	X	Y	Res.	T.
Lower right (mm)	-19	-0	19	116	-27	-11	29	86	-18	+2	19	101	-26	-6	27	72
	-20	-3	20	115	-27	-15	31	82					-30	-8	31	89
Lower left (mm)	-3	-4	5	116	19	5	20	86	-10	-7	12	101	9	-2	10	72
	-6	-4	7	115	20	4	20	82					9	-3	10	89
Upper right (mm)	-31	-2	31	116	-40	-14	42	86	-29	-1	29	101	-35	-4	35	72
	-31	-2	31	115	-36	-11	37	82					-41	-6	41	89
Upper left (mm)	NA	NA	15	116	-2	-3	3	86	NA	NA	20	101	-13	-4	13	72
	NA	NA	18	115	-3	-5	6	82					-17	-2	17	89
Rodpot (mm)	-18	/	/	116	/	/	/	/	-21	/	/	101	/	/	/	/
	-19	/	/	115	/	/	/	/					/	/	/	/

Res.: Resultant (mm)

T.: Time (ms)

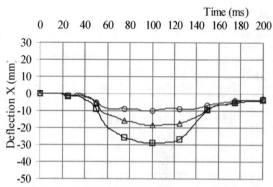
Figure A2: Hybrid III X axis deflection- 4 kN+AB restraint.

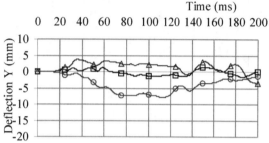
Figure A3: Hybrid III Y axis deflection- 4 kN+AB restraint.

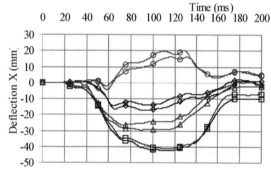
Figure A4: THOR X axis deflection- 4 kN+AB restraint.

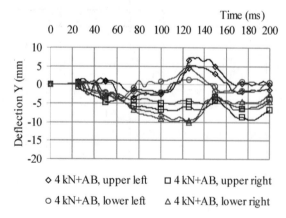
◇ 4 kN+AB, upper left   □ 4 kN+AB, upper right
○ 4 kN+AB, lower left   △ 4 kN+AB, lower right

Figure A5: THOR Y axis deflection- 4 kN+AB restraint.

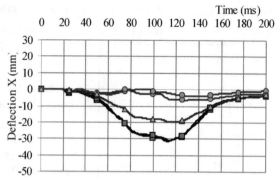
Figure A6: Hybrid III X axis deflection- 6 kN restraint.

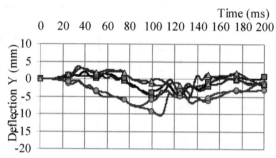
Figure A7: Hybrid III Y axis deflection- 6 kN restraint.

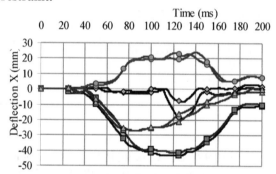
Figure A8: THOR X axis deflection- 6 kN restraint.

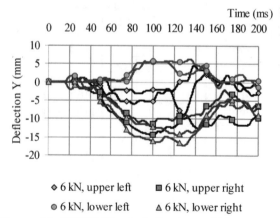
◇ 6 kN, upper left   ■ 6 kN, upper right
● 6 kN, lower left   △ 6 kN, lower right

Figure A9: THOR Y axis deflection- 6 kN restraint.

V. LABORATORY PERFORMANCE OF AIR BAGS – OCCUPANT INJURIES

POSSIBLE EFFECTS OF AIRBAG INFLATION ON A STANDING CHILD

BERTIL ALDMAN, M.D., Department of Traffic Safety, Chalmers University of Technology,

ÅKE ANDERSSON, M.D. and OLOV SAXMARK, Eng, AB Volvo, Göteborg, Sweden

IN AN AUTOMOTIVE COLLISION where the velocity of a car is rapidly changed, an empty space between the occupant and the interior of the vehicle will result in a difference in velocity between the occupant and the car in the direction of travel. The deceleration of the vehicle and the distance of unrestrained relative motion will determine the occupant's impact velocity against the vehicle interior at the end of travel, as shown in Figure 1. Since such impacts often result in occupant injury, restraint systems are used to control this relative motion. The basic principle for such systems is the interposition of load-carrying, deformable elements between the vehicle interior and the occupant.

In some restraint systems these elements are in contact with the occupant's body already during normal driving. This type of restraint system does to some extent control the occupant's posture and thereby also the spatial relation between the human body and the restraint system during the collision sequence.

One characteristic of the airbag restraint system is that the deformable element - the airbag itself - is not inflated during normal driving, but remains stored, usually in the instrument panel. This is an important characteristic of this system, because it leaves a seated occupant more freedom to move and change position than do most other systems. In order to obtain this advantage, however, the system must be activated at an early stage of a collision. Thus the airbag must inflate and be placed in position between the vehicle interior and the occupant very quickly, as shown in Figure 2. The spatial relation between the occupant and the restraint cannot be controlled in the same manner in airbag restraint systems as it can in most other systems.

The current requirements of US FMVSS 208 take into consideration only the normally seated "grown up" car passenger. It may seem quite possible that a grown up passenger only very seldom is out of position to such an extent that this would cause any serious problems. If, however, the occupant is not only out-of-position but also has other body dimensions and weight that differ too much from that of an adult occupant, these problems may become more intricate, Figure 3. A child standing by the dashboard in front of the seat during an accident would fit this description quite well and the standing child problem has been discussed at some length in the literature. Patrick and Nyquist 1972 found that, in the airbag system that they studied, there was a risk for head and internal organ injuries to an out-of-position child during static inflation.

Wu, Tang and Petrof in 1973 considered the impact of the folded portion of a deploying airbag to the chest of a standing child to be the main problem, Figure 4. Static airbag deployment tests, however, indicate that a child substitute in this position can be vigorously pushed away by a deploying airbag, .Figure 5. The potential danger of airbag inflation to a standing child, therefore, would probably be related to the inflation time sequence, Figure 6. Some variations of the inflation pulse are possible without hazardous increase of the inflation time. This has for instance been pointed out by Lundström et al. 1974.

SCOPE

The scope of this investigation was to study the influence of the three different airbag inflation pulses, shown in Figure 7, on the kinematics of and the possible mechanisms of injury to a living body in the standing child position.

The possibility that the child may be injured by being impacted by the unopened portion of the airbag has already been mentioned. This would occur at an early stage of inflation and could explain the high initial peak of acceleration recorded with child dummies.

An exact analysis of the forces exerted on the body of a standing child during the inflation period is difficult, since the shape of the airbag is quite irregular and easily influenced by contact with the child body at an early stage of deployment. However, if a child dummy is well centered in front of the airbag outlet area, it will be violently pushed away by the inflating airbag. It seems quite possible,

Fig. 1 - Unrestrained occupant

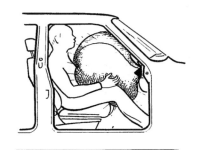

Fig. 2 - Occupant and fully inflated airbag

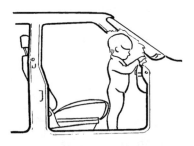

Fig. 3 - Unrestrained standing child

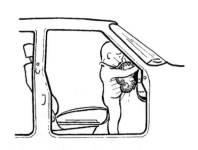

Fig. 4 - Initial stage of deploying airbag

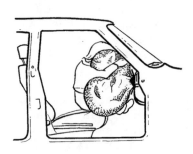

Fig. 5 - Late stage of deploying airbag

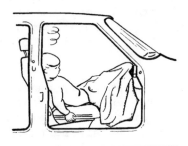

Fig. 6 - Deployed airbag

therefore, that the acceleration of the entire child body and the simultaneous deformation of that body during a large part of the inflation sequence, could be damaging to susceptible body structures of a child.

After having been accelerated by the deploying airbag, the child's body has acquired a certain speed relative to the vehicle and will eventually impact the interior of the car. This impact may well result in injuries if the impact velocity is high enough and the orientation of the body is unfavourable.

These three possible mechanisms of injury to a standing child, 1) the momentum of the unopened airbag, 2) the acceleration and deformation during inflation and 3) the impact of the body on interior car structures after deployment, are all related to the airbag inflation time sequence and should be observed, therefore, in experiments with different inflation pulses. At a distance of 10-15 cm (4-6 in.) from the airbag outlet area the deploying airbag has gained some momentum, but has still a small volume. This was therefore judged to be the most dangerous position for a standing child.

MATERIALS AND METHODS

Previous tests with airbag systems used in this investigation indicate that such systems give adequate protection to a normally seated 50th percentile male front seat passenger in simulated frontal car-barrier collisions at 48.2 km/h (30 mph).

The systems consist mainly of a package mounted in the usual glove compartment space in the dashboard. This package includes a high pressure bottle with the energy sources for the deployment of the airbag, and the folded airbag in a container. The folded airbag is kept in place in the container by a special cover, which opens in a predetermined way, when the airbag is pressurized.

The gas flow into the airbag is principally controlled by and distributed through a diffusor (Figure 8). The energy sources are compressed gas and a chemical gas generator. The latter begins to produce gas at the same moment as the compressed gas is let into the airbag through the diffusor. Two simultaneously ignited detonators open the passage from the bottle into the diffusor.

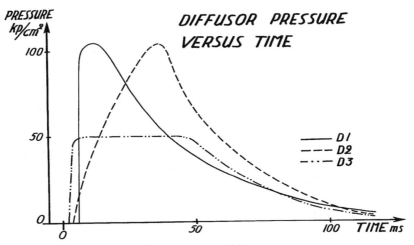

Fig. 7 - Diffusor pressure versus time

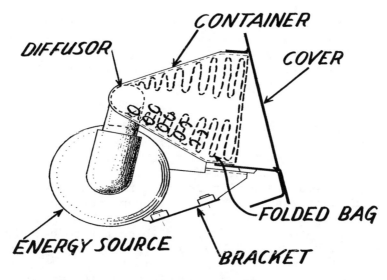

Fig. 8 - Airbag-passenger system assembly

The volume of a fully deployed airbag is 190 litres (6.7 ft^3).

In production vehicles these systems also have two sensors (acceleration detectors) which close the circuit through the detonators on and above a certain g-level, applied during a certain length of time. The tests in this paper were performed on a 12 in. Hyge sled and the sensors were replaced by a timing device.

Three different airbag inflation pulses, Figure 7, were used and their influence was studied on living bodies of approximately the same size and stage of biological development as children 3-6 years of age. Ordinary domestic pigs weighing 14-15 kilograms were used as test animals. After 24 hours of starvation the animals were anesthesized by intraperitoneal injections of 0.5-0.8 g thiopentan sodium which induced sleep and relaxation within 15-30 minutes without serious respiratory depression. The following test program was carried out.

Airbags were deployed with test animals at a distance of about 10-15 cm (4-6 in) from the center of the airbag outlet area in a stationary car and in a car body during rearward acceleration on a test sled. The dynamic tests were made using the lowest vehicle pulse (12 g peak, 17.5 mph or 28 km/h), Figure 9, which would trigger an airbag. Tests were made with and without actual airbag deployment.

In two of the eight static tests, the animals were suspended in a forward facing position with their heads and front legs resting on the instrument panel and the apex of the heart at a distance of 10 cm (4 in.) from the airbag outlet. This posture of the test animal deflected the airbag downwards in such a way that it was trapped in front of the seat cushion. In order to avoid this influence, it was necessary to suspend the other test animals in a straight vertical position. There is one difference in shape between the thoracic part of a pig and that of a child which had to be taken into account. The sagittal diameter cf the pig's thorax is slightly longer than the transverse diameter, while the opposite is the case in a child's thorax. Since this could possibly influence the tolerance, it was decided to suspend all the pigs with their right side towards and at a distance of 10-15 cm (4-6 in.) from the airbag outlet. Their vertical positions were adjusted so that the apex of the heart was at the level of the horizontal center line of the airbag outlet, Figure 11. The suspension included a loop of straps stuck to the animal by means of adhesive tape. The

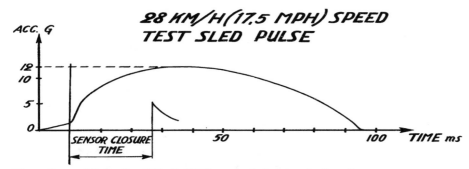

Fig. 9 - 28 km/h (17,5 MPH) speed test sled pulse

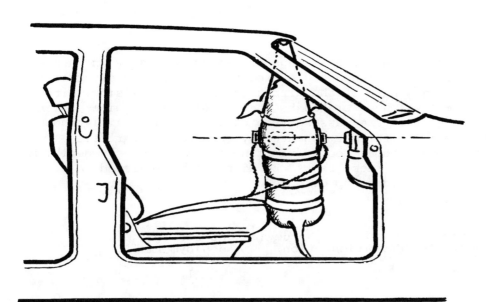

Fig. 11 - Pig in airbag test position

strap loop was fastened to a plastic bar which in turn was fixed by means of paper strings to a series of holes drilled in the roof of the car body. During the tests these strings were easily cut by the sharpedged roof sheet metal.

Each test was covered by three high-speed cameras and the films were analyzed. In order to compare the results of different tests, still pictures were made from every tenth millisecond of these films.

Animals which survived the test were sacrificed after 30 minutes. An autopsy was made on each animal, still pictures were taken of the injuries and the animals were grouped in accordance with the Abbreviated Injury Scale (AIS) as described by States 1969.

RESULTS

At an early stage of deployment, the airbags proved to be easily influenced by the presence of a test object and could expand in almost any direction. In such tests, where the major part of the airbag expanded to the side, either right or left the animal was accelerated to the opposite side quite violently and one was even ejected from the passenger compartment. An initial tendency to expand laterally would be further enhanced by this particular test arrangement in a car body without doors and without simultaneously deploying the driver airbag. In some of these tests the airbag also deviated in a vertical direction which could be up or down.

Table 1.

Airbag deployment	Lateral deviation	Vertical deviation	No deviation	No airbag deployment
Fatal (6)	2	2	3	1
Critical (5)			2	
Serious (4)	1	4	1	1
Severe (3)	2			2
No injury (0)	3			
Total number of animals	8	6	6	4
Total AIS	22	28	32	16
Average AIS	2.75	4.7	5.3	4

The airbag deviated laterally in eight cases, vertically in six cases, deployed without deviation in six cases and was purposely not inflated in four cases. To judge from the injuries as listed in Table 1 (average AIS), deviation in the vertical plane is almost as serious as a direct hit, while deviation to the side results in less injury.

In the test series (Tables 2 and 3) numbers 1-9 were static airbag tests with inflation pulse D1. These were all performed in a stationary car. Numbers 9-11, 16-21 and 30-33 were dynamic airbag tests performed in a car body mounted on a rearward accelerating test sled. The car body

Table 2

Number	Facing Pos	Restraint	Bag- +) deviation	Injury	AIS Rating	
1 stat	Forward	Airbag	D1 v	Contusion: heart	Serious	4
2 "	"	"	" v	Contusion: heart	Serious	4
3 "	Lateral	"	" s	-	No injury	0
4 "	"	"	" n	Contusion: heart, abdominal organs	Fatal	6
5 "	"	"	" v	Contusion: heart, lung	Serious	4
6 "	"	"	" v	Laceration: liver		
				Contusion: lung	Serious	4
7 "	"	"	" n	Laceration: liver	Fatal	6
8 "	"	"	" n	Laceration: liver	Critical	5
9 dyn	"	"	" s	Laceration: liver		
				Contusion: heart and lung	Fatal	7
10 "	"	"	" v	Laceration: liver and kidney		
				Fracture: ribs	Fatal	7
11 "	"	"	" n	Laceration: liver		
				Contusion: lung and heart	Fatal	7
12 dyn	Lateral	No restraint	-	Contusion: heart	Fatal	6
13 "	"	" "	-	Contusion: lung	Severe	3
14 "	"	" "	-	Contusion: lung	Severe	3

+) s = lateral deviation, v = vertical deviation, n = no deviation

Number	Facing Pos	Restraint	Bag- +) deviation	Injury	AIS Rating	
15 x)						
16 dyn	Lateral	Airbag	D2 n	Laceration: liver		
				Contusion: lung	Critical	5
17 "	"	"	D2 s	Contusion: abdominal organs	Severe	3
18 " xxx) "	"	"	" n	Cerebral injury (ejected) to-ward outside structures	Fatal	(4) 6
19 " xxx) "	"	"	" s	Overdose anesthesia	Fatal	(0) 6
20 "	"	"	" s	Laceration: liver		
				Contusion: lung	Serious	4
21 "	"	"	" s	Contusion: lung	Severe	5
22-29 x)						
30 dyn	"	"	" -	Contusion: lung, heart	Serious	4
31 dyn	Lateral	Airbag	D3 s	Laceration: liver	Fatal	6
32 "	"	"	" s	-	No injury	0
33 "	"	"	" v	Laceration: liver	Fatal	6
34 xx)						

x) tests no 15, 22-29 in this series are summarized in table 3
xx) test no 34 was static airbag inflation on the sled without human substitute
xxx) see text
+) s = laterial deviation, v = vertical deviation, n = no deviation

had a windshield, an instrument panel with one airbag on the right side and a front passenger's bucket seat. In test numbers 12-14 no attempt was made to trigger the airbag. In test 15 as well as in numbers 22-24 the airbag was not triggered and the animal was restrained in a rear-facing child seat. In tests 25, 26 and 29 a child dummy (Alderson VIP 6C) and in 27 and 28 (Alderson VIP 3C) in a rearfacing child seat replaced the test animals. Number 34 was a static test in the car body without an animal or a dummy. It was performed to check the normal inflation of the airbag. In tests 7 and 8 the front seat was filled with soft, energy-absorbing material in an attempt to exclude injury from impact against the surroundings. These two animals were injured, one fatally and one critically, both with large tears in the liver as the main injury. Pig number 18 was ejected and died from head injuries which were considered to be the result of the impact on the floor. Since these injuries were of no interest to the test conditions, they were not allowed to influence the AIS rating, which, due to lung contusions and subcapsular liver hemorrhage was judged as AIS 4. Number 19 died from an overdose of anesthetic, it had no injuries and was consequently considered as uninjured AIS 0. In test number 30, the airbag was cut open by sharp sheet metal which by accident had been left uncovered and this test has been referred, therefore, to the group where no airbag was deployed.

The first moment after impact, at which airbags with different inflation pulses were able to accelerate the test animal, was judged from still pictures from the film. Airbags with inflation pulse D_1 began to influence the animals at an earlier stage (25 milli-

Table 3

Number	Test object	Facing Pos	Restraint			Result	
15	pig	Rearwards	Child safety seat			No injury	
22	"	"	"	"	"	"	"
23	"	"	"	"	"	"	"
24	"	"	"	"	"	"	"
25	dummy	"	"	"	"	SI 200 HIC 539	
26	"	"	"	"	"	SI 264 HIC 337	
27	"	"	"	"	"	SI 252 HIC 208	
28	"	"	"	"	"	SI 316 HIC 220	
29	"	"	"	"	"	SI 312 HIC 328	

Additional tests were performed with Alderson VIP-3C dummy in a standing child position in front of the instrument panel. Following injury criteria were received.

Number	Test object	Facing Pos	Restraint	Pulse	Result
1	dummy	Forwards	Airbag	D2	SI 384 HIC 319
2	"	"	"	D2	SI 1104 HIC 894
3	"	"	"	D3	SI 584 HIC 413
4	"	"	"	D3	SI 416 HIC 413

seconds) than airbags with inflation pulses D_2 and D_3 (40 milliseconds). As mentioned earlier, the airbag is easily influenced by the position of the test animal. The comparison between the three inflation pulses has been made only from those tests where the animals were well centered and the airbag did not deviate to the side.

The injuries to the liver were mainly ruptures of different size located close to structures by which this organ is suspended in the body. For this reason this type of injury was considered to be due to displacement of the liver caused by body deformation rather than to a blow from some folded part of the bag (Hellström 1966). In the myocardium and in the lungs two types of injuries were found. In a few cases there were fairly large bleedings as can be seen as a result of direct impact. In several cases, however, there were petechial bleedings on the inner side of the thoracic cage under the pleura and in the myocardium. This type of bleedings have been described in experimental air blasts (Clemedson 1949). In the six tests where the animals were well centered in front of the airbag those two types of injury were distributed in the following manner on the three types of inflation pulses. In Table 4, P denotes pressure pulse or air blast effects and D displacement or impact injuries.

Table 4

Inflation pulse:	D_1	D_2	D_3
Severity and type of injury:	fatal (P)	critical (P+D)	fatal (D)
	fatal (D)	serious (P+D)	
	critical (D)		

The number of cases is of course too small to draw any conclusions about the effect of the different inflation pulses on type and severity of injuries.

The distribution of injury severity on all of the 24 animals in the airbag material is shown in Table 5.

Three animals remained uninjured, one in each of the inflation pulses tested. This seems to be mainly due to the deviation of the airbags which was of the same order in these three cases.

The anatomical distribution of 34 injuries of different types is shown in Table 6.

Table 5

Description	AIS	Number of animals
No injury	0	3
Severe	3	4
Serious	4	7
Critical	5	2
Fatal	6-7	8

Table 6

Type of injury	Pressure wave (P)	Displacement (D)	Other
Heart	6		
Lungs	10		
Thorax	2		
Liver		10	
Spleen		1	
Kidney		2	
Mesenterium	2		
Head			1

The main cause of death was heart injury in 3 cases, liver injury in 4 cases and asphyxia in one case.

With the exception of the three animals which escaped injury when the airbags deviated considerably to the side during inflation all the animals in the airbag test series were injured. The severity of the injuries were rated from severe (AIS 3) to fatal (AIS 6). This includes the four tests where the airbags were not deployed although the peak acceleration of the test sled was not more than 12 g and the velocity change 28 km/h (17.5 mph). Test number 15 was made in order to exclude the possibility that the animals were too sensitive when anesthesized and placed in a vertical position like the one used. In this test the animal was strapped into a rear-facing child seat, Figure 12, and subjected to the same vehicle acceleration pulse as was used in the airbag tests. This animal revealed no injury at all at the autopsy. As this had also been seen occasionally in other tests it was decided to subject three more animals to accelerations in rear-facing child seats but with acceleration pulses simulating a frontal car-barrier impact test at 48.2 km/h (30 mph) with a peak of 24 g, Figure 10.

As all these animals seemed to be completely uninjured by the tests, it was decided to let them live and to study their further development. As soon as they had recovered completely from anesthesia they were brought back to the sty and kept there together with one animal of the same litter which had been neither anesthesized nor accelerated. These four animals were observed for several weeks. They had a completely normal development including a normal weight increase over these weeks.

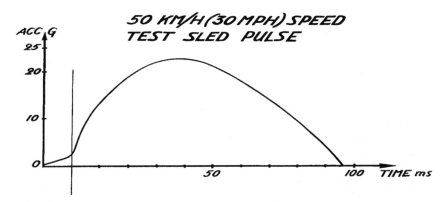

Fig. 10 - 50 km/h (30 MPH) speed test sled pulse

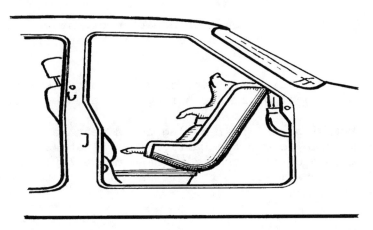

Fig. 12 - Pig in child restraint test position

SUMMARY AND CONCLUSIONS

In this investigation the influence of three different inflation pulses on the risk of injury to a standing child was inferentially studied. Pigs weighing 14-15 kilograms were used as test animals. When anesthesized and kept in an upright position in a rear-facing child seat these animals were able to withstand a simulated car-barrier impact at 48.2 km/h (30 mph) without any sign of injury. When anesthesized in the same way and suspended in a vertical position with their right side 10-15 cm (4-6 in.) from the airbag outlet area in the instrument panel and subjected to an acceleration pulse with a peak of 12 g and a velocity change of 28 km/h (17.5 mph) these animals were injured whether or not the airbags were inflated with any of the inflation pulses. Injury severity was rated from severe (AIS 3) to fatal (AIS 6) in the Abbreviated Injury Scale. Two types of injuries were seen: Tears in the liver and large bleedings in the heart or lungs as seen in impacts with displacement of the organs due to body deformation and multiple petechial bleedings mainly under the pleurae and the endocardium as seen in experimental air blasts.

Our conclusions from these results are:

- that an out-of-position passenger of the size of a child 3-6 years of age could be injured by airbag inflation even at the triggering level of vehicle acceleration (12 g),

- that injuries could be due to not only deformation of the body and displacement of internal organs, but also to the air blast effect close to the airbag outlet,

- that injuries could also result from impact on the instrument panel at the same level of acceleration (12 g) if the airbag is prevented from deploying,

-       that proper restraining in a rear-facing child seat is apparently safe during collisions comparable to vehicle-barrier impacts at 48,2 km/h (30 mph), and - that the airbag during its early phase of deployment is so easily influenced in its deployment direction that it is difficult to obtain reproduceable results.

REFERENCES

PATRICK, L.M. and NYQVIST, G.W. 1972: Airbag Effects on the Out-of Position Child. 2nd International Conference on Passive Restraints, Detroit, Michigan, May 22-25, 1972. SAE Transactions 720942.

WU, H., TANG, S.C. and PETROF, R.C. 1973: Interaction Dynamics of an Inflating Air Bag and a Standing Child. Automobile Engineering Meeting, Detroit, Michigan, May 14-18, 1973. SAE Paper 730604.

LUNDSTRÖM, L.C., WILSON, R.A. and SMITH, G.R. 1974: Relating Air Cushion Performance to Human Factors and Tolerance Levels. Fifth International Technical Conference on Experimental Safety Vehicles, London, England, June 6, 1974.

STATES, J.D. 1969: The Abbreviated and the Comprehensive Research Injury Scales. Proceedings of Thirteenth Stapp Car Crash Conference. December 2-4, 1969. SAE N Y USA.

HELLSTRÖM, G. 1965: Closed liver injury. A clinical and experimental study. Almqvist & Wiksell, Uppsala, Sweden.

CLEMEDSON, C-J 1949: An experimental study on air blast injury. Acta Physiol. Scand. 18 (Suppl. 61): 1-200.

841656

A Biomechanical Analysis of Head, Neck, and Torso Injuries to Child Surrogates Due to Sudden Torso Acceleration

Priya Prasad and Roger P. Daniel
Ford Motor Co.

ABSTRACT

This paper reports on the injuries to the head, neck and thorax of fifteen child surrogates, subjected to varying levels of sudden acceleration. Measured response data in the child surrogate tests and in matched tests with a three-year-old child test dummy are compared to the observed child surrogates injury levels to develop preliminary tolerance data for the child surrogate. The data are compared with already published data in the literature.

ALDMAN, et.al. of Volvo (1)* and Mertz, et al. of GM (2) have studied the effects of airbag inflation on out-of-position children by using piglets as surrogates for the human child. This paper reports on another study involving fifteen paired dynamic airbag tests, conducted at Southwest Research Institute, utilizing 12-15 week-old piglets** and a three-year-old child test dummy as the test subjects. The piglet test conditions and test locations within the vehicle buck were chosen to provide a broad spectrum of forces, moments and accelerations to the head, neck and torso as indicated by tests with the child dummy previously conducted at Ford and SwRI.

———————————————

*Numbers in parentheses refer to references at end of paper.

**Southwest Research Institute (SwRI) Animal Medicine Program and Facilities are approved by the American Association for Accreditation of Laboratory Animal Care and the United States Department of Agriculture. Approval by these organizations is based on guidelines of the Animal Welfare Act of 1970 and the Guide for Care and Use of Laboratory Animals. All protocols involving animal care and testing are subjected to a review by the SwRI Animal Care Committee.

TEST PROTOCOL -- The pre- and post-impact protocol for these tests is essentially the same as described by Mertz, et al. (2), and hence will not be repeated in this paper.

TERMINOLOGY -- All references to injury severity on the piglet will use the terms specified in The Abbreviated Injury Scale (AIS), 1980 Edition (3). These terms are generally accepted in the biomechanical literature for increasing injury level. They are:

Minor	} Non-	Severe	} Threat
Moderate	} dangerous	Critical	} to
Serious	} to life	Maximum (fatal)	} life

Eppinger, et.al. (4) has shown that the probability of fatality from an injury ranked "Serious" (AIS-3) is 0.4%, 4% for a "Severe" (AIS-4) injury, 28% for a "Critical" (AIS-5) injury and, by definition, 100% for "Maximum" (AIS-6) injuries. Additional injuries rapidly increase the likelihood of fatality. For example, two AIS-5 injuries raise the likelihood of fatality from 28% to 86%.

PIGLET-HUMAN ANATOMICAL COMPARISON -- Judgment of biomechanical factors was utilized to select the most appropriate surrogate for a child, given the spectrum of potential trauma sites. After consideration of many factors, the piglet was judged to be the most appropriate surrogate. However, for any given forcing function, no absolute interpretation is possible as to what injury an equivalent sized human child would experience compared to a piglet.

An assessment of the differences between the major organs of the piglet and the human child, with respect to injury potential, was made with the support of SwRI biomedical personnel when possible. Figure 1 depicts a schematic of the relative size and placement of the internal organs of a piglet and a human child superimposed on piglet and dummy templates.

469

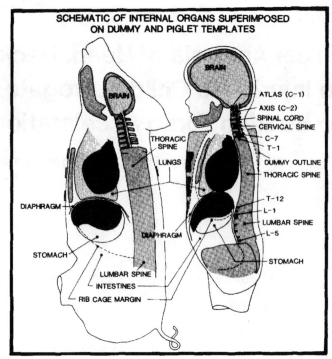

Figure 1 - Schematic of relative size and location of the major organs of a piglet and child

The brain of a piglet is protected by a massive, rigid bony structure and is approximately 10% of the mass (or volume) of a comparable sized human's brain (5,6) - 100 vs. 1000 grams. Studies on biological scaling effects (7,8,9) indicate that substantially more translational and/or angular acceleration is required to injure a small brain than one ten times more massive.

The head mass of the tested piglets was approximately 1.6-2.0 kg*, about 65-70% of the mass of the three-year-old child test dummy's head (2.6 kg). For an equal input force to the head, the acceleration of the piglet's head will be proportionately greater than the acceleration of the child dummy's head.

Although the cervical vertebral column of a piglet is covered with considerably more soft tissue than a child's neck vertebrae, the piglet's cervical spine is considered to be quite childlike in size and developmental state. Differences in the shapes in the child's and the piglet's heads are expected to cause differences in neck loading from airbag-to-head contact.

The piglet's thoracic-abdominal organ masses are judged very comparable to those of a three-year-old human child. However, the thorax of a piglet is far more rigid than that of a child. The piglet's rib cage in cross-section is shaped more like an egg standing on end so that initial sternal deflection <u>increases</u> the intra-thoracic volume. The cross-section of a child's thorax is

*Weighed at necropsy from piglets with a normal (99 g) and exceptionally large (133 g) brain.

more like an egg lying on its side so that initial sternal deflection <u>decreases</u> the intra-thoracic volume. The slanting ribs in humans also makes the human chest easier to deflect. Due to these differences in chest stiffness and shape, the primary internal injury mechanism is probably different between a child and a piglet.

The abdomen of a piglet is larger than the abdomen of an equivalent-sized human and particularly the liver and spleen are better protected by a longer and more rigid lower rib cage.

DUMMY AND PIGLET INSTRUMENTATION -- The following four possible injury mechanisms were postulated to occur during airbag development. Where possible, instrumentation was employed to measure these parameters.

1. The short duration, low mass "slap" translational acceleration effects of the airbag to the front surface of the chest and abdomen;

2. The longer duration, higher mass "punch" translational acceleration effects of the airbag to the head, chest and abdomen (i.e., HIC, spinal acceleration, etc.);

3. The tensile and shear loads, and the moment applied to the neck as the airbag deployed; and

4. The angular acceleration of the head.

The dummy represented a three-year-old child in size and weight, and was a Ford-built version of the GM child dummy (10) for inflatable restraint evaluation, but with a different head to reduce ringing. The dummy had three triaxial accelerometer arrays -- at the head c.g., on the upper spine and on the lower spine -- to measure head and torso accelerations. The head contained two other vertically oriented accelerometers to measure head angular acceleration in the mid-sagittal plane. A neck load cell measured neck axial force, neck shear force and neck moment about the lateral axis, all at the location of the occipital condyles (the head-neck attachment point). Linear accelerometers were mounted within three low mass "spools" spaced along the vertical centerline of the dummy's Ensolite™ foam torso. These spools have relative freedom to float in the foam, permitting measurement of the bag "slap" accelerations to the torso front surface. These instrumentation locations are schematically shown in Figures 2 and 3.

The instrumentation on the piglets was necessarily different from that in the dummy. A triaxial accelerometer array was attached to the piglet's snout and an additional accelerometer was attached further out on the snout for measurement of head angular acceleration. The location of these accelerometers was 75 to 100 mm

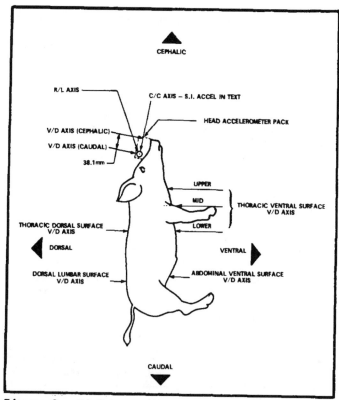

Figure 2 - Piglet surface instrumentation

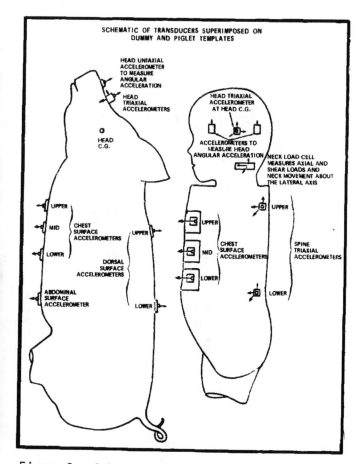

Figure 3 - Schematic of child and piglet instrumentation

from the piglet's head c.g. and thus the accelerations obtained from the piglet cannot be directly compared with the dummy's head c.g. accelerations. Three miniature accelerometers were also adhered to the flesh over the piglet's sternum to measure front chest surface "slap" accelerations. A fourth surface accelerometer in some later tests was adhered to the piglet's abdomen to measure accelerations that could possibly cause abdominal intrusion. The vertical area along the piglet's sternum covered by the sternal accelerometers approximated the area occupied by the top two spools of the dummy. The piglet's spinal acceleration was measured by one or two accelerometers adhered to the flesh just to one side of the piglet's spine; one behind the heart and one approximately opposite the abdominal surface accelerometer. These instrumentation locations also are shown in Figures 2 and 3. In addition, the piglet was instrumented with EKG leads to measure heart electrical response and with a catheter placed via the femoral artery just below the heart to measure blood pressure. Respiration was assessed visually in most cases, although it was discernible from one of the EKG traces.

RESULTS OF THE TESTS -- A broad spectrum of injuries to the head, neck, thorax, and abdomen of the piglets were observed in the 15 tests. There were four fatal neck injuries and one fatal brain stem injury. There was one critical abdominal injury due to a ruptured gall bladder. There were two severe thoracic injuries - both to the lungs. In five cases of where the piglets lived, transient arrhythmia of the heart was observed.

High speed movies of the tests indicated that the upper torsos of the piglets generally were accelerated rapidly due to airbag interaction, and the head lagged torso motion unless impact was directly to the head. Analysis of piglet spinal acceleration data showed that average accelerations over 50G's lasting more than 8 ms were associated with severe to fatal neck injuries (Figure 4).

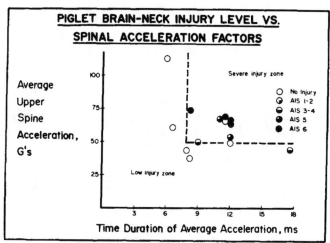

Figure 4 - Piglet upper spine acceleration factors vs. observed brain - neck injury

Analysis of dummy spinal acceleration data also shows a trend towards increasing neck loads with increasing spinal accelerations (Figure 5), implying that neck injuries are, in great part, associated with the effects of -Gx torso accelerations. The neck injuries also may be aggravated by local interaction between the head and the airbag.

The following sections of this paper describe the injuries to the various body regions of the piglet and the associations inferred with measured kinematic parameters on both the piglets and the dummy in matched tests.

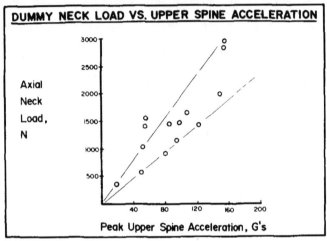

Figure 5 - Relationship between child dummy axial neck load and dummy upper spine acceleration

HEAD AND NECK

Head and neck injuries of equal or greater severity than AIS-3 were observed in seven out of the fifteen piglets tested. These injuries are described in Table 1 and all the head and neck data are given in Table 2. The head injuries ranged from a minimal extradural hemorrhage in the medial aspect of the left occipital lobe in Test 4 to a fatal severe contusion with extensive associated leptomeningial hemmorrhage of the brain stem in Test 5. The neck injuries ranged from severe hemorrhage in the joint space of the atlas (C-1) - occipital joint in Tests 4 and 5, to fatal three-fourth transection of the ventral aspect of the spinal cord at C-1 level with associated severe subdural hemorrhage surrounding the cervical cord from the brain-stem to the C-4 level in Test 10. A vertebral body fracture of the dorsal caudal crest of C-1 was the only cervical spine fracture detected in this test series. The most severe type of neck injury involved the spinal cord.

It can be seen from Tables 1 and 2 that except for one isolated case (Test 7), significant brain injuries were accompanied by severe to fatal neck injuries. Based on observation of the high speed movies and accelerometer data, the injury in Test 7 was judged to have been produced due to a severe head impact against the vehicle body during rebound. The association of severe to fatal neck injuries with brain injuries was also observed by Mertz et al. (2) in GM's test series with piglets and baboons.

Another significant result in this currently reported test series is the predominance of brain stem hemorrhages without significant injury elsewhere in the brain. Hence, the brain injury patterns do not follow the centripital pattern due to head rotational effects as suggested by Ommaya (11). This does not discredit the centripetal (rotational) theory, but infers that other injury mechanisms to the brain stem were predominant in this test series. The presence of spinal cord injuries at the C-1 level, apical ligament ruptures and hemorrhages into the Atlanto-Occipital and the C1-C2 joints suggest the presence of high tensile and bending stresses at the Atlanto-Occipital and the C1-C2 joints.

Unterharnscheidt (12) analyzed tests of Rhesus monkeys under -Gx acceleration. The torso acceleration of the piglets in this test series also were in the -Gx direction. He suggested the following three injury mechanisms either occurring simultaneously or consecutively:
1. Localized crushing of the ventral aspect of the spinal cord in the area of the lower medulla and C-1, caused by excessive movement of the tip of the odontoid process (dens) of the atlas (C-1).
2. Excessive stretch of the upper spinal cord and medulla oblongata to which the cord merges.
3. A shearing action between the rim of the foramen magnum and the arches of the atlas as the neck is both stretched and rotated relative to the head.

Spinal cord injury by the odontoid process as a possible mechansim for concussion of helmeted pilots undergoing a combination of +Gz and -Gx accelerations during crashes has been advanced by King et al. (13) based on the results of a mathematical model developed by Prasad and King (14).

The effect of abrupt head displacement relative to the torso was studied by Friede (15,16,17) in the early 1960's. It was shown to produce the same symptoms as those caused by a blow to the head, thus indicating that sudden neck stretch can lead to brain stem injuries.

Static neck stretch tests on a 12.2 kg eight-year-old baboon performed by Lenox et.al. (18) have shown interruption of Somatosensory Average Evoked Responses at 595N. This interruption was hypothesized as being due to blockage of neural transmission at the spino-medullary junction. Death followed at 1040N and neck struc-

TABLE 1

Test	Head and Neck Injuries
1	No injuries
2	No injuries
3	No injuries
4	Brain: Extradural hemorrhage, medial aspect of left occipital lobe, minimal - AIS-3. Neck: Severe hemorrhage in joint space at Atlas/occipital joint - AIS-4.
5	Brain Stem: Severe contusion with extensive, associated leptomeningial hemorrhage. No respiration post-impact. No Palberal reflex. VER's flat. AIS-6. Neck: Blood in joint space at Atlas/occip., Atlas Axis. Bilateral hemorrhage in Sternocleidomastoid muscles, Contused Thyroid. AIS-4.
6	Brain: Slight latency increase left VER one hour post-impact - AIS-2.
7	Brain: Subdural hemorrhage, left, ventral cerebral cortex, moderate with mild underlying contusion judged to be during rebound.
8	No injuries
9	Neck: Atlanto-occipital junction - Severe hemorrhage into joint space; partially torn apical ligaments; moderate hemorrhage into dorso-lateral aspect of joint capsular tissue Atlas/axis junction; Bilateral epidural hemorrhage at C-6 level, moderate to severe in spinal cord.
10	Brain: Severe subdural hemorrhage in caudal aspect of brain, ventral aspect of brain stem and over cerebellum. Neck: 3/4 transection ventral aspect of cord at C-1 level. Severe subdural hemorrhage surrounding cervical cord from brain stem to C-4 level. Severe hemorrhage into joint space at Atlanto-occipital junction. At Atlas partial tears of muscles originating at C3-C4 and inserting into Atlas.
11	Brain: Massive subdural hemorrhage caudal brain, from Optic Chiasm to brain stem ventrally. Neck: Massive subdural hemorrhage ventral aspect of cord from brain stem to C3. Partial transection of cord with dorsal and ventral tears at C-1 level. Atlanto-occipital junction - very severe hemorrhage into joint space. Ventral dura torn with subdural hemorrhage.
12	Brain: Massive subdural hemmorrhage - caudal brain, from optic chiasm to ventral brain stem; dorsal cerbellum, bilaterally; intercerebral hemisphere fissure, caudal aspect and into third ventricle. Neck: Severe subdural hemorrhage C1-C2, moderate subdural hemorrhage C3-C7; Atlanto-occipital junction - complete rupture Apical ligament; severe hemorrhage in joint space. Atlas/axis joint fluid tinged with blood. C1-C2 junction: Severe hemorrhage into joint space C1 vertebral body fracture at the Dorsal caudal crest.
13	No injuries
14	Brain: Massive subdural hemorrhage caudal brain from optic chiasma to brain stem; and ventral and dorsal brain stem - AIS-5. Neck: Subdural hemorrhage severe, circumferentially from brain stem to C3 level, ruptured apical ligament complete. Severe hemorrhage into joint space at Atlanto-occipital junction. Massive hemorrhage beneath muscles overlying ventro-lateral aspect of C1-C2 vertebral bodies.
15	No injuries

473

Table 2 - Piglet and Dummy Head and Neck Test Responses

Test No.	Type of Test	Head mass, grams	Brain mass, grams	HIC	Duration, ms	Ave. G's	Normalized Average Acc. G's	As Tested r/s²	Normalized r/s²	Duration, ms	SI Acc. G's (1)	Est. Neck Load: F = ma (N) (1)	Axial N(1)	Shear N	Moment N.m	Brain	Neck
1	Dummy	2600	-	346	181	-	-	7500	7500	3	40	1020	960	630	15.8	-	-
	Piglet	1596	100	799	65	109	109	2200	2200	30	50	790	-	-	-	0	0
2	Dummy	2600	-	476	165	-	-	8100	8100	4	55	1410	1660	1460	25.4	-	-
								4600	4600	5							
	Piglet	1453	91	1291	66	52	54	3600	3833	20	70	1000	-	-	-	0	0
3	Dummy	2600	-	442	130	-	-	5400	5400	3	65	1660	1460	1120	23.2	-	-
								6100	6100	5							
	Piglet	1229	77	1086	133	37	40	3900	4642	20	50	610	-	-	-	0	0
4	Dummy	2600	-	548	8	-	-	8100	8100	4	56	1440	1480	940	37.3	-	-
								3300	3300	7							
	Piglet	1596	100	5083	21	142	142	6500	6500	10	180	2830	-	-	-	4	4
5	Dummy	2600	-	1228	-	-	-	5800	5800	12	111	2850	2960	1340	42.4	-	-
								12500	12500	5							
	Piglet	1628	102	4886	70	87	85	12200	12040	6	150	2410	-	-	-	6	4
6	Dummy	2600	-	306	187	-	-	3800	3800	5	31	800	1050	740	17.0	-	-
	Piglet	1469	92	691	26	59	61	3000	3172	10	60	890	-	-	-	2	0
7	Dummy	2600	-	884	58	-	-	3600	3600	18	35	890	560	850	25.4	-	-
	Piglet	2054	133	5000*	6.5	226	206	2200	1819	25	105	2130	-	-	-	5(Reb.)	0
8	Dummy	2600	-	1325	52	-	-	7100	7100	10	60	1540	1570	800	37.3	-	-
	Piglet	1628	102	7893	60	112	111	5000	4934	15	190	3050	-	-	-	0	0
9	Dummy	2600	-	1570	57	-	-	7000	7000	12	78	2000	1925	540	46.3	-	-
								8000	8000	6							
	Piglet	1632	99	8226	66	109	110	6700	6745	12	140	2260	-	-	-	0	4
10	Dummy	2600	-	1337	56	-	-	9300	9300	8	67	1720	2820	1030	47.5	-	-
	Piglet	1723	108	6144	66	97	95	4200	3989	20	200	3400	-	-	-	5	6
11	Dummy	2600	-	1570	57	-	-	8000	8000	6	78	2000	1925	540	46.3	-	-
	Piglet	1500	94	8033	62	111	113	10000	10421	13	100	1480	-	-	-	5	6
12	Dummy	2600	-	1270	57	-	-	5000	5000	27	63	1610	1430	670	21.5/33.9	-	-
								7000	7000	5							
	Piglet	1771	111	7866	66	107	104	13000	12126	10	120	2100	-	-	-	5	6
13	Dummy	2600	-	63	68	-	-	2700	2700	30	-	-	400	560	11.3	-	-
	Piglet	1676	105	56	199	10	10	1100	1064	20	Low	-	-	-	-	0	0
14	Dummy	2600	-	1129	56	-	-	5000	5000	24	46	1180	1160	630	29.4	-	-
								7000	7000	6							
	Piglet	1596	100	14450	61	141	141	23000	23000	5	180	2830	-	-	-	5	6
15	Dummy	2600	-	488	6	-	-	8300	8300	6	51	1300	900(2)	900	50.9	-	-
	Piglet	1532	96	6778	9	224	227	12000	12331	12	300	4540	-	-	-	0	0

*Estimated from rebound   (1) at time of bag interaction. Some whip loads higher.   (2) Compression

tural failure took place at 1170N. These results are in agreement with those of Sances et. al. (19).

We thus believe that the primary cause of the brain stem injury in this series of piglet tests was sudden stretch of the neck.

NECK INJURIES -- Neck injuries can be caused by excessive shear forces, tensile forces, bending moments and/or torsion in the neck. These forces and moments can act alone or combine to produce damaging stresses in the neck structure. These parameters could not be directly measured in the piglet tests; thus, it is not possible to directly estimate shear forces and bending moments, or to make any estimate of torsion generated in the neck of the piglets.

Although the existence of the above injury producing mechanisms cannot be determined with certainty from the piglet tests, torn apical ligaments in the neck and cervical spinal cord hemorrhages strongly suggest the existence of high neck tensile loads, however caused. Some idea of tensile loads in the neck at the atlanto-occipital joint (the skull and neck interface) can be obtained by examining the superior-inferior (S.I.) (vertical) (Figure 2) accelerations of the head in the piglet tests, since tests with dummies indicate a good correlation between neck axial forces (measured by a neck transducer) and head S.I. accelerations multiplied by the dummy head mass (Figure 6). The result of

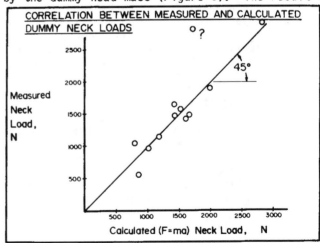

Figure 6 - Correlation between measured and calculated dummy neck axial force

Test 10 was rejected as possible gain error in the instrumentation.

In order to detect any correlation between S.I. accelerations of the piglet head, and neck injuries, the average S.I. accelerations were estimated during the high acceleration phase of the tests. These accelerations are shown in Table 2. It can be seen from the table that in seven out of ten cases in which 100G S.I. acceleration was exceeded, severe to fatal (AIS 4-6) neck injuries occurred. One of the cases of no injury involved a 300G S.I. acceleration for 3 ms. This was a case of direct impact to the snout where the accelerometers are mounted, and the data is thus considered questionable.

Based on the above data it is hypothesized that the threshold for severe-to-fatal piglet neck injuries due to S.I. acceleration is approximately 100G.

To get an estimate of the tensile neck loads in the piglet tests, the average S.I. accelerations were multiplied by the estimated head mass of the piglet in the test and are shown in Table 2. The piglet's head mass was estimated by assuming a linear relationship between the piglet's brain mass (measured during necropsy) and the head mass. The proportionality factor was determined from two necropsies where both the piglet's head and the brain masses were measured.

It can be seen from the table that there is a case of fatal neck injury with 1480N estimated tensile load for 11 ms time duration. The remaining neck injuries occurred above an estimated 2100N tensile load for time durations between 3 to 6 ms. Therefore, some time-dependence of tolerable neck loads is indicated, but there are no data between 6 and 11 ms durations. As a result, a conservative estimate of 1480N tensile load in the neck is the anticipated level at which severe-to-fatal neck injuries would be expected for piglets in this specific test condition.

HEAD RESPONSES AND INJURIES -- The association of the translational and angular accelerations (as measured by a snout-mounted accelerometer array) with piglet head injuries were investigated. These responses reflect the external forces and moments on the head generated through the neck by the interaction between the head and the deploying airbag.

The translational acceleration of the skull during impact is considered by many to be the mechanism causing brain injuries, due to the presence of shear strains or pressure gradients in the cranial cavity. Others consider the angular acceleration of the skull as the mechanism causing brain injuries due to the presence of shear strains in the brain. Although the effect of either of the above mechanisms cannot be isolated from the test data analyzed herein, the two mechanisms will be discussed separately in the following sections.

Head Translational Acceleration -- In order to study the association of head translational accelerations with piglet brain injuries, the average resultant linear accelerations of the piglet head during the HIC duration were calculated. These average accelerations were further normalized to a brain mass of 100g --the average brain mass of the piglets in these tests -- by the following relationship (6):

Normalized Average Acceleration =

$$A_{test} \left[\frac{100}{\text{Brain Mass}} \right]^{1/3}$$

where A_{test} = Average acceleration during the HIC duration in the test

The resulting normalized average accelerations are shown in Table 2. In the majority of the tests the HIC durations are over 60 ms, which is in the long duration end of the Wayne State University curve for cerebral tolerance to translational acceleration (20). This end of the curve is an extrapolation from short duration animal and human cadaver data.

Three piglet tests resulted in HIC durations of less than 20 ms. Two of these tests were injury producing; the third non-injurious. Data from this third test may be questionable due to an impact directly on the piglet's snout. This could have resulted in local vibrations and accelerations in the snout-mounted accelerometer array, thus recording accelerations which would not be experienced by the rest of the head.

All "critical" (AIS-5) piglet brain injuries observed in this test program occurred in tests producing a piglet HIC at or above the 4240 level.

In order to obtain more data in the low injury range, the piglet-air cushion test data of Mertz et al. (2) were analyzed in the same manner as in the current study. The composite result of the pooled data is shown in Figure 7. It can be seen from the figure that all but one AIS-3 or greater brain injuries are associated with HIC levels above 4240.

Head Angular Acceleration -- It has been shown by Ommaya (7) that as the brain mass decreases, the tolerance of the brain to angular acceleration increases. The brain mass of the piglets used in the two series of tests varied from 77-133g with an average of 100g. To account for this variation in brain mass, the observed average angular accelerations in the piglet tests were normalized to a 100g brain mass using the relationship below. These normalized values are shown in Table 2.

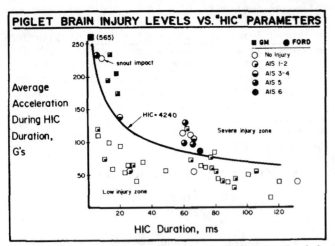

Figure 7 - Association between severity of piglet brain injury and piglet HIC parameters

$$\alpha n = \alpha a \left[\frac{100}{M}\right]^{2/3} \text{ -- from Reference 7}$$

where αn = normalized average* angular acceleration
 αa = test average angular acceleration
 M = Mass of the piglet brain in grams

The association between the angular acceleration parameters (average acceleration and time duration) and piglet brain injuries is shown in Figure 8. It can be seen from this figure that in 80% of the tests in which 10000 r/s² average angular acceleration was exceeded, critical or fatal brain injuries occurred. This is in agreement with Ommaya's (7) results with Rhesus monkeys having brain masses approximately that of the piglets in the tests herein analyzed.

* As obtained from the slope of the angular velocity curve.

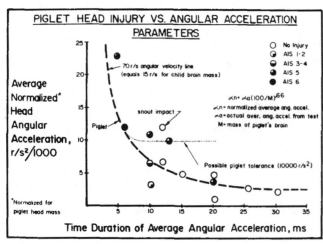

Figure 8 - Association between piglet angular acceleration parameters and piglet brain injury

However, there is one case of a critical brain injury with an average normalized angular acceleration of only about 4000 r/s² for 20 ms time duration. Therefore, some dependency on time duration for tolerable angular acceleration may exist, but, due to insufficient injury data in longer duration impacts (greater than 15 ms), firm conclusions cannot be drawn.

From Figure 8 it appears that the relationship between angular acceleration and its duration is such that an angular velocity change of 70 r/s must take place before the threshold of severe brain injury due to angular parameters is exceeded. This angular velocity change is approximately half that predicted by Ommaya for adult Rhesus monkeys (7). The difference might well reflect the effects of differences in age, neck strength, and/or species.

DUMMY HEAD AND NECK RESPONSE PARAMETERS IN IDENTICAL IMPACT ENVIRONMENT -- For each animal test, a similar test using a three-year-old child dummy (10) was conducted. An effort to correlate dummy responses with animal injuries was made. In some cases, the dummy kinematics appeared to be different from that of the animal's. As a result, engineering judgment was used in selectively disregarding some or all of the dummy response parameters in some tests. The following sections describe the results of this exercise.

Neck Axial Forces and Moments -- The peak axial forces developed in the child dummy and the neck injuries to piglets in matched tests during airbag deployment are shown in Table 3 and Figure 9. Also shown in the figure are the results of tests

TABLE 3
RANK ORDER OF NECK PARAMETERS

Neck Axial Force, N	Neck Injury Rating (AIS)	Neck Moment, N.m	Neck Injury Rating-AIS
400	0	11.3	0
560	0	15.8	0
900	0	17.0	0
960	0	23.2	0
1050	0	25.4	0
1160	6	25.4	0
1430	6	29.4	6
1460	0	33.9	6
1480	4	37.3	4
1570	0	37.3	0
1650	0	42.4	4
1925	4	46.3	4
1925	6	46.3	6
2820	6	47.5	6
2960	4	50.9	0

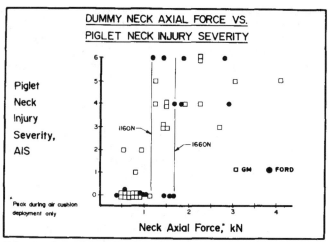

Figure 9 - Association between child dummy neck axial load and piglet neck injury severity in matched tests

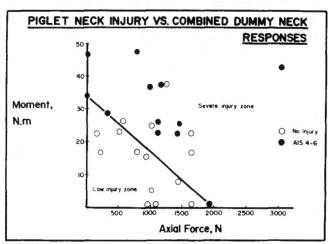

Figure 11 - Association between child dummy neck moment and axial force and piglet neck injury severity in matched tests

reported by Mertz et al. (2). The data of the Ford series is in good agreement with Mertz's series. In the range of dummy neck axial forces from 1160N to 1660N, ten out of thirteen piglets in matched tests developed significant neck injuries. Above dummy neck axial force of 1925N, all piglets had neck injuries.

The association of dummy neck moments with piglet neck injuries in matched tests during airbag deployment also is shown in Table 3 as well as in Figure 10. It can be seen from the figure that in the range of neck moments of 29N.m to 51N.m, seven out of nine piglets had significant neck injuries. The highest bending moment developed in this series, 51 N.m, was not associated with a neck injury to the piglet in the matched series. This lack of injury may be due to the fact that the 51N.m bending moment was associated with a compressive 900N axial force in the dummy test. This suggests the need to combine axial forces with moments for a composite neck injury indicator, indicating the complexity of the interactions involved in producing neck injury. Figure 11 shows an attempt to combine these two parameters. This was done using the force and moment-time traces of the neck during the time of airbag deployment and recording the combined peaks at a given time as shown in Table 4. As a result of the above process, several data points were obtained from each test.

It was assumed that a "constant stress line" resulting from combined neck axial force and bending moment is a straight line as shown by Belytschko, et.al. (21) for vertebral bodies. Based on the data of this series, a straight line joining two points (2000N, 0N.m and 0N, 34N.m) appears to delineate the no neck injury zone and the severe injury zone. Since bending moment data in Mertz's tests were not reported, further verification of the plotted equal stress line could not be carried out.

Dummy Head Translation Acceleration -- For the purpose of comparing dummy head responses with piglet tests, Tests 1, 2, 6 and 13 were rejected based on large differences in HIC durations (> 120 ms). Test 7 was rejected because the piglet head injury was judged to be during rebound at which time the dummy head did not contact the same surface as the piglet's head. It can be seen from Table 2 that the lowest dummy HIC associated with a piglet brain injury in an identical test is 1129 (Test 14) and the highest dummy HIC associated with no brain injury in a piglet is 1570 in Test 9. This test is interesting because Test 11 was a duplicate of Test 9 and the only replicate animal test in this series. In Test 11, the piglet had a critical (AIS-5) brain injury and a fatal neck injury, whereas in Test 9, the piglet had only a severe (AIS-4) neck injury. Comparisons of the piglet head responses in the two tests show that the HIC parameters were fairly close (8226 to 8033 for 66 ms), but the case of no brain injury resulted in an angular acceleration of 6700 r/s^2 compared to 10000

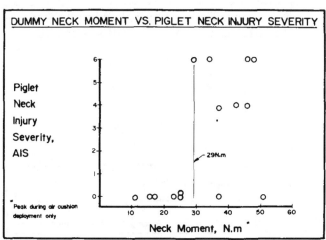

Figure 10 - Association between child dummy neck moment readings and piglet neck injury severity in matched tests

TABLE 4
DUMMY NECK PARAMETERS AND PIGLET NECK INJURY RATING

Test Number	Neck Axial Tensile Force (N)	Neck Moment, N.m	Injury Rating (AIS)
1	960 784	15.8 17.0	0
2	1030 1660	25.4 22.6	0
3	1460 506	7.9 23.2	0
4	1480 1164	25.4 37.3	4
5	1000 2960	36.7 42.4	4
6	224 1030	17.0 5.6	0
7	560	26.4	0
8	1570 1254	17.0 37.8 0	0
9 11	1925 0	0 46.3	4 6
10	760 2820	47.5 0	6
12	1430 0 0	-22.6 33.9 6.8	6
14	1160 313	-22.6 29.4	6
15	900 (Compression)	50.9	0

tests, compared to only three AIS-6 injuries in Mertz's 43 tests. As a result, Mertz's analysis reflected a lower probability of AIS 5 injury at a HIC level of 1470. It is believed that the results of Ford's series may be more representaive for the following reasons.

The average piglet head mass in the current series of tests is approximately 61% of the dummy head. As a result, under identical test conditions, the dummy head translational accelerations are expected to be 61% of those seen in matched piglet tests. Because of the 2.5 exponent in the HIC calculation, this lower acceleration would translate into dummy HIC's approximately 29% of those in piglet tests. Based on piglet acceleration data analysis previously described, HIC's of 4240 were associated with severe to fatal brain injuries. As a result of the above 29% factor, HIC's above 1230 should be observed in dummy tests where the paired piglet had a severe brain injury. This is indeed the case, since the lowest HIC's observed in the paired dummy tests producing piglet brain injuries were 1129 to 1228.

Dummy Head Angular Acceleration -- The association of dummy angular acceleration parameters with piglet brain injuries in matched tests is shown in Figure 12. A line approximately demarcating the low injury zone from the severe injury zone has been drawn. It appears that, for time durations shorter than 7 ms, dummy head angular accelerations of about 7000 r/s² are associated with severe piglet brain injuries. Beyond 7 ms time durations, 5000 r/s² appears to be associated with severe brain injuries. It is felt that more data are required to verify this curve.

THORACIC INJURIES

The severity of thoracic injuries ranged from moderate to severe (AIS-2 to AIS-4). No rib fractures were noted. The most severe injury in

r/s² for the case of the critical injury. This may be an indication of the role of angular acceleration parameters on brain injuries or it may indicate animal-to-animal differences.

The results of this series show a marked difference from the results of Mertz's paired test (22) which showed only a 1% probability of AIS-3 or greater brain injuries to piglets at dummy HIC's of approximately 1470. The main difference is probably due to the greater number (seven) of AIS-4 to AIS-6 injuries in our 15

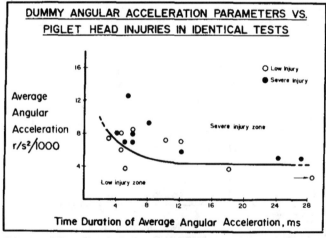

Figure 12 - Association between child dummy head angular acceleration parameters and severity of piglet head injury in matched tests

the thoracic region was to the lungs in Test 10 involving parenchymal hemorrhages and scattered, multiple contusions over the entire lung. Subendocardial hemorrhages in the heart were observed in a number of cases, including Test 13 in which the airbag was not inflated. As a result, it is believed by the authors that these small internal hemorrhages of the heart probably are artifacts extraneous to the air bag deployment event, such as the euthanising process. Transient heart arrhythmia was observed in five tests. These were rated as AIS-4 by the SwRI cardiologist.* A complete description of all the thoracic injuries is given in Table 5 at the end of the paper.

Kinematic Responses of the Piglets -- In general, the piglet torsos were accelerated suddenly with average accelerations, as measured by accelerometers mounted on the spinous process of T-6, ranging from 38 to 113G's for 6 to 18 ms. This initial phase of torso acceleration was followed by low level, long duration spinal accelerations. It is assumed here that the resulting

---

*Arrhythmia is not rated in the AIS-80 dictionary.

thoracic injuries were incurred during the initial phase of torso acceleration. Table 6 gives the thoracic and abdominal acceleration repsonses of the dummy and piglets in the matched tests.

Association of Measured Kinematic Parameters with Thoracic Injuries -- Piglet upper spine 3 ms cutoff accelerations and the resulting piglet thoracic injury AIS rating show no apparent relationship. However, in matched dummy tests, no AIS-3 or greater piglet thoracic injuries were observed when dummy upper spine accelerations were below 50G's, as seen in Figure 13. The lung injuries may well be associated with "air blast" effects as reported by Aldman, et. al. (1). We have no way to make this determination.

The correlation of heart arrhythmia with short duration, high sternal accelerations ("slap") has been reported by Viano and Artinian (23). In the current series, transient arrhythmia was initiated at piglet sternal accelerations of about 1200G's as indicated by surface accelerometer mounted at the mid-sternal location (Figure 14). The susceptibility of the piglet to arrhythmia also may be an effect of the unnatural posture of the piglets in this test series (24).

Table 6 - Piglet and Dummy Thorax and Abdominal Test Responses

Test No.	Type of Test	Piglet Age, Weeks	Piglet Mass, kg	Chest - Abdominal Surface Acceration								Arrhy-thmia, AIS	Thoracic Spine				Blood Pressure		Injury AIS	
				Upper		Mid		Lower		Abdomen			Upper		Lower		Peak, mm Hg	Dura-tion, ms	Thorax	Abdomen
				G's	Vel, m/s	G's	Vel, m/s	G's	Vel, m/s	G's	Vel, m/s		Pk/3ms G's	Vel, m/s	Pk/3ms G's	Vel, m/s				
1	Dummy	–	15.0	1390	10.7	700	8.2	300	4.6	–	–	–	81/60 @ 37	–	28/18 @ 36	–	–	–	–	–
	Piglet	10	14.5	2890	8.5	1580	3.7	770	2.7	–	–	4	133/91 @ 38	4.3	–	–	205	10	3	0
2	Dummy	–	15.0	960	11.3	370	7.3	110	3.4	–	–	–	107/87 @ 33	–	40/31 @ 34	–	–	–	–	–
	Piglet	10	14.5	2150	13.7	1650	6.1	550	4.9	–	–	4	113/93 @ 37	3.0	–	–	245	10	3	0
3	Dummy	–	15.0	535	6.4	105	2.7	85	1.5	–	–	–	55/42 @ 35	–	30/28 @ 72	–	–	–	–	–
	Piglet	10	15.0	1820	6.1	1950	6.7	470	4.6	–	–	4	119/106 @ 32	7.6	–	–	385	15	2	0
4	Dummy	–	15.0	250	8.5	235	7.3	250	4.9	–	–	–	98/87 @ 38	–	46/38 @ 32	–	–	–	–	–
	Piglet	10	15.0	920	5.2	1200	4.0	320	5.2	–	–	4	129/107 @ 36	7.6	–	–	305	15	3	2
5	Dummy	–	15.0	600	10.7	320	8.8	640	8.2	–	–	–	156/135 @ 32	–	84/76 @ 31	–	–	–	–	–
	Piglet	10	14.1	2000	9.4	1050	6.7	370	9.1	–	–	*	200/153 @ 32	7.9	–	–	595	25	2	3
6	Dummy	–	15.0	520	9.1	1130	11.6	1730	10.7	–	–	–	53/46 @ 41	–	44/35 @ 42	–	–	–	–	–
	Piglet	10	14.3	1050	8.2	2310	6.7	2770	8.2	–	–	4	118/90 @ 37	3.4	–	–	435	5	4	3
7	Dummy	–	15.0	170	4.9	90	4.0	50	3.4	–	–	–	51/43 @ 39	9.1	26/21	4.0	–	–	–	–
	Piglet	14	15.7	350	3.0	150	3.7	340	5.8	–	–	0	184/145 @ 32	6.7	–	–	370 Est. 400	20 40	2	0
8	Dummy	–	15.0	200	6.4	240	5.2	150	4.0	–	–	–	57/52 @ 32	6.1	35/29 @ 38	3.0	–	–	–	–
	Piglet	12	14.5	580	2.4	840	2.4	1900	4.9	–	–	0	65/50 @ 32	6.1	–	–	230 Est.	25	3	2
9	Dummy	–	15.0	400	11.3	280	8.5	310	8.5	–	–	–	150/68 @ 34	7.6	90/48 @ 34	4.6	–	–	–	–
	Piglet	12	14.5	700	3.7	410	3.0	500	2.4	–	–	0	110/81 @ 34	4.6	–	–	560	30	3	0
10	Dummy	–	15.0	390	12.8	290	11.3	280	7.6	–	–	–	155/92 @ 34	10.0	100/65 @ 34	6.7	–	–	–	–
	Piglet	12	15.9	780	2.4	550	3.0	1000	3.7	–	–	*	108/84 @ 32	7.6	–	–	760 Est.	25	4	5
11	Dummy	–	15.0	410	11.3	280	8.5	310	8.5	–	–	–	150/68 @ 34	7.6	90/48 @ 34	4.6	–	–	–	–
	Piglet	12	14.0	570	3.4	390	6.1	1330	8.2	370	3.0	*	77/74 @ 28	6.1	33/22 @ 35	1.8	400	30	2	3
12	Dummy	–	15.0	690	12.5	420	12.2	450	9.8	–	–	–	121/102 @ 37	6.7	70/61 @ 34	4.6	–	–	–	–
	Piglet	12	16.8	800	6.1	1180	6.1	1780	7.3	870	2.4	*	127/97 @ 33	7.6	60/53 @ 38	3.7	570	30	3	3
13	Dummy	–	15.0	Low	Low	70	3.4	Low	Low	–	–	–	20/18 @ 54	6.1	9/8 @ 55	6.1	–	–	–	–
	Piglet	12	16.8	180	1.8	110	0.6	250	1.5	350	2.7	0	21/19 @ 74	4.9	44/42 @ 75	4.6	200	25	2	0
14	Dummy	–	15.0	550	11.6	610	11.9	610	11.6	–	–	–	95/88 @ 36	8.5	65/58 @ 33	5.2	–	–	–	–
	Piglet	12	14.3	970	8.5	2170	4.9	1800	7.9	2860	9.8	*	111/82 @ 30	6.1	73/59 @ 36	4.6	500	50	3	2
15	Dummy	–	15.0	120	3.0	50	1.5	20	1.2	–	–	–	52/49 @ 34	4.0	16/15 @ 24	1.2	–	–	–	–
	Piglet	12	15.0	830	2.4	130	1.8	60	1.5	80	2.1	0	38/32 @ 29	2.4	22/20 @ 53	2.4	105, 250	10 30	3	2

* Piglet died minutes after impact

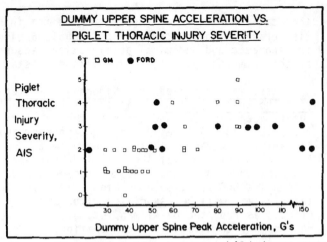

Figure 13 - Association between child dummy upper spine acceleration and piglet thoracic injury in matched tests

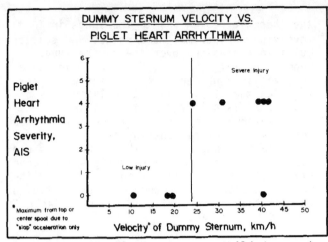

Figure 15 - Association between child dummy sternum (spool) peak velocity vs. piglet heart arrhythmia in matched tests

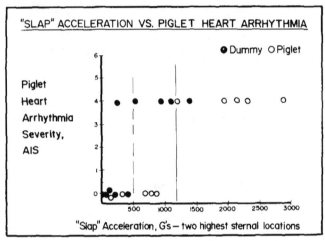

Figure 14 - Association between child dummy and piglet thoracic "slap" accelerations and piglet heart arrhythmia in matched tests

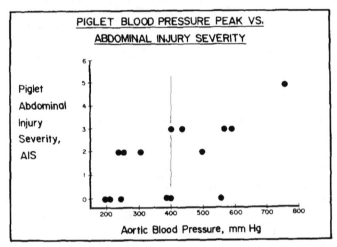

Figure 16 - Association between piglet abdominal injuries and piglet peak blood pressure as measured in the descending aorta

In paired dummy tests "slap" accelerations over 500G's (as measured at either the top or middle dummy chest surface spool) generally indicated arrhythmia in the piglets (Figure 14). Also, all arrhythmias occurred above 24 km/h maximum rate of chest compression as shown in Figure 15. The maximum rate of chest compression is the first integral of the first spike of the dummy chest spool acceleration.

ABDOMINAL INJURIES

The abdominal injuries in this series ranged from none to one critical ruptured gall bladder injury in Test 10. Partial tears of the falciform ligament were noticed in Tests 6, 10, 11 and 12. Tearing between various lobes of the liver and contusions were observed in Tests 4, 5, 8, 10, 11 and 12. These injuries to the liver could be partially acceleration-induced and partially abdominal pressure-induced due to local intrusion of the airbag in the abdominal area. Some idea of abdominal compression can be obtained by examining the blood pressure measured in the aorta of the piglets at the level of the lower end of the sternum. Figure 16 shows the association of abdominal injuries with the measured blood pressure. It can be seen that all serious to critical injuries occurred above 400 mm Hg blood pressure. There is some correlation between maximum rate of compression of the lower sternum and blood pressure as shown in Figure 17.

Association of Abdominal Injuries with Measured Kinematic Parameters -- In general, the results of this series of tests agree with those of Mertz et.al. (22) although the statistical analysis results performed by Mertz would change due to the additional data. Significant (AIS-3) abdominal injuries to piglets were observed in the current series when lower spine peak resultant acceleration exceeded 40G's in the paired child dummy tests, which agreed with the data in Mertz's paper. However, there is one case of 90G lower spine acceleration not associated with any abdominal injury, as opposed to

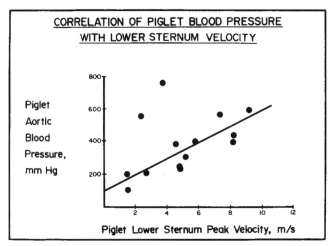

Figure 17 - Correlation between piglet lower sternum peak velocity and piglet blood pressure measured in the descending aorta

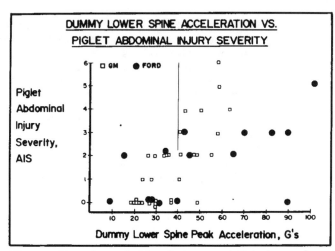

Figure 18 - Association between child dummy lower spine acceleration and piglet abdominal injury severity in matched tests

approximately 55G's in the above G.M. series. The composite data from both series are shown in Figure 18.

When the maximum rate of abdominal compression measured in dummy tests of this series is compared with that reported by Mertz et.al. (22), AIS-3 injuries are associated with about a 17 km/h abdominal compression velocity as opposed to about 20 km/h in the G.M. series. The only AIS-5 abdominal injury occurred at about 24 km/h, which is in close agreement with the reported Mertz data.

CONCLUSIONS

1. Piglet brain and neck injury mechanisms appear similar to those found in Rhesus monkeys subjected to -Gx acceleration. Thus, the stresses generated at the atlanto-occipital joints are important. The tensile loads generated at the atlanto-occipital joint of test subjects can be estimated by the S-I accelerations of the head, multiplied by the head mass. For the piglet, tensile loads in the range of 1480N for 11 ms and 2100N for 3 to 6 ms are associated with severe to fatal (AIS-4 to 6) neck injuries. Axial tensile forces and moments measured by load cells at the head/neck junction in matched child dummy tests, can explain the piglet neck injuries. A suggested curve combining the effects of neck axial tensile forces and moments has been developed.

2. Piglet translational head acceleration data show an association of HIC's above 4240 with severe to fatal (AIS-4 to 6) brain injuries. Piglet head angular rotation data suggest that 70 r/s angular velocity change in durations less than 20 ms are associated with severe to fatal (AIS-4 to 6) brain injuries. These head response parameters and their associations with injuries are useful only in test conditions and with surrogates similar to those of this series of tests.

3. There were no apparent trends found between piglet spinal accelerations and thoracic injury. However, no AIS-3 or greater thoracic injuries were found in this series of tests, or with the Mertz tests, when the upper spine acceleration of the child dummy in matched tests was less than 50G's.

4. It can be postulated that most of the small internal hemorrhages found in the hearts of most of the piglets were due to extraneous inputs such as the method of sacrificing the animals. These injuries were also found in the test where the airbag was not deployed.

5. It can be postulated that piglet lung injuries during the tests may be associated at least in part with the "air blast" effect as the air bag deploys. No instrumentation was available to measure air blast effects.

6. Piglet transient heart arrhythmia appears to be associated with piglet mid-sternal short duration ("slap") accelerations greater than 1000 to 1200G's. In the matched dummy tests, piglet transient heart arrhythmia is associated with top or center spool "slap" accelerations greater than about 500G's. The spool velocity, i.e., the area under the very short duration "slap" acceleration-time curve, also appears to delineate piglet heart arrhythmia when the velocity exceeds about 24 km/h.

7. Piglet peak blood pressure during impact (measured in the descending aorta behind the lower sternum) is associated with lower sternum velocity. Also, blood pressure peaks over 400 mm Hg were associated with AIS-3 or greater abdominal injury. Piglet abdominal injury of AIS-3 or greater also was associated with dummy lower spine accelerations greater than 40G's. The rate of compression of the dummy abdomen may help predict piglet abdominal injuries.

REFERENCES

1. Aldman, B., A. Anderson, O. Saxmark, "Possible Effects of Air Bag Inflation on a Standing Child," Proceedings of the 1980 IRCOBI Conference.

2. Mertz, H. J., G. D. Driscoll, J. B. Lenox, G. W. Nyquist, D. A. Weber, "Responses of Animals Exposed to Deployment of Various Passenger Inflatable Restraint System Concepts for a Variety of Collision Severities and Animal Positions," Proceedings of the 9th International Technical Conference on Experimental Safety Vehicles, Kyoto, Japan, Nov. 1982.

3. States, John D. and Committee on Injury Scaling, "The Abbreviated Injury Scale, 1980 Revision," American Association for Automotive Medicine, Morton Grove, Ill. 60053.

4. Eppinger, R. H., R. M. Morgan, J. H. Marcus, "Side Impact Data Analysis," Proceedings of the 9th International Technical Conference on Experimental Safety Vehicles, 1982, pg 246.

5. Francis, Carl O., "Introduction to Human Anatomy," C. V. Mosby Press, St. Louis, pg 232.

6. Sturtz, G., "Biomechanical Data of Children," Proceedings of 24th Stapp Car Crash Conference, 1980.

7. Ommaya, A. K., "Head Injury Mechanisms, Final Report," DOT-HS-800-859, 1973.

8. Hodgson, V. R. and L. M. Thomas, "Acceleration Induced Shear Strain in a Monkey Brain Hemisection," Proceedings of 23rd Stapp Car Crash Conference, 1979.

9. Roberts, V. L., V. R. Hodgson and L. M. Thomas, "Fluid Pressure Gradients Caused by Impact to the Human Skull," ASME, 66-HUF-1.

10. Wolanin, M. J., H. J. Mertz, R. S. Nyznyk, J. H. Vincent, "Description and Basis of a Three-Year-Old Child Dummy for Evaluating Passenger Inflatable Restraint Concepts," Proceedings of the 9th International Technical Conference on Experimental Safety Vehicles, Kyoto, Japan, Nov. 1982.

11. Ommaya, A. K., "Biomechanics of Head Injuries: Experimental Aspects," Biomechanics of Trauma, Appleton-Centruy-Crofts, Publishers

12. Unterharnscheidt, F., "Neuropathology of Rhesus Monkeys Undergoing $-G_x$ Impact Acceleration," Impact Injury to the Human Head, C. C. Thomas, Publisher, 1983.

13. King, A. I., S. S. Nakhla and N. K. Mital, "Simulation of Head and Neck Response to $-G_x$ and $-G_z$ Impacts," AGARD Conference Proceedings No. 253, Paper No. A7, 1978.

14. Prasad, P. and A. I. King, "An Experimentally Validated Dynamic Model of the Spine," Journal of Applied Mechanics, 41:546-550, 1974.

15. Friede, R. L., "Specific Cord Damage at the Atlas Level as a Pathogenic Mechanism in Cerebral Concussion," Journal of Neuropathol Experimental Neurol, 19266, 1960.

16. Friede, R. L., "Experimental Acceleration Concussion, Pathology and Mechanics," Arch Neurol, 4:449, 1961.

17. Friede, R. L., "The Pathology and Mechanics of Experimental Cerebral Concussion," Wright Air Development Division Report 61-256, 1961.

18. Lenox, J. B., R. L. Stalnaker, C. D. White, G. T. Moore, O. M. Anderson, R. R. Schleicher, H. H. Peel, S. S. Martin, G. D. Driscoll, H. W. Huntington, K. D. Carey, M. P. Hafner, A. X. Ommaya, "Development of Neck Injury Tolerance Criteria in Human Surrogates. 1: Static Tensile Loading in the Baboon Neck: Preliminary Observations," Proceedings of the Ninth International Technical Conference on Experimental Safety Vehicles, 1982.

19. Sances, A., R. Weber, J. Myklebust, J. Cusick, S. Larson, P. Walsh, T. Christoffel, C. Houterman, C. Ewing, D. Thomas, B. Saltzberg, "The Evoked Potential: An Experimental Method for Biomechanical Analysis of Brain and Spinal Injury," Proceedings of 24th Stapp Car Crash Conference, 1980.

20. Patrick, L. M., H. R. Lissner and E. S. Gurdjian, "Survivial by Design-Head Protection," Proceedings of 7th Stapp Car Crash Conference, 1963.

21. Belytschko, T., L. Schwer, A. Schultz, "A Model for Analytic Investigation of Three-Dimensional Head-Spine Dynamics," AMRL-TR-76-10 for Aerospace Medical Research Lab, Wright Patterson Air Force Base, Ohio, 4-19-76.

22. Mertz, H. J. and D. A. Weber, "Interpretations of the Impact Responses of a Three-Year-Old Child Dummy Relative to Child Injury Potential," Proceedings of the 9th International Technical Conference on Experimental Safety Vehicle, Kyoto, Japan, Nov. 1982.

23. Viano, and Artinian, C. G., "Myocardial Conducting System Dysfunction," Journal of Trauma, Vol. 18, No. 6, 1978.

24. Pope, M. E., C. K. Kroell, D. C. Viano, C. Y. Warner, S. D. Allen, "Postural Influences on Thoracic Impact," Proceedings of 23rd Stapp Car Crash Conference, 1979.

TABLE 5

THORACIC AND ABDOMINAL INJURIES

Test
No.

1   Heart: Four beat burst of idioventricular rhythm immediately after impact - AIS-4. Moderate to severe subendocardial internal hemmorhage over all left ventricular papillary muscles - AIS-3.
Lungs: Markedly hemorrhagic bulla; scattered subplural internal hemorrhage, moderate - AIS-3.
Hilar Lymph Node: Hemorrhage, mild to moderate - AIS-2
Regional AIS Rating: AIS-4

2   Heart: Four beat burst of idioventricular rhythm immediately post-impact - AIS-4.
Lungs: Subpleural internal hemorrhage, right apical lobe, mild - AIS-2.
Regional Thorax Rating: AIS-4

3   Heart: Subendocardial internal hemorrhage, moderate in extent, mild to moderate in severity: AIS-2. Three beat burst of idioventricular rhythm immediately followed by a premature atrial contraction: AIS-4.
Regional Thorax Rating: AIS-4

4   Heart: Subendocardial internal hemorrhage and petechiae, minimal: AIS-2. Six beats post-impact, an 18-beat sequence indicative of a conduction abnormality: AIS-4.
Lungs: Scattered hemorrhagic areas, mild to moderate: AIS-3.

Lymph Nodes: Hilar nodes mildly contused: AIS-2.
Overall Thorax Rating: AIS-4
Abdomen: Mild contusions of liver: AIS-2.

5   Heart: Marked sinus tachycardia immediately post-impact, followed by sinus bradycardia (may be entirely a result of brain stem injury): AIS-2.
Lungs: Several 6mm diameter bullae, hemorrhagic, minimal to moderate - AIS-2.
Abdomen and Liver: Laceration, mild-moderate, 6mm deep, near junction of right and middle lobes (35cc of free blood in the abdominal cavity).
Regional Abdominal Rating: AIS-3

6   Heart: No atrial activity for 10 seconds with rhythm probably idioventricular. Ventricular depolarization remained prolonged for 60 seconds: AIS-4.
Lungs: Parenchymal hemorrhages, severe in degree, moderate in extent: AIS-3.
Thymus: Scattered petechiae, mild to moderate: AIS-2.
Lymph Node: T8-9 periaortic lymph node mildly hemorrhagic: AIS-2.
Regional Thorax Rating: AIS-4
Abdomen: Peritoneal Cavity: 50 ml clotted and unclotted blood (source not found): AIS-3.
Regional Abdominal Rating: AIS-3

7   Heart: Subendocardial left ventricular internal hemorrhage, moderate in severity and extent: AIS-2. Transient, nonspecific ST-T wave changes: AIS-2.
Regional Thorax Rating: AIS-2

8   Lungs: Moderate subpleural hemorrhage covering 2/3 of intermediate lobe surface: AIS-3.
Regional Thorax Rating: AIS-3
Abdomen: Liver: Very slight tear at junction of left lateral and medial lobes: AIS-2.

9   Lungs: Contusions, moderate in extent, moderate to severe in severity on intermediate and diaphragmatic lobes: AIS-3.
Spinal Column: Subperiosteal hemorrhage, ventral aspect, T-14 vertebral body; undisplaced: AIS-3. Separation of disk just caudal to T-14 vertebral body: AIS-3.
Spinal Cord: Epidural hemorrhage, T-14 level, left side, moderate: AIS-3.
Regional Thorax Rating: AIS-3.

10 Heart: Three subendocardial internal hemorrhages, left ventricle, moderate; scattered petechiae, left ventricle, mild: AIS-2
Lungs: Scattered, multiple contusions over entire lung, moderate to severe; scattered hemorrhagic bullae and areas of severe, frank parenchymal hemorrhages: AIS-4.

Regional Thorax Rating: AIS-4
Abdomen: Liver: Two small parenchymal tears, one associated with slight subcapsular hemorrhage: AIS-2.
Gall Bladder: Ruptured distally; walls are hemorrhagic; peritoneal fluid was bile-stained: AIS-5.

11 Lymph Nodes: Moderately hemorrhagic: AIS-2.
Abdomen: Falciform Ligament: Partial tear at insertion into liver: AIS-2.
Liver: 12mm long tear between right and middle lobes; 10cc clotted blood in peritoneal cavity: AIS-3.

12 Heart: Subendocardial internal hemorrhage, left ventricle, covering 33% of surface area, severe in degree, less than 1mm in depth: AIS-2.
Lungs: Parenchymal hemorrhage, extensive, severe in degree, apical lobes, especially on vertical surfaces; also five severe hemorrhagic foci, largest 12 mm in diameter, ventral aspect, intermediate lobes: AIS-3.
Regional Thorax Rating: AIS-3.
Abdomen: Falciform Ligament: Partially torn and hemorrhagic at point of origin at diaphragm: AIS-2.
Liver: Slight tear between right and left medial lobes: AIS-3.
Lymph Nodes: Hemorrhagic, near hilum of cephalad mesenteric artery: AIS-2.
Regional Abdominal Rating: AIS-3.

13 Heart: Few subendocardial internal hemorrhages, left ventricle, mild, less than 1mm in depth: AIS-2.

14 Heart: Scattered subendocardial internal hemorrhages, left ventricle, few in number, mild in severity, less than 1mm in depth: AIS-2.
Lungs: Subpleural hemorrhages, severe, right apical lobe, involving less than 1/10 of parenchyma; moderate focus of hemorrhage, right diaphragmatic lobe: AIS-3.
Regional Thorax Rating: AIS-3
Abdomen: Falciform Ligament: Severe hemorrhage and tearing of diaphragm at origin of falciform ligament: AIS-2.

15 Heart: Subendocardial petechiae, left ventricle, very mild.
Vertebra, Last Thoracic: Epiphysial separation: AIS-3.
Abdomen: Mild hemorrhage into muscles surrounding last thoracic and L-1 vertebrae: AIS-2.

ACKNOWLEDGEMENT

The authors wish to acknowledge the expert use of the word processor by Ms. Janis Redinger in the preparation of this manuscript.

902324

Assessment of Air Bag Deployment Loads

John Horsch, Ian Lau, Dennis Andrzejak, David Viano, and John Melvin
General Motors Research Laboratories
Biomedical Science Dept.
Warren, MI

Jeff Pearson, David Cok, and Greg Miller
General Motors Current Product Engineering
General Motors Technical Center
Warren, MI

ABSTRACT

A study of air bag deployments has indicated that some occupant injury was "unexpected" and might have been related to loading by the inflating bag. Laboratory studies have found "high" loads on surrogates when they are out of a normal seating position and in the path and against an inflating air bag (out-of-position). The current study evaluated laboratory methods for assessing the significance of deployment loads and the interaction mechanics for the situation of an occupant located near or against a steering wheel mounted air bag. Analysis of the field relevance of the results must consider not only factors relating to the assessment of injury risk, but also exposure frequency.

The highest responses for the head, neck, or torso were with that body region aligned with and against the air bag module. The risk of severe injury was low for the head and neck, but high when the torso was against and fully covering the air bag module. Torso injury was related to swelling of the module before bag break-out from the module. Reducing inflator output provided only minimal reduction in torso injury risk. Providing pressure relief by an alternate bag break-out path or preventing full coverage of the module reduced the severity of torso loads.

THE RISK OF INJURY by an inflating air bag has been a consideration for passenger and driver systems [1-6]. Laboratory studies indicate that "high" loads can be developed on unrestrained surrogates when they were placed close to the air bag module and in the path of the inflating bag. Analysis of car crashes in which air bags have deployed suggest that loading by the inflating air bag has influenced injury for both driver and right-front occupants [7].

A previous laboratory study of a steering wheel mounted air bag demonstrated a substantial load on the Hybrid III dummy with the chest near or against the air bag at the initiation of inflation [1]. Maximum chest compression was recommended as the best of the test responses to assess injury risk. However, the chest "bottomed" in many tests which prevented precise injury assessment. Additionally,

"high" rates of chest compression were noted. Since that time, improved understanding of the significance of the rate of compression has lead to the development of the Viscous response [8-10] which considers both the amplitude and velocity of compression at each point in time. Because maximum VC occurs before maximum compression, VC is not sensitive to the chest "bottoming".

We estimated the Viscous response from the data reported in [1], as shown in Figure 1. The indicated amplitudes of VC for test conditions with the chest initially against the module are several times the recommended assessment value of 1.0 m/s. The characteristics of the air bag, the spacing from the air bag module at the initiation of inflation, and the interaction environment all influenced VC estimated from that study. On this basis, the perceived injury risk in out-of-position laboratory tests depends on the system tested, the test environment, and the alignment and position of the surrogate at the time of bag inflation. Additionally, the impact response of the surrogate, and the injury tolerance and/or assessment instrumentation and criterion are important.

This study emphasized developing an improved understanding of driver loading by an inflating air bag using a Hybrid III dummy and a physiological model. Laboratory tests placed unrestrained surrogates near or against the air bag module at the initiation of inflation. The study investigated the air bag deployment mechanics that influence the severity of loading in an out-of-position environment. Important questions relating to field relevance are not addressed in this study.

The laboratory tests with an out-of-position surrogate can not be readily extrapolated to real world performance. Important issues that make extrapolation difficult include: 1) the infrequent but unknown frequency for out-of-position exposures; 2) the crash conditions and occupant alignments with the steering assembly if out-of-position exposures occur; 3) the relevance of the laboratory test environment to possible real world out-of-position exposures; and 4) the appropriateness of the injury assessment technology in terms of the surrogates and injury criteria for out-of-position exposures. Thus the results of our laboratory tests should not be directly assumed to represent real world performance without considering field related issues.

Because of limited inflatable restraint field experience, it is likely that significant time will be needed to determine the actual occurrence of conditions that are "infrequent" or "rare". Instead of waiting for field experience to define the magnitude and thus priority of out-of-position exposures, we have continued studies initially reported in 1979 [1], to further understand out-of-position exposures and make continuous improvement of inflatable restraints. These studies included investigation of assessing risk of injury over a range of alignments and body regions and determining the loading mechanics for an out-of-position test surrogate.

METHODOLOGY

If real world out-of-position exposures occur, they will likely be associated with a range of crash severities and occupant alignments with the air bag module. A driver might be out-of-position at the time of inflation due to slumping or reaching for something; or a "minor" crash prior to a more severe impact sufficient to cause inflation. Potential reasons for a range of occupant alignments include: occupant size; seat adjustment; steering column or wheel tilt adjustment; precrash position; crash induced kinematics; and inflation timing. Figure 2 demonstrates a range of alignments due to steering column tilt position. Wheel orientation could vary due to precrash and crash induced rotations of the steering wheel. Belt restraint use will influence not only alignments of the occupant with the steering assembly, but also should reduce the potential for being out-of-position. Our study considered potential alignments for drivers not using the belt restraint.

The Hybrid III was selected for our study because of its human-like frontal impact response for blunt frontal sternal loading and available injury assessment instrumentation for the head, neck, and thorax [11]. The Hybrid III thorax was developed for frontal blunt loading up to 6.7 m/s impact velocity. Since that time, blunt frontal thoracic impacts at 13.4 m/s indicated similar compression and average compressive force to that of human cadavers if the "muscle tensing" built in to the Hybrid III is considered [12]. However the Hybrid III was not directly designed for air bag loading. Additionally, the Hybrid III did not have abdominal injury assessment capability. Thus a physiological surrogate was also used in our studies.

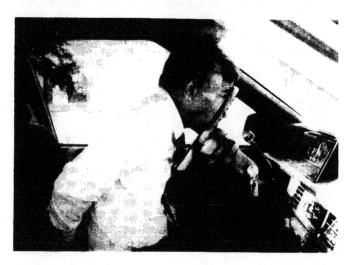

Figure 1. Hybrid III VC response estimated from a previous out-of-position of driver air bag study [1], compared with injury risk recommendations [8-10]. Thoracic injuries are indicated in the response region on the plot from studies on which VC is based.

Figure 2. Influence of column tilt adjustment on the alignment of the air bag module to a 163 cm tall occupant leaning against the steering assembly with the knees against the knee bolster.

Anesthetized swine* of 51 ± 6 kg provided a physiological model to study loading of the thorax and abdomen. The selection considered the similar mid-torso size and mass to human adults and previous use of swine for thoracic and abdominal injury studies [13-16]. An animal model was chosen instead of a human cadaver, based on expectations that injury could include soft tissue and functional injuries [8-10] due to compression velocities as high as 25 m/s as observed in the previous tests with the Hybrid III dummy [1].

TEST ENVIRONMENT - The Hybrid III dummy was tested with a range of alignments relative to the air bag module. These alignments ranged from the head centered on the module to the chest centered on the module and with various separations from the module, as shown in Figure 3 and listed in Table 1. The most representative alignment to assess torso loading was chosen with the air bag module centered on the sternum of the Hybrid III. The Hybrid III chest compression response and the association of this response with injury risk is based on blunt impact centered on the sternum. Fifteen tests with swine were conducted with the air bag module centered on the sternum. In two other tests, the module was centered at the xiphoid and 50 mm below the xiphoid to provide direct upper abdominal loading. The steering wheels and air bag module were aligned as for "straight-ahead" steering, or 90° wheel orientation from that position, Figure 4.

A static environment was used for most of our tests. For these tests, the only energy input was from the inflatable restraint. The stationary test environment was based on the use of the swine model, to minimize kinematic differences between the types of subjects, and to improve the interpretation of instrumentation such as force and acceleration responses.

The static test fixture shown in Figure 3, consisted of the air bag module and steering wheel assembly rigidly supported by a load cell, which was fixed on a bed-plate. The vertical alignment of the fixture allowed the swine, supported in a canvas sling, to be in a natural, spine horizontal position and simplified positioning relative to the test fixture. The sling was adjusted at each end to maintain the spine in a natural alignment which resulted in the wheel and module supporting 0.22-0.24 kN preload. The dummy was placed in a kneeling position. Nominally 0.5 kN of the upper body weight was supported by the steering wheel and air bag module for the sternum centered on the module. This greater preload was primarily due to the greater rigidity of the dummy spine.

---

* The rational and experimental protocol for the use of an animal model in this program have been reviewed by the Research Laboratories' Animal Research Committee. The research follows procedures outlined in publications by the U.S. Department of Health, Education, and Welfare, "Guide for the Care and Use of Laboratory Animals," or the U.S. DHEW National Institutes of Health (NIH), "Guidelines for the Use of Experimental Animals," and complies with U.S. Department of Agriculture (USDA) regulations as specified in the Laboratory Animal Welfare Act (PL 89-544), as amended in 1970 (PL91-579), 1976 (PL 94-270) and 1985 (PL 99-158).

The acceleration due to gravity in this test is low compared with potential exposures in real crashes. The absence of a compressible element such as an axial compression column and the complete covering of the air bag module surface by the test subject likely increase exposure severity. The lack of crash energy and the minimal forces and acceleration compared with an occupant in a car crash may tend to reduce the exposure severity for the test situation [1].

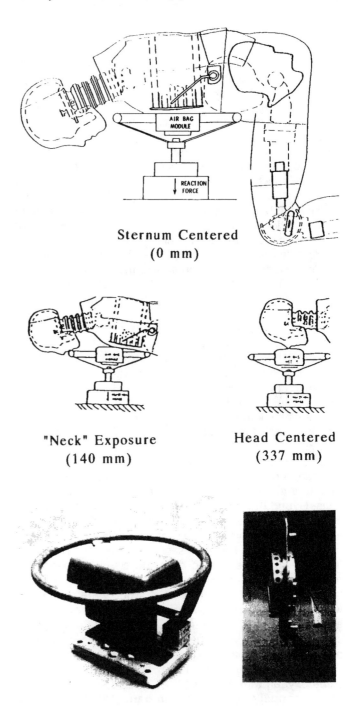

Figure 3. Test environment showing positioning of the Hybrid III dummy, the test fixture with rigid wheel, and the installation of the air bag pressure measurement tube.

Sled tests were conducted in which the initial thoracic alignment of the Hybrid III dummy with the steering wheel was comparable to that used in the static test environment with the sternum centered on the air bag module. The steering wheel was mounted on a compressible column with the column axis location similar to that in a vehicle. The 42 km/h sled tests had a 12g "square-wave" acceleration pulse of about 125 ms duration. The air bag inflation was initiated 35 ms after the start of sled acceleration. The purpose of the sled tests were to compare the static exposure with a dynamic exposure having the air bag module and wheel assembly mounted on a compressible steering column.

TEST HARDWARE - A range of inflator characteristics were used in these tests. Inflators are referred to in this report by their "kpa" and "slope" performance (the maximum pressure and rate of pressure increase when inflated into a one cubic foot tank). One cubic foot tank test data for the "350-15 kpa" inflator is provided in Figure 5. The air bag and module components are shown in Figure 6. The top of the module was nominally flush with the top of the steering wheel rim plane. Two types of steering wheel were used for the tests: 1) a steering wheel "insert" with the rim covered by a slit "rubber hose"; or 2) a "rigid" reusable wheel made from steel plates and tubing which did not exhibit plastic deformation in our experiments.

Compressed gas inflation was used for the tests having a range of longitudinal alignments of the Hybrid III dummy relative to the air bag module. In these tests, the dummy responses were slightly greater than those with a "350-15" inflator, Table 1. The compressed gas was released by a fast acting valve, resulting in a more repeatable exposure at a lower cost and without products of combustion [17].

INSTRUMENTATION - The test fixture instrumentation included a load cell which measured the reaction force of the steering wheel on the bed-plate (or steering column in the sled tests) in the direction of the steering wheel attachment. Internal cushion pressure was measured by a pressure transducer located external from the bag with a short tube ending over the center of the inflator inside of the air bag, Figure 3. These measurements helped define the mechanics of restricted inflation and provided comparisons of exposure severity between the animal and Hybrid III tests. Hybrid III thoracic instrumentation included chest compression, sternal acceleration, and spine triaxial acceleration. Instrumentation specific for the swine subjects is discussed in the following section. Processing of Hybrid III chest compression to determine VC followed the recommendations of Lau [9].

ANESTHETIZED SWINE - The swine were restrained without excitement by injection of ketamine (20 mg/Kg, IM) and acepromazine (200 ug/Kg, IM), then induced to a surgical plane of anesthesia with a mixture of nitrous oxide/oxygen, 2:1, and Halothane, 2.5%, and maintained at the surgical plane with a 1:1 mixture of nitrous oxide/oxygen with 0.5 to 1.5% Halothane. They breathed spontaneously in dorsal recumbancy on a V-shaped support for preparatory surgery.

A tracheotomy was performed and the swine was intubated through the trachael incision with a standard cuffed, eight french, endotracheal tube. The mixture of inhalation gases and oxygen was connected to the endotracheal tube which was sutured in place with a respirometer to record the respiratory cycle. The manubrium and xiphoid were palpated and the length of the sternum was divided into three sections. The sternum was exposed by cauterization of the skin and subcutaneous layers at two points: 1/3 and 2/3 the distance between the manubrium and the xiphoid. Uniaxial accelerometers, which had previously been potted in 5 mm thick by 29 mm diameter silicone disks, were placed

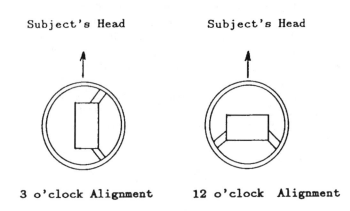

Figure 4. Air bag module rotation alignments.

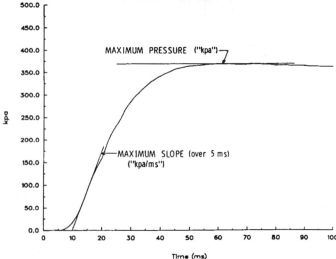

Figure 5. One cubic foot tank and test data for the "350-15 kpa" inflator.

against the sternum with the sensitive axis normal to the ventral surface. The skin and subcutaneous layers were then sutured together over the accelerometers.

A catheter tip pressure transducer was inserted through an incision in the right lateral groin into the femoral artery, and guided into the dorsal aorta to the level of the mid-thorax. ECG leads (5 leads) were attached to the limbs through the subcutaneous tissue. The leads were connected in parallel to a neo-natal physiologic monitor, and to a Holter ECG tape recorder to monitor cardiac parameters during the testing. The animal was then rotated 180° to rest in ventral recumbency.

The thoracic spinous processes transverse to the manubrium and xiphoid were identified. The skin and subcutaneous layers were exposed by cauterizing to the level of the first muscle layer. Uniaxial accelerometers, previously potted in silicone disks, were placed inside the incisions over the top of the spinous processes with the sensitive axis normal to the dorsal surface. The skin and subcutaneous layers were then sutured together over the accelerometers.

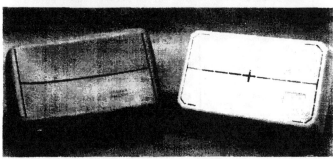

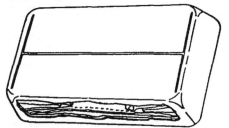

Figure 6. Air bag module components shown without the inflator and a sketch of the modified module.

The swine was transported to the test area and supported in the spine horizontal position in a canvas sling over the test fixture, Figure 3. The sling was attached to a supporting fixture by belts originating at the four out-board tethering points. The horizontal spine in the test position approximates the natural body orientation of the swine. Thus physiologic functions were not compromised as a result of the test position.

Test preparations required about 45 minutes. Immediately prior to the test, the swine was disconnected from the inhalation gases, and breathed room air for less than 1 minute. Reconnection was made immediately after the test. The swine was observed and monitored under anesthesia for 30 minutes without clinical intervention.

RESULTS

Table 1 provides responses of the Hybrid III dummy as a function of longitudinal position and separation from the air bag module. Table 2 provides test hardware, subject alignment, a summary of the observed injuries, and an AIS rating of each injury (1985 [18]) for all tests with anesthetized swine. These tests were conducted with the air bag module against the torso, 15 tests with the module centered on the sternum, one with the module center located at the xiphoid and one with the module center located on the abdomen 50 mm below the xiphoid. Table 3 provides test responses for comparable exposure conditions with the Hybrid III and swine subjects, all with the torso against the air bag module and centered on the sternum. Typical test responses are provided in Figures 7-12.

We were not able to directly measure chest compression in the swine. "Double" integration of sternal and spinal accelerations had large "drifts" and were judged as too inconsistent to accurately determine values for maximum compression or VC in our tests. This judgement was based on the large variations of VC that ranged from much less to much greater than that measured in similar tests with the Hybrid III dummy.

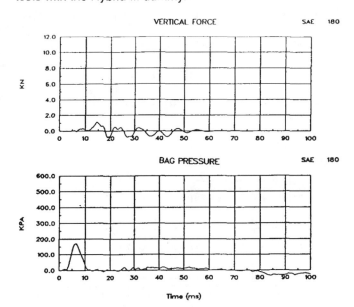

Figure 7. Test responses with no surrogate and "350-15 kpa" inflator.

TABLE 1. Influence of longitudinal alignment of the Hybrid III dummy on body region loading. Compressed gas inflation.

Longitudinal* Position (mm)	Separation (mm)	Inflator **	THORAX Compression Max (mm)	NECK Force (kN) Shear	Tension	Moment My (NM)	HEAD HIC
Head Centered	0	Comp Gas	0	1.1	0.6	99	187
337	50	Comp Gas	0	0.7	0.6	35	57
216	0	Comp Gas	9	2.6	2.4	141	102
190	0	Comp Gas	11	2.8	2.2	157	96
140	0	"350-15"	18	2.3	2.1	132	89
	0	Comp Gas	22	3.3	2.2	169	93
	25	Comp Gas	12	1.1	2.0	72	97
	50	Comp Gas	15	0.2	1.6	36	64
	0 Neck Skin	Comp Gas	23	1.4	1.9	64	84
64	0	Comp Gas	33	2.2	1.7	111	54
	50	Comp Gas	21	0.6	1.2	027	33
Chest Centered	0	Comp Gas	84	0.6	0.9	37	45
0	50	Comp Gas	25	0.2	0.6	29	18

*Longitudinal dimension from center of sternum to center of air bag module.
** Stored compressed gas used to simulate inflator. Responses with "350-15" inflator given for reference.

TABLE 2. Injury summary for swine exposures. Sternum centered on the air bag module except as noted.

Inflator	Wheel	Alignment	Module	Swine Weight (Kg)	Skeletal Injury Description	AIS	Soft Tissue Injury Above Diaphram Description	AIS	Below Diaphram Description	AIS	Injury Severity
"240-11"	Insert	12 O'clock	Std	44.4	None*	0	None*	0	None*	0	0.0
"240-11"	Insert	12 O'clock	Std	48.6	None*	0	None*	0	None*	0	0.0
"240-11"	Insert	3 O'clock	Std	53.6	None*	0	Heart Perforation Rt Atrial Appendage	5	None*	0	5.0
"240-11"	Insert	3 O'clock	Std	50.5	1 Rib Fracture	1	2 Heart Perforations Rt Atrial Appendage	5	None*	0	5.1
"240-11"	Rigid	3 O'clock	Std	54.5	10 Rib Fractures with Hemomediastinum* (Right 3-7; Left 2-6)	4	None*	0	Spleen Laceration	5	6.4
"350-15"	Rigid	3 O'clock	Std	57.1	9 Rib Fractures with Hemomediastinum* (Right 3-7; Left 3-6)	4	None*	0	Spleen Laceration Liver Laceration	5 2	6.7
"350-15"	Rigid	3 O'clock	Std	48.8	17 Rib Fractures with Hemomediastinum (Right x-y; Left s-t)	4	2 Heart Perforations	5	Spleen Laceration Liver Hematoma	4 3	8.1
"350-15"	Rigid	3 O'clock at Xiphoid	Std	47.9	1 Rib Fracture	1	Heart Contusion	3	Liver Laceration Spleen Laceration	5 4	7.1
"360-20"	Insert	3 O'clock	Std	48.6	8 Rib Fractures with Hemomediastinum* (Right 2-5; Left 2-5)	4	None*		None*	0	4.0
"360-20"	Insert	3 O'clock	Std	52.3	3 Rib Fractures with Hemothorax	3	Heart Contusion Brachiophelic Artery Laceration	4 3	None*	0	5.8
"360-20"	Rigid	3 O'clock	Std	53.6	7 Rib Fractures with hemothorax (Right 4-5; Left 4-7)	4	Heart Contusion Brachiophelic Artery Laceration	4 3	Liver Laceration	4	7.6
"360-20"	Rigid	3 O'clock 50mm below Xiphoid	Std	58.6	5 Rib Fractures (Right 4; Left 4-7)	3	None*	0	Spleen Laceration Liver Laceration Gall Bladder Rupture	3 3 3	6.0
"500-30"	Insert	3 O'clock	Std	49.6	None*	0	4 Heart Contusions	4	Inferior Vena Cava 2 Lacerations Spleen Laceration	4 3	7.6
"240-11"	Rigid	3 O'clock	Mod	51.8	None*	0	None*	0	None*	0	0.0
"240-11"	Rigid	3 O'clock	Mod	47.7	None*	0	None*	0	None*	0	0.0
"350-15"	Rigid	3 O'clock	Mod	49.0	None*	0	None*	0	Spleen Laceration	3	3.0
"350-15"	Rigid	3 O'clock	Mod	47.2	None*	0	None*	0	Spleen Laceration	5	5.0

INJURY SEVERITY: square root of the sum of AIS squared for all injuries (heart multiple injuries considered single injury)
NONE: no skeletal fractures; no organ contusive injury; isolated petechiae may be present on the lung, mediastium, peritoneum.
MODIFIED MODULE: 6 and 12 o'clock sides cut out of cover and container, fold modified.
INFLATOR: measured as the maximum pressure and rate of pressure rise in a one cubic foot tank discharge.

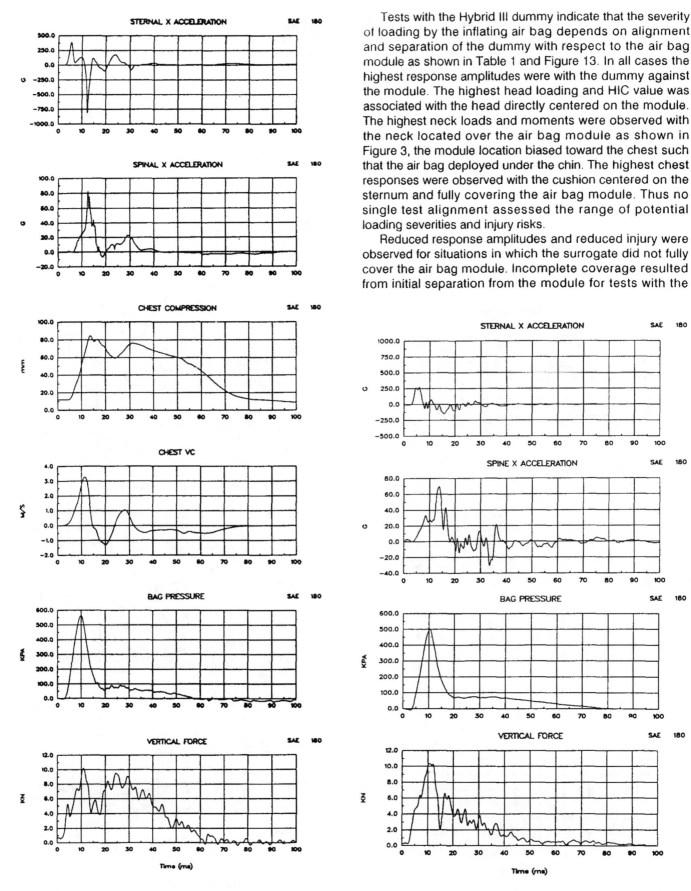

Tests with the Hybrid III dummy indicate that the severity of loading by the inflating air bag depends on alignment and separation of the dummy with respect to the air bag module as shown in Table 1 and Figure 13. In all cases the highest response amplitudes were with the dummy against the module. The highest head loading and HIC value was associated with the head directly centered on the module. The highest neck loads and moments were observed with the neck located over the air bag module as shown in Figure 3, the module location biased toward the chest such that the air bag deployed under the chin. The highest chest responses were observed with the cushion centered on the sternum and fully covering the air bag module. Thus no single test alignment assessed the range of potential loading severities and injury risks.

Reduced response amplitudes and reduced injury were observed for situations in which the surrogate did not fully cover the air bag module. Incomplete coverage resulted from initial separation from the module for tests with the

Figure 8. Test responses for the Hybrid III sternum centered on the module; "350-15 kpa" inflator.

Figure 9. Test responses for swine sternum centered on the module with the wheel at 3 o'clock; "350-15 kpa" inflator.

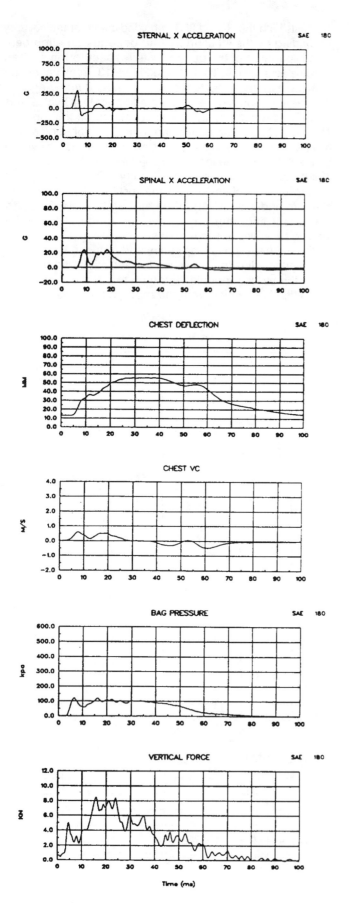

Figure 10. Test responses for Hybrid III sternum centered on the modified module (6 and 12 o'clock sides cut out); "350-15 kpa" inflator

Hybrid III (Table 1) and from the narrowness of the swine chest which could not fully cover the module in the straight ahead wheel alignment (12 o'clock, Table 2). Tests with the modified module also exhibited reduced injury and response amplitudes. Modifying the module by cutting out the sides provided an alternate break-out path for the bag from the module when the normal break-out path was blocked by the test subject. Peak air bag pressures were reduced from 550 kpa to 140 kpa due to the alternate break-out path for this out-of-position test situation.

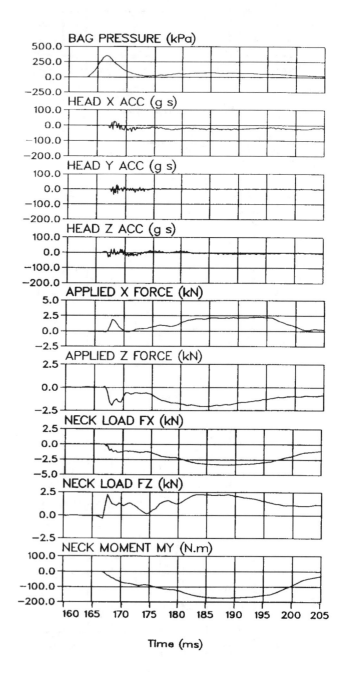

Figure 11. Test responses for Hybrid III dummy against the air bag module, the center of the module 140 mm above the center of the sternum. Compressed gas inflation.

NECK AND HEAD LOADING - The highest head acceleration of 129 g and HIC of 187 was observed with the head centered on and against the module. The peak head acceleration and HIC decreased as the head was positioned further from the module, either longitudinally or by spacing from the module. The risk of head injury in our static tests, based on HIC, appears low even for the situation of the head directly against the module.

The largest values of neck tension, shear force and neck extension bending moment were observed with the air bag module aligned with the neck region. This alignment involved the upper chest against the lower half of the module and the dummy's mouth touching the upper wheel rim. Spacing the dummy 25 mm away from the air bag module greatly reduced the neck responses, with further reduction observed for a 50 mm spacing. Comparison of the neck loads to recommended injury assessment reference values (IARVs) [19] indicates test responses were near or exceeded the values in some situations. For the test situation having the highest response magnitude, the extension bending moment was almost 300% of the IARV of 57 Nm, the neck shear force was near the IARV of 3.1 kN, and the neck tension force was 67% of the IARV of 3.3 kN, Table 1.

The anthropometry of the Hybrid III neck region is not human-like because of a lack of covering for the mechanical neck structure. As a result the neck diameter is smaller than the human neck. The smaller neck diameter in relation to the correctly sized head, and the open space behind the jaw causes Hybrid III to present an unrealistically surface for the membrane forces of an inflating air bag to react against. An appropriate neck skin, which fills in this space, provides resistance to the air bag forces without excessive compression, and also does not influence the neck flexion and extension bending properties of Hybrid III, has yet to be developed and universally accepted for air bag testing.

A prototype neck skin, fabricated from "firm" closed-cell foam appeared to satisfy the first two characteristics and not influence the extension bending response. This was used to investigate the importance of a neck skin in the test with the highest neck shear and moment responses. As shown in Table 1 and Figure 13, the use of the neck skin reduced the extension bending moment by 62%, the neck shear force by 58%, and the neck tension force by 14%. Additional research is needed to define an appropriate neck skin and the resulting interaction of the head and neck with an inflating air bag.

TORSO LOADING - The magnitude of loading and observed injury depended on several parameters. The amount of coverage of the module by the test subject strongly influenced the severity of loading and the magnitude of responses and injury. Incomplete coverage of the module resulted from tests with separation between the Hybrid III chest and the air bag module and from the "narrow" chest of the swine which could not fully cover the air bag module with the 12 o'clock module alignment. Test responses and injury were also less for the modified module due to an alternate break-out path for the air bag when the normal path was blocked by the test subject. The following sections provide results for test with the unmodified air bag module fully covered by the test subject. The other test situations are discussed later.

<u>Anesthetized Swine</u> - Severe to critical injuries were observed for all tests with the swine's torso fully covering the unmodified air bag module. The injuries are listed in Table 2 and included both skeletal and soft tissue injuries to the thorax and upper abdominal organs. Most of the subjects had multiple injuries. Although no AIS = 6 injuries were observed, 5 of the subjects died during the 30 minute observation period after the exposure. All the subjects with heart perforations died within minutes from cardiac tamponade (which prevented the heart from filling). The subject with two lacerations of the inferior vena cava also died due to exsanguination. The cardiac perforations and lacerations of the inferior vena cava were cataloged as 5 and 4 respectively based on the AIS scale. Had

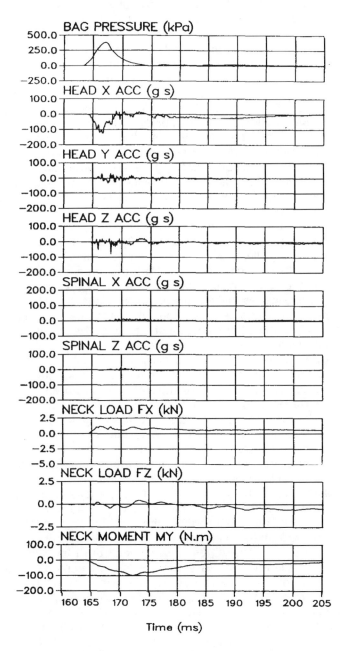

Figure 12. Test responses for the Hybrid III dummy with the head centered on the air bag module; compressed gas inflation

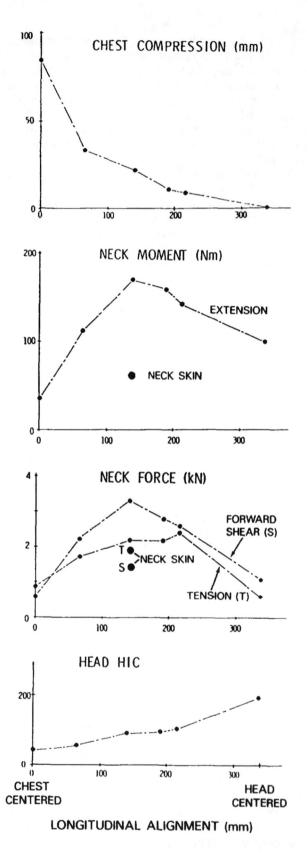

Figure 13. Influence of longitudinal alignment of Hybrid III dummy with respect to the air bag module for the condition of the dummy initially against the module. Longitudinal reference is with the sternum centered on the module. Tests conducted with out a neck skin except for test noted at 140 mm. Compressed gas inflation.

resuscitation been attempted, survival might have been possible.

Maximum AIS (MAIS) has been used as a measure of the overall severity of injury [18], but this considers only the most severe injury in terms of AIS ranking. Multiple injuries increase the severity in terms of threat-to-life [20] and other scales such as the Injury Severity Score (ISS) have been used to assess threat-to-life for situations which include multiple injuries [18,20]. The ISS considers the most severe injury in six body regions, squaring the highest AIS value for the three regions with the highest AIS. For our test situation with a load directed to one body region, the ISS does not provide insight pertaining to the increased threat to life resulting from multiple injuries in the same body region [20].

Because MAIS or ISS did not fully reflect the severity and extent of multiple tissue and skeletal injury for our tests, a Cumulative Injury Severity ranking was used which treated injury severity similar to a method reported by Sacco, et. al. to predict survival rates of 2569 injured patients [20]. They computed "the square root of the sum of squares of the AIS values of the injuries" in a body region. By this method they concluded, for example, that two AIS 4 injuries in a given body region is more severe in terms of threat-to-life than a single AIS 5 injury; stating, "These rankings so obtained agreed with rankings of multiple injuries provided by several experienced trauma surgeons." Patients with AIS 6 injuries were not included in their study since an AIS of 6 always is fatal and thus no chance of survival.

A Cumulative Injury Severity rating, based on the method discussed by Sacco [20], sums the potential interacting effects of all injuries, Table 2. The AIS severity of each discrete injury was squared, the squares summed, and the square root of the sum taken. In several cases, multiple injuries to the same organ were observed. For this situation, a judgment was made to treat this as a single or a multiple injury based on the character of the injuries and their locations. Multiple heart contusions and perforations were considered as a single injury; the multiple contusions probably having a similar influence as a single large contusion, a single perforation likely would also have resulted in cardiac tamponade. The two lacerations of the vena cava were considered as two AIS 4 injuries based on separate locations and that the severity would have been increased if the lacerations would have combined. The Cumulative Injury Severity rating method indicates that higher inflator performance ("kpa" and/or "slope") was associated with a greater risk of fatality and greater tissue damage for the swine, Figure 14. The interaction force tended to be greater with higher performance inflators, ranging from 8.0 to 20.4 kN.

Hybrid III - The VC response ranged from 1.8 to 3.5 m/s for tests in which the chest fully covered the unmodified air bag module. These values are above the 1.0 m/s recommended to represent a 25% risk of an AIS = >4 thoracic injury [8-10] which are consistent with the injuries of the swine in similar exposures. The relationship between the exposure severity in mechanical terms (loading) and the risk of injury to a subject exposed to that loading is non-linear. The risk of an AIS = >4 thoracic injury as a function of VC from [9] is shown in Figure 15. For loading

TABLE 3. Summary of test responses. Sternum centered on air bag module except as noted.

Subject	Inflator	Wheel	Alignment	Module	Chest Compression VC (m/s)	Max (mm)	Chest Acceleration (x) 3ms (g)	Max (g)	Maximum Force (kN)	Max Air Bag Pressure (kpa)
Hyb III	"240-11"	Insert	12 O'clock	Std	2.2	78	23	32	7.9	---
		Insert	12 O'clock	Std	2.1	71	24	30	8.3	---
		Insert	12 O'clock	Std	1.8	68	21	25	7.6	---
		Insert	3 O'clock	Std	1.9	71	30	48	7.0	---
Swine	"240-11"	Insert	12 O'clock	Std					8.0	---
		Insert	3 O'clock	Std					10.4	---
		Insert	3 O'clock	Std			NA		9.9	---
		Rigid	3 O'clock	Std					9.9	524
		Rigid	3 O'clock	Mod					2.3*	92
		Rigid	3 O'clock	Mod					2.6*	102
Hyb III	"350-15"	Rigid	12 O'clock	Std	3.2	84	35	92	8.8	539
		Rigid	12 O'clock	Std	3.3	84	39	83	10.2	559
		Rigid	12 O'clock	Mod	0.6	36	20	24	5.0*	119
Swine	"350-15"	Rigid	3 O'clock	Std					10.3	502
		Rigid	3 O'clock	Std					11.9	547
		Rigid	Xiphoid	Std			NA		9.7	496
		Rigid	3 O'clock	Mod					3.0*	---
		Rigid	3 O'clock	Mod					3.1*	105
Hyb III	"360-20"	Insert	12 O'clock	Std	3.0	74	50	91	12.9	---
Swine	"360-20"	Insert	3 O'clock	Std					16.3	---
		Rigid	3 O'clock	Std			NA		20.4	---
		Rigid	Abdominal	Std					17.6	524
Hyb III	"500-30"	Insert	12 O'clock	Std	3.5	Max	39	147	14.2	---
Swine	"500-30"	Insert	3 O'clock	Std			NA		18.9	---

```
INFLATOR: measured as the maximum pressure and rate of pressure rise in a one cubic foot tank discharge.
MODIFIED MODULE: 6 and 12 o'clock sides cut out of module, modified fold.
Data filtered class 180; VC follows recommendations [9].
* Maximum force during punch-out phase, force higher during membrane phase
--- Tests conducted before air bag pressure measurement was developed
Accelerations not provided for swine because accelerometer mounting not validated.
```

with injury risks near 0% or 100%, changing the loading severity does not significantly change the risk of injury. The relationship between loading and risk is important for a proper interpretation of our test results because many of the exposures were at loading severities beyond the level representing an estimate of nearly a 100% risk of severe injury. The biofidelity of the surrogate is an important issue relating test responses to injury risk, and will be discussed later for our exposure situation.

The higher performance ("kpa") inflators tended to cause more severe loading in terms of chest VC (Figure 14). Peak chest compression, reaction force and chest accelerations are provided in Table 3. Thoracic spine 3-ms acceleration ranged from 21 to 50 g's. Reaction force ranged from 7.0 to 14.2 kN. Although inflator performance influenced the severity of loading, the lowest performance inflator tested still resulted in severe injury for the swine and VC responses well above 1.0 m/s for the Hybrid III dummy

SLED TESTS - Driver side air bag restraint systems can have deformable elements such as an axial compression column supporting the air bag. The force levels observed in our tests without a compressible element suggest that an EA steering column could compress before the time of maximum VC. On this basis, column compression could act to reduce the severity of the loading.

Five sled tests were conducted in which the Hybrid III

thorax was positioned against the steering assembly. The steering assembly consisted of an axial compression column, steering wheel, and air bag module. The thorax to wheel alignment was similar to that in the static environment, the rib cage centered on the air bag module and in contact with the module. These tests used a 12-g "square-wave" sled pulse with a 35 ms delay in deploying the air bag. Compression of the Hybrid III chest ranged from 12 to 23 mm, the load on the steering column at the time of initiation of inflation ranged from 2 to 3 kN. In the static environment with a 1g acceleration, 12 mm of chest compression and 0.6 kN preload were typical. The column fully compressed in all sled tests with about 10 mm of column compression at the time of maximum VC.

A comparison of the chest VC for the Hybrid III dummy between the static and dynamic sled environment is provided in Figure 16. Each point represents the VC response for a pair of similar air bag modules, one tested in the static environment and one tested in the sled environment. The air bag modules for the comparison of static with dynamic test environments were selected to provide a range of VC responses and represent a range of experimental module designs. These comparisons indicate similar thoracic compression between the static and sled environments for the air bag modules tested.

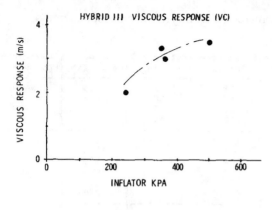
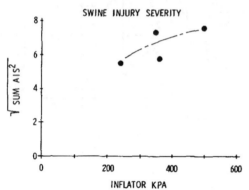
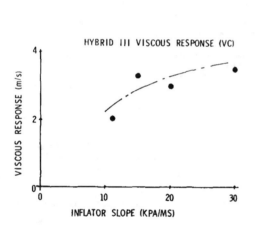
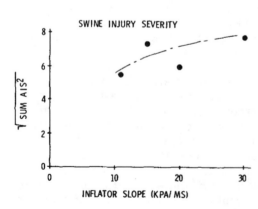

Figure 14. Hybrid III VC response and swine injury severity plotted vs inflator "kpa" and "slope" for the sternum centered on the air bag module. Inflator "kpa" and "slope" determined in one cubic foot tank test.

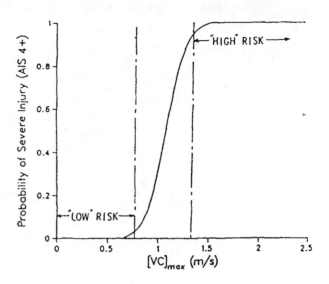

Figure 15. Relationship of injury risk (probability) to the measured loading (VC) [9]. Change of VC in the "low" or "high" risk regions does not significantly change injury risk.

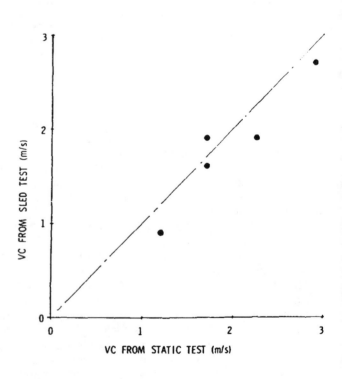

Figure 16. Comparison of out-of-position VC response between the static test environment and sled test for the sternum centered with the air bag module and the Hybrid III dummy against the module.

DISCUSSION

Loading of an out-of-position test subject by the air bag in our experiments is related to internal bag pressure, not an impact by a mass. As long as there is greater available bag volume than gas generated, there will be minimal internal bag pressure and thus minimal load on the subject. There were two primary times during an out-of-position inflation when the bag available volume is less than the volume of generated gas, referred to as "punch-out" and "membrane-force" phases, Figure 17. The significance in terms of injury risk for punch-out and membrane forces depends on the body region.

Punch-out refers to the initial pressurization and swelling of the module before the bag escaped from the module. In this phase, the module acts somewhat as a "cylinder" with a "piston" area similar to that of the module. The punch-out phase ends as the bag breaks-out of the module, rapidly adding volume and "unloading" pressure in the bag.

Membrane force refers to loading after the bag escapes from the module and the bag is acting on the subject. The force is a result of bag pressure acting on the subject and usually includes tension forces from air bag wrap-around. Membrane force can be as great as punch-out force but differ from punch-out by much larger loading areas and lower bag pressures.

PUNCH-OUT PHASE - Initially the bag is folded inside of the module cover and the available volume is constrained by the internal volume of the module. As gas initially flows into the folded bag, the internal pressure increases and the module swells, increasing stresses on the module cover until the cover opens and the bag can escape from the module.

<u>Unrestricted Inflation</u> - For an unrestricted inflation (no subject blocking the opening), the module cover tore along the designed tear pattern releasing the air bag in an axial direction. At this time the released bag added considerable available volume and the pressure dropped to low levels as the bag filled. Peak bag pressures of about 170 kpa were associated with break-out for unrestricted inflations of the air bag modules. Test responses for an unrestricted inflation are shown in Figure 7.

<u>Restricted Inflation</u> - For an out-of-position inflation with the subject fully covering the air bag module, the gas flow acted to swell the module and compress the subject. However the bag could not break-out of the module at the same internal bag pressure as for the unrestricted situation because the test subject was restricting that break-out path. Thus the pressure increased above the unrestricted condition, continuing to rise until the bag found a break-out path causing a significant free volume which resulted in rapid decrease of the bag pressure, shown in Figures 8 and 9. The first sighting of the bag beginning to break out of the module corresponded in time with the peak interaction force for both the dummy and animal subjects and was the approximate time of maximum VC (but not maximum chest compression or spine acceleration). The break-out pressure with the torso fully covering the module was nearly independent of inflator or test subject for tests with the "240", "350", and "360 kpa" inflators, Figure 18. The similar peak pressure is controlled primarily by the module design, not inflator characteristics. Tests with the "500 kpa" inflator had not yet incorporated measurement of bag pressure. Analysis of the interaction force and other tests using stored gas for inflation suggests that for high performance ("kpa") inflators such as the "500-30 kpa" inflator, bag breakout pressures are greater than typical for other inflators tested, indicating rate effects on module break-out strength.

<u>Thoracic Loading Mechanism</u> - The test results for the Hybrid III chest covering the air bag module indicated that the maximum VC and load were near the time of bag breakout from the module and just after peak internal pressure, Figure 8. A modified module was developed to evaluate the hypothesis that the break-out pressure of the bag from the module cover is an important parameter related to the severity of a restricted inflation for chest loading. The module cover was modified by cutting out the 6 and 12 o'clock sides (Figure 6) and the bag fold modified to promote bag side break-out at lower pressure. These modifications resulted in a peak bag pressure of 140 kpa needed for the bag to break-out of the modified module compared with 550 kpa for the unmodified module. The modifications resulted in a greatly reduced loading and injury assessment amplitudes for tests with the Hybrid III dummy and in greatly reduced loading and injury to swine subjects. Table 3 provides a comparison of test response

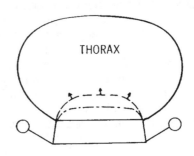

PUNCH-OUT PHASE
BAG INSIDE MODULE

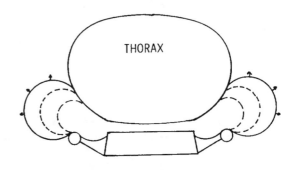

MEMBRANE LOADING PHASE
BAG OUTSIDE MODULE

Figure 17. Schematic of "punch-out" and "membrane" loading for a fully restricted inflation.

amplitudes. The "break-out" pressure of the bag from the air bag module was a major parameter influencing the severity of loading to test subjects that were against and completely covering the air bag module at the initiation of inflation. The specific modifications used to demonstrate this principle did not consider the many requirements of an air bag module; other methods to control bag break-out pressure would be required for a practical application.

Partial Restriction - The amount of restriction of bag break-out by the out-of-position subject determined the break-out path, the maximum bag pressure, and the loading of the subject. A high risk of injury based on swine tests or VC response in Hybrid III tests was associated only with the situation that the subject was almost fully covering and thus restricting the bag break-out. The head or the neck could not fully restrict bag break-out and thus punch-out did not cause "high" loads to these body regions. Axial spacing of the Hybrid III dummy chest from the module of 12 and 25 mm resulted in greatly reduced thoracic responses, reaction force, and reduced bag internal pressure developed for the bag to break-out of the module, Figure 19.

Neither of the surrogates were judged to have interacted with the test environment in a completely "human-like" manner. A limitation of these surrogates is the shape of the chest and thus the potential to restrict the air bag break-out from the module. The swine's chest is "V" shaped and could cover only about 2/3 of the top of the air bag module with the 12 o'clock alignment but could fully cover the top surface of the module with the "3 o'clock" alignment as shown in Figure 20. The partial coverage of the module surface compared with full coverage resulted in no injury (2/2 subjects) compared with critical injury (3/3 subjects) for the fully covered condition. The "240-11 kpa" inflator was the only one used for this comparison of module coverage. The reaction force was also less for the 12 o'clock alignment indicating that the reduced injury was due in part

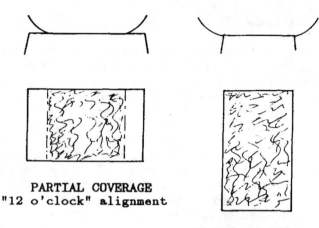

Figure 19. Influence of initial separation between the air bag module and chest for tests with the Hybrid III dummy and "350-15 kpa" inflators.

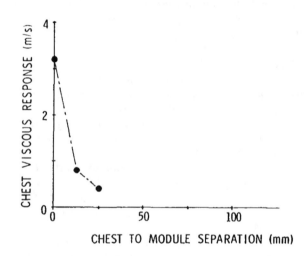

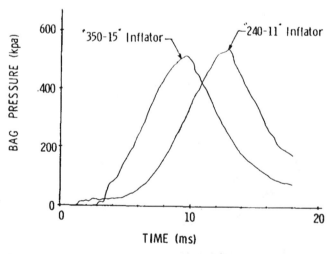

Figure 18. Air bag pressure as a function of time for "350-15" and "240-11 kpa" inflators with swine subjects. Maximum air bag break-out pressure appears to be a function of the module, not gas flow characteristics for this range of inflators.

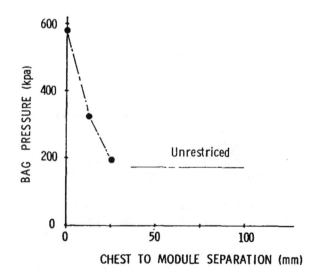

Figure 20. Influence of the "V" shape of the swine's chest on coverage of the air bag module.

to a reduced loading severity caused by less restriction of the air bag break-out of the module. In contrast, the Hybrid III chest more closely simulates the shape of a driver's chest. It could nearly cover the air bag module for either the 12 or 3 o'clock module alignment. Dummy and force responses were similar for both wheel alignments.

The swine's chest is more compliant in covering the air bag module prior to inflation than that of the Hybrid III. Thus the swine in the "3 o'clock" alignment conformed with the module surface better than the Hybrid III dummy in either the 12 or 3 o'clock alignment. The Hybrid III dummy did not provide the same degree of restriction based on the bag break-out path. Additionally, the reaction forces were greater for the swine than for the Hybrid III dummy, indicating a greater loaded area and restriction of air bag break-out from the module for tests with the swine.

The Hybrid III chest shape is much closer to that of a 50th percentile male than that of the swine. On that basis, its interaction with the deploying module might best represent the interaction for a driver. However the Hybrid III chest does not cover the air bag module as a 50th percentile male driver, Figure 21, and is statically stiff compared with the human chest, both superficially and in terms of the "spring-rate" of the rib cage, Figure 22. Thus we judged that the Hybrid III dummy chest is less likely to conform when just touching or under load to fully restrict the module as a human would.

<u>Hybrid III Positioning</u> - The Hybrid III dummy is limited in the range of possible body alignments compared with car occupants because it is not flexible due to its "tensed" posture. For example, the spatial relationships relative to the steering system are quite different between the Hybrid III dummy and humans for the condition of the knees against the bolster and then leaning against the steering assembly, Figure 23. The dummy is less flexible and adjustable than a human due to such factors as a rigid thoracic spine and general stiffness to local loading. Although many joints such as the elbow can be set at any initial angle within the range of motion, other joints such as the neck, lumbar spine, and shoulder tend to have a unique initial position which is not constrained in the human. Thus the Hybrid III dummy can not achieve the same range of out-of-position alignments as humans.

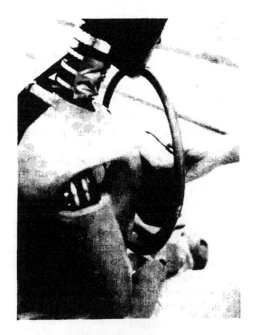

Figure 21. Curvature of the Hybrid III chest precludes full coverage and thus restriction of the air bag module at low load conditions. This contrasts with human subjects which easily fully cover the module surface. Subject in the photo has slightly less than 50th percentile male dimensions for height, weight, and chest circumference.

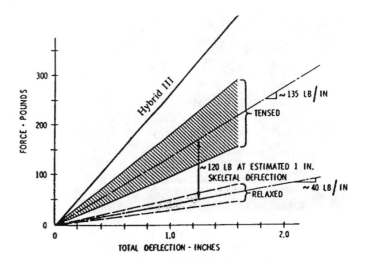

Figure 22. "Comparison of thoracic AP force-deflection characteristics for male volunteers in relaxed and tensed conditions". Reproduced from Lobdell [28] with quasistatic stiffness of Hybrid III chest loaded by a 152 mm diameter disk added to the plot.

Judgments need to be made relative to several important issues when using the Hybrid III dummy to assess the significance of out-of-position exposures for car occupants. The dummy might not achieve a position close to or against the air bag module. Even when the dummy is close to the module, the shape and stiffness of the Hybrid III chest appears less likely to fully restrict the air bag break-out than that of humans. Additionally, the bag loading on the dummy might not be directed to best utilize the injury assessment capability of the dummy. In some situations it might be important to adjust the test configuration to account for limitations of the dummy and thus provide test responses more relevant for assessing injury risk to actual car occupants. Adjusting the dummy to a position similar to that of persons in the car might be more realistic than accepting the dummy's "natural" position as a simulation of a driver's potential positioning.

MEMBRANE LOADING PHASE - Loading of the test subject can occur after the bag breaks out of the module and interacts with the subject. A subject in the path of the inflating air bag decreases the volume of the bag and with sufficient volume of generated gas, internal bag pressure increases. This bag pressure results in membrane loading of the test subject and can cause "high" loads on all body regions. For the same force on the test subject, membrane loading compared to punch-out is associated with much lower bag pressure and greater loaded area on the test subject.

163 cm
68 kg

193 cm
102 kg

Hybrid III

Figure 23. Range of positions for the Hybrid III dummy relative to the steering assembly is much more limited than for humans and in some situations prevents the Hybrid III from being close to the module while humans can easily achieve an "against" posture. Condition shown is with column tilt full-up; knees against the knee bolster, and subject leaned against the steering wheel.

Torso Exposures - A second load peak occurred after the punch-out load with the thorax against the air bag module, Figure 8. The magnitude of the membrane induced forces were as great as punch-out forces for some test situations, but differed in terms of loaded area on the subject and internal bag pressure. The punch-out phase indicates load areas of 160-200 square cm (force/pressure) for a fully restricted inflation. In contrast, the membrane loading suggests load areas of 650-1000 square cm.

The bag pressure during membrane loading is due to a greater gas volume than available bag volume, the bag volume reduced by the test subject, steering wheel, and possibly incomplete deployment of the bag from the module. Tests with the Hybrid III dummy indicated that the risk of severe thoracic injury was associated only with the punch-out phase in our test situation. The forces associated with membrane loading were distributed on the torso while the forces during punch-out were directed only on the sternum and rib cage and thus during punch-out, the full load acts to compress the chest as contrasted with membrane loading in which only part of the load acts to compress the chest.

Abdominal Injury - Abdominal injury was associated with either punch-out or membrane loading, depending on the test situation. The Hybrid III dummy does not have abdominal injury assessment, thus tests with the swine provide the basis for analysis. With the abdominal region contacting the air bag module, severe abdominal injuries occurred, likely during punch-out due to the high loads and smaller loaded area. However injury might also be associated with membrane loading. The abdominal region is only partially protected by the rib-cage and does not have shoulders to share loading. Thus, the abdomen is likely more easily compressed than the thorax for both punch-out and membrane loading.

For the mid-sternal alignment used for most tests, the punch-out loading was close to but not directly on the upper abdominal region; the liver and spleen lacerations probably occurred during the membrane loading phase. This judgment is based on two observations: 1) abdominal injury is strongly associated with rigid steering wheels; and 2) abdominal compression occurred only in the membrane phase for the modified air bag module with the 6 and 12 o'clock sides removed.

The strong association of abdominal injury with the rigid simulated steering wheel (4 of 4 subjects) and the lack of abdominal injury with the steering wheel insert (1 of 5 subjects) suggests that abdominal injury for the module centered on the sternum was associated with membrane loading. Since the air bag module is the only reaction surface during punch-out, there should be little difference in injury potential due to a rigid vs insert wheel during punch-out. However the wheel rim and spokes become reaction surfaces for the bag during membrane loading and thus deformation of the wheel insert likely reduced abdominal loading during the membrane loading phase.

Accelerometers were implanted on the sternum for most tests. Although they were not consistent for determining compression by double integration, they did provide good measures of the time of compression. Accelerometers were also located below the sternum for the tests having the 6 and 12 o'clock sides of the air bag module removed. The integration of the sternal and abdominal accelerometers indicated that the chest compressed but that no abdominal compression occurred during punch-out. Later during the onset of membrane loading, both the chest and abdomen compressed, leading to the judgment that the spleen injury probably occurred during the membrane loading and not during punch-out.

Our test situation probably emphasized abdominal injury based on several factors: 1) the alignment of the air bag module centered on the sternum was closer to the abdominal region than in tests more nearly simulating a typical vehicle test environment (this observation should also consider that real world exposures could have a range of alignments due to variations in occupant size, wheel tilt position, and crash induced column motions); 2) the 3 o'clock alignment of the module in tests with the swine covers more body length than the 12 o'clock alignment, locating part of the expanding module closer to the abdominal region for the same module center alignment (real world crash situations might have any wheel rotational alignment); and 3) the occurrence of abdominal injury is strongly related to tests with the rigid wheel as previously discussed. Thus steering wheel characteristics might be important for membrane induced injury risk, the rigid wheel used in our tests is not representative of air bag steering wheels.

Neck Loading - The largest neck loads occurred during membrane loading as the bag wrapped around the neck and chin. The peaks occurred at the same time as the second bag pressure peak shown in Figure 11. The first pressure peak was associated with bag break-out from the module. Examination of the head acceleration data indicates that the head x acceleration remains significant during the neck loading while the z acceleration is much lower and lags the corresponding z axis neck force peak.

The computed external applied z direction force acting on the head is close in magnitude to the z axis neck force, while the corresponding x axis neck shear force is much larger than the computed externally applied x direction force. This indicates that a portion of the neck shear force comes from the inertial loading of the head on the neck. The major contributions to the neck loads in both tension and shear, however, are due to the membrane loads on the head/neck area. The x applied load is acting on the head in the positive or forward direction and thus adds to the inertial forces of the head acting on the neck. This action presumably comes from the bag pushing forward on the chin area of the dummy and might be over emphasized by the construction of the Hybrid III chin. This behavior also appears to produce unrealistically large extension moments. The use of an appropriate neck skin with the Hybrid III appears to have the potential to produce what we believe to be more realistic loading of the neck during membrane loading.

The head remained in a "straight-ahead" position during the "high" neck loading and is not in hyperextension of the head/neck for which the 57 Nm IARV was recommended. Thus the significance of moment loading is not established under this particular loading condition. If the moment response is discounted, the neck shear and tension force indicate a low risk of injury in the test situation. The responses with the neck skin also suggest a low risk of neck injury for our test situation even if the moment is not discounted.

The appropriateness of the IARVs for air bag loading and the biofidelity of the Hybrid III neck responses for these loading situations has not been established. The use of an appropriate neck skin appears important. The coincidence of peak neck loads and moments is unlike lap-shoulder belt restraints, where the peak neck tension and shear forces are reached at different times. Because of the coincidence of maximum loads, it might be appropriate to consider the combined effects of tension, shear, and bending loads rather than assess each individually. Our preliminary results are unable to address these issues.

THORACIC INJURY ASSESSMENT - Dummies are not "injured" and injury assessment must be based on measured responses. The biofidelity of the dummy response in the loading environment is important. The Hybrid III dummy was chosen for thoracic exposures based on recommendations that it is the best available dummy to assess the injury risk for frontal impact [11,12,22]. However this does not assure that responses for our experiments are biofidelic.

Biofidelity of Hybrid III Thorax - A car occupant can be injured at any body location while the dummy can assess injury risk or the magnitude of loading only at discreet and incomplete locations. Thoracic injury was studied with the air bag module centered on the sternum, similar to the alignment for calibration of the Hybrid III chest impact response and the condition for which the relationship of compression to injury has been extensively evaluated. The relevance of the chest compression response to determine the magnitude of loading for substantially different alignments with the chest is unknown, but this is likely an important issue. The reduction of chest compression (and VC) for loading not at the center of the chest should not be automatically assumed to represent lower injury risk, but may represent a reduced ability to assess the severity of the loading and the risk of injury at the point of loading. The relationship of loading vs injury risk likely varies as different parts of the torso are exposed.

The Hybrid III chest compression response is similar to that of human cadavers (adjusted for "muscle-tensing") at 4.3 and 6.7 m/s impact velocity in blunt frontal impacts [11].

Additionally, the Hybrid III chest compression was shown to be similar to that of human cadavers for blunt frontal impacts at 13.4 m/s with a 4.25 kg mass which resulted in 60 mm of chest compression [12].

We observed chest compressive velocities in the range of 6 to 14 m/s for fully restricted inflations. The maximum compression velocity observed with a VC of 1.8 m/s was 6.5 m/s and a maximum chest compression of 71 mm. Sternal acceleration during compression was less than 10% of that reported for the 13.4 m/s blunt impacts. The estimated loaded area of 160-200 square cm (force/pressure) suggests similar loaded areas between the blunt impacts and the out-of-position thoracic exposures which used a 152 mm diameter flat disk. Thus the Hybrid III chest appears to be an appropriate test device for air bag punch-out loading exposures centered on the sternum for VC responses up to and above 1.0 m/s. It is less important for the response to be biofidelic at VC responses well above that associated with nearly a 100% risk of critical injury.

Injury Risk Assessment - The VC response derived from chest compression was selected as the primary response for thoracic injury assessment with the Hybrid III dummy because VC is the best indicator of injury risk for blunt loading at compressive velocities above 3 m/s [8-10]. Our analysis has demonstrated that the loaded area during punch-out is at least as large as that for the experiments on which VC is based. We observed chest compression velocities as high as 14 m/s. Thus we recommend VC as the best thoracic response to judge the severity of thoracic loading by the deploying air cushion.

The results of our study are consistent with recommendations of VC = 1.0 m/s [8-10] for a "threshold" assessment value. Hybrid III test conditions with a chest VC greater than 1.0 m/s resulted in severe injury to the anesthetized swine. Situations in which the Hybrid III chest VC was less than 1.0 m/s resulted in no chest injury, Figure 24. However this study is not directly structured to determine a tolerance level because: 1) the exposures are clustered well above or below a Hybrid III VC response of 1.0 m/s; 2) we are not able to estimate compression response for the swine; and 3) the swine had a higher load than the Hybrid III dummy.

As previously discussed, the swine provided a greater coverage and thus restriction of the deploying module resulting in greater forces, ranging from 7.0 to 14.2 kN for the Hybrid III compared with 8.0 to 20.4 kN for the animal subjects. These force amplitudes are well above those observed in blunt sternal impacts to human cadavers that resulted in severe injury (1.6 to 7.9 kN for AIS = 4-6 injury)[21].

Peak chest compression of the Hybrid III ranged from 68 mm to bottoming the chest (>84 mm) for exposure conditions resulting in severe to critical injuries in the animal model. Chest acceleration was less than 60 g (3 ms level) for all exposures, ranging from 24 to 39 g's in the Hybrid III dummy for exposure conditions resulting in severe to critical injuries in the animal model. As found for blunt impact to the chest of human cadaver and animal subjects where the impact is applied only to the rib cage [23], severe injury is associated with less than 60 g's and generally occurs before peak spinal acceleration or chest compression [24].

INFLUENCE OF TEST METHOD - The test conditions can strongly influence the level of loading to the test subject and thus the risk of injury. Our test environment does not fully simulate any potential real world exposure. The static test environment provided a means to compare responses of the surrogates in an objective environment which could minimize differences between surrogates and enhance interpretation of the underlying mechanics of restricted air bag inflations. As such, projection of loading and injury risk to real world performance must consider important differences between the test situation and the range of potential car crash exposures.

The responses in any test exposure represent the magnitude of loading for that specific situation. The interpretation of the influence of a test parameter should not be strictly in terms of the loading magnitude and associated injury risk for the specific test situation, but should consider the potential influence in real world situations [25].

If our test results were directly interpreted as representing real world performance, one might conclude that inflator output performance will not influence injury risk to out-of-position occupants since severe injury was found for all inflators tested. This might not adequately link laboratory tests to real world expectations because it does not consider important differences. Assume that we had chosen to conduct the tests with a slightly different alignment, a few millimeters away from contact. The loading severity as a function of separation, Figure 19, suggests that a lesser loading would be associated with each inflator than for the fully restricted test situation. A hypothetical relationship is shown in Figure 25 for three levels of separation that reflects possible trends. At full restriction, all levels of inflator performance ("kpa") resulted in nearly 100% risk of severe injury. For a "small" separation in the hypothetical situation (Figure 25), severe injury is not certain at "low" levels of inflator performance. At a "moderate" separation, the hypothetical risk of injury is

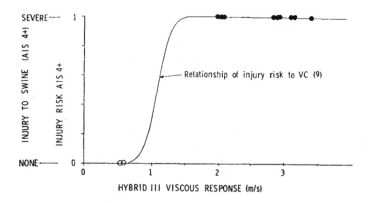

Figure 24. Comparison of Hybrid III chest VC response to swine thoracic injury in similar test exposures with the sternum centered on the air bag module, consistent with recommendations by Viano and Lau [8-10] of a VC = 1.0 m/s as an injury assessment value

low to moderate for all inflator performances. Real world exposures would likely include a range of restrictions. This hypothetical example leads to the perception that lower performance inflators will be associated with less out-of-position injury when compared to the same module with higher inflator performance, the difference being primarly in partially restricted inflations. Clearly any strategy to reduce inflator performance ("kpa") must also consider the effect it would have on restraint for the in-position occupant.

Real world out-of-position exposures will likely be dynamic -- a crash will be part of the environment. As such, accelerations, forces, and energy resulting from the crash will influence the interaction of the out-of-position unrestrained driver with the inflating air bag. In general these forces, energies, and accelerations can increase restriction of bag break-out and increase the severity of the interaction between the driver and steering system. A previous study has shown that sled acceleration increases the loading severity for a fully restricted module when both situations have an axial compression column, and that impact by the test dummy further increases the severity [1]. On this basis, the test environment is probably less severe with respect to the dynamics of car crashes.

The sled tests showed similar Hybrid III chest compression VC response to those observed in the static test, Figure 16. This should not be interpreted that this similarity is valid for all possible sled test conditions or real world situations. However the similarity does demonstrate that the static test exposure falls within the range of relevant sled test situations.

OUT-OF-POSITION INJURY DISTRIBUTION - Injury in car crashes depends not only on human tolerance, but also on the frequency and severity of exposures [25-28]. Two parameters influence injury to a drivers against an inflating air bag: 1) the risk of injury for an out-of-position exposure (the probability is zero below the minimum crash severity which will trigger inflation); and 2) the probability of a driver being out-of-position.

There are a variety of situations that might result in the driver interacting with the inflating air bag. Although our study did not address this issue, we believe that exposures modeled in our experiments will be "infrequent" or "rare". This judgment is based on the field experience of the 1970's GM air bag system [7] and the relatively "narrow" range of test alignments resulting in "high" biomechanical responses. Belt restraint should act to reduce the probability of an out-of-position exposure. Field experience will be the ultimate method to objectively address exposure frequency for out-of-position exposures.

Injury Distribution Based on NASS Data - For frontal crashes, the risk of severe injury increases as ΔV increases, Figure 26(a), as reported by Malliaris [27]. However the distribution of injured drivers is much broader than just severe crashes and much of the injury is associated with crash severities having relatively low injury risk on a per crash basis. The distribution of injury is the product of injury risk and exposure frequency, resulting in injury being distributed over a wide range of ΔV. At lower crash severities, injury is characterized by a very low risk per exposed occupant but there are many occupants exposed.

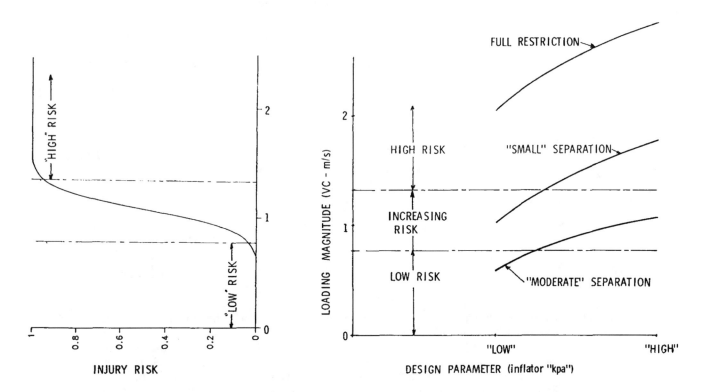

Figure 25. <u>Hypothetical</u> influence of a design parameter (inflator "kpa") on loading (VC) for three levels of the degree of restriction of air bag break-out (separation). Even though the design parameter does not change the injury risk for full restriction, the design parameter could influence injury risk for other exposure situations (partial restriction).

For severe crashes, there is a higher risk but a relatively low exposure rate. Among the reasons for injuries in low severity crashes are occupants with lower than average tolerance to impact, such as older occupants.

Even though out-of-position exposures with air bags will probably have a different severity distribution than represented by the NASS frontal data, use of the NASS distribution is helpful. Assuming that the probability of an out-of-position occurrence is related to NASS frontal crash data by a fraction "P", then the ΔV distribution of out-of-position exposures will be similar to that of NASS frontal crashes, except reduced by the constant factor P, Figure 26(b). The crash severity threshold for air bag deployment is an important factor for the exposure distribution since below that crash severity, no inflations and thus no out-of-position exposures will occur. For purposes of discussion, a simplifying assumption is made that the crash severity threshold for inflation can be related to NASS ΔV. In the hypothetical example shown in Figure 26(b), two severities for the threshold of inflation are illustrated, one at 17 km/h and another at 25 km/h.

The number of out-of-position exposures is strongly related to the inflation threshold, Figure 26(b), because there are many more "lower" than "higher" severity

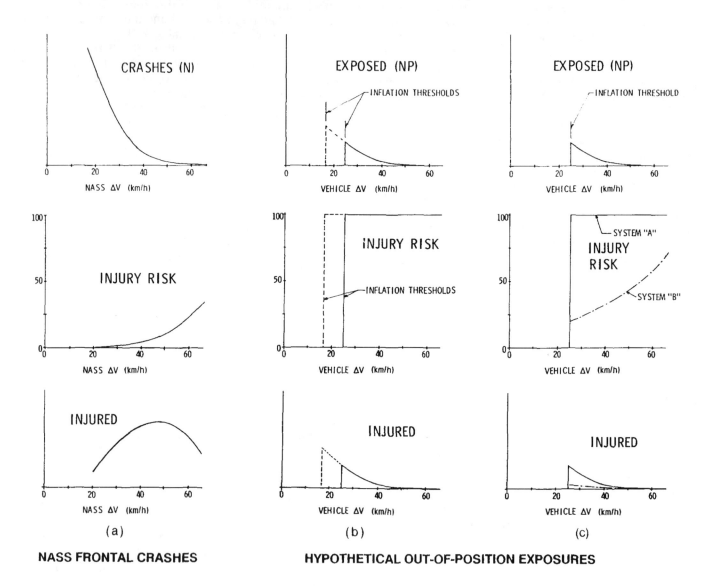

Figure 26. Factors influencing out-of-position injuries based on NASS frontal crash exposure (a). A <u>hypothetical</u> distribution of out-of-position exposures and injury is shown as a function of crash severity (ΔV) assuming: 1) the probability "P" of a restricted deployment is independent of ΔV is similar to a fraction (P, where P<<<N) of the NASS distribution; and 2) the crash inflation threshold can be determined by vehicle ΔV, with thresholds at 17 and 25 km/h indicated. Injury risk is shown for a system having in (b) nearly 100 % risk of severe injury when fully restricted and in (c) a hypothetical system having less than a 100% risk. The number of drivers injured in this analysis depends on the product of the fraction of out-of-position exposures P, the risk of injury associated with an exposure, and the inflation threshold.

crashes. There are more exposures between the 17 km/h and 25 km/h than above 25 km/h.

Injury risk in our tests with a fully restricted air bag deployment was nearly 100% even when the only energy was from the inflating air bag. Thus injury would differ greatly from that in the NASS frontal crash distribution. For a hypothetical system that results in an almost 100% risk of severe injury to a driver fully restricting the air bag break-out from the module, the injury risk to such a positioned driver is 100% for any crash severity above the inflation threshold, Figure 26(b). A system having reduced loading might have high risk of injury only in very severe crashes where the combination of forces influences injury, indicated schematically in Figure 26(c).

This discussion, based on several simplifying assumptions, suggests that out-of-position injury will occur in low severity crashes because of high incidence. It also suggests that the threshold crash severity to deploy an air bag is a significant factor. Because of the possible bias of out-of-position exposures toward crash severities just above the trigger threshold, selection of test severities for out-of-position evaluation should consider these factors. How well our discussion relates to real world exposure depends on how well the simplifying assumptions in the discussion relate to real world issues.

SUMMARY

"High" loads were developed on unrestrained test subjects near or against a steering wheel mounted air bag at the initiation of inflation. Two distinct loading phases could occur, "punch-out" and "membrane". Punch-out loading was associated with swelling of the air bag module. Unloading occurred when the air bag broke out of the module cover. Membrane loading occurred during the subject's interaction with the pressurized bag. Our test exposures ranged from the air bag aligned with the head to aligned with the chest or abdomen. The greatest risk of injury was associated with the torso against and fully covering the air bag module at the initiation of inflation during punch-out.

With the torso against the air bag module, a primary factor relating to the loading severity was the increase of bag pressure needed for the bag to escape from the module. When the test subject blocked the normal path, a pressure of 550 kpa was needed for the bag to break-out as compared to 170 kpa with unrestricted inflation. An alternate break-out path representing a pressure relief mechanism was used to demonstrate that preventing "high" internal pressures during punch-out reduces the severity of surrogate loads.

A Hybrid III dummy chest VC of 1.0 m/s was consistent with critical thoracic injuries to the swine in similar test conditions. The risk of abdominal injury could not be assessed in the Hybrid III dummy because it lacked a measurement capability. Tests with the swine indicated a risk of critical injury with the air bag aligned on the upper abdominal region. Interaction of the inflating air bag with the Hybrid III neck and chin was judged to be unrealistic due to the small neck size and open area under the chin. A neck skin greatly reduced neck loading and indicated minimal injury risk.

Our static test environment provided similar torso loading and responses to a more complex sled test. The static test has advantages in terms of cost, time, and repeatability to conduct a detailed study of out-of-position torso responses for steering wheel mounted air bags.

Injury distribution in real world crashes depends on risk of injury and the frequency of exposure. We studied issues related to severity of loading and risk of injury irrespective of crash type, severity, or frequency of exposure.

ACKNOWLEDGEMENT

The authors appreciate the technical contributions of Joseph McCleary, Gerald Horn, Todd Townsend, Brian Frantz, Howard Bender, and Bob Jones.

REFERENCES

1. J. Horsch and C. Culver, "A Study of Driver Interactions with an Inflating Air Cushion", SAE Paper No 791029, 23rd Stapp Car Crash Conference, October, 1979.

2. L. Patrick and G. Nyguist, "Airbag Effects on the Out-of-Position Child", SAE Paper No 720442, 2nd International Conference on Passive Restraints, 1972.

3. H. Wu, S. Tang, and R. Petrof, "Interaction Dynamics of an Inflating Air Bag and a Standing Child," SAE Paper No 730604, May, 1973.

4. B. Aldman, A. Anderson, and O. Saxmark, "Possible Effects of Air Bag Inflation on a Standing Child," 18th Conference of the American Association for Automotive Medicine, 1974.

5. H. Takeda and S. Kobayashi, "Injuries to Children From Airbag Deployment", SAE Paper No 806030, 8th Experimental Vehicle Safety Conference, 1980.

6. H. Mertz, G. Driscoll, J. Lenox, G. Nyguist, and D. Weber, "Response of Animals Exposed to Deployment of Various Passenger Inflatable Restraint System Concepts for a Variety of Collision Severities and Animal Positions", Paper No 826047, Ninth International Technical Conference on Experimental Safety Vehicles, November, 1982.

7. H. Mertz, "Restraint Performance of the 1973-76 GM Air Cushion Restraint System", SAE Paper No 880400, February, 1988.

8. D. Viano and I. Lau, "A Viscous Tolerance Criterion for Soft Tissue Injury Assessment." J Biomechanics 21(5):387-399, 1988.

9. I. Lau and D. Viano, "The Viscous Criterion: Bases and Applications of an Injury Severity Index for Soft Tissues." 30th Stapp Car Crash Conference, SAE Paper No 861882, October, 1986

10. D. Viano, "Cause and Control of Automotive Trauma." Bull NY Acad Med, Second Series, 64(5):376-421, June, 1988.

11. J. Foster, J. Kortge, and M. Wolanin, "Hybrid III-A Biomechanically-Based Crash Test Dummy," SAE Paper No 770938, 21st Stapp Car Crash Conference, October, 1977.

12. J. Horsch and D. Schneider, "Biofidelity of the Hybrid III Thorax in High-Velocity Frontal Impact", SAE Paper No 880718, February, 1988.

13. C. Kroell, et.al., "Interrelationship of Velocity and Chest Compression in Blunt Thoracic Injury", SAE Paper 811016, 25th Stapp Car Crash Conference, 1981.

14. J. Horsch, et.al., "Mechanism of Abdominal Injury by Steering Wheel Loading", SAE Paper No 851724, 29th Stapp Car Crash Conference, 1985.

15. I. Lau, et.al., "Biomechanics of Liver Injury by Steering Wheel Loading", The Journal of Trauma, Vol. 27, No 3, 1987.

16. D. Viano and C. Warner, "Thoracic Impact Response of Live Porcine Subjects", SAE Paper 760823, 1976.

17. T. Chan, et.al., "Exposure Characterization of Aerosols and Carbon Monoxide from Supplemental Inflatable Restraint (Automotive Air Bag) Systems", J. Aerosol Sci., Vol. 20, No. 6, pp 657-665, 1989.

18. Abbreviated Injury Scale (AIS), Publisher: AAAM Morton Grove, Il., 1985

19. H. Mertz, "Injury Assessment Values Used to Evaluate Hybrid III Response Measurements", February 1984, GM submission to NHTSA Docket 74-14, Notice 32, Enclosure 2 of attachment I of USG 2284 Part III, March 22, 1984.

20. W. Sacco, et. al., "Progress Toward a New Injury Severity Characterization: Severity Profiles", Computers in Biology and Medicine, December, 1988.

21. C. Kroell, et. al., "Impact Tolerance and Response of the Human Thorax II", SAE Paper No 741187, 1974.

22. J. Horsch and D. Viano, "Influence of the Surrogate in Laboratory Evaluation of Energy-Absorbing Steering Systems", Twenty-Eighth Stapp Car Crash Conference, October, 1984, SAE Paper 841660.

23. R. Neathery, et. al., "Prediction of Thoracic Injury from Dummy Responses", Nineteenth Stapp Car Crash Conference, November, 1975, SAE Paper 751151.

24. I. Lau and D. Viano, "How and When Blunt Injury Occurs: Implications to Frontal and Side Impact Protection", Thirty Second Stapp Car Crash Conference, PP. 81-100, SAE Paper 881714, October, 1989.

25. J. Horsch, "Evaluation of Occupant Protection from Responses Measured in Laboratory Tests", SAE Paper No 870222, February, 1987.

26. A. Malliaris, "A Review of Car Occupant Risks, Exposure, and Casualties on U.S. Roads in the 1980's", Contract Report CR-87/01/BI, January, 1987.

27. A. Malliaris, "Occupant Injury Rates in Car Crashes", Contract Report CR-87/03/BI, July, 1987.

28. A. Malliaris, "National Estimates of Car Occupant Injury Rates and Injury Counts", Contract Report CR-88/13/BI, October, 1988.

29. T. Lobdell, C. Kroell, D. Schneider, W. Hering, and A. Nahum, "Impact Response of the Human Thorax", Proceedings of the Symposium on Human Impact Response, General Motors Research Laboratories, Warren Michigan, October 2-3, 1972. Plenum Press.

922510

Investigation of Airbag-Induced Skin Abrasions

Copyright _1992_
Society of Automotive Engineers, Inc.

Matthew P. Reed and Lawrence W. Schneider
University of Michigan Transportation Research institute

Richard E. Burney
The University of Michigan

ABSTRACT

Static deployments of driver-side **airbags** into the legs of human subjects were used to investigate the effects of inflator capacity, **internal airbag** tethering, **airbag** fabric, and the distance from the module on **airbag-induced** skin abrasion. Abrasion mechanisms were described by measurements of **airbag** fabric velocity and target surface pressure. **Airbag** fabric kinematics resulting in three distinct abrasion **patterns** were identified. For all cases, abrasions were found to be caused primarily by **high-velocity fabric impact rather** than scraping associated with lateral fabric motion. Use of higher-capacity inflators increased abrasion severity, and untethered **airbags** produced more severe abrasions than tethered **airbags** at distances greater than the length of the tether. Abrasion severity decreased as the distance increased from 225 to 450 mm. Use of a finer-weave **airbag** fabric in place of a coarser-weave fabric did not decrease the severity of abrasion.

INTRODUCTION

Studies of automobile crashes in which a driver-side **airbag** was deployed have demonstrated that **airbags** can significantly reduce the incidence of severe injuries to the head and chest, particularly if a three-point belt system is also used. However, minor injuries induced by **airbags** have been reported. Digges, Roberts, and Morris (1)* cited field investigation data indicating that some vehicle occupants have complained of abrasions to the neck and face and minor burns to the hands following **airbag** deployment. In a more recent study by **Huelke**, Roberts, and Moore (2), similar injuries were reported, including abrasions to the face, neck, and forearms. Often these injuries were found to have occurred in low-speed crashes and with female drivers of small stature.

Airbag-induced abrasions have also been reported in laboratory tests in which human volunteers participated in dynamic tests of **airbag** systems. Smith, **Gulash,** and Baker (3)

tested a driver-side **airbag** system with young male volunteers on a laboratory sled over a range of impact severities. Although no serious injuries were reported, **the** volunteers sustained minor abrasions, **ecchymosis,** and **erythema** on the face, forearms, and hands. These skin injuries occurred even though the volunteers' faces and hands were coated with cold cream prior to testing to protect against abrasion. It should be noted that these tests were performed in the early 1970s with **airbag** systems that may have differed in important ways from current designs.

Research to quantify the effects of deploying **airbag** fabric on skin was conducted by Kikuchi *et al.* (4) with an **animal** model. Using compressed air to deploy an unfolded **airbag,** skin and eye injuries were quantified for a range of deployment velocities. The authors recommended design guidelines for inflation velocity and **internal airbag** pressure to reduce the potential for skin and eye injury. The **airbag** system used in the **Kikuchi** study differed considerably from those currently employed, and the use of an unfolded **airbag** produced **airbag** fabric kinematics substantially different from those of a folded **airbag** deployed from a steering wheel module.

The present study was conducted to address the problem of **airbag-induced** skin abrasion with **airbag** systems similar to those currently used in vehicles. In the initial phase of testing, **driver-**side **airbags** were deployed into the anterior surface of the legs of human subjects. Abrasion incidence and severity were evaluated with respect to several design and deployment parameters. In a second phase of testing, quantitative measurements were made of **airbag** fabric velocity and of surface pressures generated by **airbag** fabric impacts with instrumented fixtures. These data were used to describe the mechanisms of **airbag-induced** skin abrasion. It should be noted that the term "abrasion" used in this paper refers to any traumatic partial-thickness injury of the skin caused by either perpendicular impact or **scraping, and** should not be interpreted to imply a scraping mechanism alone.

METHODS

DEPLOYMENTS INTO LEGS OF HUMAN VOLUNTEERS – For the purposes of this study, static laboratory deployments with the target surface stationary relative

---

* Numbers in parentheses signify references listed at the end of the paper.

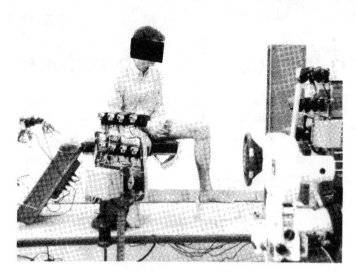

Figure 1. Laboratory for static deployment testing of driver-side airbags.

Figure 2. Subject's leg positioned for testing.

to the **airbag** module were considered to provide an adequate representation of the crash situation. Although it is reasonable to assume that, in actual vehicle crashes, the occupant will have developed a forward velocity relative to the steering wheel when the **airbag** deploys, preliminary **airbag** fabric velocity measurements indicated that the occupant velocity was likely to be a small fraction of the **airbag** fabric velocity during the deployment.

A laboratory was established to meet the needs of static **airbag** testing, as shown in Figure 1. A **9.5-mm-thick** steel plate bolted to a steel **frame** provided the platform for an **airbag**/steering-wheel support structure and an adjustable seat for the human subject. A specially designed fixture for photo documentation of skin condition was located behind the seat so that the subject could remain seated while being positioned for post-test photos. Lighting and facilities for high-speed filming of deployments were provided, and an exhaust vent system was installed to remove **airbag** inflation gases from the laboratory.

A total of 33 deployments were conducted using male subjects ranging in age from 20 to 40 years, with a median age of 23 years. Informed consent of each subject was obtained prior to testing.* Photos of abrasions resulting from deployments into the legs of the investigators were used to illustrate possible injury levels. A physician was present during testing to assess injury levels and to provide treatment to subjects who received abrasions.

Prior to testing, the anterior surfaces of the subject's legs were shaved with an electric razor to eliminate possible effects of hair on abrasion and to facilitate **pre**- and post-test evaluation of the skin condition. The skin was also targeted along the medial, anterior edge of the tibia with colored marking pens. During skin contact by the deploying **airbag**, these marks transferred to the **airbag** fabric, providing information on the areas of **airbag** fabric contacting the leg and the type of contact made (e.g., stamping or smearing). A color photo of the subject's leg was taken to record the condition of the skin prior to the test.

As shown in Figure 2, the subject was seated on the platform so that one leg was positioned vertically in front of the **airbag** module at a specified distance from the **undeployed** module cover. The subject's leg was centered both vertically and horizontally relative to the **airbag** module with the subject's foot resting on a **100-mm-thick** block of balsa wood so that the leg could easily flex at the knee during the deployment. The subject was supplied with ear plugs and instructed to remain in position until after the deployment. When the subject indicated he was ready, the **airbag** was deployed while a high-speed camera recorded the event at a nominal speed of 3000 or **6000** frames per second.

Following the deployment, the subject rotated on the seat to place his leg in the photographic fixture on the back of the test platform, as illustrated in Figure 3. Post-test photos of the subject's leg were taken at intervals of one, five, fifteen, and

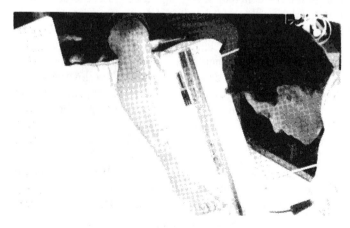

Figure 3. Photographic fixture for post-test recording of skin condition.

---

* The rights, welfare, and informed consent of the volunteer subjects who participated in this study were observed under guidelines established by the U.S. Department of Health, **Education, and Welfare Policy** (now Health and Human Services) on **Protection** of Human Subjects and accomplished under medical research design protocol standards approved by the Committee to Review Grants for Clinical Research and Investigation Involving Human Beings, Medical School, The University of Michigan.

thirty minutes after the deployment. First-aid treatment was provided to the subjects when necessary, and, in several instances, follow-up photos were taken days, weeks, or months after the test to document the healing and cosmetic appearance of the injuries.

AIRBAG TEST CONFIGURATIONS – The effects of four factors on the incidence and severity of abrasion were investigated: coarseness of fabric fibers (denier), inflator capacity, internal tethering, and distance from the module to the target surface. All tests were performed with 60-liter airbags constructed of woven nylon fabric coated internally with neoprene or silicone rubber.

Table 1 shows the matrix of test conditions for the four design and deployment factors that were investigated. These conditions were selected to represent a reasonable range of the factors and do not necessarily represent levels and combinations of factors used in production airbag systems.

Fabric: The 840-denier fabric is subjectively rougher than the 420-denier fabric because it is woven with thicker fibers.

Inflator: The inflator capacity is determined by pressure levels obtained in a standard test procedure involving deployment into a sealed tank (5). The higher-capacity inflator produces a larger volume of gas than the lower-capacity inflator.

Tether: Airbag tethering is accomplished by sewing a yoke of airbag fabric inside of the airbag between the front and back of the airbag, thereby limiting the maximum excursion of the center of the airbag to about 300 mm.

Distance: The distance between the airbag module and the target surface was measured perpendicular to the plane of the steering wheel from the center of the wheel to the skin surface.

All airbags tested were folded into the airbag module with an "accordion" fold technique in which the top and bottom of the airbag are folded together first, followed by the sides. After observation of the airbag fabric kinematics produced by this fold technique, the steering wheel was turned ninety-degrees clockwise in all tests to align the predominant airbag fabric motion during the early phases of the deployment with the long axis of the subject's leg.

QUANTITATIVE MEASUREMENT OF AIRBAG FABRIC ACTIONS – Abrasion events were assumed to be described by the airbag fabric velocity relative to the skin immediately before and during contact with the skin, and by the pressure exerted on the skin by the airbag during contact. Initially, shear stresses developed between the airbag fabric and the skin were believed to play an important role in the abrasion events, but later findings indicated that abrasion incidence and severity could be effectively predicted without consideration of shear stress.

Fabric velocities were measured by manually digitizing high-speed films taken of deployments. Selected airbags were marked with a 25-mm colored grid prior to folding to provide targets for digitization. Although the fabric velocity varies widely throughout the deployment envelope, one velocity measurement of interest is the leading-edge velocity, which describes the horizontal speed of the airbag deployed from a module mounted in a vertical plane. The leading-edge velocity was obtained by tracking the horizontal displacement of the point on the airbag furthest from the airbag module in each film frame.

Target surface pressures were measured by means of an Instrumented Leg Form (ILF) developed for this application, shown in Figure 4. Piezoelectric load cells (Kistler 903A) behind disks mounted flush with the front of the ILF were monitored at 33 kHz during deployments with the ILF positioned in a manner similar to the legs of human subjects. Surface pressure data collected with the ILF indicated the presence of a short-duration,

Table 1
Matrix of Tests with Human Subjects

Airbag Fabric (denier)	Inflator Capacity	Distance (mm) 200	225	250	275	300	325	350	375	400
420	Lower		●○ ●○ ●●			●○ ●○				
420	Higher		●○ ●○	●○	●○	●○ ●○ ○○	●○	●○		○
840	Lower		○			○				
840	Higher		○			○				

● Tethered
○ Untethered

Figure 4. Instrumented Leg Form (ILF) for measuring target surface pressure.

high-pressure impulse during the initial impact of the airbag fabric. Since the frequency response limitations of the load cells in the ILF precluded accurate measurement of the magnitude of this initial pressure impulse, Fuji Prescale film, a pressure-sensitive material, was placed over the front of the ILF and used to record peak surface pressures.

Prescale film produces a graduated color change related to peak applied pressure. The medium-sensitivity film used in these tests is effective for measuring peak pressures from 100 to 500 kg/cm^2 (10 to 50 MPa). The image on the film was evaluated by digital analysis with a Microtek 600 ZS scanner and Macintosh IIci microcomputer. The digital image was related to peak pressure by reference to a calibration curve developed by applying 1-ms-duration pressure pulses of known magnitude to samples of the film. Figure 5 shows the resulting calibration curve relating mean pixel density to applied pressure. The Prescale film was also applied to the surface of a subject's skin in two tests to determine if surface pressure data measured on the ILF were representative of surface pressures on human skin during abrasion events.

Table 2
Abrasion Rating System (ARS) Category Definitions

ARS Category	Description of Injury
1	No abrasion sufficiently severe to cause bleeding or seeping of fluid 24 hours or less after insult. Erythema or transient discoloration of the skin.
2	Bleeding, weeping of fluid, or scab (eschar) formation occurring more than 30 minutes but less than 24 hours after insult.
3	Superficial partial thickness skin abrasion, characterized by damage to the upper dermal layer and fine, punctate (pin-point) bleeding from small vessels.
4	Deep partial thickness abrasion, characterized by damage to the lower layers of dermis and coarser bleeding.
5	Full thickness abrasion, extending through the dermis into the subcutaneous tissue in places.

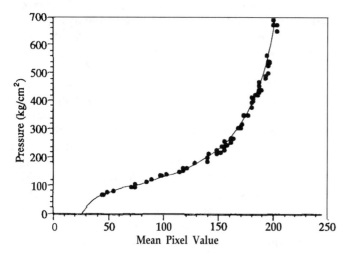

Figure 5. Prescale calibration curve relating pixel density to applied pressure.

RESULTS

ABRASION SEVERITY REPORTING SYSTEMS – Skin abrasions causing immediate surface bleeding were observed in twenty-five of thirty-three tests with human subjects. Abrasion severity was quantified in two ways. An Abrasion Rating System (ARS) was used to categorize the injuries by depth, which is related to the subsequent healing and cosmetic effects of the injury. A second system based on the area of injury provided better resolution among the different test conditions and a means of ranking the injuries from all tests for use in the analysis and presentation of the injury data.

The ARS was developed empirically through examination and treatment of the subjects in this study. Under the ARS, each test is assigned a severity level from 1 to 5 corresponding to the depth of injury observed, according to the guidelines in Table 2. This system considers only the maximum depth of injury and does not take into account other potentially important variables such as the area of injury. Figure 6 shows a schematic cross-section of human skin, indicating the relative depths of tissue damage corresponding to the various ARS levels. Maximum injury depth was selected as the most important parameter for quantifying the severity of injury because of the association between injury depth and recovery time, the potential for infection, and the potential for scarring. Each of these important considerations in the prognosis of the injury increases with increasing depth of injury. Additionally, the area of injury was found to be reasonably well correlated with the greatest depth of injury. That is, tests for which a greater maximum depth of injury was observed also exhibited a larger area of injury.

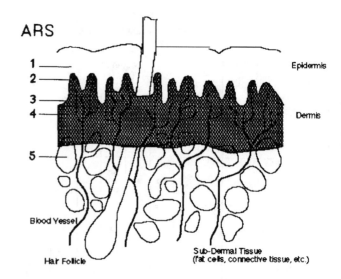

Figure 6. Schematic cross-section of human skin showing relative depths of ARS abrasion categories.

The **ARS** has several advantages for application to **airbag** abrasion injuries:

1. The scale may be applied by personnel who have not had experience with other **airbag-induced** abrasions.
2. The scale may be used to compare **airbag** injury severities among different researchers and studies.
3. **Airbag-induced** abrasion severities can be evaluated relative to abrasions caused by other types of trauma.
4. The scale is objectively determined to a large extent (i.e., bleeding and scab formation are readily identified) and contains categories **defined** without **reference** to other injuries.
5. The **category definitions** have medical relevance and contain information which can be used to assess the clinical significance of the injuries (e.g., the type of treatment required and the potential for scarring).

One disadvantage of the **ARS** is the relative coarseness of the scale. Almost all of the tests in this study were classified as **ARS** 3 or 4. **ARS** 1 and 2 injuries are **confined** to the epidermis and have little or no clinical significance. **ARS** 2 injuries are not reported in this study because no follow-up was performed for subjects with less than **ARS** 3 injuries. **ARS** 5 injuries involve damage to the full **dermal** thickness and were not observed in this study.

The **ARS** scale also gives no information about the area of injury, making comparisons between the injuries produced by different **airbag** module configurations difficult. To facilitate these evaluations, a second injury ranking system was developed. Injuries from the thirty-three tests performed with human subjects were ranked from 1 (least injury) to 33 (greatest injury) on the basis of visual examination of photographs of the injuries and the notes and observations of the investigators. The primary criterion for these rankings was the area of injury, although, as noted before, the depth of injury was fairly well correlated with the area of injury for these tests. In eight of the thirty-three deployments, no skin injury was noted **(ARS 1)**. The injuries from these tests were assigned a ranking of 4.5 (average of ranks 1 through **8)**, and the least severe injury with **ARS** 3 was assigned a ranking of 9, following the ranking technique of **Wilcoxon** (6). In relation to the **ARS** levels, abrasions with severity rankings from 25 to 33 were rated **ARS** 4, while those ranked 9 to 24 were rated **ARS** 3.

Although these rankings were performed by the investigators, who were aware of the **airbag** configurations associated with each injury, the range of injuries was large enough that the rankings were quite easily performed. Another observer might rank the injuries slightly differently, but the effects of the **airbag** design and deployment parameters are **sufficiently** clear that conclusions drawn from the data would not be substantially different. Figure 7 shows post-test photos from three tests along with their abrasion severity rankings.

EFFECTS OF DESIGN AND DEPLOYMENT FACTORS – Figures 8 and 9 show the results of the human subject tests by **airbag** configuration using the abrasion severity ranking system described above. The higher-capacity inflator produced more severe abrasions than the lower-capacity inflator. This effect is particularly apparent at the **300-mm** distance (Figure **9)**, where no abrasions were observed with the lower-capacity inflator and the higher-capacity inflator produced abrasion with both tethered and untethered **airbags**. Tethering strongly reduced the incidence and severity of abrasions for

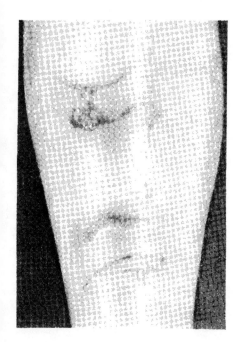
(a)
Wing Abrasion
Distance: 225 mm
420 D Untethered
Lower-Capacity Inflator
ARS 4 Rank 27

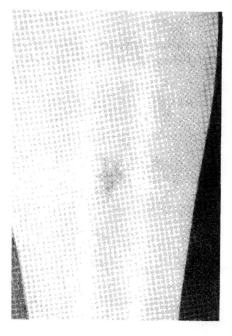
(b)
Stamp Abrasion
Distance: 300 mm
420 D Tethered
Higher-Capacity Inflator
ARS 3 Rank 13

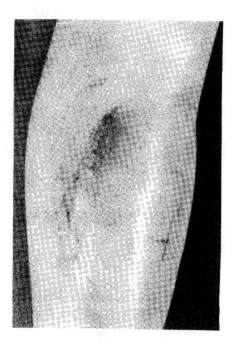
(c)
Slap Abrasion
Distance: 350 mm
420 D Untethered
Higher-Capacity Inflator
ARS 4 Rank 30

Figure 7. Post-test photos from three tests, with **ARS** levels, injury rankings, and test configurations.

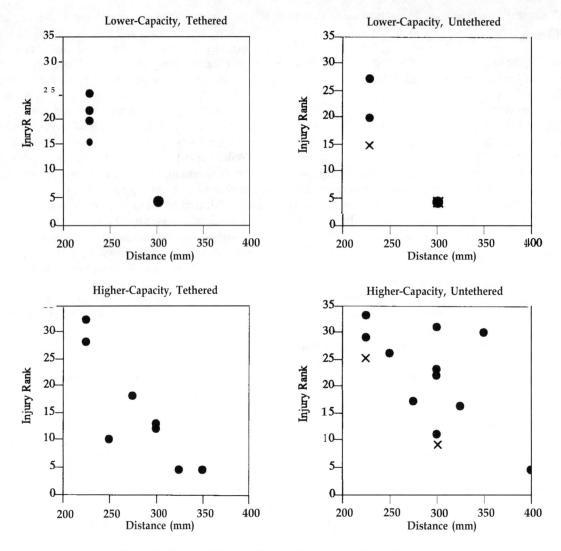

Figure 8. Injury rankings by distance, inflator capacity, and tether.
● = 420 D; ✗ = 840 D; Larger symbols denote two tests with 420-D airbags.

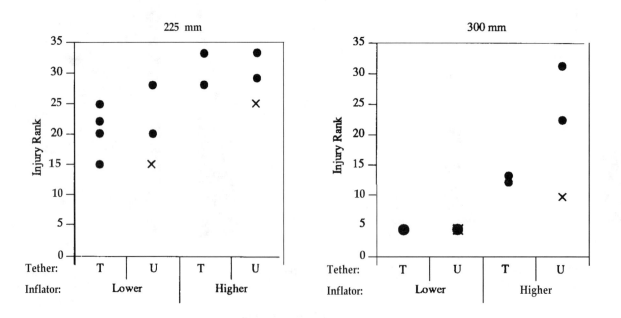

Figure 9. Injury rankings by airbag module configuration for deployments at 225 and 300 mm.
● = 420 D; ✗ = 840 D;   airbags.

512

airbags with higher-capacity inflators at distances greater than 250 mm, which corresponds to the length of the internal tethering straps.* For airbags with the lower-capacity inflators, the tethering effect was not apparent because no abrasions were produced with either tethered or untethered airbags with lower-capacity inflators at distances greater than 250 mm.

The distance from the airbag module to the skin surface was also an important factor. For airbags with lower-capacity inflators, and for tethered airbags with higher-capacity inflators, abrasion severity decreased consistently as the distance was increased from 225 to 350 mm. However, with untethered airbags and higher-capacity inflators, relatively severe abrasions were observed throughout the airbag excursion range. The only test with the higher-capacity, untethered airbag configuration that did not result in abrasion was conducted with the skin surface 400 mm from the airbag module, which is approximately at the outer edge of the deployment envelope.

The abrasion results indicate that the effect of airbag fabric is different from what was expected. It was hypothesized that coarser-weave airbag fabric (840-denier) would produce more severe abrasions than finer-weave fabric (420-denier). However, as shown in Table 3, airbags with 840-denier fabric produced less severe abrasions than airbags with 420-denier fabric in comparable deployments.

Table 3
Comparison of Injury Rankings with 420-D and 840-D Airbags

Distance	Inflator Capacity	420-D	840-D
225 mm	Lower	20 27	15
	Higher	29 33	25
300 mm	Lower	4.5 4.5	4.5
	Higher	31 23 22 11	9

ABRASION TYPES – Three distinct patterns of skin abrasion were noted under different deployment conditions. "Wing" abrasions, illustrated in Figure 7(a), occurred at distances less than 250 mm with all airbag configurations. The name derives from the sweeping action of two folds of airbag fabric that are at the leading edge of the airbag during the early phases of the deployment. The "wings" are illustrated in Figure 10, which shows a frame from the high-speed film of the deployment that produced the abrasion in Figure 7(a). Because the steering wheel has been turned 90 degrees clockwise, the

---

*Although the internal tethers are approximately 250 mm long, the center of the tethered airbag was able to deploy to about 300 mm from the module, due to the arrangement of the tether straps. Consequently, abrasions were observed up to 300 mm from the module for some tethered airbag module configurations.

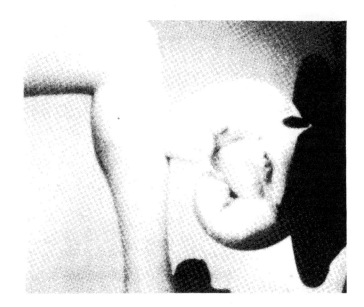

Figure 10. Frame from high-speed film showing airbag "wings."

sweeping wing motion that is vertical in these tests would be horizontal with a neutral-oriented steering wheel. One identifying characteristic of wing abrasions is the pattern of abrasion due to skin contact by the airbag seam, seen in Figure 7(a) as a string of dots.

"Stamp" abrasions were produced by tethered airbags with higher-capacity inflators at a distance of 300 mm, illustrated in Figure 7(b). These abrasions are characterized by a relatively intact epidermis with damage and bleeding in the dermis, and were relatively less severe than the other types. Stamp abrasions occurred only at the outer edge of the tethered airbag deployment envelope, and were associated with airbag fabric contact that was directed perpendicular to the skin. Although high-speed films provided the primary means of identifying airbag fabric motions during abrasion events, the pattern of target markings that transferred from the subject's skin to the airbag fabric during the deployment supports these observations.

"Slap" abrasions, illustrated in Figure 7(c), were caused by untethered airbags with higher-capacity inflators at distances from 300 to 350 mm. The severity of these abrasions was similar to wing abrasions, although more variability in severity was observed than with wing abrasions. Slap abrasions were associated with the untethered airbag fabric wrapping around the side of the subject's leg, producing areas of stamp-like abrasions along with areas where the epidermis was removed in the manner seen with wing abrasions.

AIRBAG FABRIC VELOCITY MEASUREMENTS – Figure 11 shows leading-edge velocities for the four different 420-denier airbag configurations, along with a composite plot obtained by visually approximating the mean velocity trends. Data were collected only for distances up to 225 mm from the airbag module, both because the primary interest was with the wing abrasion mechanism and because the airbag kinematics become more complex and variable beyond 225 mm, making consistent velocity measurements difficult. Two aspects of these data are particularly important First, the leading-edge velocities peak approximately 200 mm from the airbag module and drop rapidly thereafter, reflecting the one-dimensional nature of the

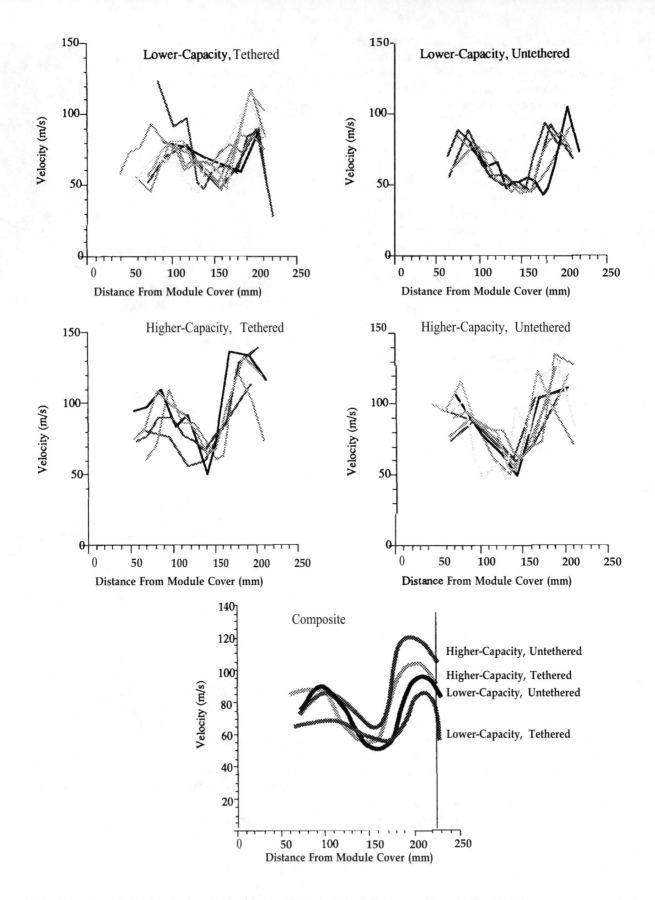

Figure 11. Leading-edge velocity data for four 420-D airbag configurations, by inflator capacity and tether. Each line represents one deployment. Composites were developed by visual approximation of trends for each configuration.

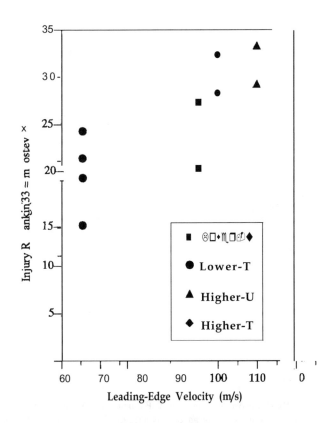

Figure 12. Leading-edge velocity at 225 mm and injury rankings for tests at 225 mm with four airbag module configurations. (Inflator Capacity - Tethered/Untethered)

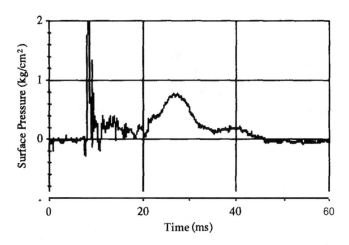

Figure 13. Surface pressure data (unfiltered) from a sensor on the ILF during airbag deployment, Horizontal axis indicates time elapsed from inflator trigger.

leading-edge velocity measurement and the largely two-dimensional sweeping motion of the airbag "wings." Second, the fabric velocities of airbags with higher-capacity inflators are generally greater than with lower-capacity inflators. Higher-capacity inflators were also associated with more severe abrasions, suggesting a positive relationship between fabric velocity and abrasion severity (Figure 12).

SURFACE PRESSURE MEASUREMENTS: INSTRUMENTED LEG FORM (ILF) – Data from the ILF indicated that the airbag fabric pressure on the target surface could be divided into two phases. The initial impact of the airbag fabric with the ILF produced a high-magnitude, short-duration pressure spike, followed by much lower pressure levels during the remainder of the deployment. Figure 13 shows the surface pressure time history for an area of the ILF where an airbag fabric "wing" initially impacted. The plot has been scaled to reveal the relatively low surface pressure that prevailed following the initial fabric impact. Characteristic differences between airbag configurations in the pressure time histories were observed, but surface pressures generally remained below 1 kg/cm^2 following the initial fabric impact.

SURFACE PRESSURE MEASUREMENTS: PRESCALE FILM – Although the ILF was effective for monitoring surface pressure during most of the deployment, the frequency response of the system did not allow accurate measurement of the magnitude of the initial peak surface pressure during fabric impact. To properly characterize this first phase of the surface pressure time history, Prescale film was applied over the surface of the ILF. After a deployment, the pattern of peak surface pressure on the ILF numerical peak pressure distribution for each test was obtained by digital image analysis using the calibration previously described.

Because of the graphic nature of the Prescale film, important aspects of the surface pressure data could be observed prior to digital analysis. Most notably, the patterns of peak surface pressure on the ILF closely matched the patterns of skin abrasion caused by similar deployments, including the patterns of abrasion "dots" associated with the airbag seam that were seen with wing abrasions. Further, the intensity of the Prescale image, which corresponds to the magnitude of the applied pressure, correlated well with the severity of the abrasions. Using the digital analysis system, areas for which the Prescale film indicated a peak surface pressure on the ILF greater than 175 kg/cm^2 were found to correspond to areas of skin abrasion in similar deployments. Figures 14, 15, and 16 show matched sets of Prescale images, the same images showing only areas for which the surface pressure on the ILF exceeded 175 kg/cm^2, and post-test photos of skin abrasions from deployments with the same airbag configuration at the same distance. These figures are examples of the wing, stamp, and slap abrasion types discussed above. Circular patterns in the Prescale images correspond to the ILF load cell locations.

Because the ILF is rigid compared with the more compliant skin and subcutaneous tissue, pressure data from the ILF are not necessarily representative of the surface pressure levels present on the skin surface during abrasion events. However, in two tests in which Prescale film was interposed between the subject's skin and the airbag, the Prescale film indicated surface pressures that were approximately 80% of the levels measured on the ILF for similar deployments. No further tests to verify these pressure levels were conducted because of difficulty in applying the stiff Prescale film to the irregular and compliant surface of the leg in such a manner that the film would not be damaged during testing.

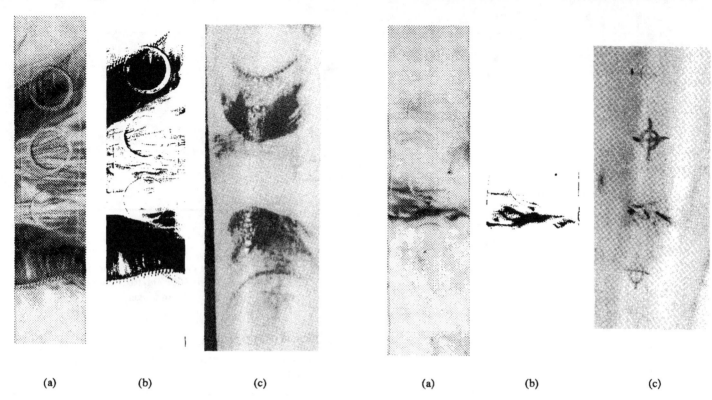

(a)   (b)   (c)   (a)   (b)   (c)

Figure 14. Relationship between ILF surface pressure patterns and abrasion patterns for untethered airbags with higher-capacity inflators at 225 mm: (a) Prescale image; (b) Digital analysis of image showing only areas exceeding 175 kg/cm2; (c) Skin abrasion.

Figure 15. Relationship between ILF surface pressure patterns and abrasion patterns for tethered airbags with higher-capacity inflators at 300 mm: (a) Prescale image; (b) Digital analysis of image showing only areas exceeding 175 kg/cm2; (c) Skin abrasion.

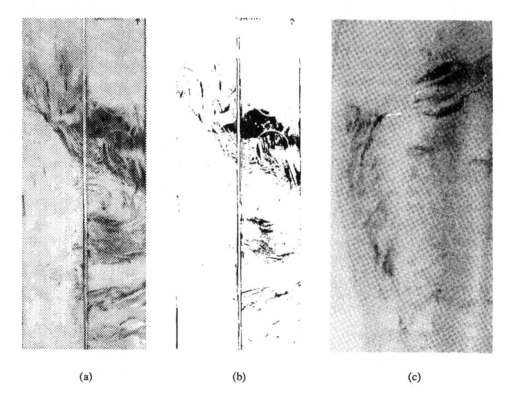

(a)   (b)   (c)

Figure 16. Relationship between ILF surface pressure patterns and abrasion patterns for untethered airbags with higher-capacity inflators at 300 mm: (a) Prescale image; (b) Digital analysis of image showing only areas exceeding 175 kg/cm2; (c) Skin abrasion.

SUMMARY AND DISCUSSION

Static testing of several **airbag configurations** with human subjects demonstrated the effects of four **airbag** design and deployment factors on skin abrasions.

- Higher-capacity inflators produced more severe abrasions.
- Tethering reduced or eliminated abrasions for **airbags** with higher-capacity inflators at distances greater than the length of the tether.
- Abrasion severity decreased with increasing distance from the module for all configurations except untethered **airbags** with higher-capacity inflators, which produced relatively severe abrasions throughout the range of **airbag** fabric excursion.
- Contrary to expectations, coarser-weave **airbag** fabric did not produce more severe abrasion than **finer-weave** fabric. However, this effect was confounded by the necessity of using different module covers for the two **airbag** fabrics, which may have altered the kinematics.

Quantitative assessment of **airbag** fabric motions that produced injury indicates that the primary cause of abrasion is the initial **airbag** fabric impact with the skin. **Airbags** that produced more severe abrasions were found to exhibit higher fabric impact velocities resulting in higher and more widely distributed peak surface pressures. The **patterns** of peak surface pressure on the **ILF** were found to be similar to the abrasion **patterns** for all three types of abrasion. This observation has important implications for the description of **airbag-induced** abrasion mechanisms.

During testing of human subjects, wing abrasions were initially believed to be caused by the scraping action of the **airbag** fabric sweeping rapidly over the skin surface. This hypothesis is consistent with a view of "abrasion" being caused by the shearing or scraping between two surfaces. However, quantitative evidence contradicts this hypothesis for **airbag-induced** abrasions, and suggests an **alternative** explanation of the abrasion mechanism.

First, high-speed films of deployments demonstrate that, during wing abrasion events, **airbag** fabric sweeps over a much larger **area of** the skin surface than is injured, indicating that **high-** velocity movement of the **airbag** fabric against the skin is not sufficient to cause abrasion. Second, data from the **ILF** and Prescale analyses indicate that surface pressures are far higher during the initial fabric impact than subsequently, and that the **pattern** of the peak pressure matches the **pattern** of abrasion. During initial impact of **airbag** fabric with the **ILF,** pressures exceeding 175 **kg/cm²** were measured on the **ILF.** In contrast, during the remainder of the deployment, as **the airbag** fabric continued to sweep across the skin, surface pressures were generally below 1 **kg/cm².** For wing abrasions, high surface pressures caused by high-velocity fabric impact were therefore found to be necessary for abrasion to occur.

Third, results of tests using tethered **airbags** and **higher-** capacity inflators with the skin or **ILF** located 300 mm from the **airbag** module indicate that high surface pressure is also sufficient **to cause** abrasion. As shown previously in Figures **7(b)** and 15, a stamping action by a tethered **airbag** at the limit of its

excursion can produce abrasion even in the absence of substantial lateral fabric motion (scraping). This type of abrasion, an "imprint" abrasion, is histologically equivalent to a more typical "scraping" abrasion, except that the damaged **dermal** and **epidermal** layers remain in place temporarily until they slough and are shed during the healing process.

Further evidence to support this view of the role of surface pressure in the etiology of these abrasions came from similar tests performed with untethered **airbags.** With human skin, untethered **airbags** with higher-capacity inflators produced slap abrasions at distances beyond 300 mm, characterized by a combination of intact and sheared epidermis found in stamp and wing abrasions, respectively. The peak surface pressure **patterns** measured for these test conditions again closely matched the abrasion **patterns** (see Figure 16).

These **findings** indicate that **airbag-induced** skin abrasions are caused primarily by high-velocity **airbag** fabric impacting perpendicular to the skin surface, although lateral fabric motion during impact events can remove the damaged epidermis. In the tests in which apiece of Prescale **film** was interposed between the **airbag** and the surface of the skin, thereby shielding the skin from scraping, abrasions were observed **underneath** the film in the areas of initial fabric impact. However, these abrasions were more typical of stamp abrasions, with relatively intact epidermis, than of wing abrasions, which would have been expected for the experimental configuration if the film *were* not interposed. These results **confirm** that the sweeping **airbag** fabric motion can act to remove the damaged upper layers of skin, although the overall depth of the abrasion is apparently similar in the absence of fabric scraping. Further, although the actual surface pressures on skin during abrasion events are probably lower than those measured on the **ILF** because of the compliance of the skin and subcutaneous tissues, the **patterns** of peak pressure are believed to be the same, based on the strong association between the Prescale images and injury **patterns.**

This view of **airbag-induced** abrasion mechanisms is consistent with the effects of the design and deployment factors discussed previously. Higher-capacity inflators produce more severe abrasions because of higher fabric velocities that result in greater surface pressures during skin impact, Tethering reduces the excursion range of the **airbag,** limiting the envelope in which high-velocity fabric impacts are possible. Because perpendicularly directed impacts, rather than lateral fabric motions, are the primary determinants of abrasion severity, **airbag** fabric roughness is not likely to have a strong effect on abrasion severity.

ACKNOWLEDGMENTS

The research described in this paper was sponsored by Honda Research and Development.

REFERENCES

1. **Digges, K.H.,** Roberts, **V.,** and Morris, **J.** (1989) *Residual injuries to occupants protected by restraint systems.* **SAE** Paper no. 891974. Society of Automotive Engineers, **Warrendale,** PA.

2. **Huelke, D.F.**, Roberts, **J.V.**, and Moore, **J.L.** (1992) Air bags in crashes: clinical studies from field investigations. To be published in *Proc. of the 13th International Technical Conference on Experimental Safety Vehicles.* U.S. Department of Transportation, National Highway Traffic Safety Administration, Washington, **D.C.**

3. Smith, **G.R., Gulash, E.C.**, and Baker, **R.G.** (1974) *Human volunteer and anthropomorphic dummy tests* of *General Motors driver air cushion system.* **SAE** Paper no. **740578.** Society of Automotive Engineers, **Warrendale,** PA.

4. **Kikuchi,** A., **Horii, M.**, Kawai, A., Kawai, **S., Komaki, Y.**, and **Matsuno, M.** (1975) Injury to eye and facial skin (rabbit) on impact with inflating air bag. *Proc. 2nd International Conference on* the *Biomechanics* of Serious *Trauma,* pp. 289-296. Edited by **J.P. Cotte** and **M.M. Presle. IRCOBI, Bron,** France.

5. Society of Automotive Engineers. (1991) Draft of 52238, **Airbag** inflator ballistic tank test procedure. Memo to members of the **SAE** Inflatable Restraints Standards Committee. Society of Automotive Engineers, **Warrendale,** PA.

6. Box, **G.E.P.** (1978) *Statistics for experimenters.* John Wiley and Sons, New York.

933119

Assessment of Air Bag Deployment Loads with the Small Female Hybrid III Dummy

John W. Melvin, John D. Horsch, Joseph D. McCleary,
Laura C. Wideman, Jack L. Jensen, and Michael J. Wolanin
General Motors Corp.

ABSTRACT

This study is an extension of previous work on driver air bag deployment loads which used the mid-size male Hybrid III dummy. Both small female and mid-size male Hybrid III dummies were tested with a range of near-positions relative to the air bag module. These alignments ranged from the head centered on the module to the chest centered on the module and with various separations and lateral shifts from the module. For both sized dummies the severity of the loading from the air bag depended on alignment and separation of the dummy with respect to the air bag module. No single alignment provided high responses for all body regions, indicating that one test at a typical alignment cannot simultaneously determine the potential for injury risk for the head, neck, and torso.

Based on comparisons with their respective injury assessment reference values, the risk of chest injury appeared similar for both sized dummies. The results suggest that the small female dummy has a greater risk of neck injury over a greater range of alignments in these exposures than does the mid-size male dummy and appears to be an important tool for assessment of neck injury risk in near-position air bag inflation tests. There are insufficient field accident data on near-position injuries to allow the results of this study to be extended to the present driving population. However, the occurrence of such injuries is expected to be infrequent.

INTRODUCTION

THE RISK OF INJURY from an inflating air bag has been a consideration for both passenger and driver air bag systems [1-7]. These laboratory studies indicate that "high" loads can be developed on unrestrained surrogates when they were placed close to the air bag module and in the path of the inflating bag. This configuration is called an out-of-position condition by many studies, but may more accurately be referred to as a near-position condition. Guidelines have recently been established for laboratory evaluation of these types of interactions [8]. The energy introduced into the passenger compartment by the air bag inflation process is, of necessity, significant in order to provide timely restraint in high velocity crashes. This energy can result in a class of injuries, denoted as inflation-induced-injuries, when the occupant is near the inflating air bag. Analysis of car crashes in which air bags have deployed suggest that loading by the inflating bag has influenced injury for both driver and right-front passengers [9].

A previous laboratory study [2] of a steering wheel mounted air bag with a mid-size (50th percentile) male Hybrid III dummy reported complete compression (bottoming) of the chest when the chest was against the air bag module at the time of inflation. The indicated amplitude of the viscous criterion for chest injury, VC, was several times the recommended assessment value of 1.0 m/s [10]. The geometric and inflation characteristics of the air bag restraint, the spacing from the module at the time of inflation, and the interaction environment all influenced the thoracic responses. Accordingly, the perceived injury risk in near-position laboratory tests depends on the system tested, the test environment, and the alignment and position of the surrogate at the time of bag inflation. Additionally, the impact response of the surrogate, and the injury tolerance and/or assessment instrumentation and criteria are important.

A recent study by Horsch, et al [1] emphasized developing an improved understanding of near-position driver loading by an inflating air bag using both a mid-size male Hybrid III dummy and physiological surrogates. Both types of surrogates were positioned, unrestrained, near or against the air bag module at the initiation of inflation in static tests. Additional sled tests, with the dummy initially leaning against the steering wheel, found dummy responses similar to those in the static tests due to compensating effects and because much of the test energy was associated with the air bag inflation. Various alignments of the module with respect to the surrogates, based on the possibility of a wide range of occupant alignments were used. The alignments ranged from the upper abdominal region to the head centered on the module. The highest responses for the head, neck, or torso were with that body region aligned with and against the air bag module. The risk of severe injury was judged to be "low" for the head and neck, based on neck loads and head acceleration of the Hybrid III dummy. The risk of injury was judged to be "high" when the torso was against and fully covering the air bag module, with both the Hybrid III dummy and the physiological surrogates. The study investigated the air bag deployment mechanics that influence the severity of loading in a near-position environment. High loads could be developed in both the "punchout" and "membrane" loading phases (see the results section of this paper for an explanation of these two types of loading). Which phase was most important depended on the body region. Inflator output was shown to influence the magnitude of loading. However, inflator output had minimal influence on torso injury risk with the torso against the module at the initiation of inflation, because severe injury was observed for the practical range of inflator outputs used with this test condition.

Smaller drivers tend to adjust the seat forward compared with larger drivers and, on this basis, would be more likely to be displaced close to the air bag module from unusual pre-inflation collision events. Thus, occupant size might influence both alignment and frequency related to near-position exposures. The recently developed [11] small (5th percentile) female Hybrid III dummy provided a small driver surrogate to extend the study by Horsch et al [1] to investigate various near-position parameters. The small female Hybrid III dummy was selected as being the most appropriate "small" driver surrogate, based on its humanlike mechanical response and injury assessment scaling from the mid-size male Hybrid III dummy [12] and available injury assessment instrumentation for the head, neck, and thorax. This study compares responses of the small female and mid-size male Hybrid III dummies in near-position exposures with similar air bag modules and test methods.

Because of limited air bag field experience, it is likely that significant time will be needed to determine the potential for occurrence of conditions that might be "infrequent" or "rare". Thus, as in the previous study [1], important questions relating to field relevance are not addressed in this study. Important issues that make extrapolation difficult include: 1) the unknown frequency for near-position exposures; 2) the crash conditions and occupant alignments with the steering assembly if near-position exposures occur; 3) the relevance of the laboratory test environment to possible real-world near-position exposures; and 4) the appropriateness of the injury assessment technology in terms of the surrogate and injury criteria for near-position exposures. Thus, the results of our laboratory tests should not be directly assumed to represent real world performance without considering these and other field related issues.

METHODOLOGY

Real-world near-position exposures will likely be associated with a range of crash severities and occupant positions relative to the air bag module. A driver might be in a near-position at the time of inflation due to slumping from illness, drowsiness, or substance abuse; or due to a "minor" crash which could displace a driver prior to a more severe impact sufficient to cause inflation. Potential variables affecting the range of occupant positions at the time of inflation include: occupant size; seat adjustment; steering column or wheel tilt adjustment; seated posture; crash induced kinematics; and inflation timing. Steering wheel orientation could vary due to precrash and crash induced rotations of the steering wheel. Belt restraint use will influence not only position of the occupant relative to the steering wheel, but also would be expected to significantly reduce the potential for being in a near-position. This study considered possible alignments for drivers not using the belt restraint.

TEST ENVIRONMENT - Both the small female and the mid-size male Hybrid III dummies were tested with a range of positions relative to the air bag module. These alignments ranged from the head centered on the module to the chest centered on the module and with various separations from the module. A static environment with a vertically oriented steering assembly was used for our tests. The dummies were positioned horizontally over the steering wheel and module assembly. Figure 1 shows the various longitudinal body positions that were used for the small female Hybrid III dummy. The steering wheel was rigidly supported by a load cell, which was fixed on a bed-plate. The steering wheel and air bag module were aligned with respect to the dummy for straight-ahead steering. The dummy was placed in a kneeling position, with the fixture supporting some of the weight of the dummy's horizontal upper torso for the case of being against the module. The dummy was supported by straps to allow clearance between the dummy and air bag module for tests having a separation between the dummy and module. Padding was placed around the test site to prevent significant impact with the floor after the test exposure. Responses from the secondary impact(s) are not included in the tabulations.

The choice of a static environment was based on its simplicity and ease of obtaining the desired initial alignment of the dummy with the air bag module and because the previous study found similarity between static and sled experiments for chest exposures [1]. However, static and dynamic exposures should not be considered equivalent. The absence of compressible elements such as the steering column and vehicle structure influence the severity of the exposure. For static tests, the only energy input is from the inflating air bag. The lack of crash energy and the minimal pre-inflation inertial forces compared to those in a car crash sufficient to deploy an air bag will influence the exposure severity for the static test situation. A previous study that used compressible steering columns and did not have a static 1G preload of the dummy on the module found higher responses in dynamic tests in comparison to static tests [2].

DUMMIES - The small female Hybrid III dummy used in this study was recently developed by a task force of government, university, and industry experts [11]. Its specifications were intended to produce the same level of biofidelity and measurement capacity as the original mid-size male Hybrid III dummy. Major body segment lengths and weights were determined from anthropometry data on the U.S. adult population [13]. Geometric and mass scaling methods were used to define the

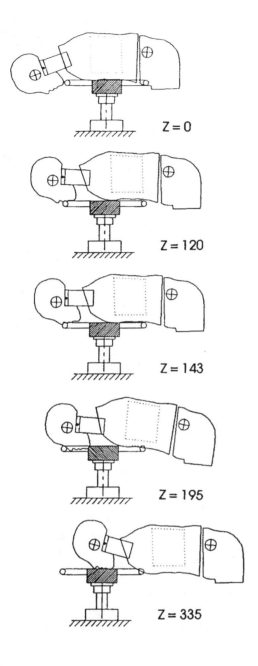

Figure 1: Static test configurations used with the small female dummy.

shape in the neck region of the Hybrid III designs. This region was left uncovered in the original Hybrid III design to avoid interference with the motion of the neck which might influence head/neck response in forward and rearward inertial motion. However, the combination of a small diameter neck and exposed horizontal surfaces at the bottom of the head and jaw provides an unnaturally large reaction surface for membrane loading from an air bag. Ideally, a neckskin for near-position air bag testing should provide an anatomical neck surface shape without introducing load paths around the load cell at the head/neck junction and also without influencing neck bending response. The neckskin must be substantial enough to resist crushing by the air bag and retain its general shape under the applied pressures. At this time, no dummy neck covering has been developed which can provide all these features. The neckskin used with the mid-size male dummy in this test series (shown in Figure 2) was a commercially available single piece item (Part No. ARL 7799, First Technology Safety Systems) composed of molded urethane foam covered with vinyl and had a total part weight of 238 grams. This weight neckskin produced neck forces and moments similar to those obtained with the prototype neckskin reported in [1]. Neckskins with heavier vinyl coatings significantly reduced the load cell forces in comparable exposures. This was thought to be due to shielding of the load cell by transfer of some of the applied load through the stiffer vinyl covering, around the load cell, and into the head, and was judged to be unrealistic.

appropriate scaled-down dimensions of the individual dummy component structures and their responses. Scaled injury assessment reference values (IARVs), based on the assumption of equal failure stress, have also been developed [12] for use with the dummy. This scaling produced IARVs with lower forces and deflections and higher accelerations in comparison to the mid-size male IARVs.

Modifications had to be made to the Hybrid III mid-size male and small female dummies used for these tests. The modification for both dummies was the addition of a neckskin which filled in the missing humanlike

Figure 2: Commercial neckskin configuration used with the mid-size male dummy.

There was no commercially available neckskin for the small female dummy, so a prototype covering (shown in Figure 3) was made in a manner similar to that used for the mid-size male in the previous study [1]. The neck was wrapped with two layers of foam rubber that formed a cylindrical structure that was open at the rear of the neck and secured around the neck by large rubber bands. This arrangement allowed unimpeded rearward bending while maintaining a uniform front surface of the chin/neck junction during rearward motion. The layer of material against the neck was a very firm rubber foam (Rubatex R-1800-FS vinyl-nitrile) 6 mm thick. The outer layer was an 8 mm thick soft foam (Rubatex G-231-N Neoprene). The outer layer extended up behind the chin of the dummy to provide continuity of the neck front surface as the head rotated rearward. The chin cavity in front of the neckskin was filled with a fitted rigid foam plug (Dow H-115 Styrofoam) to prevent unnatural interaction with the air bag surface. Adequate clearance was maintained between the plug and the neck structure and load cell to prevent interference during neck motion. Having two separate parts, the neckskin proper and the chin filler, eliminates the possibility of a load path around the neck load cell.

TEST HARDWARE - Compressed gas inflation was used for these tests. The compressed gas was released and then shut-off from two independent tanks by fast-acting valves. This method of inflation provided an adjustable and more repeatable inflation source at a lower cost and without heat or other inflator products in comparison with pyrotechnic inflators. Gas density was controlled by mixing inert gases of different densities to simulate, in part, the flow characteristics of high temperature pyrotechnic inflation gases. The inflation characteristics in this study are referred to by their maximum gage pressure (in kPa) and maximum rate of pressure increase (in kPa/msec and determined by averaging over a 5 msec interval) produced when discharged into a 0.0283 cubic meter (one cubic foot) volume tank initially at atmospheric pressure. Thus, a 350-15 inflator produces a maximum tank pressure of 350 kPa with a maximum rate of pressure increase of 15 kPa/msec.

The generic air bag and module components had some differences from those used in the previous study by Horsch et al [1]. The inflator simulated a 400-21 output compared with a 350-15 output for many of the tests previously reported [1]. The generic modules used in the current study represent a later generation. Among the differences are a modified tear pattern on the covers and a reduced width of the module. The steering wheel was a rigid, reuseable wheel made from steel plates and tubing which did not exhibit plastic deformation in our experiments. The module was mounted flush with the wheel rim.

Instrumentation included a load cell to measure the reaction force between the steering wheel and bedplate, and internal air bag pressure through a short tube ending over the center of the inflator inside of the bag. Typical dummy instrumentation included chest compression and tri-axial acceleration, force and moment transducers for the neck, and head tri-axial acceleration.

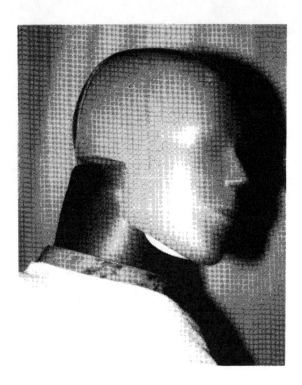

Figure 3: Prototype neckskin used with the small female dummy.

RESULTS

Table 1 provides responses of the small female Hybrid III dummy as a function of longitudinal position and separation from the air bag module. Table 2 provides similar information for the mid-size male Hybrid III dummy. The longitudinal (z) dimension (which would be vertical in the vehicle environment) to span the range from the head centered to the chest centered on the module was 335 mm for the female dummy and 400 mm for the male dummy, with zero representing the sternum centered on the module for both dummies. Tests were conducted at each extreme of this range and somewhat arbitrary locations in between, as listed in Tables 1 and 2, with the dummy against the steering wheel/module assembly and its midsagittal plane centered on the module. At several of these locations, additional tests were conducted with either a lateral shift of the dummy of 75 mm or the dummy raised either 13 or 25 mm from the against position, Table 3. Dimensions should be considered approximate due to variation of back angles among the longitudinal locations used to achieve contact with the module and due to deformation of the dummy, including bending of the neck due to preload.

The severity of the loading depended on alignment and separation of the dummy with respect to the air bag module. The highest response amplitudes for a particular body region tended to be with that body region aligned with and nearest the air bag module, as shown in Tables 1 and 2. It should be noted that the neck region could not actually contact the module when the chest and/or head is also contacting the module and/or

Table 1. Response Summary for Small Female Hybrid III Dummy Dummy Centered on and Against Module								
Longitudinal Alignment (Z) (mm)[1]	**Head**		**Neck**			**Chest**		
	Peak	HIC	Fz	Fx	My	c (3ms)	Comp	VC
	(G)		(kN)	(kN)	(Nm)	(G)	(mm)	(m/s)
0	38	48	2.00	-0.65	-16.3	30.8	58	1.94
120	53	218	2.55	-2.75	-105.7	42.9	40	0.60
143	62	255	2.80	-2.50	-95.2	25.7	38	0.38
195	72	453	3.45	-2.10	-74.0	36.4	11	0.10
335	142	526	1.65	1.85	-57.2[2]	23.1	_[3]	_[3]

[1] Center of sternum to center of module; Z = 0; Sternum centered on module; Z = 335; Head centered on module
[2] Flexion was greater, but less as percent of IARV
[3] Slight extension of sternum

Table 2. Response Summary for Mid-Size Male Hybrid III Dummy Dummy Centered on and Against Module								
Longitudinal Alignment (Z) (mm)[1]	**Head**		**Neck**			**Chest**		
	Peak	HIC	Fz	Fx	My	Acc (3ms)	Comp	VC
	(G)		(kN)	(kN)	(N.m)	(G)	(mm)	(m/s)
0	30	38	1.20	-0.25	-9.3[2]	27.0	69	2.3
140	70	141	4.40	-1.70	-17.6	24.4	62	0.75
215	60	389	2.55	-0.80	29.4	23.6	15	0.20
305	73	433	2.25	-0.95	-7.6	19.4	7	0.05
400	103	204	1.60	1.20	61.1	14.6	_[3]	_[3]

[1] Center of sternum to center of module; Z = 0; Sternum centered on module; Z = 400; Head centered on module
[2] Flexion was greater, but less as a percent of IARV
[3] Slight extension of sternum

Table 3. Influence of Separation/Lateral Alignments								
Dummy	**Z[1]**	**X, Y[2]**	**Head**	**Chest**		**Neck**		
			HIC	Comp	VC	Fz	Fx	My
	(mm)	(mm)		(mm)	(m/s)	(kN)	(kN)	(N.m)
Small	0	0, 0	48	58	1.94	2.00	-0.65	-16.3
Small	0	13, 0	52	47	1.18	1.35	-0.50	-17.1
Small	0	0, 75	48	40	0.93	1.20	-0.25	-15.9
Small	120	0, 0	218	40	0.60	2.55	-2.75	-106
Small	120	25, 0	85	33.5	0.15	1.80	-0.70	-28.5
Small	143	0, 0	255	38	0.38	2.80	-2.50	-95.2
Small	143	25, 0	177	33.5	0.12	2.45	-1.25	-59.0
Small	143	0, 75	85	19	0.13	2.18	-1.85	-77.6
Small	195	0, 0	453	11	0.10	3.45	-2.10	-74.0
Small	195	25, 0	213	7.5	0.08	2.00	0.75	-22.4
Mid	0	0, 0	38	69	2.3	1.20	-0.25	-9.3
Mid	0	13, 0	39	66.5	0.95	0.75	-0.60	-10
Mid	0	0, 75	29	53	1.2	0.85	-0.50	-10
Mid	140	0, 0	141	62	0.75	4.40	-1.70	-17.6
Mid	140	13, 0	116	53	0.2	3.75	-1.45	-25
Mid	140	0, 75	51	22	0.1	2.49	-1.00	-17

[1] Center of sternum to center of module; Z = 0; sternum centered on module
[2] X is separation from module; Y is lateral alignment of dummy midsagittal plane with module center

steering wheel rim. Thus, in those cases, the dummy is against the assembly, but the neck is not in actual contact with any part of the assembly. As observed in the previous study [1], no single alignment provided high responses for all body regions. On this basis, a single typical alignment cannot simultaneously determine the potential for injury risk for the head, neck, and torso.

Separation of the dummy from the module of either 13 or 25 mm or a lateral shift of 75 mm resulted in reductions of most responses, shown in Table 3, and a reduction of all responses that had high values when centered and against the module. These results are consistent with the findings of the previous study [1] that high loads are associated only with the surrogate very close to or against the module at the initiation of inflation.

High loads for all body regions could be associated with either punchout or membrane loading. Punchout and membrane loading were discussed in [1], and these mechanisms are indicated in Figure 4. Punchout is associated with rapid swelling of the module before much of the air bag escapes from the module. Membrane loading is associated with the inflating air bag as it wraps around the near-position dummy and then develops tension in the bag fabric as the bag becomes full. As previously demonstrated [1], loading in both the punchout and membrane loading phases is due to internal bag pressure and not inertial force from the bag. Punchout was most important for the chest, while punchout and membrane loading were found to be important for the neck.

An example of punchout loading of the chest of the small female dummy is shown by the time-history in Figure 5 where the dummy was aligned with its chest centered and against the module (Z = 0). As was typical of the responses of the mid-size dummy in this study and in the previous study [1], the viscous criterion reached its maximum slightly before maximum chest deflection. Note that this peak is reached during the high bag pressure phase of the deployment, when the bag is opening and exiting the module cover. Figure 6 shows a time-history for a punchout loading of the head/neck of the small female dummy aligned against the module at Z = 195 mm. The maximum neck loads and head accelerations occur during the high bag pressure phase of the deployment. Also shown are the applied head loads, which are calculated using the head accelerations and the neck loads. Note that the applied axial load is much higher than the neck shear load (denoted FX on the plot) while the applied vertical load is similar to the neck tension (denoted FZ on the plot). This indicates that the inertial reaction of the head influences the neck shear loads during punchout, but it does not influence the neck tension. In fact, the peak neck tension occurs during a period when the vertical accelerations are low in value and are changing direction. Figure 7 shows a time-history for a membrane loading of the head/neck of the small female dummy aligned against the module at Z = 143 mm. The maximum neck loads occur during the lower pressure filling phase of the air bag deployment. The neck tension (denoted FZ) is very close in magnitude to the applied vertical load

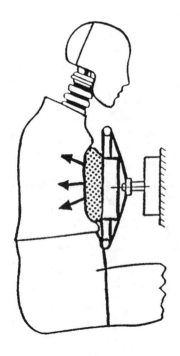

PUNCHOUT

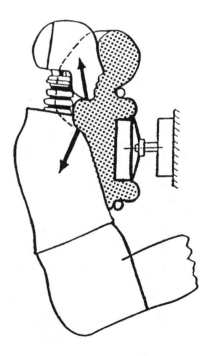

MEMBRANE

Figure 4: Driver airbag loading mechanisms (shaded regions indicate airbag shape at the time of maximum load).

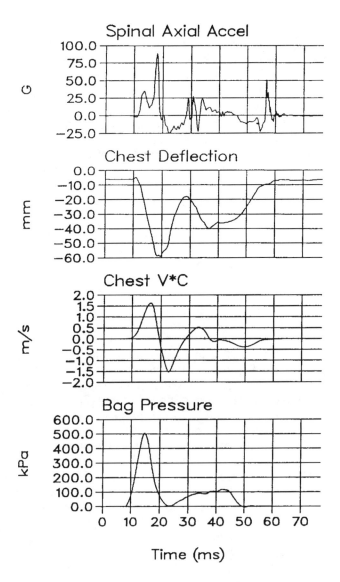

Figure 5: Small female dummy chest response data for punchout loading with the chest centered and against the air bag module (Z = 0 mm alignment). Data filtered at SAE Channel Class 180, except VC at Channel Class 60.

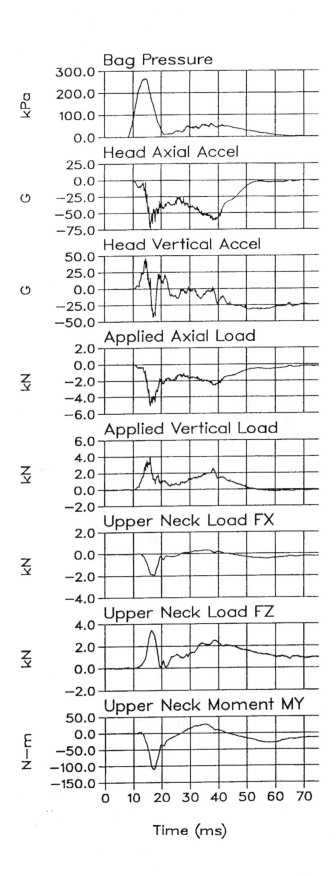

Figure 6: Small female dummy head/neck response data for punchout loading (Z = 195 mm alignment). Data filtered at SAE Channel Class 1000.

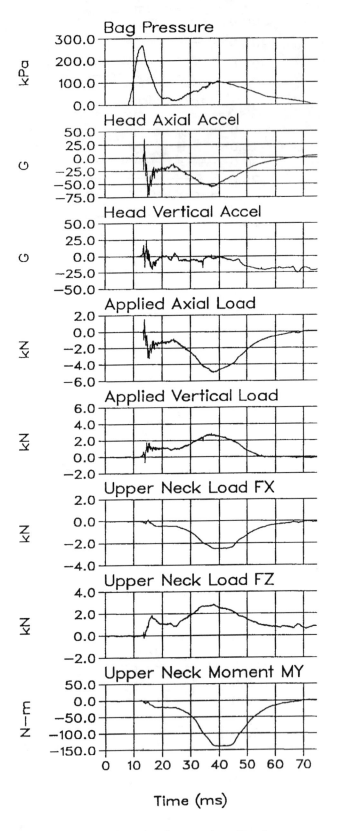

Figure 7: Small female dummy head/neck response data for membrane loading (Z = 143 mm alignment). Data filtered at SAE Channel Class 1000.

throughout the test, since the vertical accelerations are very low during the membrane loading phase. The neck shear load (denoted FX) is reduced in comparison to the applied axial load from the air bag due to the inertial reaction of the head. The neck extension moment (denoted MY) exhibits a waveform similar to that of the shear load, while the neck tension reaches its peak values somewhat earlier in the event.

Various Injury Assessment Reference Values (IARVs) have been recommended, and are in general use, for analyzing head, neck, and chest data from the mid-size male H-III dummy. An example of an IARV is a Head Injury Criterion (HIC) of 1000. IARVs are also available for the small female dummy based on scaling from the mid-size male IARVs [12]. IARVs in use for various body regions are not uniform in terms of injury risk, populations on which the injury risk was determined, methods of analysis of the injury data, etc. Selected measured responses from this study are compared as ratios of the associated IARVs for the two dummies in Table 4 and shown graphically as percentages of the IARVs in Figure 8 for the small female dummy and Figure 9 for the mid-size male dummy. The IARVs used to compare responses are listed in Table 4, and are those provided in the literature [12]. Importantly, the use of these IARVs in this paper is to provide perspective on the level of responses measured in the test exposures, but does not necessarily represent a recommendation for the particular IARV.

The comparisons of the response amplitudes for each body region with respect to the appropriate IARV show some similarities and some differences between the small female and mid-size male dummies. For both dummies, the head HIC responses are well below the IARV, suggesting a low risk of injury for the head for all test situations when judged by the HIC method. The viscous responses for the chest are approximately twice the IARV with the chest centered on and against the module for both dummies, suggesting a high risk of injury for the chest for that test exposure. The mid-size male H-III dummy had neck tension, F_z, over the IARV in one longitudinal alignment. In contrast, the small female H-III had neck tensions, shears, and moments over the IARVs in multiple longitudinal alignments, separations and lateral shifts, suggesting that the smaller dummy has a greater risk of neck injury over a greater range of alignments in these exposures than does the larger dummy. On this basis, tests with both sizes of dummies suggest that the risk of severe injury for the head, based on the HIC, is low for all test conditions and that the risk of severe injury for the chest is high with the chest against the module. However, the overall risk of severe neck injury is much greater as determined with the small female H-III dummy in comparison to that determined with the mid-size H-III dummy.

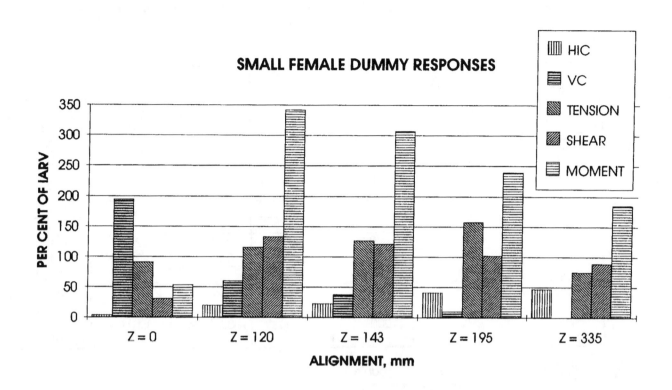

Figure 8: Comparison of small female dummy responses with respect to its Injury Assessment Reference Values (IARV's).

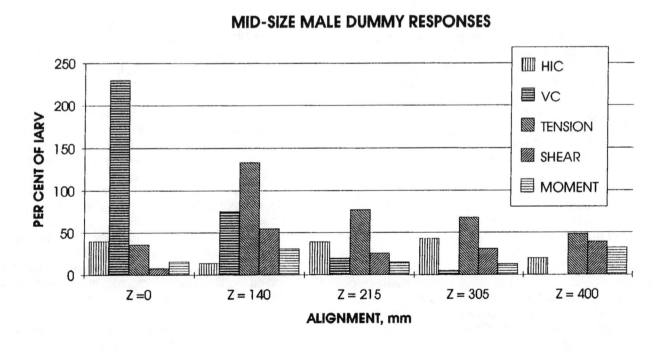

Figure 9: Comparison of mid-size dummy responses with respect to its Injury Assessment Reference Values (IARV's).

Table 4. Comparison of Selected Responses with Injury Asssessment Reference Values (IARV)
Dummy Centered on and Against Module

Dummy	Longitudinal Alignment (Z) (mm)[1]	HIC/IARV	Chest VC/ IARV	Neck		
				Fz/IARV	Fx/IARV	My/IARV
Small	0	0.04	1.94	0.91	0.31	0.53
Small	120	0.20	0.60	1.16	1.33	3.42
Small	143	0.23	0.38	1.27	1.21	3.07
Small	195	0.41	0.10	1.57	1.02	2.39
Small	335	0.47	0.00	0.75	0.89	1.85
Mid	0	0.04	2.30	0.36	0.08	0.16
Mid	140	0.14	0.75	1.33	0.55	0.31
Mid	215	0.39	0.20	0.77	0.26	0.15
Mid	305	0.43	0.05	0.68	0.31	0.13
Mid	400	0.20	0.00	0.48	0.39	0.32
Small	IARV	1113	1.0 m/s	2.20 kN	2.068 kN	+104 N.m -31 N.m
Mid	IARV	1000	1.0 m/s	3.30 kN	3.10 kN	+190 N.m -57 N.m

1 Center of sternum to center of module
Z = 0; sternum centered on module
Z = 335 Small; Head centered on module
Z = 400 Mid; Head centered on module

DISCUSSION

The mid-size male is an important sized dummy for crash testing because it is closest in size to most occupants and most development testing of restraint systems involves this size of dummy. However, since car occupants represent a range of sizes, other sizes of dummies are available. Because of limited field experience with air bags, we cannot determine which near-position situations, if any, might be most important. Horsch et al [1] speculated that, based on the frequency distribution of frontal crash severity, most identified near-position air bag exposures would be just above the threshold for system triggering, that is, in relatively minor crashes. It could be further speculated that "smaller" drivers will adjust the seat forward and thus drive in a location closer to the steering wheel than for a "large" driver. On this basis, a small female would require less displacement relative to the wheel to be in a near-position than a large male. Additionally, the small female stature represents a greater portion of car occupants than does a large male stature. All persons exceeding the stature of a small female were, at one time, the stature of a small female, although this is more likely an issue with passenger air bag systems.

The small female Hybrid III dummy is statically stiff compared with a human, a feature generally similar to that of the mid-size male Hybrid III dummy as reported in [1]. Because of this stiffness and "single" position of

articulations such as the neck, the dummies cannot, in general, achieve the "closeness" to the module that a human can [1]. For evaluation of near-position performance, it might be better to use a human to determine position relative to the steering wheel and module assembly and then position the dummy to best represent this "human-like" position instead of assuming that the dummy will always achieve a "human-like" position.

The test results for both chest compression and the viscous criterion(VC) with the small female H-III dummy indicate a high risk of serious chest injury for the condition of the sternum against the air bag module at the initiation of inflation, similar to the results with the mid-size H-III dummy. Both dummies indicated the greatest injury risk for the chest is due to punchout loading. The chest of the mid-size H-III dummy can more fully cover the air bag module, a factor shown to increase injury risk [1]. The chest acceleration levels (3 ms interval) in all the tests for both dummies were well below the IARVs, indicating that chest compression and viscous response are more appropriate parameters to discern the injury risk for this type of chest loading.

Head acceleration and HIC were in general greater for the small female than the mid-size H-III dummy. However, HIC's were well below the IARV's even for the "worst-case" alignments, with 527 being the highest value observed for the small female and 390 for the mid-size male. However, the peak acceleration associated with the high HIC value for the small female dummy was 142 G and may indicate that the exposure is more severe than shown by the HIC method.

Neck forces and moments tended to be as high or higher for the small female when compared with the mid-size male values. When compared with respect to their IARVs, the relative responses tended to be much greater for the small female dummy, with more responses over the IARVs and over a wider range of alignments. This indicates a higher risk of injury in a particular configuration and a high risk of injury over a wider range of alignments. Thus, the small female H-III dummy appears to be an important tool for assessment of neck injury risk in near-position air bag inflation tests. However, the fidelity of the dummy for neck loading by an air bag has yet to be established. This work has also shown the importance of providing a realistic neckskin to minimize artifactual head/neck/air bag interactions in Hybrid III type dummies.

As discussed in our previous paper [1], there is still not sufficient real world experience to put the results of this laboratory study into perspective. These tests placed two different sizes of dummies in close proximity to a driver air bag prior to inflation. In the field, the occurence of such a position, followed immediately by air bag inflation, has been postulated to be infrequent or rare [1]. As the U.S. vehicle population grows to 100 per cent implementation of air bags in both driver and passenger positions, even rare events could begin to occur in appreciable numbers. This should not diminish the excellent safety record of the air bag [14] but, rather, should serve as an impetus to make them even better by engineering air bag systems that minimize the near-position effects while maintaining improved occupant protection.

SUMMARY AND CONCLUSIONS

This study has compared the responses of the small female Hybrid III dummy with those of the mid-size male Hybrid III dummy when exposed to a deploying driver air bag. For both sized dummies the severity of the loading from the air bag depended on alignment and separation of the dummy with respect to the air bag module. The highest response amplitudes for a particular body region tended to be with that body region aligned with and nearest the air bag module. Separation of a dummy from the module by 13 to 25 mm or a lateral shift of 75 mm reduced many, but not all, responses to below injury assessment reference values. No single alignment provided high responses for all body regions, indicating that one test at a typical alignment cannot simultaneously determine the potential for injury risk for the head, neck, and torso. There are insufficient field accident data on near-position injuries to allow the results of this study to be extended to the present driving population, however, the occurrence of such injuries is expected to be infrequent.

Comparing the response amplitudes for each body region with respect to the appropriate IARVs shows some similarities and some differences between the small female and mid-size male dummies. For both dummies, the head HIC responses were well below the

IARV, suggesting a low risk of injury for the head for all test situations when judged by the HIC method, although the female dummy had high peak head accelerations in one configuration. The viscous responses for the chest were approximately twice the IARV with the chest centered on and against the module for both dummies, suggesting a high risk of injury for the chest for that test exposure. The mid-size male H-III dummy had peak neck tension over the IARV in one longitudinal alignment. In contrast, the small female H-III had neck tensions, shears, and moments over the IARVs in multiple longitudinal alignments, with and without separations and lateral shifts. This suggests that the small female dummy has a greater risk of neck injury over a greater range of alignments in these exposures than does the mid-size male dummy.

The small female H-III dummy appears to be an important tool for assessment of neck injury risk in near-position air bag inflation tests. This work has shown the importance of providing a realistic neckskin to minimize artifactual head/neck/air bag interactions in Hybrid III type dummies. There is a need for a neckskin that can be used for general air bag tests as well as for near-position tests for all Hybrid III type dummies.

ACKNOWLEDGEMENTS

The authors would like to express their appreciation and gratitude to Brian Frantz, Frank Wood, Dennis Andrzejak, and Janis Georgen for their contributions to the performance of this study and in the preparation of this paper.

REFERENCES

1. Horsch, J., Lau, I., Andrzejak, D., Viano, D., Melvin, J., Pearson, J., Cok, D., and Miller, G. "Assessment of Air Bag Deployment Loads." In Proceedings of the 34th Stapp Car Crash Conference, pp. 267-288, SAE Paper #902324, Society of Automotive Engineers, Warrendale, PA, 1990.

2. Horsch, J. and Culver, C. "A Study of Driver Interactions with an Inflating Air Cushion." In Proceedings of the 23rd Stapp Car Crash Conference, pp. 797-823, SAE Paper #791029, Society of Automotive Engineers, Warrendale, PA, 1979.

3. Patrick, L. and Nyquist, G. "Airbag Effects on the Out-of-Position Child." SAE Paper #720442. 2nd International Conference on Passive Restraints, Detroit, May 22-25, 1972, Society of Automotive Engineers, Warrendale, PA, 1972.

4. Wu, H., Tang, S., and Petrof, R., "Interaction Dynamics of an Inflating Air Bag and a Standing Child." SAE Paper #730604, Automobile Engineering Meeting, Detroit, May 14-18, 1973, Society of Automotive Engineers, Warrendale, PA, 1973.

5. Aldman, B., Anderson, A., and Saxmark, O. "Possible Effects of Air Bag Inflation on a Standing Child." In Proceedings of the 18th Conference of the Association for the Advancement of Automotive Medicine, pp. 15-29, Association for the Advancement of Automotive Medicine, Des Plaines, IL, 1974.

6. Takeda, H. and Kobayashi, S. "Injuries to Children from Airbag Deployment." SAE Paper #806030, 8th International Technical Conference on Experimental Safety Vehicles, Society of Automotive Engineers, Warrendale, PA, 1980.

7. Mertz, H., Driscoll, G., Lenox, J., Nyquist, G., and Weber, D., "Response of Animals Exposed to Deployment of Various Passenger Inflatable Restraint System Concepts for a Variety of Collision Severities and Animal Positions." SAE Paper #826047, 9th International Technical Conference on Experimental Safety Vehicles, Society of Automotive Engineers, Warrendale, PA, 1982.

8. SAE, Guidelines for Evaluating Out-of-Position Vehicle Occupant Interactions with Deploying Airbags. SAE Information Report J1980. SAE Human Biomechanics and Simulation Standards Committee, Warrendale, PA, November 1990.

9. Mertz, H. "Restraint Performance of the 1973-76 GM Air Cushion Restraint System." SAE Paper #880400, Society of Automotive Engineers, Warrendale, PA, 1988.

10. Lau, I. and Viano, D. "The Viscous Criterion: Bases and Applications of an Injury Severity Index for Soft Tissues." In Proceedings of the 30th Stapp Car Crash Conference, pp. 123-142, SAE Paper #861882, Society of Automotive Engineers, Warrendale, PA, 1986.

11. Mertz, H., Irwin, A., Melvin, J., Stalnaker, R., and Beebe, M., "Size, Weight and Biomechanical Impact Response Requirements for Adult Size Small Female and Large Male Dummies." SAE Paper #890756. Society of Automotive Engineers, Warrendale, PA, 1989.

12. Mertz, H., "Anthropomorphic Test Devices." In Accidental Injury: Biomechanics and Prevention, Chapter 4. Nahum, A. and Melvin, J., Editors, New York, Springer-Verlag, 1993.

13. Schneider, L., Robbins, D., Pflug, M., and Snyder, R., Development of Anthropometrically Based Design Specifications for an Advanced Adult Anthropomorphic Dummy Family. Volume 1, NHTSA Contract No. DTNH22-80-C-07502, Washington, D.C., 1983.

14. Zador, P.L. and Ciccone, M.A. "Automobile Driver Fatalities in Frontal Impacts: Air Bags Compared with Manual Belts." American Journal of Public Health 83:661-666, 1993.

THORACIC RESPONSE AND TRAUMA OF OUT-OF-POSITION DRIVERS RESULTING FROM AIR BAG DEPLOYMENT

J. R. Crandall
S. M. Duma
C. R. Bass
W. D. Pilkey
University of Virginia
Charlottesville, Virginia

S. M. Kuppa
Conrad Technologies, Inc.
Washington, D.C.

N. Khaewpong
R. Eppinger
National Highway Traffic Safety Administration
Washington, D.C.

ABSTRACT

A case review of air bag-induced fatalities for drivers in low-speed crashes indicated that those at highest risk were small females in close proximity to the air bag at the time of deployment. To approximate these conditions in the laboratory environment, the Hybrid III 5th percentile dummy and seven small female cadavers were instrumented and tested as out-of-position drivers in static air bag deployment tests. Tank test pressure profiles were used to identify more aggressive and less aggressive air bags for use in the static deployments. For comparison, a prototype dual-stage system allowing staged air bag deployment with varied peak inflator pressures and onset rates was also tested. In the out-of-position tests, the chest was positioned against the air bag module in an effort to recreate a worst-case loading environment for the thorax. Rib fractures were the most common cadaver injury and correlated well with maximum chest compression. The Viscous Criteria exceeded 1.0 m/s in nearly all of the out-of-position tests but did not correlate well with the level of observed injury severity, which was largely determined by the number of rib fractures. The results suggest that the pressure onset rate of the inflator is more important than peak pressure in determining the severity of out-of-position injuries and should be given primary consideration in inflator depowering efforts. The prototype dual-stage design provided an effective method of varying pressure onset rates and peak pressures to study their combined effect on out-of-position driver response and injury.

**41ST ANNUAL PROCEEDINGS
ASSOCIATION FOR THE ADVANCEMENT OF AUTOMOTIVE MEDICINE
November 10-11,1997, Orlando, Florida**

The air bag has proven effective in reducing fatalities in frontal crashes with estimates ranging from decreases of 11% to 30% depending on the size of the vehicle (IIHS, 1995; Kahane, 1996). At the same time, some air bag designs can cause fatalities when front-seat passengers are in close proximity to the deploying air bag. Due to the increasing number of air bag-induced injuries and fatalities, the National Highway Traffic Safety Administration (NHTSA) has recently amended Federal Motor Vehicle Safety Standard No. 208, Occupant Crash Protection, to provide vehicle manufacturers more flexibility in designing less aggressive airbags. This action is viewed as an interim solution while advanced air bags, capable of adjusting deployment rates based on occupant, restraint, and crash information, are under development.

The objective of this study was to use dummies and cadavers in out-of-position tests to examine the relationship between thoracic response or injury and air bag peak pressures, pressure onset rates, and module designs. Two production systems were evaluated relative to a dual-stage system intended for future implementation with advanced air bag sensor technology.

BACKGROUND

The term out-of-position (OOP) occupant refers to a situation in which the occupant is located unusually close to the air bag module at the time of inflation. OOP is typically used to refer to cases where an unbelted occupant is slumping or unconscious, and in close proximity to the air bag due to minor impacts prior to a more severe impact in which the air bag deployed. The first significant OOP study of thoracic loading by the deploying air bag was conducted by Horsch et al. (1979) who used a Hybrid III 50th percentile male dummy in static deployment tests. The authors determined that chest deflection was the best indicator to use for evaluation of this type of thoracic loading. An interaction pattern was noted between the air bag and the out-of-position occupant with an initial reaction as the bag breaks out of the module and a second reaction as the bag becomes fully inflated around the occupant. A second paper by Horsch et al. (1990) presented static and dynamic tests with the Hybrid III 50th percentile male and static experiments with anesthetized swine. Using data from these tests, the Viscous Criteria (VC) was identified as the best indicator of thoracic injury. The dummy maximum chest deflection was shown to vary greatly with any increase in separation or adjustment in the vertical alignment relative to the thorax centered on the module. This suggests a nonlinear sensitivity to initial occupant position and indicates that the worst case condition places the occupant's sternum in direct contact with the air bag module.

Horsch et al. (1990) refined the concept of two distinct occupant loading mechanisms from the deploying air bag. The initial loading provided the maximum VC and was referred to as 'punch-out,' since it resembled a cylinder loading the chest. The second reaction was termed membrane loading' and was the result of bag pressure acting on the occupant as well as tension forces created from the bag wrapping around the occupant. The peak thoracic reaction in the punch-out phase was shown to be primarily related to the break pressure, or internal pressure when the air bag ruptures the module and begins to inflate. It was suggested that the break pressure was determined by the module design rather than the inflator properties.

A study comparing the thoracic response during air bag deployment of the Hybrid III 5th percentile female dummy relative to the Hybrid III 50th percentile male dummy was conducted by Melvin et al. (1993). They correlated punch-out and membrane loading with the peak loads recorded by the dummies. Given the same testing conditions, the chest accelerations were slightly higher in the small female due to the lighter dummy mass while the VC and maximum chest compression were slightly lower in the small female. The authors suggested that since the 50th percentile male dummy more completely covers the module, the punch-out force is higher and thus experiences a larger VC and deflection compared to the 5th percentile female dummy.

A new air bag folding pattern, the P-fold, was presented by Malczyk et al. (1995) as a means to lower the air bag loading of the OOP occupant. This design enables the air bag to expand radially as well as longitudinally during deployment. Compared to a typical air bag folding pattern, the P-fold showed an 18% reduction in VC and a 2% decrease in the thoracic deflection and chest acceleration when tested with the Hybrid III chest against the deploying air bag module.

Comparisons between static and dynamic tests of thoracic loading by the deploying air bag have been performed in several studies. Using the 50th percentile male HIII dummy, Horsch et al. (1990) found no difference in chest compression or VC values between static deployments and those conducted with sled tests. Chest accelerations, however, appear to be sensitive to the deployment condition with static chest acceleration values comprising only 60% (Hayano et al., 1994) to 78% (Horsch et al., 1979) of the dynamic values. Presumably, this difference is a combined contribution from both the increased deceleration of the occupant from the sled pulse and from the increased aggressivity of the air bag due to increased inertial resistance during the initial deployment phase (i.e., break out).

METHODOLOGY

CASE REVIEWS - Real world cases of air bag-induced driver fatalities were investigated to identify the occupants' physical characteristics, gender, seating positions, and injury patterns. Sixteen (n=16) crashes where driver fatalities were attributed to the air bag deployment were reviewed from NHTSA's Special Crash Investigation files. In the crashes where passengers were present but there was no passenger side air bag, the passengers experienced minor injuries while the driver died of injuries attributed to air bag deployment.

All sixteen air bag related driver fatalities were low severity crashes (i.e., average delta-V of 20 ±3 km/h) with little interior intrusion. Thirteen (81%) of the fatal cases involved female drivers with an average height of 158 cm. The drivers were typically unbelted with the seat in the forwardmost position due to short stature of the driver. The driver, in some cases, was presumed to be slumping over the steering wheel due to unconsciousness or was positioned near the steering wheel at the time of air bag deployment due to pre-impact braking or to a previous minor impact.

The case reviews identified two injury mechanisms that resulted in fatal injuries to the driver due to air bag deployment. For mechanism 1, the drivers' head and neck were placed in close proximity to the air bag module at the time of deployment such that the force of air bag deployment and the force from the module cover flaps impinged on the neck, face, and upper torso of the driver. Typical injuries sustained by the driver include brain stem laceration (generally at the junction of the pons), subdural and subarachnoid hemorrhage, and basilar skull fracture. For mechanism 2, the driver's chest was placed in close proximity to the air bag module at the time of deployment. The force of the deploying air bag and the forces from the module cover flaps were applied to the thorax. The blunt force (pressure loading) from the deploying air bag and concentrated loading from the cover flaps caused multiple rib fractures as well as soft tissue injuries that included liver, abdominal aorta, and spleen lacerations.

Of the sixteen cases reviewed, five cases were a combination of mechanism 1 and mechanism 2, three cases were mechanism 1 only, and eight cases were mechanism 2 only. All sixteen (100%) cases involved thoraco-abdominal trauma with 10 cases (63%) having these injuries as the most severe trauma. Due to the frequency of mechanism 2 as an injury mechanism and the likelihood of small females being in close proximity to the air bag, the experimental testing presented in this paper concentrates on thoracic loading of the out-of-position small female driver.

AIR BAGS - Kuppa et al. (1997) evaluated less-aggressive and more-aggressive production driver-side air bags with a suspended mass in front of a deploying air bag. Based on their test results, a production less-aggressive (P-LA) and a production more-aggressive (P-MA) air bag that used conventional inflator technology were identified and included in this study. For comparison with the production air bags, a prototype dual stage driver-side hybrid air bag (D) system was used. All air bags used a Leporello (i.e., accordion type) folding scheme and an H-pattern module tear seam (Table 1).

Table 1. Air bag characteristics.

Air bag Characteristic	Figure Ref.	P-MA	P-LA	Dual-Stage
Year of Production	-	1991	1992	Prototype
Steering wheel outer diameter (cm)	a	38.1	38.1	37.8
Steering wheel rim thickness (cm)	b	2.7	2.5	2.5
Local module height (cm)	c	0.3	1.0	0.8 -1.2
Top of rim to seam (cm)	d	25.9	23.8	22.9
Vertical height of module (cm)	e	17.8	15.2	15.2
Horizontal height of module (cm)	f	20.3	17.1	17.5 @ seam
Top of module to seam (cm)	g	13.8	9.1	9.5
Flap thickness at seam (cm)	-	0.3	0.4	0.3
Deployed vertical diameter (cm)	-	68.6	69.9	68.6
Deployed horizontal diameter (cm)	-	68.6	63.5	66.0
Number of tethers	-	4	2	2
Length of tethers (cm)	-	26.7	27.9	29.2

The dual stage air bag can be combined with advanced sensor technology to discern input parameters such as occupant size and position, the crash severity, and the use of restraint systems. By varying the initiation times of the second stage, a wide range of air bag pressures and onset rates can be obtained to optimize the restraint system for a specific crash environment. The hybrid inflator of the dual-stage system uses a pressurized stored gas mixture, primarily inert, in combination with gas and heat-producing pyrotechnics to provide the gas to inflate the air cushion.

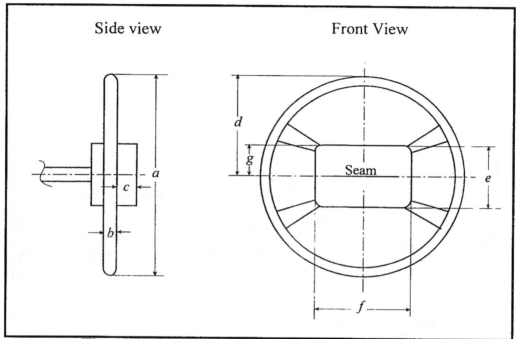

Figure 1. Steering wheel and air bag dimensions.

Inflators were characterized by both the peak pressure and the maximum time derivative of the pressure, known as the pressure onset rate, calculated using moving 10 ms linear curve fits of the pressure data. The air bags were classified using the notation "peak pressure in kPa X peak pressure onset rate in kPa/ms" such that the P-LA inflator with a peak pressure of 151 kPa and a peak pressure onset rate of 5.1 kPa/ms was identified as a "151X5.1" inflator.

An objective of this study was to provide data on the effects of peak pressure and pressure onset rate on the likelihood and severity of injury sustained by small out-of-position female drivers. The dual-stage system allowed us to investigate these parameters while other factors (e.g., tear seam design, flap properties, module location relative to the steering wheel) were held constant. Tank test pressure profiles for the dual-stage system along with those of the production air bag units are provided in Figure 2. The more aggressive dual-stage pressure profile was generated by firing the primary and secondary stages of the dual stage system simultaneously at time 0 ms (D-PS-0). The pressure onset rate of the less aggressive profile was simulated by using only the primary stage (D-P). For higher peak pressures with the same onset rate, the primary stage was fired and followed by the secondary stage after a 20 ms delay (D-PS-20). Comparable to the production air bags in terms of occupant protection, the D-PS-0 and D-PS-20 systems can provide sub-threshold injury response values for the Hybrid III 50th percentile male dummy in sled tests simulating a FMVSS 208 barrier test at 48 km/h. While the primary phase alone may not be capable of sufficiently protecting an unbelted occupant in a 48 km/h frontal crash, it is intended to prevent cosmetic injuries to the face and head of the occupant in lower speed crashes. In addition, the primary phase allows deployment of the air bag early in the impact event and provides for subsequent deployment of the secondary phase once the severity of the crash is sensed.

The differences in inflator technology between the production systems and dual-stage hybrid system can produce differences in air bag inflation characteristics due to the gas temperature and the variety of the gases involved. Wang (1991) showed that differences between stored gas and pyrotechnic air bags exist late in the deployment phase but are not evident during the initial deployment phase of concern in this study.

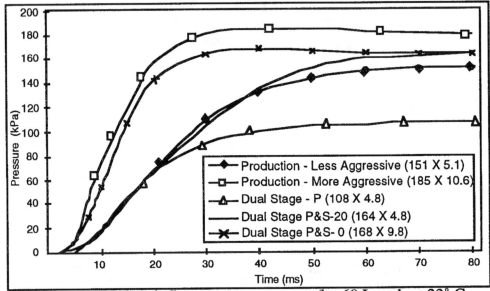

Figure 2. Air bag inflator pressure curves for 60 L tank at 22° C.

TEST SETUP - The air bag tests were conducted in the static rather than dynamic condition based on the assumption that most air bag-induced thoracic trauma could be correlated with either compression (rib fractures) or VC (soft tissue injuries). Neither of these injury criteria has exhibited differences between the static and dynamic deployment tests in previous studies. While chest acceleration has demonstrated higher values for dynamic testing, previous research has shown that most injuries in out-of-position cases are related to chest compression and the rate of compression. Furthermore, the literature shows a nonlinear sensitivity to occupant position and static testing permits a precise positioning of the occupant that cannot be achieved with dynamic testing.

The out-of-position occupant tests were conducted in a mid-size vehicle test fixture. The windshield, windshield header, and parts of the simulated "A" pillar and door frame top were removed in order to ensure minimal interference of these parts with occupant kinematics and to provide an unobstructed camera view from the driver's side. A new steering wheel was used in each test and was mounted with a 12 o'clock alignment of the top of the wheel. The steering column was clamped to prevent stroking in reaction to the forces associated with the deployment of the air bag.

The tests were photographed with three high-speed motion picture film cameras and a high-speed solid-state video camera. The views included two offboard driver's side views (2500 fps and 1000 fps) for analysis, an offboard passenger side view (1000 fps) to provide an overall view of the event, and an oblique onboard overhead view (1000 fps) to examine air bag and head interactions.

Electronic signals from sensors mounted on the test fixture and within the test subject were recorded and converted to digital data by a 128-channel DSP Technology, model TRAQ P data acquisition system. The data-collection process was controlled by DSP technology IMPAX 3.0 software. The post-test processing and conversion of the digital data were done on personal computers, using custom software.

OCCUPANT - The 5th percentile female Hybrid III was used in this study because of its biofidelity of frontal impact response, comparable anthropometry to the drivers fatally injured in real-world crashes, and available torso injury assessment instrumentation. The eight cadavers used in this test series were chosen to be as similar as possible to the occupants most frequently injured in the real-world cases and the 5th percentile female Hybrid III. Cadavers were acquired through the Virginia State Anatomical Board with the permission of the family obtained to conduct biomechanics research. All test procedures were approved by the Human Use Review Panel (HURP) of the National Highway Traffic Safety Administration and an institutional review board. Screening of blood for Hepatitis A, B, C and HIV was conducted with each cadaver prior to their being accepted into the research program. The cadavers were tested either in the fresh condition or preserved using a custom embalming technique (Crandall et al., 1994). To simulate living conditions, pulmonary and cardiovascular pressurization was performed prior to testing. Anthropometry and vital statistics for the dummy and cadavers used in this series are summarized in Appendix A.

A grounded copper mesh shielding vest was used around the thorax, between the inner and outer clothing layers, to shield the chestbands and internal instrumentation from possible electrical interference caused by static electrical discharges from the inflating air bag. A foam neck shield was installed around the Hybrid III neck to simulate a realistic neck diameter and to reduce entrapment of the air bag beneath the chin.

INSTRUMENTATION - The 5th percentile female Hybrid III was instrumented to acquire thoracic response during air bag deployment. The dummy instrumentation package included accelerometers (Endevco 7264A) and chestbands (EPIDM, Robert A. Denton Inc.). A triax of accelerometers was mounted at the chest center of gravity and at the upper spine. Accelerometers on the second and fifth ribs recorded lateral acceleration of the rib cage while accelerometers at the lower and upper sternum recorded anterioposterior accelerations. The raw acceleration data were processed by subtracting initial offset values and were filtered to SAE CFC-60 prior to calculating resultants. Although included in the instrumentation package, the sternum potentiometer failed repeatedly by sliding out of the allotted slot due to the severity of the air bag deployments and was not included in the data analysis.

The cadaver instrumentation package was developed to provide comparable response measurements to those of the Hybrid III. Triaxial accelerometers (Endevco 7267A) were mounted at the upper (T1) and lower spine (L2). Uniaxial accelerometers were mounted at the upper and lower sternum as well as on the lateral fourth and eighth ribs.

Dynamic deformation data for the upper and lower thorax were determined using chestbands, a non-invasive device designed for the measurement of cross-sectional contours of the chest during an impact event (Eppinger, 1989). For the dummy tests, the chestbands were placed at the level of the second (upper band) and fifth (lower band) ribs. For the cadavers, chestbands were placed at the level of the lateral fourth and eighth ribs. Output from the chestbands consisted of local curvature data from each chestband strain gauge. Chest deformation contours were derived from this processed data using a variant of **RBANDPC** (Chi, 1990) developed by the Chi Associates. From this position data, local gauge and sternum velocity were obtained using a four point finite-difference approximation that is further filtered to **SAE CFC-180**. Static verification data and measurements were taken before the dynamic test event to validate static chestband contours.

Air bag reaction forces and contact forces between the occupant and air bag were estimated from a five-axis load cell located at the end of the steering column. For film analysis, crosshair-type photographic targets were attached to the occupant at anatomical landmarks, and their coordinates, relative to the reference target on the test fixture, were measured and noted.

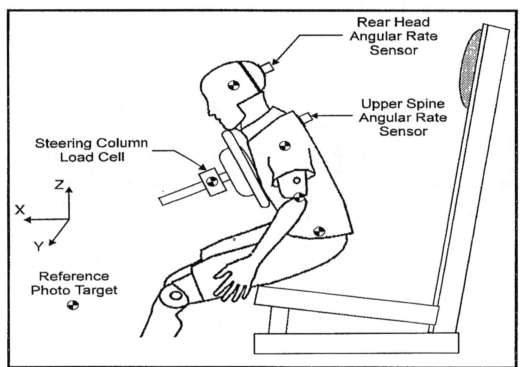

Figure 3. ISO-2 driver position for OOP occupant testing.

OCCUPANT POSITIONING - The occupant was positioned with the chest against the air bag module to maximize thoracic loading by the deploying air bag (Figure 3). The positioning followed guidelines established by the International Standards Organization (ISO) for out-of position testing and used the recommended ISO-2 driver position (ISO, 1992). This position was not intended to reflect the precise position of the occupants injured in the analysis of fatality cases, but was rather intended as a worst-case loading condition for the chest. To achieve this position, the chin was placed on top of the upper rim of the steering wheel, without hanging the head over the rim, and the chest was placed in direct contact with the air bag module. If both of these positions were not possible at the same time, the head was allowed to shift forward until the chest was in contact with the air bag module. In the case of the cadaver tests, the distance from the jugular notch to the center of the air bag module was used as a reference distance and was maintained in all tests. Head and torso rotations about the local anterior-posterior and inferior-superior axes were maintained at zero. This arrangement was achieved by adding padding under the occupant and holding the head in place with frangible masking tape. The legs were placed in front while the arms were positioned at the side of the occupant.

RESULTS AND DISCUSSION

The cadaver responses were normalized to the standard anthropometry of the 5th percentile female using the procedures of Eppinger et al. (1984). This procedure assumes that the mass density and modulus of elasticity are constant between test subjects. The scaling relationship based on occupant mass (M) in kg is shown in equation (1) using the mass of the 5th percentile female in the denominator.

$$\lambda = \left(\frac{M_{cadaver}}{46.3} \right)^{\frac{1}{3}} \qquad (1)$$

The scaled test parameters, denoted with subscript s, can then be expressed in terms of the initial measured parameters, denoted with subscript i, and the scaling factor λ (Equations 2 to 6).

Velocity: $$V_s = V_i \qquad (2)$$

Acceleration: $$A_s = \frac{A_i}{\lambda} \qquad (3)$$

Length: $$L_s = \lambda \times L_i \qquad (4)$$

Time: $$T_s = \lambda \times T_i \qquad (5)$$

Force: $$F_s = \lambda^2 \times F_i \qquad (6)$$

A gross estimate of air bag aggressivity can be made from steering column reaction forces and moments, which provide a measure of air bag deployment forces combined with inertial resistance of the out-of-position occupant. Figure 4 shows that the peak dummy and cadaver axial column forces are comparable and that there is good repeatability for a given air bag system. Steering column reaction forces varied by 10% or less in repeated tests of the P-MA air bag and in tests of the D-P and D-PS-20 air bags that had the same pressure onset rates. The data show that the dual stage system (D-P and D-PS-20) provided the lowest forces, followed by the P-LA system, while the P-MA had the highest forces. Figure 4 also demonstrates a good correlation between the steering column reaction forces and the acceleration response of the subject, suggesting that the inertial forces of the occupant retarding air bag deployment are the primary resistive forces.

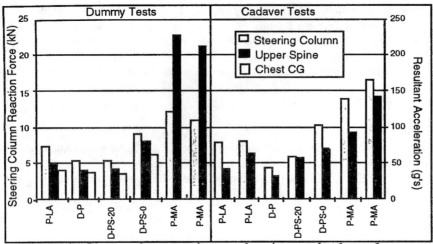

Figure 4. Chest and upper spine accelerations and column forces.

Correlation of film and sensor data indicated that peak resultant upper spine and chest accelerations occurred at the end of the punch-out phase of air bag deployment for the more aggressive air bags and at the beginning of the membrane loading for the less aggressive air bag systems. Figure 5 depicts the sternum and upper spine accelerations for a cadaver test with the P-MA air bag and illustrates the temporal relationship between different peak occupant responses.

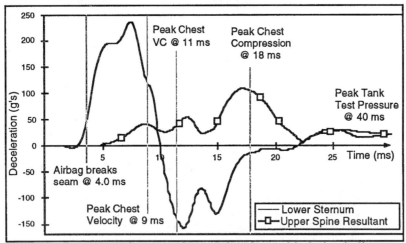

Figure 5. Sternum and upper spine accelerations with timing indicators.

Since the maximum resultant accelerations occur prior to full inflation of the air bag and the attainment of peak inflator pressure, the accelerations should be more sensitive to the onset rate than to the peak pressure. A linear regression of the data indicates good correlation ($r = 0.80$) between the peak chest acceleration of the dummy and the onset rate determined in tank tests. In particular, the dummy tests with D-P and D-PS-20 systems have resultant chest accelerations of 32.4 g's and 31.5 g's respectively. In tank tests with the dual-stage inflators, both configurations had a pressure onset rate of 4.8 kPa/ms while the D-P configuration had a peak pressure of 104 kPa and the D-PS-20 had a peak pressure of 164 kPa. Conversely, the D-PS-20 and D-PS-0 configurations had the same peak pressures but the D-PS-0 had twice the pressure onset rate at 9.8 kPa/ms which produced a considerably higher peak resultant chest acceleration of 57.8 g's. The chest acceleration injury assessment reference values (IARV) of 54 g's (Mertz, 1993) for the Hybrid III 5th percentile female thorax was exceeded only in tests of the D-PS-0 system. However, the cadaver spinal acceleration results suggest that had the sensors not failed in the P-MA dummy tests, these two tests would have also exceeded the IARV.

Sternal accelerations (anterioposterior direction) varied considerably among air bags and showed little correlation with the tank test peak pressure and onset rate results (Table 2). Since the peak sternal accelerations occur within 10 ms after deployment, the maximum sternal acceleration is more closely related to punch-out characteristics of the air bag than are peak pressures which occur at later times. This may indicate that inflator pressure onset rates and module design, such as module cover placement and dimensions, dominate the thoracic response in the early interaction between deploying air bag and occupant.

Table 2. Lower sternum accelerations.

Test		Maximum		Test		Maximum	
Number	Air bag	Value	Time	Number	Air bag	Value	Time
ATD. 381	P-LA	-244	6.1	CAD.386	P-LA	-336	5.1
ATD. 383	P-MA	-248	8.3	CAD.387	P-MA	-207	7.4
ATD. 384	P-MA	-247	9.3	CAD.388	P-LA	-273	5.3
ATD. 417	D-P	-576	6.5	CAD.422	D-P	-534	6.5
ATD. 420	D-PS-20	-738	6.7	CAD.423	D-PS-20	-612	5.5
ATD. 418	D-PS-0	-609	5.7	CAD.421	D-PS-0	-557	4.4

Maximum chest compressions and the sternal Viscous Criteria (VC) (Lau and Viano, 1986) were calculated from both the chestband and accelerometer data (Figures 5 and 6). Chestband data included the compression of the subcutaneous chest tissue and generally provided higher deflections and VC than did the accelerometer data. The accelerometer -based VC and compression values were determined from double integration of the sternal and spinal accelerometers. Although this approach is sensitive to rotations of the spine and sternum, it was felt that during the initial interaction of the deploying air bag with the occupant these effects were negligible.

The air bags produced high rates of sternal compression ranging from 7.0 m/s (P-LA) to 22.8 m/s (P-MA) which led to relatively high VC values. By definition, the maximum VC occurred earlier than the peak chest compression with the largest temporal differences evident in the less aggressive air bag systems. For the dual-stage system with 20 ms delay of the secondary stage (D-PS-20), maximum compressions and VC occurred prior to firing of the secondary stage and were similar to those obtained from the primary stage (D-P) tests. The P-LA system provided responses comparable to those with the D-P air bag while the P-MA air bag produced the highest chest compressions and VC values.

Injury assessment reference values for the Hybrid III 5th percentile female thorax include a sternal deflection of 53 mm and a maximum VC of 1.0 m/s for distributed loading (Mertz, 1993). Using accelerometer data at the lower sternum, none of the dummy or cadaver tests exceeded the compression criteria. VC values based on integration of the accelerometer data were above 1.0 m/s for the dual-stage and P-MA air bags at the lower sternum. The data collected from chestbands, however, included compression of the subcutaneous tissue and generally provided higher estimates of the injury criteria than those calculated from accelerometer data. Using chestband data, all deployments exceeded the compression criteria and had a maximum VC greater than 1 m/s.

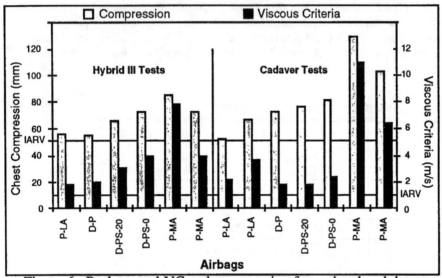

Figure 6. Peak sternal VC and compression from chestband data.

There was considerable variability in the VC calculations even for repeated testing of the same air bag with the dummy. The variability in VC results was attributed to positioning rather than inflator variability based on repeated tank tests with the same inflator. Preliminary dummy testing with the dual-stage air bags showed extreme sensitivity to initial position of the occupant with VC variations in excess of 50% for position adjustments as small as 2.0 cm.

The injury criteria calculated for the dummy tests showed better correlation with the pressure onset rate data than the peak pressure data. Linear regression analysis using the peak inflator pressure from tank tests determined correlation coefficients of r=0.56, r=0.58, and r=0.70 for the chest deflection, chest VC, and upper spine acceleration data. For the limited dummy chest acceleration data, a correlation coefficient of r=0.51 was found. Similar analysis using the peak onset rate provided correlation coefficients of r=0.71, r=0.80, r=0.87, r=0.99 for the chest deflection, chest VC, upper spine acceleration, and dummy chest acceleration data.

Given the inherent variation in cadaveric data, the Hybrid III 5th percentile female dummy provided a reasonable level of biofidelity. The injury criteria calculated for dummy and cadaver showed comparable peak values despite the fact that the dummy thorax has been adjusted to account for a baseline level of thoracic muscle tensing. While the cadavers did provide better coverage of the air bag module during initial position due to their increased compliance and degrees of freedom, analysis of the high speed film showed comparable interaction patterns between the deploying air bags and occupants.

After the air bag deployment tests with cadavers, post-test radiology and detailed necropsy were performed. The anterior breast plate was removed and the three-dimensional location of all rib and sternal fractures was recorded relative to the jugular notch. All injuries were documented, photographed, and coded using the AIS-90 scale (Appendix A). The injury patterns for the out-of-position cadaver tests were similar to those observed in real-world air bag-induced fatalities with the exception that no heart or aortic injuries were noted. Rib fractures were the most common injury and were the primary determinant for assessing the maximum AIS value (Figure 7). In addition to the number of rib fractures, the severity of the damage was estimated by totaling the number of the displaced and comminuted fractures (Figure 7). Sternal fractures accompanied rib fractures in all but the P-LA tests despite the fact that these tests had some of the highest recorded sternal accelerations. Liver lacerations occurred frequently but were mostly superficial tears of the capsule.

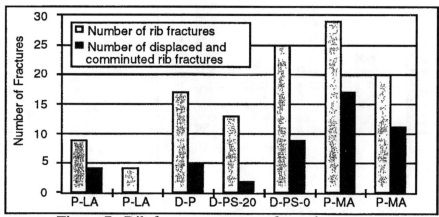

Figure 7. Rib fracture summary for cadaver tests.

Linear regression analysis of calculated injury criteria and observed injury showed maximum sternal deflection (r=0.82) correlated better with the number of rib fractures than with sternal acceleration (r = 0.06), velocity of chest compression (r = 0.49), upper spine acceleration (0.38), or VC (r = 0.56). While VC showed poor correlation with the number of rib fractures, it should be noted that VC is intended as an indicator of soft tissue injury and is not expected to correlate with hard tissue trauma. Analysis of the relationship between maximum AIS and calculated engineering parameters was obviated by the limited number of samples and the fact that all injury AIS scores were between 3 and 5. Although chest compression exhibited the best correlation with AIS scores (r=0.73), all tests except those with the PMA air bags resulted in cadaver chest compressions within 2.5 cm of each other despite AIS scores ranging from 3 to 5.

The investigation of air bag inflator attributes with measured cadaver injury criteria showed a good correlation between the onset rate and the maximum chest deflection (r = 0.86), and a moderate correlation with VC (r = 0.69). For the production systems, however, the onset rate and peak pressure could not be decoupled and significant differences existed among the module designs. Therefore, a more accurate estimation of inflator property effects should be based on the dual-stage design with different inflation attributes for the primary and secondary stages. For the dual-stage system, there was no correlation between peak inflator pressure measured in tank tests and maximum VC (r = 0.31) or chest deflection (r = 0.37) The maximum chest deflection and VC showed slightly better correlation, r = 0.58 and r = 0.64 respectively, with the inflator pressure onset rate.

While cadaver peak upper spine acceleration showed the best correlation with peak pressure (r = 0.87) and pressure onset rate (r = 0.81) of any measured response parameters, it was felt that the injuries observed in these tests were not indicative of acceleration-based mechanisms but were rather associated with the chest compression and rate of compression. This observation is further supported by the lack of correlation between observed injury and the measured sternal and spinal accelerations.

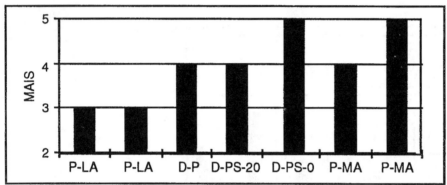

Figure 8. Maximum AIS values (MAIS) for the cadaver tests.

Although the D-P, D-PS-20, and P-LA inflators had similar pressure onset rates, tests with the P-LA air bag resulted in less severe injuries than those with the dual-stage air bag. Given the limited number of tests with each airbag system, it is not possible to determine with certainty whether these results are related to cadaver variability or to air bag parameters. However, analysis of the high speed film identified several differences in the air bag deployment kinematics that could be responsible for the differences in injury severity. Specifically, the P-LA air bag deployed outward and radially before unfolding away from the occupant through the upper opening of the steering wheel between the air bag module and steering wheel rim. This resulted in expansion of a significant portion of the air bag behind the plane of the steering wheel and away from the occupant. Meanwhile, the dual-stage air bags always remained between the steering wheel and occupant during deployment.

The influence of age on the severity of injury could not be determined from the limited data set given the number of air bag variables. In repeated cadaver tests with the P-LA and P-MA air bags, the younger occupant experienced fewer rib fractures with the P-LA air bag while in tests with the P-MA air bag the younger cadaver sustained more rib fractures. To provide a general estimate of skeletal quality for the cadavers, Computed Tomography (CT) scans were taken of the upper extremity and a cortical bone mineral density (BMD) was determined. The average cortical BMD for the cadavers was 1.95 g/cm^3(standard deviation 0.099 g/cm^3) and all values were above or within a 95% confidence interval of the average BMD for older healthy females (1.86 g/cm^3) as determined by Ruegsegger et al. (1991).

CONCLUSIONS

At the initiation of this study, production air bags and a prototype dual-stage air bag system were classified in terms of their peak pressure and pressure onset rate results from tank tests. The Hybrid III 5th percentile female dummy and seven small female cadavers were used in static deployment tests of these air bags in which the chest of the occupant was placed in direct contact with the air bag module. Rib fractures occurred in all cadaver tests and were the primary determinant of injury severity for AIS ratings. Injury severity showed better correlation with maximum chest compression than with chest velocity, VC, or accelerations. Furthermore, the test data showed that the observed injury (i.e., number of rib fractures) and the calculated engineering parameters (i.e., chest compression, velocity, VC) exhibited better correlations with peak pressure onset rate than with the peak pressure of the inflator. While the instance of injury could not be determined from the test instrumentation, the maximum chest compressions and VC occurred early in the deployment phase of the air bag for the ISO-2 test conditions. In particular, all peak response values with the dual stage delayed secondary inflator condition (D-PS-20) occurred prior to initiation of the secondary inflator at 20 ms. This suggests that air bag designs to mitigate out-of-position injuries to the thorax should focus on altering the early stages of deployment prior to full air bag inflation.

Testing with the dual-stage system allowed independent assessment of the effects of inflator peak pressure and peak pressure onset rate on the response and injury outcome. As anticipated from the overall correlation between observed injury and pressure onset rates, the injury criteria and injuries with the dual-stage air bag with delayed secondary inflation (D-PS-20) were comparable to those obtained with the primary stage inflator (D-P) which had the same pressure onset rate. In addition, the D-PS-20 air bag produced lower response parameters and less injury than the D-PS-0 air bag which had comparable inflator peak pressures but higher pressure onset rates. These results suggest that the ability to independently vary the inflator peak pressure and pressure onset rates can reduce out-of-position injury potential while still theoretically providing levels of protection equivalent to current air bag designs for normally positioned occupants during crashes.

While the dual-stage inflator design appears to have great potential in reducing out-of-position occupant injuries, the test data indicate that other air bag module and deployment factors can be as important as inflator characteristics in determining occupant response and injury. The D-PS-20 and D-P air bags had equivalent or lower peak pressure onset rates and peak pressures than the P-LA air bag but resulted in more severe injuries to the occupant. Examination of the high speed film confirms that the interaction of the air bag module cover with the chest and the unfolding pattern of the air bag are significantly different between the P-LA and dual-stage air bag systems. Similarly, the P-MA air bag resulted in more severe injuries than the D-PS-O system even though the inflators had comparable peak pressure and pressure onset rate characteristics. The more severe response and injuries associated with the P-MA air bag were attributed to the large asymmetric flap of the module cover and its prolonged interaction with the occupant during air bag deployment. These results suggest that designs with smaller module cover flaps, such as air bag systems with multiple tear seams, should reduce the likelihood of injury. The P-LA air bag deployment kinematics also suggest that multiple tear seam patterns would have the added benefit of allowing the air bag to follow a path of least resistance when the occupant is not covering the entire module.

In summary, the data suggest that decreasing the peak inflator pressure may not sufficiently reduce the incidence of out-of-position occupant injuries unless concomitant decreases are made in the pressure onset rates. Inflator systems that are capable of independently altering the inflator pressure and pressure onset rate, such as the dual-stage air bag system, show promise in mitigating injuries for the out-of-position occupant. Finally, changes in the air bag module design should accompany inflator modifications in an effort to optimize the design of the air bag system for mitigation of out-of-position injuries. Additional cadaver and dummy testing is required to determine the exact relative contributions of changes in inflator and module designs and cannot be determined from the limited number of tests in this series. Likewise, it is recognized that this study addresses only one aspect of the multifaceted air bag design problem, the OOP occupant situation as characterized by static testing, and does not purport to provide evaluations, comparisons, or recommendations regarding optimizing the overall safety performance of any of the systems studied.

ACKNOWLEDGMENTS

The opinions, findings, and conclusions expressed in this paper are those of the authors and not necessarily those of the National Highway Traffic Safety Administration. The authors would like to thank TRW for providing the dual-stage hybrid air bags. The authors would also like to thank Christine Räisänen of the University of Göteborg for her assistance in editing the manuscript.

REFERENCES

ISO Technical Report (ISO DTR 10982 - Draft), Test Procedures for Evaluation Out-of-Position Vehicle Occupant Interactions with Deploying Air bags, April, 6, 1990.

J. T. Wang, Are Tank Pressure Curves Sufficient to Discriminate Air bag Inflators?, SAE Paper No. 910808, Society of Automotive Engineers, Warrendale, PA, 1991.

Kuppa, S. M., Olson, M. B., Yeiser, C.W., Taylor, L.M.; Morgan, R. M.; Eppinger, R. H.; RAID-An Investigative Tool to Study Air bag/ Upper Extremity Interactions, SAE 970399, 1997.

Crandall, J.R.; Preservation of Human Surrogates for Biomechanical Studies, PhD Dissertation, University of Virginia, 1994.

Eppinger, R. H.; Marcus, J.H.; Morgan, R.M.; Development of Dummy and Injury Index for NHTSA's Thoracic Side Impact Protection Research Program, Proc. 27th Stapp Car Crash Conference, Paper 840885, Society of Automotive Engineers, Warrendale, PA. 1984.

Eppinger, R. H.; On the Development of a Deformation Measurement System and Its Application Toward Developing Mechanically Based Injury Indices, Proc. 33rd Stapp Car Crash Conference, Society of Automotive Engineers, Warrendale, PA, 1989.

Hayano, K.; Ono, K.; Matsuoka, F.; Test Procedures for Evaluating Out-of-Position Vehicle Occupant Interactions with Deployed Air bags, Proc. 14th ESV Conference, Paper 94-S 1-O-19, 1994.

Horsch, J.D.; Culver, C.C.; A Study of Driver Interactions with an Inflating Air Cushion, Proc. 23rd Stapp Car Crash Conference Paper 791029, Society of Automotive Engineers, Warrendale, PA, 1979.

Horsch, J.; Lau, I.; Andrzejak, D., D. Viano et al.; Assessment of Air Bag Deployment Loads, Proc. 34th Stapp Car Crash Conference, Paper 902324, Society of Automotive Engineers, Warrendale, PA, 1990.

Insurance Institute for Highway Safety, Status Report, Vol. 30, No. 3, March 18, 1995.

Kahane, C. J., Fatality Reduction by Air Bags - Analyses of Accident Data through Early 1996, NHTSA Technical Report DOT HS 808 470, National Technical Information Service, Springfield, Virginia, August, 1996.

Lau, I. V.; Viano, D. C.; The Viscous Criterion - Bases and Applications of an Injury Severity Index for Soft Tissues, Proc. 30th Stapp Car Crash Conference, SAE Paper 861882, Society of Automotive Engineers, Warrendale, PA, 1986.

Malczyk, A.; Adomeit, H.; The Air bag Folding Pattern as a Means for Injury Reduction of Out-of-Position Occupants, Proc. 39th Stapp Car Crash Conference, Paper 952704, Society of Automotive Engineers, Warrendale, PA, 1995.

Melvin, J.W.; Horsch, J.D.; McCleary, J.D.; Wideman, L.C.; Jensen, J. L.; Wolanin, M. J.; Assessment of Air bag Deployment Loads with the Small Female Hybrid III Dummy, Proc. 37th Stapp Car Crash Conference Paper 933119, Society of Automotive Engineers, Warrendale, PA, 1993.

Mertz, H. J.; Anthropometric Test Devices in Accidental Injury: Biomechanics and Prevention by A. M. Nahum and J. W. Melvin, Springer-Verlag, New York, NY, 1993.

Ruegsegger, P.; Durand, E.; Dambacher, M.; Differential Effects of Aging and Disease on Trabecular and Compact Density of the Radius, Bone, 12:99, 1991.

Appendix A: Occupant Information

TEST NUMBER	ATD.381-384, ATD.417-420	CAD.386	CAD.387	CAD.388	CAD.421	CAD.422	CAD.423	CAD.424
Air bags	Not Applicable	P-LA	P-MA	P-LA	D-PS-0	D-P	D-PS-20	P-MA
GENERAL INFORMATION								
Surrogate	5th% Hybrid III	Cadaver	Cadaver	Cadaver	Cadaver	Cadaver	Cadaver	Cadaver
Gender	Female	Female	Female	Female	Female	Female	Female	Female
Age at Death		61	45	34	68	67	51	55
Cause of Death	Not Applicable	Bronchial Carcinoma	Respiratory Failure	Myocardial Infarction	Respiratory Failure	Respiratory Failure	Myocardial Infarction	Respiratory Failure
BMD (g/cm^3)		2.02	Not Available	2.05	1.86	1.95	2.03	1.81
ANTHROPOMETRY								
Height (cm)	150	160	161	156	158	164	168	164
Weight (kg)	46.3	52.2	71.8	61.2	51.8	60.2	54.9	49.5
Scaling Factor (λ)	1.00	1.04	1.16	1.10	1.04	1.09	1.06	1.02
INJURY INFORMATION								
Rib Fractures	Not Applicable	9	29	4	25	17	13	20
Sternum Fractures	Not Applicable	0	2	0	2	2	1	1
Liver Lacerations	Not Applicable	No	Yes	Yes	No	No	Yes	Yes
CHEST RESPONSE INFORMATION (scaled values)								
Chest Compression (cm)		5.41	13.0	6.69	7.99	7.43	7.73	10.1
Chest VC (m/s)		2.4	10.5	3.9	2.4	1.8	1.8	6.2
Upper Spine Acceleration (g's)		43.2	90.7	63.8	67.5	31.9	54.1	141

980636

Evaluation of 5th Percentile Female Hybrid III Thoracic Biofidelity during Out-of-Position Tests with a Driver Air Bag

J.R. Crandall, C.R. Bass, S.M. Duma
University of Virginia

S.M. Kuppa
Conrad Technologies Inc.

Copyright © 1988 Society of Automotive Engineers, Inc.

ABSTRACT

This paper evaluates the biofidelity of the Hybrid III 5th percentile female dummy relative to seven small female cadavers tested as out-of-position drivers in static air bag deployment tests. In the out-of-position tests, the chest was positioned against the air bag module in an effort to recreate a worst-case loading environment for the thorax. Two pre-depowered production air bags and a prototype dual-stage air bag were evaluated. Thoracic accelerometers and chestbands were used to compare chest compression, velocity, acceleration, and Viscous Criteria. A statistical comparison of dummy and cadaver results indicate acceptable biofidelity of the Hybrid III dummy with significant differences observed only in the Viscous Criteria.

INTRODUCTION

The Hybrid III has demonstrated biofidelic behavior under a variety of frontal impact conditions in evaluations of belt and air bag restraint systems (Mertz, 1993). Out-of-position (OOP) testing with air bags, however, introduces loading behavior not commonly observed in vehicle frontal crash tests with rapid, distributed loads applied to the anterior chest of the dummy. A number of studies have been conducted with the Hybrid III 50th percentile male and 5th percentile female dummies to examine air bag interaction with the out-of-position occupant (Horsch et al., 1979; Malczyk et al., 1995; Melvin et al., 1993). Previous assessments of the Hybrid III subject realitve to other surrogates, however, have been limited to comparison of the 50th percentile male dummy response to that of anesthetized swine (Horsch, 1990). This testing focused primarily on determining an appropriate injury criteria rather than evaluating the dummy biofidelity. In an effort to evaluate the biofidelity of the Hybrid III dummy relative to human response, the Hybrid III 5th percentile female dummy and seven cadavers were used in out-of-position tests with three air bags of differing inflator peak pressures, inflator onset rates, and air bag module designs.

The seven cadavers used in this test series were chosen to be as similar as possible to the occupants most frequently injured in the real-world cases (Kleinberger et al., 1997; Crandall et al., 1997) and the 5th percentile female Hybrid III. Anthropometry and vital statistics for the dummy and cadavers used in this series are summarized in Appendix A. Cadavers were acquired through the Virginia State Anatomical Board with the permission of the family obtained to conduct biomechanics research. All test procedures were approved by the Human Use Review Panel (HURP) of the National Highway Traffic Safety Administration and an institutional review board. Screening of blood for Hepatitis A, B, C and HIV was conducted with each cadaver prior to their being accepted into the research program. The cadavers were tested either in the fresh condition or preserved using a custom embalming technique (Crandall, 1994). To simulate living conditions, pulmonary and cardiovascular pressurization was performed prior to testing.

METHODOLOGY

AIR BAGS - Three air bags were chosen for comparison of dummy and cadaver response in order to cover a range of air bag aggressivities. All air bag inflators were characterized using a 60L tank test at 22°C (Table 1 and 2). Based on the tank test results, the production air bags were characterized as production less-aggressive (P-LA) and production more-aggressive (P-MA). These inflators used conventional inflator technology and were developed prior to modifications in Federal Motor Vehicle Standard 208 allowing manufacturers to depower. For comparison with the production air bags, a prototype dual stage driver-side hybrid air bag (D) system was used. The hybrid inflator of the dual-stage system uses a pressurized stored gas mixture, primarily inert, in combination with gas and heat producing pyrotechnics to provide the gas to inflate the air cushion. The dual-stage system permitted variation of the initiation times of the second stage and the ability to produce a wide range of air bag pressures and onset rates. All air bags used a Leporello (i.e., accordion type) folding scheme and an H-pattern module tear seam.

Table 1.   Production air bag inflator characteristics.

Air Bag Characteristic	P-MA	P-LA
Year of Production	1991	1992
Peak Pressure (kPa)	185	151
Peak Pressure Onset Rate (kPa/ms)	10.6	5.1

Table 2.   Dual stage hybrid air bag inflator characteristics.

Air Bag Characteristic	D-P	D-PS-0	D-PS-20
Year of Production	1997	1997	1997
Peak Pressure (kPa)	108	168	164
Peak Pressure Onset Rate (kPa/ms)	4.8	9.8	4.8

The differences in inflator technology between the production systems and dual-stage hybrid system can produce differences in air bag inflation characteristics due to the gas temperature and the variety of the gases involved. Wang (1991) showed that differences between stored gas and pyrotechnic air bags exist late in the deployment phase but are not evident during the initial deployment phase of concern in this study.

TEST SETUP – The air bag tests were conducted in the static rather than dynamic condition to permit precise positioning of the occupant prior to air bag deployment. The tests were conducted in a mid-size vehicle test fixture. The windshield, windshield header, and parts of the simulated "A" pillar and door frame top were removed in order to ensure minimal interference of these parts with occupant kinematics and to provide an unobstructed camera view from the driver's side. A new steering wheel was used in each test and was mounted with a 12 o'clock alignment of the top of the wheel. The steering column was clamped to prevent stroking in reaction to the forces associated with the deployment of the air bag.

The tests were photographed with three high-speed motion picture film cameras and a high-speed solid-state video camera. The views included two offboard driver's side views (2500 fps and 1000 fps) for analysis, an off-board passenger side view (1000 fps) to provide an overall view of the event, and an oblique onboard overhead view (1000 fps) to examine air bag and head interactions.

Electronic signals from sensors mounted on the test fixture and within the test subject were recorded and converted to digital data by a 128-channel DSP Technology, model TRAQ P data acquisition system. The data-collection process was controlled by DSP technology IMPAX 3.0 software. The post-test processing and conversion of

the digital data were done on an IBM-compatible Pentium personal computers, using custom software.

INSTRUMENTATION – The 5th percentile female Hybrid III was instrumented to acquire thoracic response during air bag deployment. The dummy instrumentation package included accelerometers (Endevco 7264A) and chestbands (EPIDM, Robert A. Denton Inc.). A triax of accelerometers was mounted at the chest center of gravity and at the upper spine. To assess whole-body rotations, an angular rate sensors (ATA, Inc. Model ARS 04E) was also mounted at the upper spine. Accelerometers on the second and fifth ribs recorded lateral acceleration of the rib cage while accelerometers at the lower and upper sternum recorded anterioposterior accelerations. The raw acceleration data was processed by subtracting initial offset values and was filtered to SAE CFC-60 prior to calculating resultants. Although included in the instrumentation package, the sternum potentiometer failed repeatedly by sliding out of the allotted slot due to the severity of the air bag deployments and was not included in the data analysis.

The cadaver instrumentation package was developed to provide comparable response measurements to those of the Hybrid III. Triaxial accelerometers (Endevco 7267A) were mounted at the upper (T1 vertebra) and lower spine (L2 vertebra). It was not possible to mount an accelerometer triax at the equivalent cadaver chest center of gravity location due to interference with the chestbands and potential interaction during rebound and contact with the seatback. Uniaxial accelerometers were mounted at the upper and lower sternum as well as on the lateral fourth and eighth ribs.

Dynamic deformation data for the upper and lower thorax were determined using chestbands, a non-invasive device designed for the measurement of cross-sectional contours of the chest during an impact event (Eppinger, 1989). For the dummy tests, the chestbands were placed at the level of the second (upper band) and fifth (lower band) ribs. For the cadavers, chestbands were placed at the level of the lateral fourth and eighth ribs. Output from the chestbands consisted of local curvature data from each chestband strain gauge. Chest deformation contours were derived from this processed data using a variant of RBANDPC (Chi, 1990) developed by the Chi Associates. From this position data, local gauge and sternum velocity were obtained using a four point finite-difference approximation that is further filtered to SAE CFC-180. Static verification data and measurements were taken before the dynamic test event to validate static chestband contours.

OCCUPANT POSITIONING – The occupant was positioned with the chest against the air bag module to maximize thoracic loading by the deploying air bag (Figure 1). The positioning followed guidelines established by the International Standards Organization (ISO) for out-of-position testing and used the recommended ISO-2 driver position (ISO, 1992). This position was intended as a worst-case loading condition for the chest. To achieve this position, the chin was placed on top of the upper rim of the steering wheel, without hanging the head over the rim, and the chest was placed in direct contact with the air bag module. If both of these positions were not possible at the same time, the head was allowed to shift forward until the chest was in contact with the air bag module. In the case of the cadaver tests, the distance from the jugular notch to the center of the air bag module was used as a reference distance and was maintained in all tests. Head and torso rotations about the local anterior-posterior and inferior-superior axes were maintained at zero. This arrangement was achieved by adding padding under the occupant and holding the head in place with masking tape. The legs were placed in front while the arms were positioned at the side of the occupant.

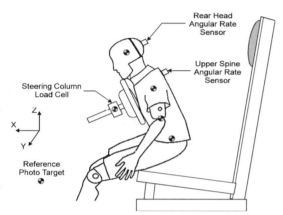

Figure 1. ISO-2 driver position for OOP occupant testing.

POST-TEST PROCEDURES – After the tests with cadavers, post-test radiology and a detailed necropsy were performed. The anterior breast plate was removed and the three-dimensional location of all rib and sternal fractures was recorded relative to the jugular notch. All injuries were documented, photographed, and coded using the AIS-90 scale (Appendix A).

RESULTS AND DISCUSSION

The injury patterns for the out-of-position cadaver tests were similar to those observed in real-world air bag-induced fatalities (Kleinberger et al., 1997; Crandall et al., 1997) but no heart or aortic injuries were noted. The reason for the lack of vascular or heart injuries is unclear given the severity of the interaction between the deploying air bag and cadaver chest and the pressurization of the cadaver cardiovascular system.

Ideally, comparison of the dummy and cadaver biofidelity would be made using only tests in which only minor injury occurred. However, rib fractures occurred in all tests and were the primary determinant for assessing the maximum AIS value (Figure 2). Rib fractures were counted as the total number of fractures rather than the number of fractured ribs such that multiple fractures per rib were possible. Sternal fractures accompanied rib fractures in all but the P-LA tests despite the fact that these tests had some of the highest recorded sternal accelerations. Liver lacerations occurred frequently but were mostly superficial tears of the capsule.

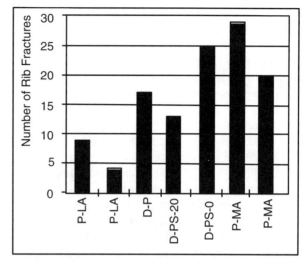

Figure 2. Rib fracture summary for cadaver tests.

The cadaver responses were normalized to the standard anthropometry of the 5th percentile female using the procedures of Eppinger et al. (1984). This procedure assumes that the mass density and modulus of elasticity are constant between test subjects. The scaling relationship based on occupant mass (M) in kg is shown in equation (1) using the mass of the 5th percentile female in the denominator.

$$\lambda = \left(\frac{M_{cadaver}}{46.3}\right)^{\frac{1}{3}}$$

(Eq. 1)

The scaled test parameters, denoted with subscript s, can then be expressed in terms of the initial measured parameters, denoted with subscript i, and the scaling factor λ (Equations 2 to 6).

Velocity:

$$V_s = V_i$$

(Eq. 2)

Acceleration:

$$A_s = \frac{A_i}{\lambda}$$

(Eq. 3)

Length:

$$L_s = \lambda \times L_i$$

(Eq. 4)

Time:

$$T_s = \lambda \times T_i$$

(Eq. 5)

Force:

$$F_s = \lambda^2 \times F_i$$ (Eq. 6)

The dummy upper spine acceleration data show that the dual stage system (D-P and D-PS-20) provided the lowest accelerations, followed by the P-LA system, while the P-MA had the highest accelerations (Figure 3). A similar trend can be observed with the cadaver data. The chest acceleration injury assessment reference value (IARV) of 54 g's (Mertz, 1993) for the Hybrid III 5th percentile female thorax was exceeded only in tests of the D-PS-0 system. However, the cadaver results suggest that had the sensors not failed in the P-MA dummy tests, these two tests would have also exceeded the IARV. A t-test at the 95% confidence level was conducted to determine whether the differences in the cadaver and dummy upper spine accelerations were significant. The t-test failed to reject the hypothesis that no difference exists between cadaver and dummy response. This implies that within the variation of the test data, the dummy upper spine acceleration response is similar to that of the cadavers.

Table 3 indicates considerable variation in sternal accelerations (anterioposterior direction) among air bags for both dummies (ATD) and cadavers (CAD). A t-test at the 95% confidence level failed to reject the hypothesis that no difference exists between cadaver and dummy response.

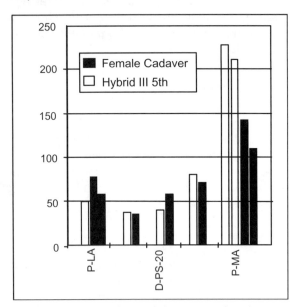

Figure 3. Comparison of dummy and cadaver upper spine accelerations.

Maximum chest compressions were calculated from both the chestband and accelerometer data (Table 4). The upper spinal anterior-posterior (x-axis) accelerometer signal was subtracted from the sternal accelerometer signal. The resulting acceleration time history was then double integrated to obtain the sternum velocity and deformation. While integration of the chest accelerometers is subject to whole body and sternal rotation, analysis of the data from an angular rate sensor mounted on the upper spine indicated insignificant whole body rotations during maximum interaction of the air bag with the occupant.

Table 3. Lower sternum accelerations.

Test Number	Ai rbag	Max. Value	Time	Test Number	Air bag	Max. Value	Time
ATD. 381	P-LA	- 244	6.1	CAD.386	P-LA	- 362	5.1
ATD. 383	P-MA	- 248	8.3	CAD.387	P-MA	- 236	7.4
ATD. 384	P-MA	<- 247	9.3	CAD.388	P-LA	- 292	5.3
ATD. 417	D-P	-576	6.5	CAD.422	D-P	-583	6.5
ATD.420	D-PS-20	-738	6.7	CAD.423	D-PS-20	-648	5.5
ATD. 418	D-PS-0	-609	5.7	CAD.421	D-PS-0	-578	4.4

Table 4. Sternal deformations from chestband (CB) and accelerometer (ACC) data (unscaled).

Parameter	Test	Airbag	Upper- CB		Lower-ACC		Lower-CB	
			Max.	Time	Max.	Time	Max.	Time
Sternum	ATD. 381	P-LA	68.0	16.0	42.0	16.4	44.0	17.4
Deflection	ATD.383	P-MA	65.0	16.6	NA	NA	106	12.6
(mm)	ATD. 384	P-MA	71.0	17.3	NA	NA	75.0	13.9
	ATD. 417	D-P	62.0	14.4	46.0	14.3	49.0	14.7
	ATD.420	D-PS-20	75.0	14.8	52.0	17.8	58.0	14.8
	ATD. 418	D-PS-0	75.0	14.3	59.7	14.5	70.0	15.3
	CAD.386	P-LA	47.0	16.7	36.5	18.0	52.0	22.9
	CAD.387	P-MA	112	13.9	56.0	17.9	73.0	12.4
	CAD.388	P-LA	61.0	27.9	33.0	19.0	57.0	23.6
	CAD.422	D-P	57	17.9	61.0	17.6	66.0	22.0
	CAD.423	D-PS-20	53	19.2	78.0	19.4	69.0	19.9
	CAD.421	D-PS-0	58.0	14.0	52.0	14.4	77.0	16.5

In general, the chestband calculated deformations including the soft tissue overlying the rib cage were consistently greater than those recorded by the accelerometers which were skeletal deformations. A correlation analysis was conducted on the dummy tests to determine whether the maximum chest deflections from chestbands were associated with the corresponding responses obtained from integration of the chest accelerometers. The results suggest that maximum chest deflection (R^2 = 0.96) obtained from chestbands showed strong correlation with the corresponding values determined from accelerometers. A t-test at the 95% confidence level was conducted to determine whether the differences in the cadaver and dummy sternal deformations were significant. The t-test failed to reject the hypothesis that no difference exists between cadaver and dummy response. This implies that within the variation of the test data, the dummy maximum compression is similar to that of the cadavers.

Injury assessment reference values for the Hybrid III 5th percentile female include a sternal deflection of 53 mm for distributed thoracic loading (Mertz, 1993). Using accelerometer data at the lower sternum, none of the dummy or cadaver tests exceeded the compression criteria (Figure 4). The data collected from chestbands, however, included compression of the subcutaneous tissue and generally provided higher estimates than those calculated from accelerometer data. Using chestband data, all deployments except one cadaver P-LA test exceeded the compression criteria.

Maximum Viscous Criteria (VC) (Lau and Viano, 1986) were calculated from both the chestband and accelerometer (accel.) data (Figure 5 and Table 5). Chestband data was differentiated to obtain velocity time history at each point on the chest circumference. The acceleration based VC was determined using the upper and lower sternal accelerometers relative to the accelerometer mounted on the upper spine. Chestband data included the compression of the subcutaneous chest tissue and generally provided higher VC values than did the accelerometer data.

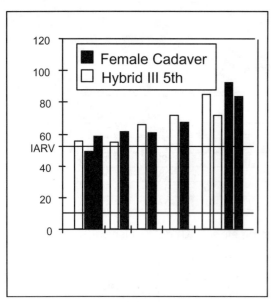

Figure 4. Peak sternal compression from chestband data (scaled).

Injury assessment reference values for the Hybrid III 5th percentile female thorax set the injury threshold at a maximum VC of 1.0 m/s for distributed loading (Mertz, 1993). VC values based on integration of the accelerometer data were above 1.0 m/s for the dual-stage and P-MA air bags at the lower sternum. However, chestband based VC exceeded the threshold in nearly every test with the exception of some with the P-LA system.

A correlation analysis was conducted on the dummy tests to determine whether the maximum chest velocity and VC obtained from chestbands were associated with the corresponding responses obtained from integration of the chest accelerometers. The results suggest that maximum chest velocity ($R^2 = 0.9$) obtained from chestbands showed strong correlation with the corresponding values determined from accelerometers but only 33% of the variation in VC obtained from chestbands was associated with the variation in VC obtained from accelerometers.

Table 5. Viscous Criteria data from chestbands and accelerometers (unscaled data)

Parameter	Test	Airbag	VC - Upper Chestband Max.	Time	VC - Upper Accel. Based Max.	Time	VC - Lower Chestband Max.	Time	VC - Lower Accel. Based Max.	Time
Sternal V*C (m/s)	ATD. 381	P-LA	1.8	10.7	NA	NA	0.7	13.0	0.92	12.9
	ATD.383	P-MA	1.9	15.3	NA	NA	7.8	9.7	NA	NA
	ATD. 384	P-MA	1.9	10.0	NA	NA	4.0	13.2	NA	NA
	ATD. 417	D-P	2.0	10.1	NA	NA	1.1	10.0	0.22	14.3
	ATD.420	D-PS-20	3.0	15.7	NA	NA	1.9	14.5	1.49	14.3
	ATD. 418	D-PS-0	3.9	12.7	NA	NA	2.4	9.7	1.90	10.8
	CAD.386	P-LA	2.4	6.6	0.37	13.1	1.0	17.6	0.46	11.8
	CAD.387	P-MA	5.8	13.4	0.2	15.0	10.5	11.8	1.27	11.0
	CAD.388	P-LA	3.9	16.7	0.19	13.9	1.1	18.9	0.36	13.2
	CAD.422	D-P	1.8	9.5	0.2	15.0	1.4	10.5	1.45	10.6
	CAD.423	D-PS-20	1.1	10.5	0.84	16.0	1.8	10.3	1.8	11.7
	CAD.421	D-PS-0	2.1	10.1	1.4	11.5	2.4	10.0	1.07	9.4

A t-test at the 95% confidence level was conducted to determine whether the differences in the cadaver and dummy VC, obtained from both chestband and accelerometer data were significant when both surrogates were evaluated under similar tests conditions (i.e., the same air bag system). The t-test failed to reject the hypothesis that no difference exists between cadaver and dummy response. This implies that within the variation of the test data, the dummy response is similar to that of the cadavers.

CONCLUSIONS

In an effort to evaluate the biofidelity of the Hybrid III 5th percentile female dummy, the dummy and seven small female cadavers were used in static out-of-position tests where the chest was placed in direct contact with the air bag module. Two production air bags and a prototype dual-stage air bags were used in the static deployment tests.

Radiographs and a detailed autopsy identified rib fractures and some sternal fractures. In order to assess biofidelity of the dummy, its responses should ideally be compared to the corresponding responses of the cadaver which experience no or minor injury such that the structural integrity of the rib cage is not compromised. However, all cadavers sustained rib fractures of AIS=3 severity. In tests with the less aggressive air bags, the cadaver sustained an average of 11 rib fractures while those of more aggressive air bag systems averaged more than 20 fractures. Accordingly, the differences between dummy and cadaver response were most similar for the less aggressive air bags.

The biofidelity evaluation of the Hybrid III 5th percentile female was conducted for sternal acceleration, upper spine resultant acceleration, sternal compression and rate of compression, and sternal VC. Sternal compressions, rates of compression, and VC derived from chestband data included the effects of the skin and soft tissue overlying the sternum, while the corresponding responses derived from accelerometer data did not.

While the chestband data generally provide higher sternal compression and velocity values, results of a correlation analysis suggested that sternal compression and rate of compression derived from chestband data were well correlated with the corresponding responses derived from accelerometer data. However, VC derived from chestband data did not correlate well with VC derived from accelerometer data. One reason for this lack of correlation may be the variability observed in VC even for repeated tests with the dummy of the same air bag system. This sensitivity emphasized the need for precise positioning during testing and for quantification of air bag inflator variability.

A t-test at the 95 percent confidence level was conducted to determine whether the differences in cadaver and dummy response (sternal and upper spine acceleration, sternal compression and rate of compression, sternal VC) under the same test conditions were significant. The test suggested that the dummy responses were comparable to that of the cadavers.

Given the inherent variation in cadaveric data, the Hybrid III 5th percentile female dummy provided a reasonable level of biofidelity. The response values and injury criteria calculated for dummy and cadaver showed comparable peak values despite the fact that the dummy thorax has been adjusted to account for a baseline level of thoracic muscle tensing and that the cadavers sustained injury in all tests. While the cadavers did provide better coverage of the air bag module during initial position due to their increased compliance and degrees of freedom, analysis of the high speed film showed comparable interaction patterns between the deploying air bags and occupants. In summary, the Hybrid III 5th percentile female dummy appears capable of representing the response of the out-of-position human occupant and showed the ability to be used for development of advanced restraint systems that mitigate air bag-induced injuries.

ACKNOWLEDGMENTS

This study was supported in part by DOT NHTSA Cooperative Agreement DTNH22-95Y-07028. The authors wish to thank Nopporn Khaewpong and Rolf Eppinger of the National Highway Traffic Safety Administration for their input and guidance. The authors would also like to thank TRW for providing the dual-stage hybrid air bags.

REFERENCES

1. ISO Technical Report (ISO DTR 10982 - Draft), Test Procedures for Evaluation Out-of-Position Vehicle Occupant Interactions with Deploying Air bags, April, 6, 1990.
2. J. T. Wang, "Are Tank Pressure Curves Sufficient to Discriminate Air bag Inflators?", SAE Paper No. 910808, Society of Automotive Engineers, Warrendale, PA, 1991.
3. Kuppa, S. M., Olson, M. B., Yeiser, C.W., Taylor, L.M.; Morgan, R. M.; Eppinger, R. H.; "RAID-An Investigative Tool to Study Air Bag/ Upper Extremity Interactions," SAE 970399, 1997.
4. Crandall, J.R.; Preservation of Human Surrogates for Bio-

mechanical Studies, PhD Dissertation, University of Virginia, 1994.
5. Crandall, J. R.; Duma, S.M.; Bass, C.R.; Pilkey, W.D.; Kuppa, S. M.; Khaewpong, N.; Eppinger, R.; "Thoracic Response and Trauma of Out-of-Position Drivers Resulting from Air bag Deployment, "Proc. 41st AAAM Conference, Orlando, Fl., November, 1997.
6. Eppinger, R. H.; Marcus, J.H.; Morgan, R.M.; "Development of Dummy and Injury Index for NHTSA's Thoracic Side Impact Protection Research Program," Paper 840885, Twenty-Seventh Stapp Car Crash Conference Proc., Society of Automotive Engineers, Warrendale, PA. 1984.
7. Eppinger, R. H.; "On the Development of a Deformation Measurement System and Its Application Toward Developing Mechanically Based Injury Indices," Proc. Thirty-Third Stapp Car Crash Conference, Society of Automotive Engineers, Warrendale, PA, 1989.
8. Horsch, J.D.; Culver, C.C.; "A Study of Driver Interactions with an Inflating Air Cushion," Proc. 23rd Stapp Car Crash Conference Paper 791029, Society of Automotive Engineers, Warrendale, PA, 1979.
9. Horsch, J.; Lau, I.; Andrzejak, D., D. Viano et al.; "Assessment of Air Bag Deployment Loads," Proc. 34th Stapp Car Crash Conference , Paper 902324, Society of Automotive Engineers, Warrendale, PA, 1990.
10. Malczyk, A.; Adomeit, H.; "The Air Bag Folding Pattern as a Means for Injury Reduction of Out-of-Position Occupants," Proc. 39th Stapp Car Crash Conference, Paper 952704, Society of Automotive Engineers, Warrendale, PA, 1995.
11. Melvin, J.W.; Horsch, J.D.; McCleary, J.D.; Wideman, L.C.; Jensen, J. L.; Wolanin, M. J.; "Assessment of Air Bag Deployment Loads with the Small Female Hybrid III Dummy," Proc. 37th Stapp Car Crash Conference Paper 933119, Society of Automotive Engineers, Warrendale, PA, 1993.
12. Kleinberger, M. and L. Summers; "Mechanisms of Injury for Adults and Children Resulting from Air Bag Interaction," Proc. 41st AAAM Conference, Orlando, Fl., November, 1997.
13. Lau, I. V.; Viano, D. C.; "The Viscous Criterion - Bases and Applications of an Injury Severity Index for Soft Tissues," SAE Paper 861882, Society of Automotive Engineers, Warrendale, PA, 1986.
14. Mertz, H. J.; "Anthropometric Test Devices" in Accidental Injury: Biomechanics and Prevention by A. M. Nahum and J. W. Melvin, Springer-Verlag, New York, NY, 1993.

APPENDIX A

TEST NUMBER	ATD.381-384, ATD.417-420	CAD.386	CAD.387	CAD.388	CAD.421	CAD.422	CAD.423	CAD.424
Air bags	Not Applicable	P-LA	P-MA	P-LA	D-PS-0	D-P	D-PS-20	P-MA
GENERAL INFORMATION								
Type	5th% Hybrid III	Cadaver	Cadaver	Cadaver	Cadaver	Cadaver	Cadaver	Cadaver
Number	VRTC 145	96-EF-58	96-EF-59	96-EF-63	96-EF-67	96-EF-66	96-EF-72	96-EF-74
Gender	Female	Female	Female	Female	Female	Female	Female	Female
Age at Death	Not Applicable	61	45	34	68	67	51	55
Cause of Death		Bronchial Carcinoma	Respiratory Failure	Myocardial Infarction	Respiratory Failure	Respiratory Failure	Myocardial Infarction	Respiratory Failure
ANTHROPOMETRY								
Height (cm)	150	160	161	156	158	164	168	164
Weight (kg)	46.3	52.2	71.8	61.2	51.8	60.2	54.9	49.5
INJURY INFORMATION								
Rib Fractures		9	29	4	25	17	13	20
Sternum Fractures	Not Applicable	0	2	0	2	2	1	1
Liver Lacerations		No	Yes	Yes	No	No	Yes	Yes

982325

Dual-Stage Inflators and OoP Occupants -A Performance Study

Axel Malczyk, Dirk Franke, Heinz-Dieter **Adomeit**
Petri AG, Engineering Center for Automotive Safety Berlin

Copyright © 1998 Society of Automotive Engineers, Inc.

ABSTRACT

Fifty-three static out-of-position tests were conducted with a "small female" dummy placed in three different positions, and with distances of 0 mm and 50 mm from the airbag.

The driver-side module with a single-stage inflator was additionally tested with inflator versions tailored to 80%, 60%, 40% and 20% of peak tank pressure, in order to simulate the first of two stages of a dual-stage inflator.

In general, biomechanical loadings decreased with less inflator propellant. Critical chest loadings were measured down to the 60%-stage. The neck extension bending moment exceeded the limit only with the 100%-charge. With distances of 50 mm, none of the threshold values were exceeded. Energy reductions of 20% between two stages did not necessarily reduce occupant loadings.

INTRODUCTION

DURING RECENT YEARS, a number of cases from the accident field have been reported in which airbags accounted for severe or even fatal injuries in low-speed crashes. The occupants affected were mostly small-statured drivers or children riding in the passenger seat who were "out-of-position" (OoP) at the moment of deployment initiation. Although most of them were found unbelted or otherwise improperly restrained, these occurences have led to the rapid introduction of a "depowering" option for airbags by NHTSA [1] as a temporary measure until September 2001. While diminuition of the inflator energy is considered to reduce airbag aggressiveness in OoP, its protective effect in severe crashes decreases as well. For this reason, the FMVSS 208 "unbelted" requirement has been revised for depowered systems. For the future, the introduction of "smart airbags" is demanded, since these are expected to provide variable inflator output for "out-of-position" and "in-position" situations.

Potential future applications of these systems are seen in tailoring of restraint system performance to other parameters such as impact velocity and occupant size under "in-position" conditions. These systems, however, require extensive and reliable input by sensors and electronics to detect crash conditions and to initiate the necessary adjustments in the system. While detailed definitions of "smart" features and performance are still under discussion, dual-stage inflators are widely considered to represent a major component of a "smart restraint system".

Dual-stage inflators have two separate chambers for solid propellant or compressed gas. They can generally be ignited separately, with a time delay, or simultaneously, and are thereby capable of producing different pressure-vs.-time histories. Depending on the ratio between the two chambers, these inflators are designated generally "X% / Y%" dual-stage inflators. "X% / Y%" combinations actually implemented range from "50% / 50%" up to "80% / 20%". The particular partition depends on the philosophy pursued in lay-out design of the airbag system.

Nevertheless, decisions on the inflator design – and therefore also on the partition of a dual-stage inflator – are necessary in an early development phase of a restraint system. Such decisions determine its tailorability in the following development process, whether for use under in-position or out-of-position conditions. Therefore, the present study aims to provide basic information on the influence of the stages of different energy levels on biomechanical loadings under out-of-position conditions. The ultimate purpose is, in turn, to facilitate decisions on inflator design for particular applications.

SCOPE OF RESEARCH PROGRAM

Although the results of depowering have been examined and discussed extensively, these measures are restricted to conventional airbag systems incorporating single-stage inflators. Broad knowledge accordingly exists only on the effect of inflator energy reductions by 20% to 35% [1;2]. Prognosis of the benefits of even greater inflator charge reductions on out-of-position occupants has proved difficult, since test results are few in number and numerical simulation based on multi-body systems does not yield reliable results for occupants in very close proximity to the airbag.

The objective of the present study is therefore to provide basic information from experiments on the **influence** of the stages of different peak pressure levels on occupant loadings under static **OoP** conditions as defined by **ISO [3]** and SAE **[4]**. Low-charged **stages** of 20% to 60% are of special interest here, because these can be assumed to be fired primarily in cases of a clearly detected **OoP** scenario. In order to keep the number of involved design parameters small, the program described here focuses at present on driver-side **airbags**. For investigation of passenger-side **airbags**, it would be necessary to consider a wider range of bag volumes and shapes, module locations, and relevant occupant sizes.

TEST SETUP

The test mock-up especially used for this study consists of two main components: an adjustable seat for accommodation of the dummy, and a test rig to support the steering wheel and the driver-side **airbag** module. The steering wheel and the **airbag** module can be adjusted horizontally and vertically, as well as in their angular position.

The steering wheel used here is made from steel tubing and simulates the shape of a four-spoke wheel. The **airbag** module is connected to the base mounting plate by three load cells to allow assessment of forces in the axial direction of the steering wheel.

DUMMY

A "small female Hybrid **III**" dummy with standard **instrumentation**, additionally equipped with a **tri-axial** upper neck load cell, served as anthropomorphic test device. In order to ensure enhanced **biofidelity** for neck and chin geometry, the dummy was fitted with a **neckskin** and a chin insert. These were prepared in close conformity to the parts described by Melvin et al. [5]. Care was taken not to alter the force-deformation behavior of the neck and head area by these measures.

AIRBAG MODULE

For the study, a baseline driver-side **airbag** module was chosen which featured state-of-the-art design, and which was representative for a great number of currently used **airbags**.

All tests took place with a driver-side **airbag** module, with an uncoated **60-liter** cushion folded according to the P-folding technique. This folding pattern has proven beneficial under **OoP** conditions, especially with regard to neck loadings [6; 7] and has become a standard feature in the **airbag** modules of various car manufacturers. The module cover in **dual-component** plastic features a horizontal split line which produces two doors of approximately the same size.

The **airbag** module additionally incorporates a diffuser made from sheet metal stamping which covers the inflator and secures the bag to the module housing. This part ensures that only one layer of cushion fabric lies on top of the **inflator**, and at the same time keeps a small gap between the inflator outlets and the fabric to reduce thermal stress. For assessment of deployment pressure in the cushion, a measuring point was located in the center of the diffuser top.

INFLATOR

Generally, dual-stage inflators are designated by the prefix "X% / Y%": e.g., "80% / 20%"-dual-stage inflator. X% and Y% stand for the size of the respective inflator stage and sum to "100%" for the entire inflator. Different definitions, however, are in use for this designation among inflator manufacturers:

For example, "**X%**" for one of the two stages may be defined as:

a) Providing **X%** of the total (i.e., 100%) space for propellant inside the respective inflator chamber

b) Providing **X%** of the total mass of propellant in the respective inflator chamber

c) Delivering **X%** of the maximum tank pressure that a **100%-reference** inflator develops in a tank test: i.e., a **60-liter** tank for driver-side inflators.

It is obvious that definitions a) and **b)** do not explicitly describe the performance and peak pressure of an X%-stage of an inflator, and that they leave room for divergence; **e.g.**, through variation of pellet sizes and shapes.

It is important to note that distribution of a given amount of propellant into two separate chambers, even if fired simultaneously, will not achieve the same pressure level as one coherent charge of the same total size.

Even definition c), which incorporates the maximum pressure at a given point in time, fails to allow for precise description of the characteristics of the pressure curve as a function of time. Earlier studies indicate that the rise rate of tank pressure appears to better characterize the aggressiveness of an inflator for the "chest against module" OoP configuration [2; 8]. These approaches propose both peak pressure and maximum rise rate as key figures for specification. **Prasad** et al. [2] have proposed supplementing this characterization by application of their so-called thrust variable, for which they found a correlation with **V*C** loading of the dummy's chest.

The driver-side inflators used for this study are based on a series-production type for 64 liters which employs solid **non-azide** propellant. Owing to the fact that dual-stage designs were not available in all of the different partitions scheduled for the test program, only one of two stages was simulated by a partially charged single-stage inflator. It may confidently be assumed that only this stage itself would be fired in an **OoP** situation, in order to achieve as little energy input on the occupant as possible.

Single-stage inflators with peak tank pressure levels of **20%, 40%, 60%,** and **80%** were prepared, compared to the known **100%-reference** inflator from series-production, applying the maximum tank pressure criterion according to definition c). Since the **series-production** inflator housing was used for all versions in this study, it was necessary to fill excess room in the combustion chamber with inert material depending on the amount of propellant used. Several **pre-tests** were conducted in **60-liter** tanks in order to tune the inflator charge to the desired tank pressure levels. Documentation took place in the form of mass-flow calculations and pressure measurements. Results from real tank tests given in Fig. 1 include peak tank pressure, maximum pressure rise rate and the rise rate at **10 ms**. The final key figure was added in order to allow comparison of inflator behavior at the point in time at which the majority of V*C maxima occurred in the tests.

METHODOLOGY

For effective study of fundamental effects with **airbags** of different energy levels, as well as analysis of their interaction with OoP, it was decided to conduct static tests. Restricting the setup to the above-described test mock-up reduces the number of influencing factors to only a few parameters. Whereas the rigidity of the steering wheel simulation and module mount entail more severe conditions in terms of energy absorption, the static test setup lacks the influence of vehicle deceleration present in a dynamic test environment. **Prasad** et al. [2] have pointed out that tests under dynamic conditions may produce loading values up to twice those measured in static tests.

The test matrix with variation of dummy positions and distances is oriented to a previous study performed to compare the influence of different bag-folding patterns under OoP conditions [7]. In accordance with ISO "Technical Report" [3] for the assessment of OoP performance of **airbags**, tests took place for "chest centered on module" and "chin on top of module", supplemented by the configuration "forehead centered on module" included in earlier ISO issues (Fig. 2). Contrary to the above-mentioned study which examined distances of 0 mm, 20 mm, 50 mm, and 100 mm, only the "against" (0 mm) and **50-mm** conditions were tested here. Testing for all the parameter combinations with dummy posture and distance from module took place separately, with inflators simulating the **20%, 40%, 60%,** and **80%-stage**; as well as with the baseline module incorporating the **100%-reference** charge.

In order to reduce the effects of random parameter deviations on results, each test was conducted twice (with the exception of a number of "chin on top of module" tests with **50 mm** distance, after the 0-mm condition had already resulted in very low loadings).

CONDUCT OF TESTING

For the tests "chest centered on the module," the axis of the steering wheel and the air bag module were horizontally positioned to pass through the center of the dummy's rib cage. For orientation of the forehead centered on the module, the steering wheel was inclined by 20 degrees to allow the head to be placed against the module. The axis passing 25 mm above the c.g. of the head was aligned with the center of the module. The "chin on module" posture was achieved by positioning the dummy's thorax parallel to the inclined steering wheel, with the tip of its chin coinciding with the horizontal line through the uppermost edge of the **airbag** cover. In the 0-mm position, the dummy's forehead touched the steering wheel rim, leaving a small gap between the chin and the module cover.

Inflator Peak Tank Pressure	Max. Pressure Rise Rate [kPa / ms]	Pressure Rise Rate at 10 ms [kPa / ms]
20%	2.26	2.16
40%	2.64	2.54
60%	5.68	5.66
80%	6.64	6.54
100%	8.88	8.1

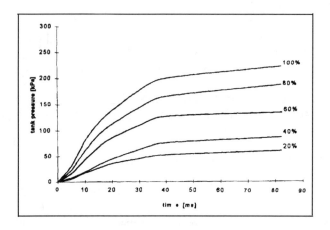

Fig. 1: Tank pressure curves and rise rates of tested inflator versions (in **60-liter** tank)

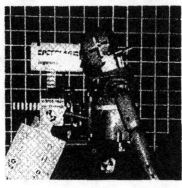 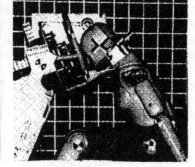 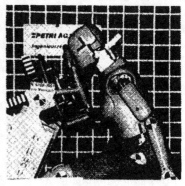

Chest centered on module   Forehead centered on module   Chin on top of module

Fig. 2 : Dummy out-of-position configurations for a distance of 0 mm

The dummy was positioned on the seat, with the height of the H-point maintained constant within all configurations. With "forehead centered on the module," constant position of the Ii-point in the longitudinal direction was likewise maintained. In order to stabilize the dummy position on the soft seat, a slab of hard foam was placed on the seating surface. In order to ensure the correct position in front of the module, it was in some cases necessary to secure the dummy at its shoulders by attaching easily-tearing adhesive paper tape to the steering wheel rim. In order to achieve as small a dummy-module separation as possible for the configuration "chest centered on the module" it was necessary in these tests to tie down the breasts of the small female with adhesive tape. Film documentation took place with two high-speed video systems (4,500 images/s) and one high-speed film camera (1,000 frames/s).

TEST RESULTS

For evaluation of occupant loading, scaled Injury Assessment Reference Values (IARV) for the small female Hybrid III [5; 9] were applied. The Viscous Criterion (V*C) values were! calculated according to the formula presented by Mertz [IO]. Bending moments measured with the tri-axial upper neck load cell were corrected to the plane through the head-neck joint.

CHEST CENTERED ON MODULE

Chest loading – For both distances from the module, sternum deflection demonstrates an immediate rise, becoming steeper with increasing inflator charge. This loading is attributable to the punch-out mechanism [5; 1 1] and coincides with occurrence of the V*C maximum. V*C maxima occurred within a time frame between 8 ms (100%-stage) and 16 ms (20%-stage) after deployment initiation, with the majority of maxima concentrated around 10 ms. For this reason, the respective pressure rise rates as indicators for inflator energy were calculated at 10 ms to allow comparison among the charge versions. Great increases in maximum rise rate resulted between 40% and 60%, and between 80% and 100%. These values coincide with sternum deflection and V*C increases in the respective tests. The latter exceeded the limit of 1 .0 m/s with the 60%, 80%, and 100%-stages with the configuration "chest against module".

In contrast to the above, the position 50 mm from the module resulted not only in halving of thorax loadings, with values remaining safely below the thresholds, but also in revealing different deployment behavior of the bag.

Corresponding to the findings of the previous OoP study [7], these results may be explained with the general character of the P-folding pattern of the bag. When it is obstructed by an OoP occupant during deployment, the bag unfolds radially under the condition that there is at least a small gap between the module cover and the dummy's body. It is only when the chest initially contacts the module that exclusively the gas forces of the deploying airbag act directly on the occupant.

The dummy thorax delayed the module doors from flipping open throughout the test configuration, whereas the upper door did not flip open with 20% and 40%-charges. This caused a smaller bag volume to escape from the module, since part of the fabric remained folded in the module. Consequently, the peak pressures in the bag were up to 50% higher than those found with a dummy/module distance of 50 mm.

Neck loading – For the neck, extension bending moment was the critical form of loading. Again, the values remained clearly below the limit of -31 Nm for the small female positioned 50 mm away from the module but exceeded the threshold at a separation of 0 mm in combination with the 100%-inflator stage. Both extension bending moment and axial tensile force increased with higher inflator charges. Their maxima occurred during the first 15 ms after airbag triggering, still without contact between the deploying bag and the

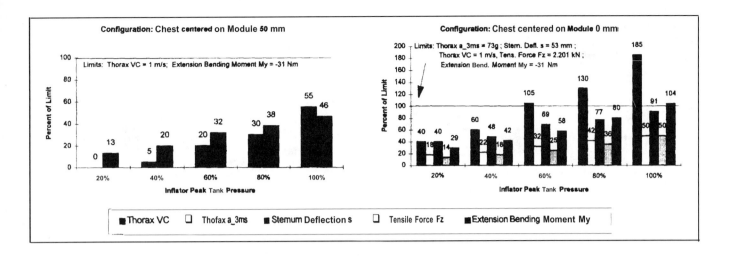

Fig.3 : Critical loadings for configuration "chest centered on module"

dummy head. These loadings may therefore be explained as coupling forces introduced by the chest being accelerated rearwards and the head remaining motionless due to its inertia. Unlike most situations in which critical neck loads are encountered due to membrane loading [5;11] through the cushion, the cause in this case for these injury mechanisms must be attributed to the punch-out effect.

The other neck loading types remained noncritical in all tests with "chest centered on module". Flexion bending moment demonstrated a slight upward tendency with increasing inflator energy.

Head loading – Head accelerations were low, with 3-ms values reaching a maximum of 22 g in the "against module" configuration.

FOREHEAD CENTERED ON MODULE

Chest loading – As could be expected, loadings of the thorax were very low due to its distance from the airbag. With 50 mm separation of the forehead from the module and charges lower than 60%, the bag did not contact the chest at all because only a partial portion of the bag was extracted from the module. Only thorax acceleration (3-ms value) reached an appreciable maximum (15 g) with the 100%-reference inflator and with the forehead resting against the cover. At the same time, sternum deflection and V*C values were negligible.

Neck loading – For both distances from the module, neck bending moments represent the major injury risk. A clear difference between the two configurations, however, is apparent. When the forehead initially contacts the module, both flexion and extension are the dominating loadings, with the baseline module reaching 55% and 62% of the scaled IARVs for flexion and extension. Flexion is introduced through punch-out loading of the head during the first milliseconds, whereas extension builds up considerably later when the cushion has achieved its full radial extent in front of the forehead. This extension phenomenon must therefore be attributed to the membrane loading effect.

In this configuration, shear force and compression force in the neck reached their maximum values throughout the entire study. With 44% and 32%, however, they remained safely below their respective limits.

At a distance of 50 mm from the module, the situation fundamentally changes. All neck injury criteria are noncritical, but extension bending moment increases considerably in the step from 80% to 100% of inflator peak pressure, and reaches 67% of its threshold value. This results in greater loading than in the 0 mm configuration. Film analysis reveals that the bag inflates almost entirely in front of the dummy's face with 50-mm separation. With direct contact on the cover it also spreads towards the thorax and distributes the loads on chest and head more equally.

Whereas neck loadings increased with higher-charged inflators throughout the range of peak pressure levels investigated, this phenomenon did not appear in general in the extension bending moment with the forehead against the module. Here, the 60%, 40%, and 20%-stages produce slightly higher extension values than those found with the 80%-stage. This may be explained by the fact that the bag did not deploy completely from the module with lower charges, but that it developed considerable pressure acting on the head. With the 80%-stage, the airbag unfolded completely in front of the chest as well which led to almost simultaneous rearward acceleration of the head and thorax. Further increase of the charge to 100% filled the bag tautly, which nearly doubled the maximum extension moment.

Head loading – Despite the position "forehead against module," the 3-ms values did not exceed 30 g; HIC values were negligibly low..

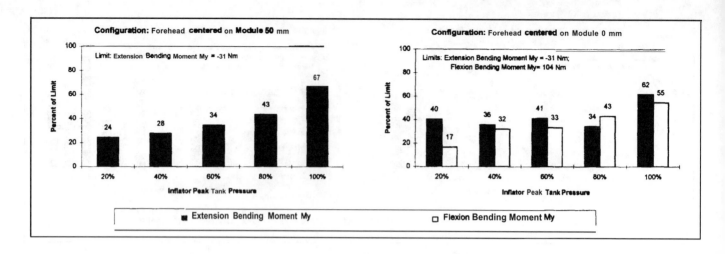

Fig. 4 : Critical loadings for configuration "forehead centered on module"

CHIN ON TOP OF MODULE

Chest loading – All thorax injury criteria remained noncritical throughout this test configuration, the highest relative loading type being sternum deflection. Nevertheless, the maximum for sternum deflection reached 20 mm with the baseline airbag remaining far below the threshold of 53 mm for the small female.

Neck loading – General increase in neck loadings becomes visible when the energy level of the airbag inflator rises. At a distance of 0 mm, neck extension is the major loading form, with maximum values of -23.4 Nm, compared with the limit of -31 Nm. Axial tensile forces range from 20% to 56% of the respective limit. Whereas flexion values were very low at 0 mm, they reached 40% of the respective IARV with 50 mm. The extension values for this distance also reached 40% of IARV.

Head loading – The highest 3-ms values were below 18 g, and HIC-36 maxima were 35.

All of the biomechanical loadings in the configuration "chin on top of module, 0 mm" were below their IARVs. Furthermore, the results for the 80% and 100%-stage and 50 mm confirmed this trend. Consequently, it was assumed that the remaining combinations with 20%, 40%, and 60%, at 50-mm distance, would produce small loading values, and they were dropped from the test matrix.

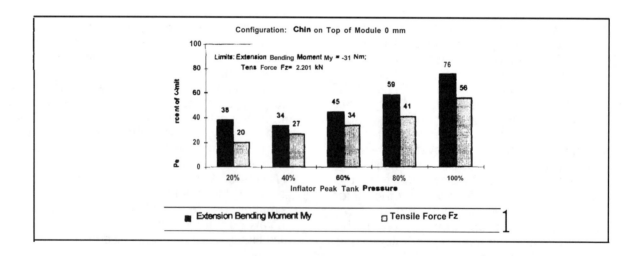

Fig. 5 : Critical loadings for configuration "chin on top of module"

SUMMARY AND CONCLUSIONS

In the present study, comparative static out-of-position (OoP) tests were conducted with driver-side airbags. The objective was to obtain information on the influence of dual-stage inflators with partitioned charges on the biomechanical loadings of an OoP occupant. Various amounts of solid non-azide propellant tailored during previous tank tests simulated the first of two inflator stages to be fired alone in case of a detected OoP situation. Peak tank pressure levels of the inflator versions ranged throughout 20%, 40%, 60%, and 80%, with percent reference to a series-production module.

The experiments were conducted on a rigid test mock-up representing the environment for the airbag and the driver. A "small female Hybrid III" dummy, additionally equipped with an upper neck load cell and neckskin, was used for assessment of injury-relevant loadings.

Testing comprised three main configurations: "chest centered on module," "forehead centered on module," and "chin on top of module," according to ISO recommendations. In addition to positioning the respective body region directly against the airbag module (distance of 0 mm), all configurations were also tested with a distance of 50 mm.

The configuration "chest centered on module" resulted in the highest biomechanical loadings in all tests conducted, both for the thorax and the neck. Biomechanical-loading results exceeded the V*C limit with O-mm distance and inflator stages of 60%, 80%, and 100%, whereas sternum deflection remained closely below the threshold for the baseline module, and thorax acceleration remained noncritical for all energy levels.

Neck extension bending moment slightly exceeded limits for the 100%-reference inflator and for the configuration of the chest against the module cover. In contrast to all other dummy/module configurations in which membrane loading by the inflating bag caused extension moments, the test results in this case must be attributed to the violent acceleration of the thorax.

With the "forehead centered on the module," only neck bending moments and tensile forces are of significance; none of these, however, exceeded the scaled IARVs. Direct positioning against the module cover resulted in both flexion and extension which represent the greatest loading in this case. At a distance of 50 mm, only extension remains relevant, due to changing bag deployment behavior under these conditions.

Whereas all loading values increased gradually as a result of higher inflator charge, this did not apply to extension moment when the forehead was placed against the module. Due to the fact that lower charges did not deploy the entire cushion from the module, and since forces therefore acted only on the head, the 20%, 40%' and 60%-stages produced higher bending moments than the 80%-stage. As in all other configurations tested, values for the head were noncritical with respect to the relevant head-injury criteria.

With the dummy's "chin on top of module," all biomechanical values likewise fell short of their respective limits. Here again, neck extension moment and tensile force delivered the relatively greatest loadings, generally rising with increasing inflator charge. Depending on the distance from the module and the resulting deployment behavior, either extension alone, or extension together with flexion, represented the dominant loading.

For all OoP configurations tested, a distance of 50 mm produced loadings which in all cases remained safely below threshold values for the small female. In view of an earlier study on the influence of bag folding patterns on OoP results, this result must be partially ascribed to the P-folding employed for the present airbag module. Under these conditions it appears that dual-stage inflators offer highest benefits for OoP drivers in the "chest against module" configuration. In this case, an 80%-charged inflator stage would suffice to observe the relevant neck extension limit, whereas a 20% or 40%-stage is required to remain below the limit for chest V*C. It is noteworthy, however, that – while the tested inflator versions were tailored to produce constant increases between "neighboring" peak tank pressures (i.e., energy levels) – this linearity was not present with their respective pressure rise rates (i.e., power levels) at the point of highest punch-out loading. Steep inclines occur especially between 40% and 60% and between 80% and 100%.

Since reduction of V*C values can also be achieved through onset fine-adjustment of the propellant, and since a gap between module and chest is possible by means of a dished steering wheel, it can be assumed that relatively slight propellant reductions – in other words, inflator stages of 60% or 80% – are realistic in order to meet all requirements for driver-side airbags under recommended ISO conditions.

Whereas reducing the inflator charge from 100% to 80% produced the best relative improvements in many cases (i.e., loading reduction with respect to inflator energy reduction), the 20%-stage did not enable significant advantages compared with the 40%-stage for the configurations "forehead centered on module" and "chin on top of module".

In view of recent discussion on the possible residual benefits of partially inflated airbags for restraint purposes, even under OoP conditions, the effects found during this study present an interesting option. As determined in tests with reduced amounts of propellant and in conjunction with the bag folding pattern, deployment of only a portion of the bag occurred – however, with considerable inflation pressure. Future concepts for adaptive airbag volume could well benefit from this design.

It must be emphasized that the objective of the present study was to provide results which are valid for a wide range of driver-side module designs. Nevertheless, significant deviations from the design presented, for instance by employing different inflator technology or radical relocation of the split line in the module cover, may also influence prospects for results. This likewise applies to the situation on the passenger-side, at which the number of influencing parameters is even greater.

Future developments to reduce injury risk for OoP occupants, however, must not sacrifice the airbag performance which is needed to provide protection for vehicle occupants in severe accidents.

ACKNOWLEDGMENTS

The authors would like to thank Dynamit Nobel GmbH for their collaboration in preparation of this study.

REFERENCES

1.   National Highway Traffic Safety Administration (NHTSA), "49 CFR Part 571, Docket No. 74-14, Notice 108", Federal Register, Vol. 62, No. 3, January 6, 1997

2.   Prasad, P.   et al., "Consideration for Belted FMVSS 208 Testing", Paper # 96-S3-O-03, 15th International Technical Conference on Enhanced Safety of Vehicles, Melbourne, 1996

3.   ISO Technical Report, "Road Vehicles – Test Procedures for Evaluating Out-of-Position Vehicle Occupant Interactions with Deploying Air Bags", ISO TR 10982, First Edition, 1996

4.   SAE Information   Report,   "Guidelines for Evaluating Out-of-Position Vehicle Occupant Interactions with Deploying Air Bags", SAE J 1980, Society of Automotive Engineers, Warrendale, PA, 1990

5.   Melvin, J.W. et al., "Assessment of Air Bag Deployment Loads with the Small Female Hybrid III Dummy", SAE-Paper # 933119, Society of Automotive Engineers, Warrendale, PA, 1993

6.   Adomeit, H.D., "The Petri Fold", in: Automotive Sourcing, Vol. 2, Issue 3, London, 1995

7.   Malczyk, A. et al., "The Airbag Folding Pattern as a Means for Injury Reduction of Out-of-Position Occupants", SAE-Paper # 952704, Society of Automotive Engineers, Warrendale, PA, 1995

8.   Nusholtz, G.S. et al., "An Evaluation of Airbag Tank-Test-Results", SAE-Paper # 980864, Society of Automotive Engineers, Warrendale, PA, 1998

9.   Mertz, H. J., "Anthropomorphic Test Devices", in "Accidental Injury", Chapter 4, Nahum, A.M., Melvin, J.W. (Editors), Springer-Verlag, New York, 1993

10.   Mertz, H.J., "V*C Formulas", ISO Document ISO/TC 22/SC 12/WG6 N 383, October 29, 1993

11.   Horsch, J.   et al., "Assessment of Air Bag Deployment Loads", SAE-Paper # 902324, Society of Automotive Engineers, Warrendale, PA, 1990

983162

Investigation into the Noise Associated with Airbag Deployment: Part II - Injury Risk Study Using a Mathematical Model of the Human Ear

Stephen W. Rouhana and Scott R. Webb
General Motors R&D Center

Vaundle C. Dunn
General Motors Powertrain Group

Copyright © 1998 Society of Automotive Engineers, Inc.

ABSTRACT

Airbag deployments are associated with loud noise of short duration, called impulse noise. Research performed in the late 1960's and early 1970's established several criteria for assessment of the risk of impulse noise-induced hearing loss for military weapons and general exposures. These criteria were modified for airbag noise in the early 1970's, but field accident statistics and experimental results with human volunteers exposed to airbags do not seem to agree with the criteria.

More recent research on impulse noise from weapons firing, in particular that of Price & Kalb of the US Army Research Laboratory, has led to development of a mathematical model of the ear. This model incorporates transfer functions which alter the incident sound pressure through various parts of the ear. It also calculates a function, called the "hazard", that is a measure of mechanical fatigue of the hair cells in the inner ear. The repeatability of the model was examined in the present study by comparing its predictive behavior for airbag noise impulses generated by nominally identical airbag systems. Calculations of potential "hazard" made by the model were also examined for reasonableness based on mechanical and biomechanical considerations. A large number of airbag noise pulses were examined using the model.

The results provide some counter-intuitive insights into the mechanism of noise-induced hearing loss from deployment of airbag systems. They also indicate that, based upon testing of feline subjects (which are believed to be a good indicator of the risk to the more susceptible segment of the human population), there could be a risk of temporary and possibly permanent threshold shifts in approximately sixty seven percent of the 1990-1995 model year vehicles from 19 manufacturers which were tested and assessed using the human ear model. A statistical estimate of the risk for the human population has yet to be quantified, but work is in progress to do so.

Work is also underway to develop a pre-production airbag component test for deployment noise that component suppliers can incorporate into the design and development process. Once injury risk curves and test protocols are established, it is recommended that the ARL Human Ear Model be utilized by the automotive community as the assessment method of choice.

INTRODUCTION

Field accident studies continue to demonstrate the effectiveness of airbags in reducing injuries and fatalities [1,2][1]. Yet, the automotive industry is continuously working to reduce the potential side effects. Airbag deployments are associated with loud noise, called impulse noise. Impulse noises are typically characterized by very short rise times (microseconds to several milliseconds), very large amplitudes (150 to 170 dB) and very short durations (on the order of a few to several hundred milliseconds). Some research was performed in the late 1960's and early 1970's to ascertain the risk of impulse noise-induced hearing loss for military and general exposures and several criteria were established [3]. These criteria were modified for airbag noise in the early 1970's [4], but field accident statistics and experimental results with human volunteers exposed to airbags do not seem to agree with the criteria [5]. More recent research on impulse noise from weapons firing, in particular that of Price & Kalb of the US Army Research Laboratory, has led to development of a mathematical model of the feline and human ears [6,7].

Some work has also been done to physically characterize the noise associated with early airbag systems and inflators [8,9,10] and with more recent systems [11]. The previous work established an understanding of the physical

---

1. Numbers in brackets designate references at the end of the paper.

phenomena contributing to the airbag noise pulse, ways to measure it, and typical sound pressure levels in vehicles of the early to mid-1990's.

The purpose of this work was to understand predictions made by the human ear model (hereafter, referred to as the "ear model" as opposed to "feline ear model" when appropriate). One aspect of this study relates to how well-behaved the model is (does it give similar results for nominally identical airbags). A second aspect considers whether the predictions made seem reasonable given what is known about the biomechanics of the ear and the physics of the impulse noise associated with airbag deployment.

Previous work has shown that the noise associated with airbag deployment inside a vehicle is 150-170 dB and is composed of two main noise pulses superimposed on one another [8-11]. In a typical automotive environment with the doors closed and the windows up, the pulse consists of high frequency noise superimposed on a very large low frequency carrier wave. Most of the energy in the pulse is in the low frequency range, with the peak in the energy spectrum typically occurring at 3-5 Hz. This portion of the noise appears to be caused by the piston-like effect of the bag deploying from its module which compresses the air inside the vehicle (which can be considered a closed volume on the time scales of typical deployments). The high frequency noise is mainly white noise associated with the gas generation, gas jetting through the inflator manifold, bag flapping and pleat popping.

Because noise and the ear are not subjects normally discussed in Stapp Car Crash Conference Proceedings, the authors have included some background material which may be of use in understanding the conclusions of this work. The reader familiar with these topics may wish to skip ahead. The various sections of the paper have been labeled to allow the reader to move more easily through the paper.

THE EAR – The human ear is a complex and fascinating organ. As shown in Figure 1, it can be viewed as consisting of an outer, middle and inner portion. The outer portion of the ear consists of the auricle or pinna and the ear canal. The pinna collects sounds or vibrations of air molecules and funnels them into the ear canal. The middle portion consists of the tympanic membrane or eardrum, the three bones of the middle ear (the malleus, incus and stapes or hammer, anvil and stirrup), and several soft connective tissues (e.g., the suspensory ligaments, the tensor tympani muscle, and the stapedius muscle). The middle ear transforms the vibrations of the air into mechanical motion of bones via the eardrum and serves as an impedance matching device to improve the flow of energy into the cochlea. The stirrup is attached to the inner ear or cochlea which is a snail shaped organ containing three fluid-filled chambers (Figure 2). The stirrup is held in the oval window of the cochlea by the annular ligament so that whenever the stapes vibrates with a piston-like motion a pressure wave characteristic of that vibration is set up in the fluid. As the pressure wave travels along inside the cochlea, the basilar membrane responds with a wavelike oscillation. Several thousand cells, called hair cells, are found within the organ of Corti which rests on the basilar membrane. When these cells are distorted by the wave motion they cause the auditory nerve to which they are connected to send signals to the brain where they are interpreted as sounds. The basilar membrane's physical properties are smoothly graded and cause it to act like a frequency analyzer with the result that hair cells near the base of the cochlea (near the oval window) respond to the high frequencies and those at the apex of the cochlea respond to the low frequencies.

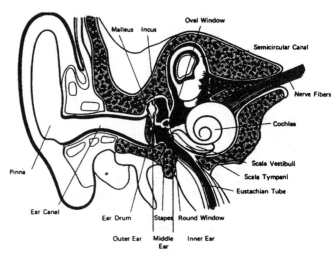

Figure 1. The anatomy of the ear [from 12].

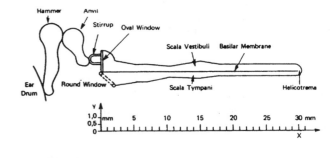

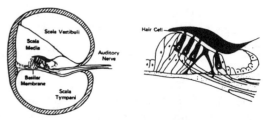

Figure 2. *Top* - Schematic representation of the middle and inner ear with the cochlea unrolled. *Bottom left* - Cross-section of the cochlea. *Bottom right* – Magnified view of the hair cells on the basilar membrane [All images from 13].

The ear typically responds to frequencies from 20 Hz to 20 kHz. The dynamic range is typically cited as going from 0 to 140 deciBels. (20 µPa to 1 kPa) although impulse noises from gunshots and airbags may go 40 dB higher. This represents an incredible dynamic range of a billion to 1.

A human subject's hearing is measured by playing pure tones of specific amplitudes and frequencies to a subject in a quiet room. The minimum amplitude at which a sound is perceived by the subject at a given frequency establishes the subject's threshold for that frequency. A threshold shift is defined as a change in a subject's hearing threshold after a specified exposure to noise [14]. Threshold shifts may be temporary (TTS) or permanent (PTS) depending on a number of factors. Every TTS has a possibility of becoming a PTS depending on the susceptibility of the subject. The definition is somewhat an exercise in semantics since a PTS is a TTS that does not go away (defined loosely as a TTS that lasts longer than six months).

The site of the physiological changes that are the cause of hearing loss is the cells within the cochlea, the ear drum and middle ear being relatively more robust. The actual mechanism of hearing loss is dependent on the type of noise exposure experienced by the subject. While the details are beyond the scope of this work, it is useful to think of the hearing loss mechanism from continuous noise as more like a disease process and that from intense impulse noise as more like a traumatic process. In fact, the mechanism for noise-induced hearing loss resulting from continuous noises at levels common in manufacturing settings appears to be biochemical or physico-chemical in nature [15] – the ear gets "tired out". The mechanism operative with intense impulse noise exposures appears to be mechanical or physical trauma to the cellular structures of the inner ear. These different mechanisms are the result of differing levels of excitation and have vastly different implications for repair and recovery of hearing acuity. In particular, continuous noise-induced hearing loss typically takes a long time to develop and the threshold shifts following individual exposures can recover upon removal of the noise source. In contrast, noise-induced hearing loss from impulse noise can occur from a single exposure and, depending on the TTS and other factors, has a higher risk of being permanent.

PREVIOUS CRITERIA – Efforts to develop damage risk criteria for noise-induced hearing loss began in the early 1930's, but were mainly focused on occupational types of continuous noise. Most of the work on impulse noise damage risk criteria has taken place since 1960 [16]. The damage risk criteria are based on noise-induced TTS and PTS.

The damage risk criteria in existence today for examining the hazard from impulse noise were developed in the late 1960's and early 1970's. One is given in a report from Working Group 57 of the Committee on Hearing and Bio-Acoustics (CHABA) of the National Academy of Sciences-National Research Council which was published in 1968 and was based on the proposal of Coles et al [17]. The second criterion is based on an analysis and adjustment of the CHABA criterion specifically for airbag impulses. This analysis was performed under contract for the National Highway Traffic Safety Administration, by Allen et al. [4] and will hereafter be referred to as the BBN method (the consulting firm was Bolt, Beranek and Newman). A third criterion, MIL-STD-1474B(MI) [18] was also based on the CHABA report and was established for military materiel acquisition. It specifies maximum sound pressure levels allowable for military weapons and systems and requires the use of hearing protection equipment for military personnel exposed to noises with peak pressures above 140 dB.

The original damage risk criterion for impulse noise [17] drew on some of the thinking behind a criterion for steady state noise developed by Working Group 46 of CHABA and published in 1966 [19]. This criterion was established from epidemiological review of hearing losses occurring in industrial settings and from studies of threshold shifts done in research laboratories. These studies attempted to correlate measured hearing losses with the exposures that individuals received in terms of sound pressure levels (SPL), duration and frequency of the noise involved. They were necessarily restricted to empirical and statistical analyses. The 1966 steady state criterion was based on the assumption that a permanent threshold shift (PTS) from continuous noise, which is eventually produced after years of exposure to the noise, is approximately equal to the temporary threshold shift (TTS) shown by a normal ear after a single day's exposure. The threshold shift value used for comparison was the TTS_2 which is the TTS measured 2 minutes after the cessation of the exposure [3]. The acceptable exposures in the CHABA criterion for impulse noise were also established to ensure that the PTS would not exceed 10 dB at 1000 Hz and below, 15 dB at 2000 Hz and 20 dB at 3000 Hz and above for 75 percent of the population exposed. These values were chosen to protect a person's ability to hear and recognize speech, most of which is transmitted between 500 Hz and 4 kHz.

CHABA – The 1968 CHABA criterion, the first specifically for impulse noise, "corrected" the Coles et al. [17] recommendations in the following manner:

a. lowered the allowable pressure by 5 dB to use normal incidence as the baseline (noise wave directed toward the eardrum). It may be increased by 5dB to allow for grazing incidence (noise wave directed parallel to the eardrum),

b. lowered the allowable pressure by another 5 dB to account for protecting 95 percent of the population exposed rather than 75 percent,

c. established a maximum acceptable peak SPL value of 179 dB with a 25 microsecond duration as a limit for a single pulse at grazing incidence to an unprotected ear,

d. accounted for reflex contraction of the middle ear muscles which takes place around 100-200 milliseconds after the onset of a pulse and reduces the effect of noise entering the ear canal subsequently by as much as 10 to 20 dB,

e. established a correction factor for single impulse exposures (rather than up to 100 impulses) by permitting a 10 dB increase in pressure for the single pulse.

The method of analyzing the noise signal was also specified in that paper. It utilized the concept devised by researchers in the field called A and B-duration analysis. The A-duration is determined from the pressure-time history by measuring the time interval between the time the low frequency portion of the pressure wave first rises above the baseline ambient pressure to the time it first returns to zero (Figure 3). The B-duration is defined as the total time the envelope of pressure fluctuations is within 20 dB of the peak pressure level including reflections (Figure 3). The A-duration was hypothesized to reflect the amount of energy emitted by a noise source and the B-duration was an indication of the emitted energy and its interactions with its surroundings [17].

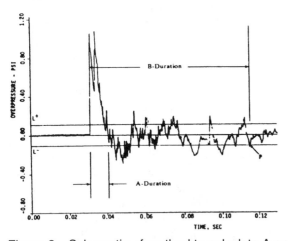

Figure 3. Schematic of method to calculate A and B-durations [modified from 18].

BBN – In 1971, Allen et al. [4] proposed modifications to the CHABA criterion to apply it to airbag noise risk assessment. In their report they used a similar A and B-duration analysis with a different frequency cutoff than Coles et al [17]. They separated low and high frequency noise into frequencies less than 300 Hz and those greater than 300 Hz. They based this separation on the original CHABA criterion indications that noise below about 300 Hz had significantly less potential risk of hearing damage. Additionally, in the absence of evidence to the contrary, they treated the low and high frequency noise independently, as if there was no interaction.

They accepted the Coles recommendations regarding increasing the tolerable pressure for a given A or B-duration but added 15 dB rather than 10 dB to account for a single pulse expected over the life of most individuals exposed to airbags [20]. They did not accept the recommendation to increase the tolerable pressure by 5 dB for grazing incidence because they felt in-vehicle exposures would be at normal incidence. They increased the tolerable pressure for a given A-duration by 7 dB to account for the longer rise times of the pressures from airbags compared with the noises in the exposures upon which the CHABA criterion was based. Their proposed damage risk criterion for airbag noise is shown in 4. Systems with in-vehicles measurements that fall below the lowest curve are expected to have less than 5% risk of noise-induced PTS. This threshold shift is defined as 10 dB at frequencies less than or equal to 1000 Hz, 15 dB at 2000 Hz and 20 dB at frequencies greater than or equal to 3000 Hz.

NAS-NRC 1992 Report – In 1992 the CHABA committee published a report reviewing the 1968 criterion in light of over 20 years of additional data [21]. They concluded that "The 1968 criterion should not be used for low-frequency impulses such as airbags, sonic booms, rapid pressurization, etc.". While the authors of this paper agree with that comment, since there was no other criterion available for airbag noise to date, we performed some analyses using the BBN version of the CHABA criterion as will be discussed below.

THE EAR MODEL – The foundation for a new way to assess risk of impulse noise-induced hearing loss began to be laid over 30 years ago with the publication of a series of papers by Price [22,23]. Since then a systematic program of research examined, among other things: the effects of frequency content on a critical level in the ear for assessment of noise hazard; relative hazard of weapons impulses; hazard from intense low frequency and mid-range impulses; effects of age on susceptibility to noise-induced hearing loss; implications of an upper limit to stapes displacement; and transformation functions of the external ear in response to impulsive stimulation [24-33]. This research lead to the conclusion that existing damage risk criteria, which were initially established for continuous noise and formulated empirically, did not adequately model the physical processes which were responsible for impulse noise-induced hearing loss.

To rectify the situation a mathematical model of the feline ear was developed as a representative of mammalian ears which are highly similar in their structure. It incorporated what was known about the anatomical and mechanical properties of the ear in the most minute detail available [6,7,34]. This model was validated for airbag noise impulses by Price et al. [35] (see also [36]). Since the introduction of the feline ear model, a parallel human ear model has been developed and both models have undergone continuous validation testing with the latest experimental data predominantly from military exposures with human volunteers [37,38,39].

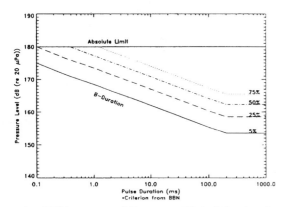

Figure 4. BBN proposed Damage Risk Criterion for airbag noise.

Details of the model have been published previously so only a brief description will be presented here [6,7,34]. Price and Kalb attempted to "relate the various elements of the model to specific physiological entities and to give them physically realistic values, to allow for maximum physical insight". In addition, "the model had the goal of beginning with the pressure history in a free field and passing the energy through the outer, middle and inner ears to provide an estimate of the [basilar membrane] displacement within the inner ear". To meet this goal the authors integrated models of the head, outer ear, middle ear and inner ear, some of which were developed by other authors, and they allowed for non-linear elements within the models.

The ear was modeled as electroacoustic elements with inductors representing mass, capacitors representing acoustic compliance and resistors representing acoustic resistances. As evidence that the model is reproducing the physical responses of the ear that can be measured, it includes graphs of the following transfer functions: free-field pressure to eardrum pressure; external ear radiation impedance; eardrum acoustic input impedance; eardrum pressure to stapes volume velocity; free-field pressure to stapes volume velocity; and stapes displacement to basilar membrane displacement.

The constants in the model were chosen to match experimental data obtained from a large number of investigators. There is excellent agreement of the transfer functions calculated by the model with actual values measured in those experiments. This agreement can be seen by plotting the transfer functions from one of the menus provided while using the model. All transfer functions utilized are available for the user to see given the parameters in the initial setup file.

Of particular significance in the development of the model was the inclusion of a nonlinear representation of the stapes. Price and Kalb note that previous research showed that "at very high intensities (above 130 dB) the middle ear becomes observably nonlinear". They note that von Bekesy showed that the stapes changes mode of vibration such that at high intensities less energy is transmitted to the cochlea [32]. In addition, they note that Yamamoto found an absolute limit to the displacement of the stapes which was uniform across various species including cadaver ears [32]. They hypothesized that the nonlinearity was due to the annular ligament, which holds the stapes in place at the oval window entrance to the cochlea. At low intensities, the annular ligament exerts a linear restoring force on the stapes and can displace easily up to 5-6 μm. At high intensities, the restoring force is nonlinear, reaching an asymptote at about 20 μm, stopping the stapes from moving. Under some circumstances, this nonlinearity appears to limit the transmission of higher frequency noise to the inner ear (as will be discussed later).

MECHANISM OF INJURY UTILIZED BY THE MODEL – At high levels the basic damage mechanism is thought to be similar to mechanical fatigue. Therefore, the hazard algorithm identifies the peaks of the upward flexes which put the critical tissues in tension where they are most likely to fail. Using the calculated stapes displacement as the driving input to the cochlea, the model calculates the displacement history of the basilar membrane for the duration of the input waveform. Hazard is then calculated at 23 locations along the basilar membrane (a little less than 1/3 octave intervals). The peak displacement in microns is squared and the values for all peaks during the impulse are summed at each location (Equation 1)[40]. This is similar to mechanical fatigue in general engineering which, in its simplest form, is calculated as the amplitude of vibration times the number of cycles of vibration. The values arrived at are named Auditory Damage Units (ADU).

$$\text{ADU (at a given location)} = \Sigma S^2 \quad \text{(Eq. 1)}$$

where,
ADU = Auditory Damage Units
S = peak basilar membrane displacement at a location (in microns)

Risk of Threshold Shift – The relationship between ADU and predicted threshold shift is provided by Equation 2 [40]. This formula was determined by fitting the data from many different impulse noise exposures with living biological ears.

As shown in Figure 5, the correlation between the model's calculation of ADU and the actual mean threshold shift measured in groups of feline ears in a dozen different experiments was very high (r= 0.94) [41].

$$\text{CTS} = 26.6 \times \text{Ln ADU} - 140.1 \quad \text{(Eq. 2)}$$

where, CTS = Compound Threshold Shift (the threshold shift measured immediately post-test)

For the feline ear 60-80% of CTS typically becomes permanent. The feline subject is believed to be more sensitive to risk of threshold shift than a human subject exposed to the same pressure-time history. Nevertheless, it is has been suggested that the average feline subject is like a susceptible human [36].

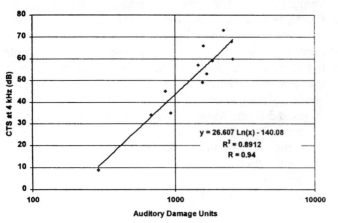

Figure 5. Correlation between experimentally measured threshold shift and auditory damage units predicted by the model. [Price, personal communication, 1998].

For the ears tested it would appear that an exposure of less than 200 ADU is essentially safe, between 200 and 500 there is an increasing risk of some permanent loss and above 500 ADU some permanent loss of hearing is likely. The correlation between ADU and changes in hearing has been supported by data from histological analyses of the ears in experiments with airbag noise [42]. As an upper limit, 2967 ADU would predict a 70 dB threshold shift and imply an almost total loss of hair cells at that location in the cochlea. In the case where the model calculates a larger number of ADU for an impulse, the design engineer can still interpret the exposure as being "higher risk" even if the ADU may not have a precise physiologic meaning.

One particularly useful feature of the model is the ability to see the development of the hazard as a function of time. This is possible because the model has a **movie** feature which plays back the calculation of hazard-time history on the same screen as the pressure-time history. This feature gives invaluable insight as to when, in time, the major portion of the hazard develops. This knowledge, in turn, has helped to understand the underlying mechanism of noise-induced hearing loss predicted by the model. This will be discussed further below as the data are presented.

METHODS – An analysis was performed using the January 1998 version of the Human Ear model, data from a parameter study of airbag noise published by Rouhana et al. [11] and some additional data generated for this part of the study. The purpose of the analysis was to assess the repeatability of the model's hazard predictions and to gain an understanding of the predictions it makes to ascertain their reasonableness and usefulness in sorting real airbag systems. The measurements of free-field pressure as a function of time were obtained using the SIRDIN system described previously [11]. In brief, this system consists of Dytran‰ pressure transducers mounted to brackets which are bolted to the back caps of mid-sized male Hybrid III dummy heads. The dummies are typically seated in the driver and right front passenger locations of the vehicles tested. The brackets position the transducers approximately 125 mm from the Tragion of the inboard ears of the dummy (that is, if a dummy had a Tragion) and approximately at grazing incidence to a sound pressure wave travelling longitudinally in a vehicle away from an airbag. The transducers are wrapped in foam for vibration isolation and press fit into a tube at the end of the brackets. The sensitive surfaces of the pressure transducers are at the level of the Tragion and "pointed" down. Unless otherwise stated, the windows and doors of the vehicle were closed.

The pressure-time data were acquired using an RS Electronics EGAA data acquisition board set to sample 16k points at 50,000 samples per second for a total time window of 327 milliseconds. A pre-trigger window was usually used and set to 500 samples. The ASCII data files were uploaded to a VAX 6220 system for analysis by the SIRDIN software which scales the data and calculates the A and B-durations by the MIL-STD-1474B method and the method used by BBN. The pressure-time data were also used as the input to the Ear Model using a software interface designed by ARL called "Import". The peak sound pressure level (SPL) in decibels (dB) is calculated using Equation 3. The peak SPL is determined relative to the standard threshold of hearing (20 µPa) using the relationship

$$SPL(dB) \equiv 20 * Log_{10} \frac{P_{measured}}{P_0} \quad \text{(Eq. 3)}$$

where,
$P_{measured}$ = measured peak pressure in Pascals
P_0 = threshold of hearing for a person with very good hearing (20 µPa)

Where enough experimental data exists, statistical comparisons of SPLs and Hazards under different test conditions have been made using the STATGRAPHICS software package (Version 7.1). Where a small number of tests at a certain condition were performed, observational analyses are reported.

RESULTS

REPEATABILITY – Two different vehicles were used as the platforms for the repeatability tests. The first vehicle was a standard size pickup truck with an interior volume of 1.8 m^3 (63 ft^3), 2 doors and full frame windows. This vehicle served as the test bed for two series of nominally identical experimental airbag deployments. The first series involved 14 tests with only passenger airbags (no driver airbags). Typical pressure-time histories from the driver and passenger side pressure transducers are given in Figures 6a and b, respectively. A clear low frequency wave is visible with higher frequency oscillations apparently superimposed as discussed earlier.

The hazards calculated using the human ear model analysis for these 14 experiments are shown in Figures 7a and b for the driver and passenger side pressure transducers, respectively. The hazard is plotted as a function of basilar membrane location. The location on the basilar membrane corresponds roughly to peaks in frequency of the noise to which the hair cells at that location respond. To facilitate comparison, these graphs are plotted on the same scale as those from the other repeatability studies presented later. This obviously minimizes differences between hazard curves with low values, but allows the reader to gain perspective since values as low as those shown in this figure are most likely well below the injury range. A mean and standard deviation for the peak SPLs and peak Hazards are given in Table 1.

Table 1.   Results of First Repeatability Series (n=14)

	Driver Side	Passenger Side
Mean Pk. SPL (dB)	168	168
± S.D. Pk. SPL	0.7	0.8
Mean Hazard (ADU)	54	42
± S.D. Hazard	18	20

Several observations are immediately obvious from these figures and tables. First, the standard deviation of the SPLs is very small. Second, the hazards range from 33 to 90 ADUs on the driver side and from 26 to 103 ADUs on the passenger side, for nominally identical airbags. Third, while most of the hazards calculated by the model fall into an envelope, there is one outlier which is almost twice the hazard of the mean hazard on the passenger side.

A different airbag system was tested in the second series of 5 nominally identical tests in the same vehicle. A typical pressure-time history for these systems is shown in Figure 8. This series showed similar results in terms of repeatability. However, the pressure-time histories from the airbag systems in this series showed more high frequency content than the systems in the first series and this appears to be responsible for much higher calculated hazards (Figure 9). The mean and standard deviation of the SPLs and hazards are given in Table 2. The driver side hazard ranged from 231 to 302 ADUs and the passenger side hazard ranged from 215 to 379 ADUs. Again, on the passenger side a single test stood out from the rest with nearly twice the hazard.

Table 2.   Results of Second Repeatability Series (n=5)

	Driver Side	Passenger Side
Mean Pk. SPL (dB)	169	169
± S.D. Pk. SPL	0.3	0.7
Mean Hazard (ADU)	265	265
± S.D. Hazard	26	77

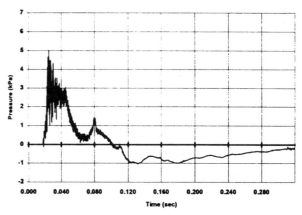

Figure 6a. Pressure-time history for the driver position in a typical passenger-only airbag test from the first repeatability series.

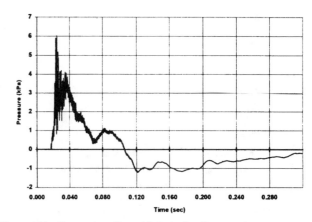

Figure 6b. Pressure-time history for the passenger position in a typical passenger-only airbag test from the first repeatability series.

Table 3.   Results of Third Repeatability Series (n=13)

DRIVER-ONLY AIRBAG SYSTEM (n=5)		
	Driver Side	Passenger Side
Mean Pk. SPL (dB)	161	160
± S.D. Pk. SPL	1.3	0.9
Mean Hazard (ADU)	296	45
± S.D. Hazard	134	20
PASSENGER-ONLY AIRBAG SYSTEM (n=5)		
	Driver Side	Passenger Side
Mean Pk. SPL (dB)	168	170
± S.D. Pk. SPL	0.4	1.2
Mean Hazard (ADU)	286	431
± S.D. Hazard	98	105
DUAL AIRBAG SYSTEM (n=3)		
	Driver Side	Passenger Side
Mean Pk. SPL (dB)	170	172
± S.D. Pk. SPL	0.3	0.9
Mean Hazard (ADU)	173	236
± S.D. Hazard	21	31

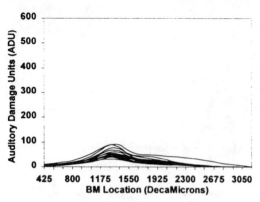

Figure 7a. Hazards calculated by the human ear model as a function of basilar membrane (BM) location for the driver position in the first repeatability series.

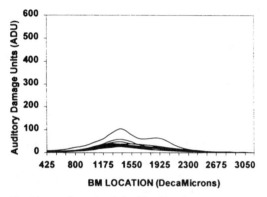

Figure 7b. Hazards calculated by the human ear model as a function of basilar membrane (BM) location for the passenger position in the first repeatability series.

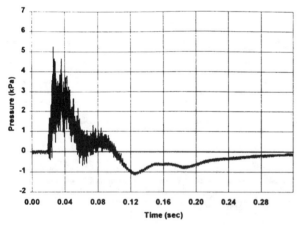

Figure 8a. Pressure-time history for the driver position in a test from the second repeatability series.

The second vehicle used as a test platform was a passenger car with an interior volume of 2.3 m^3 (82 ft^3), 2 doors and no window frames. Thirteen tests were performed in this vehicle with different configurations of airbags. Five tests were done with only a driver airbag (no passenger airbag), five tests were done with only a passenger airbag (no driver airbag), and three tests were done with both airbags (a dual airbag system). A typical pressure-time history for each test condition is shown in Figures 10a - f. The plots of the hazards are shown in Figures 11a - f for the driver and passenger pressure transducer locations. The mean and standard deviation of the SPLs and hazards, respectively, from each test condition are given in Table 3. Again, the range of hazards appears to be large, although there is less variability in the dual airbag tests.

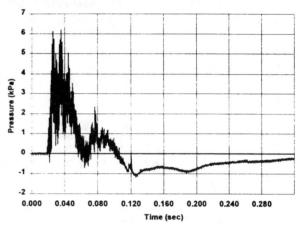

Figure 8b. Pressure-time history for the passenger position in a test from the second repeatability series.

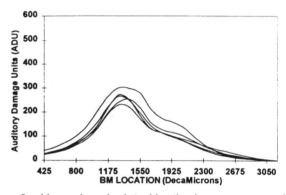

Figure 9a. Hazards calculated by the human ear model as a function of basilar membrane (BM) location for the driver position in the second repeatability series.

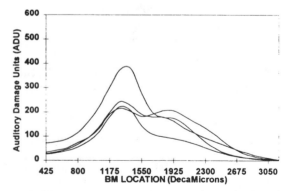

Figure 9b. Hazard calculated by the human ear model as a function of basilar membrane (BM) location for the passenger position in the second repeatability series.

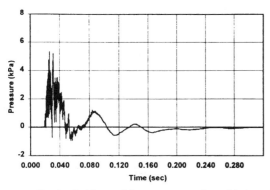

Figure 10a. Typical driver side pressure-time history for the driver-only airbag system used in the third repeatability series.

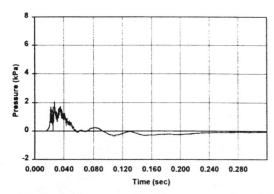

Figure 10b. Typical passenger side pressure-time history for the driver-only airbag system used in the third repeatability series.

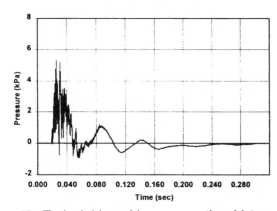

Figure 10c. Typical driver side pressure-time history for the passenger-only airbag system used in the third repeatability series.

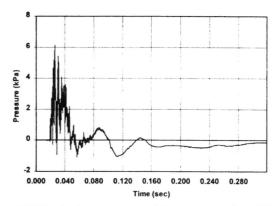

Figure 10d. Typical passenger side pressure-time history for the passenger-only airbag system used in the third repeatability series.

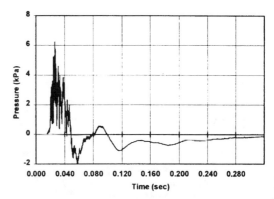

Figure 10e. Typical driver side pressure-time history for the dual airbag system used in the third repeatability series.

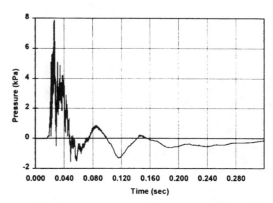

Figure 10f. Typical passenger side pressure-time history for the dual airbag system used in the third repeatability series.

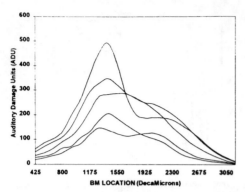

Figure 11a. Hazards calculated by the human ear model as a function of basilar membrane (BM) location for the driver position with the driver-only airbag systems in the third repeatability series.

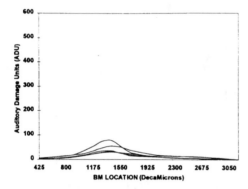

Figure 11b. Hazards calculated by the human ear model as a function of basilar membrane (BM) location for the passenger position with the driver-only airbag systems in the third repeatability series.

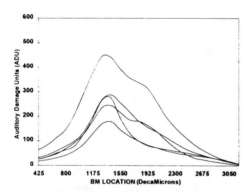

Figure 11c. Hazards calculated by the human ear model as a function of basilar membrane (BM) location for the driver position with the passenger-only airbag systems in the third repeatability series.

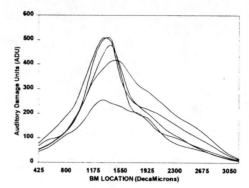

Figure 11d. Hazards calculated by the human ear model as a function of basilar membrane (BM) location for the passenger position with the passenger-only airbag systems in the third repeatability series.

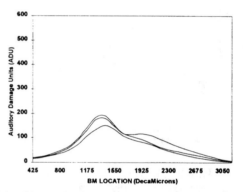

Figure 11e. Hazards calculated by the human ear model as a function of basilar membrane (BM) location for the driver position with the dual airbag systems in the third repeatability series.

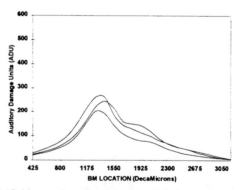

Figure 11f. Hazards calculated by the human ear model as a function of basilar membrane (BM) location for the passenger position with the dual airbag systems in the third repeatability series.

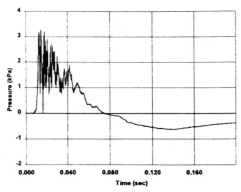

Figure 12a. Pressure-time history for driver-only airbag system in vehicle with roof.

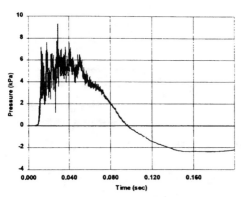

Figure 12b. Pressure-time history for dual airbag system in vehicle with roof.

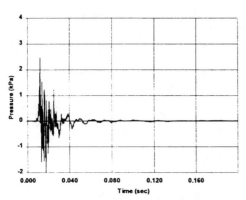

Figure 12c. Pressure-time history for driver-only airbag system in vehicle with NO roof.

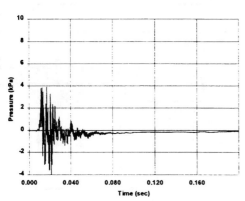

Figure 12d. Pressure-time history for dual airbag system in vehicle with NO roof.

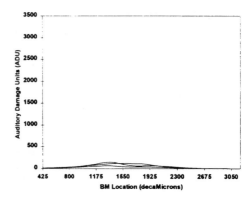

Figure 13a. Predicted hazard for 12a.

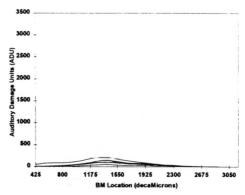

Figure 13b. Predicted hazard for 12b.

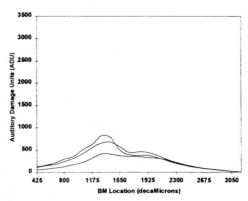

Figure 13c. Predicted hazard for 12c (NO Roof).

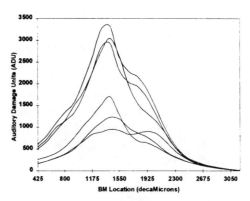

Figure 13d. Predicted hazard for 12d (NO Roof).

HAZARD PREDICTION – A number of tests were done to evaluate the risk of hearing loss associated with airbag deployment in an open passenger compartment versus in a closed compartment. For comparison, the risk of hearing loss was calculated using the ear model and the BBN method for all of the tests in this series. Three different vehicles were used in these tests.

In the first case, one test of a dual airbag system in a 1.8 m^3 (63 ft^3) convertible and one in a coupe version of the same vehicle were compared. The peak SPLs and hazards from these two tests are given in Table 4.

Table 4. Open vs Closed Compartment with Dual Airbag Systems (n=2)

CLOSED HARD TOP COUPE

	Driver Side	Passenger Side
Peak SPL (dB)	171	174
Hazard (ADU)	451	443
BBN High Freq (dB)	166	165
B-Duration (ms)	38	53
BBN Risk	50%	45%

OPEN CONVERTIBLE

	Driver Side	Passenger Side
Peak SPL (dB)	163	165
Hazard (ADU)	1530	2510
BBN High Freq (dB)	162	164
B-Duration (ms)	50	43
BBN Risk	25%	30%

In the second case, three driver-only and seven dual airbag systems were deployed in a 1.6 m^3 (55 ft^3) hard top vehicle with the windows closed. Then, the roof of the vehicle was sawed off and six dual and three driver-only airbag systems were deployed without the roof and with the windows down (open). The pressure-time histories for these tests are plotted in Figures 12a-d and the hazard predictions are plotted in Figures 13a-d. To conserve space, only the data for the passenger side is shown since the driver side data is shows the same behavior. These tests were performed outside the laboratory and away from any sound-reflecting structures to approximate free-field conditions. The mean and standard deviations of the peak SPLs and hazards are given in Table 5.

Table 5. Effect of Open vs Closed Compartment

DRIVER-ONLY AIRBAG SYSTEMS

CLOSED (n=3)

	Driver Side	Passenger Side
Mean Peak SPL (dB)	169	166
± S.D. Peak SPL	1.2	1.3
Mean Hazard (ADU)	485	109
± S.D. Hazard	40	39
Mean BBN High Freq. (dB)	164	158
± BBN High Freq. S.D.	1.2	3.0
Mean B-Duration (ms)	25	29
± B-Duration S.D.	1.0	5.1
BBN Risk	20%	<5%

OPEN (n=3)

	Driver Side	Passenger Side
Mean Peak SPL (dB)	167	163
± S.D. Peak SPL	1.9	2.7
Mean Hazard (ADU)	1051	642
± S.D. Hazard	315	207
Mean BBN High Freq. (dB)	163	158
± BBN High Freq. S.D.	2.6	2.5
Mean B-Duration (ms)	23.8	21.4
± B-Duration S.D.	4.4	2.4
BBN Risk	20%	<5%

DUAL AIRBAG SYSTEMS

CLOSED (n=7)

	Driver Side	Passenger Side
Mean Peak SPL (dB)	174	173
± S.D. Peak SPL	1.2	0.48
Mean Hazard (ADU)	213	127
± S.D. Hazard	59	57
Mean BBN High Freq. (dB)	166	164
± BBN High Freq. S.D.	2.5	0.65
Mean B-Duration (ms)	47	68
± B-Duration S.D.	10.7	5.5
BBN Risk	50%	30%

OPEN (n=6)

	Driver Side	Passenger Side
Mean Peak SPL (dB)	169	164
± S.D. Peak SPL	2.8	4.3
Mean Hazard (ADU)	1692	2198
± S.D. Hazard	674	1027
Mean BBN High Freq. (dB)	164	161
± BBN High Freq. S.D.	1.4	2.8
Mean B-Duration (ms)	37	51
± B-Duration S.D.	6.6	4.7
BBN Risk	25%	20%

In the third case, three dual airbag systems were deployed in a 1.8 m³ (63 ft³) two door hard top vehicle with the doors open and the windows down. These were compared to four dual airbag tests in the same vehicle with the doors and windows closed, and five passenger-only airbag tests in the same vehicle with the doors and windows closed and taped shut. The passenger-only tests with the taped doors were used in an attempt to keep the peak SPL constant between the doors taped versus doors not taped tests. When the doors are not taped they tend to bow outward during deployment and reduce the peak SPL [11]. The taping involved complete covering of all seams around the doors and windows with industrial duct tape which served to mechanically hold the doors and windows in place. This procedure was very effective at preventing any escape of pressure from the passenger compartment by completely preventing separation of the door and window frames from the body of the vehicle. The results show that we did obtain approximately the same peak SPLs in the passenger-only tests with taped doors and the dual airbag tests with the free doors. The mean SPLs and hazards are shown in Table 6.

OBSERVATIONAL ANALYSES – A design of experiments was performed to examine the effect of a number of variables on the risk of noise-induced hearing loss. The airbag systems tested for this series were only passenger bags (no driver bags) and all contained nominally identical inflators. The parameters studied included: material weave, compartment sealing, bag material, deployment door and bag fold. For efficiency, the conditions were chosen such that a single test under each set of conditions could be compared to tests of several other conditions. With only a single test performed at each exposure only observational analysis is possible. The results are shown in Table 7. Since there appears to be significant variability in results with nominally identical airbag systems, these observations can be considered interesting but not conclusive.

In an effort to get an understanding of the spectrum of the risk of noise-induced hearing loss in the field, deployments were performed in more than 35 vehicles from different manufacturers. The peak SPL and hazard as a function of vehicle volume are shown in Figures 14 and 15, respectively, for the dual airbag systems tested (the six driver-only airbag systems showed the same trends as the dual airbag systems). The results show that a reasonably diverse assortment of systems exist in the field. Peak SPLs and hazards range from 161 dB to 170 dB and 30 ADU to 1241 ADU for driver-only airbag systems and from 157 dB to 176 dB and 56 ADU to 2510 ADU for dual airbag systems.

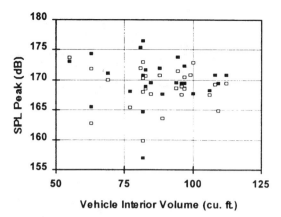

Figure 14. Peak SPL for dual airbag vehicles tested. Black squares indicate pressure at driver ear and white squares indicate pressure at passenger ear.

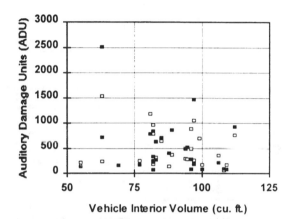

Figure 15. Hazard for vehicles shown in Figure 14.

Table 6. Effect of Sealing Passenger Compartment:

| | DUAL AIRBAG SYSTEM COMPARTMENT OPEN (N=3) ||
	Driver Side	Passenger Side
Mean Peak SPL (dB)	168	167
± S.D. Peak SPL	0.04	0.14
Mean Hazard (ADU)	2014	2069
± S.D. Hazard	625	647
Mean BBN High Freq. (dB)	166	165
± BBN High Freq. S.D.	0.92	4.1
Mean B-Duration (ms)	55	52
± B-Duration S.D.	4.8	5.8
BBN Risk	50%	45%

DUAL AIRBAG SYSTEM COMPARTMENT CLOSED & FREE (N=4)		
	Driver Side	Passenger Side
Mean Peak SPL (dB)	172	171
± S.D. Peak SPL	1.7	0.5
Mean Hazard (ADU)	**866**	**441**
± S.D. Hazard	500	98
Mean BBN High Freq. (dB)	167	164
± BBN High Freq. S.D.	2.4	2.2
Mean B-Duration (ms)	56	60
± B-Duration S.D.	9.4	6.5
BBN Risk	55%	40%

PASSENGER SYSTEM COMPARTMENT CLOSED & TAPED (N=5)		
	Driver Side	Passenger Side
Mean Peak SPL (dB)	170	173
± S.D. Peak SPL	0.93	3.2
Mean Hazard (ADU)	**143**	**252**
± S.D. Hazard	62	87
Mean BBN High Freq. (dB)	160	163
± BBN High Freq. S.D.	1.3	3.1
Mean B-Duration (ms)	69	62
± B-Duration S.D.	2.1	6.4
BBN Risk	20%	30%

Table 7. Summary of Observational Analyses

Variable	Setting	Dr SPL (dB)	Pass SPL (dB)	Driver Hazard (ADU)	Pass Hazard (ADU)
Bag	Bag	167	169	216	326
	No Bag	171	177	937	3709
Material Weave	420 D	167	168	204	199
	630 D	167	169	224	389
	840 D	166	169	219	370
Module Cover	Cover	167	166	204	199
	No Cover	168	168	153	284
Cover Fold	Cover Fold	167	168	204	199
	No cover No fold	167	170	220	440
Vehicle Doors	Closed Free	167	169	224	384
	Closed Taped	169	170	75	136

DISCUSSION

REPEATABILITY – As the results showed, while the peak SPLs appear to be quite consistent for nominally identical airbag systems (inflator, manifold, bag and cover) the hazards predicted by the Ear Model showed considerably more variability. A number of possible sources of variability were considered. Since some tests were performed on different days, temperature, humidity and/or barometric pressure variations were considered possible factors leading to variability. Since the repeatability tests were performed indoors, the temperature and relative humidity were reasonably well controlled. Nevertheless, as a standard practice all three parameters were recorded for each test. There was no correlation between any of these parameters and the peak SPL or peak hazard.

After considerable thought on the matter it appears that the variability in hazard may be related to the airbag system itself. A comparison of the passenger side pressure-time histories for the first repeatability series shows a subtle, but important difference between the curves. This is best illustrated by comparing the outlier, which has the highest hazard (103 ADU), to the test with the lowest hazard (26 ADU), as seen in Figure 16. The time scales in these plots have been expanded to facilitate the comparison. What can be observed is that the test with the higher hazard has much larger amplitude high frequency fluctuations superimposed on the low frequency portion of the pressure pulse. It's important to note that the tests with the highest and lowest hazards had similar peak SPLs (within 2 dB). This also appeared to hold true for the other tests in this series with higher hazards.

This makes sense, if the ear model and its hypothesized injury mechanism are correct, because higher frequency oscillations would drive the ear into more cycles of vibration. Therefore, the variability in the hazard rating appears to be a function of the airbag system and not the ear model. A carefully controlled parameter study of the airbag systems would be required to ascertain which components within are primarily associated with the variability.

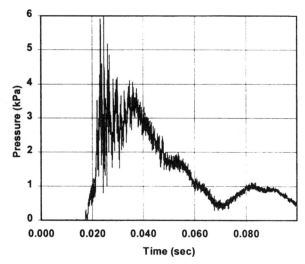

Figure 16a. Pressure-time history for test with highest hazard (103 ADU) in repeatability series one.

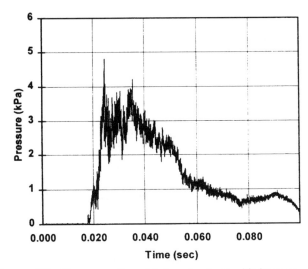

Figure 16b. Pressure-time history for test with lowest hazard (26 ADU) in repeatability series one.

It is interesting to note that, in the third repeatability series, the hazard was highest for the passenger-only airbag tests (at the passenger pressure transducer location). The hazard was lowest at the passenger side location in the driver-only airbag tests. This matches intuition since the intra-vehicular pressure is lowest for this system and the passenger side is removed from the predominant direction of the deployment (and hence from the principal direction of travel of the pressure wave). The next lowest hazard was for driver and then the passenger locations of the dual airbag system tests. It is not intuitively obvious why two airbags deploying would have less hazard than a single airbag deploying. This will be discussed below.

HAZARD PREDICTION – Two results relative to hazard prediction are immediately apparent from inspection of Tables 4-6. First, in general, the ear model and BBN methods of analysis yield a very different rank order for the hazard predictions under the different test conditions. Second and quite interestingly, the ear model predicts a greater risk of noise-induced hearing loss for the "no-roof" condition than for the closed compartment. This result is not intuitive and required further analysis for understanding.

The movie that was generated during the analysis using the ear model (see description of the ear model) was reviewed in detail for most of the test data in our files. It became clear that most of the hazard in an exposure occurs when high frequency oscillations are present and the low frequency pressure carrier wave is absent or approaches the baseline. This is the case when the roof is off of the vehicle and when the doors and windows are open. As Figure 12 shows, when the roof of a vehicle is removed, the majority of the low frequency portion of the wave disappears and the high frequency pressure oscillations are all that remain. With no low frequency pressure holding the stapes and annular ligament in tension, they are more likely to be able to transmit high frequency oscillations caused by the high frequency sound pressure. This might, in turn, drive the basilar membrane through a large number of cycles of vibration leading to mechanical fatigue.

The opposite effect is also understandable after a little explanation. When the roof is on, the low frequency part of the wave is much larger in magnitude than anything the ear could be prepared for by evolution. This pressure acts on the non-linear part of the ear, the stapes. It appears that the stapes is pushed to its limit, which is restricted by the annular ligament [6,32], and effectively acts to clip the amount of noise transmitted to the inner ear. When the low frequency carrier wave is removed, the clipping does not occur and the higher frequency oscillations drive through to the inner ear. This may be more understandable by envisioning a force applied to mass on a spring and held constant while stretching the spring to near its elastic limit. With the spring extended and the force held constant it would be more difficult to set up a high frequency vibration of the mass than if the spring was in its unextended state. This effect showed up numerous times throughout this analysis.

It is important to note that while the ear model predicts that open vehicles have the highest risk of noise-induced hearing loss (Figure 13), the BBN method generally predicts that open vehicles have the lowest risk of noise-induced hearing loss (Table 5). Thus, the two methods give very opposing views of the same event.

Likewise, the third series of tests (doors taped versus free) adds additional information. If having the doors closed reduces the risk, then we reasoned that taping them closed, to prevent the bowing mentioned previously, might reduce the risk even further. This effect is shown in the hazard results from these tests. In fact, the lowest hazards predicted by the ear model were for the tests with the doors closed and taped. The reason for this behavior can be understood from the following discussion.

As seen in comparing Figures 6, 8, 10 and 12, the airbag pressure-time history can vary in terms of the total width (time) of the main positive pressure pulse. When the doors are free to bow, a pressure relief valve of sorts exists, which allows the low frequency carrier wave to drop near ambient pressure. This pressure drop then reduces the tension on the annular ligament and allows more of the higher frequency noise to enter the inner ear. In contrast, when the doors are taped closed, the amplitude of the low frequency carrier wave is maximized which increases the peak SPL. But, that also pushes the stapes and annular ligament to their displacement limit which helps to reduce the higher frequency oscillations that can get to the inner ear and thus reduces the amplitude and number of basilar membrane oscillations. Then, with reduced amplitude and number of oscillations, the hazard is also reduced.

SENSE CHECK – Another purpose of this study is to check to see that the hazard predictions made by the model are reasonable from a common sense and scientific standpoint. At first glance, the effect of closed versus open compartment appears to contradict common sense. One might ask how the hazard could be greater, for example, in a convertible with the top down when the pressure is so much lower than in a hard top version of the same vehicle. And, in fact, the BBN method of calculating risk indicates that the risk is much less in the convertible based largely on the lower SPL magnitude and quick pressure wave decay. The ear model again gives us a fundamental understanding of the process wherein the high amplitude lower frequency portion of the pressure wave prevents much of the higher frequency noise from entering the inner ear. With the top down, the low frequency wave is absent and only large amplitude high frequency waves are present. These waves have little preventing them from entering the inner ear and driving the basilar membrane into enough oscillations for failure. Thus, while intuition implies it would be beneficial to lower the pressure in the vehicle to reduce the risk of hearing loss, doing so could have exactly the opposite effect. This highlights the difficulty with making changes without understanding the fundamental processes governing a system's behavior such as the ear.

Sommer and Nixon [43] described this exact phenomenon in a 1973 report of studies performed on ten human subjects exposed to isolated components of simulated airbag noise. They found that exposure to only the low frequency component (165 dB with a rise time of 65 ms and duration of 960 milliseconds) gave no TTS. Exposure to a high frequency noise burst (153 dB at 350 Hz to 3 kHz with a rise time of 25 milliseconds and a duration of 400 milliseconds) gave the highest TTS. Note that the higher TTS occurred with a peak pressure that was 12 dB lower! Exposure to the high and low frequency noise simultaneously gave a lower threshold shift than the high frequency by itself. They hypothesized that the "positive pressure pulse component appears to reduce the effectiveness of the high frequency burst in producing TTS". The agreement of the ear model with these results from human subject experiments is strong support for the method and algorithm used in its development.

A similar analysis could be done when reviewing the tests of the driver-only, passenger-only and dual airbag systems. In those tests, the hazard was lower for the dual airbag system than for the passenger-only airbag tests. The hazard was lowest for the passenger location in the driver-only airbag system. The hazard was similar for the driver in the driver-only and passenger-only systems and for the passenger in the passenger-only system. In the case of the dual airbag system, the combination of bags caused a larger low frequency curve than either system alone. The higher frequency noise appears to have less of a chance of superimposing. This again appears to have helped clip the high frequency portions of the pressure wave even though the overall peak SPLs were higher. This indicates that staged inflation of multiple airbags to reduce the risk of hearing loss may be less effective than timing them to inflate together to increase the lower frequency portion of the pressure-time history. It again highlights the importance of understanding the fundamental processes at play in order to effectively attack a complex problem.

There was a large difference in hazard predicted for the driver versus right front passenger in the driver-only airbag tests. The passenger in the driver-only system probably benefited from the lower peak pressure of a driver airbag, the distance from the deploying bag and the angle to the direction of wavefront propagation from the driver bag.

OBSERVATIONAL ANALYSIS – While the use of a single test per condition prevents statistical analysis of these tests, there were interesting results. The greatest difference in hazard was noted for the presence of bag material. When no airbag fabric was present in the system, the hazard increased by a factor of ten compared to a standard inflator and bag combination. In contrast, taping the vehicle doors to prevent bowing reduced the hazard by approximately a factor of three. The presence of a deployment door appeared to reduce the hazard by about 50%. This may be because most of the hazard builds up before breakout of the bag from the airbag module.

The data of hazard versus vehicle volume shows that there is a broad spectrum of risk. While there appeared to be a correlation between SPL and vehicle volume in our previous study [11], such a relationship was not apparent here. Each vehicle in the field study had a different airbag system in it which was tailored to the vehicle. So, in addition to the vehicle volume, the airbag size, inflator output and other parameters may have, and most likely did, vary from vehicle to vehicle. Thus, it should not be surprising that, with additional data, the trend observed previously is not seen.

FIELD EXPOSURE – Anecdotal reports of noise-induced hearing loss associated with airbags have recently appeared [44-46]. The fact that there have not been a plethora of reports may indicate that the risk is very low. However, it may also be that these injuries have not been recognized. Or, the hearing losses may be transient.

An example of a case of transient hearing loss was obtained by the authors from a 26 year old female who was in a collision in a compact vehicle with a driver-only airbag system. This individual was in excellent health and had a routine OSHA audiogram on file from 5 years prior to the collision. She reported no incidence of acoustic trauma during the 5 year period following the audiogram. At the time of the collision, the individual was stopped at a traffic light looking out the left side window. Her vehicle was struck at high speed from the rear with enough energy to propel it into the vehicle in front of hers. The frontal impact was severe enough to cause the driver airbag to deploy, but the subject had no interior contact other than the airbag and seat back. A post-collision audiogram, taken 17 days after the collision, showed a 10 dB threshold shift at 3 kHz in the left ear and a 15 dB shift at 8 kHz in the right ear, both relative to her previous audiogram.

Because she was looking out the window at the time of the deployment, it is likely that her head was turned toward the left. This would orient the right ear towards the airbag and to some extent would orient the left ear in the shadow of the head. That could account for the greater shift in the right ear than the left [see 36].

A second post-collision audiogram was taken 6 days after the first. This second audiogram showed a resolution of the threshold shifts at 3 and 8 kHz, but also showed a new 10 dB shift at 6 kHz in the right ear. Price and Kalb [36] have shown that threshold shifts in feline subjects can increase with time after a noise exposure. That may have occurred in this instance too. Of particular significance we note that the individual involved was not aware that she had a threshold shift until the audiogram was performed. Thus, this anecdotal case shows that individuals may sustain a noise-induced hearing loss without knowing it, and that at least in this individual there was some resolution or healing that reduced the magnitude of the initial loss. It is quite likely that the initial loss was larger immediately post-collision than that measured 17 days later.

CRASH NOISE WITH NO AIRBAGS – The possibility of impulse-noise induced hearing loss associated with airbags raises some more interesting questions. That is, how loud is a crash when there are no airbags? And, is it possible that one could sustain a hearing loss from the crash itself? This was examined briefly in 1988 by one of the authors (SWR). The sound pressure level was measured in a single 49.4 km/hour (30 mph) frontal barrier crash of a mid-sized van. The instrumentation was a B&K 4136 microphone and related amplifiers. Therefore, it had a low frequency limitation as described by Rouhana et al. [11]. In addition, the data acquisition equipment available at the time was limited to a 2500 Hz sampling rate. Thus, the results of this test should be considered cautiously since the actual peak sound pressure level was most likely higher than that recorded (like the Nixon data [5]).

The peak deceleration of the vehicle in this test was 45 g. As such, the microphone may have been affected by acceleration of the vehicle. However, it was oriented such that the longitudinal vehicle axis (the principal acceleration direction) was perpendicular to the body of the microphone and the instrumentation mounts were padded to reduce the overall acceleration environment. In this configuration, the acceleration sensitivity should have been minimized.

As shown in Figure 17, the overall peak sound pressure level measured in this test was 258 Pa (142 dB). The low frequency peak SPL was 142 dB with an A-duration of 250 milliseconds. The high frequency peak SPL was 145 dB with a B-duration of 50.3 milliseconds. Thus, according to the BBN analysis method, these values suggest that fewer than 5% of the subjects exposed to this noise would experience noise-induced threshold shifts (the BBN method of analysis indicates little risk of TTS or PTS from the noise of this crash itself).

The hazard predicted by the ear model for this test could not normally be determined because the Import program does not accept data sampled at less than 10 kHz. To obtain an order of magnitude estimate of the hazard from this crash, the data was digitally interpolated to fill in the data between the actual sampled points and bring it up to a simulated 10 kHz sampling rate. The interpolated data was then run through the ear model which yielded a predicted hazard of 104 ADU. Thus, the interpolated data also indicates a very low risk of threshold shifts from the crash noise in the absence of airbags. Clearly this result must be viewed cautiously since a significant portion of the actual data is missing. It would therefore be interesting to redo this experiment with the more modern data acquisition hardware available today, such as crash recorders, to more accurately determine the hazard calculated by the ear model.

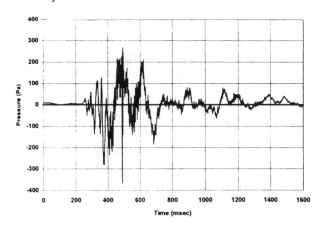

Figure 17. Pressure-time history for noise from a barrier crash of a vehicle with no airbags.

Table 8. Ear Model Results for 33 Vehicles from 18 Manufacturers (Model Years 1990-1995)

ADU Range	Dual Airbag Systems	Driver-only Systems
<250	9 (33%)	2 (33%)
250-500	6 (22%)	2 (33%)
>500	12 (44%)	2 (33%)
Total	27 (100%)	6 (100%)

EAR MODEL – Statistical calibration of the ear model to determine percent risk of noise-induced hearing loss for the human population is in process by the US Army Research Laboratory. As discussed previously, a preliminary assessment by Price et al.[40], suggests that hazards below approximately 200 ADUs have a very low risk of impulse noise-induced hearing loss. About 33% of the

1990-1995 model year vehicles tested had hazards less than or equal to 250 ADUs (Table 8). More work is necessary to define the risk in the 200 to 500 ADU range, in terms of percent of the population exposed, as is available for typical injury assessment reference values in use by automotive companies today [47].

FUTURE WORK – The Airbag Systems Task Force of the SAE has been addressing the issue of airbag noise since approximately 1995 in the hopes of developing an assessment technique for a recommended practice. A work item was also agreed to by ISO TC 22, Subcommittee 10, Working Group 3 [48,49]. In addition, a NATO Research Study Group has been evaluating the model for use as a NATO standard to assess risk of impulse noise-induced hearing loss. The Study Group's conclusions are expected to be presented in October of 1998.

In addition to selecting the injury criterion, much work still needs to be done before an SAE or ISO Standard for assessment of impulse noise-induced hearing loss can be written. The importance of testing inside of a vehicle has been shown by this work and others [8,10,11]. Yet a vehicle with the interior intended for production is not usually available when the airbag system design and validation must be done. Thus, the supplier community needs a way of assessing the hazard to a vehicle that hasn't been built. Work has begun on evaluating the correlation of tests performed in an acoustic chamber with those done inside vehicles in an attempt to develop a pre-production component test for airbag noise [50]. Additionally, the use of a human head form with a proper acoustic impedance and several other issues are still being explored by the SAE Task Force [48,49].

CONCLUSIONS

- The Army Research Lab Human Ear Model gave calculations for risk of noise-induced hearing loss that appear to be repeatable. Where differences in calculated hazard were large, a rational explanation was possible based on the biomechanics of the ear and the assumptions made by the model.

- A sense check of the calculated hazards given by the model, based on measurements made in an open compartment versus closed compartment, gave a non-intuitive result. Again, however, the result was understandable in terms of the biomechanics of the ear and the assumptions made in the model. This result also agreed with previous data from tests with human subjects.

- It appears that sealing the passenger compartment may be more effective at reducing the risk of airbag noise-induced hearing loss than venting it.

- There was a wide spectrum of calculated hazards in tests of more than 33 vehicles from 19 manufacturers from model years 1990 to 1995. Based on a preliminary correlation of the risk of noise-induced hearing loss with the calculated hazard for tests using feline

subjects (which are believed to be a good indicator of the risk to the more susceptible segment of the human population), approximately 67% of the systems tested may have the potential to be associated with noise-induced TTS or PTS upon airbag deployment. The risk for humans has yet to be quantified.

Further correlation with human data is necessary to more rigorously establish the validity of the model for humans and the statistical distribution of risk as a function of Auditory Damage Units. This correlation is being performed by ARL. Absent unexpected results, it is proposed that the Ear Model become the standard metric by which airbag noise-induced hearing loss is evaluated in the automotive environment.

ACKNOWLEDGMENTS

Many people in addition to the authors have contributed in significant ways to this work. The authors would specifically like to acknowledge the contributions of: Dawn M. Barnes, Howard Bender, Tai L. Chan, Kenneth Desaele, Joseph D. McCleary, Meghan C. Moreau, David B. Salva, Francis D. Wood and Robert G. Wooley. The authors are also grateful to Drs. G. Richard Price and Joel T. Kalb for their helpful comments on this manuscript. They reviewed this work and concurred that the model is being used appropriately.

This paper is dedicated to the memory of Howard Bender, a valued friend and colleague, whose smile and good cheer will be forever missed.

REFERENCES

1. Zador, P. and Ciccone, M.: "Automobile Driver Fatalities in Frontal Impacts: Air Bags Compared with Manual Belts", Am. J. Public Health, Vol. 83(5), May 1993.
2. Blower, D. and Campbell, K.L.: "Comparison of Occupant Restraints Based on Injury-Producing Contact Rates", Proceedings of the 38th Stapp Car Crash Conference, SAE Technical Paper Number 942219, 1994.
3. Ward, W.D., Ed.: "Proposed Damage-Risk Criterion for Impulse Noise (Gunfire)", NAS-NRC Committee on Hearing, Bioacoustics, and Biomechanics, Report of Working Group 57, 1968.
4. Allen, C.H., Bruce, R.D., Dietrich, C.W., and Pearsons, K.S. (BBN): "Noise and Inflatable Restraint Systems", BBN Report Number 2020, DOT Final Report, DOT-HS-006-1-006, 1971.
5. Nixon, C.W.: "Human Auditory Response to an Air Bag Inflation Noise", Final Report, DOT Contract Number P.O. 9-1-1151, 1969.
6. Price, G.R. and Kalb, J.T.: "Mathematical Model of the Effect of Limited Stapes Displacement on Hazard from Intense Sounds", J. Acoustical Society of America, Supplement 1, Volume 80, Fall 1986.
7. Price, G.R. and Kalb, J.T.: "Insights into hazard from intense impulses from a mathematical model of the ear", J. Acoustical Society of America, Volume 90 (1), 1991.
8. Hickling, R.: "The Noise of the Automotive Safety Air Cushion", Noise Control Engineering, Vol. 6 (3), May-June, 1976.
9. Sommer, H.C.: "Description and Use of a Measurement System for Air Bag Acoustic Transient Data Acquisition and Analysis", Final Report, DOT IA-0-1-2160, Aerospace Med-

ical Research Laboratory, Wright-Patterson Air Force Base Number AMRL-TR-73-8, March, 1973.

10. Posey, J. and Hickling, R.: "Noise Mechanisms in the Inflation of the Automotive Safety Air Bag", Presentation at the 85th Annual Meeting of the Acoustical Society of America, GM Research Publication 1411, 1973.

11. Rouhana, S.W., Webb, S.R., Wooley, R.G., McCleary, J.D., Wood, F.D., and Salva D.B.: "Investigation into the Noise Associated with Air Bag Deployment: Part I – Measurement Technique and Parameter Study", Proceedings of the 38th Stapp Car Crash Conference, SAE Technical Paper Number 942218, 1994.

12. Hassall, J.R. and Zaveri, K.: "Acoustic Noise Measurements", Figure 3.1, Page 41, Brüel & Kjaer, 1988.

13. Hassall, J.R. and Zaveri, K.: "Acoustic Noise Measurements", Figures 3.2 and 3.3, Pages 42 and 43, Brüel & Kjaer, 1988.

14. Melnick, W.: "Temporary and Permanent Threshold Shift", in Noise and Audiology, D. M. Lipscomb, ed., University Park Press, Baltimore, 1978.

15. Durrant, J.D.: "Anatomic and Physiologic Correlates of the Effects of Noise on Hearing", in Noise and Audiology, D. M. Lipscomb, ed., University Park Press, Baltimore, 1978.

16. Hodge, D.C. and Price, G.R.: "Hearing Damage Risk Criteria", in Noise and Audiology, D. M. Lipscomb, ed., University Park Press, Baltimore, 1978.

17. Coles, R.R.A., Garinther, G.R., Hodge, D.C., and Rice, C.G.: "Hazardous Exposure to Impulse Noise", J. Acoustical Society of America, Volume 43 (2), 1968.

18. United States Department of Defense: "Noise Limits for Army Materiel - Military Standard 1474B", 1979.

19. Kryter, K.D., Ward, W.D., Miller, J.D., and Eldredge, D.H.: "Hazardous Exposure to Intermittent and Steady-State Noise", Report of CHABA Working Group 46 to the NAS, J. Acoustical Society of America, Volume 39 (3), 1966.

20. McRobert, H. and Ward, W.D.: "Damage-Risk Criteria: The trading relation between intensity and the number of nonreverberant impulses", J. Acoustical Society of America, Volume 53 (5), 1973.

21. Report Committee on Hearing, Bioacoustics and Biomechanics: "Hazardous Exposure to Impulse Noise", National Academy of Sciences, National Academy Press, 1992.

22. Price, G.R.: "Functional Changes in the Ear Produced by High-Intensity Sound. I. 5.0-kHz Stimulation", J. Acoustical Society of America, Volume 44 (6), 1968.

23. Price, G.R.: "Functional Changes in the Ear Produced by High-Intensity Sound. II. 500-Hz Stimulation", J. Acoustical Society of America, Volume 51 (2, Part 2), 1972.

24. Price, G.R.: "Implications of a Critical Level in the Ear for Assessment of Noise Hazard at High Intensities", J. Acoustical Society of America, Volume 69 (1), 1981.

25. Price, G.R.: "Loss of Auditory Sensitivity Following Exposure to Spectrally Narrow Impulses", J. Acoustical Society of America, Volume 66 (2), 1979.

26. Price, G.R.: "Relative Hazard of Weapons Impulses", J. Acoustical Society of America, Volume 73 (2), 1983.

27. Price, G.R, Kim, H.N., Lim, D.J., and Dunn, D.: "Hazard From Weapons Impulses: Histological and Electrophysiological Evidence", J. Acoustical Society of America, Volume 85 (3), 1989.

28. Price, G.R. and Lim, D.J.: "Susceptibility to Intense Impulses", J. Acoustical Society of America, Supplement 1, Volume 74, Fall 1983.

29. Price, G.R.: "Hazard From Intense Low-Frequency Acoustic Impulses", J. Acoustical Society of America, Volume 80 (4), 1986.

30. Price, G.R. and Wansack, S.: "Hazard From an Intense Midrange Impulse", J. Acoustical Society of America, Volume 86 (6), 1989.

31. Price, G.R.: "Age as a Factor in Susceptibility to Hearing Loss: Young versus Adult Ears", J. Acoustical Society of America, Volume 60 (4), 1976.

32. Price, G.R.: "Upper Limit to Stapes Displacement: Implications for Hearing Loss", J. Acoustical Society of America, Volume 56 (1), 1974.

33. Price, G.R.: "Transformation Function of the External Ear in Response to Impulsive Stimulation", J. Acoustical Society of America, Volume 56 (1), 1974.

34. Kalb, J.T. and Price, G.R.: "Mathematical Model of the Ear's Response to Weapons Impulses", in Proceedings of the Third Conference on Weapon Launch Noise Blast Overpressure, Special Publication BRL-SP-66, U.S. Army Ballistics Research Lab, Aberdeen Proving Ground, MD, 1987.

35. Price, G.R., Rouhana, S.W., and Kalb, J.T.: "Hearing Hazard From the Noise of Air Bag Deployment", J. Acoustical Society of America, Volume 99 (4, Part 2), 1996.

36. Price, G.R. and Kalb, J.T.: "Auditory Hazard From Airbag Noise Exposure", Submitted for Publication to J. Acoustical Society of America, 1998.

37. Johnson, D. L.: "Blast overpressure studies with animals and men: A walk-up study", USAARL Contract Report No. CR-94-2, U.S.Army Aeromedical Res. Lab., Ft. Rucker, AL, 1994.

38. Patterson, J. H. and Johnson, D. L.: "Temporary threshold shifts produced by high intensity free field impulse noise in humans wearing hearing protection", USAARL Report No. 94-46, 24pp., U.S.Army Aeromedical Res. Lab., Ft. Rucker, AL, 1994.

39. Johnson, D. L.: "Blast overpressure studies", USAARL Contract Report No. CR-98-03, U.S.Army Aeromedical Res. Lab., Ft. Rucker, AL, 1998.

40. Price, G.R. and Kalb, J.T.: "Welcome to the ARL Auditory Hazard Assessment Algorithm – Human (AHAAH) Version 1,0", Ear Model Users Manual, 1998.

41. Price, G. R. and Kalb, J. T.: "Modeling auditory hazard from impulses with large low-frequency components", J. Acoust. Soc. Am., 99, 2464, 1996.

42. Mattox, D.E., Lou, W., Kalb, J. T. and Price, G. R.: "Histologic changes of the cochlea after airbag deployment" In Abstracts of the 20th Midwinter Meeting of the Association for Research in Otolaryngology, St. Petersburg, FL, 3797, p. 200, 1997.

43. Sommer, H.C. and Nixon, C.W.: "Primary Components of Simulated Air Bag Noise and Their Relative Effects on Human Hearing", Final Report, DOT Contract Number IA-0-1-2160, 1973.

44. Saunders, J.E., Slattery, W.H., Luxford, W.M.: "Automobile Airbag Impulse Noise: Otologic Symptoms in Six Patients", Presented at the American Academy of Otolaryngology – Head Neck Surgery Annual Meeting, September 19, 1995.

45. Kramer, M.B., Shattuck, T.G., Charnock, D.R.: "Traumatic Hearing Loss Following Air-Bag Inflation", New England J. of Medicine, Volume 337 (8), 1997.

46. McFeely, W.J., Bojrab, D.I., Davis, K.G., and Hegyi, D.: "Otologic Injuries Secondary to Airbag Deployment", Poster Presentation at the American Academy of Otolaryngology – Head Neck Surgery Foundation Annual Meeting, September 13-16, 1998.

47. Mertz, H.J.: "Anthropomorphic Test Devices", in Accidental Injury: Biomechanics and Prevention, A.M. Nahum and J.W. Melvin, Eds., Springer-Verlag, New York, 1993.

48. Rouhana, S.W.: "Airbag Noise: Research, Description and Status", Presented to SAE Airbag Systems Task Force, May 13, 1997.

49. Rouhana, S.W.: "Issues to be Considered in Drafting a Standard Measurement Procedure for Assessing Airbag Noise", Presented to SAE Airbag Systems Task Force, May 13, 1997.

50. Ochs, J.: "Airbag Inflation Noise – Evaluation and Analysis", Presented to SAE Airbag Systems Task Force, May 13, 1997.

1999-01-0764

Deployment of Air Bags into the Thorax of an Out-of-Position Dummy

C. R. Bass, J. R. Crandall, J. R. Bolton and W. D. Pilkey
University of Virginia

N. Khaewpong and E. Sun
National Highway Traffic Safety Administration

ABSTRACT

The air bag has proven effective in reducing fatalities in frontal crashes with estimated decreases ranging from 11% to 30% depending on the size of the vehicle **[IIHS-1995, Kahane-1996]**. At the same time, some air bag designs have caused fatalities when front-seat passengers have been in close proximity to the deploying air bag **[Kleinberger-1997]**. The objective of this study was to develop an accurate and repeatable out-of-position test fixture to study the deployment of air bags into out-of-position occupants. Tests were performed with a 5th percentile female Hybrid III dummy and studied air bag loading on the thorax using draft ISO-2 out-of-position (OOP) occupant positioning. Two different interpretations of the ISO-2 positioning were used in this study. The first, termed Nominal ISO-2, placed the chin on the steering wheel with the spine parallel to the steering wheel. The second, termed Chest On Module, relaxed the chin positioning criterion while maintaining the spine parallel to the steering wheel. In this position, the chest was forced against the air bag module with a 100 N force. All tests were performed in one of four nominal positions with respect to the steering wheel plane. The reference position had the center of the sternum on the center of the module. Variations from the reference position examined included the occupant sternum displaced 2 cm and 4 cm vertically in the steering wheel plane and displaced 4 cm left. Tests were performed using a production 'depowered' air bag from the current automobile fleet. To minimize the effect of external conditions, tests were performed in a climate-controlled room.

Dummy positioning on the test fixture was found to be repeatable to within 0.3 cm on all axes. This variation was within the dimensional similarity of the two available 5th percentile female Hybrid III dummies. A large variation in occupant response was found with a very small change in effective distance from the sternum to the air bag module. Nearly 50% variation in peak chest center-of-gravity resultant acceleration was found when moving from the sternum pressed on the air bag module to the

sternum effectively being 2 cm from the module. In addition, large variations in occupant response were found with vertical and horizontal displacements of the occupant with respect to the air bag module center. Also, a qualitative change in air bag deployment was found on changing the horizontal position by 4 cm to the left. These variations have significant implications for expected response from in-vehicle out-of-position dummy tests.

INTRODUCTION AND OBJECTIVES

Although the use of air bag systems as supplemental restraints has significantly decreased the overall risk of fatality in automobile collisions, there is evidence of increased risk of severe and fatal injuries to out-of-position small female and child occupants. For example, surveys of crash data **[NHTSA–1998]** indicate that small female occupants are most likely to be injured by driver-side air bags owing to their stature and proximity to the steering wheel and air bag module. Recently, automobile manufacturers have begun 'depowering' driver-side and passenger-side air bags as the result of concerns for these small female and child occupants.

The tests described in this report were designed to evaluate the relationship between air bag inflation aggressivity and occupant injury and response for a production depowered driver-side air bag. Depowered driver-side air bags were deployed into the chest of a 5th percentile female Hybrid III dummy positioned against the cover of the air bag module in an effort to simulate worst-case chest loading. The occupant, air bags, and test conditions for the tests are summarized in **Table 1**. The tests assessed the ability of chestband contours, thoracic accelerations, and head accelerations to predict injury and to quantify air bag aggressivity in terms of measured occupant response parameters.

This study presents the results and analysis of nineteen air bag deployment tests with out-of-position (OOP) occupants positioned according to the ISO-2 standard.

The tests may be considered as two subseries. The first subseries is a study of the effect of skin conditions on the dummy response. These tests, OOPS3.1-OOPS3.6, comprise three different skin conditions repeated twice. These tests are 1) skin on with breast forms on in OOPS3.1-OOPS3.2, 2) skin on with breast forms off in OOPS3.3-OOPS3.4, and 3) skin off in tests OOPS3.5-OOPS3.6.

The second subseries, OOPS3.7-OOPS3.19, is a lateral and vertical positioning study. Positions investigated include the dummy thorax centered on the air bag module, vertically displaced 2 cm and 4 cm upwards with respect to the steering wheel center, and laterally displaced 4 cm left with respect to the steering wheel module.

Goals of this research include evaluation of the position sensitivity of out-of-position deployment into a 5th percentile female Hybrid III dummy thorax as a part of a larger series in which occupant response and injury are being evaluated for a variety of occupant initial positions and air bag systems. Based upon data from these tests, a risk function that correlates observed cadaver injury with cadaver and dummy response is being developed using existing and future tests. The ultimate objective of this research is to develop a relationship between air bag inflation characteristics, occupant position, and cadaver injury.

BACKGROUND

The term out-of-position (OOP) occupant refers to a situation in which the occupant is located close to the air bag module at the time of inflation. Driver OOP cases typically involve an unbelted occupant who is in close proximity to the air bag owing to minor impacts prior to a more severe impact in which the air bag deployed. Horsch *et al.* [Horsch-1979] conducted an early significant study of OOP thoracic loading using a 50th percentile male Hybrid III dummy in static deployment tests. The authors determined that chest deflection was the best indicator for evaluating this type of thoracic loading. An interaction pattern was noted between the air bag and the out-of-position occupant with an initial reaction as the bag breaks out of the module and a second reaction as the bag becomes fully inflated around the occupant.

Subsequently, Horsch *et al.* [Horsch-1990] presented static and dynamic tests with the 50th percentile male Hybrid III dummy and static experiments with anesthetized swine. Using data from these tests, the Viscous Criteria (V*C) [Lau-1986] was identified as the best indicator of thoracic injury. The authors also noted the dummy maximum chest deflection significantly decreased with any increase in separation or adjustment in the vertical alignment relative to the thorax centered on the module. This suggests a nonlinear sensitivity to initial

Table 1.   Out-of-Position Test Matrix

Study	OOPS Test	Skin	Horizontal Position[1]	Vertical Position[1]	Comments
Skin Study	3.1	On with breasts	Center	Center	Nominal ISO-2
	3.2	On with breasts	Center	Center	Nominal ISO-2
	3.3	On without breasts	Center	Center	Nominal ISO-2
	3.4	On without breasts	Center	Center	Nominal ISO-2
	3.5	Off	Center	Center	Nominal ISO-2
	3.6	Off	Center	Center	Nominal ISO-2
Position Study	3.7	On with breasts	Center	Center	Chest On Module, No chestbands
	3.8	On with breasts	Center	Center	Chest On Module
	3.9	On with breasts	Center	Center	Chest On Module
	3.10	On with breasts	Center	Center	Chest On Module
	3.11	On with breasts	Center	4 cm up	Chest On Module
	3.12	On with breasts	Center	4 cm up	Chest On Module
	3.13	On with breasts	4 cm left	Center	Chest On Module, air bag tear
	3.14	On with breasts	4 cm left	Center	Chest On Module, air bag tear
	3.15	On with breasts	4 cm left	Center	Chest On Module
	3.16	On with breasts	4 cm left	Center	Chest On Module
	3.17	On with breasts	Center	2 cm up	Chest On Module
	3.18	On with breasts	Center	2 cm up	Chest On Module
	3.19	On with breasts	Center	2 cm up	Chest On Module

NOTES:
[1] Horizontal and Vertical Position refer to the dummy sternum center positions relative to the steering wheel center in the local axis system tangential to the steering wheel plane.

occupant position and indicates that the worst case condition places the occupant's sternum in direct contact with the air bag module. Horsch *et al.* [Horsch-1990] refined the concept of two distinct occupant loading mechanisms from the deploying air bag. The initial loading phase provided the maximum Viscous Criteria and was referred to as 'punch-out,' since it resembled a cylinder punching the chest. The second reaction was termed 'membrane loading' and was the result of air bag pressure acting on the occupant as well as tension forces created from the bag wrapping around the occupant. The peak thoracic reaction in the punch-out phase was shown to be primarily related to the break pressure, or internal pressure when the air bag ruptures the module and begins to inflate. It was suggested that the module design rather than the inflator properties determined the break pressure. Lau *et al.* [Lau-1993] used anesthetized swine to study torso injury mechanisms during air bag deployments. They found the cumulative injury severity was only marginally related to the gas output of the inflator. In addition, they determined that module alignment and position relative to the test surrogate significantly affected the injury severity.

Melvin *et al.* [Melvin-1993] conducted a study comparing the thoracic response during air bag deployment of the 5th percentile female Hybrid III dummy relative to the 50th percentile male Hybrid III dummy. They correlated punch-out and membrane loading with the peak thoracic responses recorded by the dummies. Given comparable testing conditions, the chest accelerations were slightly higher in the small female due to the lighter dummy mass while the V*C and maximum chest compression were slightly lower in the small female. The authors suggested that the 50th percentile male dummy covers the module more completely than the smaller female dummy, resulting in higher punch-out forces and larger V*C and deflection values. Comparisons between static and dynamic tests of thoracic loading by the deploying air bag have been performed in several studies. Using the 50th percentile male Hybrid III dummy, Horsch *et al.* [Horsch-1990] found no difference in chest compression or V*C values between static deployments and those conducted with sled tests. Chest accelerations, however, appear to be sensitive to the deployment condition with static chest acceleration values comprising only 60% [Hayano-1994] to 78% [Horsch-1979] of the dynamic values. Presumably, this difference is a combined contribution from both the increased deceleration of the occupant from the sled pulse and from the increased aggressivity of the air bag owing to increased inertial resistance during the initial deployment phase (i.e., break out).

TEST DESCRIPTION

The static out-of-position tests were conducted in the UVa Automobile Safety Laboratory's out-of-position (OOP) test fixture shown in *Figure 1*. The fixture incorporates a rigid dummy seat with an adjustable pelvis locating pin in the occupant centerline. The steering

wheel angle is 25 degrees (±1 degree) from laboratory vertical. The steering wheel is attached to a five-axis steering column load cell mounted to a plate with horizontal and vertical adjustment positions. These positions allow variation of steering wheel position relative to the locating pin in increments of ±2 cm and ±4 cm from the vertical center position and 2 cm and 4 cm left of the horizontal center position. The center position of the steering wheel was selected to align the center of the sternum with the center of the air bag module in a coordinate system aligned with the dummy seat. Ganged hydraulic adjustment pistons were used to move the steering wheel into position in the local X-axis. Finally, a plywood backboard, cushioned with foam and padding, was positioned behind the bench to limit the rearward translation of the occupant following interaction with the air bag.

The air bag used for the test series is a 'depowered' driver side air bag used in a 1998 model year mid-sized sedan. Air bag, air bag module and steering wheel parameters are shown in *Table 2*. The air bag incorporates a pyrotechnic inflator with a tank test (60 L tank) peak pressure of 142 ± 3 kPa and a maximum pressure rise (10 ms average) of 4.8 ± 0.3 kPa/ms. The horizontal tear seam is located approximately 55 mm below the center of the steering wheel. Vents are placed on the rear of the deployed air bag oriented away from the occupant. Air bag tears originating in these vents occurred in two of the tests in which the occupant was placed off of the vertical steering wheel centerline (OOPS3.13 and OOPS3.14). These tests were not used in subsequent analysis, though response data is presented for comparison.

Table 2. Air Bag, Air Bag Module, and Steering Wheel Parameters

Inflator Type	Sodium Azide
Tank Test Inflator Performance – (60 L Tank – Six Test Average)	
Peak pressure	142 ± 3 kPa
Maximum Pressure Onset Rate (max. 10 ms average)	4.8 ± 0.3 kPa/ms
Air Bag Volume	42.2 L
Number of Vents	2
Vent Area	177 mm^2
Air Bag Folding Pattern	Accordion
Module Dimensions	
Width (maximum)	237 mm
Width (minimum)	172 mm
Height	166 mm
Tear Seam Orientation	Horizontal
Steering Wheel Diameter	380 mm

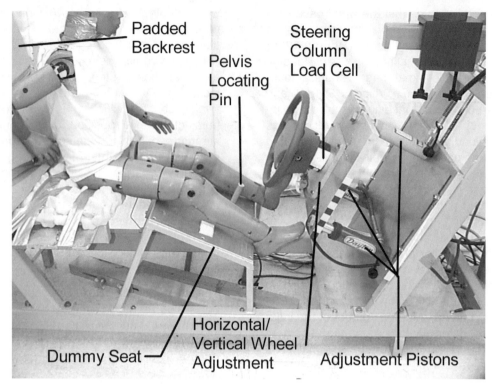

Figure 1. Out-of-Position Test Fixture

The 5th percentile female Hybrid III was instrumented to acquire head, neck, and thoracic response during air bag deployment. The dummy instrumentation package included accelerometers at the conventional head and chest center-of-gravity locations. In addition, magnetohydrodynamic (MHD) angular rate sensors and accelerometers were attached to the posterior head and upper spine in locations analogous to those of cadaveric subjects tested in previous out-of-position tests. Data from the angular rate sensors allowed neck motion to be estimated from relative motion of the head and upper spine. Accelerometers on the second and fifth ribs recorded lateral acceleration of the rib cage. Upper and lower load cells recorded neck forces and moments. Chestbands were placed at the second and fifth ribs to determine local thoracic loading by the air bag system. A sternal accelerometer and upper and lower sternum CRUX devices also recorded response of the anterior chest. In addition, contact forces between the occupant and the air bag were estimated from a five-axis load cell located at the upper end of the steering column.

In addition to the sensors discussed above, CRUX displacement sensors were used in the 5th percentile female Hybrid III thorax. Development of the CRUX displacement measurement unit was originally conducted by GESAC, Inc., as a component of the NHTSA THOR advanced frontal dummy development program. The CRUX concept provides continuous 3D measurement of sternal displacement with respect to the dummy spine, and has been designed to function reliably at sternal velocities anticipated in OOP air bag environments. The CRUX design is comprised of a two-bar linkage, with three rotational degrees of freedom that are continuously measured by three precision rotary potentiometers. Using software calculations, these angular measurements are converted to X, Y, and Z sternal displacements with respect to the CRUX base coordinate system (which is attached to the dummy spine). The CRUX design in schematic form is shown in **Figure 2**, and a CRUX mechanical assembly as currently installed as standard equipment in the 50th percentile male THOR dummy is presented in **Figure 3**. The two CRUX assemblies utilized in the 5th percentile female Hybrid III dummy tests conducted at the University of Virginia were custom modified versions of these THOR units. The units were modified to adapt the CRUX devices into the limited available thoracic volume in the 5th percentile female Hybrid III. The CRUX units were located at the level of the first rib (upper CRUX unit) and the sixth rib (lower CRUX unit). To facilitate mounting, the upper CRUX unit was located 1.1 cm left of the sternum centerline and the lower CRUX unit was located 1.1 cm right of the sternum centerline.

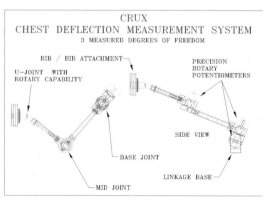

Figure 2. CRUX Schematic (Courtesy GESAC)

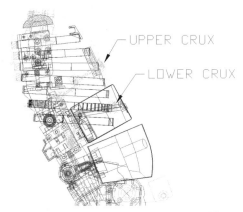

Figure 3. CRUX Installed in THOR Dummy (Courtesy GESAC)

TEST PROCEDURE

To provide a fixed position for the dummy, the occupant was initially attached to a rigid dummy-positioning arm as shown in **Figure 5**. Since repeatable dummy response is crucial to the usefulness of such research, a preliminary investigation was performed to determine the repeatability of response of the dummy ribs under quasi-static deformation under various conditions. For these tests, a 2.54 cm diameter pin was screwed into the steering column load cell at the steering wheel center, temporarily replacing the steering wheel and air bag module. The pin was pressed into the horizontal and vertical center of the dummy sternum using the hydraulic cylinders supporting the test fixture. Displacements of the pin were measured using a Faro arm measurement device at the steering wheel fixture reference location. Results of the pin press into the dummy thorax for selected tests are plotted in **Figure 4**.

The tests with skin performed before OOPS3.2 – OOPS3.4 showed very similar slopes for all cases. The thickness of the chestbands and associated tape resulted in an approximately 10 mm offset before OOPS3.2 and OOPS3.4. Testing delays between the three tests with skin were approximately an hour and approximately 24 hours. In addition, the deformations performed without skin both before testing started and after tests OOPS3.1, OOPS3.5, and OOPS3.6 all show similar force/deflection slopes. The delay between the earliest and latest tests was approximately 36 hours. Subsequent tests resulted in negligible variation of slope from those seen below. This indicates a negligible change in occupant rib stiffness over the quasistatic conditions tested and suggests that subject rib repeatability is acceptable under the test conditions used.

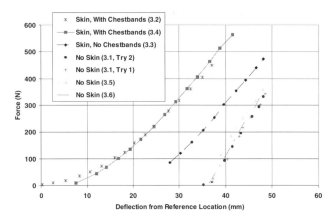

Figure 4. Force/Deflection Profiles for Mid-Thorax Quasistatic Compression with 2.54 cm Diameter Pin

After the occupant was positioned relative to the steering wheel, the steering wheel and the air bag module were moved into test position. Following this, the rigid positioning arm was carefully removed so as not to disturb the dummy position.

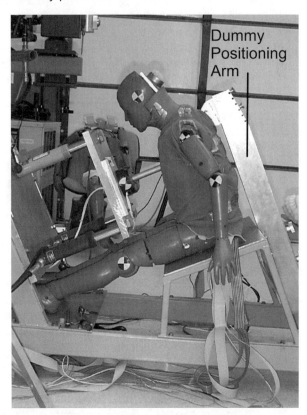

Figure 5. Dummy on Rigid Positioning Arm

OCCUPANT POSITIONING

The positioning guidelines for the ISO-2 tests were derived from the ISO DTR 10982 draft standard **[ISO-1998]**. The 5th percentile female Hybrid III occupant used in this test series was positioned according to this procedure. There is some ambiguity in the interpretation of ISO-2 positioning. The presumed intention of the ISO-2 position is to investigate maximum thoracic loading of an out-of-position occupant. To achieve this position, the chin is placed just on top of the upper rim of the steering wheel, without hanging the head over the rim, and the chest is placed in direct contact with the air bag module with the spine parallel to the plane of the steering wheel. However, the breast forms of the 5th percentile female Hybrid III dummy allow several different interpretations of this position depending on the maintenance of the chin position and spine angle, and the force of the chest on the air bag module. Consequently, this study employed two different positioning criteria:

1. **Nominal ISO-2**:

   For **OOPS3.1 – 3.6**: The position was ISO-2 with chin on wheel but not hanging over the wheel. The chest is in close proximity to the air bag with the dummy spine parallel to the plane of the steering wheel. In subsequent text, this is termed "**Nominal ISO-2**" as shown in *Figure 6*.

   Using the 5th percentile female Hybrid III dummy, this test condition places the dummy skin with breasts on the air bag module, but not deformed by the air bag module. Tests OOPS3.1-3.6 were a study of occupant response using different dummy skin conditions and the Nominal ISO-2 position with the chin and spine positions the same for all tests. This implied that the thorax contact point moved away from the air bag, over 5 cm from the tests using the dummy skin with breasts to the dummy with no skin as is discussed below.

2. **Chest On Module**:

   For **OOPS3.7 - 3.19**: The ISO-2 position was relaxed for tests OOPS3.7-3.19 to put the chest in direct contact with the air bag module with some nominal force that allowed static equilibrium to be maintained between the dummy chest and the air bag module. In subsequent text, this is termed "**Chest On Module**" as shown in *Figure 7*.

   The Chest On Module positioning implies that the dummy breasts will be deformed by the air bag module. To ensure this, the dummy was positioned with the thorax pressed against the air bag module with a force of 100 N as measured by the steering column load cell. In this position, the spine angle was maintained parallel to the plane of the steering wheel.

Occupant positioning was accomplished using an adjustable pin located between the thighs of the dummy, resting against the occupant pelvis. The chin was placed on the top of the steering wheel. For the Nominal ISO-2 position, the chin was placed in the middle of the steering wheel, not hanging over. For the Chest On Module position, the chin was allowed to hang over the steering wheel if necessary; for this position, a small wedge of foam padding was used to prevent the occupant chin from catching on the steering wheel under air bag deployment. For both positions, the spine angle was 25 degrees (±1 degrees) from vertical (i.e. parallel to the steering wheel).

Figure 6. Nominal ISO-2 Positioning - OOPS3.1 - OOPS3.6

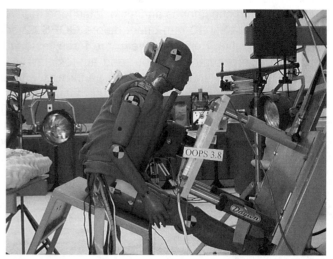

Figure 7. Chest On Module Positioning - OOPS3.7 - OOPS3.19

For runs OOPS3.1 – OOPS3.6, the positioning accuracy based on fixed bolt locations was within ±0.37 cm in local X, ±0.29 cm in local Y, and ±0.2 cm in local Z. For runs OOPS3.7 – OOPS3.19, the positioning accuracy based on bolt locations was ±0.36 cm in local X, ±0.5 cm in local Y, and ±0.32 cm in local Z. The intrinsic accuracy of the Faro Arm measurement device may be estimated using measurements made to a fixed reference location. For a representative location, positioning accuracy was ±0.06 cm in local X, ±0.23 cm in local Y, and ±1.3 cm in local Z.

To minimize the effect of temperature variation on the response of the test dummy (c.f. [Saul-1984]), tests were performed in a climate-controlled room. The dummy was placed in the room ten hours before the tests to allow the mechanical structures to equilibrate with the ambient temperature. Less than 3 °C temperature variation was seen over this time. The average test-time temperature for all tests was 28.3 °C (±1.1 °C standard deviation) with an average relative humidity of 65.5 % (±3.5 % standard deviation).

THORACIC RESPONSE

Analysis of the high-speed motion pictures identified several qualitative and quantitative differences between the air bag deployments under different conditions. A generic deployment begins with air bag squib activation. After activation, the air bag was first observed breaking through the seam of the module at approximately 4 ms.

During deployments with the dummy sternum centered on the air bag module (OOPS3.1-3.10), the air bag expanded into the occupant's chest until contacting the dummy chin at approximately 13 ms, and the occupant pelvis at 18 ms. The dummy thorax began rotating at approximately 12 ms. The dummy head began rotating downward at approximately 17 ms, and began translating rearward at approximately 24 ms. The dummy continued a rotational motion essentially centered at the pelvis for the rest of the air bag deployment. Owing to the proximity of the dummy thorax and the air bag module, the air bag was forced away from the occupant through the upper opening of the steering wheel (the space between the air bag module and the upper steering wheel rim) between approximately 14 ms and 75 ms. In OOPS3.7-3.10, a portion of the air bag did not escape the upper rim of the steering wheel before 100 ms after air bag initiation. Viewed from the top, the air bag symmetrically deploys outside the shoulders at approximately 17 ms.

For the air bag deployments with the occupant sternum center moved upward relative to the steering wheel (OOPS3.11-3.12, OOPS3.17-OOPS3.19), the air bag deploys towards the dummy abdomen and initially proceeds downward. The air bag reaches the pelvis at approximately 15 ms and reaches the chin at approximately 25 ms, later than that seen in the centered tests.

Thorax rotation occurs at approximately 13 ms. Head rotation downward occurs at approximately 15 ms with head motion rearward at approximately 25 ms. This behavior is similar to that seen in the centered tests. These tests, however, have qualitatively more pelvis translation than those with the sternum centered on the air bag module as the result of lower center of pressure of air bag contact. From the top, the air bag deploys symmetrically as in the centered tests.

For the air bag deployments with the occupant sternum center moved laterally relative to the steering wheel (OOPS3.13-OOPS3.16), we see a substantially different deployment. The air bag begins deploying to the right of the dummy at approximately 8 ms. Subsequently, the air bag deploys toward the occupant chin and into the upper right side steering wheel opening at approximately 12 ms. By 18 ms, the air bag is deployed only to the right of the dummy centerline. The succeeding deployment moves the dummy thorax to the left, while the head rotates towards the right as the result of steering wheel deformation. After approximately 50 ms, the occupant head begins rotation to the left under the influence of the air bag deployment. The air bag fully inflates at approximately 60 ms with the dummy moved to the dummy left. Differences in occupant loading and kinematics can be attributed to the differences in deployment patterns between occupant positions.

All of the sensors used for thoracic displacement, chestbands, CRUX units, and accelerometers were mounted at approximately the first and the sixth ribs in the dummy thorax. These sternal displacements are shown in **Table 3**. For most tests with skin with breasts (OOPS3.1-3.2, OOPS3.8, OOPS3.11-OOPS3.12, OOPS3.17, OOPS3.19), the upper chestband data shows much larger deflections than the CRUX or accelerometer data. This is attributable to the thickness and the compliance of the external breast forms. The tests with skin with no breasts (OOPS3.3, OOPS3.4) showed chestband deflection maximum values that were also higher than the CRUX or the accelerometer data. This is attributable to the deformation of the dummy skin.

The tests with no skin (OOPS3.5, OOPS3.6) show CRUX and chestband data that were consistent for upper sternal positions. In addition, the upper CRUX data is consistent for all groups of tests excepting the tests with air bag tears. All groups but one show a variation of less than a millisecond between peaks and a variation of less than 10% from high to low values.

The tests with the dummy thorax offset 4 cm left of the air bag module are very sensitive to details of the air bag deployment. In OOPS3.16, the exterior chestband shows smaller sternum deformation than the internal CRUX as the result of significant lateral deformation of the dummy ribs.

Table 3.   Occupant Thoracic Response Data

OOPS TEST	Position Vert., Horiz., Chest	Skin	Chest CG Resultant Acc. (g @ ms)	Upper Sternum Displacement (mm @ ms)		Lower Sternum Displacement (mm @ ms)
				Chestband	Crux	Chestband
3.1	Center, Center, Nom ISO-2	On	32.0@13.1	59.3@14.2	44.9@14.4	43.5@15.3
3.2	Center, Center, Nom ISO-2	On	33.6@13.5	60.1@14.6	47.5@14.2	50.3@15.3
3.3	Center, Center, Nom ISO-2	On/no breasts	28.1@13.6	48.5@15.0	42.8@14.6	40.9@16.0
3.4	Center, Center, Nom ISO-2	On/no breasts	28.5@13.8	51.7@13.6	43.9@14.8	41.8@12.9
3.5	Center, Center, Nom ISO-2	Off	22.6@11.0	33.1@13.2	31.7@14.1	24.8@13.5
3.6	Center, Center, Nom ISO-2	Off	22.7@11.3	27.5@12.7	29.5@13.6	26.9@13.9
3.7	Center, Center, Chest On Module	On	51.4@13.3	NA	59.2@14.6	NA
3.8	Center, Center, Chest On Module	On	60.8@13.3	74.1@14.3	58.6@14.6	59.4@15.0
3.9	Center, Center, Chest On Module	On	60.5@13.4	52.9@14.5	57.6@14.3	55.4@14.2
3.10	Center, Center, Chest On Module	On	61.8@13.8	59.4@15.6	59.4@14.9	56.0@15.7
3.11	Up 4 cm, Center, Chest On Module	On	40.1@12.8	44.5@15.9	38.9@32.2	71.9@12.4
3.12	Up 4 cm, Center, Chest On Module	On	38.9@12.9	46.3@19.7	39.0@21.2	72.7@12.2
3.13	Center,Left 4 cm, Chest On Module	On	25.1@13.2	74.8@14.7	36.9@16.1	51.3@15.4
3.14	Center,Left 4 cm, Chest On Module	On	19.0@12.5	38.3@14.0	24.7@15.1	34.2@14.3
3.15	Center,Left 4 cm, Chest On Module	On	31.8@12.3	44.4@16.4	37.0@15.8	45.4@16.1
3.16	Center,Left 4 cm, Chest On Module	On	30.7@12.6	30.1@17.0	36.4@16.1	38.9@15.6
3.17	Up 2 cm, Center, Chest On Module	On	50.7@12.7	52.3@14.9	49.5@14.3	79.0@12.7
3.18	Up 2 cm, Center, Chest On Module	On	56.2@12.3	56.6@12.4	52.0@14.3	77.6@12.6
3.19	Up 2 cm, Center, Chest On Module	On	50.7@13.5	60.8@13.7	48.9@15.1	73.8@13.7

EFFECT OF OCCUPANT SKIN/DISTANCE FROM AIR BAG MODULE

As the result of the consistent "Nominal ISO-2" positioning, the study of skin conditions effectively became a study of the effect of distance from the thorax to the air bag module. In the "Nominal ISO-2" condition, the dummy skin with breasts rested on the module with the breasts undeformed.   Maintaining the same chin and spine positions meant that the dummy thorax effectively moved away from the air bag module when the breasts were removed from the 5th percentile female Hybrid III

skin. The effective distance from module to thorax was further increased in the tests with no dummy skin. Effective skin thickness for the conditions tested are shown *Table 4*. The distance to the air bag module is defined as zero for the skin with breasts deformed by the air bag module ("Skin on Module" case).   It was further observed that the compliance of the dummy breasts is significantly variable among dummy skins, as the breasts are simply conical forms glued onto the dummy skin. Two dummy skins with breasts were available for examination. The four individual breasts have qualitatively different stiffnesses.

Table 4. Effective Sternum Distance to Air Bag Module Cover

Skin	Skin Thickness (mm)	Effective Distance from Module (mm)
Skin with breasts – Nominal ISO-2 (undeformed)	53	18
Skin with breasts – Chest On Module (deformed by module)	35	0
Skin without breasts – Nominal ISO-2	24	29
No Skin – Nominal ISO-2	0	53

The variation of peak chest resultant acceleration with effective distance from the air bag module is shown in **Figure 8**. There is over a 60% decrease in peak acceleration as the effective distance from the air bag module increased from zero (skin with breasts – Chest On Module) to 53 mm (no skin – Nominal ISO-2). The variation in peak chest CG resultant is approximately exponential in effective thorax distance to the air bag module. Similar results are seen in the dummy sensor time histories. As an example, the upper sternum acceleration time histories for the four cases are plotted in **Figure 9**. They are qualitatively similar for all cases. However, the acceleration peaks are largest for the skin with breasts in the "Chest On Module" position, with the thorax effectively closer to the module.

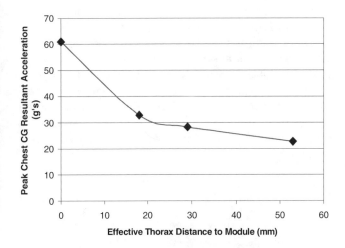

Figure 8. Peak Chest CG Resultant Acceleration vs. Effective Distance from Sternum to Module (Average of All Tests at a Given Distance)

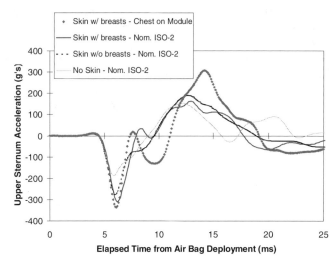

Figure 9. Time History of Upper Sternal Accelerometer - Different Skin Conditions (Representative Acceleration Time Histories)

EFFECT OF OCCUPANT POSITION

The chest accelerometer data shows substantial differences in response when the occupant is moved 4 cm up or 4 cm to the left as shown in **Figure 10**. An intermediate level of response is seen in the 2 cm up case. It is clear that the chest center-of-gravity acceleration resultant peaks group well within a test condition, and that OOPS3.13 and OOPS3.14 did not inflate properly owing to the air bag tears. Differences in peak chest center-of-gravity acceleration from the tests in the center position to the tests with the dummy offset 4 cm left are substantial, nearly 50% in peak acceleration.

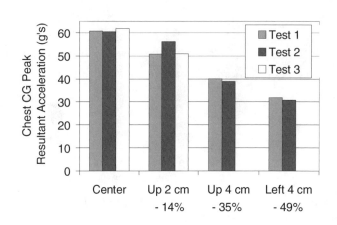

Figure 10. Acceleration Response Position Sensitivity for Repeated Tests

Typical sensitivity of the chest CG resultant acceleration for tests under the same conditions is graphed in **Figure 11**. For the dummy thorax positioned 4 cm up from the center position, the two available time histories show qualitatively and quantitatively similar behavior with a variation of less than 4% in peak acceleration. This is well within the manufacturer's suggested air bag deployment variation of 10%.

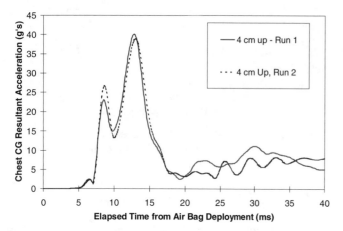

Figure 11. Chest CG Resultant Acceleration Sensitivity to Repeated Tests

Position behavior similar to the chest center-of-gravity peak acceleration is seen in the thoracic viscous response. As shown in **Figure 12**, the chest V*C decreases 16% at 2 cm up from the dummy sternum centered on the air bag module. In addition, the chest V*C decreases 60% at 4 cm up and 68% at 4 cm left.

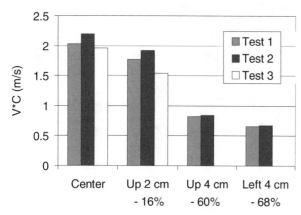

Figure 12. Position Sensitivity of Upper Crux V*C for Repeated Tests

As seen in the chest acceleration and V*C peaks, the chest center-of-gravity resultant acceleration time history shows quantitatively different peaks while showing qualitatively similar behavior. In **Figure 13**, the peaks vary over 30% while the timing of the peaks varies less than 3 ms. The initial peak in the 4 cm up case is exacerbated by enhanced initial acceleration caused by details of the air bag/thorax interaction in that position. In the 4 cm up case, the air bag deploys into the lower thorax/abdominal region of the dummy compared with the upper thorax in the centered case. This allows the air bag room for earlier deployment and forces the top of the steering wheel into the upper thorax at approximately 8 ms, causing the early peak.

As seen in **Table 5**, there is a significant variation between V*C calculated using CRUX data and chestband data owing to the compliance of the dummy skin. However, upper CRUX V*C data shows very good repeatability within test conditions. The variation in V*C among test conditions is shown in **Figure 12**. The variation is substantial, with over 16% variation from the center position to 2 cm up and 60% variation in upper sternum V*C from the center position to 4 cm up.

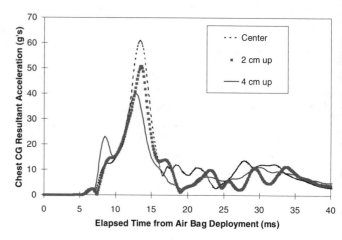

Figure 13. Chest CG Resultant Acceleration Vertical Position Sensitivity

The upper sternum CRUX deformation shows significant variation between tests in the center position and tests with the dummy thorax moved up 2 cm and 4 cm as shown in **Figure 14**. There is a 14% decrease in upper sternum deformation for a position 2 cm above center and a 33% decrease for a position 4 cm above center. In addition, there is over a 37% difference in dummy upper sternum deformation between the dummy in the center position and 4 cm to the left of center. Similar variations are seen in the chestband data, though variations are exacerbated by the non-biofidelic compliance in the dummy breasts.

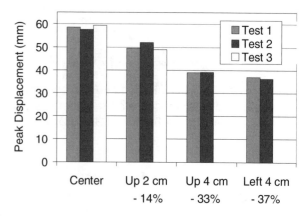

Figure 14. Position Sensitivity of Upper Sternum Deflection Measured by Crux Unit for Repeated Tests

Table 5. Thoracic V*C using CRUX Units and Chestbands

OOPS TEST	Position Vert., Horiz.	Upper Sternum V*C (m/s @ ms) Chestband	Upper Sternum V*C (m/s @ ms) CRUX	Lower Sternum V*C (m/s @ ms) Chestband
3.1	Center, Center	1.3@10.7	1.36@12.4	0.8@9.9
3.2	Center, Center	1.1@9.0	1.67@11.8	1.0@13.0
3.3	Center, Center	1.0@8.7	1.11@12.2	0.8@8.9
3.4	Center, Center	1.3@9.9	1.08@12.2	0.9@9.0
3.5	Center, Center	1.3@5.9	0.62@12.1	0.4@7.5
3.6	Center, Center	1.2@5.7	0.58@9.9	1.0@7.3
3.7	Center, Center	NA	2.01@12.1	NA
3.8	Center, Center	3.5@9.3	1.96@11.9	2.4@10.6
3.9	Center, Center	2.7@8.6	2.03@11.8	1.6@12.0
3.10	Center, Center	1.8@8.9	2.20@13.4	2.1@13.1
3.11	Up 4 cm, Center	1.3@11.1	0.82@12.2	4.1@11.2
3.12	Up 4 cm, Center	1.9@12.7	0.84@12.4	4.6@10.8
3.13	Center, Left 4 cm	3.3@11.5	0.69@10.5	1.2@11.8
3.14	Center, Left 4 cm	1.0@9.9	0.54@10.4	0.7@10.8
3.15	Center, Left 4 cm	1.5@8.1	0.65@10.1	1.7@9.4
3.16	Center, Left 4 cm	0.7@8.1	0.67@10.3	1.5@14.2
3.17	Up 2 cm, Center	1.5@11.1	1.77@12.4	5.2@10.9
3.18	Up 2 cm, Center	1.8@11.0	1.92@12.0	4.6@10.9
3.19	Up 2 cm, Center	2.4@12.3	1.54@13.3	3.8@11.5

COMPARISON WITH PROPOSED INJURY CRITERIA

As expected for these predominantly thoracic out-of-position tests, the head center-of-gravity accelerations were well below proposed injury tolerance values **[Kleinberger-1998]**. Head injury criterion (HIC) values for all tests peak within 20 ms of air bag deployment. As seen in *Figure 15*, the HIC values are all substantially lower than the proposed 5th percentile female Hybrid III dummy head injury tolerance value of 1000. This is consistent with the purpose of the draft ISO-2 test to preferentially load the subject thorax.

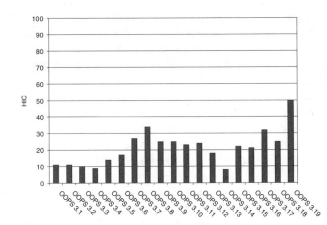

Figure 15. Head Impact Criterion for All Tests

Injury response parameters for the thorax are shown in *Figure 16*. In this figure, the chest center of gravity resultant acceleration and maximum sternal displacement using the upper CRUX unit are compared with a newly proposed injury tolerance, the Combined Thoracic Index (CTI) **[Kleinberger-1998]**. The CTI is defined as

$$CTI = \frac{A_{max}}{A_{int}} + \frac{D_{max}}{D_{int}}$$

where A_{max} and D_{max} are the maximum chest center of gravity resultant acceleration and sternal displacement respectively. For the 5th percentile female Hybrid III, the critical values are A_{int}=85 g's and D_{int}= 83 mm. CTI values larger than one are deemed injurious with a 25% risk of AIS≥3 injury. In addition, the proposed injury criterion retains the current FMVSS-208 chest injury acceleration and sternal displacement tolerances scaled for the 5th percentile female Hybrid III, taken to be 60 g's and 62 mm respectively.

In *Figure 16*, a clear delineation of injury results is seen between the Nominal ISO-2 position and the Chest On Module position. None of the Nominal ISO-2 tests saw CTI values that were greater than one. In contrast, all of the Chest On Module tests with offsets ≤2 cm had CTI values that are deemed injurious using the 5th percentile female dummy. This is the result of the increased effective distance from the air bag module of the Nominal ISO-

2 tests as compared with the Chest On Module tests. For the tests where the dummy was offset 4 cm up, the lower chestband displacements showed substantially larger values than those seen in the upper sternum since the air bag was centered nearer the lower sternum. If these displacement values were used to calculate the CTI, these tests would also show CTI values greater than the proposed injury tolerance. For the tests that were offset 4 cm left, substantial decreases in all response parameters were seen indicating that the dummy did not sustain injurious loads.

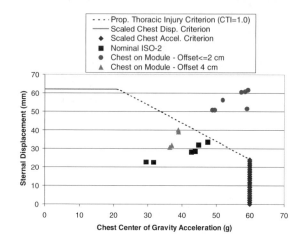

Figure 16. Sternal Displacement vs. Chest CG Acceleration Compared with Combined Thoracic Index (CTI)

A proposed neck injury tolerance, the N_{ij} criterion, is defined as **[Kleinberger-1998]**

$$N_{ij} = \frac{F_z}{F_{int}} + \frac{M_Y}{M_{int}}$$

where F_z and M_y are the neck axial force and flexion/extension moment respectively referred to the occipital condyles. For the 5th percentile female Hybrid III, the critical values are:

F_{int}=3200 N in tension/compression,
M_{int}= 210 N-m in flexion,
M_{int}= 60 N-m in extension.

N_{ij} values larger than 1.4 are deemed injurious with a 30% risk of AIS≥3 injury.

Owing to electrical noise, a number of neck load and neck moment sensors recorded questionable data during the test series. However, two tests, the sternum-centered case of OOPS3.9 and the sternum up 4 cm case of OOPS3.11, are expected to be the limiting cases for neck injury. These tests have good data on all neck load cell channels. The peak neck axial load and flexion/extension moments from these tests are compared in **Figure 17**. Both tests show N_{ij} values well below the proposed injury threshold during the time of primary air bag/occupant interaction investigated in this study.

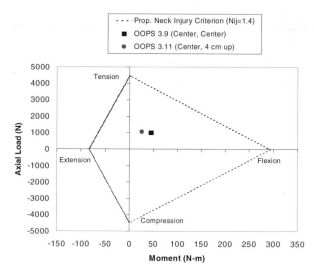

Figure 17. Neck Axial Load vs. Neck Moment Chest Compared with N_{ij} Proposed Injury Criterion

CRUX PERFORMANCE

Two CRUX units were initially installed in the 5th percentile female Hybrid III dummy used in this testing. However, the lower CRUX unit malfunctioned owing to a loose potentiometer in the CRUX housing. The upper CRUX unit performed well, however, especially considering the limited volume available for its articulated arms in the 5th percentile female Hybrid III thorax. The upper CRUX unit measured displacements of nearly 60 mm without evidence of binding or bottoming. The upper CRUX unit demonstrated generally repeatable results as shown in both thorax displacements in **Table 3** and sternal V*C in **Table 5**. There was a maximum of 6% difference in thorax displacement in repeated tests. This variation was well within the expected 10% variation owing to manufacturing differences in the air bag inflators.

A comparison of the dynamic sternal deformations measured using the upper chestband and using the upper CRUX unit is shown in **Figure 18** and **Figure 19**. In **Figure 18**, the effect of the compliance of the dummy skin is apparent in the Nominal ISO-2 position (OOPS3.1). Since the chestband is an external thoracic measuring device and the CRUX unit measures internal rib deformation, the deformation of the chestband occurs earlier and with a greater slope than the deformation of the upper CRUX unit. This provides evidence that the difference between the chestband and the CRUX unit is the result of dynamic deformation of the skin. Thereafter, when the rib cage begins deformation, the slopes of the deformations of the chestband and the upper CRUX unit are similar. In addition, peak deformations in the chestband and the upper CRUX unit occur at similar times. Similarly, as seen in **Figure 19**, the proximity of the air bag to the chest in the Chest On Module position in OOPS3.8 results in a more rapid deformation of the compliant dummy skin into the ribs. Again, the upper CRUX unit and the chestband show similar behavior and peak deformations. So, the upper CRUX unit demonstrated dynamic performance that is consistent with chestband measurements in tests where the chestband measurement accuracy is not compromised by conditions that do not affect the upper CRUX unit.

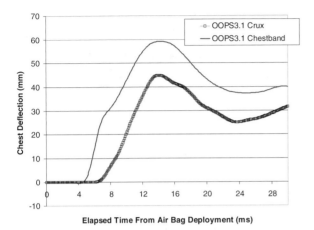

Figure 18. CRUX Thoracic Deformation vs. Chestband Sternal Deformation – OOPS3.1 - Nominal ISO-2 Position

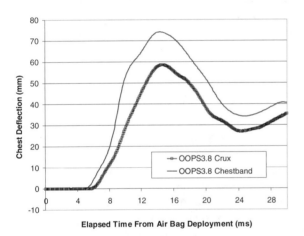

Figure 19. CRUX Thoracic Deformation vs. Chestband Sternal Deformation – OOPS3.8 – Chest On Module Position

In addition to the CRUX unit sternal deformations in the local X direction discussed above, the CRUX units are capable of measuring deflections in the local Y and local Z directions at the attachment point. In the tests with the dummy centered on the air bag module (OOPS3.1-3.12, OOPS3.17-3.19), the local Y displacement measured by the upper CRUX unit is less than 6 mm. In contrast, the tests with the dummy offset to the left of the steering wheel (OOPS3.13-3.16) show significantly larger deformations of greater than 11 mm. The timing of the maximum displacement in Y is consistent for similar test conditions for all runs. These local Y deformations, as expected, are still far smaller than the X deformations for all tests.

The local Z deformation measured by the upper CRUX unit is consistent for all tests with the dummy centered on the air bag module (OOPS3.1-OOPS3.10). In addition, the tests with the dummy sternum offset 4 cm to the left of the air bag module (OOPS3.13-OOPS3.16) and those with the dummy sternum offset 2 cm up the air bag module (OOPS3.17-OOPS3.19) showed similar Z displacements. This results from qualitatively similar Z location of air bag contact with the dummy thorax in except OOPS3.14. In OOPS3.14, a rupture of the air bag led to an anomalous deployment that affected the upper CRUX measurement. In contrast, the tests with the dummy sternum center offset 4 cm up relative to the air bag center showed much smaller displacements as a consequence of the substantially decreased interaction of the upper thorax with the deploying air bag.

CHESTBAND PERFORMANCE

Two chestbands were used in all tests except OOPS3.7. Chestbands are designed to measure external thoracic contours. So, the response of the chestbands generally includes the compliance and stiffness of both the skin and the ribs. The CRUX units, in contrast, measure the local deformation of the internal ribs. So, the chestband deformations are expected to be generally larger than the CRUX deformations. Also, since the chestband contours are calculated using all the chestband gauge curvature information, the deformation contours are sensitive to the loss of chestband gauge and gauge distribution.

Generally, the chestbands performed repeatably among similar test conditions in this study as seen in **Table 4**. Normalized by a chest deformation of 7.6 cm, the maximum sternal deformation taken from chestband contours shows approximately 1% variation among similar test conditions in OOPS3.1-3.2. There was approximately 4% difference between chestband maximum sternal deformations in OOPS3.3-OOPS3.4, and less than 5% variation in OOPS3.5-OOPS3.6. Further, there was less than 3% difference between chestband maximum sternal deformations in tests OOPS3.11 and OOPS3.12. Finally, there was approximately 11% variation in maximum sternal deformation inferred from chestbands in tests OOPS3.17-OOPS3.19.

However, maximum upper CRUX unit deformations were larger than chestband sternal deformations in tests OOPS3.6, OOPS3.9, and OOPS3.16. In addition, the maximum CRUX unit deformations in test OOPS3.10 was similar to that calculated from the chestband contours. For test OOPS3.6 with no dummy skin, the CRUX maximum sternal deformations are expected to be similar to the chestband deformations. There is less than 6% difference between the two measurements, normalized by upper CRUX unit maximum deformation. This is within the combined accuracy of the chestband and CRUX units. In addition, the upper CRUX unit was located approximately 20 mm from the centerline of the dummy sternum. The maximum sternal deformation calculated using the chestband in this region was 30.4 mm. This value is less than 4% different from the maximum sternal deformation measured by the upper CRUX unit.

In the centered Chest On Module tests, OOPS3.9 and OOPS3.10, the upper chestband maximum sternal deformation seen in *Table 3* was commensurate with deformation measured by the upper CRUX unit though compliance in the dummy skin would suggest that the chestband deformations would be larger. However, during these tests, chestband gauges were lost in regions of high curvature leaving gaps that significantly decrease the accuracy of the sternum position.

In OOPS3.16 with the sternum center positioned 4 cm to the left of the air bag center, there was evidence of the upper chestband sliding with the skin. The calculated chestband contour deflected sharply at the lateral chest and bulged near the sternum at the time of maximum deformation measured by the upper CRUX unit. This apparent sliding made the chestband sternal calculations unreliable as the computed position of the sternum may not maintain a fixed relationship with the physical sternum.

So, the chestband measurements were generally repeatable within known caveats in the use of chestbands as a contour measuring device. Among these are loss of accuracy with loss of chestband gauges in regions of high local curvature and the potential for compliance of overlying tissue to complicate the interpretation of chestband measurements when used for measuring the deformation of underlying rib structure.

CONCLUSIONS

An out-of-vehicle occupant fixture for out-of-position tests was developed that allows accurate positioning, within 0.5 cm in all axes for fixed structural points on the dummy. This fixture permits the detailed investigation of the effect of positioning on out-of-position occupants under air bag deployment into the occupant thorax. Positioning repeatability was within the dimensional variation of approximately 1 cm between the spines of two 5th percentile female Hybrid III dummies available for testing at ASL.

As expected, head and neck injury tolerance values were not generally exceeded for this testing as the ISO-2 position involves predominantly air bag/thorax interactions. In addition, though the current study used a depowered air bag from the current automobile fleet, the use of a worst-case positioning (Chest On Module position) resulted in sternal deflection, chest CG acceleration, and viscous criterion values that generally exceeded accepted and proposed injury tolerances.

The thoracic response under out-of-position air bag deployment was not found to be sensitive to the presence of chestbands used to measure thoracic contours. Upper sternal displacement varied less than 1% between tests with and without chestbands. Peak chest CG resultant acceleration varied approximately 10% among the tests. In addition, two CRUX units were installed within the 5th percentile female Hybrid III dummy to measure upper and lower thoracic deformations. Though the lower CRUX unit malfunctioned, the upper CRUX performed well giving repeatable and accurate results that compared well with other thoracic instrumentation.

Sensitivity to distance from the air bag module was found to be substantial. There was over a 60% decrease in peak acceleration as the effective distance from the air bag module increases from zero (skin with breasts – Chest On Module) to 53 mm (no skin – Nominal ISO-2). The variation in peak chest CG resultant was approximately exponential in effective thorax distance to air bag module. Similar results were seen in the dummy sensor time histories. It was further observed that the compliance of the dummy breasts was significantly variable among dummy skins, as the breasts are simply conical forms glued onto the dummy skin. Two dummy skins with breasts were available for examination. The four individual breasts had qualitatively different stiffnesses.

Also, a large variation in occupant response was found when moving a 5th percentile female Hybrid III dummy thorax from a center position to 2 cm and 4 cm up the steering wheel. Variation in response of up to 30% was seen in chest CG resultant acceleration. In addition, there was a variation of up to 34% in upper sternum deflection among the three positions and up to 60% variation in upper sternal V*C among the positions tested.

Additional testing was performed with the occupant thorax centered vertically while offset 4 cm horizontally to the left of center. This occupant position resulted in a qualitatively different deployment of the air bag. In this deployment, the air bag opened asymmetrically to the right of the occupant producing large differences in occupant response relative to a centered occupant. Over 30% variation in peak chest CG resultant acceleration was seen. Also, there was over 35% difference in upper sternum deflection between the two cases and nearly 70% difference in upper sternal V*C.

This large variation in occupant response for small differences in position may have implications for sled and vehicle testing of out-of-position occupants. Variations of ±2 cm in occupant center-of-gravity position are common in occupant positioning in sled testing at ASL. This is also likely to be the case for any positioning not using a dedicated positioning fixture.

In addition, use of different interpretations of 'ISO-2' positioning may cause large variations in occupant response under air bag deployment as a nearly 50% reduction in peak chest CG resultant acceleration is seen from positioning with dummy breasts deformed against the module to positioning with dummy breasts just touching the module. This variation may be exacerbated by significant qualitative differences in the compliance of breast forms for the 5th percentile female Hybrid III dummy.

ACKNOWEDGEMENTS

This study was conducted under Cooperative Agreement No. DTNH22-93Y-07028 with the National Highway Traffic Safety Administration (NHTSA) of the U.S. Department of Transportation. The authors gratefully acknowledge TRW Automotive for support of this project.

REFERENCES

1. **[Hayano-1994]** K. Hayano, K. Ono, and F. Matsuoka, "Test Procedures for Evaluating Out-of-Position Vehicle Occupant Interactions with Deployed Air Bags", *Proceedings of the 14th ESV Conference*, Paper 94-S1-O-19, pp. 245-259, 1994.

2. **[Horsch-1979]** J. Horsch and C. Culver, "A Study of Driver Interactions with an Inflating Air Cushion, *Proceedings of the 23rd Stapp Car Crash Conference*, Paper 791029, Society of Automotive Engineers, Warrendale, PA, pp. 799-815, 1979.

3. **[Horsch-1990]** J. Horsch, I.V. Lau, D. Andrzejak, and D.C.Viano, "Assessment of Air Bag Deployment Loads", *Proceedings of the 34th Stapp Car Crash Conference*, Paper 902324, Society of Automotive Engineers, Warrendale, PA, pp. 267-288, 1990.

4. **[IIHS-95]** Insurance Institute for Highway Safety, Status Report, Vol. 30, No. 3, Insurance Institute for Highway Safety, Arlington, VA, 1995.

5. **[ISO-1995]** ISO DTR 10982, "Road Vehicles - Test Procedures For Evaluating Out-Of-Position Vehicle/Occupant Interactions With Deploying Air Bags," International Standards Organization, November 21, 1995.

6. **[Kahane-1996]** C.J. Kahane, "Fatality Reduction by Air Bags - Analyses of Accident Data through Early 1996", NHTSA Technical Report DOT HS 808 470, National Technical Information Service, Springfield, Virginia, 1996.

7. **[Kleinberger-1997]** M. Kleinberger and L. Summers, "Mechanisms of Injuries for Adults and Children Resulting from Air Bag Interaction", *Proceedings of the 41st Annual Conference of the Association for the Advancement of Automotive Medicine*, the Association for the Advancement of Automotive Medicine, Des Plaines, IL, pp. 405-420, 1997.

8. **[Kleinberger-1998]** M. Kleinberger, E. Sun, R. Eppinger, S. Kuppa, and R. Saul, "Development of Improved Injury Criteria for the Assessment of Advanced Automotive Restraint Systems", NHTSA Document @www.nhtsa.dot.gov, September 1998.

9. **[Lau-1986]** I.V. Lau and D.C. Viano, The Viscous Criterion - Bases and Applications of an Injury Severity Index for Soft Tissues, *Proceedings of the 30th Stapp Car Crash Conference*, SAE Paper 861882, Society of Automotive Engineers, Warrendale, PA, pp. 123-142, 1986.

10. **[Melvin-1993]** J.W. Melvin, J.D. Horsch, J.D. McCleary, L.C. Wideman, J.L. Jensen, and M.J. Wolanin, "Assessment of Air Bag Deployment Loads with the Small Female Hybrid III Dummy", *Proceedings of the 37th Stapp Car Crash Conference*, Paper 933119, Society of Automotive Engineers, Warrendale, PA. pp. 121-132, 1993.

11. **[NHTSA-1998]** National Highway Traffic Safety Administration, Special Crash Investigation Summary, April 1, 1998.

12. **[Saul-1984]** R. Saul, "State-of-the-Art Dummy Selection", Vol. 1, Report SRL-29, National Highway Traffic Administration, 1984.

2001-01-0179

Air Bag Loading on In-Position Hybrid III Dummy Neck

Jian Kang, Venkatesh Agaram, Guy Nusholtz and Gregory Kostyniuk
DaimlerChrysler Corp.

Copyright © 2001 Society of Automotive Engineers, Inc.

ABSTRACT

The Hybrid III family of dummies is used to estimate the response of an occupant during a crash. One recent area of interest is the response of the neck during air bag loading. The biomechanical response of the Hybrid III dummy's neck was based on inertial loading during crash events, when the dummy is restrained by a seat belt and/or seat back. Contact loading resulting from an air bag was not considered when the Hybrid III dummy was designed. This paper considers the effect of air bag loading on the 5th percentile female Hybrid III dummies. The response of the neck is presented in comparison to currently accepted biomechanical corridors.

The Hybrid III dummy neck was designed with primary emphasis on appropriate flexion and extension responses using the corridors proposed by Mertz and Patrick. They formulated the mechanical performance requirements of the neck as the relationship between the moment at the occipital condyles and the rotation of the head relative to the torso. The deformation of the neck was due to the motion of the head relative to torso, when the torso was restrained either by the seat belt or the seat back. Air bag loading is significantly different from belt or seatback loading, because there is more than one load path on the head. During air bag loading, the dummy neck shows significantly different responses from those seen in the Mertz and Patrick tests. The neck experiences second mode bending during air bag loading as opposed to the first mode bending seen in seat belt and/or seat back loading.

The Hybrid III 5th percentile female dummy is frequently used to estimate the response of small stature occupants to air bags. This paper examines the Hybrid III 5th percentile female dummy's neck responses due to air bag loading. The neck responses were found to be highly dependent on how the air bag interacts with the dummy. Three modes of air bag-neck interaction were observed and studied: air bag directly loading the head, air bag trapped under the chin during the deployment process and air bag trapped behind the jaw of the dummy head. This paper also presents the results of some experimental modifications to the head/neck

design of the Hybrid III 5th percentile female dummy to prevent air bag entrapment.

INTRODUCTION

The members of the Hybrid III dummy family are used as human surrogates in automotive frontal impact tests to assess the performance of different occupant restraint systems. Although frontal air bags have been mandated as supplemental restraints in automobiles, the evaluation of the human-air bag interaction is not straightforward. Most current efforts to study air bag-occupant interaction use the Hybrid III family of dummies and the associated injury risk assessment criteria. One of the areas of concern in these studies is the head/neck response of the Hybrid III due to interaction with a deploying air bag. The neck of the Hybrid III dummy needs to have sufficiently human-like response characteristics to be useful in evaluating the human response to deploying air bags. Problems associated with the Hybrid III dummy head/neck design and the neck response during air bag loading make it difficult to accurately assess the response of the human subjects. One issue is the significant departure, in terms of anthropometric characteristics of the head/neck region, between the Hybrid III and the human occupant [1]. The exposed horizontal surface in the chin-jaw region and the near vertical cavity between the jaw and the neck, provide unrealistic reaction surfaces for loading due to an inflating air bag, resulting in unrealistic neck deformation.

The Hybrid III dummy neck was designed with emphasis on appropriate flexion and extension response with respect to the corridors proposed by Mertz and Patrick [2]. They formulated the mechanical performance requirements of the neck as the relationship between the moment at the occipital condyles and the rotation of the head relative to the torso. These corridors are the only existing estimates for the response of the cervical spine in bending. Neck response in these corridors is primarily due to inertial loading of the head. During impact, the seatbelts or the seatback, restrain the motion of the torso, while the neck deforms like a cantilever beam, in first mode bending, due to the motion of the head. However, as the results in this paper indicate, air bag loading can result in a significantly different neck deformation from that seen in the Mertz and Patrick [2] tests. The Hybrid III dummy's neck can experience second mode bending during air bag loading, especially when the air bag is trapped in the chin-jaw region or in the jaw-neck cavity. Since the neck response of human subjects due to interaction with deploying air bags is not known, the possibility of occurrence of second mode bending of the Hybrid III dummy neck can not be supported *a priori* as biofidelic. Moreover, the loads leading to second mode bending in the Hybrid III dummy neck may not be consistent with those leading to second mode bending in live humans.

An investigation of the response of small stature occupants to air bags, by using the Hybrid III 5th percentile female dummy, was conducted to better understand the modes of interaction between the air bag and the dummy neck. The neck response was found to be highly dependent on the nature of the air bag's interaction with the dummy's head/neck region. There are essentially three modes of interaction between the air bag and the dummy's head/neck region for the same test set-up (described below). The response varies significantly from test to test. Consequently the system response, i.e., the dummy-air bag interaction, could be considered chaotic with three distinct interaction modes, but no way to predict which interaction mode would occur in a given test. In the first interaction mode, the air bag directly loads the head. In the second interaction mode, the air bag is trapped in the chin-jaw region during the deployment process. In the third interaction mode, the air bag is trapped behind the dummy's jaw in the jaw-neck cavity. In all three cases, moment distributions resulting in the second bending mode are generated, which may not represent the response of a human subject. Modifications to the neck region of the Hybrid III 5th percentile female dummy to prevent one or both of the last two air bag-neck interaction modes were investigated. The modifications to the Hybrid III 5th percentile female dummy were for experimental purposes only. While these modifications were instrumental in demonstrating the factors associated with different neck deformation modes, they should not be considered as design recommendations for the Hybrid III 5th percentile female or any other Hybrid III dummy at this time.

TEST METHODS

A series of static air bag deployment tests were conducted to investigate the loading of the head/neck region of the Hybrid III 5th percentile female dummy due to air bag deployment. A typical test setup is shown in Figure 1. The dummy was placed, leaning towards the instrument panel, in a full-forward passenger seat. The choice of the static environment was based on its simplicity and ease of obtaining the desired initial alignment of the dummy relative to the air bag module

Figure 1. Typical Test Setup

The study was limited to frontal passenger air bags, and the seat belts were not used. Two different air bag systems were studied, a "standard" system and a "depowered" system (20% reduction in the slope of the "standard" system tank pressure-time relationship). To increase the likelihood of the air bag being caught in the chin-jaw region and/or behind the jaw, in the jaw-neck cavity, the seat was raised two inches from its standard height and the head was rotated approximately 20 degrees forward. While it is recognized that this test setup is artificially contrived, and is unlikely to occur in a vehicle crash, it is a useful test condition to illustrate the different air bag-neck interaction modes that could occur in a dynamic crash test. The primary focus of this investigation is on in-position tests such as in vehicle crashes or sled tests with the dummy seated nominally. The dummy's position however was chosen to enhance the probability of air bag entrapment in the neck-jaw cavity.

A standard Hybrid III 5th percentile female dummy, with a TMJ head skin, and a SAE neck shield[3] were used for the baseline tests. The head skin referred to as "modified" in Ref.3 is referred to as TMJ head skin in this paper. The TMJ skin was intended to prevent the air bag entrapment in the chin-jaw region as well as in the jaw-neck cavity. This was addressed by including a thin membrane of skin, which spanned the front part of the bottom section of the jaw and a second thin membrane, which spanned the area behind the jaw. The neck shield was a thin, "mouse pad like" material, which was wrapped around the neck. The head skin, in both the chin-jaw area and in the cavity behind the jaw, was painted with chalks of different colors. The colored chalks transferred to the air bag were checked to determine whether the air bag was entrapped under the chin or behind the jaw. The dummy was instrumented with upper and lower neck load cells (Models 1716 and 2150, respectively, Denton, Inc.). Head, chest and pelvic accelerations were measured using linear accelerometers (Model 7264-2000, Endevco, Inc.). The transducer data were processed according to SAE J211. The three-dimensional motion of the head and the chest were computed using the method developed by Nusholtz, et al. [4]. High-speed video cameras and film cameras, capable of 1000 frames per second, were used to monitor the head/neck and air bag interactions from the lateral direction and the oblique posterior direction above the head.

The head external loads resulting from air bag impact were calculated using the head accelerations and the upper neck loads. Figure 2 shows a free-body diagram of the dummy head with internal forces (F_N and M), external force (F_E), and acceleration (ma). The neck reaction force and moment are measured with the upper neck load cell, and the acceleration is measured with the accelerometer located at the center of gravity (CG) of the head. The solution of the dynamic equilibrium equations permits an estimation of the unknown external loads, (F_E) applied to the dummy's head.

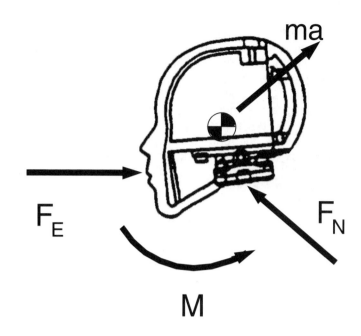

Figure 2. Free Body Diagram of Dummy Head.

Table 1. The Average Values of the Peak Response Without Dummy Modification

(SAE Sign Conv.)

Air Bag Head/Neck Interaction Mode	Statistic	Upper Neck Responses Fx (N)	Fz (N)	Mcc (Nm)	Head Res. Acc. (g)	HIC$_{15}$	Chest Res. Acc. (g)	Chest Deflec. (mm)	Head External Forces Fx (N)	Fz (N)
Air Bag Loads Head Directly (Mode 1, 10 tests)	AVE.	846	579	65	64	83	31	18	-2732	1045
	S.D.	276	167	21	14	50	9	4	668	749
Air Bag Caught Under Chin (Mode 2, 6 tests)	AVE.	-2217	3551	-121	91	266	36	23	-2842	-4606
	S.D.	509	808	39	11	122	8	4	748	1636
Air Bag Entrapped Behind Jaw (Mode 3, 9 tests)	AVE.	-5133	4446	-206	95	373	40	30	4090	-4897
	S.D.	373	619	21	14	261	5	2	843	1358

RESULTS

A summary of the test results for the unmodified dummy is shown in Table 1. The peaks in the time-histories of the upper neck loads, the head resultant acceleration, the 15ms limit HIC, the chest resultant acceleration and the chest deflections are included. Due to the large number of tests conducted, only the average values and the standard deviations are presented. The interactions between the air bag and the occupant, and the head/neck kinematics are shown in Figures 3-5. In these figures, the sequence from top to bottom shows snap shots as time progresses. The first picture in each set shows the air bag just making contact with the face, and parts of the air bag approaching the neck. The second picture shows the extent to which the air bag interacts or does not interact (mode 1) with the neck. The third picture corresponds to the air bag in the initial phases of moving away from the neck. The fourth phase corresponds to the time just before the air bag is completely free of interaction with the neck.

There appear to be three modes of air bag-dummy interaction: interaction mode 1, the neck loads are generated primarily from the air bag loading the front of the head, interaction mode 2, the neck loads are generated primarily from the air bag trapped under the chin, and interaction mode 3, the neck loads are generated primarily from the air bag trapped behind the jaw. The air bag contacted the head at approximately the same time in all three interaction modes (15 ms after air bag firing.) The loads are generated when the air bag is filled and generalized membrane tension occurs. The head and chest started to accelerate, and the forces in the neck started to develop immediately after the contact. Time history data are plotted for three typical representative tests in Figure 6-12. They are not mean time-histories obtained by averaging the results of a set of tests. The component of the external force applied to the head, which would be the major contributor to the shear loads measured by the upper neck load cell, is called as the head external shear force in Figure 7. The component of the external force applied to the head, which would be the major contributor to the axial loads sensed by the upper neck load cell are referred to as head external axial force in Figure 9. The positive directions of the two external force components are consistent with the local frame of reference located at the CG of the dummy's head. The head/neck response in each mode of interaction is described in the following sections.

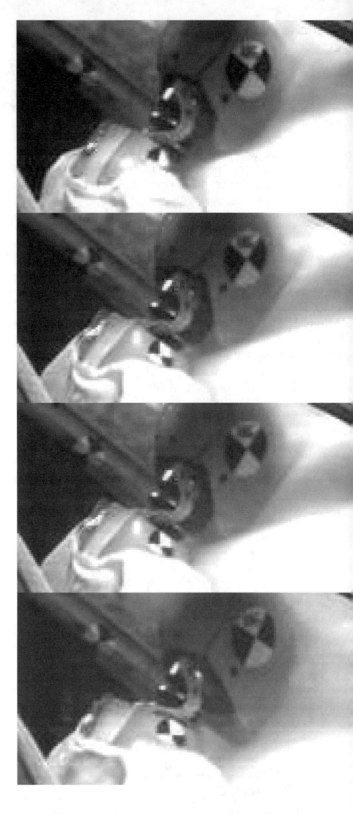

Figure 3. Air Bag Loading the Head Directly.

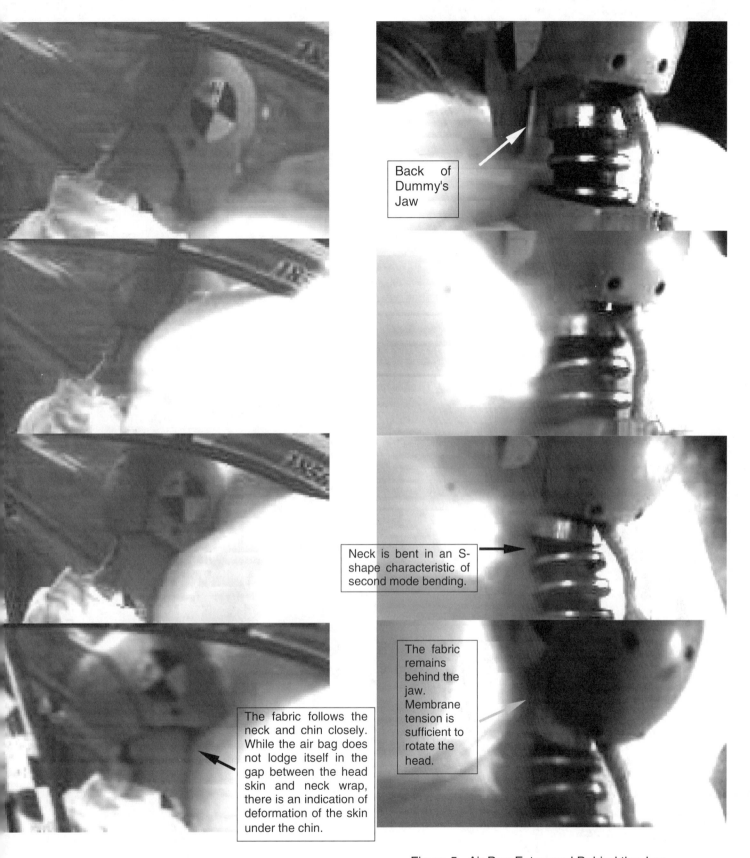

Figure 4. Air Bag Loading the Head Directly.

Figure 5. Air Bag Entrapped Behind the Jaw.

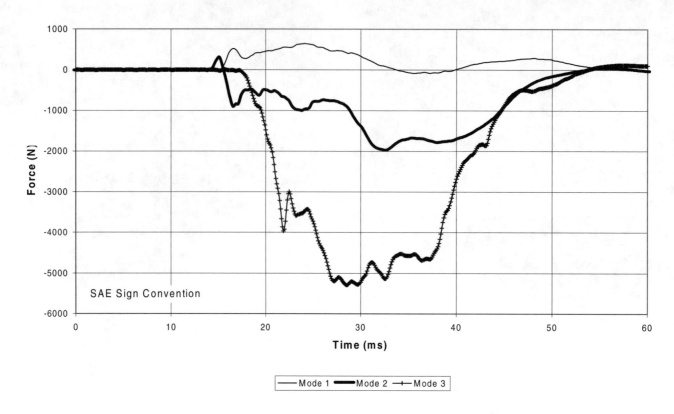

Figure 6. Upper Neck Shear Force.

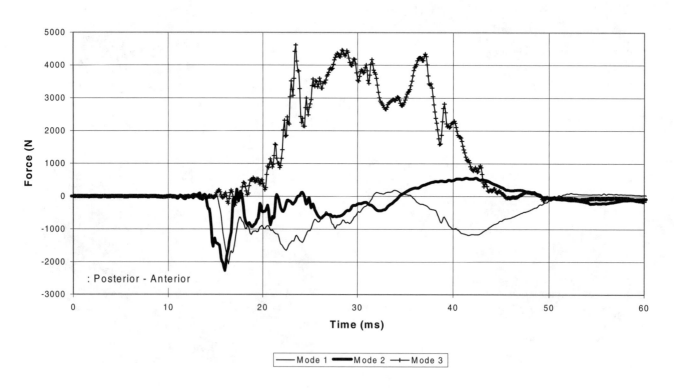

Figure 7. Head External Shear Force.

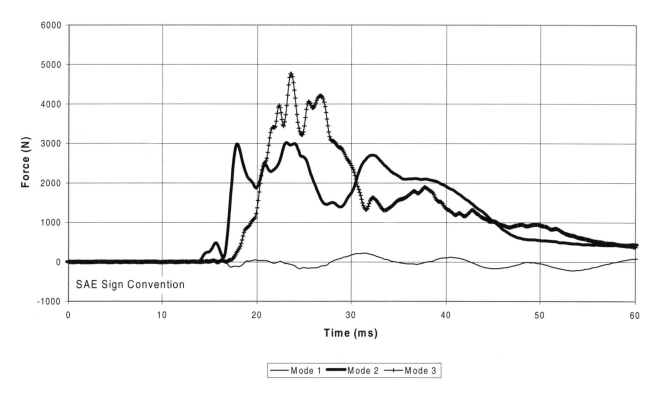

Figure 8. Upper Neck Axial Force.

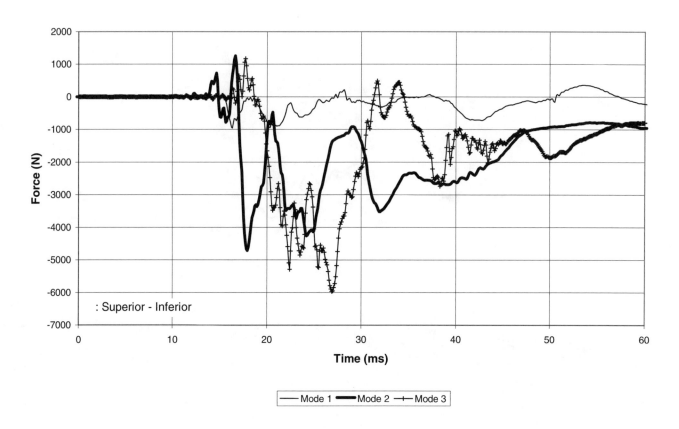

Figure 9. Head External Axial Force.

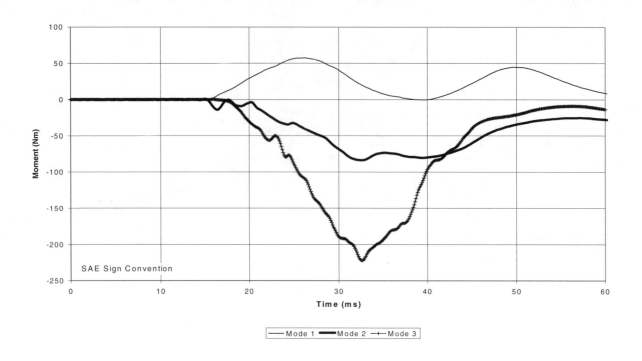

Figure 10. Upper Neck Bending Moment.

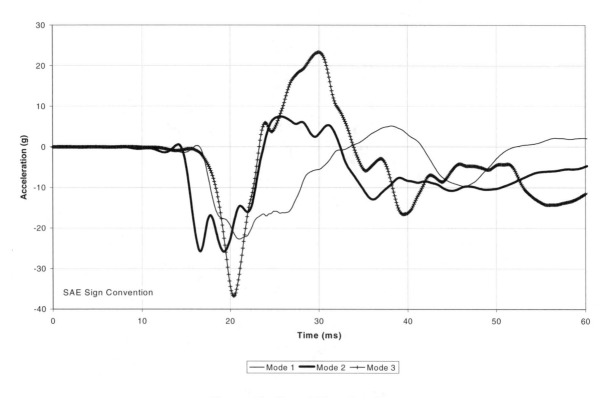

Figure 11. Chest X Acceleration.

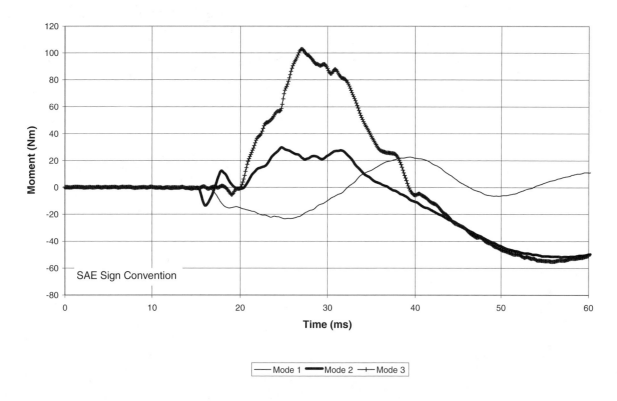

Figure 12. Lower Neck Bending Moment.

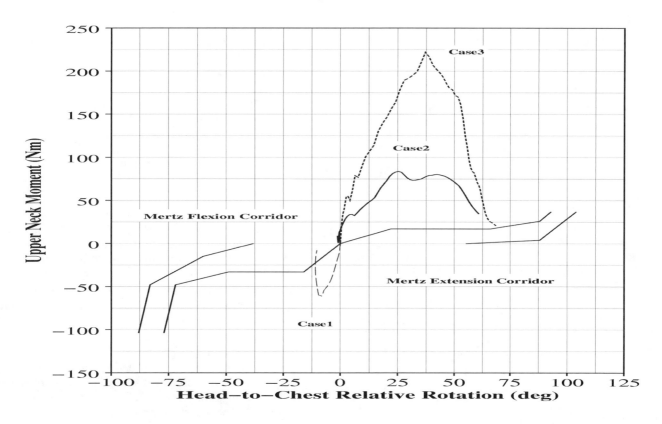

Figure 13. Upper Neck Moment vs. Head-to-Chest Rotation.

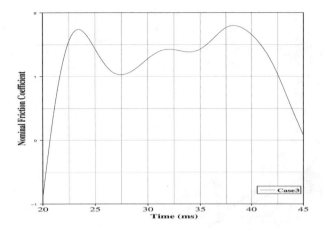

Figure 14. Estimate of the Friction Coefficient.

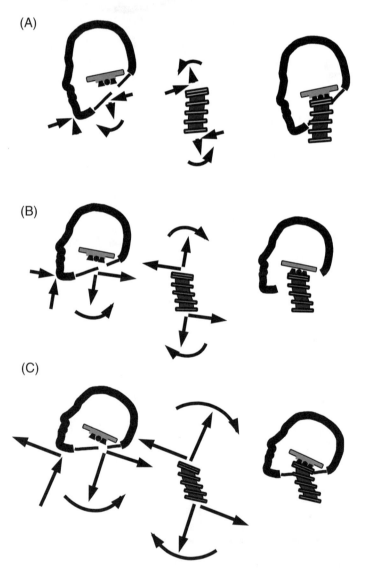

Figure 15. Free Body Diagrams of the Head and Neck, Showing the Deformed Shape of the Neck. (A) Mode 1: Airbag Loading the Head Directly, (B) Mode 2: Airbag Caught under the Chin, (C) Mode 3: Airbag Caught behind the Jaw

AIR BAG-NECK INTERACTION MODE 1

In the first air bag-neck interaction mode, the air bag directly loads the head (Figure 3), leading to a flexion moment at the neck. The neck shear is positive (Figure 6), which implies that the head is pushed rearwards relative to the neck, as a result of membrane tension. The head external shear is in the anterior-posterior direction (Figure 7) confirming the rearward pressure on the head by the air bag. The neck axial force is insignificant in magnitude and changes from compression to tension (Figure 8). The head external axial force is of small magnitude in the superior-inferior direction (Figure 9). The applied external loads are both larger than the corresponding neck forces. This indicates that, in this head/neck loading-pattern, the neck loads are primarily from the air bag loading of the head with only small loads of the air bag directly on the neck. This implies that the air bag may have formed a bridge between the chin and the chest, minimizing the load on the neck directly from the air bag. The upper neck moment is in pure flexion (Figure 10), which is an indication that the center of air bag pressure on the dummy's head is beneath the CG of the head. The torso is accelerated rearward by the deploying air bag (Figure 11). The dependence of the upper neck moment on the head rotation relative to the chest is compared to the flexion corridor proposed by Mertz, et al., [5] for Hybrid III 5th percentile female dummy (Figure 13). In this mode of air bag-neck interaction (Case 1, Figure 13), the moment-rotation relationship falls outside the corridor almost immediately after the head rotation starts, indicating that significant neck-moments can occur with little head rotation. The free body diagrams of the head and neck during air bag loading are shown in Figure 15(A). The combined upper (Figure 10) and lower neck bending moments (Figure 12) cause the neck to flex into a reflected S-shape in second mode bending.

AIR BAG-NECK INTERACTION MODE 2

In the second air bag-neck interaction mode, the air bag contacts the head under the chin (Figure 4). The bag is trapped under the chin during the deployment. The neck shear changes to negative (Figure 6), which implies that the head is pulled forward relative to the neck shortly after the initiation of the air bag-dummy interaction and membrane tension has developed. However, the head external shear is insignificant in magnitude and changes direction from anterior-posterior in the beginning to posterior-anterior in the latter part (Figure 7). This implies that a major portion of neck shear comes from the inertial loading of the head on the neck, the direct loading of the air bag on the neck, or a combination of both. Significant tension load is present in the neck (Figure 8). The head external axial force is in the inferior-superior direction (Figure 9) and causes the head to pull on the neck. The external axial load is close in magnitude to the neck tension. This indicates that the contribution to the neck tension is primarily due to the air bag loads on the head/neck area and not due to the inertial loading of the head. The upper neck moment is pure extension in nature (Figure 10). The chest acceleration is mainly in the rearward direction (Figure 11). The upper neck moment as a function of head rotation relative to the chest is compared to the extension corridor proposed by Mertz, et al., [5] (Figure 13, Case 2). The test data again falls outside the corridor, even at very small head-rotation values. The free body diagrams of the head and neck during air bag loading are shown in Figure 15(B). The forces and moments again cause a second mode bending in the neck. However, in this mode, the deformed shape of the neck is S-shaped as opposed to the reflected S-shape seen in air bag-neck interaction Mode 1 (according to the moments in Figures 10, 12).

AIR BAG-NECK INTERACTION MODE 3

In the third air bag-neck interaction mode, the air bag contacts the head below the chin (Figure 5). The fabric is entrapped in the hollow area between the neck and the jaw. As the bag continues to inflate, pressure is built up within the entrapped portion of the air bag and membrane tension develops. The air bag pulls the head forward and upward, possibly pushing on the neck at the same time. The resulting neck shear is negative to a higher degree than in Mode 2 (Figure 6), which implies that the head is pulled forward relative to the neck. The head external shear maintains the posterior-anterior direction (Figure 7) during the whole event confirming the forward pulling of the head. The neck shear is larger than the external shear load. This indicates that a portion of the neck shear comes from the inertial loading of the head on the neck. However, the major contribution to the neck shear is due to the air bag loads on the head/neck area. This could be due, either to the membrane tension in the deploying air bag in front of the dummy, pulling on the air bag material, trapped in the jaw-neck cavity, or the pressure of the trapped air bag material pushing against the neck and the jaw, or both. Elementary estimates of the forces generated by the pressure of the trapped air bag in the jaw-neck cavity indicate that they are probably negligible compared to the high shear loads experienced by the upper neck load cell. Consequently, the forward pulling of the dummy head due to the membrane tension in the deploying air bag, in front of the dummy, is considered as dominant. Tension is evident in the neck (Figure 8). The head external axial force is in the inferior-superior direction (Figure 9) pushing the head upward. The external axial load is close in magnitude to the neck tension. This indicates that the contribution to the neck tension is due to the air bag loads on the head/neck area. The upper neck moment is extension in nature (Figure 10). The chest acceleration changes from the anterior-posterior direction to the posterior-anterior direction (Figure 11). The response of upper neck moment as a function of head rotation relative to the chest is compared to the extension corridor (Figure 13, Case 3). The observed response falls outside the corridor starting with very small values of rotation. This change is due to the air bag being trapped behind the jaw and forcing the chest forward while the head rotates. If the air bag is trapped behind the jaw, the head external shear load is in the posterior-anterior direction and causes high neck shear. The friction between the air bag fabric and the material of the dummy skin, and the force from the trapped air bag behind the jaw contribute to the head external shear force.

The magnitude of the nominal friction coefficient during air bag contact, in this mode of air bag-neck interaction, was estimated (Figure 14). This was done by dividing the external shear load by the external axial load. The peak ratio is approximately 1.8. This value is significantly higher than that would be observed in a friction bench test. This indicates that significant external shear load is due to the membrane tension in the deploying air bag in front of the dummy pulling on the air bag material trapped in the jaw-neck cavity rather than strictly from sliding friction. The entrapped air bag causes high neck shear that is only seen in this mode of air bag-head/neck interaction. The free body diagrams of the head and neck during air bag loading are shown in Figure 15(C). The neck deformed into S-shape, similar to air bag-neck interaction Mode 2, but with a larger curvature. The upper and the lower neck load cell moment outputs confirm this (Figures 10, 12.)

MODIFICATIONS TO THE HYBRID III 5TH PERCENTILE FEMALE DUMMY NECK

The SAE recommended TMJ head skin and neck shield did not prevent or in any way hamper the air bag from becoming trapped in the chin-jaw region or in the jaw-neck cavity. In some cases, it appeared to make it worse because the air bag would inflate behind the thin section of the head-skin near the chin and the jaw. The additional surface area allowed greater friction, slowing the air bag's release from being trapped. To accomplish what was intended by the SAE recommended TMJ head-skin and neck shield, with respect to this artifact of air bag/head/neck interactions (air bag snagging under the chin and behind the jaw), the 5th percentile female dummy neck was modified using two different approaches.

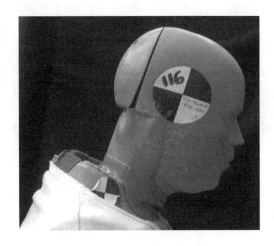

Figure 16. Modified Head/Neck Skin

The first approach used a modified head/neck skin. Using neck parts from the Hybrid II 50th percentile male dummy, additional skin and rubber and a head skin from the Hybrid III 5th percentile female dummy, a neck surface was formed that extended from the jaw to the upper torso (Figure 16). This modification prevented the air bag from being caught in the chin-jaw region or in the jaw-neck cavity. This modification produced insignificant change in the results of the pendulum extension test. The neck moment versus head rotation response satisfied the current dummy-neck-response specifications of the SAE. However, the flexion response was compromised due to interference to the bending of the neck by the additional neck skin from the Hybrid II. Eleven static air bag tests were conducted using this modified neck. All resulted in neck responses similar to air bag-neck interaction Mode 1 (Table 2).

The second approach added a pair of aluminum extensions to the jaw (Figure 17). This design also eliminated the air bag from becoming caught behind the jaw, but not from under the chin. This modification did not affect either the flexion or the extension response in the standard pendulum calibration tests. Eight static air bag tests were conducted using this modified neck design (Table 2). Five tests (62.5%) resulted in neck loads similar to air bag-neck interaction Mode 1 and three tests (37.5%) resulted in neck loads similar to air bag-neck interaction Mode 2. The Mode 3 type of air bag-neck interaction did not occur in tests using the aluminum jaw extensions. These modifications were considered for investigative purposes only and to address the issue of the air bag being trapped under and behind the jaw. They do not constitute a recommended design change for the Hybrid III 5th percentile female

Table 2. Peak Response in Airbag Tests with Neck Structure Modifications
(SAE Sign Conv.)

Airbag Head/Neck Interaction Modes	Airbag System	Head Skin	Modification	Statistic	Upper Neck Responses Fx (N)	Fz (N)	Moc (Nm)	Head Res. Acc. (g)	HIC$_{15}$	Chest Res. Acc. (g)	Chest Deflec. (mm)	Head External Forces Fx (N)	Fz (N)
Air Bag Loads the Head Directly Mode 1 (11 tests)	Standard	Standard	Skin	Ave.	1015	1241	71	98	181	42	22	-4166	786
				S.D.	153	531	11	12	38	5	4	636	1666
Air Bag Loads the Head Directly Mode 1 (5 tests)	Depowered	TMJ	Alum	Ave.	695	493	59	55	53	28	16	-2338	986
				S.D.	109	65	15	6	10	3	2	403	243
Air Bag Caught Under the Chin Mode 2 (3 tests)	Depowered	TMJ	Alum	Ave.	-2620	3856	-154	100	201	34	22	-2484	-4772
				S.D.	140	339	23	6	2	8	1	515	1478

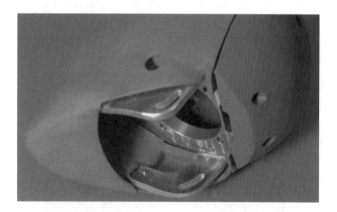

Figure 17. A Pair of Aluminum Patches Added to the Head.

dummy at this time.

Two different air bag systems were used in this study. One was a standard system and the other was a depowered system as described earlier. All the three typical air bag-head/neck interaction modes were observed with both air bag systems. The average values and standard deviations of occupant response were obtained for these two different air bag systems for the three typical head/neck and air bag interaction modes (Table 3). These two different air bags produced similar neck loads in interaction Mode 1; generally, for the depowered air bag system, higher in the Fx and lower in the Fz for interaction Modes 2 and 3 and higher for the extension moment. However, none of these comparisons are statistically significant.

Table 3. The Average Values of the Peak Response of Two Airbag Systems

(SAE Sign Conv.)

Air Bag Interaction Mode	Airbag System	Statistic	Upper Neck Responses			Head Res. Acc.	HIC_{15}	Chest Res. Acc.	Chest Deflec.	Head External Forces	
			Fx	Fz	Moc	Res. Acc.		Res. Acc.	Deflec.	Fx	Fz
			(N)	(N)	(Nm)	(g)		(g)	(mm)	(N)	(N)
Air Bag Loads Head Directly Mode.1	Standard Airbag	Ave.	1030	1119	73	98	189	41	21	-4164	813
	(12 tests)	S.D.	149	389	11	17	35	6	3	765	1491
	Depowered Airbag	Ave.	801	591	61	63	75	31	18	-2681	967
	(9 tests)	S.D.	251	172	19	15	45	9	5	687	751
Air Bag Caught Under Chin Mode.2	Standard Airbag	Ave.	-2031	3739	-92	80	379	40	27	-2805	-5454
	(2 tests)	S.D.	89	1010	11	6	189	13	6	755	1634
	Depowered Airbag	Ave.	-2311	3457	-136	96	210	35	21	-2860	-4182
	(4 tests)	S.D.	628	845	41	9	17	6	1	861	1688
Air Bag Entrapped Behind Jaw Mode.3	Standard Airbag	Ave.	-4638	5240	-192	115	789	45	31	2673	-6940
	(2 tests)	S.D.	52	551	11	3	307	3	1	109	714
	Depowered Airbag	Ave.	-5275	4220	-210	90	254	39	29	4494	-4313
	(7 tests)	S.D.	283	437	22	11	32	5	2	291	765

DISCUSSION

The flexion and extension response corridors developed by Mertz and Patrick [2] are the only biofidelity basis for Hybrid III dummy neck. The human volunteers, whose response was used to devise these extension corridors all have some degree of neck tension. Some tension is due to normal muscle activity holding the head in place. Additional tension would result from anticipation of impact. Under conditions of no neck tension i.e. cadaver response, the head would be expected to translate (due to inertial loading of the head) before showing significant rotation. The Hybrid III dummy neck is designed to deform in first mode bending in order for the moment-rotation curves to remain within the corridors. The plateau portion represents the maximum moment that the neck muscles can generate in resisting head motion before appreciable head rotation occurs. The initial bending stiffness for the 5th percentile female is 2.06 Nm/degree for flexion and 0.77 Nm/degree for extension.

After reaching a certain point, the neck muscle yields and the head keeps rotating without an increase in the bending moment. When the normal articular voluntary range of motion of the neck is reached, the action of the neck ligaments and/or passive stretch of the neck muscles, increases the bending resistance of the neck.

The lower portion of the corridors reflects the elastic behavior of the ligaments and muscles as well as energy dissipation of the muscles during rebound. These corridors represent the neck response in the particular cases of restraint with either the seatbelt or the seatback. However, they were not developed for evaluating air bag loading. By basing the design of the Hybrid III dummy neck on corridors restricted to first mode bending, it is not clear if the dummy's head/neck response in case of loading by the air bag is biofidelic or not.

The upper neck moments obtained from these series of air bag tests were plotted as a function of head rotation relative to the chest and compared to the corridors developed by Mertz and Patrick. Figure 13 shows significant differences for both the flexion mode and the extension modes. The upper neck loads build up very quickly once the air bag contacts the head. For air bag-neck interaction Mode 1, the initial stiffness of neck flexion was 7.2 Nm/degree, as compared to 2.06 Nm/degree for the flexion corridor.

For air bag-neck interaction Modes 2 and 3, the initial extension stiffness was 3.5 Nm/degree and 6.5 Nm/degree, respectively. The observed extension stiffness in these two air bag-neck interaction modes is an order of magnitude higher than the value from the extension corridor (0.77 Nm/degree). The dummy neck showed much stiffer responses in both flexion and extension during air bag loading than would be exhibited in seatbelt loading or seatback loading.

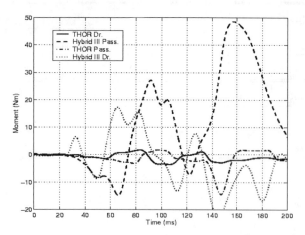

Figure 18. Neck Moment Comparison of THOR Dummy and Hybrid III Dummy [10].

The differences in head/neck response in the air bag tests and the biomechanics corridors could be due to the difference in the bending modes that occur in the two kinds of tests. Recall that the response corridors were solely generated due to the restraint by the belt or the back of the seat depending upon whether it is a forward or a rear impact respectively. Head motion loads the neck like a cantilever beam with a mass at the free end. No external forces on the head, such as those due to a deploying air bag, are present. In this case, the neck response is primarily due to head inertial loading, and the neck experiences first mode bending. During air bag loading, significant loads are applied directly to the head, which causes the dummy neck to experience second mode bending. The neck deforms in either a reflected S-shape or an S-shape due to combined upper and lower neck bending moments (Figures 10, 12 & 15). Thus the corridors based on first mode bending are not appropriate for other loading conditions such as air bag applications. The corridors do not reflect the response in other modes of neck deformation, e.g., the second mode of bending.

In live occupants, the order of the neck bending mode would depend on (1) the loads applied to the head and neck, (2) the geometric structure of the neck vertebrae and ligaments, and (3) the effects of passive and active muscle tone. Additional biomechanical data are needed to address the unique characteristics of neck muscle loading, head angular position and multiple load paths to the head and neck in the air bag testing environment. In case of compressive loading of the neck combined with shear loading, it would be possible to envisage a second mode bending in a live occupant. However, in case of air bag trapping under the chin of a live occupant, the air bag membrane forces would lead to tensile forces in the muscle pairs. The head would rotate and align the muscle pairs with the neck loads, resisting them by generating tensile stresses in the muscle pairs. During this process it is unlikely that moments of the kind observed in the Hybrid III dummy would be generated. As the live occupant's head continues to rotate, the air bag would be allowed to escape before large moments around the occipital condyles are likely to be generated. Conversely, the head rotation in the Hybrid III dummy, in cases where the air bag is trapped in the jaw-neck cavity, is resisted by generating local moments in the neck near the occipital condyle through the nodding blocks. This is possible because the Hybrid III neck is essentially a beam like structure, which is different from the load resisting structural system in a human neck that depends on muscle pairs to perform the same task.

An important factor, which could contribute to the observed head/neck response seen in the air bag tests is the effective stiffness of the occipital condyle joint in the dummy's neck. The human occipital condyle joint appears to have considerable laxity, which allows it to experience significant rotation before it can sustain a substantial moment across the joint [6]. However, the current Hybrid III neck exhibits considerable bending resistance at its occipital condyle joint. This lack of compliance may allow large moments to be transmitted to the neck by the head without significant relative motion. In a human subject, motion and resistance to motion of the neck is accomplished through muscle pairs, which are attached to the skull, the individual vertebra, and the torso. These muscle pairs respond in various group actions to produce the desired movement of the head and neck. The muscle tones are simulated in the dummy through a pair of rubber nodding blocks and four rubber neck-discs. Nightingale, et al., [7] studied the effects of upper neck joint stiffness on measured moments in the Hybrid III dummy during air bag loading using MADYMO occupant simulations. The standard dummy model was modified to simulate the axial and rotational stiffness of the ligamentous human cervical spine (no muscles). They found that decreasing the rotational stiffness had a dramatic effect on the extension moment.

The laxity of the human occipital condyle joint was studied by V. Goel, et al. [8]. They characterized the effect of muscle pairs on the motion and resistance to motion of the head and neck until the limit of range of motion of the occipital condyle is reached. The Hybrid III dummy response appears to capture the global moment (which includes the moment at the occipital condyle as well as the moment supported by the muscle pairs, the vertebrae and the ligaments) for the first mode bending. However, it has no feature, which captures the local bending moment at the occipital condyle alone. Therefore, the joint laxity and moment are not represented in the Hybrid III. To capture the global moment and head motion for the first mode bending, nodding blocks are used at the dummy's head-neck interface. These nodding blocks are designed to ensure that the Hybrid III neck response stays within the Mertz et al., [5] corridors in first mode bending. In the human neck, different bending modes are possible depending upon (1) the loads applied to the head and neck, (2) the geometric structure of the neck vertebrae and ligaments, (3) the degree of passive and active muscle tone, and (4) the bending resistance of the occipital condyle. Since in the Hybrid III, the nodding blocks coupled with the neck represent neither the geometric load path produced by muscle pairs, vertebrae and ligaments nor the response of the occipital condyle joint, it is unlikely that it will give good insights into the local moment at the occipital condyle.

By comparison, the NHTSA advanced dummy, THOR, allows significant head rotation without substantial moment in the occipital condyle joint [9]. Comparing the neck response of THOR dummy and Hybrid III dummy in vehicle crashes, the magnitude of the bending moment at the occipital condyle joint in THOR dummy was approximately 1/6 of the Hybrid III for both driver and passenger (Figure 18). This is one possible solution to the neck artifacts seen in the Hybrid III. However, the THOR is a new dummy that has not been evaluated thoroughly. Some amount of trapping of the air bag under the chin could happen in a live human. However, the large moments with small angles would not. The head would rotate and the air bag would free itself before large moments would develop.

To eliminate the effect associated with the air bag trapping under the chin or behind the jaw, an experimental modification was made to the Hybrid III 5th percentile female dummy's head skin using parts from the jaw/neck of a Hybrid II 50th percentile dummy's head skin. With this modification, all responses to the air bag loading were similar to the air bag-neck interaction Mode 1. It is possible that the neck-skin (part of Hybrid II 50th dummy) could be forced up against the neck and form an additional load path reducing the tension force measured by the neck load cell. If this were the case, then the bending mode would still be in extension with a possible reduction in magnitude. Instead, the mode is primarily in flexion-compression, which cannot be explained by the hypothetical load path. In addition, response of the dummy in interaction Mode 1 (the air bag loading the head) without the experimental modifications are slightly less than the response with the modifications in terms of forces and moments, although the results are not statistically significant (Table.1, 2). This is believed to be the result of the dynamics of the air bag. At the end of the deployment process, the gas forces will drive the air bag into a maximum volume/minimum pressure situation for a given air bag system. When this happens, the air bag as a result of membrane forces will tend to "bridge the gap" between the head and the chest. In case of head skin without modification this would result in very little force on the neck, positioning the center of pressure below the center of mass of the head, producing a flexion response. In case of the modified head skin, some forces could be transmitted through the neck, resulting in higher

measurements at the load cell, although not changing considerably, the location of the CG, and thus the flexion response. The experimental modifications discussed in this paper were primarily for studying the influence of design changes on the air bag-neck interaction modes. These modifications do not constitute recommended design changes for the Hybrid III 5th percentile female. Far more detailed investigation and design effort would be necessary for that.

A standard system and a depowered air bag system were both used in this test series. All the three typical air bag-head/neck interaction modes were observed with both air bags. Both air bags produced similar neck loads in air bag-neck interaction Mode 1, but the depowered air bag created higher neck extension moments in air bag-neck interaction Modes 2 and 3. While this indicates that the neck response is not necessarily a function of air bag power for depowering in this range, differences in response are expected for more extensively depowered systems. Due to the limited sample size, no conclusion can be drawn regarding the effects of the air bag power on the head/neck response.

CONCLUSIONS

This paper reports a limited study that analyzed the response of the Hybrid III neck to air bag loading. Only in-position occupants were addressed and the results may not be appropriate for occupants that are on or near the air bag model, i.e., out-of-position occupants. The study focused on one area of air bag loading under very specific and artificial conditions. More investigation is needed before results can be generalized. However, the results indicate that changes may be needed in the design of the Hybrid III family of dummies to accurately represent the response of human subjects subjected to air bag loading.

The test series presented in this paper indicates that the dummy head/neck response depends strongly on the manner in which the air bag interacts with the dummy's head. If the air bag is not trapped in the chin-jaw region or behind the jaw, in the jaw-neck cavity, the head is pushed rearward, resulting in neck flexion. If the air bag is caught in the chin-jaw region, the head is pulled forward and upward w.r.t the neck, subjecting the neck to extension and tension to some degree. If the air bag is entrapped behind the jaw in the jaw-neck cavity, the head is again dragged forward and upward, subjecting the neck to extension and tension to a higher degree. The SAE recommended TMJ head skin and the neck shield do not prevent or impede air bags from being trapped behind the jaw or under the chin. The TMJ skin was coated with colored chalk in two regions of the interior surface, not accessible to the air bag without reaching inside. The vertical surface behind the chin was coated with one color and the horizontal surface of the skin spanning the jaw lines was coated with another color. In cases showing high extension-tension, accompanied by shear, indicating the head being pulled forward w.r.t the neck, the marks of the colored chalk from inside the chin jaw region were observed on the air bag. The colored chalk marks on the air bag also indicated the extent to which the air bag had entered the head skin behind the chin and the jaw lines. However, two different head/neck structure modifications developed in this study demonstrate experimental means of preventing the air bag from being trapped in the chin-jaw region or in the jaw-neck cavity. They have not been studied extensively and do not appear to be applicable for general use at this time.

Even if interactions due to the air bag being trapped behind the jaw or under the chin are prevented, the dummy head/neck response due to interaction with a deploying air bag may be different than that would be expected in a live human occupant. The influence of neck vertebrae, ligaments and muscle tone on the bending of neck is not known. Additional biomechanical data related to bending behavior of human neck, subjected to airbag loads is needed.

Air bag tests result in pronounced neck bending moments in the dummy with little head rotation relative to the torso, a phenomenon which is unlikely to occur in live human subjects. In addition, the global neck moments in the Hybrid III will not necessarily give good insights into the local moments at the occipital condyle for conditions other than the first bending mode of the neck.

In the tests presented in this paper, the dummy neck was subject to second mode bending during air bag deployment. Currently available biomechanical flexion and extension response corridors of Hybrid III dummy neck do not appear applicable to the neck response in air bag tests. Biomechanical data are needed to address the neck responses during air bag loading to accurately assess the injury potential to human subjects. In addition, biomechanical data are needed to improve the mechanical characterization of the human chin/neck area so that dummies, which more accurately characterize air bag interaction, can be designed.

REFERENCES

1. Melvin, J., Horsch, J., McCleary, J., Wideman, L., Jensen, J., and Wolanin, M., "Assessment of Air Bag Deployment Loads with the Small Female Hybrid III Dummy". Proceedings of the 37th Stapp Car Crash Conference, pp. 121-132, SAE Paper 933119, 1993.
2. Mertz, H. and Patrick, L., "Strength and Response of the Human Neck". Proceedings of the 15th Stapp Car Crash Conference, SAE paper 710855, 1971.
3. Morris, C. R., Zuby, D. S. and Lund, A. K., "Measuring Air Bag Injury Risk to Out-of-Position Occupants", paper 98-S5-O-08, Proceedings of the 16th International Technical Conference on the Enhanced Safety of Vehicles, May 31 – June 4, 1998, Windsor, Canada.

4. Nusholtz, G., Wu, J., and Kaiker, P., "Passenger Air Bag Study Using Geometric Analysis of Rigid Body Motion". Experimental Mechanics, September, 1991.
5. Mertz, H., Irwin, A., Melvin, J., Stalnaker, R., and Beebe, M., "Size, Weight and Biomechanical Impact Response Requirements for Adult Size Small Female and Large Male Dummy". SAE paper 890756, 1989.
6. Snyder, R., Chaffin, D., and Foust, D., "Bioengineering Study of Basic Physical Measurements Related to Susceptibility to Cervical Hyperextension-Hyperflexion Injury". UM-HSRI-BI-75-6 (Research Report), 1975.
7. Nightingale, R., Kleinberger, M., and Myers, B., "Effects of Upper Neck Joint Stiffness on Measured Moments in the Hybrid III Dummy During Air Bag Loading". DTNH22-94-Y-07133 (Research Report), 1994.
8. Goel, V.K., Clark, C.R., Gallaes, K. and Liu, Y.K., "Moment-Rotation Relationships of the Ligamentous Occipital-Atlanto-Axial Complex". J. Biomechanics, Vol.21, No. 8, pp. 673-680, 1988.
9. Xu, L., et al., "Comparative Performance Evaluation of THOR and Hybrid III". SAE paper 2000-01-0161, 2000.

VI. FIELD PERFORMANCE
OF AIR BAGS

880400

Restraint Performance of the 1973-76 GM Air Cushion Restraint System

Harold J. Mertz
Safety & Crashworthiness Systems
General Motors Corp.

COPYRIGHT _1988_
SOCIETY OF AUTOMOTIVE
ENGINEERS, INC.

ABSTRACT

Case reviews are given of deployment accidents of the GM 1973-76 air cushion restraint system where the occupant injury was AIS 3 or greater. Many of these injuries occurred in frontal accidents of minor to moderate collision severity where there was no intrusion or distortion of the occupant compartment. Dummy and animal test results are noted that indicate that these types of injuries could have occurred if the occupant was near the air cushion module at the time of cushion deployment. An analysis is given that indicates that for frontal accidents a restraint effectiveness of 50 percent in mitigating AIS 3 or greater injuries might be achieved if an air cushion system can be designed which would not seriously injure out-of-position occupants while still providing restraint for normally seated occupants.

GENERAL MOTORS EQUIPPED 11,321 vehicles with driver and front seat passenger air cushion restraint systems (ACRS) during the 1973 through 1976 model years. The first cars equipped with the ACRS were 1000 1973 Chevrolet Impalas. This initial effort was called the Field Trial Program and the cars were identified as FTP cars. During the 1974 through 1976 model years, 10,321 Buicks, Oldsmobiles and Cadillacs were produced. For these cars, the ACRS was a customer option available on Buick Le Sabre, Electra and Riviera; the Oldsmobile 88, 98 and Toronado; and the Cadillac Deville, Brougham, Fleetwood and Eldorado. The ACRS option was priced on the basis of a projected high sale volume. This level of customer demand never developed and the ACRS option was cancelled at the end of the 1976 model year.

DESCRIPTION OF THE 1973-76 GM ACRS

The 1973-76 GM ACRS consisted of driver and passenger (center and right front) systems, two crash sensors, a large capacitor to provide ener-

gy to deploy the driver and passenger systems in the event that battery power was lost during the collision event, and a diagnostic system to provide the driver with information about the ACRS readiness to deploy. The driver air cushion module was mounted to a specially designed steering wheel and energy absorbing column. A fixed knee bolster was used to provide lower torso and leg restraint. The passenger air cushion module was mounted in the lower part of the instrument panel in front of the right front passenger seating position. It was designed to restrain both the right front and center front passengers, either singularly or together. Lower torso and leg restraint was provided by a separate high pressure, low volume knee cushion which was deployed inside of the larger volume, lower pressure head/torso cushion. Manual lap belts were provided for each front seat occupant position. A crash recorder was installed to provide data to determine (i) if the crash was below the deployment threshold, (ii) if the crash started before the cushion was deployed, and (iii) if the crash severity exceeded the severity of a 30 mph frontal, rigid barrier collision. Detailed descriptions of the various components of the 1973-76 GM ACRS are given in papers by Campbell(1)*, Klove and Oglesby(2), and Louckes, et al(3).

DEVELOPMENT AND VALIDATION OF THE 1973-76 GM ACRS

The 1973-76 GM ACRS was subjected to extensive development and validation test programs. A variety of sled tests was conducted to assure that the ACRS would provide restraint for different occupant sizes (5th percentile female, 50th percentile male and 95 percentile male), different seat positions (full forward, mid and full rear), different combinations of front seat

---

* Numbers in parentheses refer to papers listed under References

passengers (right front, right and center front, and center front) and different inflator environmental temperatures (180° F, 72° F and -30° F).

Human volunteer sled tests were conducted with both the driver system at Southwest Research Institute(4) and the passenger system at Holloman Air Force Base(5). In both test programs, the volunteers were seated in normal occupant posture and the cushions were deployed before the occupants had moved significantly forward relative to the body buck in response to the sled acceleration. Tests were conducted with increasing simulated collision severity with the most severe for each system being the 30 mph frontal, rigid barrier collision simulation. The restraint performance for the center front occupant was not evaluated with human volunteers. No significant injuries occurred. Only minor abrasions were noted.

For an early version of the passenger system (a pre FTP system), a series of static deployment tests was conducted to evaluate the effect of the interaction between the deploying cushion and the occupant. In these tests, a volunteer was seated on the seat and leaned forward, in various degrees, toward the instrument panel. The passenger cushion was deployed without the sled being accelerated. The test series was terminated after the volunteer experienced a slight concussion in a test where he was leaning forward with a torso angle of 29 degrees past vertical(5).

An animal test program to evaluate the effect of the interactions between the deploying passenger cushion and children who may be near the instrument panel at time of deployment was conducted at Wayne State University(6). Anesthetized baboons were placed in various positions near the instrument panel and various passenger system concepts were deployed while the body buck remained stationary. Test results indicated that significant injuries to the animals could be produced if the cushion was inflated too rapidly. Based on these results, the variable rate inflator concept used in the 1973-76 GM ACRS passenger system was developed. Subsequent anesthetized baboon and chimpanzee tests of the 1973-76 GM passenger ACRS produced no significant injuries for the animal positions and gas inflation rates evaluated. (Note: The results of the animal tests of the 1973-76 GM passenger ACRS are not described in the Patrick and Nyquist(6) paper.)

A series of cadaver tests was conducted at Wayne State University to evaluate the effect of the interaction between the deploying trim cover panel of the passenger air cushion module and the legs of front seat occupants. The tests were conducted in a stationary body buck with the cadavers seated in a normal seating posture. Their legs were aligned squared to the passenger module trim cover panel and the cushion was deployed. No lower extremity fractures occurred with the 1973-76 GM passenger ACRS.

To assess the efficacy of the crash sensing system, a variety of car-to-car and car-to-obstacle collisions was staged. These tests and their results are described in a paper by Wilson and Piepho(7). The tests included 30 mph and 40 mph frontal rigid barrier impacts; right and left 30 mph, 30 degree angle rigid barrier impacts; 30 mph offset rigid pole impacts; 30 mph frontal impact with a bumper underride; and car-to-car impacts involving front of ACRS car to side of car, front of ACRS car to rear of car and a multiple impact scenario. In the majority of these tests, the Hybrid II dummy responses were well below the FMVSS 208 limits. The main exception was the 40 mph, rigid barrier tests where none of the Hybrid II dummies (driver, center front, right front) met all the limits. This test result indicated that in the more severe accidents, the 1973-76 GM ACRS would not mitigate all serious occupant injury.

The results for FMVSS 208, 30 mph, frontal rigid barrier tests are given in Table 1. Note that the 1973-76 GM ACRS values were well below the compliance limits defined by FMVSS 208 for passive restraint systems. These results, plus the results obtained from the human volunteer, animal, cadaver, dummy sled tests and full scale vehicle tests, indicated that the 1973-76 GM ACRS should have provided reasonably effective occupant restraint for a variety of real-world, frontal accident configurations and severities.

RESTRAINT EFFECTIVENESS OF THE 1973-76 GM ACRS

A note was put in the owner's manual of all vehicles equipped with the 1973-76 GM ACRS requesting that GM be notified of any accident involving the vehicle. A toll-free telephone number was given for the purpose of communicating this information. In addition, agents of GM Motors Insurance Corporation (MIC) and GM car dealers were asked to report the occurrence of accidents involving the ACRS vehicles. For each deployment accident reported, an investigator from the GM Field Accident Review group was sent to inspect the vehicle. Photographs were taken of the exterior and interior damage. Police reports were reviewed for accident descriptions and occupant injuries. If agreeable, occupants of the ACRS vehicles were interviewed and medical records reviewed. The severities of the injuries were rated according to the AAAM Abbreviated Injury Scale (AIS). For each reported accident, a report was written summarizing the pertinent observations.

A study was done by Pursel et al(8) to estimate the restraint effectiveness of the 1973-76 GM ACRS in mitigating AIS 2 and greater and AIS 3 and greater injuries. For each reported deployment accident, a search was done of the GM Motors Insurance Company (MIC) files for comparable accidents involving non-air cushion restrained occupants. Comparable MIC accident cases were selected on the basis of collision condition and severity, exterior vehicle damage, and occupant sex, age, and seating position. Percent occurrences of AIS = 0 or 1, AIS = 2, AIS = 3 or 4 and AIS = 5 or 6 were calculated for the ACRS deployment cases that were matched and for the corresponding non-air cushion restraint cases. The ACRS effectiveness was ob-

620

Table 1 - Results of FMVSS 208, 30 mph,
Frontal Barrier Tests of the 1973-76
GM ACRS Using Hybrid II Dummies(7)

	HEAD HIC	CHEST ACC. (3 ms, G)	LT. FEMUR (1b)	RT. FEMUR (1b)
Test C-3352 **(Lap Belts Not Used)**				
Driver	340	40	1420	1270
C. Front	320	45	470	1210
Rt. Front	360	44	930	1110
Test C-3353 **(Lap Belts Not Used)**				
Driver	310	36	1330	1180
C. Front	370	44	540	1190
Rt. Front	490	45	660	1170
Test C-3321 **(Lap Belts Not Used)**				
Driver	590	50	1250	1230
R. Front	260	40	910	1010
Test C-3094 **(Lap Belts Used)**				
Driver	370	47	1230	500
R. Front	430	42	660	760

Table 2 - Estimates of Percent Effectiveness
of the 1973-76 GM ACRS in Mitigating
AIS 2 or Greater and AIS 3 or Greater
Injuries in Frontal Deployment
Accidents

	MITIGATING AIS = 2 OR GREATER INJURIES	MITIGATING AIS = 3 OR GREATER INJURIES
1973-76 GM Driver ACRS	18%	21%
1973-76 GM Passenger ACRS	-34%	16%
1973-76 GM Combined ACRS	6%	18%

tained by subtracting the percent of deployment accident occupants that experienced a given AIS level from the percent of non-deployment accident occupants that experienced the same AIS level. This difference was divided by the latter percentage to give the ACRS effectiveness. The combined effectiveness estimates of the driver and passenger system for mitigating AIS 2 or greater injuries and AIS 3 or greater injuries were 6 and 18 percent, respectively. Effectiveness estimates for the driver and passenger systems separately were calculated using the same procedure. All effectiveness estimates of the GM 1973-76 ACRS are given in Table 2. Note these effectiveness estimates are for frontal, deployment type accidents and do not assess the ACRS effectiveness in non-deployment type accidents such as roll overs, side impacts and rear end collisions. If these non-deployment type accidents were included, the overall effectiveness would be lower.

The effectiveness of both the driver and passenger systems in mitigating AIS 3 or greater injuries was less than expected when compared to the "no significant injuries" results obtained in the human volunteer programs. The negative 34 percent effectiveness for mitigating AIS 2 or greater injuries for the passenger system was quite disturbing since the implication is that overall the ACRS passengers experience more AIS 2 or greater injuries than their matched case, non-air cushion restrained counterparts.

MODERATE TO SEVERE DEPLOYMENT ACCIDENTS WHERE THE OCCUPANT WAS PROTECTED

A case-by-case study was done of each ACRS deployment accident. At the time of that review (December 1980), there were 216 deployment accidents in the 1973-76 GM ACRS file (FTP cases through AC-3020 and 1974-76 cases through LA350). These accidents involved 216 drivers, 86 right front occupants and 11 center front occupants. Of the 97 front seat passengers, thirteen (13/97 = 13.4%) were children whose ages were ten years or less.

Based on analyses of collision descriptions and photographs of exterior and interior damage, each deployment accident was classified in terms of collision severity and interior intrusion/-distortion severity. Three collision severity classes were used: minor, moderate and severe. An accident was judged as minor collision severity if the exterior damaged involved only front-end sheet metal (hood and/or fenders), grille and/or the bumper. For an accident to be judged as moderate collision severity, some rearward displacement of the engine and/or wheels had to occur and the estimated speed of the accident had to be greater than 15 mph. If there was extensive damage to the vehicle and the accident speed was estimated to be greater than 25 mph, then the accident was classified as severe. Two classifications of interior intrusion/distortion were used: significant and non-significant. If the interior intrusion/distortion occurred in the vicinity of a front seat

occupant and if it was difficult to conceive how any air cushion could provide protection from the intrusion/distortion, then the accident was judged as having significant occupant compartment intrusion and/or distortion.

The majority of the 216 deployment accidents were classified as minor. None of these accidents had significant occupant interior intrusion or distortion. Several of the severe and moderate collision accidents did have significant occupant interior intrusion/distortion. In these accidents it was difficult to conceive how any air cushion system could have mitigated the injuries experienced by the occupants.

There were 14 accidents where the collision severities were classified as moderate or severe, there was no significant occupant interior intrusion or distortion, and the occupant injuries caused by the restraint forces applied by the ACRS were of minor consequence. A summary of the accident conditions, collision severities and occupant injuries for these accidents is given in Table 3. Note that with the exception of one driver injury, the injury severities for these ACRS occupants were AIS 2 or less with the majority being AIS = 1. The exception was a driver who experienced a bruised kidney which was rated as AIS = 3. The kidney injury could have been caused by the vertical loading associated with the car going into a 10 foot deep ravine. Even if the kidney injury occurred during the principal front collision event, the restraint performance of the 1973-76 driver ACRS would be judged as effective because the matched case drivers all experienced AIS = 4, 5 or 6 injuries. These 14 accidents demonstrate the occupant restraint potential of an air cushion system in moderate to severe frontal accidents when there is no significant intrusion or distortion of the interior in the vicinity of the front seat occupants.

1973-76 GM ACRS DEPLOYMENT ACCIDENTS WITH UNEXPECTED OCCUPANT INJURIES

DEPLOYMENT ACCIDENTS WITH AIS 3 OR GREATER INJURIES - Twenty-three front seat occupants (23/313 = 7.3%) experienced AIS 3 or greater injuries: fourteen drivers (14/216 = 6.5%) and nine front seat passengers (9/97 = 9.3%), three of whom were children. Pertinent information concerning the collision conditions and occupant injuries is given in Table 4. There were five fatalities, four drivers and a one-month-old child who was lying on the seat cushion prior to the accident. Four occupants, all right front passengers, experienced AIS = 4 injuries, all related to leg fractures. Ten drivers and four passengers experienced AIS = 3 injuries. These injuries consisted of fractures to the ribs, arms, legs, and lumbar vertebra; a concussion; a persistent neurological impairment and a bruised kidney.

An analysis of the accident data indicated that in ten of these twenty-three cases there was either significant intrusion/distortion of the interior in the vicinity of the occupant

TABLE 3 - SUMMARY OF GM 1973-76 ACRS DEPLOYMENT ACCIDENTS OF MODERATE TO SEVERE COLLISION SEVERITY, NO SIGNIFICANT OCCUPANT INTERIOR INTRUSION OR DISTORTION, AND RESTRAINT PROTECTION PROVIDED BY THE ACRS

CASE NO.	COLLISION INFORMATION			OCCUPANT INFORMATION				
	SEVERITY	DESCRIPTION	POS.	AGE	SEX	AIS	INJURY DESCRIPTION
LA 350	SEVERE	HEAD-ON WITH SECOND CAR	DR	16	M	3	BRUISED KIDNEY
AC 356	SEVERE	UNDERBODY TO RAILROAD TRACKS FOLLOWED BY RT. FRONT POLE IMPACT	DR	14	F	1	BROKEN NOSE, CUT LIP
			RF	14	F	1	ABRASION TO HAND
LA 36	SEVERE	PARTIAL HEAD-ON WITH SNOW PLOW, BIASED TO DRIVER'S SIDE	DR	41	M	1	BRUISED NOSE, CHEST, KNEE
			RF	35	F	1	FACIAL BRUISES
LA 250	SEVERE	PARTIAL HEAD-ON WITH SECOND CAR, BIASED TO DRIVER'S SIDE	DR	36	F	1	BRUISED KNEES, FOOT, ELBOW
			RF	21	F	1	BRUISED TIBIAS
LA 96	SEVERE	PARTIAL HEAD-ON WITH BUS	DR	62	M	2	BROKEN RIB, BRUISED TIBIA
AC 3012	MODERATE	UNDERBODY TO RAILROAD TRACKS, LT. FRONT POLE IMPACT	DR	18	M	1	DAZED, FACIAL BRUISES
LA 265	MODERATE	LT. FRONT TREE IMPACT	DR	69	F	2	SPRAINED ELBOW, KNEE BRUISES.
LA 307	MODERATE	LT. FRONT IMPACT TO CONCRETE TRAFFIC SIGNAL BASE	DR	21	M	1	BRUISED KNEE, CHIN ABRASION
			RF	20	F	2	SPRAINED ANKLE, BRUISES
AC 1504	MODERATE	PARTIAL HEAD-ON WITH SECOND CAR, BIASED TO DRIVER'S SIDE	DR	18	M	1	CUT ON BACK OF HEAD
			RF	19	M	1	DAZED
LA 85	MODERATE	LT. FRONT IMPACT TO TREE	DR	16	M	1	BRUISED KNEE
			RF	15	M	1	BRUISED KNEES
LA 275	MODERATE	HEAD-ON WITH SECOND CAR	DR	38	F	1	BRUISED CHEST, KNEES, ARM, HAND
			RF	38	F	2	SPRAINED KNEE
LA 174	MODERATE	IMPACT TO SIDE OF CAR	DR	20	M	1	FACIAL BRUISES
			RF	18	M	0	NO INJURIES NOTED
LA 234	MODERATE	PARTIAL HEAD-ON WITH CAR	DR	49	M	1	BRUISED CHEST, CHIN, ARMS, HANDS
LA 235	MODERATE	IMPACT TO SIDE OF CAR FOLLOWED BY POLE IMPACT	DR	68	M	1	SCALP CUT, ARM ABRASIONS

TABLE 4 - SUMMARY OF FRONT SEAT OCCUPANTS WITH AIS 3 OR GREATER INJURIES
IN DEPLOYMENT ACCIDENTS INVOLVING THE 1973-76 GM ACRS

CASE NO.	COLLISION DATA		OCCUPANT INFORMATION					INJURY EXPECTED
	SEVERITY	SIGNIFICANT INTRUSION	POS.	AGE	SEX	AIS	MOST SEVERE INJURY	
LA 130	SEVERE	YES	DR	29	M	6	SEVERE HEAD AND CHEST INJURIES	YES
LA 303	SEVERE	YES	DR	29	M	6	SEVERE CHEST INJURIES	YES
LA 343	SEVERE	YES	DR	81	M	6	MASSIVE HEAD INJURIES	YES
LA 324	SEVERE	YES	DR	49	M	3	FRACTURED ARM AND FEMUR	YES
LA 350	SEVERE	NO	DR	16	M	3	BRUISED KIDNEY	YES
AC 1291	MODERATE	YES	DR	39	M	3	FRACTURED RIBS AND ANKLE	YES
LA 23	MODERATE	YES	DR	42	M	3	FRACTURED RIBS	YES
LA 46	MODERATE	NO	DR	58	M	3	FRACTURED LUMBAR VERTEBRA	YES
LA 128	MODERATE	NO	DR	41	M	6	POSSIBLE CHEST/NECK INJURY	NO
LA 8	MODERATE	NO	DR	46	M	3	FRACTURED RIBS	NO
LA 74	MINOR	NO	DR	45	M	3	FRACTURED RIBS	NO
LA 126	MINOR	NO	DR	24	F	3	UNCONSCIOUS FOR 25 MINUTES	NO
LA 152	MINOR	NO	DR	46	F	3	FRACTURED ARM	NO
LA 165	MINOR	NO	DR	56	F	3	FRACTURED CLAVICLE AND RIBS	NO
LA 180	MODERATE	YES	RF	10	M	4	FRACTURED FEMUR	YES
LA 53	MODERATE	NO	RF	65	F	3	FRACTURE OF LUMBAR VERTEBRA	YES
AC 316	MODERATE	NO	RF	1 Mo	M	6	SEVERE HEAD INJURY	NO
LA 173	MODERATE	NO	RF	34	F	4	FRACTURED FEMUR AND ANKLE	NO
LA 195	MODERATE	NO	RF	40	F	4	FRACTURES OF BOTH LEGS	NO
LA 294	MINOR	NO	RF	59	F	4	FRACTURES OF BOTH LEGS	NO
LA 310	MODERATE	NO	RF	21	F	3	FRACTURES OF TIBIA AND FIBULA	NO
LA 246	MODERATE	NO	RF	4	M	3	PERSISTENT NUMBNESS LEFT SIDE	NO
LA 170	MINOR	NO	RF	75	M	3	FRACTURED TIBIA	NO

and/or the injuries were produced by non-frontal collision forces. In these cases it was judged that no air cushion restraint system could have mitigated these injuries. These cases, involving eight drivers and two passengers, are classified as injury expected in Table 4. For the thirteen other cases, the collision severities were minor or moderate. The collision forces were primarily frontal and there was no significant intrusion or distortion of the interior. Since an air cushion system is expected to provide occupant protection in such accidents, these injuries were more severe than expected. These thirteen cases, involving six drivers (6/216 = 2.8%) and seven front seat passengers (7/97 = 7.2%, two of whom were children), are classified as occupant injuries more severe than expected in Table 4. It is these occupant injuries which reduced the effectiveness of the 1973-76 GM ACRS for AIS 3 or greater injuries.

SUMMARY OF AIS 2 OR GREATER INJURIES THAT WERE MORE SEVERE THAN EXPECTED - The remainder of the deployment accident cases were reviewed to identify all occupants who experienced neurological problems; significant thoracic organ injuries including rib fractures; fractures of the hand, arms, legs, pelvis and vertebrae; and significant abdominal organ injuries that were classified as AIS 2. In each case that was identified, the accident data were reviewed in the same manner as was described for the AIS 3 or greater cases and a judgment was made as to whether or not the injury was more severe than expected. A summary of all AIS 2 or greater injuries which were judged as more severe than expected is given in Table 5 including those injuries noted in Table 4. Twenty drivers (20/216 = 9.3%) and seventeen passengers (17/97 = 17.5%) experienced injuries that appeared to be more severe than expected. These injuries were the principal reason why the effectiveness estimates of the 1973-76 GM ACRS noted in Table 2 were lower than expected.

POSSIBLE CAUSES OF UNEXPECTED INJURIES

By definition, the only injuries noted in Table 5 were to occupants involved in accidents of minor to moderate collision severity where the collision force was primarily frontal and where there was no significant intrusion or distortion of the car interior in the vicinity of the occupant. In such accidents, only minor injuries to the occupants were expected provided the occupant was not near the air cushion module at the time of deployment. This expectation is based on the fact that human volunteers experienced either no injuries or only minor abrasions in similar simulated collision environments while being restrained by the ACRS(4, 5). On the other hand, if the occupant was near the air cushion module at the time of deployment, significant forces would be developed between the deploying cushion and the occupant. Human volunteer and animal evaluation of such out-of-position occupant/cushion interactions indicated the potential for significant injuries(5, 6). Anal-

ysis of the accident data for cases noted in Table 5 indicated that both the fatally injured driver and the one-month-old passenger may have been close to the air cushion module at the time of deployment. The legs of the passengers with leg fractures were likely close to the passenger air cushion module at the time of deployment since the module is located in the lower part of the instrument panel, directly in front and close to the legs of the passenger. Consequently, it was hypothesized that most of the injuries noted in Table 5 were due to the occupant being close to the air cushion module at the time of deployment and interacting with the deploying cushion. To investigate the nature of these interactions a series of dummy and animal tests was conducted. The following is a summary of these tests.

DRIVER CHEST INJURIES - Horsch and Culver(9) conducted a series of simulated frontal collisions of moderate severity (23 mph, 14 G) sled tests of the 1973-76 driver ACRS using a Hybrid III dummy. The dummy was seated in a normal driving posture and was allowed to slide forward on the seat in response to the simulated collision pulse. The air cushion was deployed when various spacings between the dummy's chest and the cover of the driver's inflator module were reached. These spacings included the condition where the driver's chest had impacted the module cover and was compressing the column when the air cushion was deployed. For comparison purposes, two tests were conducted where the dummy was leaning forward with its chest against the module prior to subjecting it to the sled pulse. In one of these tests, the air cushion was deployed. In the other test, the cushion was not deployed.

Pertinent results from these tests are given in Table 6. When the dummy was initially in a normal driver's posture and the cushion was deployed prior to the dummy moving close to the module (Test 1), the maximum chest deflection experienced by the dummy was only 0.7 inch. However, in those tests where the dummy's chest was against the module at deployment (Tests 4, 5 and 6), the dummy's sternum bottomed out on its spine box and a maximum chest compression of 3.3 inches was recorded in each instance. According to Neathery et al(10), this level of thoracic compression is indicative of life threatening thoracic injury.

The added effect of deploying the air cushion when the chest is against the module is demonstrated by comparing the results of Tests 6 and 7. For these tests, the dummy was leaned forward with its chest against the module prior to subjecting it to the sled acceleration. The air cushion was deployed in Test 6 and the dummy's sternum was bottomed out on its spine (3.3 inches). In Test 7, the cushion was not deployed and the maximum chest deflection was only 1 inch. These results clearly demonstrate that the 1973-76 driver ACRS had the potential to produce the various thoracic lesions noted in Table 5 if the drivers were near the inflator module when the cushion was deployed. Although

Table 5 - Summary of Injuries to 1973-76 GM ACRS Occupants
That Were More Severe Than Expected

Body Region	Driver Injuries	Passenger Injuries
Head/Neck	3 Drivers Concussed 1 AIS = 3, Unconscious 25 M. 2 AIS = 2	1-Month-Old Baby With Subdural Hematoma, AIS = 6 4-Year-Old With Persistent Numbness, AIS = 3
Arm/Hand	9 Drivers With Fractures (9/216 = 4.2%)	6 Passengers With Fractures (6/97 = 6.2%)
Thorax	Possible Fatal Chest Injury, No Autopsy 5 Drivers With Rib Fractures (3 AIS = 3, 2 AIS = 2)	None
Abdomen	None	None
Leg/Pelvis	2 Drivers With Fractures (2/216 = 1.1%)	9 Passengers With Fractures (9/97 = 9.3%)

Horsch and Culver did not evaluate the potential for head injuries, it is not difficult to envision the potential to produce head/neck injuries if the head is close to the module when the cushion is deployed.

CHILD INJURIES - Anesthetized pig tests conducted by Chambers University for Volvo[11] indicated that fatal liver injuries could be produced if the pig was oriented with the region of the abdomen containing the liver aligned with the path of the deploying cushion. These results suggested that the animal positions used by Patrick and Nyquist[6] to evaluate the out-of-position child injury concern for the 1973-76 GM ACRS may not have been the most critical positions. To evaluate this possibility, a second series of animal sled tests was conducted using anesthetized pigs and baboons[12]. In these tests, the animals were positioned near the 1973-76 GM ACRS passenger inflator module with either their head, chest or abdomen placed in the path of the deploying cushion. The sled was subjected to a variety of simulated collision pulses and the cushion was deployed. Life threatening injuries to the brain, cervical spine, heart and liver were produced. These results suggest that children who were near the 1973-76 GM ACRS passenger air cushion module when it was deployed could have experienced life threatening injuries as well. These results are entirely consistent with the injuries experi-

enced by the two children noted in Table 5 and suggest that the children may have been close to the air cushion module when the cushion was deployed.

PASSENGER LEG FRACTURES - The relative high frequency of passenger leg fractures (9/97 = 9.3%) which occurred in minor to moderate collisions without significant occupant compartment intrusion was quite unexpected because the axial compressive femur loads that were measured in the 30 mph, frontal barrier tests (Table 1) were quite low, ranging from 470 to 1210 pounds.

Seven of the nine fractures were to the tibia suggesting that axial compressive femur load may not be the most appropriate response measurement for evaluating the restraint potential of the 1973-76 GM ACRS deploying passenger knee cushion. A pair of specially instrumented lower legs for the Hybrid III dummy were developed by Nyquist and Denton[13]. Since the majority of the fractures were to the tibial plateau and malleolus, these legs were instrumented to provide measurements of the internal medial and lateral loading of the knee and ankle joints. The shafts of the legs were instrumented to provide measurement of the bending moment at two cross sections and the axial compressive load. Sled tests were conducted using a Hybrid III dummy equipped with these instrumented lower legs. Since none of the passenger leg fractures noted in Table 5 occurred in a se-

Table 6 - Summary of Sled Tests Conducted to Investigate Out-of-Position Driver Interactions With the 1973-76 Driver ACRS. Moderate Collision Severity Pulse (23 mph, 14 G) and Hybrid III Dummy Used(9).

Test No.	Dummy Initial Position	Distance Between Chest and Module at Deployment (in)	Thoracic Responses			Column Stroke (in)
			Chest Compression		Spine Acc. (3 ms, G)	
			Max. Defl. (in)	Max. Rate (ft/s)		
1	Normal	10.5	0.7	3	32	3.2
2	Normal	1.7	2.3	49	64	6.0[5]
3	Normal	0.8	3.1	52	75	6.0[5]
4	Normal	0	3.3[4]	75	113	6.0[5]
5	Normal	Impacting[1]	3.3[4]	49	93	6.0[5]
6	Chest on Module	0[2]	3.3[4]	56	66	4.8
7	Chest on Module	0[3]	1.0	7	23	3.0

Notes:

1. In Test 5, cushion deployed 7 ms after chest contacted module.

2. In Test 6, cushion deployed 9 ms after beginning of sled pulse which is same timing as Test 1.

3. In Test 7, cushion was not deployed.

4. In Tests 4, 5 and 6, Hybrid III sternum bottomed out on spine box.

5. Maximum column stroke available in test fixture was 6 inches.

vere collision environment, a moderate collision severity (18 mph, 10 G) was chosen and only the low level of the passenger inflator was activated. Two different leg positions were evaluated: knees squared to the inflator module cover and knees rotated inboard with the right leg extended and closest to the module. The pertinent leg loads measured in these tests are summarized in Table 7. With the knees squared to the module cover all the leg loads, including the femur loads, were quite low indicating a low potential for leg fracture. In contrast, with the knees rotated inboard, the lateral compressive internal knee load of the right leg (the one closest to the module) was 1607 lb. Cadaver fracture load data of Hirsch and Sullivan(14) which was analyzed by Mertz(15) indicate that internal compressive loads of 900 lb between the tibial plateau and femoral condyle may fracture the tibial plateau if the load is biased medially or laterally. Clearly the results of the inboard facing leg test demonstrate the impor-

tance of leg orientation in evaluating the efficacy of deploying knee cushions and that many of the passenger tibial plateau fractures noted in Table 5 could have been caused by the legs not being squared to the inflator module cover when the air cushion was deployed.

HAND AND ARM FRACTURES - No tests were conducted to investigate possible causes for the hand and arm fractures noted in Table 5. However, it is easy to deduce that many of these fractures may have occurred if the cushion impacted the hand, or if the hand was flung hard against the car interior by the deploying air cushion.

RESTRAINT EFFECTIVENESS OF AN AIR CUSHION SYSTEM DESIGNED TO REDUCE THE SEVERITY OF CUSHION FORCES APPLIED TO OUT-OF-POSITION OCCUPANTS

If a driver and passenger air cushion system can be designed which would not injure out-of-position occupants while still providing re-

Table 7 - Summary of Peak Femur Loads and Lateral and
Medial Internal Knee and Ankle Loads for Two
Positions of the Legs, Moderate Severity Sled
Pulse (18 mph, 10 G), and Low Level Deployment
of the 1973-76 GM Passenger ACRS.

Leg Position	Femur Forces (lb) L/R	Knee Forces (lb)		Ankle Forces (lb)	
		LT. LAT. LT. MED.	RT. LAT. RT. MED.	LT. LAT. LT. MED.	RT. LAT. RT. MED.
Knees Squared To Module	452	259	479	265	319
	470	308	281	324	297
Knees Rotated Inboard, RT Leg Extended and Closest to Module	72	486 T	1607	104	339
	387	517	1310 T	145	299

Note:

1. All loads are peak compressive forces except those followed by a T which
    indicates tension.

straint for normally seated occupants, then the restraint effectiveness of that system would be greater than the effectiveness of the 1973-76 GM ACRS. For example, if the 1973-76 GM ACRS had been designed so that the unexpected AIS 3 or greater injuries noted in Table 4 were AIS 2 or less, then the restraint effectiveness in frontal deployment accidents of such a system would have been 50% instead of 18%. The individual effectiveness estimates for such a driver system and passenger system would have been 41% and 71% instead of 21% and 16%, respectively. These restraint effectiveness estimates were calculated using the same data and method used to calculate the 1973-76 GM ACRS effectiveness estimates except the injury severity classification for all the AIS 3 or greater unexpected injuries were changed to AIS = 2. The results of this analysis indicate that a sizeable gain in restraint effectiveness can be achieved if air cushion systems are designed to reduce the severity of cushion forces applied to out-of-position occupants.

Some methods to achieve this objective are discussed in papers by Klove and Oglesby(2) and Horsch and Culver(9). Mertz(16) notes that the tradeoff between inflating the cushion fast enough to provide restraint protection in the 30 mph, frontal barrier crash test required by FMVSS 208, but slow enough so as to not to seriously injure out-of-position occupants is the design dilemma of air cushion systems. Much more effort is needed to address this concern (17, 18, 19, 20) if the restraint effectiveness

potential of the air cushion system concept is to be realized.

SUMMARY

General Motors equipped 11,321 vehicles with driver and front seat passenger air cushion systems during the 1973 through 1976 model years. Based on an analysis of field accident data, Pursel et al estimated the restraint effectiveness of the 1973-76 GM ACRS to be 18% in mitigating AIS 3 or greater injuries when compared to unrestrained occupant injuries that occurred in comparable accidents. A case by case review of AIS 3 or greater injuries indicated that many of the injuries occurred in frontal accidents of minor to moderate collision severity where there was no intrusion or distortion of the occupant compartment in the vicinity of the occupant. These injuries were classified as unexpected and appear to be due to the occupant being near the air cushion module at the time of deployment. Dummy and animal tests of the 1973-76 GM ACRS were conducted and confirmed the fact that the types and severities of the unexpected injuries could be produced if the occupant was near the ACRS module at the time of deployment. It was estimated that a restraint effectiveness of 50% in mitigating AIS 3 or greater injuries in frontal deployment accidents could be achieved if an air cushion system could be developed that reduced the cushion forces applied to out-of-position occupants.

REFERENCES

1. Campbell, D. D., "Air Cushion Restraint Systems Development and Vehicle Application", SAE 720407, Second International Conference on Passive Restraints, May, 1972.

2. Klove, E. H. and Oglesby, "Special Problems and Considerations in the Development of Air Cushion Restraint Systems", SAE 720411, Second International Conference on Passive Restraints, May, 1972.

3. Louckes, T. N., Slifka, R. J., Powell, T. C., and Dunford, S. G., "General Motors Driver Air Cushion Restraint System", SAE 730605, May, 1973.

4. Smith, G. R., Gulash, E. C., and Baker, R. G., "Human Volunteer and Anthropomorphic Dummy Tests of General Motors Driver Air Cushion System", SAE 740578, 1974.

5. Smith, G. R., Hurite, S. S., Yanik, A. J., and Greer, C. R., "Human Volunteer Testing of GM Air Cushions", SAE 720443, Second International Conference on Passive Restraints, May, 1972.

6. Patrick, L. M. and Nyquist, G. W., "Airbag Effects on the Out-of-Position Child", SAE 720442, Second International Conference on Passive Restraints, May, 1972.

7. Wilson, R. A. and Piepho, L. L., "Crash Testing the General Motors Air Cushion", Fifth International Technical Conference on Experimental Safety Vehicles, London, England, July, 1974.

8. Pursel, H. D., Bryant R. W., Scheel, J. W. and Yanik, A. J., "Matching Case Methodology for Measuring Restraint Effectiveness", SAE 780415, February, 1978.

9. Horsch, J. D. and Culver, C. C., "A Study of Driver Interactions With an Inflating Air Cushion", SAE 791029, Twenty-Third Stapp Car Crash Conference, October, 1979.

10. Neathery, R. F., Kroell, C. K. and Mertz, H. J., "Prediction of Thoracic Injury From Dummy Responses", SAE 751151, Nineteenth Stapp Car Crash Conference, November, 1975.

11. Aldman, B., Andersson, A., and Saxmark, O., "Possible Effects of Airbag Inflation on a Standing Child", Proceedings of the International Meeting on Biomechanics of Trauma in Children, Lyons, France, 1974.

12. Mertz, H. J., Driscoll, G. D., Lenox, J. B., Nyquist, G. W., and Weber, D. A., "Responses of Animals Exposed to Deployment of Various Passenger Inflatable Restraint System Concepts for a Variety of Collision Severities and Animal Positions", Proceedings of the Ninth International Technical Conference on Experimental Safety Vehicles, Kyoto, Japan, November, 1982.

13. Nyquist, G. W. and Denton, R. A., "Crash Test Dummy Lower Leg Instrumentation for Axial Force and Bending Moment", Transactions of the Instrument Society of America, Vol. 18, No. 3, 1979.

14. Hirsch, G. and Sullivan, L., "Experimental Knee Joint Fractures -- A Preliminary Report", ACTA Orthopaedica Scandinavica, Vol. 36, 1965.

15. Mertz, H. J., "Injury Assessment Values Used to Evaluate Hybrid III Response Measurements", ISO/TC22/SC12/WG5, Document N123, May, 1984.

16. Mertz, H. J., and Marquardt, J. F., "Small Car Air Cushion Performance Considerations", SAE 851199, 1985.

17. Prasad, P. and Daniel, R. P., "A Biomechanical Analysis of Head, Neck, and Torso Injuries to Child Surrogates Due to Sudden Torso Acceleration", SAE 841656, Twenty-Eighth Stapp Car Crash Conference, November, 1984.

18. Montalvo, F., Bryant R. W. and Mertz, H. J., "Possible Positions and Postures of Unrestrained Front-Seat Children at Instant of Collision", Proceedings of the Ninth International Technical Conference on Experimental Safety Vehicles, Kyoto, Japan, November, 1982.

19. Wolanin, M. J., Mertz, H. J., Nyznyk, R. S., and Vincent, J. H., "Description and Basis of a Three-Year-Old Child Dummy for Evaluating Passenger Inflatable Restraint Concepts", Proceedings of the Ninth International Technical Conference on Experimental Safety Vehicles, Kyoto, Japan, November, 1982.

20. Mertz, H. J. and Weber, D. A., "Interpretations of the Impact Responses of a Three-Year-Old Child Dummy Relative to Child Injury Potential", Proceedings of the Ninth International Technical Conference on Experimental Safety Vehicles, Kyoto, Japan, November, 1982.

910901

Effectiveness of Safety Belts and Airbags in Preventing Fatal Injury

David C. Viano
Biomedical Science Dept.
General Motors Research Laboratories
Warren, MI

ABSTRACT

Airbags and safety belts are now viewed as complements for occupant protection in a crash. There is also a view that no single solution exists to ensure safety and that a system of protective technologies is needed to maximize safety in the wide variety of real automotive crashes. This paper compares the fatality prevention effectiveness, and biomechanical principles of occupant restraint systems. It focuses on the effectiveness of various systems in preventing fatal injury assuming the restraint is available and used. While lap-shoulder belts provide the greatest safety, airbags protect both belted and unbelted occupants.

RESTRAINT EFFECTIVENESS

Estimates of Effectiveness: Laboratory tests involving dummy injury assessment provide an objective evaluation of safety systems, but not an accurate estimate of restraint effectiveness in saving lives and preventing injury. In part, this is due to the limited type of crash testing conducted in relation to the wide range of real world crashes and the evolution of test dummies and injury criteria in simulating the responses of real occupants. While significant improvements have been made in the biofidelity of dummies, understanding of biomechanics, and validity of injury criteria that enable laboratory tests to better predict restraint effectiveness and better related to real-world safety performance, the most objective assessment of the effectiveness of occupant restraints is by analysis of real-world crashes.

Early epidemiologic studies of motor vehicle injury dealt with fleet evaluations of interior safety features, including the energy absorbing steering system, high penetration resistant windshields, and side-guard door beams. The first comprehensive study of lap-shoulder belts was conducted by Bohlin (1967) in Sweden and showed impressive injury and fatality prevention. Subsequent studies in various other countries have substantiated the earlier levels of safety belt effectiveness. However, variability in the underlying data and analysis approaches has led to a relatively wide range in the estimation of effectiveness in preventing fatality and serious injury.

The 1980's saw the introduction of sophisticated new statistical methods for epidemiologic analyses of field accident data. Evans (1986a) developed the double pair comparison procedure to isolate the effectiveness of belt use from other confounding factors in automotive crashes. Variations in the method have been used by Partyka, Kahane and others to determine the safety of lap-shoulder belts, rear lap belts and child safety seats. Evans (1991) has extended the approach to investigate the effects of alcohol use, occupant age, seating position, and direction of crash on fatality risks. The methodology has enabled accurate quantification of factors influencing crash injuries.

Statistical Analysis of Restraints in Fatal Crashes: Evans (1986a) developed the double paired comparison method to determine the effectiveness of occupant restraint as a function of seating position and crash direction. The method compares the number of fatalities to either of two occupants under two conditions, such as restrained or unrestrained, driver or passenger, and ejected or non-ejected. One of the occupants serves a normalizing or exposure estimating role for the frequency of fatality of the other under a particular crash situation. This forms the basis for the estimation of restraint effectiveness.

For example, the effectiveness of safety belt use by the right-front passenger (RFP) is determined by the double pair comparison using two sets of fatal crashes. The first set consists of a restrained RFP and an unrestrained driver, at least one of whom is killed. From the numbers of RFP and driver fatalities, a restrained RFP to unrestrained driver fatality ratio is calculated. From the second set of data on unrestrained RFPs and unrestrained drivers, an unrestrained RFP to unrestrained driver fatality ratio is similarly estimated. When the first fatality ratio is divided by the second, it gives the probability that a restrained RFP is killed compared to that of an unrestrained RFP in actual traffic accidents. This is the estimate of restraint system effectiveness.

Restraint Effectiveness: Determining the fatality prevention effectiveness of lap-shoulder belts was one of the first applications of the double pair comparison (Evans 1986b). Lap-shoulder belts were shown to be (41 ± 4)% effective in preventing fatality for front-seated occupants. They are (42 ± 4)% effective for drivers and (39 ± 4)% for right-front passengers (Table 1). The overall effectiveness was estimated at 43% by combining the average from Evans with those of NHTSA and others estimating 40-50% effectiveness.

Subsequently, the effectiveness of lap-shoulder belts was determined as a function of the direction of impact (Evans 1988c), including contributions from reducing ejection (Evans and Frick 1989). This type of analysis (Figure 1) helps differentiate two essential components of occupant protection with safety belt use. One is protection against ejection and is primarily due to the lap portion of the belt system. The other is mitigation of interior impact and is largely contributed by upper body restraint from the shoulder harness.

The relative safety contribution by reducing ejection with belt use significantly depends on the type of crash and is highest in primarily rollover accidents. Safety belt use is most effective in preventing driver fatality in crashes where rollover is the first harmful event and where occupant containment is a key feature of safety performance. The lap-shoulder belt system is least effective in preventing driver fatality in left-side impacts as the principal point of vehicle impact and deformation of the occupant compartment are critical factors in increasing fatality risk for an occupant.

The difference between overall lap-shoulder belt effectiveness and ejection reduction was

used by Evans (1988c) to determine the safety benefit of a driver and passenger airbag system. The analysis inferred the fatality reduction effectiveness of the airbag only system as a component of impact mitigation and estimated an (18 ± 4)% effectiveness for the unbelted driver and (13 ± 4)% for the unbelted right-front passenger. This level compares favorably with the results of an expert judgment of fatality prevention potential of airbag restraints (Wilson and Savage 1973).

In a further study, Evans (1988a) determined an overall (18 ± 9)% effectiveness of lap-belt use by rear-seated occupants. Most of the safety benefit is from anti-ejection since this level compares favorably with the 17-19% effectiveness of lap-shoulder belts in preventing ejection. A major component of lap-belt effectiveness is occupant containment in the vehicle. For primarily frontal crashes, the effectiveness of lap-belt use by rear seated occupants has a larger variability and is relatively low for impact mitigation.

Lap-shoulder belts for rear outboard occupants are available in most new passenger vehicles and will increase in the vehicle fleet. However, there isn't sufficient crash injury data to conduct a statistical analysis of effectiveness. It is possible to use the understandings of restraint effectiveness in other seating positions to make a judgment about the level of impact mitigation and ejection prevention. Lap-shoulder belts are estimated to be 27% effective in preventing fatal injury in rear seats. This is essentially due to impact mitigation improvements.

Table 2 summarizes the available estimates of belt restraint and airbag effectiveness in preventing occupant fatalities. The results show the substantial effectiveness of lap-shoulder belts. They are clearly the principal safety feature in protecting occupants in severe crashes. Since airbags and other interior components do not hold the occupant in a seating position, they have lower overall effectiveness because of a much lower effectiveness in reducing ejection. Their contribution to overall crash protection is essentially limited to impact mitigation which is about half the overall benefit of lap-shoulder belt use.

The overall safety benefit of the combination of lap-shoulder belt use and airbag has not been determined from field accident data. However, it is possible to estimate the effectiveness by considering the current safety effectiveness studies and the frequency of unsurvivable crashes. Based on analysis of fatal crashes to unbelted occupants, Huelke, Sherman and Murphy (1979) estimated that approximately 50% of the crash fatalities are unpreventable by currently available occupant restraints. The limit, in part, reflects the

Table 1

Effectiveness of Occupant Protection Systems
(Adapted from Evans with additional estimates added)

Driver

Safety System	Impact Mitigation	Ejection Prevention	Overall Effectiveness
Lap-Shoulder Belt	(23 ± 4)%	(19 ± 1)%	(42 ± 4)%
Airbag	(18 ± 4)%	-	(18 ± 4)%
EA Steering System	6%	-	(6 ± 3)%
Lap-Shoulder Belt (Plus Air Bag)	[27%]	[19%]	[46%] *

Right Front Passenger

Safety System	Impact Mitigation	Ejection Prevention	Overall Effectiveness
Lap-Shoulder Belt	(22 ± 4)%	(17 ± 1)%	(39 ± 4)%
Airbag	(13 ± 4)%	-	(13 ± 4)%
Friendly Interior	[<6%]	-	[<6%]
Lap-Shoulder Belt (Plus Airbag)	[26%]	[17%]	[43%]

Rear-Seat Passenger

Safety System	Impact Mitigation	Ejection Prevention	Overall Effectiveness
Unbelted Rear Versus Front-Seat Position	-	-	(26 ± 2)%
Lap-Belt	(1 ± 9)%	(17 ± 1)%	(18 ± 9)%
Lap-Shoulder Belt	[10%]	[17%]	[27%]

*[] New estimates of restraint effectiveness based on judgement.

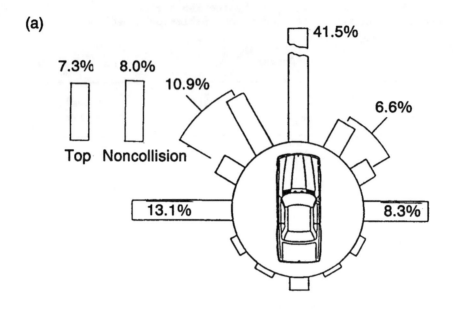

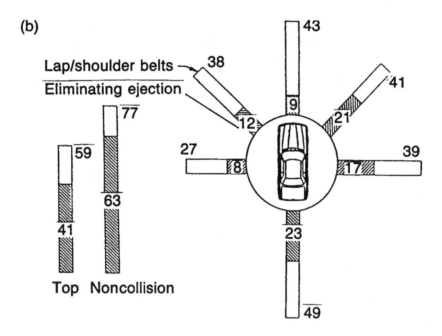

Figure 1: (a) Distribution of driver deaths by principal impact point and (b) effectiveness of lap/shoulder belts in preventing driver fatalities. The fraction of fatalities prevented by eliminating ejection is shown in the hashed portion of the bar according to impact direction. For example, in frontal (12 o'clock) crashes, lap/shoulder belts prevent 43% of driver fatalities; 9% of this is due to eliminating ejection, so that 34% is due to interior impact reduction (from Evans 1988c with permission).

Table 2

Effectiveness of Occupant Restraints

	Driver	Right Front Passenger	Rear Passenger
Lap-Shoulder Belts	(42 ± 4)%	(39 ± 4)%	[27%]
Airbag Only	(18 ± 4)%	(13 ± 4)%	--
Belts and Airbag	[46%]	[43%]	--
Lap Belt	--	--	(18 ± 9)%

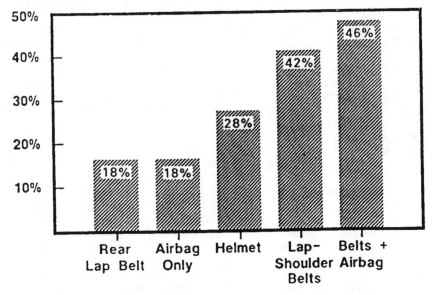

Figure 2: Effectiveness of safety devices in preventing fatal driver injury in severe motor vehicle crashes and motorcyclist injury with helmet use (from Viano 1990 with permission).

severity of many fatal crashes which may involve extreme vehicle damage and forces on the passenger compartment, unusual crash configurations and causes of death, and unique situations associated with particular seating positions and crash dynamics.

By comparing the level of unpreventable fatalities with the overall protective effect of lap-shoulder belts, Viano (1988a) found a maximum of 7% additional fatality prevention may occur with a combination of restraints or additional supplemental features. Based on a similar review of fatalities to lap-shoulder belted front-seated occupants, a potential 3-5% additional fatality prevention is estimated with a combination of lap-shoulder belt and airbag system.

Using a 4% additional benefit to safety belt wearers from airbags, there is an overall effectiveness of 46% for the belted driver with supplemental airbag (Figure 2) and 43% for the right-front passenger. However, this estimate is based on expert judgment and not statistical methods applied to crash injury data.

The 4% increment in effectiveness with the airbag supplementing lap-shoulder belt use is consistent with a 5% additional benefit determined by NHTSA (1984). The fact that a lap-shoulder belt and airbag are only 43% to 46% effective in preventing fatality underscores that absolute protection is not achievable by occupant restraints and that injury and fatality will continue to occur to belt wearers despite a significant overall net safety gain by restraint usage, even if augmented by airbags. This is because of accident severity, configuration, and limits of human tolerance.

Fatality risk depends on the particular occupant seating position, the principal impact point and the proximity of passenger compartment crush. Evans and Frick (1988) have shown (Figure 3) more than a 7 to 1 increase in fatality risk for a right-front passenger in right-side (nearside) versus a left-side (farside) impact.

A better appreciation of restraint effectiveness estimates may be possible by a fuller understanding of the biomechanics of restraint systems in reducing impact forces and controlling occupant kinematics over the wide variety of real-world crash. The following section expands on a recent review by Viano (1988b) and addresses key features of occupant restraint performance in frontal and rollover accidents.

RESTRAINT BIOMECHANICS AND PERFORMANCE

Driver Airbag: During the development of crash protection systems for automobile occupants in the early 1960's, many concepts for energy absorption and load distribution were conceived and evaluated. A driver airbag took advantage of rapid filling of a concealed bag to provide a cushion in front of an occupant in a crash (see review of SAE publications in Viano 1988c). This provided a large area to gradually decelerate the driver. However, it was necessary to provide lower torso restraint through either a lap-belt or knee bolster to prevent the driver from submarining the airbag (Figure 4a and b) and experiencing higher forces on interior impact in a frontal crash.

Although compressed gas in a cylinder was one of the early concepts, the eventual production system took advantage of a relatively small mass (70-100 g) of sodium azide. The chemical can rapidly inflate a 80-100 l driver airbag by ignition and conversion to harmless nitrogen gas when sensors detected a severe frontal crash. The airbag is stored inconspicuously in the interior and deployed only during a severe crash, sensors are located in the engine compartment to detect rapid decelerations of the front-end during a crash and an electronic system is used to monitor and initiate deployment of the airbag.

Production airbag systems were available from General Motors in 1974. However, the customer demand never developed for the safety option so the program was cancelled in 1976 after selling only 10,321 vehicles. Mertz (1988) recently published an analysis of the field performance of the airbag based on the work of Pursel et al (1978) using comparable non-airbag crashes. The driver airbag system was found to be 21% effective in preventing AIS 3+ injuries (16% for the passenger system), but was -34% ineffective for AIS 2+ passenger injury. However, the matched crash analysis only allows the determination of effectiveness in deployment accidents. The overall effectiveness of a driver airbag in typical crashes would be approximately 9% in preventing AIS 3+ driver injuries, assuming deployment accidents represent 45% of all severe crashes.

Many of the airbag inflation injuries occurred in moderate severity crashes without distortion of the occupant compartment. They were considered "unexpected" and caused by the occupant being near the cushion at the time of deployment. Horsch and Culver (1979) and more recently Horsch et al (1990) have observed significant forces adjacent to an airbag if the normal path of inflation is blocked by an occupant. Thus, the deployment, per se, of an airbag has the potential to seriously injury or kill, independent of crash severity.

The use of lap-shoulder belts helps minimize the risk of airbag injury, but there is a potential consequence of public misperception of driver airbag protection. Some car occupants may not recognize that the steering wheel airbag is only a supplement, and not an alternative, to the primary occupant protection system in the vehicle, the lap-shoulder belts.

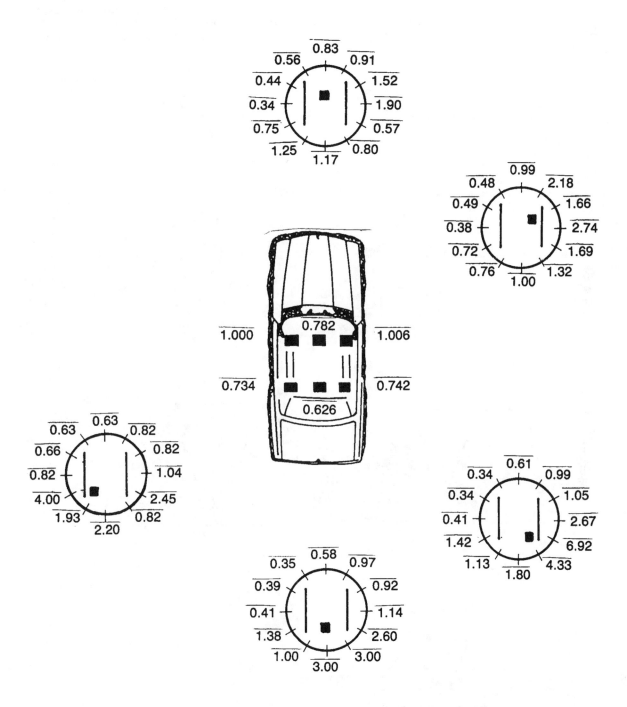

Figure 3: Relative fatality risk to passengers in different seating positions in relation to that of the driver set at 1.000 and fatality risk for passengers in various car seating positions relative to the principal direction of impact (based on Evans and Frick 1988 with permission).

If drivers fail to buckle up, they significantly reduce their driving safety, particularly in side impact and rollover crashes where the lack of belt restraint in the seat may lead to interior impact or ejection. Evans (1989) calculated a 41% increase in fatality risk by drivers ceasing to wear the lap-shoulder belt because they have a supplemental bag system. Fortunately, current belt use surveys are not finding a lower rate of belt use in airbag equipped vehicles.

Lap-belts: An early use of the lap-belt was to complement crash protection of the chest by a driver airbag or the energy absorbing (EA) steering system, another 1960's interior safety concept. The belt controlled forward excursion of the lower torso (Figure 4c) by restraining the pelvis and maintained an upright posture of the occupant (Horsch, Peterson and Viano, 1982).

In the absence of pelvic restraint, loads would be applied through the knees and seat pan to restrain the occupant and control the upright posture of the driver. These loads are necessary to take advantage of the cushioning affect of the airbag or EA steering column on the upper body. The lap-belt also complemented early designs for passenger protection by a padded dash board and found their way into rear seating positions to supplement padded seat backs (Figure 4d).

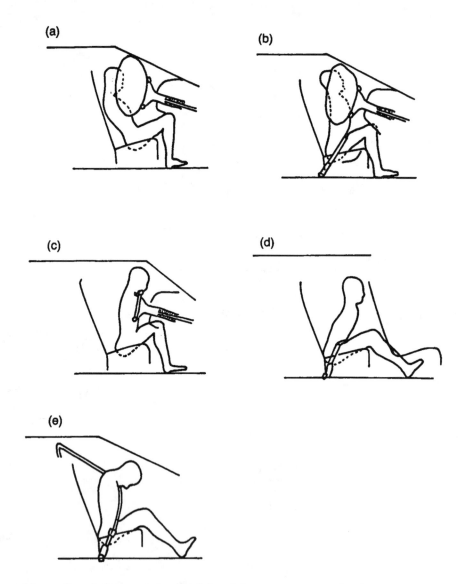

Figure 4: Kinematics of a driver restrained by (a) a steering wheel airbag where loads act on the upper body but not on the lower extremity, (b) the combination of lap belt and airbag giving pelvic and upper body restraint, (c) the energy absorbing steering system and knee bolster, and passenger restrained by (d) a lap belt in the rear seat and (e) a lap-shoulder belt in the right-front seating position.

Although much of the development work on restraints has involved frontal barrier and sled testing, safety engineers recognized the importance of the lap-belt in preventing ejection, which had been identified as the leading cause of fatality in the mid-1950's by crash investigations. Thus, occupant containment within the vehicle during a crash was a principal benefit of lap-belt usage. Since lap-belt restraint was only one part of the systems engineering for occupant protection by friendly interior components, development work focused on assuring that the various safety components worked together as a system in a crash.

During testing, it also became apparent that lap-belt use also reduced the potential for rear-seat occupant loading on the front-seat back. This is particularly important for belt restrained front-seat occupants, since Park (1987) has shown that unrestrained rear-seat occupants increase the fatality risk of belted front-seat occupants by (4 ± 2)% because of the additional load applied on the restrained occupant. This decreases the safety effectiveness for a belted front-seat driver from 42% to 40% and a belted right-front passenger from 39% to 37%, a significant increment in crash injury risk.

With the advent of safety belt use in passenger cars in the late 1950's and early 60's, physicians started reporting on the injury patterns of belt restrained victims (Kulowski and Rost, 1956). The typical pattern of upper body injury to unrestrained occupants was replaced by belt related injury to abdominal organs and tissues. These injuries led to the phrase "seat belt syndrome" (Garrett and Braunstein 1962, Fish and Wright, 1965) as the new injury patterns in motor vehicle crashes received attention due to concentrated forces on the lower abdominal region for lap-belt wearers. In many cases improper belt wearing was identified as a cause of abdominal loading and injury, although other factors may play a role. Since belt use modifies injury patterns and it is not possible to prevent all injury to occupants in crashes, some in the public are suspicious about the safety effectiveness of belts, in spite of now over-whelming evidence of benefit.

Lap-Shoulder Belts: As safety engineers pursued continued improvements in crash protection, the combination of a lap and diagonal shoulder belt gained rapid acceptance (Bohlin 1967). Lap-shoulder belts provide occupant restraint during a crash by routing safety belts over the boney structures of the pelvis and shoulder. This takes advantage of a relatively high tolerance to impact forces for these regions of the skeleton and avoids concentrating load on the more compliant abdominal and thoracic regions. Control of occupant kinematics helps insure maximum protection by belt restraints.

The fundamentals of a high quality belt restraint system involve kinematic controls (Adomeit and Heger 1975, Adomeit 1977, 1979, Viano and Arepally, 1990) which maintain the lap portion of the belt low on the pelvis through adequate seat cushion support. This minimizes pelvic rotation and reduces the tendency for the lap-belt to slide off the illium and directly load the abdomen. Forward rotation of the upper torso of slightly greater than 90° upright posture (Figure 4e) directs a major portion of the upper torso restraint into the shoulder and upper chest region. This reduces loads on the more compliant areas of the lower rib cage. Each of the kinematic controls helps direct forces onto the skeletal structures above and below the center of gravity of the torso thus balancing the restraint and keeping loads away from more compliant body regions vulnerable to injury.

There is another component of quality restraint. Biomechanical responses related to occupant protection assessment need to be evaluated and kept below human tolerance for the severity of the crash test (Horsch 1987). Although there is a rich history in the use of chest acceleration as a measure of injury risk and the current standards require less than 60 g's for 3 ms duration, chest acceleration is an inadequate measure of injury risk.

Recent evaluations by Groesch et al (1986) have demonstrated that chest acceleration is not a logical measure of restraint effectiveness in real-world crashes or an adequate indicator of fundamentally different restraint configurations. The current evidence points to deformation of the body and its organs as the cause of injury, and the tolerance to deformation depends on the velocity of loading. The faster the deformation, the lower the tolerance to compression (normalized deformation). This type of injury is related to the Viscous response which is a cause of soft tissue injury and a measure of energy dissipated during rapid compression (Viano and Lau 1988).

With a significant increase in safety belt use after state passage of mandatory wearing laws and a greater health consciousness in America, Orsay et al (1988) has seen reductions in hospital admissions and injury severities for lap-shoulder belted occupants, even as higher relative numbers of belted victims are being treated.

Injury patterns associated with safety belt wearing are becoming better understood (Denis et al 1983, Arajarvi, Santavirta and Tolonen, 1987, Banerjee 1989). The potential for improper use has also continued with the advent of lap-shoulder belt systems, particularly placement of the shoulder harness under the arm (States et al 1987) and wearing of the lap-belt high on the abdomen with poor seating posture.

An improvement in passenger safety has been made by adding a shoulder belt to the out-board rear-seat lap-belt. This further improves the safety of rear-seat occupants, which is otherwise similar to belted front-seat occupants because of the inherently greater safety of rear seating positions (Evans 1988a). That is, the restraint effectiveness of lap-shoulder belt use by front-seat occupants is similar to lap-belt use by rear-seat occupants (39% effectiveness for lap-belted rear-seat occupants versus 43% for a lap-shoulder belted driver and 39% for the right-front passenger). Thus, combining the shoulder harness to the rear-seat lap-belt should increase effectiveness by mitigating interior contacts of the upper body. This should increase the safety of belted rear-seat occupants to an estimated level of 27% with an overall gain in safety over those belted in the front seat.

An analysis of rollover crashes by Huelke, Compton and Studer (1985) determined that safety belt use virtually eliminated the risk of paralyzing cervical injury and ejection. This observation is consistent with Evans' (1986b) finding that belt restraints are most effective in rollover crashes, particularly where rollover is the first harm event. In such crashes, lap-shoulder belts are (82 ± 5)% effective in preventing driver fatalities. Sixty-four percent (64%) of the overall effectiveness is by preventing ejection with the remaining 18% effectiveness by reducing forces due to interior impacts.

In contrast, unrestrained occupants in rollovers are subjected to a series of impacts and potential ejection during a rollover (Viano 1990). In a simulation of experimental rollovers by Robbins and Viano (1984), the complex kinematics of an unbelted occupant were studied using several crash scenarios, including complete occupant containment, and subsequent driver or passenger door opening (Figure 5). The potential for serious injury by interior impacts or ejection were apparent in the rollover sequences for the unrestrained driver.

PERCEPTIONS OF OCCUPANT RESTRAINTS

Interestingly, the overall fatality prevention effectiveness of a driver airbag without safety belt use and lap-belt use in the rear is similar. Both systems provide an 18% reduction in fatal crash injury risk. This level of effectiveness is far greater than built-in safety devices, such as the EA steering column at 6% effectiveness, and high penetration resistant windshields, side door-beams, and interior padding which have lower effectiveness levels. Therefore, airbags and rear-seat lap-belts are effective automotive safety features and are second only to front-seat lap-shoulder belts which have a 41% safety effectiveness. However, this is the effectiveness when the

airbag is available and the lap belt is used. It also does not consider the effects of occupancy rates on overall injury prevention.

Driver airbags and rear-seat lap-belts provide protection in different crash types. Airbag effectiveness is essentially due to interior impact mitigation in primarily frontal crashes. It has minimal effectiveness in lateral or rollover crashes. In contrast, rear-seat lap-belt effectiveness is essentially due to ejection prevention in primarily non-frontal crashes such as rollovers, since it is less effective in reducing impact forces in frontal crashes. Rear-seat lap-belts are currently available in virtually all passenger cars and only require buckling by an occupant to be effective. On the other hand, driver airbags are available in much fewer vehicles but provide crash protection independent of any action by the occupant.

Both systems have the possibility of modifying injury in particular situations. The high energy release of an airbag may injure an occupant against the system at the instant of deployment. Blocking the path of deployment increases pressures in the cushion during gas generation and develops high forces on the occupant. Since the force occurs with high velocity, there is a risk of injury by a Viscous mechanism.

Placing the lap-belt on the abdomen as compared to correctly on the pelvis does not direct restraining loads through the skeleton in severe frontal crashes. Abdominal loading and deformation occurs as forces in the lap-belt restrain forward excursion of the lower body. This situation, and another in which the lap-belt slips above the pelvis during a crash, may result in abdominal organ injury and hemorrhage by submarining pelvic restraint.

Since there is a general lack of information on the technical aspects, performance, and efficacy of safety systems, the public has developed an exaggerated perception of the safety effectiveness of a driver airbag and insecurities about rear-seat lap-belt use. Many falsely believe that airbags are safer than safety belts. The facts about safety systems need to be accurately covered in the news and conveyed to the public to gain understanding.

In addition, there should be a better awareness that safety devices cannot provide full or complete protection, function in a specific range of crash types, and may be associated with injury in particular crash situations. Misleading information or rumor about either safety benefits of crash protection components or injury risks, which may be infrequent in comparison to overall safety benefit, may cause occupants to reduce their overall

640

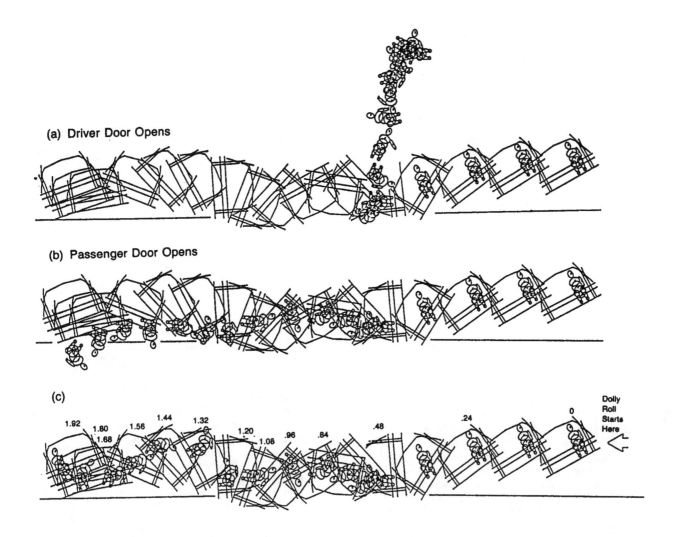

Figure 5: Kinematics of an unrestrained driver in a rollover crash in which (a) the driver door opens and the occupant ejects and is thrown upward, (b) the passenger door opens, the driver ejects and is crushed between the rolling car and the road, and (c) the occupant is contained in the vehicle (from Robbins and Viano 1984 with permission).

driving safety by failing to properly use inherently safe technologies to protect themselves and their families.

REFERENCES

1. Adomeit, D. and Heger, A., "Motion Sequence Criteria and Design Proposals for Restraint Devices in Order to Avoid Unfavorable Biomechanic Conditions and Submarining." In Proceedings of the 19th Stapp Car Crash Conference, 139-166, SAE Technical Paper #751146, Society of Automotive Engineers, Warrendale, PA, 1975.

2. Adomeit, D., "Evaluation Methods for the Biomechanical Quality of Restraint Systems During Frontal Impact." In Proceedings of the 21th Stapp Car Crash Conference, 911-932, SAE Technical Paper #770936, Society of Automotive Engineers, Warrendale, PA, 1977.

3. Adomeit, D., "Seat Design -- A Significant Factor for Safety Belt Effectiveness." In Proceedings of the 23rd Stapp Car Crash Conference, 39-68, SAE Technical Paper #791004, Society of Automotive Engineers, Warrendale, PA, 1979.

4. Arajarvi, E., Santavirta, S. and Tolonen, J., "Abdominal Injuries Sustained in Severe Traffic Accidents by Seat Belt Wearers." Journal of Trauma 27(4):393-397, 1987.

5. Banerjee, A., "Seat Belts and Injury Patterns: Evolution and Present Perspectives." Postgraduate Medical Journal, 65:199-204, 1989.

6. Bohlin, N.I., "A Statistical Analysis of 28,000 Accident Cases with Emphasis on Occupant Restraint Value." In Proceedings of the 11th Annual Stapp Car Crash Confer-

ence, University of California, Los Angeles, CA, October 10-11, 1967. SAE Technical Paper #670925, Warrendale, PA, Society of Automotive Engineers, 455-478, 1967.

7. Denis, R., Allard, M., Atlas, H. and Farkouh, E., "Changing Trends with Abdominal Injury in Seat Belt Wearers." Journal of Trauma 23(11):1007-1008, 1983.

8. Evans, L., "The Effectiveness of Safety Belts in Preventing Fatalities." Accident Analysis and Prevention 18:229:241, 1986b.

9. Evans, L., "Double Pair Comparison -- A New Method to Determine How Occupant Characteristics Affect Fatality in Risk in Traffic Crashes." Accident Analysis and Prevention 18:217-227, 1986a.

10. Evans, L., "Passive Compared to Active Approaches to Reducing Occupant Fatalities." GM Research Publication GMR-6596, Experimental Safety Vehicles paper #ESV 89-5B-0-005, Goteborg, Sweden, May, 1989.

11. Evans, L. and Frick, M.C., "Potential Fatality Reductions Through Eliminating Occupant Ejection from Cars." Accident Analysis and Prevention, 22(2):169-182, 1989.

12. Evans, L., "Rear Seat Restraint System Effectiveness in Preventing Fatalities." Accident Analysis and Prevention 20:129-136, 1988a.

13. Evans, L., "Occupant Protection Device Effectiveness in Preventing Fatalities." In Proceedings of the 11th International Technical Conference on Experimental Safety Vehicles, Washington, DC, May 12-15, 1987, U.S. Department of Transportation, National Highway Traffic Safety Administration, DOT HS 807 233, pp. 220-227, 1988b.

14. Evans, L., "Restraint Effectiveness Occupant Ejection from Cars, and Fatality Reductions." General Motors Research Report #GMR-6398, General Motors Research Laboratories, Warren, MI, 1988c.

15. Evans, L. and Frick, M., "Seating Position in Cars and Fatality Risk." American Journal of Public Health, 78:1456-1458, 1988.

16. Evans, L., Traffic Safety and the Driver. Van Nostrad Reinhold, January, 1991.

17. Fish, J. and Wright, R.H., "The Seat Belt Syndrome -- Does It Exist?" Journal of Trauma, 5:746-750, 1965.

18. Garrett, J.W. and Braunstein, P.W., "The Seat Belt Syndrome." Journal of Trauma 2:220-238, 1962.

19. Groesch, L., Katz, E., Marwitz, H., Kassing, L., "New Measurement Methods to Assess the Improved Injury Protection of Airbag Systems." In Proceedings of the 30th Annual Conference of the American Association for Automotive Medicine, Association for the Advancement of Automotive Medicine; Des Plains, IL, 235-246, 1986.

20. Horsch, J.D. and Culver, C.C., "A Study of Driver Interactions with an Inflating Air Cushion." In Proceedings of the 23rd Stapp Car Crash Conference, SAE Technical Paper #791029, Society of Automotive Engineers, Warrendale, PA, 1979.

21. Horsch, J.D., Petersen, K.R. and Viano, D.C., "Laboratory Study of Factors Influencing the Performance of Energy Absorbing Steering Systems." SAE Transactions, Vol. 91, 1982, SAE Technical Paper #820475, in SAE Special Publication 507, Occupant Interaction with the Energy Absorbing Steering System, pp. 51-63, 1982.

22. Horsch, J.D., "Evaluation of Occupant Protection From Responses Measured in Laboratory Tests." SAE International Congress and Exposition, Detroit, MI, February 23-27, 1987. SAE Paper #870222. Society of Automotive Engineers, Warrendale, PA, 1987.

23. Horsch, J., Lau, I., Andrzejak, D., Viano, D., Melvin, J., Pearson, J., Cok, D., Miller, G., "Assessment of Airbag Deployment Loads." In the Proceedings of the 34th Stapp Car Crash Conference, pp. 276-288, SAE Technical Paper #902324, November 1990.

24. Huelke, D.F., Sherman, H.W., Murphy, M.J., "Effectiveness of Current and Future Restraint Systems in Fatal and Serious Injury Automobile Crashes." SAE Technical Paper #790323. Warrendale, PA, Society of Automotive Engineers, 1979.

25. Huelke, D.F., Compton, C. and Studer, R., "Injury Severity, Ejection, and Occupant Contacts in Passenger Car Rollover Crashes." SAE Paper #850336, Society of Automotive Engineers, Warrendale, PA, 1985.

26. Kulowski, J. and Rost, W.B., "Intra-Abdominal Injuries from Safety Belt in Auto Accident." Archives of Surgery, 73:970-971, 1956.

27. Mertz, H.J., "Restraint Performance of the 1973-76 GM Air Cushion Restraint System." SAE Technical Paper #880400, 1988:61-72, Society of Automotive Engineers, Warrendale, PA.

28. National Highway Traffic Safety Administration, "FMVSS 208 Regulatory Impact Analysis." Department of Transportation, 1984.

29. Orsay, E.M., Turnbull, T.L., Dunne, M., Barrett, J., Langenberg, P., and Orsay, C.P., "Prospective Study of the Effect of Safety Belts on Morbidity and Health Care Costs in Motor-Vehicle Accidents." Journal of American Medical Association 260(24):3598-3603, 1988.

30. Park, S., "The Influence of Rear-Seat Occupants on Front-Seat Occupant Fatalities: The Unbelted Case." General Motors Research Laboratories Research Publication GMR-5664, January 8, 1987.

31. Pursel, H.D., Bryant, R.W., Scheel, J.W., and Yanik, A.J., "Matching Case Methodology for Measuring Restraint Effectiveness." SAE Technical Paper #780415, Society of Automotive Engineers, Warrendale, PA, February, 1978.

32. Robbins, D.H., and Viano, D.C., "MVMA-2D Modeling of Occupant Kinematics in Rollovers." SAE Transactions, Vol. 93, 1984. In Mathematical Simulation of Occupant and Vehicle Kinematics, (P-146), 65-77, SAE Technical Paper #840860, Society of Automotive Engineers, Warrendale, PA, 1984.

33. States, J.D., Huelke, D.F., Dance, M. and Green, R.N., "Fatal Injuries Caused by Underarm Use of Shoulder Belts." Journal of Trauma 27(7):740-, 1987.

34. Viano, D.C., "Limits and Challenges of Crash Protection." Accident Analysis and Prevention, 20(6):421-429, 1988a.

35. Viano, D.C., "Cause and Control of Automotive Trauma." Bulletin of the New York Academy of Medicine, Second Series, 64(5):376-421, 1988b.

36. Viano, D.C., (editor), SAE Passenger Car Inflatable Restraint Systems: A Compendium of Published Safety Research, SAE Progress in Technology PT-31, Society of Automotive Engineers, Warrendale, PA, 1988c.

37. Viano, D.C. and Lau, I.V., "A Viscous Tolerance Criterion for Soft Tissue Injury Assessment." Journal of Biomechanics, 21(5):387-399, 1988.

38. Viano, D.C., Arepally, S., "Assessing the Safety Performance of Occupant Restraint Systems." In the Proceedings of the 34th Stapp Car Crash Conference, pp. 301-328, SAE Technical Paper #902328, November, 1990.

39. Viano, D.C., "Cause and Control of Spinal Cord Injury in Automotive Crashes." G. Heiner Sell Lecture. Submitted to Paraplegia, 1990 and available until publication as GMR-6885, 12/7/89.

40. Viano, D.C., "Testimony Before the United States Senate Committee on Environment and Public Works Subcommittee on Water Resources, Transportation, and Infrastructure Concerning Senate Bill S.1007 The National Highway Fatality and Injury Reduction Act of 1989." October 17, 1989.

41. Wilson, R. A. and Savage, C.M., "Restraint System Effectiveness - A Study of Fatal Accidents." In the Proceedings of Automotive Safety Engineering Seminar, Warren, MI, General Motors Corporation, Automotive Safety Engineering, Environmental Activities Staff; 27-39. 1973.

940714

Airbag Protected Crash Victims - The Challenge of Identifying Occult Injuries

Jeffrey S. Augenstein
University of Miami/Jackson Memorial Hospital

Elana B. Perdeck
University of Miami/Jackson Memorial Hospital

Kennerly H. Digges
George Washington University

James E. Stratton
Arvin Caslspan Corp.

Louis V. Lombardo
National Highway Traffic Safety Administration

A. C. Malliaris
Data Link, Inc.

Patricia M. Byers, Diego B. Nunez, Jr., Gregory A. Zych, Jonathan L. Andron, A. Kevin Craythorne, and Carla Verga
University of Miami/Jackson Memorial Hospital

ABSTRACT

A multidisciplinary, automobile crash investigation team at the Jackson Memorial Hospital/Ryder Trauma Center in Miami, Florida, is conducting a detailed medical and engineering study. The focus is restrained (seatbelts and/or air bag) occupants involved in frontal crashes, who have also been severely injured. More than 60 crashes have been included in the study to date.

Analysis of the initial data indicates that restraint systems are working to reduce many of the head and chest injuries which unrestrained occupants suffer.

However, internal injuries among air bag-protected occupants may be unrecognized in the field providing new challenges in triage and injury diagnosis. In other cases, survival in extremely high severity crashes presents trauma management challenges due to the extent and complexity of the multiple injuries which result. The paper provides case examples to illustrate types of chest and abdominal injuries associated with air bag cases. Three types of cases are presented: (1) Jackson Study Involving Occult Chest /Abdominal Injury, (2) National Accident Sampling System (NASS) Special Crash Investigation (SCI) and (3) Jackson Study Involving Crash Severities Greater than 45 MPH.

To assist in recognizing the extent of injuries to occupants protected by air bags, it is suggested that additional evidence from the crash scene be used in the triage criteria. For the occult chest/abdominal cases observed in the Jackson study, deformation of the steering system was the vehicle characteristic most frequently observed.

The challenges of recognizing injuries to air bag-protected occupants are discussed. The presence of steering wheel deformation may be a sufficient signal of caution to justify transporting the injured victim to a Level 1 or 2 trauma center so that a close examination for occult injuries can be made.

INTRODUCTION

Air bags first became available in passenger automobiles about twenty years ago. During the present decade, the penetration of these devices in the new car fleet will approach one hundred percent. It is estimated that by the year 2000 more than five hundred thousand air bag deployments will occur each year. Yet, there is little medical literature detailing the injury patterns of air bag-protected occupants involved in crashes [1]. Anecdotal reports exist particularly on minor injuries associated with air bag deployments such as abrasions of the cornea of the eye and lacerations of the face [2].

Because air bags are less than 100% effective, injuries will still occurr. The medical community will be challenged by occupants who avoid head, chest or abdominal trauma but experience injuries of their un(air bag)-protected extremities. Of similar concern, occupants may sustain trauma to the chest or abdominal organs through impacts with the air bag or through collisions with internal contents like the steering wheel. The mechanisms of injury may include overwhelming the air bag's energy management capabilities due to very high crash forces or multiple collisions wherein the air bag is deflated during some of the collisions. In any event, the injury patterns in air bag-protected occupants must be discerned so that treatment strategies can be optimized and hopefully injury countermeasures can be developed.

This paper summarizes the detailed investigation of a number of air bag-protected individuals involved in crashes whose injuries warranted admission to a trauma center.

METHODS AND PROCEDURES

A research protocol for addressing seriously injured, restrained occupants of frontal automobile collisions was implemented at the University of Miami/Jackson Memorial Medical Center in Miami, Florida beginning July, 1991. In August, 1992, all trauma services were moved into a newly constructed, dedicated trauma care facility named the Ryder Trauma Center. The Center includes unique capabilities to perform injury research. In particular, a computer system has been developed and implemented throughout the Ryder Trauma Center to address the challenge of acquiring and analyzing the enormous amount of information associated with injuries. The **CARE** system is designed to be integrated with the clinical, administrative, research and educational components of the center. Innovative technologies such as radio-linked terminals, computer-compatible cameras and multimedia electronic displays are incorporated into the system [3]. The research data elements include reconstructions of the automobile crash scene and the vehicular damage, description of the patient's clinical course and outcome status and definition of the economic implications. The database encompasses electronic images such as x-rays, digitized voice and video segments in addition to conventional elements.

Every automobile crash-related admission to the trauma center was considered for possible inclusion in the study. Once a patient was determined, on a preliminary basis, to meet the study criteria, a number of parallel activities began:

- A crash study researcher attempted to interview the patient (and perhaps other occupants of the crash vehicle), scene-responding Emergency Medical Services (EMS) personnel and the investigating police. If it were concluded that the study requirements had been met, the patient was then asked (after being briefed on the nature of the study and his/her rights) to sign a form consenting to participate in the study.

- If the patient consented, the crash investigator sought to evaluate the vehicle and the scene. This individual is an experienced National Accident Sampling System (NASS) investigator. Vehicle location was typically contained in the police report. Every effort was made to have access to such data sources within 24 hours of the crash since the evidence at the accident scene, as well as that in and on the vehicle, deteriorated within a short time after the crash.

- Once the accident vehicle became available, the second phase of data collection began. That is, the NASS investigator confirmed whether or not the collision was frontal and whether or not the subject had been restrained. The physical evidence of air bag deployment and/or seat belt/child-restraint engagement in the position of the subject was crucial for establishing that study restraint criteria had been met.

It should be noted that an extremely useful source of information was photographs or videos taken at the scene. In several cases, representatives of the police, the EMS, local television stations and even amateur videographers provided useful documentation. In a few cases, it established restraint status.

- Once a patient consented to participate in the study and the crash investigator determined that the direction of force and the restraint status was appropriate, the remaining vehicle and scene data as well as pertinent clinical and economic data were collected. The crash investigator attempted to establish occupant/vehicle contact points as well as the energy characteristics of the crash following NASS reference standards [4].

- The extent of the injuries, the therapeutic interventions used to correct them, the post-therapeutic course (both within the hospital and after discharge) as well as the economic, social and psychological implications of the subject's injuries were then assessed through medical record review and interviews. A major analytical approach of the study was the multidisciplinary review, during which injury mechanisms were defined. The goal of these meetings was to incorporate the judgments of general and specialty surgeons, forensic pathologists, trauma nurses, the crash investigator, automotive safety engineers, police accident investigation specialists and EMS personnel. The various data elements and evidence from all phases of the investigation were presented to them for discussion and review in their role as an expert panel.

DESCRIPTION OF DATA COLLECTED TO DATE

The study to date includes 62 injured occupants. All occupants were protected by some type of restraint system. Table 1 shows the distribution of restraint systems

TABLE 1
Restraint Distribution

TYPE OF RESTRAINT	NUMBER
LAP & SHOULDER BELT	39
SHOULDER BELT ONLY	6
AIR BAG ONLY	10
LAP & SHOULDER BELT, AIR BAG	7

One of the criteria for admission to this study is that the vehicle occupant was admitted to the Ryder Trauma Center. In most cases, the victim was suspected on admission of having one or more severe injuries. The cases collected are suitable for developing hypotheses regarding the cause and mechanisms of the injuries and possible injury reduction countermeasures. However, population-based crash studies are required to assess the overall effectiveness of the safety belt and air bag systems. Many such studies have been reported in the literature. The National Highway Traffic Safety Administration (NHTSA) recently reported that safety belts were about 45% effective in reducing moderate to critical injuries. In 1992, a total of 5,226 lives were saved by safety belts. Air bags in conjunction with safety belts are reported to be between 55%-68% effective in reducing moderate to critical injuries. Through 1992, air bags and/or safety belts have saved an estimated 558 lives in air bag equipped cars, and reduced nearly 40,000 moderate to critical injuries.[5].

A general observation of the injured occupants in the study to date confirms other findings that air bags and seat belt systems are doing a good job of reducing the severity of injuries. The severity and extent of the head and chest trauma are generally much less than has been observed for the unrestrained population which made up most of the motor vehicle trauma admissions in years past.

This reduced extent of trauma still carries with it the need for careful examination and diagnosis of the residual trauma. Especially for air bag cases, the physiological response of the victim at the crash site does not consistently predict the gravity of their trauma.

The analysis to follow will address the residual injuries observed in air bag cases and will focus on the occult (not immediately obvious) chest and abdominal injuries observed to date.

The 17 occupants protected by air bags who were entered into the study to date had a total of 101 injuries. The distribution of injuries by body region is shown in Table 2. This table also shows the Harm distribution. In the Harm accounting procedure, weighing factors are applied to the injury data, with each injury described according to its threat to life and weighted in proportion to its monetary cost. The procedure was pioneered by Malliaris [6,7]. The monetary weighing factors are based on cost data developed by Miller [8].

TABLE 2
Injury Distribution in Air Bag Equipped Cars
17 Cases

Region	# of Injuries	Harm %
Head/Neck	28	22%
Chest/Abdomen	22	38%
Lower Extremity	35	36%
Upper Extremity	16	4%
TOTAL	**101**	**100%**

Table 2 shows that head/neck injuries constitute 28% of the injuries, but only 22% of the Harm. In contrast, chest/abdominal injuries comprise 22% of the injuries, but 38% of the Harm. This result suggests that the chest/abdominal injuries are generally more serious than the other injuries for the cases collected to date.

The most common vehicle feature which was associated with chest/abdominal injuries was a bent steering wheel or a compressed steering column. Ninety-five percent of the Harm to the chest/abdomen region was associated with a steering wheel/column deformation.

CASE STUDIES-In the sections to follow, nine selected air bag cases will be summarized.

Four of the 17 air bag cases in the Jackson study involved occult chest/abdominal injuries. Among these cases, three involved multiple impacts, at a relatively moderate delta V. The fourth was a high severity (37 MPH delta V) impact. All occupants survived.

In addition to the Jackson cases, three cases from the NASS-SCI on-site air bag deployment investigation fleet are presented. These three cases are all moderate delta V (less than 25 MPH) frontal impacts into trees or poles. In all three cases, the occupant died of chest/abdominal injuries.

JACKSON OCCULT INJURY CASES
Case # 92-004

FIGURE 1

A 1992 Lincoln Town Car (Fig. 1) was traveling on a two lane road through a swampy region at a speed of 60 MPH when the left front tire blew out. The posted speed was 45 MPH. The time was 8:42 PM, the weather was rainy and the road was wet. The Lincoln swerved left of center, across the path of a 1982 Mercury Marquis. The two vehicles impacted left-headlight to left-headlight, with a contact width of 8". The Lincoln, directed to the left by this impact, continued along the left shoulder until it sustained a centerline impact with a concrete utility pole, approximately 350' from the first impact. The displacement measurements were as follows: Maximum vehicle crush - 30", Left "A" pillar intrusion - 3.25", Left instrument panel intrusion - 3", Steering wheel deformation - 2". The estimated delta

V for the car-to-car impact was 19 MPH and for the pole the impact was 22 MPH.

The driver was a healthy 83-year-old retired male who weighed 200 lbs. and was 5' 10" tall. He was married to a young woman and was the father of a then 4-year-old child. Protection was provided by an air bag and manual lap and shoulder belts. He sustained the following injuries:

- AIS-5   Hematoma, subdural bilateral
- AIS-1   Laceration, ear left
- AIS-2   Fracture, ribs left (3)
- AIS-4   Rupture, spleen with hemorrhage
- AIS-2   Laceration, descending colon mesentery
- AIS-1   Laceration, elbow left
- AIS-2   Fracture, navicular left
- AIS-2   Fracture, acetabulum with dislocation femur left
- AIS-2   Fracture, ankle left

The EMS personnel at the scene of the crash found that the driver did not meet any of the objective criteria which would have mandated transport to a trauma center (trauma center admission criteria) [9]. Because of suspicion of severe injury, EMS transported the patient by helicopter to a Level 1 trauma center. The patient had a multi-week stormy hospital course wherein he underwent multiple surgical procedures. He was discharged to a nursing home, where he remains today.

Case #92-023

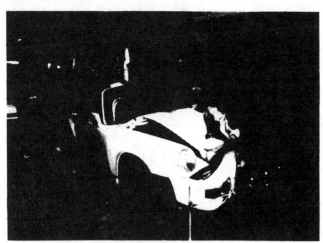

FIGURE 2

A 1990 Porsche Carrera (Fig. 2) impacted the rear of a 1985 Pontiac Fiero which was parked for a tire change in the fast lane of a busy three lane divided highway. The travel speed prior to impact was at the speed limit, about 55 MPH and the delta V is estimated at 22 MPH. The time was 3:30 PM, the weather was clear and the surface was dry. A van traveling ahead of the Porsche blocked the view of the parked Pontiac until the last instant, preventing timely evasive action. The Porsche sustained two impacts. The major impact was with the rear of the parked Pontiac. A second impact occurred with the left front fender of the Porsche impacting the concrete median barrier at a low delta V. The Porsche sustained a maximum frontal crush of 28.5" and a left toe pan intrusion of 4.5". The steering wheel was deformed 1". The driver was a 50-year-old male businessman. He was 5' 11" tall and weighed 220 lbs. He was protected by an air bag and manual belt. His injuries were:

- AIS-2   Fracture, ribs left (3-5)
- AIS-2   Laceration, liver
- AIS-1   Contusion, antecubital fossae (bilateral)
- AIS-2   Fracture, acetabulum left

At the scene, the patient did not meet trauma center admission criteria and was transported to an emergency room, not a trauma center. After a number of hours in that facility, the patient's condition deteriorated and he was transferred to a trauma center. He had a complex hospital course during which he underwent numerous surgical procedures. During his hospital care he developed a number of complications. A year after the crash the patient was unable to resume his pre-injury activities largely due to psychological problems.

Case #92-006

FIGURE 3

At 5:00 AM, a 1992 Honda Civic DX (Fig. 3) approached a right angle intersection in a residential area. The vehicle was traveling at 40 MPH, 10 MPH above the posted speed limit. Unable to negotiate the 90 degree turn to the left, the driver locked the brakes and proceeded straight ahead, leaving 45 feet of skid marks. The vehicle crossed a 5.0" high curb and a sidewalk, crashed through a wood fence and continued another 20' before impacting a cement block stucco house. The left side of the car went through the sliding glass door but the right side impacted the cement block exterior wall and the inside bedroom wall which was parallel to the direction of travel (Fig. 4).

FIGURE 4

The vehicle continued into the house striking an occupied bed in the master bedroom and coming to rest inside the house. The maximum vehicle crush was 23.0", the right toe pan intrusion was 8.0" and the steering wheel deformation was 4.0". The estimated delta V was 23 MPH. The driver was a 34-year-old male bartender who weighed 134 lbs. and was 5' 6" tall. His BAC was .079, measured 12 hours after the crash. He was wearing the manual belt system and the air bag deployed. He sustained a single injury:
- AIS-2   Laceration, liver with hematoma

The patient did not meet trauma center admission criteria and was transported to the nearest hospital, not a trauma center. He developed severe abdominal pain, suggestive of intra-abdominal injury, a number of hours later. He was then transferred to a Level 1 trauma center. The liver injury was treated non-surgically. The patient was discharged home a few days after admission and has done well.

Case #92-017

FIGURE 5

A 1991 Mercury Grand Marquis (Fig. 5) was traveling northbound on a two lane road. The posted speed for this area is 50 MPH. At 8:24 PM, the weather was cloudy and the road surface was dry when a 1989 Chevrolet pick-up truck had to brake suddenly to avoid striking a slower moving vehicle in the same lane. The pick-up truck swerved to the left into the northbound lanes. The Mercury struck the pick-up truck in the front end with the front bumper. After the initial contact, both vehicles rotated and sideslapped. The Mercury came to rest facing in a northeasterly direction. There was 65.0" of direct contact across the front bumper of the Mercury and the maximum extent of crush for this impact was 37.75". The vehicle sustained a left toe pan intrusion of 8.0" and a left instrument panel intrusion of 4.0". The steering wheel deformation was 3.0". The delta V was calculated to be 37 MPH.

The driver was a 63-year-old male engineer who weighed 175 lbs. and was 5' 10" tall. Protection was provided by an air bag as well as lap and shoulder belts. He sustained the following injuries:
- AIS-3   Fracture, ribs left (6-8) right (1)
- AIS-3   Contusion, lungs bilateral
- AIS-4   Contusion, cardiac
- AIS-2   Fracture, calcaneus right

This crash occurred in an area not served by a trauma center. The patient was transported from the scene to a local hospital. His injuries led to cardiac and respiratory insufficiency. These problems were not adequately treated. A number of days after the crash he was transferred to a Level 1 trauma center in a critically-ill state. The patient had a multi-week critical course which included a number of operations. He was discharged home and within two months had resumed all his pre-injury activities.

NASS SPECIAL CRASH INVESTIGATION OF FATALITIES INVOLVING CHEST/ABDOMINAL INJURIES IN LOW SEVERITY CRASHES

Case #91-12

A 1990 Dodge Shadow impacted a 10" diameter utility pole on the left side of a rural two lane state route where the speed limit was 35 MPH. The impact was at 12 o'clock and at the centerline of the Dodge's bumper. It produced a maximum of 14.5" of bumper crush. The resulting delta V of 14.4 MPH deployed the air bag. The steering wheel was bent .25" and the shear capsule was compressed 1.6".

The driver was a 36-year-old female, 5' 3" tall, weighing 112 lbs. She had a history of epilepsy. She was wearing her 3-point belt system. Her injuries were:
- AIS-3   Fractures, ribs bilateral
- AIS-3   Rupture, spleen
- AIS-5   Rupture, abdominal aorta

The patient was reportedly unconscious at the scene and she was transported to a local hospital.

Shortly after admission, the patient's condition deteriorated. She was taken to the operating room wherein she died presumably of the aortic rupture.

Case # 92-4

The driver of a 1986 Ford Tempo lost control of the car during a coughing attack. The Tempo traversed a 5" curb and impacted a wooden utility pole located on the right side of an urban street. The driver air bag deployed. The maximum depth of bumper crush was 14.8" which produced a delta V of 17 MPH. The steering wheel was deformed 3.3" forward.

The driver was a 57-year-old female, 5' 6" tall, weighing 150 lbs. Her seat adjustment was 3" from the full forward position. Her injuries were:

- AIS-3   Rupture, spleen
- AIS-1   Contusion, abdominal wall
- AIS-1   Contusion, breast bilateral
- AIS-1   Abrasion, chin
- AIS-1   Laceration, lower lip

The crash victim was conscious at the scene and reported no pain. She was transported to a local hospital where on admission she was reportedly stable. Four hours following the crash she expired. There was concern by the case reviewers that the injuries to the spleen were not initially recognized.

Case #91-10

A 1991 Pontiac Firebird ran through a T-intersection in a residential area and impacted an 18" diameter tree with its center front. The estimated impact speed was 21 MPH with a delta V of 19 MPH. The vehicle sustained a maximum depth of crush of 22" from the 12 o'clock direction of force impact. The steering wheel was deformed 2" forward and the steering column shear capsule was separated 3".

The driver was a 46-year-old male, 5' 11" tall and weighed 230 lbs. He was wearing manual lap and shoulder belts. He temporarily lost consciousness as he approached the T-intersection. The impact with the tree caused the air bag to deploy. The driver sustained the following injuries:

- AIS-4   Fractures, ribs bilateral (2-9)
- AIS-4   Contusion, heart
- AIS-1   Ecchymosis, legs, distal to the knees
- AIS-1   Abrasion, chin
- AIS-1   Abrasion, forehead with ecchymosis

The driver remained conscious post-crash and was transported to a local hospital where he expired several hours following the crash. The apparent cause of death was the contusion of the heart. However, the apparent loss of consciousness prior to the crash opens the question of a heart attack at that time. Acute and/or chronic heart disease could have exacerbated the problem of the contusion, according to the case reviewers.

JACKSON HIGH CRASH SEVERITY CASES

Case #93-001

A 1992 Honda Civic DX (Fig. 6) was traveling northbound at 12:43 AM. The road surface was dry, the weather conditions were clear and the estimated speed was 40 MPH, the same as the posted speed. As the Honda entered a right hand curve, the vehicle crossed the centerline into the path of a 1993 Mitsubishi Starion. The vehicles struck head-on in the southbound lane. After the impact with the Mitsubishi, the Honda struck and mounted the west barrier curb. The Honda continued traveling west and struck a bridge rail with the left front bumper. After this impact, the Honda rotated clockwise and came to rest on the west sidewalk. There was 54" of direct contact on the front bumper of the Honda. The maximum extent of crush was 42.25". In addition, there was a brake pedal intrusion of 14.5" and a left toe panel intrusion of 12.5". The delta V for the Honda was 55 MPH.

FIGURE 6

The Honda driver was a 40-year-old male factory worker who weighed 155 lbs. and was 5' 8" tall. He was protected by an air bag and lap and shoulder belts. His BAC was .200. He sustained the following injuries:

- AIS-1   Laceration, forehead (minor)
- AIS-1   Abrasion, neck left
- AIS-1   Contusion, chest upper bilateral
- AIS-3   Fracture, ribs right (2-4) with pneumothorax
- AIS-1   Abrasion, forearm left
- AIS-1   Abrasion, upper thigh right
- AIS-1   Abrasion, knee left
- AIS-2   Fracture, calcaneus right
- AIS-2   Fx/Dislocation, tarsal/metatarsal joint right
- AIS-2   Fracture, metatarsals right (2-4)
- AIS-2   Fracture, metatarsals left (2-4)
- AIS-2   Fracture, medial malleolus left

The patient met trauma center admission criteria and was transported directly to a Level 1 trauma center from the scene. The hospital course was less than two weeks during which he underwent multiple orthopedic surgeries. He was discharged home with outpatient rehabilitation. Follow-up revealed that the patient was walking with support but was still unable to wear his right shoe. He returned to work approximately three months post-crash.

Case #91-002

FIGURE 7

A 1991 Volvo 740 Turbo (Fig. 7) was traveling westbound in the second lane at an estimated speed of 60 MPH, exceeding the posted speed by 15 MPH. At 11:55 AM, the weather was clear and the road was dry when the driver reportedly "blacked out". The vehicle continued forward crossing the eastbound lanes and struck a 5.5" barrier curb with the left front wheel. It then mounted the curb and continued in a westerly direction across a 7' concrete sidewalk and grass lawn. It continued approximately 75' striking an aluminum traffic signal control box which measured 2'x3'x7'. The control box was uprooted from the impact and came to rest 95' west and 57' north of its point of origin. The Volvo continued forward another 10' and struck an 18" square concrete utility pole. The direct contact of the front bumper was 36" and the maximum extent of crush was 47.7". The vehicle sustained a right "A" pillar intrusion of 5.5" and a right dash intrusion of 4.0". The delta V was calculated to be 55 MPH.

The driver was a 39-year-old male businessman who weighed 323 lbs. and was 5' 11" tall. Protection was provided by an air bag only. He sustained the following injuries:

    AIS-3   Hemorrhage, subarachnoid
    AIS-2   Laceration, scalp major
    AIS-2   Fracture, C7 vertebra
    AIS-1   Contusion, chest
    AIS-2   Laceration, mesentery small bowel
    AIS-4   Avulsion, small bowel
    AIS-2   Contusion, colon
    AIS-1   Abrasion, forearm right
    AIS-3   Fracture, femur right open

The patient did not meet trauma center admission criteria at the crash scene. He was transported to the Level 1 trauma center because it was the nearest facility. Although initially stable, within an hour of admission the patient's condition began to deteriorate significantly. He was taken to the operating room for treatment of seemingly correctable problems. During the abdominal operations, he developed severe problems including cardiac and respiratory failure. The cause of these problems is unclear. He died less than 24 hours later of multiple organ system failure.

Table 3 summarizes the Jackson air bag occult injury cases. Note steering column deformation ("DEF") occurred in all these cases.

TABLE 3
AIR BAG CASES-OCCULT INTERNAL INJURIES

CASE #	CAR	DEF	INJURY	TRIAGED
92-004	92-Lincoln	2.0"	4-Spleen	Trauma Ctr EMS Susp.
92-023	90-Porsche	1.0"	2-Liver	Hospital
92-006	92-Honda	4.0"	2-Liver	Hospital
92-017	91-Mercury	3.0"	3-Lung	Hospital

Table 4 summaries the NASS data for air bag cases. Note that steering wheel deformation was reported in 23 cases. Of those 15 admitted to the hospital, nine had AIS-3 or greater injuries.

TABLE 4
NASS DATA 1989-1991- AIR BAG DEPLOYMENT FRONTAL CRASHES

135	Cases of air bag deployment
626	Injuries to occupants
23	Cases with reported steering wheel deformation
9	Occupants suffered AIS-3+ injuries

DISCUSSION

OCCULT INJURIES-The air bag may have ushered in a new era in injury management. Strategies for dealing with severely injured people emanate from military approaches. The methods of assessing injuries on the field, stabilizing life threatening derangements in physiology and anatomy, rapidly transporting patients to hospitals wherein definitive diagnostic and therapeutic interventions can be provided have continued to improve throughout the history of American military conflicts. The concept of "triage" is fundamental to these processes. It involves the allocation of resources to injured individuals based on the severity of the injuries and the likelihood of survival.

Injuries which occur in military situations are typically related to the penetration of one or more body parts by bullets and shrapnel. The presence of a life- or limb-threatening injury is obvious by the degree of physiological compromise, such as shock, and/or the entry location of the projectile and its presumed path in the body. In automobile related injuries the damaging forces are typically blunt. At the scene of a crash the presence of an internal injury, such as a laceration of the liver, is suggested if there is external evidence of contact (over the area of the liver) such as bruising or pain, and/or if there is physiological abnormality such as shock.

The clinician dealing with blunt trauma must always maintain a high level of suspicion of injury; the failure to recognize injuries can lead to loss of life or limb. Decisions on whether a trauma victim needs to be taken to a hospital and the level of trauma care expertise of the receiving hospital is based on triage criteria mandated typically by state governments. The most capable trauma centers are designated Level 1 or 2 according to the American College of Surgeons Committee on Trauma (ACSCOT) standards [9].

ACSCOT and other organizations have developed criteria that largely are the basis for local trauma system's criteria. These are typically based on physiological abnormalities such as systolic blood pressure below 90 millimeters of mercury or Glasgow Coma Score below 13. Location of injury is also included in some systems' criteria, such as bullet wound to the abdomen. Another group of criteria is mechanism of injury such as a fall of more than two building stories or ejection from a vehicle in a crash. The last group of criteria is called "index of suspicion". These are based on the opinion of EMS personnel at the scene. Even though other objective criteria are not present, the patient does not "look right" and therefore should be evaluated in a trauma center. The criteria which are utilized in Dade County, Florida, wherein the Ryder Trauma Center is located, are listed below.

ADULT TRAUMA CRITERIA
USED IN DADE COUNTY, FLORIDA
- Systolic BP ≤ 90
- Respiratory rate <10 or >29 BPM
- Glasgow Coma Scale ≤ 13
- Penetrating injury to head, neck, chest, abdomen or groin
- Paralysis
- Second or third degree burns ≥ 15% TBSA
- Amputation proximal to wrist or ankle
- Ejection from motor vehicle

Most studies have shown that the combination of criteria that are utilized by individual trauma systems work well [10]. Quality of care analyses, which are mandated for trauma systems, typically show very few cases where injured individuals did not receive adequate treatment. One way of evaluating the quality of care in a system is to evaluate deaths and determine if any were preventable. Most well organized trauma systems have low preventable death rates [11].

The air bag may affect the ability to successfully evaluate crash-involved occupants utilizing existing criteria. The air bag-protected occupant typically will not have physiologically compromising head, chest or abdominal injuries nor the facial lacerations, bruises and fractures that often make crash victims "look" seriously injured. Personnel at the scene may not send occupants, who later turn out to have serious injuries, to trauma centers.

In most cases, the injuries to occupants protected by air bags are less severe than to occupants without airbags, but the problem is that the residual injuries may not be recognized by EMS or emergency room personnel. It is not that the air bag contributes to injury, but that it changes a very serious, obvious injury into a less serious, but less obvious injury. If this less obvious injury is not treated, it may become a serious problem.

What kinds of injuries may not be initially obvious but become life threatening? Any chest or abdominal injury wherein the immediate physiological implications are minimal fall into this category, such as contusions and lacerations of the lungs, aorta and heart. In the abdomen tears of the solid organs, in particular of the liver and spleen, can initially cause limited bleeding with little blood pressure loss and minimal abdominal pain. The mesenteries which connect the bowel components to their central blood supplies can be torn with limited initial bleeding. The bowel can also be torn. Bleeding is less problematic with bowel injuries; contamination of the abdominal cavity with irritating and often infection producing components is the problem. Continued loss of blood and/or contamination of the abdomen typically leads to very serious problems. Even minutes of uncorrected shock and contamination can lead to death or failure of other organ systems such as the lungs or the immune system.

How can the air bag contribute to serious abdominal and chest injuries which are not initially noticeable? There appears to be three mechanisms:

- The air bag may contact the chest or abdomen directly during deployment with enough force to injure internal organs. This appears to be a rare event. It occurs mainly when the occupant is in close proximity to the bag as it deploys.

- The occupant's position and size, and/or the velocity of the crash may exceed the air bag's protective capability allowing the occupant to contact an internal component of the automobile, particularly the steering wheel. The examples to date of this include large occupants and crash velocities in excess of 35 miles per hour.

- There may be multiple collisions wherein the air bag has partially or completely deflated after the initial external collision. Thus, the occupant can contact internal objects such as the steering wheel.

The problem is that the occupant may not meet existing trauma center admission criteria after one of these types of events. However, the individual may have sustained some chest or abdominal internal injury as previously described. It appears that suspicion of injury must occur if these potentially injured occupants are to receive a full evaluation in the hospital. In present trauma centers the determination that an individual has an internal abdominal or chest injury is based on a combination of available historical information, physical examination and diagnostic tests. The latter include: x-rays (which are particularly useful in evaluating lung injuries), ultrasound (to examine the heart for internal injury and/or bleeding into its surrounding sac), CAT scanning (which is useful mainly for abdominal injuries and in a limited fashion chest injuries), angiography (which is particularly useful for evaluating the most potentially life-threatening occult injury--partial tear of the thoracic aorta) and diagnostic peritoneal lavage (which is extremely useful for discovering bleeding in the abdominal cavity).

Not all injuries require surgery. Today many lacerations of the liver and spleen are treated without surgery. The diagnosis of these non-operated injuries is typically made via CAT scan. The stable patient can be observed in an intensive care unit. The immediate capability of performing surgery if the patient deteriorates is a necessary requirement of this strategy. Non-operative therapy works in part because the patients are kept at bed rest. At rest the injured organs, especially the liver, typically heal rapidly. When patients with these injuries are sent home from the crash scene, because they appear uninjured, their normal activities can further stress the injured organs causing life-threatening bleeding.

Dealing with injury in the era of the air bag is dramatically different than in the setting of the battlefield. The injuries in the air bag setting will often not be obvious and the receiving hospital may evaluate many patients who turn out to be injury-free. In the military scenario the injuries are obvious and hospitals are for the clearly life-threatened patient.

The challenge is to find reliable indicators of potential injury that are available at the crash scene. Deformation of the steering wheel in frontal collisions appears, in the small number of cases studied to date by the Jackson investigation team, to be a useful indicator of possible abdominal or thoracic injury. As Table 4 indicates, all four of the cases wherein occult abdominal injury occurred to air bag-protected drivers, there was steering wheel deformation. Based on these observations a Research Note was developed by NHTSA suggesting that EMS evaluate the steering wheel in these cases [12]. The Note recommends that deformation of the steering wheel be considered as a

criterion for further evaluation, in a hospital setting, of the stable, air bag-protected driver involved in a frontal crash.

Other potential sources of information about possible internal abdominal or chest injuries in air bag-protected crash occupants are the amount of deformation to the exterior of the automobile and/or intrusion of internal components. Occupant stature and position at the crash moment may turn out to be very predictive, particularly if the occupant is out of position or on top of the air bag as it deploys. Crashes which include multiple collisions may correlate highly with injuries.

Continued, detailed study of air bag-involved crashes wherein occult injuries occur is necessary. Traditional analyses of real-life crashes, such as NASS, do not provide precise definition of the clinical course. For example, the timeline from the crash to injury discovery and treatment is of real importance. Hospital-based studies involving multidisciplinary teams such as the ones ongoing at the Ryder Trauma Center and the Maryland Institute for Emergency Medical Services are necessary to provide these data [13].

SUMMARY

The number of air bag-involved crashes wherein difficult triage decisions need to be made is unknown at this time. As previously stated, it is estimated by NHTSA that by the year 2000 more than five hundred thousand air bag deployments will occur yearly. To avoid missing injuries, triage strategies err on the side of transporting patients, who may have serious injuries, to trauma centers for evaluation. Applying this to air bag-protected occupants, there may be many admissions to trauma centers for work-up of occult injuries. On the other side of the equation, admissions of severely injured occupants may decrease as a function of air bag protection.

Determinations need to be made as to whether evaluations of initially stable crash victims, with suspicion of major injury, can be performed in the most sophisticated trauma care arenas (Level 1 or 2 trauma centers). They have the expertise and equipment to work-up these patients and treat their spectrum of injuries. At this time, however, these centers concentrate on treating individuals with obvious threat to life or limb and are often functioning at or above capacity [14]. In the air bag era, the role of the trauma center may need to be broadly expanded to include the evaluation of stable patients who may have incurred occult injuries.

REFERENCES

1. Zador, Paul L.; Ciccone, Michael A. (Insurance Institute for Highway Safety, Arlington, VA). Automobile Driver Fatalities in Frontal Impacts: Air Bag Compared

With Manual Belt. American Journal of Public Health. 1993 May; 83(5): 661-66.

2. Rosenblatt, Marc; Freilich, Benjamin; Kirsh, David. Air Bags: Trade-Offs. New England Journal of Medicine. 1991 Nov 21; 325(21): 1518-1519. ISSN: 0028-4793.

3. Augenstein, J. S.; Peterson, E. A. Computerization: Solution to Problems in the Input, Manipulation, and Storage of Intensive Care Unit Data. Textbook of Critical Care. 3rd ed. Philadelphia, Pennsylvania: W. B. Sanders-Society of Critical Care in Medicine; 1990: Chapter 3. (Shoemaker, W. C.; Ayres, S. M.; Holbrook, P. [and others]).

4. NASS. NHTSA. MDE Users Manual. Revision 1990 ed.; U. S. DOT; 1990 Jan.

5. U.S. Department of Transportation, National Highway Traffic Safety Administration, Effectiveness of Occupant Protection Systems and Their Use, Report to Congress, January, 1993, DOT HS 808 019, NRD-31; also, "Final Regulatory Evaluation, FMVSS No. 208, Mandatory Air Bag installation", NHTSA, June 1993; also, U.S. DOT NHTSA "Motor Vehicle Fatalities in 1992 Fall to 30-Year Low", Press release Tuesday June 22, 1993; also U.S. Department of Transportation, National Highway Traffic Safety Administration, Plans and Policy Office of Regulatory Analysis, Final Regulatory Evaluation, FMVSS No. 208, Mandatory Air Bag Installation, June, 1993.

6. Malliaris, A. A Search for Priorities in Crash Protection. SAE. ; 1982; 820242.

7. Malliaris, A. Harm Causation and Ranking in Car Crashes. SAE. ; 1985; 850090.

8. Miller, T. [and others]. The Cost of Highway Crashes. FHWA Publication. : FHWA; 1991 Oct; FHWA-RD 91-055.

9. Committee on Trauma. Committee on Trauma: Hospital and Prehospital Resources for the Care of the Injured Patient. Bulletin of the American College of Surgeons. 1986 Oct; 71: 4-12.

10. MacKenzie, Ellen J.; Steinwachs, Donald M.; Ramzy, Ameen I. (John Hopkins University, School of Hygiene & Public Health). Evaluating Performance of Statewide Regionalized Systems of Trauma Care. Journal of Trauma. 1990 Jun; 30(6): 681-8.

11. Kreis, D. J.; Placencia, G.; Augenstein, D. Preventable Trauma Deaths: Dade County Florida. Journal of Trauma. 1986; 26: 649-654.

12. Lombardo, Louis V.; Ryan, Susan D.; "Detection of Internal Injuries in Drivers Protected by Air Bags", Research Note, U.S. Department of Transportation, National Highway Traffic Safety Administration, August 1993.

13. Siegel, J. H.; Mason-Gonzalez, S.; Dischinger, P.; Cushing, B.; Read, K.; Robinson, R.; Smialek, J.; Heatfield, B.; Hill, W.; Bents, F.; Jackson, J.; Livingston, D.; Clark, C. Safety Belt Restraints and Compartment Intrusions in Frontal and Lateral Motor Vehicle Crashes: Mechanisms of Injuries, Complications, and Acute Care Costs. Journal of Trauma. 1993 May; 34(5): 736-759.

14. Hoff, William S.; Tinkoff, Glen H.; Lucke, Joseph F.; Lehr, Shannon (Department of Surgery, Division of Trauma, Lehigh Valley Hospital Center, Allentown, PA). Impact of Minimal Injuries on a Level I Trauma Center. Journal of Trauma. 1992; 33(3): 408-412; ISSN: 0022-5282.

940716

Upper Extremity Injuries Related to Air Bag Deployments

Donald F. Huelke, Jamie L. Moore, and Timothy W. Compton
University of Michigan Transportation Research Institute

Jonathan Samuels and Robert S. Levine
Wayne State Univ.

ABSTRACT

From our crash investigations of air bag equipped passenger cars, a subset of upper extremity injuries are presented that are related to air bag deployments. Minor hand, wrist or forearm injuries-contusions, abrasions, and sprains are not uncommonly reported. Infrequently, hand fractures have been sustained and, in isolated cases, fractures of the forearm bones or of the thumb and/or adjacent hand. The close proximity of the forearm or hand to the air bag module door is related to most of the fractures identified. Steering wheel air bag deployments can fling the hand-forearm into the instrument panel, rearview mirror or windshield as indicated by contact scuffs or tissue debris or the star burst (spider web) pattern of windshield breakage in front of the steering wheel.

INTRODUCTION

The air bag should be considered as part of the preventative medicine armamentariun but it is not the ultimate polio vaccine for traffic medicine. The injury reducing potential and the lifesaving benefits of the air bag, the supplemental restraint now commonly found in the steering wheels of most new passenger cars, have been documented in the medical and engineering literature (1-14). However, case reports of injuries from the air bag, particularly corneal/scleral abrasions, facial erythema, contusions, or abrasions have also been noted (15-25).

This report documents the variety of injuries to the upper extremities related to air bag deployments, ranging from forearm and hand erythema, contusions, lacerations, or thermal burns, to fractures or dislocations of the thumb, wrist, and forearm.

These cases were investigated by members of the field investigation team from the University of Michigan Transportation Research Institute (UMTRI) and some from the National Highway Traffic Safety Administration (NHTSA) Special Crash Investigation Section.

MATERIALS AND METHODS

At the University of Michigan Transportation Research Institute (UMTRI) there has been an ongoing field accident research program to determine the causes of injuries and deaths in automobile crashes since 1961, with investigations primarily in Ann Arbor, Michigan and the surrounding county. In the mid-1980's investigations of air bag crashes began, with a gradual increase in frequency as more new cars were equipped with air bags. Since 1988 UMTRI's program has expanded nationwide in search of frontal crashes with air bag deployments. More than 1900 air bag notifications have been received from various sources nationwide. Car tear-down, repair, or lack of occupant injury information has precluded detailed study of many of these notifications. Over 250 air bag crashes have been investigated in detail as of December, 1993.

The National Highway Traffic Safety Administration (NHTSA) has also been conducting a special study of crashes involving automobiles equipped with air bags. To date they have investigated about 1200 air bag collisions with little overlap of cases between NHTSA and UMTRI. NHTSA contributed to this study by providing case information from their air bag deployment investigations.

In each case the investigators document vehicle, environmental, and occupant data including make, model, and year of the car, the amount of car crash damage, age, height, weight, and detailed injuries of the occupant(s), as well as information about the crash. From these investigations selected cases are presented here in capsule format. Injury contact data are the objective opinions of the crash investigators and rarely based on occupant statements.

RESULTS

ERYTHEMA, CONTUSIONS, ABRASIONS AND BURNS - Following an air bag deployment, the affected occupant frequently complains of various minor upper extremity injuries such as discoloration's, abrasions, or contusions. We have identified these minor injuries in about 50 air bag cases collected by UMTRI. Such injuries are caused by the deploying air bag impacting the skin. Several examples follow:

Case 1 - A 1991 Cadillac Eldorado driven by a 30-year-old unrestrained female sustained extensive right front damage when it struck a 12 cm diameter tree (Fig. 1). The driver, height-163 cm, mass-64 kg, had her seat positioned forward. She sustained contusions of both anterior forearms, had contusions of both temples, and complained of a headache, all from contact with the air bag. Contusions of the knees and cervical strain were also reported. (UM-2909-91)

Fig. 1 Frontal collision, 1991 Cadillac Eldorado. Minor injuries sustained including contusions of forearms. Similar injuries in Cases 2 and 3.

Case 2 - A 1992 Honda Accord LX sedan struck a guardrail and then was hit by a tractor-trailer unit. Frontal damage from the guardrail impact was severe. Damage to the left rear from the truck was minor. The 35-year-old lap-shoulder belted male driver, height-180 cm, mass-72 kg, had the seat at mid position. His right wrist and thumb were contacted by the air bag as it deployed, resulting in an abrasion on the posterior right thumb and a contusion of the anterior right wrist. The driver also had cervical strain and abrasions of the left knee and left leg. (UM-2999-92)

Case 3 - On a slippery, snow-covered road a 1992 Plymouth Sundance hit the rear of a parked car. Minor front right damage was noted. The air bag deployed. The 39-year-old lap-shoulder belted female driver, height-165 cm, weight-55 kg, sustained a sore right anterior forearm muscle when her forearm was "thrown off" the steering wheel by air bag deployment. Her extremity did not impact anything within the car after bag deployment. No other injuries were reported. (UM-3118-93)

Such forearm injuries as described above have also been noted to front right passengers who interacted with the deploying passenger side air bag (see cases 17, 19 and 20)

Case 4 - A 1992 Dodge Grand Caravan, driven by a 52-year-old lap-shoulder belted male, height-183 cm, mass-84 kg, crossed the center line and was struck head-on by another car (Fig 2). The driver sustained minor injuries including lacerations of the knees and ankles, a contusion of the chest from the shoulder portion of the restraint system and multiple small lacerations of both the anterior right and left forearms from air bag contact. (UM-2996-92)

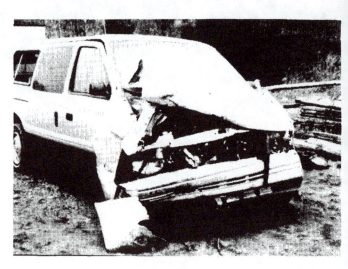

Fig. 2 Forearm lacerations from air bag contact in this 1992 Dodge Grand Caravan frontal crash.

There have been reports of individuals experiencing burns on their hands from the hot gases generated during air bag deployment. Two similar cases in our files are summarized below:

Case 5 - A 1990 Infiniti M30 rear-ended a stopped car. The Infiniti had moderate right front damage. The 37-year-old lap-shoulder belted female driver, height-160 cm, mass-45 kg, apparently had both of her hands on the steering wheel spokes at the time of the crash. She sustained burns on both hands from the air bag gas vented through the openings in the back of the bag. These injuries were noted as first degree burns on the dorsal left wrist and of the dorsal left hand between thumb and index finger, second degree burns in the same location on the right hand, and multiple burns of the right fingertips. (UM-NAB-022)

Case 6 - A 1990 Plymouth Sundance struck the rear of a car stopped in traffic. The Sundance had very minor front damage. The car was equipped with an early design untethered air bag (see discussion). The 65-year-old female driver, height-163 cm, mass-72 kg, was wearing her 3-point restraint. She sustained an abrasion beneath her chin and a slight contusion of her left lateral forearm from contact with the deploying air bag. The air bag vent gas melted her polyester glove at the base of her left thumb, beneath which was a minor burn. (UM-2854-90)

FLAILING EXTREMITY INJURIES

Occasionally, injuries result from the air bag flinging an upper extremity towards car interior surfaces, primarily to the windshield or instrument panel.

Case 7 - A 1985 Ford Tempo, attempting a left turn, struck the left side of an oncoming car. The Tempo was moderately damaged in the left front (Fig. 3). The 36-year-old male driver, height-175 cm, mass-78 kg, was wearing his 3-point restraint. His hand was thrown into the windshield by the deploying air bag, causing a contusion and slight sprain of the left wrist. The windshield was cracked from the hand impact. (Fig. 4) (UM-FMC-003)

Fig. 3 Frontal crash, 1985 Ford Tempo

Fig. 5 Frontal crash into tree, 1991 Dodge Shadow

Fig. 4 Windshield cracked from hand impact due to air bag deployment.

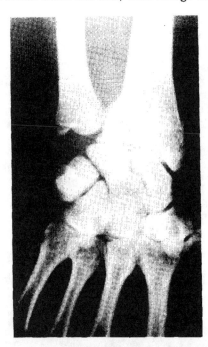

Fig. 6 Impact fracture right distal radius when the extremity was flung into the instrument panel by the deploying air bag.

Case 8 - A 1991 Dodge Shadow went off the right side of the road and struck a tree (Fig. 5). The 25-year-old unrestrained male driver, height-180 cm, mass-98 kg, did not have any facial injuries from the air bag. Both forearm bones were fractured including an impacted fracture of the right distal radius and fractures of the adjacent ulna and of the base of the second metacarpal from impacting the instrument panel when his hand was flung from the steering wheel by the deploying air bag (Fig. 6). His right femoral neck and mid shaft fractures from impacting the console would have been prevented if the available 3-point restraint had been worn. (UM-3098-93)

Case 9 - On a wet, slippery expressway an out of control car crossed the grassy median and struck a 1990 Plymouth Acclaim in the left side (Fig. 7). In this side impact the air bag deployed. The 28-year-old female driver, height-171 cm, mass-90 kg, was wearing her 3-point restraint system. She sustained lacerations of the spleen and contusions of the lower rib cage from the interior of the driver's door, abrasions and contusions of the knee and pretibial area from the lower instrument panel and, due to air bag deployment, her right hand was flung and impacted the upper instrument panel, producing fractures of the scaphoid (wrist bone) and base of the fifth metacarpal. She also had an abrasion of the left forearm from the air bag. (UM-3078-92)

Fig. 7 Side swipe type of crash. Wrist and hand fractures from instrument panel impact due to deploying air bag.

Case 10 - A 55-year-old male driver, height-173 cm, mass-61 kg, was wearing a 3-point restraint system. While driving on an expressway he fell asleep and his 1993 Infiniti J30 went off the road and struck an earthen embankment (Fig. 8). The air bag deployed in this frontal crash. The driver sustained a small laceration of his right wrist from his wristwatch and displaced transverse fractures of the right mid radius and distal ulna and fracture of the right distal radius (from steering wheel contact due to forearm entrapment by the air bag) (Fig. 9). No other injuries were reported. (UM-NAB-039)

Fig. 8 1993 Infiniti J30 impacted an embankment. Forearm injuries are shown in Fig. 9

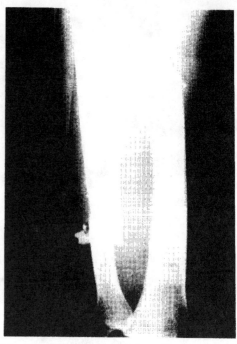

Fig. 9 Displaced fractures of right radius and ulna from steering wheel-air bag entrapment

Case 11 - While attempting to turn left, a 1990 Lincoln Continental struck a van head-on. Damage to the front of the Continental was moderate (Fig. 10). The 33-year-old lap-shoulder belted female driver, height-166 cm, mass-57 kg, sustained a fracture and abrasions of the right wrist and abrasions of the left wrist. Air bag deployment caused the wrist abrasions and propelled her right hand into the center instrument panel producing a wrist fracture. (UM-FMA-036)

Fig. 10 1990 Lincoln Continental with frontal damage. Air bag deployment flung the driver's right hand to the instrument panel producing a wrist fracture.

Case 12: - A van applied its brakes and was rear-ended by a 1992 Lincoln Continental, 4-door sedan. Neither driver nor front passenger wore the available lap-shoulder belts. Both air bags deployed. The frontal damage to the Continental was minor (Fig. 11). The 52-year-old male driver, height-185 cm,

inor (Fig. 11). The 52-year-old male driver, height-185 cm, mass-95 kg, sustained abrasions and contusions of the right anterior forearm and abdomen from the steering wheel air bag. The front right 51 year-old male, height-175 cm, mass-84 kg, had a contusion of the anterior aspect of the right forearm, a sprain of the joints of the third finger and three overlying lacerations of the left hand from contact with the passenger side air bag. (UM-FMA-057)

Fig. 11 1992 Lincoln Continental. Both driver and front passenger sustained forearm abrasions and hand injuries from the air bag.

Case 13 - A 1989 Dodge Daytona struck the front of an incoming vehicle that had crossed the center line. Left front corner impact damage to the Daytona was very minor. The 9-year-old lap-shoulder belted female driver, height-170 cm, mass-65 kg, had her seat at mid position. The driver's left hand was propelled into the windshield by air bag deployment as indicated by the star burst pattern on the windshield (Fig. 12). The driver's injuries included a bruise of the left hand, from the windshield impact, a cervical spine strain, and bruises of the chest and right hip from the belt restraint webbing. (UM-CCA-006)

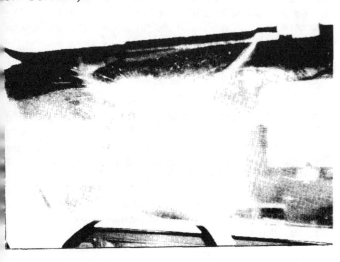

Fig. 12 Windshield star burst fracture due to driver's hand impact.

Case 14 - A 1991 Dodge Caravan SE hit the side of an oncoming car that had turned in front of it. Impact was to the Caravan's left front, producing minor damage. The 34-year-old male driver, height-178 cm, mass-86 kg, was wearing his 3-point restraint. When the air bag deployed his hand was forced upwards from the steering wheel to strike the upper left area of the windshield producing a spider web impact pattern. The driver sustained pain and muscle strain of the left hand as a result of impacting the windshield. No other injuries were reported. (UM-CCA-074)

Case 15 - A 1990 Lincoln Town Car Signature Series was traveling on a rural road when it hit the side of a farm tractor that had failed to stop (Fig. 13). Moderate impact damage was located in the left front of the car. The 76-year-old male driver, height-183 cm, mass-82 kg, was lap-shoulder belted. The air bag deployment flung his left hand into the windshield (Fig. 14). As a result of this impact he sustained a contusion and multiple lacerations on the dorsum of his left hand as well as a sprain of the left hand. (UM-FMA-044)

Fig. 13 1990 Lincoln Town Car driver sustained minor injuries.

Fig. 14 Hand impact producing windshield fracture. Contusions and multiple lacerations on dorsum of left hand with wrist sprain were sustained.

Case 16 - A 1991 Dodge Spirit driven by a 62-year-old 3-point restrained male, height-170 cm, mass-66 kg, failed to stop at a red light and struck the left front of another car (Fig. 15). The two cars then rotated together, striking again. The initial impact to the right front of the Spirit caused moderate damage while the second impact resulted in minor damage to the right side. The driver's hands and forearms were contacted by the deploying air bag. His right hand was forced off of the steering wheel and struck the rear view mirror. The mirror rotated and hit and cracked the windshield. The driver sustained a contusion of his posterior right hand from this rear view mirror impact, and air bag induced abrasions of the anterior forearms and an abrasion between the right first finger and thumb. The driver also had contusions of both hips from the lap portion of the 3-point restraint. (UM-3048-92)

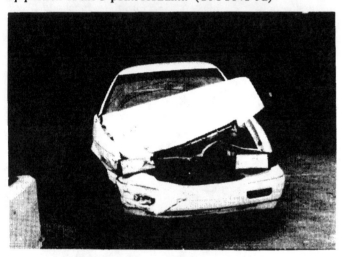

Fig. 15 1991 Dodge Spirit driver's hand injury sustained by impacting the rear view mirror.

AIR BAG MODULE COVER AND DEPLOYING AIR BAG INJURIES

During every air bag deployment the steering wheel air bag module cover is flung open with the considerable force necessary to achieve timely full deployment. We have seen several thumb injuries produced by the opening of the module covers or by the deploying air bag.

Case 17 - A 1990 Lincoln Town Car struck the right front of a car which disregarded a stop sign. The 42-year-old male driver, height-180 cm, mass 59-kg, was wearing his lap-shoulder restraint and had his seat at mid position. Air bag deployment resulted in contusions of the driver's left thumb and his anterior left upper arm. The forceful opening of the air bag module was responsible for the thumb contusion, as evidenced by the marks on the left side of the module cover (Fig. 16). The driver also sustained muscle strain across the center of his chest from the shoulder belt. The right front lap-shoulder belted passenger, a 42-year-old female, height-157 cm, mass-59 kg, with her seat in the mid position, sustained a contusion to her left anterior upper arm from the passenger air bag, as well as a contusion and muscle strain of her left hip from the seat belt buckle. (UM-FMA-045)

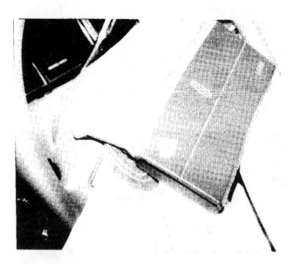

Fig. 16 Scuff on upper left of air bag module cover from driver's left thumb.

Case 18 - A 1990 Infiniti Q45 sustained front end damage when it rear-ended a Chevrolet S-10 Blazer stopped at a traffic light. The 56-year-old lap-shoulder belted male driver, height-180 cm, mass-105 kg, only sustained a sprain of the left thumb, caused by the deploying air bag. (UM-NAB-004)

Case 19 - A 1989 Lincoln Continental, traveling at about 55 mph on a rural road, struck a large deer. The center front bumper, grille area, and hood were damaged. The 67-year-old male driver, height-170 cm, mass-82 kg, sustained an abrasion across his left chest from the shoulder belt portion of the 3-point restraint. Air bag deployment resulted in a left thumb sprain and an abrasion of his right forearm. The restrained 67-year-old female right front passenger, height-173 cm, mass-72 kg, sustained a contusion of her right wrist from air bag contact. (UM-FMA-028)

Case 20 - A 1993 Lincoln Continental rear-ended a stopped car. Damage was minor. The driver sustained minor injuries. The lap-shoulder belted front passenger, a 66-year-old female, height-157 cm, mass-61 kg, had a 5 cm tear of the skin of the web between the left thumb and hand with tearing of the underlying adductor pollicis muscle. The metacarpophalangeal joint of the thumb was dislocated with the ulnar side of the capsular ligament being torn.

In a number of cases we have noted upper extremity fractures which are related to air bag deployment. (see also cases 9 and 10).

Case 21 - A 1993 Ford Taurus was turning left when hit in the right front area. Damage to the car was minor. The male driver, 59-years old, height-170 cm, mass-54 kg, was wearing his 3-point restraint. No occupant contact marks were identified in the car. Air bag deployment produced comminuted mid-shaft fractures of the right radius and ulna (Fig. 17). His right forearm and/or hand was forced into his face producing a nasal fracture, and a fracture of the right maxillary sinus (Fig.18). Chip fractures of two lower front teeth were also sustained (UM-FMA-055).

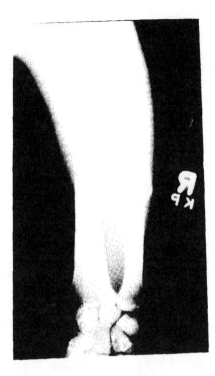

Fig. 17 Driver forearm fractures produced in a left turning maneuver. Forearm was on the air bag module cover at deployment.

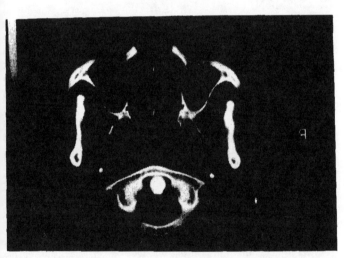

Fig. 18 Fracture of the right cheek (arrow) from driver's hand impact due to air bag deployment.

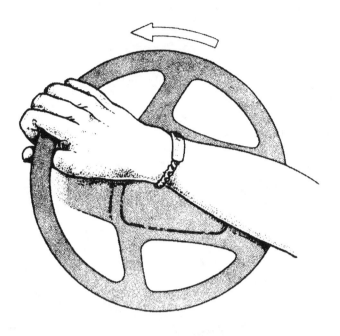

Fig. 19 Position of right forearm at time of air bag deployment producing injuries shown in Fig. 17.

Case 22 - A stopped car was struck by a 1992 Volvo 960 4-door sedan. Damage to the Volvo was negligible. The 37-year-old lap-shoulder belted female driver, height-163 cm, mass-60 kg, sustained an upper chest contusion due to shoulder belt loading. The air bag module cover tore her light weight jacket below the elbow and produced a comminuted displaced Monteggia fracture of the right proximal ulna and a dislocation of the radial head (Fig. 20). The driver subsequently developed a compartment syndrome. The driver stated that her hand struck her face during the air bag deployment. No facial injuries were reported. (CS 92-18)

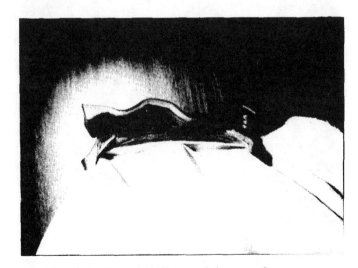

Fig. 20 Deformed air bag module cover from upper extremity contact.

Case 23 - A 1991 Mercury Grand Marquis, in a left turning maneuver, was struck along the left front side by another car (Fig. 21). On impact the upper air bag module flap contacted the mid right forearm of the 73-year-old lap-shoulder belted female driver, height-150 cm, mass-57 kg (Fig. 22). From impact by the air bag module cover and the contact of the deploying air bag she sustained multiple segmental open fractures of the right radius and ulna, with fractures of both the proximal and distal ulna as well as the olecranon. In addition, a circumferential degloving laceration (about 340 degrees) involving the skin and subcutaneous tissue was sustained. The right proximal fifth phalanx was also fractured, probably from the inflating air bag. The air bag flung her right upper extremity upward as indicated by tissue and blood spatters on the roof liner, side pillar and rear seat area. (CS 92-22)

Fig. 21 Left side front damage, 1991 Mercury Grand Marquis.

Fig. 22 Upper air bag module cover deformed by contacting the driver's right upper extremity.

Case 24 - A 75-year-old unrestrained female driver, height-163 cm, mass-41 kg, claims she passed out when driving in a left curve on a two lane road. Her car, a 1992 Mercury Grand Marquis LS, went straight and struck a 22 inch diameter tree at a change of velocity of 12 mph. The upper air bag module cover impacted her right forearm producing a nearly circumferential laceration of the mid right forearm and comminuted fractures of the distal radius and ulna. Eyeglass damage was due to face/air bag contact or by right arm/face impact. A cervical spine fracture was also diagnosed. (CS 93-01)

Case 25 - A 1991 Mercury Grand Marquis, 4 door, attempting to stop on an icy roadway slid across the oncoming lane, went off-road, and struck a tree at a relatively low speed. The 50-year-old restrained female driver, height-168 cm, mass-89 kg, sustained an open comminuted fracture of the distal right radius and ulna, and a contusion of the right abdomen. (DS 92-AB-10).

Case 26 - A 1992 Nissan NX 1600, 3 door hatchback was on an entrance ramp approaching a thruway, sustained minimal frontal damage when it struck the rear of a stopped vehicle. The 44-year-old restrained female driver, height-168 cm, mass-53 kg, sustained multiple fractures of her right radius and ulna. Her forearm was over the steering wheel at the time of air bag deployment. (NC 93-02)

Case 27 - A 1990 Ford Taurus, 4 door was attempting a left turn when it was struck by an on-coming car. The 52-year-old restrained female driver, height-155 cm, mass-59 kg, sustained fractures of the right ulnar (mid shaft), right olecranon and a dislocation of the head of the right radius from the air bag module cover. (IN 93-06)

DISCUSSION

In many automobile crashes that we and others have investigated, deployments of current generation air bags have been shown to reduce the frequency of fatalities and serious injuries to the head, face, and torso (12,13,14). In order to reduce the risk of serious and life threatening injuries, air bags must rapidly inflate with considerable force. Nevertheless, injuries related to air bag deployment are most often minor. There is strong evidence that, on balance, today's air bags decrease the frequency of significant injuries and are an effective supplemental restraint system. In a systematic study of accidents involving air bag deployments about 30 percent of drivers reported sustaining some sort of an air bag injury to the head, usually erythemia or minor skin abrasions to the face, chin, or neck. Additionally, abrasions to the upper chest or hands, wrists and forearms were reported (4). Other air bag related injuries include contusions, lacerations, of the forearm, neck, and face, as well as occasional eye trauma, usually corneal/scleral abrasions independent of the severity of the collision. This paper documents, for the first time, a subset of upper extremity injuries related to air bag deployments.

At initial deployment the air bag fabric can reach a velocity in the range of 160-320 kmph. Reed and Schneider have demonstrated that contact injury occurs at the point of initial air bag impact with the skin, rather than when the smooth fabric of the inflated air bag sweeps across the skin surface (26,27). New folding patterns and lower-mass-uncoated fabrics may reduce the frequency of these minor fabric impact injuries in the future, but due to the proximity of the driver's upper extremities to the steering wheel air bag module and the necessity to deploy the air bag rapidly, these more common superficial extremity injuries may not be capable of being completely eliminated.

Hand and wrist burns, produced by the venting of the air bag gas, are being noted less often with current vehicle modules than with earlier models. Some air bag designs had the exhaust vents at the nine and three o'clock positions on the back of the air bag. This resulted in more frequent minor burns than are seen in some other designs, where the vents are located at eleven and one o'clock positions. Still, in crashes that result in an air bag deployment, the driver's hands can be forced off of the steering wheel towards the exhaust vents by either inertial forces or by the deploying air bag itself.

Another area of improved design is the tethering of the central portion of the air bag. The early untethered air bags bulge out in the center when fully inflated. These untethered air bags were more likely to cause the minor facial injuries because these air bags project farthest in the center, where the face contact is most often occurred. For several years tethered driver-side air bags have been installed. These bags have circles of stitching around the center of the air bag's front surface where the internal tethering strips are anchored. The tethering fabric inside the air bag restricts the excursion of the center of the air bag during deployment. The differences between tethered and untethered air bags and the various vent designs mentioned above indicate how crucial it is to include the vehicle year, make, and model when discussing occupant injuries in order to determine the type of air bag involved. Seat position is also important, for the near proximity of the front occupant to the air bag module may explain some of the air bag contact injuries.

In six of our investigations, drivers have noted that the air bag contacted their watch bands or bracelets, causing minor lacerations or contusions of the wrist or forearm. In several cases the deploying air bag caught the band of the wrist watch, ripping it off, and the broken wrist band was the cause of the wrist or forearm injury.

To benefit the front seat occupants the air bag must deploy before significant forward occupant movement has taken place. Air bag deployment occurs in less than 60 ms in which time the sensors identify the crash pulse, and the air bag is fully inflated. Therefore, the air bag(s) must be deployed with significant force in order to break through the air bag module cover and fully inflate in this short time frame. In one air bag module design, deployment flings the cover flap outward and upward at 360 kmph. If the driver has his hand or arm across the steering wheel hub as the module cover opens, it is possible that a contusion, laceration, or fracture of the forearm may be produced. Recently, a case involving the traumatic avulsion of a 23-year-old female driver's left thumb as a result of contact by the air bag module cover was reported (25). The injuries included a laceration at the base of the first metacarpal with deep transection of all thenar (thumb) muscles, except the tendon of the flexor pollicis longus, and an open fracture of the first metacarpal with dislocation at the metacarpalphalangeal joint.

Alternate module cover designs may reduce the potential for injury due to contact between the opening module cover and the driver's hands or forearms. Modifying designs with smaller flaps of material moving at lower velocities may reduce the potential for lacerations and fractures. Further research is needed to determine the optimal design to reduce the incidence and severity of these hand, wrist and forearm injuries within the deployment envelope while preserving the life-saving performance of air bag systems.

In certain accidents involving deceleration of the car before the major impact, as with pre crash braking or a minor collision preceding a significant impact, the occupants may move forward before the air bag sensor system has triggered, especially if the lap-shoulder restraint is not worn (8, page 38). The air bag deployment is then much more likely to cause injuries. The greatest risk of injury as demonstrated in testing with crash dummies and anesthetized swine occurs when a vehicle occupant's torso is against and fully covering the air bag module at the time of deployment (28,29).

The air bag will occasionally cause the hand/forearm to be flung towards the windshield, instrument panel, or rear view mirror resulting in injuries including minor lacerations, contusions of the posterior hand, wrist, and/or forearm, and occasionally sprains or fractures of the hand. The star burst or spider web pattern of windshield damage, previously noted to be produced by the head of an unbelted occupant, is now seen with air bag induced upper extremity flailing into the windshield. In fact, in almost all instances a lap-shoulder belted driver's head cannot reach the windshield over a deployed air bag, and rarely is there windshield contact by the head of even unbelted drivers in frontal crashes where the steering wheel air bag deployed. Thus, in frontal accidents involving air bag equipped cars, the star burst windshield damage is now most likely caused by air bag induced upper extremity impact.

In the review of the cases presented, minor head, face, or torso injuries were often noted, although facial injuries from hand/wrist impact caused by air bag deployment have also been observed. Only in very high speed crashes have more extensive torso or head injuries been observed, usually due to collapsing car structures and associated high-velocity torso loading.

The effectiveness of the passenger side air bag in frontal crashes is not known, because few cars are so equipped at this writing, and the lower occupancy rate of the front passenger seat makes data on passenger side air bag more difficult to obtain.

Acknowledgment:

To Richard Reed, NHTSA Special Crash Investigation Section, our thanks for his cooperation in providing some of the air bag cases presented in this paper.

REFERENCES

1. Huelke DF, Roberts JV, Moore JL. Air Bags in Crashes Clinical Case Studies from Field Investigations. *XIII ESV Conf*, pp 140-148, Paris, France, 1991.
2. Huelke DF, Moore JL, Ostrom M. Air Bag Injuries and Occupant Protection. *J Trauma*, 33:894-898, 1992.
3. Huelke DF, Moore JL. Field Investigations of the Performance of Air Bag Deployments in Frontal Collisions. *Proc 36th Am Assn for Auto Med*, pp 43-57, Des Plaines, IL, 1992.
4. *Insurance Special Report.*, Driver Injury Experience in 1990 Models Equipped with Air Bags or Automatic Belts. Highway Loss Data Institute, Arlington, VA, Report No. 38, 1991.
5. Jagger J, Vernberg K, Jane JA. Air Bags: Reducing the Toll of Brain Trauma. *Neurosurg*, 20:815-817, 1987.

6. Marsh JC. Supplemental Air Bag Restraint Systems: Consumer Education and Experience. SAE No. 930646, *Soc Auto Eng Cong and Expo*, Detroit, MI, 1993.

7. Viano DC. Effectiveness of Safety Belts and Airbags in Preventing Fatal Injury. *Proc Frontal Crash Safety Technologies for the 90's*, pp 159-171, 1991.

8. Huelke DF, Sherman HW, Murphy MJ, et al. Effectiveness of Current and Future Restraint Systems in Fatal and Serious Injury Automobile Crashes. SAE No. 790323, *Soc Auto Eng Cong and Expo*, Detroit, MI, 1979.

9. Evans L. Traffic Safety and the Driver. New York: Van Nostrand Reinhold, 1991.

10. Digges KH, Roberts V, Morris J. Residual Injuries to Occupant Protected by Restraint Systems. SAE No. 891974, *1989 Passenger Car Meeting & Expo*, Soc of Auto Eng, Warrendale, PA, 1989.

11. Backaitis SH, Roberts JV. Occupant Injury Patterns in Crashes with Airbag Equipped Government Sponsored Cars. *Proc 31st Stapp Car Crash Conf*, pp 251-266, Warrendale, PA, 1987.

12. Zador PL, Ciccone MA. Driver Fatalities in Frontal Impacts: Comparisons Between Cars with Air Bags and Manual Belts, *Am J of Pub Health*, 83: 661-666, 1993.

13. O'Neill B, Lund AK. The Effectiveness of Air Bag in Preventing Driver Fatalities in the United States. *Presented at: Canadian Assoc Road Safety Professionals Intnl Conf on Air Bags and Seat Belts: Evaluations and Implications for Public Policy*, Oct. 1992.

14. NHTSA. Evaluation of the Effectiveness of Occupant Protection. *Federal Motor Vehicle Safety Standard 208*, Interim Report, Washington D.C. 74 pp, June, 1992.

15. Blair G, Larkin GL. Airbag-Mediated Facial Trauma. *4th Intl Conf on Emerg Med*, Washington, D.C, May, 1992.

16. Braude LS. Protective Eyewear Needed With Driver's Side Air Bag? *Arch Opthalmol*; 110:1201, 1992.

17. Ingraham HJ, Perry HD, Donnenfeld ED. Air-Bag Keratitis. *New Eng J Med*, 324:1599-1600, 1991.

18. Larkin GL. Airbag-Mediated Corneal Injury. *Am J Emerg Med*, 9:444-446, 1991.

19. Lubeck D, Greene JS. Corneal Injuries. *Opthal Emerg & Ocular Trauma*, 6:73-94, 1986.

20. Mishler KE. Hyphema Caused by Air Bag. *Arch Opthalmol*, 109:1635, 1991.

21. Rimmer S, Shuler JD. Severe Ocular Trauma From a Driver's-Side Air Bag. *Arch Opthalmol*, 109:774, 1991.

22. Rosenblatt M, Freilich B, Kirsch D. Air Bags: Trade-Offs. *New Eng J Med*, 325:1518-1519, 1991.

23. Steinmann R. A 40 Year Old Woman With an Air Bag-Mediated Injury. *J Nursing*, 18:308-310, 1991.

24. Lancaster G L, DeFrance JH, Borruso JJ. Unusual Heart Injury Caused by Air Bag Deployment, Lack of Safety Belt . *New Eng J Med*, 328:357-358, 1993.

25. Smock WS. Scientific Poster "Traumatic Avulsion of the First Digit Secondary to Air Bag Deployment." *Proc 36th Am Assn for Auto Med*, p 444, Des Plaines, IL, 1992.

26. Reed MP, Schneider LW, Burney RE. Investigation of Airbag-Induced Skin Abrasion. *Proc 36th Stapp Car Crash Conf*, pp 1-12, Warrendale, PA, 1992.

27. Reed MP, Schneider LW. A Laboratory Technique for Assessing the Skin Abrasion Potential of Airbags. *Frontal Impact Protection: Seat Belts and Air Bags*. Soc Auto Eng Cong and Expo, SAE No 930644, Detroit, MI, 1993.

28. Lau IV. Mechanism of Injury from Air Bag Deployment Loads. *Acc Anal and Prev*, 25:29-45, 1993.

29. Horsch J, Lau I, Andrzejak D, et al. Assessment of Air Bag Deployment Loads. *Proc 34th Stapp Car Crash Conf*, pp 267-288, Warrendale, PA, 1990.

940802

Survey of Airbag Involved Accidents
An Analysis of Collision Characteristics,
System Effectiveness and Injuries

John V. Werner and Wayne W. Sorenson
State Farm Mutual Automobile Insurance Co.

ABSTRACT

In May, 1989, State Farm began to collect field data on crashes involving airbag deployments. The data include crash configurations and locations, driver actions, deployment thresholds and rates, repair cost contributions, and injury profiles.

The effectiveness of airbags in reducing injuries is described, including efforts by one manufacturer to reduce abrasion and contusion injuries, which are among the most common minor injuries in airbag deployments. Analysis shows that airbag systems are working as intended; that is, in frontal crashes where moderate to severe injuries to the face and torso would otherwise be likely.

CURRENTLY, AIRBAGS are being introduced into the U.S. fleet of passenger cars at a very rapid rate. This infusion began in earnest with the introduction of a driver side airbag by Mercedes in 1984. Although airbags occurred in a few makes and models in the mid-eighties, the watershed event leading to their enormous popularity occurred when Chrysler introduced driver side airbags as standard equipment in several of their most popular models in 1988.

The effectiveness of airbags in reducing serious injuries and fatalities has often been described in individual, frequently dramatic, crash events and much anecdotal evidence of their effectiveness has been provided by crash survivors. Even though the airbag has become institutionalized both in law and in conventional wisdom, the responsibility remains to measure the airbag's efficacy, optimize its performance, and minimize its costs.

Any doubt about the life-saving potential of airbag technology has been removed by Insurance Institute for Highway Safety statistical studies (1,2). Field investigations of a large number of airbag deployments can provide meaningful baseline documentation for airbag performance.

This paper helps address the foregoing questions of efficacy, performance, and costs by studying crash data of a large insurer. The State Farm policyholder group presently incurs about 30 crashes per day in which airbags deploy. Analyses of many of these crashes reveal that airbags have, indeed, had a salutary effect in terms of reducing serious injuries. The analyses also reveal that significant variations occur among various designs and suggest that airbag technology is improving as experience with the airbags is accruing.

Some suggestions for which airbag designs are superior also derive from the analyses. Questions about the cost-effectiveness of airbags are also addressed by contrasting the safety benefits with the cost to the buying public of airbags.

Modern passenger cars include a variety of passive safety mechanisms; e.g., head restraints, collapsible steering columns, and safety glass. However, in the judgment of the authors, no other single passive safety device has had as dramatic effect on reducing the human cost of automobile fatalities and serious injuries as has the airbag.

DATA COLLECTION

There are three major data sources which form the basis for analysis and discussion in this paper:

1. State Farm Mutual Automobile Insurance Company's electronically prepared damage repair estimates.
2. State Farm survey of claims involving crashes with an airbag deployment and a comparison group of claims involving crashes with belt-only restraint.
3. Survey of claims involving fatal crashes with a driver airbag deployment.

State Farm estimators were required to complete detailed damage estimates on all vehicles with an airbag deployment, even if the crash resulted in a total loss. An airbag deployment was identified for survey from damage estimates indicating an airbag module replacement. Only State Farm policyholders were included in this study. The claim representative was contacted to provide initial information regarding the policyholder involved accident. This initial information included an accident description; age and sex of the insured driver and right front passenger; driver and right front passenger injury descriptions for the head, torso, and extremities; belt usage; unusual circumstances; and vehicle repair status.

In the typical claim settlement process, claim representatives complete injury evaluations, conduct interviews with the insured and witnesses, obtain police reports, and determine contributing causes for the crash. Doctor reports of injury descriptions and treatment are obtained to substantiate medical payments.

Telephone interviews with the insured driver were conducted by State Farm Research staff to obtain supportive accident and injury description information. The insureds were asked to provide qualitative information regarding their experience. Up to six injuries each were recorded for the driver and right front occupant. The Abbreviated Injury Scale (AIS) and Occupant Injury Classification (OIC) coding were used to document injuries (3).

DATA LIMITATIONS

Automobile insurance data systems have not been used extensively in the United States as a source for traffic safety information. Insurance data systems are maintained for the operation of the business of insurance and as a result, contrary to popular misconceptions, they do not contain much of the detail found in accident research data files maintained by government agencies and universities.

The limitations in insurance data systems can be overcome by implementing a special data collection process. State Farm surveyed crashes resulting in airbag deployments, beginning in May, 1989. The survey involved the investigating claim representative and the insured for obtaining accident and injury relevant data.

The State Farm electronic damage estimate data represent moderate to severe collision damage. Where the damage to the vehicle was extensive and resulted in an "obvious" total loss, the State Farm estimator may not have completed an electronic damage estimate indicating an airbag deployment. The effect of this possibility was minimized by requiring estimators to write damage estimates for all vehicle damages resulting in airbag deployments.

The accuracy of survey results depends on the completeness of claim representative responses and the dependability of the insured driver's accident and injury descriptions. Every attempt was made to elicit complete and detailed responses from the insured. Unless there was an exceptional situation, no vehicle inspection was completed. The severity of the accident was obtained indirectly from the dollar amount of damage. Despite the recognized data constraints, a great deal of information has been distilled from these data-- information which is complementary to that derived from engineering studies

and laboratory testing. These data provide basic information on large numbers of vehicles, and, therefore, can be used to evaluate trends in crash configurations and resultant injuries.

CRASHES RESULTING IN AIRBAG DEPLOYMENT

Table 1 shows the collision types resulting in airbag deployments. The most common crash configuration resulting in an airbag deployment was insured-front into claimant-rear; e.g., the insured driver followed too close or was inattentive to the driver action in front. The second most common collision type was a single vehicle crash. Single vehicle crashes usually involved a driver who did not maintain control of the vehicle and struck a fixed object.

Table 1
Collision Types
Resulting in Airbag Deployments
(n=2,415)

Collision Type	Percent
Vehicle-to-Vehicle	
Front-Rear	
Rear End	28
Angle Rear	2
Front-Front	
Angle Front	19
Head-On	6
Front-Side	
Front into Side	13
Angle Side	6
Single Vehicle	26

An airbag is designed to inflate in a frontal or near-frontal impact of 13-23 kph (8-14 mph) barrier equivalent or above (4). The vehicle's initial damage location was compared to the most severe area of damage. Vehicle damage locations were determined from the accident description and a review of the damage estimate. Table 2 shows that when the initial damage location was to the front of the vehicle, a low percent had subsequent more severe secondary impacts. When the angle of impact incidence increased from front-center, the likelihood of a more serious secondary impact increased marginally.

Table 2
Most Severe Damage Compared to Initial Damage
All Collision Types - Airbag Deployed
(n=2,323)

Initial Damage Location Insured Vehicle	Percent Crashes Where Frontal Damage Was Most Severe
Front-Center	98
Angle Front-Bumper	88
Angle Front-Fender	84

Table 3 shows vehicle damage locations to both the insured and the initially involved claimant vehicle, in two or more vehicle collisions. Eleven percent of the airbag deployment crashes involved initial impacts to the side (7%) or rear (4%) of the insured's vehicle.

Table 3
Vehicle Damage Locations
Insured Vehicle Airbag Deployment
Two or More Vehicle Crashes
(n=1,582)

Initial Damage Location Insured Vehicle	Initial Damage Location Claimant	Percent Crashes
Front	Rear	42
Front	Front	27
Front	Side	20
Side	All Locations	7
Rear	All Locations	4

Most crashes occur in congested areas. Analysis of survey results determined that most accidents involving airbag deployments occurred in congested areas. Table 4 shows that local city streets and intersections are the most common locations for crashes resulting in airbag deployments.

Table 4
Accident Location
When Airbag Deployed
(n=1,959)

Location	Percent
Local City Street	33
Four-Way Stop	22
Divided Highway	13
State Route	11
Two-Way Stop	9
Interstate	9
Other	3

Many different objects were struck in single vehicle crashes resulting in airbag deployments. Table 5 shows that posts and trees accounted for 40 percent of the objects struck for this collision type. The vehicle undercarriage contacted medians and curbs in 6.4 percent of these single vehicle crashes. Out of 2,818 deployment crashes investigated, there have been very few instances of what might be considered inadvertent deployments. One case involved a single point sensor and control unit located in the passenger compartment that received multiple inputs from a loose drive shaft impacting the floor pan area. Other cases were related to improper servicing or the absence of required servicing to the airbag system, prior to the crash.

Table 5
Objects Struck
Single Vehicle Crashes
Resulting in Airbag Deployment
(n=641)

Object Struck	Percent
Post	24.5
Tree	15.6
Embankment/Ditch	13.4
Guard Rail	12.2
Building/Wall	11.4
Deer/Animal	7.6
Divider/Median	3.6
Curb	2.8
Culvert	1.4
Rock/Boulder	1.1
Hydrant	1.1
Fence	0.9
Other	4.4

The survey determined drivers' actions leading to the crash event. Table 6 shows that insured drivers were turning, swerving, or changing lanes in 16 percent of all crashes leading to an airbag deployment. These driver actions may contribute to forearm or hand interactions with the deploying airbag.

Table 6
Insured Driver Action
Leading to Crash Event With Airbag Deployment
All Collision Types
(n=2,728)

Driver Action	Percent
Going Straight	62.5
Lost Control	17.0
Swerving	5.8
Turning in Traffic	4.9
Turning at Intersection	3.4
Stopped or Slowed	3.4
Changing Lanes/Passing	2.1
Other	0.9

DEPLOYMENT RATES AND THRESHOLDS

DEPLOYMENT RATES - Automobile manufacturers, airbag system component suppliers, and the vehicle owner are very interested in the reliability of their airbag systems in discriminating crash events that should result in deployment. This study has shown an expected diverse crash environment.

State Farm has determined the deployment rate per 1,000 collision coverage claims(*), for most makes and models with airbags. The deployment rate was calculated for driver side deployment events. A

---

* Collision coverage provides physical damage coverage for the insured's vehicle in the event of a crash.

deployment was known to occur if the airbag module was quoted for replacement on the damage estimate.

The probability of deployment in any given year P_y can be calculated by multiplying the probability of a collision coverage claim P_c by the probability of a deployment P_d when the crash event occurs.

$$P_c \ X \ P_d = P_y \qquad (1)$$

Table 7 shows the range in deployments per 1,000 insured vehicles in one year (IVY) for several representative 1992 model year vehicles. Appendix A contains a more comprehensive itemization of vehicles.

deployments are expected for every 161 million kilometers (100 million miles) traveled.

When restricted to frontal impacts only, State Farm has determined that there is an industry average 95 driver side deployments per 1,000 frontal crashes. Approximately 60 percent of all collision coverage claims involve initial impacts to the front area of the vehicle.

Ninety-five deployments per 1,000 frontal crashes can be compared to the probability of injury projected by the U.S. Department of Transportation in their 1988 and 1991 General Estimates System (5,6). Table 8 shows the G.E.S. estimated maximum injury count in passenger cars when the maximum damage area is

Table 7
Collision Claim Frequency and Deployment Rates
Selected 1992 Model Year Vehicles
Driver Side Deployments

Manufacturer Vehicle	Collision Claims per 1,000 IVY* P_c	Deployments per 1,000 Claims P_d	Deployments per 1,000 IVY P_y
Mercedes			
190 Class('89-'93)	89	28	2.5
300 Class	82	9	0.7
General Motors			
Geo Metro Cv.	119	91	10.8
Chev. Camaro	110	62	6.8
Olds 88	76	29	2.2
Buick Roadmaster	80	10	0.8
Chrysler			
Dodge Shadow	113	118	13.3
Plymouth Sundance	108	86	9.3
Concorde/Intrepid/Vision	45	49	2.2
Minivans	53	48	2.5
Jeep G. Cherokee	58	25	1.4
Ford			
Probe('93)	121	90	10.9
Taurus	69	55	3.8
Aerostar	62	38	2.4
Honda			
Civic	109	38	4.1
Nissan			
Sentra	87	117	10.2
Toyota			
Camry	80	55	4.4
All Makes and Models (1990 to 1993 MYs)	**70**	**56**	**4.0**

* IVY Insured Vehicle Years.

Based on State Farm insureds' experience, there are an average 70 collision claims per 1,000 insured vehicle years. There are an average 56 driver side deployments per 1,000 collision coverage involved crashes (all impact directions). Applying EQ (1) results in an average of four deployments per 1,000 IVYs. Assuming that a vehicle is driven an average 24,140 km (15,000 miles) in one year, approximately 27

front. The maximum injury in the car was serious or fatal in 35 per 1,000 frontal crashes. The maximum injury in the car was minor or moderate in 190 per 1,000 frontal crashes. The total of these two estimates is 225 per 1,000 frontal crashes. State Farm's estimate of 95 deployments per 1,000 frontal crashes for all vehicles appears to approximate the government's estimate of the proportion of crashes with the potential to cause

Table 8
U.S. Department of Transportation
NHTSA General Estimates System - 1988 & 1991
Maximum Damage Area - Front
Maximum Injury in Car

Total Crashes (000)	No Injury (000)	Minor or Moderate Injury (000)	Serious or Fatal Injury (000)
1988			
3,336	2,597 (78%)	621 (19%)	118 (3.5%)
1991			
3,272	2,523 (77%)	626 (19%)	123 (3.8%)

moderate or more serious injury.

DEPLOYMENT THRESHOLD - A complex set of factors must be considered when engineering an airbag system. These may include vehicle stiffness, crash zone extent, vehicle mass, the expected crash environment, and the manufacturer's knowledge of the car buyer's expectations for occupant protection.

State Farm determined the probability of driver side deployment by level of severity for several passenger cars. The level of crash severity was indirectly determined by dividing the range of repair costs into ten equal increments. Although repair cost is a less-than-perfect surrogate for measuring crash energy, it does provide an overall view of deployment consistency within levels of damage and also standardizes the level of damage for vehicle-to-vehicle comparisons.

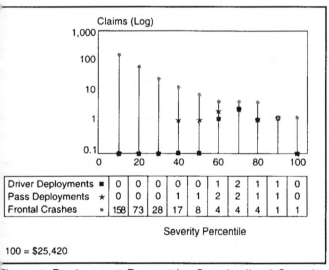

Figure 1 Deployment Percent by Standardized Severity Level - Mercedes 300.

Figure 1 shows the number of deployments by standardized severity level for the Mercedes 300 Class, 1991-1993 model year experience combined. The Mercedes airbag system is designed to detect belt usage and is set to a higher threshold for belted occupants. Above the 50th percentile severity, 36 percent of the frontal crashes resulted in a driver side deployment.

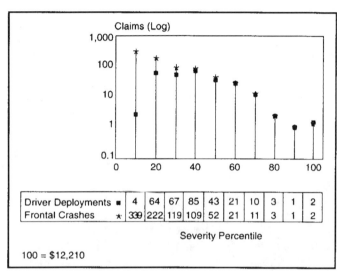

Figure 2 Deployment Percent by Standardized Severity Level-Dodge Spirit.

Figure 2 shows the number of deployments by standardized severity level for the Dodge Spirit. Above the 50th percentile, virtually every frontal crash resulted in a deployment. Appendix B contains a more comprehensive itemization of vehicle profiles.

Figure 3 shows an overall profile of deployments by

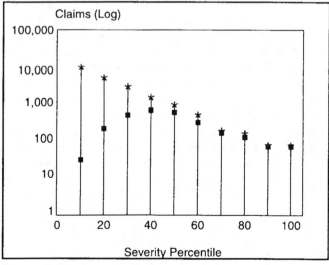

Figure 3 Deployment Percent by Standardized Severity Level - Twenty-Nine Representative Vehicles.

standardized severity level for a representative mix of airbag installed vehicles. For the vehicles examined, airbag system designs are consistently detecting severe crashes necessitating airbag deployments.

SERVICE AND REPAIR COSTS

When an airbag deploys, automobile manufacturers require that some components are replaced, regardless of damage. Manufacturers also recommend that the insurance appraiser and the repairer inspect other system components and replace them if damaged. State Farm has prepared an internal guide for appraisers to assist them in determining manufacturer recommendations for all model years of airbag installed vehicles. This guide is based upon the Inter-Industry Conference on Auto Collision Repair (I-CAR) research of manufacturer service guides for 1993 model year vehicles (7).

Table 9
Percent Vehicles Totaled After Airbag Deployment
Selected Vehicles in First Year of Operation
and All Vehicles

Vehicle	Percent Totaled
Toyota Celica	51
Acura Legend	33
Ford Mustang	30
Lincoln Continental	23
Dodge Daytona	28
Cadillac DeVille	21
LeBaron Coupe	25
Plymouth Spirit	16
All Vehicles	30

Appendix C contains representative airbag system replacement costs. The replacement costs include costs associated with manufacturer mandatory part replacement recommendations and forward mounted sensors. The replacement costs shown in Appendix C

Table 10
Marginal Additional Premium Cost per Policy
$3,000 Airbag System Replacement Cost
by Age of Vehicle

Vehicle Age	Average Deployments Per 1,000 IVY*	Percent Repaired*	Repaired Per 1,000 IVY	Marginal Additional Premium Cost Per Policy
0-1	4.0	70	2.8	$8.4
1-2	4.0	59	2.4	7.2
2-3	4.0	56	2.2	6.6
3-4	4.0	40	1.6	4.8
4-5	4.0	21	0.8	2.4
5-6	4.0	10	0.4	1.2
6-7	4.0	<10	<0.4	<1.2

* Average for all airbag installed vehicles.

REPAIR COSTS - The survey determined how often vehicle crash damage, with an airbag deployment, resulted in a total loss. One might assume that crash damage serious enough to cause a deployment would be unrepairable. During the first few years of operation, vehicles retain relatively high market values and therefore are repairable. Table 9 shows that the percent vehicles totaled varies by make and model. The observed differences in total loss percent can be expected because of variations in the vehicles' market values and deployment thresholds.

vary from $720 to $5,435. These cost estimates do not include other potentially damaged components related to the airbag system -- such as the windshield, instrument panel, or wiring. A review of damage estimates shows that most manufacturers are designing passenger side systems which limit the amount of secondary damage to the instrument panel upon deployment. Top deploying passenger side airbags cause more frequent windshield replacements.

The chance that a crash will result in an insurance total loss increases with the age of a vehicle. State Farm estimates that virtually all crashes with airbag deployments will result in total losses when the vehicle reaches seven years of age. The average market value of a seven year old vehicle totaled by State Farm is currently $4,000. The marginal additional auto insurance premium cost per policy for airbag replacement can be estimated, by age of vehicle, knowing the deployment rate per 1,000 IVYs and the total loss rate. Table 10 shows the marginal additional premium cost for a $3,000 airbag system replacement cost.

INJURIES

Up to six injuries were coded for both the driver and right front occupant from claim representative provided injury descriptions. This information was supplemented by injury descriptions obtained by telephone interview with the insured driver. Injury type and severity were coded using the Abbreviated Injury Scale (AIS). In addition, OIC body region and lesion coding was used to clarify injury location for abrasion, laceration, and contusion injuries.

SIZE CLASS OF CAR - Table 11 presents injury experience for cars grouped by size class. The distribution of car size differs between airbag and belt equipped cars. To control for this difference, comparisons of data standardized for size are also presented, using a technique developed by the Insurance Institute for Highway Safety (8). Data are presented for all moderate and severe injuries combined; i.e., AIS severity codes 3, 4, 5, and 6 are combined to represent severe injuries. Injury data represent the maximum AIS severity.

The standardized results suggest that the frequency of overall injury is slightly greater for crashes involving airbag deployments. However, drivers involved in airbag deployment crashes were 35 percent less likely to suffer moderate and severe injuries. Drivers of small

Table 11
Injury Experience of Drivers in Frontal Crashes
With More Than $5,000 Damage by Size Class* of Car
Belt-Only Compared to Vehicles With Airbags That Deployed

	Small		Midsized		Large		Standardized	
	Belt	Airbag	Belt	Airbag	Belt	Airbag	Belt	Airbag
Crashes	30	434	152	721	68	499	250	1,654
Avg Damage Amount	$8,313	$8,564	$8,263	$8,630	$10,882	$9,981	—	—
Drivers Injured**	600	654	592	634	632	655	606	645
Drivers w/ Moderate & Severe Injuries**	100	83	119	61	74	60	101	66

* Small - Wheelbase ≤ 251 cm (99 in)..24%
   Midsized - Wheelbase > 251 cm (99 in) and ≤ 277 cm (109in) 46%
   Large - Wheelbase > 277 cm (109in) 30%
** Per 1,000 frontal crashes.

The number of crashes with airbag deployments investigated was 2,818. Included in this dataset were 1,654 frontal crashes resulting in more than $5,000 in repair cost. The number of crashes involving vehicles with belt-only restraint was 417. Two-hundred and fifty of these crashes were frontal with damages more than $5,000. As shown in Appendix B, $5,000 represents a level of repair cost that is likely to result in an airbag deployment.

The sample size of crashes involving vehicles with belt-only restraints may be considered marginally adequate for detailed comparisons to the airbag deployment injury experience; however, the data are believed to be satisfactory to provide indications of major injury pattern differences.

cars with airbag deployments were 17 percent less likely to have moderate or severe injuries than drivers of cars with belts.

BELT USAGE - The drivers of belt-only vehicles reported an average 89 percent belt use. This belt use rate was nearly identical to that reported by drivers of airbag vehicles. Table 12 shows the injury experience of belted and non-belted drivers. Belted drivers of airbag cars with deployments were 23 percent less likely to have moderate or severe injuries than unbelted drivers. This result reinforces the fact that the airbag is an important supplemental restraint system for the lap-shoulder belted front seat occupant.

Table 12
Injury Experience of Drivers in Frontal Crashes
With More Than $5,000 Damage by Reported Belt Use

	Crashes	Drivers Injured*	Drivers with Moderate and Severe Injuries*
Airbag Vehicles With Deployments			
Not Belted	186	680	98
Belted	1,167	733	75
Belt-Only Vehicles			
Not Belted	23	757	221
Belted	194	641	98

* Per 1,000 frontal crashes. Standardized for size class of car.

Extreme caution is advised for anyone trying to extol the benefits of airbags for unbelted occupants. The airbag works best when the occupant is held in position, directly in front of the deploying airbag. In one surveyed case, the right front of an insured's vehicle was struck with moderate severity (see Fig. 4). The unbelted male driver moved right and forward, rotating upward and over the deployed airbag. He was fatally injured when his head contacted the top area of the windshield.

LOCATION OF DRIVER INJURIES - Table 13 shows driver injuries separated by body region and lesion. Because some drivers sustained injuries to multiple body locations, the number of injuries per 1,000 frontal crashes in Table 11 is less than the total number of injuries in Table 13.

Because detailed injury experience is presented, standardization by size class was not practical. The results shown in Table 11 indicate a consistent trend between airbag deployment crash results and belt-only crash results, by car size class of car. Therefore, the results shown in Table 13 should reliably indicate where major differences in injury patterns occur.

Drivers of cars with airbag deployments have lower rates of potentially more serious head and chest injuries such as head concussions and rib fractures; however, they have higher rates of abrasion, laceration, and contusion injuries to the face, arms, and wrist. Complete recoveries from these minor injuries typically occurred within three weeks. A few finger and wrist fractures were observed. One serious multiple fracture to the forearm and elbow has been documented. The injury resulted from a deploying airbag contacting the driver's forearm in a turning maneuver. This study was not able to determine the injury causing mechanics; however, investigating claim representatives have commented that the deploying airbag can force the hand and arm into contact with the windshield and other unforgiving interior objects.

Infrequent eyeball injuries have been observed to include abrasions, contusions, and lacerations to the cornea. In 1,654 severe frontal crashes, this study has documented one retinal detachment, one sclera laceration, and one cornea laceration.

Drivers of cars with belt-only restraint had higher rates of injury to the neck and shoulder; i.e., injury types that are typically associated with belt use (strains and contusions). The airbag appears to provide additional upper-body support to limit belt-induced injuries. Although the rate of abdomen injuries was less for the driver in airbag deployment crashes, it is not clear how the airbag might have helped, unless additional upper-body support was also a contributing factor. The knee bolster may also be a contributing factor for reducing abdominal injuries. Appendix D contains the detailed AIS injury description and severity results.

Figure 4 Unbelted Driver With Airbag Deployment and Fatal Injury.

Table 13
Driver Injuries by Body Region and Lesion
Crashes with More than $5,000 Damage

Body Region Lesion	Injuries per 1,000 Frontal Crashes		Body Region Lesion	Injuries per 1,000 Frontal Crashes	
	Airbag Deployed (n=1,654)	Belt-Only (n=250)		Airbag Deployed (n=1,654)	Belt-Only (n=250)
Head-Skull			**Upper Arms**		
Concussion	44	80	Contusion	69	36
Contusion	26	72	Abrasion	40	—
Laceration	10	24	Laceration	15	—
Abrasion	4	—	Fracture	—	—
Fracture	1	—	Unk Lesion	6	—
Unk Lesion	12	—	**Upper Limbs**		
Face			Contusion	—	4
Contusion	72	60	Fracture	18	16
Abrasion	71	12	**Elbow**		
Laceration	57	52	Contusion	12	4
Fracture	13	28	Laceration	2	—
Dislocation	4	—	Abrasion	2	—
Unk Lesion	4	1	Sprain	2	—
Neck/Spine			Unk Lesion	1	—
Strain	141	196	**Forearm**		
Abrasion	12	—	Contusion	12	4
Contusion	8	16	Abrasion/Burn	10	—
Laceration	1	—	Laceration	3	8
Fracture	1	4	**Wrist-Hand**		
Dislocation	1	8	Abrasion/Burn	37	—
Unk Lesion	8	—	Contusion	43	16
Chest			Sprain	25	8
Contusion	114	144	Fracture	13	8
Fracture	11	28	Laceration	16	4
Abrasion	8	4	Dislocation	4	—
Laceration	3	8	Unk Lesion	6	—
Strain	—	4	**Knee**		
Unk Lesion	22	16	Contusion	65	80
Back			Sprain	7	8
Strain	113	104	Laceration	4	16
Fracture	2	4	Abrasion	3	4
Dislocation	2	8	Fracture	2	—
Unk Lesion	6	8	Unk Lesion	5	4
Abdomen			**Lower Leg**		
Contusion	15	28	Contusion	35	36
Laceration	—	12	Fracture	5	12
Unk Lesion	2	—	Laceration	4	—
Pelvis-Hip			Abrasion	4	4
Contusion	7	4	Unk Lesion	2	4
Sprain	2	—	**Ankle-Foot**		
Dislocation	1	—	Sprain	15	4
Shoulder			Contusion	7	4
Contusion	22	48	Fracture	5	8
Sprain	18	20	Dislocation	2	—
Dislocation	2	—	Laceration	2	—
Fracture	2	—	Unk Lesion	2	8
Detachment	1	—			
Abrasion	1	—			
Laceration	1	—			

Airbag Design Considerations - Auto manufacturers are continuing to refine their airbag system designs in response to field reported experience. Table 14 shows some known differences between a 1st and 2nd generation design for one manufacturer's airbag system. These design changes were made in an effort to reduce injuries attributed to the deploying airbag.

Table 14
Airbag System Design Differences
One Manufacturer's Design Changes

	1st Generation	2nd Generation
Airbag Inflator (1 Cu. foot tank)	350KPA	320KPA
Airbag Fabric	Nylon-840 Denier	Nylon-420 Denier
Coating	Chloroprene Rubber (3 coats)	Silicone Rubber (1 coat)
Tether	No	Yes

Laboratory studies have indicated that tethered airbags may help reduce abrasion severity (9,10). Figure 5 shows a side view of an untethered Acura Legend airbag at maximum extension (peak displacement = 51cm (20in)). Figure 6 shows a side view of a tethered Chrysler 5th Avenue airbag at maximum extension (peak displacement = 30cm (12in)). A silhouette of the Hybrid III 5th percentile female dummy was positioned 38cm (15in) away from the steering wheel hub.

Figure 5 Untethered Acura Legend Airbag at Maximum Displacement.

State Farm field survey data was used to evaluate the effectiveness of these design changes. Table 15 shows the results of this analysis. The combination of design changes has been very effective in reducing abrasion, contusion, and laceration injuries, particularly to the face.

Figure 6 Tethered Chrysler 5th Avenue at Maximum Displacement.

Table 15
Injury Experience of Drivers in Frontal Crashes
1st Generation Compared to 2nd Generation
Airbag Deployed

	Injuries per 1,000 Frontal Crashes	
	1st Generation (n=80)	2nd Generation (n=46)
Injuries		
Abrasion/Burn		
All Body Regions	213	43
Face	75	22
Contusion		
All Body Regions	538	457
Face	138	43
Laceration		
All Body Regions	150	109
Face	63	22

DRIVER AGE, SEX, AND STATURE - The survey mix of drivers by sex, age, and stature involved in airbag deployment crashes is shown in Table 16. A comparison of injuries showed that the aged and females were more likely to have injuries. Short females <160cm (63 inches) were 135 percent more likely to have moderate or severe injuries than taller females, for crashes involving airbag deployments. The aged (>59) were 137 percent more likely to have moderate or severe injuries than younger drivers, for crashes involving airbag deployments. This study was not able to compare these stature and age related results to belt-only vehicle crash experience due to sample size limitations; however, the aged are known to be more susceptible to injuries in severe crashes. Also, one might expect short individuals to be more prone to injury in severe crashes due to the expected close proximity of the steering wheel rim.

PAIRED COMPARISON - A right front passenger was present an average twenty-eight percent of all airbag crashes. There were 211 frontal crashes with more than $5,000 damage, a driver-side deployment,

Table 16
Driver Age, Sex, and Stature
Crashes With Airbag Deployment

Sex	Percent	Age	Percent	Stature	Percent
Male	47	<60	87	<160cm	12
Female	53	≥60	13	≥160cm	88

and a right front passenger present with no airbag available. Table 17 shows the injury experience of the driver-passenger pair, standardized for driver sex. Male drivers were present in 58 percent of the paired crash events. Male right front passengers were present in 35 percent of the paired crash events. Crashes with right front passengers less than 16 years of age were excluded from the paired analysis.

Table 17
Injury Experience of Right Front Passengers
In Frontal Crashes With More Than $5,000 Damage
Driver Airbag Deployed - No Passenger Airbag

	Driver	Male	Female	Standardized for Driver Sex
Crashes	211	74	137	211
Injured*	706	527	613	563
Moderate and Severe Injuries*	62	54	101	74

* Per 1,000 frontal crashes.

Although the space in front of the right front passenger might be considered an additional margin for safety when compared to the more limited space in front of the driver, this paired analysis shows that drivers were 16 percent less likely to have moderate or severe injuries than right front passengers with belt-only restraint available. Right front passengers reported an average 80 percent belt use rate.

FATALITIES WITH AIRBAG DEPLOYMENTS - Periodically, all personal injury coverage claims were surveyed to determine the occurrence of fatalities to drivers of airbag equipped vehicles. The data collection methodology was similar to that used for the airbag survey, except that all communication was restricted to the State Farm claim representative investigating the loss.

This research was conducted in an attempt to determine the occurrence of fatal injuries, where the airbag may have been a contributing factor. The National Highway Traffic Safety Administration has reported that among drivers protected by airbags, serious internal injuries may be present but not externally apparent (11). The NHTSA study stated that blunt trauma, regardless of the restraint system, can cause internal injuries that are often survivable if treated in time, but can be fatal is not detected and treated appropriately and promptly. The case studies researched by the NHTSA included a variety of circumstances to include severe crashes, multiple impacts, and unbelted drivers that sustained multiple injuries. The lower rim of the steering wheel was a contributing factor in several cases.

Figure 7 Example of Vehicle Damage Associated with Massive Crush and Airbag Deployment.

Table 18 shows the distribution of crash types that involved fatalities to State Farm insured drivers, with an airbag deployment. Figure 7 is a photograph of the vehicle crash damage that was typical of the massive crush observed in this fatality survey.

Table 18
Collision Types Involving Fatalities
to Drivers of Airbag Equipped Vehicles
Airbag Deployed

Collision Type	Cases	Percent
Massive Frontal Crush or Rollover	15	38
Massive Side Intrusion	10	25
Multiple Impacts - High Speeds Involved	6	15
Mitigating Circumstances - Strokes	4	10
Potentially Survivable - No Belts Used	3	7
Potentially Survivable - Age a Factor	2	5

The two cases of potentially survivable frontal crashes involved belted elderly women, 78 and 79 years of age. The frontal impact forces involved appeared to be survivable for a more youthful and healthy individual. It is likely that these two individuals could have sustained serious injury from belt loading alone, in the absence of an airbag deployment.

Very close proximity to the deploying airbag can be a factor contributing to additional chest loads. The additional risk appears to apply primarily to out-of-position drivers, slumping or leaning over the steering wheel. Based on the survey results presented in this paper, the frequency of this possibility appears to be negligible; however, the fact that a few cases have been identified should reinforce the efforts made by the vehicle safety research community to advocate restraint designs minimizing such events.

SUMMARY

This study has identified several characteristics of crashes and injuries related to airbag deployments. Among the findings:

1. State Farm estimates a fleet average of 27 airbag deployments per 161 million km (100 million miles) and 95 deployments per 1,000 frontal crashes.
2. Airbags work. Moderate and severe injuries were 35% less likely for crashes involving airbag deployments, even though the overall injury rate was slightly greater. In a paired comparison, drivers in airbag-deployment crashes were 16% less likely to receive moderate or severe injuries than were right front passengers protected only by belts.
3. Seatbelt usage is still important. Belted drivers of airbag cars with deployments were 23% less likely to receive moderate or severe injuries than those who were not belted.
4. Drivers of cars with airbag deployments have lower rates of potentially serious head and chest injuries, but higher rates of abrasion, laceration, and contusion injuries to the face, arms, and wrist.
5. A second generation airbag system design incorporating a tethered bag was shown to produce a marked reduction in abrasions, contusions, and lacerations.
6. A study of 40 driver-fatalities involving airbag deployments showed that massive crush was involved in 63% of the cases. Mitigating circumstances such as nonuse of belts, strokes, multiple impacts, and age applied to the remaining cases.
7. Deployment rates by standardized crash severity level vary widely among makes and models.
8. For vehicles at least seven years old, virtually all airbag-deployment crashes will result in total losses under collision insurance coverage. This varies by make and model and is a function of several factors including deployment threshold, vehicle value, and part prices.

ACKNOWLEDGEMENT

The authors appreciate contributions made by State Farm field staff in providing claim file reviews and survey responses. Also, the authors appreciate the contributions made by Shawn Bill Ensenberger, Lisa Craver, and Renee Bacon in collection, organization and processing of the airbag survey data.

REFERENCES

1. Paul L. Zador and Michael A. Ciccone, "Driver Fatalities in Frontal Impacts: Comparisons Between Cars with Air Bags and Manual Belts", Insurance Institute for Highway Safety Research Report, Arlington, VA, October 1991.
2. Brian O'Neill and Adrian K. Lund, "The Effectiveness of Air Bags in Preventing Driver Fatalities in the United States", Insurance Institute for Highway Safety Research Report, Arlington, Va, October 1992.
3. J. C. Marsh, "An Occupant Injury Classification Procedure Incorporating the Abbreviated Injury Scale", Proceedings of the International Accident Investigation Workshop-Pilot Study on Road Safety for Committee, NATO, Brussels, Belgium, June 28-29, 1973, pp 143-162.
4. "The Air Bag Systems in Your Car-What the Layperson Needs to Know about Air Bag Systems", An SAE Information Brochure, June 25 1992, p 6.
5. U.S. Department of Transportation, National Highway Traffic Safety Administration General Estimates System 1988, August 1990, p 54.
6. U.S. Department of Transportation, National Highway Traffic Safety Administration General Estimates System 1991, August 1993, p 68.
7. "Passive Restraints Update-Part 1, 1993 Model Supplemental Restraint System Part Replacement Recommendations", Inter-Industry Conference on Auto Collision Repair, I-CAR Advantage, Volume VI, No. 3, May-June 1993, pp 10-11.
8. "Driver Injury Experience in 1990 Models Equipped with Air Bags or Automatic Belts", Highway Loss Data Institute Special Report A-38, Arlington, VA, October 1991.
9. M. P. Reed, L. W. Schneider, and R. E. Burney, "Investigation of Airbag-Induced Skin Abrasions", SAE Paper No. 922510, Society of Automotive Engineers, Warrendale, PA.
10. "Air Bag Deployment Characteristics", U.S. Department of Transportation, National Highway Traffic Safety Administration, Report No. DOT HS 807-869, February 1992.
11. Louis V. Lombardo and Susan D. Ryan, "Cases of Air Bag Crashes with Life Threatening Occupant Injuries-Detection of Internal Injuries in Drivers Protected by Air Bags", U.S. Department of Transportation, National Highway Traffic Safety Administration, November 1993.

Appendix A
Collision Coverage[1] Claim Frequency and Deployment Rates
Vehicles with Airbags[2]

Manufacturer Vehicle	Claims per 1,000 IVY[3]	Deployments per 1,000 Claims[4]	Deployments per 1,000 IVY
BMW			
BMW 7 Series ('89-'93)	87	50	4.4
BMW 3 Series ('90-'93)	102	34	3.5
BMW 5 Series ('89-'93)	81	34	2.8
Mercedes			
190 ('89-'93)	89	28	2.5
SL	58	35	2.0
300 Series('89)	82	20	1.6
300 Series('92)	82	9	0.7
Porsche	73	42	3.1
Saab			
9000 ('89-'93)	99	39	3.9
900 ('90-'93)	94	32	3.0
Volvo			
940 ('91-'93)	75	67	5.0
740 ('89-'92)	91	44	4.0
240 ('90-'93)	72	45	3.2
VW			
Cabriolet ('90-'92)	113	53	6.0
General Motors			
Geo Metro Conv.	119	91	10.8
Beretta	113	74	8.3
Camaro	110	62	6.8
Geo Storm	137	39	5.3
Corsica	90	57	5.1
Saturn SL ('93)	80	61	4.9
Saturn SC ('93)	97	42	4.1
Eldorado	72	41	3.0
98	80	37	3.0
Corvette ('90-'93)	46	64	2.9
Le Sabre	67	36	2.4
DeVille/Fleetwood	78	28	2.2
Seville	70	32	2.2
88	76	29	2.2
Park Avenue	74	29	2.1
Bonneville	62	33	2.0
Caprice	83	19	1.6
Roadmaster	80	10	0.8
Chrysler			
Shadow	113	118	13.3
Spirit	73	169	12.3
Sundance	108	86	9.3
Daytona	114	67	7.6
Le Baron	84	79	6.6
Dynasty	61	92	5.6
Stealth	102	48	4.9
Acclaim	75	54	4.1
5th Avenue	48	81	3.9
Caravan, etc.[5]	53	48	2.5
Concorde, etc.[5]	45	49	2.2
G. Cherokee	58	25	1.4

Appendix A (continued)

Manufacturer Vehicle	Claims per 1,000 IVY[3]	Deployments per 1,000 Claims[4]	Deployments per 1,000 IVY
Ford			
Probe ('93)	121	90	10.9
Mustang	100	89	8.9
Capri ('91)	111	46	5.1
Continental	67	71	4.8
Town Car	65	68	4.4
Sable	70	57	4.0
Taurus	69	55	3.8
Grand Marquis	61	54	3.3
Crown Victoria	61	50	3.1
Aerostar	62	38	2.4
Acura			
Vigor	73	29	2.1
Legend	89	22	1.9
Honda			
Prelude	115	50	5.8
Civic	109	38	4.1
Accord	80	30	2.4
Isuzu			
Stylus ('91)	117	50	5.9
Mazda			
MX-6 ('93)	112	83	9.3
929	96	69	6.6
626 ('93)	89	47	4.2
Miata	78	51	4.0
Mitsubishi			
3000	108	49	5.3
Diamante	84	23	1.9
Nissan			
NX ('91 - '92)	127	146	18.7
Sentra ('93)	87	117	10.2
Infiniti J30 ('93)	105	40	4.2
Altima ('93)	83	46	3.8
Maxima	80	46	3.7
Infiniti Q45 ('90-'93)	93	39	3.6
Infiniti M30 ('90-'92)	99	32	3.2
300ZX ('91 - '93)	87	4	0.3
Toyota			
MR2 ('91 - '93)	99	77	7.6
Tercel ('93)	95	65	6.2
Paseo ('93)	87	64	5.6
Lexus ES250 ('90-'91)	88	64	5.6
Camry	80	55	4.4
Celica	118	32	3.8
Lexus ES300	80	39	3.1
Previa ('93)	59	51	3.0
Lexus LS400 ('90-'92)	84	32	2.7
Lexus SC	109	24	2.6
Corolla	87	26	2.3

[1] The coverage under which people insure their own vehicle against loss caused by collision.
[2] Not all vehicles with airbags listed due to sample size limitations.
[3] IVY, Insured Vehicle Year
[4] 1992 model year driver side deployment rate, unless otherwise specified.
[5] All minivans combined. Concorde, Intrepid, and Vision combined.

Appendix B
Number of Driver-Side Deployments by Standardized Severity Level
Selected Vehicles with an Airbag[1]

Vehicle	Standardized Severity Level									
	10%	20%	30%	40%	50%	60%	70%	80%	90%	100%
Dodge Caravan (100=$19,117)										
Driver Deployments	2	10	38	21	7	5	1	0	0	1
Frontal Crashes	343	125	82	31	12	8	1	1	0	1
Dodge Dynasty (100=$10,271)										
Driver Deployments	0	13	31	32	45	29	16	12	4	2
Frontal Crashes	210	214	90	56	57	32	17	14	5	3
Jeep G. Cherokee (100=$13,341)										
Driver Deployments	0	0	1	1	0	4	2	1	0	1
Frontal Crashes	65	34	20	6	1	5	3	3	0	1
Plymouth Acclaim (100=$11,058)										
Driver Deployments	7	51	79	82	81	40	13	4	2	1
Frontal Crashes	368	271	153	109	91	43	13	4	2	1
Ford Aerostar (100=$11,944)										
Driver Deployments	0	0	0	5	9	9	9	3	3	0
Frontal Crashes	165	121	54	40	18	15	11	6	4	1
Ford Mustang (100=$16,025)										
Driver Deployments	1	25	65	95	65	24	11	6	0	1
Frontal Crashes	694	406	185	149	76	29	14	6	2	2
Ford Taurus (100=$13,887)										
Driver Deployments	0	2	25	89	102	61	17	14	2	1
Frontal Crashes	906	688	359	203	136	73	19	16	2	1
Lincoln Continental (100=$16,809)										
Driver Deployments	0	1	6	20	11	11	2	1	0	1
Frontal Crashes	195	116	45	40	15	11	3	2	0	1
Cadillac DeVille/Fleetwood (100=$24,535)										
Driver Deployments	0	8	38	45	10	9	1	1	1	0
Frontal Crashes	772	381	132	72	14	13	1	1	1	1
Chevrolet Camaro (100=$14,049)										
Driver Deployments	0	1	17	43	23	19	5	6	5	3
Frontal Crashes	541	319	125	99	44	29	10	8	6	3
Chevrolet Corsica (100=$11,231)										
Driver Deployments	0	2	9	23	37	29	20	9	0	3
Frontal Crashes	372	262	117	74	64	37	23	10	1	4
Geo Storm (100=$17,854)										
Driver Deployments	0	3	32	56	46	11	5	1	0	1
Frontal Crashes	926	407	206	104	62	13	6	1	0	1
Oldsmobile 88 (100=$12,707)										
Driver Deployments	0	0	0	0	5	3	5	6	3	3
Frontal Crashes	111	74	38	27	17	6	7	8	3	4
Saturn SL (100=$13,423)										
Driver Deployments	0	1	12	11	12	10	0	0	1	0
Frontal Crashes	173	106	412	21	14	14	4	1	1	1
Acura Legend (100=$24,868)										
Driver Deployments	2	4	8	7	10	3	4	4	1	1
Frontal Crashes	547	151	66	28	15	3	6	4	1	1
Honda Accord (100=$20,491)										
Driver Deployments	3	20	35	25	18	19	8	2	0	0
Frontal Crashes	1,299	491	179	61	30	23	9	3	0	1
Honda Civic (100=$11,901)										
Driver Deployments	1	2	14	13	24	20	20	11	3	3
Frontal Crashes	683	361	187	83	62	30	26	20	5	4
Mazda Miata (100=$14,316)										
Driver Deployments	1	1	10	16	13	9	9	4	0	1
Frontal Crashes	290	115	55	26	23	10	9	4	0	1

Appendix B (continued)

Vehicle	Standardized Severity Level 10%	20%	30%	40%	50%	60%	70%	80%	90%	100%
Nissan Maxima (100=$13,186)										
Driver Deployments	0	1	1	0	1	5	2	2	2	2
Frontal Crashes	110	58	34	17	6	10	4	2	2	2
Infiniti Q45 (100=$18,373)										
Driver Deployments	0	0	0	0	4	3	0	1	1	2
Frontal Crashes	73	53	11	11	14	4	1	1	1	2
Lexus LS 400 (100=$33,859)										
Driver Deployments	0	3	5	2	7	0	1	1	0	1
Frontal Crashes	294	92	36	11	7	1	1	1	0	1
Toyota Camry (100=$18,112)										
Driver Deployments	0	7	23	20	47	25	21	7	4	2
Frontal Crashes	789	340	140	53	60	92	25	7	5	2
Audi 80/90 (100=$13,952)										
Driver Deployments	0	0	1	1	1	1	1	0	0	1
Frontal Crashes	30	16	13	7	6	3	1	0	0	1
BMW 3 Series (100=$22,581)										
Driver Deployments	1	1	2	10	4	8	6	5	1	3
Frontal Crashes	204	96	69	34	16	11	7	10	3	3
BMW 5 Series (100=$24,592)										
Driver Deployments	0	0	1	1	2	0	2	3	3	2
Frontal Crashes	95	53	28	16	5	3	4	3	3	2
Saab 900 (100=$19,372)										
Driver Deployments	2	0	1	2	3	5	1	1	3	1
Frontal Crashes	83	42	34	14	6	8	2	1	3	1
Volvo 740 (100=$15,930)										
Driver Deployments	0	2	2	3	2	6	2	5	0	1
Frontal Crashes	143	80	41	36	17	10	7	10	1	1

[1] Model years 1990 and later airbag installed vehicles, as applicable based on VIN.

Appendix C
Representative 1993 MY Airbag System Replacement Costs
OEM Replacement Recommendations and Forward Mounted Sensors

Manufacturer	Model	Driver Side Module	Passenger Side Module	Sensors	Other Parts	Labor (Hrs)[1]	Total Cost
Acura	Legend	$ 774	$1,010	$305x2	—	2.5	$2,469
	Vigor	774	—	305x2	—	1.7	1,435
Audi	100	878	1,483	839	$ 904[2]	8.8	4,368
	90	878	—	839	—	0.8	1,741
BMW	850Ci	1,262	1,915	192x2	1,574[3]	8.9	5,402
	325	1,262	—	192x2	1,134[3]	5.4	2,942
Buick	LeSabre	540	—	135x2	—	2.2	876
Cadillac	Seville	540	540	135x2	—	1.6	1,398
	DeVille	540	—	135x2	—	2.7	891
Chevrolet	Camaro	373	540	135x1	—	1.9	1,105
	Beretta	540	—	135x1	—	1.5	720
Dodge	Intrepid	414	414	77x2	101[4]	5.6	1,251
	Shadow	466	—	77x2	101[4]	1.9	778
Ford	Taurus	579	773	55x1	—	1.5	1,452
	Mustang	579	—	69x2	—	1.2	753
Infiniti	J30	1,021	1,327	119x1	670[5]	2.6	3,215
Lexus	GS300	1,248	2,606	107x2	561[6]	3.6	4,737
	ES300	927	—	171x2	554[6]	1.8	1,877
Lincoln	Continental	577	573	55x2	—	1.8	1,314
Mazda	929	735	1,684	342x2	—[7]	4.5	3,238
	626	744	—	186X2	—[7]	0.9	1,143
Mercedes	300 Class	1,100	2,730	1,050	480[8]	2.5	5,435
Mitsubishi	Diamante	793	—	321x2	1,773[9]	6.7	3,409
Nissan	Maxima	1,118	—	221x2	736[10]	5.0	2,446
Porsche	928	767	1,161	64x2	529[11]	4.6	2,723
Saab	900	904	—	119x2	1,078[12]	2.7	2,301
Saturn	All	703	—	63x2	160[13]	2.4	1,061
Toyota	Camry	868	—	162x2	778[14]	2.6	2,048
Volvo	850	1,113	946	895	1,412[15]	9.5	4,651
	240	1,113	—	879	178[16]	1.5	2,215

[1] Labor rate of $30/hr used in total cost calculation.
[2] Dash panel.
[3] Control/Diagnostic module, contact ring, belt tensioners, and belts.
[4] Clock spring.
[5] Control module. Pre-tensioner cost not shown.
[6] Center airbag sensor assembly.
[7] Clock spring cost not available.
[8] ETR belt assemblies.
[9] SRS diagnostic unit, clock spring, steering wheel, steering column, intermediate joint.
[10] Diagnostic module.
[11] Control unit and steering wheel contact.
[12] SRS control unit. Coil spring unit and passenger seat belt tensioner costs not shown.
[13] SIR coil assembly.
[14] Center airbag sensor assembly.
[15] Contact reel, steering wheel, seat belt tensioner, seat belts, windshield, dashboard cover panel, igniter leads.
[16] Seat belt tensioner and seat belts. Seat belt tensioner cost not shown.

Appendix D
Driver Injuries by AIS Injury Description
Crashes With More Than $5,000 Damage
Injuries Per 1,000 Frontal Crashes

	Airbag Deployed (1,654 Crashes)	Belt-Only (250 Crashes)		Airbag Deployed (1,654 Crashes)	Belt-Only (250 Crashes)
External			**Thorax (cont.)**		
10201.1 Contusion, NSF	334	236	52401.2 Thoracic duct, laceration	< 1	—
10101.1 Abrasion, minor	144	8	52519.3 Rib fracture, ≥ 4	< 1	—
10301.1 Laceration, NSF	71	52	51405.3 Lung laceration, NFS	—	4
10202.1 Contusion, minor	29	144	51410.3 Lung laceration, segmental	—	4
10302.1 Laceration, minor	26	36	**Abdomen**		
10701.1 Burn, NFS	24	—	60605.4 Artery, splenic	< 1	—
10303.2 Laceration, deep	17	4	61702.2 Kidney, contusion NFS	< 1	—
10102.1 Abrasion, NSF	16	36	61802.2 Liver, contusion NFS	< 1	—
10203.2 Contusion, major	6	12	61902.2 Spleen, contusion NFS	< 1	—
10103.2 Abrasion, major	2	—	60102.2 Abdominal wall, laceration	—	4
Cranium and Brain			60103.3 Abdominal wall, lac. with tissue loss	—	4
20601.1 C-1, minor	22	28	60802.1 Abdominal injury, hematuria	—	4
20637.2 C-2, concussion	14	44	60804.3 Retroperitoneal injury	—	4
20606.2 C-1,< 1 hr.	2	4	61701.2 Kidney, NFS	—	4
20602.2 C-1, unk hrs.	1	4	61805.2 Liver, laceration NFS	—	4
20604.2 C-1, amnesia	1	—	**Cervical Spine**		
20626.5 C-1, > 24 hrs.	< 1	—	70101.1 Strain, acute	140	196
20636.5 C-2, > 24 hrs.	< 1	—	70501.2 Dislocation	< 1	8
20704.2 Vault, NFS	< 1	—	70601.2 Fracture, NFS	—	4
20408.4 Cerebellum, subdural NFS	< 1	—	**Thoracic Spine**		
20501.3 Cerebrum, NFS	< 1	—	73101.1 Strain	48	68
20504.4 Cerebrum, laceration	< 1	—	73501.2 Dislocation	1	4
Face			**Lumbar Spine**		
32401.1 Nasal, contusion	8	8	76101.1 Strain	65	36
32603.1 Teeth, fracture	7	12	76701.2 Fracture, NFS	< 1	—
32703.2 Temp. joint	3	—	76707.2 Vertebral body, NFS	< 1	4
32402.1 Nasal, fracture, NFS	2	12	76801.2 Nerve Root, NFS	< 1	—
32201.1 Mandible, fracture, NFS	2	4	76601.2 Dislocation, NFS	—	4
31102.1 Cornea, abrasion	2	4	**Upper Extremity**		
32701.1 Temp. joint, NFS	1	—	82003.1 Wrist sprain	24	8
32702.1 Temp. joint, sprain	1	—	81802.1 Shoulder, contusion	19	28
32602.1 Teeth, dislocation	1	—	82101.2 Fracture, NFS	18	16
31103.1 Cornea, contusion	< 1	—	81806.1 Shoulder, sprain	17	20
31104.2 Cornea, laceration	< 1	—	81502.1 Elbow, contusion	12	4
31401.2 Optic Nerve, NFS	< 1	—	82202.1 Finger, fracture	9	—
31502.2 Retinal detachment	< 1	—	82201.1 Finger, NFS	4	16
31601.2 Sclera laceration	< 1	—	82002.1 Wrist, contusion	10	4
Thorax			81801.1 Shoulder, NFS	3	—
52501.1 Rib cage, contusion	21	32	82005.3 Wrist, dislocation	2	—
52508.2 Rib fracture, > 1 NFS	3	4	81902.1 Sternoclavicular joint, contusion	2	4
52503.1 Rib fracture, 1	2	4	82102.2 Carpal/Metacarpal fracture	2	8
52602.2 Sternum, fracture	2	4	81508.1 Elbow sprain	2	—
52509.2 Rib fracture, 2-3	2	8			
50105.3 Asthmatic attack	2	—			
51802.3 Pericardium, NFS	1	—			
52502.1 Rib fracture, NFS	1	4			
52601.1 Sternum, contusion	1	4			
50201.4 Aorta, NFS	< 1	—			
51101.3 Vein, NFS	< 1	—			
51705.6 Myocarduim, complex	< 1	—			

Appendix D (continued)

	Airbag Deployed (1,654 Crashes)	Belt-Only (250 Crashes)
Upper Extremity (cont.)		
82001.1 Wrist, NFS	2	—
82301.2 Radius, NFS	2	—
82401.2 Ulna, NFS	2	—
82601.2 Clavicle fracture	2	—
Lower Extremity		
91802.1 Knee, contusion	65	76
91703.1 Ankle, sprain	14	4
91803.2 Knee, sprain	7	8
91702.1 Ankle, contusion	3	4
92401.2 Tibia fracture, NFS	3	4
91801.1 Knee, NFS	2	—
91705.3 Ankle, dislocation	2	—
92701.2 Patella, fracture	2	—
91902.3 Hip, dislocation	1	—
92201.2 Calcaneus fracture	1	—
92801.2 Pelvis fracture, NFS	1	—
91102.3 Knee Collateral ligament laceration	<1	8
91706.3 Ankle crush	—	4
91803.2 Knee sprain	—	8
92001.2 Foot fracture, NFS	—	4
92402.2 Tibia fracture, closed	—	4
90103.3 Crush, below knee (whole area)	< 1	—
90104.4 Crush, above knee (whole area)	< 1	—
91503.1 Foot joint, dislocation	< 1	—
91701.1 Ankle, NFS	< 1	—
91706.3 Ankle, cursh	< 1	—
91805.3 knee, dislocation	< 1	—
92101.2 Metatarsal fracture	< 1	—
92301.1 Toe, NFS	< 1	—
92302.1 Toe, fracture	< 1	—
92502.2 Fibula, peroneal nerve	< 1	—
92503.2 Fibula, fracture	< 1	—
93001.3 Sacroiliac fracture	< 1	—

950866

"An Overview of Air Bag Deployments and Related Injuries. Case Studies and a Review of the Literature."

Donald F. Huelke
University of Michigan
Transportation Research Institute

ABSTRACT

This is an overview of air bag injuries, a review of the literature and descriptions of air bag related injuries to the various body areas. Some unusual injuries are also included. The cases presented are from the author's files and one from the Special Studies Division of NHTSA.

INTRODUCTION

THIS PAPER PRESENTS AN OVERVIEW of the injuries seen in the various body areas to car occupants who were in frontal crashes when the air bag deployed. This review article is augmented by some cases from the authors files of the unusual or unexpected air bag related injuries.

MATERIALS AND METHODS

THE LITERATURE ON AIR BAG INJURIES was reviewed as well as the data on air bag crashes from the author's previous publications, additional cases from the University of Michigan Transportation Research Institute (UMTRI) field accident investigations, and those of the Special Crash Investigations Section of NHTSA.[1-3] Most of the injury data are of drivers, for there are few right front passengers in the cars equipped with passenger side air bags.

HEAD AND FACE - From a review of the literature, it is apparent that most of the injuries to the face are either errythemia, abrasions, contusions, and much less often, ocular injuries [1, 2, 4-23]. The medical literature on the abrasions, lacerations or chemical keratitis to the eye, or facial fractures, adequately describe the injury [24, 25]. Often however, the mechanism of injury, when presented, is speculative at best . Kikuchi, et al, using rabbits, found that the inflating air bag produced ocular injuries [26]. Fukagawa, et al, found corneal epithelial cell loss due to contact with the inflating air bag [27].

The mechanisms for facial abrasions and errythemia due to air bag contact has been well described by Reed, et al [28, 29]. Abrasions were noted with untethered air bags within 350 mm from the module to the subject. Abrasions with a typical tethered air bag occurred at less than 300 mm (approximately 275 mm). In general, the closer the subject to the air bag module the more extensive were the abrasions because of the air bag fabric velocity. They showed that improved fold techniques reduces fabric velocity and thus reduced the extent of the abrasion.

Recently a case of a pipe stem perforating the eye has been reported [30]. Also, perforations of the eyeball from eye glass frames or lenses have been identified [31].

There are cases of individuals striking themselves in the face when the airbag flings the hand towards the face. In one case, the hand of the driver caused nasal and anterior face fractures (see case no. 14). Scott, et al, presented a case of a 64 year old lap-shoulder belted female driver of a 1992 Mercury Grand Marquis that struck a guard rail [25]. It was opined that the air bag contact caused an ocular injury, periorbital abrasions and ecchymosis, with nasal bone and right maxillary fractures . More probably the fractures were sustained by a "fist slap" and not by direct air bag contact. Air bag related hearing deficit has been reported [32]. The temporomandibular joint has been injured allegedly by air bag contact [33].

THE NECK - Neck injuries are almost always minor and consist of pain in the posterior area. Such injuries are not uncommon in air bag deployments. Whether the neck injury, i.e., cervical sprain, is due to the drivers head flexing over the lap-shoulder belt, which is known to be related to cervical sprains, or whether it is due to the interaction of the head and the air bag, is unknown. Numerous cases of cervical strain involving air bags, irrespective of lap-shoulder belt use and conversely, with lap-shoulder belts and no air bags, has been identified at UMTRI.

For drivers, there are few cases reported in the literature of the more serious cervical spine injury. An elderly unbelted female driver was diagnosed some weeks after the crash as having a small chip fracture of one of the cervical vertebra without associated neurological symptoms (34). An upper cervical fracture was sustained by a 40 year old lap-shoulder belted male driver of a Mercedes-Benz in a head on car to car crash (35, 36). A fracture of the C2 body, extending through both pedicles and the left lateral mass, were identified. Blacksin reviewed the hospital records of seven drivers in air bag deployment crashes (23). In one case a lap-shoulder belted 23 year old female driver sustained a C1-C2 posterior arch fracture. In another, an unbelted 18 year old male had a C4-C5 disk herniation, cord contusion, and a C4 end plate fracture following a high speed crash into a wall. High cervical fractures with cord injuries have been observed in experimental laboratory thorax impacts of predeployed and nonvented air bags. (37)

Infrequently, there are some anterior neck (throat) abrasions or errythemia when this under-chin area is contacted by the air bag (2, 38). No significant air bag related injuries in the throat area have been identified.

Case 1 A 1993 Plymouth Sundance Duster was struck head-on by a 1995 Chrysler LeBaron that crossed the centerline (Fig. 1). The driver of the Plymouth, an unbelted 38 year old female, (165 cm, 91 kg) was dead at the scene. She sustained a partial transection of the spinal cord at the base of the skull with complete dislocation of the atlas to occipital joint. Also, multiple bilateral rib fractures with lung contusion were identified at autopsy (UM-3221).

Figure 2

THE THORAX - In general, injuries to the thorax are related to the shoulder belt in frontal crashes and infrequently minor injuries about the upper thorax from the air bag have been observed (39). Thoracic injuries of any significance from air bag contact are extremely rare and are, for the most part, found only in the higher speed collisions where both the steering assembly and air bag can be related to intrathoracic injury or rib fractures (23). An air bag rupture of the right atrium has been reported as well as of the aortic valve (40, 41). Augenstein, et al, indicated that occult torso injuries may occur in survivors of high speed crashes without exterior signs of torso trauma (42). The response of asthmatics to air bag aerosols and gasses has also been studied (43).

The following cases, from the literature and the author's files, exemplify some of the thoracic injuries related to torso interaction with the air bag.

Case 3 - An unrestrained 31 year old male sustained nasal, orbital, and xyphoid fractures along with a T3 compression fracture and fractures of the adjacent posterior ribs at T2 and T3 (9).

Case 4 - A 22 year old unbelted female driver sustained a rupture of the right atrium from air bag deployment following a 10-15 mph collision with a parked car. The make, model, or year of the car was not given, nor the drivers height and weight(11).

Below are two cases of drivers who sustained more than minor chest injuries from air bag/torso interaction.

Case 5 - A 1990 Ford Taurus, driven by a 40 year-old unrestrained male (173 cm, 91 kg), went off the road and struck a tree. The driver was dead at the scene. The autopsy indicated elevated blood alcohol level as well as non prescription drug ingestion. The front passenger indicated that the driver slumped over the steering wheel prior to the car leaving the road. (Fig 3) Autopsy revealed multiple bilateral rib fractures, a sternal fracture, bilateral lung lacerations, multiple epicardial and endocardial contusions and superficial lacerations of the liver. (UM-3216)

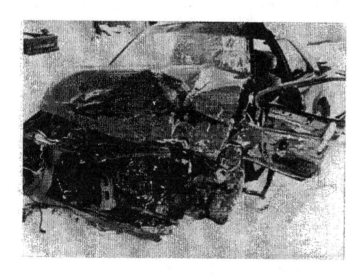

Figure 1

Case 2 A 1994 Oldsmobile 98 was struck head-on by a 1991 Aerostar van (Fig. 2). The 49 year old female lap-shoulder belted driver died, sustaining a complete separation of the C1-C2 vertebrae. In addition she had a laceration of the right ventricular wall, and of the thoracic aorta (UM 3286).

Figure 3

Case 6 - A 1991 Plymouth Acclaim 4 dr. struck the right side of a 1985 Chevrolet Silverado 4 x 4 pick up truck that turned left in front of the Acclaim (Fig 4). The unrestrained 30 year old unrestrained driver (175 cm, 82 kg) sustained three right rib fractures and a sternal fracture from contact with the air bag (UM-CCA-062).

Figure 4

THE ABDOMEN - Air bag related intraabdominal organ injuries have been rarely reported. Infrequently errythemia on the drivers anterior abdominal wall, usually somewhat circular, has been related to contact with the air bag.

THE BACK - Other than the cervical spine, the infrequent complaint of aches or pain in the mid and low back area cannot be related solely to the air bag, for such injuries have been noted in lap-shoulder belted occupants (without air bags) and are mainly related to the active restraint rather than to the air bag-occupant interaction. Rarely a thoracic fracture has been identified in a non-belted occupant who interacted with the air bag (see case #3 above).

THE UPPER EXTREMITY - Air bag related injuries of the upper extremity rarely have been reported to the shoulder or arm [39]. However, a number of injuries to the forearm, wrist, hand or elbow are related to the air bag module door or to the deploying air bag [12, 15, 34, 44, 45]. Dermal burns reportedly from air bag powder residue has been identified, [45] but is rarely found.

Not infrequently, abrasions, contusions, and errythema of the forearm are reported, especially on the anterior (palmar) surface. The deploying air bag can slide along the forearm, or slap the forearm, causing these superficial skin injuries. At times the deploying air bag can snag on a watch band, which in turn scuffs or lacerates the skin. Infrequently we have seen drivers who have localized burns on the dorsal side of the base of the thumb, or adjacent wrist, due to exhaust gases directly contacting the skin and not from air bag contact.

The flailing upper extremity has recently been identified in field accident investigations (34). In these crashes the upper limb is flung by the inflating air bag and impacts the instrument panel, rear view mirror or windshield, causing contusions, abrasions, lacerations and/or fractures of the fingers, hand or wrist, or distal forearm.

Near amputation of the thumb or laceration of the web between the thumb and index finger from air bag contact have been reported [34, 46].

If the forearm is close to the air bag module door at the time of deployment, forearm fractures, or tearing of the forearm skin, are due to either the high velocity of the deploying air bag, or the direct contact of the forearm with the door of the air bag module.

Case 7 - A 1991 Mercury Grand Marquis, in a left turning maneuver, was struck along the left front side by another car (Fig. 5). On impact the upper air bag module flap contacted the mid right forearm of the 73-year-old lap-shoulder belted female driver, (150 cm, 57 kg) (Fig. 6). From impact by the air bag module cover and the contact of the deploying air bag she sustained multiple segmental open fractures of the right radius and ulna, with fractures of both the proximal and distal ulna as well as the olecranon. In addition, a circumferential degloving laceration (about 340 degrees) involving the skin and subcutaneous tissue was sustained. The right proximal fifth phalanx was also fractured, probably from the inflating air bag. The air bag flung her right upper extremity upward as indicated by tissue and blood spatters on the roof liner, side pillar and rear seat area. (CS 92-22)

vent ports which were facing her on initial bag unfolding. (Fig 8), (CCA-067).

Figure 5

Figure 7

Case 9 -This 1991 Chrysler New Yorker Fifth Avenue, driven by a 68 year old lap-shoulder belted driver (170 cm, 75 kg) attempted a left turn when struck in the left front corner of her car by another car. Car damage was minor. (Fig 9). The air bag deployed and melted portions of her blouse and bra She sustained minor injuries consisting of abrasions to the chin and lower face, right forearm, and wrist and contusion to the left forearm, wrist and portions of the thumb. All injuries were air bag induced (UM-CCA-068).

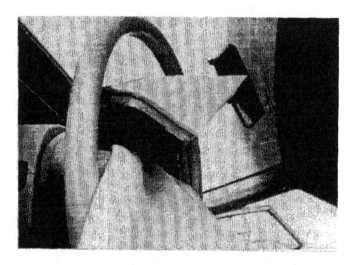

Figure 6

THE LOWER EXTREMITY - In that the lower extremities are not in the path of the deploying air bag, it is not surprising that air bag related injuries have not been reported in the medical literature or identified in field accident investigations.

THE UNUSUAL AND THE BAZAAR

Case 8 - A 1991 Plymouth Acclaim, driven by a lap-shoulder belted 68 year old female driver (163 cm, 64 kg) was traveling slowly and made a left turn and struck a light pole in a parking lot. The damage to the left front corner of the vehicle was minor (Fig 7). The air bag deployed. The driver sustained facial errythema and abrasions. Her polyester tank top and bra were melted, due to direct contact with the

Figure 8

Case 10 - A 1993 Dodge Intrepid, driven by a 41 year old lap-shoulder belted male, (171 cm, 77 kg) went off the road and struck a tree. The air bag caused his jaw to snap

688

shut and in so doing chipped and fractured two teeth (CCA-104).

Case 11 - A 1992 Dodge Spirit driven by a lap-shoulder belted 32 year old male driver (178 cm, 86 kg) was struck head on by a 1990 Chevrolet Corsica The lap-shoulder belted driver of the Corsica died. In this high speed collision, there was extensive exterior damage to the car, as well as significant compromise of occupant space within the vehicle (Fig. 10). The driver sustained multiple extensive injuries to his lower extremities and also intraabdominal injuries. Other air bag related injuries included abrasions to the posterior, abrasions of the left forehead, right anterior elbow, and a slight concussion. His most severe injuries were to the lower extremities. In this crash, estimated to be a closing velocity of approximately 130 mph, the air bag saved his life (UM-3133).

Case 12 - Minor damage was sustained to this 1992 Plymouth Sundance America, driven by a 39 year old lap-shoulder belted female (165 cm, 55 kg). The front of her wool sweater showed a centrally located scorch pattern and the underlying clothing was also scorched. Again, this was probably due to the hot gases of the air bag venting from the exhaust ports rather than air bag contact..

Case 13 - A 1988 BMW 750 skidded and struck a guard rail. The lap-shoulder belted 51 year old driver (185 cm, 87 kg) sustained a right cheek abrasion and a ruptured right eyeball. His pipe and pipe tobacco was found in the car. The pipe stem was broken and is believed to have caused the perforation of the eyeball [30].

Case 14 - A 1993 Ford Taurus was turning left when hit in the right front area. Damage to the car was minor. The driver, a 59-year old male (170 cm, 54 kg), was wearing his 3-point restraint. No occupant contact marks were identified in the car. Air bag deployment produced comminuted mid-shaft fractures of the right radius and ulna (Fig. 11). His right forearm and/or hand was forced into his face producing a nasal fracture, and a fracture of the right maxillary sinus (Fig. 12). Chip fractures of two lower front teeth were also sustained (UM-3113).

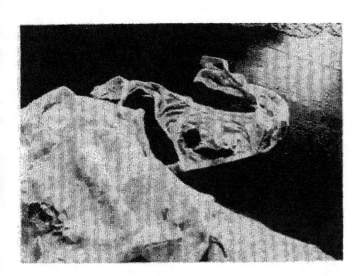

Figure 9

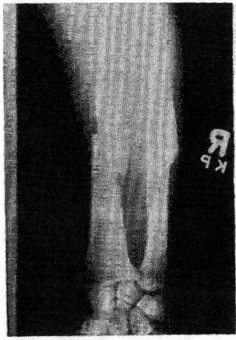

Figure 11

Figure 10

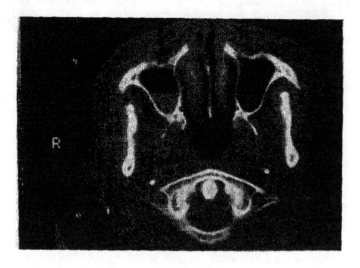

Figure 12

Case 15 - A 1994 Volvo struck the rear side of a passenger car at low speed. Both air bags deployed. The 60 year old lap-shoulder belted driver did not sustain any injuries. The 59 year old front seat lap-shoulder belted passenger (188 cm, 89 kg), had her left foot on the instrument panel over the air bag module door. Post crash her left foot was deviated medially. She had multiple comminuted fractures of the medial malleolus and the dome of the talus (Fig. 13).

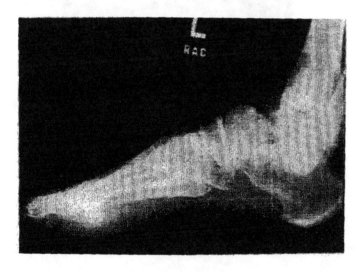

Figure 13

THE FRONT RIGHT PASSENGER

There are but few cases of passenger side air bag deployments with a front right passenger in the car at the time of the deployment. Several injury cases are presented below.

Case 16 - A 1992 Lincoln Continental struck the rear of a stopped van. The grill of the Lincoln was broken-no other frontal damage was sustained. The unrestrained 51 year old male front passenger sustained three laceration of the left 3rd finger and "jammed" the joints of that finger. A year post crash he still could not bend the finger (FMA-057).

Case 17 - A 1993 Lincoln Continental rear-ended a stopped car. Damage was minor. The driver sustained minor injuries. The lap-shoulder belted front passenger, a 66-year-old female, height 157 cm, mass 61 kg, had a 5 cm tear of the skin of the web between the left thumb and hand with tearing of the underlying muscle. The metacarpophalangeal joint of the thumb was dislocated with the ulnar side of the capsular ligament being torn.

DISCUSSION

"There is no polio vaccine for traffic medicine." Any and all structures within the vehicle can cause injury. It is well known that the lap-shoulder belts are effective in reducing the frequency and severity of serious injuries and fatalities, yet there have been injuries attributed to the lap-shoulder belt. Similarly, the HPR windshield is an excellent safety device that reduces the incidence of significant lacerations to the face of an unrestrained occupant, yet it may be the cause of injuries as well. Similarly, the air bag is known to be a significant adjunct to the lap-shoulder belt in reducing the frequency of serious injuries and death. However, as shown in this paper, the air bag can be related to injuries that are usually minor in nature.

Uniquely, however the upper extremity injuries, involving lacerations, fractures, and dislocations are specific to the air bag for, in general, these types of injuries are relatively infrequently noted to lap-shoulder belted drivers who do not have a steering wheel air bag. The strange and unusual cases should be of interest to the reader, showing the variety of accident injury scenarios that exist in the real world.

The passenger side air bag is yet to be discussed in detail because of the lack of crashes with front seat passengers. When further data are available, information will be presented.

All in all, what has been discussed in this paper are the "worst case scenarios" that we are seeing in the field investigations of air bag crashes. These are the outliers, the unique type of cases, and should not be construed as being typical occurrences. In our studies we now have over 350 air bag crashes investigated and the injuries described to the upper extremities, cervical spine and the flailing extremity are relatively infrequent.

REFERENCES

1. Huelke DF, Moore JL, Ostrom M. Air Bag Injuries and Occupant Protection. *J Trauma*, 33:894-898, 1992.
2. Huelke DF, Moore JL. Field Investigations of the Performance of Air Bag Deployments in Frontal Collisions. *Acc Anal and Prev*, 25:717-730, 1993.
3. Huelke DF, Moore JL, Compton TW. Upper Extremity Injuries Related to Air Bag Deployments. *J Trauma*, pp 894-898, 1992.
4. Backaitis SH, Roberts JV. Occupant Injury Patterns in Crashes with Airbag Equipped Government Sponsored

Cars. *Proc 31st Stapp Car Crash Conf,* pp 251-257, Warrendale, PA, 1987.

5.   Blair G, Larkin GL. Airbag-Mediated Facial Trauma. *4th Intl Conf on Emerg Med,* Washington, DC, May, 1992.

6.   Braude LS. Protective Eyewear Needed With Driver's Side Air Bag? *Arch Opthalmol*; 110:1201, 1992.

7.   Ingraham HJ, Perry HD, Donnenfeld ED. Air-Bag Keratitis. *New Eng J Med*, 324:1599-1600, 1991.

8.   Larkin GL. Airbag-Mediated Corneal Injury. *Am J Emerg Med*, 9:444-446, 1991.

9.   Lubeck D, Greene JS. Corneal Injuries. *Opthal Emerg & Ocular Trauma*, 6:73-94, 1986.

10.   Mishler KE. Hyphema Caused by Air Bag. *Arch Opthalmol*, 109:1635, 1991.

11.   Rimmer S, Shuler JD. Severe Ocular Trauma From a Driver's-Side Air Bag. *Arch Opthalmol*, 109:774, 1991.

12.   Mertz HJ. Restraint Performance of the 1973-76 GM Air Cushion Restraint System. In Automatic Occupant Protection Systems, SP-736 Paper No 88044, SAE, Detroit, Feb. 1988.

13.   Digges KH, Roberts V, Morris J. Residual Injuries to Occupant Protected by Restraint Systems. SAE Paper No 891974, Soc of Auto Eng, Warrendale, PA, 1989.

14.   Conover K. Chemical Burn from Automotive Air Bag. *Ann Emerg Med*. 21:6, 1992.

15.   Han DP. Retinal Detachment Caused by Air Bag Injury. *Arch Opthalmol*, 111:1317-1318, 1993.

16 .   Rosenblatt M, Freilich B, Kirsch D. "Air Bag-Asociated Ocular Injury." *Arch Opthalmol*, Vol III, pp 1318, 1993.

17.   Whitacre MM, Pilchard WA. Air Bag Injury Producing Retinal Dialysis and Detachment. *Arch Opthalmol*, 111:1320, 1993.

18.   Campbell JK. Automobile Air Bag Eye Injuries. *Neb Med J*, pp 306-307, 1993.

19.   Lesher MP, Durrie DS, Stiles MC, et al. Corneal Edema, Hyphema, and Angle Recession After Air Bag Inflation *Arch Opthalmol*, 111:1320-1322, 1993.

20.   Kuhn F, Morris R, Witherspoon CD, et al. Ocular Injuries in Motor Vehicle Crashes, *Opthal*, 100:1280, 1993.

21.   Anonymous. Study of Injuries from Air Bags Yields Concern, Not Alarm. *Acc Reconstr J*, 6:12, 61,1994.

22.   Driver PJ, Cahswell LF, Yeatts, R. Airbag-associated Bilateral Hyphemas and Angle Recission, *Am J Opthalmol* 118:250-1, 1994

23.   Blacksin MF. Patterns of Fracture After Air Bag Deployment., *J Trauma*, 35:840-843, 1993.

24.   Smally AJ, Binzer,A, Dolin S, et al. Alkaline Chemical Keratitis: Eye Injury from Airbags. *Ann Emer Med*, 21:1400-1402, 1992.

25.   Scott I, John G, Stark WJ. Airbag-Associated Ocular Injury and Periorbital Fractures. *Arch Opthalmol* , 111:25, 1993.

26.   Kikucki A, Horii M, Kawai A, et al. Injury to Eye and Facial Skin (rabbit) on Impact with Inflating Air Bag., *Biomech of Serious Trauma*, 2nd Intl Conf Proc, pp 240, 289-296, 1975.

27.   Fukagawa K, Tsubota, K, Kimura C, et al. Corneal Endothelial Cell Loss Induced by Air Bags.*Opthal.*, 100:1819-1823, 1993.

28.   Reed MP, Schneider LW, Burney RE. Investigation of Airbag-Induced Skin Abrasion. *Proc 36th Stapp Car Crash Conf,* pp 1-12, Warrendale, PA, 1992.

29.   Reed MP, Schneider LW. A Laboratory Technique for Assessing the Skin Abrasion Potential of Airbags. *Frontal Impact Protection: Seat Belts and Air Bags.* Soc Auto Eng Cong and Expo, SAE Paper No 930644, Detroit, MI, 1993.

30.   Walz F, Mackay M, Gloor B. Air bag and eye perforation by tobacco pipe., *J Trauma*, 1995.

31.   Gault J, Vichnin, M, Jaeger E, et al. Ocular Injuries Associated with Eyeglass Wear and Air Bag Inflation, J Trauma, 1995.

32.   Beckerman B, Elberger S. Air Bag Ear, Letter to Editor, *Annals of Emerg Med*, pp 195-196, 1991.

33.   Garcia R. Air Bat Implicated in Temporomandibular Joint Injury. *Cranio, 12:125-27, 1994*

34.   Huelke DF. Upper Extremity Injuries Related to Air Bag Deployments. SAE Paper No 940716, Soc of Auto Eng, Warrendale, PA, 1994.

35.   Steinmann R. A 40 Year Old Woman With an Air Bag-Mediated Injury. *J Nursing*, 18:308-310, 1991.

36.   Traynelis VC, Gold M. Cervical Spine Injury in an Air Bag-Equipped Vehicle. *J Spinal Disorders*, 1:60-61, 1993.

37.   Cheng T, Yang, KH, Levine RS, et al. Injuries to the Cervical Spine Caused by a Distributed Frontal Load to the Chest. Proc Stapp Car Crash Conf, pp 1-40, 1982.

38.   Shebar ET, Laurenzano Letter to the Editor, *N Eng J Med*, 325:1519, 1991.

39.   Hoel EN. An Unusual Airbag-Mediated Injury. *J Emerg Nursing*, 19:6, 1993.

40.   Lancaster GI, DeFrance, JH., Borruso, JJ. Air-Bag Associated Rupture of the Right Atrium. *New England J. Med*, 358, 1993.

41.   Reiland-Smith J, Weintraub RM, Selke FW. Traumatic Aortic Valve Injury Sustained Despite the Deployment of an Automobile Air Bag. *Chest*, 103:1603, 1993.

42.   Augenstein JS, Digges KH, Lombordo LV, et al. Air bag Protected Crash Victims-The Challenge of Identifying Occult Injuries. SAE Paper No 940714, Soc. Auto Eng, Warrendale, PA, 1994.

43.   Gross KB, Haidar AH, Bashama CH, et al. Acute Pulmonary Response of Asthmatics to Aerosols and Gasses Generated by Airbag Deployment. *Am J Resp & Critical Car Med*, 150:408-14, 1994.

44.   Roth T, Meredith P. Hand Injuries from Inflation of an Air Bag Security System., *J Hand Surgery (British and European Volume)*, 18B:510-522, 1993.

45.   Swanson-Bierman B, Mrvos R, Dean BS, et al. Air Bags: Lifesaving With Toxic Potential? *Am J Emerg Med* , 11:38-9, 1993.

46.   Smock,WS. Air Bag Module Cover Injuries. *J Trauma*, 19995.

Acknowledgment:

Thanks to Richard Reed of the Special Crash Investigation Section of NHTSA for providing case materials for review.

960659

Air Bag Field Performance and Injury Patterns

A. C. Malliaris, J. H. DeBlois
DeBlois Associates, Inc.

K. H. Digges
The George Washington Univ.

Copyright 1996 Society of Automotive Engineers, Inc.

ABSTRACT

This investigation addresses and evaluates: (a) the frequency of air bag deployments in comparison with belt only protection or no restraint, as a function of calendar year and model year; (b) injury and harm rates as a function of crash severity and restraint use and type; (c) restraint effectiveness in reducing fatalities and injuries as a function of restraint, crash severity, type of impact, and car size; (d) the confounding effects of crash severity; (e) injury patterns by injured body region, injuring contact, and injury severity; and (f) variation of injury patterns as a function of crash severity. It is found that restraints, irrespective of type, appear to be more effective at higher injury severities; the lowest casualty rates, and highest effectiveness values, are associated with the use of an air bag plus safety belt, or safety belt without air bag deployment; the air bag even without the concurrent use of a safety belt appears to offer a certain crash protection. Significant crash severity differences exist among the populations protected by various restraint types. These differences are such as to inflate the effectiveness of belt use only, and depress the effectiveness of the air bag, whether or not with belt use. When controlled for crash severity, the effectiveness of restraints in reducing fatalities, by comparison to no restraint, is found to be in all car crashes:

(56.3 +/- 13.2)%, (25.4 +/- 11.8)%, and (48.5 +/- 5.5)%,

for the air bag plus belt, air bag only, and belt only, respectively. The corresponding values in frontal crashes are:

(76.3 +/- 11.8)%, (19.4 +/- 17.8)%, and (51.6 +/- 7.5)%.

It is further found that the air bag plus belt reduces very significantly brain, spinal cord, facial, and abdominal injuries all across the board, at the expense of minor skin and flesh injuries. However the relative importance of a body-region/injuring-contact pair may vary by a large factor, depending on the crash severity range and restraint type under consideration.

INTRODUCTION

The air bag is already one of the primary means of occupant crash protection in the field. This, plus a continuing progress in accident and exposure data collection and analysis, suggest the appropriateness and timeliness of a comprehensive evaluation of air bag field performance, in comparison with the no restraint, or other restraint conditions.

DATA SOURCES

The findings of this investigation are based on two primary sources of national coverage: (i) the field crash experience contained in the records of the National Highway Traffic Safety Administration's (NHTSA's) National Accident Sampling System/Crashworthiness Data System (NASS/CDS) 1988-1994; and (ii) the fatal accident experience contained in NHTSA's Fatal Accident Reporting System (FARS) 1991-1994. Selection of the late 1980 and early 1990 years is primarily dictated by the relatively recent introduction of air bags as a means for crash protection. The NASS/CDS file has been selected as the primary source of data concerning accidents of virtually all severities (all towaway crashes). In addition to non-fatal accident coverage, this file provides significant resolution of crash, vehicle, occupant, and injury data not currently available in other sources.

POPULATIONS AT ISSUE

The data sources addressed in this investigation could, and will eventually, provide data covering all primary classes of restraint use and type. This could be provided for the three primary occupant seating positions: Driver, Outboard Front Passenger, and All Other; and for all four principal categories of light vehicles: cars, pickups, vans, and multipurpose vehicles (MPV).

However, on the basis of the data available to date, that cover calendar years 1988 to 1994, the populations at issue will be limited to car drivers only. The reason for this is that, except for car drivers, all other populations are either not represented, or represented by a very meager sample.

For this reason the balance of this investigation will be limited to car drivers, to cars, and to car crashes. Also, the condition of "unknown restraint use or type" will not be given any further consideration, as such a condition represents a small fraction of the crash involved car occupants.

The annual nationwide incidence of car drivers involved in towaway crashes is estimated at:

1,900,000 per year with a standard error of 25,000
This is a NASS/CDS19 88-94 average. The computation of the standard error reported above and elsewhere in the investigation is discussed in Appendix A.

MIX OF OCCUPANT RESTRAINTS VERSUS CALENDAR YEAR

The portion of car driver air bag deployments grew from about 0.1% in calendar year (CY) 1988 to about 20% in CY 1994. After redistributing a small fraction of unknowns, the mix of restraints that provide protection to crash involved car drivers is shown in Figure 1, as a function of calendar year.

The sharp rise of air bag deployments for crash protection is evident in this figure. The portion of such deployments reaches about 20% of all crash involved car drivers in 1994, when the "with belt" and "no belt" conditions are combined.

This rise in drivers protected by air bag deployments takes place at the expense of unrestrained drivers, that decline from about 35% in CY 1988 to about 20% in CY 1994. In the same period, the portion of drivers restrained by safety belt only remains relatively unchanged.

MIX OF OCCUPANT RESTRAINTS VERSUS MODEL YEAR

The sharp increase of air bag deployments in crash protection, as shown above, is the result of the cumulative and increasing influx of air bag equipped cars into the fleet on the road. More impressive trends are evident on a model year (MY) basis, as may be seen in Figure 2, where the towaway car driver restraint mix is resolved by model year. The portion of air bag deployments is seen to increase sharply as the model years progress from '87 to '94, reaching values nearly 80% by MY '94, when air bag deployments are addressed irrespective of belt use.

It is also important to note that, in the same model year range, the sharp increase of air bag deployments occurs in conjunction with a concurrent decline of unrestrained drivers (apparently a ten-fold decline), and a concurrent but modest decline of drivers restrained by belt only.

USE OF SAFETY BELTS IN AIR BAG DEPLOYMENTS

Air bag deployment is not always associated with safety belt use. Figures 1 and 2 show the split between belt users and non-users. As is evident, this split varies significantly by calendar year and model year.

However, among all air bag deployments, the portion without belt use appears to be declining, from levels between 30% and 40% in pre-1990 to levels about 10% in 1994.

It is important to note that unbelted car drivers, even under air bag deployment, not only miss the extra protection provided by a belt, but are also exposed to more severe crash conditions. This will be discussed in detail later in this investigation.

FATALITY AND MOST SEVERE INJURY RATES

Fatality rates and most severe injury rates for survivors are addressed next, as a function of restraint use and type. These rates are calculated per 100 towaway crash involved car drivers.

Conversion to absolute numbers may be accomplished by recalling that the annual nationwide incidence of car drivers involved in towaway crashes is estimated at about: 1,900,000 per year with a standard error of 25,000.

The results of these calculations are summarized in Figure 3 for fatality rates, and in Figures 4 to 8 for the most severe injury rate of survivors, injured at maximum AIS (MAIS) = 5 ,4, 3, 2, and 1, respectively.

In each of these illustrations, the injury rate is resolved by the use and type of restraint used by car drivers: no restraint, safety belt only, air bag only, and air bag plus belt. In all instances the injury rates are accompanied by their corresponding standard error.

OVERALL HARM RATE

The concept of harm is applied in order to consolidate all casualties: fatalities and injured survivors by severity into a single descriptor. The way used in this investigation to measure harm is by assigning an equivalent fatality estimate to the most severe injuries of car driver survivors, and then adding this to the actual fatality count. A more detailed description of the harm estimation procedure is presented in Appendix B.

Results obtained on the basis of the cited approach are summarized in Figure 9, where an overall harm rate to car drivers is presented, in fatalities (actual + equivalent) per 100 crash involved car drivers, as a function of restraint use and type.

Contrary to an earlier notion in the safety community, the equivalent fatalities assigned to injured survivors is comparable to or exceeds the actual fatalities experienced by the same general population. This is evident in the illustration of Figure 10, where actual fatality rates are compared side by side with overall harm rates.

EFFECTIVENESS OF OCCUPANT RESTRAINTS IN TOWAWAY CRASHES

The casualty rate results summarized and illustrated in Figures 3 to 10 can be further and more quantitatively evaluated by computing a restraint effectiveness. Using the cited results, we compute the effectiveness of three restraint types: safety belt, air bag -- no belt, and air bag plus belt. In each case the computation is made by reference to the no restraint state, following the simple formula:

$$e = (Rwo - Rw) / Rwo \qquad (1)$$

where:

Rwo is the injury rate without a restraint use, and
Rw is the injury rate with use of the restraint at issue.

The results of these computations and the associated standard errors are shown below in Figure 11 for fatality reduction effectiveness; in Figure 12 for reduction of most severe injuries of survivors; and in Figure 13 for overall harm reduction.

CONCLUSIONS ABOUT CASUALTY RATES AND RESTRAINT EFFECTIVENESS

A sequential review of the results summarized and illustrated in Figures 3 to 13 leads to the following conclusions:

(i) the difference between injury rates for restrained car

drivers, in comparison with unrestrained drivers, is progressively decreasing as the injury severity decreases. Note the difference between Figure 3 or 4 and Figure 8;

(ii) alternatively it may be stated that restraints, irrespective of type, appear to be more effective at higher injury severities, see Figure 12;

(iii) the lowest casualty rates, and highest effectiveness values, are associated with the use of an air bag plus safety belt, or safety belt without air bag deployment;

(iv) the air bag even without the concurrent use of a safety belt appears to offer crash protection. This is clear at the higher severities, but is masked by uncertainties at the lower severities.

These conclusions must be further evaluated and modified when several confounding effects are taken into account. For example, it appears that drivers using a safety belt, whether with an air bag deployment or not, are favored by a lower severity exposure. Conversely, drivers not using a safety belt are exposed to more severe conditions, irrespective of air bag deployment. These confounding effects are addressed and evaluated next.

CONFOUNDING EFFECTS DUE TO CRASH SEVERITY DIFFERENCES

An early warning, concerning crash severity differences among the car driver populations protected by different restraint types, appears in Figure 14 which illustrates the portion of car drivers that are intoxicated in fatal crashes, as a function of restraint use and type. It is evident in the results of this figure that drivers protected by air bag only are as exposed to the severe conditions of driving while intoxicated, as the unrestrained drivers are.

More concrete indications of significant crash severity differences, by restraint type, are found in the analysis of delta V distributions in all towaway crashes, as shown in Figure 15. In this figure it is evident that drivers protected by safety belts only are quite privileged by lower frequencies at high severities, while the opposite is true for drivers protected by an air bag deployment, without the benefit of a safety belt use. The restraint conditions of "no restraint" or "air bag plus belt" are in between the cited extremes.

A higher resolution display of these results at the higher crash severities is given in Figure 16, where the portions of crashes at higher crash severities are displayed as a function of restraint use and type. A similar display is given in Figure 17 where all crashes at delta v values under 10 mph are excluded from the distributions, since air bags are very infrequently deployed in this range.

Given that exposure to a lower frequency of high crash severities is a privilege, Figures 15 to 17 indicate clearly how privileged is the population of drivers protected by belt use only; and how underprivileged are the populations protected by air bag deployment. It should be noted that the scale in the cited figures are logarithmic, and that relatively small differences in crash severity frequency, at high crash severities, makes a large difference in casualty rates.

CASUALTY AND HARM RATES AS A FUNCTION OF CRASH SEVERITY

A rigorous control of the confounding effects addressed above is not yet possible, as the sample sizes available in the NASS/CDS

are still small and the resulting statistical variabilities large. However a rudimentary remedy is available when casualty and harm rates are resolved by crash severity range.

This is done in Figures 18 to 21, where injured driver rates and harm to injured drivers are displayed as a function of crash severity, for each of the restraint types under consideration.

CONCLUSIONS ABOUT THE CONFOUNDING EFFECTS OF CRASH SEVERITY

Significant crash severity differences exist among the populations protected by various restraint types. These differences are such as to inflate the effectiveness of belt use only, and depress the effectiveness of the air bag, whether or not with belt use. These considerations should be kept in mind when interpreting the restraint effectiveness values illustrated in Figures 11 or 13.

An informative evaluation, that minimizes confounding effects, is discussed next in connection with the method followed in the determination of effectiveness from the FARS records.

EFFECTIVENESS IN FATAL ACCIDENTS UNDER CONTROLLED EXPOSURE

Independent estimates of restraint effectiveness in fatality reduction may be obtained from the FARS data. These data are sufficiently voluminous to allow an effectiveness computation to be made on the basis of matched driver/passenger pairs, exposed to identical crash conditions, using a method detailed in reference 1. Thus in such computations, the confounding effects of crash severity differences between restrained and unrestrained occupants would be minimized.

The results of the cited computations are summarized in Table I below, and are illustrated in Figure 22. In these cases, effectiveness refers to fatality reduction irrespective of impact mode or of any other consideration.

Table I. Restraint Effectiveness by Restraint Type in All Fatal Crashes

Impact Mode	Safety Belt	Air Bag - No Belt	Air Bag with Belt
All	48.5 +/- 5.5	25.4 +/- 11.8	56.3 +/- 13.2

It is important to note how the effectiveness of the safety belt assumes a value about 48% in this evaluation, as opposed to a value around 80%, as shown in Figure 11. The most likely reason for this lowering of the effectiveness in the analysis of the FARS data is the application of the matched pair methodology that minimizes crash severity differences between unrestrained and restrained drivers.

It is also noteworthy that, under controlled crash severity conditions, the effectiveness of the air bag/no belt restraint remains in the range 20% to 30%, primarily because no major crash severity differences are expected between unbelted populations, irrespective of air bag deployment.

Based on the findings of Table I, the air bag plus belt restraint might have a certain advantage over the belt only restraint, but this is not yet strongly evident due to large errors. Specifically the air bag plus belt restraint, when evaluated by reference to the belt only restraint, shows an effectiveness of:

(13.3 +/- 12.4)%,

in fatality prevention.

RESTRAINT EFFECTIVENESS AS A FUNCTION OF IMPACT MODE

Estimates of restraint effectiveness in fatality reduction, as a function of impact mode have been obtained from the FARS 1991-1993 data, using the matched pair methodology, and thus controlling for crash conditions as much as possible. The results of these determinations are shown in Table II below and are illustrated in Figure 23.

Table II. Restraint Effectiveness by Restraint Type
in Fatal Crashes as a Function of Impact Mode

Impact Mode	Safety Belt		Air Bag - No Belt		Air Bag with Belt	
Frontal	51.6 +/- 7.5		19.4 +/- 17.8		76.3 +/- 11.8	
Nonfrontal	45.3	8.1	30.6	15.8	26.5	29.6
All	48.5	5.5	25.4	11.8	56.3	13.2

These results show a fairly clear and substantial advantage of the air bag plus belt restraint in frontal impacts over all other impacts. The results for other restraints are inconclusive.

RESTRAINT EFFECTIVENESS AS A FUNCTION OF CAR SIZE

Furthermore, estimates of restraint effectiveness in fatality reduction, as a function of car wheelbase or curb weight, have been obtained from the FARS 1991-1993 data, using the matched pair methodology, and thus controlling for crash conditions as much as possible. The results of these determinations are shown in Table III below, and Figure 24.

Table III. Restraint Effectiveness by Restraint Type in Fatal
Crashes as a Function of Car Wheelbase

Wheelbase Inches	Safety Belt		Air Bag - No Belt		Air Bag with Belt	
< 105	50.8 +/- 6.3		21.0 +/- 15.4		74.6 +/- 11.6	
> 105	44.1	11.7	34.3	18.3	31.3	30.2
All	48.5	5.5	25.4	11.8	56.3	13.2

A review of this table and of the illustration in Figure 24 shows no significant dependence of restraint effectiveness on car size, except perhaps in the case of air bag plus belt restraint.

INJURED BODY REGIONS AND SYSTEM/ORGANS

Major changes occurred in NASS 1993 concerning injury coding and classification, as a result of implementing in this file the new AIS 90 Occupant Injury Classification system. Injured body regions and system/organs are the most profoundly affected.

Thus in expanding the NASS 1988-1992 pool to include NASS 1993 and 1994, several issues must be resolved concerning:

(a) interpretation of new codes consistent with the older codes;

(b) differences due to legitimate changes in classification; and

(c) clerical misclassification.

At this time, without further work and experience, a fully detailed mapping of new into old, or old into new, body region and system/organ codes is not yet satisfactory. Specifically, a large fraction of codes are not categorically mapped one way or another, thus leaving a large fraction of unknowns.

In order to circumvent this difficulty, and also in order to retain ample samples, we have selected a total of 10 categories for body regions and system organs, as follows:

1. brain, irrespective of any other qualifications;
2. spinal cord, cervical, thoracic, or lumbar spine;
3. other head, excluding brain & cervical spine;
4. face, excluding integumentary;
5. neck and back, excluding spine and integumentary;
6. chest and contents of thoracic cavity;
7. abdomen and contents of abdominal cavity;
8. upper extremities & shoulders;
9. lower extremities & pelvis; and
10. integumentary (skin & flesh), irrespective of other qualifications.

These ten categories are mutually exclusive and can be assembled commonly from either pre-93 or post-92 or both NASS records. They are extensively used in the analyses and evaluations that follow. Occasionally, one or more of these categories may be further resolved, if the sample sizes are statistically adequate.

FREQUENCY OF INJURED BODY REGIONS AND SYSTEM/ORGANS

Figure 25 shows the distribution of the most severe injuries of car drivers, among the ten cited categories of body regions and system/organs, for the entire 1988-1994 NASS pool, irrespective of injury severity. This is done primarily for the most important restraint system: air bag plus belt in comparison with the no restraint condition.

A similar comparison is made in Figure 26, concerning harm (actual plus equivalent fatalities) assigned to the most severely injured body regions. The frequency of all injuries, as opposed to the most severe injury per car driver are shown in Figure 27 in the same fashion.

More comprehensive comparisons are made in Figures 28 to 32. All four occupant protection systems are compared in each of these five figures, one body region at a time. The descriptor used in the comparisons is the harm frequency assigned to the most severe injury, expressed in equiv fatalities, as % of the total in each class of restraint.

Note that in the displays shown in Figures 25 to 32, the use of % of the total in each restraint class serves the purpose of underscoring the body regions that may be over- or under-represented in that class. However, beyond such injured body region over- or under-representation, there is the issue of injury frequency differences to be taken into account as a function of restraint class.

Injury frequencies, expressed per 100 crash involved drivers, have been addressed in Figures 18 to 21. It is evident in these figures, for example, that in air bag deployments the number of injured drivers per 100 crash involved is often significantly lower than that corresponding to the no restraint condition. This would have the effect of further strengthening any advantage of air bag deployments observed in Figures 25 to 32.

CONCLUSIONS ABOUT FREQUENCIES OF INJURED BODY REGIONS

A review of Figures 25 to 32 leads to the following conclusions:

(i) the air bag plus belt, i.e. the expectedly most common occupant protection system of the future, reduces very significantly brain, spinal cord, facial, and abdominal injuries all across the board, at the expense of minor skin and flesh injuries.

(ii) for other body regions: other head, neck and back, chest, and extremities, the results are mixed and could go either way, depending on uncertainties, but also depending on what specific descriptors are at issue: frequency of or harm assigned to injuries, and whether per 100 injuries or per 100 crash involved drivers; also depending on whether the most severe or all injuries per injured are addressed.

(iii) a further review of these figures reveals that the findings cited above hold in comparisons of the air bag plus belt protection against any of the other three protection systems under consideration.

INJURING CONTACTS

A total of 17 individual components or car areas have been selected, from nearly 100 available, for the evaluation of the most frequent and most harmful injuring contacts of towaway car drivers. They are shown below with the abbreviations used in later tables:

Steering Assembly	(SteerA)
Dash	(Dash)
Windshield	(Wndshld)
A Pillars	(Pillrs)
Lower Side Interior	(LwrSIn)
Upper Side Interior	(UprSIn)
Roof Rail	(Rail)
Roof	(Roof)
Floor	(Floor)
Seat Back	(SeatBk)
Head Rest	(HdRest)
Seat Belt	(StBelt)
Flying Glass	(FGlass)
Non Contacts	(NonCon)
Exterior	(Exter)
Air Bag, and	(Air Bg)
All Other	(AllOthr)

The frequency of and harm assigned to injuries from each of these injuring contacts is evaluated below in conjunction with the corresponding injured body region, selected from the ten regions discussed earlier in the investigation as shown below, including appropriate abbreviations:

Brain	(Brain)
Spinal Cord	(SpCord)
Other Head	(OthrHd)
Face	(Face)
Neck and Back	(N & B)
Chest	(Chest)
Abdomen	(Abdmn)
Upper Extremities	(UprXtr)
Lower Extremities	(LwrXtr)
Integumentary	(Integu)

INJURING-CONTACT/INJURED-BODY-REGION PAIRS

In investigating the frequency and harm fraction of Injuring-Contact/Injured-Body-Region Pairs we:

(a) address all injuries, by straight as well as by harm weighted frequency;

(b) resolve the results by crash severity in three delta V ranges: 1-15 mph, 16-30 mph, and 31-Hi mph; and

(c) emphasize the air bag plus belt restraint in comparison with the belt only and no restraint conditions.

Because the number of pairs at issue is very large, we tabulate the leading pairs, arbitrarily cutting off the tabulation when the fractions reach values under 1%. Under the cited provisions, our findings are tabulated in the following six tables:

Table IV. Frequency of Injuries at Delta V: 1 to 15 mph
Table V. Frequency of Injuries at Delta V: 16 to 30 mph
Table VI. Frequency of Injuries at Delta V: 31 to Hi mph
Table VII. Harm Weighted Frequency, Delta V: 1 to 15 mph
Table VIII. Harm Weighted Frequency, Delta V: 16 to 30 mph
Table IX. Harm Weighted Frequency, Delta V: 31 to Hi mph

Table IV. Frequency of Injuries Assigned to Specific Body-Region/Injuring-Contact Pairs in <u>Delta V 1-15 mph</u>

Body Region	Injuring Contact	Air Bag & Belt	Belt Only	No Restraint
		Car Driver Injuries, Percent of Total in Each Column		
Integu	Dash	14.22	15.10	17.71
Integu	Air Bg	13.89	.	.
Integu	StBelt	12.08	11.66	0.05
Integu	SteerA	11.30	9.61	13.77
N & B	AllOth	7.57	12.51	4.03
Integu	UprSIn	7.02	3.62	2.64
Integu	Floor	6.59	2.54	1.84
Integu	LwrSIn	6.52	6.20	6.86
UprXtr	SteerA	2.59	0.98	0.89
Integu	Wndsld	1.91	3.02	14.25
Integu	AllOth	1.48	1.29	1.10
Integu	Unknwn	1.40	6.80	7.13
Integu	FGlass	1.12	1.71	3.08
N & B	SeatBk	1.10	1.15	1.60
All Other		11.21		
Total		100.00		

697

Table V. Frequency of Injuries Assigned to Specific Body-Region/Injuring-Contact Pairs in Delta V 16-30 mph

Body Region	Injuring Contact	Car Driver Injuries, Percent of Total in Each Column		
		Air Bag & Belt	Belt Only	No Restraint
Integu	Air Bg	23.46	.	0.01
Integu	StBelt	12.54	14.86	0.06
Integu	Dash	10.58	16.91	21.24
N & B	AllOth	7.51	8.13	2.37
Integu	SteerA	5.94	11.51	11.70
Integu	Wndsld	5.73	4.00	19.71
Integu	LwrSIn	4.50	5.20	4.36
Integu	Unknwn	4.00	5.58	5.45
UprXtr	SteerA	2.48	1.20	1.42
LwrXtr	Dash	2.47	0.79	0.93
Chest	StBelt	2.17	0.77	.
Integu	FGlass	1.76	2.45	1.63
UprXtr	Air Bg	1.66	.	.
Chest	LwrSIn	1.56	0.66	0.73
Integu	Floor	1.44	2.58	3.08
N & B	StBelt	1.33	0.59	.
Integu	UprSIn	1.03	1.68	1.08
All Other		9.84		
Total		100.00		

Table VI. Frequency of Injuries Assigned to Specific Body-Region/Injuring-Contact Pairs in Delta V 31-Hi mph

Body Region	Injuring Contact	Car Driver Injuries, Percent of Total in Each Column		
		Air Bag & Belt	Belt Only	No Restraint
Integu	Air Bg	13.24	.	.
LwrXtr	Floor	9.12	2.88	2.89
Integu	Dash	8.49	14.16	16.30
Chest	LwrSIn	5.66	0.77	1.29
Integu	FGlass	5.51	1.96	1.17
N & B	SeatBk	5.37	0.55	0.31
Integu	LwrSIn	5.09	2.31	3.95
Integu	StBelt	4.83	16.85	0.12
Integu	Unknwn	4.12	4.53	7.64
LwrXtr	Dash	4.06	2.69	4.09
Abdmn	LwrSIn	3.91	0.44	0.61
LwrXtr	LwrSIn	3.71	0.55	1.07
Integu	SteerA	2.27	14.24	9.60
Chest	StBelt	1.82	1.82	.
UprXtr	SteerA	1.71	1.46	0.71
Abdmn	SeatBk	1.31	0.18	0.06
Brain	Roof	1.31	0.01	0.22
Chest	SeatBk	1.31	0.05	0.06
Integu	Roof	1.31	0.22	0.24
Integu	AllOth	1.31	0.72	0.86
Integu	Floor	1.25	1.99	1.97
OthrHd	UprSIn	1.06	0.03	0.01
All Other		12.23		
Total		100.00		

Table VII. Harm Weighted Frequency of Injuries Assigned to Body-Region/Injuring-Contact Pairs in Delta V 1-15 mph

Body Region	Injuring Contact	Harm Weighted Driver Injuries, Percent of Total in Each Column		
		Air Bag & Belt	Belt Only	No Restraint
Integu	Dash	9.75	7.27	5.88
Integu	Air Bg	9.52	.	.
Integu	StBelt	8.28	5.53	0.02
Integu	SteerA	7.75	6.20	5.53
N & B	AllOth	7.08	6.91	1.50
Integu	AllOth	6.32	0.55	0.34
UprXtr	SteerA	5.74	2.09	2.67
Brain	UprSIn	5.48	3.81	2.07
Integu	UprSIn	4.82	1.84	0.80
Integu	Floor	4.52	1.19	0.82
Integu	LwrSIn	4.47	5.13	2.39
UprXtr	LwrSIn	2.33	0.78	0.71
LwrXtr	Floor	2.23	1.45	1.59
LwrXtr	LwrSIn	1.71	2.40	0.71
Face	LwrSIn	1.42	.	0.01
Integu	Wndsld	1.31	1.43	4.95
UprXtr	Air Bg	1.28	.	.
UprXtr	StBelt	1.25	0.70	.
LwrXtr	Dash	1.07	4.70	2.68
All Other		13.67		
Total		100.00		

Table VIII. Harm Weighted Frequency of Injuries Assigned to Body-Region/Injuring-Contact Pairs in Delta V 16-30 mph

Body Region	Injuring Contact	Harm Weighted Driver Injuries, Percent of Total in Each Column		
		Air Bag & Belt	Belt Only	No Restraint
UprXtr	Air Bg	16.27	.	.
LwrXtr	Dash	15.35	4.00	7.92
Chest	LwrSIn	12.11	5.07	6.55
UprXtr	SteerA	10.92	3.45	3.07
UprXtr	LwrSIn	7.08	1.47	1.23
Integu	Air Bg	5.28	.	0.00
Chest	Air Bg	4.34	.	.
LwrXtr	LwrSIn	3.42	3.60	1.99
Integu	StBelt	2.76	4.41	0.01
LwrXtr	StBelt	2.76	0.47	.
Integu	Dash	2.38	4.60	3.50
N & B	AllOth	2.38	2.59	0.50
Integu	SteerA	1.30	3.34	2.17
Integu	Wndsld	1.29	1.75	4.02
UprXtr	Dash	1.20	0.75	1.30
LwrXtr	Floor	1.09	10.06	4.79
All Other		10.07		
Total		100.00		

Table IX. Harm Weighted Frequency of Injuries Assigned to Body-Region/Injuring-Contact Pairs in Delta V 31-Hi mph

Harm Weighted Driver Injuries, Percent of Total in Each Column

Body Region	Injuring Contact	Air Bag & Belt	Belt Only	No Restraint
LwrXtr	Dash	19.88	11.16	15.92
Brain	HdRest	15.10	1.24	0.80
UprXtr	SteerA	6.91	1.40	1.23
UprXtr	Rail	6.59	.	.
LwrXtr	LwrSIn	6.04	1.70	2.68
Brain	Exter	5.52	4.66	0.09
Brain	Pillrs	4.67	4.92	4.66
Chest	LwrSIn	4.11	5.96	3.17
LwrXtr	Floor	4.05	5.66	5.11
Brain	Roof	3.76	0.03	0.69
N & B	SeatBk	2.55	0.09	0.03
Integu	Roof	2.15	0.01	0.01
Abdmn	StBelt	1.64	0.89	.
UprXtr	AllOth	1.62	0.16	0.01
UprXtr	LwrSIn	1.58	1.05	0.93
Abdmn	LwrSIn	1.35	0.49	0.85
Integu	Air Bg	1.22	.	.
Chest	AllOth	1.21	0.13	0.09
OthrHd	HdRest	1.05	.	.
All Other		9.00		
Total		100.00		

It should be noted that the above tables emphasize the frequency of injuries sustained under the restraint: air bag plus belt, as the projected most prominent restraint in current and future car fleets. Frequencies under restraint conditions: belt only and no restraint are presented for comparative purposes in all entries made for the air bag plus belt.

FURTHER PERSPECTIVE

The frequency distributions presented above, expressed as percent of all injuries in each specified class, have the purpose of flagging body-region/injuring-contact pairs that may require attention. These distributions, however, should be further interpreted in conjunction with the count of injuries sustained per 100 crash involved occupants.

The count of injuries per 100 crash involved occupants varies rather sharply, as a function of crash severity (delta V) range, and of restraint type, as may be seen in Tables X and XI.

Table X. Injury Count by Crash Severity and Restraint Type

Restraint	Injuries per 100 Crash Involved Car Drivers		
	1-15 mph	16-30 mph	31-Hi mph
No Restraint	152 +/- 6	352 +/- 9	519 +/- 16
Belt Only	96 3	195 5	428 16
Air Bag & Belt	134 15	137 11	372 31

Table XI. Harm Weighted Injury Count by Crash Severity and Restraint Type

Restraint	Harm Weighted Injuries per 100 Crash Involved Car Drivers		
	1-15 mph	16-30 mph	31-Hi mph
No Restraint	0.70 +/- 0.07	3.41 +/- 0.27	9.46 +/- 0.69
Belt Only	0.31 0.02	1.24 0.08	5.62 0.52
Air Bag & Belt	0.27 0.03	0.83 0.19	5.33 0.82

It is readily seen that the relative importance of a body-region/injuring-contact pair may vary by a large factor, depending on the crash severity range and restraint type under consideration.

CONCLUSIONS CONCERNING PAIRS OF BODY-REGIONS/INJURING CONTACTS

A review of Tables IV to IX under the further perspective discussed above leads to the following conclusions concerning crash involved car driver injuries under the air bag plus belt restraint:

(i) considering frequency of injuries irrespective of injury severity, i.e. without harm weighing, the great majority of injuries are skin and flesh injuries (integumentary), by contact with the air bag and several other car components;

(ii) the cited conclusion holds true irrespective of crash severity;

(iii) vital body parts such as the brain, the spinal cord, the face (skeletal), other head (skeletal) and the abdomen appear to have very low frequencies, also irrespective of crash severity;

(iv) the chest, upper extremities, lower extremities, and neck & back appear to have noticeable frequencies, but still well below the integumentary frequencies.

(v) the cited conclusions hold for injury frequencies irrespective of injury severities.

These conclusions are modified when injury severities are taken into account, i.e. when harm weighted frequencies are addressed. Specifically:

(vi) although skin and flesh (integumentary) injuries still dominate the low delta v range, other pairs appear dominant in higher delta v ranges, such as: the upper extremities (as injured by the air bag, steering assembly, and lower side interior); the lower extremities (as injured by the dash); and the chest (as injured by the lower side interior and the air bag).

(vii) the harm weighted frequency of brain injuries becomes noticeable at high delta v values, but other vital body parts are absent.

ACKNOWLEDGEMENT

The authors, solely responsible for the analyses, evaluations, and conclusions presented herein, acknowledge thankfully that this investigation has been supported by the Federal Highway Administration's/National Highway Traffic Safety Administration's National Crash Analysis Center at The George Washington University.

REFERENCES

1. Evans, L., "Double Pair Comparison", Accident Analysis and Prevention 18: 217-227; 1986.

2. Miller, T., "The Comprehensive Cost of Motor Vehicle Injuries by Body Region and AIS Severity", 35th Annual Proceedings of the AAAM, 1991.

APPENDIX A: ESTIMATION OF REPORTED ERRORS

When dealing with analyses of data from the NASS, all standard errors in this investigation are estimated with the help of statistical procedures prescribed in "Survey Data Analysis" (SUDAAN) software, Research Triangle Institute, Research Triangle Park, North Carolina, 1992. Such procedures are applicable in the analysis of data from multi-stage sample designs, like that of the NASS.

These procedures are readily applied in the computation of standard errors that accompany the estimation of simple counts and the estimation of count proportions appearing in injury rate estimation.

When injury rates or other similar results are further processed, as for example in the estimation of restraint effectiveness via relation (1) in the text, the standard errors are compounded through an application of a general relation for the variance of a ratio, namely:

$$[Var(A/B) / (A/B)**2] = [Var(A) / A**2] + [Var(B) / B**2]$$

where:
Var here stands for standard error squared.

A different procedure is used for the estimation of errors of results, whether concerning counts or rates, when the data at stake are provided by a census file, such as the FARS. In the absence of any distortions, the fatality or any other counts obtained from the FARS obey Poisson statistics. This essentially means that the variance of each of these counts is given by the count itself, and that the standard error of each count equals the square root of the count.

The application of this error computation procedure is cumbersome but straightforward when fatality rates and restraint effectiveness are estimated according to the matched pair methodology, as is the case in this investigation.

APPENDIX B: HARM ESTIMATION

The concept of harm is applied in order to consolidate all casualties: fatalities and injured survivors by severity into a single descriptor. One way used in this investigation to measure harm is by assigning an equivalent fatality estimate to the most serious injury of a car driver. At this time, the equivalent fatality assignments recommended by the NHTSA's Planning Office and used in this investigation are shown in Table B-1.

Table B-1. Equivalent Fatality Estimates

AIS Severity of Most Severe Injury:	Equivalent Fatality:
Survivor @ MAIS=1	.0024
Survivor @ MAIS=2	.0411
Survivor @ MAIS=3	.1528
Survivor @ MAIS=4	.3882
Survivor @ MAIS=5	.8100
Fatality	1.0000

Further resolution appears in order, since it is known that at each injury severity, the consequences of the most severe injury differ substantially according to the body region that sustains the said injury. In the absence of other alternatives, we adopt the comprehensive costs per injury as a function of severity and injured body region, as developed in reference 2.

We recast these costs in nondimensional factors, used as weighing multipliers of the equivalent fatality data shown in Table B-1. This is done by normalizing the cost of each body region's most severe injury by the cost corresponding to all body regions, at the same severity. The resulting adjustment factors are shown in Table B-2.

Table B-2. Adjustment Factors by Body Region and AIS

Body Region	AIS	Adjustment	Body Region	AIS	Adjustment
All	1	1.00	N & B	1	0.89
All	2	1.00	N & B	2	0.28
All	3	1.00	N & B	3	0.82
All	4	1.00	N & B	4	0.66
All	5	1.00	N & B	5	0.30
Brain	1	1.25	Chest	1	0.89
Brain	2	1.69	Chest	2	0.28
Brain	3	1.08	Chest	3	0.82
Brain	4	1.34	Chest	4	0.66
Brain	5	1.27	Chest	5	0.30
SpCord	3	0.87	Abdmn	1	0.89
SpCord	4	1.47	Abdmn	2	0.28
SpCord	5	1.32	Abdmn	3	0.82
OthrHd	1	1.42	Abdmn	4	0.66
OthrHd	2	0.48	Abdmn	5	0.30
OthrHd	3	0.81	UprXtr	1	1.42
OthrHd	4	0.91	UprXtr	2	0.60
OthrHd	5	0.64	UprXtr	3	0.75
Face	1	1.42	LwrXtr	1	0.89
Face	2	0.48	LwrXtr	2	0.57
Face	3	0.81	LwrXtr	3	1.18
Face	4	0.91	LwrXtr	4	0.47
Face	5	0.64			

In essence then, the harm sustained by each casualty population or subpopulation is estimated by summing up the counts of fatalities and injured survivors (according to most severe injury), each weighed according to injury severity and injured body region, as shown in Tables B-1 and B-2 above. The body regions shown above are defined as in the text.

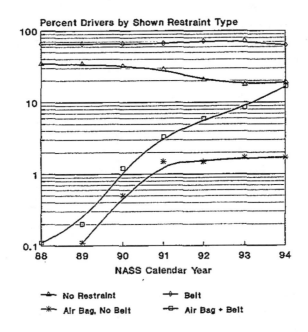

Fig. 1. Percent of Towaway Crash Involved Car Drivers, by Restraint Type in Use, v. Calendar Year

The NASS/CDS 1988-1994

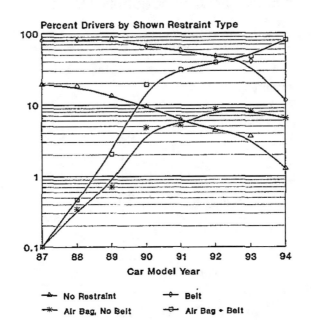

Fig. 2. Percent of Towaway Crash Involved Car Drivers, by Restraint Type in Use, v. Car Model Year

The NASS/CDS 1988-1994

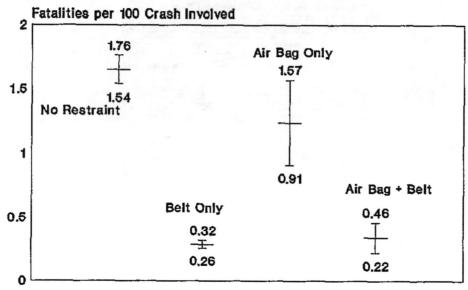

Fig. 3. Fatality Rate of Car Drivers per 100 Involved in Towaway Crashes, as a Function of Restraint

The NASS/CDS 1988-1994;

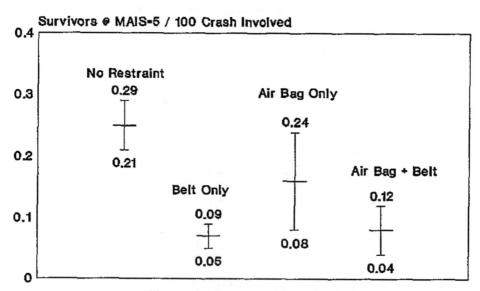

Fig. 4. Rate of Surviving Car Drivers @ MAIS=5 per 100 Involved in Towaway Crashes, as a Function of Restraint

The NASS/CDS 1988-1994;

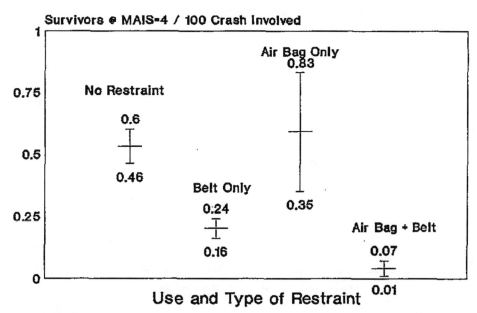

Fig. 5. Rate of Surviving Car Drivers @ MAIS=4 per 100 Involved in Towaway Crashes, as a Function of Restraint

The NASS/CDS 1988-1994;

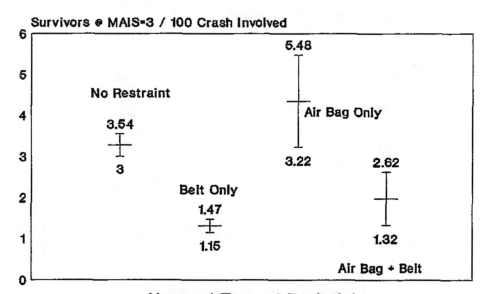

Fig. 6. Rate of Surviving Car Drivers @ MAIS=3 per 100 Involved in Towaway Crashes, as a Function of Restraint

The NASS/CDS 1988-1994;

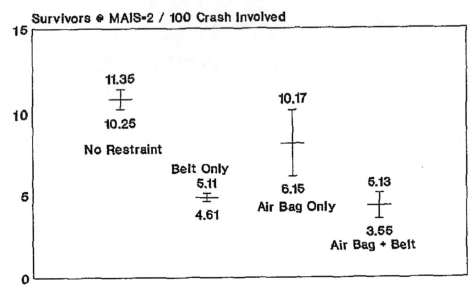

Fig. 7. Rate of Surviving Car Drivers @ MAIS=2 per 100 Involved in Towaway Crashes, as a Function of Restraint

The NASS/CDS 1988-1994;

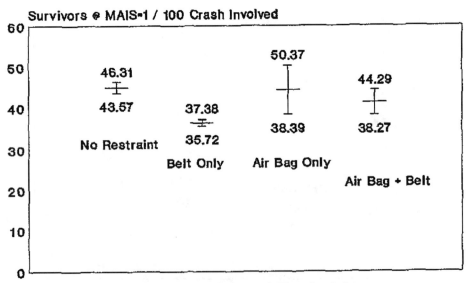

Fig. 8. Rate of Surviving Car Drivers @ MAIS=1 per 100 Involved in Towaway Crashes, as a Function of Restraint

The NASS/CDS 1988-1994;

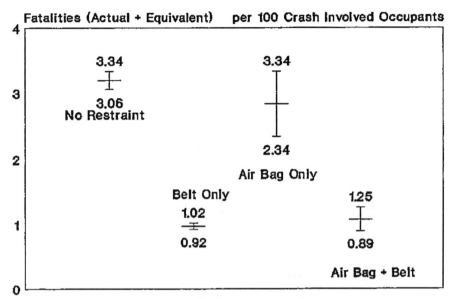

Fig. 9. Rate of Harm to Car Drivers per 100 Involved in Towaway Crashes, as a Function of Restraint

The NASS/CDS 1988-1994;

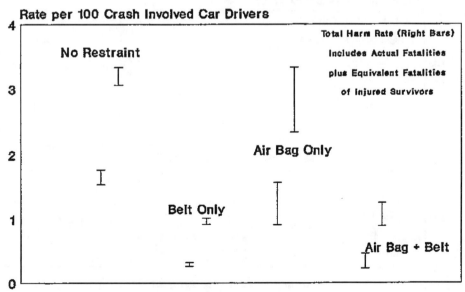

Fig. 10 Fatality Rate (Left Bars) and Total Harm Rate of Car Drivers as a Function of Restraint Use and Type

The NASS/CDS 1988-1994;

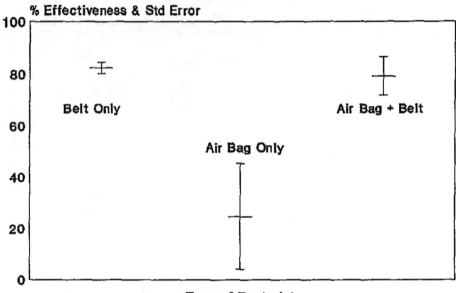

Fig. 11. Restraint Effectiveness
Re: Fatality Reduction among Car Drivers
in Towaway Crashes

The NASS/CDS 1988-1994;

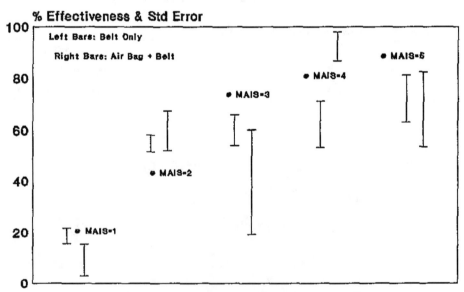

Fig. 12. Restraint Effectiveness
Re: Most Severe Injury Reduction among
Car Driver Survivors in Towaway Crashes

The NASS/CDS 1988-1994;

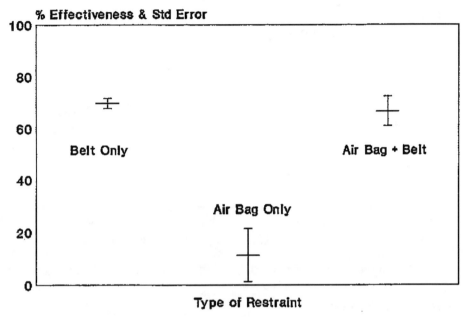

Fig. 13. Restraint Effectiveness Re: Reduction of Total Harm to Car Drivers in Towaway Crashes

The NASS/CDS 1988-1994;

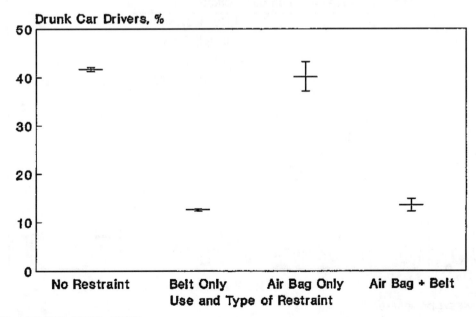

Fig. 14. Portion of Car Drivers that Are Drunk in Fatal Accidents as a Function of Restraint Use and Type

The FARS 1990-1993

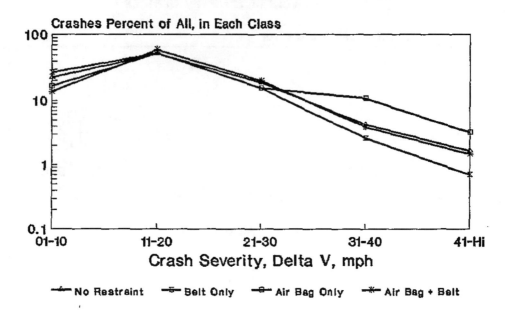

Fig. 15. Distribution of Car Crashes among Crash Severities, Delta V, as a Function of Restraint Use and Type

The NASS/CDS 1988-1994

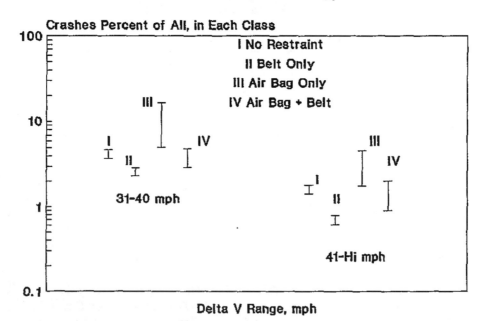

Fig. 16 Portion of Car Crashes at High Crash Severities, as a Function of Restraint Use and Type

NASS/CDS 1988-94;

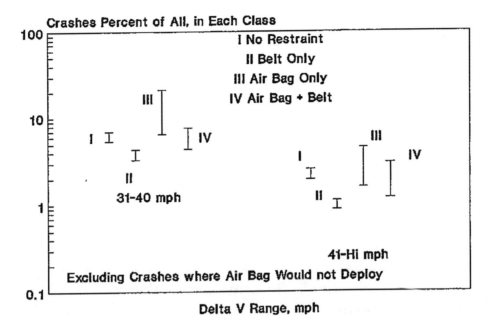

Fig. 17 Portion of Car Crashes at High Crash Severities, as a Function of Restraint Use and Type

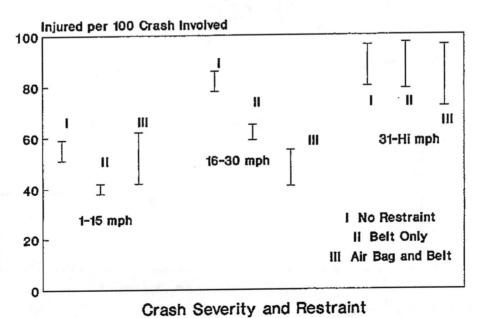

Fig. 18 Injured Car Drivers per 100 Crash Involved, as a Function of Crash Severity and Restraint

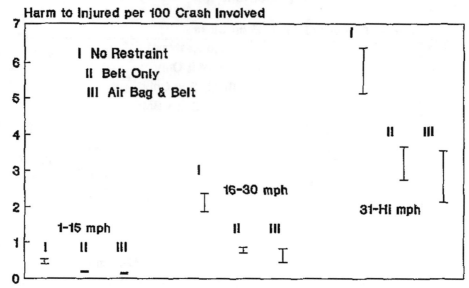

Fig. 19 Harm to Injured Car Drivers per 100 Crash Involved, as a Function of Crash Severity and Restraint

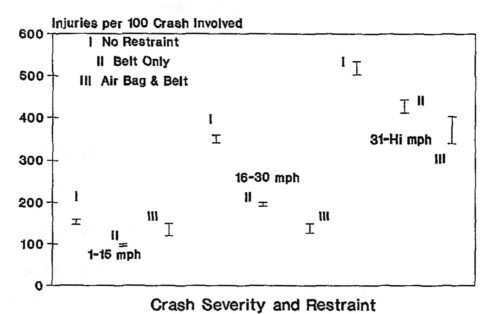

Fig. 20 Injuries of Car Drivers per 100 Crash Involved, as a Function of Crash Severity and Restraint

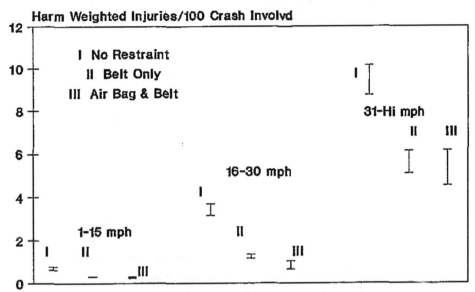

Fig. 21 Harm Weighted Injuries of Car Drivers per 100 Crash Involved, as a Function of Crash Severity and Restraint

The NAS/CDS 1988-1994

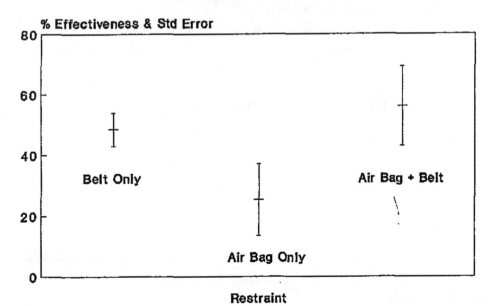

Fig. 22. Restraint Effectiveness in Reducing Car Driver Fatalities; All Impacts; All Car Sizes

The FARS 1991-1993;
Matched Pairs of Front Seat Occupants

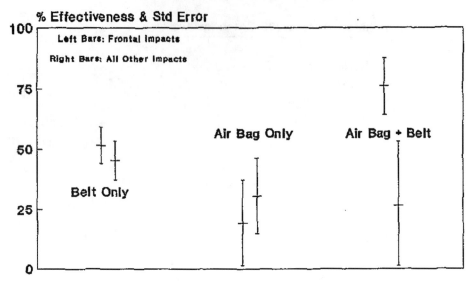

Fig. 23. Restraint Effectiveness in Reducing Car Driver Fatalities; All Car Sizes by Impact Type

The FARS 1991-1993;
Matched Pairs of Front Seat Occupants

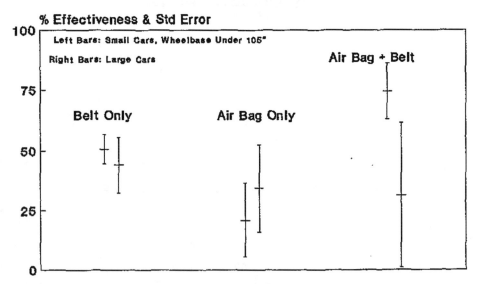

Fig. 24 Restraint Effectiveness in Reducing Car Driver Fatalities; All Impacts by Car Size

The FARS 1991-1993;
Matched Pairs of Front Seat Occupants

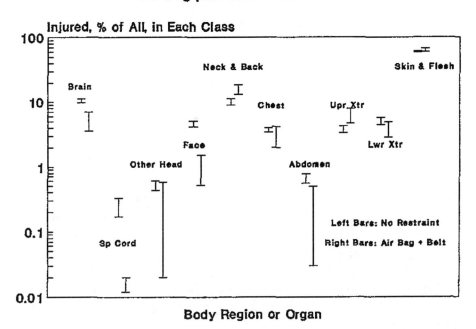

Fig. 25 Frequency of Most Severe Injury by Body Region of Car Drivers; Air Bag plus Belt v. No Restraint

NASS/CDS 1988-93;

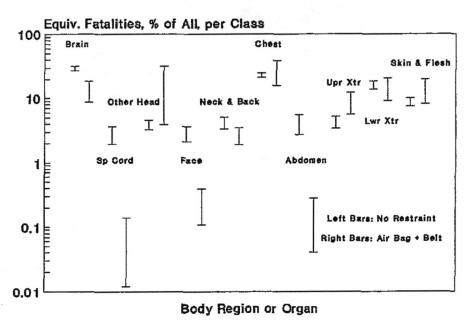

Fig. 26 Harm Due to Most Severe Injury by Body Region of Car Drivers; Air Bag plus Belt v. No Restraint

NASS/CDS 1988-93;

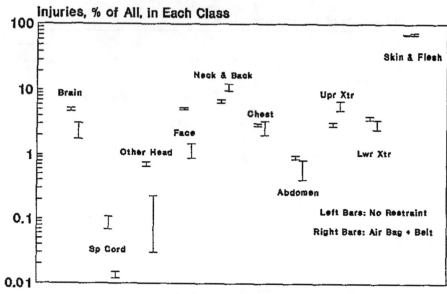

Fig. 27 Frequency of Car Driver Injuries by Body Region; Air Bag plus Belt v. No Restraint

NASS/CDS 1988-93;

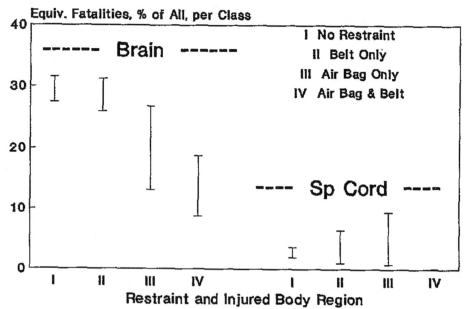

Fig. 28 Harm Due to Most Severe Injury to Shown Body Regions of Car Drivers, as a Function of Use and Type of Restraint

NASS/CDS 1988-1993

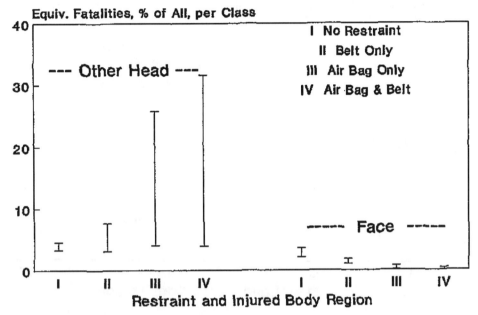

Fig. 29 Harm Due to Most Severe Injury to Shown Body Regions of Car Drivers, as a Function of Use and Type of Restraint

NASS/CDS 1988-1993

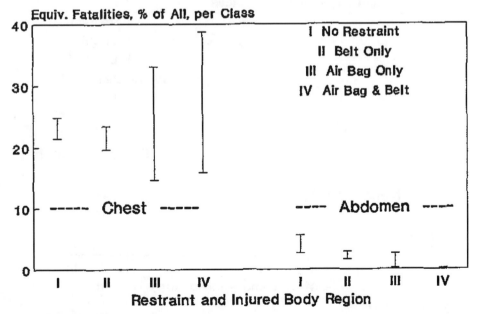

Fig. 30 Harm Due to Most Severe Injury to Shown Body Regions of Car Drivers, as a Function of Use and Type of Restraint

NASS/CDS 1988-1993

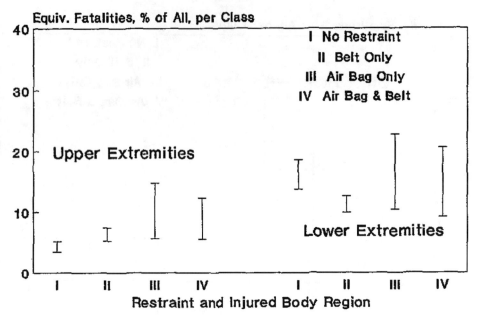

Fig. 31 Harm Due to Most Severe Injury to Shown Body Regions of Car Drivers, as a Function of Use and Type of Restraint

NASS/CDS 1988-1993

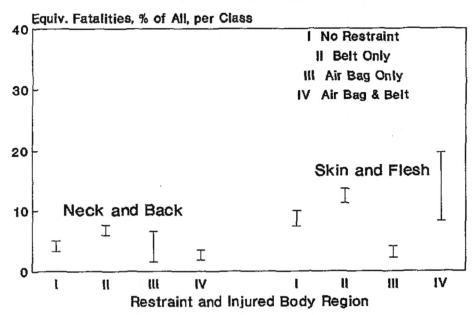

Fig. 32 Harm Due to Most Severe Injury to Shown Body Regions of Car Drivers, as a Function of Use and Type of Restraint

NASS/CDS 1988-1993

960664

Air Bag Deployment Frequency and Injury Risks

John V. Werner and Steve F. Roberson
State Farm Mutual Automobile Insurance Co.

Susan A. Ferguson
Insurance Institute for Highway Safety

Kennerly H. Digges
George Washington University

ABSTRACT

Automobile insurance claims were examined to determine the air bag deployment frequencies of cars with significant frontal crash damage. Air bag deployment frequencies were found to vary greatly by car model. Two popular midsize cars -- the Dodge Neon and Honda Civic -- were studied in detail to better understand the differences in the performance of different air bag systems in low severity crashes (delta V ≤ 15 mph). The Neon had a greater frequency of low speed air bag deployments than the Civic, which in turn resulted in a greater likelihood of air bag-induced minor injuries, in particular upper extremity injuries for females. Differences in air bag leading edge speed and excursion distance may also contribute to the different likelihood of injuries. The comparisons provide no information on the overall performance -- in particular the protection offered in high speed crashes -- but they do illustrate the importance of both nominal and actual deployment characteristics in low speed crashes. An important goal for future air bag designs should be to minimize deployments in low speed crashes, without compromising protection in the more important high speed crashes.

INTRODUCTION

Air bags are very effective in reducing serious injuries and fatalities in frontal crashes (1, 2, 3, 4, 5). Recent studies have estimated about a 20 percent reduction in driver fatalities in frontal crashes of air bag-equipped vehicles compared with vehicles equipped only with seat belts (1, 3, 4). Despite their effectiveness in preventing deaths and serious injuries, injuries sometimes result when the vehicle occupant contacts the deploying air bag. People who are close to the air bag when it begins to deploy are particularly vulnerable. This group includes unbelted occupants who move forward early in the crash or during precrash braking. It also includes people, often women, who sit close to the steering wheel.

Studies of data from the National Accident Sampling System (NASS) and from insurance claims indicate that air bag injuries are predominantly minor, and involve abrasions and contusions to the head, neck, face, and upper extremities. Serious injuries to drivers from air bag deployments are rare (6, 7, 8, 9, 10). However, in a few isolated cases, the air bag is suspected as the source of serious or fatal injuries (6, 7, 8, 11, 12, 13).

Several researchers have investigated the forces produced by inflating air bags to determine the mechanism of air bag-induced injuries. Tests with human volunteers showed that air bag-induced skin abrasions may be caused by the impacting fabric slapping the skin at high speed (14). Tests also have been conducted to evaluate the air bag injury potential for out-of-position occupants using 5th percentile female Hybrid III dummies and 50th percentile male dummies placed directly against the air bag module (15, 16). The results indicate that serious chest and neck injuries can occur for occupants located very close to the air bag at the time of deployment, even in crashes of low to moderate severity. Some researchers have expressed concern about the large number of air bag deployments in low speed crashes in which air bags may not be needed to mitigate injury, particularly for belted occupants (17). Air bags are designed to deploy in frontal crashes at barrier equivalent speeds of about 8 to 14 mph. However, studies have found a tremendous variation in deployment rates and deployment thresholds among different makes and models (5).

To better understand and illustrate the differences that exist among different air bag designs, the crash experience of two popular midsize cars -- the Dodge Neon and Honda Civic -- was studied. State Farm claims data for vehicle damage repair costs, air bag deployment frequencies, and injury claims were compared for the two models. In addition, data from NASS (1989-1994) were used to fit estimated delta V distributions to the insurance dollar damage amounts. The comparisons focus on the performance of the air bag systems in low severity crashes (delta V of 15 mph or less) and the likelihood of injuries in these crashes.

The study does not address the <u>overall</u> performance of the air bag systems studied, and it is important to recognize that differences identified in these comparisons of low severity crashes have no implications for the more important higher speed crash protection offered by the different air bag designs.

METHOD

The following four sources of information form the basis for the analysis and discussion in this paper:

1. State Farm Mutual Automobile Insurance Company's electronically prepared damage repair estimates for towed vehicles.
2. State Farm survey of claims involving frontal crashes for the 1995 Dodge/Plymouth Neon and the 1992-1995 Honda Civic.
3. The 1989-1994 NASS database, which contains detailed investigations of a sample of all crashes, involving towed vehicles, in the United States. Weighting factors permit the prediction of national estimates from this sample.
4. The leading edge speed and excursion profile of driver air bag fabric in the 1995 Dodge/Plymouth Neon and the 1993 Honda Civic.

DEPLOYMENT FREQUENCY - State Farm estimators complete detailed damage estimates on all severely damaged vehicles. Estimators are required to record primary and secondary impact locations. Only towed vehicles with frontal damage were included in the study. Multiple impacts were eliminated by excluding repair estimates with secondary impact locations. Vehicles with air bag deployments were identified by analyzing the repair estimate for driver air bag module replacement. Deployment frequency was calculated by dividing the number of driver air bag replacements by the total number of towed vehicles with frontal crash damage.

CLAIM SURVEY - Two popular midsize car models, the Plymouth/Dodge Neon and the Honda Civic, with similar wheelbase, weight and average cost to repair (Table 1) were selected for the claim survey. These cars have similar repair cost severity distributions, but a large difference in their air bag deployment frequencies - the Honda Civic with an estimated deployment frequency of 20 per 100 frontal crashes, and the Plymouth/Dodge Neon with an estimated 50 deployments per 100 frontal crashes. The car with a higher overall frequency of deployment was expected to have a larger number of deployments at low crash severities compared to the car with a low deployment frequency. Appendix A shows the driver air bag deployment frequency for a wide range of vehicles.

Because most insurance claim investigations do not require crash severity or delta V to be determined as part of the claim settlement process, the "delta V" distribution of the vehicle crashes represented by the insurance claims data was approximated by fitting the

rank ordered cumulative distribution of dollar damages for these two cars to the cumulative delta V distribution determined from a NASS (1989-1994) analysis of midsize cars (Table 2). The two models chosen for this comparison are similar in their average repair costs (see Table 1). The repair costs were adjusted to remove air bag system related expenses. This was done to control for the known difference in repair costs due to the higher replacement frequency of air bag components for the Neon. This approach assumes that the dollar amount of the repair increases with crash severity. Researchers looked at the number of components damaged as repair costs increased and found that, as expected, the number of components repaired increased systematically with higher repair costs (Figure 1). It is estimated that about 70 percent of the crashes of these two vehicles occur at delta Vs of $\leq$ 15 mph.

Table 1
Civic and Neon Comparisons

	Wheelbase	Length	Weight	Average Repair Cost*
Neon	104.1"	171.8"	2,338 lbs	$3,549
Civic	103.2"	173.0"	2,313 lbs	$3,708

*Average repair cost is less cost of air bag components.

Claim files for both models were surveyed only for those crashes that according to the delta V distribution estimated from NASS were at speeds of 15 mph or less. State Farm claim representatives were contacted to provide information regarding policyholder involved accidents. Information tabulated included an accident description, age and gender of the insured driver, driver injury descriptions, verification of air bag deployment status, and belt use of the injured driver. The Abbreviated Injury Scale (AIS) and Occupant Injury Classification (OIC) coding were used to document the injuries to insured drivers that sought treatment resulting in claim payments.

Table 2
Cumulative Delta V Distribution of Midsize Towed Passenger Cars With Frontal Damage
NASS 1989-1994 - Weighted

Delta V (mph)	Percent Midsize Cars	Neon Crashes	Civic Crashes
0 - 5	0.7	3	5
6 - 10	32.2	122	207
11 - 15	69.7	263	448
16 - 20	89.1	336	572
21 - 25	96.0	362	616
26 - 30	98.3	370	631
31 - 35	99.3	374	638
Over 35	100.0	377	643

In the typical claim settlement process, claim representatives complete injury evaluations, conduct interviews with the insured driver and witnesses, obtain police reports, and determine contributing causes for the crash. Physicians reports of injury descriptions and treatment are obtained to substantiate medical payments.

NASS ANALYSES - Analyses were conducted using NASS 1989-1994 data to determine the distribution of injuries by body region in 1989 or newer model vehicles for comparison with the State Farm data. To match the criteria used in the insurance data analyses, only qualifying vehicles involved in frontal crashes (10 - 2 o'clock, with no secondary impacts, at known delta Vs) were included.

FABRIC SPEEDS AND EXCURSION - Research has shown that differences in air bag deployment characteristics, for example the air bag fabric leading edge speed, may affect the potential for minor injury from the air bag (14). To allow a comparison of the air bag deployment characteristics of the Neon and Civic driver air bags, a 1995 Dodge Neon driver air bag was deployed statically and filmed. The excursion history for the leading edge of the air bag fabric for the Neon was generated by analysis of the film. The excursion history of the Honda Civic air bag was available from a previous study of leading edge deployment speeds of production air bags (18). Detailed methodology can be obtained from that paper.

RESULTS

REPAIR COST DISTRIBUTION - Figure 1 depicts the cumulative percentage of Neons and Civics repaired by damage severity percentile for frontal crashes. Severity percentile was determined by dividing each car's repair cost range into eight equal repair dollar intervals. A Kolmogorov-Smirnov test was used to compare the equivalency of the two cumulative repair cost distributions. The test results support the assumption that the repair cost distributions are equal (KSA [1020] = 0.697, $p = 0.716$).

A logistic regression model (Equation 1) was used to determine the damage severity at which the probability of an air bag deployment reached 50 percent. P is the probability of deployment and x is the damage severity (in dollars). For p = 0.5, the damage severity is given by Equation 2. The derived value is defined for this study as the air bag deployment threshold, that is, the dollar point in the distribution where the probability of deployment is 50 percent. The adjusted repair cost, where the probability of deployment is 50 percent is $3,479 for Neons (95 percent CI = $3,223 to $3,757). Civics have a 50 percent probability of air bag deployment at a $6,908 adjusted repair cost (95 percent CI = $6,452 to $7,484).

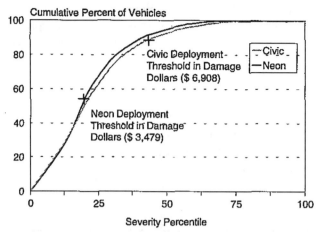

Note: 100th severity percentile represents $16,198 for the Neon and $15,893 for the Civic.

Figure 1. Cumulative Neon and Civic collision claim severity distributions for frontal impacts with towing.

An estimate of delta V for the derived damage severity deployment thresholds was obtained by fitting the rank ordered adjusted repair cost distribution for Neons and Civics to the NASS delta V distribution shown in Table 2. Interpolation within the 5 mph interval was used to determine the delta V for deployment threshold.

$$\log(p/(1-p)) = b_0 + b_1 x \quad (1)$$

$$x(p=.5) = -b_0/b_1 \quad (2)$$

The Neon 50 percent deployment threshold was estimated to be 13 mph delta V compared with 20 mph delta V for the Civic (see Figure 2).

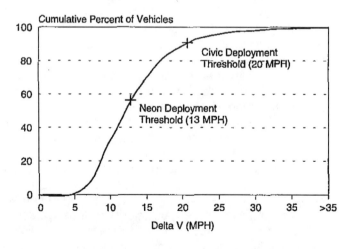

Figure 2. Cumulative NASS delta V distribution for towed midsize cars with frontal impacts.

DEPLOYMENT FREQUENCIES - Figure 3 shows air bag deployments per 100 frontal crashes by delta V, for the comparison vehicles. The deployment rates are lower than expected for estimated delta Vs above 15 mph. The 50 percent deployment threshold estimates for the comparison are surprisingly high in light of the fact that the typical nominal all-fire threshold is around 15 mph. The reason for these results is not clear, but there are a number of possible contributing factors. It is likely that the insurance claims data include underride and other crash types where delta V would be difficult to determine. As previously stated, NASS data used in this study include crashes with known delta Vs. Therefore, claims data may include crashes, which although expensive to repair, may be of lower severity. These higher repair costs may skew the delta V distribution. It is also possible that the insurance claims data, even though only crashes involving towing were included, may contain crashes of lower severity than those reported in NASS (which samples only police-reported towaway crashes).

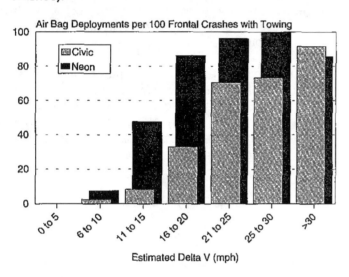

Figure 3. Air bag deployment rate by estimated delta V, frontal crashes with towing.

However, even if the estimated delta V distribution for both vehicles is higher than expected, the data still show a higher deployment rate for the Neon within most delta V intervals. An analysis of deployment frequencies in the 0 to 15 mph delta V range shows that there are significantly more deployments for Neons (OR = 6.9; 95 percent CI = 4.24 to 11.11). About 43 percent of all Neon deployments occurred at or below a delta V of 15 mph compared to 20 percent for the Civic.

INJURIES - The following analyses compare injury rates in the two cars for low severity crashes (at or below 15 mph). As shown in Table 3, about 20 percent of Neon drivers report an injury in these crashes compared to 14 percent of Civic drivers. The relative risk of injury for Neon drivers compared to Civic drivers is 1.53 (95 percent CI = 1.00 to 2.30).

Table 3
Driver Injury by Vehicle
Frontal Crashes with Towing
Crash Severity 15 mph or Less Delta V

	Driver Injury No	Yes	Total Crashes
Neon	211 (80.2%)	52 (19.8%)	263 (100%)
Civic	386 (86.2%)	62 (13.8%)	448 (100%)

Table 4 shows the maximum injury severity (MAIS) for the Neon and Civic drivers. Overall, injury severity was low for these drivers. Of the 711 Neon and Civic drivers, 703 (98.9 percent) had injuries of MAIS-1 severity or less; 7 (1.0 percent) had MAIS-2 level injuries; and 1 (0.1 percent) had a MAIS-3 injury. Very few MAIS-2 or greater injuries were associated with crashes in this estimated severity range. The insurance data revealed claims for one temporal mandibular dislocation, two arm fractures, one knee sprain, one asthmatic attack due to stress, and one instance of two AIS 2 injuries to the same driver--an elbow fracture and a cerebral concussion. It should be noted that the injury claims data do not include untreated injuries. Thus, there may be additional drivers who suffered minor injuries (AIS 1) who would not be counted here because they did not seek treatment for their injuries. Many minor abrasions and contusions heal without requiring treatment.

Table 4
Drivers Injured by Maximum AIS
Frontal Crashes with Towing
Crash Severity 15 mph or Less Delta V

MAIS	Neon Count	Percent	Civic Count	Percent	Total Count	Percent
0	211	80.2	386	86.2	597	84.0
1	47	17.9	59	13.2	106	14.9
2	5	1.9	2	0.4	7	1.0
3	0	0.0	1	0.2	1	0.1
>3	0	0.0	0	0.0	0	0.0
Total	263	100.0	448	100.0	711	100.0

Injuries by Gender - The injury data were also analyzed by gender of driver (Table 5). The proportion of male to female drivers involved in crashes below 15 mph delta V did not differ significantly for the two vehicles ($x^2[1] = .175$, $p = .676$), so overall injury risk can be compared without having to control for gender differences. Comparing across vehicles, the relative risk of injury to female drivers for Neons compared to Civics was 1.4. However, this relative risk was not significant (95 percent CI = 0.89 to 2.22). For male drivers, the relative risk of injury was 1.6 for Neons compared to Civics, again not significant (95 percent CI = 0.61 to 4.08).

Table 5
Driver Injury Status by Gender
Frontal Crashes with Towing
Crash Severity 15 mph or Less Delta V

Driver Gender	Driver Injury No	Yes	Total Crashes
Neon			
Female	116 (73.0%)	43 (27.0%)	159 (100%)
Male	66 (88.0%)	9 (12.0%)	75 (100%)
Civic			
Female	196 (79.0%)	52 (21.0%)	248 (100%)
Male	116 (92.1%)	10 (7.9%)	126 (100%)

Within a given vehicle, a comparison of injury risk differences by gender, shows that Neon female drivers had significantly higher odds of injury when compared to Neon male drivers (OR = 2.7; 95 percent CL = 1.25 to 5.92). In Civics, female drivers had a relative risk of injury equal to 3.1 when compared to male drivers (95 percent CI = 1.52 to 6.29).

Injuries by Air Bag Deployment Status - Table 6 shows injury rates by car make and deployment status. Neon drivers in air bag deployment crashes had 2.1 times higher odds of injury than Civic drivers. However, this difference was not significant (95 percent CI = 0.78 to 5.56). For crashes in which the air bag did not deploy, the relative risk of injury for drivers was 0.7 for Neons, again not significant (95 percent CI = 0.41 to 1.25).

Table 6.
Driver Injury Status by Deploy Status
Frontal Crashes with Towing
Crash Severity 15 mph or Less Delta V

Deploy Status	Driver Injury No	Yes	Total Crashes
Neon			
No	169 (90.4%)	18 (9.6%)	187 (100%)
Yes	42 (55.3%)	34 (44.7%)	76 (100%)
Civic			
No	368 (87.0%)	55 (13.0%)	423 (100%)
Yes	18 (72.0%)	7 (28.0%)	25 (100%)

The odds of injury for Neon drivers in air bag deployment crashes was higher than for Neon drivers in non-deployment crashes (OR = 7.6; 95 percent CI = 3.91 to 14.76). Likewise, Civic drivers had higher odds of injury for air bag deployment crashes compared to non-deployment crashes (OR = 2.6; 95 percent CI = 1.04 to 6.52). However, it should be noted that crashes in which an air bag deployment occurs typically are of higher severity.

Location of Driver Injuries - Table 7 shows driver injury rates by body region and car model for estimated delta Vs of 15 mph or less. The injury rate (injuries per 1,000 frontal crashes with towing) includes multiple injuries per injured driver. Lower extremity injuries are reported in the other body area category. As previously stated, most injuries were minor.

Table 7
Driver Injuries per 1,000 Frontal Crashes with Towing
by Car Model
15 mph or Less Delta V

Body Region	Neon	Civic
Head	38	22
Face	20	13
Upper Extremity		
Elbow	4	--
Upper arms	46	6
Wrist-Hand	81	17
Forearm	24	2
Total Upper Extremity	155	25
Chest	30	
Neck/Back		
Neck/C. Spine	80	80
Shoulder	34	20
Back/T. Spine	53	71
Total Neck/Back	167	171
Other Body Areas	47	34
Total	457	296

Multiple injuries per injured driver possible.

Neon Drivers had higher rates of injury to the upper extremities, including the upper arms, wrists, hands, and forearms than Civic drivers. The rate of head and facial injuries to Neon drivers is marginally higher than the Civic facial injury rate. Injuries to other body regions are fairly similar between the two cars. The overall injury rate for Neon drivers was about 1.5 times the injury rate for Civic drivers.

Table 8 shows driver injury rates by gender and car model for different body regions. Female Neon drivers had the highest overall injury rate in this comparison. However, both Neon and Civic female drivers had substantially higher numbers of injuries compared with males, to all body regions

Table 9 shows driver injuries by air bag deployment status and car model. The majority of non-deployments occurred at estimated crash speeds below 10 mph delta V, thus it is not surprising that injury rates are lower in the non-deployment crashes. Neon drivers involved in air bag deployment crashes had the highest injury rate in this comparison. Higher driver injury rates to the upper extremities are associated with air bag deployment crashes, with the wrist-hand area having the highest injury rate among body regions, followed in magnitude by injuries to the upper arms.

Table 8
Driver Injuries per 1,000 Frontal Crashes with Towing by Gender and Car Model
Delta V Less Than 15 mph

Body Region	Neon Male	Neon Female	Civic Male	Civic Female
Head	12	51	7	30
Face	--	30	--	20
Upper Extremity				
Elbow	--	6	--	--
Upper Arms	24	56	--	10
Wrist/Hand	72	85	7	23
Forearm	24	23	--	3
Total Upper Extremity	120	170	7	36
Chest	24	34	14	40
Neck/Back				
Neck/C. Spine	12	112	40	101
Shoulder	12	45	--	30
Back/T. Spine	12	73	40	88
Total Neck/Back	36	230	80	219
Other Body Areas	--	69	14	47
Total	192	584	122	392

Multiple injuries per injured driver possible.

Table 9
Driver Injuries per 1,000 Frontal Crashes with Towing by Air Bag Deploy Status and Car Model
Delta V Less Than 15 mph

Body Region	Neon Deploy Status Yes	Neon Deploy Status No	Civic Deploy Status Yes	Civic Deploy Status No
Head	52	32	40	21
Face	39	10	40	12
Upper Extremity				
Elbow	13	--	--	--
Upper Arms	144	5	40	5
Wrist/Hand	248	11	120	11
Forearm	78	--	40	--
Total Upper Extremity	483	16	200	16
Chest	66	16	160	24
Neck/Back				
Neck/C. Spine	145	53	80	80
Shoulder	66	21	80	16
Back/T. Spine	79	43	80	71
Total Neck/Back	290	117	240	167
Other Body Areas	65	36	40	35
Total	995	227	720	275

Multiple injuries per injured driver possible.

NASS ANALYSES - Analyses of NASS crash data were conducted to provide comparisons of the insurance claims data with a national sample of crashes. It should be noted that any differences found may be related to the wider distribution of vehicle makes and models and model years present in the NASS data. Because of sample size it is not possible to provide analysis at the level of vehicle make. However, to the extent that similarities are found, it can point to the generalizability of these findings.

Distribution of Air Bag Deployment Speeds in Frontal Crashes - Table 10 shows the distribution, by delta V, of air bag deployments in frontal crashes of 1990 and newer model vehicles, using NASS 1990-1994 data. About 60 percent of all air bag deployments in NASS occur at estimated delta Vs of 15 mph or less. This is somewhat higher than the insurance-based estimates of 43 percent of all Neon deployments that occur at delta Vs of 15 mph or less, and much higher than that found for the Civic. The NASS analysis, however, includes a much wider range of vehicle makes, models, and model years.

Table 11 shows the MAIS for drivers in frontal crashes at delta Vs of 15 mph or less. As found in the insurance data, the vast majority of drivers in low speed crashes sustain no or only minor injuries. However, more drivers in the NASS sample sustained minor injuries. This is likely related to the data collection methods. NASS includes self-reported injury which may or may not have been treated. The insurance data is based on injury claims which would include only injuries for which treatment is sought.

Table 10
Distribution of Air Bag Deployments in Frontal Crashes by Total Delta V (NASS 1990-1994)

Delta V	Raw Number	Percent Weighted
< 10 mph	39	12
10-15 mph	187	51
16-19 mph	106	23
≥ 20 mph	164	14
Total	496	100

Table 11
Driver Injuries in Frontal Crashes
by Maximum Injury Severity
Delta V 15 mph or Less

MAIS	Air Bag-equipped Cars Percent	Belt Only Cars Percent
0	55	57
1	39	37
2	5	3
≥3	1	3
Unknown	<1	<1

Table 12 shows driver injuries by body region in frontal crashes (10-2 o'clock direction of force with no secondary impact) of air bag-equipped and non air bag-equipped vehicles at delta Vs of 15 mph or less. Because air bags deployed relatively infrequently in the Civic crashes, driver injuries in cars equipped only with automatic or manual belts are presented for comparison to the Civic claims data.

Table 12
Driver Injuries per 1,000 Frontal Crashes with Towing
by Body Region for Air Bag-equipped and
Non Air Bag-equipped Cars
NASS 1989-94, 1989 and Newer Model Vehicles
Delta V 15 mph or Less

Body Region	Air Bag-equipped Cars	Belt Only Cars
Head	18	45
Face	373	134
Upper Extremity	502	141
Chest	158	183
Neck/Back	183	255
Lower Extremity	137	243
Abdomen	32	34
Unknown	19	33
Total	1,422	1,068

Multiple injuries per injured driver possible.

Generally, the rate of injury per 1,000 towed vehicles in NASS is much higher than the rate of injury estimated from insurance claims. However, as noted above, this may in part be related to differences in the way injury data are collected. The insurance claims data show that the Neon, with the higher proportion of air bag deployments, had a much higher rate of injury to the upper extremities and the overwhelming majority of these injuries were sustained in crashes in which air bags deployed. Likewise, in the NASS data, injuries to the upper extremities are much higher in vehicles equipped with air bags than in vehicles with belts only (Table 12). Others have documented higher rates of upper extremity injuries with air bag deployments (6, 7, 10), and it is probable that the higher rate of injuries to the upper extremities in the Neon is related to the higher rate of air bag deployments. One particularly notable finding is that the facial injury rate per 1,000 towable crashes at estimated delta Vs 15 mph or less in NASS is higher for air bag-equipped cars than for belt only cars. It should be noted, however, that these facial injuries are most often abrasions and contusions that heal without treatment.

Driver Air Bag Fabric Leading Edge Speeds and Excursions - Both the Neon and the Civic air bags are tethered and constructed of relatively lightweight materials. The maximum excursion of the air bag fabric leading edge is also quite similar for the two bags (see Table 13). However, the maximum leading edge speed of the 1995 Neon air bag is much higher than that of the Civic (see also Figures 4 and 5). There is an increased potential for abrasive type injuries for higher speed air bags, particularly if they reach speeds in excess of 230 km/h (18) . However, it should be noted that the Neon air bag underwent a design change for the 1996 model year.

Table 13
Air Bag Maximum Leading Edge Speeds and
Maximum Excursions

Model	Tethered	Maximum Excursions (mm)	Maximum Leading Edge Speed (km/h)
Honda Civic	Yes	328	227
Dodge Neon	Yes	334	325

DISCUSSION

Air bags are designed to protect unbelted and belted occupants in frontal crashes. Unbelted occupants are more vulnerable to injury, and consequently dictate air bag deployment thresholds for those car models that do not have different deployment thresholds for belted and unbelted occupants. The results of this study confirm earlier findings (5) that the rate of deployment as well as the deployment threshold can vary widely between different vehicle makes and models. In this study, not surprisingly, drivers in the vehicle with more low speed deployments (delta V ≤ 15 mph) were injured more often than drivers in the vehicle with few deployments. Clearly, designs that can eliminate some of the unneeded deployments in relatively low speed crashes without sacrificing the performance in the much less frequent but more important high speed crashes, would reduce the number of air bag-induced injuries. The comparative performance of the Neon and Civic in higher severity

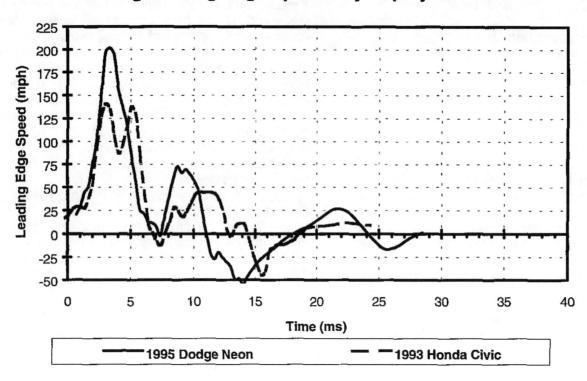

Figure 4. Air bag leading edge speeds by deployment time for the Neon and Civic air bags.

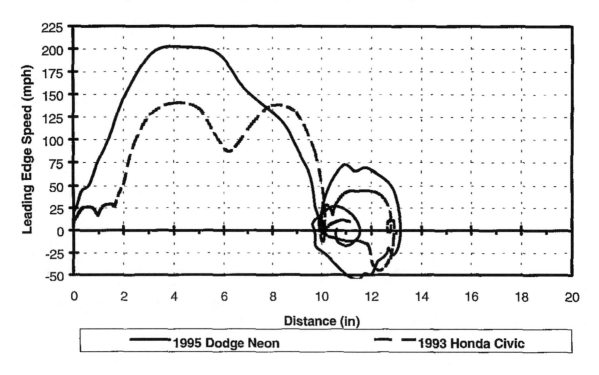

Figure 5. Air bag leading edge speeds by deployment distance for the Neon and Civic air bags.

crashes has not been considered and may be different from the low crash speed performance.

The results of this study confirm earlier findings that upper extremity injuries to drivers are the most frequent air bag-induced injury (6, 7, 10, 19). Moreover, these findings indicate that the rate of upper extremity injuries is higher the higher the rate of deployment. The rate of facial injuries in these air bag-equipped vehicles, however, seems low compared with the findings of the NASS analysis. No definitive explanation can be offered for this finding, but a number of possibilities exist. Drivers may be sustaining minor facial injuries for which they do not seek treatment, thus these injuries do not appear in the insurance claims data yet appear when self-reported injuries are included. It is also possible that the incidence of facial injuries may actually be lower in more recent model vehicles with driver air bags that are tethered as was the case with both of the air bags studied. Tethered bags have a much lower excursion distance from the module into the occupant compartment (18), and it has been shown that abrasive type injuries decrease as distance from the air bag increases (14). Further analyses would be required to distinguish between these possibilities.

A new finding, however, is that females appear to be sustaining more injuries, particularly to the upper extremities, than males in these low severity crashes. Previous NHTSA crash investigations of air bag-related injuries in low to moderate severity crashes have indicated that females may more often sustain serious arm fractures. However, these in-depth crash investigations are not based on a statistically representative sample of crashes so cannot be considered generalized to the population of crashes. Nevertheless, female drivers, particularly older females, may be more susceptible to arm injuries because of fragility. Females are often shorter than males and tend to sit closer to the steering wheel (20). Proximity to the air bag at the time of deployment is clearly a risk factor. Less aggressive air bag designs and air bag systems currently under development that take into account occupant size, weight, and proximity to the air bag, so called "smart" air bags, may help to alleviate this problem.

Because of the much lower belt use rates in the United States compared to other countries such as Canada and Europe, and the increased vulnerability to injury of unbelted occupants, it is unlikely that deployment thresholds will be changed. However, sensor technology improvements should result in fewer deployments below the nominal thresholds. Furthermore, some manufacturers already have incorporated dual deployment thresholds in their vehicles using electronic sensors which, based on belt use information, have higher nominal thresholds for belted occupants. Nominal thresholds for unbelted occupants are typically around 10-12 mph for unbelted occupants compared to 16 mph for belted occupants. Based on the average speeds at which air bags currently deploy, about half of all deployments for belted

occupants could be eliminated if dual deployment thresholds were available in all passenger cars (21).

The higher rate of injury in low severity crashes in the Neon compared with the Civic could be due to a number of factors. The Neon experiences many more air bag deployments in frontal crashes at delta Vs of 15 mph and less, and studies have shown (8) that about 40 percent of all deployments result in at least one air bag-related injury. However, there may also be variations in these mainly minor injuries based on air bag deployment characteristics. Air bag-induced skin abrasions are thought to be caused primarily by high velocity air bag fabric (14) and the Neon air bag also has a higher leading edge speed. However, the injury potential of an air bag can be reduced by changes in deployment characteristics. One manufacturer added tethers, used a lighter air bag fabric and decreased inflator output to greatly reduce leading edge speed. A study of insurance claims found a marked reduction in the rate of minor facial injuries for the redesigned air bag (5).

CONCLUSIONS

A comparison of low severity (delta V $\leq$ 15 mph) air bag deployments in two popular midsize cars -- the Dodge Neon and the Honda Civic -- illustrates the substantial differences in the deployment characteristics that exist among current air bag designs and how such differences affect the likelihood of air bag induced injuries. The results indicate that low speed air bag deployments frequently produce injuries to the upper extremities, especially for female drivers. Facial injuries, although reported less often in insurance claims than in detailed crash investigations, are also frequent in low severity air bag deployment crashes. Because low severity crashes are much more common than high severity crashes, relatively minor changes in both the nominal and actual air bag deployment thresholds can substantially change the numbers of air bag deployments.

The present study provides comparisons between two different air bag systems, indicating large differences in the likelihood of air bag induced minor injuries in low speed crashes. It is important, however, that these results be considered only in the context of low speed crashes and not be used to infer differences in the overall performance of the systems studied. A better understanding of when air bags do, and do not, deploy in the complete range of real-world crash configurations and severities is essential if future air bag designs are to be improved by minimizing deployments in low speed crashes, without compromising protection in the more important high speed crashes.

Acknowledgement

The authors appreciate the statistical consulting from Michael A. Greene, IIHS, and Randall C. Overton, State Farm. Also, the authors acknowledge the contributions from Lisa Craver and Linda Hoover in the collection of the survey data.

References

1. Ferguson, S.A., Lund, A.K. and Greene, M.A. 1995. Driver fatalities in 1985-1994 air bag cars. Arlington, VA. Insurance Institute for Highway Safety.

2. Highway Loss Data Institute: Driver Injury Experience in 1990 Models Equipped with Air Bags or Automatic Belts. A-38. Arlington, VA. Highway Loss Data Institute, 1991.

3. Kahane, C.J., Fatality reduction by automatic occupant protection in the United States. Paper 94-S5-0-08. Presented at the 14th International Technical Conference on Enhanced Safety of Vehicles, Munich, Germany, 1994.

4. Lund, A.K. and Ferguson, S.A. 1994. Driver fatalities in 1985-1993 cars with airbags. *The Journal of Trauma Injury, Infection, and Critical Care*, Vol. 38, No. 4, April 1995.

5. Werner, J.S. and Sorenson, W.W., Survey of airbag involved accidents. An analysis of collision characteristics, system effectiveness and injuries. S*afety Technology.* SP-1041. Paper 940802. Detroit, SAE, 1994, pp 1-19.

6. Huelke, D.F., Moore, J.L.and Compton, T.W., et al. Upper extremity injuries related to air bag deployments. *In-Depth Accident Investigation: Trauma Team Findings in Late Model Vehicle Collisions.* SP-1042. Paper 940716. Detroit, SAE, 1994, pp 55-64.

7. Huelke, D.F., Moore, J.L., Compton, T.W., Samuels, J., and Levine, R.S. Upper extremity injuries related to airbag deployments. *The Journal of Trauma Injury, Infection, and Critical Care.* Vol. 38, No. 4, April 1995.

8. Insurance Institute for Highway Safety. *Special issue: Air Bag Effectiveness, Status Report* 30(3): 1995.

9. Malliaris, A., Digges, K., and DeBlois, H. Injury patterns of car occupants under air bag deployment. Paper 950867. SAE, Detroit, 1995.

10. Stucki, S.L., Ragland, C., Hennessey, B., and Hollowell, W.T., NHTSA's improved frontal protection research program. *Issues in Automotive Safety Technology.* SP-1072. Paper 950497, SAE, Detroit, 1995, pp 279-291.

11. Augenstein, J.S., Digges, K.H., Lombardo, L.V., et al. Airbag protected crash victims--The challenge of identifying occult injuries. *In-Depth Accident Investigation: Trauma Team Findings in Late Model Vehicle Collisions.* SAE, Detroit, Paper 940714, 1994, pp. 45-54.

12. Augenstein, J.S., Digges, K.H., Lombardo, L.V., et al. Chest and abdominal injuries suffered by restrained occupants. *Advances in Occupant Protection Technologies for the Mid-Nineties.* SP-1077, Paper 950657, SAE, Detroit, 1995, pp 37-44.

13. Lancaster, G.I., DeFrance, J.H., and Borruso, J.J. Air bag-associated rupture of the right atrium. New England Journal of Medicine 328:358, 1993.

14. Reed, M., Schneider, L., and Burney, R. Investigation of airbag induced skin abrasion injuries. Paper 922510, SAE, Detroit, 1992.

15. Horsch, J., Llau, I., Andrzejak, D., Viano, D., Pearson, J., Cok, D., and Miller, G. Assessment of air bag deployment loads. Paper 902324, SAE, Detroit, 1990.

16. Melvin, J., Horsch, J., McCleary, J., Wideman, L., Jensen, J., and Wolanin, M. Assessment of air bag deployment loads with the small female hybrid III dummy. Paper 933119, SAE, Detroit, 1993.

17. Dalmotas, D.J., German, A., and Hurley, R.M. Air bag deployments involving restrained occupants. Detroit, Paer 950868, SAE, Detroit, 1995.

18. Powell, M.R. and Lund, A.K. Leading edge deployment speed of production air bags. Detroit, SAE, No. 950870, 1995.

19. Ferguson, S.A., Reinfurt, D.W., and Williams, A.F., 1995. Survey of passenger and driver attitudes in air bag deployment crashes. Arlington, VA: Insurance Institute for Highway Safety.

20. Parkin, S., Mackay, G.M., and Cooper, A. 1993. How drivers sit in cars. Proceedings of the 37th Annual Association for the Advancement of Automotive Medicine Conference. San Antonio, TX:375-388.

21. Insurance Institute for Highway Safety. Comments to Docket No.74-14 on January 22, 1996.

Appendix A
Driver Air Bag Deployment Frequency
Frontal Crashes With Towing
Selected Vehicles

Model	Deployment Frequency (per 100 Frontal Crashes)
Large Cars, Four-Door Models	
Dodge Intrepid	19
Buick LeSabre	25
Pontiac Bonneville	27
Midsize Cars, Four-Door Models	
Buick Century	17
Honda Accord	19
Honda Civic	20
Chevrolet Corsica	21
Saturn SL	27
Toyota Camry	28
Nissan Maxima	30
Ford Taurus	30
Mazda 626	36
Dodge/Plymouth Neon	50
Dodge Spirit	54
Plymouth Acclaim	55
Small Cars, Four-Door Models	
Toyota Corolla	23
Chevrolet Prizm	24
Large Cars, Two-Door Models	
Ford Thunderbird	27
Mid Size Cars, Two-Door Models	
Acura Integra	14
Honda Accord	19
Honda Civic	19
Pontiac Grand Am	19
Chevrolet Beretta	23
Mazda MX-6	34
Ford Probe	37
Toyota Celica	39

Model	Deployment Frequency (per 100 Frontal Crashes)
Small Cars, Two-Door Models	
Saturn SC	30
Toyota Tercel	35
Plymouth Sundance	42
Ford Escort	44
Dodge Shadow	50
Midsize Cars, Sports Models	
Ford Mustang	40
Large Cars, Luxury Models	
Cadillac DeVille	14
Large Cars, Station Wagons, Vans	
Ford Aerostar	27
Plymouth Voyager/Dodge Caravan	30
Small Cars, Station Wagons, Vans	
Ford Escort	42
Intermediate Utility Vehicles	
Jeep Grand Cherokee	15
Standard Pickups	
Ford F-150	27

SUPPLEMENTAL RESTRAINT SYSTEMS:
FRIEND OR FOE TO BELTED OCCUPANTS?

D.J. Dalmotas, R.M. Hurley, A. German
Road Safety and Motor Vehicle Regulation Directorate
Transport Canada

ABSTRACT

In North America, air bag inflation characteristics are determined largely by the unbelted test requirements of US FMVSS 208. Air bag deployment thresholds are set by vehicle manufacturers primarily to prevent facial fracture to unbelted drivers in low speed collisions. This paper examines the extent to which these design practices may be at odds with the protection needs of belted occupants in jurisdictions such as Canada where the seat belt wearing rate is close to 95 percent. Drawing on data compiled in Canada and the US, the field performance of current air bag systems is examined. Analyses of Canadian and US data for both belted drivers (Canada and US) and unbelted drivers (US only) have been performed using an injury/harm model. US injury trends pertaining to belted drivers are in close agreement with the findings of the Canadian field accident studies. Supplementary air bag systems significantly reduce the risk of severe head and facial injuries among belted drivers. However, these benefits are being negated by air bag-induced injuries, most notably to the face in moderate and low speed collisions, and to the upper extremities at all collision severities. Unless current air bag design practices are changed, the widespread use of these systems is unlikely to produce an overall benefit in jurisdictions with high seat belt wearing rates. In particular, these findings require consideration of some specific countermeasures if a satisfactory level of air bag performance is to be achieved in Canada.

SUPPLEMENTAL RESTRAINT SYSTEMS (SRS), in the form of air bags, are now common in the front outboard seating positions of passenger cars and many multipurpose passenger vehicles. As of September 1997, the fitment of such devices will be mandatory for all passenger cars sold in the United States. By September 1998, this requirement will apply also to all light-duty vehicles (e.g. vans, pickup trucks). A similar hardware requirement has not been introduced in Canada. Starting in 1997, occupant restraint systems in Canada will be subjected to dynamic testing. To date, the associated performance requirements have only been satisfied consistently by vehicles fitted with supplementary air bag systems [Welbourne, 1992; Dalmotas and Welbourne, 1991]. Consequently, it is anticipated that most manufacturers will elect to meet the Canadian requirements through the fitment of supplementary air bags.

40th ANNUAL PROCEEDINGS
ASSOCIATION FOR THE ADVANCEMENT OF AUTOMOTIVE MEDICINE
October 7-9, 1996, Vancouver, British Columbia

Assessments of vehicle performance in the near future in Canada and the US will be based on responses measured on two 50th percentile male Hybrid III dummies seated in the two front outboard seating positions in a 48 km/h frontal barrier crash test. The US limits include a requirement that the chest deflection not exceed 76.2 mm and the resultant chest acceleration not exceed 60 g when the vehicle is tested with the dummies belted as well as with the dummies unbelted. In contrast, only a single belted test is proposed for Canada. Assessment of chest response in the Canadian test is confined to chest deflection which must not exceed 50 mm.

The imposition of the unbelted test in the US to regulate the performance of a safety device now universally marketed as "supplemental" to the manual three-point seat belt continues to be a subject of debate and controversy. It has been argued that the US unbelted test forces air bag designs which are overly aggressive to occupants, particularly drivers. The rapid rise in the US in the number of infants and young children killed by passenger-side air bags has greatly increased the intensity of this debate.

A related limitation shared by both the US and proposed Canadian requirements is that each front outboard seating position is tested with a dummy of 50th percentile male dimensions in one specific seating posture. The performance levels reflected in the test may not be indicative of the levels of protection likely to be afforded to individuals of different stature. Of particular concern are possible adverse air bag-occupant interactions if the seat is located forward of the middle position. Evidence from laboratory testing shows that the proximity of an occupant to the air bag module has a strong influence on the response of the neck and the chest [Melvin et al, 1993; Horsch et al, 1990].

In addition, the intervention of the air bag can be expected to introduce a variety of new injury mechanisms such as facial injuries from "bag slap", upper extremity fractures either directly from the deploying air bag module or from arm flailing, and thermal burns to the face and arms [Huelke et al, 1994; Marsh, 1993].

In order to investigate the field performance of air bags systems in Canada, in the fall of 1993, Transport Canada initiated a directed study to document the injury experience of occupants involved in crashes resulting in the deployment of an air bag. The present paper summarizes the findings of Phase I of this study as they pertain to belted drivers. The number of right front passengers exposed to air bag deployments contained in the database is still too small to support meaningful analyses. Additional analyses were carried out using US field accident data. These findings are compared with the Canadian results.

METHODS

The Air Cushion Restraint Study (ACRS) was conducted between October 1993 and May 1995. Details of the methodology have been described previously [Dalmotas, German, and Welbourne, 1995]. Eight collision investigation teams, based in universities across Canada, selected crashes where an air bag deployed, irrespective of injury outcome. The cases selected form a convenience sample believed to be free of intentional biases.

As a follow-up to ACRS, additional air bag deployment crashes are currently being researched where a further criterion is that an occupant protected by an air bag must be transported to hospital as a result of the collision. Our specific air bag studies are being supplemented by a programme of special collision investigations which includes incidents where air bags deploy in low severity crashes, yet produce high consequences in terms of occupant injuries. While analyses conducted for the present paper using Canadian data are based exclusively on the ACRS cases, a number of individual case studies are presented which are drawn from both our air bag studies and special collision investigation programme.

RESULTS

The final ACRS database consists of 409 case vehicles containing 635 occupants. Frontal collisions accounted for almost 87% of SRS deployments, followed by side impacts (11.5%), and top and undercarriage impacts (1.7%).

All drivers experienced air bag deployment, as did 36 of the 138 right front passengers. Of the air bag exposed occupants, 93% were restrained by the lap and torso seat belt (380 drivers and 32 passengers). In terms of gender, 66% (250) of belted drivers were males and 34% (130) were females. The belted driver sample ranged in age from 16 to 84 years. Of the total belted drivers, 65% (246) were injured. Of the belted male drivers, 58% (145) sustained an injury; while 78% (101) of belted females sustained injury. How injury experience varies as a function of gender will be explored in greater detail later in the paper.

For the belted driver subset, estimates of collision severity, as defined by the total change in vehicle velocity during impact (delta-V), were available for 304 of the case vehicles, while estimates of equivalent barrier speed (EBS) were made for 359 case vehicles. Table 1 shows that about 73% of cases involved low speed collisions with an EBS or a delta-V of 25 km/h or less. It is also worth noting that EBS or delta-V values in excess of 40 km/h accounted for less than 5% of the air bag deployments.

Table 1
Cumulative Distributions of Case Vehicles
by Estimated Equivalent Barrier Speed and Delta-V

Velocity Change	EBS Cumulative		Delta-V Cumulative	
km/h	%	n	%	n
00 - 15	20.9	75	24.0	73
16 - 20	53.5	192	52.3	159
21 - 25	74.4	267	72.7	221
26 - 30	85.2	306	83.6	254
31 - 35	92.8	333	90.8	276
36 - 40	95.0	341	93.8	285
≥ 41	100.0	359	100.0	304

As shown in Table 2, 91.3% of the individual injuries were of minor severity (AIS 1), as measured by the Abbreviated Injury Scale [AAAM, 1990]. The most frequently injured body region was the upper extremity, followed by the face, and the lower extremity. Injuries of moderate or greater severity (AIS $\geq$ 2) accounted for 8.4% of individual injuries and typically consisted of head injuries involving a loss of consciousness (LOC), followed by fractures of the upper extremities, and fractures of the lower extremities. The severity of two individual injuries (0.3%) was unknown.

Table 2
Distribution of Individual Injuries Sustained by Belted Drivers
(n=246 injured) in Collisions Which Resulted in the Deployment
of the Driver-Side Air Bag System

Body Region	Anatomic Structure	AIS 1 %	AIS 2 %	AIS≥3 %	Total %
Head	Organs	-	-	1.09	1.09
	Head-LOC	0.31	0.78	0.31	1.40
	Skeletal	-	-	0.47	0.47
	Skin	0.93	0.16	-	1.09
Face	Organs	0.78	-	-	0.78
	Skeletal	0.62	0.16	-	0.78
	Skin	24.30	-	-	24.30
Neck	Skin	2.34	-	-	2.34
Thorax	Organs	-	0.31	0.62	1.09
	Skeletal	1.09	0.31	0.31	1.71
	Skin	7.79	-	-	7.79
Abdomen	Organs	-	0.62	-	0.62
	Skin	2.34	-	-	2.34
Spine	Organs	6.07	-	-	6.07
	Skeletal	-	0.31	-	0.31
	Skin	0.16	-	-	0.16
Upper Extremity	Organs	0.93	-	-	0.93
	Skeletal	2.34	1.25	0.47	4.05
	Skin	23.68	-	-	23.68
Lower Extremity	Organs	0.47	-	-	0.47
	Skeletal	1.56	0.93	0.31	2.80
	Skin	15.26	-	-	15.26
Total %		91.30	4.80	3.60	100.00
Total n		586	31	23	640

The findings noted above with respect to the incidence of upper extremity and facial injury are consistent with other studies which have examined the field performance of air bag systems. Huelke et al (1994), using the University of Michigan Transportation Research Institute collision database, also observed that upper extremity injuries are a common consequence of air bag deployments. As in the present study, the injuries documented ranged from AIS 1 contusions, abrasions and burns, to AIS 3 fractures of the forearm. A University of Virginia study [Crandall et al, 1994] also showed an elevated rate of facial abrasion among belted drivers in air bag collisions. The University of Virginia study also showed that drivers protected by both the seat belt and the air bag experienced a higher incidence of facial abrasions than occupants protected only by seat belts.

In addition to the relatively low representation of head injuries in the present sample, another striking feature of the injury distribution was the low incidence (less than 2%) of facial injury at the AIS 2 and AIS 3 levels. Historically, such facial injuries accounted for over 20% of all AIS $\geq$ 2 injuries observed among belted drivers [Dalmotas, 1980].

Another notable feature in the present study was the overrepresentation of AIS $\geq$ 2 injuries in very low speed impacts. Table 3 shows that 27% of all maximum AIS $\geq$ 2 cases were sustained in crashes for which the EBS was under 20 km/h.

Table 3
Distribution of Belted Drivers with Air Bag Deployment
as a Function of MAIS and EBS

EBS (km/h)	No Injury (%)	MAIS 1 (%)	MAIS $\geq$2 (%)	Total (%)
0 - 19 km/h	56.3	44.4	26.9	47.4
Over 19 km/h	43.8	55.6	73.1	52.7
Total	100.0	100.0	100.0	100.0
(n)	(128)	(205)	(26)	(359)

Additional analyses were carried out drawing on US field data compiled under the National Accident Sampling System (NASS) - Crashworthiness Data System in order to compare directly the injury experience of belted drivers as a function of SRS fitment. The methodology and findings are discussed in detail in a separate publication [Dalmotas et al, 1996].

A breakdown of the harm distribution among belted drivers represented in the NASS data set as a function of body region and SRS fitment is provided in Table 4. While the mean level of harm to the head in the SRS sample can be seen to be substantially lower than that in the non-SRS sample, these head injury reduction benefits are offset by an increased level of overall harm to the face and upper extremities. This trade-off is entirely consistent with the Canadian findings presented in this paper.

Table 4
Mean Level of Harm by Body Region Grouping

Body Region Grouping	Mean Level of Harm*		% Change in Mean Harm Given SRS Deployment
	Belted, No SRS Deployment	Belted, SRS Deployment	
Head	0.0026	0.0009	-65.4%
Face, Upper Extremities	0.0038	0.0065	71.1%
Other	0.0068	0.0056	-17.6%
All	0.0133	0.0130	-2.3%
N (Weighted) n (Unweighted)	1,521,829 3,173	266,206 666	

* Expressed in terms of equivalent fatality units.

The mean harm levels for the two belted driver populations as a function of collision severity, represented by delta-V, are summarized in Table 5. Here we can see that belted drivers in the SRS sample showed a greatly reduced mean harm level at collision severities in excess of 39 km/h, but showed higher overall mean harm rates in the under 24 km/h and in the 24-39 km/h delta-V intervals. Consequently, it can be seen that the benefits achieved at higher collision severities by the intervention of the air bag are being negated by deployments in low and moderate speed collisions. This trade-off is again consistent with the Canadian findings.

Table 5
Mean Level of Harm as a Function of Delta-V

Collision Severity Grouping	Mean Level of Harm		% Change in Mean Harm Given SRS Fitment
	Belted, No SRS Deployment	Belted, SRS Deployment	
Under 24 km/h	0.0063	0.0112	77.8%
24-39 km/h	0.0215	0.0270	25.6%
40+ km/h	0.1287	0.0944	-26.7%
N (Weighted) n (Unweighted)	1,521,829 3,173	266,206 666	

How the injury experience of belted drivers varies as a function of gender was also examined. These results are summarized in Table 6. As would be predicted on the basis of the Canadian experience, belted females in the SRS sample showed higher injury and harm rates than belted males in the SRS sample. They also showed consistently higher injury and harm rates than belted females in the non-SRS sample. The differences are most pronounced with respect to the incidence of AIS $\geq$ 3 injuries which, in turn, have been linked to the elevated AIS 3 injury rate to the upper extremities among belted females in low severity crashes (<24 km/h) in the SRS sample. It is interesting to note that, given SRS deployment, the AIS 3 injury rate in low severity collisions was twice that observed in higher severity collisions. This was not unexpected since maximum opportunity for direct injury from the module cover as well as from arm flailing would be expected to occur during turning manoeuvres when the forearm is centered over the module. This is far more likely to occur at lower driving speeds. Similar analysis using AIS 2 or greater also showed that the rate of upper extremity injury in low speed collisions was twice that observed in higher speed collisions.

Table 6
Overall Injury Risk and Harm Rates
Among Belted Female and Male Drivers as a Function of SRS Fitment

Performance Measure	Belted, No SRS	Belted, SRS Fitted	% Change Given SRS Fitted
Male:			
Injury Probability AIS >= 1	37.8%	39.9%	5.5%
Injury Probability AIS >= 2	6.5%	5.7%	-12.9%
Injury Probability AIS >= 3	1.1%	1.3%	10.4%
Mean Number of Injuries	1.038	0.957	-7.8%
Mean Level of Harm	0.0112	0.0099	-11.6%
Female			
Injury Probability AIS >= 1	50.7%	57.4%	13.2%
Injury Probability AIS >= 2	7.6%	8.2%	8.5%
Injury Probability AIS >= 3	1.7%	2.5%	41.1%
Mean Number of Injuries	1.521	1.937	27.4%
Mean Level of Harm	0.0152	0.0166	9.2%

The collective findings presented above show close agreement with a recent US study which examined the relationship between injury risk and deployment frequency using both insurance claim frequency data and NASS data [Werner et al, 1996]. The study compared the injury experience of two popular midsize car models, one with a higher frequency of air bag deployment (50 per 100 frontal crashes) and one with a low frequency of deployment (20 per 100). The study observed that the former produced a 50% higher injury rate in low speed collisions. The body regions most likely to be adversely affected were the face and the upper extremities. The authors also observed that females showed a higher rate of bag induced injury in low speed collisions than males, particularly with respect to the upper extremities.

CANADIAN CASE STUDIES - Case studies illustrating the level of added occupant protection afforded to fully-restrained occupants by air bag systems in moderate to severe frontal crashes have been reviewed in previous papers [Dalmotas et al, 1995; Dalmotas, Hurley, and German, 1995]. At lower collision severities, injury outcome can be adversely affected by deployment of the air bag. The severity of such bag-induced injuries can vary from minor skin abrasions to fatal injury, depending on situational factors such as the posture of the occupant at the instant of deployment. This variation in injury outcome is illustrated for drivers in the following three case studies. All of these cases involve the same vehicle model. In each case the driver is a short-statured female, and the seat is positioned forward of its mid position. All three case vehicles sustained only minor front-end damage.

ASFS-1641: The driver of a 1994 Plymouth Sundance arrived at work, drove slowly through a narrow entranceway, and turned sharply to the right, intending to leave the vehicle in a parking space adjacent to a building. The right front bumper of the vehicle made minor contact (12FRLN1) with the concrete wall of the building and the driver's air bag deployed. Several scrapes were evident on the face of the bumper and there was slight deformation of the bumper cover immediately below the right headlight. The direct damage extended 29 cm across the face of the bumper, commencing at the right front corner, with a maximum crush of 2 cm. The vehicle was driveable following the collision.

 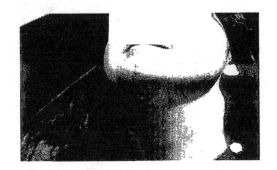

The belted driver (female/24 years) was 165 cm (5'5") tall with a mass of 64 kg (140 lb). Her seat was adjusted forward of the middle and the seat back was slightly reclined. As the air bag deployed, it struck the driver under the chin, producing a 5 cm x 5 cm abrasion (290202.1,8).

ACRS-1118: Apparently under the influence of alcohol and prescription drugs, the operator of a 1993 Plymouth Sundance fell asleep and drove through a red traffic light at low speed. The front of the case vehicle struck the side of a 1989 Ford Thunderbird. The collision produced only minor damage to the front of the Sundance (12FYEW1).

The belted driver (female/32 years) had the seat positioned fully forward. The driver was unconscious at the collision scene and, on arrival at hospital, was responsive only to pain (160899.3,0). She sustained a chipped upper molar (251404.1,8), and minor contusions and abrasions to the arms and chest. Contact to the driver's neck by the deploying air bag resulted in an abrasion, and swelling which closed off the airway (442699.3,4). She was admitted to hospital and was placed on a respirator for three days.

ASF2-1802: The case vehicle, a 1994 Plymouth Sundance, drifted off the right side of the roadway striking a wooden utility pole. Vehicle damage was minimal and was concentrated on the vehicle's front bumper and grille (12FRENI), with a maximum crush of 18 cm.

The driver (female/58 years) was fully restrained, and had her seat located fully forward with the seat back upright. The crash occurred directly outside a hospital where she was immediately examined, but was pronounced dead. The autopsy indicated minor contusions to the right upper arm, left arm and to the left chest. Death resulted from a 0.6 cm tear to the root of the left main pulmonary artery with cardiac tamponade (421006.3,4). The pathologist noted that there was no pre-crash degeneration of the arterial tissue. The driver's blood alcohol content was found to be at the legal limit. This fact, in combination with the shallow angle of departure from the roadway, the lack of evasive action, and the injury pattern to the thorax, leads to the conclusion that the driver had fallen asleep and was slumped directly on top of the air bag module at the time of collision.

As noted earlier, upper extremity injury is a common occurrence in air bag deployments. Such injuries, however, are not confined to drivers, as is illustrated by the following case study:

ACR2-1102: The driver of a 1995 Oldsmobile 98 failed to negotiate a left turn, ran off the right side of the roadway and struck a multi-stemmed tree. Only minor damage resulted; the maximum crush was 21 cm, measured at the bumper (12FLENI).

The driver (male 75 years) and the right-front passenger (female/78 years) were both fully restrained, and both front air bags deployed. The front seats were adjusted forward of the mid-seating positions.
The driver was unconscious on admission to hospital, and suffered from amnesia and confusion when he recovered (160204.3,0). He sustained fractures to the right ulna (753200.2,1), the eighth rib on the right side (450212.1,1), and the nose, in addition to multiple minor contusions, lacerations, and abrasions.

The right-front passenger received a fracture of the left ulna (753202.2,2), a hemorrhage to the right cornea (240604.1,1), and multiple minor contusions, lacerations, and abrasions.

Of particular concern has been the incidence of fatalities to small children occupying the right-front seat where an air bag has deployed. As of June, 1996, twenty-one such cases had been reported in the United States, involving unrestrained or improperly restrained children, and infants located in rear-facing infant carriers. A recent Canadian case illustrates the vulnerability of small children to air bag deployment in extremely minor collisions:

<u>ACR3-1314</u>: The case vehicle, a 1995 Hyundai Accent, was travelling southbound on an urban arterial. A 1992 Honda Civic, travelling ahead of the case vehicle, stopped in traffic. The driver of the Accent braked; however, the front of his vehicle struck the rear of the Civic. Minor damage resulted to the front bumper and hood of the Accent (12FDEW1). The corresponding direct damage to the rear of the Civic was limited to the rear bumper assembly, with induced damage to the spare wheel well (06BDLWI).

The driver (male/35 years) of the Accent was 178 cm (5' 10") tall, with a mass of 113 kg (250 lb). He was using the available three-point seat belt correctly, with his seat positioned fully rearward, when the driver's air bag deployed. He sustained only a minor contusion to his left hand (890402.1,2).

The right-front passenger (male/4 years) was 107 cm (3'6") tall and weighed 18 kg (40 lb). The driver, the child's father, had buckled the passenger's seat belt prior to commencing the trip, with the torso portion of the seat belt behind the child's back. The passenger's seat was adjusted to the mid-position for its range of travel. The driver stated that the child was leaning forward, he thought to play with the radio controls, just prior to the crash. As a result of the deployment of the rightfront air bag, the child received a large abrasion to the right side of the neck and face, and a thermal burn to the right cheek. Complete dislocation of the spine, at C1 and the base of the cranium, was accompanied by complete transection of the spinal cord (640276.6,6), and a large haematoma in the region of C 1-C7. The right atrium of the heart was bruised, and portions of skin were avulsed from the little finger and wrist of the right hand. In addition, there was head contact with the floor-mounted transmission shift lever resulting in a contusion (7 cm x 4 cm) to the occipital region.

DISCUSSION

Historically, safety devices found to be effective in reducing the risk of fatal injury were also found to be effective in preventing non-fatal injuries. Available evidence suggests this is unlikely to prove true for air bag systems given their current design and deployment characteristics. The ability of air bag systems to substantially reduce the risk of serious head injuries among belted drivers is demonstrated clearly in both Canadian and US collision databases. Since head injury is a leading cause of fatal injury, it is not surprising that previous analyses of mass accident data have observed that air bags are highly effective in preventing fatal injury. However, the most recent NHTSA report to Congress on the effectiveness of occupant protection systems also showed that current air bag systems provide no added benefit in terms of reducing the likelihood of moderate or greater severity injuries, either to belted or unbelted drivers [NHTSA, 1996]. While it is evident from the findings presented in the present study that air bags provide added protection against certain forms of non-fatal injury such as facial fracture, there is clearly a substantial increase in the risk of injury to other body regions, most notably to the upper extremities. Furthermore, it can be seen that the safety benefits achieved at higher collision severities are being negated by the high incidence of bag-induced injury in low and moderate collision severities.

Since the benefits and drawbacks of current air bag systems are drawn largely from mutually exclusive segments of the collision severity spectrum, a significant improvement in the overall level of protection afforded belted occupants by air bags could be achieved by increasing the deployment threshold. In jurisdictions such as Canada which have high levels of seat belt usage, the vast majority of air bag deployments in low speed collisions serve no useful purpose. In such collisions, injury outcome is either unchanged or adversely affected. While a majority of air bag-related injuries are AIS 1 facial and AIS 1-3 upper extremity injuries, they can include AIS≥3 injuries to other body regions when the occupant is close to the deploying air bag.

An alternative approach to increasing deployment thresholds is to reduce the aggressiveness of current air bag systems to levels consistent with those required to provide adequate protection to the belted driving population. Reducing air bag inflation rates and bag pressures, in combination with seat belt enhancements such as belt pre-tensioners, affords an attractive means of substantially improving belted protection in low and moderate severity crashes without compromising protection in high speed collisions. The fitment of less energetic air bag systems in combination with seat belt pre-tensioners is becoming increasingly more common outside of North America. Vehicle design and air bag fitment practices in Canada continue to be strongly influenced by US regulatory practices, most notably the US unbelted test requirements which promote aggressive air bag designs and provide no incentive to industry to incorporate enhanced seat belt technologies such as pre-tensioners.

In the future, it may prove possible to reconcile differences in occupant protection policies in Canada and in the US through the use of "smart" air bag technologies which incorporate features such as the capability to alter the air bag deployment threshold and inflation characteristics as a function of whether or not the occupant is belted. In the interim, Transport Canada is encouraging manufacturers to voluntarily optimize their occupant restraint systems to protect belted occupants. In the short term, this might be achieved by raising the deployment threshold or by "de-powering" air bag systems.

In addition, a number of possible regulatory options to greater ensure that occupants of short stature are not adversely affected by air bag systems are being evaluated by Transport Canada. One such option is to add a second belted 48 km/h test using 5th percentile female Hybrid III dummies with the seats set in the full forward position. Another option is to introduce a low speed offset frontal crash test, again using 5th percentile dummies with the seats set in the full forward position. The impact speed could be set to correspond closely to what would constitute, in a general sense, an appropriate air bag deployment threshold for a belted occupant. Vehicle manufacturers would then have the option of satisfying the associated performance requirements either through deployment of the air bag system or by other means as deemed appropriate. Such a test condition would encourage consideration of countermeasures such as belt pre-tensioners, compliant steering wheel rims, padded hubs and padded A-pillars to provide low speed protection. The use of air bags could then be limited to higher severity collisions where the benefits in terms of improved protection against serious head and facial injury can be more reasonably expected to outweigh the risk of any bag-induced injury.

It is unlikely that the aggressivity of air bag systems can be reduced to the point that they would pose no risk to certain segments of the population such as small children who are unrestrained or incorrectly restrained, elderly occupants in very close proximity to the air bag, or any occupant in direct contact with the air bag module (e.g. draped over the steering wheel, asleep). The use of proximity sensors to detect such conditions and to automatically deactivate the air bag system represents the best long term solution to this problem. In the short term, it is critical that the general public be made aware of the potential hazards associated with air bags, of the importance of ensuring that small child occupants travel in the rear seat, and of the importance of consistent and proper restraint use by all occupants.

In summary, while air bags provide added protection against fatal injury, current designs are still far from "belt-friendly", a situation which will only be rectified when they are actually designed to protect the belted rather than the unbelted.

ACKNOWLEDGMENTS AND DISCLAIMER

The authors would like to acknowledge the contributions of the collision investigation teams at the Technical University of Nova Scotia, University of New Brunswick, École Polytechnique de Montréal, Ryerson Polytechnic University, University of Western Ontario, University of Saskatchewan, University of Calgary, and the University of British Columbia in providing the field data for this study. The assistance afforded by Floyd Dempsey in the creation and maintenance of the collision databases is similarly acknowledged.

The conclusions reached, and opinions expressed, in this paper are solely the responsibility of the authors. Unless otherwise stated, they do not necessarily represent the official policy of Transport Canada.

REFERENCES

Association for the Advancement of Automotive Medicine (AAAM). The Abbreviated Injury Scale, Des Plaines, IL 60018, USA, 1990.

Crandall, J.R., Kuhlmann, T.P., Martin, P.G., Pilkey, W.D., and Neeman, T. Differing patterns of head and facial injury with air bag and/or belt restrained drivers in frontal collisions: Advances in Occupant Restraint Technologies. Proceedings of Joint AAAM/IRCOBI Special Session; Lyon, France; September 22, 1994.

Dalmotas, D.J. Mechanisms of injury to vehicle occupants restrained by three-point seat belts. SAE 801311, 1980.

Dalmotas, D.J., German, A., Hendrick, B.E., and Hurley, R.M. Air bag deployments: The Canadian experience. J.Trauma 38(4): 476-481, April, 1995.

Dalmotas, D.J., German, A., Hurley, R.M., and Digges, K. Air bag deployment crashes in Canada. Enhanced Safety of Vehicles Conference, 96-S1-0-05, Melbourne, Australia, May 1996.

Dalmotas, D.J., German, A., and Welbourne, E.R. Directed studies: A focused approach to collision investigation. Proc. Canadian Multidisciplinary Road Safety Conference IX, pp. 13-23, M6ntreal, Quebec, Canada, 1995.

Dalmotas, D.J., Hurley, R.M., and German, A. Air bag deployments involving restrained occupants. SAE 950868, 1995.

Dalmotas, D.J. and Welbourne, E.R. Improving the protection of restrained front seat occupants in frontal crashes. Proc. 13th International Technical Conference on Experimental Safety Vehicles, Paris, France, November 4-7, 1991.

Horsch, J., Lau, I., Andrzejak, D., Viano, D., Melvin, J., Pearson, J., Cok, D., and Miller, G. Assessment of air bag deployment loads. SAE 902324, 1990.

Huelke, D.F., Moore, J.L., Compton, T.W., Samuels, J., and Levine, R.S. Upper extremity injuries related to air bag deployments. Advances in Occupant Restraint Technologies. Proceedings of Joint AAAM/IRCOBI Special Session; Lyon, France; September 22, 1994.

Marsh, J.C. Supplemental air bag restraint systems: Consumer education and experience. SAE 930646, 1993.

Melvin, J., Horsch, J., McCleary, J., Wideman, L., Jensen, J., and Wolanin, M. Assessment of air bag deployment loads with the small Hybrid III dummy. SAE 933119, 1993.

National Highway Traffic Safety Administration (NHTSA). Second Report to Congress: Effectiveness of occupant protection systems and their use, Washington, DC, USA, February, 1996.

Welbourne, E.R. Specifying performance requirements to reduce the risk of closed head injury. Proc. International Conference on Air Bags and Seat Belts: Evaluation and Implications for Public Policy, Montreal, Quebec, Canada, October 18-20, 1992.

Werner, J.V., Roberson, S.F., Ferguson, and Digges, K.H. Air bag deployment frequency and injury risks. SAE 960664, 1996.

970491

Injury Risks in Cars with Different Air Bag Deployment Rates

John V. Werner and Steve F. Roberson
State Farm Mutual Automobile Insurance Co.

Susan A. Ferguson
Insurance Institute for Highway Safety

Kennerly H. Digges
George Washington University

ABSTRACT

Automobile insurance claims of two popular midsize cars with different air bag deployment frequencies -- the Dodge/Plymouth Neon and Honda Civic -- were examined to determine performance in higher severity crashes (the upper 30 percent of crashes ranked by adjusted repair cost). Previously, it was found that drivers sustained more, mainly minor, injuries in the Neon which had a higher deployment frequency in low speed crashes. This study examined, for these two cars, whether there was any trade-off associated with a higher deployment threshold. It was found that even at higher speeds, the Neon had a greater frequency of air bag deployments, which in turn resulted in a greater likelihood of driver injury. Once again upper extremity injuries were most prevalent for Neon drivers and were highest for female drivers. At the same time, there was little evidence that driver protection was compromised in the Civic in the more important high speed crashes. The limited evidence from this study suggests that efforts should be made to reduce the number of air bag deployments in low speed crashes.

INTRODUCTION

Air bags can be designed to deploy at various speeds, but the assumption is that deployments should not occur below the injury threshold. However, there is growing concern about the large numbers of air bag deployments in crashes where speeds are slow enough that air bags are not needed to provide protection, particularly for belted occupants (1, 2, 3). This is particularly troubling because air bags can cause injury even in very low severity crashes. Air bag-related injuries are mostly minor, but in rare instances they involve serious and fatal injury to the head and chest to people who are close to them when they deploy (2).

Air bags are designed to deploy in frontal crashes at barrier equivalent speeds of about 10-12 mph. However, many of them are occurring at speeds lower than this

nominal threshold depending on sensor type and location as well as crash type. Analyses of data from the National Highway Traffic Safety Administration (NHTSA) National Accident Sampling System (NASS 1990-1994) reveals that about a third of all deployments occur in frontal crashes at longitudinal delta Vs of less than 10 mph (2). Forty-two percent occur between 10 and 15 mph. There is also evidence of tremendous variation in deployment frequencies and deployment thresholds among different makes and models (3,4). Moreover, a recent study found, in a comparison of low severity crashes in two popular midsize cars, that drivers in the vehicle with more low speed deployments were injured more often than drivers in the vehicle with fewer deployments; albeit these were mostly minor injuries (3).

It has been suggested that the solution to the high number of low speed deployments is to raise the deployment threshold. A recent petition to NHTSA (5) recommended that Federal Motor Vehicle Safety Standard 208 be amended to set a 12 mph threshold below which the air bag cannot deploy. There are, however, some concerns that need to be addressed. If deployment thresholds are higher there is the possibility that injury rates will increase for unbelted occupants who no longer will be protected by an air bag in some crashes. There is also the concern that higher deployment thresholds may result in later deployments. This could have the effect of increasing injuries because occupants have the potential to be closer to the air bag at the time of deployment.

The previous study using insurance claims data for the 1995 Dodge/Plymouth Neon and the 1992-1995 Honda Civic found higher injury rates for the vehicle with higher deployment frequencies in low speed crashes (3). However, these comparisons provided no information on the overall performance of these two vehicles. That is, there was no analysis of the protection offered at higher speeds. This paper provides a comparison of injury likelihood in the same make/model vehicles, but in crashes of higher severity -- the crashes that were not included in the previous paper.

METHOD

The following sources of information form the basis for the analysis and discussion in this paper:

1. State Farm Mutual Automobile Insurance Company's electronically prepared damage repair estimates for towed vehicles.

2. State Farm survey of claims involving frontal crashes for the 1995 Dodge/Plymouth Neon and the 1992-1995 Honda Civic.

3. The leading edge speed and excursion profile of driver air bag fabric in the 1995 Dodge/Plymouth Neon, the 1996 Neon, and the 1993 Honda Civic.

DEPLOYMENT FREQUENCY - A 1996 study described the use of State Farm damage repair estimates to identify towed vehicles with frontal damage (3). Deployment status was determined by analyzing the damage estimate for driver air bag module replacement. The Plymouth/Dodge Neon and the Honda Civic have similar repair cost severity distributions, but a large difference in their air bag deployment frequencies - the Honda Civic with an estimated deployment frequency of 20 per 100 frontal crashes, and the Plymouth/Dodge Neon with an estimated 50 deployments per 100 frontal crashes.

CLAIM SURVEY - Because insurance claim investigators do not require crash severity or delta V to be determined as part of the claim settlement process, the "delta V" distribution of the vehicle crashes represented by insurance data was approximated by fitting the rank ordered cumulative distribution of dollar damages for these two cars to the cumulative delta V distribution determined from a analysis of midsized cars in the National Accident Sampling System (NASS 1989-1994, 3). The repair costs were adjusted to remove air bag system related expenses. Claim files representing the upper 30 percent of the cumulative distribution of crashes, rank ordered by adjusted repair cost, were selected for this study.

In the absence of crush measurements for insurance data, the correlation between the rank ordered repair cost distribution and the NASS delta V distribution is not exact; however, a review of the photographs of vehicle damage and a review of the damage estimates show a systematic increase in the extent of unibody structural damage. The authors are confident that the claim files selected for this study are representative of the moderate to serious crashes seen in NASS.

Claim files were surveyed to provide information regarding policyholder involved crashes. Information tabulated included a crash description, age and gender of the insured driver, driver injury descriptions, and verification of air bag deployment status. The Abbreviated Injury Scale (AIS) and Occupant Injury Classification (OIC) coding were used to document the injuries to insured drivers that sought treatment resulting in claim payments.

RESULTS

REPAIR COST DISTRIBUTION - Figure 1 depicts the cumulative percentage of 1995 Neons and 1992-1995 Civics repaired by damage severity percentile for frontal crashes. Severity percentile was determined by dividing each car's repair cost range into eight equal repair dollar intervals. A Kolmogorov-Smirnov test was used to compare the equivalency of the two cumulative repair cost distributions. The test results support the assumption that the repair cost distributions are equal for 1995 Neons and 1992-1995 Civics (KSA [1020] = 0.697, $p = 0.716$).

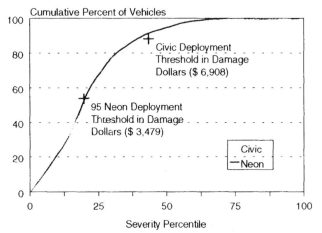

Note: 100th severity percentile represents $16,198 for the Neon and $15,893 for the Civic.

Figure 1. Cumulative Neon and Civic collision claim severity distributions for frontal impacts with towing.

Cases included in these analyses were those in the upper 30 percent of the cumulative distribution of crashes rank ordered by adjusted repair costs. This portion of the distribution includes crashes judged to be of moderate to high severity. For 1995 Neons, these crashes represent damages in excess of $4,264 in adjusted repair costs. For Civics, crashes with more than $4,461 in damages were included.

$$\log(p/(1-p)) = b_0 + b_1 x \quad (1)$$

$$x(p=.5) = -b_0/b_1 \quad (2)$$

A logistic regression model (Equation 1) was used to determine the damage severity at which the probability of an air bag deployment reached 50 percent. P is the probability of deployment and x is the damage severity (in dollars). For $p = 0.5$, the damage severity is given by Equation 2. The derived value is defined for this study as the air bag deployment threshold, that is, the dollar point in the distribution where the probability of deployment is 50 percent. The adjusted repair cost, where the probability of deployment is 50 percent is $3,479 for Neons (95 percent CI = $3,223 to $3,757). Civics have a 50 percent probability of air bag deployment at a $6,908 adjusted repair cost (95 percent CI = $6,452 to $7,484).

DEPLOYMENT FREQUENCIES - Figure 2 shows air bag deployments per 100 frontal crashes by damage severity, for the comparison vehicles. The Civic has an estimated deployment frequency of 48 per 100 moderate to serious, frontal crashes, and the Neon has an estimated 89 deployments per 100 moderate to serious, frontal crashes.

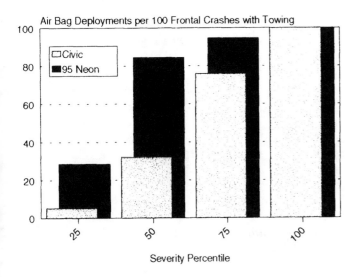

Figure 2. Air bag deployment rate by severity, frontal crashes with towing.

An analysis of the odds of deployment for moderate to serious crashes shows that there are significantly more deployments per 100 frontal crashes for Neons than for Civics (OR = 9.17; 95 percent CI = 4.72 to 17.54).

INJURIES - The following analyses compare injury rates in the two cars for moderate to serious crashes (the upper 30 percent of crashes ranked by adjusted repair amount). As shown in Table 1, 50 percent of Neon drivers report an injury in these crashes compared to 35 percent of Civic drivers. The relative risk of injury for Neon drivers compared to Civic drivers is 1.83 (95 percent CI = 1.16 to 2.96).

Table 1
Driver Injury by Car Model
for Moderate to Serious Crashes

	Driver Injury No	Driver Injury Yes	Total Crashes
Neon	57 (50.0%)	57 (50.0%)	114 (100%)
Civic	126 (64.6%)	69 (35.4%)	195 (100%)

Table 2 shows the maximum injury severity (MAIS) for the Neon and Civic drivers. Overall, injury severity was low for these drivers. Of the 309 Neon and Civic drivers, 292 (94.5 percent) had either minor injury or no injuries (MAIS-1); 15 (4.9 percent) had moderate injuries (MAIS-2); and 2 (0.6 percent) had a MAIS-3 injury. Very few MAIS-2 or greater injuries were associated with crashes in this crash severity range.

Appendix A shows the MAIS-2 or greater injuries, with driver age, gender, reported seatbelt use, and deployment status, rank ordered by the cumulative distribution percentile of crash severity.

Compared to the overall female-to-male driver ratio (2:1), females represented a higher proportion of the MAIS 2+ injuries (3.5:1). The odds of MAIS-2+ injury for female drivers was higher than for MAIS-2+ injured male drivers (OR = 1.81; 95 percent CI = 0.58 to 5.68). However, this relative risk was not significant.

Table 2
Drivers Injured by Maximum AIS
for Moderate to Serious Crashes

MAIS	Neon Count	Neon Percent	Civic Count	Civic Percent	Total Count	Total Percent
0	57	50.0	126	64.6	183	59.2
1	51	44.7	58	29.7	109	35.3
2	6	5.3	9	4.7	15	4.9
3	0	0.0	2	1.0	2	0.6
> 3	0	0.0	0	0.0	0	0.0
Total	114	100.0	195	100.0	309	100.0

Injuries by Gender - The injury data were also analyzed by gender of driver (Table 3). The proportion of male to female drivers involved in these crashes did not differ significantly for the two vehicles ($x^2[1]$ = .032, p = .857), so overall injury risk can be compared without having to control for gender differences. Comparing across vehicles, the relative risk of injury to female drivers for Neons compared to Civics was 1.62. However, this relative risk was not significant (95 percent CI = 0.87 to 3.01). For male drivers, the relative risk of injury was 1.54 for Neons compared to Civics, again not significant (95 percent CI = 0.64 to 3.72).

Table 3
Driver Injury Status by Gender
for Moderate to Serious Crashes

Driver Gender	Driver Injury No	Driver Injury Yes	Total Crashes
Neon			
Female	26 (38.2%)	42 (61.8%)	68 (100%)
Male	19 (54.3%)	16 (45.7%)	35 (100%)
Civic			
Female	53 (50.5%)	52 (49.5%)	105 (100%)
Male	33 (64.7%)	18 (35.3%)	51 (100%)

Within a given vehicle, a comparison of injury risk differences by gender, shows that Neon female drivers had higher odds of injury when compared to Neon male drivers (OR = 1.92; 95 percent CI = 0.84 to 4.39). In Civics, female drivers had a relative risk of injury equal to 1.83 when compared to male drivers (95 percent CI = 0.92 to 3.66) with neither result being statistically significant.

Injuries by Air Bag Deployment Status - Table 4 shows injury rates by car make and deployment status. In air bag deployment crashes Neon drivers had 1.45 times higher odds of injury than Civic drivers. However, this difference was not significant (95 percent CI = 0.83 to 2.54). For crashes in which the air bag did not deploy, the relative risk of injury for drivers was 0.87 for Neons, again not significant (95 percent CI = 0.22 to 3.45).

Table 4
Driver Injury Status by Deploy Status for Moderate to Serious Crashes

Deploy Status	Driver Injury No	Yes	Total Crashes
Neon			
No	9 (75.0%)	3 (25.0%)	12 (100%)
Yes	47 (46.1%)	55 (53.9%)	102 (100%)
Civic			
No	73 (72.3%)	28 (27.7%)	101 (100%)
Yes	52 (55.3%)	42 (44.7%)	94 (100%)

The odds of injury for Neon drivers in air bag deployment crashes was higher than for Neon drivers in non-deployment crashes (OR = 3.51; 95 percent CI = .898 to 13.73). Likewise, Civic drivers had higher odds of injury for air bag deployment crashes compared to non-deployment crashes (OR = 2.11; 95 percent CI = 1.16 to 3.82). However, it should be noted that crashes in which an air bag deployment occurs, typically are of higher severity and more injuries could be expected.

Location of Driver Injuries - Table 5 shows driver injury rates for "High" severity crashes (the upper 30 percent of crashes ranked by adjusted repair amount) and "Low" severity crashes (the lower 70 percent of crashes ranked by adjusted repair amount) by body region and car model. The injury rate (injuries per 1,000 frontal crashes with towing) includes multiple injuries per injured driver.

As might be expected, drivers in high severity crashes had higher rates of injury to all body regions, with the exception of head injuries to Neon drivers. Neon drivers had higher rates of injury to the upper extremities, including the upper arms, wrists, and hands for both high and low severity crashes. The rate of predominately minor facial injuries to Civic drivers is substantially higher than the Neon facial injury rate for the high crash severity category (103 compared to 35). The rate of facial injuries to Neon drivers is marginally higher than the Civic facial injury rate for the low crash severity category (20 compared to 13). The rate of lower extremity injuries for high severity crashes has a five-to-six fold increase over the rate of lower extremity

injuries for low severity crashes. Neon and Civic drivers have similar lower extremity injury rates.

Table 6 shows driver injury rates by gender and car model for different body regions. Female drivers had generally higher injury rates compared with male drivers. Female Neon drivers had the highest overall injury rate in this comparison, with substantially higher injury rates to upper extremities compared with all other car model and gender categories. However, the rate of head, facial, and chest injuries to Civic female drivers is higher than for Neon female drivers.

Table 5
Driver Injuries per 1,000 Frontal Crashes with Towing by Car Model

Body Region	Neon Crash Severity High	Low	Civic Crash Severity High	Low
Head	26	38	41	22
Face	35	20	103	13
Upper Extremity				
Elbow	9	4	15	--
Upper arms	149	46	31	6
Wrist-Hand	228	81	56	17
Forearm	--	24	--	2
Total Upper Extremity	386	155	102	25
Chest	70	30	97	31
Neck/Back				
Neck/C. Spine	184	80	159	80
Shoulder	70	34	51	20
Back/T. Spine	105	53	138	71
Total Neck/Back	359	167	348	171
Lower Extremity				
Knee	88	19	113	22
Lower Leg	44	8	15	6
Ankle-Foot	61	8	36	--
Pelvic-Hip	9	4	26	2
Total Lower Extremity	202	39	190	30
Other Body Areas	--	--	10	--
Unknown	44	8	10	4
Total	1,122	457	901	296

Multiple injuries per injured driver possible.

Table 7 shows driver injuries by air bag deployment status and car model. There were very few claims to review for the Neon non-deployment classification, because of the high deployment frequency for Neons in these moderate to serious crashes. Neon drivers involved in air bag deployment crashes had higher rates of injury to the upper extremities compared with Civic drivers involved in air bag deployment crashes.

Table 6
Driver Injuries per 1,000 Frontal Crashes with Towing
by Gender and Car Model
Moderate to Serious Crashes

| Body Region | Neon | | Civic | |
	Male	Female	Male	Female
Head	57	15	38	58
Face	57	29	58	163
Upper Extremity				
Elbow	--	15	19	19
Upper Arms	29	234	--	58
Wrist/Hand	143	309	58	77
Total Upper Extremity	172	558	77	154
Chest	114	59	58	154
Neck/Back				
Neck/C. Spine	200	206	154	221
Shoulder	86	74	58	67
Back/T. Spine	143	103	212	154
Total Neck/Back	429	383	424	442
Lower Extremity				
Knee	86	103	77	173
Lower Leg	--	74	--	29
Ankle-Foot	29	88	--	67
Pelvic-Hip	29	--	58	19
Total Lower Extremity	144	265	135	288
Other Body Areas	--	--	--	19
Unknown	57	44	19	10
Total	1,093	1,353	809	1,288

Multiple injuries per injured driver possible.

The rate of facial and chest injuries to Civic drivers involved in air bag deployment crashes was higher than the Neon facial and chest injury rate for air bag deployment crashes. Civic air bag deployments occur at the upper range of the crash severity distribution and likely reflect a more serious crash, on average, compared with Neon air bag deployment crashes.

Driver Air Bag Fabric Leading Edge Speeds and Excursions - As noted in the previous paper (3) both the Neon and the Civic air bags are tethered and constructed of relatively lightweight materials. The maximum excursion of the air bag fabric leading edge is also quite similar for the two bags (see Table 8). However, the maximum leading edge speed of the 1995 Neon air bag is much higher than that of the Civic (see also Figures 3 and 4). There is an increased potential

for abrasive type injuries for higher speed air bags, particularly if they reach speeds in excess of 230 km/h (6).

Table 7
Driver Injuries per 1,000 Frontal Crashes with Towing
by Air Bag Deploy Status and Car Model
Moderate to Serious Crashes

| Body Region | Neon Deploy Status | | Civic Deploy Status | |
	Yes	No	Yes	No
Head	29	--	21	58
Face	39	--	149	58
Upper Extremity				
Elbow	10	--	11	19
Upper Arms	167	--	43	19
Wrist/Hand	255	--	106	10
Total Upper Extremity	432	--	160	48
Chest	69	83	138	58
Neck/Back				
Neck/C. Spine	186	167	149	163
Shoulder	78	--	64	38
Back/T. Spine	108	83	128	144
Total Neck/Back	372	250	341	345
Lower Extremity				
Knee	98	--	138	87
Lower Leg	49	--	32	--
Ankle-Foot	69	--	43	29
Pelvic-Hip	10	--	53	--
Total Lower Extremity	226	--	266	116
Other Body Areas	--	--	21	--
Unknown	49	--	21	--
Total	1,216	333	1,117	683

Multiple injuries per injured driver possible.

However, it should be noted that the Neon air bag underwent a design change for the 1996 model year and leading edge speeds in the 1996 Neon air bag now are lower than either the Honda Civic, or the 1995 Neon.

Table 8
Air Bag Maximum Leading Edge Speeds and
Maximum Excursions

Model	Tethered	Maximum Excursions (mm)	Maximum Edge Speed (km/h)
Honda Civic	Yes	328	227
1995 Dodge Neon	Yes	334	325
1996 Dodge Neon	Yes	274	158

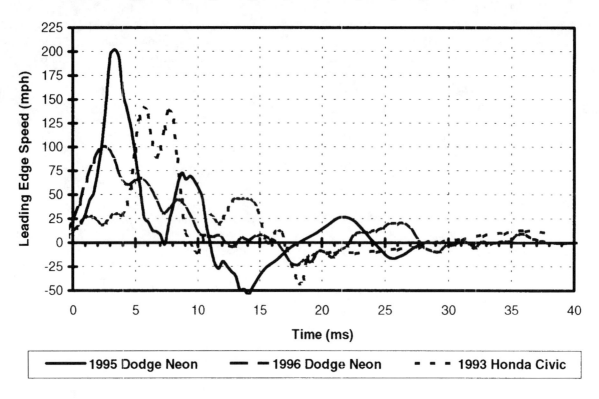

Figure 3. Air bag leading edge speeds by deployment time for the Neon and Civic air bags.

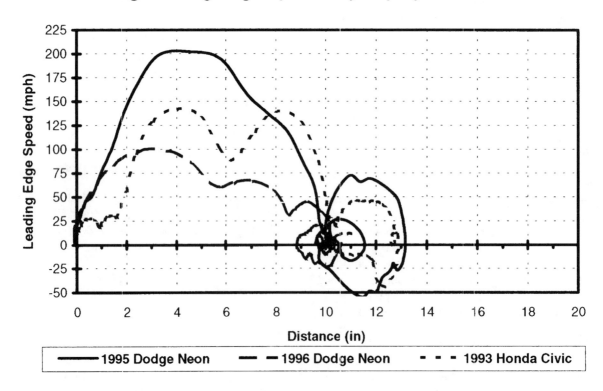

Figure 4. Air bag leading edge speeds by deployment distance for the Neon and Civic air bags.

DISCUSSION

The comparative performance of two similar mid-sized cars in moderate to serious crashes was considered in this paper to supplement results from a previous paper examining injury likelihood for the same two vehicles in low severity crashes. The previous study determined that drivers in the vehicle with more low speed deployments were injured more often than drivers in the vehicle with fewer deployments.

For the moderate to serious crashes surveyed in this study, one vehicle had substantially more deployments than the other vehicle. One might expect more moderate and serious injuries to drivers of the vehicle with fewer deployments in these moderate to serious crashes if air bags are failing to deploy when needed to prevent injury. This was not observed. There were very few moderate and greater injuries to the drivers of both vehicles and the proportion of these injuries was very similar; moreover, drivers in the vehicle with more deployments were injured more often overall than drivers in the vehicle with fewer deployments.

A comparison of the MAIS-2 and greater injury patterns, by body region, between vehicles, revealed no clear differences. Although drivers in both vehicles suffered head and chest injuries in crashes in which the air bag did not deploy, these injuries were predominantly minor. The absence of serious injuries to drivers of the vehicle with fewer deployments in moderate to serious crashes is encouraging. However, because of the limited number of cases available, and the absence of any very high speed crashes, further research is needed to confirm that in the Civic, and more generally, in other vehicles with fewer low speed deployments, that there is no injury trade-off when setting deployment thresholds higher.

The level of belt use is an important consideration for interpreting these results. Unbelted occupants are more vulnerable to injury. However, only self-reported belt use is available for drivers in this study and in the Civic and Neon, about 75 percent of the injured drivers reported that they used their belt. High belt use decreases the likelihood of observing serious injuries in non-deployment, moderate to serious crashes.

The results of this study indicate that the rate of upper extremity injury is higher, the higher the rate of deployment, regardless of crash severity. In addition, female drivers appear to be sustaining more injuries, particularly to the upper extremities, than males in these moderate to serious crashes. Proximity to the air bag at the time of deployment is clearly a risk factor. Less aggressive air bag designs and air bag systems currently under development that take into account occupant size, weight, and proximity to the air bag, may help alleviate this problem.

The authors' previous study outlined two factors that could be contributing to the higher rate of injury to Neon drivers compared with Civic drivers. Results of this study indicate that these same factors apply to the interpretation of the higher rate of Neon driver injury compared with Civic drivers in moderate to serious crashes. The 1995 Neon experiences many more air bag deployments in moderate to serious frontal crashes and the air bag fabric has a higher leading edge speed compared with the Civic. Both of these air bag performance characteristics appear to be changed for the 1996 Neon. The estimated deployment frequency for the 1996 Neon is 35 per 100 frontal crashes compared with 50 deployments per 100 frontal crashes for the 1995 Neon and 20 deployments per 100 frontal crashes for the Civic. Also, the leading edge speed in the 1996 Neon air bag is lower than either the Honda Civic, or the 1995 Neon. Results of this study and the previous related study suggest that these changes will result in lowering the injury rate to drivers of 1996 Neons involved in frontal crashes.

CONCLUSIONS

Auto manufacturers, the NHTSA, air bag system component suppliers, auto safety researchers, and auto safety advocates are currently challenged to find effective solutions to unintended injuries from air bag deployments. Smart air bag technology is expected to address the issues associated with the most serious of these unintended air bag induced injuries in the longer term. However, air bag system designs that optimize the air bag deployment threshold, using current and available technology, can partially address these concerns by reducing the number of unneeded deployments. For example, dual deployment thresholds that are set to deploy the air bag at different crash severities based on belt use could reduce unneeded deployments for belted occupants. Serious consideration should be given to raising the deployment threshold for unbelted occupants where the vehicle's air bag deployment threshold results in large numbers of low speed deployments. One concern in raising the unbelted threshold is that facial injuries, such as fractures to facial bones, could increase. This study, although limited in scope, found no evidence of such a trade-off.

The higher rate of mainly minor injuries for Neon drivers could be related both to the higher frequency of air bag deployment as well as the higher leading edge speed of the 1995 Neon air bag. Air bag design characteristics associated with the number of deployments and air bag leading edge speed were changed for the 1996 Neon. These changes will likely reduce the air bag induced injuries in low to moderate severity crashes, while maintaining air bag performance in higher speed crashes.

ACKNOWLEDGEMENTS

The authors acknowledge the contributions from Karon Bakel and Lisa Craver in the collection of the survey data.

REFERENCES

1. Dalmotas, D.J., Hurley, J., German, A., and Digges, K. D., 1996. Air bag deployment crashes in Canada. Proceedings of the 18th International Technical Conference on Experimental Safety Vehicles, Melbourne, Australia, Paper 96-S1-O-05.

2. Ferguson, S. A., Update on airbag performance in the United States: Benefits and Problems Proceedings of the Airbag 2000 Conference in Karlsruhe, Germany, November 26, 1996.

3. Werner, J.V., Roberson, S.F., Ferguson, S.A., Digges, K.D. Air bag deployment and injury risks. Paper 960664, SAE, Detroit, 1996.

4. Werner, J.S. and Sorenson, W.W., Survey of airbag involved accidents. An analysis of collision characteristics, system effectiveness and injuries. Safety Technology. SP-1041. Paper 940802. Detroit, SAE, 1994, pp 1-19.

5. Center for Auto Safety, Petition to the National Highway Traffic Safety Administration. November 8, 1996.

6. Powell, M.R. and Lund, A.K. Leading edge deployment speeds of production air bags. Detroit, SAE, No. 950870, 1995.

Appendix A
MAIS-2 or Greater Injuries
by Vehicle
Moderate to Serious Crashes

Vehicle	Distribution Percentile	MAIS	Deploy Status	Gender	Age	Belt Use	Injury
Civic	76	2	No	Female	19	No	Concussion
	82	3	No	Female	26	Yes	Ankle Fracture
	85	2	No	Female	35	Yes	Sternum Fracture
	86	2	No	Female	Unknown	Unknown	Scalp Laceration
	90	2	Yes	Female	Unknown	Unknown	Patella Fracture
	93	3	Yes	Male	Unknown	Unknown	Hip Dislocation
	93	2	Yes	Male	45	Yes	C-2 Fracture
	94	2	Yes	Female	52	Yes	Patella Fracture
	95	2	Yes	Female	48	Yes	Knee Sprain
	95	2	Yes	Female	34	Yes	Kidney Contusion
	98	2	Yes	Female	Unknown	Unknown	Face - Not Further Specified
Neon	75	2	No	Male	Elderly	Unknown	2 Rib Fractures
	80	2	Yes	Female	26	Unknown	Hand Laceration
	87	2	Yes	Female	Unknown	Unknown	Upper Extremity Fracture
	95	2	Yes	Male	71	Yes	Scalp Laceration
	97	2	Yes	Female	Unknown	Unknown	Lower Extremity Fracture
	100	2	Yes	Female	20	Yes	Multiple Lower Extremity Fracture

**Mechanisms of Injuries for Adults and Children
Resulting from Airbag Interaction**

Michael Kleinberger and Lori Summers
National Highway Traffic Safety Administration
Office of Crashworthiness Research
Washington, DC

ABSTRACT

As of April 15, 1997, the National Highway Traffic Safety Administration (NHTSA) has documented a total of 63 fatalities of automotive occupants attributed to airbag deployment. This paper summarizes the serious and fatal injuries sustained by both children and adults as a result of this airbag interaction. It is important to realize that the cases discussed in this paper represent a very small percentage of the total number of airbag deployments.

Injury patterns are identified and primary mechanisms proposed for each pattern. Three distinct injury patterns are discussed for adults, and two patterns for children. Biomechanical differences between adults and children are discussed with respect to the resulting injuries.

Primary injury mechanisms are proposed based on information collected from a number of different sources. Analytical modeling was used, along with documented occupant contact points within the vehicle interior, to predict occupant kinematics and timing with respect to airbag deployment. Experimental testing with anthropomorphic test dummies provided valuable information on the magnitude and direction of loads delivered by the deploying airbag to an out-of-position occupant. All of this information, along with biomechanical expertise and existing injury tolerance data, were used to develop the proposed injury mechanisms.

As of April 15, 1997, the NHTSA has documented a total of 63 fatalities of automotive occupants attributed to airbag deployment. Thirty-eight cases involved children in the passenger seat, while the remaining 25 cases involved adults (22 drivers and 3 passengers). Of the 38 child cases, 9 involved infants in rear-facing seats, and the other 29 cases involved forward-facing children, almost all of whom were either unrestrained or improperly restrained. In addition to the 63 fatal cases, another 16 cases have been documented (all children) where serious non-fatal injuries

41ST ANNUAL PROCEEDINGS
ASSOCIATION FOR THE ADVANCEMENT OF AUTOMOTIVE MEDICINE
November 10-11,1997, Orlando, Florida

resulted from occupant-airbag interaction, bringing the total of serious and fatal injuries up to 79 cases. In all cases, the victim remained inside the vehicle during the crash event.

For each documented case of airbag-induced injuries, crash investigations were conducted, detailing the vehicle dynamics, occupant anthropometry, and resulting injuries based on autopsy and other medical reports. Any evidence of contact between the occupant and vehicle was carefully noted, such as tissue/fluid transfer, broken glass, or other structural damage to the occupant compartment. Each injury sustained by the occupant was matched to a particular contact within the vehicle. Final written reports have been completed for 41 of the 79 cases, which comprises the knowledge base upon which this paper was written.

CHILD TRAUMA RESULTING FROM AIRBAG INTERACTION

Thirty-eight cases involved children in the right front passenger seating position. Nine cases involved infants in rear-facing child seats and the remaining 29 cases involved forward-facing children, almost all of whom were either unrestrained or improperly restrained by the available seat belt system. Twenty-five written reports were reviewed and summarized for this paper. In all cases, the crash investigators have identified the passenger airbag and/or cover flap as the source of the injuries. Little to no intrusion of the occupant compartment was reported, with an average estimated change in velocity (delta-V) of 19.2 kph (12 mph) with a range from 11.2-33.6 kph (7-21 mph).

Among the crashes resulting in the 25 child cases reviewed, only one driver sustained any significant injuries (AIS > 1), mostly caused by submarining in a 2-point torso belt without the lap belt. This case was the most severe crash, with a delta-V of 33.6 kph (21 mph) in an offset frontal configuration to the left side of the vehicle. Seven drivers were identified as small females with heights not exceeding 1.60 meters (63 inches), who sustained only minor injuries consisting mostly of burns and contusions. Three of these small females were restrained by the available seat belt systems, while the other four were not.

Eight of the 25 reviewed child cases involved infants in rear-facing child safety seats; the other 17 cases involved forward-facing children. Thirteen of the 17 forward-facing children were not restrained by the available seat belt system. Three children were wearing only the lap belt portion of the available 3-point belt system, and one child was seated in a booster seat restrained by the available 3-point belt. Four of the eight rear-facing cases were fatalities resulting from brain injuries associated with skull fractures. All except for two of the forward-facing cases were fatalities resulting from either brain or cervical injuries. One of these nonfatal cases was the child seated in a booster seat, and the other was wearing only the lap belt portion of the available 3-point restraint. Injuries sustained for the reviewed child cases are summarized in Appendix A.

754

MECHANISMS OF INJURY FOR REAR-FACING INFANTS - All the infants were properly restrained in an appropriate rear-facing infant seat, and all but one of these seats were secured to the vehicle. In one case (CA9602), the child seat was not properly secured, which may have contributed to the severity of the injuries.

The crash scenario for airbag involvement with rear-facing infant seats is similar for all cases. During deployment, the passenger airbag or cover flap impacts the back of the rear-facing infant seat, typically with sufficient force to crack or break the plastic shell. The shell in turn accelerates rearward and impacts the child's head, causing skull fractures and associated subdural and subarachnoid hemorrhages. Skull fractures typically occur bilaterally in the parietal region. The most likely mechanism of these fractures is that, if the child is facing straight back, the infant seat back will impact the occipital region of the child's head. This load will compress the occipital bone inward, causing the parietal bones to bend outward. This bending of the skull appears to cause the bilateral fractures. Cervical injuries are difficult to diagnose in infants due to their immature skeleton and lax joints. In addition, most of the cases involved only a cursory pathologic evaluation in the upper cervical region since fatal brain injuries had already been documented. It is therefore possible that cervical injuries existed in these cases but were not reported.

MECHANISMS OF INJURY FOR FORWARD-FACING CHILDREN - Occupant kinematics for forward-facing children assumes the children are either unrestrained or improperly restrained by the available seat belt system. Almost all of the cases involved unrestrained children with pre-impact braking, which caused the child to move forward within close proximity of the stored airbag prior to deployment. It is not clear whether contact is made with the instrument panel prior to deployment.

The airbag deploys into the area of the upper chest, neck, and face of the child. Rapid translation and rotation of the skull causes a number of cervical spine injuries; most notable is a dislocation of the atlanto-occipital joint with contusion or laceration of the brain stem or rostral (upper) spinal cord. Other cervical injuries include C2/C3 fractures and subluxations resulting in fatal spinal cord injuries. Closed head injuries, such as subdural and subarachnoid hemorrhages and intraventricular bleeding, were common but skull fractures were typically not observed. Closed head injuries are consistent with large and rapid rotations of the head produced by a distributed force. Mandibular fractures and avulsed teeth have also been reported as a result of airbag or cover flap impact with the chin and face. Thoracic injuries such as lung contusions and atrial hemorrhages have also been reported.

PRE-DEPLOYMENT CHILD PASSENGER KINEMATICS - To determine the most likely child occupant position prior to airbag deployment, analytical simulations were conducted using the Madymo occupant modeling software. Models of the P-3 3-year-old and P-6 6-year-old child dummies were initially placed in typical seating positions for unrestrained children, and were subjected to pre-impact braking conditions. Standard occupant databases, developed, validated, and distributed by TNO-Madymo, were used for this study. The airbag was not inflated for these simulations since the purpose was to determine the occupant's position prior to airbag deployment. In real-world crashes, it is typically not possible to determine the initial seating position of the child occupant. Thus, these simulations can not be directly compared with any specific case, but are indicative only of the general motion of a typical child occupant.

Figure 1 illustrates the results of an analytical modeling simulation conducted to predict occupant kinematics for a 3-year-old child during pre-impact braking. The P3 child dummy was initially positioned on the passenger seat such that its legs were hanging over the seat ledge. A rectangular waveform with a peak magnitude of 0.7G was used to simulate hard-braking on dry pavement [Hitchcock, 1980; Kaleps, 1982]. Simulation results show that a 3-year-old child translates forward towards the instrument panel, while pivoting about the hip and waist. Due to its height and initial seating position, the 3-year-old child's head comes in contact with the instrument panel at approximately 320 msec after the initiation of braking while his buttocks remain in contact with the seat. This position is comparable to what is commonly referred to as ISO position 2 for a child [ISO, 1997].

Figure 2 illustrates the results of an analytical modeling simulation for a 6-year-old child during pre-impact braking. Similar initial positioning and braking were applied; however physical differences between the two child dummies resulted in different kinematic responses. The 6-year-old child sits up higher and has longer legs than the 3-year-old. The 6-year-old's feet contact the floor early in the event at approximately 30 msec, whereas the 3-year-old's feet contact the floor very late at over 200 msec. The 6-year-old's buttocks is initially further away from the seat ledge and sinks deeper into the seat cushion, providing a greater resistive force. The 6-year-old child translates forward towards the instrument panel and contacts the knee bolster at approximately 540 msec after the initiation of braking. The 6-year-old child does not pivot as much about the hip and waist due to increased flexural stiffness in the model. This results in the child's upper torso being flush against the instrument panel, which is comparable to what is commonly referred to as ISO position 1 for a child.

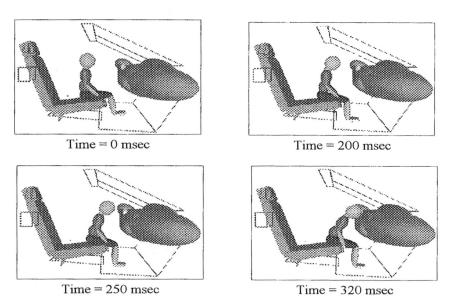

Figure 1. Pre-impact Braking Simulation with 3-Year-Old Child.

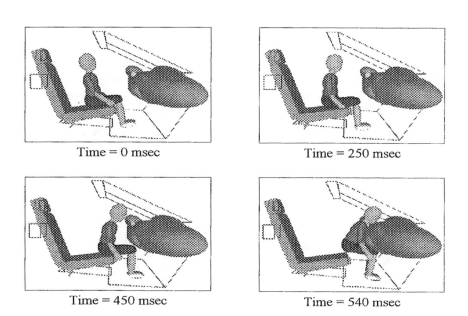

Figure 2. Pre-impact Braking Simulation with 6-Year-Old Child.

INJURY ASSESSMENT WITH ANTHROPOMORPHIC CHILD TEST DUMMIES - As has already been stated, the primary injury mechanisms differ between rear-facing and forward-facing child occupants. Rear-facing occupants typically sustain skull fractures resulting from a direct blow to the back of the skull. Relevant measurements related to these injuries are the translational accelerations of the skull. It must be noted that since the infant skull cannot be assumed rigid, the efficacy of the Head Injury Criterion (HIC) to predict these injuries is questionable. In fact, the suggested cause of fracture is the bending stresses created by skull deformation.

Forward-facing child occupants typically sustain atlanto-occipital dislocation with brain stem or spinal cord laceration. Skull fractures are typically not observed since the airbags are quite effective at distributing the load over a relatively large area of the child's body. Closed head injuries are most closely associated with rotational accelerations and velocities measured at the head center of gravity.

Newly developed neck injury criteria, referred to as N(ij), combine the effects of tension, shear, and extension moment measured at the occipital condyles, and are expected to provide a better predictor of craniocervical injuries. For the case of tension-extension injuries, the distractive load (F_{dist}) is defined as the resultant of the tensile (F_z) and shear (F_x) loads. This is normalized with respect to a critical value for distraction (F_{crit}). Extension moment (M_{ext}) is similarly normalized with respect to a critical value for extension (M_{crit}). Critical values for calculating the N(ij) are uniquely defined for each specific dummy. The normalized neck tension-extension criteria can be written as the sum of these two normalized loads.

$$N_{TE} = (F_{dist}/F_{crit}) + (M_{ext}/M_{crit}) \qquad (1)$$
$$N_{TE} \leq= 1 \qquad (2)$$

Table 1 shows Injury Assessment Reference Values (IARV) which have been used to evaluate injury risk [Klinich, 1996]. These IARV are preliminary values, not yet accepted by the automotive safety community, and as such some contradictions may exist between values for children and small female occupants.

For the forward-facing, unbelted 3-year-old child passenger following pre-impact braking, the most likely initial position at the time of airbag deployment is up against the instrument panel with the head positioned just above the airbag module. Experimental data for this initial position (ISO 2) was selected from a series of static airbag deployments conducted with the Hybrid III 3-year-old child dummy. The most likely initial position for a 6-year-old child at the time of deployment is up against the instrument panel with the chest positioned in front of the airbag module. Data for this initial position (ISO 1) was selected from a series of static airbag deployments conducted with the Hybrid III 6-year-old child dummy. For both child dummies, an additional single axis accelerometer was mounted on the skull in the mid-sagittal plane to enable calculation of rotational acceleration and velocity. Maximum dummy measurements recorded during these tests are shown in Table 1 along with the recommended IARV.

Table 1. Static Out-of-Position Airbag Deployment Test Data.

	Head		Neck			Chest
	HIC	Ang. Vel. (rad/s)	Tension (N)	Ext. Moment (Nm)	N(ij)	Chest Res. Acc. (g's)
Hybrid III 3-Year-Old Child Dummy						
IARV*	1000	34	2500	30	1.0	60
Max. Measured Value	3511	122	2999	111	4.77	103
Hybrid III 6-Year-Old Child Dummy						
IARV*	1000	33	3000	35	1.0	60
Max. Measured Value	2470	62	8596	175	6.85	121

* Preliminary values only; not accepted by automotive safety community.

HIC and head angular velocities exceeded the preliminary IARV by more than a factor of 3 for the 3-year-old (double for the 6-year-old), showing a high risk potential for head injuries. Neck injury measurements, such as tension load, extension moment, and N(ij), greatly exceeded the IARV, showing an extremely high potential for cervical injuries. Weak muscle development and lax ligaments in the cervical region makes the child susceptible to atlanto-occipital dislocation with spinal cord laceration, which was found to be the leading cause of death among the reviewed cases. Thoracic injury measurements were roughly double the recommended IARV, showing a high risk of thoracic injury. Although rib fractures were not reported, possibly due to the compliance of the child's rib cage, contusions of the heart and lungs were common. Mandibular fractures and avulsed teeth were also quite common for the forward-facing child passengers.

ADULT TRAUMA RESULTING FROM AIRBAG INTERACTION

As of April 15, 1997 the NHTSA has identified 22 cases in which the driver's airbag has been attributed with causing fatal injuries during deployment. These cases often involve short statured women who are sitting in close proximity to the steering wheel and airbag module. In addition to the 22 drivers, three adult passengers have also been fatally injured by a deploying passenger airbag. Sixteen of the 25 adult fatality cases have been reviewed.

Airbag related injuries in adults can be associated with three primary injury mechanisms. The first involves airbag contact with the face or chin of the driver causing basilar skull fracture with associated brainstem lacerations and/or subdural and subarachnoid hemorrhages. The second mechanism involves multiple rib fracture, usually bilateral, with associated lacerations of the underlying thoracic and abdominal organs. Frequently lacerated organs include the heart, spleen, liver, and aorta. The third injury mechanism is not as common as the first two, but involves cardiac and pulmonary contusions and hemorrhages without rib fractures. A summary of the injuries sustained in the reviewed adult cases is presented in Appendix B.

Internal injuries sustained by occupants with airbag restraints are often difficult to diagnose with the absence of external indications. Liver lacerations are a good example of occult injuries that may be present as a result of airbag interaction. To assist in recognizing the extent of injuries to occupants protected by airbags, it is suggested that additional evidence from the crash scene be used in the triage criteria. Potentially important crash parameters include steering wheel deformation, amount of external vehicle deformation and/or intrusion of internal components, seating position, and seat belt use, particularly if the occupant is in close proximity to the airbag as it deploys. The presence of steering wheel deformation in a low-speed crash with little to no external vehicle deformation, for example, may be a sufficient warning to justify transporting the occupant to a level 1 or 2 trauma center so that a close examination for occult injuries can be made [Augenstein, 1994].

PRE-DEPLOYMENT DRIVER KINEMATICS - In order to predict the most likely driver position prior to the initiation of air bag deployment, analytical crash reconstructions were conducted. Since many of the fatally injured drivers were unbelted and short statured, an unbelted 5th percentile female driver in a forward seating position was chosen for these simulations. A rectangular waveform with a peak magnitude of 0.7G was again used to simulate hard-braking on dry pavement.

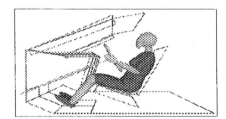 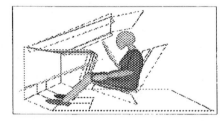

Figure 3. Pre-impact Braking Simulation with Unbelted, 5th Percentile Female Driver.

Figure 3 illustrates the initial and final positions of a small female driver subjected to pre-impact braking. The driver translates forward until contact is made with the knee bolster and lower rim of the steering wheel. She then rotates about the waist, placing her head in close proximity to the air bag module. This position is comparable to what is commonly referred to as ISO position 1 for the adult driver [ISO, 1997].

INJURY ASSESSMENT USING ANTHROPOMORPHIC SMALL FEMALE TEST DUMMIES - As has already been stated, there are several primary injuries that cause adult driver fatalities. Basilar skull fractures and upper cervical injuries are most closely associated with neck tension loads and extension moments calculated at the occipital condyles (mechanism 1) [Hopper, 1994; Myers, 1995]. Also important to this crash scenario are closed head injuries, which can be related to large and rapid rotations of the head. Rotational accelerations and velocities measured at the head center of gravity are most closely associated with these injuries. Thoracic injuries can be associated with the Viscous Criterion (V*C), which gives an indication of thoracic injury risk resulting from a wide range of loading rates. Slower loading rates with large rib displacement can cause fractures with associated soft tissue laceration (mechanism 2). Very high loading rates without large rib displacement can cause thoracic internal organ injury in the absence of rib fracture (mechanism 3). Table 2 shows Injury Assessment Reference Values (IARV) for the small female dummy which have been used to evaluate injury risk [Mertz, 1994].

Table 2. Out-of-Position Test Data Using Hybrid-III 5th Percentile Female Dummy.

	Head		Neck			Chest	
	HIC	Ang. Vel. (rad/s)	Ten-sion (N)	Ext. Mom. (Nm)	N(ij)	Chest Res. Acc. (g's)	V*C (m/s)
IARV*	1113	30	2201	31	1.0	73	1.0
Max. Measured Value	189	NA	2759	84	6.1	61	2.3

* Preliminary values only; not accepted by automotive safety community.

Small statured females are considered to be at greatest risk of airbag induced injuries for two reasons. First, being small statured, they tend to sit relatively close to the steering wheel and stored airbag. Pre-impact braking will bring them even closer to the airbag module, especially if they are unbelted. Second, women in general have more slender bones with lower density than men, and therefore have a lower tolerance to applied external forces. Combining their proximity to the airbag module with lower bone strength makes small females more susceptible to airbag loading than other adults. It is important to realize, however, that the cases discussed in this paper represent a very small percentage of the total number of airbag deployments.

An unbelted small female driver, following pre-impact braking, is likely to be positioned very close to the steering wheel with her chin just above the airbag module. Experimental data for this position (adult ISO 1) was selected from a series of static airbag deployments conducted with the Hybrid III 5th-percentile small female dummy. Maximum dummy measurements recorded during these tests are presented in Table 2.

HIC measurements were very low for the adult dummy out-of-position tests, which might explain the absence of calvarial fractures in real world airbag deployments. Subdural and subarachnoid hemorrhages have been documented in these cases, which are typically associated with the rotation of the head. Unfortunately, dummy head rotational accelerations and velocities were not measured in these experiments; thus a high injury risk could not be verified for this loading scenario. Maximum tension loads were only slightly above the recommended IARV, but the maximum extension moments were more than double the IARV. The combined effect of tension and extension loading, as presented by the N(ij) parameter, would predict a high risk of upper cervical injury or basilar skull fractures. Compared to the immature pediatric cervical spine, the musculature and ligaments of the adult cervical region are quite strong. In fact, the ligaments and muscles surrounding the foramen magnum can be stronger than the skull base itself Tensile loading of the cervical region in the adult can often result in basilar skull fracture, rather than the atlanto-occipital dislocation without fracture as seen in children [Hopper, 1994; Sances, 1981]. Viscous Criterion measurements in the thorax were more than double the IARV, which would indicate a high risk of thoracic injury. This injury could present itself as either rib fractures with associated organ laceration, or as internal organ ruptures or contusions in the absence of rib fractures.

CONCLUSIONS

Pre-impact braking, combined with lack of belt usage, can place both drivers and child passengers at increased risk of sustaining airbag induced injuries. Although this paper summarizes 63 documented cases of airbag induced fatalities, it should be clearly understood that this represents a very small percentage of all airbag deployments. Preliminary estimates indicate that less than 4 fatalities occur for every 100,000 airbag deployments. In contrast to the 63 fatalities, an estimated 1900 lives have been saved by airbags over roughly the same time period.

Experimental static out-of-position airbag deployment tests using anthropomorphic test dummies were conducted using initial positions predicted by analytical modeling. Measurements taken from the dummy's head, neck, and chest provide scientific evidence that airbags can impart considerable loads to an occupant positioned within the airbag deployment region. Neck loads, in particular, were found to greatly exceed recommended tolerance levels.

Among the cases examined, injury patterns were identified for various occupants. Infants in rear-facing child seats typically sustain bilateral skull fractures with associated subdural and subarachnoid hemorrhages. Forward-facing children typically sustain upper cervical and closed head injuries. Atlanto-occipital dislocations with cord transection, as well as subdural and subarachnoid hemorrhages, were commonly reported. Driver injuries typically include basilar skull fractures with associated brainstem lacerations and/or subdural and subarachnoid hemorrhages. Other common driver injuries include multiple rib fractures with associated thoracic or abdominal organ lacerations.

ACKNOWLEDGMENTS

The authors would like to express their thanks to a number of individuals who have greatly contributed to this paper through the highquality professional performance of their normal work assignments. We would like to thank the members of NHTSA's Special Crash Investigation Program, especially Lee Franklin and Chip Chidester, for collecting and providing the crash investigation data used for this study. We would also like to thank Glen Rains of NHTSA's Vehicle Research and Test Center for providing the analyses of occupant responses from the out-of-position airbag deployment tests conducted at this facility.

REFERENCES

Augenstein J, et al. Injuries sustained by air bag occupants in frontal crashes. Proceedings of the Fourteenth International Technical Conference on Enhanced Safety of Vehicles, pp. 641-648, 1994.

Hitchcock RJ and Nash CE. Protection of children and adults in crashes of cars with automobile restraints. 8th International Technical Conference on Experimental Safety Vehicles, Wolfsburg, 1980.

Hopper RH, McElhaney JH, and Myers BS. Mandibular and Basilar Skull Fracture Tolerance. Proceedings of the 38th Stapp Car Crash Conference, SAE Paper No. 942213, pp. 123-32, 1994.

ISO. Road vehicles - Test procedures for evaluating out-of-position vehicle occupant interactions with deploying air bags. International Standards Organization, Draft Technical Report 10982, 1997.

Kaleps I and Marcus JH. Predictions of child motion during panic braking and impact. SAE Paper No. 821166, 1982.

Klinich KD, et al. Techniques for developing child dummy protection reference values. National Highway Traffic Safety Administration Docket Report #74-14 Notice 97 Item 69, 1996.

Mertz H. Anthropomorphic test devices. In Hybrid III: The First Human-Like Crash Test Dummy. Society of Automotive Engineers, Warrendale, PA, SAE-PT44, pp. 387-405, 1994.

Myers BS and Winkelstein BA. Epidemiology, classification, mechanism, and tolerance of human cervical spine injuries. Critical Reviews in Biomedical Engineering, vol. 23, issues 5&6, pp. 307-409, 1995.

Sances A, et al. Experimental studies of brain and neck injury. SAE Paper No. 811032, 1981.

APPENDIX A

Summary of Airbag-Related Trauma Cases in Children

Case #*, (Model)	Age, Sex (Ht, Wt)	Restraint System	Major Injuries
Rear-Facing Infants in Right Front Passenger Seat			
DS9423 ('94 Corolla)	3 mos., female (20 in, 13 lbs) (0.51 m, 5.9 kg)	Rear-facing infant seat	- bilateral parietal skull fractures, - frontal and bilateral temporal subdural hematomas, - bilateral parietal and temporal lobe contusions
CA9516 ('95 Escort)	22 days, female (24 in, 11 lbs) (0.61 m, 5 kg)	Rear-facing infant seat	- bilateral parietal and occipital skull fracture - brain stem contusion - massive subdural and subarachnoid hemorrhage - subgaleal hematoma - basilar skull fracture
DS9519 ('94 Camry)	5 mos., female (27 in, 20 lbs) (0.69 m, 9.1 kg)	Rear-facing infant seat	- left temporo-occipital and right occipital skull fractures - subdural and subarachnoid hemorrhages - subgaleal hematoma
CA9522 ('95 Saturn)	4 mos., male (24 in, 22 lbs) (0.61 m, 10 kg)	Rear-facing infant seat	- severe brain edema - tonsillar herniation - subdural and epidural hemorrhage - right middle fossa fracture thru sphenoid wing and floor - depressed right parietal and temporal skull fractures with brain matter extruding from the fracture sites - comminuted left parietal skull fracture - left subdural hemorrhage - rt. temporal/frontal lobe brain laceration - left temporal and right occipital lobe contusions - bilateral hemothorax
DS9522 ('95 Escort)	6 mos , male (24 in, 22 lbs) (0.61 m, 10 kg)	Rear-facing infant seat	- 4 mm linear left parietal skull fracture - possible small subdural hematoma
DS9525 ('94 Aspire)	3 mos., female (25 in, 15 lbs) (0.64 m, 6.8 kg)	Rear-facing infant seat	- occipital skull fracture - subdural hematoma
IN9521 ('96 Caravan)	7 weeks, male (22 in, 9 lbs) (0.56 m, 4.1 kg)	Rear-facing infant seat	- bilateral nondisplaced parietal skull fractures - bilateral subdural hematomas - right occipital white matter injury - right lateral intraventricular hemorrhage - frontal subarachnoid hemorrhage
CA9602 ('95 Isuzu Trooper)	4 mos., male (24 in, 20 lbs) (0.61m, 9.1 kg)	Rear-facing infant seat (not attached properly)	- basilar skull fracture - epidural, subdural, and subarachnoid hemorrhage
Forward-Facing Children in Right Front Passenger Seat			
CA9307 ('93 Volvo 850)	6 years, female (44 in, 51 lbs) (1.12 m, 23.2 kg)	Unrestrained	- uncal and cerebellar tonsillar herniation - pontine contusion - frontal and superior sagittal subgaleal hematoma - subarachnoid hemorrhage
DS9420 ('94 Caravan)	4 years, female (41 in, 35 lbs) (1.04 m, 15.9 kg)	Unrestrained	- left side depressed and open basilar skull fracture - two avulsed teeth
CA9443R ('94 Mustang)	4 years, male (44 in, 54 lbs) (1.12 m, 24.5 kg)	Unrestrained	- C2/C3 cervical subluxation - NO MEDICAL REPORT AVAILABLE
IN9508 ('95 Voyager)	9 years, male (55 in, 65 lbs) (1.40 m, 29 5 kg)	Lap belt only	- atlanto-occipital dislocation - bilateral subdural and subarachnoid hemorrhages - diffuse frontal subgaleal hemorrhage

* Underlined case numbers are fatalities.

APPENDIX A (cont.)

CA9515R ('93 Lexus)	7 years, female	Unrestrained	- atlanto-occipital dislocation - C2/C3 cervical fracture with spinal cord contusion - mandibular fracture, avulsed teeth - atrial hemorrhage with lung contusions
CA9520 ('94 Camaro)	5 years, male	Unrestrained	- subgaleal hemorrhages under left side of scalp - 9 cm coronal suture diastasis - bilateral cranial vault compression injury - cortical brain hemorrhages - subdural and subarachnoid hemorrhage - compressed brain ventricles - disruption of spinal cord just below medulla - complete separation of intervertebral disc at C2/C3 level - C1/C2 dislocation with spinal cord involvement - stretching laceration of inferior vena cava
CA9521 ('95 Caravan)	7 years, female (51 in, 55 lbs) (1 30 m, 25.0 kg)	Lap belt only	- intraventricular bleeding at fourth ventricle - subarachnoid hemorrhage - probable C2 fracture with C2/C3 subluxation
IN9518 ('95 Jaguar XJS)	3 years, male (41 in, 37 lbs) (1 04 m, 16 8 kg)	Booster seat w/ 3-point belt	- brain stem shear injury - bilateral cerebral shearing of cortical gray-white junctions - right temporal lobe contusion - rt temporal lobe subdural and subarachnoid hemorrhage - diffuse right brain swelling - concussion and coma - right nasal bone fracture and unstable mandible
IN9520R ('95 Contour)	5 years, female (42 in, 45 lbs) (1.07 m, 20.5 kg)	Unrestrained	- brain stem laceration at pontomedullary junction - atlanto-occipital dislocation - dens fracture, C1/C2 dislocation, transection of spinal cord - bilateral subarachnoid hemorrhage and cerebral contusions - mandibular fracture - liver laceration and lung contusion
CA9523R ('93 Intrepid)	5 years, female (45 in, 45 lbs) (1.14 m, 20.5 kg)	Unrestrained	- closed head injury with lack of cerebral blood flow - severe bilateral cerebral edema - bilateral pulmonary contusions
ID9501R ('95 Camry)	6 years, female	Unrestrained	- fatal closed head injuries - NO AUTOPSY WAS PERFORMED
CA9601 ('95 Caravan)	9 years, male (54 in, 90 lbs) (1 37 m, 40 9 kg)	Unrestrained	- fracture dislocation of C1 with spinal cord transection - right petrous temporal bone fracture - subarachnoid and epidural hemorrhages - sternal fracture - left pulmonary contusion
DS9605 ('95 Metro)	3 years, female (39 in, 29 lbs) (0 99 m, 13.2 kg)	Unrestrained (sitting on lap of RF pass.)	- comminuted, depressed left frontal bone skull fracture with associated cortical contusions - subarachnoid and subgaleal hemorrhage - coronal suture diastasis - C4 vertebral fracture with spinal cord contusion - atlanto-occipital subluxation with cord compression - contusions of right lung, heart, and thymus
DS9609 ('95 Grand Prix)	4 years, male (43 in, 40 lbs) (1.09 m, 18.2 kg)	Lap belt only	- subdural hematoma - right humeral fracture
IN9618 ('95 Lumina)	5 years, male (? in, 42 lbs) (? m, 19 1 kg)	Unrestrained	- atlanto-occipital dislocation - concussion - avulsed teeth
IN9619 ('95 Caravan)	4 years, male (48 in, 50 lbs) (1.22 m, 22.7 kg)	Unrestrained	- concussion - upper cervical injury - presumed basilar skull fracture (hemotympanum)
IN9624(R) ('96 Neon)	6 years, male (50 in, 54 lbs) (1 27 m, 24.5 kg)	Unrestrained	- upper cervical fracture dislocation with severed spinal cord - subarachnoid hemorrhage, concussion, and brain edema - bilateral temporomandibular fracture dislocation

* Underlined case numbers are fatalities

APPENDIX B

Summary of Airbag-Related Trauma Cases in Adults

Adults in Driver's Seat				
Case #* **(Vehicle)**	**Age, Sex** **(Height)**	**Belt** **Usage**	**Injury** **Mecha-** **nism******	**Major Injuries**
CA9109 ('91 Taurus)	79 yr., female (62 inches) (1.57 m)	yes mis-use	2	- bilateral rib fractures w/ hemothoraces - anterior pericardial and right atrial laceration - small splenic laceration
CA9112 ('90 Shadow)	36 yr., female (64 inches) (1.63 m)	yes slumped	1 2	- extensive subarachnoid hemorrhage - bilateral rib fractures - splenic rupture - abdominal aortic rupture - cardiac contusion
NC9208R ('90 Taurus)	64 yr., female (56 inches) (1.42 m)	?	2	- bilateral rib fractures w/ hemothoraces - bilateral pulmonary contusions - aortic tear - thyroid cartilage fracture
CA9308 ('91 Corsica)	37 yr., female (62 inches) (1.57 m)	none	1	- brainstem laceration - basilar skull fracture propagating into occipital bone
CA9309 ('90 Taurus)	71 yr., female (62 inches) (1.57 m)	none	1 2	- brainstem laceration - subdural and subarachnoid hemorrhage over cerebellum - multiple bilateral rib fractures - aortic and pericardial laceration - liver laceration - mesenteric laceration - cervical vertebral fracture - right shoulder fracture
CA9405R ('92 Corsica)	76 yr., female (62 inches) (1.57 m)	none	1 2	- ring-type basilar skull fracture - complete atlanto-occipital dislocation - multiple bilateral rib fractures
DS9422 ('92 Corsica)	74 yr., female (59 inches) (1.50 m)	yes	3	- cardiac and ascending aortic rupture - cardiac contusion
IN9505 ('90 Eldorado)	51 yr., female (64 inches) (1.63 m)	none	1 2	- C1/C2 cervical fracture dislocation - brainstem laceration - subarachnoid hemorrhage - sternal and multiple rib fractures - aortic and liver lacerations
IN9506 ('90 Continental)	38 yr., female (62 inches) (1.57 m)	none	1 2	- hinge-type basilar skull fracture involving bilateral temporal bones - transverse left parietal and occipital bone fractures - medullo-pontine transection - bilateral internal carotid artery laceration - left rib fractures - cardiac laceration - mediastinal contusion - bilateral pulmonary contusions
DS9523 ('89 Daytona)	17 yr., female (59 inches) (1.5 m)	none	1 3	- basilar skull fracture thru cavernous sinuses between petrous bones - intrapulmonary hemorrhage
IN9519R ('90 Taurus)	36 yr., female (65 inches) (1.65 m)	none	3	- cardiac lacerations thru right ventricle - pulmonary contusion - concussion

* Underlined case numbers are fatalities.
** Injury Mechanism =   (1) **basilar skull fracture with associated brain injuries**
                                     (2) **rib fracture with associated thoracic and abdominal injury**
                                       (3) **thoracic and abdominal injury without rib fracture**

APPENDIX B (cont.)

IN9609R ('94 Camry)	46 yr , female (63 inches) (1 60 m)	none	2 3	- concussion - atlanto-occipital dislocation with 1 cm of distraction - cardiac arrest - liver contusion
CA9110 ('91 Firebird)	46 yr., male (72 inches) (1.83 m)	yes slumped	2	- bilateral rib fractures - myocardial contusion
NC9307R ('91 Corsica)	78 yr., male (70 inches) (1.78 m)	?	2	- multiple bilateral rib fractures - cardiac laceration
CA9502 ('94 F150)	56 yr., male (70 inches) (1 78 m)	none	2	- sternal and bilateral rib fractures - aortic transection - liver laceration - pulmonary contusion
Adults in Right Front Passenger Seat				
IN9608 ('94 Caravan)	98 yr , female (63 inches) (1.60 m)	yes	1	- right subdural hematoma w/ brainstem herniation - rt. frontal cerebral contusion w/ edema and asymmetric ventricles - open comminuted right distal ulnar and radial fractures - ulnar nerve and artery injuries

* **Underlined case numbers are fatalities.**
** **Injury Mechanism = (1) basilar skull fracture with associated brain injuries**
(2) rib fracture with associated thoracic and abdominal injury
(3) thoracic and abdominal injury without rib fracture

Driver Airbag Effectiveness by Severity of the Crash

American Journal of Public Health, October 2000, Volume 90, Number 10.

Copyright, 2000, American Public Health Association

Driver Air Bag Effectiveness by Severity of the Crash

Maria Segui-Gomez, MD, ScD

ABSTRACT

Objectives. This analysis provided effectiveness estimates of the driver-side air bag while controlling for severity of the crash and other potential confounders.

Methods. Data were from the National Automotive Sampling System (1993–1996). Injury severity was described on the basis of the Abbreviated Injury Scale, Injury Severity Score, Functional Capacity Index, and survival. Ordinal, linear, and logistic multivariate regression methods were used.

Results. Air bag deployment in frontal or near-frontal crashes decreases the probability of having severe and fatal injuries (e.g., Abbreviated Injury Scale score of 4–6), including those causing a long-lasting high degree of functional limitation. However, air bag deployment in low-severity crashes increases the probability that a driver (particularly a woman) will sustain injuries of Abbreviated Injury Scale level 1 to 3. Air bag deployment exerts a net injurious effect in low-severity crashes and a net protective effect in high-severity crashes. The level of crash severity at which air bags are protective is higher for female than for male drivers.

Conclusions. Air bag improvement should minimize the injuries induced by their deployment. One possibility is to raise their deployment level so that they deploy only in more severe crashes. (*Am J Public Health.* 2000;90:1575–1581)

Air bags have existed for more than 40 years and have always been surrounded by controversy.[1,2] Unlike most safety devices, a deploying air bag increases the amount of energy being released during the crash and, hence, potentially increases the frequency and severity of injuries sustained by the driver.[3,4] Frontal driver-side air bags, which have been mandated in all new passenger vehicles sold in the United States since model year 1997,[5] were available in many automobile makes and models before regulation took effect. (Note that this article refers to frontal driver-side air bags only, not to frontal passenger-side air bags or to side air bags for either the driver or any other occupant.) In fact, in 1999, almost half of all passenger cars in the US fleet were equipped with a frontal driver-side air bag (National Highway Traffic Safety Administration, written communication, May 1998), and over the next 15 years or so, this ratio will increase to nearly 100%.

Air bags (also called *supplemental restraint systems*) are supposed to protect the occupants in combination with lap and shoulder safety belts. Yet, the air bag systems sold in the United States were designed to meet a federal performance standard that requires that in an experimental frontal crash at approximately 48 km/h, the forces on an unbelted 50th percentile male dummy's head, chest, and thighs do not exceed a specified level.[5] A typical air bag system consists of one or more sensors that detect the longitudinal velocity change of the vehicle during the crash, an electronic unit that monitors the system, and a module that houses the inflator and the bag,[6] but the design varies across makes, models, and years.[7]

Since the early 1970s, researchers have evaluated the performance of air bag systems. Most evaluations have focused on the air bag's effect on preventing fatalities. Initial estimates of the percentage reduction in fatalities due to air bags plus safety belts, based on expert judgment or experimental data or both, ranged from 18% to 55%.[5,8–15] More recent effectiveness estimates, computed with real-world fatal crash data (and in some cases computed with the double pair comparison method[16]), suggest an approximately 19% reduction in fatality risk among belted drivers.[17–19] The only estimates available to date regarding air bag effectiveness in nonfatal injuries indicate reductions in moderate to serious nonfatal injuries to the head, face, and upper torso among occupants in frontal crashes,[18,20–23] but the magnitude of these benefits varies greatly from study to study. More complete reviews of the reported effectiveness estimates are available elsewhere.[24,25]

In contrast to these encouraging findings, several studies have linked air bags to the causation of injuries. To date, air bags have been linked to the deaths of 57 drivers (National Highway Traffic Safety Association, National Center for Statistics and Analysis [NCSA] online. Available at: http://www.nhtsa.dot.gov. Accessed December 1, 1999) and to numerous nonfatal injuries of varying degrees of severity, including corneal abrasions, aortic rupture, lung contusions, abdominal injuries, and open fractures of the forearm.[26–35]

It has been suggested that air bag–related injuries may be associated with specific design features, such as the amount of energy released by the deploying air bag; the speed of inflation; and the volume, shape, or folding pattern of the bag.[36,37] It has also been suggested that these air bag–induced injuries are more likely in female drivers.[21] In addition, air bag–induced injuries are the most serious injuries reported in relatively low-speed crashes,[22]

At the time of the study, Maria Segui-Gomez was with the Risk Analysis and Injury Control and Research Center, Harvard School of Public Health, Boston, Mass.

Requests for reprints should be sent to Maria Segui-Gomez, MD, ScD, Center for Injury Research and Policy, Johns Hopkins School of Hygiene and Public Health, 624 N Broadway St, Room 543, Baltimore, MD 21205-1996 (e-mail: mseguigo@jhsph.edu).

This article was accepted February 3, 2000.

TABLE 1—Selected Driver, Vehicle, and Crash Characteristics in Frontal and Near-Frontal Crashes With Known Longitudinal Delta V: National Automotive Sampling System Crashworthiness Data System Calendar Years 1993–1996, Passenger Cars of Model Years 1986–1997

	All (N = 5003), %	Air Bag Deployed (n = 1095), %	Air Bag Not Deployed (n = 3908), %
Driver			
Age, y			
13–24	30.6	28.5	31.1
25–54	51.8	53.5	51.3
55–64	6.2	7.1	6.0
≥65	10.8	10.2	11.0
Missing	0.6	0.7	0.6
Sex			
Male	49.3	50.9	48.8
Female	50.6	49.0	51.0
Missing	0.1	0.1	0.2
Injury severity[a]			
MAIS			
0	40.4	37.0	41.4
1	32.9	34.4	32.5
2	11.3	12.2	11.1
3	8.4	10.9	7.7
4	1.7	1.5	1.8
5	1.1	1.2	1.0
6	4.2	2.8	4.5
ISS			
0	40.4	37.0	41.4
1–3	32.9	34.4	32.5
4–8	9.3	10.6	8.8
9–15	8.1	10.0	7.6
16–24	3.1	2.9	3.1
25–75	6.2	5.1	6.6
MFCI			
0	83.7	79.6	84.9
1–20	4.1	6.5	3.4
21–40	2.8	3.9	2.5
41–60	3.7	6.2	3.0
61–80	1.1	0.6	1.2
81–100	4.6	3.2	5.0
Vehicle			
Air bag present	34.0	100.0	15.5
Crash			
Air bag deployed	21.9	100.0	0.0
Delta V, km/h			
<12	17.5	10.6	19.4
12–23	42.8	45.6	42.0
24–31	19.2	21.5	18.6
32–39	9.7	10.8	9.4
≥40	10.8	11.5	10.6
Seat belt used			
Yes	66.0	69.9	64.9
Missing	5.6	5.5	5.7

Note. MAIS = Maximum Abbreviated Injury Scale; ISS = Injury Severity Score; MFCI = Maximum Functional Capacity Index.

[a] Injuries selected for analyses include up to 14 of the most severe injuries per driver reported from autopsy, hospital, emergency room, and medical records. Fatalities were considered as MAIS = 6, ISS = 75, and MFCI = 100 (excludes drivers who died of causes unrelated to the crash). Drivers with no injuries were coded as MAIS = 0, ISS = 0, and MFCI = 0.

which raises the question of whether these air bags are deploying unnecessarily in some crashes.

When is the crash serious enough to warrant air bag deployment? Current air bags are designed to deploy at crashes between 12 and 26 km/h, although the precise threshold varies by make, model, and, in some cases, the restraint use of the occupant.[7] Lower deployment levels imply higher air bag deployment rates, which have been associated with a higher incidence of air bag–related injuries.[38,39] In low-speed crashes, the injuries induced by the deploying air bag may be more serious than injuries that would otherwise have occurred, whereas in higher-speed crashes, air bag deployment may actually prevent the driver from sustaining more severe injuries.

The goal of this analysis is to provide net effectiveness estimates of the driver-side air bag in preventing fatal and nonfatal injuries in frontal and near-frontal crashes by severity of the crash, while controlling for characteristics known to influence the frequency and severity of injuries, such as age and sex of the driver,[17,40–43] vehicle size and mass,[44,45] and safety belt use.[46–48]

Methods

Outcome Measures

The effectiveness of air bags in reducing injury severity was evaluated with the Abbreviated Injury Scale, the Injury Severity Score, the Functional Capacity Index, and survival.

The Abbreviated Injury Scale is a consensus-derived, anatomically based system that classifies each injury on an ordinal scale that ranges from 1 (minor injury) to 6 (virtually unsurvivable).[49] The most severe injury sustained by an occupant is referred to as the maximum Abbreviated Injury Scale score. Drivers who died were assigned a maximum Abbreviated Injury Scale score of 6, and those with no injuries were assigned a maximum Abbreviated Injury Scale score of 0.

The Injury Severity Score constitutes the most common method of computing overall severity when a patient has multiple injuries. It is defined as the sum of the squares of the highest Abbreviated Injury Scale scores in 3 different body regions (or is assigned the value of 75 if at least 1 Abbreviated Injury Scale score of 6 is reported).[50] Thus, the Injury Severity Score is an ordinal scale that ranges from 0 to 75, although it is often treated by data analysts as a continuous scale.

The Functional Capacity Index is the first preference-based multiattribute score system that reflects the predicted extent of functional limitation 1 year postinjury.[51] Each injury has its own Functional Capacity Index. The Functional Capacity Index is a continuous score that ranges from 0 (no limitation) to 100 (maximum limitation). For our analyses, patients with no injuries were coded as having a Functional Capacity Index of 0, fatalities were coded as a Functional Capacity Index of 100, and the remaining patients were classified by the functional limitation of the injury leading to the maximum (i.e., worst) functional limitation.

The scores from these severity measures generated the outcome (or dependent) vari-

Driver Air Bag Effectiveness

TABLE 2—Multivariate Regression Models—Adjusted [a] Air Bag Deployment–Related Coefficients, Variances, and Covariances in Frontal and Near-Frontal Crashes With Known Longitudinal Delta V (n = 4697): National Automotive Sampling System Crashworthiness Data System Calendar Years 1993–1996, Passenger Cars of Model Years 1986–1997

Regression Technique	Outcome Variable	Coefficients			Variances			Covariances		
		Air Bag Deployment (Main Term)	Air Bag Deployment × Sex (Interaction Term)	Air Bag Deployment × Delta V (Interaction Term)	Air Bag Deployment	Air Bag Deployment × Sex	Air Bag Deployment × Delta V	Air Bag Deployment, and Air Bag Deployment × Sex	Air Bag Deployment and Air Bag Deployment × Delta V	Air Bag Deployment × Sex and Air Bag Deployment × Delta V
Logistic	Fatality	0.54	NA	NA	0.051	...	...	...	...	...
Ordinal	3-level MAIS	0.56	0.47	0.01	0.022	0.017	0.00003	0.0069	0.0006	0.00005
	7-level MAIS	0.58	0.44	0.01	0.027	0.020	0.00003	0.0089	0.0007	0.00005
Linear	ISS	2.83	2.29	0.10	1.220	0.955	0.0014	0.423	0.032	0.0023
	MFCI	5.32	4.18	0.13	2.875	2.251	0.0032	0.996	0.075	0.0055

Note. Three-level Maximum Abbreviated Injury Scale (MAIS) categorizes injuries into "none" (MAIS=0), "minor" (MAIS=1–3), and "severe" (MAIS=4–6); 7-level MAIS uses the entire Abbreviated Injury Scale (AIS) for severity categorization (i.e., 0, 1, 2, 3, 4, 5, and 6). ISS=Injury Severity Score; MFCI=Maximum Functional Capacity Index. Injuries selected for analyses include up to 14 injuries with the highest AIS reported from autopsy, hospital, emergency room, and medical records. Fatalities were considered as MAIS=6, ISS=75, and MFCI=100 (excludes drivers who died of causes unrelated to the crash). Drivers with no injuries were considered as MAIS=0, ISS=0, and MFCI=0. NA=not applicable (i.e., the term did not reach statistical significance).

[a] Controlling for longitudinal Delta V, safety belt use, age, and sex of driver.

---

ables: a dichotomous variable indicating whether the driver died (i.e., maximum Abbreviated Injury Scale score of 6), 2 ordinal variables indicating the maximum Abbreviated Injury Scale level of the driver, and 2 continuous variables with the Injury Severity Score and the maximum Functional Capacity Index. The 2 maximum Abbreviated Injury Scale–related ordinal variables were a 3-level maximum Abbreviated Injury Scale that categorized the drivers' injuries into "none" (Maximum Abbreviated Injury Scale score of 0); "minor" (maximum Abbreviated Injury Scale score of 1, 2, or 3); and "severe" (Maximum Abbreviated Injury Scale score of 4 or 6) and a 7-level maximum Abbreviated Injury Scale (i.e., 0, 1, 2, 3, 4, 5, and 6). The use of these ordinal variables allowed for a more comprehensive evaluation of the air bag effect on each injury severity level.

Data

The National Automotive Sampling System Crashworthiness Data System (NASS CDS), formerly the National Accident Sampling System, for calendar years 1993 through 1996 was used. The NASS CDS is a stratified sample of police-reported crashes involving passenger vehicles in which at least 1 of the vehicles is towed away from the scene because of damage from the crash. Trained crash investigators complete an extensive question-naire of data elements that describe the crash, the vehicles, the occupants, and their injuries. About 5000 crashes are investigated per year.

Inclusion criteria for this analysis entailed being the operator of a passenger car of model years 1986 through 1997 in a frontal or near-frontal crash during 1993 through 1996 for which the severity of the crash was known. For each driver, we analyzed injuries reported in the autopsy, the hospital, the emergency room, or the medical records. Some drivers were not injured, whereas others had 1 or more injuries. For each driver, we included up to 14 injuries with the highest Abbreviated Injury Scale scores (these 14 injuries accounted for 97% of all reported injuries and included the most severe ones). Drivers who died because of causes not related to the crash were not included.

The analyzed NASS CDS data included driver characteristics (age, sex, and height), crash consequences (number, type, and severity of the injuries), crash circumstances (severity of the crash, direction of crash, air bag deployment, safety belt use), and vehicle characteristics (air bag presence, wheelbase, curb weight, and the Vehicle Identification Number [VIN]). The severity of the crash was defined on the basis of the longitudinal component of the maximum velocity change incurred by the vehicle during the crash (the so-called longitudinal Delta V—herein, Delta V). Crash investigators calculate the Delta V with a computerized algorithm that is solely based on vehicle deformation as measured in the postcrash investigation. The vehicle identification number was decoded with VINDICATOR[52] to corroborate the information about air bag presence and restraint systems. Vehicles were characterized as equipped with a frontal driver-side air bag when either the VIN or the NASS CDS data indicated so. Crashes were classified as frontal or near-frontal if the direction of force of the primary or secondary impacts was within the 10:00 to 2:00 range.[17,31] A computerized mapping algorithm (M. Waltz, MS, National Center for Statistics and Analysis, National Highway Traffic Safety Administration, written communication, November 1997) was used to assign Functional Capacity Index scores to the injuries sustained by the drivers. Stata[53] and Microsoft Excel[54] were used for data management and statistical analysis.

Analyses

We conducted a descriptive analysis of driver demographics, injuries, vehicles, and crash characteristics. Univariate and multivariate regression techniques were then used to evaluate (1) the effect of air bag deployment on injury frequency and severity; (2) the association between injury severity and several personal, vehicle, and crash characteristics; and (3) the possible confounding and effect modification between air bag deployment and personal, vehicle, and crash characteristics.

Independent variables for inclusion in the multivariate regression were those that had significant or quasi-significant coefficients ($P<.25$) in the univariate regressions (i.e., driver's sex, age, and height; seat belt use; vehicles' wheelbase; and crash severity) and a dummy variable indicating whether the air bag deployed. For each dependent variable, models were built systematically and included 2, 3, or more independent variables and the interaction terms between air bag deployment and each of the covariates (e.g., air bag deployment and crash severity). More complex models were evaluated with log likelihood ratios[55] or the residuals sum of square test.[56] The independent variables that retained statistical significance ($P<.1$) in most (if not all) of the regression models (and thus were included in the final models) were air bag deployment, crash severity, the driver's sex and age, and safety belt use. In the final models, we also included the 2 terms reflecting the interaction between air bag deployment and driver's sex and air bag deployment and Delta V when these terms achieved statistical significance ($P<.1$).

The air bag deployment coefficients in the final multivariate regression models reflected the point estimates of air bag deployment effectiveness, while controlling for severity of the crash, age and sex of the driver, and his or her seat belt use. "Effectiveness" was defined as a decrease or increase in (1) the probability of sustaining injuries of different severity levels (i.e., when evaluating maximum Abbreviated Injury Scale), (2) the overall severity of the injuries sustained (i.e., when evaluating Injury Severity Score), and (3) the functional limitations associated with those injuries (i.e., when evaluating maximum Functional Capacity Index). Statistically significant interaction coefficients indicate whether the air bag effectiveness varies across circumstances. For example, air bags could be more protective for male than for female drivers or for more severe than for less severe crashes. In the models that have a significant interaction term, the air bag effectiveness estimates must be reported for the different circumstances that were used in defining the interaction term.[55,57]

Results

Of the 13 092 drivers in the NASS CDS from 1993 to 1996 in passenger cars of model years no earlier than 1986, 6409 (49.0%) had a known Delta V. A comparison of drivers with and without known Delta Vs found differences both in the proportion of cases with missing data and in the proportion of (near) frontal crashes. Drivers with unknown Delta Vs were more likely than drivers with known Delta Vs to have missing information for variables such as age, sex, air bag presence, air bag deployment, seat belt use, vehicle size or mass, and direction of crash. For those cases with known direction of crash, frontal crashes were more common among drivers with known Delta Vs (86.9%) than among drivers with unknown Delta Vs (81.5%) ($P=.0005$). However, no statistically significant differences were found regarding the distribution of any other variables, including the maximum Abbreviated Injury Scale, Injury Severity Score, and maximum Functional Capacity Index (data not shown).

Of the 6409 drivers with known Delta Vs, 11 died of causes unrelated to the crash, 655 had missing information about the direction of crash, and 740 were in nonfrontal crashes. Hence, 5003 drivers met the study's inclusion criteria; their personal, vehicle, and crash characteristics are summarized in Table 1.

Among these 5003 drivers, 208 (4.2%) died as a consequence of the crash. No injuries were reported for 2023 drivers, including 187 of the drivers who died. Among the drivers who had at least 1 injury, 518 sustained only 1 injury each, and the other 2441 drivers had a total of 10 055 injuries.

The NASS CDS indicated that 1545 vehicles were equipped with a frontal driver-side air bag, whereas the decoding of the vehicle identification number identified 1580 such vehicles. Agreement between the 2 sources occurred in 1424 cases, whereas at least 1 of the data sources indicated the presence of an air bag in 1701 cases (34.0%). Air bag deployment occurred in 1095 cases (21.9% of the total or 64.4% of the air bag–equipped vehicles).

The logistic multivariate regression confirmed that air bag deployment was associated with a statistically significant decrease in the probability of fatal injuries (odds ratio [OR] = 0.58, 95% confidence interval [CI] = 0.37, 0.90). This protective effect did not differ by sex of the driver (Table 2).

In the ordinal and linear multivariate regression models, both the interaction between air bag deployment and Delta V and the interaction between air bag deployment and driver's sex were statistically significant (Table 2). Air bag deployment per se *increases* the overall injury severity and functional limitations as measured by the 3- and 7-level Maximum Abbreviated Injury Scale, the Injury Severity Score, and the maximum Functional Capacity Index. In contrast, the interaction terms have *protective* effects. As a consequence, (1) air bag deployment at low Delta V induces (more severe) injuries, particularly among female drivers; (2)

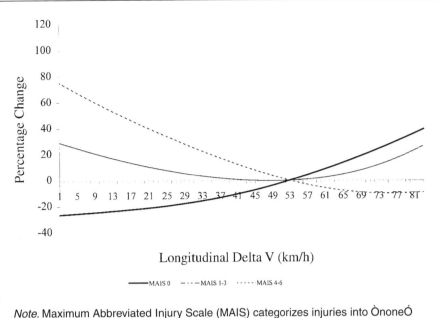

FIGURE 1A— Percentage change (with control for age and seat belt use) in the probability of female drivers (n = 2532) sustaining injuries, by injury severity level (as defined by maximum Abbreviated Injury Scale) in frontal or near-frontal crashes (with known longitudinal Delta V): National Automotive Sampling System Crashworthiness Data System calendar years 1993–1996, passenger cars of model years 1986–1997.

Note. Maximum Abbreviated Injury Scale (MAIS) categorizes injuries into "none" (MAIS = 0), "minor" (MAIS = 1–3), and "severe" (MAIS = 4–6). Fatalities were considered as MAIS = 6 (excludes drivers who died of causes not related to the crash).

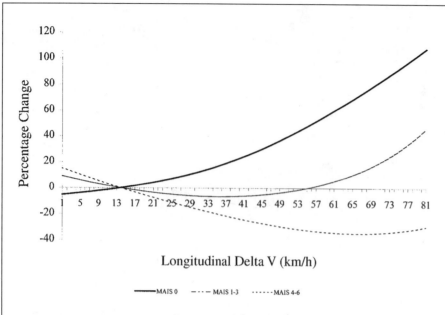

FIGURE 1B—Percentage change (controlling for age and seat belt use) in the probability of male drivers (n=2466) sustaining injuries, by injury severity level (as defined by Maximum Abbreviated Injury Scale) in frontal or near-frontal crashes (with known longitudinal Delta V): National Automotive Sampling System Crashworthiness Data System calendar years 1993–1996, passenger cars of model years 1986–1997.

Note. Maximum Abbreviated Injury Scale (MAIS) categorizes injuries into "none" (MAIS = 0), "minor" (MAIS = 1–3), and "severe" (MAIS = 4–6). Fatalities were considered as MAIS = 6 (excludes drivers who died of causes not related to the crash).

at higher Delta Vs, the air bag's protective effect becomes large enough to offset its injurious effect, and the air bag deployment net effect becomes protective; and (3) the Delta V at which the net effect of air bag deployment becomes protective differs between female and male drivers.

Figures 1A and 1B illustrate net air bag effectiveness at each crash severity level as the percentage change in the probability that female and male drivers will sustain no injury (i.e., Maximum Abbreviated Injury Scale score of 0) or injuries of different severity (Maximum Abbreviated Injury Scale score of 1–3 or 4–6), after controlling for age and seat belt use. For example, the net air bag effect on "severe" injuries among female drivers ranged from a 10% protective effect in crashes with Delta Vs around 70 km/h to a 70% increase in crashes with Delta Vs below 3 km/h. The effect on "severe" injuries among male drivers ranged from a 35% protective effect in crashes with Delta Vs around 64 km/h to a 10% increase in crashes with Delta Vs below 5 km/h. Overall, the Delta V at which the air bag deployment changes from injurious to protective occurs at 52.0 km/h for female and 12.9 km/h for male drivers.

The effect of air bag deployment evaluated with the 7-level Maximum Abbreviated Injury Scale, Injury Severity Score, and Functional Capacity Index as the outcome measures produced results consistent with the findings just described, although the precise Delta V at which air bag deployment became protective varied depending on the outcome measure evaluated (Table 3). The lowest crossover points occurred when Injury Severity Score was the analyzed outcome. Air bag deployment among male drivers had a net protective effect in crashes with Delta Vs at or above 5.2 km/h, whereas the protective effects for female drivers occurred only in more severe crashes (at or above 27.5 km/h).

All other covariates included in the final multivariate regression models had statistically significant effects in the anticipated directions (e.g., more severe injuries were associated with higher Delta Vs, older drivers, and no seat belt use) (data not shown).

Discussion

The results show that crashes with low Delta V result in an overall injurious effect of air bags largely caused by an increase in injuries of female drivers. The air bag's detrimental effect is offset in crashes at higher Delta Vs, when air bags become protective and prevent all drivers from having more severe injuries (e.g., Maximum Abbreviated Injury Scale score of 4 or higher), from having injuries associated with more functional limitations, and from dying.

Among female drivers, air bag deployment in low-severity crashes increases the probability of "minor" injury (i.e., Maximum Abbreviated Injury Scale score of 1, 2, or 3), the

TABLE 3—Severity of Crash Above Which Air Bag Deployment Exerts a Net Protective Effect (Adjusted[a] Point Estimates and 95% Confidence Intervals [CIs]) in Frontal and Near-Frontal Crashes With Known Longitudinal Delta V (n = 4697): National Automotive Sampling System Crashworthiness Data System Calendar Years 1993–1996, Passenger Cars of Model Years 1986–1997

	All Drivers		Female Drivers		Male Drivers	
Outcome Variable	Point Estimate, km/h	95% CI	Point Estimate, km/h	95% CI	Point Estimate, km/h	95% CI
3-level MAIS	32.8	0.0, 74.8	52.0	0.0, 127.6	12.9	0.0, 53.9
7-level MAIS	36.5	0.0, 85.6	62.2	0.0, 157.8	9.7	0.0, 53.5
ISS	16.6	0.0, 39.2	27.5	0.0, 65.6	5.2	0.0, 30.2
MFCI	25.3	0.0, 58.2	41.1	0.0, 98.8	8.8	0.0, 41.9

Note. MAIS = Maximum Abbreviated Injury Scale: 3-level MAIS categorizes injuries into "none" (MAIS = 0), "minor" (MAIS = 1–3), and "severe" (MAIS = 4–6), whereas 7-level MAIS uses each of the Abbreviated Injury Scale levels for severity categorization (i.e., 0, 1, 2, 3, 4, 5, and 6); ISS = Injury Severity Score; MFCI = maximum Functional Capacity Index. 95% confidence interval truncated at 0 km/h.
[a]Controlling for safety belt use, age, and sex of driver.

overall injury severity (as indicated by the Injury Severity Score), and the functional limitations (as indicated by the maximum Functional Capacity Index score).

The strength of these findings resides in the breadth of dependent variables evaluated, which includes minor injuries. Previous researchers predominantly evaluated fatality reduction effectiveness or the effectiveness in reducing the most serious nonfatal injuries (e.g., Maximum Abbreviated Injury Scale score of 3 or higher).[17,18,31,58,59] The estimates are consistent with those presented in the technical literature, including the often reported 19% fatality reduction estimate,[17] which is within the 95% confidence interval of the fatality reduction estimate reported here.

This analysis offers new data on the differential effect of air bags across the severity of crashes. The statistical significance of the interaction terms between air bag deployment and Delta Vs confirms the hypothesis that air bags (in vehicles of model year 1997 or older) deploy in low-severity crashes in which the risk of incurring any injuries is actually exacerbated by the deployment of the bag.[22,60] These induced injuries (more frequently, Maximum Abbreviated Injury Scale score of 1, 2, or 3) are disproportionately borne by female drivers, although male drivers are also at higher risk for injuries. The Delta Vs above which air bag deployment is beneficial are higher than the deployment levels reported by most manufacturers (much higher in the case of female drivers). The confidence estimates around the Delta Vs above which air bag deployment has a net protective effect are very large because of the limited sample size of real-world crashes available for analysis. It is unlikely that analysis of future years of NASS CDS data will permit greater precision in these switch points because of the current changes in air bag design.

Our analyses had some peculiarities that are worth noting. Research conducted to date has evaluated the effectiveness of air bag presence (regardless of actual deployment), instead of the effectiveness of air bag deployment.[17–23,31] In our data, 606 additional drivers had an air bag that did not deploy during the crash. To evaluate the robustness of our findings, we ran the final regression models with air bag presence instead of air bag deployment as the variable of interest. The resulting effectiveness estimates were comparable (i.e., had the same direction and similar magnitude) to those reported here, although the statistical significance of some estimates was lost.

Despite the relevance of our findings, one should exercise caution when interpreting them for several reasons. First, the findings describe the aggregate effects of air bags available in all makes, models, and years included in the data set. Obviously, air bag systems have design and performance differences, which may change the effectiveness of any particular system. Furthermore, the recent federal regulation allowing for depowering of the air bags (i.e., reducing the speed and/or volume of inflation) may change the effectiveness of the system.[37] Two factors prevent us from performing a more refined analysis of specific air bag systems: (1) specific air bag design parameters across models and years are proprietary information not available to researchers, and (2) even if this information became available, the NASS CDS may not have enough cases to allow statistically significant findings in analyses by air bag design.

Second, our air bag effectiveness estimates were based on crashes for which the Delta Vs of the crashed passenger car were known. The NASS CDS data set has about half as many crashes without such information. This limits the external validity of the findings. Delta Vs are most often missing because the algorithm used by crash investigators in their computation cannot be used when the crash involves a rollover, other nonhorizontal forces, sideswipe, severe override, and overlapping damage or when data about the crash or vehicle are insufficient.[61] In fact, direction of impact was the only statistically significant difference between drivers with known and unknown Delta Vs in our sample.

Finally, the NASS CDS data are subject to some measurement error, particularly regarding the severity of the crash. The validity of the estimates for Delta V values lower than 40 km/h is questionable, because the algorithm is calibrated at approximately 48 km/h (M. Finkelstein, MA, oral communication, June 1998), and it tends to underestimate the actual impact speed, especially in nonfrontal crashes.[62,63] However imperfect this measure might be, it is the only proxy for crash severity that is available for this type of analysis. Although the exact degree of error present in the NASS CDS data is unknown, qualitative evaluations included in the NASS CDS data set indicate that for 75% of the 5003 crashes, the computed Delta Vs fit the crash description. In 17%, the Delta Vs appear reasonable, and in 8%, the Delta Vs appear high or low (4% of each). The effect of these measurement errors on our findings is difficult to evaluate, but our estimates of Delta Vs at which air bags become protective are probably underestimates of the actual Delta Vs at which this happens.

In conclusion, our results showed that in frontal or near-frontal crashes, air bag deployment is effective in reducing the most severe and fatal injuries, whereas in low-severity crashes, air bag deployment induces injuries of Maximum Abbreviated Injury Scale level 1 to 3 (predominantly among females). Raising the crash severity level at which air bags are designed to deploy should be considered an injury prevention strategy in conjunction with other changes in air bag design. However, one should be cautious in interpreting this recommendation. With imperfect sensor technology, raising the air bag deployment threshold means that (in some crashes) it will take longer for the air bag system to recognize that a crash is severe enough to justify air bag deployment. This may be a particular problem in off-frontal crashes, when the sensors might take longer to acknowledge the existence of a crash because of their positioning.[64,65] Late deployment may injure drivers who have moved forward into the air bag's deployment zone. Use of crush-zone sensors and/or more advanced sensor technology may help alleviate this problem.

Acknowledgments

This research was partially supported by the Harvard Center for Risk Analysis and the Harvard Injury Control Research Center (Centers for Disease Control and Prevention grant R49/CCR115279-01).

Preliminary findings were presented at the 4th World Conference on Injury Prevention and Control, Amsterdam, the Netherlands, May 17–20, 1998.

The author is greatly indebted to David Wypij for his assistance in the design of the statistical analysis, Marie Waltz for providing the computerized algorithm to transform Abbreviated Injury Scale codes into Functional Capacity Index scores, and John Graham for his encouragement to undertake this project as well as his guidance and support while completing it. John Graham, Ellen MacKenzie, Milton Weinstein, and 3 anonymous reviewers provided helpful comments on previous drafts of the paper.

References

1. Campbell DD. *Air Cushion Restraint Systems Development and Vehicle Application.* Warrendale, Pa: Society of Automotive Engineers; 1972. SAE Paper No. 720407.
2. Graham JD. *Auto Safety: Assessing America's Performance.* Dover, Mass: Auburn Publishing House; 1989.
3. King AI, King HY. Research in biomechanics of occupant protection. *J Trauma.* 1995;38:570–576.
4. Struble DE. *Airbag Technology: What It Is and How It Came to Be.* Warrendale, Pa: Society of Automotive Engineers; 1998. SAE Paper No. 98648.
5. *Amendment to Federal Motor Vehicle Safety Standard No. 208: Passenger Car Front Seat Occupant Protection. Federal Regulatory Impact Analysis.* Washington, DC: US Dept of Transportation, National Highway Traffic Safety Administration; 1984.
6. Breed DS. How airbags work (design, deployment criteria, cost perspectives). *Chronic Dis Can.* 1992;14(suppl 4):S70–S78.
7. Lund AK. Letter to Mr. W. Boehly. Arlington, Va: Insurance Institute for Highway Safety; 1995.
8. Mertz HJ. *Restraint Performance of the 1973–1976 GM Air Cushion Restraint System.* Warrendale, Pa: Society of Automotive Engineers; 1988. SAE Paper No. 880400.
9. Goldmuntz L, Gates HP. *Comparative Effectiveness of Occupant Restraint Systems: Response*

to Passive Restraint Hearing Docket No. 74-14 and 37-77. Washington, DC: Economics and Science Planning Inc; 1977.

10. Grush ES, Henson SE, Rittlerling OR. *Restraint System Effectiveness.* Detroit, Mich: Ford Motor Co Automotive Safety Affairs Office; 1971. Report No. S-71-40.

11. Wilson R, Savage C. *Restraint System Effectiveness: A Study of Fatal Accidents.* Warrendale, Pa: Society of Automotive Engineers; 1973.

12. Graham JD, Henrion M. A probabilistic analysis of the passive restraint question. *Risk Anal.* 1984; 4:25–40.

13. *Analysis of Proposed Changes to Passenger Car Requirements of FMVSS208.* Washington, DC: US Dept of Transportation, National Highway Traffic Safety Administration; 1974. Docket 74-14-N01-104.

14. Cook C. *Injury and Fatality Rates, With and Without Air Bags.* Washington, DC: US Dept of Transportation, National Highway Traffic Safety Administration; 1979.

15. Mohan D, Zador PL, O'Neill B, Ginsberg M. Air bags and lap/shoulder belts—a comparison of their effectiveness in real world frontal crashes. In: Proceedings of the 20th annual meeting of the American Association for Automotive Medicine; November 1–3, 1976; Morton Grove, Ill.

16. Evans L. Double pair comparison—a new method to determine how occupant characteristics affect fatality risk in traffic crashes. *Accid Anal Prev.* 1986;18:217–227.

17. Kahane C. *Fatality Reduction by Airbags: Analyses of Accident Data Through Early 1996.* Washington, DC: US Dept of Transportation, National Highway Traffic Safety Administration; 1996.

18. Ferguson SA. *Update on Airbag Performance in the United States: Benefits and Problems.* Arlington, Va: Insurance Institute for Highway Safety; 1996.

19. Lund AK, Ferguson SA, Powell MR. *Fatalities in Airbag-Equipped Cars: A Review of 1989–1993 NASS Cases.* Warrendale, Pa: Society of Automotive Engineers; 1996. SAE Paper No. 960661.

20. *Automobile Occupant Crash Protection.* Washington, DC: US Dept of Transportation, National Highway Traffic Safety Administration; 1980. Progress Report No. 3.

21. Werner JV, Sorenson WW. *Survey of Airbag-Involved Accidents: An Analysis of Collision Characteristics, System Effectiveness and Injuries.* Warrendale, Pa: Society of Automotive Engineers; 1994. SAE Paper No. 94082.

22. Dalmotas DJ, German A, Hendrick BE, Hurley J. Airbag deployments: the Canadian experience. *J Trauma.* 1995;38:476–481.

23. Langwieder K, Hummel TA, Muller CB. Experience with airbag-equipped cars in real-life accidents in Germany. Paper presented at: 15th International Technical Conference on the Enhanced Safety of Vehicles; May 13–16, 1996; Melbourne, Australia. SAE Paper No. 96-S1-O-04.

24. Graham JD, Thompson KM, Goldie SJ, Segui-Gomez M, Weinstein MC. The cost-effectiveness of airbags by seating position. *JAMA.* 1997;278: 1418–1425.

25. Thompson KM, Segui-Gomez M, Graham JD. Validating analytical judgments: the case of airbags' lifesaving effectiveness. *Reliability Eng Syst Safety.* 1999;66:57–68.

26. Ingram HJ, Perry HD, Donnenfeld ED. Airbag keratitis. *N Engl J Med.* 1991;324:1599–1600.

27. Reiland-Smith J, Weintraub RM, Sellke FW. Traumatic aortic valve injury sustained despite the deployment of an automobile airbag. *Chest.* 1993; 103:1603.

28. Huelke DF, Moore JL, Ostrom M. Air bag injuries and occupant protection. *J Trauma.* 1992; 33:894–898.

29. Huelke DF. *An Overview of Airbag Deployment and Related Injuries: Case Studies and a Review of the Literature.* Warrendale, Pa: Society of Automotive Engineers; 1995. SAE Paper No. 950866.

30. Lombardo LV, Ryan SD. *Detection of Internal Injuries in Drivers Protected With Airbags* [research note]. Washington, DC: US Dept of Transportation, National Highway Traffic Safety Administration; 1993.

31. *Effectiveness of Occupant Protection Systems and Their Use: Fourth Report to Congress.* Washington, DC: US Dept of Transportation, National Highway Traffic Safety Administration; 1999.

32. Libertiny GZ. *Airbag Effectiveness—Trading Major Injuries for Minor Ones.* Warrendale, Pa: Society of Automotive Engineers; 1995. SAE Paper No. 950871.

33. Malliaris AC, Digges KH, DeBlois JH. *Evaluation of Airbag Field Performance.* Warrendale, Pa: Society of Automotive Engineers; 1995. SAE Paper No. 950869.

34. Malliaris AC, Digges KH, DeBlois JH. *Injury Patterns of Car Occupants Under Air Bag Deployment.* Warrendale, Pa: Society of Automotive Engineers; 1995. SAE Paper No. 950867.

35. *Driver Injury Experience in 1990 Models Equipped With Air Bags or Automatic Belts.* Arlington, Va: Highway Loss Data Institute; 1991.

36. *Passenger-Side Airbags: Which Perform the Best?* Washington, DC: Public Citizen and Center for Auto Safety; 1997.

37. *Actions to Reduce the Adverse Effects of Airbags: Depowering, Final Regulatory Evaluation.* Washington, DC: US Dept of Transportation, National Highway Traffic Safety Administration; 1997.

38. Werner JV, Roberson F, Ferguson SA, Digges KH. *Airbag Deployment Frequency and Injury Risks.* Warrendale, Pa: Society of Automotive Engineers; 1996. SAE Paper No. 960664.

39. Werner JV, Robertson SF, Ferguson SA, Digges KH. *Injury Risks in Cars With Different Airbag Deployment Rates.* Warrendale, Pa: Society of Automotive Engineers; 1997. SAE Paper No. 970491.

40. Evans L. Risk of fatality from physical trauma versus sex and age. *J Trauma.* 1988;28:368–378.

41. Huelke DF. *Does Stature Influence Driver Injuries in Airbag Deployment Crashes? Analysis of UMTRI Crash Investigations.* Warrendale, Pa: Society of Automotive Engineers; 1998. SAE Paper No. 980640.

42. Kleinberger M, Simmons L. Mechanism of injuries for adults and children resulting from airbag interaction. In: Proceedings of the Association for the Advancement of Automotive Medicine 41st Annual Meeting; November 10–11, 1997; Orlando, Fla.

43. Ludstrom LC. Relating air cushion performance to human factors and tolerance levels. Paper presented at: 5th International Conference on Experimental Safety Vehicles; 1974; London, England.

44. Evans L, Frick MC. Car size or car mass: which has greater influence on fatality risks? *Am J Public Health.* 1992;82:1105–1112.

45. Pflug JA. *Air Cushion Systems for Full-Sized Cars.* Warrendale, Pa: Society of Automotive Engineers; 1972. SAE Paper No. 720408.

46. Zador PL, Ciccone MA. Automobile driver fatalities in frontal impacts: air bags compared with manual belts. *Am J Public Health.* 1993;83: 661–666.

47. Campbell BJ. Safety belt injury reduction related to crash severity and front seat position. *J Trauma.* 1987;27:733–739.

48. Zinke DT. *Small Car Front Seat Passenger Inflatable Restraint System, Volume 1: Interim Results.* Washington, DC: US Dept of Transportation, National Highway Traffic Safety Administration; 1981.

49. *The Abbreviated Injury Scale, 1990 Rev.* Des Plaines, Ill: Association for the Advancement of Automotive Medicine; 1990.

50. Baker SP, O'Neill B. The injury severity score: an update. *J Trauma.* 1976;16:882–885.

51. MacKenzie EJ, Damiano AM, Miller TS, Luchter S. The development of the Functional Capacity Index. *J Trauma.* 1996;41:799–807.

52. *Vindicator 96* [computer program]. Arlington, Va: Highway Loss Data Institute; 1996.

53. *Stata Statistical Software, Release 5.0* [computer program]. College Station; Tex: Stata Corp; 1997.

54. *Microsoft Excel, Version 97 SR-1* [computer program]. Redmond, Wash: Microsoft Corp; 1997.

55. Hosmer DW, Lemeshow S. *Applied Logistic Regression.* New York, NY: John Wiley & Sons; 1989.

56. Seber GAF. *Linear Regression Analysis.* New York, NY: John Wiley & Sons; 1977.

57. Rice JA. *Mathematical Statistics and Data Analysis.* Belmont, Calif: Wadsworth Inc, Wadsworth & Brooks/Cole Advance Books & Software; 1988.

58. Barry S, Ginpil S, O'Neill T. The effectiveness of air bags. *Accid Anal Prev.* 1999;31:781–787.

59. Deery HA, Morris AP, Fildes BN, et al. Airbag technology in Australian passenger cars: preliminary results from real world investigations. *J Crash Prev Inj Control.* 1999;1:121–128.

60. Otte D. *Review of Airbag Effectiveness in Real Life Accidents: Demands for Positioning and Optimal Deployment of Airbag Systems.* Warrendale, Pa: Society of Automotive Engineers; 1995. SAE Paper No. 952701.

61. National Accident Sampling System Crashworthiness Data System. *Data Collection, Coding, and Editing Manual.* Washington, DC: US Dept of Transportation, National Highway Traffic Safety Administration, National Center for Statistics and Analysis; 1993–1996.

62. Stucki SL, Fessahaie O. *Comparison of Measured Velocity Change in Frontal Crash Tests to NASS Computed Velocity Change.* Warrendale, Pa: Society of Automotive Engineers; 1998. SAE Paper No. 980649.

63. O'Neill B, Preuss CA, Nolan JM. Relationships between computed delta V and impact speeds for offset crash tests. Paper presented at: 15th International Technical Conference on the Enhanced Safety of Vehicles; May 13–16, 1996; Melbourne, Australia. SAE Paper No. 96-S9-0-11.

64. Breed DS, Sanders T, Castell V. *A Critique of Single Point Sensing.* Warrendale, Pa: Society of Automotive Engineers; 1992. SAE Paper No. 920124.

65. Klove EH, Oglesby RN. *Special Problems and Considerations in the Development of Air Cushion Restraint Systems.* Warrendale, Pa: Society of Automotive Engineers; 1972. SAE Paper No. 720411.

2001-01-0156

Analysis of Driver Fatalities in Frontal Crashes of Airbag-Equipped Vehicles in 1990-98 NASS/CDS

David S. Zuby and Susan A. Ferguson
Insurance Institute for Highway Safety

Michael X. Cammisa
Association of International Automobile Manufacturers, Inc.

Copyright © 2001 Society of Automotive Engineers, Inc.

ABSTRACT

This study, which is an extension of an earlier study, examined an additional 64 frontal crashes of airbag-equipped vehicles in the 1997-98 National Automotive Sampling System Crashworthiness Data System (NASS/ CDS) in which the driver died. The principal cause of death in each case was determined based on an examination of the publicly available case materials, which primarily consisted of the crash narrative, the injury/source summary, and photographs of the crashed vehicle. Results were consistent with the earlier analyses of the 1990-96 NASS/CDS files. In the combined data set (1990-98), gross deformation of the occupant compartment was the leading cause (42 percent) of driver deaths in these 116 frontal crashes. The force of the deploying airbag (16 percent) and ejection from the vehicle (13 percent) also accounted for significant portions of the driver deaths in these frontal crashes. There continues to be little or no evidence that airbags deploy with too little energy. In contrast to previous analyses, there were 5 crashes in 1997-98 in which the driver's fatal injuries may have resulted from bottoming the airbag. However, there were equally plausible, alternative explanations for these fatal injuries, including the possibility that they were caused by the airbag itself in 4 of the 5 cases.

INTRODUCTION

In May 2000, the National Highway Traffic Safety Administration (NHTSA) published a final interim rule that will require manufacturers to install advanced airbags beginning with the 2004 model year (49 CFR Part 552, May 12, 2000). The regulation has two aims. The first and foremost is to reduce the risks that airbags pose to out-of-position (OOP) occupants, particularly children and small adults. To this end, the regulation includes a battery of static airbag deployment tests using dummies representing a range of people from infants to small adults in a variety of near-the-airbag positions. The regulation offers safety system de-

signers the option to demonstrate that the airbag will not deploy when a child occupant is seated in the airbag-protected seat, will not deploy if the seat occupant is in a dangerous position, or that a close-range deployment has a low risk of causing injury to an OOP occupant. Secondly, the standard aims to improve the level of crash protection offered by current airbag technology. To accomplish this aim, regulators added crash tests with dummies representing small adult occupants and, after a phase-in period starting September 1, 2007, will raise the test speed for rigid barrier frontal crash tests involving the 50th percentile male belted dummy from 30 to 35 mi/h. A new test configuration, a 25 mi/h (40 km/h) frontal offset crash with a dummy representing the 5th percentile adult female, will also be required. The goal of this test is to improve crash sensing technology in the low-acceleration crashes near the threshold at which airbags typically deploy. Crash tests have shown that airbags often deploy very late in similar scenarios, putting occupants at greater risk of being close to airbags when they deploy. The agency also lowered the rigid barrier test speed from 30 to 25 mi/h for tests using unbelted dummies. This was the most controversial provision of the final rule.

The basis for the debate about test speeds for crash tests using unbelted dummies is whether higher test speeds are needed to ensure that airbags have sufficient power to prevent occupant collisions with vehicle interiors in the most severe crashes. It is argued that if airbag energy is reduced as a result of reduced test speed, some unbelted drivers may die because of insufficient airbag energy. The other side of the argument is that the powerful airbags needed to restrain unbelted dummies in ways that allow injury criteria to be met have been shown to cause injuries and deaths to vulnerable car occupants in some cases. The higher the test speed, the more energy an airbag must have to catch and decelerate an unbelted dummy. In addition, higher test speeds shorten the time available to deploy the airbag. Both of these factors lead to energetic deployments that can cause injuries to people near the airbags regardless of the crash severity. Since 1998, when airbags were first de-powered as the result of a change in Federal Motor Vehicle Safety Standard (FMVSS) 208 that allowed manufacturers to use 30 mi/h sled tests with unbelted dummies, no cases involving insufficiently powerful airbags have been identified (see Insurance Institute for Highway Safety (IIHS) [1]* for methodology on identifying such crashes). Furthermore, prior studies have clearly shown that unbelted drivers are dying not because of insufficiently powered airbags, but either because intrusion is so massive that no airbag could help, because drivers are ejected, or because the airbag inflation force is so powerful it contributed to the drivers' fatal injuries [2-3].

The debate over the appropriate test speed to ensure protection for unbelted occupants has been hampered by a lack of recognition that airbags cause injuries to OOP occupants in high-speed crashes as well as in the low-speed events that have received much public attention. Many high-speed crashes, like less severe crashes, include some form of preimpact maneuvering that can cause unbelted occupants to move near the airbags before they deploy. Furthermore, alcohol impairment, illness, and falling asleep behind the wheel can result in drivers being out of position. Examples of such scenarios from the NASS/CDS have been described elsewhere [2-3].

Since 1996, IIHS has reviewed NASS/CDS cases of frontal crashes of airbag-equipped cars that were fatal to drivers as part of an effort to better understand how drivers are dying in these crashes. Researchers found that catastrophic collapse of the occupant compartment was the leading cause (51 percent) of driver fatalities [2]. The force of the airbag deployment (17 percent) and ejection (13 percent) were the second and third most common cause for the drivers' fatal injuries. No examples of deaths caused by insufficiently powered airbags were found. The present study analyzed additional frontal crashes from NASS/CDS (1997-98) in which a driver died in an airbag-equipped passenger vehicle.

METHODS

The NASS/CDS Statistical Analysis System (SAS) database for calendar years 1989-98 was queried to identify all cases in which a driver death occurred in a model year 1990 or newer airbag-equipped vehicle involved in a crash in which the area most severely impacted was frontal (GAD1=F) and the primary direction of force (DOF1) was between 10 and 2 o'clock, inclusive. Also included were frontal crashes in which vertical or lateral shifting of a vehicle's end structure occurred. This is indicated in the database by incrementing the DOF1 variable: 00 = no shift, 20 = end shifted up, 40 = end shifted down, 60 = end shifted right, 80 = end shifted left (reference: page EV-23 of the National Automotive Sampling

System 1997 Crashworthiness Data System Data Collection, Coding, and Editing Manual for a more detailed explanation). Frontal crashes (GAD1=F) with unknown DOF1 also were included. Cases in which a driver's death was classified as "fatal-ruled disease" (as identified by the TREATMNT variable) were not included in this analysis, because the performance of the airbag would not be expected to have influenced the outcome.

Publicly available NASS case materials were reviewed for each of the 127 fatalities that were identified using the search criteria. Case materials for 1997-98 calendar years were downloaded from the NASS/CDS section of NHTSA's website (www-nass.nhtsa.dot.gov/nass). Printed copies of cases for the earlier calendar years were obtained from NHTSA through Zimmerman and Associates. Both the electronic and printed case materials included data collection forms describing the crash events, interior and exterior vehicle damage, and injury information including the likely source for each injury as determined by the NASS investigator. Photographs of the vehicles involved and the crash site were included for most cases.

The causes of the drivers' fatal injuries in these cases were classified into 1 of 7 categories (Table 1) based on a careful review of the case materials. In most cases, the NASS-assigned injury sources for the driver's most severe injuries were the basis for this determination. Thus, our review was an accounting of the circumstances under which the injury contact occurred. For example, if a driver's severe chest injuries were attributed to the airbag by the NASS investigator, then our review focused on deciding whether the case circumstances suggested that the driver may have been near the airbag as it was deploying or whether the crash circumstances suggested the driver was normally seated and the injuries were caused by high restraining forces. Some cases included notes describing the investigator's assessment. In cases that did not state the investigator's conclusions, the time of day that the crash occurred, preimpact maneuvers, driver size and age, alcohol and drug use, and seat belt use were among the factors considered in trying to decide whether an occupant may have been near the airbag at the time of deployment. Sometimes, the nature of the injuries was used to help classify a particular case. For example, if the driver's most severe injury was a fracture/dislocation in the upper cervical spine and there was no evidence of a head contact with hard surface, then the force of the deploying airbag was suspected. Regardless of the NASS estimated delta V, if the vehicle was so utterly destroyed that there was no remaining survival space for the driver, then the case was classified with the intrusion category. While such a retrospective examination of the records of a sometimes inconclusive investigation cannot with certainty determine the exact cause of a driver's death, the 7 categories are useful for better understanding how such fatalities could be prevented.

---

*Numbers in brackets designate references at the end of the paper.

Table 1. Cause-of-Death Classifications

Classification	Definition
Airbag	The driver's most severe or fatal injuries were typically attributed to the airbag by the NASS investigator, and our review of the case materials found evidence that the driver likely was near the airbag when it deployed. Thus, energy from the deploying airbag either directly or indirectly caused these injuries.
Ejection	The NASS investigator indicated the driver was ejected or partially ejected from the vehicle and his/her most severe injuries were attributed to contact with something outside his/her own vehicle.
Intrusion	The crash was so severe that there was little or no survival space around the driver's position. Alternately, the NASS investigator attributed the most severe or fatal injuries to a component that was contacted because it intruded into the driver's space.
Interior Surface	The NASS investigator attributed the driver's most severe or fatal injuries to impact with an intruding or non-intruding interior vehicle surface. Intrusion may have been present in cases with this classification, but it was not extensive enough to significantly compromise driver survival space.
Non-deployment	The driver's most severe or fatal injuries would most likely have been prevented or ameliorated if the airbag had deployed. Crash conditions indicated that a deployment was expected.
Other	Some agent, other than those described by the main categories, most likely caused the driver's most severe or fatal injuries.
Unknown	The driver's most severe or fatal injuries could not be explained readily by the available information or, there were multiple and equally plausible explanations and no way to choose among them.

Despite having electronically filtered out cases that were classified in NASS as "fatal-ruled disease," 9 more cases in which the fatality was related to a medical condition rather than injuries sustained from the impact were identified upon review. In addition, there were 2 cases in which the fatal injuries were attributed to fire. Performance of the airbag would not be expected to have influenced the outcome of these 11 cases (Table 2), therefore they were not included in the remaining summary tables and figures. A total of 116 cases remained for further analysis to determine if airbag performance directly or indirectly contributed to the fatal injuries.

Table 2. Cases Dropped from Analysis Because Death was Due to Disease or Fire

Calendar Year	PSU	Case ID	Reason for Exclusion
1995	45	082A	Heart failure before crash
1996	72	026B	Heart attack before crash
1996	11	036A	Fire
1997	4	018A	Arteriosclerotic heart disease
1997	11	146A	Fire
1997	12	099A	Heart attack before crash
1997	45	042B	Cardiac arrest; only minor injuries
1998	45	020A	Arteriosclerotic heart disease
1998	45	152E	Stroke
1998	72	117B	Disease, unspecified
1998	79	004J	Congestive heart disease

RESULTS

The additional cases for the 1997-98 calendar years accounted for more than half (54 percent) of the 116 cases in the combined data set. Figure 1 shows the distribution of the frontal crashes included in the study across the 9 calendar years. There were no frontal crashes of model year 1990 and newer cars meeting the criteria for this study in the 1989 NASS/CDS file.

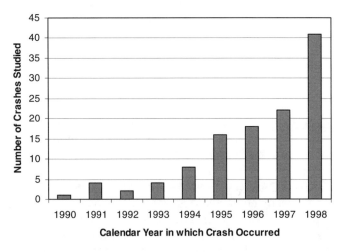

Figure 1. Distribution of Studied Crashes by Year in Which Crash Occurred.

These crashes had estimated velocity changes (delta Vs) ranging from 13 to 100 km/h (8 to 60 mi/h), but delta V was unknown in 37 cases. Forty-one (35 percent) of the drivers were believed to be using their seat belts at the time of the crash, but belt use could not be determined in 6 of the cases. Airbags deployed in all but 8 crashes (7 percent), and airbag deployment could not be ascertained in one case in which the vehicle burned after crashing.

Figure 2 shows the distribution of the 116 cases by the 7 cause-of-death categories. Intrusion-related injuries were the most prevalent cause of driver death, accounting for 49 fatalities (42 percent). Airbag-related injuries accounted for 18 deaths (16 percent), ejection for 15 deaths (13 percent), and contact with interior vehicle surfaces accounted for 7 deaths (6 percent). One of the fatal crashes (1995 079-002A), with a delta V of 31 mi/h, involved a 1995 Ford F-150 pickup in which the driver airbag did not deploy. The unbelted driver's fatal head injuries may have been prevented if the airbag had deployed, but there was no information in the case file indicating why it did not. Four cases (3 percent) were classified as "other." One case (1998 43-262A) involved a deer that, after being struck by the case vehicle, crashed through the windshield and caused the driver's injuries. The drivers' most serious injuries in the other 3 cases (1995 45-100A, 1997 79-109B, and 1997 82-186B) appeared to have been caused by the seat belt. A definite cause of death could not be determined in 22 of the cases ("unknown" category). The appendix lists all 116 cases grouped by cause-of-death category and includes information on crash severity (delta V), seat belt use, and the most critically injured body region.

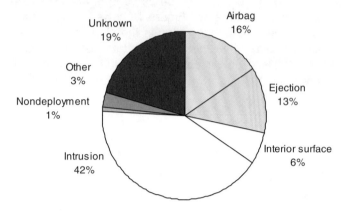

Figure 2. Causes of Death to Drivers of Airbag-Equipped Vehicles in Frontal Crashes (1990-98 NASS/CDS).

AIRBAG CASES – Table 3 summarizes the driver, vehicle, and crash characteristics for the airbag-related fatal crashes. The airbag was identified as the injury source by the NASS investigator in all of the 1997-98 cases and some of the earlier cases in this category. A number of factors were used to determine if the death was caused by the airbag. Some of the airbag cases were identified by a very low delta V, because in such crashes it seemed unlikely that the driver would have died had the airbag not deployed. Because a shorter person is more likely to drive in a position seated close to the steering wheel, stature was also an important factor. The lack of seat belt restraint was also considered important, because an unbelted driver is more likely to have moved near the airbag before it deployed due to preimpact braking or other precrash maneuvers. If the driver's survival space was maintained, if there was no evidence of forceful occupant contacts with parts of the vehicle interior other than the airbag, and if the drivers' injuries were also consistent with the kinds of injuries typically seen among occupants close to airbags, then a determination was made that the injuries were airbag induced regardless of estimated delta V.

A few crashes in this category had higher delta Vs, and may have been fatal despite the airbag deployment, but one injury or more was attributed by the NASS investigator to the deploying airbag. In case 1998 09-087, for example, the 73-year-old female driver of a Mercury Tracer, which crashed into a telephone pole, died with fatal injuries recorded for both the head (brain stem laceration, AIS 6) and chest (multiple lacerations of the heart, AIS 6). The delta V of the crash was 59 km/h (37 mi/h), and the heart lacerations and concomitant chest injuries were attributed to forces applied by the automatic shoulder belt. The fatal brain stem injury and concomitant cervical fractures were attributed to the deploying airbag. While this old and probably frail driver likely would have died without the neck and brain stem injuries, a younger driver in the same situation may have survived the high belt loading but would have been at the same risk of dying from neck and brain injuries. Other cases with higher delta Vs included circumstances that suggested the driver likely was near the airbag when it deployed. Delta Vs for all the airbag crashes ranged from 26 to 61 km/h (16 to 38 mi/h).

The higher severity crashes tended to have only moderate levels of intrusion on the driver side. In case 1998 09-087B, for example, the vehicle had a delta V of 59 km/h (37 mi/h) and 9-17 cm of intrusion in the footwell area, but intrusion of the instrument panel and steering column were not indicated. Similarly in case 1998 43-088B, the vehicle had a delta V of 54 km/h (34 mi/h), damage was offset to

Table 3. Airbag-Related Fatal Crashes

Case ID	Gender	Age	Height (cm)	Weight (kg)	Critically Injured Body Region	Belt Use	Delta V (km/h)	Vehicle
1993 05-125A	Female	62	157	81	Chest	Belted	46	1992 Toyota Camry
1993 06-006A	Male	64	170	73	Chest	Unbelted	46	1990 Plymouth Acclaim
1993 08-133A	Male	58	183	85	Chest	Unbelted	—	1992 Chrysler Fifth Ave
1994 11-150A	Female	23	160	56	Chest	Unbelted	61	1992 Plymouth Sundance
1995 09-167A	Male	30	183	93	Head	Unbelted	—	1995 Infiniti J30
1995 02-140A	Male	66	168	93	Chest	—	46	1993 Chrysler Town & Country
1996 41-024A	Male	79	175	82	Chest	Belted	26	1995 Ford Escort
1996 08-100A	Male	85	—	—	Chest	—	37	1995 Nissan Maxima
1997 04-146A	Female	84	—	—	Chest	Lap only	30	1996 Geo Metro
1997 72-103A	Female	62	163	66	Chest	Unbelted	48	1996 Ford Taurus
1997 41-157J	Male	77	—	—	Head	—	35	1996 Geo Prizm
1998 02-148A	Female	30	163	70	Head	Belted	—	1995 Mitsubishi Galant
1998 09-087B	Female	73	163	110	Head/chest	Belted	59	1994 Mercury Tracer
1998 43-088B	Female	64	157	66	Chest	Belted	54	1992 Honda Accord
1998 04-004B	Female	71	157	64	Chest	Unbelted	—	1993 Buick Century
1998 11-018B	Female	42	155	59	Head/chest	Unbelted	48	1991 Plymouth Voyager
1998 41-076B	Female	80	155	59	Chest	Unbelted	27	1993 Honda Accord
1998 48-090B	Female	79	147	46	Chest	Unbelted	—	1991 Ford Crown Victoria

the passenger side, and no intrusion was noted in the driver's space. Half of the drivers (9 of 18) in these crashes were not using their seat belts at the time of the crash, but seat belt use could not be determined in 3 of the crashes. Most of these drivers were shorter than the average heights of 162 cm for females and 180 cm for males [4], and most of the drivers believed to be fatally injured by their airbags (13 of 18) were older than 60 years.

INTRUSION CASES – Intrusion was judged to be the cause of the driver's fatal injuries when pictures or the NASS investigator's measurements indicated that the driver's survival space collapsed to a volume not large enough to contain the driver or when the driver's critical injuries were caused by contact with a specific intruding component. Consequently, the cases in this category tended to have relatively high delta Vs [mean = 59 km/h (37 mi/h)]. There were also a large number of cases (17 of 48) for which delta V could not be determined. Typically, delta V was unknown because damage to the vehicle was so extensive that measurement was rendered impossible. For example, in case 1998 73-061A, a 1995 Dodge W-series truck (Figure 3) crashed into a Peterbilt

Figure 3.   Subject Vehicle from Case 1998 073-061A, 1995 Dodge W-Series Truck.

Figure 4.   Subject Vehicle from Case 1997 04-104, 1995 Pontiac Bonneville.

tractor trailer then fell off a bridge. Four crashes in the intrusion category were less drastic. Nevertheless, the driver's critical injuries could only be attributed to an intruding component. In case 1997 04-104A, for example, a 1995 Pontiac Bonneville (Figure 4) crashed into the side of a Ford Bronco, which rolled on top of the Bonneville and pushed the roof and windshield header into the occupant compartment, causing fatal head injuries to the seat belted driver.

Four of the cases involved drivers whose critical injuries resulted from intrusion caused by an impact other than the principal impact. Three of these cases (1997 43-180A, 1998 45-170A, and 1998 49-176A) involved rollovers subsequent to a frontal impact, and one case (1998 81-092A) involved a side impact with a tree subsequent to front impacts with two fences and other smaller trees.

EJECTION AND INTERIOR SURFACE CASES – All of the cases in these two categories involved drivers who were not using seat belts at the time of the crash. Deaths attributed to contact with an interior surface involved vehicles with intact driver survival spaces; the injury sources were assigned by the NASS investigator on the basis of contact evidence, and the contact was judged likely to occur regardless of intrusion. This category was used only when the principal direction of force and the resulting driver kinematics indicated that the fatal contact would have occurred regardless of the presence of the airbag (i.e., the airbag would not be expected to be between where the driver started and where the fatal contact occurred). All of the vehicles in this category except one had asymmetric frontal damage patterns, and all but one had oblique principal directions of force.

UNKNOWN CATEGORY – Among the 23 cases in the unknown category, most did not have enough information to make a cause-of-death determination. Nine of these cases had no information about the driver's injuries or contained injury information that did not adequately account for the driver's death, so the critically injured body region could not be identified. Some cases were put in this category despite thorough documentation in the NASS/CDS record, because a single most likely cause of death could not be determined. Table 4 lists the cases that were unknown because the drivers' deaths could be explained by more than one plausible mechanism. Included among these 9 cases are 3 (1992 79-021A, 1997 49-097A, and 1998 45-044A) in which the driver's fatal chest injuries likely were caused by the steering wheel because the steering column moved and the airbag was not in position to protect the driver. The driver in one case (1998 49-151A) may have died of severe chest injuries caused by forces applied by the shoulder belt webbing. The remaining 5 are crashes in which the driver's critical injuries may have resulted from contact with the steering wheel subsequent to pushing through the airbag. There was intrusion in all 5 of these crashes, which might have contributed to the driver's critical injuries, but

Table 4.   Cases with Unknown Cause of Death because of Competing Possibilities

Calendar Year	PSU	Case ID	Critically Injured Body Region	Belt Use	Delta V (km/h)	Possible Cause-of-Death Categories
1992	79	021A	Chest	No	74	Steering wheel, airbag
1997	49	097A	Chest	No	96	Steering wheel, airbag
1998	04	064A	Head/chest	No	88	Airbag, steering wheel through airbag, steering wheel
1998	05	012A	Head/chest	No	34	Instrument panel, steering wheel through airbag
1998	06	137A	Head	No	74	Airbag, steering wheel through airbag, instrument panel
1998	11	135J	Head	Yes	62	Airbag, steering wheel through airbag, intrusion
1998	49	151A	Chest	Yes	85	Seat belt, steering wheel
1998	45	044A	Chest	No	Unknown	Steering wheel, instrument panel, airbag
1998	72	073B	Head/chest/abdomen	No	Unknown	Steering wheel through airbag, airbag, steering wheel

none was characterized by a total loss of survival space. Also, four of the possible bottomed-out airbag cases had circumstances that indicated the drivers might have been near the airbags when they deployed, so an airbag-related death could not be ruled out. All 10 cases with multiple possible cause-of-death scenarios are presented in more detail below.

Case 1992 79-021A – This crash had been characterized previously as an airbag-related death [3]. It involved a 1991 Dodge Grand Caravan that crashed into the rear end of a parked 1958 Chevrolet Bel Air. The estimated delta V was 74 km/h (46 mi/h), and the 35-year-old un-belted male driver died of chest injuries (crushed sternum with concomitant tear in the wall of the right ventricle) that the NASS investigator attributed to contact with the steering wheel. No intrusion of any kind was mentioned in the NASS file, and the steering wheel was reported as not deformed. However, a comparison of the steering wheel's postcrash position with that in a similar vehicle crashed under similar conditions at IIHS's Vehicle Research Center indicated that the column probably rotated upward and forward during the crash. In the crash test, the dummy's chest contacted the steering wheel rim without the airbag intervening, because the rotation of the column occurred early in the crash. Although the crash test now suggests contact with the steering wheel was the most likely cause of the driver's fatal injuries, the possibility remains that the airbag caused the driver's death. According to the NASS file, the driver's blood alcohol concentration was 0.16 percent, so it is possible that he had fallen asleep under the influence of the alcohol and was near the airbag when it deployed.

Case 1997 49-097A – This crash involved a 1997 Ford Escort that sideswiped the right side of a car stopped in the curb lane then jumped the curb and crashed into a light pole. The 71-year-old female driver who was not using her seat belt died of lacerations to the heart and aorta (AIS 6) with accompanying rib fractures. The NASS records also indicated a basilar skull fracture (AIS 3) and many other broken bones. The estimated delta V for the Escort was 97 km/h, but the investigator's notes indicated that this estimate was high. While considerable intrusion was noted (29 cm measured at the instrument panel), it is possible that the initial sideswipe impact deployed the airbag so that it was not available for the more

severe crash with the pole. It is also possible that the initial events caused the driver to be out of position and near the airbag when it deployed. Abrasions on the face, which the investigator attributed to the airbag, and the basilar skull fracture is a typical airbag inflation injury pattern but also might have been caused by head impact against some other part of the car's interior. Measurements recorded by the investigator also indicate that the steering wheel moved laterally 20 cm, so it is possible that the column moved before the unbelted driver contacted the airbag. Thus she would not have the full advantage of its load-spreading benefits.

Case 1998 04-064A – This crash involved a 1995 Toyota Camry that ran off the road and crashed into a tree with an estimated delta V of 88 km/h (55 mi/h). The damaged area was offset to the passenger side of the car. The 22-year-old male driver (183 cm tall, 98 kg), was not using his seat belt and died of head (brainstem laceration) or chest (aorta tear) injuries, both of which the NASS investigator attributed to contact with the steering wheel. Both injuries are common among drivers believed to be killed by their airbags, but both injuries are also reported in nonairbag crashes. Shatsky et al. [5] and Got et al. [6] reported causing brainstem lacerations in experimental head impacts with human surrogates, and both authors indicated that the injury is observed clinically. A review of clinical studies of injuries caused by motor vehicle crashes indicated that aorta tears account for 10-20 percent of automobile crash-related deaths [7].

No steering column intrusion is noted in the documentation of this case, but photographs show that it is rotated upward somewhat. Consequently, the driver may have contacted the steering wheel without the airbag intervening, as in case 1992 79-021A, or he may have contacted the steering wheel after pushing through the airbag. Another possibility, however, is that the airbag itself caused the driver's fatal injuries. The NASS report indicated that his blood alcohol concentration was 0.121 percent, so it is possible that the driver fell asleep and was near the airbag prior to the crash.

Case 1998 05-012A – This crash involved a 1995 Ford Escort that yawed across the centerline and was struck at the front of the right side with an oblique principal direction of force. Although photographs of the damage suggest this was a side crash, it was selected for this study on the basis of the NASS-assigned collision deformation classification. The 32-year-old male driver was not using his seat belt and died of multiple injuries to the head, chest, and abdomen. The NASS investigator attributed the head injuries (cerebellum hematoma, AIS 4, and basilar skull fracture, AIS 3) to contact with the center instrument panel. The oblique direction of force, which was from the right of center, and the fact that deformation of the vehicle showed that right side structures were pushed toward the driver suggest the driver airbag likely could not have prevented this contact. The case was assigned to the unknown category because the driver's chest and abdomen injuries (bilateral rib cage fractures, AIS 4, and lacerated liver, AIS 4) were attributed to the steering wheel by the investigator. This could have occurred by one of two means — either movement of the steering column during the crash exposed an unprotected portion of the rim to the driver's chest or the unbelted driver pushed through the airbag then contacted the steering wheel rim.

Case 1998 06-137A – This crash involved a 1997 Pontiac Grand Am that had a left frontal offset crash into a tree on the side of the road opposite the lane in which the car was originally traveling. After impacting the tree and rotating counterclockwise, the car slid sideways and struck a utility pole with the right rear quarter. The 47-year-old male driver (183 cm tall, 156 kg) was not using his seat belt and died of severe head injuries (bilateral cerebrum hematoma, diffuse axonal injury, basilar skull fracture), which the NASS investigator attributed to contact with the roof on the passenger side of the car. Contact evidence and the likely driver kinematics during a secondary right side impact against a utility pole could support this possibility. However, it seems more likely that the driver sustained his fatal injuries during the initial impact or rebound following the initial impact by one of three mechanisms. The driver may have fallen asleep and consequently was near the airbag when it deployed. This possibility is suggested by the crash circumstances; the crash occurred early in the morning (8:05 a.m.) on the opposite side of the road with no preceding crash avoidance maneuver. Alternatively, this relatively tall and heavy driver, if he was normally seated, may have impacted the windshield header or A-pillar despite the airbag deployment, which may not have been strong enough or timely enough to prevent such impact, or during rebound from the airbag. Finally, the driver's head may have impacted the steering wheel rim, which was bent, after pushing through the airbag.

Case 1998 11-135J – This crash involved a 1998 Dodge Dakota pickup that ran off the road and crashed into a tree. The damaged portion of the vehicle's front was offset to the passenger side, and the estimated delta V was 63 km/h (39 mi/h). The 16-year-old male driver (178 cm tall, 66 kg) was using his seat belt at the time of the crash and died of head injuries (diffuse axonal injury, brainstem com-

pression and others) that the NASS investigator attributed to a noncontact mechanism. There is no explanation in the case materials of why the investigator did not attribute the driver's head injuries to some part of the vehicle interior. The steering column and instrument panel were moved 10 cm rearward into the driver's space, and intrusion of the toepan measured 27 cm. Consequently, intrusion likely contributed to the driver's injuries. Comparison with an IIHS 40 mi/h offset frontal crash test of the same model pickup suggests the driver's head injuries may have been caused by contact with the steering wheel through the airbag. In this test, in which rearward steering column intrusion was half that in the NASS crash, the driver dummy recorded high neck forces (3.9 kN) and high head accelerations (163 g) when its head contacted the steering wheel through the airbag. Another possibility is that the driver's fatal injuries were caused by the deploying airbag. The crash occurred early in the morning, and it is possible that the driver fell asleep before the crash. Plus, the truck continued 50 m after leaving the roadway and traversed down a steep hill before impacting the tree. Consequently, despite being belted, the driver might have been near the airbag when it deployed.

Case 1998 45-044A – This was a frontal crash, offset to the passenger side, between a 1997 Ford Taurus and a 1991 Ford Aerostar van. The delta V for the Taurus was not estimated. The 45-year-old male driver of the Taurus (175 cm tall, 100 kg) was not using his seat belt during the crash and died of unspecified thoracic injuries that the NASS investigator attributed to the airbag and steering wheel. Despite massive intrusion into the front passenger space, the driver side had moderate levels of intrusion (16 cm at the toepan, 18 cm at the instrument panel near the center of the car, driver door opening appeared to be intact). It is possible that the principal direction of force was not 12 o'clock as indicated by the NASS investigator and that the driver's injuries were caused by contact with the intruding center instrument panel. Alternately, the driver may have contacted the steering wheel despite the presence of the airbag as the NASS investigator surmised. Finally, the airbag itself cannot be ruled out as a possible cause for the driver's fatal injuries. Blood test results indicated the presence of unspecified drugs, which might have caused drowsiness and could explain why the driver drifted across the centerline and into oncoming traffic.

Case 1998 49-151A – This crash involved a 1995 Ford Escort station wagon that had a right frontal offset crash with a 1988 Honda Accord while travelling the wrong way on a one-way road. The 51-year-old female driver of the Escort, who was using only the automatic shoulder belt (i.e., not manual lap belt), died of severe chest injuries that included bilateral rib fractures (AIS 5) and a major contusion of the heart (AIS 4). The investigator attributed these injuries to the steering wheel, which was deformed in the crash, but high shoulder belt forces may also have caused the injuries. If the steering wheel was the agent that caused her chest injuries, then likely the intruding steering column moved in a way that prevented the airbag from cushioning the driver's chest from this contact.

Case 1998 72-073B – This crash involved a 1992 Chevrolet Caprice that crashed head on into a 1989 Ford Crown Victoria. The 60-year-old male driver of the Caprice was not using his seat belt during the crash and died with head, chest, and abdomen injuries. The chest and abdomen injuries were attributed to the steering wheel, which was deformed and intruded into the occupant compartment. Thus, the driver may have contacted the steering wheel rim without protection of the airbag because rotation of the steering column moved the airbag. It is also possible that the driver contacted the steering wheel through the airbag. The investigator attributed the driver's head injuries to contact with the airbag module flap. If this assessment is correct, then it seems likely that the unbelted driver was too near the steering wheel before the airbag deployed for it to provide much protection.

SUMMARY

In spite of a large increase in the number of cases available for analysis, the results of this update of driver fatalities in frontal crashes of airbag-equipped vehicles are consistent with previous analyses [2-3]. As before, the catastrophic collapse of the occupant compartment was identified as the leading cause of driver deaths, accounting for 42 percent of the fatalities. The force of deploying airbags and ejection from the vehicle also accounted for a large proportion of the drivers' fatal injuries (16 percent and 13 percent, respectively). In contrast with earlier findings, there were 5 cases in which bottoming out the airbag could not be ruled out as a possible explanation for the driver's fatal injuries. However, even if bottoming out the airbag was the cause of these drivers' injuries, the intrusion noted in all 5 cases likely played a role. Furthermore, the precrash circumstances in 4 of these cases also suggested the possibility that these drivers were near their airbags and may have been injured by them when they deployed.

Many of these frontal crashes were simply not survivable because the driver's survival space was obliterated. However, some other deaths are the result of a driver impacting intruding vehicle components in an otherwise intact occupant compartment. In addition, there were a number of cases in this sample of frontal crashes in which the steering wheel apparently rotated upward or intruded rearward toward the driver. When the steering wheel rotates upward, the driver can impact the stiff rim of the steering wheel causing serious injuries to the chest and abdomen. Furthermore, intrusion of the steering wheel into the occupant compartment can compromise the protection the airbag affords. In IIHS frontal offset crashes at 64 km/h (40 mi/h), large movement of the steering column has been associated with the dummy's head striking the steering wheel through the airbag and

uncontrolled sideways movement of the driver dummy during rebound. Airbags and seat belts can protect occupants most effectively if the safety cage remains intact and the steering column is stable.

Airbags continue to kill drivers in both high- and low-speed crashes. Four of the 18 airbag-related deaths analyzed were at estimated delta Vs lower than 25 mi/h, but in 9 cases delta Vs were 29 mi/h and higher (4 were unknown). In the 1998 model year, many manufacturers depowered airbags as the result of temporary changes in FMVSS 208 allowing them to certify frontal crash performance for unbelted occupants using a 30 mi/h sled test. There were 10 1998 and 1999 model year vehicles included in this analysis, too few to judge whether airbag-related deaths have decreased as a result of depowering. Of the 10 deaths in late-model vehicles, 6 were the result of intrusion, 1 from impact with an interior surface, 1 was ejected, and 2 were of unknown causes in which bottoming out of the airbag could not be ruled out. In the next few years, more and more vehicles will be equipped with advanced airbags incorporating such features as dual deployment thresholds and dual-stage deployments that should reduce airbag injuries substantially. It will be important to continue crash investigations of late model airbag vehicles to see whether the benefits are realized and to ensure that airbags continue to perform well in protecting occupants in frontal crashes.

Evidence that first generation airbags deploy with too little power for some severe crashes remains scant. There were a few cases in which it was possible that the driver may have bottomed out the airbag. However, in each of these cases there were competing and sometimes equally plausible alternative explanations for the fatal injuries. Crashes are complicated events, and it is sometimes difficult after the fact to reliably determine the cause of death. In some cases crash reconstruction may shed light on the most feasible explanation if the crash circumstances can be reliably replicated (see the Dodge Caravan case discussed earlier). These cases point to the need to continue monitoring airbag performance. In cases where bottoming out is a possible explanation, further investigation may be warranted. However, even if there were convincing evidence these drivers died because their airbags had insufficient power, drivers overall were much more likely to die because their airbags were too powerful and actually contributed to their fatal injuries (18 vs. 5 possible cases).

ACKNOWLEDGMENT

This work was supported by the Insurance Institute for Highway Safety.

REFERENCES

1. Insurance Institute for Highway Safety. 1998. Looking for evidence of inadequate airbag energy: methodology for evaluation. Arlington, VA.

2. Cammisa, M.X.; Reed, R.T.; Ferguson, S.A.; and Lund, A.K. 2000. Driver fatalities in frontal crashes of airbag-equipped vehicles: a review of 1989-96 NASS cases. SAE Technical Paper Series 2000-01-1003. Warrendale, PA: Society of Automotive Engineers.

3. Lund, A.K.; Ferguson, S.A.; and Powell, M.R. 1996. Fatalities in air bag-equipped cars: a review of 1989-93 NASS cases. SAE Technical Paper Series 960661. Warrendale, PA: Society of Automotive Engineers.

4. U.S. Department of Health and Human Services. 1987. Vital and health statistics; anthropometric reference data and the prevalence of overweight, United States, 1976-80, pp. 24-45, tables 14-15. Washington, DC: U.S. Government Printing Office.

5. Shatsky, S.A. et al. 1974. Traumatic distortions of primate head and chest: correlation of biomechanical, radiological, and pathological data. *Proceedings of the 18th Stapp Car Crash Conference*, 351-82. Warrendale, PA: Society of Automotive Engineers.

6. Got, C. et al. 1978. Results of experimental head impacts on cadavers: the various data obtained and their relations to some measured parameters. *Proceedings of the 22nd Stapp Car Crash Conference*, 55-100. Warrendale, PA: Society of Automotive Engineers.

7. Viano, D.C. 1983. Biomechanics of non-penetrating aortic trauma: A review. *Proceedings of the 27th Stapp Car Crash Conference*, 109-14. Warrendale, PA: Society of Automotive Engineers.

Appendix – All NASS Cases Grouped by Cause-of-Death Category

Calendar Year	PSU	Case ID	Cause of Death	Body Region of Critical Injury	Crash Event	Belt Use	Delta V (km/h)	Airbag Deployment	Model Year	Make	Model
1993	5	125A	Airbag	Chest	Single	Belted	47	Yes	1992	Toyota	Camry
1993	6	006A	Airbag	Chest	Single	Unbelted	47	Yes	1990	Plymouth	Acclaim
1993	8	133A	Airbag	Chest	Multiple	Unbelted	Unknown	Yes	1992	Chrysler	Fifth Ave
1994	11	150A	Airbag	Chest	Single	Unbelted	61	Yes	1992	Plymouth	Sundance
1995	2	140A	Airbag	Chest	Multiple	Unknown	47	Yes	1993	Chrysler	Town & Country
1995	9	167A	Airbag	Head	Multiple	Unbelted	Unknown	Yes	1995	Infiniti	J30
1996	8	100A	Airbag	Chest	Single	Unknown	37	Yes	1995	Nissan	Maxima
1996	41	024A	Airbag	Chest	Multiple	Belted	26	Yes	1995	Ford	Escort
1997	4	146A	Airbag	Chest	Single	Lap only	31	Yes	1996	Geo	Metro
1997	41	157J	Airbag	Head	Single	Unknown	35	Yes	1996	Geo	Prizm
1997	72	103A	Airbag	Chest	Single	Unbelted	48	Yes	1996	Ford	Taurus
1998	2	148A	Airbag	Head	Multiple	Belted	Unknown	Yes	1995	Mitsubishi	Galant
1998	4	004B	Airbag	Head/chest	Multiple	Unbelted	Unknown	Yes	1993	Buick	Century
1998	9	087B	Airbag	Head/chest	Single	Belted	60	Yes	1994	Mercury	Tracer
1998	11	018B	Airbag	Head/chest	Multiple	Unbelted	48	Yes	1991	Plymouth	Voyager
1998	41	076B	Airbag	Chest	Single	Unbelted	27	Yes	1993	Honda	Accord
1998	43	088B	Airbag	Chest	Single	Belted	55	Yes	1992	Honda	Accord
1998	48	090B	Airbag	Chest	Multiple	Unbelted	Unknown	Yes	1991	Ford	Crown Victoria
1992	43	210A	Ejection	Head	Multiple	Unbelted	58	Yes	1992	Chevrolet	Camaro
1992	48	122A	Ejection	Head	Multiple	Unbelted	21	Yes	1991	BMW	318i
1994	74	125A	Ejection	Head	Multiple	Unbelted	74	Yes	1991	Geo	Metro
1995	4	029A	Ejection	Head	Multiple	Unbelted	34	Yes	1993	Dodge	Shadow
1995	49	135A	Ejection	Head	Multiple	Unbelted	34	Yes	1995	Ford	F-150
1995	73	117A	Ejection	Head	Single	Unbelted	35	Yes	1995	GMC	Sierra
1996	49	195A	Ejection	Head	Multiple	Belted	18	Yes	1994	Mitsubishi	Mirage
1997	11	068B	Ejection	Head	Multiple	Unbelted	Unknown	Yes	1990	Dodge	Daytona
1997	72	105A	Ejection	Head	Multiple	Unbelted	26	Yes	1993	Chevrolet/Geo	Nova/Prizm
1997	72	126B	Ejection	Head	Multiple	Unbelted	Unknown	Yes	1992	Chrysler	LeBaron
1997	73	127A	Ejection	Spine	Multiple	Unbelted	23	No	1996	Chevrolet	Blazer
1998	5	045B	Ejection	Head	Multiple	Unbelted	Unknown	Yes	1994	Chevrolet	Camaro
1998	12	186B	Ejection	Spine	Single	Unbelted	Unknown	Yes	1990	Plymouth	Horizon
1998	49	046A	Ejection	Head	Multiple	Unbelted	Unknown	Yes	1996	Ford	Ranger
1998	49	176A	Ejection	Head	Multiple	Unbelted	Unknown	Yes	1998	Ford	Mustang
1994	6	021A	Interior surface	Abdomen	Single	Unbelted	Unknown	Yes	1993	Buick	LeSabre
1995	12	072A	Interior surface	NA	Single	Unbelted	19	Yes	1994	Cadillac	DeVille
1997	11	108B	Interior surface	Chest	Single	Unbelted	40	Yes	1993	Mazda	626
1997	75	188B	Interior surface	Head	Single	Unbelted	40	Yes	1991	Dodge	Dynasty
1998	45	165J	Interior surface	Head	Multiple	Unbelted	Unknown	No	1996	Honda	Accord
1998	72	139B	Interior surface	Head	Single	Unbelted	Unknown	Yes	1991	Chevrolet	Corsica
1998	73	106A	Interior surface	Head	Multiple	Unbelted	40	Yes	1998	Lincoln	Continental
1990	11	111A	Intrusion	Unknown	Multiple	Belted	Unknown	Yes	1990	Ford	Taurus
1991	75	023A	Intrusion	Chest	Single	Unbelted	63	Yes	1992	Ford	Tempo
1991	79	139A	Intrusion	Chest	Multiple	Unbelted	48	Yes	1990	Chrysler	Lebaron
1991	82	057A	Intrusion	Chest	Single	Unbelted	95	Yes	1990	Geo	Storm
1993	11	112A	Intrusion	Head	Single	Belted	Unknown	Yes	1990	Mercury	Sable
1994	12	025B	Intrusion	Unknown	Multiple	Belted	Unknown	Yes	1991	Dodge	Caravan
1994	12	064A	Intrusion	Head	Single	Unbelted	Unknown	Yes	1994	Cadillac	Eldorado
1994	41	199A	Intrusion	Head	Single	Belted	Unknown	Yes	1993	Toyota	Camry

continued on next page

Calendar Year	PSU	Case ID	Cause of Death	Body Region of Critical Injury	Crash Event	Belt Use	Delta V (km/h)	Airbag Deployment	Model Year	Make	Model
1994	72	082A	Intrusion	Lower extremity	Multiple	Unbelted	92	Yes	1994	Ford	Mustang
1994	74	157A	Intrusion	Chest	Single	Unbelted	76	Yes	1993	Dodge	Intrepid
1995	12	163A	Intrusion	Head	Single	Unbelted	55	Yes	1995	Saturn	SL
1995	13	149A	Intrusion	Head	Single	Unbelted	50	Yes	1992	Dodge	Caravan
1995	13	208A	Intrusion	Chest	Multiple	Unbelted	74	Yes	1995	Nissan	Maxima
1995	41	069A	Intrusion	Chest	Single	Unbelted	68	Yes	1994	Dodge	Stealth
1995	49	213A	Intrusion	Chest	Single	Unbelted	85	Yes	1992	Chevrolet	Corvette
1995	75	103A	Intrusion	Unknown	Single	Belted	47	Yes	1995	Chevrolet	Astro
1996	8	084A	Intrusion	Chest	Single	Unbelted	34	Yes	1994	Ford	Aspire
1996	9	010B	Intrusion	Chest	Single	Belted	77	Yes	1992	Dodge	Caravan
1996	11	192A	Intrusion	Head	Single	Belted	63	Yes	1994	Ford	Aspire
1996	43	069J	Intrusion	Head	Multiple	Belted	42	Yes	1994	Chevrolet	Astro
1996	45	053B	Intrusion	Head	Single	Belted	Unknown	No	1990	Chrysler	New Yorker
1996	74	033A	Intrusion	Head	Multiple	Belted	76	Yes	1995	Honda	Accord
1996	74	144B	Intrusion	Chest	Multiple	Unknown	90	Yes	1995	GMC	Sonoma
1996	75	098A	Intrusion	Chest	Multiple	Unbelted	47	Yes	1995	Plymouth	Voyager
1996	75	158A	Intrusion	Unknown	Multiple	Belted	13	Yes	1996	GMC	Suburban
1996	78	122A	Intrusion	Unknown	Single	Unbelted	61	Yes	1996	Ford	Mustang
1996	81	014A	Intrusion	Head	Multiple	Unbelted	34	Yes	1995	Chevrolet	pickup
1997	4	104A	Intrusion	Chest	Single	Belted	23	Yes	1995	Pontiac	Bonneville
1997	11	139J	Intrusion	Head	Single	Belted	Unknown	Yes	1995	Ford	Probe
1997	43	180A	Intrusion	Head	Multiple	Belted	43	Yes	1995	Toyota	Tacoma
1997	72	099A	Intrusion	Chest	Single	Unbelted	43	Yes	1995	Toyota	Tercel
1997	73	061A	Intrusion	Head	Multiple	Unbelted	Unknown	Yes	1995	Dodge	W-series truck
1997	73	115B	Intrusion	Head	Multiple	Belted	Unknown	Yes	1990	Dodge	Dynasty
1997	78	009A	Intrusion	Head/chest	Single	Belted	47	Yes	1996	Mitsubishi	Gallant
1998	2	119A	Intrusion	Head	Multiple	Belted	64	Yes	1995	Ford	Windstar
1998	2	154A	Intrusion	Abdomen	Single	Belted	43	Yes	1995	Toyota	Avalon
1998	8	124A	Intrusion	Head/chest	Single	Unbelted	63	Yes	1995	Ford	Taurus
1998	9	053A	Intrusion	Head	Single	Belted	60	Yes	1998	Saturn	SL
1998	9	144A	Intrusion	Head	Single	Belted	81	Yes	1998	Chevrolet	Malibu
1998	41	004A	Intrusion	Head	Single	Unknown	Unknown	Yes	1998	Mercury	Tracer
1998	45	151B	Intrusion	Unknown	Single	Unknown	Unknown	Yes	1992	Plymouth	Acclaim
1998	49	058A	Intrusion	Chest	Multiple	Unbelted	Unknown	Yes	1997	Ford	F-series
1998	72	064A	Intrusion	Head	Single	Unbelted	Unknown	Yes	1996	Isuzu	Rodeo
1998	72	073B	Intrusion	Chest/abdomen	Single	Unbelted	76	Yes	1992	Chevrolet	Caprice
1998	73	123B	Intrusion	Head	Single	Unbelted	Unknown	Yes	1993	Ford	Taurus
1998	74	111B	Intrusion	Head	Multiple	Belted	Unknown	Yes	1993	Jeep	Cherokee
1998	79	046A	Intrusion	Head	Multiple	Unbelted	68	Yes	1998	Chevrolet /Geo	Metro
1998	45	170A	Intrusion; related to rollover	Head	Multiple	Belted	Unknown	Yes	1995	Jeep	Cherokee
1998	81	092A	Intrusion; related to side impact	Head/chest	Multiple	Belted	Unknown	Yes	1999	Chevrolet	Blazer
1995	79	002A	Nondeployment	Head	Single	Unbelted	50	No	1995	Ford	F-150
1998	43	262A	Other	NA	Multiple	Belted	Unknown	Yes	1997	Chevrolet	G-series van
1995	45	100A	Other (seat belt)	Chest/abdomen	Single	Belted	64	Yes	1994	Ford	F-150
1997	79	109B	Other (seat belt)	Chest	Single	Shoulder	39	Yes	1993	Nissan	Altima
1997	82	186B	Other (seat belt)	Chest	Single	Belted	47	Yes	1992	Volvo	240 series
1991	79	021A	Unknown	Chest	Single	Unbelted	74	Yes	1991	Dodge	Caravan
1995	79	501A	Unknown	Abdomen	Multiple	Unbelted	60	Yes	1995	Dodge	Neon
1996	12	081B	Unknown	Unknown	Multiple	Unbelted	42	Yes	1992	Plymouth	Sundance

continued on next page

Calendar Year	PSU	Case ID	Cause of Death	Body Region of Critical Injury	Crash Event	Belt Use	Delta V (km/h)	Airbag Deployment	Model Year	Make	Model
1996	13	113B	Unknown	Head	Multiple	Unbelted	13	Yes	1992	Plymouth	Sundance
1996	78	024A	Unknown	Head	Multiple	Belted	21	Yes	1994	Buick	LeSabre
1997	12	065A	Unknown	Unknown	Single	Unbelted	Unknown	Yes	1997	Dodge	Avenger
1997	12	167A	Unknown	Unknown	Multiple	Unbelted	35	Yes	1997	Chevrolet	S-10
1997	12	203A	Unknown	Unknown	Multiple	Unbelted	Unknown	Yes	1997	Pontiac	Grand Am
1997	49	097A	Unknown	Chest	Multiple	Unbelted	97	Yes	1997	Ford	Escort
1998	4	064A	Unknown	Head/chest	Single	Unbelted	89	Yes	1995	Toyota	Camry
1998	5	012A	Unknown	Head/chest	Single	Unbelted	34	Yes	1995	Ford	Escort
1998	5	115B	Unknown	Unknown	Single	Unbelted	89	Yes	1994	Chevrolet	Beretta
1998	6	137A	Unknown	Head	Multiple	Unbelted	47	Yes	1997	Pontiac	Grand Am
1998	6	147A	Unknown	Unknown	Single	Unbelted	61	Yes	1997	Honda	Civic
1998	11	135J	Unknown	Head	Multiple	Belted	63	Yes	1998	Dodge	Dakota
1998	11	196A	Unknown	Chest	Multiple	Unbelted	Unknown	Unknown	1996	Ford	F-150
1998	11	214B	Unknown	NA	Multiple	Belted	19	No	1994	Buick	LeSabre
1998	12	040A	Unknown	Unknown	Single	Belted	18	No	1995	Chevrolet	S-10
1998	12	234A	Unknown	Unknown	Single	Unbelted	55	Yes	1999	Pontiac	Grand Prix
1998	45	044A	Unknown	Chest	Multiple	Unbelted	Unknown	Yes	1997	Ford	Taurus
1998	49	151A	Unknown	Chest	Multiple	Shoulder	85	Yes	1995	Ford	Escort
1998	78	083A	Unknown	Unknown	Multiple	Belted	Unknown	Yes	1995	Chevrolet	Lumina

AIR BAG CRASH INVESTIGATIONS

Augustus "Chip" B. Chidester
Thomas A. Roston
National Highway Traffic Safety Administration
United States of America
Paper Number 246

ABSTRACT

The performance of air bags, as an occupant protection system, is of high interest to the National Highway Traffic Safety Administration (NHTSA or Agency). Since 1972, the NHTSA has operated a Special Crash Investigations (SCI) program which provides in-depth crash investigation data on new and rapidly changing occupant protection technologies in real world crashes. The Agency uses these in-depth data to evaluate vehicle safety systems and form a basis for rulemaking actions. The data are also used by the automotive industry and other organizations to evaluate the performance of motor vehicle occupant protection systems such as air bags.

This paper presents information from NHTSA's SCI program concerning crash investigations on air bag equipped vehicles. The paper focus is on data collection and some general findings in air bag crash investigations including: air bag related fatal and life threatening injuries; side air bags; redesigned air bags and advanced air bags.

BACKGROUND

The NHTSA performs research and develops safety programs and standards in an effort to reduce the toll of deaths, injuries, and property damage from traffic crashes. In-depth field investigations on crashes with an air bag deployment are conducted in the SCI program under the auspices of the National Center for Statistics and Analysis (NCSA). SCI cases are an anecdotal data set used to examine and evaluate the latest safety systems. These investigations play a vital role by providing data relative to real world events. The objective of the SCI crash investigation is to provide detailed data for the analysis of air bag deployments. These in-depth investigations include the documentation of crash circumstances, the identification of injury mechanisms, the evaluation of safety countermeasure effectiveness, and the early detection of design and functional problems relative to air bags and vehicle occupants.

SPECIAL CRASH INVESTIGATIONS

The SCI investigators follow up the on-site investigations by interviewing crash victims and other involved parties and by reviewing medical records to determine the nature, cause, and severity of the injuries. Each investigation provides extensive information about pertinent pre-crash, crash, and post-crash events involving the occupants, vehicles, rescue procedures, and environmental factors that may have contributed to the event's occurrence and/or resulting severity. Included in each report are analyses and determinations of occupant kinematics and vehicle dynamics.

From 1972 to 1990, the SCI program investigated all crashes reported to NHTSA that involved an air bag equipped vehicle. However, due to the rapid growth in the number of air bag equipped vehicles present in the marketplace after 1990, the SCI program shifted from investigating all air bag vehicle crashes to investigating only air bag related special interest cases. These cases involve such issues as air bag related serious and fatal injuries, air bag success stories, interaction between air bags and child safety seats, air bag non-deployment crashes, inadvertent air bag deployments, front right passenger air bag performance, side air bag performance, redesigned air bag systems, and the effectiveness of advanced occupant protection systems. These SCI air bag cases have been utilized by the agency and the automotive safety community to acquire knowledge in real world performance of new and emerging air bag systems and have been instrumental in influencing improvements to new generations in air bag technologies.

HISTORY OF AIR BAG FATALITIES

In 1991, the SCI program investigated and confirmed the first allegation of a driver air bag related fatality (CA9109). At that time, the SCI was tasked with locating, investigating, confirming and reporting air bag related life threatening and fatal injury cases. In 1993, the first air bag deployment related child fatality (CA9307) was confirmed. The first air bag related fatality cases were not sampled in the National

Automotive Sampling System (NASS) Crashworthiness Data System (CDS) until 1997.

In 1996, the SCI program was significantly expanded in order to perform investigations of all air bag related life threatening or fatal injuries. In October 1996, NHTSA began publishing summary tables for each confirmed air bag related fatality and seriously injured occupant. The tables are available through the NHTSA website at http://www.nhtsa.dot.gov/people/ncsa /scireps.html. These summary tables contain basic information about serious injuries and fatalities related to air bag deployments in low speed crashes to:

(1) infants in rear facing child safety seats (RFCSS);
(2) children not in RFCSS;
(3) drivers; and
(4) adult passengers

NHTSA has defined children as occupants 12 years of age and under. Serious injury has been defined as a level sufficient to be a threat to life. The injuries that are considered a threat to life have a significant effect on mortality. Low speed crashes have been defined as those with a speed change less than 25 miles per hour.

To be fatally injured by an air bag, the deployment energy of the air bag must be imparted to the occupant. For the deployment energy to be imparted to the occupant, he/she must be in the path of the deploying air bag. In low speed crashes the occupant is most typically out-of-position (OOP) and in the path of a deploying air bag in one of the following two scenarios:

5. The occupant's initial seating position will place them in the air bag deployment path. Initial positioning may include: small or short-stature occupants seated in close proximity to the air bag, as well as occupants that fall asleep, have passed out or are leaning into the air bag deployment path. This scenario includes both belted and unbelted occupants.

6. The occupant is repositioned to a location within the air bag deployment path just prior to deployment by a pre-impact or at-impact event. The event that repositions the occupant into the deployment path includes a number of factors such as: pre-impact braking, multiple closely spaced near deployment events, running off the road or long crash pulses which also result in late deployments. Unbelted or improperly belted occupants are more likely to become out-of-position in this scenario.

In an effort to create as close to a census of air bag related fatalities as possible in the SCI, the Fatality Analysis Reporting System(FARS) is queried for possible cases. This process is performed annually and has been ongoing since 1992. In 2000, this process was significantly improved. The FARS coding of confirmed air bag related fatality cases was reviewed in an effort to upgrade the criteria used to locate potential cases. A number of additional cases were identified and investigated. As a result, the SCI files contain a near census of low speed air bag related fatalities. Information regarding non-fatal crashes provides valuable engineering information regarding occupant/air bag interactions, however there is no scientific method in place to ensure that these crashes are thoroughly sampled. Since including non-fatal injuries in the calculations could produce risk estimates that are inconsistent with actual trends, only fatality counts were used in these estimates. The SCI continues to monitor the FARS, NASS, law enforcement community and media for potential cases.

SCI cases are divided into two categories unconfirmed and confirmed. Unconfirmed cases are crashes under active investigation where the air bag is suspected of being the injury mechanism. The unconfirmed case fatal counts were initially reported to alleviate a false sense of improvement in declining confirmed case counts. Confirmed case counts typically lag approximately six months from initiation to confirmation. The primary reason for the lag time is medical record acquisition for injured occupants.

For unconfirmed cases, there is always the possibility that the investigation, when completed, will not support a conclusion of an air bag-related injury or fatality. However, since 1997, the SCI headquarters team has pre-screened the notifications submitted. As a result, approximately 90% of the unconfirmed cases are eventually confirmed.

The SCI has confirmed, as of January 1, 2001, 172 air bag related fatalities as noted in Figure 1. One hundred and two air bag related fatalities were children. Eighty three were children not in a rear facing child safety seat (RFCSS), 19 were infants in rear facing child safety seats. There have been 63 adult drivers and 7 adult front right passenger fatalities.

Chidester, pg. 2

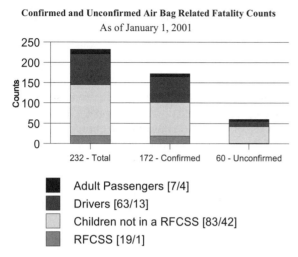

Figure 1. Air Bag Related Fatality Counts by Occupant Type Reported in NHTSA Tables.

Beginning in May 1998, NHTSA began reporting unconfirmed air bag related fatal injury counts in their monthly reports. As of January 1, 2001, NHTSA's SCI program had a total of 232 cases (172 confirmed and 60 unconfirmed) where the deployment of the driver or passenger air bag resulted in a fatal injury to an occupant in a low speed crash.

AIR BAG RELATED CHILD OCCUPANT FATALITIES

Of the 232 fatal cases reported (Figure 1), 145 are children (102 confirmed and 43 unconfirmed). One hundred and twenty five (83 confirmed and 42 unconfirmed) are children not in a RFCSS. Two of the 145 were children who were fatality injured by the driver air bag. Twenty (19 confirmed and 1 unconfirmed)) were children in a RFCSS.

Figure 2 presents the data for 12-month production periods for passenger air bags. The data was calculated by dividing the count of children fatally injured by a deploying air bag for each 12-month production period by the total number of registered vehicles with passenger air bags during that same interval. Each 12-month production period was aligned with the vehicle production year, September 1 through August 31.

NHTSA and its partners, (manufacturers, insurance companies and other organizations) have committed a high volume of public education resources in an effort to prevent air bag related injuries and fatalities, especially to children. This media attention appears to be having a positive effect on reducing child fatality cases. Despite the emergence of more than 15 million vehicles equipped with passenger air bags annually, there has been a significant down trend (see Figure 2) in the air bag related child fatality rate when normalized for vehicles equipped with passenger air bags.

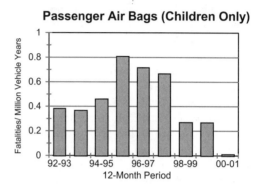

Figure 2. Normalized - Child Fatally Data by 12-Month Vehicle Sales Periods.

NHTSA issued a rulemaking action in March of 1997 that allowed automobile manufacturers to expediently reduce the force at which their air bags deployed. A number of manufacturers began installing these reduced power air bags in their 1998 model year vehicles. NHTSA refers to these reduced power as redesigned air bags.

NHTSA has investigated 10 cases (4 confirmed 6 unconfirmed) where the deployment of a redesigned passenger air bag resulted in a child fatality. In all cases the unbelted child was out-of-position and struck by the deploying redesigned air bag, resulting in a fatal

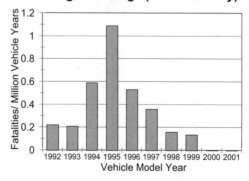

Figure 3. Children Fatally Injured by an Air Bag by Vehicle Model Year

Chidester, pg. 3

head or neck injury. As of January 1, 2001, NHTSA has not been notified of any crashes involving a 2000 or 2001model year vehicle where a a child passenger air bag related fatality is suspected (see Figure 3.)

CHILD PASSENGER KINEMATICS AND INJURY MECHANISMS

The following provides a discussion of the child passenger kinematics and injury mechanisms associated with front right passenger air bag induced injuries. In all cases, the crash investigators have identified the passenger air bag and/or air bag cover flap as the source of the critical-to-fatal injuries. Little or no intrusion of the occupant compartment was reported, and the cases have a speed changes less than 25 miles per hour. Given the level of the crash severities involved in most of the cases investigated, one would not expect that these children would have sustained life threatening or fatal injuries in the absence of an air bag.

The child air bag problem is most logically broken down into two distinct situations: infants in RFCSS or children not in a RFCSS located in the front right passenger seat position. A discussion of the injury mechanisms for each group are provided below.

Rear-facing Infant Injuries

Of the 20 infant fatalities, 11 were restrained in an appropriate infant seat, and the seat was secured by the seat belt in the front right seating position of a passenger air bag equipped vehicle. However, this is not considered properly restrained, since a RFCSS should never be placed in the front seat of a vehicle equipped with a passenger air bag. The only exception is for vehicles with no back seat that are equipped with an air bag on/off switch.

In the remaining nine cases, three were in RFCSS being held on the lap of the front right occupant and six were either not properly secured in the RFCSS or the RFCSS was not secured with the vehicle's seat belt. In all 20 cases, the vehicle's driver and/or other adult passengers ignored the warning labels located on the sun visor and/or on the child safety seat and placed the infant in the front right seating position.

The crash scenario for air bag involvement with rear-facing infant seats is similar for all cases. Upon impact, the deploying passenger air bag interacts violently with the back of the rear-facing infant seat, typically with sufficient force to crack or break the plastic shell. The force and rapid acceleration of this impact are carried through the rear facing child safety seat and into the child's head, typically causing skull fractures and associated brain injuries.

Children NOT in a RFCSS Injuries

As noted earlier, there are a total of 125 children, 83 are confirmed and 42 are unconfirmed as fatally injured by a deploying air bag. The confirmed cases contain information on the restraint use and injury information. The two children fatally injured by a deploying driver air bag will be excluded from this analysis. This analysis is based on the 81 confirmed children not in a RFCSS fatally injured by the deploying passenger air bag. With the exception of six, the children were either unrestrained or improperly restrained by the available seat belt system (see Figure 4.)

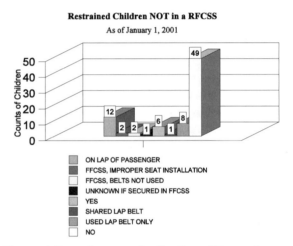

Figure 4. Restraint usage for Confirmed Forward Facing Children Fatally Injured by a Passenger Air Bag

Six children were restrained by the lap and shoulder belts. NHTSA recommends the proper restraint for all children 12 years of age and under is properly restrained in a rear seating position. For Five of the six belted children the proper restraint for their physical dimensions would have been a child safety seat. The sixth child was leaning into the path of the deploying passenger air bag.

Sixty-nine of the 81 fatalities involved pre-impact braking which caused the child to move forward into close proximity of the stored air bag. Occupant contact with the instrument panel prior to deployment has been confirmed for some cases by the identification of tissue, fluid, and/or clothing transfers on the air bag cover flap and/or instrument panel.

In the vast majority of cases, upon impact, the air bag deploys into the out-of-position child's chest, neck, and face resulting in a rapid translation and extension of the air bag under the chin against the neck and then wrapping upward from ear to ear. The occupant's head is effectively lifted upward off the neck resulting in an atlanto-occipital fracture (C1-C2) and/or a transection of the spinal cord, and/or brain stem injuries. Axonal diffuse brain injuries, consistent with rapid movements of the head are also commonly reported. Skull fractures were typically not observed.

There appears to be a correlation between restraint usage and the injury patterns (see Table 1). When the child passenger has any form of positive indication of restraint usage the head appears to have a higher propensity for injury. When there is no indication of restraint usage the neck appears to be the most likely source of the fatal injury.

Table 1 Children NOT in a RFCSS Passengers Confirmed as Fatally Injured by the Air Bag As of January 1, 2001			
Any Type of Restraint Usage	Head	Head & Neck	Neck
Positive	10	4	3
Negative	17	10	37

In Figure 5, the cumulative percentage of forward facing children indicates the majority of children involved in air bag related fatalities are less than seven years of age.

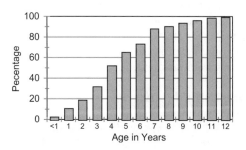

Figure 5. Cumulative Age in Percent of Children Not in a RFCSS.

AIR BAG RELATED ADULT FATALITIES

As of January 1, 2001 the Agency had a total of 87 cases (70 confirmed and 17 unconfirmed) in which the deployment of the driver or passenger air bag resulted in a fatal injury to an adult occupant in a low speed crash (see Table 2). In Table 2 the unconfirmed counts are in brackets.

Table 2
Adults Fatally Injured by the Air Bag
By Crash Year
As of January 1, 2001

YEAR	ADULT DRIVERS	ADULT PASSENGERS	TOTALS
1990	1	0	1
1991	4	0	4
1992	3	0	3
1993	4	0	4
1994	7	0	7
1995	5	0	5
1996	7	2	9
1997	16 [5]	4 [1]	20[6]
1998	10 [1]	0 [2]	10[3]
1999	2 [4]	0	2[4]
2000	4 [3]	1 [1]	5[4]
TOTAL	63[13]	7[4]	70[17]

Figures 6 and 7 present the data for 12-month production periods for adults fatally injured by a driver or passenger air bag. The data were calculated by dividing the count of adults fatally injured by a deploying air bag for each the emergence of more than 15 million vehicles equipped with driver and passenger air bags annually, there has been a significant downward trend (see Figures 7 and 8) in the adult air bag related fatality rate when normalized for vehicles equipped with air bags. 12-month production period by the total number of registered vehicles with driver or passenger air bags during that same interval. Each 12-month production period was aligned with the vehicle production year.

NHTSA and its partners (manufacturers, insurance companies and other organizations) have committed a high volume of public education resources in an effort to prevent air bag related injuries and fatalities, especially to short statured adults. This media attention appears to be having a positive effect on reducing adult driver fatality cases.

Chidester, pg. 5

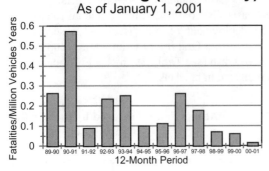

Figure 6. Adult Driver Air Bag Fatalities Normalized by 12-month Vehicle Sales Period.

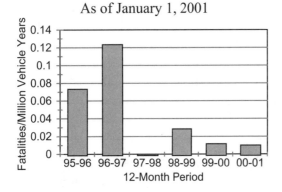

Figure 7. Adult Passenger Air Bag Fatalities Normalized by 12-month Vehicle Sales Period.

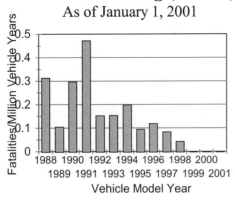

Figure 8. Normalized Adult Driver Fatally Data by Vehicle Model Year.

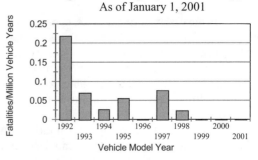

Figure 9. Normalized Adult Passenger Fatally Data by Vehicle Model Year.

Figures 8 and 9 present the data for adult air bag related fatalities by vehicle model years for driver and passenger air bags. There has been a general decrease in the fatality rate for the driver and passenger air bag over the last few years; and, as of January 1, 2001, there have been no adult driver or passenger air bag fatalities in model year 1999 vehicles and later.

DRIVER AND ADULT PASSENGER KINEMATICS AND INJURY MECHANISMS

The air bag injury patterns for drivers and adult passengers are uniquely different due to the location, size and shape of the air bags. The driver's air bag, because of its location in the steering wheel hub, is situated in close proximity to the occupant seating position. The front right passenger air bag is typically mounted in the instrument panel and is much larger in size. The larger air bag is needed to fill the space between the instrument panel and the right front passenger and, if present, a front middle passenger.

In all cases, the crash investigators have identified either the driver or passenger air bag and/or air bag cover flap as the source of the fatal injuries. Little to no intrusion of the occupant compartment was reported. Given the level of the crash severities involved, fatal injuries would not be expected.

Adult Driver Injuries

As discussed earlier, to be fatally injured by a driver air bag, the deployment energy of the air bag must be imparted to the out-of-position occupant. Typically, out of position includes small statured occupants seated directly in the path of a deploying air bag or occupants

Chidester, pg. 6

who become out-of-position into the deployment path during a pre-impact event.

The driver air bag injury patterns are directly affected by the occupants at-deployment positioning in relationship to the body region exposed to the deploying air bag. When the occupant is seated with his or her chest in the deployment path, the most common fatal-to-life threatening injuries include: Multiple bilateral rib fractures, flail chest, lung contusions, fractured sternum, laceration of the myocardium/pericardium, or aorta laceration/tear. In addition, some short stature drivers have received neck extension fractures at the atlanto-occipital joint (fracture at C1-C2) with and without spinal cord involvement. Head injuries are typically diffuse axonal brain injuries, brain stem injuries and basilar skull fractures resulting from rapid acceleration of the head from interaction with the inflating driver air bag.

Table 3
Adults Fatally Injured by the Air Bag By Height and Restraint Usage As of January 1, 2001

Height in Inches	Any Type of Restraint Usage Positive	Any Type of Restraint Usage Negative/Unknown	Totals
47		1	1
56		1	1
58	1		
59	1	2	3
60	1	2	3
61	1	2	3
62	4	7	11
63	3	2	5
64	5	8	13
65		5	5
66	1	4	5
67		2	2
68		2	2
69	1		
70		1	1
71		2	2
72	2	1	3
UNK	1		
Total	21	42	63

The 63 confirmed air bag fatality cases contain restraint and injury mechanism information. Twenty-one of the drivers fatally injured by a deploying driver air bag were either restrained or improperly restrained by the available safety belt system. In all but four of these cases the driver was 64" or less in height (see table 3). NHTSA recommends that drivers keep approximately 10 inches of free space between their breastbone and the air bag.

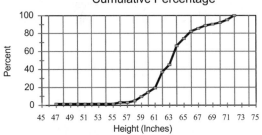

Figure 10. Cumulative Percentage of Confirmed Driver Air Bag Related Fatalities by Height.

In Figure 10, the cumulative percentage of drivers fatally injured by an air bag indicates the majority (80%) are 66 inches or less in height.

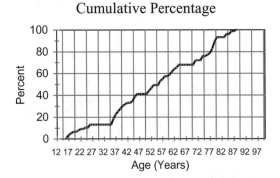

Figure 11. Cumulative Percentage of Adult Drivers in Years

Figure 11, shows the cumulative percentage of adults fatally injured by a deploying driver air bag. Sixty percent of the drivers fatally injured by a deploying driver air bag are over 50 years of age.

Adult Passenger Injuries

The injury patterns for adult passengers are similar to those seen in the forward facing children. The passenger air bags are typically much larger to afford crash protection to the front right occupant. Some of these air bags inflate over an extremely large area in an effort to protect the front middle occupied positions. Typically the front right passenger air bag requires more inflation volume, thereby creating a high potential for injury when an occupant is in close proximity to the air bag

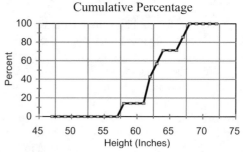

Figure 13. Cumulative Percentage of Confirmed Passenger Air Bag Related Fatalities by Height

In Figure 13, the cumulative percentage of passengers fatally injured by an air bag indicates the majority (71%) are 64 inches or less in height.

In Figure 14, the cumulative percentage of adult passengers fatally injured by a deploying air bag indicates the majority of adults involved are more than 57 years of age.

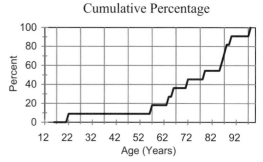

Figure 14. Cumulative Percentage of Adult Passengers in Years

Figure 13. Adults Confirmed as Fatally Injured by a Passenger Air Bag by Restraint Usage.

deployment path.

Adult passenger air bag fatalities are primarily unbelted occupants (see Figure 13). Most are placed out-of-position into the deployment path by a number of factors including pre-impact braking, multiple closely spaced near deployment events, running off the road or long crash pulses which also result in late deployments.

Upon impact, the air bag deploys into the out-of-position adult passenger's neck and head. As the air bag expands, it results in the rapid translation and extension of the air bag under the chin against the neck and then wrapping upward from ear to ear. The occupant's head is effectively lifted upward off the neck resulting in an atlanto-occipital joint fracture (C1-C2) and a transection of the spinal cord, and probable brain stem injuries. Diffuse axonal brain injuries are also commonly reported, but skull fractures were typically not observed. These head injuries are consistent with rapid movements of the head.

SIDE AIR BAGS

As of January 1, 2001 the SCI has 48 side air bag cases. In February 2000, the side air bag case selection criteria was modified from any crash involving the deployment of a side air bag and/or side curtain to only crashes where one of these devices deployed into an occupied seating position.

Table 4 Number of Cases by Impact Plane As of January 1, 2001	
Impact Plane	Number of Cases
Left	26
Right	11
Rollover	5
Front	4
No Impact	2

As noted in table 4, the distribution of impact planes is as follows: Left side 26 cases, Right side 11 cases, frontal plane four cases, Multiple plane (roll overs) five cases and two cases where there was no impact. The two non impact cases both involved seat mounted mechanical sensors that were activated when a foreign object (a hockey stick and a hand bag) were placed between the door and the side of the seat.

Table 5 Number of Cases by Type of Side Air Bag As of January 1, 2001		
Type of Side Air Bag	Protected Body Region	Number of Cases
Seat mounted	Thorax	22
Door Mounted	Thorax	19
Seat Mounted	Head and Thorax	2
Inflatable Curtain	Head	1
Inflatable Tubular Structure	Head	4

Table 5 includes a case count of the types of side air bags investigated. The majority of the investigations of side air bags have been for the protection of the thorax only. In all cases the air bag provided an increase in occupant protection. However, most noteworthy was the occupant protection noted in the seven cases with head protection, including at least one fatality due to a thorax injury, no reported life threatening head injuries.

There are 10 drivers and three right front passengers with an injury code of AIS-3 or higher. All but one of these injuries have been attributed to an interior component, the ground or the other vehicle due to a high impact speed, ejection or massive intrusion into the occupant compartment. In the single exception, the driver received an AIS-3 thorax injury attributed to the air bag cover flap in a door mounted thorax air bag system.

REDESIGNED AIR BAGS

In March of 1997, NHTSA issued a rulemaking action that allowed automobile manufacturers to reduce the force at which their air bags deployed. Most of the manufacturers began introducing redesigned air bags into their fleets beginning with 1998 model year vehicles. In order to determine how changes in the air bag affected occupants in real world crashes the NCSA initiated the Redesigned Air Bag Special Study (RABSS) in October, 1997 to collect data on crashes involving these redesigned air bags.

The objective of this study was to collect data on crashes of high interest (children, out of position occupants, high damage severity, and multiple injured occupants) involving vehicles equipped with a redesigned air bag system in which the air bag has deployed.

As of January 1, 2001, the RABSS has a total of 472 cases. Of these 472 cases, 100 are still under active investigation, 131 are undergoing an agency review process and 237 are available to the public. Also included in the reporting system for redesigned air bags are 178 cases listed as PARTNERS cases. These cases are provided by the manufacturers, insurance companies and other organizations.

From October 1997 to January 1998, the SCI was selecting any case with a redesigned air bag deployment, requiring the screening of thousands of police crash reports involving 1998 and newer vehicle. However due to the overwhelming volume of cases and limited resources available, In January 1998, a minimum case selection criteria for inclusion in the study was

Chidester, pg. 9

initiated. The case selection used from January 1998 to January 2000 was based on a minimum criteria and their interest to NHTSA. All cases that met the minium criteria were reviewed, however only cases that met a selection factor based on their interest to NHTSA were selected. The following is the minimum criteria and NHTSA interest priority list:

Minimum Criteria for RABSS Selection

- 1998 model year vehicle equipped with a redesigned air bag.
- The crash configuration must be an impact where the air bag is designed to protect the occupants (e.g. 11 to 1 o'clock PDOF). Do not include side or back plane impacts. In addition, exclude rollovers with or without ejection.
- The 1998 model year vehicle must have a complete exterior and interior vehicle inspection and be towed due to damage.
- An occupant must be seated in a position in which a redesigned air bag has deployed.

If minimum criteria is met, then select cases were selected based on their interest to NHTSA in the following priority order.

1. A child under 13 years old is seated in a position with a redesigned air bag deployment.
2. An occupant protected by a deploying redesigned air bag receives fatal injuries.
3. The driver and/or front right occupant protected by a deploying redesigned air bag that are transported to a medical facility for treatment of injuries.
4. Crashes of high severity, regardless of the injury level. These include crashes where the delta V would be greater than 24 MPH. These may be identified by specific information on the police traffic crash report.

From July 1998 to December 1999 the NASS CDS selected redesigned air bag cases out-of-sample. These cases are anecdotal only (not weighted) and have an 800 series case number.

In an effort to focus the redesigned air bag investigations toward true case of interest to the Agency, the following case selection criteria has been established for cases selected after January 2000:

- The crash configuration is a frontal and a child is in a seat position where a redesigned air bag has deployed.

- The crash configuration is a frontal, is not a

rollover and an adult in a seat position protected by a redesigned air bag receives fatal injuries.

- Any crash involving a 2000 model year or newer passenger car or light truck equipped with "smart" or advanced air bag system. These systems include but are not limited to multi-stage inflators, systems with the capability to detect out of position occupants etc.

- Any crash involving a 2000 model year or newer passenger car or light truck equipped with an Event Data Recording device and a deployed driver or passenger redesigned air bag.

NHTSA is currently analyzing this data to determine the effectiveness of redesigned air bags in real world crashes. The conclusions from this data analysis will be published by the Agency at a later date.

ADVANCED OCCUPANT PROTECTION

In May 2000, an interim final rule for FMVSS No. 208 was issued by NHTSA that amends the occupant crash protection standard to require that future air bags be designed to create less risk of serious air bag-induced injuries than current air bags, particularly for small women and young children; and provide improved frontal crash protection for all occupants, by means that include advanced air bag technology. To achieve these goals, it adds a wide variety of new requirements, test procedures, and injury criteria, using an assortment of new dummies. It replaces the sled test with a rigid barrier crash test for assessing the protection of unbelted occupants.

The interim final rule is also a performance based standard that does require manufacturers to specifically incorporate advanced sensors, inflators, or suppression technologies.

A few automobile manufacturers began introducing advanced occupant protection systems into their fleets beginning with 2000 model year vehicles. In an effort to determine how changes in the occupant protection systems affected occupants in real world crashes the NCSA initiated the Advanced Occupant Protection Special Study (AOPSS) in September, 1999 to collect data on crashes involving 2000 model year vehicles equipped with advanced occupant protection systems.

Chidester, pg. 10

The objective of the AOPSS is to provide data that will assess the real world performance of new occupant protection technologies. For this study, advanced occupant protection may include one or more of the following: seat belt sensors, weight sensors, seat position sensors, multi-stage inflators, systems that may provide automatic air bag suppression, rollover sensors and event data recorders.

Due to the few number of vehicles currently available with an advanced occupant protection system the case selection criteria has been very general. Also because the air bag deployment thresholds have been significantly raised, air bag deployment is no longer considered. The minimum criteria for AOPSS case selection are:

1. >2000 model year vehicle equipped with an advanced occupant protection system.
- The crash configuration must be an impact where the advanced restraint system is designed to protect the occupants and the vehicle is towed due to damage
- Back plane impacts and rollovers are excluded
- Side impact crashes are included only if the vehicle is equipped with inflatable side impact protection and that protection deployed into a occupied seating position

As of January 1, 2001 the AOPSS program has a total of 35 cases. Thirty-one cases are still being actively investigated. Four have been submitted for agency review. The Agency will begin publishing tables with the advanced occupant protection system data in 2001.

The Agency is sharing the field data with the automobile manufacturers regarding the technology surrounding these new occupant protection devices. This collaborative effort combines the talents of crash investigators, engineers, and designers, which enable all interested parties to perform case-by-case evaluation of the real world performance of these advanced technologies. Particularly noteworthy is the technical analysis of the event data recorder (EDR) output . The EDR data has provided invaluable information relating to occupant status, severity assessment, and deployment control in researching crashes with advanced occupant protection systems.

CONCLUSIONS

NHTSA and its partners, (manufacturers, insurance companies and other organizations) have committed a high volume of public education resources in an effort to prevent air bag related injuries and fatalities, especially to children. This media attention appears to be having a positive effect on reducing child fatality cases.

NHTSA's rulemaking changes beginning in March of 1997 have had a positive effect on reducing air bag related fatalities. As of January 1, 2001, there have been no adult driver or adult passenger air bag fatalities in 1999 and newer model year vehicles. In addition, there have been no 2000 or 2001 model year vehicles involved in a child passenger air bag related fatality. However, the SCI program will continue to monitor, confirm and report air bag related fatality data.

In all crashes investigated with a side air bag deployment, the air bag provided an increase in occupant protection. Most noteworthy is that in the seven cases with head protection, including at least one fatality due to a thorax injury, there have been no life threatening head injuries reported.

AIR BAG CRASH INVESTIGATION DATA AVAILABILITY

The NHTSA has a number of methods in which the air bag crash data is distributed. Beginning in 2001, summary tables will be published quarterly on the National Highway Traffic Safety Administration's (NHTSA) Internet web site at the following web address:

http://www.nhtsa.dot.gov/people/ncsa/sci.html

Copies of the summary tables can also be obtained by calling one of the following telephone numbers and requesting "Air Bag Fatality Reports for Special Crash Investigations"

Toll Free 800-934-8517
Local Number 202-366-4198

Special Crash Investigations cases will be available from the National Automotive Sampling System's web page beginning June 2001.

Http://www-nass.nhtsa.dot.gov/nass

Copies of completed hard copy SCI reports listed as available on the summary tables can be obtained at the address below. The reports contain images and accordingly there is a cost associated with reproduction of the crash report.

Chidester, pg. 11

Marjorie Saccoccio, DTS-44
DOT/Volpe National Transportation Systems Center
Kendall Square
Cambridge, MA 02142
USA

Completed SCI reports can be reviewed at the hard copy storage facility. There is a nominal cost for case retrieval and handling.

Acknowledgment of thanks are due to the Special Crash Investigators at Veridian Corporation, Indiana University, and Dynamic Science, Inc.

REFERENCES

1. "Selecting and Using the Most Appropriate Car Safety Seats for Growing Children: Guidelines for Counseling Parents", Pediatrics Volume 97 No 5 May 1996, American Academy of Pediatrics, Committee on Injury and Poison Prevention

2. "Monthly Counts for Air Bag Related Fatalities and Seriously Injured Persons", January 2001, Special Crash Investigation, U. S. Department of Transportation, National Highway Traffic Safety Administration

3. "Redesigned Air Bag Summary", January 2001, Special Crash Investigation, U. S. Department of Transportation, National Highway Traffic Safety Administration

4. "Side Air Bag Summary", January 2001, Special Crash Investigation, U. S. Department of Transportation, National Highway Traffic Safety Administration

5. "Advanced Air Bag Summary", January 2001, Special Crash Investigation, U. S. Department of Transportation, National Highway Traffic Safety Administration

6. "Air Bag Crash Investigations", Chidester, Augustus., U. S. Department of Transportation, National Highway Traffic Safety Administration

2002-01-0186

Performance of Depowered Air Bags in Real World Crashes

J. Augenstein, E. Perdeck and J. Stratton
William Lehman Injury Research Center, University of Miami School of Medicine

K. Digges and J. Steps
The National Crash Analysis Center, George Washington University

Copyright © 2002 Society of Automotive Engineers

ABSTRACT

During the period 1992 through 2000, the William Lehman Injury Research Center collected crash and injury data on 141 drivers and 41 right front passengers in frontal crashes with air bag deployment. Among these cases were twenty-eight cases with depowered air bags. The paper compares the crash characteristics for injured occupants in vehicles with 1st generation and depowered air bags.

The population with 1st generation air bags contains unexpected fatalities among as well as fatalities at low delta-V's. To date, these populations are absent among the fatally injured occupants of vehicles with depowered air bags. The depowered cases include both belted and unbelted survivors at crash severities above 40 mph delta-V. The maximum injury in these severe crashes was AIS 3 with no evidence of unsatisfactory air bag performance. However, serious internal chest injuries were observed in two cases with unrestrained drivers at crash severities of 19 and 24 mph.

INTRODUCTION

The Lehman Injury Research Center air bag data represents a near census of the seriously and fatally injured air bag protected occupants in Miami and Southern Florida. The database of frontal crashes involving 1st generation air bags contains 154 occupants, 45 of whom were fatally injured.

In the database of 1st generation air bags, 13 of the 45 fatalities were seniors (over 65) and 4 were small children. Eight of the fatalities were at speeds below 15 mph.

The depowered air bag database contains 28 occupants, 4 of whom were fatally injured. In the depowered cases, there have been no child fatalities and no fatalities as speeds below 20 mph. Three of the four fatalities were in severe crashes with delta-V greater than 40 mph and with massive intrusion of the occupant compartment.

Among the depowered cases there were four survivors of extremely severe crashes – greater than 40 mph. Two occupants were belt restrained and two were unrestrained. The chest injuries were of remarkably low severity and no head injuries were sustained to either occupant. These cases suggest that air bags as currently depowered provide protection in high severity crashes to both restrained and unrestrained occupants. These cases are summarized in the sections to follow.

There were two cases of severe injuries to unrestrained occupants at moderate crash severities – 19 and 24 mph. In these cases, the occupant may have been close to the deploying bag. The case summaries follow.

COMPARISON OF FATAL CRASHES

The distribution of crash severity for the fatally injured drivers with 1st generation and depowered air bags is shown in Table 1. The table shows the number of fatalities in each delta-V (mph) increment. The 1st generation air bag fatalities are fairly uniformly distributed. There are a relatively large number of fatalities in the lower speed ranges for the 1st generation air bags.

Table 1. Distribution of Driver Fatalities by Delta-V

Delta-V	1st Generation	Depowered
0-15	5	0
16-20	3	0
21-25	5	1
26-30	5	0
31-35	3	0
36-40	5	0
40+	6	1
Total	32	2

Table 2. Distribution of Driver Fatalities by Age

Age	1st Generation	Depowered
15-20	3	0
21-30	5	0
31-40	3	1
41-50	6	1
51-60	5	0
61-70	5	0
71+	5	0
Total	32	2

The age distribution of the fatally injured drivers with 1st generation and depowered air bags is shown in Table 2. There are a relatively large number of fatally injured older occupants among the fatally injured group with 1st generation air bags.

The distribution of crash severity for the fatally injured right front passengers with 1st

Table 3. The 1st generation air bag fatalities are fairly uniformly distributed. There are a relatively large number of fatalities in the lowest speed range (0-10 mph) for the 1st generation air bags.

Table 3. Distribution of Passenger Fatalities by Delta-V

Delta-V	1st Generation	Depowered
0-10	3	0
11-20	1	0
21-30	4	0
31-40	2	0
40+	3	2
Total	13	2

The age distribution of the fatally injured right front passengers with 1st generation and depowered air bags is shown in Table 4. There are a relatively large number of fatally injured children among the fatally injured group with 1st generation air bags. To date no injuries to children have been observed in the depowered Lehman Center cases.

Table 4. Distribution of Passenger Fatalities by Age

Age	1st Generation	Depowered
0-3	4	0
21-30	2	0
31-40	0	1
41-50	1	0
51-60	2	0
61-70	1	0
71+	3	1
Total	13	2

SEVERE CRASHES WITH FAVORABLE AIR BAG PERFORMANCE

Four extremely severe crashes of drivers with depowered air bags are in the Lehman Center database. These cases had a delta-V of 40 mph or greater. Two of the drivers were restrained and two were unrestrained. None had any significant head injuries.

Case D015-99

The case vehicle was a 1999 Mitsubishi Mirage, involved in a frontal offset crash with a 1992 Honda Accord. The maximum crush was 59" and the delta-V was 45 mph. There was extensive intrusion in the driver location – 17" at the toepan, and 9" at the dashboard. The driver was a 34 year old male, 6'1" tall, weighing 185 lbs., restrained by a lap and shoulder belt and a depowered air bag. The case vehicle is shown in Figure 1. The injuries were as follows:

Laceration, Ant. Right Seer, Right Lobe
    AIS-2
Fracture, Anterior Right Rib Cage, Multiple
    AIS-3
Fracture, Post. Left Lumbar Vertebra, Multiple
    AIS-2
Fracture, Anterior Shaft of Femur, Left
    AIS-3
22 minor Lacerations and Contusions
    AIS-1

Figure 1. D01599 Case Vehicle

CASE D023-00

The case vehicle was a 1998 Toyota Tacoma Pickup that impacted an embankment at 12 o'clock with a delta-V of 40 mph. The driver was a 33 year old male, 68" tall, 176 lbs. He was restrained by a 3-point seatbelt, and the air bag deployed. The case vehicle is shown in Figure 2. The injuries were as follows:

Fracture of Shaft, Right Femur   AIS-3
Fracture of Intertrochanteric Crest of Right Femur   AIS-3

There were no head or chest injuries.

Figure 2. D023-00 Case Vehicle

Case U001-99

The case vehicle was a 1999 Dodge Ram 1500 Pickup, involved in a frontal offset crash with an unyielding pole. The delta-V was 43 mph and the pole intruded into the right occupant compartment. The right front passenger was an unrestrained 29 year old female. The air bag on the passenger side had been deactivated by the on-off switch and it did not deploy. The passenger was fatally injured from multiple AIS 6 level head and chest injuries. The driver was an unrestrained 29 year old male protected by the driver air bag. He sustained no head injury or chest injures greater than AIS 1. He did sustain ulna and femur fractures. The case vehicle is shown in Figure 3.

Figure 3. Case Vehicle, U001-99

Case D022-00

The case vehicle was a 2000 Dodge Ram 1500 Pickup that impacted an unyielding pole at 1 o'clock with a delta-V of 41 mph. The unrestrained driver was a 28 year old male, 71" tall, 189 lbs. The vehicle sustained a max of 59" of crush. The vehicle is shown in Figure 4. The only injury was an AIS-3 laceration of the spleen.

Figure 4. Case Vehicle D023-00

MODERATE SEVERITY CRASHES WITH UNFAVORABLE AIR BAG PERFORMANCE

There were two cases at moderate crash severity in which severe or fatal injuries resulted.

Case D018-99

Figure 5. D01899 Case Vehicle

impacted a tree. The and the delta-V was 24 mph. There was 2" of dashboard intrusion. The driver was a 44 year old male, 5'8" tall, weighing 148 lbs. He was unrestrained and the depowered air bag deployed. The case vehicle is shown in Figure 5. The injuries were as follows:

Sternum Fracture	AIS-2
Tear of the Pericardium	AIS-2
Heart Valve Laceration	AIS-4
Ventricle Laceration	AIS-6

CASE D010-99

The case vehicle was a 1998 Honda Accord that impacted the right side of a 1982 Buick Park Avenue at the rear wheel. The delta-V was 19 mph. There was no intrusion. The driver was an unrestrained 18 year old female, 5'4" tall, weighing 130 lbs. The case vehicle is shown in Figure 6. The injuries were as follows:

Laceration, Anterior, Right Medial Segment of Liver, Left Lobe	AIS-5
Avulsion, Anterior Knee, Right	AIS-1
Multiple Contusions, Upper Chest	AIS-1

Figure 6. D01099 Case Vehicle

SEVERE FATAL CASES

The depowered air bag cases with fatalities and crash severities above 40 mph involved massive intrusion of the occupant compartment. The vehicles in these cases were a 1998 Chevrolet Cavalier that impacted a van with a 55 mph delta-V and a 1998 Pontiac Bonneville that impacted a tractor-trailer with a delta-V of 40 mph. These vehicles are shown in Figures 7 and 8.

Figure 7. Fatal Crash with Delta-V 55 mph.

complex occupant loading. One such crash is non-fatal case D016-99. This crash involved a frontal underride of a trailer followed by a severe rear impact. The air bag deployed in the frontal impact. However, the occupant compartment intrusion resulted from the rear impact that produced 39 inches of crush. The restrained 22 year old female driver suffered AIS2 rib fractures and an AIS 3 liver laceration. These were attributed to occupant compartment intrusion from the second impact. The damage from the rear impact is shown in Figure 9. The occupant compartment intrusion is shown in Figure 10.

Figure 9. D01699 Case Vehicle Rear Impact

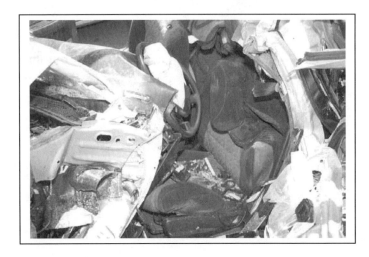

Figure 10. D01699 Case Vehicle Intrusion

Figure 8. 40+ mph Fatal Crash with Multiple Impacts

MULTIPLE IMPACT CASE

Crashes involving airbag deployment

DISCUSSION

At present, there is insufficient data on depowered air bags to make statistically significant comparisons with 1st generation air bags. However, to date the data from the William Lehman Injury Research Center contains no fatalities at low crash speeds when depowered air bags were present. There have been no reported fatalities to children.

The two of the cases with depowered air bags produced unexpected injuries to unrestrained drivers. These cases were D018-99 and D010-99, summarized earlier. In all other cases, the depowered air bags performed as expected. In all other cases with crash severity less than 40 mph there were no AIS 3+ head and only one AIS 3 chest injury. The AIS 3 chest injury involved intrusion from a rear impact (Case D016-99). In two cases with delta-V greater than 40 mph the depowered air bags performed exceptionally well. In all cases above 40 mph with fatalities, the occupant compartment suffered extensive intrusion.

CONCLUSION

In the limited number of depowered cases investigated by the William Lehman Injury Research Center, the performance of depowered air bags has been very good. High speed protection at crash severities greater than 40 mph has been observed for both restrained and unrestrained occupants. The database of depowered air bags contains no significant injuries in very low speed crashes, and no injuries to children.

Three of four fatalities occurred in crashes that were so severe that the occupant compartment was destroyed.

However, serious internal chest injuries were observed in two cases with unrestrained drivers at crash severities of 19 and 24 mph. One of these crashes produced in a fatal heart injury and the other an AIS 5 liver injury. These cases contained the only unexpected injuries among the population protected by depowered air bags.

ACKNOWLEDGEMENT

The authors would like to express appreciation to the Alliance of Automobile Manufacturers for sponsoring this research.

VII. ABSTRACTS FOR FURTHER READING

1999-01-0633, SP-1432

US and UK belted driver injuries with and without airbag deployments~A field data analysis

Parenteau, Chantal S., Shah, Minoo, Desai, Trilok
Delphi Automotive Systems
Frampton, Richard
Loughborough University of Technology

This study compares the effect of US and European airbag deployments on injury outcomes for belted drivers in frontal crashes. Driver weight, height and seat track position was also examined in relation to those outcomes. This information may help to prioritize and guide the logic for "Smart" airbags.

For this study, only airbag-equipped cars were considered. Two accident databases were used: 1) the weighted and unweighted National Accident Sampling System (NASS-CDS) from the US, calendar years 1995 to 1996, and 2) the unweighted Co-operative Crash Injury Study (CCIS) from the UK, calendar years 1992 to 1998. The parameters investigated were Injury Severity Score (ISS), Equivalent Test Speed (ETS), occupant weight, occupant height and seat location. For US drivers, the injury rate and occurrence were calculated using weighted data, while for UK drivers, the rate and occurrence were obtained using unweighted data. Because the sample size in the two databases was small, the results should only be used for trend analysis.

ETS was an important factor influencing injury rate. For UK- deployed cases, all severely injured drivers were involved in a severe crash (ETS \mg 24 mph), while about 70% of the severely injured US drivers were in crashes with an ETS \mg 18 mph. In the US, 6 out of 47 deployed cases involving a severely injured (ISS \mg 12) driver occurred in a low-speed accident (ETS \ml 14 mph). There were no severely injured US drivers in low-speed crashes without an airbag deployment. The rate to be severely injured was thus higher for deployed than undeployed cases at ETS \ml 14 mph. In the UK, there were no undeployed cases in which the driver was severely injured. Although this study does not include an analysis of injury causation, this suggests that minimizing airbag deployments at an ETS \ml 14 mph would be beneficial for the US market.

The effect of weight, height, and seat location was analyzed for belted drivers involved in a low-to-moderate crash severity (ETS \ml 24 mph). Using US-deployed data, drivers weighing less than 55 kg had the highest rate to be severely injured. However, more than 50% of US and UK drivers weigh between 56 to 86 kg. US and UK drivers weighing more than 86 kg had the highest rate to be moderately injured when the airbag deployed.

US deployed and undeployed data revealed a trend for a higher moderate injury rate in the \ml 160 cm height category than other height categories. This finding is also true for UK drivers involved with an airbag deployment. However, short-statured (\ml 160 cm) drivers accounted for less than 8% of the overall belted sample. Seat location was also found to possibly influence US driver injury responses. Though less than 8% of the belted drivers were sitting in the forward seat location, these drivers had the highest rate to be severely injured when the airbag deployed. No drivers sustained an ISS \mg 12 in the seat forward location when the airbag did not deploy. Since seat location can be used to estimate an occupant's height, seat location sensing may thus provide important input for US airbag deployment decision logic. Data were not available on seat location in the UK database. Even though the sample size was too small to yield a statistically significant analysis, the trends observed in this study can provide general insight that may be useful in developing future occupant safety countermeasures.

1999-01-0760, SP-1411

Adult front-seat passengers exposed to airbag deployments
Huelke, Donald F.
Unviersity of Michigan

In comparison to drivers exposed to steering-wheel airbag deployments in frontal crashes, there have been fewer front-seat passengers exposed to airbag deployments for 1) many of the cars in crashes did not have dual airbags, and 2) the front passenger seat is less often occupied. Of the 826 airbag crashes detailed by UMTRI crash investigators at the time of this manuscript preparation, there were 145 front-seat passengers, exposed to instrument-panel-mounted airbags. Most of these front-seat

passengers, 124, were involved in the frontal crashes. There were 92 who were 16 years of age or older, 24 were under 12 years of age and 11 young teenagers, 13-15 years of age. Of those who were 16 years or older in frontal crashes 70% had an MAIS-1 injury. None of the MAIS-2 injuries were directly related to airbag deployments. Of the AIS-3+ level injuries, about two- thirds were not airbag related.

1999-01-0763

Correlation of driver inflator predictor variables with the viscous criterion for the mid-sized male, instrumented test dummy in the chest-on-module condition
Laituri, Tony R., Prasad, Priya
Ford Motor Co.

A new inflator specification, the "inflator thrust variable," was developed to better explain measured mid-sized male, instrumented test dummy responses in the chest-on-module test condition. Specifically, controlled laboratory experiments were conducted with non-production, driver airbag modules with inflators of various outputs and gas constituents in an effort to assess their effects on a pertinent occupant response. Regression analyses showed that the inflator thrust variable is a better predictor of the observed variation in peak viscous criterion responses than either peak tank pressure or the related pressure rise rate when inflators of differing gas composition were compared.

1999-01-0764, SP-1411

Deployment of air bags into the thorax of an out-of-position dummy
Bass, C. R., Crandall, J. R., Bolton, J. R., Pilkey, W. D.
University of Virginia
Khaewpong, N., Sun, E.
National Highway Traffic Safety Administration

The air bag has proven effective in reducing fatalities in frontal crashes with estimated decreases ranging from 11% to 30% depending on the size of the vehicle (IIHS-1995, Kahane-1996). At the same time, some air bag designs have caused fatalities when front-seat passengers have been in close proximity to the deploying air bag (Kleinberger-1997). The objective of this study was to develop an accurate and repeatable out-of-position test fixture to study the deployment of air bags into out-of- position occupants. Tests were performed with a 5\ut\uh percentile female Hybrid III dummy and studied air bag loading on the thorax using draft ISO-2 out-of-position (OOP) occupant positioning. Two different interpretations of the ISO-2 positioning were used in this study. The first, termed Nominal ISO-2, placed the chin on the steering wheel with the spine parallel to the steering wheel. The second, termed Chest On Module, relaxed the chin-positioning criterion while maintaining the spine parallel to the steering wheel. In this position, the chest was forced against the air bag module with a 100 N force. All tests were performed in one of four nominal positions with respect to the steering wheel plane. The reference position had the center of the sternum on the center of the module. Variations from the reference position examined included the occupant sternum displaced 2 cm and 4 cm vertically in the steering wheel plane and displaced 4 cm left. Tests were performed using a production "depowered'air bag from the current automobile fleet. To minimize the effect of external conditions, tests were performed in a climate-controlled room.

Dummy positioning on the test fixture was found to be repeatable to within 0.3 cm on all axes. This variation was within the dimensional similarity of the two available 5\ut\uh percentile female Hybrid III dummies. A large variation in occupant response was found with a very small change in effective distance from the sternum to the air bag module. Nearly 50% variation in peak chest center-of-gravity resultant acceleration was found when moving from the sternum pressed on the air bag module to the sternum effectively being 2 cm from the module. In addition, large variations in occupant response were found with vertical and horizontal displacements of the occupant with respect to the air bag module center. Also, a qualitative change in air bag deployment was found on changing the horizontal position by 4 cm to the left. These variations have significant implications for expected response from in-vehicle,

1999-01-1063, SP-1411
Airbags and their impact on collision insurance losses
Mirnazari, Mir T.
Vehicle Information Centre of Canada
Norup, Henning M.
Insurance Information Centre of Canada

It is generally accepted that airbags have beneficial impacts on vehicle occupants, reducing serious to minor injuries and in some cases avoiding injuries. There have also been some indications of the drawback of airbags deployments in certain situations, especially for unrestrained infants. The cost associated with reinstallation of replacement airbags may, however, adversely affect the cost of insurance collision claims. This cost and its impact on collision insurance losses are the subjects of this study.
Factorial and paired comparison designs are utilized to analyze the losses in five accident years. The results of these analyses indicate an overall 4.6% increase in the collision portion of claims severity for the airbag-equipped vehicles. This increase is statistically highly significant and supported by the strength of the evidence provided by the data.

2000-01-1003, SP-1493
Driver fatalities in frontal crashes of airbag-equipped vehicles: A review of 1989-96 NASS cases
Cammisa, Michael X.
Insurance Institute for Highway Safety
Reed, Richard T.
Accident Research and Analysis
Ferguson, Susan A., Lund, Adrian K.
Insurance Institute for Highway Safety

Using data from the National Automotive Sampling System/Crashworthiness Data System (NASS/CDS) for 1995-96, this study updates previous analyses of driver fatalities in airbag-equipped vehicles in the NASS/CDS database for 1989-93 and 1989-94. A total of 59 cases of frontal crashes of airbag-equipped vehicles with driver fatalities were identified in these 8 years of NASS/CDS data, but in 9 cases the fatalities were not related to the impacts (e.g., fire, medical condition). Vehicle intrusion was the cause of the fatal injuries in 27 cases, and 7 drivers died from injuries sustained when they were either partially or totally ejected from their vehicles. There was one case in which the airbag did not deploy, although the crash conditions indicated it should have. One driver died from contact with a nonintruding vehicle surface, and the causes of the fatal injuries in 5 cases were unknown. There were no cases in which drivers died because airbags had insufficient power to prevent the fatal injuries, although in one of the vehicle-intrusion-related fatalities the airbag deployed in the first of two impacts and thus was not able to protect the driver throughout the crash sequence. There were 9 cases in which airbags likely contributed to the drivers' fatal injuries. Seven of these occurred in higher severity crashes with delta Vs greater than 30 km/h (20 mi/h), one was in a crash with a delta V of 26 km/h (16 mi/h), and one had an unknown delta V.

2000-01-SC09
Influence of airbags and seatbelt pretensioners on AIS1 neck injuries for belted occupants in frontal impacts
Kullgren, Anders, Krafft, Maria, Malm, Sigrun, Ydenius, Anders
Folksam Research

Tingvall, Claes
Monash University Accident Research Center

AIS1 neck injuries are the most frequent disabling injuries among car occupants in road traffic accidents. Although neck injury is mostly regarded as resulting from rear end collisions, almost one third of all neck injuries occur in frontal impacts. Several studies have shown the effect of airbags on injury and fatality rates. However, studies of the effect of airbags on the risk of injuries to different body regions are rare. Airbags and seatbelt pretensioners may influence especially the

risk of neck injuries.
This paper presents influence of airbags and pretensioners on reported neck injury risk in frontal impacts. Cars fitted with airbags in combination with pretensioners and cars without have been analyzed. Since 1992, approximately 150,000 vehicles on the Swedish market have been equipped with crash pulse recorders to measuring frontal impacts. This study includes results from 158 frontal impacts, where the crash pulses have been recorded using crash pulse recorders and where the status of airbag deployments was known. Only reported neck injuries, often denoted as short-term disability to the neck, were considered in this study. Injury risk functions for cars with and without airbags versus change of velocity and mean and peak accelerations were analyzed.
It was found that airbags in combination with seat belt pretensioners reduced the number of AIS1 neck injuries in frontal impacts with 41% +/- 15.2%. In impacts at a change of velocity between 1 and 30 km/h, airbags and pretensioners was found to reduce the neck injury risk with 59% +/- 18.6%.

2001-01-0155, SP-1615
The effects of belt use and driver characteristics on injury risk in frontal airbag crashes
Huelke, Donald F.

From the crash investigation files at the University of Michigan Transportation Research Institute (UMTRI), the crashes involving deployed airbags were reviewed. The total number of deployments is 898 of which 764 are frontal crashes with the principal direction of force (PDF) at 11-1 o'clock. Of the drivers in these frontal crashes 83% were using the belt restraint.
Overall, seven of ten drivers have an AIS-0 or 1 level injury as the maximum or highest injury severity level (MAIS). Of the survivors, one in six had a moderate level injury (AIS-2) as their most significant injury and one in nine had an MAIS 3 or greater injury. Fatalities are rare. There is a difference between injury severity frequencies of belted versus non-belted drivers. Three-quarters of the belted drivers had minor injuries compared to only half of those not belted. A difference was also noted at the AIS-2 level-belted versus unbelted 14% versus 23%. Of the belted drivers 10% had an AIS 3-5 injury level compared to 20% of the unbelted.
Drivers with AIS 3, 4 or 5 as their highest AIS level, whether belted or not, have the lower extremity as the body region most often injured. Belted drivers have a slightly higher frequency of upper extremity injuries.

2001-01-0159, SP-1615
Robust automated airbag module calibration
Schubert, Peter J.
Delphi Delco Electronics Systems

Increasing sophistication of electronic safety systems requires more advanced tools for design and optimization. Systems of safety products already being designed are becoming too interdependent to calibrate as stand-alone modules. Compounding this difficulty is the trend towards fewer test crashes and more sophisticated regulatory requirements. This paper presents a unified calibration approach to produce robust performance. First, the set of crash samples are extended using statistical techniques. Then an automated calibration tool using Genetic Algorithms is used to provide robust performance against deployment requirements. Finally, an expert systems is employed to ensure logical behavior. Together, these powerful methods yield calibrations which out-perform manual calibrations and can be completed in far less time.

2001-01-0164, SP-1615
Influence of air bag folding pattern on OOP-injury potential
Mao, Yiqin, Appel, Hermann
Institute of Automotive Technology

Four different air bag folding patterns are discussed in this paper.
The influence of air bag folding pattern on inflation characteristics and occupant OOP-injuries is simulated and evaluated. The results indicate

the high opening pressure and high "punch out"-, "bag slap"-, "membrane loading"-effect is proved to be a critical folding pattern for the OOP-Injuries.

This study also shows that the combination of design, test and simulation provides a comprehensive and effective tool to investigate the influence of air bag design parameters on the OOP-injury and to develop the "Advanced Air Bags."

2001-01-0179, SP-1577

Air bag loading on in-position Hybrid III dummy neck
Kang, Jian, Agaram, Venkatesh, Nusholtz, Guy, Kostyniuk, Gregory
DaimlerChrysler Corp.

The Hybrid III family of dummies is used to estimate the response of an occupant during a crash. One recent area of interest is the response of the neck during air bag loading. The biomechanical response of the Hybrid III dummy's neck was based on inertial loading during crash events, when the dummy is restrained by a seat belt and/or seat back. Contact loading resulting from an air bag was not considered when the Hybrid III dummy was designed. This paper considers the effect of air bag loading on the 5\ut\uh percentile female Hybrid III dummies. The response of the neck is presented in comparison to currently accepted biomechanical corridors.
The Hybrid III dummy neck was designed with primary emphasis on appropriate flexion and extension responses using the corridors proposed by Mertz and Patrick. They formulated the mechanical performance requirements of the neck as the relationship between the moment at the occipital condyles and the rotation of the head relative to the torso.
The deformation of the neck was due to the motion of the head relative to torso, when the torso was restrained either by the seat belt or the seat back. Air bag loading is significantly different from belt or seatback loading, because there is more than one load path on the head. During air bag loading, the dummy neck shows significantly different responses from those seen in the Mertz and Patrick tests. The neck experiences second mode bending during air bag loading as opposed to the first mode bending seen in seat belt and/or seat back loading.
The Hybrid III 5\ut\uh percentile female dummy is frequently used to estimate the response of small stature occupants to air bags. This paper examines the Hybrid III 5\ut\uh percentile female dummy's neck responses due to air bag loading. The neck responses were found to be highly dependent on how the air bag interacts with the dummy. Three modes of air bag-neck interaction were observed and studied: air bag directly loading the head, air bag trapped under the chin during the deployment process and air bag trapped behind the jaw of the dummy head. This paper also presents the results of some experimental modifications to the head/neck design of the Hybrid III 5\ut\uh percentile female dummy to prevent air bag entrapment.

2002-01-0022, SP-1665

An analysis of recent accidents involving upper extremity fractures associated with airbag deployment
Atkinson, Patrick
Kettering Univ.
Mclean, Michelle, Telehowski, Paul, Khan, Imran, Martin, Sidney
McLaren Regional Medical Center
Hariharan, Prem
Kettering Univ.
Atkinson, Theresa
IPAC, LLC

Prior experimental and field studies have demonstrated an increased risk of upper extremity fracture due the deployment of frontal airbags. The experimental studies provide valuable insight as to likely injury mechanisms; namely, increasing proximity increases the risk of forearm fracture. Still, field data is needed to validate these experimental findings. The available field data has largely been derived from direct case study analysis or a review of government accident statistics. In both cases, the datasets were comprised solely of pre-1995 era vehicles. Such data represents early generation airbag designs and there has been little additional study in this area. In addition, there has been an absence of fracture pattern analyses as a function of airbag deployment

and non-deployment. Such an analysis would help elucidate the role of the deploying airbag on upper extremity fracture in the current fleet. In the current study we analyzed the NASS database and cases admitted to our trauma service. Our analysis shows that the forearm was fractured most frequently and that airbag deployment is associated with a 2 to 3 times increase in the proportion of forearm fractures versus no bag deployment. Further, it was found that females were statistically more at risk to suffer a forearm fracture even after pair matching for airbag deployment exposure and occupant sex. The current study shows that airbag deployment increases the risk of forearm fracture as a proportion of all upper extremity fractures and that females are at an increased risk of such injury.

2002-01-0182, SP-1669

Important issues in crash severity sensing
Gioutsos, Tony
Breed Technologies, Inc.

In this paper, we describe some important aspects of crash severity sensing for "smart'airbag implementation. In particular, gray zone considerations are described in detail. Another important issue discussed is the loss of the additional front sensor used for crash severity sensing and its impact on system performance. A third issue of great importance is the second-stage (assume a dual-stage inflator) firing delay. Finally, a discussion of some inherent problems with the crash severity sensor approach are discussed as well as dynamic range/resolution problems that accelerometers can encounter when used in the crush zone of a vehicle. A summary with conclusions and recommended approaches is presented.

2002-01-0190, SP-1669

Intelligent algorithms for early detection of automotive crashes
Singh, Gautam B., Song, Haiping
Oakland Univ.

In this paper we describe the methodology for building crash pulses using accelerometer crash pulses. This procedure is applied to a set of 9 Chevrolet Cavalier crash pulses derived both from rigid and offset-deformable barrier collisions. Five (5) of these are used to train a 7-state, fully-connected HMM. The HMM model acquired from the training
pulses is next utilized for classification of the remaining four crash pulses that have been intermixed with random pulses of comparable strength. The log-likelihood of deploy pulses was around -200, whereas the log-likelihood of non-deploy pulses ranged between -2000. As the public databases provide us crash pulses only for airbag deployment events, we have tested our algorithm using randomly generated pulses as negative control. The statistical characteristics of the randomly generated pulses are made to approximate that of the true pulse. The log likelihood of the crash pulses is very different from that of random pulses, which means we can distinguish a crash pulse from the random pulse only with the first 10 milliseconds' information. The classification of the crash and non-crash pulses was thus achieved without ambiguity.

2002-01-0192, SP-1669

The computation of airbag deployment times with the help of precrash information
Theisen, Marc, Eisele, Sybille, R\arlleke, Michael
Robert Bosch GmbH

Modern airbag control units are required to compute airbag deployment times with a high degree of precision. Therefore, the crash situation has to be recognized unambiguously, i.e., the goal is to obtain precise information about the relative speed, the barrier and the position of impact. One way of achieving this aim is via the implementation of a precrash sensing system using radar sensors. With these sensors, the relative closing velocity and the time-to-impact can be measured, thereby enabling a precise analysis of the crash situation. In this paper the algorithm for the computation of the airbag deployment decision will be

700904, P-33

Evaluation of the lap belt, air bag, and Air force restraint systems during impact with living human sled subjects

Gragg, C. D., Bendixen, C. D., Clarke, Thomas D., Klopfenstein, H. S., Sprouffske, James F.

6571st Aeromedical Res. Lab.

Abrupt linear decelerations (-g\dx) were conducted with human volunteers in order to study the loading to the human anatomy while restrained with the lap belt, lap belt plus air bag, and Air Force harness systems.

Impulses and peak forces in the lap belts and peak forces in the seat pan, seat back, and foot cells were measured and compared. Each subject was compared with himself using the different systems, and the range and mean of these comparisons are shown.

The results indicated that in comparison with the lap belt only system, both the lap belt plus air bag and the Air Force harness systems significantly reduced the impulses and peak forces transmitted to the pelvis.

720407

Air cushion restraint systems development and vehicle application

Campbell, D. D.

Fisher Body Div., General Motors Corp.

The discussion presented in this paper includes: A design review of General Motors air cushion restraint system, performance in dynamic testing, and the General Motors program for production build and fleet tryout. Among the items covered in the design review are: The driver's restraint system, the front passenger's restraint system, and the sensing mechanism. The discussion of air cushion system performance testing includes: The text matrix, two significant crash configurations, the importance of variable inflation levels, and the need to continue baboon and human volunteer testing. The discussion of the General Motors pilot production build and fleet tryout program covers the purpose for gaining "on-road" air cushion system experience

720408

Air cushion systems for full-sized cars

Pflug, J. A.

Ford Motor Co.

The initial air cushion research by Ford and suppliers has primarily involved full-sized cars. Technical approaches for these cars in packaging, kinematics, sensing and deployment are presented. Also, efforts to meet the requirements of the potential restraint standard as well as to resolve "real world" problems are discussed.

720409

Development problems with inflatable restraints in small passenger vehicles

Seiffert, Ulrich W., Borenius, Gunnar H.

Volkswagenwerk AG (Germany)

This paper summarizes the work done on air bag systems under Volkswagen's restraint system development program. The paper first presents theoretical considerations as applied to small cars. Based on this, some typical design parameters for the Volkswagen air bag systems are given.

Sensor considerations and various air bag system concepts and configurations are discussed. The solutions chosen for the VW are presented, followed by some typical test data from sled tests and full-scale crash tests.

Acceptable small-car systems cannot be expected simply by scaling down large car systems, mainly because of the differences in vehicle deceleration characteristics and passenger compartment layout between large and small cars. Volkswagen's development experience over the last few years has shown that, before solving all the peripheral problems of inflatable restraints which are not covered by FMVSS 208, an extensive

development effort is required to reach the basic 30 mph performance levels attained by large car air bag systems one or two years ago.

In summary, the advantages and disadvantages of the air bag system are described. These indicate that the air bag is not clearly superior to other restraints, and Volkswagen believes that the customer should be allowed to select the type of restraint system for his vehicle

720411

Special problems and considerations in the development of air cushion restraint systems

Klove, E. H.

Fisher Body Div., General Motors Corp.

Oglesby, Robert N.

Engineering Staff, General Motors Corp.

Presented in this paper is a discussion of the details of the General Motors air cushion restraint system and of specific technical problems of system development and of implementing a production build program. The details of the General Motors system include a description of the components of the driver's and front passenger's systems, crash sensing, and "variable inflation." The discussion of specific technical problems includes performance considerations, such as: Occupant rebound, child-size occupants, out-of-position occupants, nonbarrier type crashes, and the function of the appearance cover. Also included is a discussion of the toxicity potential, noise risk, sensor development, reliability considerations, and field service requirements

720442

Airbag effects on the out-of-position child

Patrick, L. M.

Wayne State Univ.

Nyquist, Gerald W .

Wayne State Univ.

This paper describes experiments involving airbag systems. Because there is the least amount of data on the tolerance of children to impact, the out-of-position child was used in the experiments.

After careful consideration it was decided that a primate of approximately the same weight as a child be used, which would provide the most realistic evaluation of potential injuries. The animal chosen for the experimental program was the baboon.

Five distinct positions were chosen and this paper describes in detail the experimental physiological conditions and results.

720443

Human volunteer testing of GM air cushions

Smith, G. R., Hurite, S. S., Yanik, A. J.

General Motors Corp.

Greer, C. R.

St. Joseph's Hospital

From November 1970 through August 1971 an extensive program of static and dynamic air cushion inflation tests utilizing human volunteers was conducted at Holloman Air Force Base, New Mexico, sponsored by the Department of Transportation. Forty-one full cushion deployment static firings were made, with air cushion hardware and seating buck environment designed by General Motors. The static series was followed by 35 dynamic sled firings of human volunteers, beginning at 8.6 g (15.1 mph) and culminating at 21.7 g (31.5 mph). A major objective of both the static and dynamic test series was to identify changes in air cushion design found necessary to improve its protective capability for human beings. Because of the severity of cushion deployment, one modification was made following the initial static tests: The orifice diameter size of the bag inlet was reduced from 1.0 to 0.6 in to diminish the rapidity of bag inflation. This modification proved effective in the dynamic series. As far as the human element in the dynamic series is concerned, no severe injuries beyond erythema, abrasion, contusion, and blister were received. As a result of this program, and testing of baboons at Wayne State University, an improved "augmented" GM air cushion is being developed

730605

General Motors driver air cushion restraint system

Louckes, T. N.

Oldsmobile Div., General Motors Corp.

Slifka, R. J.

Delco Electronics Div., General Motors Corp.

Powell, T. C.

Saginaw Steering Gear Div., General Motors Corp.

Dunford, S. G.

Inland Manufacturing Div., General Motors Corp.

This paper describes the Oldsmobile driver air-cushion restraint system designed to meet the new federal Motor Vehicle Safety Standard 208. Described is the overall system, indicating how the components must function together to provide optimum vehicle barrier performance. Also included are discussions of the sensor system, the new steering column and mounting system, and the steering wheel and wheel-mounted air-cushion restraint module.

740576, P-53

The development of an air bag on collapsible dashpanel restraint system for right front seat occupants

Shoemaker, Norris E.

Calspan Corp.

Biss, David J.

Calspan Corp.

An air bag on collapsible dashpanel (ABCD) passive restraint system concept was researched, developed, and demonstrated at Calspan. Elements of the ABCD system are: a collapsible dashpanel which is positioned within steering wheel distance or greater of the occupant to absorb the primary portion of the kinetic energy of the occupant-vehicle interaction, and two small air bags which deploy at speeds above 20 mph, to distribute chest contact forces and control head motions. A crushable kneebar is used for lower torso restraint. The Calspan 3-D Crash Victim Simulation was used as a preliminary design tool in developing the concept. Component tests of the collapsible dashpanel were conducted on the Calspan linear accelerator impactor. Sled tests were conducted to refine the restraint system design and to evaluate the performance of the restraint system with respect to accepted injury criteria. Satisfactory restraint system performance was demonstrated for the 50 lb child at 40 mph and the 50th percentile male at 50 mph. Performance for the 95th percentile male at 45 mph was marginal. The ABCD concept was demonstrated to be a feasible passive restraint system which shows promise for improving occupant protection for the range of occupant sizes from the 50 lb child through the 95th percentile male.

740580, P-53

The efforts of the National Highway Traffic Safety Administration in the development of advanced passive protection systems and child restraint systems

Strother, Charles E., Morgan, Richard M.

National Highway Traffic Safety Administration

This report presents an overview of the occupant packaging research program within the National Highway Traffic Safety Administration. The report discusses the program's efforts to establish the feasibilies of practical methods for providing the highest levels of occupant protection. In the area of frontal impact protection, work is progressing on advanced driver air bag systems, on a bag and bolster approach to passenger protection, on the development of improved inflation techniques for inflatables and on the passive application of the air belt concept. Efforts in the other areas of side, rear, and rollover protection are discussed as are NHTSA's efforts in child restraint research

741185, P-56

Otologic hazards of airbag restraint system

RichtersII, Harry J., Stalnaker, R. L., Pugh @sJr., James E.

University of Michigan

Since the airbag passive restraint system may be in general use in early 1976, and in fact is now an option on some automobiles, its potential biomedical hazards need to be thoroughly examined. Previous investigations in this area have been extremely limited.
The objective of this study was to investigate the effects of local slap pressure of airbag deployment against the external ear and tympanic membrane and to measure its effects on subsequent hearing acuity. Adult and infant squirrel monkeys were used as experimental subjects, because the gross structure of their ear and tympanic membrane closely resembles man's. To create an adequate simulation of the airbag trauma, a small airbag was fabricated and mounted on a pneumatic impact facility. This device was designed to produce a specific velocity to determine the behavior of objects under impact conditions simulating accident kinematics. Cochlear nerve action potentials were measured in both ears of 10 subjects prior to blast, immediately postblast, and several weeks postblast. High-speed photography recorded the events of the blast, as well as the technique of recording the potential from the cochlea and the appearance of the drumhead pre- and posttrauma. No permanent hearing damage, eardrum perforation, or disruption of ossicles occurred at airbag velocities up to 100 mph and a sound intensity level of 150 dB

746031

Relating air cushion performance to human factors and tolerance levels-program development

Ludstrom, Louis C.

General Motors Corp.

The development of the crash deployed air cushion restraint system for use in passenger cars required many special programs over a number of years to be reasonably certain the cushions would perform as designed in the protection of both children and adults in a wide range of environmental exposures and accident situations. This paper will identify the major programs and summarize their findings. Details are not repeated in this paper, but will be found in referenced material. When significant, the results of continuing projects will be updated.

750389

Passive and active restraint systems~performance and benefit/cost comparison

Patrick, L. M.

Wayne State Univ.

Five different restraint systems~mandatory harness, airbag + 20% lap belt usage, airbag, passive three point harness, and torso and knee bar~are analyzed for fatality and injury reduction, benefit/cost ratio, and cost-effectiveness. The mandatory harness is superior to the others in all comparisons with approximately 100,000 lives saved over the first 10 years which is about twice as many as would be saved by the other systems. A major advantage of the mandatory harness is that practically all of the vehicles are equipped while the other systems will require 10 years for complete installation.

750395

Determination of restraint effectiveness, airbag crash test repeatability

Versace, John, Berton, Roger J.

Ford Motor Co.

Thirty-three airbag-equipped Mercury cars were crash tested at three different test laboratories in order to determine the repeatability of test results in the proposed FMVSS 208 procedure. Nine Hybrid-II crash dummies were used. There were significant inconsistencies in results from the three testing agencies; by comparison, the different dummies caused little or no difference in test results; there was a large component of

751142, P-62

Efficiency comparison between three-point belt and air bag in a subcompact vehicle

Dejeammes, M., Quincy, R.

Laboratore des Chocs Organisme National de S\aecurit\ae Routi\afre

The purpose of this paper is the comparison of the protection efficiency between three-point belt and air bag systems under various crash conditions.

Dynamic tests have been performed with subcompact vehicles (Renault R 12) in which two dummies were restrained, either by three-point belts with load limiting devices, or by air bags consisting of solid gas generators and bags including porous outlets (the driver's knees were protected by a collapsable structure).

Three types of crashes were chosen: frontal barrier crash at 50 km/h (13,9 m/s); head-on crash between two vehicles with overlap at 50 km/h (13,9 m/s); crash against a guardrail at 80 km/h (22 m/s) with 30\mD angle of incidence.

The comparison drawn from commonly used biomechanical indices shows that the three-point belt ensures a protection in each analysed crash type but it should be improved in order to reduce head deceleration. The air bag results depend on the crash type and show the problems of adaptation in a subcompact vehicle.

The frontal barrier crash tests conducted with another type of dummy reveal that the results obtained for the two restraint types depend on the dummy, so that the efficiency assessment is difficult

751170, P-62

Front passenger passive restraint for small car, high speed, frontal impacts

Romeo, David J.

Calspan Corporation

A front passenger passive restraint system has been developed which provides frontal impact protection under small car, high speed crash conditions. The system consists of an extended crushable dashpanel, a knee bar and a relatively small volume air bag. Computer simulations, static tests and sled tests have been used to develop this system for the range of occupant sizes from 6 yr. child to 95th percentile adult for crash speeds to 50 mph. This paper reviews these efforts and presents observations regarding not only the performance of the system but those concerned with production feasibility and consumer acceptance as well. This research was conducted under contract to the U. S. Department of Transportation, NHTSA, under Contract DOT-HS-4-00972.

770155

Air bag update~recent crash case histories.

Smith, George R.

Environmental Activities Staff General Motors Corporation

About 11,000 air cushion-equipped cars have been put on the road by General Motors, including fleet Chevrolets in the field trial program and privately owned 1974-76 Oldsmobiles, Buicks, and Cadillacs. Although these vehicles have accumulated over 400 million car miles, the total is very small when compared with the mileage being driven by the entire vehicle population. Therefore, statistically valid conclusions cannot yet be drawn about the safety benefits of the system. Nevertheless, sampling of accidents illustrates the protection provided by the air cushion in a variety of situations.

780415

Matching case methodology for measuring restraint effectiveness

Pursel, H. D., Bryant, R. W., Scheel, J. W.

General Motors Proving Ground

Yanik, A. J.

GM Technical Center

This paper describes a procedure used to evaluate the injury and fatality prevention effectiveness of automobile occupant restraint systems using field accident data. The technique involves the direct comparison of accidents involving a specific restraint with a control group of accidents with similar injury producing potential. This technique is called the Matching Case Methodology and has been initially applied by General Motors in determining air cushion effectiveness.

780893, P-77

Evaluation of air cushion and belt restraint systems in identical crash situations using dummies and cadavera

Walsh, Michael J., Kelleher, Barbara J.

Advanced Technology Center, Calspan Corporation

An experimental program is discussed wherein fresh cadavera and anthropometric test devices (ATD) were exposed to identical crash situations utilizing both belt and air cushion restraint systems. Results will include symmetric and one-half offset frontal full size car-to-car tests conducted on the Calspan Vehicle Experimental Research Facility (VERF) at 60 MPH closing speed. Data obtained include head and chest triaxial accelerations from externally located sensors and thoracic accelerations from implanted sensors on the cadavera, normally measured internal triaxial head and chest accelerations and femur loads on the ATDs and belt loads for both cadavera and ATDs.

Osteologic data allows comparison between the cadavera regarding their relative skeletal quality.

Results of the study allow comparisons of the restraint systems effectiveness with respect to cadaver vs. cadaver and cadaver vs. ATD based upon autopsy evaluations and acceleration measurements.

780896, P-77

Bolster impacts to the knee and tibia of human cadavers and an anthropomorphic dummy

Viano, David C., Culver, Clyde C., Haut, Roger C.

Biomedical Science Department, General Motors Research

Melvin, John W., Bender, Max, Culver, Roger H.

Highway Safety Research Institute, University of Michigan

Levine, Robert S.

Wayne State Univ.

Knee bolsters on the lower instrument panel have been designed to control occupant kinematics during sudden deceleration. However, a wide variability in car occupant anthropometry and choice of seating posture indicates that lower-extremity contacts with the impingement bolster could predominantly load the flexed leg through the knee (acting through the femur) or through the tibia (acting through the knee joint). Potential injuries associated with these types of primary loading may vary significantly and an understanding of potential trauma mechanisms is important for proper occupant restraint. Impacts of the bolster panel against the knee or lower leg were simulated in 10 human cadaver and anthropomorphic dummy tests and the following aspects were assessed: 1) biomechanical response for lower-extremity impacts, 2) potential mechanisms of skeletal and ligamentous trauma, 3) differences between human cadavers and an anthropomorphic test dummy response, and 4) knee-joint ligament failure characteristics in isolated knee-joint tests.

Knee impacts with a 55.9 kg bolster covered mass at 6.0 m/s resulted in frequent avulsion fractures of the posterior cruciate ligament at its osseous attachment to the tibia with peak contact loads of 7.02 kN (7.74 kN peak dummy femur load). In this study, analysis of high-speed movies and radiographs indicated that the bolster loaded against the tibial tuberosity early in the event, translated the tibia posteriorly and resulted in a stretching of the posterior cruciate ligament. Lower-leg impacts produced tibial/fibular fractures or knee-joint ligament failures with peak bolster contact loads of 5.15 kN (4.21 kN peak dummy femur load). Isolated knee-joint tests indicated complete failure of the ligament after 2.26 cm of relative posterior tibial subluxation and a resistive load of 2.48 kN. However, the absolute values of the maximally tolerated loads may be significantly influenced by the deficiencies of the cadaver model and cannot be directly extrapolated for real-life situations. Since the lower extremity of the dummy cannot accommodate translatory motion at the knee joint and the skeletal mass of the dummy significantly exceeds that of the human, substantial kinematic and

790321
Advanced restraint system concepts
Reidelbach, W., Scholz, H.
Daimler-Benz AG (Stuttgart/Germany)

The seat belt pretensioner designed to eliminate belt slack in lap/shoulder belt systems with emergency locking retractors, is described as well as the current Mercedes-Benz passive restraint system which consists of air bags deployed by means of solid propellant gas generators, knee bolsters, and an electronic crash sensor with dual level triggering function.
Combinations of lap/shoulder belts, air bags, and pretensioners are presented and lead to the conclusion that an optimal restraint system would not be in compliance with current regulations.

790323
Effectiveness of current and future restraint systems in fatal and serious injury automobile crashes. Data from on-scene field accident investigations
Huelke, D. F.
University of Michigan Medical School
Sherman, Harold W., Murphy, Michael J., Kaplan, Richard J., Flora, Jerry D.
Highway Safety Research Institute, University of Michigan

Data from 101 front seat automobile occupant fatality crashes that the authors had investigated were reviewed along with 70 front seat automobile occupants who had the more severe (AIS 3, 4, or 5) level injuries who did not die. The effectiveness of the lap belt alone, lap-shoulder belt, air bag alone, air bag with lap belt, and the passive shoulder belt were made. The estimates reveal that none of the restraints would have prevented 42 to 51 of the fatalities. The air bag with lap belt, and the lap-shoulder belt system, have the highest effectiveness for reducing fatalities (AB+LB, 34%; LB+SH, 32%). The air bag with lap belt has an effectiveness of 68% in reducing the more serious injuries with the lap-shoulder belt nearly as equal (64%). NHTSA's fatality reduction estimates are excessively high and overly optimistic compared to ours, but theirs are noticeably lower for serious injury reduction than are ours. Comparisons with other restraint effectiveness studies are also made

791029, P-82
A study of driver interactions with an inflating air cushion
Horsch, John D., Culver, Clyde C.
Biomedical Science Department, General Motors Research

Conceptually, a steering wheel mounted air cushion is inflated before the upper torso of the driver significantly interacts with the cushion. However, this might not be the case for some seating postures or vehicle crash environments which could cause the driver to significantly interact with an inflating cushion.
These experiments utilized several environments to study the interaction between an inflating driver air cushion and mechanical surrogates. In these laboratory environments, the measured responses of mechanical surrogates increased with diminishing distance between the surrogate's sternum and the steering wheel mounted air cushion

800293
Driver and passenger air bag unit assemblies
Romeo, David J.
Talley Industries of Arizona, Inc., Mesa, AZ

The air bag restraint system or automatic restraint system is specifically designed to control forces and deceleration to the human body during an automobile accident. The three basic components comprising the air bag restraint system include: 1) the crash sensor(s), the diagnostic package for determining operability status, and the driver and front passenger air bag assemblies.
The information presented is limited to a description and the

operational features of the driver and passenger unit assemblies; parts which are located in the steering wheel and instrument panel respectively. Both assemblies consist of three basic components: 1) the inflator, 2) the module, and 3) the air bag.

800294
The development of air cushion restraint systems for small car front seat occupants
Zinke, D. Theodore
Minicars, Inc.

This paper presents progress, to date, on development and evaluation of three front seat airbag restraint systems for small cars. The subject vehicles were a Chevrolet Chevette, for which a passenger restraint system was developed, and a Dodge Omni, for which both driver and passenger systems were developed. The systems were primarily evolved during a series of sled tests and evaluated in vehicle barrier impacts.

821165, P-113
Unrestrained, front seat, child surrogate trajectories produced by hard braking
Stalnaker, R. L., Klusmeyer, L. F., Peel, H. H., White, C. D.
Southwest Research Institute
Smith, G. R., Mertz, H. J.
General Motors Corp.

This paper describes a study to determine the influence of preimpact vehicle braking on the positions and postures of unrestrained, children in the front seat at the time of collision.
Anesthetized baboons were used as child surrogates. The unrestrained animals were placed in various initial sitting, kneeling, and standing positions typically assumed by children while traveling in automobiles. Tests were conducted with various front seat positions and seat covering materials. Measurements were made of pertinent vehicle dynamics and surrogate kinematics during the hard braking event. For each initial condition evaluated, a photosequence is given showing typical positions and postures of the surrogate during the braking event.
The principal conclusion of the study is that the kinematic response of a child will be somewhere between that of a child dummy (probably more representative of a "braced" child) and that of an anesthetized baboon (probably more representative of a sleeping or relaxed child). The anesthetized animal pivoted about his hips and tumbled forward in response to the braking, whereas the child dummy maintained his seat position while sliding forward.

826045
Possible positions and postures of unrestrained front-seat children at instant of collision
Montalvo, F., Bryant, R. W., Mertz, H. J.
General Motors Corp.

The development of any front-seat passenger inflatable restraint system should consider the added deployment forces applied to occupants who may be close to the instrument panel at the time of deployment. A major factor influencing occupant position and posture is the effect of preimpact braking which often occurs prior to the collision. Most susceptible to preimpact braking are small, unrestrained children who can move off the front edge of the seat during hard braking and be near or against the instrument panel in a variety of different positions and postures at the instant of collision. Based on an analysis of hard-braking tests conducted with anesthetized baboons and child dummies, 13 positions are identified as being representative of the expected child positions. An estimate is given that, of the 149 small children (infants through four years old) expected to be unrestrained front-seat occupants in collisions of sufficient severity to deploy an inflatable restraint system per million car-years of exposure, 51 of them will be in one of these 13 positions near the instrument panel at the instant of collision.

826047

Responses of animals exposed to deployment of various passenger inflatable restraint system concepts for a variety of collision severities and animal positions

Mertz, H. J.
General Motors Corp.
Driscoll, G. D., Lenox, J. B.
Southwest Research Institute
Nyquist, G.W., Weber, D. A.
General Motors Corp.

This paper summarizes the results of tests conducted with anesthetized animals that were exposed to a wide range of passenger inflatable restraint cushion forces for a variety of impact sled - simulated accident conditions. The test configurations and inflatable restraint system concepts were selected to produce a broad spectrum of injury types and severities to the major organs of the head, neck and torso of the animals. These data were needed to interpret the significance of the responses of an instrumented child dummy that was being used to evaluate child injury potential of the passenger inflatable restraint system being developed by General Motors Corporation. Injuries ranging from no injury to fatal were observed for the head, neck and abdomen regions. Thoracic injuries ranged from no injury to critical, survival uncertain. Graphs are presented that show associations between the severity of the animal injuries by body region and selected measured animal responses and restraint system- accident characteristics. Caution must be used in interpreting the significance of these injuries relative to the expected performance of passenger inflatable restraint systems since aggressive restraint system concepts and accident conditions were selected for some tests in order to produce severe injuries.

851199

Small car air cushion performance considerations

Mertz, Harold J., Marquardt, James F.
Current Engineering and Manufacturing Services Staff, General

A critical performance issue in the development of any air cushion restraint system is the dichotomy that exists between the inflation rate required to meet the 30 mph frontal, rigid barrier restraint performance requirements and the effect that this parameter has on increasing the risk of deployment-induced injuries to out-of-position occupants. In general, small cars experience greater vehicle deceleration levels than large vehicles in FMVSS 208, 30 mph frontal, rigid barrier tests due to tighter packaging of their front-end components. In order to meet the FMVSS 208 performance requirements for such cars, the small car air cushion must be thicker and inflated faster than the large car air cushion. Such air cushion technology will increase the risk of life-threatening, deployment-induced injuries to out-of-position occupants of the small car. A harm reduction analysis is done that indicates that a greater benefit can be derived from installing a large car air cushion technology in a small car than from installing the same air cushion technology in a large car, even though the FMVSS 208 restraint performance requirements are not met when installed in the small car. This statement is true for any occupant restraint system (belts, passive interiors) identically applied to both the large and small car since small car occupants will always be exposed to more severe collision severities due to the car's lighter mass and reduced front-end crush space. Based on the analyses presented, it is suggested that performance requirements addressing deployment-induced injury concerns of out-of-position occupants be added to FMVSS 208 for evaluating air cushion restraints and that the FMVSS 208, 30 mph frontal, rigid barrier test conditions be changed to a frontal test of equal collision severity for all car sizes

851201

Can we develop less expensive airbags?

Breed, Allen
Breed Corp.

An all-mechanical self-contained airbag has been developed that avoids the complexities of electrical systems and thus is less expensive

and easier to install. The system, based on conventional military fuzing technology, evolved from computer math models that found a sensor response curve which would provide timely airbag deployment while detecting the crash pulse outside the crush zone.

851247

Chest injury criteria for combined restraint systems

Gr\arsch, Lothar
Daimler-Benz AG

The expected improved performance of a combined restraint system where an airbag supplements the conventional safety belt was not reflected in reduced g-values on the dummy's chest. However, by the distribution of force over the wider area of the airbag and the corresponding reduction of the specific pressure exerted by the three point belt, improved occupant protection is actually produced. Therefore, measurable quantities other than acceleration should be selected to evaluate the risk of chest injury, such as belt load or chest deflection.

A new method to measure the deflection of dummy ribs with strain gauges has been developed. The resulting data indicate a significantly reduced chest deflection when a combined system is used

856015

Supplemental driver airbag system ~ Ford Motor Company Tempo and

Topaz vehicles

Maugh, Roger E.
Ford Motor Co.

This paper describes Ford's supplemental airbag system that is available to commercial fleet users in the United States on Tempo and Topaz car lines. Several key components of the system were based on designs used in a 1981 experimental program. Design considerations in adapting that experimental system to a smaller production vehicle are presented.

In my discussion, I would like to describe the supplemental driver-side airbag system currently being sold as a commercial fleet option in the United States on Ford's Tempo and Topaz compact front-wheel- drive car lines. This system went into production early this year. It was developed in part on the basis of experience we gained with an experimental design originally developed for the full-size 1981 Lincoln Town Car. The Town Car is a large luxury car, approximately 43 in (1,088 mm) longer and 1,540 lb (700 kg) heavier than the Tempo/Topaz. Since the focus of our panel discussion is on airbag technology and optimizing airbag system performance, I would like to describe a few of the changes Ford made during a nearly 2-year program to adapt the experimental Lincoln-based airbag system to the much smaller Tempo/Topaz vehicles

Automobile design

872216

P-202

Occupant injury patterns in crashes with air bag equipped government

sponsored cars

Backaitis, Stanley H.
National Highway Traffic Safety Administration
Roberts, J. Vernon
National Highway Traffic Safety Administration

In 1983, the National Highway Traffic Safety Administration (NHTSA) initiated two air bag vehicle fleet programs. The objective was to demonstrate that both original equipment and retrofit air bag systems operate in vehicles as intended. As of July 1, 1987, the two fleets together have accumulated over 200 million miles. Data are presented for 112 crashes involving air bag deployment in these government sponsored fleet vehicles in service between 1984 and July 1, 1987.

Of the 112 drivers involved in the crashes, 103 sustained either no injury or only minor (AIS 1)[1] injuries. Of the nine remaining cases, six were AIS 2 and three AIS 3. To date, the limited data indicate that the air bag deployed as expected in all frontal crashes severe enough to

, in collisions in which the air bag did not deploy, the

crashes were of such low severity that no actuation was expected and none took place. In one case the crash was so severe (95 mph delta V and complete override by a logging truck) that deployment did not occur because the electrical system was destroyed before the crash sensors could initiate air bag deployment. This accident was judged to be unsurvivable regardless of the type of restraint system used to protect the occupant.

902324, P-236
Assessment of air bag deployment loads
Horsch, John, Lau, Ian, Andrzejak, Dennis, Viano, David C., Melvin, John
Biomedical Science Dept., General Motors Research Labs.
Pearson, Jeff, Cok, David, Miller, Greg
General Motors Current Product Engrg.

A study of air bag deployments has indicated that some occupant injury was "unexpected" and might have been related to loading by the inflating bag. Laboratory studies have found "high" loads on surrogates when they are out of a normal seating position and in the path and against an inflating air bag (out-of-position). The current study evaluated laboratory methods for assessing the significance of deployment loads and the interaction mechanics for the situation of an occupant located near or against a steering wheel mounted air bag. Analysis of the field relevance of the results must consider not only factors relating to the assessment of injury risk, but also exposure frequency.
The highest responses for the head, neck, or torso were with that body region aligned with and against the air bag module. The risk of severe injury was low for the head and neck, but high when the torso was against and fully covering the air bag module. Torso injury was related to swelling of the module before bag break-out from the module. Reducing inflator output provided only minimal reduction in torso injury risk. Providing pressure relief by an alternate bag break-out path or preventing full coverage of the module reduced the severity of torso loads.

910808, SP-852
Are tank pressure curves sufficient to discriminate airbag inflators
Wang, J. T.
Engrg. Mechanics Dept., General Motors Research Labs.

The validity of the current practice of using tank pressure curves to discriminate airbag inflators is evaluated. Sled test results of two inflators, which have similar tank pressure curves, are first compared. Significant differences in airbag performance have been observed which suggests that the two inflators are not similar. The gas dynamics of airbag inflation are then reviewed to develop theories to explain the phenomenon. The dual-pressure method, which was previously developed for modeling airbag inflators, is found to be useful in this analysis. The analysis clearly shows that the inequality is due to the difference in gas temperature among inflators. We find that the higher the gas temperature the faster the gas venting and leaking will be. This is why different airbag performance is obtained from inflators which have similar tank pressure curves. Based on both theoretical and experimental evidence, we conclude that knowing the tank pressure curve alone is insufficient to determine the capability of an inflator for both driver and passenger airbag systems, as long as vents and/or leaks exist in the systems. The inability to discriminate among inflators early in the design, could lead to late changes in the design of the restraint system when the actual performance of the airbag is measured during sled or barrier tests. Finally, four new methods to discriminate inflators are discussed.

920124, SP-906
A critique of single point sensing
Breed, David S., Sanders, W. Thomas, Castelli, Vittorio
Automotive Technologies International, Inc.

In two previous SAE papers by the authors, supporting analysis was presented showing the difficulty in achieving a timely response to real-crash events using a single point sensor mounted in the non-crush zone of the vehicle (tunnel, cowl, etc.). The analysis demonstrated the propensity to deploy the air bag(s) late during certain of these events. If a vehicle occupant was not wearing a safety belt, the deceleration forces of the crash could place the occupant out of position and resting against the air bag when it was deployed. In another SAE paper by H. J. Mertz et al, the authors demonstrated that animals, used as surrogates for humans, could be injured if positioned against an air bag at the time of deployment.
Arguments are presented here to show that there is insufficient information in the crash pulse as sensed in the non-crush zone to deploy an air bag in time for the unbelted occupant. It is, therefore, not possible to create an algorithm for an electronic sensor, based on the crash pulse information in the non-crush zone alone, which will initiate air bag deployment in time for all cases. Therefore, sensing in the crush zone is required.

922523, P-261
Crash injury prevention: a case study of fatal crashes of lap-shoulder belted occupants
Viano, David C.
General Motors Research and Environmental Staff

A case study was conducted of 123 crashes involving 144 fatally injured lap-shoulder belted front-seat occupants. The crashes occurred throughout the United States in 1985-86 and involved 97 driver and 47 right-front passenger deaths in new vehicles. A judgment was made by consensus of a safety panel on the potential for saving the victim's life by the addition of safety technology. Supplemental airbags provided the greatest potential for improving the life-saving effectiveness of current lap-shoulder belts. Overall, airbags may have prevented 12% of the belted occupant fatalities and 27% of the deaths in frontal crashes. The benefit of supplemental airbags was greater for the right-front passenger, in part, because of more females and occupants over 60 years of age in that seating position. A majority (68%) of the belted fatalities were judged unpreventable by reasonable restraint or vehicle mofidications. This level is indicative of the extreme severity of many of the fatal crashes involving extensive vehicle damage and forces on the occupant compartment, unusual crash configurations and causes of death, and unique situations related to seating position and crash dynamics. Using published levels of belt effectiveness, 50% of all fatalities may not be preventable by the use of lap-shoulder belts, supplemental airbags, and practicable changes in crashworthiness systems. The study addressed only the potential benefit of supplemental-restraint components and airbags and did not consider possible adverse effects which might occur in real-world crashes

940714, SP-1042
Airbag protected crash victims~The challenge of identifying occult injuries
Augenstein, Jeffrey S., Perdeck, Elana B., Byers, Patricia M., NunezsJr., Diego B., Zych, Gergory A., Andron, Jonathan, Craythorne, A. Kevin, Verga, Carla
University of Miami
Digges, Kennerly H.
George Washington Univ.
Lombardo, Louis V.
National Highway Traffic Safety Administration
Stratton, James E.
Arvin Caslspan Corp.
Malliaris, A. C.
Data Link, Inc.

A multidisciplinary, automobile crash investigation team at the Jackson Memorial Hospital/Ryder Trauma Center in Miami, Florida, is conducting detailed medical and engineering study. The focus is restrained (seatbelts and/or airbag) occupants involved in frontal crashes, who have also been severely injured. More than 60 crashes have been included in the study to date.
Analysis of the initial data indicates that restraint systems are

However, internal injuries among airbag-protected occupants may be unrecognized in the field providing new challenges in triage and injury diagnosis. In other cases, survival in extremely high severity crashes presents trauma management challenges due to the extent and complexity of the multiple injuries which result. The paper provides case examples to illustrate types of chest and abdominal injuries associated with airbag cases. Three types of cases are presented: (1) Jackson Study Involving Occult Chest/Abdominal Injury, (2) National Accident Sampling System (NASS) Special Crash Investigation (SCI) and (3) Jackson Study Involving Crash Severities Greater than 45 MPH.

To assist in recognizing the extent of injuries to occupants protected by airbags, it is suggested that additional evidence from the crash scene be used in the triage criteria. For the occult chest/abdominal cases observed in the Jackson study, deformation of the steering system was the vehicle characteristic most frequently observed.

The challenges of recognizing injuries to airbag-protected occupants are discussed. The presence of steering wheel deformation may be a sufficient signal of caution to justify transporting the injured victim to a Level 1 or 2 trauma center so that a close examination for occult injuries can be made.

940716

Upper extremity injuries related to air bag deployments
Huelke, Donald F., Moore, Jamie L., Compton, Timothy W.
University of Michigan
Samuels, Jonathan, Levine, Robert S.
Wayne State Univ.

From our crash investigations of air bag equipped passengers cars, a subset of upper extremity injuries are presented that are related to air bag deployments. Minor hand, wrist or forearm injuries-contusions, abrasions, and sprains are not uncommonly reported. Infrequently, hand fractures have been sustained and, in isolated cases, fractures of the forearm bones or of the thumb and/or adjacent hand. The close proximity of the forearm or hand to the air bag module door is related to most of the fractures identified. Steering wheel air bag deployments can fling the hand-forearm into the instrument panel, rearview mirror or windshield as indicated by contact scuffs or tissue debris or the star burst (spider web) pattern of windshield breakage in front of the steering wheel.

942206, P-279

Thoracic trauma assessment formulations for restrained drivers in simulated frontal impacts
Morgan, Richard M., Eppinger, Rolf H., Haffner, Mark P.
NHTSA
Yoganandan, Narayan, Pintar, Frank A., Sances @sJr, Anthony
Medical College of Wisconsin
Crandall, Jeff R., Pilkey, Walter D., Klopp, Gregory S.
University of Virginia
Kallieris, Dimitrios, Miltner, Erich, Mattern, Rainer
University of Heidelberg
Kuppa, Shashi M., Sharpless, Cheryl L.
Conrad Technologies, Inc.

Sixty-three simulated frontal impacts using cadaveric specimens were performed to examine and quantify the performance of various contemporary automotive restraint systems. Test specimens were instrumented with accelerometers and chest bands to characterize their mechanical responses during the impact. The resulting thoracic injury severity was determined using detailed autopsy and was classified using the Abbreviated Injury Scale.

The ability of various mechanical parameters and combinations of parameters to assess the observed injury severities was examined and resulted in the observation that belt restraint systems generally had higher injury rates than air bag restraint systems for the same level of mechanical responses. To provide better injury evaluations from observed mechanical parameters without prior knowledge of what restraint system was being used, a dichotomous process was developed. The process first determines, based on multiple chest deformation patterns, whether the restraint system is behaving belt like or bag like and then uses either the separately developed belt or bag criteria, to estimate injury severity. This two-step process had shown improved predictive capabilities for this data set.

942217, P-279

Laboratory investigations and mathematical modeling of airbag-induced skin burns
Reed, Matthew P., Schneider, Lawrence W., Burney, Richard E.
University of Michigan

Although driver-side airbag systems provide protection against serious head and chest injuries in frontal impacts, injuries produced by the airbag itself have also been reported. Most of these injuries are relatively minor, and consist primarily of skin abrasions and burns. Previous investigations have addressed the mechanisms of airbag-induced skin abrasion. In the current research, laboratory studies related to the potential for thermal burns due to high-temperature airbag exhaust gas were conducted. A laboratory apparatus was constructed to produce a 10-mm- diameter jet of hot air that was directed onto the leg skin of human volunteers in time-controlled pulses. Skin burns were produced in 70 of 183 exposures conducted using air temperatures ranging from 350 to 550oC, air velocities from 50 to 90 m/s, and exposure durations from 50 to 300 ms. A mathematical model of heat transfer to the skin and burn injury was developed, along with an empirical description of the threshold for partial- thickness skin burn as a function of gas velocity, gas temperature, and exposure duration. The mathematical burn injury model was combined with a lumped-parameter gas-dynamics model of airbag inflation to demonstrate the application of the skin thermal tolerance data to prediction of airbag-induced skin burn.

942218, P-279

Investigation into the noise associated with air bag deployment: Part I~Measurement technique and parameter study
Rouhana, Stephen W., Webb, Scott R., Wooley, Robert G., McCleary, Joseph D., Wood, Francis D., Salva, David B.
General Motors

High-amplitude, short-duration noise is called impulse noise. A large body of literature on impulse noise has been developed primarily by military researchers for multiple exposures such as those caused by weapons firing. Some research into the impulse noise associated with air bag deployments was performed in the late 1960's and early 1970's to ascertain the risk of hearing loss. Several criteria for risk of noise-induced hearing loss were proposed and much was learned about the sources of the noise. Unfortunately, the instrumentation used to measure the noise in many of those studies lacked adequate low frequency response characteristics. Perhaps more importantly, results from experiments with human volunteers do not seem to agree with the proposed criteria. For this study, a new system consisting of commercially available pressure transducers and microphones was assembled and a new software package was developed. This system allows analysis of the pressure-time data using two analysis methods and criteria proposed in the early 1970's. A series of experiments using this system was run over a four year period to investigate the parameters that affect the impulse noise associated with a deploying air bag. Some observations are presented and conclusions drawn from the data

942219, P-279

Comparison of occupant restraints based on injury-producing contact rates
Blower, Daniel, Campbell, Kenneth L.
Transportation Research Institute, University of Michigan

The objective of this analysis is to evaluate the effectiveness of restraints in preventing injury-producing contacts of specific body regions, such as the head or chest, with specific interior components. In order to make comparisons by restraint use, an injury rate is calculated as the number of injury-producing contacts per hundred involved occupants. Data, including the Occupant Injury Classification (OIC), are Accident Sampling System (NASS) Crashworthi-

vehicle drivers in towaway, frontal impacts. Injury-producing contact rates are compared for four restraint configurations: unrestrained, three-point belted, driver airbag alone, and driver airbag plus three-point belt. For each restraint configuration, contact rates are compared by three categories of injury severity, AIS 1, AIS 2, and AIS 3-6, body region injured, and contact area producing the injury. The three point belt provides substantial reductions in driver injury rates for head/face and torso contacts with the glazing, pillar/rails, and steering assembly. The addition of the driver airbag to the three-point belt appears to offer further reductions in these injury rates. The driver airbag alone did not show similar reductions, although sample size was very limited. Also, the injury rate for airbag contacts is more than three times the rate for belt contacts. The effects of occupant age, gender, and stature are identified as areas for further study.

950865

Regulatory history of automatic crash protection in FMVSS 208
Kratzke, Stephen R.
National Highway Traffic Safety Administration

This paper summarizes the regulatory history of the automatic crash protection requirements in Federal Motor Vehicle Safety Standard 208. It is intended to give the reader an overview of the regulatory history involved in Standard 208, from its beginning in 1968 as a requirement that passenger cars be equipped with seat belts to its present requirement that, as of 1998, all passenger cars and light trucks must be equipped with airbags. It also discusses and summarizes the various court cases that have challenged different aspects of the automatic crash protection requirements.

950867

Injury patterns of car occupants under airbag deployment
Malliaris, A. C., Digges, K. H., DeBlois, J. H.
DeBlois Associates, Inc.

This investigation addresses and evaluates the injury patterns of car occupants as a function of airbag deployment with or without belt use. A companion paper provides a comprehensive evaluation of airbag field performance, in comparison with the no restraint, or other restraint conditions, concerning: cars, exposure, occupants, restraints, and protection provided by the restraints.
The findings reported in this investigation are based exclusively on the data contained in the records of NHTSA's NASS/CDS 1988-1992. The investigations focuses on car drivers. Because of the relatively late and limited introduction of airbags, all other light vehicle populations are either not represented or represented by a very meager sample in the sources cited above

950868

Air bag deployments involving restrained occupants
Dalmotas, Dainius J., Hurley, Regina M., German, Alan
Transport Canada

As a consequence of various federal and provincial initiatives to promote the use of seat belts in Canada, the wearing rate of seat belts among front outboard passenger car occupants is now estimated at 90 percent. Accordingly, the vast majority of air bag deployments in Canada involve restrained occupants. In order to gain a better understanding of the field performance of air bag systems, Transport Canada recently initiated an in-depth study of motor vehicle collisions involving air bag deployments. To date, investigations have been completed on 242 such collisions. While the preliminary data suggest that supplementary air bag systems provide considerable added protection against serious head injuries in moderate and high severity frontal crashes, they also suggest that, in low severity crashes, deployment of an airbag system may expose belted occupants to unnecessary injury risk from the air bag itself. A more favorable trade-off can be achieved by increasing the deployment threshold of the airbag system for belted occupants.

950869

Evaluation of airbag field performance
Malliaris, A. C., Digges, K. H., DeBlois, J. H.
DeBlois Associates, Inc.

This investigation encompasses a comprehensive evaluation of airbag field performance, in comparison with the no restraint, or other restraint conditions. The paper at hand addresses: cars, exposure, occupants, restraints, and protection provided by the restraints. A companion paper addresses the injury patterns of car occupants. The findings of the investigation are based on two primary sources of national coverage: (i) the field crash experience contained in the records of NHTSA'a NASS/CD 1988-1992, and (ii) the fatal accident experience contained in NHTSA's FARS 1991- 1993. The investigation focuses on car drivers. Because of the relatively late and limited introduction of airbags, all other light vehicle populations are either not represented or represented by a very meager sample in the sources cited above

950870

Leading-edge deployment speed of production airbags
Powell, Michael R., Lund, Adrian K.
Insurance Institute for Highway Safety

Airbags have proven to be effective in preventing deaths and serious injuries; however, in some instances when an occupant contacts an airbag while it is still deploying, injuries may result from this contact. Most of these are minor injuries, such as skin abrasions, which are believed to be caused by the contact pressure created by the deploying airbag surface. To assess the relative potential of different airbag designs to cause skin abrasions, a series of static deployment tests were conducted to measure the leading-edge speed of driver-side airbags from several 1993 model cars. The results of the tests indicate that airbags exhibit a wide range of leading-edge speeds and that, in some cases, maximum leading-edge speed is a highly variable characteristic among airbags from the same model car. Maximum leading-edge speeds ranged from 171 to 328 km/h. Comparison of speed profiles to an approximate abrasion reference value showed that many of the tested airbags are traveling at a speed great enough to cause abrasions as far away as 286 mm from the steering wheel; however, some airbags never reached this reference speed.

950871

Air bag effectiveness~Trading major injuries for minor ones
Libertiny, George Z.
Ford Motor Co.

A number of articles in professional journals and newspapers have questioned the effectiveness of supplemental air bags. The articles maintain that insurance injury claims have increased after adoption of air bag systems and thus air bags provide no net benefit. The insurance industry itself disagreed with this interpretation of its own data, but their protest is not highly publicized.
The object of this paper is to shed light on the effectiveness of air bags using the National Accident Sampling System (NASS) of the National Highway Traffic Safety Administration (NHTSA). Our results indicated that the air bags are doing what they were designed to do, that is, to decrease the severity of injuries in major accidents. On the other hand, air bags sometimes might cause minor injuries, when no injury would have occurred without air bags. Since minor accidents are more frequent than major accidents, the frequency of injuries might increase when air bags are introduced. This does not mean that air bags are ineffective; on the contrary, the air bags decrease the severity of major injuries in exchange for increasing the number of minor injuries, surely a desirable design tradeoff.

950886, SP-1077

The effect of limiting shoulder belt load with airbag restraint
Mertz, Harold J., Williamson, James E., Vander Lugt, Donald A.
General Motors Corp.

The dilemma of using a shoulder belt force limiter with a 3- point belt system is selecting a limit load that will balance the reduced risk of significant thoracic injury due to the shoulder belt loading of the chest against the increased risk of significant head injury due to the greater upper torso motion allowed by the shoulder belt load limiter. However, with the use of airbags, this dilemma is more manageable since it only occurs for non-deploy accidents where the risk of significant head injury is low even for the unbelted occupant. A study was done using a validated occupant dynamics model of the Hybrid III dummy to investigate the effects that a prescribed set of shoulder belt force limits had on head and thoracic responses for 48 and 56 km/h barrier simulations with driver airbag deployment and for threshold crash severity simulations with no airbag deployment. For the belt/bag/occupant configuration that was evaluated, the analysis gives an optimum shoulder belt load limit that minimizes the dummy responses in the 56 km/h simulation without significantly increasing the risk of significant head injury in the non-deploy simulations.

952700, P-299

On the synergism of the driver air bag and the 3-point belt in frontal collisions
Keenan, Lori
Conrad Technologies, Inc.
Kallieris, Dimitrios, Rizzetti, Andreas, Mattern, Rainer
University of Heidelberg
Morgan, Richard, Eppinger, Rolf
National Highway Traffic Safety Administration

The number of passenger vehicles with combined 3-point belt/driver air bag restraint systems is steadily increasing. To investigate the effectiveness of this restraint combination, 48 kph frontal collisions were performed with human cadavers. Each cadaver's thorax was instrumented with a 12-accelerometer array and two chest bands. The results show, that by using a combined standard 3-point belt (6% elongation)/driver air bag, the thoracic injury pattern remained located under the shoulder belt. The same observation was found when belts with 16% elongation were used in combination with the driver air bag. Chest contours derived from the chest bands showed high local compression and deformation of the chest along the shoulder belt path, and suggest the mechanism for the thoracic injuries. On the other hand, in tests where the air bag was the only available torso restraint, forces were distributed uniformly over the front of the chest; high local compression/deformation and injuries were reduced.
This study asks if it is possible to obtain both the thoracic injury mitigating benefits of an air bag only restraint and the all-impact-direction benefits of the belt from a combination restraint system by adding a force limiter to the shoulder belt. For this reason, tests with force limiters were performed. Initially, the investigation was carried out with Hybrid III dummies using two different levels of force limiters: 4 kN and 5 kN. The force limiter with the level of 4 kN showed, through examination of the chest band contours, a more bag-like uniform compression of the chest, with the belt effect only slightly pronounced. The chest compressions were 4 to 8 cm, and the resultant spinal accelerations were 30 to 40 g's.
By using the same restraint combination and force limiter, comparable vertebral accelerations and chest compressions were measured in cadaver testing. No injuries were found in the cervical spine and only an AIS 1 was observed in the thorax, for the age range of 60 to 65 years. The results also suggest that when a driver air bag is combined with a 3-point belt system that limits the torso belt loop load to 4 kN, additional injury mitigation benefits for both the cervical spine and the thorax are obtained in frontal collisions.
Analytical simulations were also conducted using different size occupants in both the baseline and the optimized belt/air bag restraint and in other crash conditions. These simulations suggest the harmonized belt/air bag also improves safety performance for other than mid-sized male occupants and does not adversely affect the performance of the restraint system in other prevalent crash configurations

952701, P-299

Review of the air bag effectiveness in real life accidents~Demands for positioning and optimal deployment of air bag systems
Otte, D.
Medical University Hannover

In Europe there is less experience with air bag deployments in accidents than the U.S. scene. Within the continuing in-depth investigations at the Medical University Hannover, 41 accidents with air bag deployments could be documented. A detailed description of selected accidents is given in the study. The injuries of the occupants are described and the injury mechanisms caused by the air bag deployment are discussed. In the study, the distribution of injury severities are compared for accidents with seat belt system only correlating to those with air bag additionally.
The air bag-fitted cases demonstrate that the protective effect was essentially based on using the seat belt. Especially for less load level some injuries occurred which can called as air bag specific injuries. These are some hematoma in the thorax and face region as well as burn injuries, but a higher risk for distortions of the cervical spine AIS 1 could be seen.
These results lead to the supposition that in case of air bag - seat belt systems the current level of deployment is too low. It is the proposal of the study to use the effectiveness of the seat belt system only up to delta-v 35 or 40 km/h and the air bag effectiveness should be aimed at above this delta-v level. The benefit of an air bag could be established with higher delta-v.
A second part of the study quantifies the benefit of an airbag for frontal and lateral collisions.

952702, P-299

How airbags and seat belts work together in frontal crashes
Deng, Yih-Charng
General Motors R & D Center

This study examines the combined effects of the passenger airbag and the seat belt on the occupant impact response. It was found that while an airbag is beneficial in reducing unbelted occupant injury, its restraint force is in general additive to that of the belts in a 30 MPH barrier impact and tends to increase belted occupant response numbers. A number of possible design strategies were discussed and the inherent performance trade-offs among various impact conditions were illustrated. Concepts for two types of Adaptive Restraint System (ARS) are discussed which might achieve even greater levels of occupant protection for both belted and unbelted occupants. For a belted occupant, these ARS designs have embedded logic to determine when and how to use an airbag and/or a belt under various impact conditions. These ARS designs try to utilize the combined strengths of the airbag and the seat belt systems. Possible design strategies for these systems were also discussed.

960660, SP-1144

Injuries sustained by air bag protected drivers
Augenstein, Jeffrey S., Perdeck, Elana B.
University of Miami
Digges, Kennerly H.
George Washington Univ.
Lombardo, Louis V.
National Highway Traffic Safety Administration
Malliaris, A. C.
Data Link, Inc.

The William Lehman Injury Research Center has conducted multi-disciplinary investigations of fifty crashes involving drivers protected by air bags. In all cases, serious injuries were suspected. Nine cases involved fatal injuries. These cases are not representative of crashes in general. However, when used in conjunction with NASS/CDS they provide insight into the most severe injuries in crashes of vehicles

A comparison with data from the National Accident Sampling System; Crashworthiness Data System (NASS/CDS) shows that head injury and abdominal injury make up a larger fraction in the Lehman data than in NASS/CDS. Examination of fatal cases indicates that head injuries are frequently caused by intruding structure or by unfavorable occupant kinematics among the unrestrained population. The more precise injury examination provided to patients admitted to a Level I Trauma Center may contribute to the higher proportion of abdominal injuries observed at the Lehman Center.

Serious upper extremity injuries are rare in the Lehman data, but contribute a large harm fraction in NASS/CDS. The upper extremity injuries in NASS/CDS occur predominately at the lower crash severities. Both NASS/CDS and Lehman data indicate that lower extremity injuries are an increasing harm factor among occupants in severe crashes. The Lehman data suggests that air bags are remarkably effective at high severity frontal crashes. However, in some rare events, the protection at low crash severity may be less than without the air bag.

960661

Fatalities in airbag-equipped cars: a review of 1989-93 NASS cases

Lund, Adrian K., Ferguson, Susan A., Powell, Michael R.
Insurance Institute for Highway Safety

A review of 39 driver fatalities in 1990-93 cars with airbags from the National Accident Sampling System indicated most of these fatalities were due to causes unrelated to frontal airbag performance. Two-thirds occurred in side-impact or rollover crashes, in which airbag effectiveness is limited; of 15 frontal crash fatalities, 6 died of causes unrelated to the frontal impact and 5 in cars with severe intrusion. The remaining 4 fatalities, 3 of whom were unbelted, were in moderate to high severity crashes which could have been survivable; however, the deploying airbags, instead of protecting, probably contributed to the fatal injuries. A similar review of 12 fatalities of unbelted drivers in cars without airbags revealed 3 could have been prevented by airbags, but 4 were in crashes that could have put them in position to be injured by the airbag. These results suggest that reducing deployment energy would improve airbag effectiveness in relatively severe crashes as well as low severity crashes, even for unbelted drivers. No examples of fatal driver injuries from airbags in low severity crashes were found.

970129
SP-1231

Chest injury risks to drivers for alternative air bag inflation rates

Digges, Kennerly
George Washington Univ.
Noureddine, Ahmad
George Washington Univ.
Bedewi, Nabih E.
George Washington Univ.

While the present air bag systems have been shown to be highly effective in high severity crashes, undesirable side effects have been reported in some low severity events. The inflation rate of the air bag during deployment has been cited as a factor which induces injuries. A rapid air bag deployment rate is advantageous to provide protection to occupants in severe crashes. On the other hand, air bag aggresivity associated with the high inflation rate can increase injuries in the lower severity crashes. The injury producing forces from the air bag increase as the occupant position becomes closer to the bag at the time of deployment.

This paper describes the results of an analytical study to evaluate chest injury measures for reduced inflation rates of a Taurus type air bag in a variety of crash modes. A detailed nonlinear finite element model of an unfolding air bag and a 50\ut\uh percentile male Hybrid III dummy are used in conjunction with a test buck to simulate frontal crashes. The model is calibrated for a 30 mph crash, and then used to evaluate the effects of chest loading for other crash modes and air bag flow rates. Different acceleration time histories are introduced to model

high-speed (35 mph) crashes and low-speed (16 mph) pole impacts. A series of parametric simulations are performed for the unbelted dummy by varying the deployment time and the gas flow rate.

A barrier crash at 20 mph was simulated with three variations in flow rate, and occupant position at inflation. The results suggest that a flow rate greater than 60% is desirable to prevent steering wheel contact by the dummy.

Simulations of different crashes indicated that a 75% flow rate reduced chest injury measures in the 35 mph offset crash, and at all cases simulated at severities of 30 mph and below. The 75% flow rate increased chest injury measures in the 35 mph barrier crash.

The simulation of a 16 mph pole crash with late deployment indicated that chest injury readings for this case were higher than in the 35 mph barrier crash. The 75% flow rate reduced chest injury measures in this case by 25%.

Reducing the air bag flow rate by about 75% significantly reduced the chest injury risk to a normally seated, unbelted, 50\ut\uh percentile male dummy in moderate and low severity crashes. A reduction in the margin of distance between the chest and steering wheel also resulted. However, the 75% flow rate prevented contact with the steering wheel in all cases simulated.

970490, SP-1226

Facial, periorbital and ocular injuries related to steering-wheel airbag deployments

Huelke, Donald F., Schneider, Lawrence W., Reed, Matthew P., Gilbert, Ryan J.
University of Michigan

To determine the frequency of facial injuries from steering-wheel airbag deployments, 540 consecutive steering-wheel airbag deployments, investigated by the University of Michigan Transportation Research Institute (UMTRI) personnel, were reviewed. About 1 in 3 drivers sustain an injury to the face. Injuries to the area surrounding the eye (periorbital) or to the eyeball (ocular) rarely occur. The frequencies of facial or ocular injuries are the same for belted and unbelted drivers. Drivers of short stature had a higher frequency of facial injury. Females sustained ocular injuries more frequently than males. Untethered airbags were not overly involved in drivers with an ocular injury. No specific make or model car was overly represented in the ocular injury cases.

970493, SP-1226

Upper extremity injuries from steering-wheel airbag deployments

Huelke, Donald F., Gilbert, Ryan, Schneider, Lawrence W.
University of Michigan

In a review of 540 crashes in which the steering-wheel airbag deployed, 38% of the drivers sustained some level of upper extremity injury. The majority of these were AIS-1 injuries including abrasions, contusions and small lacerations. In 18 crashes the drivers sustained AIS-2 or 3 level upper extremity injuries, including fractures of the radius and/or ulna, or of the metacarpal bones, all related to airbag deployments. It was determined that six drivers sustained the fracture(s) directly from the deploying airbag or the airbag module cover. The remaining 12 drivers had fractures from the extremity being flung into interior vehicle structures, usually the instrument panel. Most drivers were taller than 170 cm and, of the 18 drivers, 10 were males.

970776, SP-1226

Adaptive airbag-belt-restraints~An analysis of biomechanical benefits

Adomeit, H.-D., Wils, O., Heym, A.
PETRI AG Engineering Center for Automotive Safety

Current restraint systems with pretensioner, belt force limiter and airbag are not only designed for a crash test such as FMVSS 208. Automobile manufacturers carry out test configurations over and above these parameters to check the effectiveness of restraint systems, for , frontal crash with 40% offset against a deformable

for all new European vehicles as of October 1998. However it is not possible to cover the entire spectrum of accident configurations found in real-world accident occurrences.

In the development of restraint systems with reference to aspects of occupant biomechanics, the possibilities on the testing side for reproducing reality are particularly limited. Aside from the 50%-HYBRID III dummy as occupant of average size and weight, only the two extremes of the occupant, 5% female dummy and 95% male dummy, are used. This paper defines the concept of adaptability of the restraint system, that is the adaptability of its protection performance to various different possible initial parameters such as type and severity of accident, and the occupancy/car interior/constellation. The adaptation requirement for the restraint system is determined in an occupancy/car interior model with simulation for individual crash occupancy combinations in a frontal crash. With the help of selected crash configurations with high adaptation requirements on the restraint system (RS), the potential of such restraint systems with regard to the increase in occupant protection is demonstrated. The question of the value of adaptability should be discussed in comparison to a restraint system which has been alternatively further optimized by primary measures.

973295, P-316

Children as front seat passengers exposed to airbag deployments
Huelke, Donald F.
University of Michigan

From a review of 722 "airbag" crashes investigated by UMTRI personnel, there were 117 front seat passengers exposed to passenger-side airbag deployments. The majority of these were 16 years of age or older (90 of the 117), with 20 passengers, 11 years of age or younger. Two cases, both fatalities, will not be described, for these crashes have been investigated also by NHTSA's Special Crash Investigation Program personnel. The crashes and injuries of these 18 passengers are described. In this group there were 13 children who had MAIS-1-level injuries and two sustained an MAIS-2 injury. One child was without injury, and two had MAIS-5-level injuries. The description of the crash and of the major injuries sustained are detailed in each of the case capsule descriptions. Most of the children were properly restrained

973296, P-316

Airbag-induced injury mechanisms for infants in rear-facing child restraints
Augenstein, Jeffrey S., Perdeck, Elana, Williamson, Jami, Stratton, James
University of Miami
Digges, Kennerly
George Washington Univ.
Lombardo, Louis
National Highway Traffic Safety Administration

The National Highway Traffic Safety Administration (NHTSA) Special Crash Investigations database contains twelve completed cases of child fatalities in rearward-facing child seats caused by deploying airbags. Three of these are now available for examination. An additional two cases were investigated by the William Lehman Injury Research Center at the University of Miami School of Medicine. These five cases are examined to evaluate crash environment, injury mechanisms, and circumstances which caused the child to be in front of the passenger-side airbag.

Four of the cases were crashes with impacts with the side of other cars with crash severities less than 15 mph. The predominate injury mechanism was brain and skull injury from a blow transmitted to the rear of the head through the child seat back. In one case, the force to the head was transmitted downward, directly from airbag contact. In this case, the infant showed no sign of injury at the scene, and there was no damage or displacement of the rearward-facing child seat. The child died 24 hours later of brain injuries.

Examination of the circumstances surrounding the crashes indicates that many opportunities were missed for providing warning of the danger of airbag injuries to infants in rear-facing child safety seats.

Inaccurate, incomplete, and inadequate warnings existed regarding the use of rearward-facing child safety seats during the time period in which the crashes in this investigation occurred (1995-1996).

973297, P-316

Nonfatal airbag deployments involving child passengers
Gotschall, Catherine S., Eichelberger, Martin R., Morrissey, J. Ren\ae, Better, Allison I., Reardon, Jean
Children's National Medical Center
Bents, Francis
Dynamic Science, Inc.

This study describes nonfatal interactions between children and front passenger airbags. Twelve surviving children aged 12 and younger who experienced passenger airbag deployments were identified from hospital emergency department records and police accident reports. Data from crash reconstructions and medical records were analyzed to determine mechanisms responsible for each injury sustained. Four of six children in safety seats and one of six children restrained in lap/shoulder belts sustained significant brain injuries. It is recommended that infant seats and forward-facing safety seats be placed in the rear seats of vehicles equipped with passenger airbags.

973323, P-315

Response and vulnerability of the upper arm through side air bag deployment
Kallieries, Dimitrios, Rizzetti, Andreas, Mattern, Rainer
University of Heidelberg
Jost, Stefan, Priemer, Peter, Unger, Michiel
AlliedSignal

The number of passenger cars equipped with side air bags is steadily increasing. With the aim of investigating the mechanical responses and the injuries of the arm under the influence of a side air bag, tests in probably higher injury risk configurations with dummies and cadavers were performed.

The air bag was installed at the outer side of the seat back, with the subject seated in the driver or front passenger seat of a passenger car. During the inflation of the air bag, the left or right forearm of the subject was positioned on the arm rest while the upper arm made contact with the seat back edge. The volume of the thorax air bag was 15 liters and for the thorax-head air bag, 28 liters. The dummy was instrumented at the thorax c.g., shoulder, elbow and wrist with triaxial accelerometers. In the cadaver, triaxial accelerations in three orthogonal directions were measured at the upper and the lower humerus, the upper radius and the lower radius and the first thoracic vertebrae. Bending tests on three humeri were performed to characterize the mechanical properties of the bone.

Two high-speed cameras with a frame rate of 1000 or 2000 frames/sec were used in order to record the deployment phase as well as the arm movement.

During the post-test film review, it was determined that the interaction of the Hybrid III dummy with the Side Air Bag (SAB) was different from that of the cadaver. In all of the dummy tests, a clockwise rotation of the forearm between 20\mD and 90\mD occurred with a simultaneous slip-through of the air bag between the arm and the vehicle B-pillar/door trim. The general motion of the cadaver's arm was different to the dummy due to the different behavior of the airbag.

The duration of the time histories for the resultant accelerations varied between 10 and 15 ms. For the cadavers, the highest acceleration values were measured at the humerus with a maxima between 193 and 334 g, the lowest acceleration were measured at the Th1 with a maximum between 13 and 26 g. The dummies showed the highest values at the elbow and the wrist. In cadaver tests acceleration values were clearly than in the dummy tests.

The maximum bending moments for the three humeri were between 130 and 148 Nm. In only one case did the exposed humerus fracture, due to cushion fabric failure and the brittle condition of the bone. In a second case, a front rupture of the shoulder joint capsule with subluxation of the joint (arthritic joint) was observed.

The results indicate that even old people can endure the inflation of a side air bag without suffering humerus fractures

973324, P-315

The interaction of air bags with upper extremities

Bass, C. R., Duma, S. M., Crandall, J. R., Morris, R., Martin, P., Pilkey, W. D., Hurwitz, S.
University of Virginia

Khaewpong, N., Eppinger, R.
National Highway Traffic Safety Administration

Sun, E.
Conrad Technologies

Recently there has been a greater awareness of the increased risk of certain injuries associated with air bag deployment, especially the risks to small occupants, often women. These injuries include serious eye and upper extremity injuries and even fatalities. This study investigates the interaction of a deploying air bag with cadaveric upper extremities in a typical driving posture; testing concentrates on female occupants. The goals of this investigation are to determine the risk of upper extremity injury caused by primary contact with a deploying air bag and to elucidate the mechanisms of these upper extremity injuries. Five air bags were used that are representative of a wide range of air bag "aggressivities" in the current automobile fleet. This air bag "aggressivity" was quantified using the response of a dummy forearm under air bag deployment.

For this study, sixteen excised cadaveric upper extremities were mounted to a four-degree-of-freedom universal joint that functioned as a shoulder. The upper extremity position was a "natural" driving posture in a one-hand turn crossover maneuver. In this posture, the forearm was pronated with the humerus normal to the plane of the steering wheel and the forearm positioned directly on the air bag module so that the distal third of the forearm was over the air bag tear seam. This position represents a "worst case" for deployment of an air bag into a forearm. Injuries sustained during testing included ulna nightstick fractures and multiple fractures. A nightstick fracture is a diaphaseal ulna fracture unaccompanied by a radius fracture. The results indicate that the air bag/flap interaction may be significant in the production of some injuries, but is not necessary to produce peak moments in the forearm. These primary contact injuries occurred within 13 ms of the deployment near the site of interaction with the flap. In addition, these results show that the risk of injury increases with increasing air bag aggressivity. The testing also suggests that there is an aggressivity among those air bags installed in the current automobile fleet below which the risk of forearm injury for occupants is low. Further, the results imply that there is a forearm bone strength above which the risk of injury is low even for an aggressive air bag deployment. Results also suggest that humerus position, forearm pronation angle, and forearm position relative to the air bag module affect the risk of injury from air bag deployment.

Comparison tests were performed with the SAE 5\ut\uh Percentile Female Instrumented Arm using four of the air bags tested in the cadaveric series. These tests were performed using the SAE arm mounted on a Hybrid III dummy and the universal joint test fixture. Repeated tests revealed no significant difference between the forearm response on the fixture and on the dummy. Resulting forearm moments were used to develop logistic injury risk functions for ulna and ulna/radius fractures of female forearms in the worst case position under air bag deployment. For 50% risk of ulna or ulna/radius fractures, the dummy forearm moment is 61 N-m (+/- 13 N-m standard deviation). For 50% risk of fracture of both the ulna and the radius, the dummy forearm moment is 91 N-m (+/- 14 N-m standard deviation).

973325, P-315

Biomechanical investigation of airbag-induced, upper-extremity injuries

Hardy, Warren N., Schneider, Lawrence W., Reed, Matthew P., Ricci, Leda L.
Transportation Research Institute, University of Michigan

The factors that influence airbag-induced, upper-extremity injuries sustained by drivers were investigated in this study. Seven unembalmed human cadavers were used in nineteen direct-forearm-interaction static deployments. A single horizontal-tear-seam airbag module and two different inflators were used. Spacing between the instrumented forearm and the airbag module was varied from 10 cm to direct contact in some tests. Forearm-bone instrumentation included triaxial accelerometry, crack detection gages, and film targets. Internal airbag pressure was also measured. The observed injuries were largely transverse, oblique, and wedge fractures of the ulna or radius, or both, similar to those reported in field investigations. Tears of the elbow joint capsule were also found, both with and without fracture of the forearm.

Forearm fracture occurrence was analyzed with respect to time of airbag deployment, distal forearm speed, airbag-module-to-forearm spacing, bone mineral content, and upper-extremity mass. Forearm fractures occurred within 5 to 10 ms of the onset of airbag pressure, as indicated by crack detection gage output. Time of fracture was found to coincide with local reductions in resultant distal forearm acceleration, and local plateaus of distal forearm speed. Peak distal forearm speed ranged from 9 to 21 m/s, and decreased approximately 30 percent when initial airbag-module-to-forearm spacing was increased from direct contact to 2.5 cm. A distinct division between fracture and nonfracture cases was found at about 15.2 m/s peak, and 11.7 m/s average, distal forearm speed. Distal forearm speed, fracture incidence, and fracture severity were typically reduced as initial airbag-module-to-forearm spacing was increased. Additionally, under direct-contact conditions, fractures occurred in bones having mineral contents less than 1.03 g/cm, but not in bones having greater mineral contents. Both distal forearm speed and mineral content were found to be strongly related to the mass of the upper extremity. Therefore, the potential for forearm fracture was also found to be strongly related to the mass of the upper extremity. The results of this study suggest that airbag-module-to-forearm spacing has a substantial influence on airbag-induced, upper-extremity injuries sustained by drivers. Increased initial spacing between the airbag module and the forearm reduces the incidence and severity of forearm fractures caused by direct interaction with the airbag. Furthermore, increased initial spacing reduces the speeds ultimately achieved by the distal forearm, thereby potentially mitigating flailing injuries. The results also suggest that a simple airbag agressivity-assessment tool for the prediction of forearm fracture might be based upon peak or average distal forearm speed of a biofidelic, surrogate arm.

980637, SP-1333

Effect of occupant position and airbag inflation parameters on driver injury measures

Digges, Kennerly, Noureddine, Ahmad, Eskandarian, Azim, Bedewi, Nabih E.
George Washington Univ.

This paper investigates the effects of driver airbag inflation characteristics, airbag relative position, airbag-to-dummy relative velocity, and steering column characterist using a finite element model of a vehicle, airbag, and Hybrid III 50% male dummy. Simulation is conducted in a static test environment using a validated finite element model. Several static simulation tests are performed where the airbag module's position is mounted in a rigid steering wheel and the vertical and horizontal distances are varied relative to the dummy. Three vertical alignments are used: one position corresponds to the head centered on module, another position corresponds to the neck centered on module, and the third position centers the chest on the module. Horizontal alignments vary from 0 mm to 50 mm to 100 mm. All of these tests are simulated using a typical pre-1998 type inflation curve (mass flow rate of gas entering the bag). Simulation tests are then repeated using two other airbag inflation curves. One curve is a simple 75% reduction of baseline and another that has a shorter but wider and more delayed peak that is thought to produce smaller bag pressure. Inertial effects are then explored along with steering column resistance to better simulate real accident environment.

Results indicate the benefit of keeping a minimum distance of 50 mm (2 inches) between airbag and dummy. Results also indicate benefits of less aggressive inflation. A milder inflation curve that supplies the same amount of energy but takes longer to peak may also result in reduced inflation-induced injuries

980640, SP-1333

Does stature influence driver injuries in airbag deployment crashes?~Analysis of UMTRI crash investigations
Huelke, Donald F.
Transportation Research Institute, University of Michigan

At the University of Michigan Transportation Research Institute (UMTRI), 763 crashes involving steering-wheel airbag deployments have been investigated in detail (as of 12/1/97). A subset of only frontal crashes, in which the steering-wheel airbag deployed, and stature was known, was formed (636 drivers). In these crashes, there were 201 "short" stature drivers, 165 cm or less in height (32% of all drivers). The vast majority of all drivers were lap-shoulder belted. Of these drivers, 69% sustained no injuries or an AIS-1 level injury. Of the shorter drivers there were 40 MAIS-2 level injuries and 15 who survived with an MAIS injury level of 3, 4, or 5. These higher level injuries were usually found in only one body area. Details of the injury locations and contacts are presented. Data on the taller drivers (435) were similarly tabulated. Of the taller drivers (\mG 168 cm), 74% had a MAIS-0 or 1 injury. Of taller drivers with the MAIS-3,4, or 5 injuries, the majority (70%) had such injuries unrelated to the deployment of the airbag. Of all the MAIS-2+ injured drivers, short or tall, 57% had such injuries unrelated to airbag deployments. The lower extremity was the body area most often involved, followed by the brain and upper extremity injuries.

980643, SP-1333

Estimation of occupant position from probability manifolds of airbag fire-times
Nusholtz, Guy S., Xu, Lan, Kostyniuk, Gregory
Chrysler Corp.

This paper outlines a method for estimating the probablistic nature of airbag crash sensor response and its effect on occupant position. Probability surfaces of airbag fire times are constructed for the impact velocities from 0 to 40 mph. These probability surfaces are obtained by using both frontal offset deformable barrier and frontal rigid barrier crash data. Another probability surface of displacement is constructed to estimate the occupant displacement time history before airbag deployment. This probability surface is constructed by using the initial occupant seating position data and the vehicle impact velocity and deceleration data. In addition, the probability of airbag firing at a given crash velocity is estimated from NASS-CDS, frontal offset and rigid barrier crash data. Finally, a calculus is constructed to estimate the effects of different: fire/no-fire probability, occupant position when the airbag fires, the number of occupants in contact with the airbag module when it fires. However, it should be mentioned that all work presented in this paper is preliminary and may be modified in the future. The purpose of this paper is to outline the method used to estimate the probabilistic nature of airbag fire times and then use the estimated firing times to model its effect on the occupant.

980646, SP-1333

An innovative approach to adaptive airbag modules
Ryan, Shawn
Delphi Interior & Lighting Systems

An airbag module with adaptive capability can be achieved by several methods. Separate inflators, a single hybrid inflator with multiple heaters, or a dual-stage pyrotechnic inflator can all be used to control the inflation level of an airbag. In addition, Pyrotechnically Actuated Venting (PAV) offers an innovative method for regulating airbag inflation energy. PAV allows a controlled amount of gas to be vented out of the module before it enters the cushion. This paper discusses several methods for achieving a variable output airbag module. Static and sled tests were conducted to evaluate PAV in comparison to other adaptive restraint modules.

980650, SP-1333

Recent regulatory history of airbags
Stocke, James E.
JES Engineering

The recent experience in this country with serious and fatal injuries to children and some adults caused by airbags in moderate and low-speed crashes has generated considerable activity at the National Highway Traffic Safety Administration (NHTSA) to address this situation. The recent federal regulatory history on airbags, beginning in August of 1996 to the present, provides a look at this portion of NHTSA's efforts to reduce the serious side effects of airbag deployments. This paper will discuss notices of proposed rulemaking and final rules relating to new warning labels, the depowering of airbags and the deactivation of airbags

980905, SP-1333

Airbags~Legions of fable~Consumer perceptions and concerns
Frame, Phil, MacPherson, Rebecca
National Highway Traffic Safety Administration

This paper discusses the consumer and news media perceptions about airbags that had to be taken into account by the National Highway Traffic Safety Administration in making rulemaking decisions in 1997. Addressing these perceptions was a major concern as the agency made preparations to allow identifiable groups of people at risk from an airbag deployment to have on-off switches installed in their vehicles.

982325, P-335

Dual-stage inflators and OoP occupants~A performance study
Malczyk, Axel, Franke, Dirk, Adomeit, Heinz-Dieter
Petri AG

Fifty-three static out-of-position tests were conducted with a "small female' dummy placed in three different positions, and with distances of 0 mm and 50 mm from the airbag.
The driver-side module with a single-stage inflator was additionally tested with inflator versions tailored to 80%, 60%, 40% and 20% of peak tank pressure, in order to simulate the first of two stages of a dual-stage inflator.
In general, biomechanical loadings decreased with less inflator propellant. Critical chest loadings were measured down to the 60%-stage. The neck extension bending moment exceeded the limit only with the 100%-charge. With distances of 50 mm, none of the threshold values were exceeded. Energy reductions of 20% between two stages did not necessarily reduce occupant loadings

983149, P-337

Response and tolerance of the human forearm to impact loading
Pintar, Frank A., Yogananda, Narayan, Eppinger, Rolf H.
National Highway Traffic Safety Administration

With the widespread use of supplemental restraint systems (airbags), occasional rare injuries have occurred because of the force associated with these systems upon deployment. Recent case studies have demonstrated forearm fractures associated with airbag deployment. The present study was conducted to determine the tolerance of the human forearm under a dynamic bending mode. A total of 30 human cadaver forearm specimens were tested using three-point bending protocol to failure at 3.3 m/s and 7.6 m/s velocities. Results indicated significantly (p<0.01) greater biomechanical parameters associated with males compared to females. The bending tolerance of the human forearm, however, was found to be most highly correlated to bone mineral density, bone area, and forearm mass. Thus, any occupant with lower bone mineral density and lower forearm geometry/mass is at higher risk. The mean failure bending moment for all specimens was 94 Nm. The smaller sized occupant with lower bone mineral density, however, has one-half of this tolerance (approximately 45 Nm). The present investigation offers quantitative information regarding tolerance of the human forearm which may be useful for design of injury-mitigating devices.

986096/98-S5-O-08

Measuring airbag injury risks to out-of-position occupants
Morris, Christina R., Zuby, David S., Lund, Adrian K.
Insurance Institute for Highway Safety

Real-world crash experience has shown the need to reduce the risk of injury from inflating airbags to out-of-position occupants including small adult drivers. The small (5th percentile) female Hybrid III dummy positioned very close to the steering wheel/airbag assembly is the primary means currently available for assessing potential risk of severe chest and head/neck injury to out-of-position drivers. However, researchers have identified shortcomings with this dummy in reliably assessing potential interactions between the head/neck and the deploying airbag. Several noncrash airbag deployment tests in two late-model vehicles were conducted with a small female Hybrid III dummy. The dummy's spine had been modified to permit the upper torso to rotate forward without the buttocks leaving the driver seat in order to better simulate at-risk positions that a belted driver could achieve. Tests with the standard Hybrid III head and neck, even though supplemented with a foam neck shield as recommended by Melvin et al., (1993), confirmed nonbiofidelic interaction between the dummy and deploying airbag. Several modifications to the dummy's head skin and neck shield were tested to determine whether any gave more repeatable and biofidelic results. None of the head skin/neck shield configurations tested were found to provide a reliably biofidelic indication of airbag inflation injury risk.

986159

98-S7-O-12

The interaction of air bags with upper extremity test devices
Bass, C. R., Duma, S. M., Crandall, J. R., George, S.
University of Virginia
Kuppa, S.
Conrad Technologies Inc.
Khaewpong, N., Sun, E., Eppinger, R.
National Highway Traffic Safety Administration

This study examines and compares the response of two upper extremity test devices under driver-side air bag deployment to contribute to the development of dummy surrogates for the investigation of primary contact forearm injuries during air bag deployments. The first of these test devices, the SAE 5th Percentile Female Arm (SAE Arm), is an anthropomorphic representation of a small female forearm and upper arm that is instrumented with load cells, accelerometers and potentiometers to enable the determination of upper extremity kinematics and dynamics. The second, the Research Arm Injury Device (RAID), is a simple beam test device designed for detailed investigation of moments and accelerations resulting from close contact in the initial stages of air bag deployment. The RAID includes strain gauges distributed along its length to measure the distribution of moment applied by the air bag deployment. The study used four air bags representing a wide range of aggressivities in the current automobile fleet. The upper extremity position was a "natural" driving posture when turning left with one hand across the steering wheel. The forearm was positioned directly on the air bag module with the forearm oriented perpendicular to the air bag module tear seam. For the SAE Arm, the humerus was oriented normal to the steering wheel. Tests with the SAE Arm were performed both with the arm attached to a 5th Percentile Female Hybrid III dummy and with the arm mounted to a universal joint test fixture. The RAID was mounted to an articulated test fixture. In addition to the dynamic tests, a detailed comparison of the inertial properties of each of the test devices with the inertial properties of a typical small female was performed. Forearm response from both test devices confirmed the levels of air bag aggressivity determined using previous cadaveric injury results. In addition, logistic risk functions for forearm fracture were developed using existing cadaver studies and the moment response of each test device. These risk functions indicate that for 50% risk of ulna or ulna/radius fractures, the SAE arm peak forearm moment is 61 N-m (\mP 13 N-m standard deviation) while the RAID peak forearm moment is 373 N-m (\mP 83 N-m standard deviation). For 50% risk of fracture of both the ulna and the radius, the SAE arm peak dummy forearm moment is 91 N-m

473 N-m (\mP 60 N-m standard deviation).

99SC03

Three-year-old child out-of-position side airbag studies
Pintar, Frank A., Yoganandan, Narayan
Medical College of Wisconsin and VAMC
Maltese, Matthew R., Samaha, Randa R., Eppinger, Rolf H.
US DOT NHTSA

A series of twenty-nine tests was completed by conducting static deployment of side airbag systems to an out-of-position Hybrid III three-year-old dummy. Mock-ups (bucks) of vehicle occupant compartments were constructed. The dummy was placed in one of four possible positions for both door- and seat-mounted side airbag systems. When data from each type of position test were combined for the various injury parameters it was noted that the head injury criteria (HIC) were maximized for head and neck tests, and the chest injury parameters were maximized for the chest tests. For the neck injury parameters, however, all of the test positions produced high values for at least one of the parameters. The study concluded the following. Static out-of-position child dummy side airbag testing is one possible method to evaluate the potential for injury for worst-case scenarios. The outcome of these tests are sensitive to preposition and various measurements should be made to reproduce the tests. For the same make and model of airbag, different deployment characteristics were observed and, therefore, duplicate tests should be run for repeatability in future research programs. For the airbags tested, these data suggest that the kind of tests necessary to evaluate head, neck, and chest injury potential may be limited to two different positions: a head-neck position, and a chest position. Other airbag designs may require evaluation of additional positions. The side airbags evaluated in this series appear to have the potential for injuring the out-of-position child under worst-case scenarios; the less aggressive airbags reduced this potential.

99SC27

The influence of occupant and vehicle characteristics on risk of pediatric air bag injury
Arbogast, K. B., Durbin, D. R., Resh, B. F., Winston, F. K.
Children's Hospital of Pennsylvania

A case-comparison study was conducted of children between one and twelve years of age exposed to passenger air bag (PAB) deployment. Cases were children fatally injured by PAB exposure and were investigated by the Special Crash Investigation Program of NHTSA. For comparison, children exposed to PABs, but suffering minor injury were identified through the Partners for Child Passenger Safety (PCPS) Study, a system utilizing insurance claims data for crashes involving children. The crash severity as measured by Delta V was not significantly different between the two groups. Restraint status in conjunction with pre-impact braking highly influenced injury outcome indicating the importance of pre-crash positioning as a risk factor in child exposure to PAB deployment. Other related variables such as child size and age reinforced the importance of restraint. No vehicle characteristics or interior vehicle space measurements were significantly different between the two groups. Current vehicle designs cannot be differentiated with respect to their potential for producing serious child injuries due to PAB deployment. Researchers must continue to examine the entire spectrum of occupant injury severity in order to fully understand injury risk and ensure that future design of vehicles considers the safety of all occupants, including children.

VIII. BIBLIOGRAPHY FOR FURTHER READING

Bibliography

"Actions to Reduce the Adverse Effects of Air Bags, FMVSS No. 208 Depowering," Final Regulatory Evaluation, Office of Regulatory Analysis, Plans and Policy, February, 1997.

"Effectiveness of Occupant Protection Systems and Their Use", Fourth Report to Congress, Department of Transportation Report DOT HS 808 537, May, 1999.

"Effectiveness of Occupant Protection Systems and Their Use", Report to the Congress, Department of Transportation Report DOT HS 808 019, January, 1993.

"Effectiveness of Occupant Protection Systems and Their Use", Second Report to the Congress, Department of Transportation Report DOT HS 808 389, February, 1996.

"Effectiveness of Occupant Protection Systems and Their Use", Third Report to Congress, Department of Transportation Report DOT HS 808 537, December, 1996.

"FMVSS 208, Air Bag On-Off Switches", Final Regulatory Evaluation, Office of Regulatory Analysis, Plans and Policy, November, 1997.

"Ford Reports - Safety Features Save Lives", Traffic Safety and Highway Improvement Department, Ford Motor Company, 1957.

Aboud, G., Repp, J., Sirola, C., Husby, H. (1993) Inadvertent air bag sensor testing for off-road vehicles. Paper Number 933020, Society of Automotive Engineers, Warrendale, Pennsylvania.

Adams SL, Petri RW. (1996) Injury with spontaneous deployment of an automobile air bag. Academic Emergency Medicine 3(2):179-80.

Adams, B., Labib, M. (1999) Fundamental of a stored liquefied gas inflator. Paper Number 940710, Society of Automotive Engineers, Warrendale, Pennsylvania.

Adams, J. (1983) Public safety legislation and the risk compensation hypothesis; the example of motorcycle helmet legislation. Environment and Planning C: Government and Policy 1:193-203.

Adams, J. (1985) Smeed's law, seat belts, and the emperor's new clothes. In: Evans, L. Schwing, R. (eds.) Human Behavior and Traffic Safety, pp. 193-248. Plenum Press, New York, NY.

Advanced Notice of Proposed RuleMaking on FMVSS 208, 34 Federal Register 11148, July 2, 1969.

Adomeit D (1979) Seat design – A significant factor for safety belt effectiveness. Paper Number 791004, Society of Automotive Engineers, Warrendale, PA.

Adomeit D, Heger A. (1975) Motion sequence criteria and design proposals for restraint devices in order to avoid unfavorable biomechanic conditions and submarining. Paper Number 751146, Society of Automotive Engineers, Warrendale, PA.

Adomeit, D. (1977) Evaluation methods for the biomechanical quality of restraint systems during frontal loading. Paper Number 770936, Society of Automotive Engineers, Warrendale, PA.

Agaram, V., Kang, J., Nusholtz, G., Kostyniuk, G. (2001) Hybrid III dummy neck response to air bag loading. Paper Number 469, 17th Conference on the Enhanced Safety of Vehicles (ESV), National Highway Traffic Safety Administration, Washington, DC.

Air Bag Fleet Retrofit Program Crash Tests, NHTSA Report DTNH22-82-A-17148, September, 1983.

Alam M, Bickers DR. (2002) Airbag trauma induced cutaneous fistulae in a heart transplant patient. Journal of the American Academy of Dermatology. 47(2 Pt 2):S175-6.

Alcala, E., Valverde, P. Moreno, J. (2001) Advanced driver airbag system. Paper Number 465, 17th Conference on the Enhanced Safety of Vehicles (ESV), National Highway Traffic Safety Administration, Washington, DC.

Aldman B., Andersson A., Saxmark O. (1974) Possible effects of airbag inflation on a standing child. Proceedings of the International Research Council on the Biomechanics of Impact (IRCOBI) Conference, Lyon, France.

Allen, J. (1992) Power-rate crash sensing method for safety device actuation. Paper Number 920478, Society of Automotive Engineers, Warrendale, Pennsylvania.

Angel CA, Ehlers RA. (2001) Images in clinical medicine. Atloido-occipital dislocation in a small child after air-bag deployment. New England Journal of Medicine 345(17):1256.

Arbogast, K., Durbin, D., Resh, B., Winston, F. (1999) The influence of occupant and vehicle characteristics on risk of pediatric air bag injury. Paper Number 99SC27, Society of Automotive Engineers, Warrendale, Pennsylvania.

Asaria RH, Zaman A, Sullivan PM. (1999) Retinitis sclopeteria associated with airbag inflation. British Journal of Ophthalmology 83(9):1094-5.

Atkinson, P., Atkinson, T., Haut, R., Eusebi, C., Maripudi, V., Hill, T., Sambatur, K. (1999) Development of injury criteria for human surrogates to address current trends in knee-to-instrument panel injuries. Paper Number 993146, Society of Automotive Engineers, Warrendale, Pennsylvania.

Atkinson, P., Hariaharan, P., Mari-Gowda, S., Telehowski, P., Martin, S., Van Hoof, J., Bir, C., Atkinson, T. (2002) An Under-Hand Steering Wheel Grasp Produces Significant Injury Risk to the Upper Extremity During Air bag Deployment. Proc. 46th Annual Scientific Conference of the Association of the Advancement of Automotive Medicine, Tempe, AZ.

Augenstein JS, Digges KH, Lombardo LV, Perdeck EB, Stratton JE, Malliaris AC, Quigley CV, Craythorne AK, Young PE. (1995) Occult abdominal injuries to airbag-protected crash victims: a challenge to trauma systems. The Journal of Trauma 38(4):502-8.

Augenstein, J., et al. (1994) Injuries sustained by air bag occupants in frontal crashes. Paper Number 94-S4-O-18, 14th Conference on the Enhanced Safety of Vehicles (ESV), National Highway Traffic Safety Administration, Washington, DC.

Augenstein, J., Perceck, E, Bowne, J., Stratton, J., Horton, T., Singer, M., Rao, A., Digges, K., Malliaris, A. (1999) Injury patterns among belted drivers protected by air bags in 30 to 35 mph crashes. Paper Number 1999-01-1062, Society of Automotive Engineers, Warrendale, Pennsylvania.

Augenstein, J., Perdeck, El, Digges, K., Lombardo, L., Malliaris, A. (1996) Injuries sustained by air bag protected drivers. Paper Number 960660, 1996, Society of Automotive Engineers, Warrendale, PA.

Avanessian, H., Ridella, S., Mani, S., Krishnaswamy, P. (1982) An Analytical Model to Study the Infant/Seat Airbag Interaction. Paper Number 920126, Society of Automotive Engineers, Warrendale, Pennsylvania.

Bailey H, Perez N, Blank-Reid C, Kaplan LJ. (2000) Atlanto-occipital dislocation: an unusual lethal airbag injury. Journal Emergency Medicine 18(2):215-9.

Baker RS, Flowers CW Jr, Singh P, Smith A, Casey R. (1996) Corneoscleral laceration caused by air-bag trauma. American Journal of Ophthalmology 121(6):709-11.

Baker, M., Frank, R., Puhl, L., Dabbish, E., Danielsen, M. (1990) Sensing and systems aspects of fault tolerant electronics applied to vehicle systems. Paper Number 901123, Society of Automotive Engineers, Warrendale, Pennsylvania.

Ball DC, Bouchard CS. (2001) Ocular morbidity associated with airbag deployment: a report of seven cases and a review of the literature. Cornea. 20(2):159-63. Review.

Bandstra RA, Carbone LS. (2001) Unusual basal skull fracture in a vehicle equipped with an air bag. American Journal of Forensic Medcine and Pathology 22(3):253-5.

Bandstra RA, Meissner U, Huelke D. (2001) Re.: Severe head injury caused by airbag displacement. The Journal of Trauma 50(6):1160-1.

Barron L. (1998) More on 'the invisible disability'. Postgrad Med 104(5):49; discussion 50.

Barry, S., Ginpil, S., O'Neill, T. (1999) The effectiveness of air bags. Accident Analysis and Prevention, 31:781-7.

Baruchin AM, Jakim I, Rosenberg L, Nahlieli O. (1999) On burn injuries related to airbag deployment. Burns 25(1):49-52.

Bauer, J., Hamade, T. (1995) Part design and material selection for single shot injection molded passenger airbag deployment doors. Paper Number 950349, Society of Automotive Engineers, Warrendale, Pennsylvania.

Baur, P., Lange, W., Messner, G., Rauscher, S., Pieske, O. (2000) Comparison of real world side impact/rollover collisions with and without thorax airbag/head protection system: a first field experience study. Proceedings of the Association for the Advancement of Automotive Medicine, Des Plaines, IL.

Beckerman B, Sama A. (1995) Air bag "tattoo," a lasting impression. Journal of Emergerncy Medicine 13(5):680-2.

830

Bedell JR, Malik V. (1997) Facial nerve paresis involving passenger airbag deployment: a case report. Journal of Emergerncy Medicine15(4):475-6.

Begeman, P., Pratima, K., Prasad, P. (1999) Bending strength of the human cadaveric forearm due to lateral loads. Paper Number 99SC24, Society of Automotive Engineers, Warrendale, Pennsylvania.

Benecke JE Jr. Automobile airbag impulse noise.) Otolaryngology and Head and Neck Surgery 1999 Jul;121(1):166.

Berg, F., Grandel, J., Grzelak, R., Schmall, G. (1996) Crash Tests using Passenger Cars fitted with Air bags and a Simulated Out of Position Passenger. Proc. International Research Conference on the Biomechanics of Impact (IRCOBI), Dublin, Ireland.

Berg, F., Schmitt, B., Epple, J., Mattern, J. (1997) Dummy-Loadings Caused by an Air bag in Simulated Out-of-Position Situations. Proc. International Research Conference on the Biomechanics of Impact (IRCOBI). Hanover, Germany.

Berg, F., Schmitt, B., Epple, J., Mattern, R., Kallieris, D. (1998) Results of Full-scale Crash Tests, Stationary Tests and Sled Tests to Analyze the Effects of Air bags on Passengers with or without Seat Belts in the Standard Seating Position and OOP. Paper Number 98-S5-O-10, Proc. 16th ESV Conference.

Berkowitz, A. (2001) Evaluation of the effect of air bags in multi-year light truck NCAP tests. Paper Number 392, 17th Conference on the Enhanced Safety of Vehicles (ESV), National Highway Traffic Safety Administration, Washington, DC.

Berliner, J., Athey, J., Baayoun, E., Byrnes, K., Elhagediab, A., Hultman, R., Jensen, J., Kim, A., Kostyniuk, G., Mertz, H., Prest, J., Rouhana, S., Scherer, R., Xu, L. (2000) Comparative evaluation of the Q3 and hybrid III 3-year-old dummies in biofidelity and static out-of-position airbag tests. Paper Number 2000-01-SC03, Society of Automotive Engineers, Warrendale, Pennsylvania.

Bhavsar AR, Chen TC, Goldstein DA. (1997) Corneoscleral laceration associated with passenger-side airbag inflation. British Journal of Ophthalmology 81(6):514-5.

Billen, K., Federspiel, L., Schockmel, P., Serban, B. (1999) Occupant classification system for smart restraint systems. Paper Number 1999-01-0761, Society of Automotive Engineers, Warrendale, Pennsylvania.

Bischoff, M., Fendt, G., Zechmair, D. (1997) Different approaches for air bag triggering units using a firing bus. Paper Number 971049, Society of Automotive Engineers, Warrendale, Pennsylvania.

Biss, D., Fitzpatrick, M., Zinke, T., Strother, C., and Kirchoff, G. (1980) A Systems Analysis Approach to Airbag Design and Development. 8th Conference on the Enhanced Safety of Vehicles (ESV), National Highway Traffic Safety Administration, Washington, DC.

Blacksin MF. (1993) Patterns of fracture after air bag deployment. The Journal of Trauma 35(6):840-3.

BMA (British Medical Association) (1987) Living with Risk. Wiley Medical Publication, John Wiley & Sons.

Boggess, B., Sieveka, E., Crandall, J., Pilkey, W., Duma, S. (2001) Interaction of the hand and wrist with a door handgrip during static side air bag deployment: Simulation study using the CVS/ATB multi-body program. Paper Number 2001-01-0170, Society of Automotive Engineers, Warrendale, Pennsylvania.

Bohlin, N., (1967) Statistical analysis of 28,000 vehicle accident cases, with emphasis on occupant restraint value. SAE 670925, 11th Stapp Car Crash Conference, Society of Automotive Engineers, Inc., Warrendale, Pennsylvania, USA.

Boldin C, Peicha G, Passler JM, Hauser H, Riccabona M. (2002) Inferior thyroid artery injury due to airbag deployment. Injury 33(3):283-4.

Bondar, R., Haas, J., Shrewsburg, J. (1994) New material for supplemental inflatable restraint doors. Paper Number 940649, Society of Automotive Engineers, Warrendale, Pennsylvania.

Boran, C., Boran, E., McConnell, D. (1999) Expanding restraint sensing system discrimination. Paper Number 1999-01-1067, Society of Automotive Engineers, Warrendale, Pennsylvania.

Bourke GJ. (1996) Airbags and fatal injuries to children. Lancet. 1996 347(9001):560.

Brantman, R., Breed Driver Air Bag Fleet Demonstration Program, NHTSA Contract DTNH-22-84-C-07077, Copy in Docket No. 74-14-GR-505, 1984-85.

Brantman, R., Breed, D. (1985) Use of computer simulation in evaluating airbag system performance. Paper Number 851188, Society of Automotive Engineers, Warrendale, Pennsylvania.

Braude LS. (1995) Passenger side airbag ocular injury while wearing sunglasses. British Journal of Ophthalmology 79(4):391.

Breed, D. (1985) The Breed All-mechanical Airbag Module. Paper 856014, Society of Automotive Engineers, Warrendale, Pennsylvania.

Breed, D. and Castelli, V. (1989) Trends in sensing frontal impacts. Paper 890750, Society of Automotive Engineers, Warrendale, Pennsylvania.

Breed, D., Castelli, V. Shokoohi, F. (1990) Are barrier crashes sufficient for evaluating air bag sensor performance. Paper Number 900548, Society of Automotive Engineers, Warrendale, Pennsylvania.

Breed, D., L. Summers, J. Carlson, Koyzreff, M. (2001) Development of an Occupant Position Sensor System to Improve Frontal Crash Protection. Paper Number. 325, 17th Conference on the Enhanced Safety of Vehicles (ESV), National Highway Traffic Safety Administration, Washington, DC.

Breed, D., Summers, L., Carlson, J., Koyzreff, M. (2001) Development of an Occupant Position Sensor System to Improve Frontal Crash Protection. Paper Number 325, 17th Conference on the Enhanced Safety of Vehicles (ESV), National Highway Traffic Safety Administration, Washington, DC.

Brison RJ. (1997) The accidental patient. Canadian Medical Association Journal 1997 157(12):1661-2. Review.

Brown CR. (1997) TMJ injuries from direct trauma. Practical Periodontics and Aesthetic Dentistry 9(5):581-2.

Brown CR. (1998) Dental injuries as a result of air bag deployment. Practical Periodontics and Aesthetic Dentistry 10(7):856, 859.

Brown DK, Roe EJ, Henry TE. (1995) A fatality associated with the deployment of an automobile airbag. The Journal of Trauma 39(6):1204-6.

Browne, A., Stephenson, D., Kiel, W. (1993) Using seat mounted accelerometers to differentiate between normal seated passengers and infants in infant seats. Paper Number 933092, Society of Automotive Engineers, Warrendale, Pennsylvania.

Brujis, W., de Coo, P., Ashmore, R., Giles, A. (1992) Airbag simulations with the MADYMO fem module. Paper Number 920121, Society of Automotive Engineers, Warrendale, Pennsylvania.

Buckley, G., Setchfield, N., Frampton, R. (1999) Two case reports of possible noise trauma after inflation of air bags in low speed car crashes. British Medical Journal 318:499-500.

Burst, H., Tautenhahn, U., Wierschem, F. (1988) The airbag-system in the Porsche 944. Paper Number 880613, Society of Automotive Engineers, Warrendale, Pennsylvania.

Burton JL. Air-bag injury. Journal of Accident and Emergency Medicine 1994 Mar;11(1):60.

Butler, P., Krier, H., Faigle, E., Semchena, J., Thompson, R. (1992) Numerical simulation of passenger-side automotive airbag inflators. Paper Number 920848, Society of Automotive Engineers, Warrendale, Pennsylvania.

Cacciatori M, Bell RW, Habib NE. Blow-out fracture of the orbit associated with inflation of an airbag: a case report. Br J Oral Maxillofac Surg. 1997 Aug;35(4):241-2.

Calspan/Chrysler Research Safety Vehicle: Final Technical Report, Volume 1, Executive Summary, Report No. DOT HS-805-322, April, 1980.

Campbell JK. (1993) Automobile air bag eye injuries. Nebraska Medical Journal 78(9):306-7.

Carter, R. (1972) Passive Protection at 50 Miles Per Hour. National Highway Traffic Safety Administration. DOT HS-810-197.

Centers for Disease Control and Prevention. (1996) Update: fatal air bag-related injuries to children--United States, 1993-1996. MMWR 45(49):1073-6.

Centers for Disease Control and Prevention. (1995) Air-bag-associated fatal injuries to infants and children riding in front passenger seats-- United States. MMWR 44(45):845-7.

Centers for Disease Control and Prevention. (1995) Air-bag-associated fatal injuries to infants and children riding in front passenger seats-- United States. Journal of the American Medical Association 274(22):1752-3.

Chen JF, Lee ST. (2001) An adult bilateral atlantoaxial rotary dislocation caused by airbag inflation. The Journal of Trauma 2001 51(3):572-6.

Cheng, R., Yang, K., Levine, R., King, A., Morgan, R. (1982) Injuries to the cervical spine caused by a distributed frontal load to the chest. Paper Number 821155, Society of Automotive Engineers, Warrendale, Pennsylvania.

Chialant D, Damji KF. (2000) Ultrasound biomicroscopy in diagnosis of a cyclodialysis cleft in a patient with corneal edema and hypotony after an air bag injury. Canadian Journal of Ophthalmology 35(3):148-50.

Chidester, A., Roston, T. (2001) Air bag crash investigations. Paper Number 246, 17th Conference on the Enhanced Safety of Vehicles (ESV), National Highway Traffic Safety Administration, Washington, DC.

Childester, A. Rutland, K. (1998) Airbag crash investigations. Paper Number. 98-S6-0-02, 16th Conference on the Enhanced Safety of Vehicles (ESV), National Highway Traffic Safety Administration, Washington, DC.

Chou, C., Lev, A., Lenardon, D. (1980) MVMA-2D Air Bag/Steering Wheel Assembly Simulation Model. Paper 800298, Society of Automotive Engineers, Warrendale, Pennsylvania.

Clark, C., Blechschmidt, C. (1965) Human Transportation Fatalities and Protection Against Rear and Side Crash Loads by the Airstop Restraint. Paper Number 650952, Society of Automotive Engineers, Warrendale, Pennsylvania.

Clark, C., Blechschmidt, D., Gordon, F. (1964) Impact Protection with the 'AirStop' Restraint System. Paper Number 640845, Society of Automotive Engineers, Warrendale, Pennsylvania.

Clark, C., Young, W. (1994) Airbag bumpers inflated just before the crash. Paper Number 941051, Society of Automotive Engineers, Warrendale, Pennsylvania.

Clark, C., Young, W. (1995) Car crash theory and tests of airbag bumper systems. Paper Number 950156, Society of Automotive Engineers, Warrendale, Pennsylvania.

Clement, D., Reid-Harnisch, S. (1999) Technological Evolution of the Airbag Safe Infant Seat. Paper Number 1999-01-0084, Society of Automotive Engineers, Warrendale, Pennsylvania.

Conlee, J. (1997) Passenger side air bag system for open interior architecture. Paper Number 970721, Society of Automotive Engineers, Warrendale, Pennsylvania.

Consumer Warning Issued by the NHTSA re the Use of Rear-Facing CRS's in the Front Seat of Air Bag Equipped Vehicles, 1991.

Cooper JT, Balding LE, Jordan FB. (1998) Airbag mediated death of a two-year-old child wearing a shoulder/lap belt. Journal of the Forensic Science Society 43(5):1077-81.

Cooper, G.J. et al. (1982) The Biomechanical response of the Thorax to Non-penetrating trauma with Particular Reference to Cardiac Injuries. The Journal of Trauma 22(12):994-1008, 1982.

Corazza M, Bacilieri S, Morandi P. (2000) Airbag dermatitis. Contact Dermatitis 42(6):367-8.

Crandall, J. R., Bass, C.R., Pilkey, W. D., Morgan, R. M., Eppinger, R. H., Miller, H. J., Sikorski, J. (1996) An Evaluation of Thoracic Response and Injury with Belt, Airbag, and Constant Force Retractor Restraints," Proc. NATO Crashworthiness of Transportation Systems: Structural Impact and Occupant Protection, Portugal, July 1996.

Crandall, J., Bass, Duma, S., Kuppa, S. (1998) Evaluation of 5th percentile female Hybrid III thoracic biofidelity during out-of-position tests with a driver airbag. Paper Number 980636, Society of Automotive Engineers, Warrendale, Pennsylvania.

Crandall, J., Duma, S., Bass, C., Pilkey, W., Kuppa, S., Khaewpong, N., Eppinger, R. (1999) Thoracic Response and Trauma in Air Bag Deployment Tests with Out-of-Position Small Female Surrogates. Journal of Crash Prevention and Injury Control 1(2): 101-112.

Crandall, J., et al. (1994) A Comparison of two and three point restraint systems. Proc. Joint Session of AAAM and International Research Council on the Biomechanics of Impact Conference, Lyon, France.

Crandall, J., Kuhlmann, T., Martin, P., Pilkey, W., Neeman, T. (1994) Differing patterns of head and facial injury with air bag and/or belt restrained drivers in frontal collisions. Proc. Advances in Occupant Restraint Technologies: Joint AAAM-IRCOBI Special Session, Lyon, France.

Crandall, J., Martin, P., Pilkey, W. (1996) Variability of head injury criteria with the Hybrid III dummy. Paper Number 960094, Society of Automotive Engineers, Warrendale, Pennsylvania.

Crouch, E. (1993) Evolution of airbag components and materials. Paper Number 932912, Society of Automotive Engineers, Warrendale, Pennsylvania.

Cullinan, M., Merriman, T. (2001) Oesophageal rupture resulting from air bag deployment during a motor vehicle accident. The Australian and New Zealand Journal of Surgery 71(9):554-5.

Cummings, P., McKnight, B., Rivara, F., Grossman, D. (2002) Association of driver air bags with driver fatality: a match cohort study. BMJ 324:1119-22.

Cunningham CD 3rd, Weber PC, Cure J. (2000) Neurotologic complications associated with deployment of airbags. Otolaryngology and Head and Neck Surgery 123(5):637-9.

Cunningham K, Brown TD, Gradwell E, Nee, PA. (2000) Airbag associated fatal head injury: case report and review of the literature on airbag injuries. Journal of Accident and Emergency Medicine 17(2):139-42.

Dalmotas DJ, German A, Hendrick BE, Hurley RM. (1995) Airbag deployments: the Canadian experience. The Journal of Trauma 38(4):476-81.

Dalmotas, D., German, A., Tylko, S. (2001) The crash and field performance of side-mounted airbag systems. Paper Number 442, 17th Conference on the Enhanced Safety of Vehicles (ESV), National Highway Traffic Safety Administration, Washington, DC.

Dalmotas, D., Hurley, R., German, A. (1996) Supplemental restraint systems: friend or foe to belted occupants? Proceedings of the Association for the Advancement of Automotive Medicine, pp. 63-75, Des Plaines, IL.

Dalrymple, G. (1996) Effects of assistive steering devices on air bag deployment. Paper Number 960223, Society of Automotive Engineers, Warrendale, Pennsylvania.

Danne P. Serious injuries from airbags. (2001) The Australian and New Zealand Journal of Surgery 71(9):507-8. Review.

Daris, F., Bell, S. (1997) Magnesium passenger airbag housing for Chrysler Minivan. Paper Number 970329, Society of Automotive Engineers, Warrendale, Pennsylvania.

Dejeammes, M. and Quincy, R. (1975) Efficiency comparison between three-point belt and air bag in a subcompact vehicle. Paper 751142, Society of Automotive Engineers, Warrendale, Pennsylvania.

Deng, Y., Wang, J. Peng, J., Kulkarni, S. (1998) An analytical study of side airbag designs for TTI reduction in a large size car. Paper Number 982322, Society of Automotive Engineers, Warrendale, Pennsylvania.

Dickson, K., Nuhammad, A. (1990) Airbag restraint system design by crash simulation modeling and design of experiments. Paper Number 901718, Society of Automotive Engineers, Warrendale, Pennsylvania.

Diller, R. (1991) Electronic sensing of automobile crashes for airbag deployment. Paper Number 910276, Society of Automotive Engineers, Warrendale, Pennsylvania.

Dischinger, P., Ho, S., Kerns, T., Brennan, P. (1996) Patterns of injury in frontal collisions with and without airbags. Proceedings of the International Research Council on the Biomechanics of Impact (IRCOBI) Conference.

Domb, E., Kowalick, J. (1998) Applying TRIZ to develop new design for the future of the airbag. Paper Number 980647, Society of Automotive Engineers, Warrendale, Pennsylvania.

Driver PJ, Cashwell LF, Yeatts RP. (1994) Airbag-associated bilateral hyphemas and angle recession. American Journal of Ophthalmology 118(2):250-1.

Dubois J, Stewart E. (1998) Ocular injuries from air bag deployment. Journal of Ophthalmic Nursing and Technology 17(4):147-50.

Duma SM, Crandall JR. (2000) Eye injuries from airbags with seamless module covers. The Journal of Trauma 48(4):786-9. Review.

Duma SM, Kress TA, Porta DJ, Martin PG, Simmons RJ, Alexander CL. (1997) An experimental study of airbag impact to the orbit using an instrumented Hybrid III headform. Biomedical Sciences Instrumentation 33:59-64.

Duma SM, Kress TA, Porta DJ, Simmons RJ, Alexander CL, Woods CD. (1997) Airbag-induced eye injuries: experiments with in situ cadaver eyes. Biomedical Sciences Instrumentation 33:106-11.

Duma SM, Kress TA, Porta DJ, Woods CD, Snider JN, Fuller PM, Simmons RJ. (1996) Airbag-induced eye injuries: a report of 25 cases. The Journal of Trauma 41(1):114-9.

Duma SM, Schreiber PH, McMaster JD, Crandall JR, Bass CR, Pilkey WD. (1999) Dynamic injury tolerances for long bones of the female upper extremity. Journal of Anatomy 194 (Pt 3):463-71.

Duma, S., Rudd, R., Kress, T., Porta, D. (1997) A Pneumatic airbag deployment system for experimental testing. Paper Number 970124, Society of Automotive Engineers, Warrendale, Pennsylvania.

Duncan MA, Dowd N, Rawluk D, Cunningham AJ. (2000) Traumatic bilateral internal carotid artery dissection following airbag deployment in a patient with fibromuscular dysplasia. British Journal of Anaesthics 85(3):476-8.

Dunn JA, Williams MG. Occult ascending aortic rupture in the presence of an air bag. Annals of Thoracic Surgery 1996 62(2):577-8.

Durbin, D., Kallan, M., Elliott, M., Arbogast, K., Cornejo, R., Winston, F. (2002) Risk of injury to restrained children from passenger airbags. Proc. Annual Scientific Conference of the Association for the Advancement of Automotive Medicine, AAAM, Des Plaines, IL.

Edwards, W. R. (1994) An effectiveness analysis of Chrysler driver airbags after five years exposure. Paper Number 94-S4-O-09, 14th Conference on the Enhanced Safety of Vehicles (ESV), National Highway Traffic Safety Administration, Washington, DC.

Enouen, S., Guenther, D., Saul, R., MacLaughlin, T. (1984) Comparison of models simulating occupant response with air bags. Paper Number 840451, Society of Automotive Engineers, Warrendale, PA.

Epperly NA, Still JT, Law E, Deppe SA, Friedman B. (1997) Supraglottic and subglottic airway injury due to deployment and rupture of an automobile airbag. American Surgeon 63(11):979-81.

Eppinger, R., et al. (1999) Development of improved injury criteria for the assessment of advanced automotive restraint systems – II. National Highway Traffic Safety Administration, U.S. Department of Transportation.

Esser, R., "Air Bag Demonstration Program", Report RD 583-2. 1983/1984.

Evaluation of the Hybrid III dummy interactions with air bag in frontal crash by finite element simulation. (1995) Lin, T., Wawa, C., Khalil, T. Paper Number 952705, Society of Automotive Engineers, Warrendale, Pennsylvania.

Evans, L. (1986) The effectiveness of safety belts in preventing fatalities. Accident Analysis and Prevention 18:229-241.

Evans, L. (1989) Passive compared to active approaches to reducing occupant fatalities. 12th Conference on the Enhanced Safety of Vehicles (ESV), National Highway Traffic Safety Administration, Washington, DC.

Evans, L. (1989) Passive Compared to Active Approaches to Reducing Occupant Fatalities. 12th Conference on the Enhanced Safety of Vehicles (ESV), National Highway Traffic Safety Administration, Washington, DC.

Evans, L. (1997) Air bags - Even the smartest technology is a dumb, dangerous mandate. Detroit Free Press, 16 June.

Evans, L. (1997) Air bags have been oversold. Plain Dealer, 16 November, 1997.

Evans, L. (1989) Airbag effectiveness in preventing fatalities predicted according to type of crash, driver age, and blood alcohol concentration. Proc. 33rd Annual Scientific Conference of the Association for the Advancement of Automotive Medicine, pp. 307-322, Des Plaines, IL.

Evans, L. (1991) Traffic Safety and the Driver. Van Nostrand Reinhold, New York, NY

Evans, L. (1999) Transportation Safety. In Handbook of Transportation Science, R.W. Hall Editor, Kluwer Academic Publishers, Norwell, MA.

Federal Register (1977) U.S. Federal Register, Vol. 42, No. 128, Part 571 – Federal Motor Vehicle Standards: Occupant protection systems, Docket No. 75-14, Notice 10, pages 34289-34305.

Federal Register, Docket 2000-7013-1. Final Rule on Advanced Air Bags. May 5, 2000.

Federal Register, Docket 74-14. NPRM on Advanced Air Bags.

Federal Register, Volume 56, Final Rule to Amend FMVSS 201 to Allow Top-Mounted Passenger Air Bag Systems, June, 1991.

Federal Register, Volume 58, Final Rule to Mandate Air Bags (rather than Passive Restraints) in Vehicles, September 2, 1993.

Federal Register, Volume 59, Final Rule to Amend FMVSS 213 to Require Warning Labels on Rear-Facing CRS's Re Placement in the Front Seat of Air Bag Equipped Vehicles, February 16, 1994.

Federal Register, Volume 60, Number 99, Final Rule Allowing Cutoff Switches for Certain Vehicles, May 23, 1995.

Federal Register, Volume 61, Number 152. Docket 74-14; Notice 100. Warning Label NPRM for Air Bags. August,8, 1996.

Federal Register, Volume 61, Number 230, Docket 74-14; Notice 103. Warning Label Final Rule. November 27, 1996.

Federal Register, Volume 62, Number 3, Docket 74-14; Notice 107. NPRM; Air Bag Deactivation. January 6, 1997.

Federal Register, Volume 62, Number 3, Docket 74-14; Notice 108. NPRM; Air Bag Deactivation. January 6, 1997.

Federal Register, Volume 62, Number 3, Docket 74-14; Notice 109. January 6, 1997.

Federal Register, Volume 62, Number 53, Docket 74-14. Depowering Final Rule. March 19, 1997.

Federal Register, Volume 62, Number 62, Docket 74-14. Final Rule on Cutoff Switches, November 18, 1997.

Ferguson, S., Braver, E., Greene, M., Lund, A. (1996) Initial estimates of reductions in deaths in frontal crashes among right-front passengers in vehicles equipped with passenger airbags. The Chronicle of ADTSEA, 44(4).

Fildes, B., Cameron, M., Vulcan, A., Digges, K. (1992b) Injury mitigation for a range of vehicle safety measures. Proceedings of the Association for the Advancement of Automotive Medicine, pp. 93-108, Des Plaines, IL.

Fildes, B., Cameron, M., Vulcan, A., Digges, K., Taylor, D. (1992a) Airbag and facebag benefits and costs. Proceedings of the International Research Council on the Biomechanics of Impact (IRCOBI) Conference.

Fildes, B., Deery, H., Vulcan, P. (1997) A preliminary analysis of the effectiveness of airbag technology in reducing seatbelt injuries. Proceedings of the International Research Council on the Biomechanics of Impact (IRCOBI) Conference.

Film Submitted to the FMVSS 208 Docket showing unrestrained NHTSA Representative in the Right Front Passenger Seat subjected to panic braking at 5, 10, 20, and 40 mph - Various occupant prebrading situations. Anon, 1971.

Final Regulatory Impact Analysis Amendment to FMVSS 208, "Passenger Car Front Seat Occupant Protection", NHTSA, Plans and Programs, Office of Planning and Analysis, 11 July 1984.

Fischer, K. (1995) Passenger airbag system tailoring algorithm. Paper Number 950874, Society of Automotive Engineers, Warrendale, Pennsylvania.

Fitzpatrick, M., "Development of Advanced Passive Restraint System for Sub-Compact Car Drivers", Final Report under NHTSA Contract DOT-HS-113-3-742, 1974.

Fitzpatrick, M., "Development of the DEPLOY Computerized Math Model of a Deploying Air Bag", Final Report for U.S. DOT, NHTSA, Office of Passenger Vehicle Research, Contract No. DTNH22-80-C-07120, Oct., 1980.

Fitzpatrick, M., "RSV - Phase II, Comprehensive Technical Results, Section 4.3.3 System Operation - Out of Position Child", November, 1977.

Fitzpatrick, M., Pinto front passenger air bag final report, Minicars Inc., 1975.

FMVSS No. 208 Air Bag Cutoff Device, Final Regulatory Evaluation, Office of Regulatory Analysis, Plans and Policy, November, 1996.

FMVSS No. 208, Advanced Air Bags, Final Economic Assessment, Office of Regulatory Analysis & Evaluation, Plans and Policy, May, 2000.

Foley E, Helm TN. (2000) Air bag injury and the dermatologist. Cutis 66(4):251-2.

Foret-Bruno, J., Trosseille, X., Le Coz, J-Y, Bendjellal, F., Steyer, C., Phalempin, T., Villeforce, D., Dandres, P., Got, C. (1998) Thoracic injury risk in frontal car crashes with occupants restrained with a belt load limiter. Paper 983166, Society of Automotive Engineers, Warrendale, PA.

Foret-Bruno, J-Y., Trosseille, X., Page, Y., Huere, J-F, Le Coz, J-Y., Bendjellal, F., Diboine, A., Phalempin, T., Villeforceix, D., Baudrit, P., Guillemot, H., Coltat, J-C (2001) Comparison of thoracic injury risk in frontal car crashes for occupants restrained without belt load limiters and those restrained with 6 kN and 4 kN belt load limiters. Stapp Car Crash Journal 45:205-224.

Frampton, R., Sferco, R., Welsh, R., Kirk, A., Fay, P. (2000) Effectiveness of airbag restraints in frontal crashes – what European studies tell us. Proceedings of the International Research Council on the Biomechanics of Impact (IRCOBI) Conference.

Freedman EL, Safran MR, Meals RA. (1995) Automotive airbag-related upper extremity injuries: a report of three cases. The Journal of Trauma 38(4):577-81.

Frey, S. M., "History of Air Bag Development", Paper Presented at the International Conference on Passive Restraints, May, 1970.

Fukagawa K, Tsubota K, Kimura C, Hata S, Mashita T, Sugimoto T, Oguchi Y. (1993) Corneal endothelial cell loss induced by air bags. Ophthalmology 100(12):1819-23.

Fusco A, Kelly K, Winslow J. (2001) Uterine rupture in a motor vehicle crash with airbag deployment. The Journal of Trauma 51(6):1192-4.

Gallup, D. (1991) Gas damped sensing of automobile crashes for airbag deployment. Paper Number 910275, Society of Automotive Engineers, Warrendale, Pennsylvania.

Garcia R Jr. (1994) Air bag implicated in temporomandibular joint injury. Cranio 12(2):125-7.

Gault JA, Vichnin MC, Jaeger EA, Jeffers JB. (1995) Ocular injuries associated with eyeglass wear and airbag inflation. The Journal of Trauma 38(4):494-7.

Geggel HS, Griggs PB, Freeman MI. (1996) Irreversible bullous keratopathy after air bag trauma. The CLAO Journal 22(2):148-50.

German, A., Dalmotas, D., Hurley, R. (1998) Air bag collision performance in a restrained occupant population. Paper Number 98-S5-O-04, 16th Conference on the Enhanced Safety of Vehicles (ESV), National Highway Traffic Safety Administration, Washington, DC.

Ghafouri A, Burgess SK, Hrdlicka ZK, Zagelbaum BM. (1997) Air bag-related ocular trauma. American Journal of Emergency Medicine 15(4):389-92.

Giguere JF, St-Vil D, Turmel A, Di Lorenzo M, Pothel C, Manseau S, Mercier C. (1998) Airbags and children: a spectrum of C-spine injuries. Journal of Pediatric Surgery 33(6):811-6.

Gimovsky ML, Nunez G, Beck P. Fetal heart rate monitoring casebook. (2000) Airbag-associated rupture of membranes: evaluation of trauma in pregnancy. Journal of Perinatology 20(4):270-3.

Gioutsos, T., Gillis, Ed. (1994) Testing techniques for electronic single point sensing systems. Paper Number 940803, Society of Automotive Engineers, Warrendale, Pennsylvania.

Gioutsos, T., Tabar, D. (1999) Determination of crash severity using a ball-in-tube and accelerometer sensing system (BASS) Paper Number 1999-01-1326, Society of Automotive Engineers, Warrendale, Pennsylvania.

Glass RJ, Segui-Gomez M, Graham JD. (2000) Child passenger safety: decisions about seating location, airbag exposure, and restraint use. Risk Analysis 20(4):521-7.

Goch, S., Krause, T., Gillespie, A. (1990) Inflatable restraint system design considerations. Paper Number 901122, Society of Automotive Engineers, Warrendale, Pennsylvania.

Goldberg MA, Valluri S, Pepose JS. (1995) Air bag-related corneal rupture after radial keratotomy. American Journal of Ophthalmology 120(6):800-2.

Goldstein DA. (2001) Airbag-related injuries in children: a MEDLINE search tip. Ophthalmology. 108(6):1008-9.

Goor, P., Goor, D. (1993) Occupant integral self adjusting quasi intelligent (pre-programmed) inflatable restraint systems using forces and cushioning to dynamically enhance protection. Paper Number 930241, Society of Automotive Engineers, Warrendale, Pennsylvania.

Goral, T., Nagasawamy, V., Schlenke, R. (1999) A predictive design methodology for active top pads during airbag deployment. Paper Number 1999-01-0688, Society of Automotive Engineers, Warrendale, Pennsylvania.

Gossman W, June RA, Wallace D. (1999) Fatal atlanto-occipital dislocation secondary to airbag deployment. American Journal of Emergency Medicine 17(7):741-2.

Gottesman M, Sanderov B, Ortiz O. (2002) Carotid artery dissection and stroke caused by airbag injury. American Journal of Emergency Medicine 20(4):372-4.

Graefe, H., Krummheuer, W., Siejak, V. (1990) Computer simulation of static deployment tests for airbags, air permeability of uncoated fabrics and steady state measurements of the rate of volume flow through airbags. Paper Number 901750, Society of Automotive Engineers, Warrendale, Pennsylvania.

Graham J., Corso, P., Morris, J., Segui-Gomez, M., Weinstein, M. (1998a) Evaluating the Cost-Effectiveness of Clinical and Public Health Measures. Annual Review of Public Health, 19:125-152.

Graham J., Goldie, S., Segui-Gomez, M., Thompson, K., Nelson, T., Glass, R., Simpson, A., Woerner, L. (1998b) Reducing Risks to Children in Vehicles with Passenger Air Bags. Pediatrics 102(1).

Graham, J. (1997a) Statement before the National Transportation Safety Board, Supplemental Restraint Panel, Washington, DC.

Graham, J. (1997b) Statement before the National Transportation Safety Board, Effectiveness Panel, Washington, DC.

Graham, J. et al. (1997) The cost-effectiveness of air bags by seating position. Journal of the American Medical Association 278:1418-1425.

Grisoni ER, Pillai SB, Volsko TA, Mutabagani K, Garcia V, Haley K, Schweer L, Marsh E, Cooney D. (2000) Pediatric airbag injuries: the Ohio experience. Journal of Pediatric Surgery 35(2):160-2; discussion 163.

Groenenboom, P, Lasry, D., Subbian, T. Narwani, G. (1993) A diffusive gas jet model in PAM-SAFE for airbag inflation. Paper Number 930238, Society of Automotive Engineers, Warrendale, Pennsylvania.

Grösch, L. (1985)Chest injury criteria for combined restraint systems. Paper Number 851247, Society of Automotive Engineers, Warrendale, Pennsylvania.

Grösch, L., Katz, E., Kassing, L., Marwitz, H., Zeidler, F. (1987) New measurement methods to assess the improved protection potential of airbag systems. Paper Number 870333, Society of Automotive Engineers, Warrendale, Pennsylvania.

Gross, K., Haidar, A., Basha, M., Chan, T., Gwizdala, C., Wooley, R., Popovich, J. (1994) Acute pulmonary response of asthmatics to aerosols and gases generated by air bag deployment. American Journal of Respiratory and Critical Care Medicine 150:408-14.

Gross, K., Kelly, N., Reddy, S., Shah, N., Grain, T. (1999) Assessment of human response to non-azide air bag effluents. Paper Number 99SC26, Society of Automotive Engineers, Warrendale, Pennsylvania.

Gross, K., Koets, M., D'Arcy, J., Chan, T., Wooley, R., Basha, M. (1995) Mechanism of induction of asthmatic attacks initiated by the particles generated by air bag system deployment. The Journal of Trauma 38:521-7.

Hahm, S. Ki,, D., Lee, M. (1998) Airbag depowering for a compact vehicle through MADYMO simulation and sled test. Paper Number 980904, Society of Automotive Engineers, Warrendale, Pennsylvania.

Hallock GG. Mechanisms of burn injury secondary to airbag deployment. Annals of Plastic Surgery 39(2):111-3.

Han, D.P. (1993) Retinal detachment caused by air bag injury. Archives of Ophthalmology 111(10):1317-8.

Hanna KM, Weiman DS, Pate JW, Wolf BA, Fabian TC. (1997) Aortic valve injury secondary to blunt trauma from an air bag. Tennessee Medicine 90(5):195-6.

Hansen TP, Nielsen AL, Thomsen TK, Knudsen PJ. (1999) Avulsion of the occipital bone--an airbag-specific injury. Lancet 353(9162):1409-10.

Hardy, W. and Schneider, L. (1998) Investigation of inertial factors involved in airbag-induced forearm fractures. Paper Number 98-S7-O-13, 16th Conference on the Enhanced Safety of Vehicles (ESV), National Highway Traffic Safety Administration, Washington, DC.

Hardy, W., Schneider, L., Reed, M. (1998) Comparison of airbag-aggressively predictors in relation to forearm fractures. Paper Number 980856, Society of Automotive Engineers, Warrendale, Pennsylvania.

Hart RA, Mayberry JC, Herzberg AM. (2000) Acute cervical spinal cord injury secondary to air bag deployment without proper use of lap or shoulder harnesses. Journal of Spinal Disorders 13(1):36-8.

Hassan, J. and Nuscholtz, G. (2000) Development of a combined thoracic injury criterion – a revisit. Paper Number 2000-01-0158, Society of Automotive Engineers, Warrendale, Pennsylvania.

Hayano, K., Ono, K., Matuoka, F. (1994) Test Procedures for Evaluating Out-of-Position Vehicle Occupant Interactions with Deployed Air bags. Paper Number 94-S1-O-19, Proc. 14th ESV Conference.

He, S. (1997) Advanced finite element analysis in the structural design of airbag modules. Paper Number 970773, Society of Automotive Engineers, Warrendale, Pennsylvania.

Heimbach D. (2000) Full-thickness burn to the hand from an automobile airbag. Journal of Burn Care Rehabilitation21(3):288-9.

Helleman, H., Brantman, R., Nakhla, S. (1996) Seat and airbag design to mitigate frontal crash lower limb injuries Paper Number 960503, Society of Automotive Engineers, Warrendale, Pennsylvania.

Henderson, M. (1987) Living with risk : the British Medical Association Guide. John Wiley & Sons, Chichester, New York.

Hendrickx I, Mancini LL, Guizzardi M, Monti M. (2002) Burn injury secondary to air bag deployment. Journal of the American Academy of Dermatology 46(2 Suppl Case Reports):S25-6.

Hendrix, T., Kelley, J., Piper, L. (1990) Mechanical versus accelerometer based sensing for supplemental inflatable restraint systems. Paper Number 901121, Society of Automotive Engineers, Warrendale, Pennsylvania.

Henriksson, M., Hallén. A., Höwing, Lundell, B. (1997) Passenger airbag status indication awareness study. Paper Number 970276, Society of Automotive Engineers, Warrendale, Pennsylvania.

Hersman, M., Farrington, S., Tansey, B. (1999) Coverstock materials for instrument panel with top-mounted invisible airbag doors. Paper Number 1999-01-1018, Society of Automotive Engineers, Warrendale, Pennsylvania.

Highway Loss Data Institute (1991) Driver injury experience in 1990 models equipped with air bags or automatic belts. Highway Loss Data Institute, Arlington, VA.

Hinger, J. and Clyde, H. (2001) Advanced air bag systems and occupant protection: recent modifications to FMVSS 208. Paper Number 2001-01-0157, Society of Automotive Engineers, Inc., Warrendale, Pennsylvania, USA.

Hinger, J., Clyde, H. (2001) Advanced air bag systems and occupant protection: Recent modifications to FMVSS 208. Paper Number 2001-01-0157, Society of Automotive Engineers, Warrendale, Pennsylvania.

Hoel EN. (1993) An unusual air bag-mediated injury. Journal of Emergency Nursing 19(1):6.

Hoffer, G., Millner, E., Peterson, S. (1994) Are drivers of airbag-equipped cars more aggressive?: a test of the Peltzman hypothesis. The Journal of Law and Economics 38: 251-264.

Hoffman, R., Pickett, A., Ulrich, D., Haug, E., Lasry, D., Clinkemaillie, J. (1989) A finite element approach to occupant simulation--The PAM-CRASH airbag model. Paper Number 890754, Society of Automotive Engineers, Warrendale, Pennsylvania.

Hoffmann, R., Ulrich, D., Protard, J., Wester, H., Jaehn, N., Scharnhorst, T. (1990) Finite element analysis of occupant restraint system interaction with PAM-CRASH. Paper Number 902325, Society of Automotive Engineers, Warrendale, Pennsylvania.

Hollands CM, Winston FK, Stafford PW, Lau HT. (1996) Lethal airbag injury in an infant. Pediatric Emergency Care 12(3):201-2.

Hollands, C., Winston, F., Stafford, P., Shochat, S. (1996) Severe Head Injury Caused By Air Bag Deployment. The Journal of Trauma 41(5):920-2.

Hollowell, W., Stucki, S. (1996) Improving occupant protection systems in frontal crashes. Paper Number 960665, Society of Automotive Engineers, Warrendale, Pennsylvania.

Hollowell, W., Summers, L., Prasad, A., Narwani, G., Ato, T. (2001) Performance evaluation of dual stage passenger air bag systems. Paper Number 234, 17th Conference on the Enhanced Safety of Vehicles (ESV), National Highway Traffic Safety Administration, Washington, DC.

Hopkins, J. B. (1973) Anticipatory Sensors for Collision Avoidance and Crash Safety Prediction as Applied to Vehicle Safety Research.

Horsch, J.D., Culver, C.C. (1979) A Study of Driver Interactions With an Inflating Air Cushion. Paper Number 791029, Society of Automotive Engineers, Warrendale, Pennsylvania.

Horsch, J., Lau, I., Andrzejak, D., Viano, D., Melvin, J., Pearson, J., Cok, D., Miller, G. (1990) Assessment of air bag deployment loads. Paper Number 902324, Society of Automotive Engineers, Warrendale, Pennsylvania.

Hou, J., Tomas, J., Sparke, L. (1995) Optimization of driver-side airbag and restraint system by occupant dynamics simulation. Paper Number 952703, Society of Automotive Engineers, Warrendale, Pennsylvania.

Huebner CJ, Reed MP. (1998) Airbag-induced fracture in a patient with osteoporosis. The Journal of Trauma 45(2):416-8.

Huelke DF, Moore JL, Compton TW, Rouhana SW, Kileny PR. (1999) Hearing loss and automobile airbag deployments. Accident Analysis and Prevention 31(6):789-92.

Huelke DF, Moore JL, Compton TW, Samuels J, Levine RS. (1995) Upper extremity injuries related to airbag deployments. The Journal of Trauma 38(4):482-8.

Huelke DF, Moore JL, Ostrom M. (1992) Air bag injuries and occupant protection. The Journal of Trauma 33(6):894-8.

Huelke, D. and Moore, J. (1993) Field investigations of the performance of air bag deployments in frontal collisions. Accident Analysis and Prevention 25(6):717-30.

Huelke, D., Compton, C., Studer, R. (1985) Injury severity, ejection, and occupant contacts in passenger car rollover crashes. Paper Number 850336, Society of Automotive Engineers, Warrendale, Pennsylvania.

Huere, J. et al. Airbag efficiency in frontal real world accidents. Paper Number 2001-S1-O-193, 17th Conference on the Enhanced Safety of Vehicles (ESV), National Highway Traffic Safety Administration, Washington, DC.

Huff GF, Bagwell SP, Bachman D. (1998) Airbag injuries in infants and children: a case report and review of the literature. Pediatrics 102(1):e2. Review.

Humayun, S., Tekelly, J., Bapu, G. (1995) CAE analysis of an airbag mounting structure in an instrument panel. Paper Number 950639, Society of Automotive Engineers, Warrendale, Pennsylvania.

Hunt L. (1995) Ocular injuries from driver's air bag. Insight 20(1):18-9.

Hussain SS. (1999) Noise trauma after inflation of air bags in low speed car crashes. Article was unclear and possibly misleading. British Medical Journal 318(7195):1421.

Hussey, B., Rink, L. (1995) Predictive techniques for airbag inflator exit properties. Paper Number 950344, Society of Automotive Engineers, Warrendale, Pennsylvania.

Hwang YS, Lai CC, Yang KJ, Chen TL. (2001) A rapid and successful treatment for airbag-related traumatic macular hole. Chang Gung Medical Journal 24(8):530-5.

Igarashi, M, Atsumi, M. (1985) An analysis of 3 pt. belted occupant impact dynamics in frontal collision and its application. Paper Number 850436, Society of Automotive Engineers, Warrendale, Pennsylvania.

Ito, K., Ishikawa, M., Sakamoto, K., Shiga, I., Sakane, K., Kondoh, Y., Miya, M., Iwai, Y. (1995) A drive-side airbag system using a mechanical firing microminiature sensor. Paper Number 950346, Society of Automotive Engineers, Warrendale, Pennsylvania.

Jagger, J., Vernberg, K. (1987) Jane, J. Air bags: reducing the toll of brain trauma. Neurosurgery 20(5):815-7.

James, M., Smith, G., Kent, R., Perl, T. R. (1999) Residual crush and delta-v as postcollision evaluators of airbag system performance. Proceedings of the 32nd International Symposium on Automotive Technology and Automation, Vienna, Austria.

Janovicz, M., Winkelbauer, D. (1998) Airbag sensor system evaluation methods. Paper Number 982357, Society of Automotive Engineers, Warrendale, Pennsylvania.

Jernigan, M. et al. (2001) Eye injury patterns in automobile crashes: the effect of frontal airbags. Proceedings of the Association for the Advancement of Automotive Medicine, Des Plaines, IL.

Jinno, K., Ofuji, M., Saito, T., Sekido, S. (1997) Occupant sensing utilizing perturbation of electric fields. Paper Number 971051, Society of Automotive Engineers, Warrendale, Pennsylvania.

Johnston, K., Klinich, K., Rhule, D. (1997) Assessing Arm Injury Potential from Deploying Airbags. Paper Number 970400, Society of Automotive Engineers, Warrendale, Pennsylvania.

Jumbelic MI. (1995) Fatal injuries in a minor traffic collision. Journal of the Forensic Science Society 40(3):492-4.

Kahane, C. J. "Fatality Reduction by Airbags, Analysis of Accident Data through Early 1996," Department of Transportation Report DOT HS 808 470, August 1996, NHTSA, US DOT, Washington, DC.

Kallieris, D., Conte-Zerial, P., Rizzetti, A., Mattern, R. (1998) Prediction of thoracic injuries in frontal crashes. Paper Number 98-S7-O-04, 16th Conference on the Enhanced Safety of Vehicles (ESV), National Highway Traffic Safety Administration, Washington, DC.

Kallieris, D., Mattern, R., Schmidt, G., Klaus, G., (1982) Comparison on Three-Point Belt and Air Bag-Knee Bolster Systems. Injury Criteria and Injury Severity at Simulated Frontal Collisions. Proceedings of the International Research Council on the Biomechanics of Impact (IRCOBI).

Kamiji, K. and Nobuyuki, K. (2001) Study of airbag interference with out of position occupant by the computer simulation. Paper Number 374, 17th Conference on the Enhanced Safety of Vehicles (ESV), National Highway Traffic Safety Administration, Washington, DC.

Kanamoto, Miyamori, M., Jinno, K., Hirao, M. (1994) Electronic crash sensing unit for airbag. Paper Number 940624, Society of Automotive Engineers, Warrendale, Pennsylvania.

Kang, J., Wang, J. (1999) H-ISP-- A hybrid -inflator simulation program. Paper Number 1999-01-1069, Society of Automotive Engineers, Warrendale, Pennsylvania.

Karlow, J., Jakovski, J., Seymour, B. (1994) Development of a new downsized airbag system for use in passenger vehicles. Paper Number 940804, Society of Automotive Engineers, Warrendale, Pennsylvania.

Katz, E., Grosch, L., Kassing, L. (1987) Chest compression response of a modified hybrid III with different restraint systems. Paper Number 872215, Society of Automotive Engineers, Warrendale, Pennsylvania

Keller, G., Miller, J. (1999) Rear-impact air bag protection system. Paper Number 1999-01-1066, Society of Automotive Engineers, Warrendale, Pennsylvania.

Kelley, J. (1993) Sensing considerations and tradeoffs for single point sensing. Paper Number 932916, Society of Automotive Engineers, Warrendale, Pennsylvania.

Kemmerer, R., Chute, R., P. Haas, and Slack, W. (1968) Automotive Inflatable Occupant Restraint System, Parts I and II. Paper Number 680033, Society of Automotive Engineers, Warrendale, Pennsylvania.

Kent, R., Crandall, J., Bolton, J., Duma, S. (2000) Driver and right-front passenger restraint system interaction, injury potential, and thoracic injury prediction. Proceedings of the Association for the Advancement of Automotive Medicine, pp. 261-282, Des Plaines, IL.

Kent, R. and Crandall, J. (2001) Boundary condition effects on thoracic deformation response to anterior impact loading. Summer Bioengineering Conference, American Society of Mechanical Engineers, Snowbird, Utah.

Kent, R., Bolton, J., Crandall, J., Prasad, P., Nusholtz, G., Mertz, H., Kallieris, D. (2001) Restrained Hybrid III dummy-based criteria for thoracic hard tissue injury prediction. Conference of the International Research Council on the Biomechanics of Impact (IRCOBI).

Kent, R., Crandall, J., Bolton, J., Duma, S. (2002) Comparison and evaluation of contemporary restraint systems in the driver and front-passenger environments. Journal of Automobile Engineering, Proceedings of the Institution of Mechanical Engineers, Volume 215 Part D, pp. 1-13.

Kent R, Funk J, Crandall J. (2002) U.S. injury trends projected to 2012: the influence of an aging population. Proceedings of the Association for the Advancement of Automotive Medicine, Des Plaines, IL.

Kent, R., Crandall, J., Patrie, J., Fertile, J. (2002) Radiographic detection of rib fractures: a restraint-based study of occupants in car crashes. Traffic Injury and Prevention, 3(1):49-57.

Keshavaraj, R., Tock, R., Nusholtz, G. (1995) A novel blister-inflation technique for evaluating the thermal aging of airbag fabrics during deployment. Paper Number 950341, Society of Automotive Engineers, Warrendale, Pennsylvania.

Keshavaraj, R., Tock, R., Nusholtz, G. (1995) Comparison of contributions to energy dissipation produced with safety airbags. Paper Number 950340, Society of Automotive Engineers, Warrendale, Pennsylvania.

Keshavaraj, R., Tock, R., Nusholtz, G. (1995) Modeling of Biaxial deformation of airbag fabrics using artificial neural nets. Paper Number 950343, Society of Automotive Engineers, Warrendale, Pennsylvania.

Khalil, T., Wasko, R., Hallquist, J., Stillman, D. (1991) Development of a 3-dimensional finite element model of air bag deployment and interactions with an occupant using DYNA3D. Paper Number 912906, Society of Automotive Engineers, Warrendale, Pennsylvania.

Khandhadia, P., Klosinski, R., Vitek, J. (1995) Development of advanced inflator technology for automotive airbag modules. Paper Number 951062, Society of Automotive Engineers, Warrendale, Pennsylvania.

Kikuchi, A., Horii, M., Kawai, A., Kawai, S., Kamaki, Y., Matsuno, M. (1975) Injury to Eye and Facila Skin (Rabbit) on Impact with Inflating Air Bag. Proc. International Research Conference on the Biomechanics of Impact (IRCOBI), Birmingham, Alabama.

Kirchhoff R, Rasmussen SW. (1995) Forearm fracture due to the release of an automobile air bag. Acta Orthopaedica Scandinavica 66(5):483.

Kisielewicz, L., Kodama, N., Ohno, S., Uchio, E. (1998) Numerical prediction of airbag-caused injuries on eyeballs after radial keratotomy. Paper Number 980906, Society of Automotive Engineers, Warrendale, Pennsylvania.

Kitada, N., Watanabe, K. (1998) Accurate predictive algorithm for air bag expansion by fusing the conventional predictive algorithm and proximity sensor. Paper Number 980907, Society of Automotive Engineers, Warrendale, Pennsylvania.

Kithil, P. (1998) Capacitive occupant sensing. Paper Number 982292, Society of Automotive Engineers, Warrendale, Pennsylvania.

Kleinberger, M. and Summers, L. (1997) Mechanisms of injuries for adults and children resulting from airbag interaction. Proceedings of the Association for the Advancement of Automotive Medicine, Des Plaines, IL.

Klinich KD, Schneider LW, Moore JL, Pearlman MD. (2000) Investigations of crashes involving pregnant occupants. Proceedings of the Association for the Advancement of Automotive Medicine, Des Plaines, IL.

Knudsen PJ. (2002) Air bag injuries. American Journal of Forensic Medicine and Pathology 23(2):208.

Kompaˇ, K., Witte, M. (1996) The BMW seat occupancy monitoring system - A step towards ""situation appropriate airbag deployment. Paper Number 960226, Society of Automotive Engineers, Warrendale, Pennsylvania.

Konert, J., Lee, C., Bayley, G. (1996) Assessment of air bag module durability test specifications using CAE techniques. Paper Number 960972, Society of Automotive Engineers, Warrendale, Pennsylvania.

Kramer MB, Shattuck TG. (1997) Charnock DR. Traumatic hearing loss following air-bag inflation. New England Journal of Medicine 337(8):574-5.

Kress, T., Porta, D., Duma, S., Snider, J., Nino, N. (1996) A discussion of the air bag system and review of induced injuries. Paper Number 960658, Society of Automotive Engineers, Warrendale, Pennsylvania.

Kuhn F, Morris R, Witherspoon CD, Byrne JB, Brown S. (1993) Air bag: friend or foe? Archives of Ophthalmology 111(10):1333-4.

Kuhn F, Morris R, Witherspoon CD. (1995) Eye injury and the air bag. Current Opinion in Ophthalmology 6(3):38-44. Review.

Kulkarni, K., Kuschinsky, S., Thyagarajan, R. (1997) Using CAE to guide passenger airbag door design for optimal head impact performance. Paper Number 970772, Society of Automotive Engineers, Warrendale, Pennsylvania.

Kuner EH, Schlickewei W, Oltmanns D. (1996) Injury reduction by the airbag in accidents. Injury 27(3):185-8.

Kuo, C., Liang, H., Yang, W., Kang, J., Wang, J. (1998) An experimental investigation of transient heat losses to tank wall during the inflator tank test. Paper Number 982326, Society of Automotive Engineers, Warrendale, Pennsylvania.

Kuppa, S. and Eppinger, R. (1998) Development of an improved thoracic injury criterion. Paper Number 983153, Society of Automotive Engineers, Warrendale, Pennsylvania.

Kuppa, S., Yeiser, C., Oslon, M., Taylor, L. (1997) RAID - An Investigative Tool to Study Air Bag/ Upper Extremity Interactions. Paper Number 970399, Society of Automotive Engineers, Warrendale, Pennsylvania.

Lakshminarayan, V., Lasry, D. (1991) Finite element simulation of driver folded air bag deployment. Paper Number 912904, Society of Automotive Engineers, Warrendale, Pennsylvania.

Lambert, D. (1998) Direct thermal detection for front passenger seat airbag suppression. Paper Number 982293, Society of Automotive Engineers, Warrendale, Pennsylvania.

Lancaster GI, DeFrance JH, Borruso JJ. (1993) Air-bag-associated rupture of the right atrium. New England Journal of Medicine 328(5):358.

Langweider, K. and Hummel, T. (1998) The effect of airbags on injuries and accident costs. Paper Number 98-S1-W-27, 16th Conference on the Enhanced Safety of Vehicles (ESV), National Highway Traffic Safety Administration, Washington, DC.

Larkin GL. (1991) Airbag-mediated corneal injury. Am Journal of Emergency Medicine 9(5):444-6.

Lasry, D., Hoffmann, R., Protard, J. (1991) Numerical simulation of fully folded airbags and their interaction with occupants with PAM-SAFE. Paper Number 910150, Society of Automotive Engineers, Warrendale, Pennsylvania.

Lau IV, Horsch JD, Viano DC, Andrzejak DV. (1993) Mechanism of injury from air bag deployment loads. Accident Analysis and Prevention 25(1):29-45.

Lau, E., Ray, R., Cheng, L. (1997) Deployment of airbags in traffic accidents: characteristics and consequences. Proceedings of the Association for the Advancement of Automotive Medicine, Des Plaines, IL.

Lee WB, O'Halloran HS, Pearson PA, Sen HA, Reddy SH. (2001) Airbags and bilateral eye injury: five case reports and a review of the literature. Journal of Emergency Medicine 20(2):129-34. Review.

Lee, Y., Yu, C., Green, P., Chen, L., Butler, P. (1998) A study of aspiration effects in reduced-scale model airbag modules. Paper Number 982324, Society of Automotive Engineers, Warrendale, Pennsylvania.

Lemley HL, Chodosh J, Wolf TC, Bogie CP, Hawkins TC. (2000) Partial dislocation of laser in situ keratomileusis flap by air bag injury. Journal of Refractive Surgery 16(3):373-4.

Lenard, J., Frampton, R., Thomas, P. (1998) The influence of European air bags on crash injury outcomes. Paper Number 98-S5-O-01, 16th Conference on the Enhanced Safety of Vehicles (ESV), National Highway Traffic Safety Administration, Washington, DC.

Leonardis, D., Ferguson, S., Pantula, J. (1998) Survey of driver seating positions in relation to the steering wheel. Paper Number 980642, Society of Automotive Engineers, Warrendale, Pennsylvania.

Lesher MP, Durrie DS, Stiles MC. (1993) Corneal edema, hyphema, and angle recession after air bag inflation. Archives of Ophthalmology 111(10):1320-2.

Levitt, S. and Porter, J. (2001) Sample selection in the estimation of air bag and seat belt effectiveness. The Review of Economics and Statistics, 83(4):603-15.

Levy Y, Hasson O, Zeltser R, Nahlieli O. (1998) Temporomandibular joint derangement after air bag deployment: report of two cases. Journal of Oral and Maxillofacial Surgery 56(8):1000-3.

Lian, W., Narwani, G. (1998) Effect of gas thermodynamics on the aggressiveness of airbag inflator. Paper Number 982323, Society of Automotive Engineers, Warrendale, Pennsylvania.

Lim MS, Stavrou P. (2001) Severe ocular injury associated with airbag inflation. Eye 15(Pt 6):805-6.

Lissy, K., Cohen, J., Park, M., Graham, J. (2000) Cellular phones and driving: weighing the risks and benefits. Risk in Perspective 8(6), Harvard Center for Risk Analysis, Cambridge, MA.

Liter, Park, C., Kaviany, M., Wantg, J., Kang, J., Lee, Y. (1999) An experiment-based model of fabric heat transfer and its inclusion in air bag deployment simulations. Paper Number 1999-01-0437, Society of Automotive Engineers, Warrendale, Pennsylvania.

Liter, S, Burgess, D., Kaviany, M., Wang, J., Kang, J. (1998) Transient heating of airbag fabrics: experiment and modeling. Paper Number 980865, Society of Automotive Engineers, Warrendale, Pennsylvania.

Liter, S., Kaviany, M., Wang, J. (1997) Permeability and transient thermal response of airbag fabrics. Paper Number 971063, Society of Automotive Engineers, Warrendale, Pennsylvania.

Liu, J. (1993) SIR sensor closure time prediction for frontal impact using full vehicle finite element analysis. Paper Number 930643, Society of Automotive Engineers, Warrendale, Pennsylvania.

Loo, G. et al. (1996) Airbag protection versus compartment intrusion effect determines the pattern of injuries in multiple trauma motor vehicle crashes. The Journal of Trauma 41(6)935-51.

Lueder GT. Air bag-associated ocular trauma in children. Ophthalmology. 2000 107(8):1472-5.

Lund, A. and Ferguson, S. (1995) Driver fatalities in 1985-1993 cars with airbags. The Journal of Trauma 38(4):469-475.

Lundell, B. Edvardsson, M., Johansson, L., Korner, J., Pihali, S. (1995) Sipsbag - The seat-mounted side -impact airbag system. Paper Number 950878, Society of Automotive Engineers, Warrendale, Pennsylvania.

Lundy DW, Lourie GM. (1998) Two open forearm fractures after airbag deployment during low speed accidents. Clinical Orthopaedics and Related Research 351:191-5. Review.

Lupker, H., Brujis, W. (1993) Gas jet model for airbag inflators. Paper Number 930645, Society of Automotive Engineers, Warrendale, Pennsylvania.

MacVean, S., "Seat Belt Usage and Vehicle Occupancy Data", Ford Motor Company Technical Memo PRM 66-26, August 1966.

Madreperla SA, Benetz BA. (1997) Formation and treatment of a traumatic macular hole. Archives of Ophthalmology 115(9):1210-1.

Mahon, G., Masiello, M. (1992) Single point sensing and structural design of vehicles. Paper Number 920119, Society of Automotive Engineers, Warrendale, Pennsylvania.

Major MS, MacGregor A, Bumpous JM. (2000) Patterns of maxillofacial injuries as a function of automobile restraint use. Laryngoscope 110(4):608-11.

Malczyk, A., Adomeit, H. (1995) The air bag folding pattern as a means for injury reduction of out-of-position occupants. Paper Number 952704, Society of Automotive Engineers, Warrendale, Pennsylvania.

Malczyk, A., Franke, D., Adomeit, H. (1998) A Study On the Benefits of Dual-Stage Inflators Under Out-Of-Position Conditions. Proc. International Research Conference on the Biomechanics of Impact (IRCOBI), Goteborg, Sweden.

Malliaris, A., Digges, K., Deblois, J. (1997) Relationships between crash casualties and crash attributes. Paper Number 970393, Society of Automotive Engineers, Warrendale, Pennsylvania.

Manche EE, Goldberg RA, Mondino BJ. (1997) Air bag-related ocular injuries. Ophthalmic Surgery and Lasers 28(3):246-50.

Manon, G., Hensler, R. (1993) Tradeoffs encountered in evaluating crash sensing systems. Paper Number 930648, Society of Automotive Engineers, Warrendale, Pennsylvania.

Marco F, Garcia-Lopez A, Leon C, Lopez-Duran L. (1996) Bilateral Smith fracture of the radius caused by airbag deployment. The Journal of Trauma 40(4):663-4.

Marsh, J. (1993) Supplemental air bag restraint systems: consumer education and experience. Paper Number 930646, Society of Automotive Engineers, Warrendale, Pennsylvania.

Marshall KW, Koch BL, Egelhoff JC. (1998) Air bag-related deaths and serious injuries in children: injury patterns and imaging findings. AJNR American Journal of Neuroradiol 19(9):1599-607.

Martin, P., Crandall, J., Pilkey, W. (2000) Injury trends of passenger car drivers in frontal crashes in the USA. Accident Analysis and Prevention 32:541-557.

Martinez R. (1996) Improving air bags. Annals of Emergency Medicine 28(6):709-10.

Materna, P. (1992) Advances in analytical modeling of airbag inflators. Paper Number 920120, Society of Automotive Engineers, Warrendale, Pennsylvania.

Matsumoto, H., Sakakida, M. Kurimoto, K. (1990) A parametric evaluation of vehicle crash performance. Paper Number 900465, Society of Automotive Engineers, Warrendale, Pennsylvania.

Maxeiner H, Hahn M. (1997) Airbag-induced lethal cervical trauma. The Journal of Trauma 42(6):1148-51.

Mbamalu D, Banerjee A, Shankar A, Grant D. (2000) Air bag associated fatal intra-abdominal injury. Injury 31(2):121-2.

McDermott ML, Shin DH, Hughes BA, Vale S. (1995) Anterior segment trauma and air bags. Archives of Ophthalmology 113(12):1567-8.

Mcfeely WJ Jr, Bojrab DI, Davis KG, Hegyi DF. (1999) Otologic injuries caused by airbag deployment. Otolaryngology and Head and Neck Surgery 121(4):367-73.

McGovern MK, Murphy RX Jr, Okunski WJ, Wasser TE. (2000) The influence of air bags and restraining devices on extremity injuries in motor vehicle collisions. Annals of Plastic Surgery 44(5):481-5.

McKay MP, Jolly BT. (1999) A retrospective review of air bag deaths. Academic Emergency Medicine 6(7):708-14.

McKay, M., Fitzharris, M., Fildes, B. (1999) Driver injury patterns in the United States and Australia: Does beltwearing or airbag deployment make a difference? Proceedings of the Association for the Advancement of Automotive Medicine, Des Plaines, IL.

McKendrew, C., Hines, M., Litsky, A., Saul, R. (1998) Assessment of forearm injury due to a deploying driver-side air bag. Paper Number 98-S5-O-09, 16th Conference on the Enhanced Safety of Vehicles (ESV), National Highway Traffic Safety Administration, Washington, DC.

McKinley, R., Nixon, C. (1998) Human auditory response to an air bag inflation noise: has It been 30 years?," Journal of Acoustical Society of America. 104(3): 1769.

Mertz H.J., Driscoll G.D., et al. (1982) Responses of animals exposed to deployment of various passenger inflatable restraint system concepts for a variety of collision severities and animal positions. 9th Conference on the Enhanced Safety of Vehicles (ESV), National Highway Traffic Safety Administration, Washington, DC.

Mertz H.J., Weber D.A. (1982) Interpretation of the impact responses of a three-year-old child dummy relative to child injury potential. 9th Conference on the Enhanced Safety of Vehicles (ESV), National Highway Traffic Safety Administration, Washington, DC.

Michaeli-Cohen A, Neufeld M, Lazar M, Geyer O, Haddad R, Kashtan H. (1996) Bilateral corneal contusion and angle recession caused by an airbag. British Journal of Ophthalmology 80(5):487.

Michie, V. and M. Bronstad, "Evaluate Airbag Restraints for Subcompact", DOT-024-1-165, May, 1973.

Miller, H. (1995) Injury reduction with smart restraint systems. 39th Annual Proceedings of the Association for the Advancement of Automotive Medicine, pp. 527-541, Des Plaines, IL.

Miller, J. (1996) Occupant performance with constant force restraint systems. Paper Number 960502, Society of Automotive Engineers, Warrendale, Pennsylvania.

Miller, P., Gau, H. (1997) Sled testing procedure for side impact airbag development. Paper Number 970570, Society of Automotive Engineers, Warrendale, Pennsylvania.

Mohamed AA, Banerjee A. (1998) Patterns of injury associated with automobile airbag use. Postgrad Medical Journal 74(874):455-8. Review.

Mohan, D., Zador, P., O'Neill, B., Ginsburg, M. (1976) Airbags and lap/shoulder belts – a comparison of their effectiveness in real world, frontal crashes. Proceedings of the Association for the Advancement of Automotive Medicine, Des Plaines, IL.

Molia LM, Stroh E. (1996) Airbag injury during low impact collision. British Journal of Ophthalmology 80(5):487-8.

Morgan, R., Eppinger, R., Haffner, M., Yoganandan, N., Pintar, F., Sances, A., Crandall, J., Pilkey, W., Klopp, G., Dallieris, D., Miltner, E., Mattern, R., Kuppa, S., Sharpless, C. (1994) Thoracic trauma assessment formulations for restrained drivers in simulated frontal impacts. Paper Number 942206, Society of Automotive Engineers, Warrendale, Pennsylvania.

Morgenstern K, Talucci R, Kaufman MS, Samuels LE. (1998) Bilateral pneumothorax following air bag deployment. Chest 114(2):624-6.

Morrall J.F. (1986) A Review of the Record. Regulation 25-34, November/December.

Morris MS, Borja LP. (1998) Air bag deployment and hearing loss. American Family Physician 57(11):2627-8.

Morris MS, Borja LP. (1998) Noise levels associated with airbag deployment may result in occupants experiencing irreversible hearing loss. The Journal of Trauma 44(1):238-9.

Morris, A. ,Barnes, J., Fildes, B. (2001) The effectiveness of airbags in Australia as determined by in-depth crash injury research. Paper Number 167, 17th Conference on the Enhanced Safety of Vehicles (ESV), National Highway Traffic Safety Administration, Washington, DC.

Morris, A. and Fildes, B. (1998) Preliminary experience of passenger airbag deployments in Australia. Paper Number 98-S5-W-17, 16th Conference on the Enhanced Safety of Vehicles (ESV), National Highway Traffic Safety Administration, Washington, DC.

Morris, A., Barnes, J., Fildes, B. (2001) A preliminary analysis of passenger airbag effectiveness in Australia. Paper Number 169, 17th Conference on the Enhanced Safety of Vehicles (ESV), National Highway Traffic Safety Administration, Washington, DC.

Morris, A., Fildes, B., Digges, K., Dalmotas, D., Langweider, K. (1998) Injuries in primary and supplementary airbag systems. Paper Number 98-S5-O-03, 16th Conference on the Enhanced Safety of Vehicles (ESV), National Highway Traffic Safety Administration, Washington, DC.

Morris, J. (1985) Air bags for small cars. Paper Number 851200, Society of Automotive Engineers, Warrendale, Pennsylvania.

Morris, R., Duma, S., Bass, C., Sieveka, E., Pellettiere, J., Crandall, J., Pilkey, W. (1998) Analysis of humerus orientation in upper extremity experiments with a deploying airbag. Paper Number 980639, Society of Automotive Engineers, Warrendale, Pennsylvania.

Morrison AL, Chute D, Radentz S, Golle M, Troncoso JC, Smialek JE. (1998) Air bag-associated injury to a child in the front passenger seat. American Journal of Forensic Medicine and Pathology 19(3):218-22.

Mouzakes J, Koltai PJ, Kuhar S, Bernstein DS, Wing P, Salsberg E. (2001) The impact of airbags and seat belts on the incidence and severity of maxillofacial injuries in automobile accidents in New York State. Archives of Otolaryngology and Head and Neck Surgery 127(10):1189-93.

Mu, W. (2001) Driver out-of-position injuries mitigation and advanced restraint features development. 17th Conference on the Enhanced Safety of Vehicles (ESV), National Highway Traffic Safety Administration, Washington, DC.

Mueller, H., Linn, B. (1998) Smart airbag systems. Paper Number 980558, Society of Automotive Engineers, Warrendale, Pennsylvania.

Murphy RX Jr, Birmingham KL, Okunski WJ, Wasser T. (2000) The influence of airbag and restraining devices on the patterns of facial trauma in motor vehicle collisions. Plastic Reconstructive Surgery 105(2):516-20.

Musiol, J. A., Norgan-Curtiss, L. Wilkins, M. (1997) Control and Application of Intelligent Restraint Systems. Paper Number 971052, Society of Automotive Engineers, Warrendale, Pennsylvania.

Musiol, J., Norgan-Curtiss, L., Wilkins, M. (1997) Control and application of intelligent restraint systems. Paper Number 971052, Society of Automotive Engineers, Warrendale, Pennsylvania.

Nabarro M, Myers S. (2000) Airbag injuries: upper limb fractures due to airbag deployment. The Australian and New Zealand Journal of Surgery 70(5):377-9.

Nader, Ralph. (1965) Unsafe at Any Speed - The Designed-In Dangers of the American Automobile. Grossman, New York, NY.

Narwani, G., Subbian, T. (1993) Optimization of passenger airbags using occupant simulation. Paper Number 930240, Society of Automotive Engineers, Warrendale, Pennsylvania.

National Center for Statistics and Analysis, Special Crash Investigation Data on Air Bag Fatalities, as Reported on the NHTSA Website (www-nrd.nhtsa.dot.gov/pdf/nrd-30/NCSA/SCI/4Q.htm), January, 2002.

National Center for Statistics and Analysis, Special Crash Investigation Data on Airbag Fatalities. www-nrd.nhtsa.dot.gov, NHTSA, US DOT, Washington, DC.

Nelson TF, Sussman D, Graham JD. (1999) Airbags: an exploratory survey of public knowledge and attitudes. Accident Analysis and Prevention 31(4):371-9.

Newman L, Hopper C. Driver's airbags and facial injuries. British Journal of Oral and Maxillofacial Surgery 34(5):480.

NHTSA (1974) Analysis of effects of proposed changes to passenger car requirements of MVSS 208. National Highway Traffic Safety Administration, U.S. Department of Transportation, Washington DC

NHTSA (1984) "Final Regulatory Impact Analysis, Amendment to FMVSS No. 208, Passenger Car Front Seat Occupant Protection," July 11, 1984. NHTSA, Plans and Programs, Office of Planning and Analysis, US DOT, Washington, DC.

NHTSA (1984) FMVSS 208 regulatory impact analysis. National Highway Traffic Safety Administration, U.S. Department of Transportation, Washington DC

NHTSA "Effectiveness of Occupant Protection Systems and Their Use," Second Report to Congress, Department of Transportation Report DOT HS 808 389, February, NHTSA, US DOT, Washington, DC.

NHTSA "Effectiveness of Occupant Protection Systems and Their Use," Third Report to Congress, Department of Transportation Report DOT HS 808 537, December, NHTSA, US DOT, Washington, DC.

NHTSA Press Release, "Warning of Dangers to Children and Unbelted Adults Riding in the Front Passenger Seat of Air Bag Equipped Vehicles", October 27, 1995.

Nieboer, J., Wismans, J., de Coo, P. (1990) Airbag modeling techniques. Paper Number 902322, Society of Automotive Engineers, Warrendale, Pennsylvania.

Nieboer, J., Wismans, J., Fraterman, E. (1988) Status of the MADYMO 2D airbag model. Paper Number 881729, Society of Automotive Engineers, Warrendale, Pennsylvania.

Nixon, C. (1969) Human Auditory Response to an Air Bag Inflation Noise", Final Report, DOT Contract No. P.O. 9-1-1151.

Nixon, C. (1970) The Human Ear in an Air Bag Noise Environs. Proc. 14th Annual Scientific Conference of the Association of the Advancement of Automotive Medicine, Tempe, AZ.

Norden RA, Perry HD, Donnenfeld ED, Montoya C. (2000) Air bag-induced corneal flap folds after laser in situ keratomileusis. American Journal of Ophthalmology 130(2):234-5.

Nusholtz, G., Wylie, E., Wang, D. (1996) Air bag aspiration simulation. Paper Number 960499, Society of Automotive Engineers, Warrendale, Pennsylvania.

Nusholtz, G., Wang, W., Wylie, E. (1997) Flow and energy pattern in pyrotechnic airbag inflator-canister system. Paper Number 970388, Society of Automotive Engineers, Warrendale, Pennsylvania.

Nusholtz, G., Wang, D., Wylie, E. (1998) An Evaluation of Airbag Tank-test Results. Paper Number 980864, Society of Automotive Engineers, Warrendale, Pennsylvania.

Nusholtz, G., Xu, L., Mosier, R., Kostyniuk, G. (1998) Estimation of OOP from Conditional Probabilities of Air bag Fire Times and Vehicle Response. Paper Number 98-S5-O-16, Proc. 16th ESV Conference.

Nusholtz, G., Wu, J., Wang, D., Wylie, E. (1999) Energy and entropy in airbag deployment: The effect on an out-of-position occupant. Paper Number 1999-01-1071, Society of Automotive Engineers, Warrendale, Pennsylvania.

Nusholz, Shi, Y., Xu, L. (2000) Optimization of single-point frontal airbag fire threshold. Paper Number 2000-01-1009, Society of Automotive Engineers, Warrendale, Pennsylvania.

O'Connor, C., Rao, M. (1992) Development of a model of a three-year-old child dummy used in air bag applications. Paper Number 922517, Society of Automotive Engineers, Warrendale, Pennsylvania.

O'Neill, B. (2000) Airbag test requirements under proposed new rule. Statement before the Transportation Subcommittee, U.S. House of Representatives Appropriations Committee, Washington DC.

Odell C. (1996) Air bags and lower extremity trauma: implications for flight crew members. Journal of Emergency Nursing 22(2):139-41.

O'Halloran HS, Draud K, Stevens JL. (1998) Primary enucleation as a consequence of airbag injury. The Journal of Trauma 44(6):1090.

Okamoto K, Takemoto M, Okada Y. (2002) Airbag-mediated craniocervical injury in a child restrained with safety device. The Journal of Trauma 52(3):587-90.

Omura, H., Shimamura, M. (1989) Analysis of airbag inflating. Paper Number 890192, Society of Automotive Engineers, Warrendale, Pennsylvania.

Ong CF, Kumar VP. (1998) Colles fracture from air bag deployment. Injury 29(8):629-31.

Onwuzuruigbo CJ, Fulda GJ, Larned D, Hailstone D. (1996) Traumatic blindness after airbag deployment: bilateral lenticular dislocation. The Journal of Trauma 40(2):314-6.

Otte, D. (1995) Review of the air bag effectiveness in real life accidents - Demands for positioning and optimal deployment of air bag systems. Paper Number 952701, Society of Automotive Engineers, Warrendale, Pennsylvania.

Padmanaban, J., Ray, R. (1992) Performance of rear seat occupant restraint systems. Paper Number 922524, Society of Automotive Engineers, Warrendale, Pennsylvania.

Park, B., Morgan, R., Hackney, J., Lowrie, J. (1998) The effect of redesigned air bags on frontal USA NCAP. Paper Number 98-S11-O-01, 16th Conference on the Enhanced Safety of Vehicles (ESV), National Highway Traffic Safety Administration, Washington, DC.

Park, B., Mortan, R., Hackney, J., Partyka, S., Kleinberger, M., Sun, E., Smith, H., Lowrie, J. (1998) Upper neck response of the belt and air bag restrained 50th percentile hybrid III dummy in the USA's new car assessment program. Paper Number 983164, Society of Automotive Engineers, Warrendale, Pennsylvania.

Patrick, L.M., Kroell, C.K., Mertz, H.J. (1965) Forces on the human body in simulated crashes. Proc. 9th Stapp Car Crash Conference, pp. 237-260.

Patrick, L, Bohlin, N., Anderson, A. (1974) Three-point harness accident data and laboratory data comparison. Paper Number 741181, Society of Automotive Engineers, Warrendale, PA.

Pattimore, D., Thomas, P., Dave, S. (1992) Torso Injury Patterns and Mechanisms in Car Crashes: An Additional Diagnostic Tool. Injury: the British Journal of Accident Surgery 23(2): 123-126.

Pearlman JA, Au Eong KG, Kuhn F, Pieramici DJ. (2001) Airbags and eye injuries: epidemiology, spectrum of injury, and analysis of risk factors. Surv Ophthalmol 46(3):234-42. Review.

Pellettiere, J. (1999) Computational strength determination of human long bones. Paper Number 1999-01-1904, Society of Automotive Engineers, Warrendale, Pennsylvania.

Peltzman, S. (1975) The effects of automobile safety regulation. Journal of Political Economy 83:677-725.

Perdikis G, Schmitt T, Chait D, Richards AT. (2000) Blunt laryngeal fracture: another airbag injury. The Journal of Trauma 48(3):544-6.

Perez J, Palmatier T. (1996) Air bag-related fatality in a short, forward-positioned driver. Annal of Emergency Medicine 28(6):722-4.

Perez-Camarero ER, De Cortazar JL, Caudevilla CJ, Cabane JM, Boleas ME, Cilveti ML. (2000) Airbag lesions. European Journal of Emergency Medicine 7(2):160.

Peterson, S. and Hoffer,G. (1994) The impact of airbag adoption on relative personal injury and absolute collision insurance claims. J. of Consumer Research 20:657-662.

Petitejean, A., Lebarbe, M., Potier, P., Trosseille, X., Lassau, J. (2002) Laboratory reconstructions of real world frontal crash configurations using the Hybrid III and THOR dummies and PMHS. Paper Number 2002-22-0002, Society of Automotive Engineers, Warrendale, Pennsylvania.

Pewinski, W., Ventura, K., Sadek, J. (2001) A preliminary look at occupant sensing and child restraints. Paper Number 2001-01-0163, Society of Automotive Engineers, Warrendale, Pennsylvania.

Pham T, Palmieri TL, Greenhalgh DG. Sodium azide burn: a case report. JournalBurn Care Rehabil. 2001 May-Jun;22(3):246-8.

Phen, R., Dowdy, M., Ebbeler, D., Kim, E-H, Moore, N., VanZandt, T. (1998) Advanced Air Bag Technology Assessment, Final Report, April 1998, prepared for NHTSA and NASA by the Jet Propulsion Laboratory, Jet Propulsion Laboratory, California Institute of Technology, Pasadena, California.

Pintar, F., Mayer, R., Yoganandan, N., Sun, E. (2000) Child Neck strength characteristics using an animal model. Paper Number 2000-01-SC06, Society of Automotive Engineers, Warrendale, Pennsylvania.

Plank, G. R., Kleinberger, M., and Eppinger, R. H. (1994) Finite element modeling and analysis of thorax/restraint system interaction. Paper Number 94-S1-O-16, 14th Conference on the Enhanced Safety of Vehicles (ESV), National Highway Traffic Safety Administration, Washington, DC.

Plank, G., Kleinberger, M., Eppinger, R. (1998) Analytical investigation of driver thoracic response to out of position airbag deployment. Paper Number 983165, Society of Automotive Engineers, Warrendale, Pennsylvania.

Polk JD, Thomas H. (1994) Automotive airbag-induced second-degree chemical burn resulting in Staphylococcus aureus infection. Journal of the American Osteopathic Association 94(9):741-3.

Prasad, P., Laituri, T. (1996) Consideration for belted FMVSS 208 Testing. Paper Number 86-S3-O-03, Proc. 15th ESV Conference.

Price GR, Kalb JT. (1999) Auditory hazard from airbag noise exposure. Journal of the Acoustical Society of America 106(5):2629-37.

Pudpud AA, Linares M, Raffaele R. (1998) Airbag-related lower extremity burns in a pediatric patient. American Journal of Emergency Medicine 16(4):438-40.

Rains, G., Prasad, A., Summers, L., Terrell, M. (1998) Assessment of advanced air bag technology and less aggressive air bag designs through performance testing. Paper Number 98-S5-O-06, 16th Conference on the Enhanced Safety of Vehicles (ESV), National Highway Traffic Safety Administration, Washington, DC.

Reed, D. (1999) Father of the Air Bag. Automotive Engineering February 99(2).

Reed, M., Rupp, J., Hardy, W., Schneider, L. (1999) Methods for laboratory investigation of airbag-induced thermal skin burns. Paper Number 1999-01-1064, Society of Automotive Engineers, Warrendale, Pennsylvania.

Reed, M., Rupp, J., Reed, S., Schneider, L. (1999) Development of an improved airbag-induced thermal skin-burn model. Paper Number 1999-01-1065, Society of Automotive Engineers, Warrendale, Pennsylvania.

Reidelbach, W., (1985) The Daimler-Benz Supplemental Restraint System. SAE 856016, Tenth International Technical Conference on Experimental Safety Vehicles, Oxford, England, Highway Traffic Safety Administration, Washington, D.C.

Reiland-Smith J, Weintraub RM, Sellke FW. (1993) Traumatic aortic valve injury sustained despite the deployment of an automobile air bag. Chest 103(5):1603.

Reilly DA, Garner WL. Management of chemical injuries to the upper extremity. Hand Clin. 2000 May;16(2):215-24. Review.

Report on ASL/Takata Cooperative Research with NHTSA, September, 1997.

Rich, D., Kosiak, W., Manlove, G., Schwarz, D. (1997) A remotely mounted crash detection system. Paper Number 973240, Society of Automotive Engineers, Warrendale, Pennsylvania.

Richter HJ 2nd. (1976) Investigation of acoustic trauma from the airbag. Laryngoscope 86(8):1188-95.

Riling, J. (1995) Sensing and diagnostic module for airbags. Paper Number 952682, Society of Automotive Engineers, Warrendale, Pennsylvania.

Rivara FP. (1997) Motor vehicle occupant protection: have we become too complacent? Injury Prevention 3(2):78.

Robbins, D. (1989) Restraint systems computer modeling and simulation state of the art and correlation with reality. Paper Number 891976, Society of Automotive Engineers, Warrendale, Pennsylvania.

Roberts D, Pexa C, Clarkowski B, Morey M, Murphy M. (1999) Fatal laryngeal injury in an achondroplastic dwarf secondary to airbag deployment. Pediatric Emergency Care 15(4):260-1.

Roccia F, Servadio F, Gerbino G. (1999) Maxillofacial fractures following airbag deployment. Journal of Craniomaxillofacial Surgery 27(6):335-8.

Rocket Research Corporation, "Development of Improved Inflation Techniques", Final Report DOT HS-801 724, August, 1975.

Roesler JS, Kinde MR. (2000) Air bags. An imperfect and incomplete solution. Minnesota Medicine 83(8):57-60.

Romeo, D. (1975) Front passenger passive restraint for small car, high speed, frontal impacts. Paper Number 751170, Society of Automotive Engineers, Warrendale, PA.

Roselt, T., Langwieder, K., Hummel, T., Koster, H. W. (2002) Injury Patterns of Front Seat Occupants in Frontal Car Collisions with Air bags. Effectivity and Optimisation Potential of Air bags. Proc. International Research Conference on the Biomechanics of Impact (IRCOBI), Munich, Germany.

Rosenblatt MA, Freilich B, Kirsch D. (1993) Air bag-associated ocular injury. Archives of Ophthalmology 111(10):1318.

Roth T, Meredith P. (1993) Hand injuries from inflation of an air bag security system. Journal of Hand Surgery [Br] 18(4):520-2.

Roychoudhury, T., Sun, S., Hamid, M., Hanson, C. (2000) Fifth Percentile Driver Out of Position Computer Simulation. Paper Number 2000-01-1006, Society of Automotive Engineers, Warrendale, Pennsylvania.

Rozner L. (1996) Air bag-bruised face. Plastic Reconstructive Surgery 1996 97(7):1517-9.

Ruiz-Moreno JM. (1998) Air bag-associated retinal tear. European Journal of Ophthalmology 8(1):52-3.

Rutan TC, Stocco GS. (1997) Burns from airbags. Nurs Spectr (Wash DC) 7(19):11.

Rutherford, W., H. (1985) The medical effects of seat-belt legislation in the United Kingdom: a critical review of the findings. Archives of Emergency Medicine 2(4):221-3.

Ryan, S. (1998) An Innovative Approach to Adapative Airbag Modules. Paper Number 980646, Society of Automotive Engineers, Warrendale, Pennsylvania.

Sagberg F, Fosser S, Saetermo IA. (1997) An investigation of behavioural adaptation to airbags and antilock brakes among taxi drivers. Accident Analysis and Prevention 29(3):293-302.

Sama AE, Barnaby DP, Wallis KJ, Gadaleta D, Hall MH, Nelson RL, Naidich J, Ward RJ. (1995) Isolated intrathoracic injury with air bag use. Prehospital Disaster Medicine 10(3):198-201.

Sastry SM, Copeland RA Jr, Mezghebe H, Siram SM. (1995) Retinal hemorrhage secondary airbag-related ocular trauma. The Journal of Trauma 38(4):582.

Saul, R., Backaitis, S., Beebe, M., Ore, L. (1996) Hybrid III Dummy Instrumentation and Assessment of Arm Injuries During Air Bag Deployment. Paper Number 962417, Society of Automotive Engineers, Warrendale, Pennsylvania.

Saul, R., Pritz, H., McFadden, J., Backaitis, S., Hallenback, H., Rhule, D. (1998) Description and Performance of the Hybrid III Three Year Old, Six Year Old and Small Female Test Dummies in Restraint System and Out-Of-Position Air Bag Environments. Paper Number 98-S7-O-01, 16th ESV Conference.

Schroeder, G., Eidam, J. (1997) Typical Injuries Caused by Air-Bag in Out-of-Position Situations– An Experimental Study. Proc. International Research Conference on the Biomechanics of Impact (IRCOBI), Hannover, Germany.

Saunders JE, Slattery WH 3rd, Luxford WM. (1998) Automobile airbag impulse noise: otologic symptoms in six patients. Otolaryngology and Head and Neck Surgery 118(2):228-34.

Sawamura D, Umeki K. (2000) Airbag dermatitis. Journal of Dermatology 27(10):685-6.

Schelkle, E., Remensperger, R. (1991) Integrated occupant-car crash simulation with the finite element method: the PORSCHE hybrid III-dummy and airbag model. Paper Number 910654, Society of Automotive Engineers, Warrendale, Pennsylvania.

Schimkat, H, Weissner, R. (1974) A Comparison Between Volkswagen Automatic Restraint and Three-Point Automatic Belt on the Basis of Dummy and Cadaver Tests. Paper Number 741183, Society of Automotive Engineers, Warrendale, Pennsylvania.

Schimkat, H., Weissner, R., Schmidt, G. (1974) A Comparison Between Volkswagen Automatic Restraint and Three-Point Automatic Belt on the Basis of Dummy and Cadaver Tests. Paper Number 741183, Society of Automotive Engineers, Warrendale, Pennsylvania.

Scott IU, Greenfield DS, Parrish RK 2nd. (1996) Airbag-associated injury producing cyclodialysis cleft and ocular hypotony. Ophthalmic Surgery and Lasers 27(11):955-7.

Scott IU, John GR, Stark WJ. (1993) Airbag-associated ocular injury and periorbital fractures. Archives of Ophthalmology 111(1):25.

Segui-Gomez M, Graham JD. (2000) Patterns of injury among drivers hospitalized in Level-I Trauma Centers: have frontal airbags made a difference? Proceedings of the Association for the Advancement of Automotive Medicine, Des Plaines, IL.

Segui-Gomez, M. (2000) Driver air bag effectiveness by severity of the crash. American Journal of Public Health 90:1575-1581.

Shah GK, Penne R, Grand MG. (2001) Purtscher's retinopathy secondary to airbag injury. Retina 21(1):68-9.

Shah N, Weinstein A. (1997) Reflex sympathetic dystrophy syndrome following air-bag inflation. New England Journal of Medicine 337(8):574.

Sharma OP, Mousset XR. (2000) Review of tricuspid valve injury after airbag deployment: presentation of a case and discussion of mechanism of injury. The Journal of Trauma 48(1):152-6. Review.

Sharma OP. (1999) Pericardio-diaphragmatic rupture: five new cases and literature review. Journal of Emergency Medicine 1999 17(6):963-8. Review.

Shaw, G., Crandall, J., Butcher, J. (2000) Biofidelity evaluation of the THOR advanced frontal crash test dummy. Proc. International Research Conference on the Biomechanics of Impact (IRCOBI).

Shaw, G., Dalrymple, G., Ragland, C. (1999) Reducing the risk of driver injury from common steering control devices in frontal collisions. Paper Number 1999-01-0759, Society of Automotive Engineers, Warrendale, Pennsylvania.

Shaw, M. (2002) Accelerometer overload considerations for automotive airbag applications. Paper Number 2002-01-0193, Society of Automotive Engineers, Warrendale, PA.

Shelton, T. and Lago, J. (1996) National occupant protection use survey. 15th Conference on the Enhanced Safety of Vehicles (ESV), National Highway Traffic Safety Administration, Washington, DC.

Sherman, D. (1995) The Rough Road to Air Bags. Invention and Technology. Summer. 1995: 48-56.

Sherman, D. (1995) The Rough Road to Air Bags. Invention and Technology 2(1).

Sherman, S., Samuels, H., Riedel, W. (1997) A low-cost dual-axis accelerometer. Paper Number 970607, Society of Automotive Engineers, Warrendale, Pennsylvania.

Shinto, H., Ogata, K., Teraoka, F., Fukabori, M. (1991) Development of the all-mechanical air bag system. Paper Number 910149, Society of Automotive Engineers, Warrendale, Pennsylvania.

Shoemaker, N., Biss, D. (1974) The Development of an Air Bag on Collapsible Dashpanel Restraint System for Right Front Seat Occupants. Paper Number 740576, Society of Automotive Engineers, Warrendale, Pennsylvania.

Shokoohi, F. (1995) Air bag sensor fire time - Occupant performance criterion. Paper Number 950873, Society of Automotive Engineers, Warrendale, Pennsylvania.

Shokoohi, F., Sanders, T., Castelli, V., Breed, D. (1993) Characterization of the cross-axis acceleration crash sensor environment and test method. Paper Number 930651, Society of Automotive Engineers, Warrendale, Pennsylvania.

Sieveka, E., Crandall, J., Duma, S., Pilkey, W. (1999) Three-year-old child and side airbag interaction study using the CVS/ATB multi-body simulation program. Paper Number 1999-01-0756, Society of Automotive Engineers, Warrendale, Pennsylvania.

Sieveka, E., Duma, S., Pellettiere, J., Crandall, J. (1997) Multi-Body Model of Upper Extremity Interaction with Deploying Airbag. Paper Number 970398, Society of Automotive Engineers, Warrendale, Pennsylvania.

Singer HW. (1998) Potential air bag-related eye injuries require special ER attention. Journal of Ophthalmic Nursing Technology 17(1):21-2.

Skeels, P., Falzon, R. (1964) A New Laboratory Device for Simulating Vehicle Crash Conditions. Paper Number 1962-12-0014, Society of Automotive Engineers, Warrendale, Pennsylvania.

Slack, W. (1968) Automotive Inflatable Occupant Restraint System, Part II. Paper Number. 680033, Society of Automotive Engineers, Warrendale, Pennsylvania.

Smally, A.J., Binzer, A., Dolin, S., Viano, D. (1992) Alkaline chemical keratitis: eye injury from air bags. Annals of Emergency Medicine 21(11):1400-2.

Smith DP, Klein FA. (1997) Renal injury in a child with airbag deployment. The Journal of Trauma 42(2):341-2.

Smith, G., Hurite, S., Yanik, A., Greer, C. (1972) Human Volunteer Testing of GM air cushions. Paper 720443, Society of Automotive Engineers, Warrendale, Pennsylvania.

Smith, G. (1973) Field testing of 1000 air cushion equipped vehicles. Proceedings of the Association for the Advancement of Automotive Medicine, Des Plaines, IL.

Smith, R. and Moffatt, C. (1975) Accident experience in airbag-equipped cars. Proceedings of the Association for the Advancement of Automotive Medicine, Des Plaines, IL.

Smock WS, Nichols GR 2nd. (1995) Airbag module cover injuries. The Journal of Trauma 38(4):489-93.

Snyder, R., Young, J., Snow, C. (1967) Preliminary primate tests with air bag and inertia reel/inverted-Y yoke torso harness. Paper 670922, Society of Automotive Engineers, Warrendale, PA.

Southerland, G. (1970) Self-Deployed, Air Induction Inflation Systems. Paper presented at the International Conference on Passive Restraints.

Spiess, O., Marotzke, T., Zahn, M. (1997) Development Methodology of an Airbag Integrated Steering Wheel in Order to Optimize Occupant Protection Balanced Against Out-of-Position Risks. Paper Number 970777, Society of Automotive Engineers, Warrendale, Pennsylvania.

Stalnaker, Klusmeyer, Peel, White, Smith, Mertz, (1982) Unrestrained, Front Seat, Child Surrogate Trajectories Produced by Hard Braking Paper 821165, Society of Automotive Engineers, Warrendale, Pennsylvania.

Starner, A. (1998) Airborne particulates in automotive airbags. Paper Number 980645, Society of Automotive Engineers, Warrendale, Pennsylvania.

States, J., Rosenau, W. (1982) Field Performance of Volkswagen Automatic Restraint System, Progress Report.

Stein JD, Jaeger EA, Jeffers JB. (1999) Air bags and ocular injuries. Trans Am Ophthalmology Society 97:59-82; discussion 82-6.

Steiner, P., Wetzel, G. (1999) New aspects on static passenger and child seat recognition and future dynamic out-of-position detection for airbag control systems. Paper Number 1999-01-0765, Society of Automotive Engineers, Warrendale, Pennsylvania.

Stoneham MD. (1995) Bilateral first rib fractures associated with driver's air bag inflation: case report and implications for surgery. European Journal of Emergency Medicine 2(1):60-2.

Stranc MF. (1999) Eye injury resulting from the deployment of an airbag. British Journal of Plastic Surgery 1999 52(5):418.

Strother, C., and T. Zinke, "Small Car Driver Inflatable Restraint System Evaluation Vol. 4: Evolving a Low Mount Passenger Air Cushion Restraint System (ACRS) for the Vega Subcompact Vehicle", Final Report, Contract DOT-HS-6-01412, July, 1978.

Strother, C., and W. Broadhead, "Small Car Driver Inflatable Restraint System Evaluation", Final Report, Contract DOT-HS-6-01412, 1978.

Strother, C., Fitzpatrick, M., Egbert, T. (1976) Development of Advanced Restraint Systems for Minicars RSV. 6th Conference on the Enhanced Safety of Vehicles (ESV), National Highway Traffic Safety Administration, Washington, D.

Struble, D. (1998) Airbag Technology: What it is and How it Came to Be. Paper Number 980648, Society of Automotive Engineers, Warrendale, PA.

Stucki, S. and Biss, D. (2000) A study of the NASS-CDS system for injury/fatality rates of occupants in various restraints and a discussion of alternative presentation methods. Proceedings of the Association for the Advancement of Automotive Medicine, Des Plaines, IL.

Stucki, S., Fessahaie, O. (1998) Comparison of measured velocity change in frontal crash tests to NASS-computed velocity changes. Paper Number 980649, Society of Automotive Engineers, Warrendale, Pennsylvania.

Submission to Docket No. NHTSA-2000-7013-16, Concerning the NHTSA Testing of 13 late-1990's Production Air Bag Equipped Vehicles to Determine if Depowering Adversely Affected High Speed Protection, June 20, 2000.

Sullivan, L., "Child Restraint/Passenger Air Bag Interaction Analysis", NHTSA Report VRTC-87-0074, October, 1992.

Summers, L., Hollowell, W., Prasad, A. (2001) Analysis of occupant protection provided to 50th percentile male dummies sitting mid-track and 5th percentile female dummies sitting full-forward in crash tests of paired vehicles with redesigned air bag systems. Paper Number 338, 17th Conference on the Enhanced Safety of Vehicles (ESV), National Highway Traffic Safety Administration, Washington, DC.

Summers, L., Hollowell, W., Rains, G. (1998) NHTSA's advanced air bag technology research program. Paper Number 98-S5-W-29, 16th Conference on the Enhanced Safety of Vehicles (ESV), National Highway Traffic Safety Administration, Washington, DC.

Sutyak, J., Passi, V., and Hammond, J. (1997) Airbags alone compared with the combination of mechanical restraints and airbags: implications for the emergency evaluation of crash victims. Southern Medical Journal 90(9): 915-919.

Suzuki, N., Inoue, S., Nakahama, R. (1989) Determination of airbag sensor threshold level by graphic method. Paper Number 890193, Society of Automotive Engineers, Warrendale, Pennsylvania.

Swanson-Biearman B, Mrvos R, Dean BS, Krenzelok EP. (1993) Air bags: lifesaving with toxic potential? American Journal of Emergency Medicine 11(1):38-9.

Swihart, W., Lawrence IV, A. (1995) Investigation of sensor requirements and expected benefits of predictive crash sensing. Paper Number 950347, Society of Automotive Engineers, Warrendale, Pennsylvania.

Tabani, Y., Labib, M., Hensler, R. (1999) Automobile airbag inflator based on combustion of methane-oxygen mixture. Paper Number 1999-01-1070, Society of Automotive Engineers, Warrendale, Pennsylvania.

Tanavde, A., Khandelwal, H., Lasry, D., Ni, X., Haug, E., Schlosser, J., Balakrishnan, P. (1995) Airbag modeling using initial metric methodology. Paper Number 950875, Society of Automotive Engineers, Warrendale, Pennsylvania.

Tenofsky PL, Porter SW, Shaw JW. (2000) Fatal airway compromise due to retropharyngeal hematoma after airbag deployment. American Surgeon 66(7):692-4.

The National Traffic and Motor Vehicle Safety Act of 1966.

Thompson, K., Graham, J., Zellner, J. (2001) Risk-benefit analysis methods for vehicle safety devices. Paper Number 340, 17th Conference on the Enhanced Safety of Vehicles (ESV), National Highway Traffic Safety Administration, Washington, DC.

Thomson BN, Davis SM. (2001) Carotid artery dissection: another airbag injury. The Australian and New Zealand Journal of Surgery 71(9):552-3.

Thompson, K., Segui-Gomez, M., Graham, J. (2002) Validating benefit and cost estimates: the case of airbag regulation. Risk Analysis 22(4):803-11.

Traynelis VC, Gold M. (1993) Cervical spine injury in an air-bag-equipped vehicle. Journal of Spinal Disorders 6(1):60-1.

Tsuda Y, Wakiyama H, Amemiya T. (1999) Ocular injury caused by an air bag for a driver wearing eyeglasses. Japanese Journal of Ophthalmology 43(3):239-40.

Turbell, T., Lowne, R., Lundell, B., Tingvall, C. (1993) ISOFIX - A new concept of installing child restraints in cars. Paper Number 933085, Society of Automotive Engineers, Warrendale, Pennsylvania.

Tylko, S. and Dalmotas, D. (2001) Static out-of-position test methodologies: identifying a realistic worst case for small stature female drivers. Paper Number 421, 17th Conference on the Enhanced Safety of Vehicles (ESV), National Highway Traffic Safety Administration, Washington, DC.

U.S. Patent No. 2,649,311, "Safety Cushion Assembly for Automotive Vehicles", John W. Hetrick,, Inventor, August 18, 1953.

Uchio E, Kadonosono K. (2001) Are airbags a risk for patients after radial keratotomy? British Journal of Ophthalmology 85(6):640-2.

Uchio E, Ohno S, Kudoh K, Kadonosono K, Andoh K, Kisielewicz LT. (2001) Simulation of air-bag impact on post-radial keratotomy eye using finite element analysis. Journal of Cataract Refractive Surgery 27(11):1847-53.

Ulrich D, Noah EM, Fuchs P, Pallua N. (2001) Burn injuries caused by air bag deployment. Burns 27(2):196-9.

Van der Linden WJ. Dislocated fracture of the mandibular condylar process after airbag deployment: report of a case. Journal of Oral and Maxillofacial Surgery 2002 Jan;60(1):113-5.

Vezin, P., Bruyere-Garnier, K., Bermond, F., Verriest, J. (2002) Comparison of Hybrid III, THOR-α and PMHS response in frontal sled tests. Paper Number 2002-22-0001, Society of Automotive Engineers, Warrendale, Pennsylvania.

Viano, D., Artinian, C. (1978) Myocardial Conducting System Dysfunctions for Thoracic Impact. The Journal of Trauma, 18(6):452-459.

Viano, D. (1988) Limits and challenges of crash protection. Accident Analysis and Prevention 20(6):421-429.

Viano, D. (1995) Restraint effectiveness, availability and use in frontal crashes: implications to injury control. The Journal of Trauma 38(4):538-546.

Viano DC (1988) Cause and control of automotive trauma. Bulletin of the New York Academy of Medicine, Second Series 64(5):376-421.

Vichnin MC, Jaeger EA, Gault JA, Jeffers JB. (1995) Ocular injuries related to air bag inflation. Ophthalmic Surgery and Lasers 26(6):542-8.

Vitello W, Kim M, Johnson RM, Miller S. (1999) Full-thickness burn to the hand from an automobile airbag. Journal of Burn Care Rehabilitation 20(3):212-5.

Walter DP, James MR. (1996) An unusual mechanism of airbag injury. Injury 27(7):523-4.

Walz FH, Mackay M, Gloor B. (1995) Airbag deployment and eye perforation by a tobacco pipe. The Journal of Trauma 38(4):498-501.

Wang, J. (1995) On airbag inflator grade designation - A five index grading system. Paper Number 950345, Society of Automotive Engineers, Warrendale, Pennsylvania.

Wang, J., Lin, K. (1988) A CAL3D steering system impact model. Paper Number 880650, Society of Automotive Engineers, Warrendale, Pennsylvania.

Wang, J., Ngo, T. (1990) Modeling of passenger side airbags with a complex shape. Paper Number 900545, Society of Automotive Engineers, Warrendale, Pennsylvania.

Wang, J., Tung, D. (1997) A procedure for quantifying the effective leak area of a full-size airbag. Paper Number 970577, Society of Automotive Engineers, Warrendale, Pennsylvania.

Warner, C. (1979) Passive protection: promulgation or politics. Proceedings of the Association for the Advancement of Automotive Medicine, Des Plaines, IL.

Watanabe, K., Umezawa, Y. (1993) Optimal timing to trigger and airbag. Paper Number 930242, Society of Automotive Engineers, Warrendale, Pennsylvania.

Watts DD, Kokiko J. (1999) Air bags and eye injuries: assessment and treatment for ED patients. Journal of Emergency Nursing 25(6):572-4.

Weber, K. (1993) Child restraint and airbag interaction: problem and progress. Paper Number 933094, Society of Automotive Engineers, Warrendale, Pennsylvania.

Weinman SA. (1995) Automobile air bag-mediated injury: a case presentation. Journal of Emergency Nursing 21(1):84-5.

Weintraub BA. (1997) Air bag--mediated injury in the emergency department population. International Journal of Trauma Nursing 3(2):46-9. Review.

Wetzel, G., Braunwarth, H., Pulido, F., Reisacher, G. (1997) Occupant detection systems. Paper Number 971047, Society of Automotive Engineers, Warrendale, Pennsylvania.

Whitacre MM, Pilchard WA. (1993) Air bag injury producing retinal dialysis and detachment. Archives of Ophthalmology 111(10):1320.

White JE, McClafferty K, Orton RB, Tokarewicz AC, Nowak ES. (1995) Ocular alkali burn associated with automobile air-bag activation. Canadian Medical Association Journal 153(7):933-4.

White, C., Behr, L. (1990) Inflatable restraint sensing and diagnostic strategy. Paper Number 901120, Society of Automotive Engineers, Warrendale, Pennsylvania.

Williams, A., Wells, J., Lund, A. (1990) Seat belt use in cars with air bags. American Journal of Public Health 80(12):1514-1516.

Willis BK, Smith JL, Falkner LD, Vernon DD, Walker ML. (1996) Fatal air bag mediated craniocervical trauma in a child. Pediatric Neurosurgery 24(6):323-7.

Wilson, E. (1995) Investigation of improving energy absorption performance and reducing weight of passenger air bag modules using computer aided analysis. Paper Number 950338, Society of Automotive Engineers, Warrendale, Pennsylvania.

Wilson, R. and Savage, C. (1973) Restraint system effectiveness – a study of fatal accidents. Proceedings of Automotive Safety Engineering Seminar, General Motors Corporation, Warren, MI.

Winston, F., Reed, R. (1996) Air bags and children: Results of a National Highway Traffic Safety Administration special investigation into actual crashes. Paper Number 962438, Society of Automotive Engineers, Warrendale, Pennsylvania.

Xu, L., Agaram, V., Rouhana, S., Hultman, R., Kostyniuk, G., McCleary, J., Mertz, H., Nusholtz, G., Scherer, R. (2000) Repeatability evaluation of the pre-prototype NHTSA advanced dummy compared to the Hybrid III. Paper Number 2000-01-0165, Society of Automotive Engineers, Warrendale, Pennsylvania.

Yang, K., Wang, H., Wang, H., King, A. (1990) Development of a two-dimensional diver side airbag deployment algorithm. Paper Number 902323, Society of Automotive Engineers, Warrendale, Pennsylvania.

Yaremchuk K, Dobie RA. (2001) Otologic injuries from airbag deployment. Otolaryngology and Head and Neck Surgery 125(3):130-4.

Yoganandan, N., Skrade, D., Pintar, F., Reinartz, J., Sances, A. (1994) Thoracic deformation contours in a frontal impact. In: Biomechanics of Impact Injury and Injury Tolerances of the Thorax-Shoulder Complexes, Backaitis, S. (ed). pp. 765-781. Society of Automotive Engineers publication PT-45.

Zabler, E., Dukart, Hermann, T. (1998) Contactless airbag firing and signal transmission on the steering wheel with and inductive contact unit. Paper Number 981103, Society of Automotive Engineers, Warrendale, Pennsylvania.

Zabriskie NA, Hwang IP, Ramsey JF, Crandall AS. (1997) Anterior lens capsule rupture caused by air bag trauma. American Journal of Ophthalmology 123(6):832-3.

Zador, P. and Ciccone, M. (1991) Driver fatalities in frontal impacts: comparison between cars with airbags and manual lap belts. Insurance Institute for Highway Safety, Arlington, VA.

Zechmair, D., Baur, R., Wörle, E. (1997) Plastic housings for safety-critical air bag triggering units. Paper Number 970131, Society of Automotive Engineers, Warrendale, Pennsylvania.

Zhou, Q., Rouhana, S., Melvin, J. (1996) Age effects on thoracic injury tolerance. Paper Number 962421, Society of Automotive Engineers, Warrendale, Pennsylvania.

Ziegahn, K. (1998) Driving factors and future developments of airbag technology. Paper Number 980556, Society of Automotive Engineers, Warrendale, Pennsylvania.

Zinke, D. (1980) The Development of Air Cushion Restraint Systems for Small Car Front Seat Occupants. Paper Number 800294, Society of Automotive Engineers, Warrendale, Pennsylvania.

Zinke, T., "Small Car Front Seat Passenger Inflatable Restraint System - Volume I - Interim Results", Final Report, Contract DOT-HS-8-01809, April, 1981.

Zinke, T., "Small Car Front Seat Passenger Inflatable Restraint System - Volume II - Citation Air Bag Systems", Final Report, Contract DOT-HS-8-01809, April, 1981.

Ziperman, H. H. and G. R. Smith. (1975) Startle Reaction to Air-Bag Restraints. Journal of the American Medical Association 223(5):436-440.

Zuby, D. and O'Neill, B. (2001) Steering column movement in severe frontal crashes and its potential effect on airbag performance. Paper Number 136, 17th Conference on the Enhanced Safety of Vehicles (ESV), National Highway Traffic Safety Administration, Washington, DC.

Zuppichini, F., Trenchi, G., Rigo, C., Marigo, M. (1994) Unexpected deaths in airbag equipped cars: case reports. Proc. Advances in Occupant Restraint Technologies: Joint AAAM-IRCOBI Special Session, AAAM, Des Plaines, IL.

About the Contributors

Cameron R. "Dale" Bass is a Research Assistant Professor in the Department of Mechanical and Aerospace Engineering (MAE) at the University of Virginia. He received his Ph.D. in MAE from UVa in 1994. He also received a B.S. from UVa in 1984. His recent research interests include the biomechanics of high rate shocks including effects of shock on the lumbar, thoracic and cervical spine, the head, and the thoracic viscera. In addition, he has studied the biomechanics of lower extremity impact. He is a member of SAE, the American Physical Society, the American Association for the Advancement of Science, and Sigma Xi. Recognition of his work includes the 1996 AAAM Conference Best Paper Award and the Siegel Award.

Jeff R. Crandall is Associate Professor in the Department of Mechanical and Aerospace Engineering (MAE) at the University of Virginia, with a joint appointment in Biomedical Engineering, and is Director of the UVa Center for Applied Biomechanics. He received his Ph.D. in MAE from UVa in 1994. He also holds a M.E. from UVa and a B.A. in Engineering Sciences from Dartmouth. His research interests include restraint system effectiveness, protection of vulnerable road users, and lower extremity injuries. He is co-Editor of the journal Traffic Injury Prevention. He is past-president of the Association for the Advancement of Automotive Medicine and sits on numerous boards and committees, including the Board of IRCOBI and the Autoliv Technical Advisory Board. Recognition of his work includes the Lloyd L. Withrow Distinguished Speaker Award, the 1996 AAAM Best Paper Award, the System Safety Society Scientific Merit Award, and the 1996 Siegel Award for Best Paper at the Stapp Car Crash Conference.

Michael U. Fitzpatrick is a mechanical engineer who has spent the last 35 years developing air bag systems for both the federal government and automotive corporations. Mr. Fitzpatrick obtained his Bachelor of Science degree from Marquette University and his M.S. degree from Stanford University, both in Mechanical Engineering. In 1977, Mr. Fitzpatrick founded Fitzpatrick Engineering (FE) to perform air bag consulting, development, simulation and restraint system software sales and support. FE has been in continuous operation since that time. FE has clients of international scope including virtually all of the major automotive companies and their suppliers plus the federal government, vehicle testing firms, and many others. One of the major areas of specialization of FE is the development of software to aid in the development of seat belt and airbag systems. Mr. Fitzpatrick received the first "DOT Safety Award for Engineering Excellence" from DOT/NHTSA. He also received the "Bliss Award" from Mr. Ralph Nader and the "Center for Auto Safety". Here, Mr. Fitzpatrick was honored for being one of the eleven people worldwide most responsible for the invention of the air bag. He was also honored as the one person worldwide most responsible for computer simulation being applied to airbag design and development.

Charles J. Griswold, Jr. is President of C.J.Griswold,Inc. which was formed in 1987 to provide engineering analysis in automotive and aircraft crashworthiness design. He held a variety of engineering and management positions in the Fisher Body Division of General Motors Corporation, beginning in Research & Development as a Project Engineer and retiring as Manager of Safety Systems in GM Advanced Vehicle Engineering. He received a special GM Patent Award for the development of patentable designs, among them the first safety interlock door latch designed to prevent doors from opening under crash loads. He was the Engineer-in-Charge for the GM Air Restraint Program from 1970 to 1976, and currently serves as a consultant in air bag design. He is a life member of SAE, Chairman of the SAE Occupant Protection Committee, and has been an organizer for Air Bag, Biomechanics, Side Impact and Rollover technical sessions in the SAE Congress, as well as the ISATA and SAE-ATT Conferences in Europe. He also is a member of the SAE Accident Investigation & Reconstruction Committee, the SAE Impact & Rollover Committee, the Association for the Advancement of Automotive Medicine, and the International Traffic Medicine Association. He studied Engineering at Middlebury College with the U.S.Navy, Mechanical Engineering at the University of Buffalo and holds a Bachelor of Science Degree in Engineering Mechanics from Purdue University. He has done work in post-graduate engineering at University of Michigan, Northwestern University and Wayne State University.

Sherman Henson is a Product Safety Engineering professional with 36 years of experience in the automotive industry at Ford Motor Company. Thirty-one of those years were in product safety. While at Ford, he held positions in product development, computer-aided design, safety research, and vehicle safety regulations. Mr. Henson received Ford Motor Company Environmental and Safety Engineering awards for his work on Side Impact. He retired from Ford in 2001 as Manager, Impact Dynamics and Occupant Dynamics Regulations. He is currently an engineering consultant. He holds a Bachelor of Science in Mechanical Engineering from West Virginia University and a Master of Science in Mechanical Engineering from the University of Michigan. He is a member of SAE where he has been a co-organizer of the Air Bag session for the 2002 and 2003 World Congress and he is a member of the SAE Accident Reconstruction Committee. Mr. Henson was a member of the Stapp Car Crash Conference Advisory Committee during the years 1988-2000.

Joseph N. Kanianthra is the Associate Administrator for Applied Research at NHTSA involved in directing all of applied vehicle safety research related to crashworthiness, crash avoidance, driver-vehicle performance, tire research, and driver distraction research programs in the agency. Additionally, he is also responsible for heavy truck related safety research. Dr. Kanianthra holds BS, MS degrees and a Doctoral degree in Mechanical Engineering. He has been at NHTSA for over 26 years. He came to NHTSA from a distinguished career in teaching and research. At NHTSA, he has spent several years in many parts of the agency, including research and safety standards development. His past research activity included crash reconstruction, crash avoidance, and crashworthiness research. He was also involved in the agency's Research Safety Vehicle Program in the mid 1970s in the development, test, and evaluation of the Research Safety Vehicles. He has also served as a Special Assistant for Research and Technology in the Office of the Secretary of Transportation, in addition to his duties. His chief accomplishments in research and in safety standards development in NHTSA are dynamic side impact protection, roof crush injury prevention, ejection mitigation, head injury protection and frontal impact protection of occupants in light passenger vehicles, and in the area of international harmonization of research and safety standards. He has also represented NHTSA as a member of the group of experts of the U.N. ECE vehicle safety standards development group for several years. Dr. Kanianthra has received several distinguished performance awards and superior achievement awards in NHTSA and the Department of Transportation (DOT), including the Secretary's award for meritorious and technological achievement and the Presidential Rank Award for sustained superior achievements. He has over 40 technical papers and publications in professional journals, and patent to his credit. He is a member of the Society of Automotive Engineers and the American Society of Mechanical Engineers.

Richard W. Kent is a Research Assistant Professor in the Department of Mechanical and Aerospace Engineering (MAE) at the University of Virginia. He received his Ph.D. from UVa in 2002 after receiving his M.S. in Mechanical Engineering in 1997 from the University of Utah. Dr. Kent was named the SAE Doctoral Scholar for 1999-2001, and also received the NASA GSRP Fellowship to support his Ph.D. studies on restraint system performance. Prior to joining the MAE Department at UVA, Dr. Kent spent 8 years at Collision Safety Engineering in Orem, Utah, where he evaluated restraint systems and vehicle crashworthiness in the field. His current research focuses on thoracic injury and how factors such as musculature, restraint condition, aging, and load distribution influence the structural and injury response of the thorax. The validity of injury criteria and dummies for the assessment of injury risk with diverse restraint types is of particular interest. Dr. Kent is a member of SAE, the Association for the Advancement of Automotive Medicine, Sigma Xi, and ASB. He is a member of the Scientific Programming Committee of the Association for the Advancement of Automotive Medicine and the Occupant Protection Committee of SAE. He is a past organizer of the Air Bags Technical Session at the annual SAE Congress and is the current co-organizer of the Biomechanics Technical Session. His work has won several awards, including the Stapp Best Student Paper award, and he was recently named Best Young Researcher by the International Research Council on the Biomechanics of Impact (IRCOBI).

Hugo Mellander, born 1943, studied mechanical engineering in Sweden. He started employment at Volvo Cars in 1974 and became Director of Crash Safety Engineering in 1989. From more than thirty years of experience in biomechanics, development and testing of crash safety systems (including "Eurobag" and side impact air bags) he has published over thirty scientific papers. He was given the NHTSA Safety Award for Engineering Excellence in 1987 and, on behalf of Volvo, the Prince Michael Safety Award for an integrated child booster seat in 1990. Mr. Mellander is founder and President of Traffic Safety Research & Engineering. He recently served as President of AAAM (Association for the Advancement of Automotive Engineering) and is board member of the IRCOBI (International Committee on the Biomechanics of Impact).

Richard M. Morgan is a graduate of North Carolina State University with a bachelor and a master's degree in physics. After joining the National Highway Traffic Safety Administration in 1973, he worked to show the feasibility of using air bags in frontal collisions. Beginning in 1980, Mr. Morgan helped develop the side impact dummy and the injury criteria used in the US side impact standard. In the early 1990s, he established a network of leading universities to study the biomechanics of impact trauma. In 1996, Mr. Morgan assumed leadership of the New Car Assessment Program in the US.

Karl Erik Nilsson, born 1938, received a mechanical engineering degree in Sweden in 1958. After different employment and further studies, he joined Autoliv in 1967 (at that time still Lindblads Autoservice) with responsibility mainly for seat belt engineering and manufacturing. He got a SIF ("Svenska Industritjänstemannaförbundet") award to study occupant safety in the USA in 1972. In 1973, he established a cooperation with Bayern-Chemie, Germany, and joined this company one year later, first as project manager for seat belt pretensioner R&D and in 1976 also for air bag inflators. He got an over-all responsibility for the commercial programs in 1979. Mr. Nilsson assumed a new position at Morton, Automotive Safety Products, USA, as European Marketing Manager in 1991. Since the merger with Autoliv into Autoliv Inc. in 1997, he holds a position as Director, International Business Support. (He plans to retire in February of 2003).

Hansjürgen Scholz, born 1941, graduated as mechanical engineer in Stuttgart in 1969. In the same year, he joined Daimler-Benz and became involved in the air bag development. He got the cross-divisional responsibility for the air bag system development in 1971 and for all occupant restraint systems in 1972, since 1980 as Department Manager. In 1986, Mr. Scholz assumed a new position within Daimler-Benz as Head of the Test Department, which he held until his retirement in 1998. Mr. Scholz received the NHSTA Safety Award for Engineering Excellence in 1980 and in 1987, the Porsche Award of the Technical University of Vienna for his contribution to the development of the air bag system.

Charles E. Strother graduated from the University of Maryland, College of Mechanical Engineering in 1963. After spending seven years working at the Naval Ship Research and Development Center, Mr. Strother joined the Crashworthiness Research Division of what would evolve into the National Highway Traffic Safety Adminstration. At the NHTSA he planned and managed programs to develop and evaluate advanced occupant restraint systems, primarily air bag systems. In 1975, he left the NHTSA to work as a restraint engineer at Minicars, Inc. At Minicars, Mr. Strother worked on various occupant protection systems incorporated into the Research Safety Vehicle, principle among them being the driver air bag system. Later, he supervised a number of NHTSA-sponsored research programs conducted at Minicars to develop and evaluate advanced air bag systems for a variety of vehicle types. In 1981, he joined what would later become Collision Safety Engineering (CSE) as a consulting engineer, analyzing real-world automobile accidents and evaluating the role vehicle safety systems played in the injury outcomes. Also at CSE, he conducted research in the fields of automotive safety and accident reconstruction. Mr. Strother retired from CSE in 2001. During the course of his career, he authored or co-authored about 30 technical papers dealing with automotive safety or accident reconstruction. One of these papers, written in 1985, won the SAE Ralph H. Isbrandt Award for the best safety paper written that year.

Donald E. Struble holds a BS, MS, and PhD degrees from California Polytechnic State University, Stanford University, and Georgia Institute of Technology, respectively, all in engineering with an emphasis on structural mechanics. He was an Assistant Professor of Aeronautical Engineering at Cal Poly, manager of the Research Safety Vehicle program and Senior Vice President of Engineering and Research at Minicars, Inc., and President of Dynamic Science in Phoenix, Arizona. He has been working in automotive safety since 1972 and reconstructing accidents since 1983. His publications include invited presentations at two ESV conferences and papers on accident reconstruction, airbags, and other topics regarding automotive safety. He is co-holder of a patent on a side impact airbag. He has been an SAE TopTec instructor on air bags, high-speed rear impacts, and accident reconstruction. Dr. Struble is a member of SAE, AAAM, and Sigma Xi, The Scientific Research Society. Formerly Senior Engineer at Collision Safety Engineering in Phoenix, Arizona, he is now President of Struble-Welsh Engineering in San Luis Obispo, California.

David C. Viano is a specialist in injury biomechanics and impact protection in automotive crashes, sport impacts, and defense/law-enforcement actions. He is Adjunct Professor of Traffic Medicine at Chalmers University of Technology and Biomedical Engineering at Wayne State University, where he serves as the Director of the Sport Biomechanics Laboratory. He has advised and graduated five doctoral students. He is a Fellow of SAE, ASME, AIMBE, and AAAM, and serves on the Board of Directors of several scientific, medical, and charitable organizations. Dr. Viano is Editor-in-Chief of Traffic Injury Prevention. He has published over 200 scientific papers on injury biomechanics and impact protection; and, he recently published a book on the Role of the Seat in Rear Crash Safety with SAE (R-317). He has more than a dozen patents, most notably the first commercial active head restraint system to prevent whiplash and the high retention seat for rear impact safety. He is recently retired as Principal Scientist from General Motors Corporation, and now serves as a consultant to a number of organizations, including the National Football League in their efforts to understand and prevent concussion. He works through his company ProBiomechanics LLC. Dr. Viano served on the National Academy of Science committee that wrote Injury in America, and received the Award of Engineering Excellence from NHTSA/DOT as well as numerous SAE awards. Dr. Viano received his Ph.D. in Applied Mechanics from the California Institute of Technology, Pasadena, California, and Dr. Med. from the Karolinska Institute and Medical University, Stockholm, Sweden.